T0331964

Ecosystem-Based Fisheries Management

Ecosystem-Based Fisheries Management

Progress, Importance, and Impacts in the United States

Jason S. Link

Senior Scientist for Ecosystem Management, National Oceanic and Atmospheric Administration's (NOAA) National Marine Fisheries Service (NMFS), USA

Anthony R. Marshak

Program Analyst, CSS, Inc., supporting the National Oceanic and Atmospheric Administration (NOAA)'s National Centers for Coastal Ocean Science (NCCOS), USA

OXFORD
UNIVERSITY PRESS

Great Clarendon Street, Oxford, OX2 6DP,
United Kingdom

Oxford University Press is a department of the University of Oxford.
It furthers the University's objective of excellence in research, scholarship,
and education by publishing worldwide. Oxford is a registered trade mark of
Oxford University Press in the UK and in certain other countries

© 2021 U.S. Department of Commerce, U.S. Government

The moral rights of the author have been asserted

First Edition published in 2021

Impression: 1

All rights reserved. No part of this publication may be reproduced, stored in
a retrieval system, or transmitted, in any form or by any means, without the
prior permission in writing of Oxford University Press, or as expressly permitted
by law, by licence or under terms agreed with the appropriate reprographics
rights organization. Enquiries concerning reproduction outside the scope of the
above should be sent to the Rights Department, Oxford University Press, at the
address above

You must not circulate this work in any other form
and you must impose this same condition on any acquirer

Published in the United States of America by Oxford University Press
198 Madison Avenue, New York, NY 10016, United States of America

British Library Cataloguing in Publication Data
Data available

Library of Congress Control Number: 2021937971

ISBN 978–0–19–284346–3

DOI: 10.1093/oso/9780192843463.001.0001

Printed in Great Britain by
Bell & Bain Ltd., Glasgow

Links to third party websites are provided by Oxford in good faith and
for information only. Oxford disclaims any responsibility for the materials
contained in any third party website referenced in this work.

Table of Contents

Detailed Table of Contents

About the Authors

Dr. Jason S. Link is the Senior Scientist for Ecosystem Management with the National Oceanic and Atmospheric Administration's (NOAA) National Marine Fisheries Service (NMFS), USA. In this role, Dr. Link leads approaches and models to support development of ecosystem-based management plans and activities throughout the agency, serving as the agency's senior-most authority on ecosystem science. Dr. Link has written several books and book chapters on the topic of Ecosystem-based Fisheries Management, has written over 200 peer-reviewed publications, over 300 reports (~100 citable), has over 325 published abstracts, and has convened over 10 major international symposia and summits on marine ecosystem management, climate change impacts on marine ecosystems, and modeling-related topics. Dr. Link has been a champion of ecosystem science and ecosystem-based management, both as a discipline and as a practice, for resource management agencies in the U.S. and around the world, sitting on several international advisory boards and the editorial board of an international marine science journal. Dr. Link has also served on several executive/leadership boards of several community, church, foundation, and non-profit organizations unrelated to fisheries. Dr. Link holds an adjunct faculty position at the School for Marine Science and Technology at the University of Massachusetts, is a fellow of the American Institute of Fishery Research Biologists, and has received a Department of Commerce Bronze medal and the Fisheries Society of the British Isles Medal for significant advances in fisheries science. He enjoys fishing, hunting, snorkeling, weight-lifting, writing non-scientific works (under a pseudonym), helping with relief work around the world, chopping wood, and is active in his church. He lives on Cape Cod with his wife, kids, and pets, including a high-energy dog.

Dr. Anthony (Tony) R. Marshak is a Program Analyst with CSS-Inc. in support of NOAA's National Centers for Coastal Ocean Science (NCCOS), USA. Prior to this role, he worked as a Research Associate in the NMFS Office of Science and Technology where his duties included co-leading its habitat science program, organizing workshops and symposia, and collaborating with the NMFS Senior Scientist for Ecosystem Management. Dr. Marshak has conducted research on a variety of topics including coral reef fisheries ecology, climate-related range shifts, the effects of marine protected areas, and socioecological studies related to ecosystem-based management. Before joining NOAA full-time as a Sea Grant John A. Knauss Fellow in 2014, Dr. Marshak did his graduate studies at the University of South Alabama/Dauphin Island Sea Lab (Ph.D.) and University of Puerto Rico (M.S.), and led several research projects in the U.S. Caribbean during most of his twenties. In addition, he has conducted research throughout the Gulf of Mexico, in systems of the Mediterranean and Baltic Seas, and contributed to several international working groups. During his academic studies, Dr. Marshak was supported by NOAA and regularly collaborated with its researchers, including briefly working at the NMFS Galveston Laboratory after completing his undergraduate (Texas A&M University). His father is also a retired NOAA National Weather Service employee, so he has been acquainted with the agency since birth. In his spare time, he enjoys playing music solo or in bands (esp. with The Kingfish Trio, Spadefish, Narwhals of the North, and Rabbits in the Basement), diving, hiking, traveling, attending plays and concerts, and spending time with his family. He lives in Washington, DC, with his wife and sort of well-behaved dog.

Preface

Are You Kidding Me?

I (JL) smacked my forehead loudly. It was an instantaneous, reflexive reaction. Which was not a good thing to do. First of all, it hurt. Second, it left a big red mark on my head and as a bald guy, that showed pretty clearly. Third, I was giving a talk to a room of about 50 people who at that point all gave me their full and very undivided attention. And finally, I was reacting to a comment that one of the senior leaders from that region had made. As a senior leader in my organization, and more generally as a leader anywhere, expressing that level of consternation publicly the way I did is probably not wise. But it was revealing to me.

What caused my very public and loud head-smack was a comment by that leader to the effect of, "Yeah, but I don't know what EBFM is? It would be good if we could all agree on what EBFM means and why we want to do it." This statement would have been fine, except I had just spent the past 30 minutes briefing that room on the need for, the definition of, and the rationale of why we thought doing EBFM was such a good idea. Of course, that leader had been stepping in and out of the room taking phone calls and hadn't quite gotten those points. Plus, at that point this very debate had been going on in the literature for at least the past 15 years. And conservatively speaking, by that point I (JL) had probably given over 100 talks, presentations, and briefings on that exact topic during the preceding 10 years. After a decade of doing this, and after 4 years of singularly and repeatedly pushing EBFM within my organization (and for the discipline more broadly) as my primary job function, to hear such unawareness was incredibly frustrating. But again, upon reflection this was revealing to me.

Another time, I nearly smacked my forehead again, but managed to catch myself and feign an ear scratch. What caused this was while at a similar meeting in a different region, the local leader interrupted me, stating: "Why do we need this, we're already doing EBFM?" Which was also intriguing because that region had been struggling with the basics of fisheries interactions, i.e. bycatch, and thought that just doing more single-species stock assessments coupled with the odd habitat designation was sufficient to address all the issues warranted for EBFM. Which was understandable from the view that there was difficulty in just conducting the major stock assessments, but clearly missed several of the habitat and biogenic habitat loss issues. Again, frustrating to assume that business-as-usual (BAU) was equivalent to EBFM, but also revealing to me.

Frustrations aside, what these head-smack moments revealed to me was a few things. First, all my interest and energies and passion for the topic just wasn't translating into action or awareness, and we needed to figure out why. Second, the lack of importance or perceived lack of importance of the topic, as exhibited by that one leader (stepping out to take calls), was obvious. Third, maybe we had been making more progress than I was aware of or could observe. Fourth, we needed to more formally evaluate just how much progress we were making. Finally, it became clear that this issue was not so much a lack of what those leaders or others thought EBFM meant, but rather how it would actually work, how we would actually do it, and how it would help them and their staff to do their jobs better was the issue. And the point about how this would actually work truly resonated with me. As did the need to systematically evaluate our progress toward EBFM.

Reflecting on my head-smack moments, these revelations suggested that it was time to more thoroughly evaluate and codify the discipline and practice of EBFM. It was time to better explore the impediments and successes of EBFM. It was to time evaluate the progress we have been making toward EBFM. After discussing with several colleagues, one of them (an incredibly capable partner (AM)) agreed to join me to compile just such a tome. Hence this book.

Aims of This Work

The primary aim of this work was to take a step back, take stock of the discipline and practice of EBFM (as gleaned from U.S. examples), and comment on how well we are progressing toward EBFM. The rationale for doing this was to see what areas we've grown in and are succeeding at, and to identify those areas where we could improve. Although focused on U.S. fishery ecosystems, many of the lessons learned can be applicable to other marine ecosystems around the world. Ultimately, we want to see if we can identify and highlight suggestions that will lead to even further improvements of how we manage living marine resources.

Any undertaking like this is fraught with challenges. Would we ever be able to adequately capture the salient details of so many ecosystems? How do we provide a nationally consistent look at EBFM while maintaining the appropriate degree of specificity for a particular region? Would we ever be able to accurately convey the status of any specific situation regarding its progress toward EBFM, even if for a particular point in time? We knew we'd probably forget at least something, we'd likely leave something out, we'd undoubtedly overlook a particular detail or reference or bit of data, we'd potentially alienate a colleague or constituent by omitting their favorite taxa-issue-reference-process. So what to do when trying to evaluate progress for EBFM at such a large scale and scope?

Upon reflection, we simply needed to provide some context-setting (what we call the "bookend" chapters here) by way of introductory and summary and synthesis materials, and then get on with any such evaluation as best we could. Thus, really the bulk of this work is aimed at evaluating all eight U.S. territorial fishery regional ecosystems, with one chapter on international waters, by examining common measures known to track status of EBFM progress (Link & Marshak 2019; *Rev. Fish Biol. Fish.* 29:23–70). We make no claim to be perfect, entirely comprehensive, nor totally representative, but we did attempt to be as exhaustive as we could be in a reasonable period of time that we set for ourselves, as accurate (and fact-checked) as we could, as fair and unbiased as we possibly could be in what we present, and systematic in our approach. Thus, a suite of over 90 different metrics are presented for all the major U.S. fishery ecosystems that capture the many important facets of EBFM and successful fisheries management. Because of that, in some places this work can begin to be a bit dry if not even encyclopedic; to combat that, we have attempted to insert some case study text boxes in each chapter that purposefully contain a bit of levity. In some ways, this work could really almost serve as an introduction on the status and situation of U.S. marine fisheries systems in the late 2010s–early 2020s, as well as an introduction more broadly to U.S. fisheries issues in each region. But more so, examining the many measures of these socioecological systems across so many different regions, this work provides a composite snapshot at this point in time of the state of the discipline and practice of EBFM. It should provide an overarching view, at least in the U.S., of how well and in what areas EBFM is progressing, as well as those areas that remain challenges. If this book provides even part of such a composite picture, it will have succeeded in its objective.

Caveat Emptor, Caveat Mare

Kindly allow us a few caveats.

Scientists tend to be a particularly finicky group of people. As I (JL) tell my kids when arguing over whether some LEGO-built construct that fell into the bowl of green beans at the dinner table was actually Voltron or a transformer, does it really matter? (as in, why did you have LEGOs at the table in the first place? As in, will this materially make a difference tomorrow when we forget all about it? As in, don't chip your tooth when you eat your vegetables. etc.). We acknowledge that we've undoubtedly missed some details, forgotten a particular reference, glossed over some specifics, misrepresented nuances, and generally not provided as full of a treatment of any particular process or topic for a particular ecosystem. For that we apologize in advance. We attempted to mitigate this by having each chapter reviewed by a team of regional experts, which was exponentially helpful. But at one level, does knowing the specific frontal boundary location and how that impacts a specific mechanism that is driving fish recruitment dynamics for one species of fish ultimately help us grasp that sometimes we may need to more carefully consider the environment in fisheries management? Our main aim here was for accuracy versus precision, generality versus specifics, common patterns versus exceptions.

Resource managers, regulators, and policy setters tend to be a cautious group of people. As I (JL) tell my kids when nudging them to get off the couch and video game time to get outside, yes, it's hard to catch a fish, but it's easier to do so with a line in the water. We are highly sensitized to the fact that a large part of our agency regulates, or approves regulations, for our nation's fisheries. We entirely understand and appreciate the need to do so with as much caution, wisdom, transparency, fairness, and judiciousness as possible. We appreciate that misunderstandings and misinterpretations can be costly, leading to conflict, frustration, lawsuits, and even violence in some limited cases. This can understandably lead to cautious, risk-averse attitudes when approaching a topic. Concurrently, we also recognize that being entirely risk-averse could lead to missed opportunities or at best constrained discussions on how to make improvements. We recognize that the different attitudes toward risk leads to an inherent tension in a science-based, resource management organization. The last thing we want to do with this work is to inadvertently cause strife or harm to anyone in our organization, any of our partners, any of our stakeholders, or any of the interested parties in our nation's fisheries. Yet we also want to have an honest and open discussion about areas we think, as backed by supporting evidence, that could be improved. Thus, to mitigate any of these concerns we have endeavored to present all the materials in this work as openly and objectively as possible, to have those involved in fisheries management for each region review that particular work, have other reviewers from the regulatory side of our organization review parts of this work, and have been careful to present any items herein that are benchmarked as options, not requirements.

The ocean is dynamic. And then it changes even more. Trying to manage the living marine resources in such a dynamic environment, subject to what are almost chaotic if not even fractal, and certainly non-linear conditions, can seem to be impossible. Doing so at appropriate spatial and temporal scales, where the nature of the dynamics fundamentally changes with scale, adds to the challenge. Doing so with the complexities inherent to natural marine food webs, and then bio-geo-chemical connections and processes, into full marine ecosystems, and then layering in human aspects for full socioecological systems, can seem simply daunting. Upon conducting our review, we do not claim to have an exhaustive and perfect knowledge of all the processes in these and related marine ecosystems to be able to ever fully elucidate each of these conditions and situations. However, from the experience and knowledge gleaned from concerted and dedicated studies of U.S. marine ecosystems for at least the past half century, we assert that we do know enough things to make a few definitive, general statements that can facilitate EBFM. And more so, we assert that one does not need perfect knowledge of every process to begin to manage living marine resources from an ecosystem perspective. So while we acknowledge the complexity and dynamism of marine ecosystems, we do think that there have emerged enough repeatable, generalizable patterns and knowledge from which we can be able to make progress toward EBFM.

We submit this work to you the reader with humility at the scope of the task, hope that it will serve its intended purpose, trust that you will find it even mildly useful, and most of all with a desire that it will spur on even further progress toward EBFM.

Acknowledgments

To compile this work obviously took a mild bit of effort and time. But to ensure that we got this work mostly correct, we needed a large community to verify what we compiled, provide regionally specific perspectives, fact-check our information, and generally give this work a look-over. Thus, we thank Bill Arnold, Skyler Sagarese, Tauna Rankin, Reni Garcia, Trika Gerhard, Mandy Karnauskas, Robert Leaf, Chris Kelble, Todd Kellison, Kevin Craig, Jeff Buckel, Paul Rudershausen, Sarah Gaichas, Jay Odell, Rob Latour, Tom Miller, Scott Large, Mike Fogarty, Rich Bell, Jake Kritzer, Yvonne DeRenier, Toby Garfield, Chris Harvey, Kristin Marshall, Rusty Brainard, Ivor Williams, Michael Parke, Phoebe Woodworth-Jefcoats, Frank Parrish, Don Kobayashi, Erik Franklin, Anne Hollowed, Franz Mueter, Stefani Zador, Gordon Kruse, Tobey Curtis, Hassan Moustahfid, George Watters, Carrie Soltanoff, Beth Fulton, Tim Essington, Peg Brady, Howard Townsend, Wendy Morrison, and Kenric Osgood for their helpful comments and suggestions on prior versions of the various chapters of this work. We want to especially thank Kenric Osgood for his review of each of these chapters and for his assistance in shepherding them through the federal document and research product review process. We also acknowledge and thank the numerous U.S. Federal and State agency, Fishery Management Council, academic, and private industry employees and staff responsible for the collection and archiving of all datasets that were used in this work. We thank Jim Morley of East Carolina University and National Oceanic and Atmospheric Administration (NOAA) National Marine Fisheries Service staff including Karen Abrams, Lee Benaka, Brian Fredieu, Kevin Friedland, Kimberly Hyde, Todd Kellison, Michael Lewis, Alan Lowther, Sean Lucey, Rebecca Peters, Bernando Vargas-Angel, Ivor Williams, and Anthony Winbush for their assistance in providing and acquiring data. Additionally, we thank Tim Haverland and John Kennedy of NOAA Fisheries' Science Information Division, and David "Moe" Nelson (NOAA National Centers for Coastal Ocean Science) for their assistance in creating spatial frameworks in which to conduct our analyses.

Although this was an authored, not edited, tome on EBFM, we whole-heartedly acknowledge that much of the thinking herein has resulted from and been influenced by conversations with several of our colleagues over the years. To those noted above who did double duty as chapter reviewers, we thank them for such conversations. In addition to those noted above, we also thank the following for their many helpful conversations on these topics over the years: Richard Merrick, Howard Townsend, Patrick Lynch, Doug Lipton, Mark Dickey-Collas, Steve Murawski, Erik Olsen, Wes Patrick, Pamela Mace, Isaac Kaplan, John Field, Becky Shuford, Stephanie Oakes, Kevern Cochrane, Astrid Jarre, Lynne Shannon, Yunne Shin, Philippe Cury, Jörn Schmidt, Olivier Thebaud, Roger Griffis, David Smith, Tony Smith, Keith Sainsbury, Ellen Pickitch, Marc Mangel, Christine Santora, Phil Levin, Lisa Suatoni, Becky Goldberg, Lee Crockett, Charlotte Hudson, Chuck Fowler, Tim Smith, Anthony Charles, John Pope, Henry Regier, Alan Longhurst, Ed Houde, Steve Hall, Tony Pitcher, Sam Rauch, Mike Sissenwine, Brian Rothschild, Phil Mundy, Brian Wells, Jeff Polovina, Boris Worm, Rick Robbins, Rich Seagraves, Tom Hoff, Jeff Kaelin, Greg DiDomenico, John Williamson, John Pappalardo, Karen Abrams, Andy Rosenberg, Howard Browman, Eillen Soebeck, Eric Schwaab, Toby Garfield, Cisco Werner, Clay Porch, John Walter, Bern Megrey, Tom Bigford, Serge Garcia, Anna Rindorf, Anna Christine Dusseldorf, Geir Huse, Alf Hakon Hoel, Jake Rice, Alida Bundy, Stefan Nueunfeldt, Evan Howell, Kevin Friedland, John Kocik, Jay O'Reilly, Tom Noji, Jon Hare, Mike Seki, Yimin Ye, Gabriella Bianchi, Nico Gutierrez, Fabio Pranovi, Simone Libralato, Vic Adamowicz, Richard Appeldoorn, Stephen Brown, Correigh Greene, Ken Heck, Kenyon Lindeman, Patrick McConney, Tom Minello, Mark Monaco, Wendy Morrison, Will Patterson, Sean Powers, John Valentine,

Stacey Williams, Reg Watson, Fiorenza Micheli, Larry Crowder, Heather Leslie, Dave Fluharty, Ray Hilborn, Carl Walters, Andre Punt, Simon Jennings, Michel Kaiser, Manuel Barange, Villy Christensen, Daniel Pauly, Kalei Shotwell, Kerim Aydin, Jameal Samhouri, Jeremy Collie, Ellen Johannesen, and Gavin Fay. To the many we have undoubtedly and inadvertently omitted, we thank you as well.

We also acknowledge that our interactions with several working groups, workshops, tiger teams, conference committees, review panels, and task teams have helped advanced our thinking on EBFM as well. We particularly thank: the NOAA Fisheries EBFM Policy task force, NOAA Fisheries EBFM Road Map writing and review team, and Regional Action Plan (RAP) Teams, NMFS Climate Science Strategy Development Team, and Regional Implementation Plan (RIP) teams, NMFS Science Board (SB), NOAA Fisheries Next Generation Stock Assessment Improvement Plan (SAIP) writing and editorial team, NOAA Integrated Ecosystem Assessment (IEA) Program, NOAA Council of Fellows, NOAA Unified Modeling Committee, NOAA Research Council Ecosystem Indicators Working Group (RC EIWG), NOAA Fisheries and the Environment (FATE) Program, NOAA's Science Advisory Board (SAB)'s Ecosystem Science and Management Working Group (ESMWG), Atlantic Ocean Research Alliance (AORA) Ecosystem Approaches to Ocean Health and Services (EA2OHS) WG, various ICES WGs, but esp. ICES WGNARS, ICES WGSAM, ICES Regional Seas Program, ICES SCICOM, all 8 U.S. Fishery Management Councils (FMCs) and their Scientific and Statistical Committees (SSCs), the U.S. States Marine Fisheries Commissions (SMFCs), Lenfest FEP Task Force, BONUS, IUCN RLE, MYFISH, CAMEO SPMW, MSEAS, ECCWO, various PICES Ecosystem WGs, various FAO workshops, Lowell Wakefield Symposia and AK SG, Mare Frame, Mote Symposia, various IWC ecosystem model review panels, all CIE reviewers for NMFS Ecosystem Program Reviews (esp. Stephanie Oakes), EwE Symposia, AFS Habitat Columns (esp. Tom Bigford), MREP, IMBER ESSAS Regional Program, IRD's/IOC-UNESCOs Indiseas, NEMoWs, MARACOOS, NERACOOS, and other GOOS, NAFO EAF WG, Arctic Council PAME, VECTORS, CORILA, and a host of other acronyms representing groups and efforts that we can't possibly all adequately capture or even begin to remember here. For the members of all those groups listed and the many others we've been involved with but could not fit here, we thank you for your valuable interactions that helped us sharpen our thinking on EBFM.

We also thank Ian Sherman at Oxford University Press. We appreciate their agreement to take on a work such as this.

Although we thank all our colleagues and reviewers, the responsibility for this work remains ours alone. Any good signs and progress, we readily acknowledge and share with the community of EBFM practitioners and theorists. But we retain all responsibility for any errors or challenges herein.

We thank our wives (Julie Link, Sarah Stein), families, and friends for all their support and encouragement throughout this effort; our dogs seemed rather indifferent about it. Finally, as in all the major works for one of us (JL), JL asserts *Soli Deo Gloria*.

Foreword

This is the right time for this review.

In 2014, NOAA's Ecosystem Science and Management Working Group (ESMWG) assessed the progress that the U.S. had made toward use of ecosystem science in fisheries management, specifically targeting the implementation of ecosystem-based fishery management (EBFM) in the regional fishery management council (RFMC) system. This study, led by Dr. David Fluharty (Co-Chair of the ESMWG) with assistance by Dr. Tony Marshak (NOAA liaison to the ESMWG), concluded the science enterprise for fisheries management was strong, with a large amount of the effort going into stock assessments, essential fish habitat, and other mandates that are at the foundation of EBFM. Though all eight RFMCs were using EBFM science in management, the demand for and use of ecosystem science was highly variable by council region. Where RFMCs were committed to development and implementation of Fishery Ecosystem Plans, there was increased demand for and use of EBFM science. The review's "principal" recommendation was that a needs assessment should be undertaken to prioritize the ecosystem science inputs that would improve the performance of RFMCs. The report included an additional seven "primary" recommendations to NOAA Fisheries that would further expand the use of EBFM in the U.S.

NOAA's implementation of EBFM (stimulated to a certain degree by the ESMWG's review) moved forward with the agency's statement of an explicit EBFM Policy in 2016, an effort led by Dr. Jason Link. The Policy made it clear that NOAA Fisheries "strongly supports implementation of ecosystem-based fisheries management (EBFM) to better inform and enable better decisions regarding trade-offs among and between fisheries (commercial, recreational, and subsistence), aquaculture, protected species, biodiversity, and habitats. Recognizing the interconnectedness of these ecosystem components will help maintain resilient and productive ecosystems (including the human communities on which they depend), even as they respond to climate, habitat, ecological, and other environmental changes." The Policy provides for a definition of what EBFM is, then describes the benefits of EBFM and how it relates to existing living marine resource management legal authorities and requirements. Perhaps most importantly, the Policy establishes a framework of six guiding principles to enhance and accelerate the implementation of EBFM. Finally, the policy specifically tasks NOAA Fisheries' leadership to work with its "stakeholders and partners, including the Regional Fishery Management Councils, to achieve effective implementation of the EBFM policy."

Guidance for implementation of the Policy is provided through 2016's NOAA Fisheries "Ecosystem-Based Fisheries Management Road Map," developed by a team of NOAA scientists and managers led again by Dr. Link. This EBFM Road Map guides and enhances NOAA Fisheries' efforts to implement the EBFM Policy by describing recommended actions to address each of the Policy's six guiding principles for near-term work. The Road Map provides a menu of options, but is not a prescription of "must-do's." Given the breadth and magnitude of implementing EBFM, the Road Map is an initial national articulation of priorities that the agency will continue to review, revise, and build through EBFM's implementation in the U.S.

We are now well into implementation of the ESMWG recommendations, the NOAA EBFM Policy, and the associated Road Map. So, where are we? This is a time for taking stock, and to update the state of affairs since the 2014 report.

This book, coauthored by two of the key players in the development of the U.S. approach to EBFM, provides just such a timely review. As they describe it, they use a set of standard indicators with case studies to explore the progress the U.S. has made toward making EBFM an operational reality. They, however, sell their work short, as the information presented extends far beyond a review of the progress that has been made in EBFM within the U.S. EEZ. Their report actually represents a thorough examination of U.S. marine ecosystems, the living resources within them, and the social/governance systems that have developed around these ecosystems.

As I have said, this is the right time for this review, because it should further stimulate discussions at both the national and regional levels, that move implementation of EBFM forward. This is a critical time for this to occur. The ocean environment is facing unprecedented impacts from natural and anthropogenic stressors, such that a whole ecosystem view of fisheries management is the only viable path toward conservation of our living marine resources and the communities that depend upon them.

Richard Merrick, Ph.D.
Former Chief Science Advisor and Director of Scientific Programs, NOAA Fisheries

Endorsements

What leading voices in the discipline of fisheries science and management are saying about EBFM and *Ecosystem-based Fisheries Management: Progress, Importance, and Impacts in the United States*

"For the last 20 years the FAO has been advocating the use of the ecosystem approach to fisheries management, in recognition of the dynamic and interactive nature of ecosystems, the resources they harbor, and the people that make use of them. But turning concepts and principles into operational objectives is not that simple. Link and Marshak have systematically reviewed the implementation of the ecosystem approach to fisheries management in the USA, warts and all, and concluded that successful EAFM implementation is synonymous to improved management. They prove that ecosystem-based management works, and provide guidance for all to see that fisheries sustainability is a reality."

– **Manuel Barange**
Director, Fisheries Division (NFI)
Food and Agriculture Organization of the United Nations (FAO)

"In this era of increasing and competing stressors on our marine environments, ecosystem-based fisheries management (EBFM) is widely recognized as essential to sustainable oceans management, although there are recognized challenges to this interdisciplinary approach. In this very timely review of progress in EBFM across the USA, recognized experts Link and Marshak bring their enormous expertise to evaluate progress and identify key areas where advances can be made. They demonstrate that EBFM is possible, that barriers are virtual and that real challenges can be overcome."

– **Alida Bundy**
Fisheries and Oceans Canada

"My interest in ecosystem-based fisheries management of marine resources (EBFM) goes back to the early 2000s when I realized, through discussions with many colleagues, that few of us understood what it was all about conceptually, and even fewer (anyone?) what it meant operationally. Many view EBM as the only rational way to manage the oceans sustainably. Jason Link and his colleagues have been among the global leaders in defining what EBFM means and what must be done to operationalize it. Link's first book on EBFM summarized where we were in this regard ten years ago. Now, in this new book, he and Tony Marshak bring us up to date on where we are with EBFM, and where we still need to go. The case studies that they present, and the lessons learned, provide valuable information and insights for anyone grappling with operationalizing and implementing EBFM."

– **Howard Browman**
Institute of Marine Research
Ecosystem Acoustics Research Group
Austevoll Research Station, Norway
Editor-in-Chief, ICES Journal of Marine Science

"Ecosystem-based fisheries management is a holistic way to manage fisheries and marine resources. Sharing experiences from the U.S. can help communicate the benefits, and the lessons learned, in using this management approach."

– **Anne Christine Brusendorff**
International Council for the Exploration of the Sea (ICES), General Secretary

"Everyone in fisheries agrees that ecosystem-based management is essential for, even synonymous with, effective modern management. What is much less clear is how to make it work. That is why it is so important to look in detail at how the approach is used in practice – in this case in the United States – and to do so in an integrated way, covering environmental, socioeconomic, and governance aspects. Link and Marshak are well placed to accomplish this, bringing to the task both a vast range of experience and detailed current knowledge."

– **Tony Charles**
Senior Research Fellow in Environment and Sustainability
Professor, School of the Environment & School of Business
Director, Community Conservation Research Network
Saint Mary's University

"Ecosystem-based fisheries management (EBFM) is a holistic way of managing fisheries so that ecosystems are maintained in a productive, healthy, and resilient condition to provide sustainable services for human needs. Numerous studies have been conducted to develop EBFM. However, there is limited comprehensive and systematic evaluation of EBFM. Dr. Jason Link and Dr. Tony Marshak are two of the leading scientists in EBFM theory, development, and practice. They are highly qualified and in the perfect position to conduct a systematic evaluation of the progress we have made toward advancing ecosystem-based fisheries management."

– **Yong Chen**
Professor of Fisheries, School of Marine Sciences, University of Maine
Editor-in-Chief, Canadian Journal of Fisheries and Aquatic Sciences

"Are we finally there? The writing has been on the wall for decades, EBFM is coming! But only now do Link and Marshak demonstrate that we actually can walk the walk, drawing eminently upon case studies to provide lessons for making EBFM operational. We have made progress, and Link and Marshak are at the forefront of the development as demonstrated through this book."

– **Villy Christensen**
Institute for the Oceans and Fisheries
Univ. of British Columbia

"Ecosystem-based fisheries management (EBFM) is recognized as the only way to manage fisheries comprehensively and effectively but can be complex and demanding and management agencies around the world are still struggling to come to terms with the concept and practices. The United States has been a leader in the development and implementation of EBFM and this book of progress in the country and remaining challenges, by two leading experts in the field, will be a valuable and sought-after resource for scientists and practitioners throughout the world."

– **Kevern Cochrane**
Department of Ichthyology and Fisheries Science
Rhodes University, South Africa

"Moving consistently toward ecosystem-based fisheries management requires rigorous analyses such as those presented in this book. Undoubtedly this will be of great interest to all those interested in advancing toward policies that achieve a balance between the use and exploitation of marine organisms and their ecosystems."

– **Marta Coll**
Senior Researcher
Institute of Marine Science (CSIC), Barcelona, Spain

"In the global context of the Sustainable Development Goals (SDGs) of the UN Agenda 2030, EBFM is crucial when defining sustainability of marine life. This outstanding book published by two eminent scientists in the field of fisheries is reviewing case studies throughout the U.S. It provides insightful information regarding the functioning of marine ecosystems and their management. As such it will inspire the rest of the world, as EBFM is today an internationally acknowledged objective when managing marine resources."

– **Philippe Cury**
Représentant IRD auprès des Instances Européennes/IRD Representative
IRD (Institut de Recherche pour le Développement), France

"I think that an update of the status of EBFM in the USA is extremely timely. I think that the discipline is ready for an assessment and there are a number of universities, organizations and policy developers that will find such a resource extremely useful and a powerful tool."

– **Mark Dickey-Collas**
Chair of the Advisory Committee
International Council for the Exploration of the Sea (ICES)

"Link and Marshak's book will be an invaluable contribution to fisheries management. The science and practice of ecosystem-based fisheries management is evolving at a rapid pace, and wider adoption (and subsequent better management outcomes) will be greatly enhanced by synthesizing the lessons learned across the globe. Moreover, Link and Marshak are uniquely qualified to provide this synthesis—both are deeply knowledgeable in both the underlying science, but also the decision-making context."

– **Tim Essington**
Professor, School of Aquatic and Fishery Sciences
University of Washington

"This book *Ecosystem-based Fisheries Management: Progress, Importance, and Impacts in the United States* by Jason Link and Tony Marshak offers a timely assessment of the status of implementation of EBFM in the U.S., with global relevance. The authors are leaders in the national initiative to make the transition to EBFM and are eminently qualified to undertake this valuable synthesis of the U.S. experience."

– **Michael J. Fogarty**
Senior Scientist, Ecosystem Dynamics and Analysis
Northeast Fisheries Science Center
National Marine Fisheries Service

"EBFM is earning its way into U.S. fishery management as a tool to inform decisions that improve sustainability and avoid management mistakes. Marine ecosystems are dynamic and EBFM engages the fishing community, managers and scientists in seeking understanding of the multiple challenges in confronting climate change and other ecosystem drivers affecting fisheries. Link and Marshak are critically situated and capable to evaluate the progress being made in application of EBFM in the United States and to advise where further progress is possible."

– **David Fluharty**
Associate Professor (WOT)
School of Marine Affairs, University of Washington
& Former Chair, NOAA Science Advisory Board
Ecosystem Science and Management Working Group

"Ecosystem-based fisheries management has been recognized as a necessary approach to balancing exploitation and ecosystem health for more than 20 years now. There is a lot written about its mixed implementation to date so having an objective assessment of true progress is refreshing and timely. The credentials of Marshak and Link lay considerable weight to the assessment, both have been major contributors bedding down and moving practical EBFM forward in America and more widely."

– **Beth Fulton**
Senior Principal Research Scientist
Commonwealth Scientific and Industrial Research Organisation (CSIRO), Australia
Deputy Director, Centre for Marine Socioecology, UTAS
Research Program Leader, Environment and Ecosystems, Blue Economy CRC

"The ecosystem approach to fisheries was adopted in FAO in 2001 and its effective implementation has been painfully slow in most nations. A study of its implementation and of the lessons learned, two decades later, in one of the most active nations, by two highly regarded experts in the field, is extremely timely, and may serve as a benchmark for other, similar ones elsewhere."

– **Serge Garcia**
Chair, International Union for Conservation of Nature (IUCN)
Commission of Ecosystem Management (CEM)
Fisheries Expert Group (FEG)

"With the increasing, and increasingly diverse, ways that people interact with, impact, and benefit from the oceans, the need for more holistic approaches to managing these critical ecosystems is ever more pressing. EBFM has matured as a science and practice to meet these needs, yet there are still important areas for further understanding and improved practice. Jason and Tony's book comes at this key moment of reflection on what we have learned and where EBFM needs to go in the coming decade and beyond."

– **Benjamin S. Halpern**
Director, Nat'l. Center for Ecol. Anal. & Synth. (NCEAS)
University of California
Professor, Bren School of Environmental Science and Management

"Although ecosystem-based fisheries management (EBFM) has been proposed as a holistic way to achieve sustainable fisheries since the 1990s, very few countries have thus far included EBFM in their national fisheries management policies. Possibly because the science had not reached the maturity needed to show its value. However, without EBFM we will never achieve sustainable harvesting of our seas and oceans. The USA's fisheries management strategies consider EBFM as the most effective approach to achieving its management objectives, in large part due to the hard work of Jason Link and others. Jason has been at the forefront of creating the maturity and rigor needed to make EBFM a reality and when we were looking for an international reviewer for our future science brief on ecosystem modeling, he was my first choice. Jason and Tony Marshak have brought together state-of-the art case studies of EBFM in the USA."

– **Prof. Sheila JJ Heymans**
Executive Director
European Marine Board

"Ecosystem-based management fisheries management is an essential step in ensuring that fisheries are managed in a way to maintain ecosystems ability to continue to contribute to human well-being. Understanding what has been done in U.S. fisheries to move toward EBFM is an important step to refining how to implement EBFM, and there is no one better able to tell this than Link and Marshak."

– **Ray Hilborn**
Professor, School of Aquatic and Fishery Sciences
University of Washington

There has been a lot of developments in the field of ecosystem-based fisheries management in recent decades. NOAA and the United States have been at the forefront of this development. It is therefore very timely and valuable to perform a review of the state of progress in our field of ecosystem-based fisheries management in the nine major U.S. fishery ecosystem jurisdictions. Such a contribution will obviously have a great impact in the U.S., but will likely also have a global interest. And no one better to do such a synthesis work than Link and Marshak.

– **Geir Huse**
Forskningsdirektør—Marine økosystem og ressurser/
Research Director—Marine Ecosystems and Rescources
Havforskningsinsituttet/Institute of Marine Research

"We have been receiving the blessings of the ocean: fish and shellfish, but we have to not forget that the ocean is not just for humans. The ocean is the mother (source) of life on the Earth, and keeping the ocean healthy by EBFM is the foundation of a sustainable Earth."

– **Shin-ichi Ito**
Atmosphere and Ocean Research Institute,
The University of Tokyo

"Integrated approaches to management of human activities in the ocean, importantly including fishing, are essential toward improved sustainability. Dr. Jason Link and colleagues at NOAA have been at the very forefront of implementing ecosystem-based fisheries management in the USA since its inception, and have also been key facilitators of such approaches globally. Drs. Link and Marshak's work has previously highlighted how cross-disciplinary integration and systematic comparisons offer generalized insight into successful management strategies across key U.S. marine ecosystems. Their in-depth analysis of the progress achieved to date is a source of knowledge as much as a source of inspiration to everyone around the world concerned with integrated approaches to management."

– **Prof. Astrid Jarre**
South African Research Chair in Marine Ecology & Fisheries
University of Cape Town

"Ecosystem-based approaches to fisheries management are a keystone element to achieve the diverse goals that people have for ocean systems, globally. In order to make progress in implementing ecosystem-based approaches, it is critical to evaluate the efforts to date, recognizing how social and environmental factors shape both the form and the outcomes of EBFM in different places around the world. Jason Link and Tony Marshak are at the frontier of this type of assessment."

– **Heather Leslie**
Director, Darling Marine Center
University of Maine & School of Marine Sciences
co-author of *Ecosystem-Based Management for the Oceans*

"Fishing is critical to food security and the well-being of individuals and communities around the world. Conventional approaches to fisheries management can be successful at avoiding overfishing and rebuilding depleted stocks, but often fail to address the ecological, economic, and social systems that fisheries are part of. In this timely volume, two leaders in ecosystem-based fisheries management, Jason Link and Tony Marshak, conduct a comprehensive assessment of the state of EBFM in the United States. Their work highlights that EBFM is critically needed, and feasible today using existing science tools, policy instruments, and management structures. This invaluable resource highlights the many lessons learned across the U.S., and points to needed areas of improvement. This book is a must-have for any student or practitioner of modern fisheries management."

– **Phil Levin**
Lead Scientist, The Nature Conservancy Washington Field Office
& Professor-of-Practice, University of Washington
School of Environmental & Forest Sciences
School of Marine & Environmental Affairs

"The ecosystem-based approach to management has rapidly developed in the twenty-first century, where it needs to become the tool for managing human interactions with nature. But it is not a trivial task, and there is still plenty of resistance to the idea from people who are stuck in their ways, unable to grasp the idea, or unwilling to work hard enough to make it a reality. Thus, we need clear expositions of why and how to do it. Jason Link began doing so in his 2010 book, and now offers a series of next steps."

– **Marc Mangel**
Distinguished Research Professor, UC Santa Cruz
Adjunct Professor, Theoretical Ecology Group, University of Bergen
Research Affiliate Professor, Puget Sound Institute, University of Washington, Tacoma
Chair Emeritus, Board of Directors, FishWise

"Increasingly, sustainable development requires ecosystem-based fisheries management (EBFM), but the information demands can be a barrier to its development. By providing lessons learned, Link and Marshak provide an invaluable tool to regions looking to further develop and sustain a commitment to EBFM."

– **Thomas Miller**
Professor and Director
Chesapeake Biological Laboratory
University of Maryland Center for Environmental Science

"Implementing ecosystem-based fisheries management (EBFM) is nowadays more than ever of great importance to assure a sustainable use of the ocean´s resources and to achieve related UN Sustainable Development Goals (SDGs). EBFM is furthermore important to combat the important challenges of humanity such as climate change and the biodiversity crisis. Despite the EBFM concept now being about two decades old and having scientifically matured, its implementation is often lacking in many parts of the world. The U.S. is without doubt at the forefront of developing EBFM and hence a review and an evaluation of the progress that has been made here will be of great benefit for fisheries managers in other regions of the globe and the scientific community in general. Hence, I greatly appreciate the timely initiative of Link & Marshak to produce a state-of-the-art review of the efforts that have been made in the U.S. to develop and implement the EBFM concept. The authors are leading figures in the development of the EBFM concept but also in implementing it in real life fisheries management. They are without doubt the most qualified persons to conduct this enterprise. Furthermore, as a university scholar on fisheries science the published book will be the standard source for teaching EBM, complementing the first author's book on EBFM (a classic in the field on its own)."

– **Prof. Dr. Christian Möllmann**
Institute for Marine Ecosystem and Fisheries Science
Center for Earth System Research and Sustainability (CEN)
University of Hamburg

"The quest for ecosystem-based fisheries management (EBFM) requires assessing the effects of all individual fishing activities in an area on the entire ecosystem and in relation to the regional and national socio-economics and governance. EBFM is instrumental to curb overfishing and provide a basis for managing fisheries in the context of climate change and increasing ocean uses from all marine and coastal sectors. As leading scientists in the development of the scientific basis for EBFM, Link and Marshak provide a timely review of the progress toward EBFM in the United States that is of great relevance not only to the U.S., but also internationally to scientists, managers, IGOs, and NGOs developing EBFM in other regions."

– **Erik Olsen**
Head of Research
Demersal Fish Research Group
Institute of Marine Research

"Many jurisdictions that still seem to be trapped in a 1950s single-species world (e.g. parts of north America and North Atlantic ICES fisheries) are being challenged to modernize with a need to address the deep ecological and ethical issues raised by the ecosystem approach that we see cogently discussed in this important new book written by Jason Link and Tony Marshak."

– **Tony Pitcher**
Professor, University of British Columbia
Founding Editor, *Reviews in Fish Biology and Fisheries* and *Fish and Fisheries*

"All the federal regional fisheries management councils in the U.S. have been employing EBFM to varying degrees. A comparative review, looking at how this approach is employed and is performing, across all regions is timely, will be of interested to a broad audience from marine ecologist to fisheries managers, and will greatly help move this approach forward. The authors are both internationally recognized experts in EBFM and ideally situated in NOAA to conduct this important synthesis."

– **Jeff Polovina**
Senior Scientist (retired)
Pacific Islands Fisheries Science Center
NOAA Fisheries, Honolulu, Hawaii

"Ecosystem-based fisheries management has been an aspirational goal for fisheries management for decades, yet on-water progress is considered to be lacking compared to policy development. The U.S. has been a leader in the science supporting EBFM. In this new book, Jason Link and Tony Marshak provide a comprehensive evaluation of how EBFM goals are being achieved in the U.S. (and where progress is lacking), and an essential reference for researchers, teachers, and managers interested in all aspects of marine resource management."

– **Andre Punt**
Chair, School of Aquatic and Fishery Sciences
University of Washington

"Within the context of the unsustainability of the human presence on the planet, very little time remains for changing the route and reducing our impacts on the living and non-living resources. Fisheries represent a good 'case study' of the complexity that has to be considered for the implementation of effective management strategies. EBFM is the only approach for gaining the objective of sustainable marine resource exploitation. In order to move further, now we need stories of success in the implementation of EBFM for specific ecosystems to be used as best practices, to speed up the process at the global level. Given their long experience in the field, the colleagues Jason Link and Anthony Marshak are highly qualified for this task. They offer a high-quality product, useful for both the scientific community and policymakers."

– **Fabio Pranovi**
Full Professor in Ecology
Environmental Sciences, Informatics and Statistics Dept.
University Ca'Foscari of Venice

"A decade ago EBFM was a great idea, but often used as a Trojan horse for each expert to bring something of ecological interest into fisheries. Jason's first book helped bring order and structure to that situation, and was a foundation for much progress. Now he and Tony extract the lessons learned from that progress, and lay out the basis for future developments."

– **Jake Rice**
Chief Scientist—Emeritus
Fisheries and Oceans Canada

"The recent acceleration of EBFM calls for a synthesis like this book to enable accurate assessment of successes and areas of improvements. The authors are uniquely qualified to provide such a review, and importantly, this book is not just a review but a synthesis at the national level, and illustrated and grounded with real-world case studies. This book will provide a much-needed foundation and guide as EBFM enters the next phase of increased applications and implementations."

– **Kenneth Rose**
France-Merrick Professor in Sustainable Ecosystem Restoration
University of Maryland Center for Environmental Science
Horn Point Laboratory

"Fisheries managers in many parts of the world have turned in earnest to the challenges of managing not just individual fish stocks but the impacts of fisheries on ecosystems as a whole. It is none too soon. Drs. Link and Marshak, leaders in the field, present a critical evaluation of progress to date and gaps that need to be addressed. This is an important guide for practitioners and researchers alike."

– **Andrew A. Rosenberg**
Director, Center for Science and Democracy, Union of Concerned Scientists

"Ecosystem-based fisheries management is how to account for, and protect, ecosystem structure and function while the many other human benefits of fishing are also achieved. The USA has a national program to address this need and the authors have been very engaged with that effort. Here they review progress and provide interpretations from tropical to sub-Arctic fisheries and ecosystems. This is a very welcome contribution. It both highlights the successes and points to approaches to some remaining challenges. There are solutions available for many issues and there are practical options for the remaining challenges."

– **Keith Sainsbury**
Associate Professor, University of Tasmania
Senior Scientist, CSIRO (ret), Australia

"Despite the best intentions of scientists and managers alike, the enormous complexities and practicalities associated with establishing and implementing workable EBFM frameworks and actions can deter noticeable advances. In this book, we benefit enormously from the highly recognized and in-depth knowledge, experience, and wisdom of Link and Marshak, who draw on U.S. case studies to demonstrate EBFM progress to date. This is exactly what the world needs at this time—to inspire its nations to continue their EBFM journeys, in search of more sustainable fisheries and well-functioning marine ecosystems supporting these."

– **Lynne Shannon**
Chief Researcher: Ecosystem Approach to Fisheries
Department of Biological Sciences
University of Cape Town, South Africa

"Critical science thinking continues to drive comparative ecosystem-based fisheries assessment and management practices. Jason Link and Tony Marshak in their new book, *Ecosystem-Based Fisheries Management: Progress, Importance, and Impacts in the United States*, bring together under a single cover the results of an assessment of over 90 indicators for nine major United States fisheries ecosystem jurisdictions for tracking progress toward ecosystem-based fisheries management. The book provides the reader with an innovative and insightful approach to fisheries assessments from an ecosystems perspective that is at the cutting edge of this application of fisheries science in support of sustaining fisheries stocks."

– **Ken Sherman**
Adjunct Professor of Oceanography
Cultural Resources Center
Graduate School of Oceanography
University of Rhode Island
& Senior Scientist (ret), NOAA

"A one-of-a-kind book by world renowned experts, showing tangible progress in EBFM. Such evaluation is absolutely timely and necessary when new targets and indicators are being defined for the post-2020 Global Biodiversity Framework of the CBD."

– **Yunne Shin**
IRD (Institut de Recherche pour le Développement)
UMR MARBEC

"*Ecosystem-Based Fisheries Management: Progress, Importance, and Impacts in the United States*" by Jason Link and Tony Marshak is a timely review, by well-qualified authors that know what's really happening. As a former NOAA leader when ecosystem-based management became a priority and as a member of an FMC that is preparing an eFMP, progress to me is more than documents that describe ecosystems, reiterate habitat protection, address climate change, etc. Progress means plans that are consistent with the undisputed reality of trophic interactions. This means all species cannot be winners at the same time, although none should be written off. It is analogous to the NFL striving for competitive balance, but not the impossibility of all teams winning in the same season! Is EBFM making management better, or just making more work?"

– **Michael Sissenwine**
Visiting Scholar, Woods Hole Oceanographic Institution
Adjunct Professor, University of Massachusetts School of Marine Science and Technology
New England Fishery Management Council member

"It is a good time to review progress in implementing EBFM and Drs Link and Marshak are eminently qualified to do so. While there has been considerable progress, further research is required. This includes adequately dealing with cumulative impacts and also a more robust discussion on what impacts are acceptable."

– **David Smith**
Honorary Fellow, CSIRO Oceans and Atmosphere

"EBFM has been one of the key innovations in fisheries science and management during the twenty-first century, and is only increasing in importance as pressures increase on our oceans and their resources. Jason Link and his colleagues have been at the forefront of many of the key developments, and their review of the U.S. experience in applying these approaches will be of keen interest globally."

– **Tony Smith**
Honorary Fellow, CSIRO Oceans and Atmosphere
& Adjunct Professor, Institute of Marine and Antarctic Studies
University of Tasmania, Hobart, Australia

"EFBM is rapidly becoming a basic approach and standard in fisheries management, with potential to substantially impact both unintended impacts of fishing and overall catch levels. EFBM approaches have now been adopted in enough cases that it is critical to have case study reviews on progress and impacts of EFBM implementation, especially in relation to possible unexpected and unintended impacts."

– **Carl Walters**
OBC, FRSC
Professor Emeritus, University of British Columbia

"Ecosystem-based fishery management is a holistic approach to address the three challenges in fisheries: biodiversity, climate change, and sustainable development. This book's evaluation of its past practice and progress offers valuable lessons to all those who work toward global fishery sustainability."

– **Yimin Ye**
Chief, Marine and Inland Fisheries Brach (FIAF)
FAO of the United Nations

CHAPTER 1

Introduction

1.1 Background

The underlying concepts of ecosystem-based management (EBM) have been considered for over a century (Baird 1873, c.f. Smith 1994). The idea of actually operationalizing EBM has been formally discussed in natural resource management, sustainable development, and ecological science communities since the 1970s and early 1980s (e.g., Eberhardt 1977, Austin 1983, Belsky 1985, Butterworth 1986, WCED 1987, c.f., Link 2010, Robbins 2012). A decade later this was more cogently considered with respect to marine ecosystems (Larkin 1996). Applied to the marine fisheries sector, EBM morphed into ecosystem-based fisheries management (EBFM) and this has been an important driving concept for fisheries since the late 1990s (Arkema et al. 2006, Murawski & Matlock 2006, Link 2010, McLeod et al. 2014, Dolan et al. 2015, Patrick & Link 2015a).

Around that time, in 2002 one of us (JL) gave a talk at the Mote Marine Symposium where we noted, "Now we're to how," implying that we had sorted out much of the debate and it was time to get on with actually doing EBFM. But in retrospect that seems premature by at least a decade, as demonstrated by my head-smack moment noted in the preface. We wrongly assumed that the debate over what EBFM means and why one would want to do EBFM was mostly settled and that everyone agreed on what it meant. And more so, that everyone agreed that the need for it was obvious. As we enter the third decade of the twenty-first century, those "what and why" debates still continue to nibble at the edges of the discipline (Link & Browman 2014, Lidström & Johnson 2020), but we truly do see that this has become much more than just an academic debate and rather the discipline has seriously begun to wrestle with how to implement the practice of EBFM. The "whats and whys" may never be truly settled, but they have reached a point where a large enough quorum in the community of practice find them generally understood-enough to be agreed upon (Yaffee 1999, Pitcher et al. 2009, Link & Browman 2017, Marshak et al. 2017) so that working on implementation can actually occur. Certainly, we could list an entire suite of literature that captures the full extent of this debate, but to be that exhaustive would encompass an unduly large section of this work. After the decade or so of marine EBFM discussion in the 2000s, one of us (JL) attempted to advance the debate surrounding EBFM through a book (Link 2010) that sought to codify what a more operational look at the topic might entail. Suffice to say that since then, the past decade of the 2010s has seen a lot of activity regarding EBFM.

Thus, it seemed timely, useful, and appropriate to examine the progress we have made, both as a discipline and a practice, toward EBFM. The aim of this book is ultimately to assess (any) progress we have made in the implementation of EBFM. It is now very appropriate to take stock of the discipline and practice of EBFM. We do so largely from the perspective of federal scientists working in the United States system. We hope that any lessons learned here can be portable to other marine ecosystems and their associated management structures around the world, given the scale and scope of U.S. Marine jurisdictions–U.S. marine waters cover 11 large marine ecosystems, representing nearly 10% of the world's ocean surface area, spanning over 70 degrees of latitude and 100 degrees of longitude, from tropical to polar regions, and intersect with major parts of two ocean basins (Duda & Sherman 2002), and from differing living marine resource (LMR) management frameworks, capabilities, capacities, and socioeconomic drivers—making the U.S. experience relevant to most other global conditions. We want to highlight upfront that our view of this topic entails the full set of social-ecological system thinking (Walker et al. 2004,

Ecosystem-Based Fisheries Management: Progress, Importance, and Impacts in the United States. Jason S. Link and Anthony R. Marshak, Oxford University Press. © U.S. Department of Commerce, U.S. Government 2021. DOI: 10.1093/oso/9780192843463.003.0001

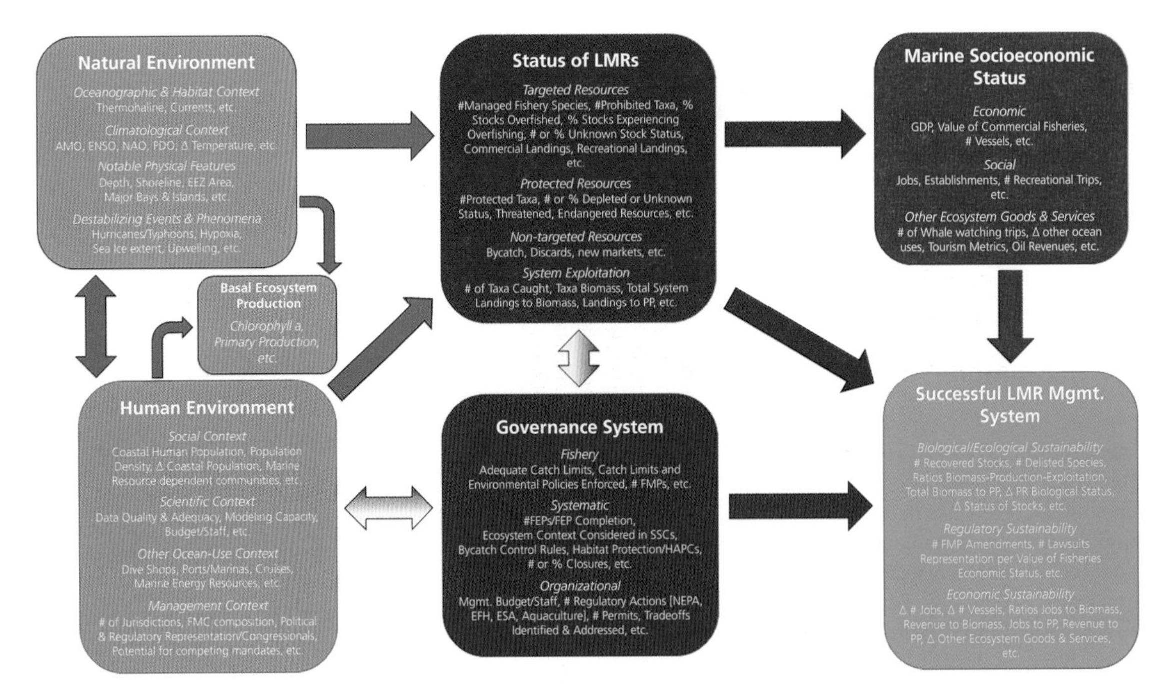

Figure 1.1 Schematic of the determinants and interconnectivity of successful Living Marine Resource (LMR) systems management criteria, presented as a coupled socio-ecological system with key facets broken into different sections, and suggested metrics in each. Our view is that successful LMR management requires EBFM and is truly synonymous with EBFM.

Adapted from Link & Marshak (2019). LMR = living marine resource; EEZ = exclusive economic zone; AMO = Atlantic Multidecadal Oscillation; ENSO = El Niño–Southern Oscillation; NAO = North Atlantic Oscillation; PDO = Pacific Decadal Oscillation; Δ = delta, i.e., change in; GDP = gross domestic product; PP = primary production; FMC = Fishery Management Council; SSC = FMC Scientific and Statistical Committee; HAPC = Habitat Area of Particular Concern; PR = protected resource (i.e., species); FMP = fishery management plan; FEP = fishery ecosystem plan; NEPA = National Environmental Policy Act; EFH = Essential Fish Habitat; ESA = Endangered Species Act.

Folke et al. 2005, De Young et al. 2008, Ostrom 2009) and is presented in a diagram that depicts the multiple facets of the issue (Fig. 1.1). Clearly there are many facets of this issue, but focusing on only the natural sciences aspect, or just the economic aspect, misses the important features required for full EBFM. We truly think this multidisciplinary view reflects some of the important, fundamental changes in knowledge and practice as EBFM has developed over the past decade. We attempt to highlight social, cultural, and economic considerations throughout this book, but acknowledge that both of us are trained as biologists and thus we likely revert to those examples more frequently than socioeconomic ones. To ameliorate this tendency, we explicitly and intentionally consider socioeconomic aspects as part of our EBFM evaluations.

1.2 Clarifying the What of EBFM

Before proceeding and to ensure all readers are on the same page, here we provide a brief synopsis of the definitions of EBFM and discuss why one would want to do EBFM. The canons of EBM have been expounded upon for decades (e.g., Grumbine 1994, Christensen et al. 1996, Griffis & Kimball 1996, Lackey 1998; Table 1.1), as a holistic approach to natural resource (fisheries) management that recognizes the physical, biological, economic, and social complexities of managing natural (living marine) resources. It really is a key, if not *the* key example of a way to approach the management of such complex social-ecological systems for the marine realm. In short, one can pose the question: How does one catch fish, preserve habitat, conserve other critters, utilize the ocean, have lots of tourists, make some cash, avoid too much bad stuff, and keep as many people as possible happy all at once? In many respects, that is another way of saying that we need to consider doing EBFM, and in so doing inadvertently describing it. In practice, it is often centered on an actual ecosystem, or a place in which we interact with the ocean and its organisms. People connect to the idea of EBFM by emphasizing that it is place-based: it starts with the place, the system in which fisheries operate. The "place" is the literal place (the area), the

Table 1.1 Various definitions of ecosystem approaches to fisheries management (EAFM), ecosystem-based fisheries management (EBFM), ecosystem-based management (EBM), and Ecosystem approaches to management (EAM). Adapted from internal reviews and documents (W. Patrick, pers. comm.; J. Haugen, pers. comm.)

EAFM	Reference	EBFM	Reference	EBM	Reference	EAM	Reference
EAFM is the inclusion of ecosystem factors into a (typically single species) stock focus to enhance our understanding of fishery dynamics and to better inform stock-focused management decisions.	Patrick and Link 2015a	EBFM recognizes the combined physical, biological, economic and social tradeoffs for managing the fisheries sector as an integrated system, specifically addresses competing objectives and cumulative impacts to optimize the yields of all fisheries in an ecosystem.	Patrick and Link 2015a	EBM is a multi-sectored approach to management that accounts for the interdependent components of ecosystems, and the fundamental importance of ecosystem structure and functioning in providing humans with a broad range of ecosystem services.	Patrick and Link 2015a	Ecosystem management means different things to different people, but the underlying concept is similar to that of the long-standing ethic of conservation.	Larkin 1996
For NOAA's purposes, an ecosystem approach to management is management that is adaptive, specified geographically, takes into account ecosystem knowledge and uncertainties, considers multiple external influences, and strives to balance diverse social objectives.	Fluharty et al. 2006	A comprehensive ecosystem-based fisheries management approach would require managers to consider all interactions that a target fish stock has with predators, competitors, and prey species; the effects of weather and climate on fisheries biology and ecology; the complex interactions between fishes and their habitat; and the effects of fishing on fish stocks and their habitat. However, the approach need not be endlessly complicated. Definition in Glossary: "Fishery management actions aimed at conserving the structure and function of marine ecosystems, in addition to conserving the resource."	EPAP 1999	Does not include a definition. However the goal of EBM should be to "maintain ecosystem health and sustainability" and the policies for EBM include: "Change the burden of proof, apply the precautionary approach, purchase "insurance" against unforeseen, adverse ecosystem impacts, learn from management experiences, make local incentives compatible with global goals, promote participation, fairness and equity in policy and management."	EPAP 1999	Many definitions of an ecosystem-based approach to natural resource and fisheries management have recurring themes. There is recognition of a broader constituency of uses and users of the marine environment (including fishing)...	Marasco et al. 2007

Continued

Table 1.1 Continued

EAFM	Reference	EBFM	Reference	EBM	Reference	EAM	Reference
		Ecosystem management is shorthand for more holistic approaches to resource management. Ecosystem management, as shorthand for more holistic approaches to resource management, is, from a fisheries management perspective, centered on multispecies interactions in the context of a variable physical and chemical environment. Broader perspectives include social, economic and political elements which are best considered pragmatically as a part of the context of fisheries management.	Larkin 1996	The concept of EBM has no universal definition or consistent application, but hierarchical context, ecological boundaries, ecological integrity, data collection, monitoring, adaptive management, interagency cooperation, organizational change, humans embedded in nature, and values are dominant themes of ecosystem management.	Brodziak and Link 2002	A strategy for the integrated management of land, water and living resources that promotes conservation and sustainable use in an equitable way.	UN CBD 2011
		EBFM is a holistic approach to maintaining ecosystem quality and sustaining associated benefits.	Brodziak and Link 2002	An ecosystem approach to management is a management that is adaptive, specified geographically, takes into account ecosystem knowledge and uncertainties, considers multiple external influences, and strives to balance diverse social objectives.	Francis et al. 2007	An ecosystem approach has management that takes into account all the delicate and complex relationships between organisms (of all sizes) and physical processes (such as currents and sea temperature) that constitute the Antarctic marine ecosystem	CCAMLR 2001
		Ecosystem-based fishery management recognizes the physical, biological, economic, and social interactions among the affected components of the ecosystem and attempts to manage fisheries to achieve a stipulated spectrum of societal goals, some of which may be in competition.	Marasco et al. 2007	An ecosystem-based strategy to manage fisheries involves using best available scientific information to promote long-term sustainability and to prevent adverse and irreversible harm to ecosystem structure and functioning by addressing how fishing activities affect biodiversity, food web interactions, and habitat.	Heltzel et al. 2011	An ecosystem approach to management (EAM) is management that is adaptive, geographically specified, takes account of ecosystem knowledge and uncertainties, considers multiple external influences, and strives to balance diverse social objectives.	Holliday and, Gautam 2005

EAFM	Reference	EBFM	Reference	EBM	Reference	EAM	Reference
		a systematic approach to fisheries management in a geographically specified area that contributes to the resilience and sustainability of the ecosystem; recognizes the physical, biological, economic, and social interactions among the affected fishery-related components of the ecosystem, including humans; and seeks to optimize benefits among a diverse set of societal goals.	NMFS 2016b	An integrated approach to resource management that considers the entire ecosystem, including humans. It requires managing ecosystems as a whole instead of separately managing their individual components or uses. EBM considers all the elements that are integral to ecosystem functions and accounts for economic and social benefits as well as environmental stewardship concerns. It also recognizes that ecosystems are not defined or constrained by political boundaries. The concept of EBM is underpinned by sound science and adaptive management as information or changing conditions present new challenges and opportunities. . . . it is an integrated approach to management that drives decisions at the ecosystem level to protect the resilience and ensure the health of the ocean, our coasts and the Great Lakes. EBM is informed by science and draws heavily on natural and social science to conserve and protect our cultural and natural heritage, sustaining diverse, productive, resilient ecosystems and the services they provide, thereby promoting the long-term health, security, and well-being of our Nation.	National Ocean Council 2013	An ecosystem approach to management (EAM) is one that provides a comprehensive framework for living resource decision-making. In contrast to individual species or single-issue management, EAM considers a wider range of relevant ecological, environmental, and human factors bearing on societal choices regarding resource use.	Murawski and Matlock 2006

Continued

Table 1.1 Continued

EAFM	Reference	EBFM	Reference	EBM	Reference	EAM	Reference
		An ecosystem approach to fisheries strives to balance diverse societal objectives, by taking into account the knowledge and uncertainties about biotic, abiotic and human components of ecosystems and their interactions and applying an integrated approach to fisheries within ecologically meaningful boundaries.	FAO 2003	An integrated management approach that recognizes the full array of interactions within an ecosystem, including humans, rather than considering single issues, species, or ecosystem services in isolation. The current and future environmental challenges facing ocean, coastal, and Great Lakes ecosystems benefit from EBM by utilizing a broad management approach that considers cumulative impacts on marine environments; an approach that works across sectors to manage species and habitats, economic activities, conflicting uses, and the sustainability of resources. EBM allows for consideration of resource tradeoffs that help protect and sustain diverse and productive ecosystems and the services they provide.	NOAA 2015	The ecosystem approach is a strategy for the integrated management of land, water and living resources that provides sustainable delivery of ecosystem services in an equitable way. EBM and EA used interchangeably as they mean generally the same thing	UNEP 2011
		Definition in Glossary: refers to personal, social, political, and management decisions that are made considering ecological information. Ecosystem-based decisions acknowledge that the environment, even in the absence of anthropogenic influence, is always changing. Ecosystem approach decisions are three-dimensional because they (1) include stakeholders,	Busch et al. 2003	In ecosystem-based management, the associated human population and economic/social systems are seen as integral parts of the ecosystem. Most importantly, ecosystem-based management is concerned with the processes of change within living systems and sustaining the services that healthy ecosystems produce. Ecosystem-based management	UNEP 2011	An ecosystem approach to management is management that is adaptive, specified geographically, takes into account ecosystem knowledge and uncertainties, considers multiple external influences, and strives to balance diverse social objectives.	NOAA 2004

EAFM	Reference	EBFM	Reference	EBM	Reference	EAM	Reference
		perspectives, and human goals, (2) consider the health and vitality of ecosystems into the indefinite future, and (3) include the larger landscape and connections among other landscapes. This approach requires attention to ecosystem integrity, interagency cooperation, spatially explicit management measures, and time-series data for multiple species and habitats. The goal of the ecosystem approach to fisheries management is to conserve natural resources and protect biodiversity while optimizing social and economic benefits and minimizing negative social and economic impacts to communities. Ecosystem goals are set with reference to the larger environment, including ecosystem parameters or environmental conditions (e.g., water quality) that limit fishery management options.		is therefore designed and executed as an adaptive, learning-based process that applies the principles of the scientific method to the processes of management. EBM is an approach that goes beyond examining single issues, species, or ecosystem functions in isolation. Instead it recognizes ecological systems for what they are: a rich mix of elements that interact with each other in important ways			
		NOAA Fisheries defines EBFM in the following manner: "Ecosystem-based fishery management recognizes the physical, biological, economic and social interactions among the affected components of the ecosystem and attempts to manage fisheries to achieve a stipulated spectrum of societal goals, some of which may be in competition.	NMFS 2014	**Author's definition:** "An [EBM] strategy for marine fisheries would be to minimize potential impacts while allowing for extraction of fish resources at levels sustainable for both the fish stock and the ecosystem." **NPFMC's draft definition:** "[EBM] is a strategy to regulate human activity towards maintaining long-term system sustainability (within the range of natural variability as we understand it) of the North Pacific."	Witherell et al. 2000	An ecosystem approach to management (EAM) is one that provides a comprehensive framework for marine, coastal, and Great Lakes resource decision making.	Levin et al. 2009

Continued

Table 1.1 Continued

EAFM	Reference	EBFM	Reference	EBM	Reference	EAM	Reference
				EBM differs from a single species or single sector approach to management by considering complex interactions between humans and the living and non-living environment over multiples scales in space and time.	Clarke and Jupiter 2010	Referred to EPAP definition.	NMFS 2009
				EBM emphasizes the maintenance or enhancement of ecological structure and function, and the benefits that healthy oceans provide to society	Link and Browman 2017	A strategy for the integrated management of land, water and living resources that promotes conservation and sustainable use in an equitable way.	UN CBD 2011
				EBM is a broad approach, involving the management of species, other natural commodities, and humans as components of the larger ecosystem.	Arkema et al. 2006	An ecosystem approach has management that takes into account all the delicate and complex relationships between organisms (of all sizes) and physical processes (such as currents and sea temperature) that constitute the Antarctic marine ecosystem	CCAMLR 2001
				EBM is an integrated approach that considers the entire ecosystem, including humans.	Leslie and McLeod 2007	An ecosystem approach to management (EAM) is management that is adaptive, geographically specified, takes account of ecosystem knowledge and uncertainties, considers multiple external influences, and strives to balance diverse social objectives.	Holliday and, Gautam 2005

EAFM	Reference	EBFM	Reference	EBM	Reference	EAM	Reference
				EBM, is an approach that goes beyond examining single issues, species, or ecosystem functions in isolation. Instead it recognizes our coasts and oceans for what they are: a rich mix of elements that interact with each other in important ways.	UNEP 2010	An ecosystem approach to management (EAM) is one that provides a comprehensive framework for living resource decision-making. In contrast to individual species or single-issue management, EAM considers a wider range of relevant ecological, environmental, and human factors bearing on societal choices regarding resource use.	Murawski and Matlock 2006
				Ecosystem management integrates scientific knowledge of ecological relationships within a complex sociopolitical and values framework toward the general goal of protecting native ecosystem integrity over the long term. NOTE- Discusses Ecosystem Management, not Ecosystem Based Management, but includes many of the EBM principles we have in EBM definitions today	Grumbine 1994	An ecosystem approach to management (EAM) is one that provides a comprehensive framework for marine, coastal, and Great Lakes resource decision making.	Levin et al. 2009
				Ecosystem management is management driven by explicit goals, executed by policies, protocols, and practices, and made adaptable by monitoring and research based on our best understanding of the ecological interactions and processes necessary to sustain ecosystem composition, structure, and function.	Christensen et al. 1996	An ecosystem approach to management is management that is adaptive, specified geographically, takes into account ecosystem knowledge and uncertainties, considers multiple external influences, and strives to balance diverse social objectives.	NOAA 2004
				Ecosystem-based management (EBM) is an integrated approach to resource management that considers the entire ecosystem, including humans. It requires managing	ORAP 2013	An ecosystem approach to management is one that is "geographically specified, adaptive, takes account of ecosystem knowledge and uncertainties, considers multiple	Barnes et al. 2005 & NOAA 2004

Continued

Table 1.1 Continued

EAFM	Reference	EBFM	Reference	EBM	Reference	EAM	Reference
				ecosystems as a whole instead of separately managing their individual components or uses. EBM considers all the elements that are integral to ecosystem functions and accounts for economic and social benefits as well as environmental stewardship concerns. It also recognizes that ecosystems are not defined or constrained by political boundaries. The concept of EBM is underpinned by sound science and adaptive management as information or changing conditions present new challenges and opportunities		external influences, and strives to balance diverse societal objectives" with implementation needing to be "incremental and collaborative." Since EAM's inception, numerous scholars, environmental managers, and conservation groups have developed EAM definitions to fit their respective needs and missions and with varying philosophies. The evolving concept of EAM is often the focus of much current debate and, as can be seen from the differences in definitions, EAM has not been uniformly defined or consistently applied by different agencies; Although the literature does not present a single agreed upon EAM definition, natural resource scientific and management communities agree that the following elements are needed for successful EAM: collaboration, adaptive management, ecological integrity, integrated data and information, and the connection between all landscape levels. NOAA is committed to moving toward an EAM of the Nation's coastal and marine ecosystem using the best current scientific information and the most reasonable adoption of EAM for NOAA's mission goals. This will require increased understanding of these complex systems as well as improved integration and collaboration.	

EAFM	Reference	EBFM	Reference	EBM	Reference	EAM	Reference
				Ecosystem-based management is an integrated approach to management that considers the entire ecosystem, including humans. The goal of ecosystem-based management is to maintain an ecosystem in a healthy, productive and resilient condition so that it can provide services humans want or need. Ecosystem-based management differs from current approaches that usually focus on a single species, sector, activity or concern; it considers the cumulative impacts of different sectors. An in-depth, inclusive definition developed by over 200 science and policy experts in the United States.	McLeod et al. 2005		
				Ecosystem-based management is an interdisciplinary approach that balances ecological, social and governance principles at appropriate temporal and spatial scales in a distinct geographical area to achieve sustainable resource use. Scientific knowledge and effective monitoring are used to acknowledge the connections, integrity and biodiversity within an ecosystem along with its dynamic nature and associated uncertainties. EBM recognizes coupled social-ecological systems with stakeholders involved in an integrated and adaptive management process where decisions reflect societal choice.	Long et al. 2015		

Continued

Table 1.1 Continued

EAFM	Reference	EBFM	Reference	EBM	Reference	EAM	Reference
				Ecosystem-based management is fundamentally about perceiving the big picture, recognizing connections, and striving to maintain the elements of ecosystems and the processes that link them.	Guerry 2005		
				Ecosystem-based management is the comprehensive, integrated management of human activities based on best available scientific and traditional knowledge about the ecosystem and its dynamics, in order to identify and take action on influences that are critical to the health of ecosystems, thereby achieving sustainable use of ecosystem goods and services and maintenance of ecosystem integrity.	Arctic Council 2013		
				In ecosystem-based management, the associated human population and economic/social systems are seen as integral parts of the ecosystem. Most importantly, ecosystem-based management is concerned with the processes of change within living systems and sustaining the services that healthy ecosystems produce. Ecosystem-based management is therefore designed and executed as an adaptive, learning-based process that applies the principles of the scientific method to the processes of management.	UNEP 2006		

EAFM	Reference	EBFM	Reference	EBM	Reference	EAM	Reference
				The comprehensive integrated management of human activities based on the best available scientific and traditional knowledge about the ecosystem and its dynamics, in order to identify and take action on influences which are critical to the health of marine ecosystems, thereby achieving sustainable use of ecosystem goods and services and maintenance of ecosystem integrity	ICES 2017		
				To support long-term sustainable management in aquatic ecosystems, strong policy integration in terms of objectives, knowledge base methods and tools, as well as engagement and knowledge exchange, is essential. The integrative nature of ecosystem-based management (EBM) shows in theory a lot of promise for supporting all of the above. Ultimately, EBM is a collaborative management approach used with the intention to restore, enhance and protect the resilience of an ecosystem so as to sustain or improve ESs and protect biodiversity, while considering nature and society, i.e. the full social-ecological system	Langhans et al. 2019		

Continued

Table 1.1 Continued

EAFM	Reference	EBFM	Reference	EBM	Reference	EAM	Reference
				U.S. ocean and coastal resources should be managed to reflect the relationships among all ecosystem components, including human and nonhuman species and the environments in which they live. Applying this principle will require defining relevant geographic management areas based on ecosystem, rather than political, boundaries.	U.S. Commission on Ocean Policy 2004		

habitats, the fish, and the oceanographic features. It is also the people, communities, and markets. And importantly, it is how all of these are linked to each other. Put another way, EBFM is a mindset for thinking of how fisheries truly work, distinct from a population or stock concept (T. Essington, pers. comm.).

We note that there are many (40+) definitions of marine-oriented EBM and EBFM (c.f. Link 2002, Arkema et al. 2006, Link 2010, Patrick & Link 2015a, Lidström & Johnson 2020; Table 1.1). Although there are several derivatives of EBFM defined in the scientific literature (e.g., Larkin 1996, Browman & Stergiou 2004, Arkema et al. 2006, McLeod et al. 2014, Lidström & Johnson 2020), significantly they have all mostly coalesced to substantively mean the same thing, just with different subtle points of emphasis (Table 1.2). Additionally, EBM has been used synonymously, and in our view sometimes erroneously, with terms like *marine spatial planning*, *integrated ocean management*, *integrated coastal zone management*, *multiple-use management*, *ocean health*, etc. Although acknowledging the similarities among them, we do not delve into those particulars nor distinctions in this work, as our primary focus is on EBFM. Suffice it to say that all of these terms and the various definitions of EBM and EBFM obviously vary in specific verbiage and particular emphasis given the context they were developed in, but most have commonly repeated aspects. Here we provide a few examples to demonstrate that point. Before we do, we need to clarify one other item.

Using a synonymous definition of EBFM (produced by the Food and Agriculture Organizations (FAO), an intergovernmental organization with representatives from 194 nations; FAO 2003), Wes Patrick (Dolan et al. 2015, Patrick & Link 2015b) summarized elements of the FAO definition of EBFM:

- Strive to balance diverse societal objectives
- Account for knowledge and uncertainty about biotic, abiotic, and human components of ecosystems
- Apply an integrated approach to fisheries
- Have ecologically meaningful boundaries

Like many of these types of works (e.g., Arkema et al. 2006, Pitcher et al. 2009, Long et al. 2015), he then benchmarked how well the U.S. was doing with these in relation to the context of Optimal Yield. Others use these aspects of EBFM definitions and principles to benchmark particular ecosystems, jurisdictions, or regions to help define EBFM and ascertain progress thereto, but more on that later. An important point of showing this FAO definition is that these are features of EBFM that are commonly expressed in many other definitions (FAO 2015, 2016). A synthesis from Patrick (pers. comm.) clearly shows how these touch on the same topics (Tables 1.1, 1.2).

Other aspects of defining EBFM can be adapted from international EBM contexts, as those based on the Convention on Biological Diversity[1] and other facets of the FAO Ecosystem Approach to Fisheries (Garcia 2000, FAO 2003, Garcia et al. 2003, Garcia & Cochrane 2005). This results in a list of 20 common EBM principles (Rudd et al. 2018; Table 1.3). These align with the previously noted synthesis (Table 1.2).

So although quite generic and broad in scope, and although (as is common for a lot of UN types of documents; WCED 1987, UNEP 2006, UNCBD 2011, etc.) a bit focused on governance structures and processes, some of these principles repeat important facets of the FAO EBFM definition (and the National Oceanic and Atmospheric Administration (NOAA) Fisheries definition; see next). This particularly includes aspects noting that there needs to be consideration of humans as part of the ecosystem; that trade-offs need to be addressed; that it is understood that there are multiple goals or objectives; and that there are multiple ecosystem components, and hence multiple disciplines, that should be considered as part of a whole system and how that system is managed (Tables 1.1, 1.2).

The Ecosystem Principles Advisory Panel (EPAP 1999) proposed eight, arguably more specific, recommendations for EBFM in the U.S. context (Table 1.4). The EPAP posited, as an overarching goal, the need to: "Maintain ecosystem health and sustainability." The EPAP principles reinforce other definitions and expressly state them in the context of U.S. fisheries. Wilkinson & Abrams (2015) benchmarked these principles against U.S. Fishery Management Council efforts, to show that some progress was being made. Again, these principles did drive some action and thinking about EBFM, and also helped to codify what EBFM might be defined as in practice.

NOAA Fisheries subsequently developed an EBFM Policy Statement (NMFS 2016a) and associated EBFM Road Map (NMFS 2016b). In those are six guiding principles for EBFM (Fig. 1.2). From that work, NOAA Fisheries committed to and acknowledged that the implementation of EBFM should reflect those six guiding principles. These principles flow from the foundational basis of science, through strategic planning, prioritization, and trade-off analyses, and into management advice, all with the ultimate aim of maintaining productive and resilient ecosystems. These guiding principles are:

[1] https://www.cbd.int/ecosystem/principles.shtml

Table 1.2 A literature review of prominent articles that define and describe the principles of ecosystem-based fisheries management (EBFM). Overall, 10 core EBFM principles were identified based on common key phrases used in these articles. Adapted from internal reviews and documents (W. Patrick, pers. comm.). Percentage refers to the proportion of these articles that contained the principle noted

EBFM Principles	Arkema et al. 2006	Barnes et al. 2005, Grumbine 1994	Broadziak & Link 2002	Christensen et al. 1996	EPAP 1998	FAO 2003	Francis et al. 2007	Hall & Mainprize 2004	Larkin 1996	Leslie & McLeod 2007	Link 2010	Pikitch et al. 2004	Pitcher et al. 2009	Ruckelshaus et al. 2008	Percentage
Identify and monitor systemic reference points that align with the goals and objectives of the fishery	x	x	x	x	x		x	x	x	x	x	x	x	x	93%
Optimize yield through interdisciplinary trade-off analysis	x	x	x		x	x	x	x		x	x	x	x	x	86%
Maintain resiliency of the ecosystem	x	x	x	x	x	x	x	x	x		x	x	x		86%
Specify long-term interdisciplinary goals & objectives	x	x	x	x	x	x		x		x	x	x	x	x	86%
Moving beyond just a single-species focus	x	x	x	x	x		x		x	x	x	x	x	x	86%
Incorporate ecosystem considerations into management decisions	x	x	x	x	x		x		x	x	x	x	x		79%
Use a precautionary approach to management	x	x		x	x	x	x	x	x		x	x		x	79%
Identify appropriate level of spatial management and coordinate with other management agencies	x	x		x	x	x		x		x	x	x		x	71%
Incentivize stakeholders to participate	x	x	x		x	x	x	x		x		x	x		71%
Promote sustainability	x	x	x	x	x	x		x	x		x	x			71%
Policy framework to implement EBFM		x			x					x	x		x	x	43%

Table 1.3 Important aspects defining ecosystem-based management (EBM), as adapted from the Convention on Biological Diversity (https://www.cbd.int/ecosystem/principles.shtml) and other facets of the FAO ecosystem approach to fisheries (Garcia et al. 2003). This results in a list of 20 common EBM principles (Rudd et al. 2018)

1. The objectives of management of (land, water and) living resources are a matter of societal choices
2. Management should be decentralized to the lowest appropriate level
3. Ecosystem managers should consider the effects (actual or potential) of their activities on adjacent and other ecosystems
4. Recognizing potential gains from management, there is usually a need to understand and manage the ecosystem in an economic context
5. Recognizing potential gains from management, there is usually a need to understand and manage the ecosystem in a social context
6. Recognizing potential gains from management, there is usually a need to understand and manage the ecosystem in a cultural context
7. In order to maintain ecosystem services, the conservation of ecosystem structure and functioning should be an objective of the ecosystem approach
8. Ecosystems must be managed within the limits of their functioning
9. The ecosystem approach should be undertaken at the appropriate spatial and temporal scales
10. Recognizing the varying temporal scales and lag-effects that characterize ecosystem processes, objectives for ecosystem management should be set for the long term
11. Management must recognize that change is inevitable
12. The ecosystem approach should seek the appropriate trade-off (balance) between, and integration of, conservation and use of marine resources (e.g., biological diversity)
13. The ecosystem approach should consider all forms of relevant information, including scientific and indigenous and local knowledge, innovations, and practices
14. The ecosystem approach should involve all relevant sectors of society and scientific disciplines
15. The interdependence between human well-being and ecosystem well-being is recognized
16. An appropriate policy, legal, and institutional framework is adopted to support the sustainable and integrated use of the resources
17. An institutional framework is utilized
18. Objectives are reconciled through prioritization and making trade-offs
19. The need to maintain the productivity of ecosystems for present and future generations is recognized, and
20. Efforts are made to establish and preserve equity in all its forms (intergenerational, intra-generational, cross-sectoral, cross-boundary, and cross-cultural), with special attention given to rights of minorities

Table 1.4 The Ecosystem Principles Advisory Panel (EPAP 1999) proposed eight recommendations for EBFM

1. Delineate the geographic extent of the ecosystem(s) that occur with [Fishery Management] Council authority, including characterization of the biological, chemical, and physical dynamics of those ecosystems and "zone" the area for alternative uses
2. Develop a conceptual model of the food web
3. Describe the habitat needs of different life-history stages for all plants and animals that represent the "significant food web" and how they are considered in conservation and management measures
4. Calculate total removals—including incidental mortality. Show how they relate to standing biomass, production, optimum yields, natural mortality, and trophic structure
5. Assess how uncertainty is characterized and what kinds of buffers against uncertainty are included in conservation and management
6. Develop indices of ecosystem health as targets for management
7. Describe available long-term monitoring data and how they are used
8. Assess the ecological, human, and institutional elements of the ecosystem, which most significantly affects fisheries, and are outside Council/U.S. Department of Commerce authority. Included should be a strategy to address those influences in order to achieve both fisheries management plan (FMP) and fishery ecosystem plan (FEP) objectives

1. Implement ecosystem-level fishery planning
2. Advance our understanding of ecosystem processes
3. Prioritize vulnerabilities and risks to ecosystems and their components
4. Explore and address trade-offs within an ecosystem
5. Incorporate ecosystem considerations into management advice
6. Maintain resilient ecosystems

Again, these reinforce what has converged in the literature on EBFM definitions (Table 1.1). As such, these six guiding principles, which are currently being applied in U.S. national and regional EBFM implementation, are part of what we evaluate throughout this book. Many of the metrics and indicators we examine can link to these guiding principles (c.f. Chapter 2).

For the context of this book, we primarily use the NOAA Fisheries definition of EBFM (NMFS 2016a) as:

> EBFM is a systematic approach to fisheries management in a geographically specified area that contributes to the resilience and sustainability of the ecosystem; recognizes the physical, biological, economic, and social interactions among the affected fishery-related components of the ecosystem, including humans; and seeks to optimize benefits among a diverse set of societal goals.

As defined in the NOAA Fisheries EBFM policy (NMFS 2016a), EBFM includes considerations of interactions among fisheries, protected species, aquaculture, habitats, and other ecosystem components, including the human communities that depend upon them and their associated ecosystem services. EBFM examines not only the broader suite of factors that impact fisheries, but also considers the potential impacts of fisheries and fished stocks on other parts of the ecosystem (e.g., on other fish species, marine mammals). "Societal goals" should consider and include any relevant economic, social, and ecological factors in the context of relating to fisheries and fishery resources. EBFM is cognizant of both human and ecological considerations.

These examples of principles, guidelines, etc. are useful to demonstrate not only how the discipline has begun to implement EBFM, but also how the discipline has defined and begun to unpack what EBFM means. One can see these principles moving from broader, almost platitudinal statements to ever-increasingly specific ideas, that although still broad, have become more amenable to actionable efforts able to be implemented. In many ways, this is the institutional "unpacking" noted by O'Boyle & Jamieson (2006). One could continue to list all sorts of guidelines, principles, pillars, and other such considerations for EBFM, as there is certainly no shortage of those in the literature. At one level, that might be overwhelming and confusing. At another level, it is a sign that the discipline has certainly wrestled with and debated the meaning and definition of EBFM. The salient point is that these definitions have largely coalesced into some common themes (Tables 1.2, 1.3). They have also become increasingly more

Figure 1.2 Ecosystem-based fisheries management (EBFM) guiding principles from the NMFS (2016a) EBFM Policy Statement. This depicts the interconnected and interdependent nature of the major EBFM guiding principles. The salient points are that we need science as the foundation of what we do, we need to start with the end in mind of that science, and we need to consider how best to achieve those stated goals as we conduct our science and management efforts.

focused on implementable, specific elements representing a suite of options for defining EBFM in practice.

1.3 Yeah, but What Level of EBFM?

One of the main challenges in defining EBFM has been linguistic uncertainty. One of us (AM) synthesized a survey of dozens of international colleagues regarding international perceptions on ecosystem approaches to management (Marshak et al. 2017). What emerged was not that there was no definition of EBM or EBFM, but that there was not a commonly utilized understanding of it in practice. We often hear phrases of "What do you mean by ecosystem?" or "What level of EBM are you talking about?" or, more commonly, "Why do I want to include all that stuff about the ecosystem when I just want to understand _*insert my favorite fish here* ?", that all belie a lack of understanding about what we are referring to when invoking EBM and EBFM.

To address this concern, one of the things that we have found helpful has been what we colloquially now term the "blue infographic of EBM" (Fig. 1.3). Within

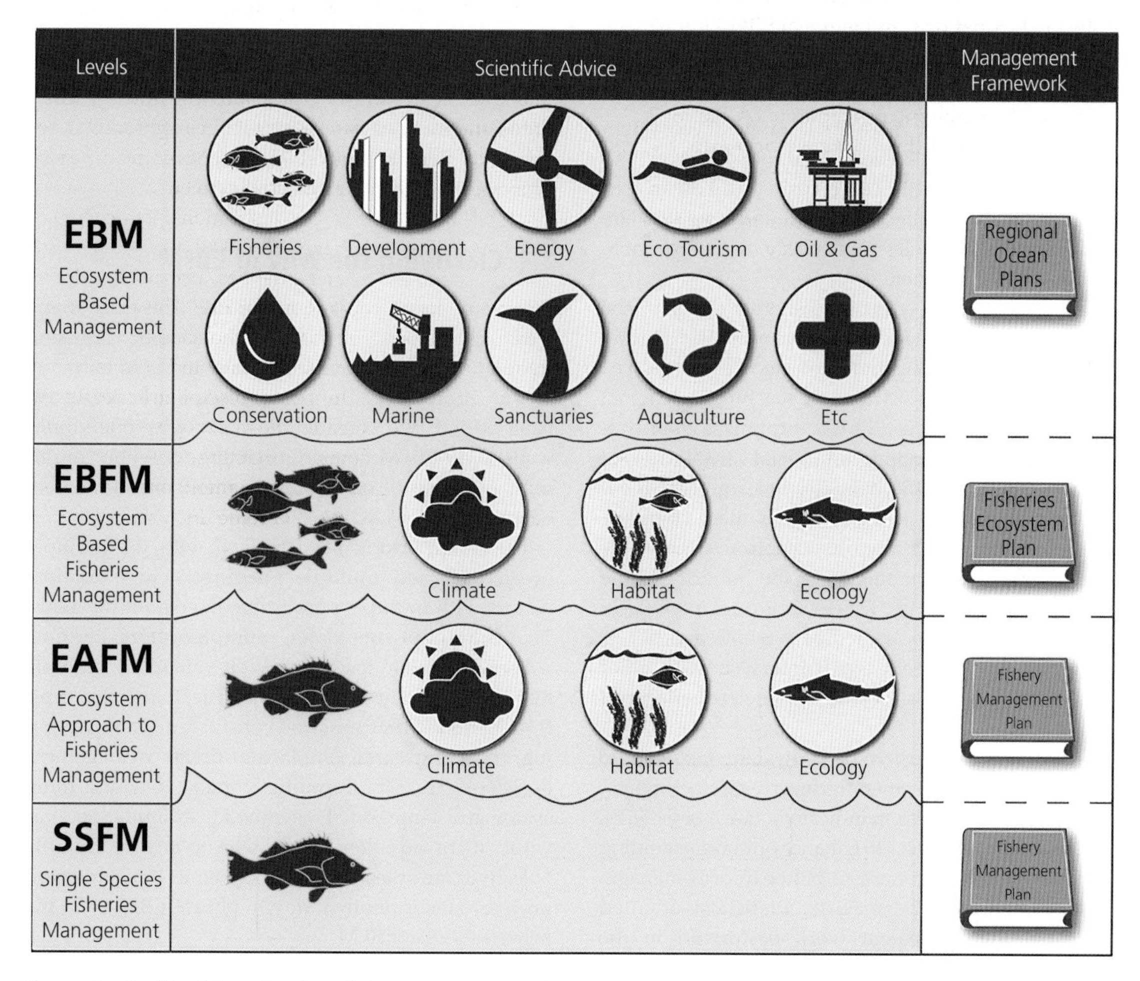

Figure 1.3 The "blue infographic of EBM" showing various levels of ecosystem management. The paradigms of ecosystem management (EM) focusing on the fisheries sector, building upwards from single-species fisheries management (SSFM), to ecosystem-based management (EBM). Scientific advice and the sectors of management build with each level, as well as the management framework. Key differences between ecosystem approaches to fisheries management (EAFM) and ecosystem-based fisheries management (EBFM) is that the later considers the trade-offs of multiple species, as opposed to a stock within a fishery, and EBFM takes a more coordinated approach to management through the use of strategic planning documents like FEPs. The considerations broaden as one moves up the levels of EM, as depicted by the elements in each circle. The delivery of the advice at each level is delivered in different types of plans. The breaks between these levels are not sharp, and information can be used at multiple levels, c.f. Table 1.5 for further details.

Adapted from Dolan et al. (2015), and discussions, Patrick & Link (2015b), Link & Browman (2014), and Link (2017).

the fisheries discipline, managers and scientists frequently describe EBFM as one level along a continuum of ecosystem approaches to management: 1) ecosystem approach to fisheries management (EAFM); 2) EBFM; and 3) EBM. This work has been presented and serially adapted by Link and Browman (2014), Dolan et al. (2015), Patrick and Link (2015b), and a published conversation with the American Fisheries Society (Link 2017). In it, we flag three levels of EBM beyond the classical approach for the fisheries sector:

- EBM is defined as a multisectoral approach to management that accounts for the interdependent components of ecosystems, and the fundamental importance of ecosystem structure and functioning in providing humans with a broad range of ecosystem services (Curtin & Prellezo 2010, Link & Browman 2014, Long et al. 2015). This focus is on all ocean uses of a particular marine ecosystem and how they are managed as a whole, particularly with respect to trade-offs among them.
- EBFM focuses solely on the fisheries sector of the ecosystem, which includes recreational fisheries, commercial fisheries, and protected resources (i.e., marine mammals, sea turtles, sea birds, etc.) that interact with fisheries. EBFM recognizes the physical, biological, economic, and social complexities of managing LMRs, the habitats that supports these species, and strategically considers these complexities and competing interests to optimize the yields of fisheries, while maintaining the integrity of the ecosystem (Link 2010). This focus is on the system of fisheries and fishes, treated as a whole and not as separate stocks, fleets, etc. It intersects with other ocean use sectors since they impact fisheries and vice versa.
- EAFM is the inclusion of ecosystem factors into analytical tools and other products (e.g., stock assessment reports, ecosystem indicators, risk assessments, stock assessments, etc.) to enhance our understanding of fishery dynamics and to better inform management decisions. The majority of fishery-oriented ecosystem management work performed in the U.S. and in other countries is considered EAFM (Pitcher et al. 2009, Skern-Mauritzen et al. 2016). The focus is largely on a stock, but includes broader considerations.
- This continuum also includes classical single-species fisheries management (SSFM). In this instance, the focus remains on a single stock and tends to omit direct consideration of other factors, even if they are assumed or subsumed in population parameters.

We assert and have found that applying EBM at appropriate levels of the hierarchy may help alleviate a lot of the confusion associated with linguistic uncertainty (Table 1.5). By identifying at which level one is speaking to, confusion about how much information is required, which parties need to be involved, and how the decision process will work, all become much clearer.

Again, by EBFM we are essentially exploring how to catch fish, preserve habitat, conserve other critters, derive energy, avoid too much bad stuff, have lots of tourists, utilize the ocean, and keep people happy all at once. Applying the appropriate level of EBM, EBFM, or EAFM to make that a reality is necessary. When presented as a broad set of overarching goals plainly stated and clarified as to how it will be approached, we have experienced general agreement by most people every time we have presented it as such.

1.4 Clarifying the Why of EBFM

Why would one want to do EBFM? There are many reasons, but one central reason emerges: assessing trade-offs. That has been the subtitle to EBFM for many of our efforts over the past decade (Link 2010). We have added two considerations for why one would want to do EBFM beyond just trade-offs—the "multi-multi" context of ocean management, and ultimately better outcomes (AORA 2017, 2019).

There is a strong need to deal with the multiple ocean uses and multiple pressures across multiple sectors with multiple parties that have multiple goals. That is, multiple mandates, multiple sectors, multiple fisheries, multiple species, multiple valuations, multiple human communities, multiple objectives, and multiple contexts that must be considered when weighing, enacting, and emplacing ocean management decisions. This "multi-multi" context is why a more systematic, integrated, synthetic examination has value, if for no other reason than to ensure multiple objectives are at least not overlooked in a management process. This is another way to phrase EBM, or in the fisheries sector, EBFM.

It has been repeatedly demonstrated that there are many potentially competing interests and objectives given the multiple jurisdictions, mandates, fleets, targeted taxa, etc. in a region of the ocean. Yet trade-offs and ecosystem considerations are still largely ignored in most locales, at least in terms of being directly considered in the management process (Levin et al. 2009, Link 2010, Patrick & Link 2015a, b). Certainly, increasing contextual information is being developed and

Table 1.5 Levels of marine ecosystem management (EM), with a description of how each level focuses on different sectors, biological components, objectives, scientific advice, implementation framework, and ideal implementation body. Single-species fisheries management (SSFM), ecosystem approaches to fisheries management (EAFM), ecosystem-based fisheries management (EBFM), ecosystem-based management (EBM), biological reference point (BRP), fishery management plan (FMP), fishery management council (FMC), ecosystem-level reference points (ELRP), National Environmental Policy Act (NEPA)

Feature	SSFM	EAFM	EBFM	EBM
Sector of focus	Fisheries	Fisheries	Fisheries	All
Focus of biological hierarchy	Stock/population	Stock/population	Community	Whole system & connected systems*
Primary analysis objective	Determine the status of stocks. Ascertain stock productivity	Determine the status of stocks. Ascertain stock productivity	Address fisheries sector LMR trade-offs. Ascertain ecosystem productivity. Identify best mix of goods and services across fisheries	Address cross-sector trade-offs. Identify best mix of goods and services across systems
Scientific advice produced	Biological reference points (BRPs)	BRPs	Ecosystem-Level Reference Points (ELRPs), which include BRPs	ELRPs
Implementation framework	Fishery management plan (FMP)	FMP	Fishery Ecosystem Plan	Regional Ocean Action Plans, NEPA
Implementation body	Fishery Management Council	FMC	FMC	Regional Ocean Council

Adapted from Dolan et al. (2015) and Link & Browman (2014).

* The components within the ecosystem as well as the broader scale community and ecosystem structure at an entire system level, i.e., the high-level systemic view.

used in many regions in the U.S. (Moffitt et al. 2016, Zador et al. 2017, Marshall et al. 2018) and elsewhere in the world (Berghöfer et al. 2008, Metcalf et al. 2009, Marshak et al. 2017, Smith et al. 2017), with successful approaches toward certain LMR management applications. Nevertheless, formal examination of the trade-offs facing fisheries systems remains an important issue to be addressed (Christensen & Walters 2004, Link 2010, White et al. 2012, Andersen et al. 2015), hence the continued calls for EBFM (Fogarty 2014, Fulton et al. 2014, Patrick & Link 2015b, Link 2018, Marshall et al. 2018). We note that there have always been trade-offs in LMR management; EBFM raises the level of trade-offs beyond those found when considering just a single stock and its population dynamics. Ignoring trade-offs, particularly in a broader ecosystem context, does mean that the trade-offs do not exist. And addressing these trade-offs seems increasingly wise to do.

Additionally, there are myriad issues facing the management of fisheries. They range from the classically persistent challenges known to population dynamics (Schaefer 1957, Hilborn & Mangel 1997, Salas et al. 2007, Levin et al. 2009, Cadrin et al. 2014, Link & Marshak 2019) to a broader array of processes that influence the dynamics of these LMRs (Keyl & Wolff 2008, Link et al. 2012, Thorson et al. 2015). Due to the recognition that business-as-usual (BAU) single-species management may not fully address the issues that impact fisheries, and certainly not the cumulative effects across multiple fisheries in a given ecosystem (Jennings & Kaiser 1998, Halpern et al. 2008, Micheli et al. 2014, Coll et al. 2016), numerous calls to implement EBFM continue (Botsford et al. 1997, Link 2002, Pikitch et al. 2004, Beddington et al. 2007, Link 2010, Fogarty 2014, Fulton et al. 2014, NMFS 2016a). Core to the calls for EBFM is a more systematic, and prioritized, consideration of all fisheries, pressures, risks, threats, objectives, and outcomes for a given marine ecosystem (Browman & Stergiou 2004, 2005, Szuwalski et al. 2015, Link 2018). Critics and proponents of EBFM alike often interpret calls to execute EBFM as a means to examine broader fisheries-related issues over the classical, single-species approach still common in fisheries management (Hall & Mainprize 2004, Hilborn 2011, Patrick & Link 2015b, Ballesteros et al. 2018).

One of the persistent challenges facing LMR management has been integration across fleets, taxa, disciplines, and even mandates (Beddington et al. 2007, Leslie & McLeod 2007, Link 2010, NMFS 2016a). There is copious stand-alone information for LMRs and marine ecosystems, but rarely is it integrated and synthesized (Fulton et al. 2014, Link & Browman 2014). Collating the disparate data from international, national, or regional perspectives is certainly challenging, but also has high value to facilitate the comparative, systematic approaches that lead to an understanding of marine ecosystems. More so, such collation helps to address a key challenge facing EBFM operationalization, namely that of relativity of all the processes affecting LMRs in any given ecosystem (Link et al. 2012, Patrick & Link 2015a). Knowing which processes, features, pressures, human activities, and taxa group responses are strongest, and cumulatively which are most prominent, requires a comprehensive, systematic examination. Factors including natural environmental features, human stressors, and basal ecosystem production are fundamental components of a given fisheries ecosystem, which ultimately influence the status of its LMRs and socioeconomics (Fig. 1.1). All are interrelated, and together with governance capacity and efficiency determine the effective LMR management strategies for a given system (Link & Marshak 2019). Until all factors are considered simultaneously, the risk is that important considerations will be missed. More so, until all fisheries objectives in a region are considered concurrently, the risk is that the solution space for some fisheries objectives will override that of others, be they competing fleets, competing taxa, competing markets, or entirely different goals.

We acknowledge that there have been notable successes in fisheries science and management, such that now many parts of the world have well-managed fisheries and reasonably healthy fish stocks (Pitcher et al. 2009, Worm et al. 2009, Hilborn et al. 2015, 2020), even though this is not globally true (Pitcher et al. 2009). Yet whether stocks are poorly or well managed, several other challenges to marine fisheries persist which have long been well chronicled (e.g., Botsford et al. 1997, Micheli 1999, Jackson et al. 2001, Pauly et al. 2002, 2003, Pikitch et al. 2004, Worm et al. 2009, Link 2010, Fogarty 2014). The list of challenges often involves the need to more clearly consider the broader impact of fisheries—on other biota, habitat, and socioeconomic systems—as well as the impacts of those broader considerations on fisheries, and all that in a systematic, comprehensive manner. In the context of improving the status of fisheries and fish stocks—in both well and poorly managed situations—numerous calls for EBFM continue to arise (Garcia 2003, Pikitch et al. 2004, Garcia & Cochrane 2005, Levin & Lubchenco 2008, Link 2010, Fogarty 2014). For "poorly managed stocks" (poor conditions, poor decisions, poor capacity, poor enforcement) and data-poor stocks, providing additional information has the potential to better elucidate those stocks, their dynamics, and their status, and any such additional information will be an improvement over the status quo. It also makes sense that we see this evolution toward exploring ecosystem approaches for many "well-managed stocks," in that there are diminishing returns to greater investment (i.e., model and data improvements) in single-species assessments and management; whereas, the potential benefit of incorporating ecosystem information can and has provided significant returns on fisheries investments.

The highly dynamic oceanographic environment and directional changes facing the oceans due to climate change is another strong reason for EBFM (Barange et al. 2014, Busch et al. 2016). The case for EBFM is made much easier given a world that recognizes the scale and scope of rapid ecological change caused by significant climate change effects on the oceans (Barange et al. 2014, Busch et al. 2016). Very simply, any projections, stock assessments, catch limits, reference points, and other LMR management advice provided that does not account for these rapidly changing conditions can ultimately be misleading (Punt et al. 2014, Lynch et al. 2018).

The complexities, challenges, need for efficiencies in a declining budget context, and increasing non-fish and environmental drivers all highlight the need for integrative coordination of LMR management in a region—e.g., EBFM (Link 2010, Fulton et al. 2014, Micheli et al. 2014, Patrick & Link 2015b). There are myriad dynamics, processes, and events occurring in any given ecosystem; this is true whether it be competing mandates, competing ocean uses, competing fishing fleets, or competing taxa (Gutiérrez et al. 2011, Dickey-Collas 2014, Samhouri et al. 2014, Szuwalski et al. 2015, Folke et al. 2016, Harvey et al. 2017). As we move forward with these increasing "multi-multi" demands facing LMR management, it is clear that a systematic, integrated approach is needed. What has been noted but warrants reiterating is that an ecosystem approach will allow us to better coordinate, prioritize LMRs at higher risk, deal with this huge set of ecological and human-dimension complexities, gain efficiencies, address multiple mandates and objectives, and explore all goals and objectives simultaneously rather than attempting to do this piecemeal, species-by-species, mandate-by-mandate, and fleet-by-fleet (Fulton et al. 2014, Ballesteros et al. 2018, Levin et al. 2018, Link 2018). Directly addressing the "multi-multi" conflicting objectives is the essence of EBFM; ignoring the "multi-multi" situation does not make the conflicts go away.

Implementing EBFM can help LMR management agencies and their partners optimize benefits among a diverse set of societal goals. These benefits are realized across multiple federal and state mandates by considering salient environmental and ecological factors that affect trust resources and by identifying trade-offs among trust resources, including fisheries, protected species, and their habitats. A rather compelling case has been made that EBFM is more efficient than conventional management (Fletcher et al. 2010, Essington et al. 2016, Levin et al. 2018, Link 2018, Rudd et al. 2018, Fulton et al. 2019). Through EBFM, fisheries managers can have a better understanding of the cumulative impact of a management action beyond just a single species. Additionally, EBFM can help communicate risks, uncertainties, and implications of management decisions across marine fisheries and a range of affected species. Better understanding, articulation, and consideration of the risks, benefits, and effectiveness of management alternatives is important for better LMR management. This is also true of the interconnectedness and trade-offs between and among management objectives, which will all ensure more transparent decision processes, outcomes, and more efficient resource use by fisheries managers.

Management advice from EBFM will also be more comprehensive and accurate, and will likely help reduce uncertainty by taking into consideration interacting elements in the ecosystem. EBFM can maintain ecosystem function and fishery sustainability, which support economic and social stability and fishing community well-being. EBFM applies the best available scientific information to improve decision-making via consideration of the holistic impact of management decisions. EBFM can also use forecasts of future ecosystem conditions and services, incorporating natural variability, anthropogenic forcing, and change in climate and ocean conditions to predict and evaluate outcomes from a range of alternative management strategies. Combined stability and efficiency outcomes for business and regulatory planning result from adopting EBFM (Link 2018).

This approach is an improvement over BAU single species or single sector approaches because it allows one to manage fisheries in marine ecosystems with many benefits (Table 1.6). We (and many others) really do think EBFM is an improved way to manage LMRs, because in empirical and simulation studies (e.g., Edwards et al. 2004, Sanchirico et al. 2008, Jin et al. 2016, Fay et al. 2019, Fulton et al. 2019) the benefits of an ecosystem-based approach truly have been demonstrated to have notably higher benefits—the

Table 1.6 The improvement of EBFM over "business-as-usual" (BAU) single-species or single sector approaches

- risk of overfishing is minimized;
- multiple populations of fishes, catches, and profits are more stable;
- overall value across all stocks is maximized;
- bureaucratic oversight and regulatory interventions are minimized;
- catch and yield are optimized;
- biomass of the resource (in aggregate) is maximized;
- stakeholder disenfranchisement and legal challenges are minimized;
- catch per unit effort is optimized; and,
- risk of ancillary ecosystem impacts is minimized.

Adapted from Rudd et al. (2018) and Link et al. (2018)

objective is for these benefits to be more broadly realized.

1.5 Clarifying the Obstacles to EBFM

EBFM has not yet been made fully operational, although the premise of this book is that there has been notable progress to that end. While currently being implemented within the U.S., EBFM actions generally have not been applied in a thorough manner at a national level or throughout U.S. regions. That is, EBFM is recognized as still needing to be more widely and fully implemented, even though it is now widely recognized as desirable (Link 2002, Pitcher et al. 2009, Patrick & Link 2015b, Marshak et al. 2017, Link 2018, Rudd et al. 2018). There is a growing broadscale consensus of what this implementation needs to be.

The call to execute EBFM is not novel. Nearly 150 years ago, Baird (1873) certainly noted factors and facets that resonate with themes noted in the modern vernacular of EBFM (Link 2010). Over time significant debate has occurred on the topic, usually centered around the relative importance of density-dependent vs. independent factors, typically sharpened as the relative importance of fishing vs. environmental or other exogenous (to the fish stock) factors that affect stock dynamics (Browman & Stergiou 2004, 2005). Events in the 1940s such as the Thompson-Burkenroad debate (c.f., Skud 1975, Smith 1994), similar discourse again in the 1950s–60s ecological literature (Andrewartha & Birch 1954, Hairston et al. 1960), and continuing to this day (Walters & Collie 1988, Rose 2000, Punt et al. 2014, Szuwalski et al. 2015) all represent snapshots of the overarching debate regarding the prominence of internal vs. external factors that shape the dynamics of fish stocks. In our view, the debate has become somewhat artificial and has remained singularly philosophical rather than having pragmatic solutions—of course, both density-dependent and independent factors influence fish stocks; of course, both fishing and the environment influence fish stocks. And, of course, the relative importance of each varies under any given set of conditions. The argument against a broader inclusion of environmental, socioeconomic, or density-independent elements really centers on:

- the analytical level of emphasis (single stock, multispecies, aggregate group, entire fisheries, or entire fishing and ecological system);
- whether including other factors in a model makes a significant difference in statistical performance or outcomes of models and management advice based thereon (Burgess et al. 2017);
- whether we will ever understand the ocean enough to model functional forms of all salient processes;
- whether governance institutions are structured to address trade-offs across stocks; or,
- concerns over limited data.

The issues of data and governance structure remain a legitimate concern, although perhaps less so than is typically assumed (Patrick & Link 2015b). We would argue that in some instances these broader EBFM considerations are not needed to manage some fisheries, albeit trade-offs in objectives among and across fisheries would still remain. We would also argue that the other objections to broader inclusion of ecosystem considerations are also valid in certain situations, but there are solutions for those concerns. We also note that the discipline has been rooted in SSFM and branched into EAFM. Some of this is due to disciplinary "siloes" and the simple challenges of speaking across departments, jargons, and perspectives. But de-emphasizing the stock emphasis and moving to EBFM of the *entire, composite system* of fisheries has remained either a conceptual or philosophical challenge for some in the practice of LMR management. We find this particularly intriguing as many of the earliest fisheries scientists and stock assessors, individuals who established much of the basis for our practice of fisheries today (e.g., Hjort 1914, Schaefer 1956, Beverton & Holt 1957, Gulland 1974, c.f. Andersen & Ursin 1977, c.f. Gulland 1983, Cushing 1988; Daan & Sissenwine 1991, Smith 1994, S. Holt, pers. comm., T. Smith, pers. comm.), all recognized the importance and the role of the ecosystem. They simply had to make simplifying assumptions of going to single species as they didn't have the knowledge available or computational capacity to deal with more than a single stock at the time. After discussing this with some of them, it is in effect the equivalent of getting stuck in what was supposed to be a temporary measure to keep things tractable. They always assumed we would return to broader considerations as more information and computing power became available.

Thus, skepticism still remains about fishery managers', scientists', and policymakers' ability to operationalize EBFM. There are many articles that discuss the daunting challenges of operationalizing EBFM (e.g., Browman & Stergiou 2004, Curtin & Prellezo 2010), and how they may differ between developed and developing countries (Pitcher et al. 2009, Tallis et al. 2010). In general, these works point to impediments such as: defining, prioritizing, and monitoring long-term ecosystem related goals and objectives (e.g., Cury et al. 2005, Ruckelshaus et al. 2008, Jennings & Rice 2011); issues with linguistic uncertainty and understanding the levels of ecosystem management as noted above (Arkema et al. 2006, Link & Browman 2014, Dolan et al. 2015, Marshak et al. 2017); developing appropriate data collection, analytical tools, and models (e.g., Hilborn 2011, Cowan et al. 2012, Walther & Möllmann 2014); and, the need for (potentially) drastically different governance structures to deal with the uncertainty and complexities of EBFM, as well as long-term planning (e.g., Leslie et al. 2008, Jennings & Rice 2011, Berkes 2012). Recent reports on developing Fishery Ecosystem Plans as a means to implement EBFM (Essington et al. 2016, Levin et al. 2018) note the repeated challenges of complexity, uncertainty, and unclear objectives, or really the perceptions thereof, along with clear solutions to overcome these challenges.

These and other issues have been around for over 20 to 30 years, many of which, if not all, have essentially been resolved in the U.S. and other developed countries (Pikitch et al. 2004, Murawski 2007, Curtin & Prellezo 2010, Cowan et al. 2012, Essington et al. 2016). There are, of course, some issues that just won't go away, such that these "myths" of EBFM impediments live on, still pervading the minds of the public, interest groups, fishery managers, scientists, and fishery policymakers who play a role in implementing EBFM (Patrick and Link 2015b). Until these myths are refuted, the operationalization of EBFM will continue to be hindered in developed countries. Previously, Murawski (2007) addressed 10 myths that "counter-revolutionists" use to circumvent or disrupt the implementation of EBM, and we find that approach an interesting tactic. In a subsequent work, Patrick & Link (2015b) adopted a similar approach to refute the myths that continue to impede the implementation of EBFM. Unlike Murawski (2007), however, who thought that these myths are primarily used to maintain status quo, Patrick & Link (2015b) rather thought they are (or have become) misconceptions about what is needed to operationalize EBFM (i.e., make it functional). Here we reiterate each of these common myths, address why they are indeed factually inaccurate today, and suggest ways to move forward. Adapted from Patrick & Link (2015b), these myths are:

- Myth 1: Marine EBM lacks universal terminology, making it difficult to implement.
- Myth 2: Ecosystem-based fisheries management lacks the governance structure and mandates to implement EBFM (i.e., there is no clear mandate for EBFM).
- Myth 3: EBFM can only be implemented in regions where we have copious data, and the corollary, doing EBFM requires models that are too complicated.
- Myth 4: EBFM always results in too conservative and restrictive advice.
- Myth 5: EBFM is a naïve pie-in-the-sky panacea for an already complex socioeconomic system.
- Myth 6: We don't have enough resources to do EBFM (i.e., capacity, capability, data, information, personnel, etc.).

Patrick & Link (2015b) go on to dispel these myths, at one point even having an official NOAA Fisheries webpage to address them. These myths have discouraged some managers from even trying EBFM and have prevented them from getting the best available information needed for resource management. Instead of viewing EBFM as a complex management process that requires an overabundance of information, it should be viewed as a framework to help managers work with the information they have and address competing objectives. As well as provide the formalized framework to re-assess a situation as more information becomes available.

Another myth that continues is the idea that "if we just did single-species management well that would be enough," implying that any other ecosystem issues would not exist (e.g., Mace 2001, 2004, Hilborn 2011; c.f. Browman & Stergiou 2004). Even if stocks are well managed, several other challenges to marine fisheries persist (e.g., Botsford et al. 1997, Micheli 1999, Jackson et al. 2001, Pauly et al. 2002, 2003, Pikitch et al. 2004, Link 2010, Fogarty 2014). And more to the point, by definition trade-offs across taxa, fleets, fisheries, and other aspects of the "multi-multi" situation we find in marine fisheries ecosystems cannot be addressed by focusing on only one stock at-a-time.

The sum of addressing these myths and many of the barriers to EBFM can be stated as this: One does not need perfect knowledge of every process to implement EBFM, but it would be good to think about more than the usual processes we typically think about. Yet given that succinct and workable statement, implementation obstacles still remain.

Why do the barriers to adopting a holistic, systems approach—EBFM—remain? We submit that there are at least four reasons.

First is simply unfamiliarity with the broad concepts and approaches of systems thinking as applied to the discipline of fisheries. That can be rectified by familiarity of works that continue to elaborate on systems-related approaches (Jackson 2003, Allen 2009, Mele et al. 2010, Wu 2013) and specifically reinforcing works specific to fisheries (Walters 1971, Walters & Hilborn 1976, Walters 1980, Charles 2001, Apollonio 2002, 2010, Link 2018). We would also encourage University fisheries departments, schools, programs, and other training fora to present this ecosystem approach to better broaden awareness of its existence and the probable benefits from its application.

Second is simply historical inertia of this (or any) discipline. Largely the theory and practice of fisheries have been developed on a stock-oriented basis, emphasizing component parts and a reductionist perspective. There is a long history of why and how this came about (Smith 1994, Finley 2011). In many ways this has been embodied in the either-or debates noted above (Andrewartha & Birch 1954, Hairston et al. 1960, Skud 1975, Walters & Collie 1988, Smith 1994, Rose 2000, Punt et al. 2014, Szuwalski et al. 2015). But as just noted, single-species approaches were used because that was all that could be done at the time; progress in the amount of knowledge, efficiency of data collection, and computing power now enables much more complete and systematic approaches of EBFM. Reasonable minds are recognizing that we need to maintain the stock-oriented perspective in many instances, but that the scale and scope of challenges facing the discipline of fisheries is forcing an evaluation of methods and means toward a broader consideration of issues. Evaluating, re-evaluating, and tweaking the underlying philosophy (and associated worldview) of how one understands phenomenon, and even approaches such understanding, is certainly difficult, but is also the hallmark of scientific progress.

The third reason is that there remains skepticism to new or innovative approaches that is beyond just resistance to change, but healthy skepticism that is typical and somewhat required in a scientific context. Until the theory, clear predictability, mathematical representations, applied outcomes, and benefits of a new or different approach such as EBFM are demonstrated, such an approach will nearly always be slow to be adopted. As a corollary, most management or governance systems tend to oppose change until they understand and become familiar with the background and behavior of the proposed changes, as well as comfortable with any new benefits (e.g., increased certainty, effectiveness, and efficiency). For instance, directing management action at one level in the EBM hierarchy above the level where effects are desired is an elementary feature (and recommendation) of hierarchy theory, but may seem counterintuitive at first for those trained to think and operate at a population level; as such, this type of thinking would need familiarization and socialization before being enacted. Related to this, at least in a management context, is the perception that EBFM means more work. Given the typical daily demands and full workloads, the thinking is it is difficult to take time to think systematically, with limited time or energy to list, let alone prioritize, a broader set of objectives (Essington et al. 2016). The reality is that most management bodies do in fact engage in these visioning, strategic planning, and systematic scoping exercises, and increasingly they are including ecosystem considerations.

The fourth reason is one that relates to the first—paralyzed inaction that stems from the recognition of overwhelming complexity in these systems. That thinking decries the limited data, difficulty in understanding, and near infeasibility of predictions associated with marine ecosystems and their fisheries. We and several others before us (von Bertalanffy 1968, Apollonnio 2002, 2010, c.f. Luhmann 2013, Capra & Luisi 2014; see references in Link 2018) clearly recognize the essentially impossible task of understanding these coupled social-ecological marine ecosystems in a mechanistic, reductionist matter for each and every component. Yet this thinking entirely misses the utility of adopting a systems approach, namely in that it frees one from having to have such detailed data and understanding for each and every component of the system. Rather, from hierarchy theory, we don't have to have that level of understanding, and in fact may actually glean further understanding of the system by adopting a systems approach that elucidates emergent properties. We are under no illusion that these barriers to EBFM will resolve any time soon. Yet we are confident that at least some readers will give EBFM thinking due consideration for their work and application.

1.6 Moving to EBFM

Objections to EBFM are important, yet they can be readily addressed. That is, any objections to EBFM are not insurmountable. And that the challenges to implement ecosystem-level approaches to management can be overcome. The benefits, need, and value of EBFM have also been recognized as desirable. Ultimately, it has been recognized that BAU in fisheries science and management needs to change. So, what has happened in the past decade? Here we briefly note some of the highlights that have occurred in fisheries management to better facilitate EBFM.

One of the key leads for LMR management in the world, especially the U.S., is NOAA and its National Marine Fisheries Service (NMFS, aka NOAA Fisheries). NOAA Fisheries has invested considerable time and effort in the later 2000s and 2010s to better implement EBFM (Table 1.7). These investments have included things ranging from new positions (personnel) to changes in policies to rearranged resources or programs to clear strategic guidance documents (Table 1.7; Merrick 2018). The NOAA Fisheries Strategic Plan 2019–2022 clearly states that to accomplish its mission will " . . . all be backed by sound science and an ecosystem-based approach to management" so the commitment to EBFM is now routinely codified. A parallel and complementary effort was occurring in the U.S. via the Lenfest Ocean Program's Fishery Ecosystem Plan (FEP) task force (Essington et al. 2016, Levin et al. 2018). Although our work here is focused on the U.S., we recognize the lessons learned (which we attempt to provide throughout the rest of this book) can have global applications. Certainly there have been similar efforts and investments around the world (c.f. the International Council for the Exploration of the Sea (ICES) largely in Europe and North Atlantic waters, the North Pacific Marine Science Organization (PICES), the Integrated Marine Biosphere Research (IMBER) project, the Commonwealth Scientific and Industrial Research Organisation (CSIRO) in Australia, the Department of Fisheries and Oceans (DFO) in Canada, the Norwegian Institute for Marine Research (IMR), the United Nations FAO, the United Nations Environmental Programme (UNEP), the Global Environment Facility (GEF), the United States Agency for International Development (USAID), etc.), and we acknowledge that collectively those represent copious, comparable efforts to generally advance EBFM. In several of those jurisdictions, there have been various EBM- and EBFM-related polices, laws, and efforts. Examples include:

- Australia's Oceans Policy (1998) and Environment Protection and Biodiversity Conservation Act (1999) are both based on EBM principles, resulting in a clear need for Ecological Risk Assessments for the Effects of Fishing (e.g., Hobday et al. 2011).
- Also in Australia, the Intergovernmental Agreement on the Environment (1990)—is an Australian subnational agreement between states to implement a coordinated Ecologically Sustainable Development (ESD) approach.
- Canada's Oceans Act (1996) and Sustainable Development Act (2008) both emphasize EBM-related principles.
- In the European Union, there is a need to codify and move toward Good Environmental Status in the Marine Strategy Framework Directive (European Parliament and Council of the European Union 2008), which has resulted in numerous ICES working groups to address many of the facets of EBM and EBFM (ICES 2017).
- In Norway, the Marine Resources Act (2009) covers all wild LMRs and mandates an EAFM.
- Egypt is exploring fisheries management in waters of the Red Sea by attempting EBFM, with a particular

Table 1.7 NOAA NMFS investments to better implement EBFM

- Establishment of and continued resourcing of the Integrated Ecosystem Assessment (IEA; Levin et al. 2009) program
- Several other cross-NOAA programs
- Joint programs across multiple federal agencies (and re-emphasis of funds)
- A new Senior Advisor, national-level position to champion the issue (Merrick 2018)
- National NOAA Fisheries programmatic reviews that had an entire year dedicated to ecosystem science—2016
- The NOAA Fisheries EBFM Policy Statement (NMFS 2016a)
- The NOAA Fisheries EBFM Road Map (NMFS 2016b)
- The NMFS Climate Science Strategy (Busch et al. 2016)
- The Updated Stock Assessment Improvement Plan, which had copious ecosystem linkages in it (Lynch et al. 2018)
- The Habitat Assessment Improvement Plan (NMFS 2010) and update (Peters et al. 2018)
- Clarification that EBFM is permissible under extant statutes (Link et al. 2018, NMFS 2016a)
- A new National EBFM Coordinator (and standing EBFM Committee)—2017
- New staff, FTEs, and rearranged programs dedicated to ecosystem analyses—2016–present
- Regional climate action plans and EBFM implementation plans as part of national and regional strategic planning processes—2017–2019
- EBFM terminology becoming a key, clear pillar in NMFS mission statements (e.g., strategic plans, guidance memoranda, etc.)—2017–present

emphasis on adopting indicator reference points (Alsolami et al. 2020).

- Korea is also adopting aspects of ecosystem approaches to management (EAM) for its fisheries, particularly by using an Integrated Fisheries Risk Analysis Method for Ecosystems (Zhang et al. 2009, 2011), with some follow-up work for other Pacific countries in the PICES context.

Here we largely focus on U.S. examples, but are fully cognizant that many of our counterparts in other locales are equally pushing forward on EBFM implementation. The critical element is that in many places we can now begin to point to specific investments, policies, and efforts to make EBFM more fully operational.

Thus, we wanted to pause, take the proverbial step back, and evaluate the progress that the U.S. has made toward marine EBFM. We did so in two ways. First, and comprising the bulk of this book, will be by evaluating the major U.S. regions and territories against common, standard criteria (Link & Marshak 2019). Others have done so by benchmarking against EPAP criteria (Wilkinson & Abrams 2015) or FAO ecosystem approach to fisheries (EAF) principles (Patrick & Link 2015a). We had originally executed such an exercise in a synthetic, comparison mode (Link & Marshak 2019), but here we expand those data from that original work to explore each region and its notable subregions in more detail. Herein we also benchmark against the EBFM Road Map action items (NMFS 2016b) and provide an update relative to the EPAP criteria. Second, is by attempting to both synthesize, relativize, and contrast all the regions of the U.S. in one chapter, and then via a chapter that will provide a high-level overview of progress made on the more detailed, technical steps in the LMR science-to-management advice process, placing this work in the context of "so what?" and what is next. We do so to ascertain not only how well each region is doing with this range of possible voluntary options for implementing EBFM, but also to identify what may or may not be appropriate for each region along with what is a common utility across them all. We also do so to get a sense of the trajectory of where EBFM is headed in the U.S.

We previously noted that we are necessarily multidisciplinary in our thinking (Fig. 1.1; Link & Marshak 2019). Our main rubric for thinking about how to implement EBFM and improve LMR management is driven by this proposed pathway:

$$PP \rightarrow B_{\text{targeted, protected spp, ecosystem}} \leftrightarrow L_{\uparrow\text{targeted spp, }\downarrow\text{bycatch}} \rightarrow \text{jobs, economic revenue.}$$

where PP is primary production, B is biomass of either targeted or protected species or an entire ecosystem, L is landings of targeted or bycaught species, all leading to the other socioeconomic factors. This operates in the context of an ecological and human system, with governance feedbacks at several of the steps (i.e., between biomass and landings, jobs, and economic revenue), implying that fundamental ecosystem features can determine the socioeconomic value of a set of fisheries in a region, as modulated by human interventions. We think this rubric is helpful in that it suggests common and standard criteria. It is also, in many ways, the path up the pyramid of the six EBFM guiding principles (Fig. 1.2). Compiling information to explore this proposed pathway to delineate the determinants of successful LMR management not only facilitates comparison across regions but should elicit common, emergent features contributing to LMR management success. Doing so will also demonstrate how much progress we have made in the implementation of EBFM. We also think this is a useful rubric as it likely reflects the reality of true limits to fisheries production (Friedland et al. 2012, Stock et al. 2017, Link & Watson 2019).

If we are being honest, even though we know that it takes a while for change, we had been viewing EBFM implementation as a threshold to be surpassed. Or as a major phase shift that needed to occur, that would somehow happen and then we could say that we were (finally) doing EBFM. Instead, we now think that it is more accurate and appropriate to note that the discipline has been making (slow) steady progress toward EBFM. After compiling this tome, we better recognize that implementation of EBFM is not a single large action but rather a series of ongoing and cumulative actions leading to comprehensive (and improved) management of LMRs. Certainly, there are challenges and difficulties to overcome, and many of the obstacles remain. But we do think that there has been demonstrable progress toward EBFM, as measured objectively by the many indicators we will present. We trust that the information shown herein will demonstrate that EBFM is well underway. Our hope is that this book helps to capture this progress for US LMR management as a whole, and spurs on additional progress toward EBFM.

1.7 References

Allen TFH. 2009. "Hierarchy theory in ecology." In: *Ecosystem Ecology—A Derivative of Encyclopedia of Ecology*. Edited by SE Jørgensen. Amsterdam: Elsevier (pp. 114–20).

Alsolami LS, Abdelaty M, Zhang CI. 2020. An ecosystem-based fisheries assessment approach and management system for the Red Sea. *Fisheries Research* 227:105551.

Andersen KH, Brander K, Ravn-Jonsen L. 2015. Trade-offs between objectives for ecosystem management of fisheries. *Ecological Applications* 25:1390–6.

Andersen KP, Ursin E. 1977. A multispecies extension to the Beverton Holt theory of fishing, with accounts of phosphorous circulation and primary production. *Medd Dan Fisk Havunders* 7:319–45.

Andrewartha HG, Birch LC. 1954. *The Distribution and Abundance of Animals*. Chicago, IL: University of Chicago Press.

Apollonio S. 2002. Hierarchical Perspectives on Marine Complexities: Searching for Systems in the Gulf of Maine. New York, NY: Columbia University Press.

Apollonio S. 2010. A testable hypothesis for ecosystem-based fisheries management. *Reviews in Fisheries Science* 18(2):183–8.

AORA (Atlantic Ocean Research Alliance). 2019. Working Group on the Ecosystem Approach to Ocean Health and Stressors (p. 42).

AORA. 2017. Working Group on the Ecosystem Approach to Ocean Health and Stressors (p. 58).

Arctic Council. 2013. *Ecosystem-Based Management in the Arctic*. Report submitted to Senior Arctic Officials by the Expert Group on Ecosystem-Based Management, as adopted May 2013. Hein Rune Skjoldal and Phil Mundy, Co-Chairs; PAME Ecosystem Approach Expert Group (p. 63).

Arkema KK, Abramson SC, Dewsbury BM. 2006. Marine ecosystem-based management: from characterization to implementation. *Frontiers in Ecology and the Environment* 4:525–32.

Austin HM. 1983. *Ecosystem Management, an Alternative to Fisheries Management*. Marine Resource Report No. 83–10. Virginia Institute of Marine Science, College of William and Mary.

Baird SF. 1873. *Report on the Condition of the Sea Fisheries of the South Coast of New England in 1871 and 1872*. Washington, DC: Government Printing Office.

Ballesteros M, Chapela R, Ramírez-Monsalve P, Raakjaer J, Hegland TJ, Nielsen KN, Laksá U, Degnbol P. 2018. Do not shoot the messenger: ICES advice for an ecosystem approach to fisheries management in the European Union. *ICES Journal of Marine Science* 75:519–30.

Barange M, Merino G, Blanchard JL, Scholtens J, Harle J, Allison EH, Allen JI, Holt J, Jennings S. 2014. Impacts of climate change on marine ecosystem production in societies dependent on fisheries. *Nature Climate Change* 4:211–16.

Barnes C, Bozzi L, McFadden K. 2005. *Exploring an Ecosystem Approach to Management: A Review of the Pertinent Literature*. Silver Spring, MD: NOAA (National Oceanic and Atmospheric Administration) Ecosystem Goal Team (p. 19).

Beddington JR, Agnew DJ, Clark CW. 2007. Current problems in the management of marine fisheries. *Science* 316:1713–16.

Berghöfer A, Wittmer H, Rauschmayer F. 2008. Stakeholder participation in ecosystem-based approaches to fisheries management: A synthesis from European research projects. *Marine Policy, Interaction Between Environment and Fisheries* 32:243–53.

Belsky MH. 1985. Management of large marine ecosystems: developing a new rule of customary international law. *San Diego Law Review* 22:733–64.

Berkes F. 2012. Implementing ecosystem-based management: evolution or revolution? *Fish and Fisheries* 13:465–76.

Beverton RJH, Holt SJ. 1957. *On the Dynamics of Exploited Fish Populations. Fisheries Investment Series 2, Vol 19*. London, UK: UK Ministry of Agriculture and Fisheries.

Botsford LW, Castilla JC, Peterson CH. 1997. The management of fisheries and marine ecosystems. *Science* 277: 509–15.

Brodziak J, Link J. 2002. Ecosystem-based fishery management: what is it and how can we do it? *Bulletin of Marine Science* 70:589–611.

Browman H, Stergiou K. 2005. Politics and socio-economics of ecosystem-based management of marine resources. *Marine Ecology Progress Series* 300:241–96.

Browman H, Stergiou K. 2004. Perspectives on ecosystem-based approaches to the management of marine resources. *Marine Ecology Progress Series* 274:269–303.

Burgess MG, Giacomini HC, Szuwalski CS, Costello C, Gaines SD. 2017. Describing ecosystem contexts with single-species models: a theoretical synthesis for fisheries. *Fish and Fisheries* 18:264–84.

Busch DS, Griffis R, Link J, Abrams K, Baker J, Brainard RE, Ford M, Hare JA, Himes-Cornell A, Hollowed A. 2016. Climate science strategy of the US national marine fisheries service. *Marine Policy* 74:58–67.

Busch WDN, Brown BL, Mayer GF. 2003. *Strategic Guidance for Implementing an Ecosystem-Based Approach to Fisheries Management*. Silver Spring, MD: United States Department of Commerce, National Oceanic and Atmospheric Administration, National Marine Fisheries Service (p. 62).

Butterworth DS. 1986. Antarctic marine ecosystem management. *Polar Record* 23(142):37–47.

Cadrin SX, Kerr LA, Mariani S. 2014. *Stock Identification Methods: Applications in Fishery Science*. Cambridge, MA: Academic Press.

Capra F, Luisi PL. 2014. *The Systems View of Life: A Unifying Vision*. Cambridge. UK: Cambridge University Press.

Charles A. 2001. *Sustainable Fishery Systems*. Oxford, UK: Wiley-Blackwell.

Christensen NL, Bartuska AM, Brown JH, Carpenter S, D'Antonio C, Francis R, Franklin JF, et al. 1996. The report of the Ecological Society of America Committee on the scientific basis for ecosystem management. *Ecological Applications* 6:665–91.

Christensen V, Walters CJ. 2004. Ecopath with Ecosim: methods, capabilities and limitations. *Ecological Modelling Placing Fisheries in their Ecosystem Context* 172:109–39.

Clarke P, Jupiter S. 2010. *Principles and Practice of Ecosystem-Based Management: A Guide for Conservation Practitioners in the Tropical Western Pacific*. Suva, Fiji: Wildlife Conservation Society (p. 80).

Coll M, Steenbeek J, Sole J, Palomera I, Christensen V. 2016. Modelling the cumulative spatial–temporal effects of environmental drivers and fishing in a NW Mediterranean marine ecosystem. *Ecological Modelling Ecopath 30 years—Modelling Ecosystem Dynamics: Beyond Boundaries with EwE* 331:100–14.

CCAMLR. 2001. *Commission for the Conservation of Antarctic Marine Living Resources*. Hobart, Tasmania: Management of the Antarctic (p. 20).

Cowan Jr JH, Rice JC, Walters CJ, Hilborn R, Essington TE, Day Jr JW, Boswell KM. 2012. Challenges for implementing an ecosystem approach to fisheries management. *Marine and Coastal Fisheries* 4:496–510.

Curtin R, Prellezo R. 2010. Understanding marine ecosystem based management: a literature review. *Marine Policy* 34:821–30.

Cury PM, Mullon C, Garcia SM, Shannon LJ. 2005. Viability theory for an ecosystem approach to fisheries. *ICES Journal of Marine Science* 62:577–84.

Cushing DH. 1988. *The Provident Sea*. Cambridge, UK: Cambridge University Press.

Daan N, Sissenwine MP. 1991. Multispecies models relevant to management of living resources. *ICES Marine Science Symposium* 193:6–11.

De Young C, Charles A, Hjort A. 2008. *Human Dimensions of the Ecosystem Approach to Fisheries: An Overview of Context, Concepts, Tools and Methods*. Fisheries Technical Paper No. 489. Rome, Italy: Food and Agriculture Organization of the United Nations (p. 152).

Dickey-Collas M. 2014. Why the complex nature of integrated ecosystem assessments requires a flexible and adaptive approach. *ICES Journal of Marine Science* 71:1174–82.

Dolan TE, Patrick WS, Link JS. 2015. Delineating the continuum of marine ecosystem-based management: a US fisheries reference point perspective. *ICES Journal of Marine Science* 73:1042–50.

Duda AM, Sherman K. 2002. A new imperative for improving management of large marine ecosystems. *Ocean and Coastal Management* 45:797–833.

Eberhardt LL. 1977. "Optimal" management policies for marine mammals. *Wildlife Society Bulletin* 5(4):162–69.

Edwards SF, Link JS, Rountree BP. 2004. Portfolio management of wild fish stocks. *Ecological Economics* 49:317–29.

Essington TE, Levin PS, Marshall KN, Koehn L, Anderson LG, Bundy A, Carothers C, et al. 2016. *Building Effective Fishery Ecosystem Plans: A Report from the Lenfest Fishery Ecosystem Task Force*. Washington, DC: Lenfest Ocean Program (p. 60).

EPAP (Ecosystem Principles Advisory Panel). 1999. *Ecosystem-based Fishery Management: A Report to Congress by the Ecosystem Principles Advisory Panel*. Washington, DC: The Service.

Fay G, DePiper GS, Steinback S, Gamble R, Link JS. 2019. Economic and ecosystem effects of fishing on the Northeast US shelf. *Frontiers in Marine Science* 6:133.

FAO. 2015. EAF-Nansen Project/FAO. Minutes of the Joint Meeting of the EAF-Nansen Project Regional Steering Committees, Casablanca, Morocco, October 24, 2014/ LFAO EAF-Nansen Project Report/ N°26. Rome, Italy: FAO (p. 21).

Fletcher WJ, Shaw J, Metcalf SJ, Gaughan DJ. 2010. An ecosystem based fisheries management framework: the efficient, regional-level planning tool for management agencies. *Marine Policy* 34:1226–38.

Fogarty MJ. 2014. The art of ecosystem-based fishery management. *Canadian Journal of Fisheries and Aquatic Sciences* 71:479–90.

Folke C, Biggs R, Norström AV, Reyers B, Rockström J. 2016. Social-ecological resilience and biosphere-based sustainability science. *Ecology and Society* 21(3):Art 41.

Folke C, Hahn T, Olsson P, Norberg J. 2005. Adaptive governance of social-ecological systems. *Annual Reviews of Environment and Resources* 30:441–73.

FAO (Food and Agriculture Organization). 2016. *State of the World's Fisheries and Aquaculture*. Rome, Italy: Food and Agricultural Organization of the United Nations.

FAO. 2003. Fisheries Management. *2: The Ecosystem Approach to Fisheries, FAO Technical Guidelines for Responsible Fisheries*. Rome, Italy: Food and Agriculture Organization of the United Nations (p. 112).

Finley C. 2011. *All the Fish in the Sea: Maximum Sustainable Yield and the Failure of Fisheries Management*. Chicago, IL: University of Chicago Press.

Fluharty D, Abbott M, Davis R, Donahue M, Madsen S, Quinn T, Rice J, Sutinen J. 2006. *The External Review of NOAA's Ecosystem Research and Science Enterprise—A Report to the NOAA Science Advisory Board. Evolving an Ecosystem Approach to Science and Management Throughout NOAA and its Partners*. Washington, DC: NOAA (p. 85).

Francis RC, Hixon MA, Clarke ME, Murawski SA, Ralston S. 2007. Ten commandments for ecosystem-based fisheries scientists. *Fisheries* 32:217–33.

Friedland KD, Stock C, Drinkwater KF, Link JS, Leaf RT, Shank BV, Rose JM, Pilskaln CH, Fogarty MJ. 2012. Pathways between primary production and fisheries yields of large marine ecosystems. *PLoS One* 7:e28945.

Fulton EA, Punt AE, Dichmont CM, Harvey CJ, Gorton R. 2019 Ecosystems say good management pays off. *Fish and Fisheries* 20:66–96.

Fulton EA, Smith ADM, Smith DC, Johnson P. 2014 An integrated approach is needed for ecosystem based fisheries management: insights from ecosystem-level management strategy evaluation. *PLoS One* 9:e84242.

Garcia SM. 2000. The FAO definition of sustainable development and the Code of Conduct for Responsible Fisheries: an analysis of the related principles, criteria and indicators. *Marine and Freshwater Research* 51:535.

Garcia SM. 2003. *The Ecosystem Approach to Fisheries: Issues, Terminology, Principles, Institutional Foundations, Implementation and Outlook*. Rome, Italy: Food & Agriculture Org.

Garcia SM, Cochrane KL. 2005. Ecosystem approach to fisheries: a review of implementation guidelines. *ICES Journal of Marine Science* 62:311–18.

Garcia SM, Zerbi A, Aliaume C, Do Chi T, Lasserre G. 2003. The Ecosystem Approach to Fisheries: Issues, Terminology, Principles, Institutional Foundations, Implementation and Outlook. FAO Fisheries Technical Paper 443. Rome, Italy: FAO (p. 71).

Griffis RB, Kimball KW. 1996. Ecosystem approaches to coastal and ocean stewardship. *Ecological Applications* 6:708–12.

Grumbine RE. 1994. What is ecosystem management? *Society for Conservation Biology* 8:27–38.

Guerry AD. 2005. Icarus and Daedalus: conceptual and tactical lessons for marine ecosystem-based management. *Frontiers in Ecology and the Environment* 3:202–11.

Gulland JA. 1974. *The Management of Marine Fisheries*. Seattle, WA: University of Washington Press.

Gulland JA. 1983. *Fish Stock Assessment: A Manual of Basic Methods*. New York, NY: Wiley.

Gutiérrez NL, Hilborn R, Defeo O. 2011. Leadership, social capital and incentives promote successful fisheries. *Nature* 470:386–9.

Hairston NG, Smith FE, Slobodkin LB. 1960. Community structure, population control, and competition. *American Naturalist* 94:421–5.

Hall SJ, Mainprize B. 2004. Towards ecosystem-based fisheries management. *Fish and Fisheries* 5(1):1–20.

Halpern BS, Walbridge S, Selkoe KA, Kappel CV, Micheli F, D'Agrosa C, Bruno JF, et al. 2008. A global map of human impact on marine ecosystems. *Science* 319:948–52.

Harvey CJ, Kelble CR, Schwing FB. 2017. Implementing "the IEA": using integrated ecosystem assessment frameworks, programs, and applications in support of operationalizing ecosystem-based management. *ICES Journal of Marine Science* 74:398–405.

Heltzel JM, Witherellm D, Wilson WJ. 2011. Ecosystem-based management for protected species in the North Pacific fisheries. *Marine Fisheries Review* 73:20–35.

Hilborn R. 2011. Future directions in ecosystem-based fisheries management: a personal perspective. *Fisheries Research* 108:235–9.

Hilborn R, Amoroso RO, Anderson CM, Baum JK, Branch TA, Costello C, De Moor CL, et al. 2020. Effective fisheries management instrumental in improving fish stock status. *Proceedings of the National Academy of Sciences* 117(4):2218–24.

Hilborn R, Fulton EA, Green BS, Hartmann K, Tracey SR, Watson RA. 2015. When is a fishery sustainable? *Canadian Journal of Fisheries and Aquatic Sciences* 72:1433–41.

Hilborn R, Mangel M. 1997. *The Ecological Detective: Confronting Models with Data*. Princeton, NJ: Princeton University Press.

Hjort J. 1914. Fluctuations in the great fisheries of Northern Europe, viewed in the light of biological research. *Rapports et Procès-Verbaux des Réunions du Conseil Permanent International pour l'Exploration de la Mer* 20:1–228.

Hobday AJ, Smith ADM, Stobutzki IC, et al. 2011. Ecological risk assessment for the effects of fishing. *Fisheries Research* 108:372–84.

Holliday MC, Gautam AB. 2005. *Developing Regional Marine Ecosystem Approaches to Management. U.S. Dept Commer NOAA Tech Memo NMFS-F/SPO-77*. Washington, DC: NOAA (p. 38).

ICES (International Council for the Exploration of the Sea). 2017. *Explaining ICES approach to ecosystem based management*. https://www.ices.dk/news-and-events/news-archive/news/Pages/Explaining-ICES-approach-to-ecosystem-based-management.aspx

Jackson JBC, Kirby MX, Berger WH, Bjorndal KA, Botsford LW, Bourque BJ, Bradbury RH, et al. 2001. Historical overfishing and the recent collapse of coastal ecosystems. *Science* 293:629–37.

Jackson M. 2003. *Systems Thinking: Creative Holism for Managers*. Chichester, UK: John Wiley & Sons Ltd.

Jennings S, Kaiser MJ. 1998. "The Effects of Fishing on Marine Ecosystems." In: *Advances in Marine Biology*. Edited by JHS Blaxter, AJ Southward, PA Tyler. Cambridge, MA: Academic Press (pp. 201–352).

Jennings S, Rice J. 2011. Towards an ecosystem approach to fisheries in Europe: a perspective on existing progress and future directions. *Fish and Fisheries* 12, 125–37.

Jin D, DePiper G, Hoagland P. 2016. Applying portfolio management to implement ecosystem-based fishery management (EBFM). *North American Journal of Fisheries Management* 36:652–69.

Keyl F, Wolff M. 2008. Environmental variability and fisheries: what can models do? *Reviews in Fish Biology and Fisheries* 18:273–99.

Lackey RT. 1998. Seven pillars of ecosystem management. Modified from a presentation given at the Symposium Ecosystem Health and Medicine: Integrating Science, Policy, and Management, Ottawa, Ontario, Canada, June 19–23, 1994. *Landscape and Urban Planning* 40:21–30.

Langhans SD, Jahnig SC, Lago M, Schmidt-Kloiber A, Hein T. 2019. The potential of ecosystem-based management to integrate biodiversity conservation and ecosystem service provision in aquatic ecosystems. *Science of the Total Environment* 672:1017–20.

Larkin PA. 1996. Concepts and issues in marine ecosystem management. *Reviews in Fish Biology and Fisheries* 6:139–64

Leslie H, Rosenberg AA, Eagle J. 2008. Is a new mandate needed for marine ecosystem-based management? *Frontiers in Ecology and the Environment* 6:43–8.

Leslie HM, McLeod KL. 2007. Confronting the challenges of implementing marine ecosystem-based management. *Frontiers in Ecology and the Environment* 5:540–8.

Levin PS, Fogarty MJ, Murawski SA, Fluharty D. 2009. Integrated ecosystem assessments: developing the scientific basis for ecosystem-based management of the ocean. *PLoS Biology* 7(1):e1000014.

Levin PS, Essington TE, Marshall KN, Koehn LE, Anderson LG, Bundy A, Carothers C, et al. 2018. Building effective fishery ecosystem plans. *Marine Policy* 92:48–57.

Levin SA, Lubchenco J. 2008. Resilience, robustness, and marine ecosystem-based management. *BioScience* 58:27–32.

Lidström S, Johnson AF. 2020. Ecosystem-based fisheries management: a perspective on the critique and development of the concept. *Fish and Fisheries* 21:216–22.

Link J. 2010. *Ecosystem-Based Fisheries Management: Confronting Tradeoffs*. Cambridge, UK: Cambridge University Press.

Link J. 2017. A conversation about NMFS' ecosystem-based fisheries management policy and road map. *Fisheries* 42:498–503.

Link JS. 2002. What does ecosystem-based fisheries management mean. *Fisheries* 27:18–21.

Link JS, Gaichas S, Miller TJ, Essington T, Bundy A, Boldt J, Drinkwater KF, Moksness E. 2012. Synthesizing lessons learned from comparing fisheries production in 13 northern hemisphere ecosystems: emergent fundamental features. *Marine Ecology Progress Series* 459:293–302.

Link JS, Browman HI. 2014. Integrating what? Levels of marine ecosystem-based assessment and management. *ICES Journal of Marine Science* 71:1170–3.

Link JS, Browman HI. 2017. Operationalizing and implementing ecosystem-based management. *ICES Journal of Marine Science* 74:379–81.

Link JS. 2018. System-level optimal yield: increased value, less risk, improved stability, and better fisheries. *Canadian Journal of Fisheries and Aquatic Sciences* 75:1–16.

Link JS, Dickey-Collas M, Rudd M, McLaughlin R, Macdonald NM, Thiele T, Ferretti J, Johannesen E, Rae M. 2018. Clarifying mandates for marine ecosystem-based management. *ICES Journal of Marine Science* 76(1):41–4.

Link JS, Marshak AR. 2019. Characterizing and comparing marine fisheries ecosystems in the United States: determinants of success in moving toward ecosystem-based fisheries management. *Reviews in Fish Biology and Fisheries* 29:23–70.

Link JS, Watson RA. 2019. Global ecosystem overfishing: clear delineation within real limits to production. *Science Advances* 5(6):eaav0474

Long RD, Charles A, Stephenson RL. 2015. Key principles of marine ecosystem-based management. *Marine Policy* 57:53–60.

Luhmann N. 2013. *Introduction to Systems Theory*. [Translated by P. Gilgen.] Malden, MA: Polity.

Lynch PD, Methot RD, Link JS. 2018. *Implementing a Next Generation Stock Assessment Enterprise. An Update to the NOAA Fisheries Stock Assessment Improvement Plan. U.S. Dep. Commer., NOAA Tech. Memo. NMFS-F/SPO-183*. Washington, DC: NOAA (p. 127).

Mace PM. 2001. A new role for MSY in single-species and ecosystem approaches to fisheries stock assessment and management. *Fish and Fisheries* 2:2–32.

Mace PM. 2004. In defense of fisheries scientists, single-species models and other scapegoats: confronting the real problems. *Marine Ecology Progress Series* 274:285–291.

Marasco RJ, Goodman D, Grimes CB, Lawson PW, Punt AE, Quinn II TJ. 2007. Ecosystem-based fisheries management: some practical suggestions. *Canadian Journal of Fisheries and Aquatic Sciences* 64(6):928–39.

Marshak AR, Link JS, Shuford R, Monaco ME, Johannesen E, Bianchi G, Anderson MR, et al. 2017. International perceptions of an integrated, multi-sectoral, ecosystem approach to management. *ICES Journal of Marine Science* 74:414–20.

Marshall KN, Levin PS, Essington TE, Koehn LE, Anderson LG, Bundy A, Carothers C, et al. 2018. Ecosystem-based fisheries management for social–ecological systems: renewing the focus in the United States with next generation fishery ecosystem plans. *Conservation Letters* 11(1):e12367.

McLeod K, Leslie H, Aburto M. 2014. *Ecosystem-Based Management for the Oceans*. Washington, DC: Island Press.

McLeod KL, Lubchenco SR, Palumbi SR, Rosenberg AA. 2005. *Scientific consensus statement on marine ecosystem-based management*. Communication Partnership for Science and the Sea (COMPASS). http://www.onlyoneplanet.com/marineEBM_ConsensusStatement.pdf

Mele C, Pels J, Polese F. 2010. A brief review of systems theories and their managerial applications. *Service Science* 2(1–2):126–35.

Merrick R. 2018. Mechanisms for science to shape US living marine resource conservation policy. *ICES Journal of Marine Science* 75:2319–24.

Metcalf SJ, Gaughan DJ, Shaw JL. 2009. *Conceptual Models for Ecosystem Based Fisheries Management (EBFM) in Western Australia*. Perth, WA: Fisheries Department of Western Australia.

Micheli F. 1999. Eutrophication, fisheries, and consumer-resource dynamics in marine pelagic ecosystems. *Science* 285:1396–8.

Micheli F, De Leo G, Butner C, Martone RG, Shester G. 2014. A risk-based framework for assessing the cumulative impact of multiple fisheries. *Biological Conservation* 176:224–35.

Moffitt EA, Punt AE, Holsman K, Aydin KY, Ianelli JN, Ortiz I. 2016. Moving towards ecosystem-based fisheries management: options for parameterizing multi-species biological reference points. *Deep Sea Research Part II Topical Studies Oceanography* 134:350–9.

Murawski SA, Matlock GC. 2006. *Ecosystem Science Capabilities Required to Support NOAA's Mission in the Year 2020. NOAA Tech Memo NMFS-F/SPO-74*. Washington, DC: NOAA (p. 100).

Murawski SA. 2007. Ten myths concerning ecosystem approaches to marine resource management. *Marine Policy* 31:681–90.

NMFS (National Marine Fisheries Service). 2009. *Report to Congress: The State of Science to Support an Ecosystem Approach to Regional Fishery Management. U.S. Dep. Commerce, NOAA Tech. Memo. NMFS-F/SPO-96*. Washington, DC: NOAA (p. 24).

NMFS. 2010. *Marine Fisheries Habitat Assessment Improvement Plan. Report of the National Marine Fisheries Service Habitat Assessment Improvement Plan Team. U.S. Dep. Commer., NOAA Tech. Memo. NMFS-F/SPO-108*. Washington, DC: NOAA (p. 115).

NMFS. 2014. *Report to the NOAA Science Advisory Board—Exploration of Ecosystem Based Fishery Management in the United States*. Washington, DC: NOAA (p. 110).

NMFS. 2016a. *Ecosystem-based Fisheries Management Policy. National Marine Fisheries Service Policy Directive 01-120*. Washington, DC: NOAA.

NMFS. 2016b. *Ecosystem-based Fisheries Management Policy. National Marine Fisheries Service Policy Directive 01-120-01*. Washington, DC: NOAA.

NOAA (National Oceanic and Atmospheric Administration). 2004. *New Priorities for the 21st Century—NOAA's Strategic Plan: Updated for FY 2005-FY 2010*. Washington, DC: NOAA (p. 28).

NOAA. 2015. Ecosystem-Based Management: NOAA EBM 101. Department of Commerce. https://ecosystems.noaa.gov/EBM101/WhatisEcosystem-BasedManagement.aspx

National Ocean Council. 2013. National Ocean Policy Implementation Plan. Washington, DC: National Ocean Council (p. 32).

O'Boyle R, Jamieson G. 2006. Observations on the implementation of ecosystem-based management: experiences on Canada's east and west coasts. *Fisheries Research* 79:1–12.

ORAP (Ocean Research Advisory Panel). 2013. *Implementing Ecosystem-Based Management. A Report to the National Ocean Council*. Washington, DC: ORAP (p. 33).

Ostrom E. 2009. A general framework for analyzing sustainability of social-ecological systems. *Science* 325:419–22.

Patrick WS, Link JS. 2015a. Hidden in plain sight: using optimum yield as a policy framework to operationalize ecosystem-based fisheries management. *Marine Policy* 62:74–81.

Patrick WS, Link JS. 2015b. Myths that continue to impede progress in ecosystem-based fisheries management. *Fisheries* 40:155–60.

Pauly D, Alder J, Bennett E, Christensen V, Tyedmers P, Watson R. 2003. The future for fisheries. *Science* 302:1359–61.

Pauly D, Christensen V, Guénette S, Pitcher TJ, Sumaila UR, Walters CJ, Watson R, Zeller D. 2002. Towards sustainability in world fisheries. *Nature* 418:689–95.

Peters R, Marshak AR, Brady MM, Brown SK, Osgood K, Greene C, Guida V, Johnson M, Kellison T, McConnaughey R, Noji T, Parke M, Rooper C, Wakefield W, Yoklavich M. 2018. Habitat Science is a Fundamental Element in an Ecosystem-Based Fisheries Management Framework: An Update to the Marine Fisheries Habitat Assessment Improvement Plan. U.S. Dept. of Commerce, NOAA. NOAA Technical Memorandum NMFS-F/SPO-181, 29p.

Pikitch EK, Santora C, Babcock EA, Bakun A, Bonfil R, Conover DO, Dayton P, et al. 2004. Ecosystem-based fishery management. *Science* 305:346–7.

Pitcher TJ, Kalikoski D, Short K, Varkey D, Pramod G. 2009. An evaluation of progress in implementing ecosystem-based management of fisheries in 33 countries. *Marine Policy* 33:223–32.

Punt AE, A'mar T, Bond NA, Butterworth DS, de Moor CL, De Oliveira JAA, Haltuch MA, Hollowed AB, Szuwalski C. 2014. Fisheries management under climate and environmental uncertainty: control rules and performance simulation. *ICES Journal of Marine Science* 71:2208–20.

Robbins K. 2012. *An Ecosystem Management Primer: History, Perceptions, and Modern Definition*. Akron, OH: Akron Law Publications.

Rose KA. 2000. Why Are quantitative relationships between environmental quality and fish populations so elusive? *Ecological Applications* 10:367–85.

Ruckelshaus M, Klinger T, Knowlton N, DeMaster DP. 2008. Marine ecosystem-based management in practice: scientific and governance challenges. *BioScience* 58:53–63.

Rudd MA, Dickey-Collas M, Ferretti J, Johannesen E, Macdonald NM, McLaughlin R, Rae M, Thiele T, Link JS. 2018. Ocean ecosystem-based management mandates and implementation in the North Atlantic. *Frontiers in Marine Science* 5:485.

Salas S, Chuenpagdee R, Seijo JC, Charles A. 2007. Challenges in the assessment and management of small-scale fisheries in Latin America and the Caribbean. *Fisheries Research* 87:5–16.

Samhouri JF, Haupt AJ, Levin PS, Link JS, Shuford R. 2014. Lessons learned from developing integrated ecosystem assessments to inform marine ecosystem-based management in the USA. *ICES Journal of Marine Science* 71: 1205–15.

Sanchirico JN, Smith MD, Lipton DW. 2008. An empirical approach to ecosystem-based fishery management. *Ecological Economics* 64:586–96.

Schaefer MB. 1956. "The scientific basis for a conservation program." In: *Papers Presented at the International Technical Conference on the Conservation of the Living Resources of the Sea*. New York, NY: United Nations Publication (pp. 15–55).

Schaefer MB. 1957. Some considerations of population dynamics and economics in relation to the management of the commercial marine fisheries. *Journal of the Fisheries Research Board of Canada* 14:669–81.

Skern-Mauritzen M, Ottersen G, Handegard NO, Huse G, Dingsør GE, Stenseth NC, Kjesbu OS. 2016. Ecosystem processes are rarely included in tactical fisheries management. *Fish and Fisheries* 17:165–75.

Skud BE. 1975. *Revised Estimates of Halibut Abundance and the Thompson-Burkenroad Debate.* Seattle, WA: International North Pacific Halibut Commission.

Smith TD. 1994. *Scaling Fisheries: The Science of Measuring the Effects of Fishing, 1855–1955.* New York, NY: Cambridge University Press.

Smith DC, Fulton EA, Apfel P, Cresswell ID, Gillanders BM, Haward M, Sainsbury KJ, Smith ADM, Vince J, Ward TM. 2017. Implementing marine ecosystem-based management: lessons from Australia. *ICES Journal of Marine Science* 74:1990–2003.

Stock CA, John JG, Rykaczewski RR, Asch RG, Cheung WWL, Dunne JP, Friedland KD, Lam VWY, Sarmiento JL, Watson RA. 2017. Reconciling fisheries catch and ocean productivity. *Proceedings of the National Academy of Sciences* 114:E1441–9.

Szuwalski CS, Vert-Pre KA, Punt AE, Branch TA, Hilborn R. 2015. Examining common assumptions about recruitment: a meta-analysis of recruitment dynamics for worldwide marine fisheries. *Fish and Fisheries* 16:633–48.

Tallis H, Levin PS, Ruckelshaus M, Lester SE, McLeod KL, Fluharty DL, Halpern BS. 2010. The many faces of ecosystem-based management: making the process work today in real places. *Marine Policy* 34:340–8.

Thorson JT, Cope JM, Kleisner KM, Samhouri JF, Shelton AO, Ward EJ. 2015. Giants' shoulders 15 years later: lessons, challenges and guidelines in fisheries meta-analysis. *Fish and Fisheries* 16:342–61.

UNCBD (United Nations Convention on Biological Diversity). 2011. *Ecosystem Approach.* http://www.cbd.int/ecosystem/

UNEP (United Nations Environment Program). 2006. *Ecosystem-based Management, Markers for Assessing Progress.* New York, NY: United Nations Publication (p. 50).

UNEP. 2010. *Marine and Coastal Ecosystem-Based Management. An Introductory Guide to Managing Oceans and Coasts Better. Regional Seas Conventions and Action Plans DEPI/RS.12/6.* New York, NY: United Nations Publication.

UNEP. 2011. *Taking Steps toward Marine and Coastal Ecosystem-Based Management—An Introductory Guide. UNEP Regional Seas Reports and Studies No. 189, DEP/1409/NA.* New York, NY: United Nations Publication (p. 67).

USCOP (U.S. Commission on Ocean Policy). 2004. *An Ocean Blueprint for the 21st Century.* Washington, DC: U.S. Commission on Ocean Policy (p. 92).

von Bertalanffy L. 1968. *General System Theory: Foundations, Development, Applications.* New York, NY: George Braziller.

Walker B, Holling CS, Carpenter SR, Kinzig A. 2004. Resilience, adaptability and transformability in social—ecological systems. *Ecology and Society* 9(2):5.

Walters CJ. 1980. "Systems principles in fisheries management." In: *Fisheries Management.* Edited by RT Lackey, LA Nielsen. New York, NY: Wiley (pp. 167–83)

Walters CJ, Hilborn R. 1976. Adaptive control of fishing systems. *Journal of the Fisheries Research Board of Canada* 33(1):145–59.

Walters CJ. 1971. "Systems ecology: the systems approach and mathematical models in ecology." In: *Fundamentals of Ecology.* Edited by EP Odum. Philadelphia, PA: Saunders (pp. 276–92).

Walters CJ, Collie JS. 1988. Is research on environmental factors useful to fisheries management? *Canadian Journal of Fisheries and Aquatic Sciences* 45:1848–54.

Walther YM, Möllmann C. 2014. Bringing integrated ecosystem assessments to real life: a scientific framework for ICES. *ICES Journal of Marine Science* 71:1183–6.

WCED (United Nations World Commission on Environment and Development). 1987. *Report of the World Commission on Environment and Development: Our Common Future. Annex A/42/427.* New York, NY: WCED (p. 247).

White C, Halpern BS, Kappel CV. 2012. Ecosystem service tradeoff analysis reveals the value of marine spatial planning for multiple ocean uses. *Proceedings of the National Academy of Sciences* 109:4696–701.

Wilkinson EB, Abrams K. 2015. Benchmarking the 1999 EPAP recommendations with existing fishery ecosystem plans. NOAA Tech. Memo. NMFS-OSF-5. Washington, DC: NOAA (p. 22).

Witherell D, Pautzke C, Fluharty D. 2000. An ecosystem-based approach for Alaska groundfish fisheries. *ICES Journal of Marine Science* 57:771–7.

Worm B, Hilborn R, Baum JK, Branch TA, Collie JS, Costello C, Fogarty MJ, et al. 2009. Rebuilding global fisheries. *Science* 325(5940):578–85.

Wu JG. 2013. "Hierarchy theory: an overview." In: *Linking Ecology and Ethics for a Changing World: Values, Philosophy, and Action.* Edited by R Rozzi, STA Pickett, C Palmer, JJ Armesto, JB Callicott. New York, NY: Springer (pp. 281–301).

Yaffee SL. 1999. Three faces of ecosystem management. *Society for Conservation Biology* 13:713–25.

Zador SG, Holsman KK, Aydin KY, Gaichas SK. 2017. Ecosystem considerations in Alaska: the value of qualitative assessments. *ICES Journal of Marine Science* 74:421–30.

Zhang CI, Kim S, Gunderson D, Marasco R, Lee JB, Park HW, Lee JH. 2009. An ecosystem-based fisheries assessment approach for Korean fisheries. *Fisheries Research* 100:26–41.

Zhang CI, Hollowed AB, Lee JB, Kim DH. 2011. An IFRAME approach for assessing impacts of climate change on fisheries. *ICES Journal of Marine Science* 68:1318–28.

CHAPTER 2

Methods for Characterizing and Examining Marine Fishery Ecosystems

2.1 Introduction

All methodologies used to examine the socioecological aspects of United States (US) marine fishery ecosystems, and current progress toward US ecosystem-based fisheries management (EBFM) (and that of its participatory regional fisheries management organizations (RFMOs)), are briefly described within this chapter. In characterizing regional marine fisheries ecosystems, a suite of geographic, environmental, managerial, fisheries, socioeconomic, and ecological criteria were examined. Ninety unique indicators characterizing each regional and subregional fisheries ecosystem related to (1) socioeconomic, (2) governance, (3) environmental forcing and major features, (4) major pressures and exogenous factors, and (5) systems ecology and fisheries were examined using extant datasets (Table 2.1). The majority of these factors were similarly examined by Link & Marshak (2019) nationally to characterize and compare US fisheries ecosystems when evaluating successful ecosystem-based strategies for Living Marine Resource (LMR) management. These indicators track major parts of the National Oceanic and Atmospheric Administration (NOAA) Fisheries EBFM Road Map's six guiding principles (NMFS 2016a; Fig. 1.2) and were selected to cover a range of features of the socioecological system, mapping to our general schema thereof (Fig. 1.1). For each U.S. region (as defined by the eight regional fishery management council jurisdictions), two to six bays, and/or island areas of geographic, cultural, and economic significance were chosen as subregions. Indicators were examined for redundancy and interdependence in correlative tests (results not shown), from which non-collinearity was generally observed among indicators from separate datasets within an indicator class (i.e., subheadings in Table 2.1 and in subsequent chapters). We also note that even in instances of collinearity, the indicators are not redundant as they refer to different facets of these components of the socioecological system (e.g., targeted and protected species; chlorophyll concentrations and [pelagic] primary production). Additionally, ratios of ecosystem indicators for production, LMR status (biomass, fisheries landings), and socioeconomic status (LMR employments and fisheries value) were developed to provide an integrated perspective. Rankings based on mean anomaly values for each indicator category were calculated among regions, subregions, and U.S. components of RFMO jurisdictions. Based upon the geographic extent, jurisdictional organizations, environmental conditions, and mandated responsibilities of each defined U.S. region or subregion (Link & Marshak 2019, Supplementary Tables 2, 3), data were compiled to examine current and historic trends. Data sources for these variables are noted in Table 2.1.

Results are presented for each of the eight defined main U.S. regions and their subregions, and for U.S. Exclusive Economic Zone (EEZ) waters within 15 RFMOs or global conservation organizations. We also present them based on corresponding fishery management councils (FMCs), Large Marine Ecosystems (LMEs), or other jurisdictional considerations in accordance with data availability and resolution, where appropriate. Given the larger geographic extent of many of these datasets, examinations were often conducted at macroresolutions for a given fisheries ecosystem. Where possible we present this information at the subregional scale when data resolution was warranted, where here we refer to region as the entire region of focus in a chapter, and smaller "subregions" representing subsets thereof, be they bays, estuaries, or parts of LMEs. Thus, not all indicators were presented for each of the eight main regions and domestic or international RFMO jurisdictions, but the majority of indicators were applicable to all U.S. regions,

Ecosystem-Based Fisheries Management: Progress, Importance, and Impacts in the United States. Jason S. Link and Anthony R. Marshak,
Oxford University Press. © U.S. Department of Commerce, U.S. Government 2021. DOI: 10.1093/oso/9780192843463.003.0002

Table 2.1 Criteria used to characterize U.S. fishery ecosystems, their subregions, and U.S. and international contributions to participatory Regional Fishery Management Organization (RFMO) jurisdictions, including data sources. Asterisks indicate indicators that were not included in comparative synthesis rankings. Bold type under the Ecosystem Indicator/Variable heading indicates particular subsets of a given indicator type that were included in comparative synthesis rankings

Criteria	Ecosystem Indicator/Variable	Data Source	URL
Socioeconomic Criteria *Social & Regional Demographics*	Human Population/**Coastal Population Density**	U.S. Census, NOAA	https://coast.noaa.gov/digitalcoast/data/demographictrends.html
Socioeconomic Status & Regional Fisheries Ecosystems	Regional Economy of Living Marine Resources (LMR; Number of Establishments, Employments) and **Percent Regional Economy Dependent on Fisheries**	National Ocean Economics Program	https://coast.noaa.gov/digitalcoast/tools/enow.html
	Regional Economy of LMRs (GDP) and **Percent Regional Economy Dependent on Fisheries**	National Ocean Economics Program	https://coast.noaa.gov/digitalcoast/tools/enow.html
	Number of Permitted Vessels	NOAA Fisheries Office of Sustainable Fisheries Council Reports to Congress (1990–2016)	https://www.fisheries.noaa.gov/national/partners/council-reports-congress
	Total Value of Commercial Fisheries	NOAA Fisheries Commercial Fisheries Statistics	https://www.fisheries.noaa.gov/national/sustainable-fisheries/commercial-fisheries-landings
	Ratio of Total LMR Employment (Jobs) to Total Biomass	National Ocean Economics Program, NOAA Fisheries, Rutgers Ocean Adapt Database	https://coast.noaa.gov/digitalcoast/tools/enow.html
	Ratio of Total Value (Revenue) of Commercial Fisheries to Total Biomass	National Ocean Economics Program, NOAA Fisheries, Rutgers Ocean Adapt Database	https://coast.noaa.gov/digitalcoast/tools/enow.html
	Ratio of Total LMR Employments (Jobs) to Total Primary Production	National Ocean Economics Program, NASA Ocean Color Web—Vertically Generalized Production Model (VGPM)	https://coast.noaa.gov/digitalcoast/tools/enow.html
	Ratio of Total Value (Revenue) of Commercial Fisheries to Total Primary Production	National Ocean Economics Program, NASA Ocean Color Web—Vertically Generalized Production Model (VGPM)	http://www.oceaneconomics.org/Market/ocean/oceanEcon.asp
Governance Criteria *Human Representative Context*	Organizations Responsible for Environmental and Fisheries Management, Decisions, Implementation, and Enforcement	NOAA Fisheries Office of Sustainable Fisheries, United Nations Food and Agriculture Organization (FAO) Fisheries & Aquaculture Department	https://www.fisheries.noaa.gov/about/office-sustainable-fisheries http://www.fao.org/fishery/en

Criteria	Ecosystem Indicator/Variable	Data Source	URL
Governance Criteria *Human Representative Context*	Number of States and Congressionals (and per mile and total value of fisheries)	U.S. Census	https://catalog.data.gov/organization/census-gov
Fishery/Systematic Context	Composition of Regional Fishery Management Council Members and Atlantic Highly Migratory Species (HMS) Advisory Panel	NOAA Fisheries Office of Sustainable Fisheries Council Reports to Congress (1990–2016)	https://www.fisheries.noaa.gov/national/partners/council-reports-congress
	Composition of Marine Mammal Scientific Review Group (SRG) Members	NOAA Fisheries Office of Protected Resources Scientific Review Committee Reports (1995–2017)	https://www.fisheries.noaa.gov/national/marine-mammal-protection/scientific-review-groups
	Number of Fishery Management Plans (FEPs) and Amendments	NOAA Fisheries	https://www.fisheries.noaa.gov/about/office-sustainable-fisheries
	Number of Fishery Ecosystem Plans (FEPs) and Amendments	NOAA Fisheries	https://www.fisheries.noaa.gov/about/office-sustainable-fisheries
	Number of Habitat Areas of Particular Concern (HAPCs), and List of National Parks, National Seashores, National Estuarine Research Reserves (NERRs), NOAA National Marine Sanctuaries, and NOAA Habitat Focus Areas	NOAA Fisheries Office of Habitat Conservation, NOAA National Estuarine Research Reserve System, NOAA Office of Marine Sanctuaries, National Park Service	http://www.habitat.noaa.gov/protection/efh/newInv/index.html; https://coast.noaa.gov/nerrs/; https://sanctuaries.noaa.gov/; https://www.nps.gov/findapark/index.htm
	Current Number and Area of Closed Areas and Locations where Fishing is Prohibited or Restricted, and **Percent of EEZ and Ocean Basin where Fishing is Permanently Prohibited**	NOAA Marine Protected Areas Inventory, International Union for the Conservation of Nature (IUCN) Protected Areas Inventory	https://marineprotectedareas.noaa.gov/dataanalysis/mpainventory/ https://www.iucn.org/theme/protected-areas/our-work/world-database-protected-areas
Organizational Context	Budget of Regional Fishery Management Council Relative to Value of Fisheries	NOAA Fisheries Office of Sustainable Fisheries, NOAA Fisheries Commercial Fisheries Statistics	https://www.fisheries.noaa.gov/about/office-sustainable-fisheries; https://www.fisheries.noaa.gov/national/sustainable-fisheries/commercial-fisheries-landings
	Cumulative Number of National Environmental Policy Act (NEPA) Actions Per Region	U.S. Environmental Protection Agency Environmental Impact Statement (EIS) Database	https://cdxnodengn.epa.gov/cdx-enepa-public/action/eis/search
	Total Number of Fisheries-Related Lawsuits per Region (and per year)	NOAA Office of General Council	http://www.gc.noaa.gov/
Status of Living Marine Resources (Targeted & Protected Species)	**Number** and List of Managed Targeted & Prohibited Species, Stocks, and Taxa	NOAA Fisheries	https://www.fisheries.noaa.gov/
	Number/Percent of Stocks Overfished, Experiencing Overfishing, and of Unknown Status	NOAA Fisheries Office of Sustainable Fisheries	https://www.fisheries.noaa.gov/national/population-assessments/fishery-stock-status-updates

Continued

Table 2.1 Continued

Criteria	Ecosystem Indicator/Variable	Data Source	URL
Governance Criteria *Status of Living Marine Resources (Targeted & Protected Species)*	**Number** and List of Managed Protected Species, Stocks, and Taxa	NOAA Fisheries Office of Protected Resources	https://www.fisheries.noaa.gov/about/office-protected-resources
	Number/**Percent** of Marine Mammal Strategic Stocks and of Unknown Population Size	NOAA Fisheries Office of Protected Resources	https://www.fisheries.noaa.gov/national/marine-mammal-protection/marine-mammal-stock-assessment-reports-region
	Number of ESA-Listed Threatened and Endangered Species	NOAA Fisheries Office of Protected Resources, U.S. Fish and Wildlife Service	https://www.fisheries.noaa.gov/species-directory/threatened-endangered https://www.fws.gov/endangered/species/us-species.html
	Number/Percent of IUCN-Listed Threatened and Endangered Species	International Union for the Conservation of Nature (IUCN) Red List of Threatened Species	https://www.iucnredlist.org/
Environmental Forcing & Major Features *Oceanographic and Climatological Context*	Major Currents and Circulation Patterns of the U.S. Exclusive Economic Zone (EEZ)*	Satellite Applications for Geoscience Education; Beletsky et al. 1999	https://cimss.ssec.wisc.edu/sage/oceanography/lesson3/images/ocean_currents2.jpg
	Annual Sea Surface Temperature	NOAA National Centers for Environmental Information Extended Reconstructed Sea Surface Temperature (ERSST)	https://www.ncdc.noaa.gov/data-access/marineocean-data/extended-reconstructed-sea-surface-temperature-ersst-v3b
	Major Climate Forcing (e.g., NAO, PDO, etc.)*	NOAA Earth System Research Laboratory	https://www.esrl.noaa.gov/psd/data/climateindices/list/
Environmental Forcing & Major Features *Notable Physical Features & Destabilizing Events and Phenomena*	Major Bays and Islands*	Google Earth	https://www.google.com/earth/
	Area, Depth, and Miles of Coastline per Region Throughout U.S. Exclusive Economic Zone (EEZ)*	General Bathymetric Chart of the Oceans (GEBCO), NOAA Office of Coast Survey U.S. Maritime Boundaries and Limits, and derived National Fish Habitat Partnership Spatial Framework from U.S. Maritime Boundaries	https://coastalscience.noaa.gov/project/national-fish-habitat-action-plan-coastal-assessment/; https://www.gebco.net/; https://nauticalcharts.noaa.gov/data/us-maritime-limits-and-boundaries.html
	Number of Hurricanes or Typhoons per decade	NOAA Office for Coastal Management Digital Coast	https://coast.noaa.gov/hurricanes/
	Annual Extent of Gulf of Mexico Bottom Water Hypoxia	Gulf of Mexico Hypoxia Database, Louisiana Universities Marine Consortium (LUMCON)	https://gulfhypoxia.net/research/shelfwide-cruises/
	Proportion of Sea Ice Extent	NOAA Earth System Research Laboratory	https://www.esrl.noaa.gov/psd/data/gridded/data.noaa.oisst.v2.highres.html
Major Pressures & Exogenous Factors *Other Ocean Use Context*	**Total Regional Ocean Economy**, including non-LMR marine sectors (Number of Establishments, Employments) and their **Percent Contributions to Regional Ocean Economy**	National Ocean Economics Program	https://coast.noaa.gov/digitalcoast/tools/enow.html
	Total Regional Ocean Economy, including non-LMR marine sectors (GDP) and their **Percent Contributions to Regional Ocean Economy**	National Ocean Economics Program	https://coast.noaa.gov/digitalcoast/tools/enow.html

Criteria	Ecosystem Indicator/Variable	Data Source	URL
Major Pressures & Exogenous Factors *Other Ocean Use Context*	Number of Tourism Indicators (Dive Shops, Major Ports/Marinas, Cruise Ship/Passenger Metrics)	Professional Association of Diving Instructors (PADI), U.S. Department of Transportation	http://apps.padi.com/scuba-diving/dive-shop-locator/; https://www.transportation.gov/
	Average Number of Oil Rigs	Baker Hughes Rotary Rig Counts	http://phx.corporate-ir.net/phoenix.zhtml?c=79687&p=irol-reportsother
	Current Number of Offshore Wind Energy Areas	Bureau of Ocean Energy Management (BOEM)	https://www.boem.gov/
Systems Ecology & Fisheries *Basal Ecosystem Production*	Mean Surface Chlorophyll a	NASA Ocean Color Web	https://oceancolor.gsfc.nasa.gov/
	Total Primary Production	NASA Ocean Color Web—Vertically Generalized Production Model (VGPM)	https://oceancolor.gsfc.nasa.gov/
System Exploitation	Number of Taxa Captured by Commercial and Recreational Fisheries	NOAA Fisheries Commercial and Recreational Fisheries Statistics, Alaska Department of Fish and Game Recreational Fisheries Statistics, Pacific Fishery Management Council Recreational Statistics, Pacific States Marine Fisheries Commission Recreational Fisheries Information Network (RecFIN)	https://www.fisheries.noaa.gov/national/sustainable-fisheries/commercial-fisheries-landings; https://www.fisheries.noaa.gov/recreational-fishing-data/recreational-fishing-data-and-statistics-queries; http://www.adfg.alaska.gov/sf/sportfishingsurvey/; https://www.pcouncil.org/; www.recfin.org
	Total Fish and/or Invertebrate Biomass	NOAA Fisheries, Rutgers Ocean Adapt Database	https://www.fisheries.noaa.gov; http://oceanadapt.rutgers.edu/
	Ratio of Total Fisheries Landings to Total Biomass	NOAA Fisheries Commercial and Recreational Fisheries Statistics, Alaska Department of Fish and Game Recreational Fisheries Statistics, Pacific Fishery Management Council Recreational Statistics, Pacific States Marine Fisheries Commission Recreational Fisheries Information Network (RecFIN), NOAA Fisheries, Rutgers Ocean Adapt Database, FAO Fisheries Statistics	https://www.fisheries.noaa.gov/national/sustainable-fisheries/commercial-fisheries-landings; https://www.fisheries.noaa.gov/recreational-fishing-data/recreational-fishing-data-and-statistics-queries; http://www.adfg.alaska.gov/sf/sportfishingsurvey/; https://www.pcouncil.org/; www.recfin.org; http://www.nmfs.noaa.gov/; http://oceanadapt.rutgers.edu/; http://www.fao.org/fishery/statistics/en
Systems Ecology & Fisheries *System Exploitation*	Ratio of Total Biomass to Total Primary Production	NOAA Fisheries, Rutgers Ocean Adapt Database, NASA Ocean Color Web—Vertically Generalized Production Model (VGPM)	https://www.fisheries.noaa.gov; http://oceanadapt.rutgers.edu/; https://oceancolor.gsfc.nasa.gov/
	Ratio of Total Fisheries Landings to Total Primary Production	NOAA Fisheries Commercial and Recreational Fisheries Statistics, Alaska Department of Fish and Game Recreational Fisheries Statistics, Pacific Fishery Management Council Recreational Statistics, Pacific States Marine Fisheries Commission Recreational Fisheries Information Network (RecFIN), NASA Ocean Color Web—Vertically Generalized Production Model (VGPM), FAO Fisheries Statistics	https://www.fisheries.noaa.gov/national/sustainable-fisheries/commercial-fisheries-landings; https://www.fisheries.noaa.gov/recreational-fishing-data/recreational-fishing-data-and-statistics-queries; http://www.adfg.alaska.gov/sf/sportfishingsurvey/; https://www.pcouncil.org/; www.recfin.org; https://oceancolor.gsfc.nasa.gov/; http://www.fao.org/fishery/statistics/en

Continued

Table 2.1 Continued

Criteria	Ecosystem Indicator/Variable	Data Source	URL
Systems Ecology & Fisheries *Targeted & Non-Targeted Resources*	Total Commercial and Recreational Fisheries Landings (and per km^2)	NOAA Fisheries Commercial and Recreational Fisheries Statistics, Alaska Department of Fish and Game Recreational Fisheries Statistics, Pacific Fishery Management Council Recreational Statistics, Pacific States Marine Fisheries Commission Recreational Fisheries Information Network (RecFIN), FAO Fisheries Statistics, NOAA Office of Coast Survey U.S. Maritime Boundaries and Limits	https://www.fisheries.noaa.gov/national/sustainable-fisheries/commercial-fisheries-landings; https://www.fisheries.noaa.gov/recreational-fishing-data/recreational-fishing-data-and-statistics-queries; http://www.adfg.alaska.gov/sf/sportfishingsurvey/; https://www.pcouncil.org/; www.recfin.org; http://www.fao.org/fishery/statistics/en; https://nauticalcharts.noaa.gov/data/us-maritime-limits-and-boundaries.html
	Total Bycatch (by weight and number of individuals reported)	NOAA Fisheries National Bycatch Report Database	https://www.fisheries.noaa.gov/resource/document/national-bycatch-report

subregions, and RFMOs. All following unspecified references to "region(al)" and "subregion(al)" refer to those occurring within U.S. waters.

Additionally, up to 68 of these indicators were examined across regions to investigate differences in U.S. regional and subregional capacities (and their contributing RFMO-level components) for elucidating the determinants of successful (and ability to execute an ecosystem approach to) LMR management. These comparisons are presented in a summary, synthesis chapter (Chapter 12). Although these data are amenable to further, multivariate statistical analysis, we did not emphasize that approach here to ensure we did not lose important context by emphasizing additional analytics, and instead emphasized narrative threads within any given region among these common metrics.

Results are presented up to 2017–2018. This decision was based on data availability and what we were able to compile for this book. We note this choice as we had to stop updating data so we could execute our analysis and interpretation. We acknowledge that in doing so we may have missed some important updates between then and the time it took to publish this work. Thus, we recognize that upon completion this work may already be partially outdated. Given this consideration, we submit that the main trajectories, trends, and patterns we provide still probably represent a fair depiction of the progress on EBFM over the past decade. We recognize future evaluations and short-term updates may change the specifics for a given set of data or a particular geography. In many ways, that is excellent and the point of this work—to provide a snapshot of progress toward EBFM in the U.S. and spur on further progress. If that progress has occurred while this work was completed, all the better. To ameliorate some of this, immediately prior to publication we re-evaluated the status of major stocks in each region and provide those updates as footnotes in each chapter.

2.2 Socioeconomic Criteria

For socioeconomic indicators, including *social and regional demographics*, trends for coastal human population and human population density were derived from NOAA Digital Coast U.S. Census decadal data available within coastal counties for the past four decades, and summed for a given ecosystem region or subregion. Coastal counties are defined by NOAA and the U.S. Census Bureau as those counties, where at least 15% of a county's total land area is located within the U.S. coastal watershed (NOAA 2013b). Additionally, national trends in the proportion of individuals living within coastal counties were examined (NOAA 2013a). *Socioeconomic status and regional fisheries ecosystems* were examined in terms of U.S. regional and subregional trends in the total number of LMR establishments (defined as a place of work in an industry with explicit ties to ocean LMRs—excluding recreational fisheries, which are accounted for in the tourism and recreation sector data), employments (defined as the number of individuals working in LMR establishments—also excluding recreational fisheries), and their associated gross domestic product (GDP) values, including their related percent contributions to the total ocean economy of a given region or subregion as defined and recorded by the National Ocean Economics Program. These values were calculated per year over approximately the past decade (since 2005). Annual trends in the number of fishing vessels were examined using data from NOAA Fisheries and the regional FMCs. Additionally, the total value of all U.S. commercially landed species as reported by NOAA Fisheries and regional FMCs was examined at the RFMO and U.S. regional and subregional scales. Although landed highly migratory species are included in NOAA Fisheries statistics for all regions, including the Western Pacific, the numbers and values are underestimated in light of the international jurisdictions of these species, records of capture beyond U.S. waters throughout their range, and indeterminate landings among international ports (Lynch et al. 2011, 2012, Punt et al. 2015, Craig et al. 2017). When applicable among RFMO jurisdictions and U.S. regions, total surveyed fish and invertebrate biomass and primary production values were related to trends in total LMR employments and commercial fisheries value (USD). Integrative ratio relationships among production, biomass, LMR employments (jobs), and LMR revenue were examined with either total biomass or total primary production as the denominator. Factors identified by Link & Marshak (2019) as "Other Ecosystem Goods and Services" were not directly included, but trends from the examined socioeconomic indicators may be inferred toward assessing changes in other ocean uses, tourism metrics, and oil revenues, as examined for exogenous factors.

2.3 Governance Criteria

To examine the *human representative managerial context*, the number of domestic and international organizations, states, and jurisdictions (including congressional representation of a given U.S. regional or subregional ecosystem or contributing U.S. component of an RFMO) was tallied for each U.S. Census Congressional district boundary over time. These counts were then standardized per mile of coastline and relative to the total annual U.S. dollar (USD) value of all commercially landed species for a given region or subregion. Increased representation may suggest higher governmental attention to issues within a given region, but can additionally lead to increased conflict and less streamlined or transparent approaches to governance when centralized or aggregated over a larger area (Pomeroy & Berkes 1997, Hilborn 2007). Therefore, it has been suggested that lower values for total representation (i.e., number of congressionals or states) and higher values for standardized representation (i.e., per mile of shoreline or fisheries value) of a given region would be more effective toward LMR management (Link & Marshak 2019). Trends in the composition of representatives serving on the eight regional U.S. FMCs, the U.S. Atlantic Highly Migratory Species (HMS) Advisory Panel, and U.S. regional marine mammal Scientific Review Groups (SRGs) were also examined.

For elucidating *fishery and systematic governance and science systems*, all regional states marine fisheries commissions and federal fishery management plans (FMPs), fishery ecosystem plans (FEPs), and fishing regulations were examined to count FMPs and FEPs. Each FMP and FEP was also examined for the total number of modifications (i.e., amendments, frameworks, motions, specifications, and addendums) it had undergone since its original release, and all values were summed per region. While low numbers of FMP modifications may reflect overall stability within a region, they may also reflect less attention to certain fisheries or stressors (Stram & Evans 2009) or a lower degree of adaptive management (Maas-Hebner et al. 2016). It has additionally been assumed that mid-level numbers (closest to cross-regional mean value; see **2.6. Synthesis**) of modifications can be most effective for successful LMR management (Link & Marshak 2019).

All state and federally protected coastal and offshore areas—including coastal national parks, national seashores and lakeshores, National Estuarine Research Reserves (NERRs), National Marine Sanctuaries (NMS), NOAA Habitat Focus Areas (HFAs), and Habitat Areas of Particular Concern (HAPCs) sets—were enumerated and summarized per region. Total ocean basin-wide, and U.S. regional and subregional number and spatial extent of named permanent and seasonal fisheries closures (i.e., those areas with "closed" or "closure" in their title), and marine protected areas where commercial and/or recreational fishing is prohibited or restricted were tallied. The percent coverage of areas where commercial and/or recreational fishing is permanently prohibited was also estimated relative to the EEZ of a given ocean basin, U.S. region, or subregion. It is important to note that many named closures are not necessarily areas where fishing is permanently prohibited, and may only contain partial fishing restrictions (Link & Marshak 2019).

In terms of *organizational context*, annual trends in the budget of a given FMC related to the total commercial value of its managed fisheries were examined. Regulatory actions were indexed as the number of U.S. National Environmental Policy Act (NEPA) Environmental Impact Statements (EISs) from 1987 to 2016 and number of fisheries management-related lawsuits from 2010 to present, which were tallied per region and subregion. The number of EIS actions is indicative of broader ecosystem uses and pressures (Link & Marshak 2019).

LMR status for targeted resources was enumerated for all managed targeted fishery species and species for which fishing or harvest is prohibited ("prohibited species") under U.S. state or federal regulations. Each managed U.S. fisheries stock was examined for its June 2017 overfishing, overfished, or unknown status as reported for NOAA's Fish Stock Sustainability Index (FSSI) and non-FSSI stocks (NMFS 2017), and totals and proportions of stocks of a given status were summarized per region and for U.S. regulated species within RFMO jurisdictions. To assess *LMR status of protected resources*, all species protected under the U.S. Endangered Species Act (ESA) and/or the U.S. Marine Mammal Protection Act (MMPA) per region ("protected species") were enumerated. Additionally, the status of all protected species, including managed marine mammal stocks (including "strategic"—those threatened, endangered, declining, and/or depleted stocks for which the level of direct human-caused mortality exceeds the potential biological removal level—and "non-strategic" stocks, and stocks of unknown status) and ESA-listed species ("threatened" or "endangered") was summed and examined proportionally per region. For global conservation organizations and associated jurisdictions (e.g., International Whaling Commission, IWC), all International Union for the Conservation of Nature (IUCN) threatened and endangered species were additionally summed and examined proportionally at the ocean basin scale.

2.4 Environmental Forcing and Major Features

To characterize environmental forcing and major features of regional fisheries ecosystems, *oceanographic and climatological properties* (and water column habitats) were examined, including the major currents, trends in sea surface temperature (SST), and climate forcing oscillations operating in each of the regions. To assist in defining their geographic extent and regional oceanographies, qualitative examination of major currents encompassing specific U.S. marine regions and U.S. EEZ components of RFMO jurisdictions were noted based upon characterizations by CIMSS (2007). Average annual trends in SST were spatially examined for defined U.S. EEZ regions and subregions using the two-degree resolution NOAA Extended Reconstructed Sea Surface Temperature (ERSST) database. Under a climatological context, trends in climate forcing oscillations (AMO—Atlantic Meridional Oscillation Index, AO—Arctic Oscillation Index, ENSO—El Niño Southern Oscillation Index and MEI—Multivariate El Niño index, NAO—North Atlantic Oscillation Index, NOI—Northern Oscillation Index, NPGO—Northern Pacific Gyre Oscillation Index, PDO—Pacific Decadal Oscillation Index) from 1948 to present were examined using NOAA Earth Systems Research Laboratory (ESRL) climate index datasets. Additionally, regional trends in temperature increase over time were calculated using the ERSST database.

Indicators for *notable geographic and bathymetric physical features* were also characterized within U.S. marine regions. Major bays and islands in a given region or subregion were documented and enumerated using Google Earth at a 500 m resolution, while total EEZ area and miles of coastline per region or subregion were calculated using NOAA Office of Coast Survey U.S. maritime boundaries and limits spatial shapefile data (NOAA 2017). Additionally, gridded bathymetric data at a 30 arc-second resolution were obtained from the General Bathymetric Chart of the Oceans (GEBCO; Carpine-Lancre et al. 2003) program, and examined and averaged over the defined EEZ area of a given U.S. region or subregion.

Destabilizing events and phenomena that were quantified included the frequency of hurricanes and typhoons, bottom water hypoxia, and sea ice extent. Total and decadal trends in typhoon and hurricane frequency per U.S. region since the 1850s were examined using the NOAA Office of Coastal Management Digital Coast hurricanes platform at a 200 nautical mile resolution for U.S. regions and a 50 nautical mile resolution for subregions. Trends in the spatial extent of the midsummer bottom water hypoxia event over time were examined for the Gulf of Mexico region from the Louisiana Universities Marine Consortium (LUMCON) hypoxia database. Hypoxia events are highly pronounced in the northern Gulf of Mexico, covering expansive areas, and comprising the largest hypoxic zone in the United States (Rabalais et al. 2002). However, it is worth noting that natural hypoxic conditions with depth are also observed in the Pacific region as related to seasonal upwellings (Connolly et al. 2010). Similarly, the proportional areal extents of sea ice throughout the New England and North Pacific EEZ regions were spatially averaged over time using ESRL high-resolution annually blended analysis data of daily ice concentrations at a one-quarter degree global grid (Banzon et al. 2016).

2.5 Major Pressures and Exogenous Factors

Other ocean use context was examined in terms of trends in total U.S. regional ocean economies and those of regional, non-LMR, other ocean uses—as defined and recorded by the National Ocean Economics Program (i.e., marine construction, marine transportation, offshore mineral extraction, ship and boat building, tourism, and recreation). As similarly conducted for LMR economies, the total number of regional ocean and non-LMR establishments (defined as a place of work in an industry with explicit ties to a specific non-LMR other ocean use), regional ocean and non-LMR employments (defined as the number of individuals working in a given non-LMR other ocean use establishment), and their associated GDP values, including their related percent contributions to the total ocean economy of a given region (as defined and recorded by the National Ocean Economics Program) were calculated. Additionally, other ocean uses were examined by characterizing tourism proxy indices, including sums of the current number of Professional Association of Diving Instructors (PADI) dive shops and Department of Transportation listed major ports and marinas in a given region or subregion. Trends in the total number of cruises and vessels, and number of cruise destination and departure port passengers per region or subregion, were also examined using Department of Transportation cruise vessel datasets for the years 2009–2011. The average number of oil rigs per region over time, and total number of currently identified Bureau of Ocean Energy Management (BOEM) offshore wind energy areas were additionally tabulated to examine marine energy trends. These provide a broader marine-based economic context for a region (Link & Marshak 2019).

2.6 Systems Ecology and Fisheries

Basal ecosystem production estimates for each U.S region and RFMO jurisdiction were measured by characterizing annual regional primary productivity (g carbon m^{-2} y^{-1}; from NASA Ocean Color Web Data SeaWiFS years 1998–2007 and MODIS-Aqua years 2008–2014, 4 km resolution), using the Behrenfield and Falkowski Vertically Generalized Production Model (VGPM) estimation method (Eppley 1972, Behrenfeld & Falkowski 1997). Primary productivity values were averaged and calculated over published U.S. and international LME areas. Additionally, to account for primary producer concentration, mean annual chlorophyll concentration was also examined spatially for defined U.S. EEZ regions and subregions over time using NASA Ocean Color Web data (SeaWiFS years 1998–2001 and MODIS-Aqua years 2002–2014, 4 km resolution; NASA 2014). Nearshore benthic production throughout vegetated habitats (e.g., macroalgae, mangrove, salt marsh, seagrass) is important in these systems, but data are less comprehensively and synoptically available as the extents of many of these areas are not well-mapped (Peters et al. 2018). Given these limitations, and that the scale of this study occurred throughout the entire U.S. EEZ and within multiple international LMEs, we did not incorporate benthic primary productivity estimates. Although performed in other studies that examine satellite data (Cannizarro & Carter 2006), no additional correction for chlorophyll or productivity values was made for optically shallow waters.

System exploitation was characterized by examining annual trends in the number of reported taxa captured by commercial and recreational fisheries per year for each U.S. region (and subregion when applicable). Additionally, annual trends in total surveyed fish and invertebrate biomass as summed from NOAA Fisheries seasonal fishery-independent surveys of demersal and pelagic species were examined for most U.S. regions (Reid et al. 1999, Stauffer 2004). Total biomass was not estimable for the U.S. Caribbean and only available for the nearshore (3–10 m) zone in the South Atlantic region, given surveying methodology constraints. When applicable, total biomass values were related to trends in total (summed) annual commercial and recreational landings tonnage (i.e., an exploitation index), and annual regional primary production. Integrative ratio relationships among production, biomass, and fisheries landings were examined with either total biomass or total primary production as the denominator. For *targeted resources*, trends in total U.S. and RFMO-wide regional commercial and recreational landings (and as standardized by EEZ area for U.S. regions and subregions; km^{-2}) reported by NOAA Fisheries, regional FMCs, and Food and Agriculture Organization (FAO) as of their 2017 database entries were additionally examined. *Non-targeted resources* were examined by calculating total bycatch per region (by weight and number of individuals) as reported by NOAA Fisheries (NMFS 2016b).

2.7 Synthesis

A subset of up to 68 ecosystem indicators and reported metrics (Table 2.1) were separately compared among (1) the eight U.S. regions of interest (n = 68), (2) U.S. subregions (n = 45), and (3) for RFMO jurisdictions (n = 55). This information is presented in a separate synthesis chapter (Chapter 12). Among the five indicator categories, a given indicator was examined based upon its current value or cumulative average value and relative standard error over time. To assess cumulative nationwide trends for indicators, the number of regions with values above the total calculated cross-regional mean value for a given indicator (i.e., anomaly) were tabulated following Link & Marshak (2019). Additionally, the number of regions for which relative standard error was greater than 10% (z-score equivalent: 1.645) were tabulated to identify the total highly variable regions for which collective dynamic trends were occurring per indicator. Tabulated values were also averaged among the five indicator categories to gage overall trends per category (Link & Marshak 2019).

Current or average values of time-series data as related to relative standard error (i.e., signal to noise ratios) per ecosystem indicator were separately ranked across U.S. regions, subregions, and RFMO jurisdictions. Rankings were averaged together and within the five indicator categories to examine comparative regional relative success (high, mid, or low) for components of LMR management. Based upon Link & Marshak (2019), it was assumed that limited and less variable natural and human stressors, in addition to higher and more stable productivity, LMR status, socioeconomic status, and governance and scientific capacity contributed more toward LMR management success in a given region. Additionally, for those natural stressors occurring within limited geographies (i.e., sea ice and hypoxia) rankings were restricted to the regions in which they occurred. Ranked signal-to-noise ratios and regression relationships among primary productivity, biomass, fisheries landings, LMR employments, and total revenue of commercial fisheries from 2005 to 2014 were also examined per region and RFMO jurisdiction, as detailed productivity and biomass values were not available at the subregional scale. Additionally, the integrative ratios noted above also provide some indication of the relative trends across these indicator categories, relationships among them, and system-level emergent features. The anomaly method noted was also applied to these ratios of indicators and ranked by region.

Here we report basic summary statistics of these indicators to elucidate and compare major patterns across U.S. regional fisheries ecosystems. We present findings largely as time series where possible or otherwise report current snapshots, comparing and ranking them across indicator categories and relative to these *ad hoc*, anomaly-based thresholds.

In addition to evaluating these metrics, EBFM progress was benchmarked in each of the eight U.S. regions of interest, and their subregions, against the Ecosystem Principles Advisory Panel (EPAP) main recommendations (EPAP 1999) and NOAA Fisheries EBFM Road Map action items (NMFS 2016a). For each element we scanned the literature and associated reports, talked with regional experts, and looked at FMC webpages and plans to ascertain if these items have been considered, and if so, the degree of any progress to delineate if there has been action on these nationally agreed upon policies. To be clear, particularly for the EBFM Road Map, it was understood that these action items were meant to suggest options for consideration and were not requirements. Conversely, any common action, or consistent omission, is informative. Thus, this synthesis was not done as a report card or grading exercise, but rather to note common and national areas of progress and impediment, respectively, toward EBFM.

2.8 References

Banzon V, Smith TM, Chin TM, Liu C, Hankins W. 2016. A long-term record of blended satellite and in situ sea-surface temperature for climate monitoring, modeling and environmental studies. *Earth System Science Data* 8:165–76.

Behrenfeld MJ, Falkowski PG. 1997. Photosynthetic rates derived from satellite-based chlorophyll concentration. *Limnology and Oceanography* 42:1–20.

Cannizzaro JP, Carder KL. 2006. Estimating chlorophyll a concentrations from remote-sensing reflectance in optically shallow waters. *Remote Sensing of Environment* 101:13–24.

Carpine-Lancre J, Fisher R, Harper B, Hunter P, Jones M, Kerr A, Laughton A, Ritchie S, Scott D, Whitmarsh M. 2003. *The History of GEBCO 1903–2003: The 100-Year Story of the General Bathymetric Chart of the Oceans*. Lemmer, NL: GITC bv.

CIMSS (Cooperative Institute for Meteorological Satellite Studies). 2007. Satellite applications for geoscience education. Oceanography-ocean currents-global ocean currents. https://cimss.ssec.wisc.edu/sage/oceanography/lesson3/images/ocean_currents2.jpg.

Connolly TP, Hickey BM, Geier SL, Cochlan WP. 2010. Processes influencing seasonal hypoxia in the northern California current system. *Journal of Geophysical Research* 115:C03021.

Craig M, Bograd S, Dewar H, Kinney M, Lee HH, Muhling B, Taylor B. 2017. *Status Review Report of Pacific Bluefin Tuna (Thunnus Orientalis). NOAA Technical Memorandum NMFS-SWFSC-587*. Washington, DC: US Department of Commerce, NOAA.

EPAP (Ecosystem Principles Advisory Panel). 1999. *Ecosystem-Based Fishery Management: A Report to Congress*. Silver Spring, MD: National Marine Fisheries Service (p. 62).

Eppley RW. 1972. Temperature and phytoplankton growth in the sea. *Fishery Bulletin* 70:1063–85.

Hilborn R. 2007. Moving to sustainability by learning from successful fisheries. *Ambio* 36:296–303.

Link JS, Marshak AR. 2019. Characterizing and comparing marine fisheries ecosystems in the United States: determinants of success in moving toward ecosystem-based fisheries management. *Reviews in Fish Biology and Fisheries* 29:23–70.

Lynch PD, Graves JE, Latour RJ. 2011. *Challenges in the Assessment and Management of Highly Migratory Bycatch Species: A Case Study of the Atlantic Marlins. Sustainable Fisheries: Multilevel Approaches to a Global Problem*. Bethesda, MD: American Fisheries Society (pp. 197–226).

Lynch PD, Shertzer KW, Latour RJ. 2012. Performance of methods used to estimate indices of abundance for highly migratory species. *Fisheries Research* 125:27–39.

Maas-Hebner KG, Schreck C, Hughes RM, Yeakley JA, Molina N. 2016. Scientifically defensible fish conservation and recovery plans: addressing diffuse threats and developing rigorous adaptive management plans. *Fisheries* 41:276–85.

NASA (National Aeronautics and Space Administration). 2014. *NASA Goddard Space Flight Center, Ocean Ecology Laboratory, Ocean Biology Processing Group. Moderate-resolution Imaging Spectroradiometer (MODIS) Aqua Data*. Greenbelt, MD: NASA OB.DAAC.

NMFS (NOAA National Marine Fisheries Service). 2016a. *NOAA Fisheries Ecosystem-Based Fisheries Management Road Map*. Department of Commerce (p. 50). https://www.st.nmfs.noaa.gov/Assets/ecosystems/ebfm/EBFM_Road_Map_final.pdf.

NMFS. 2016b. *U.S. National Bycatch Report First Edition Update 2*. Edited by LR Benaka, D Bullock, J Davis, EE Seney, H Winarsoo. Washington, DC: US Department of Commerce.

NMFS. 2017. National marine fisheries service. http://www.nmfs.noaa.gov/sfa/fisheries_eco/status_of_fisheries/archive/2017/second/q2-2017-stock-status-table.pdf.

NOAA (National Oceanic and Atmospheric Administration). 2013a. *National Coastal Population Report. Population Trends from 1970 to 2020*. A product of the NOAA State of the Coast Report Series, a publication of the National Oceanic and Atmospheric Administration, Department of Commerce, developed in partnership with the US Census Bureau. https://aamboceanservice.blob.core.windows.net/oceanservice-prod/facts/coastal-population-report.pdf.

NOAA. 2013b. *Spatial Trends in Coastal Socioeconomics (STICS): Coastal County Definitions*. https://coast.noaa.gov/htdata/SocioEconomic/NOAA_CoastalCountyDefinitions.pdf.

NOAA. 2017. *NOAA Office of Coast Survey Maritime Zones of the United States*. https://nauticalcharts.noaa.gov/data/us-maritime-limits-and-boundaries.html.

Peters R, Marshak AR, Brady MM, Brown SK, Osgood K, Greene C, Guida V, et al. 2018. *Habitat Science is a Fundamental Element in an Ecosystem-Based Fisheries Management Framework: An Update to the Marine Fisheries Habitat Assessment Improvement Plan. NOAA Technical Memorandum NMFS-F/SPO-181*. Washington, DC: US Department of Commerce, NOAA.

Pomeroy RS, Berkes F. 1997. Two to tango: the role of government in fisheries co-management. *Marine Policy* 21: 465–80.

Punt AE, Su NJ, Sun CL. 2015. Assessing billfish stocks: a review of current methods and some future directions. *Fisheries Research* 166:103–18.

Rabalais NN, Turner RE, Wiseman WJ. 2002. Gulf of Mexico hypoxia, aka "The dead zone". *Annual Review of Ecology and Systematics* 33:235–63.

Reid RN, Almeida FP, Zetlin CA. 1999. *Essential Fish Habitat Source Document: Fishery-Independent Surveys, Data Sources, and Methods. NOAA Technical Memorandum NMFS NE 122*. Washington, DC: US Department of Commerce, NOAA.

Stauffer G. 2004. *NOAA Protocols for Groundfish Bottom Trawl Surveys of the Nation's Fishery Resources. NMFS-F/SPO-65*. Washington, DC: US Department of Commerce, NOAA.

Stram DL, Evans DC. 2009. Fishery management responses to climate change in the North Pacific. *ICES Journal of Marine Science* 66:1633–9.

CHAPTER 3

The New England Region

➲ The Region in Brief

- New England contains the second-lowest number of managed taxa among U.S. marine ecosystems, including historically important groundfish species such as Atlantic cod, Haddock, Atlantic halibut, commercially valuable Atlantic sea scallop and American lobster, and federally protected Atlantic salmon.
- This region contains the highest number and percentage of stocks that are overfished or experiencing overfishing and the lowest number and percentage of stocks of unknown status in the nation.
- Human population and population density are high in this region, contributing significant pressures to marine habitats while also contributing to one of the nationally largest coastal and oceanic economies.
- New England has very high socioeconomic status, with living marine resources (LMRs) contributing substantially to its marine economy (second-highest fisheries value and proportional economic contribution of LMRs among regions).
- New England is tied with the Mid-Atlantic for the second-highest basal productivity in the U.S., contributing toward higher system biomass, landings, and LMR-dependent jobs and revenue.
- Important concentrations of offshore canyons and seamounts, in addition to inshore rocky shorelines and submerged aquatic vegetation support economically and ecologically important species throughout their life histories.
- Lower numbers of depleted marine mammal stocks and threatened/endangered species are found in this region.
- Within this region and its subregions, average sea surface temperature increases have been observed at >1.2°C since the mid-twentieth century.
- Other ocean uses and coastal tourism are less pronounced in New England. Although the region is tied nationally for second-highest in the development of offshore wind energy areas.
- Fisheries management among state and federal entities has been focused upon strict rebuilding plans for historically overfished commercially and recreationally important species, with contentious relationships (including the nationally highest number of fisheries-related lawsuits) observed among managers, stakeholders, and fishing sectors.
- While subject to high overexploitation and ecosystem shifts over time, the Gulf of Maine and offshore Georges Bank still house major national fisheries (e.g., scallop, lobster) and continue to be significant contributors to New England's LMR economy.
- Among its subregions, Narragansett Bay, Casco Bay, and Long Island Sound contribute most heavily toward regional fisheries landings, LMR revenue, and LMR employments, while human population stressors surrounding Long Island Sound and Narragansett Bay are some of the highest nationally.
- Overall, ecosystem-based fisheries management (EBFM) progress has been made at the regional and subregional level in implementing ecosystem-level planning, advancing knowledge of ecosystem principles, and examining system trade-offs.
- This ecosystem is excelling in the socioeconomic status of its LMRs, and is relatively productive, as related to the determinants of successful LMR management.

Ecosystem-Based Fisheries Management: Progress, Importance, and Impacts in the United States. Jason S. Link and Anthony R. Marshak,
Oxford University Press. © U.S. Department of Commerce, U.S. Government 2021. DOI: 10.1093/oso/9780192843463.003.0003

3.1 Introduction

Thoughts of the New England ecosystem immediately bring to mind images of major national fisheries species such as American lobster (*Homarus americanus*), Atlantic cod (*Gadus morhua*), Atlantic salmon (*Salmo salar*), sea scallop (*Placopecten magellanicus*), and other harvested bivalves, including surfclams (*Spisula solidissima*), soft-shell clams (*Mya arenaria*), and ocean quahogs (*Arctica islandica*). Sights of lobster traps, New England trawlers, lighthouses, fishing piers, historical whaling ships, and rocky shorelines come to mind when thinking about the marine environment and culture of this maritime region, as do nationally popular seafood dishes such as New England clam chowder, lobster rolls, baked cod, baskets of steamers, or fish and chips prepared with fried haddock. Sounds of foghorns, pot haulers, sea shanties, Boston classic rock bands, New England folk music, seagulls, and breaching whales punctuate its coastal and offshore environments and comprise major parts of its "wickedly" proud regional heritage. Iconic maritime tales as told by Herman Melville, descriptions of its natural settings by American Romantic era authors and poets like Emerson, Dickinson, and Thoreau, and even gothic and modern horror stories from H.P. Lovecraft and Stephen King make up the diverse tales of New England coastal communities. The natural environments of this area serve as a foundation for all of these aspects of its maritime cultural identity, which is also heavily dependent on the local living marine resources (LMRs) that are closely associated with this regional ecosystem. They serve as a major basis for its local marine economy, which is also rooted in significant tourism, marine transportation, and a prominent ship and boatbuilding industry.

There are many works that characterize the New England ecosystem (Bigelow & Schroeder 1953; Backus 1987; Sherman et al. 1996, Steneck 1997, Fogarty & Murawski 1998, Link et al. 2001, Link et al. 2002a, b, Sherman et al. 2002, Read & Brownstein 2003, Stevenson et al. 2004, Link et al. 2008a, 2008b, 2009, NEFSC 2009, Pranovi & Link 2009, Jordaan et al. 2010, Link et al. 2010a, Lucey & Nye 2010, Mountain & Kane 2010, Johnson et al. 2011, Link et al. 2011a, 2012, NEFSC 2012, Shackell et al. 2012, Nye et al. 2013, Large et al. 2015a, 2015b; Wyatt et al. 2017, Fay et al. 2019, Gaichas et al. 2019b, Link & Marshak 2019), its important and major fisheries (e.g., lobster, clams, cod, haddock, flounders, other groundfish; McFarland 1911, Royce et al. 1959, Sissenwine 1974, Lindsay & Savage 1978, Anthony 1990, Shelley et al. 1995, Garrison 2000, Garrison & Link 2000a, 2000b, Jin et al. 2002, Ames 2004, Howell 2012, Pershing et al. 2015, NMFS 2020b), its protected species (e.g., whales, seals, dolphins, salmon, sturgeons; Saunders 1981, Beland et al. 1982, Mattila et al. 1987, Payne & Selzer 1989, Gilbert & Guldager 1998, Hamilton & Mayo 1990, Orciari et al. 1994, Fullard et al. 2000, Weinrich et al. 2001, Savoy & Pacileo 2003, Kraus et al. 2005, Craddock et al. 2009, Dunton et al. 2010), and their contributions toward the marine socioeconomics of this region, especially in how humans use, enjoy, and interact with this system (Robadue & Tippie 1980, Bassett 1987, Hall-Arber et al. 2001, Bell 2009, Colburn et al. 2010, Pollnac et al. 2011). This marine region, its resources, and its goods and services have played a major role in American history, especially contributing toward the early U.S. colonial marine economy, its maritime history, and its increasing industrialization (Vickers 1994, Newell 1998, Roland et al. 2008, Weigold & Pillsbury 2014). This region also served as a major initial destination for European immigrants (Plymouth Colony) and contains sites where pivotal events in the American Revolutionary War, such as the Boston Tea Party and Battles of Lexington and Concord, occurred (Adams 1923, Demos 1965, Tourtellot 2000, Shalev 2009, Carp 2010, Raphael 2011). In addition to being an iconic setting for American maritime history, iconic groundfish fisheries that developed off New England significantly accelerated the emergent U.S. marine economy (Jensen 1967, Bolster 2008).

What follows is a brief background for this region, describing all the major considerations of this marine ecosystem. Then there is a short synopsis of how all the salient data for those factors are collected, analyzed, and used for LMR management. We then present an evaluation of key facets of ecosystem-based fisheries management (EBFM) for this region.

3.2 Background and Context

3.2.1 Habitat and Geography

The New England ecosystem is part of the Northeast Shelf Large Marine Ecosystem (Sherman et al. 1996) and is often referred to as the northeast U.S. (NEUS), Northeast Shelf (NES) or U.S. Northwest Atlantic. By area, this ecosystem makes up the third-largest component of the United States Atlantic Exclusive Economic Zone (EEZ), with the entire New England EEZ contained in this ecosystem designation. It connects to the Mid-Atlantic through continuation of the Gulf Stream north of Cape Hatteras, and with influence

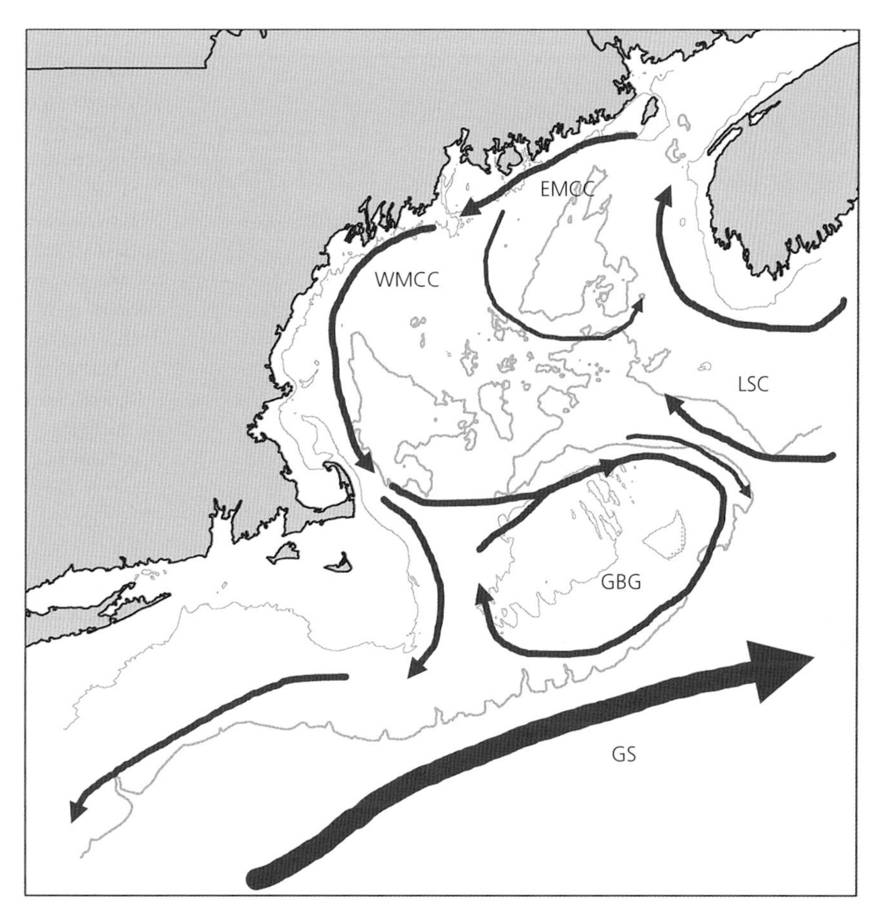

Figure 3.1 Major currents and circulation patterns throughout and encompassing the New England region and subregions. GS = Gulf Stream; LSC = Labrador Slope Current; EMCC = Eastern Maine Coastal Current; WMCC = Western Maine Coastal Current; GBG = Georges Bank Gyre.

from northern boreal waters via the Labrador Current (Fig. 3.1; also see text below in *Oceanographic and Climatological Context*).

Significant concentrations of human population, fisheries landings, and economic production are found throughout the New England region, particularly in four key subregions: **Long Island Sound**—Connecticut and New York (partially also a subset of the U.S. Mid-Atlantic region); **Narragansett Bay**—Massachusetts and Rhode Island; and **Casco** and **Penobscot Bays**—Maine (Fig. 3.2). There are other subregional features in this region, but we emphasize these four herein as exemplary, smaller-scale ecosystems.

The main features of this ecosystem include rocky intertidal habitats, coastal salt marshes, eelgrass (*Zostera marina*) habitats, tide pools, offshore sediments, rocky hardbottoms, deep corals, low to mid-range numbers of managed species, and an oceanography driven by continuation of the Gulf Stream and influence of the Labrador Current (Menge 1976, Sutcliffe et al. 1976, Menge 1978, Lubchenco 1980, Niering & Scott-Warren 1980, Brooks 1985, Collette 1986, Petraitis 1987, Townsend 1991, Heck et al. 1995, Fogarty & Murawski 1998, DeAlteris et al. 2000, Leonard 2000, Raposa & Roman 2001, Packer et al. 2003, Adamowicz & Roman 2005, Auster et al. 2005, Watling & Auster 2005, Auster 2007, Mountain & Kane 2010, Link et al. 2011a, Gawarkiewicz et al. 2012, Mountain 2012, Fratantoni et al. 2015, 2019). Many prominent groundfish and shellfish species in this region are associated with the offshore rocky habitats of the Gulf of Maine and Georges Bank, while sandy and muddy bottoms also support commercially and recreationally important bivalves and finfishes, including recently recovered skate species (Wigley 1968, Auster et al. 1995, Cargnelli 1999a, b, Carmichael et al. 2004, Link et al. 2006, Curtis & Sosebee 2015).

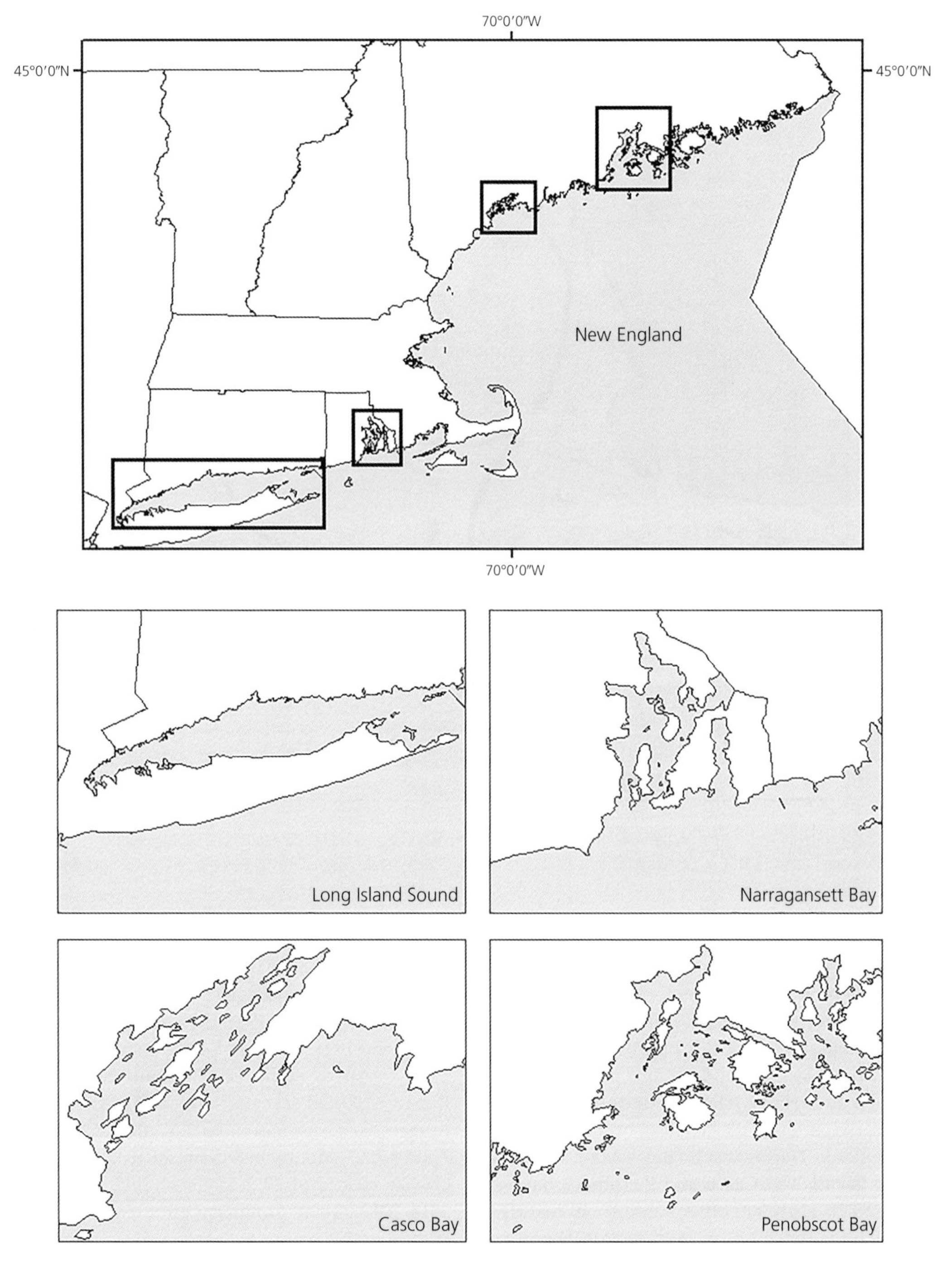

Figure 3.2 Map of the New England region and identified subregions.

3.2.2 Climate and Other Major Stressors

This marine region has also been subject to significant warming, with increasingly abundant species emerging from lower latitudes and iconic species such as lobster and cod moving into Canadian and deeper offshore waters, which may foster changes to the overall composition of this temperate ecosystem and the future of its economically significant fisheries (Frank et al. 1990, Murawski 1993, Fogarty et al. 2007, Nye et al. 2009, Lucey & Nye 2010, Pinsky & Fogarty 2012, Pinsky et al. 2013, Pershing et al. 2015, Hare et al. 2016, Kleisner et al. 2016, 2017, Morley et al. 2017, 2018, Flanagan et al. 2019).

In addition to the effects of climate-associated regional warming, the New England ecosystem is subject to other stressors that include habitat loss and degradation, coastal development, and associated increased nutrient loading and pollution, sea-level rise, species range shifts, intensely concentrated historical overfishing, and recent invasions from exotic species (Warren & Niering 1993, Auster et al. 1996, Foster 1995, Fogarty & Murawski 1998, Flanagan et al. 1999, Roman et al. 2000, Driscoll et al. 2003, Keser et al. 2003, Bromberg & Burtness 2005, Buchsbaum et al. 2005, Deegan & Buchsbaum 2005, Pederson et al. 2005, Frumhoff et al. 2007, Johnson et al. 2008, Lengyel et al. 2009, Nye et al. 2009, Valentine et al. 2009, Nettleton et al. 2013, Flanagan et al. 2019). Northward shifts in the Gulf Stream accompanied by climatic changes in the Atlantic Meridional Oscillation (AMO) and the North Atlantic Oscillation (NAO) are associated with some of the highest and above-average increases in sea surface temperature for the nation (sea surface temperature (SST); Sutton & Hodson 2007, Gawarkiewicz et al. 2012, Pershing et al. 2015). These temperature increases have led to subsequently observed ecosystem responses (Drinkwater et al. 2003, Greene et al. 2013, Friedland et al. 2019, 2020, Tableau et al. 2018). These environmental factors have also exacerbated recruitment failures for several prominent and overharvested groundfishes, further constraining stock rebuilding efforts (Drinkwater et al. 2005, Fogarty et al. 2008, Klein et al. 2017, Tableau et al. 2018). Degradation of offshore habitats by bottom-tending fishing practices such as trawling and scallop dredging have significantly affected biological communities in those areas (Auster et al. 1996, Watling & Norse 1998, Collie et al. 1997, Sparks-McConkey & Watling 2001, Watling et al. 2001, Thrush & Dayton 2002, Deegan & Buchsbaum 2005, Hiddink et al. 2017).

Nearshore habitat quality has also been reduced as a consequence of high coastal population density and development, upstream land-based activities, historical depletion of filter feeding bivalves, and with consequential losses of wetlands and eelgrass beds as a result of these stressors (Flanagan et al. 1999, Dahl 2000, Roman et al. 2000, Hughes et al. 2002, Driscoll et al. 2003, Bromberg & Burtness 2005, Gedan et al. 2011, Deegan et al. 2012, Watson et al. 2017). Eutrophication of this system, with increasing die-off of marsh vegetation is also of concern regarding the continued sustainability of this increasingly thermally stressed system (Gedan & Bertness 2009, 2010, Gedan et al. 2011, Deegan et al. 2012, Watson et al. 2017). Reductions in kelp bed coverage and losses of offshore kelp forests in the Gulf of Maine have also been observed as a result of ongoing stressors, including continued regional warming (Steneck et al. 2002, Horton & McKenzie 2009, Krumhansl et al. 2016). Additionally, inland rivers, streams, and nearshore estuaries once supported abundant Atlantic salmon, which are now listed as an endangered species (Saunders 1981, Beland et al. 1982, Parrish et al. 1998, Robinson et al. 2009). These areas also serve as habitat for diadromous American eel (*Anguila rostrata*; Busch et al. 1998, Lazzari et al. 2003) and river herrings (*Alosa* spp.), the latter of which are a significant prey species for offshore groundfish (Garrison & Link 2000a, 200b, Smith et al. 2007, Link et al. 2009b, Lake et al. 2012, Ames & Lichter 2013, McDermott et al. 2015, Willis et al. 2017). Compounding these losses are the effects of sea level rise, which while not predicted to be as intensive as in the South and Mid-Atlantic regions, are still affecting coastal habitats and its coastal communities and economies, with rates having accelerated to 2.4 mm y^{-1} in New England over the past century (Gornitz 1990, Roman et al. 1997, Orson et al. 1998, Hapke et al. 2010, Gedan et al. 2011). Within these stressed coastal and offshore habitats, shifts in community composition have been observed as a result of natural and human-derived stressors (Fogarty & Murawski 1998, Jackson et al. 2001, Deegan et al. 2002, Thrush & Dayton 2002, Bourque et al. 2008, Gedan et al. 2009, Mountain & Kane 2010, Smith et al. 2012). Ecological consequences from climate-related species range expansions are predicted to occur in New England (Kordas et al. 2011, Sorte et al. 2010, McMahan & Grabowski 2015, McMahan 2017). These particularly include increasingly abundant Scup (*Stenotomus chrysops*; Bell et al. 2015), Atlantic croaker (*Micropogonias undulatas*; Hare & Able 2007, Hare et al. 2010), and Black Sea bass (*Centropristis striata*) that have caused associated predation and aggressive competitive interactions with resident New England species (Wahle et al. 2013, Mc Mahan & Grabowski 2015, McMahan 2017). Significant shifts in the distributions of American lobster, Atlantic cod, American shad (*Alosa sapidissima*), Silver (Merluccius bilinearis) and Red (*Urophycis chuss*) hakes, and Yellowtail (*Pleuronectes ferruginea*) and Summer (*Paralichthys dentatus*) flounders toward more northern portions of this system have also been observed (Nye et al. 2009, Lucey & Nye 2010, Pinsky & Fogarty 2012, Walsh et al. 2015, Kleisner et al. 2016).

3.2.3 Exploitation

While renowned for its maritime and fisheries-based history, heavily concentrated historical overharvesting has left a significant impact on the New England ecosystem, particularly on its commercially and recreationally important fishes, invertebrates, and protected species (Brown et al. 1976, Clark & Brown 1977, Anthony 1990, Serchuk & Wigley 1992, Whitman & Sebens 1992, Anthony 1993, Shelley et al. 1995, Auster et al. 1996, Fogarty & Murawski 1998, Murawski et al. 2000, Wahle 2000, Jackson et al. 2001, Stone et al. 2004, Buchsbaum et al. 2005, Murawski 2005, Brodziak et al. 2008, Mountain & Kane 2010, Rothschild et al. 2013, Pershing et al. 2015). Although stock status in this region is among the best known nationally (Link & Marshak 2019) and some stock recovery has been observed (Safina et al. 2005, Brodziak et al. 2008a, b, NEFSC 2012), this region still contains the highest number and proportion of overfished stocks in the nation (Overholtz 2002, Stone et al. 2004, Hart & Rago 2006, Rosenberg et al. 2006, Brodziak et al. 2008a, b, Link & Marshak 2019), with significant depletion of groundfish species and shifts to a crustacean-dominant (>90%) offshore fishery (Anthony 1993, Fogarty & Murawski 1998, Healey 2000, Hollan & Maguire 2003, Acheson 2006, Alexander et al. 2009, Acheson 2011, Steneck & Wahle 2013, Wahle et al. 2013, Acheson & Gardner 2014). As of the 1960s, the arrival of factory trawlers facilitated sequential overexploitation of several species on Georges Bank, particularly Haddock, Silver hake, Atlantic sturgeon (*Acipenser* spp.), Atlantic halibut (*Hippoglossus hippoglossus*), herring (*Clupea harengus*), and mackerel (*Scomber scombrus*) (Fogarty & Murawski 1998, Fogarty et al. 2012). Regional warming and significant overfishing of Atlantic Cod in the Gulf of Maine and Georges Bank led to its collapse during the 1990s, and again during the 2010s (Kurlansky 2011, Pershing et al. 2015), with it and 11 other groundfish stocks currently remaining overfished (Link & Marshak 2019, NMFS 2019). Consequential growth and recruitment overfishing have led to smaller-sized individuals, earlier maturity times, and negative population growth for several New England groundfishes (Fogarty & Murawski 1998, Hutchings 2005, Hutchings & Baum 2005, Murawski 2005).

Regime shifts in groundfish recruitment have been observed as related to low spawning biomass, bottom-up forcing, and zooplankton dynamics (Perretti et al. 2017). Over time, a shift from a demersal to pelagic oriented community (dominated by herring and mackerel) in this iconic fisheries system has been observed in response to fishing pressure and environmental factors (Link 1999, Garrison & Link 2000a, 2000b, Link et al. 2005, Lucey & Nye 2010, Link et al. 2011a). Removal of predatory groundfish biomass, shorter generation times for small pelagic species, and potential competitive release have contributed to this transition (Fogarty & Murawski 1998, Garrison & Link 2000a, 2000b, Lucey & Nye 2010, Link et al. 2011a, Link & Auster 2013). As a result of continued overfishing and sequential depletion, Atlantic Mackerel remains overfished (Link & Marshak 2019, NMFS 2019). These effects, together with losses of higher trophic levels, cascading top-down effects, and footprints of destructive fishing practices in essential fish habitats have significantly shifted this iconic fisheries system, while exacerbating the ecological impacts of environmental stressors (Fogarty & Murawski 1998, Stevenson et al. 2004, Link et al. 2005, 2006, 2008a, b, 2010, 2011).

While advances in New England fisheries management have occurred, ecological consequences from overfishing and resistance to strict rebuilding plans by stakeholders have continued (Acheson 1980, Pollnac & Poggie 1988, Robertson & Caporossi 2004, Salz & Loomis 2004, Hartley & Robertson 2008, Holland et al. 2010, Acheson 2011, Acheson & Gardner 2011, Pollnac et al. 2015). These factors have led to a contentious environment with differing perceptions among managers, scientists, and fishers as to the actual status of fishery and protected species, concerns about economic stability, and the continuation of marine livelihoods, resulting in difficulty in enacting new or continued management actions (Hartley & Robertson 2006, Holland et al. 2010, Acheson 2011, Acheson & Gardner 2011, Pollnac et al. 2015). First-hand evidence of the consequences of historical overfishing has brought attention to the need for more holistic management strategies and more sustainable fishing practices (Roman et al. 2000, Brodziak & Link 2002, Brodziak et al. 2004, Bakun et al. 2009, Link 2010, Acheson & Gardner 2011, Link et al. 2011a, Portman et al. 2011, Jin et al. 2012, 2013, NEFSC 2012, Gaichas et al. 2019b, NMFS 2019).

Protected species populations have also been depleted in this region due to historical whaling, ongoing bycatch and entanglements with fishing gears, overfishing of their prey species, and overexploitation of New England Atlantic salmon populations (Parrish et al. 1998, Johnson et al. 2005, Kraus et al. 2005, Clapham & Link 2006, Saunders et al. 2006, Bakun et al. 2009, Knowlton et al. 2012, Reeves et al. 2013, Shoemaker 2014, Hare et al. 2019). While these factors continue, efforts toward stock rebuilding, more comprehensive understanding of this ecosystem and the implementation of holistic management practices have emerged as major priorities for its Fishery Management Council (FMC), NOAA Fisheries, the Atlantic States Marine Fisheries Commission (ASMFC), and other state-level management bodies (Sissenwine & Cohen 1991, Overholtz et al. 1995, Murawski et al. 2000, Link et al. 2002a, b, Brodziak et al. 2004, Acheson 2006, Link et al. 2008, Bakun et al. 2009, Link et al. 2011a, Jin et al. 2012, 2013, NEFSC 2012, Shelley 2012, Jin et al. 2016, NMFS 2019).

3.2.4 Invasive Species

The introduction of exotic, invasive species, including European green crab (*Carcinus maenas*), Asian shore crab (*Hemigrapsus* sanguineus), Chinese mitten crab (*Eriocheir sinensis*), dead man's fingers green algae (*Codium fragile*), European flat oyster (*Ostrea edulis*), and several species of invasive tunicates (*Ascidiella* spp., *Botryllus* spp., *Diplosoma* spp., *Didemnum* spp.; Dijkstra et al. 2007, Osman & Whitlatch 2007, Daley & Scavia 2008, Lengyel et al. 2009, Mercer et al. 2009, Carman & Grunden 2010, GMCME 2010a, Simkanin et al. 2016) has led to increasing concerns regarding their ecological effects on New England native fauna (Grosholz & Ruiz 1996, Pederson et al. 2005, Valentine et al. 2009, Bentley 2011, Griffen et al. 2012, Epifiano 2013, Wells et al. 2013, Neckles 2015). European green crabs have invaded salt marshes and seagrasses throughout New England coastal areas, with effects on eelgrass beds and anticipated effects on native fish communities (Davis et al. 1998, Neckles 2015, Matheson et al. 2016). Asian shore crabs have expanded along the U.S. Atlantic Coast, with predatory effects observed on resident sessile invertebrates (McDermott 1991, Ledesma & O'Connor 2001, Brousseau et al. 2003, Griffen et al. 2012, Heinonen & Auster 2012). They have been observed outcompeting and preying upon resident crustaceans, and are now the dominant crab species in New England rocky intertidal communities and vegetated habitats (Epifiano 2013, Peterson et al. 2014, Lord & Dalvano 2015). Invasive Chinese mitten crabs are also able to outcompete resident New England fauna, while consequences to sediment habitats have been observed as a result of their extensive large-scale burrowing (Bentley 2011). Blooms of dead man's fingers have also spread throughout the Gulf of Maine and other portions of New England nearshore habitats, causing consequences to bivalve populations and smothering other vegetation that is foundational for coastal marine habitats (Mathieson et al. 2003, Wells et al. 2013). Invasive ascidians (aka tunicates) are impacting native species in rocky intertidal, eelgrass, and even some offshore benthic communities, particularly in their ability to dominate sessile benthic communities and reduce overall species diversity in those communities (Pedeson et al. 2005, Weigle 2007, Bullard & Carman 2009, Carman et al. 2009, Mercer et al. 2009, Valentine et al. 2009, Karlson & Osman 2012). Together, all of these natural and human-induced stressors have resulted in consequences and shifts to the New England ecosystem and its associated LMRs.

3.2.5 Ecosystem-Based Management (EBM) and Multisector Considerations

In addition to these species and stressors, the New England ecosystem is also known for the trade-offs that arise across its LMRs (and their usage) and between LMR management and other ocean uses (Link 2002, Brodziak et al. 2004, Link 2010, Link et al. 2011a, NEFSC 2012, White et al. 2012, Samhouri et al. 2013, Samoteskul et al. 2014, Maxwell et al. 2015). Beyond the differential and dynamic trade-offs among regional commercial and recreational fisheries, multiple coastal and ocean uses that support the New England marine economy co-occur within areas providing habitat to many managed species, which support these local fisheries, and are foundational for marine tourism and emerging wind energy sites (Andrews & Rossi 1986, Hoagland et al. 2005, Martin & Hall-Arber 2008, Anthony et al. 2009, Link 2010, Link et al. 2011a, Portman et al. 2011, Holland et al. 2012, Bailey et al. 2014, Samoteskul et al. 2014, Link &

Blow Hards: The Trade-Offs Among Conflicting Objectives Between Offshore Windfarms and Scallop Beds

—(c.f. Chapter 8)

If you were going to put a windmill in my neighborhood to generate electricity for all of us living there, some people might not like it in their backyard, but I wouldn't mind that. But it would be downright unkind of you to try and put it in my vegetable garden.

That's essentially what has been happening with offshore wind and major fishing grounds. Energy companies have done their wind studies, have worked with the Bureau of Ocean Energy Management (BOEM; the federal U.S. agency responsible for approving such efforts) to obtain all the appropriate permits, and by all accounts are trying to both do the right thing and make a buck.

But their proposals have been to plop the wind farms right smack dab in the middle of one of the most lucrative fisheries in the country—Atlantic sea scallops (*Placopecten magellicanicus*). It's one thing to set up a wind farm, with all its attendant underwater cables, restricted areas of access, and infrastructure when there is a fishing ground for species that can move, even though that has some concerns. But scallops, although they can move (a little), tend to stick around certain locations. As in, they pile up in good habitat. Just because the mostly sandy, wave-swept bottom is in an area of high wind, or maybe even because of it, doesn't mean that scallops can easily and readily move. They are in those locations for a reason—a composite rationale based on a bunch of oceanographic conditions that make it good for their larvae to settle and that makes it good for them to filter feed on adequate food. Not too deep, not too shallow, but just right. These scallop habitats have to be in the "goldlilocks" of situations, and scallops are designed to do well in those locations. But not necessarily other locations.

And scallops on the east coast weren't doing so well a couple decades ago (Serchuck & Murawski 1997, Murawski et al. 2000, Buchsbaum et al. 2005, Hart & Rago 2006, Rosenberg et al. 2006). Various forms of area closures have led to their population recovery (NEFSC 1997, Fogarty & Murawski 1998, Murawski et al. 2000, Smith & Rago 2004, Truesdell et al. 2016). These closures have provided the breathing room for scallops to reproduce and "seed" surrounding areas. Scallops are now one of the most valuable wild-caught fisheries in the entire U.S. (NMFS 2018, Rheuban et al. 2018).

So, you have arguably one of the bigger fishery success stories, and along comes another ocean-use sector that wants to drop right into the middle of it. This truly highlights the importance of considering all ocean uses systematically, at the very least coordinating across sectors.

Fortunately, BOEM and the National Marine Fisheries Service (NMFS) recognized this situation and the National Environmental Policy Act (NEPA; related to permitting processes) flagged the situation. This highlights the value of such an overarching mandate like NEPA, one that is often underused, but one that provides the opportunity to consider all uses and impacts in an ecosystem simultaneously.

Negotiations are ongoing about the best place(s) to site these offshore windfarms. The fear by many is that as valuable as the scallop fishery is, that it is still small potatoes compared to the energy sector. But at the very least, providing a venue for all parties to talk and to acknowledge that as good as a wind farm may be, it likely isn't worth it if placed in a locale that'll mess up a half a billion-dollar industry.

References

Buchsbaum R, Pederson J, Robinson WE. 2005. *The Decline of Fisheries Resources in New England: Evaluating the Impact of Overfishing, Contamination, and Habitat Degradation*. Cambridge, MA: MIT Sea Grant College Program (p. 175).

Fogarty MJ, Murawski SA. 1998. Large-scale disturbance and the structure of marine systems: fishery impacts on Georges Bank. *Ecological Applications* 8(sp1):S6–22.

Hart DR, Rago PJ. 2006. Long-term dynamics of US Atlantic sea scallop Placopecten magellanicus populations. *North American Journal of Fisheries Management* 26(2):490–501.

Murawski SA, Brown R, Lai HL, Rago PJ, Hendrickson L. 2000. Large-scale closed areas as a fishery-management tool in temperate marine systems: the Georges Bank experience. *Bulletin of Marine Science* 66(3):775–98.

NEFSC (NOAA Northeast Fisheries Science Center). 1997. *Report of the 23rd Northeast Regional Stock Assessment Workshop (23rd SAW). Northeast Fisheries Science Center Reference Document 97-05.* Washington, DC: NOAA (p. 191).

NMFS (National Marine Fisheries Service). 2018. *Fisheries of the United States, 2017. NOAA Current Fishery Statistics No. 2017.* Washington, DC: NOAA.

Rheuban JE, Doney SC, Cooley SR, Hart DR. 2018. Projected impacts of future climate change, ocean acidification, and management on the US Atlantic sea scallop (Placopecten magellanicus) fishery. *PloS One* 13(9).

Rosenberg AA, Swasey JH, Bowman M. 2006. Rebuilding US fisheries: progress and problems. *Frontiers in Ecology and the Environment* 4(6):303–8.

Serchuck FM, Murawski SA. 1997. "The offshore molluscan resources of the northeastern coast of the United States: Surfclams, ocean quahogs, and sea scallops." In: *The History, Present Condition, and Future of Molluscan Fisheries of North and Central America and Europe. Volume 1, Atlantic and Gulf Coasts. NOAA Technical Report NMFS 127.* Edited by CL Mackenzie, VG Burrell, A Rosenfield, WL Hobart. Washington, DC: NOAA (p. 234).

Smith SJ, Rago P. 2004. Biological reference points for sea scallops (Placopecten magellanicus): the benefits and costs of being nearly sessile. *Canadian Journal of Fisheries and Aquatic Sciences* 61(8):1338–54.

Truesdell SB, Hart DR, Chen Y. 2016. Effects of spatial heterogeneity in growth and fishing effort on yield-per-recruit models: an application to the US Atlantic sea scallop fishery. *ICES Journal of Marine Science* 73(4):1062–73.

Marshak 2019). Together with LMRs, these sectors contribute heavily toward the region's marine economy, while their overlapping and interrelated activities emphasize connectivity among this fisheries ecosystem and its other ocean uses. In New England offshore habitats, the effects of wind energy siting on essential fish habitat (EFH) and accessibility to historical fishing grounds are of concern, while the attraction, production, and mortality potential of these large structures toward fishery populations are also receiving attention (Bailey et al. 2014, Samoteskul et al. 2014, Guida et al. 2017, Wyatt et al. 2017, Cruz-Marrero et al. 2019). Increasing marine traffic over the last half-century has resulted in the regular introduction of invasive species into the New England region through ship ballast waters, with increasing prevalence of exotic bivalves, crustaceans, and pathogens attributed to this source (Pederson et al. 2005, Thayer & Stahlnecker 2006, Mathieson et al. 2008a, b, GMCME 2010, Hobbs et al. 2015). While conflicting interests among commercial and recreational fisheries are observed in New England, their overall magnitude and effects have been relatively more muted compared to other regions (Ross & Biagi 1991, Shepherd & Terceiro 1994, Hall-Arber et al. 2001, Ihde et al. 2011). However, conflicts among the full spectrum of fisheries sectors within management entities in this region are among some of the most contentious in the nation (Murawski 1991, Repetto 2001, Brodziak et al. 2004, Hartley & Robertson 2006, Hilborn 2007, Acheson 2011, Acheson & Gardner 2014). Conflicts among fisheries, marine transportation, and protected species are of concern in this region, especially with the effects of gear entanglements and ship strikes on Northern Right Whales (*Eubalaena glacialis*) and other cetaceans (Kenney et al. 1997, Knowlton & Kraus 2001, Russell et al. 2001, Ward-Geiger et al. 2005, Knowlton et al. 2012, Hamilton & Kraus 2019, Howle et al. 2019, Moore 2019). Ecological consequences from fishing and climate change on marine mammal prey have also been observed, affecting trophodynamics in this region (Read & Brownstein 2003, Bakun et al. 2009, NEFSC 2012, Meyer-Gutbrod 2017, Gaichas et al. 2019b). Bycatch associated with longlining and other pelagic gear types is still of concern, with reported marine mammal bycatch in this region being the highest in the nation (Zollett & Rosenberg 2005, Read et al. 2006,

Hatch et al. 2016, Link & Marshak 2019, Orphanides 2019).

As ocean uses expand, more holistic management approaches are necessary to address resource conflicts including and beyond fishing interests (Leslie & McLeod 2007, Ehler & Douvere 2010, Nutters & da Silva 2012, Olsen et al. 2014, Wyatt et al. 2017). Concerns related to the effects of coastal and tourism-related development and LMR habitat quality, and conflicts among marine transportation and recreational interests continue to intensify. Emerging trade-offs between aquaculture and marine energy (and other ocean uses) siting in this region are becoming of increasing priority (Leslie & McLeod 2007, Ehler & Douvere 2010, Nutters & da Silva 2012, White et al. 2012, Samhouri et al. 2013, Olsen et al. 2014, Smythe 2017, Smythe & McCann 2019). Exploring, identifying, and resolving these differences in priorities and goals for the New England ecosystem is the essence of EBFM, and the basis of its EBFM implementation plan and forthcoming Fishery Ecosystem Plan (FEP; NMFS 2019).

3.3 Informational and Analytical Considerations for this Region

3.3.1 Observation Systems and Data Sources

In moving toward a more holistic management approach, data regarding the physical, biological, and socioeconomic components of the New England ecosystem are available through regional observation systems and multiple monitoring programs spanning several decades. The Northeastern Regional Association of Atlantic Coastal Ocean Observing Systems (NERACOOS) provides oceanographic and atmospheric data from multiple sensor stations throughout coastal and offshore locations (NERACOOS 2016). The system is a composite of aggregated data from several federal, academic, and regional sources, with observations from coastal and offshore sensors maintained by NOAA's National Data Buoy Center (NDBC), New England state agencies, universities and institutions, the Coastal Data Information Program (CDIP), and via the Northeast Regional Ocean Council (NROC) data portal (NROC 2020). Since 2001, physical data reported at daily to annual scales have been available at the NERACOOS portal with monitored atmospheric variables including biological (chlorophyll, animal tracking, and abundance), chemical (pH, carbon dioxide, dissolved oxygen, oxygen saturation), and physical parameters (current velocity, wind velocity, wave height, period, and direction, air and water temperature, salinity, air pressure, and sea level). This information is also complemented by underwater gliders, high-frequency radar, satellite data, and other sensors on offshore buoys. These data are being applied toward providing comprehensive information regarding the New England system, in addition to forecasting regional sea-level rise, coastal flooding, and predicting regional climatological, thermohaline, and wind and wave patterns. As seen for other ocean observing systems, NERACOOS priorities also include providing information for safe maritime practices, addressing coastal hazards, forecasting coastal and offshore climatological effects, and improving water quality (NERACOOS 2016). The Gulf of Maine Council on the Marine Environment has also collected and provided regional data on water quality, environmental indicators, and habitat delineations, as featured in its State of the Gulf of Maine Report (GMCME 2010b).

Several oceanographic and atmospheric data sources have been incorporated as part of broader understanding of the New England and greater northeast ecosystem, particularly its overall status (NEFSC 2012, Samhouri et al. 2013, Gaichas et al. 2019b), focal components, and their applicability toward regional ecosystem status reports (ESRs) and Integrated Ecosystem Assessments (IEAs; Levin et al. 2009, NEFSC 2012, Samhouri et al. 2013, DePiper et al. 2017, Gaichas et al. 2019b). Data regarding the AMO and NAO indices and their influences on satellite-derived SST have been applied toward broader understanding of climate forcing in the New England region (NEFSC 2012, Nye et al. 2014, Pershing et al. 2015, Gaichas et al. 2019b). Trends in freshwater input via precipitation and influences of other physical pressures such as position of the Gulf Stream, and wind and thermohaline dynamics, have been derived from NWS datasets and oceanographic data taken aboard NOAA northeast regional surveys characterizing the major water masses throughout the Mid-Atlantic and New England and changes in their stratification. All of this physical information has been incorporated into ESRs for this region and the broader northeastern U.S, including dynamics of the Gulf Stream, Labrador Current, and riverine input to this system (NEFSC 2012, Gaichas et al. 2018, 2019a, b).

Information regarding habitat factors and aspects of primary and secondary production are monitored via satellite information and ongoing monitoring programs throughout the New England ecosystem accounting for water quality, chlorophyll-a concentrations, phytoplankton biomass, and estimated coverage and risk assessments of nearshore vegetated habitats (Fahay et al. 1999, Packer et al. 1999, Lough 2004, NFHP 2010, NEFSC 2012, Crawford et al. 2016, NFHP 2015, NEFSC 2018, Gaichas et al. 2019b). These data also account for phytoplankton blooms. Zooplankton abundance and biomass are monitored in surveys (i.e., Marine Resources Monitoring Assessment and Prediction Program—MARMAP, Northeast Shelf Ecosystem Monitoring Program—ECOMON) from the NOAA Northeast Fisheries Science Center (NEFSC) seasonal monitoring program, which are performed multiple times per year and began in the 1970s (Sibunka & Silverman 1989, Kane 2007, Richardson et al. 2010, Politis et al. 2014).

Comprehensive databases for secondary through upper trophic level species include fishery-independent information collected from seasonal bottom trawl surveys, which account for biomass and abundance of benthic invertebrates, particularly crustaceans, oysters, and sea scallops, while other sessile benthic species have been examined through remotely operated vehicle (ROV) and "HABCAM" (Habitat Mapping Camera) surveys (Azarovitz 1981, NEFC 1988, Smith, 2002, Howland et al. 2006, Taylor et al. 2008, NEFSC 2012, Hourigan et al. 2017, NEFSC 2018, Gaichas et al. 2019b). Most states in this region have surveys (Northeast Area Monitoring and Assessment Program—NEAMAP), which provide fish abundance and associated data for more nearshore locations. They also provide fishery-independent information regarding the biomass and abundance of commercially important finfish species and composition of offshore assemblages. Their data are applied toward broader understanding of ecosystem functioning and informing assessments of stock status for New England fisheries (Reid et al. 1999, Wigley et al. 2003, McElroy et al. 2019). These surveys are also complemented by NOAA Fisheries fishery-dependent monitoring programs for commercial and recreational harvest, landings, revenue, and effort, including surveys such as the Marine Recreational Information Program (MRIP) and fisheries observation data on commercial landings, catches, and bycatch gathered by the National Observer Program (NAS 2017, NMFS 2017b). Factors including trends in average size, spatial distributions, aggregate biomass of pelagic and demersal species groupings and their ratios, predator–prey relationships and mean trophic level, condition indices, and thermal preferences represent many of the ecosystem-level metrics that have been derived from this fishery-dependent and independent survey information (NEFSC 2012, Lynch et al. 2018, NEFSC 2018b, Gaichas et al. 2019a, b). Socioeconomic information for fisheries and other uses, particularly regarding revenue and employment, are also collected by NOAA Fisheries, the National Ocean Economics Program (NOEP), and NOAA's Office for Coastal Management Economics: National Ocean Watch (ENOW) program. Monitoring programs exist for evaluating the status, recovery, and bycatch of protected species that include sea turtles and cetaceans, and several monitoring programs for marine birds, such as the USGS Breeding Bird Survey, are overseen by NOAA, the U.S. Department of Interior, and state agencies. In the New England region and the northeast ecosystem, all of these aforementioned data sources have been applied toward integrative ecosystem measurements that examine cumulative changes in these metrics over time and their interdependent relationships (Link et al. 2002, 2006, 2008a, b, NEFSC 2012, Gaichas et al. 2019b).

3.3.2 Models and Assessments

Assessments of economically important and protected species in New England are conducted by NOAA Fisheries in collaboration with the New England Fishery Management Council (NEFMC; Ferguson et al. 2017, Lynch et al. 2018). Stock assessment models are used to generate estimates of stock status with a multitude of data sources (e.g., landings, catch per unit effort, life history characteristics, survey biomass, etc.; Karp et al. 2019). Stock assessments for New England species are primarily conducted through the Stock Assessment Workshop/Stock Assessment Review Committee (SAW/SARC) process (Lynch et al. 2018), with some transboundary stocks assessed as part of the Transboundary Resources Assessment Committee (TRAC). Some assessments are also conducted in conjunction with the ASMFC, mostly through the SAW process. NOAA Fisheries also examines the effects of commercial and recreational fisheries on New England marine socioeconomics through application of the NMFS Commercial Fishing Industry Input/Output and Recreational Economic Impact Models (NMFS 2018). Data from NOAA Fisheries marine mammal stock abundance surveys are used to conduct population assessments as published in regional marine mammal stock assessment reports and peer reviewed by regional scientific review groups (SRGs; Carretta et al. 2019, Hayes et al. 2019, Muto et al. 2019). The U.S. Fish and Wildlife Service also prepares stock assessment reports for nearshore and inland threatened and protected species under its jurisdiction. Information collected on protected species is also applied toward five-year status reviews of those listed under the Endangered Species Act (ESA).

The New England region has been extensively modeled with regard to its biophysical processes, including bottom-up effects on its primary and secondary production (Franks & Chen 2001, Steele et al. 2007, Link et al. 2008a, b, 2010, Runge et al. 2010, Johnson et al. 2011, Fogarty et al. 2012). In addition, major advances in understanding trophic interactions and predatory effects on pelagic forage species (e.g., herring, mackerel, squid) have occurred in this region, as related to refining species biological reference points and describing overall system functioning (Overholtz et al. 2000, Tsou & Collie 2001, Read & Brownstein 2003, Overholtz & Link 2007, Overholtz et al. 2008, Tyrrell et al. 2008, Link et al. 2009, Moustahfid et al. 2009a, b, Garrison et al. 2010, Link et al. 2011a). Available data and species assessments have also been applied toward broader ecosystem models for the New England region, particularly the Gulf of Maine and Georges Bank. These models have identified the importance of bottom-up processes, high productivity, and small pelagic species in driving many of the dynamics of the New England fisheries ecosystem (Link et al. 2005, Link et al. 2006, Steele et al. 2008, Link et al. 2008a, b, Overholtz et al. 2008, Link et al. 2009a, 2010). Their findings have also advanced ecosystem approaches to fisheries management (EAFM) and led to more comprehensive stock assessment modeling approaches that incorporate broader ecological information (Moustahfid et al. 2009a, b, Link et al. 2011a, b, Tyrrell et al. 2011, Fogarty et al. 2012, Gaichas et al. 2016). Efforts to incorporate IEA products (e.g., ESRs) into single-species stock assessments and LMR management decisions have been ongoing for some time in this region (Townsend et al. 2008, Link et al. 2010b, Townsend et al. 2014, 2017).

Stock assessments can span a range of models, including functional group production models and catch-at-age models, which incorporate trophic information, while emerging economic modules also examine aspects of revenue, profitability, and fisheries portfolios. Extending these assessment modeling approaches to multispecies has shown some promise in this region (Link et al. 2011b). Biomass dynamic models, multispecies virtual population analyses (MSVPA) and their extended versions (MSVPA-X) have also been applied in these contexts (Tsou & Collie 2001, Garrison & Link 2004, Overholtz et al. 2008, Tyrrell et al. 2008, 2011, Gamble & Link 2009, 2012). A length-based multispecies model (Hydra) examines the effects of the environment on growth, maturity, fecundity, and recruitment, while components of dietary relationships, spatial variability, and the influences of multiple fishing fleets are also being incorporated (Gaichas et al. 2017, Townsend et al. 2017). A multispecies statistical catch-at-age model has also been developed and applied for this region (Curti et al. 2013). Current ecosystem modeling efforts throughout the New England region are closely associated with the northeast IEA program, which employs models that range from the single-species to ecosystem level (Gaichas et al. 2016, 2017, Townsend et al. 2017). These include qualitative conceptual models of Georges Bank and the Gulf of Maine that are based on submodels regarding aspects of fisheries ecology, climate, and socioeconomic components (ICES 2016, DePiper et al. 2017). Aggregate group production models have also been used to account for trade-offs between management objectives related to fisheries and marine mammals, including the effects of human and fishery interactions and trophic relationships on their biomass (Link et al. 2011a, b, Fogarty et al. 2012, Gaichas et al. 2012a, b, Gamble & Link 2012, Smith et al. 2015). Trade-offs at scales ranging from those at the species to ecosystem level have been investigated to calculate set limits on ecosystem removals based on total system productivity as applied to specific functional groups and bioeconomic portfolios (Link et al. 2008a, b, 2010a, 2011, Gaichas et al. 2012a, b, 2016). These models are also investigating the ecosystem responses of varying trophic levels to differential fishing effort (Townsend et al. 2014, 2017).

The effects of decreasing forage fish biomass on predatory fishes throughout the New England ecosystem have also been accounted for in larger models, including Ecopath with Ecosim (EwE; Polovina 1984; Pauly et al. 2000; Christensen and Walters 2004; Christensen et al. 2005) and Atlantis-based platforms (Fulton et al. 2007, 2011) for the northeastern EEZ, which also examined influences of climate change and fisheries-related metrics on species functional groups (Link et al. 2008a, b, 2010a, 2011, Nye et al. 2013,

Bite Me: The Importance of Considering Predation Mortality for Small Pelagic Fishes and Invertebrates as it Informs Stock Assessment Models and Harvest Control Rules/Biological Reference Points

—(c.f. Chapters 4, 8, and 9)

What if there were a fishery that was removing 5 to 10 times more fish than all the other fishing fleets targeting that fish combined? And what if we totally ignored that bigger fishery? There would be an outrage. There would be cries of not using the best available information. There would be concerns that we'd be missing not only important dynamics of that stock, but also be totally off on the magnitude of anything to do with that stock.

Yet, that is typically what we do when it comes to evaluating and managing forage fish.

Pick your favorite small harvested marine critter. Herring. Sardines. Mackerel, menhaden, squids, shrimp, krill, anchovies, whiting, and so on. All of them are eaten by a lot of other critters. However, for only a few of these species do we actually account for this predation in their assessments and management.

We ignore predation mortality of most forage fish (and commercially targeted invertebrates). It has clearly been shown that predators remove 5–10× more biomass than fishing fleets (Cohen et al. 1982, Bax 1991, Overholtz et al. 1991, 2000, Wysujack & Mehner 2002, Gamble & Link 2009, Moustahfid et al. 2009a, b, Tyrrell et al. 2011). The biological reference points upon which these forage fish are managed can be 2–3, even 5× off (Overholtz et al. 2000, Collie & Gislason 2001, Overholtz et al. 2008, Gamble & Link 2009, Moustahfid et al. 2009a, b, Tyrrell et al. 2011, Holsman et al. 2016). These refer to both total biomass and fishing pressure, which leads to underestimates of total mortality at a third to a half of what the stock is really experiencing.

We have the analytical tools to incorporate predation mortality (M2) into most assessments. Both adding on to single-species and via multispecies models. But these efforts are not routinely performed or utilized.

We know that predators like seabirds, marine mammals, large, highly migratory fish, and other demersal fishes all feed on and even rely on forage fish. We have gross estimates of the feeding demands of these groups. But we don't typically include them when evaluating forage fish stock dynamics.

Mostly the arguments against doing so are arcane statistical, sampling, and analytical caveats that are not defensible upon first principles. Secondarily the arguments are a

continued

Bite Me: *Continued*

dearth of data, or at least insufficient data density. So what? At the very least, some accommodation of this known source of removal needs to be included, even if as "priors" in a Bayesian sense or as risk-based policy modifiers.

Fortunately, the number of assessments in the U.S. including ecosystem considerations in general has doubled over the past 10 years (Skern-Mauritzen et al. 2016, Lynch et al. 2018, Marshall et al. 2019). But it still has only gone from 4 to 8%. More could be done.

Fortunately, a number of management bodies have begun to adopt forage limits or ecosystem component species considerations under the National Standard Guidelines (NS 1) of the Magnuson-Stevens Act to account for the pivotal role of these fish in the ecosystem.

Without accounting for this source of mortality, however that happens, for these forage fish, we will miss arguably one of the easiest opportunities and most avoidable problems facing us as we continue to attempt to adopt EBFM.

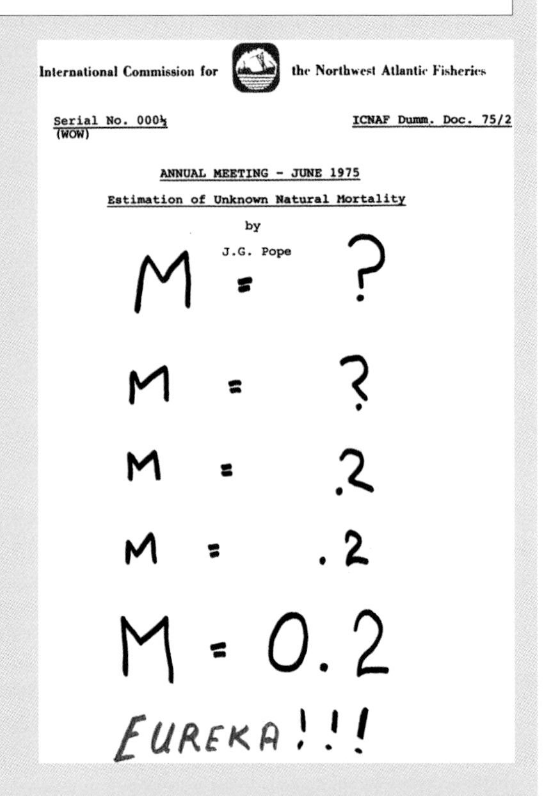
International Commission for the Northwest Atlantic Fisheries

Serial No. 000½ (WOW) ICNAF Dumm. Doc. 75/2

ANNUAL MEETING - JUNE 1975

Estimation of Unknown Natural Mortality

by

J.G. Pope

M = ?

M = ?

M = .2

M = .2

M = 0.2

EUREKA!!!

References

Bax N. 1991. A comparison of fish biomass flow to fish, fisheries and marine mammals for six marine ecosystems. *In ICES Mar Science Symposium* 193:217–24.

Cohen EB, Grosslein MD, Sissenwine MP, Steimle F, Wright WR. 1982. An energy budget for Georges Bank. *Canadian Special Publication of Fisheries and Aquatic Sciences* 59:95–107.

Collie JS, Gislason H. Biological reference points for fish stocks in a multispecies context. *Canadian Journal of Fisheries and Aquatic Sciences* 2001 Nov 1;58(11):2167–76.

Gamble RJ, Link JS. 2009. Analyzing the tradeoffs among ecological and fishing effects on an example fish community: a multispecies (fisheries) production model. *Ecological Modelling* 220(19):-2570–82.

Holsman KK, Ianelli J, Aydin K, Punt AE, Moffitt EA. 2016. A comparison of fisheries biological reference points estimated from temperature-specific multi-species and single-species climate-enhanced stock assessment models. *Deep Sea Research Part II: Topical Studies in Oceanography* 134:360–78.

Lynch PD, Methot RD, Link JS. 2018. *Implementing a Next Generation Stock Assessment Enterprise.* An update to the NOAA fisheries stock assessment improvement plan. US Dep. Commer. NOAA Tech. Memo. NMFS-F/SPO-183.127.

Marshall KN, Koehn LE, Levin PS, Essington TE, Jensen OP. 2019. Inclusion of ecosystem information in US fish stock assessments suggests progress toward ecosystem-based fisheries management. *ICES Journal of Marine Science* 76(1):1–9.

Moustahfid H, Link JS, Overholtz WJ, Tyrrell MC. 2009a. The advantage of explicitly incorporating predation mortality into age-structured stock assessment models: an application for Atlantic mackerel. *ICES Journal of Marine Science* 66(3):445–54.

Moustahfid H, Tyrrell MC, Link JS. 2009b. Accounting explicitly for predation mortality in surplus production models: an application to longfin inshore squid. *North American Journal of Fisheries Management* 29(6):1555–66.

Overholtz WJ, Jacobson LD, Link JS. An ecosystem approach for assessment advice and biological reference points for the Gulf of Maine–Georges Bank Atlantic herring complex. North American Journal of Fisheries Management. 2008 Feb;28(1):247–57.

Overholtz W, Link JS, Suslowicz LE. 2000. Consumption of important pelagic fish and squid by predatory fish in the northeastern USA shelf ecosystem with some fishery comparisons. *ICES Journal of Marine Science* 57:1147–59.

Overholtz WJ, Murawski SA, Foster KL. 1991. Impact of predatory fish, marine mammals, and seabirds on the pelagic fish ecosystem of the northeastern USA. *ICES Marine Science Symposia* 93:198–208.

Skern-Mauritzen M, Ottersen G, Handegard NO, Huse G, Dingsør GE, Stenseth NC, Kjesbu OS. 2016. Ecosystem processes are rarely included in tactical fisheries management. *Fish and Fisheries* 17(1):165–75.

Tyrrell MC, Link JS, Moustahfid H. 2011. The importance of including predation in fish population models: implications for biological reference points. *Fisheries Research* 108(1):1–8.

Wysujack K, Mehner T. 2002. Comparison of losses of planktivorous fish by predation and seine-fishing in a lake undergoing long-term biomanipulation. *Freshwater Biology* 47(12):2425–34.

Fay et al. 2017, Townsend et al. 2017). A multispecies production assessment model is also examining species interactions that account for predation, functional responses, and competition and how they vary spatially. This information has been directly applied toward a multispecies assessment for Georges Bank that accounts for trophic interactions, environmental and climate trends, and socioeconomic information (Townsend et al. 2018). More recently, a Management Strategy Evaluation (MSE) was undertaken for Atlantic herring, which relied on a multispecies modeling framework and was built on the previously mentioned studies (Deroba et al. 2019, Townsend et al. 2019). Similar efforts have been applied for Atlantic surfclam (Hoffmann et al. 2018). There are other models that inform management discussions in this region and its subregions (Kenney et al. 1995, Holland & Sutinen 1999, Holland 2000, Franks & Chen 2001, Jin et al. 2003, Runge et al. 2010, Chen et al. 2011, Nye et al. 2013, Grieve et al. 2017, Link et al. 2017, Tanaka et al. 2019), but most of these have not been directly incorporated into the SAW/SARC process noted above. However, the New England region has been among the leaders in terms of system modeling and incorporating environmental and ecosystem-based considers into its assessments, but it remains difficult to get the full ecological or environmental data integrated fully into operational stock assessments.

3.4 Evaluation of Major Facets of EBFM for this Region

To elucidate how well the New England ecosystem is doing in terms of LMR management and progress toward EBFM, here we characterize the status and trends of socioeconomic, governance, environmental, and ecological criteria pertinent toward employing an EBFM framework for this region, and examine their inter-relationships to ascertain its successful implementation. We also quantify those factors that emerge as key considerations under an ecosystem-based approach for the region and for four major subregions of interest. Ecosystem indicators for the New England region related to the (1) human environment and socioeconomic status, (2) governance system and LMR status, (3) natural environmental forcing and features, and (4) systems ecology, exploitation, and major fisheries of this system are presented and synthesized below. We explore all these facets to evaluate the strengths and weaknesses along the pathway:

$$PP \rightarrow B_{\text{targeted, protected spp, ecosystem}} \leftrightarrow L_{\uparrow\text{targeted spp, }\downarrow\text{bycatch}} \rightarrow \text{jobs, economic revenue.}$$

Where PP is primary production, B is biomass of either targeted or protected species or the entire ecosystem, L is landings of targeted or bycaught species, all leading to the other socioeconomic factors. This operates in the context of an ecological and human system, with governance feedbacks at several of the steps (i.e., between biomass and landings, jobs, and economic revenue), implying that fundamental ecosystem features can determine the socioeconomic value of fisheries in a region, as modulated by human interventions.

3.4.1 Socioeconomic Criteria

3.4.1.1 Social and Regional Demographics

In the New England region, high coastal human population values and population densities (Fig. 3.3) have persisted from 1970 to present with values currently around 9.3 million residents and ~160 people km^{-2}, respectively. This region makes up 3.1% of the United States total population (2010), while density values have risen at ~1.2 people km^{-2} year^{-1}. Human population is heavily concentrated in the Long Island Sound subregion, given its proximity to New York City (NY), Suffolk and Nassau Counties (Long Island, NY), and the cities of Bridgeport, Fairfield, New Haven, and New London (CT). High population concentrations are found near Narragansett Bay, with proximity to Providence (RI) and Bristol County (MA). Comparatively lower populations occur in areas surrounding Penobscot and Casco Bay, which includes the city of Portland (ME) and the counties of Cumberland and Sagadahoc, while three coastal counties (Hancock, Knox, Waldo) surround Penobscot Bay. Population trends have remained relatively steady throughout all New England subregions over the past 40 years, including near Long Island Sound (increasing by ~21 thousand people year^{-1}), with currently up to 9.6 million residents and 573.7 people km^{-2} in surrounding areas. Similar trends have been observed for Narragansett Bay (increasing by ~5200 people year^{-1}), which while having ~1.6 million residents also has a concentrated human population at 276.4 people km^{-2}. Increases in human population have been less pronounced in Casco (~2500 people year^{-1}) and Penobscot Bays (~1150 people year^{-1}), where ~550 thousand (77 people km^{-2}) and 133 thousand residents (11.8 people km^{-2}) are found, respectively.

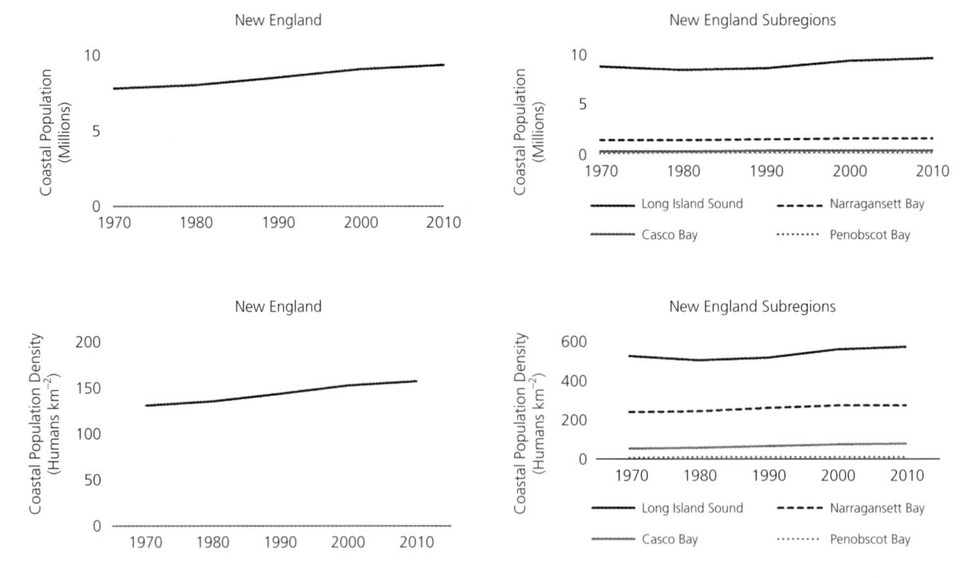

Figure 3.3 Decadal trends (1970–2010) in coastal human population (top) and population density (bottom; humans km^{-2}) for the New England region and identified subregions. Note: Values for Long Island Sound also include New York surrounding counties.

Data derived from U.S. censuses and taken from https://coast.noaa.gov/digitalcoast/data/demographictrends.html.

3.4.1.2 Socioeconomic Status and Regional Fisheries Economics

High numbers of LMR establishments and employments, and high LMR-associated Gross Domestic Product (GDP) values (~$1 billion USD), are found throughout the New England region (Fig. 3.4). Second-highest in the nation, New England contains significant proportions of LMR-based establishments, employments, and GDP value, which especially contribute toward the national ocean economy. Although they have been declining since 2005, LMR establishments continue to comprise 8–9% of regional ocean economy establishments and LMR employments have contributed ~3–4% of regional ocean employments over time. LMR-associated GDP has slowly increased over time, and contributes ~6–7% toward the regional total ocean economy. Values have remained relatively steady throughout subregions, with the highest numbers of LMR establishments and employments historically occurring in Long Island Sound counties (284.6 ± 3.5 standard error (SE) establishments; 1013.3 ± 25.3 SE employments), and GDP value historically highest within both Long Island Sound ($67.8 ± 3.2 million SE, USD) and Narragansett Bay ($67.4 ± 4.4 million SE, USD) counties. Increasing numbers of LMR establishments have been observed over the past decade in Penobscot Bay, comprising up to 23% of total ocean economy establishments. LMR establishments in the other three subregions have remained relatively steady, with establishments in Casco Bay comprising ~8–10% of total establishments and those in Long Island Sound and Narragansett Bay contributing ~2–3% to their total ocean economies. Overall, total LMR employments and GDP among subregions have also remained steady with Casco Bay and Penobscot Bay LMR employments contributing the highest percentages to their total ocean economies (~2–4%) and Casco Bay and Narragansett Bay contributing the highest percentage of LMR associated GDP (~2–6%). In more recent years, increasing contributions of LMR employments and GDP to the Penobscot Bay total ocean economy have been observed.

Since 1990, high numbers of permitted fishing vessels have operated in New England (Fig. 3.5), with values expanding up to ~15,000 during the early 1990s. Following lower reported vessel numbers in the mid to late 1990s, values have remained >10,000 since the 2000s. However, in more recent years, decreases in numbers have been observed with values currently similar to those of the early 1990s. Total revenue (year 2017 USD) of landed commercial fishery catches (Fig. 3.6) has increased over time, peaking at $1.2 billion in 2015 and currently second-highest in the nation. Among subregions, commercial catch revenues have similarly increased, with highest values observed in Narragansett (~$94 million) and Penobscot Bays ($89 million) as of 2016, and peak values for Long Island Sound (~$133 million) occurring during 1996.

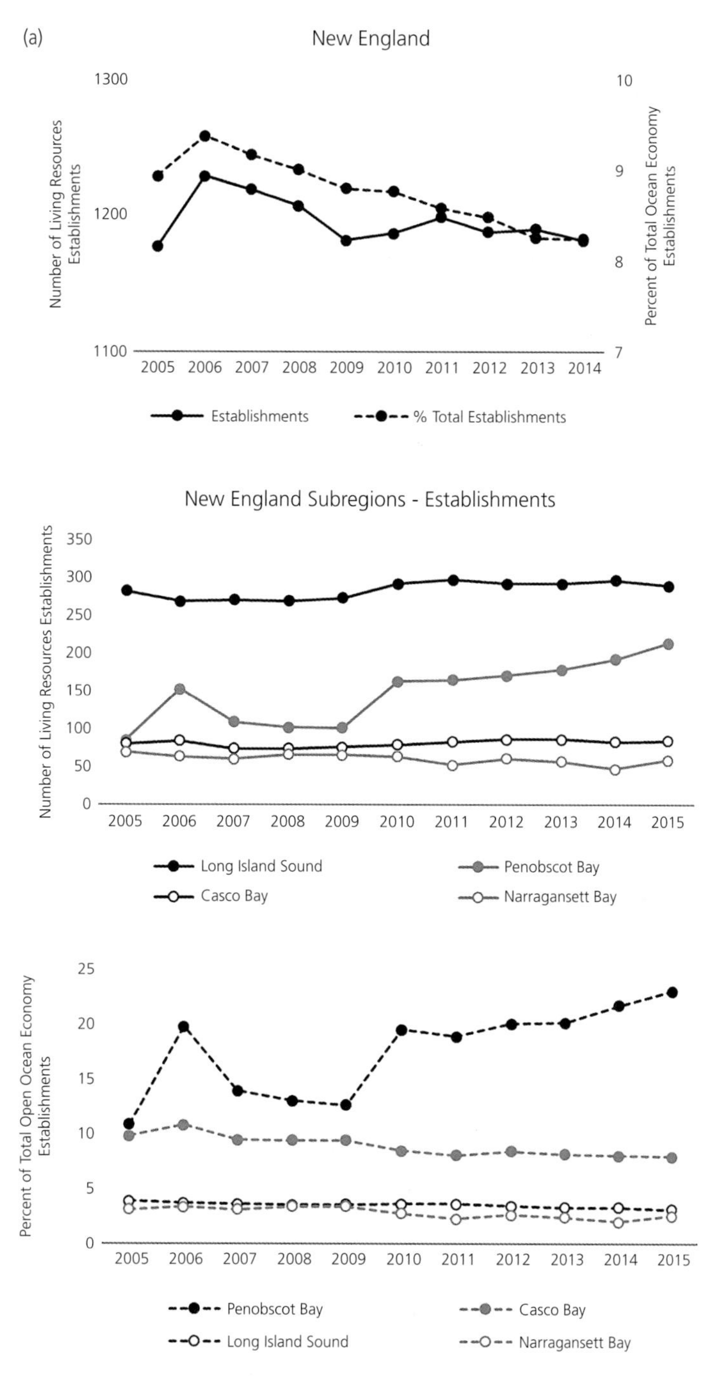

Figure 3.4a The number of living marine resources establishments and their percent contribution to total multisector oceanic economy establishments in the New England region and identified subregions (years 2005–2015).

Data derived from the National Ocean Economics Program.

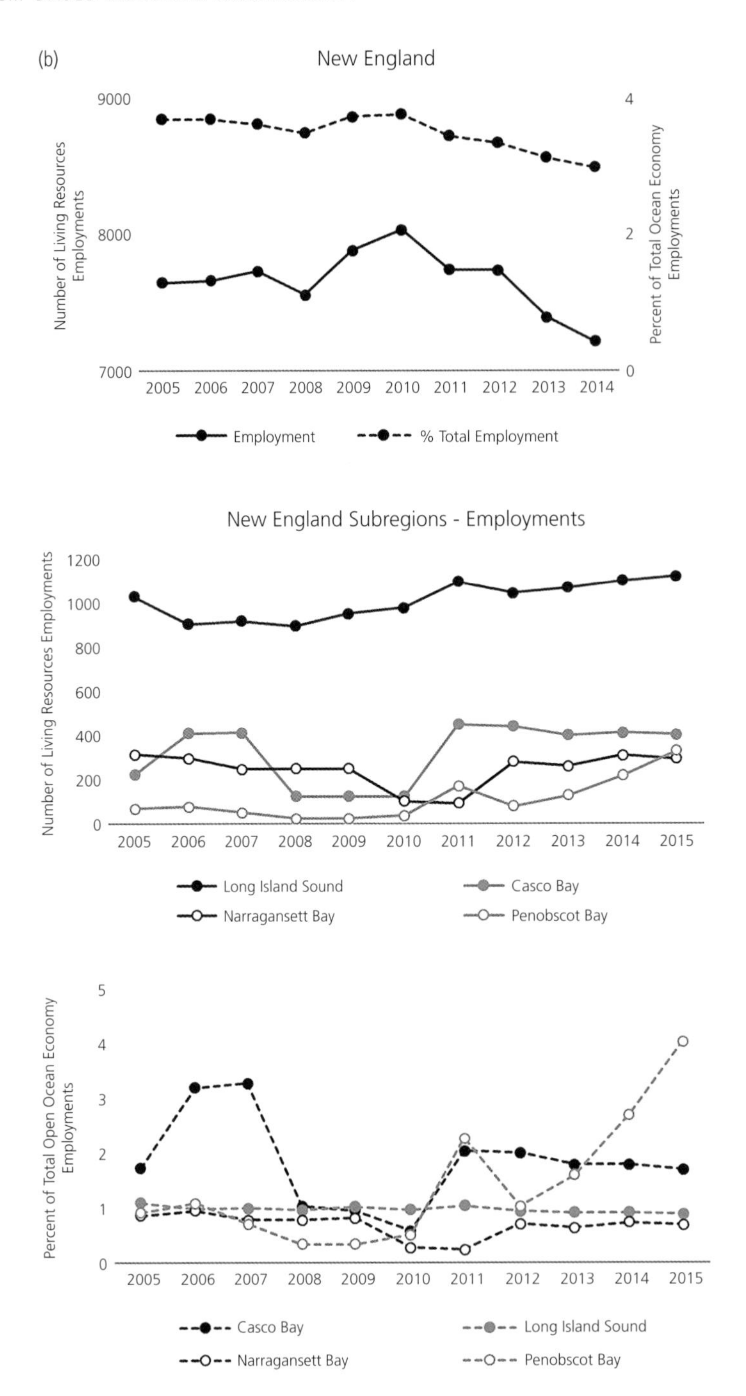

Figure 3.4b The number of living marine resources employments and their percent contribution to total multisector oceanic economy employments in the New England region and identified subregions (years 2005–2015).

Data derived from the National Ocean Economics Program.

Figure 3.4c Gross domestic product value (GDP; USD) from living marine resources revenue and percent contribution to total multisector oceanic economy GDP for the New England region and identified subregions (years 2005–2015).

Data from the National Ocean Economics Program.

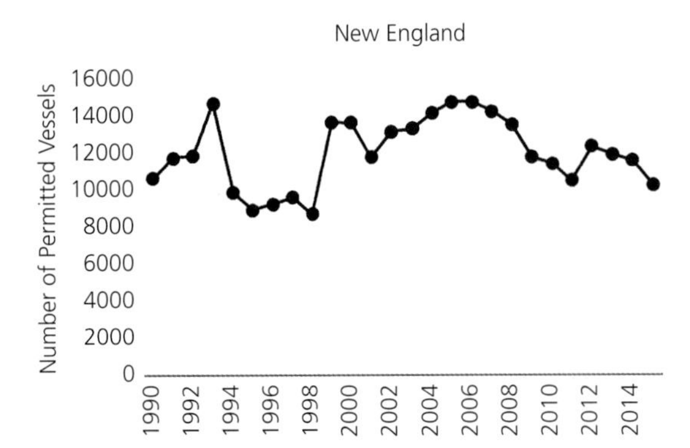

Figure 3.5 Total number of permitted vessels for the New England region over time (years 1990–2015).

Data derived from NOAA National Marine Fisheries Service Council Reports to Congress.

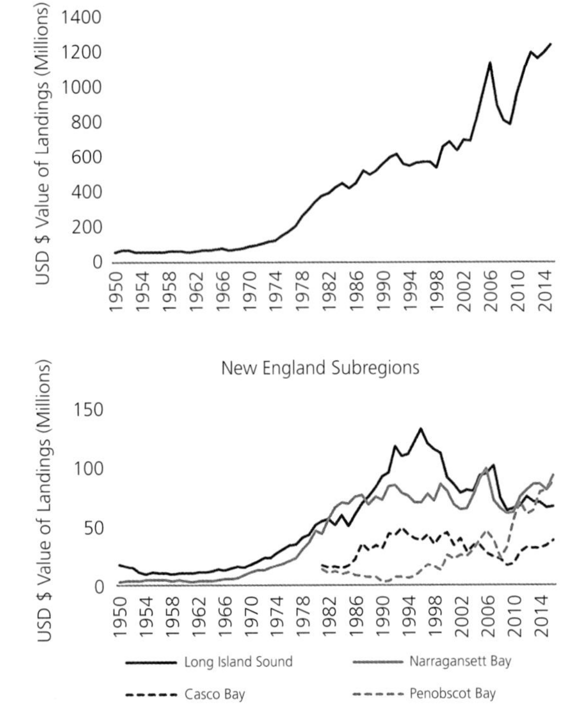

Figure 3.6 Total revenue (Year 2017 USD) of landed commercial fishery catches for the New England region (years 1950–2015) and identified subregions (years 1950–2016), over time.

Data derived from NOAA National Marine Fisheries Service commercial fisheries statistics.

Ratios of commercial fisheries revenue to total biomass (up to $140/metric ton; Fig. 3.7) and jobs to biomass (up to 1.9 jobs/thousand metric tons) are second- and third-highest nationally, respectively, in New England (tied with the Mid-Atlantic) where revenue/biomass values have increased and jobs/biomass has decreased over time. Ratios of the same variables noted above relative to primary production (metric tons wet weight y^{-1}) are respectively first- and second-highest in the nation. Jobs/productivity ratios have ranged from 13.9 to 15.0 jobs/million metric tons wet weight y^{-1} over the past decade, while ratios of total commercial landings revenue to productivity have increased steadily over time to ~$1.5/metric ton wet weight y^{-1}.

3.4.2 Governance Criteria

3.4.2.1 Human Representative Context

Twenty-two federal and state agencies and organizations operate within the New England region and are directly responsible for fisheries and environmental management (Table 3.1). Fisheries resources throughout state waters (0–3 nautical miles) are managed by individual state agencies and the ASMFC, while those in federal waters are managed by both NOAA Fisheries (Greater Atlantic Region) and the NEFMC. Casco and Penobscot Bays fall under the jurisdiction of the Maine Department of Marine Resources (DMR). Long Island Sound encompasses both New York (Mid-Atlantic federal region) and Connecticut (New England) state waters under the jurisdictions of the Connecticut Department of Energy and Environmental Protection and New York Department of Environmental Conservation—Division of Fish and Wildlife. Rhode Island and a portion of Massachusetts state waters surround Narragansett Bay, which is under the jurisdictions of the Massachusetts Department of Fish and Game—Division of Marine Fisheries (DFG) and Rhode Island Department of Environmental Management—Marine Fisheries Section (DEM). Broader efforts toward EBM and consideration of other ocean uses have also been undertaken by the Northeast Regional Planning Body in accordance with the Northeast Ocean Plan (NRPB 2016).

Nationally, New England contains the fourth-highest number of Congressional representatives, which are concentrated within five states (Fig. 3.8), and second-highest Congressional representation when standardized per mile of shoreline. Historically, composition of the NEFMC has been dominated by representatives

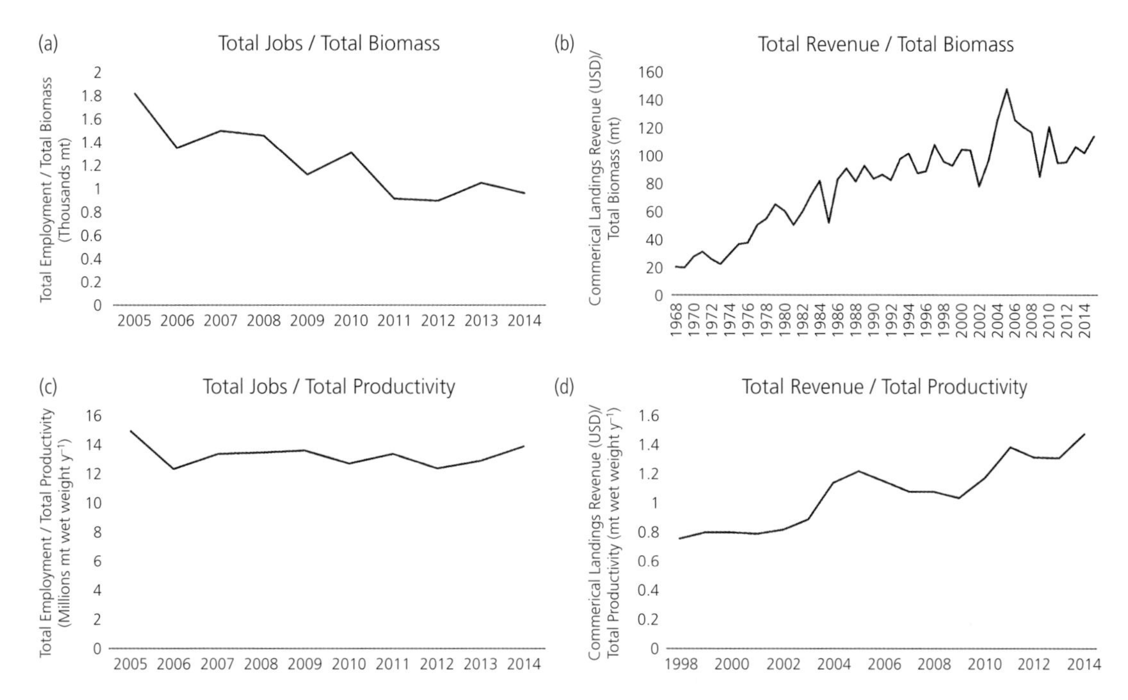

Figure 3.7 Over time, ratios of (a) total living marine resources employments to total biomass (thousands of metric tons); (b) total commercial landings revenue (USD) to total biomass (metric tons); (c) total living marine resources employments to total productivity (metric tons wet weight year^{-1}); (d) total commercial landings revenue (USD) to total productivity (metric tons wet weight year^{-1}) for the New England region.

from the commercial fishing sector, with expanding representation from the recreational sector and those of other interests in more recent years (Fig. 3.9a). The NEFMC also contains the second-highest number of appointed council members nationally, while regional representatives from New England also serve on the Atlantic Highly Migratory Species Advisory Panel. The Atlantic marine mammal SRG is responsible for advising federal agencies on the status of marine mammal stocks covered under the Marine Mammal Protection Act (MMPA), for regions including New England. Its composition and membership have remained steady over time (Fig. 3.9b), with individuals from the academic and private institutional sectors making up the majority of its representatives.

3.4.2.2 Fishery and Systematic Context

Seven New England-specific fishery management plans (FMPs) are managed by NOAA Fisheries and the NEFMC, these include plans for northeast multispecies groundfish (10 species, some with multiple stocks per species); small mesh multispecies groundfish (3 species of hake); Atlantic sea scallop; Atlantic salmon; deep sea red crab; Atlantic herring; and the Northeast skate complex (7 species; Table 3.2). Two more FMPs are shared among the NEFMC and Mid-Atlantic Fishery Management Council (MAFMC) for Monkfish and Spiny dogfish. In total, these FMPs have been subject to 191 modifications (i.e., amendments, frameworks, specifications), which is highest in the nation. The ASMFC manages 26 FMPs for state fisheries resources, including bony fishes (20 species), coastal sharks (22 species), horseshoe crab, Jonah crab, northern shrimp, and notably American lobster. In total, Atlantic state-wide FMPs have been subject to 230 modifications over time. Currently, no FEP exists for the New England region; however, one is in development (NMFS 2019).

One National Seashore (Cape Cod), seven coastal national memorials/parks or historic sites, one NOAA Habitat Focus Area (Penobscot River), four National Estuarine Research Reserves (NERRs), and one major marine sanctuary (Stellwagen Bank) are found throughout the New England region (Table 3.3). Nineteen inshore and offshore areas are identified as Habitat Areas of Particular Concern (HAPCs) in New England, including 16 offshore canyons grouped into 11 HAPCs. Twenty-one named fishing closure zones occur in New England, while fishing is prohibited in only 7 specific areas and restricted in 62 of them

Table 3.1 Federal and state agencies and organizations responsible for fisheries and environmental consulting and management in the New England region and subregions

Region	Subregion	Federal Agencies with Marine Resource Interests	Council	State Agencies & Cooperatives	Multi-Jurisdictional
New England		U.S. Department of Commerce **DoC** (NOAA Fisheries Northeast Fisheries Science Center **NEFSC**, Greater Atlantic Regional Fisheries Office **GARFO**, National Marine Sanctuary Program **NMS**); U.S. Department of Interior **DoI** (U.S. Fish and Wildlife Service **FWS**, U.S. Bureau of Ocean Energy Management **BOEM**, U.S. National Park Service **NPS,** U.S. Geological Survey **USGS**); U.S. Environmental Protection Agency **EPA**	New England Fishery Management Council **NEFMC**	Atlantic States Marine Fisheries Commission **ASMFC**; Connecticut Department of Energy and Environmental Protection; Maine Department of Marine Resources **DMR**; Massachusetts Department of Fish and Game—Division of Marine Fisheries **DFG**; New Hampshire Fish and Game; Rhode Island Department of Environmental Management – Marine Fisheries Section **DEM**	International Commission for the Conservation of Atlantic Tunas **ICCAT**; North Atlantic Salmon Conservation Organization **NASCO**; North Atlantic Fisheries Organization **NAFO**; Transboundary Management Guidance Committee **TMOC**; Transboundary Resources Assessment Committee **TRAC**
	Casco Bay	**DoC** (**NEFSC**, **GARFO**); **DoI** (**FWS**, **BOEM**, **USGS**); **EPA**	**NEFMC**	**DMR**	**NASCO; NAFO; TMOC; TRAC**
	Long Island Sound	**DoC** (**NEFSC**, **GARFO**); **DoI** (**FWS**, **BOEM**, **USGS**); **EPA**	Mid-Atlantic Fishery Management Council **MAFMC**; **NEFMC**	Connecticut Department of Energy and Environmental Protection; New York Department of Environmental Conservation—Division of Fish and Wildlife	**NAFO**
	Narragansett Bay	**DoC** (**NEFSC**, **GARFO**); **DoI** (**FWS**, **BOEM**, **NPS, USGS**); **EPA**	**NEFMC**	**DEM, DFG**	**NAFO**
	Penobscot Bay	**DoC** (**NEFSC**, **GARFO**); **DoI** (**FWS**, **BOEM**, **USGS**); **EPA**	**NEFMC**	**DMR**	**NASCO; NAFO; TMOC; TRAC**

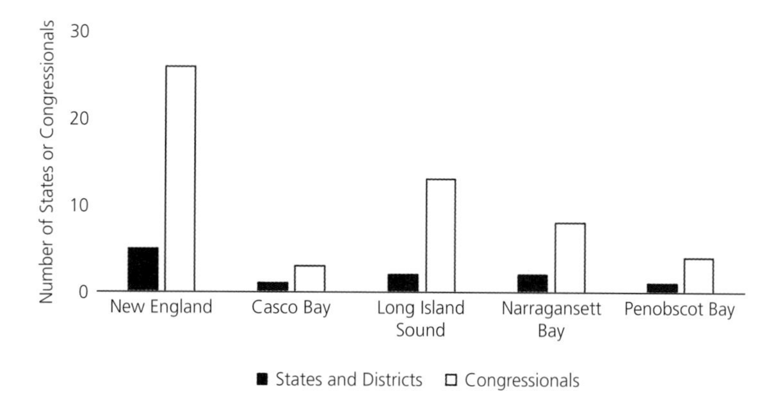

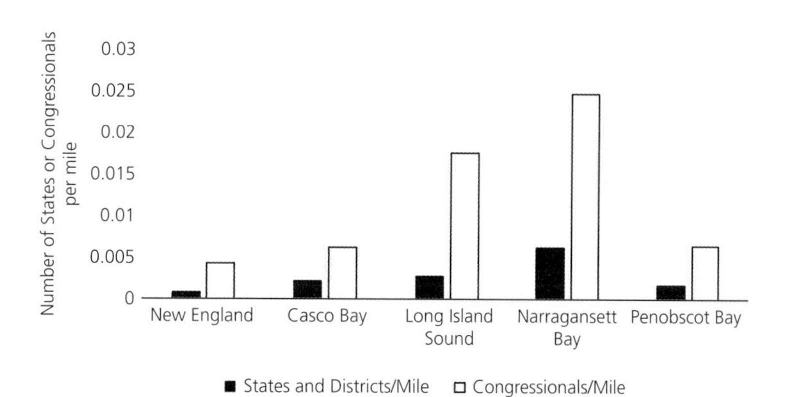

Figure 3.8 Number of states and U.S. Congressional representatives (top panel) and as standardized by miles of shoreline (bottom panel) for the New England region and its identified subregions.

(Fig. 3.10). Few of these closures occur in New England subregions, of which Penobscot Bay contains the highest number with 3 closures and 6 restricted areas. No large-scale prohibited fishing areas are identified in New England subregions, but there are many in the individual states. Very little area of New England waters contains permanent fishing prohibitions (~0.16% of the EEZ), while ~136,000 km^2 is subject to some fishing restrictions, including major areas on Georges Bank. Among subregions, the largest restricted area occurs in Long Island Sound (~3200 km^2) as part of the Southern Nearshore Trap/Pot Waters Area, while all portions of Narragansett, Casco, and Penobscot Bays are subject to fishing restrictions defined by the Northern Nearshore Trap/Pot Waters Area and northeast Gillnet Waters Area. Portions of Penobscot Bay are restricted by the Northeast Closure Area and the Gulf of Maine Rolling Closure Areas, the latter of which also encompass the waters of Casco Bay.

3.4.2.3 Organizational Context

Relative to the total commercial value of its managed fisheries, the NEFMC budget is second lowest when compared to other regions (Fig. 3.11), and makes up a very small proportion relative to the value of its fisheries. These values increased in the early 2010s and have remained relatively stable over time. Cumulative National Environmental Policy Act Environmental Impact Statement (NEPA-EIS) actions from 1987 to 2016 in New England have been relatively low compared to other regions with 38 observed over the past ~30 years (Fig. 3.12). Relatively few actions have occurred within New England subregions, with the highest cumulative value of 14 actions occurring near Long Island Sound. There are 19 fisheries-related lawsuits that have occurred in the New England region since 2010, making it the most litigious among all marine regions in the country.

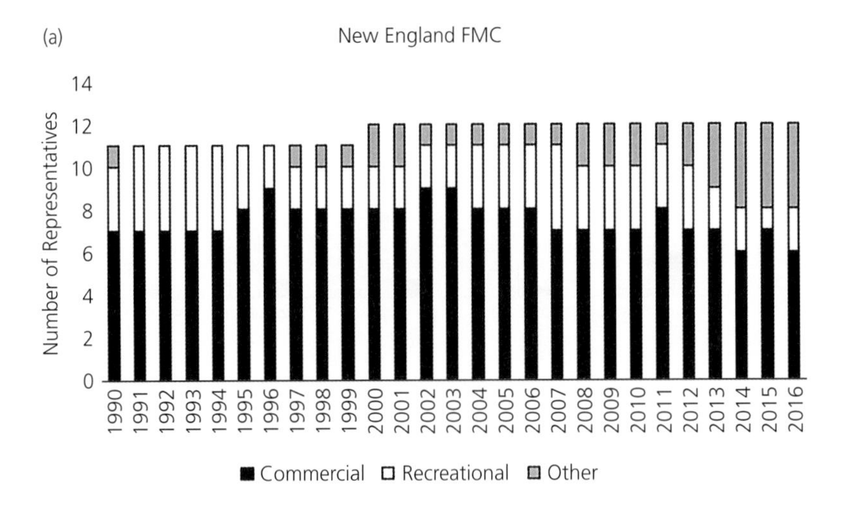

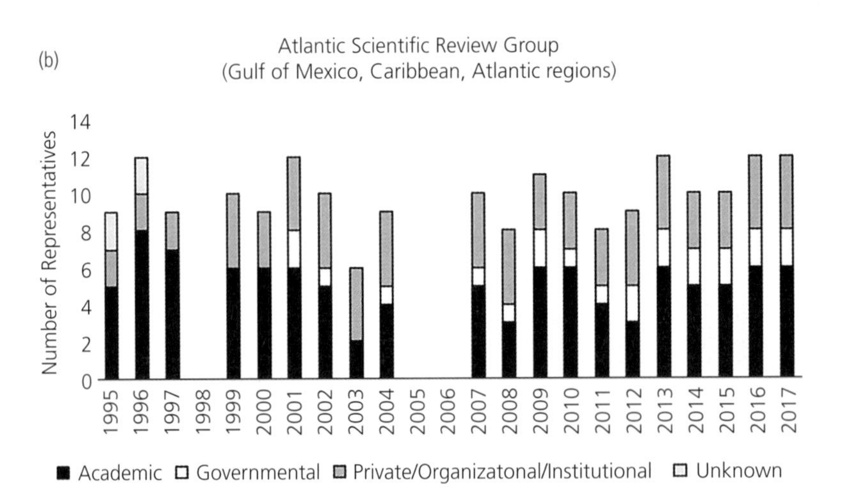

Figure 3.9 (a) Composition of the New England Fishery Management Council and the (b) Atlantic marine mammal scientific review group (data unavailable 1998, 2005–2006) over time.

Data derived from NOAA National Marine Fisheries Service Council Reports to Congress and scientific review group reports.

3.4.2.4 Status of Living Marine Resources (Targeted and Protected Species)

Nationally, the New England region contains the second-lowest number of managed species. Of the 81 federally managed species in the region, 26 are targeted fishery species covered under FMPs, while 6 federally managed bony fishes and 3 skate species, and 19 Atlantic-wide ASMFC managed shark species are prohibited from harvest. Forty-six species of marine mammals, sea turtles, and fishes are federally protected under the ESA or MMPA (Fig. 3.13, Tables 3.2, 3.4–3.5). An overlapping 68 species are also managed under the state jurisdiction of the ASMFC.

As of mid-2017, 39 stocks were federally managed in New England with 38 listed as a NOAA Fish Stock Sustainability Index (FSSI) stock and Atlantic salmon-Gulf of Maine listed as the only non-FSSI stock (Fig. 3.14; NMFS 2017a, 2020a).[1] Among these stocks, six were identified as experiencing overfishing (**Atlantic cod**—*Gadus morhua*, Georges Bank and Gulf of Maine stocks; **Yellowtail flounder**—*Pleuronectes ferruginea*, Cape Cod/Gulf of Maine, Georges Bank, S. New England/Mid-Atlantic stocks; and **Winter flounder**—*Pseudopleuronectes americanus*, Georges Bank stock), and 14 were overfished (**Atlantic cod**, Georges Bank and Gulf of Maine stocks; **Windowpane flounder**—*Scophthalmus aquosus*, Gulf of

[1] As of June 2020, only 25 stocks are listed as FSSI stocks, and all other stocks (n = 14; Atlantic salmon-Gulf of Maine; Deep sea red crab-Northwestern Atlantic; six Northeast multispecies stocks; and five Northeast skate complex stocks) classified as non-FSSI stocks.

Table 3.2 List of major managed fishery species in the New England region, under State and Federal Jurisdictions

Council/Agency	Fishery Management Plan (FMP)	FMP Modifications	Major Species/ Species Group	Scientific Name(s) or Family Name(s)
Mid-Atlantic Fishery Management Council **MAFMC**	Atlantic surf clam and ocean quahog (2 species)	2 amendments 18 amendments	Atlantic surfclams Ocean quahogs	*Spisula solidissima* *Arctica islandica*
MAFMC/ New England Fishery Management Council **NEFMC**	Monkfish	7 amendments, 9 frameworks	Monkfish	*Lophius americanus*
	Spiny dogfish	2 frameworks, 4 amendments	Spiny dogfish	*Squalus acanthias*
New England Fishery Management Council **NEFMC**	Northeast multispecies (10 species)	23 amendments, 56 frameworks	Atlantic cod Haddock	*Gadus morhua* *Melanogrammus aeglefinus*
			Atlantic pollock Ocean perch (redfish) Yellowtail flounder American plaice Winter flounder Witch flounder White hake Atlantic halibut	*Pollachius pollachius* *Sebastes norvegicus* *Pleuronectes ferruginea* *Hippoglossoides platessoides* *Pseudopleuronectes americanus* *Glyptocephalus cynoglossus* *Urophycis tenuis* *Hippoglossus hippoglossus*
	Small mesh multispecies (3 species)	5 amendments, 4 frameworks, 2 specifications	Whiting, other hakes	Silver hake *Merluccius bilinearis,* Red Hake *Urophycis chuss,* Offshore Hake *Merluccius albidus*
	Atlantic sea scallop	17 amendments, 28 frameworks	Atlantic sea scallop	*Placopecten magellanicus*
	Atlantic salmon	4 amendments	Atlantic salmon	*Salmo salar*
	Deep sea red crab	4 amendments, 1 framework, 2 specifications	Deep sea red crab	*Chaceon quinquedens*
	Atlantic herring	8 amendments, 4 frameworks, 2 specifications	Atlantic herring	*Clupea harengus*
	Northeast skate complex (7 species)	3 amendments, 3 frameworks	Skates	Winter skate *Leucoraja ocellata*, barndoor skate *Dipturis laevis*, thorny skate *Amblyraja radiata*, smooth skate *Malacoraja senta*, little skate *Leucoraja erinacea*, clearnose skate *Raja eglanteria*, rosette skate *Leucoraja garmani*

Continued

Table 3.2 Continued

Council/Agency	Fishery Management Plan (FMP)	FMP Modifications	Major Species/ Species Group	Scientific Name(s) or Family Name(s)
Atlantic States Marine Fisheries Commission **ASMFC**	American eel	4 addendum	American eel	*Anguilla rostrata*
	American lobster (co-managed with NOAA Fisheries)	3 amendments, 24 addendum, 2 technical addendum	American lobster	*Homarus americanus*
	Atlantic croaker	1 amendment, 2 addendum	Atlantic croaker	*Micropogonias undulatus*
	Atlantic herring	3 amendments, 7 addendum, 2 technical addendum	Atlantic herring	*Clupea harengus*
	Atlantic menhaden	2 amendments, 1 Revision, 1 supplement, 6 addendum, 2 technical addendum	Atlantic menhaden	*Brevoortia tyrannus*
	Atlantic striped bass	6 amendments, 1 supplement, 9 addendum	Atlantic striped bass	*Morone saxatilis*
	Atlantic sturgeon	1 amendment, 1 technical addendum, 4 addendum	Atlantic sturgeon	*Acipenser oxyrhynchus*
	Black drum		Black drum	*Pogonias cromis*
	Black sea bass	2 amendments, 27 addendum	Black sea bass	*Centropristis striata*
	Bluefish	1 amendment, 1 addendum	Bluefish	*Pomatomus saltatrix*
	Coastal sharks (22 species)	4 addendum	Aggregated large coastal	*Carcharhinus* spp., *Galeocerdo cuvier*, *Ginglymostoma cirratum*, *Negaprion brevirostris*
			Hammerhead	*Sphyrna* spp.
			Pelagic	*Alopias vulpinus*, *Carcharhinus longimanus*, *Isurus oxyrinchus*, *Lamna nasus*, *Prionace glauca*
			Blacknose	*Carcharhinus acronotus*
			Non-blacknose small coastal	*Carcharhinus isodon*, *Rhizoprionodon terraenovae*, *Sphyrna tiburo*
			Smoothhound	*Mustelus* spp.
	Cobia		Cobia	*Rachycentron canadum*

Council/Agency	Fishery Management Plan (FMP)	FMF Modifications	Major Species/ Species Group	Scientific Name(s) or Family Name(s)
Atlantic States Marine Fisheries Commission **ASMFC**	Horseshoe crab	7 addendum	Horseshoe crab	*Limulus polyphemus*
	Jonah crab	2 addendum	Jonah crab	*Cancer borealis*
	Northern shrimp	2 amendments, 1 addendum	Northern shrimp	*Pandalus borealis*
	Red drum (Atlantic coast)	2 amendments, 1 addendum	Atlantic coast red drum	*Sciaenops ocellatus*
	Scup	3 amendments, 29 addendum	Scup	*Stenotomus chrysops*
	Shad and river herring (3 species)	2 addendum, 3 amendments, 1 technical addendum	American shad and river herring	*Alosa aestivalis*, *A. pseudoharengus*, *A. sapidissima*
	Spanish mackerel	1 amendment, 1 addendum	Spanish mackerel	*Scomberomorus maculatus*
	Spiny dogfish	5 addendum	Spiny dogfish	*Squalus acanthias*
	Spot	1 addendum	Spot	*Leiostomus xanthurus*
	Spotted seatrout	1 amendment	Spotted seatrout	*Cynoscion nebulosus*
	Summer flounder	2 amendments, 28 addendum	Summer flounder	*Paralichthys dentatus*
	Tautog	6 addendum	Tautog	*Tautoga onitis*
	Weakfish	4 amendments, 1 technical addendum, 4 addendum	Weakfish	*Cynoscion regalis*
	Winter flounder	1 amendment, 5 addendum	Winter flounder	*Pseudopleuronectes americanus*

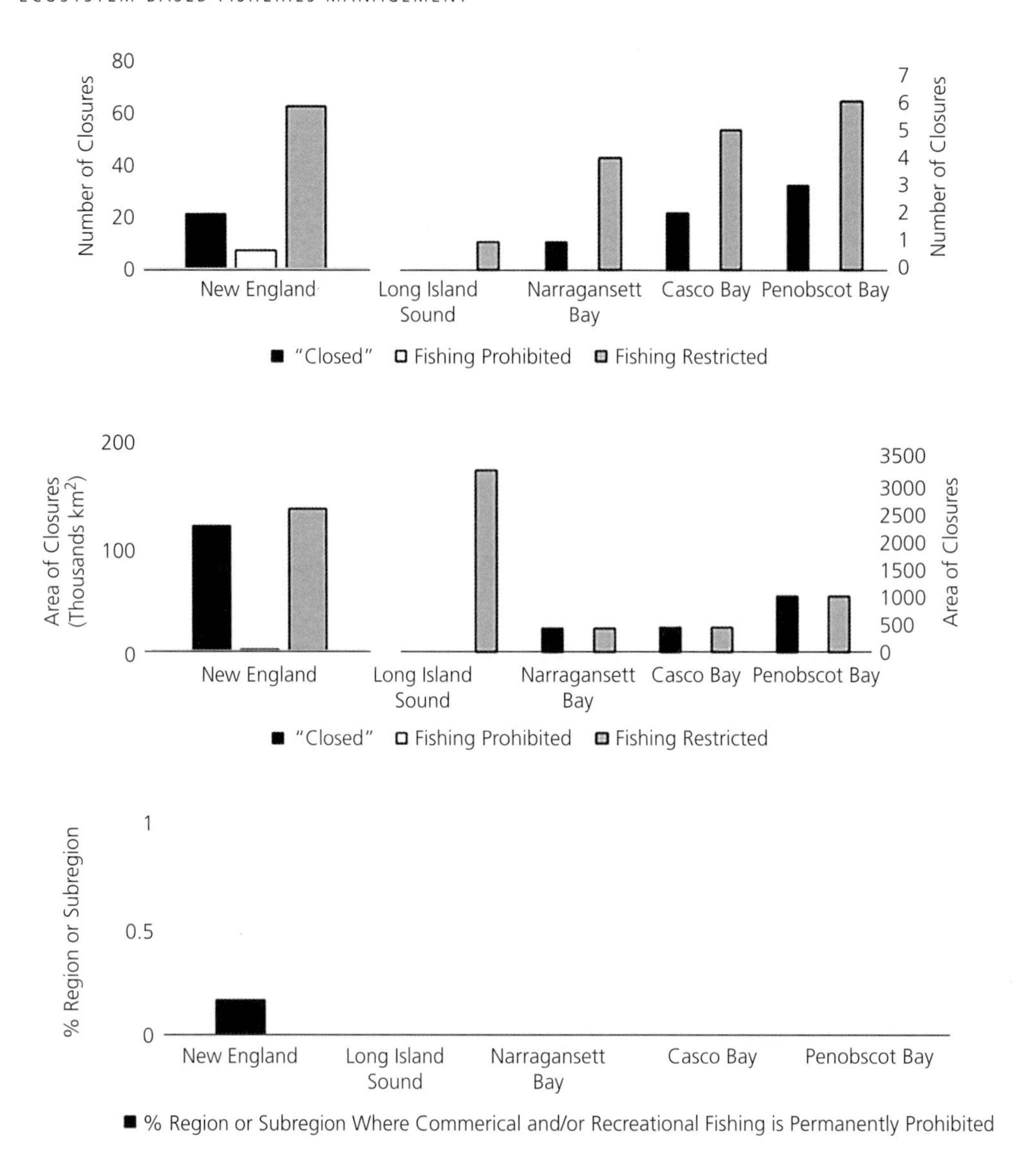

Figure 3.10 Number (top panel) and area (middle panel; km²) of named fishing closures, and prohibited or restricted fishing areas of the New England region and identified subregions, including percent region or subregion (bottom panel) where commercial and/or recreational fishing is permanently prohibited. Values for New England subregions (top and middle panels) plotted on the secondary (right) axis.

Data derived from NOAA Marine Protected Areas inventory.

Maine/Georges Bank stock; **Witch flounder**—*Glyptocephalus cynoglossus*; **Yellowtail flounder**, Cape Cod/Gulf of Maine, Georges Bank, S. New England/Mid-Atlantic stocks; **Thorny skate**—*Amblyraja radiate*, Gulf of Maine stock; **Atlantic halibut**—*Hippoglossus*; **Atlantic salmon**—*Salmo salar*; **Atlantic wolffish**—*Anarhichas lupus*; **Ocean pout**—*Zoarces americanus*; and **Winter flounder**, Georges Bank, southern New England stocks).[2] It should be noted that an increasing number are not able to have biological reference points, and hence status determination criteria

[2] As of June 2020, four New England stocks are experiencing overfishing (Atlantic cod—Georges Bank and Gulf of Maine stocks; Yellowtail flounder—Georges Bank; Red hake—*Urophycis chuss*, Southern Georges Bank/Mid-Atlantic) and 15 are overfished (Atlantic cod—Georges Bank and Gulf of Maine stocks; Windowpane flounder—Gulf of Maine/Georges Bank; Witch flounder, Yellowtail flounder—Georges Bank and S. New England/Mid-Atlantic stocks; Thorny skate—Gulf of Maine; Atlantic halibut; Atlantic salmon; Atlantic wolffish; Ocean pout; Winter flounder—Georges Bank and southern New England stocks; Red hake—Southern Georges Bank/Mid-Atlantic; and White hake—Gulf of Maine/Georges Bank).

Table 3.3 List of major bays and islands, NOAA Habitat Focus Areas (HFAs), NOAA National Estuarine Research Reserves (NERRs), NOAA National Marine Monuments and Sanctuaries (NMS), Coastal National Parks, National Seashores, and number of Habitats of Particular Concern (HAPCs) in the New England region and subregions

Region	Subregion	Major Bays	Major Islands	HFAs	NERRs	Monuments and NMS	Coastal National Parks	National Seashores	HAPCs
New England		Buzzards Bay, Cape Cod Bay, Casco Bay, Ipswich Bay, Long Island Sound, Massachusetts Bay, Merrimack River, Nahant Bay, Narragansett Bay, Penobscot Bay, Piscataqua River, Saco Bay	Isle Au Haut, Martha's Vineyard, Mt. Desert Island, Nantucket Island, Naushon Island, Prudence Island	Penobscot River, Maine HFA	Great Bay NERR, Narragansett Bay NERR, Waquoit Bay NERR, Wells NERR	Northeast Canyons and Seamounts Marine National Monument, Stellwagen Bank NMS	Acadia National Park, Boston Harbor Islands National Recreation Area, New Bedford Whaling Historical Park, Roosevelt Compobello International Park, Sagamore Hill National Historic Site, Saint Croix Island International Historic Site, Salem Maritime National Historic Site	Cape Cod National Seashore	19
	Casco Bay		Bailey Island, Bustins Island, Cliff Island, Cousins Island, Cushing Island, Great Diamond Island, Great Chebeague Island, Long Island Mackworth Island, Orr's Island, Peaks Island, Sebascodegan Island (Great Island)						
	Long Island Sound		City Island, Davids Island, Fishers Island, Hart Island, Plum Island, Shea Island, Thimble Islands				Sagamore Hill National Historic Site		
	Narragansett Bay		Aquidneck Island, Coasters Island, Conanicut Island, Cornelius Island, Despair Island, Dutch Island, Dyer Island, Hog Island, Patience Island, Prudence Island		Narragansett Bay NERR				
	Penobscot Bay		Criehaven Island, Deer Isle, Great Spruce Head Island, Isle au Haut, Islesboro Island, Little Deer Isle, Matinicus Isle, Nautilus Island, North Haven Island, Sears Island, Vinalhaven Island	Penobscot River, Maine HFA					

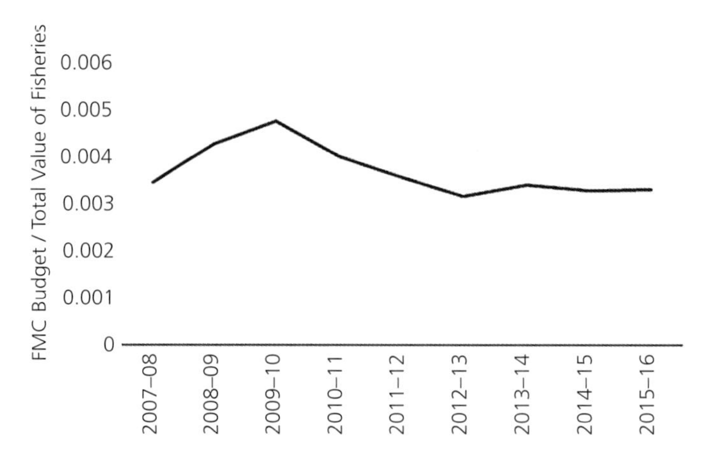

Figure 3.11 Total annual budget of the New England Fishery Management Council as compared to the total value of its marine fisheries. All values in USD.

Data from NOAA National Marine Fisheries Service Office of Sustainable Fisheries and NOAA National Marine Fisheries commercial landings database.

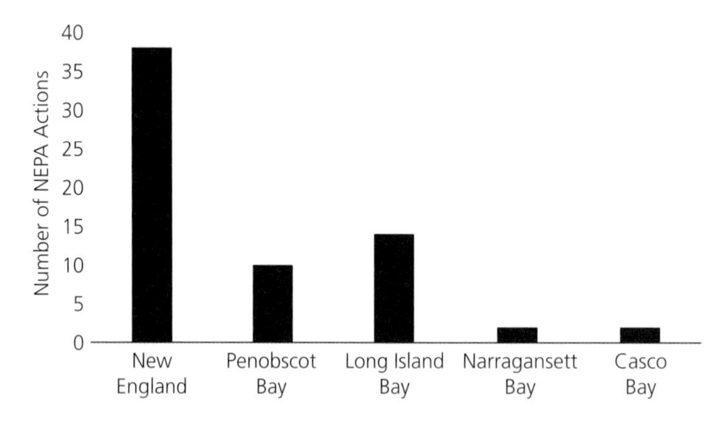

Figure 3.12 Number of National Environmental Policy Act Environmental Impact Statement (NEPA-EIS) actions from 1987 to 2016 for the New England region and identified subregions.

Data derived from U.S. Environmental Protection Agency EIS database.

established, due to challenges with stock assessments. New England has the nationally highest number and percentage (35.9%) of known overfished stocks, while also having some of the lowest numbers and percentages of stocks of unknown status among regions.[3] Only two stocks have unknown overfishing status (**Offshore hake**—*merluccius albidus*; **Witch flounder**, northwestern Atlantic Coast stock), while three have unknown overfished status (**Deep sea red crab**—*Chaceon quinquedens*; **Offshore hake**; **Winter flounder**, Gulf of Maine stock).[4]

[3] As of June 2020, New England continues to have the nationally highest number and percentage (29.4%) of known overfished stocks, and some of the lowest numbers and percentages of stocks of unknown status among regions.

[4] These values and stocks remain the same as of June 2020.

For ASMFC-managed stocks, only four of its New England stocks are considered sustainable or rebuilt as of 2020, while six are depleted (**American eel**—*Anguilla rostrata*; **American lobster**—southern New England stock; **American shad**—*Alosa sapidissima*; **Atlantic sturgeon**—*Acipenser oxyrhynchus*; **Northern shrimp**—*Pandalus borealis*; **River herring**—*Alosa* spp.), five are overfished (**Bluefish**—*Pomatomus saltatrix*; **Striped bass**—*Morone saxatilis*; **Tautog**—*Tautoga onitis* Long Island Sound stock; **Winter flounder**—*Pseudopleuronectes americanus*), and two are currently undergoing overfishing (**Striped bass**; **Tautog** Long Island Sound stock; ASMFC 2020b). The protected species of New England (also grouped with the Mid-Atlantic region) are divided into 39 independent marine mammal stocks, and 15

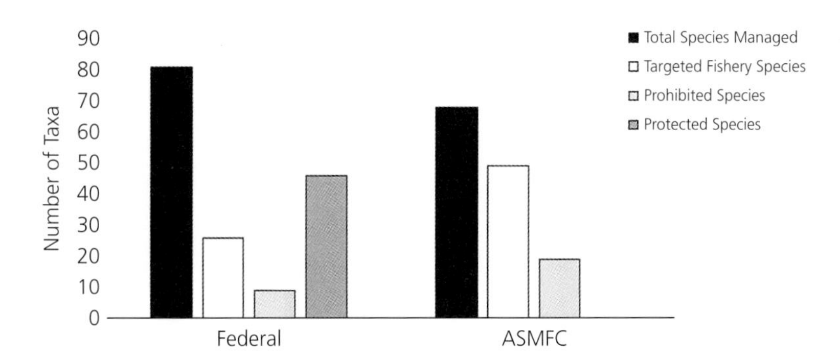

Figure 3.13 Number of managed taxa (species or families) for the New England region, including federal and state (Atlantic States Marine Fisheries Commission—ASMFC) jurisdictions.

ESA-listed distinct population segments, of which 80% are endangered (Fig. 3.15). Nationally, lower numbers of marine mammal stocks are found in New England (12% of total marine mammal stocks), with one-third listed as strategic marine mammal stocks. There are 28% of marine mammal stocks with unknown population size that occur in New England, which is tied with the Mid-Atlantic for fifth-highest in the nation.

3.4.3 Environmental Forcing and Major Features

3.4.3.1 Oceanographic and Climatological Context

Circulation patterns throughout the New England EEZ are driven primarily by components of the Gulf Stream, the inflow of the Labrador Current, and the north-south flow of the Eastern and Western Maine Coastal Currents (EMCC, WMCC; Fig. 3.1). These two coastal currents are the principal branches of the cyclonic Gulf of Maine Coastal Current (Pettigrew et al. 2005, Xue et al. 2008). Continuing along the United States Atlantic Coast and connected to Mid-Atlantic circulation patterns, northern components of the Gulf Stream serve as the major western boundary current of the United States. At the northern edge of the Gulf Stream, a sharp temperature front known as the "North Wall" occurs, where warmer Gulf Stream waters separate from those of the continental shelf and slope (Hameed et al. 2018). Latitudinal changes observed in North Wall positioning also influence New England thermal dynamics (Gawarkiewicz et al. 2012, Hameed et al. 2018). Seasonal southern counterclockwise "cold-core" and northern clockwise "warm-core" rings also develop due to the meandering of the Gulf Stream, which break off the main current and provide additional means of propagule transport along coastal regions and throughout the Atlantic EEZ (Brooks 1987). The Gulf Stream and meandering rings also influence water temperatures, productivity, and counterclockwise circulation patterns of Long Island Sound (Schmalz & Devine 2003), Narragansett Bay (Spaulding & Swanson 2008), the Gulf of Maine, and occasionally near Georges Bank (Brooks 1987, Garfield & Evans 1987, Ryan et al. 2001). These rings also tend to persist for longer timeframes with increasing distance from the Gulf Stream North Wall (Brown et al. 1986). The coastal westward flow of the WMCC and EMCC continue along the coastal regions of Penobscot and Casco Bay to diverge into offshore components of New England, especially surrounding Georges Bank (Zemeckis et al. 2014). EMCC waters are well mixed out to approximately 50 m depth, with the influence of tidal mixing extending to 100 m depth. The WMCC consists mainly of a surface trapped plume originating from the Kennebec River (Hetland & Signell 2005). The Gulf of Maine Coastal Current, including its principal branches, is a pressure gradient-driven system (Pettigrew et al. 2005). Due to increasing freshwater input, the WMCC and EMCC increase their transport in the spring and summer. Offshore veering of the EMCC limits southwest transport near Penobscot Bay (Pettigrew et al. 2005) and leads to intermittent southeastern accumulations of offshore propagules (Xue et al. 2008). Recurring intrusion of Labrador Slope Water into the Gulf of Maine (at depths 150–200 m) through the northeast Channel also has been regularly observed, as related to NAO dynamics (Mountain 2012). When more prominently occurring, these colder and fresher waters circulate in the Gulf of Maine and move southward down the coast (Petrie & Drinkwater 1993, Drinkwater et al. 1998, Mountain 2012).

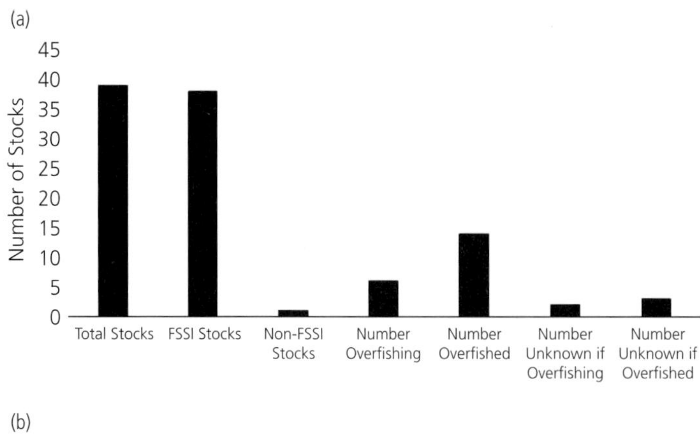

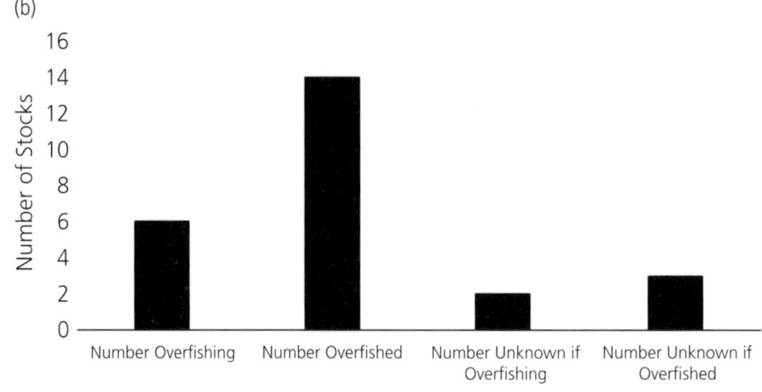

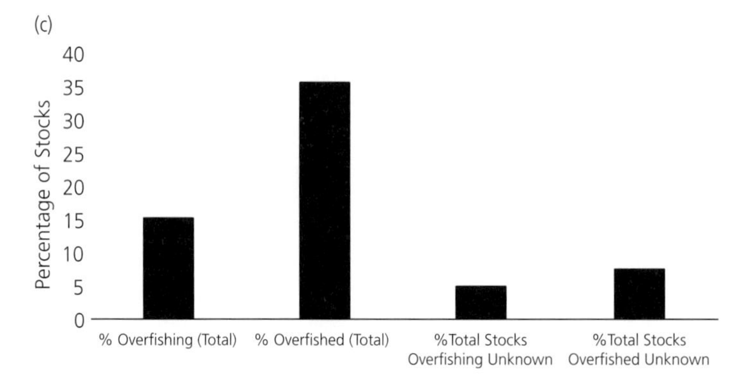

Figure 3.14 For the New England region as of June 2017 (a) Total number of managed Fish Stock Sustainability Index (FSSI) stocks and non-FSSI stocks, and breakdown of stocks experiencing overfishing, classified as overfished, and of unknown status. (b) Number of stocks experiencing overfishing, classified as overfished, and of unknown status. (c) Percent of stocks experiencing overfishing, classified as overfished, and of unknown status.

Data from NOAA National Marine Fisheries Service. Note: stocks may refer to a species, family, or complex.

Clear interannual and multidecadal patterns in average SST have been observed for the New England region (Fig. 3.16), with temperatures averaging 13.7°C ± 0.04 (SE) over time (1854–2016), and having increased by 1.2°C since the mid-twentieth century. Pershing et al. (2015) have identified the Gulf of Maine region of New England as one of the most rapidly warming areas worldwide, with SST increasing faster than 99% of the global ocean. Increasing temperatures have also been observed within other key subregions, with increases of 1.3°C since 1950 in Casco and Penobscot Bays, and 0.8 and 1.1°C in Long Island Sound and

Table 3.4 List of prohibited fishery species in the New England region, including federal and state jurisdictions

Council/Authority	Region	Prohibited Species	Scientific Name
New England Fishery Management Council **NEFMC**; Greater Atlantic Regional Fisheries Office **GARFO**	New England (9 species)	Atlantic salmon	*Salmo salar*
		Federal Red drum	*Sciaenops ocellatus*
		Ocean pout	*Zoarces americanus*
		Windowpane flounder	*Scophthalmus aquosus*
		Atlantic wolffish	*Anarhichas lupus*
		Barndoor skate	*Dipturus laevis*
		Thorny skate	*Amblyraja radiata*
		Smooth skate	*Malacoraja senta*
		Federal Striped bass	*Morone saxatilis*
Atlantic States Marine Fisheries Commission **ASMFC**	Atlantic state waters (19 species)	Coastal sharks	*Alopias superciliosus*, *Carcharhinus* spp., *Carcharias taurus*, *Carcharodon carcharias*, *Hexachus* spp., *Isurus paucus*, *Notorynchus cepedianus*, *Odontaspis noronhai*, *Rhincodon typus*, *Rhizoprionodon porosus*, *Squatina demeril*

Table 3.5 List of protected species under the Endangered Species Act (ESA) or Marine Mammal Protection Act (MMPA) in the New England region

Region	Managing Agency	Protected Species Group	Common Name	Scientific Name
New England (46 species)	New England Fishery Management Council **NEFMC**; Greater Atlantic Regional Fisheries Office **GARFO**	Whales	Blainville's beaked whale	*Mesoplodon densirostris*
			Blue whale	*Balaenoptera musculus*
			Cuvier's beaked whale	*Ziphius cavirostris*
			Dwarf sperm whale	*Kogia sima*
			False killer whale	*Pseudorca crassidens*
			Fin whale	*Balaenoptera physalus*
			Gervais' beaked whale	*Mesoplodon europaeus*
			Humpback whale	*Megaptera novaeangliae*
			Killer whale (Orca)	*Orcinus orca*
			Long-finned pilot whale	*Globicephala melas*
			Melon headed whale	*Peponocephala electra*
			Minke whale	*Balaenoptera acutorostrata*
			North Atlantic right whale	*Eubalaena glacialis*
			Northern bottlenose whale	*Hyperoodon ampullatus*
			Pygmy killer whale	*Feresa attenuata*

Continued

Table 3.5 Continued

Region	Managing Agency	Protected Species Group	Common Name	Scientific Name
New England (46 species)	New England Fishery Management Council **NEFMC**; Greater Atlantic Regional Fisheries Office **GARFO**	Whales	Pygmy sperm whale	*Kogia breviceps*
			Sei whale	*Balaenoptera borealis*
			Short-finned pilot whale	*Globicephala macrorhynchus*
			Sowerby's beaked whale	*Mesoplodon bidens*
			Sperm Whale	*Physeter macrocephalus*
			True beaked whale	*Mesoplodon mirus*
		Dolphins	Atlantic spotted dolphin	*Stenella frontalis*
			Atlantic white-sided dolphin	*Lagenorhynchus acutus*
			Common bottlenose dolphin	*Tursiops truncatus truncatus*
			Clymene dolphin	*Stenella clymene*
			Common dolphin	*Delphinus delphis*
			Fraser's dolphin	*Lagenodelphis hosei*
			Harbor porpoise	*Phocoena phocoena*
			Pantropical spotted dolphin	*Stenella attenuata*
			Risso's dolphin	*Grampus griseus*
			Rough-toothed dolphin	*Steno bredanensis*
			Spinner dolphin	*Stenella longirostris*
			Striped dolphin	*Stenella coeruleoalba*
			White-beaked dolphin	*Lagenorhynchus albirostris*
		Sea Turtles	Green sea turtle	*Chelonia mydas*
			Hawksbill sea turtle	*Eretmochelys imbricata*
			Kemp's Ridley sea turtle	*Lepidochelys kempii*
			Leatherback sea turtle	*Dermochelys coriacea*
			Loggerhead sea turtle	*Caretta caretta*
		Fishes	Atlantic salmon	*Salmo salar*
			Atlantic sturgeon	*Acipenser oxyrinchus oxyrinchus*
			Shortnose sturgeon	*Acipenser brevirostrum*
		Seals	Gray seal	*Halichoerus grypus*
			Harbor seal	*Phoca vitulina*
			Harp seal	*Phoca groenlandica*
			Hooded seal	*Crystophora cristata*

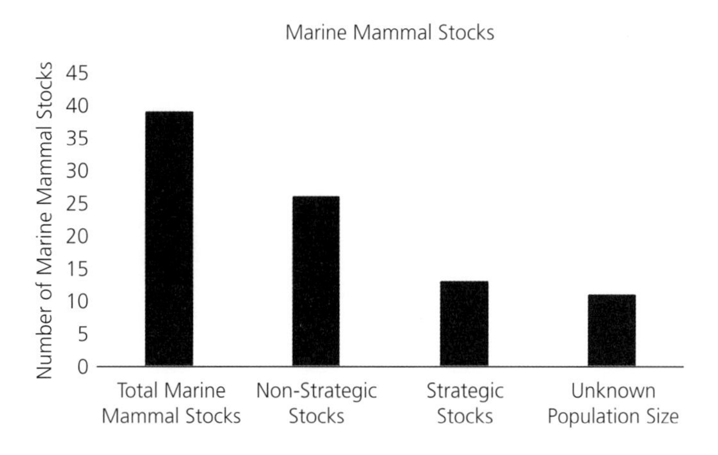

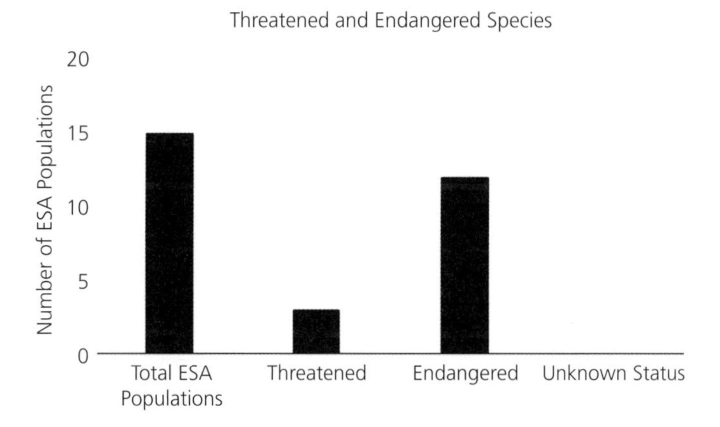

Figure 3.15 Number and status of federally protected species (marine mammal stocks, top panel; distinct population segments of species listed under the Endangered Species Act—ESA, bottom panel) occurring in the New England region.

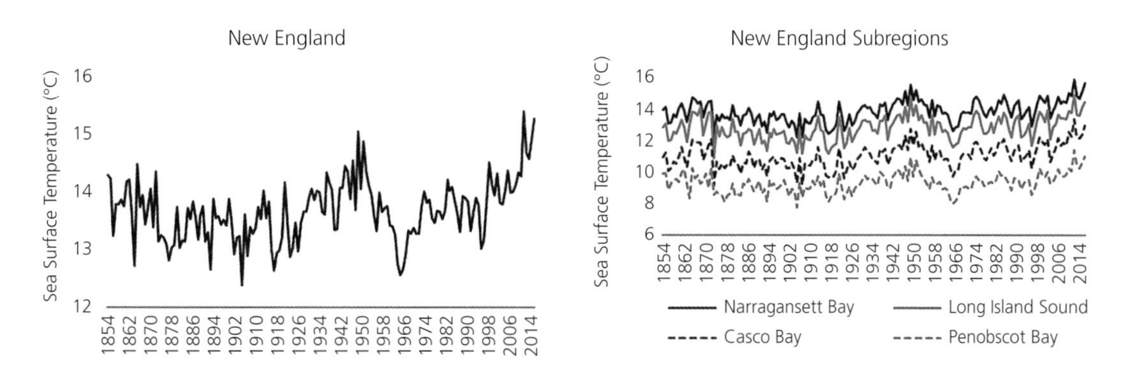

Figure 3.16 Average sea surface temperature (°C) over time (years 1854–2016) for the New England region and identified subregions.

Data derived from the NOAA Extended Reconstructed sea surface temperature dataset (https://www.ncdc.noaa.gov/).

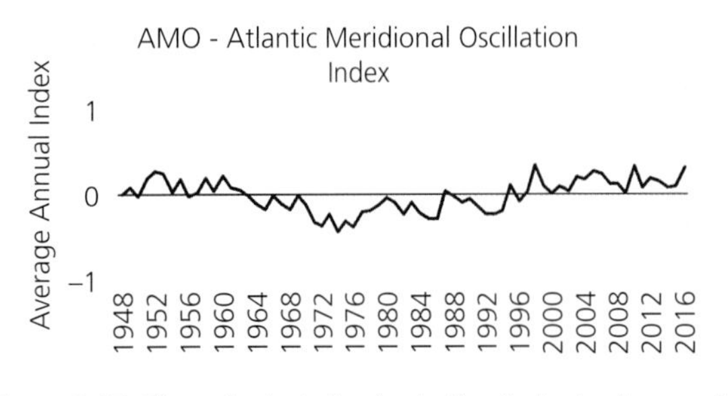

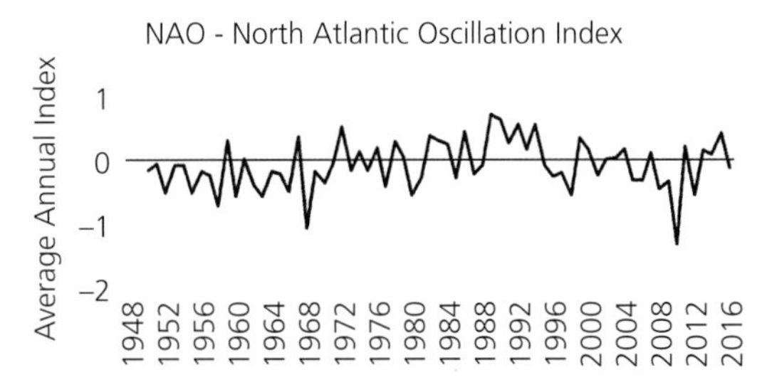

Figure 3.17 Climate forcing indices for the New England region over time (years 1948–2016).
Data derived from NOAA Earth System Research Laboratory data framework (https://www.esrl.noaa.gov/).

Narragansett Bay, respectively. Average temperatures over the 162-year time period have been highest in Narragansett Bay at 13.8°C ± 0.05 (SE) and slightly lower in Long Island Sound (12.8°C ± 0.06, SE). Although subject to the highest warming rates, average SST values are still cooler in Casco (11.0 °C ± 0.06, SE) and Penobscot Bays (9.3 °C ± 0.04, SE) than in the more southern components of New England. Coincident with these temperature observations are Atlantic basin-scale climate oscillations (Fig. 3.17), including the AMO and NAO, which exhibit decadal cycles that can influence the environmental conditions and ecologies of several U.S. regions, including New England.

3.4.3.2 Notable Physical Features and Destabilizing Events and Phenomena

Besides its four main subregions, New England also contains 12 major bays (Table 3.3), primarily off the coasts of Connecticut, Rhode Island, Massachusetts, and Maine, and 6 major islands, mostly concentrated off the coasts of Massachusetts and Maine. Proportionally, the New England region makes up 0.8% of the U.S. EEZ (~136,000 km²), 4.2% of the national shoreline (~9,900 km), while containing the shallowest average depth (~870 m) among U.S. Atlantic regions and maximum depths of ~4.4 km (Fig. 3.18). Among the four subregions, Long Island Sound comprises the largest proportion (2.4%; ~3200 km²) of the New England EEZ with ~1200 km of shoreline and comparatively the shallowest average (6.6 m) and maximum depths (35 m). Penobscot, Casco, and Narragansett Bays cumulatively make up 1.3% of the New England EEZ, among which Penobscot Bay has the greatest area (~990 km²) and longest shoreline (~995 km). Casco and Penobscot Bays both have greater depths than Long Island Sound and Narragansett Bay, with Casco Bay having an average depth of 18.4 m and deepest depths of 88 m, and Penobscot Bay the deepest among all subregions (average depth: 27.7 m; maximum depth: 100 m).

In New England, 126 hurricanes have been reported since 1850 with an average of 9.4 hurricanes per decade and peak numbers (~13–17 per decade) recorded in the 1870s, 1950s, and 2000s (Fig. 3.19). Of U.S. Atlantic regions, New England has been least prone to hurricanes. Among subregions, hurricanes have been observed occurring at much lower intensities, with the highest numbers observed in Long Island Sound (36 total; average: 2.1 per decade) and peak values of four per decade observed in Long Island Sound during the 1900s and in Casco Bay during the 1860s.

3.4.4 Major Pressures and Exogenous Factors

3.4.4.1 Other Ocean Use Context

Apart from LMR extraction, ocean uses in New England are most associated with the tourism/recreation, marine transportation, and ship and boatbuilding sectors (Fig. 3.20). Among sectors, trends for New England ocean establishments have increased over the past decades, while employments have remained relatively consistent. Tourism establishments and employments comprise 65–84% of total New England ocean establishments and 59–74% of total New England ocean employments.

Substantial increases in total ocean establishments have also been observed, while at a much lesser rate for employments, with highest values occurring in 2016 at ~15,000 establishments and ~260,000 employments. The GDP from tourism is highest among sectors, contributing 39–51% toward total New England ocean GDP. Over the past couple of decades, especially during the 2010s, ocean GDP has also increased, with the most recent estimated values at ~$19.3 billion. Overall, the New England region contributes toward 5% of the U.S. total ocean GDP, which is third-lowest nationally. Establishments and employments in the

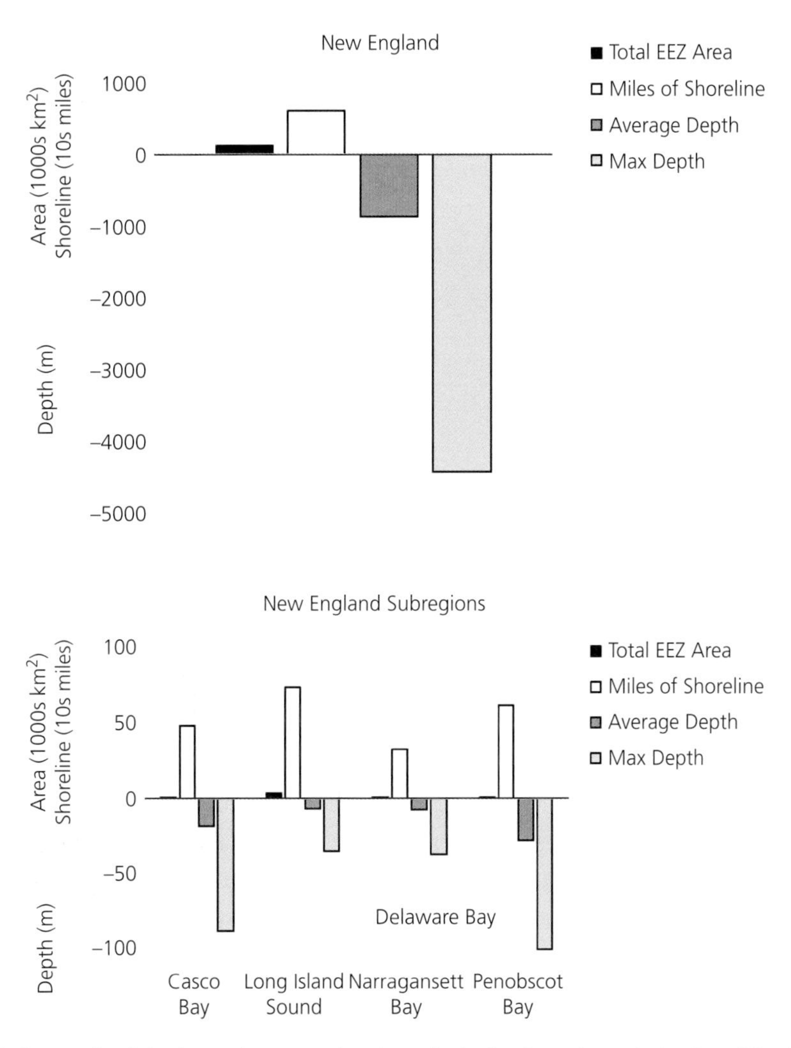

Figure 3.18 Total shelf area, miles of shoreline, and average and maximum depth of marine regions and subregions of New England. EEZ = Exclusive Economic Zone

New England marine transportation sector contribute 6–7% toward total U.S. ocean establishments and ocean employments. New England also contributes toward 11% of U.S. marine tourism establishments and ~8–9% of national marine tourism employments and GDP, all of which are mid-range nationally. This region is also mid-range in its contributions (~7%) to U.S. marine transportation establishments, employments, and GDP. Contributions from New England ship and boatbuilding make up a significant proportion of national ship and boat production (12% of U.S. ship and boatbuilding establishments, 17% of U.S. ship and boatbuilding employments, and 15% of U.S. ship and boatbuilding GDP). Overall, New England is above average in terms of its marine socioeconomic status, especially as associated with its LMR-based economy and increasing total ocean GDP.

The degree of tourism, as measured by number of dive shops, number of major ports and marinas, and number of cruise ship departure passengers is mid-range nationally for New England (Fig. 3.21), with comparatively lower numbers of dive shops and marinas, and a moderate number of major ports. There are few dive shops, ports, or marinas found throughout the key subregions, with most concentrated near Long Island Sound (14 dive shops, 4 major ports, and 3 major marinas). Cruise destination passengers are prominent in the New England region (third-highest

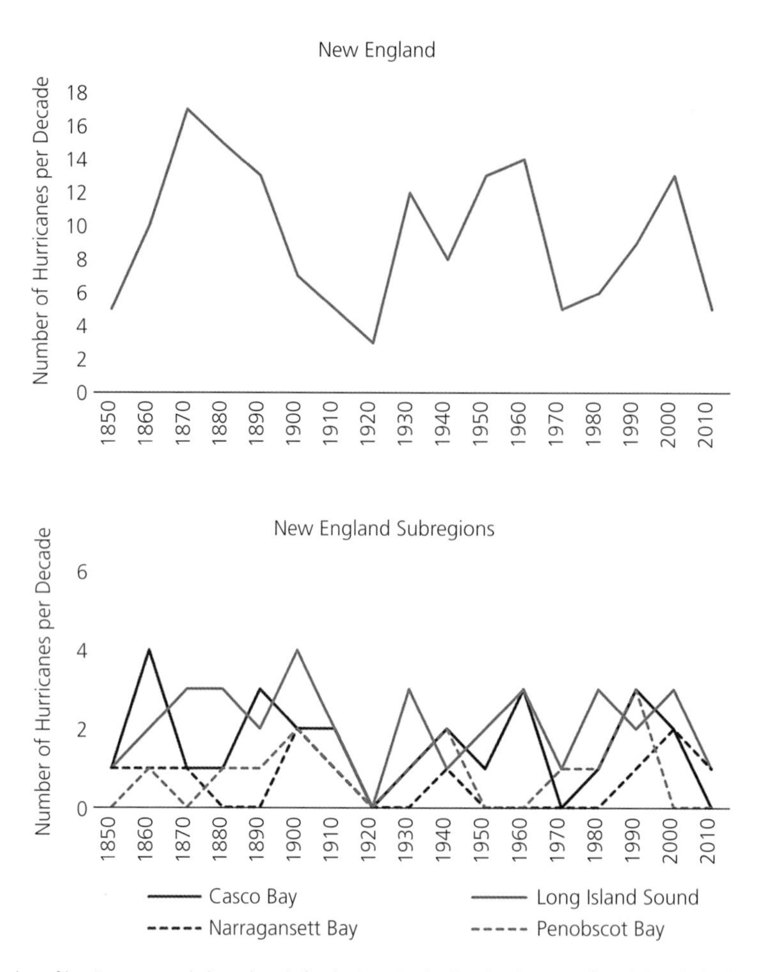

Figure 3.19 Total number of hurricanes recorded per decade for the New England region (top panel) and identified subregions (bottom panel) since year 1850.

Data derived from NOAA Office for Coastal Management (https://coast.noaa.gov/hurricanes/).

nationally), with lower numbers of departure passengers annually observed for the region. While no active offshore oil rigs are found throughout the U.S. Atlantic continental shelf, New England has emerged as a leading area for offshore wind energy production with six established U.S. Bureau of Ocean Energy Management (BOEM) wind energy sites (tied with the South Atlantic for second-highest nationally).

3.4.5 Systems Ecology and Fisheries

3.4.5.1 Basal Ecosystem Production

Since 2002, mean surface chlorophyll values throughout the New England EEZ have been observed at ~1.59 ± 0.04 (SE) mg m^{-3} (Fig. 3.22). Comparatively, nearshore key subregions subject to higher freshwater input and nutrient loading average up to 12.1 ± 0.7 (SE) mg m^{-3} in Penobscot Bay, and 9.3 ± 0.9 (SE) to 9.5 ± 0.2 (SE) mg m^{-3} in Casco Bay and Long Island Sound, respectively. Lower average values (5.6 ± 0.7, SE) are found in Narragansett Bay. Regular interannual oscillations in chlorophyll concentration have been observed throughout Casco, Narragansett, and Penobscot Bays over the past 1.5 decades. As of 2002, average annual primary productivity within the U.S. Northeastern (Mid-Atlantic-New England) EEZ has remained relatively steady between 250 and 300 g C m^{-2} $year^{-1}$ (second-highest nationally).

3.4.5.2. System Exploitation

The fifth-highest numbers of landed taxa have been captured in the New England region over time, with values continuing to increase since the 1980s and peaking at 138 reported taxa in 2004 (Fig. 3.23). Among

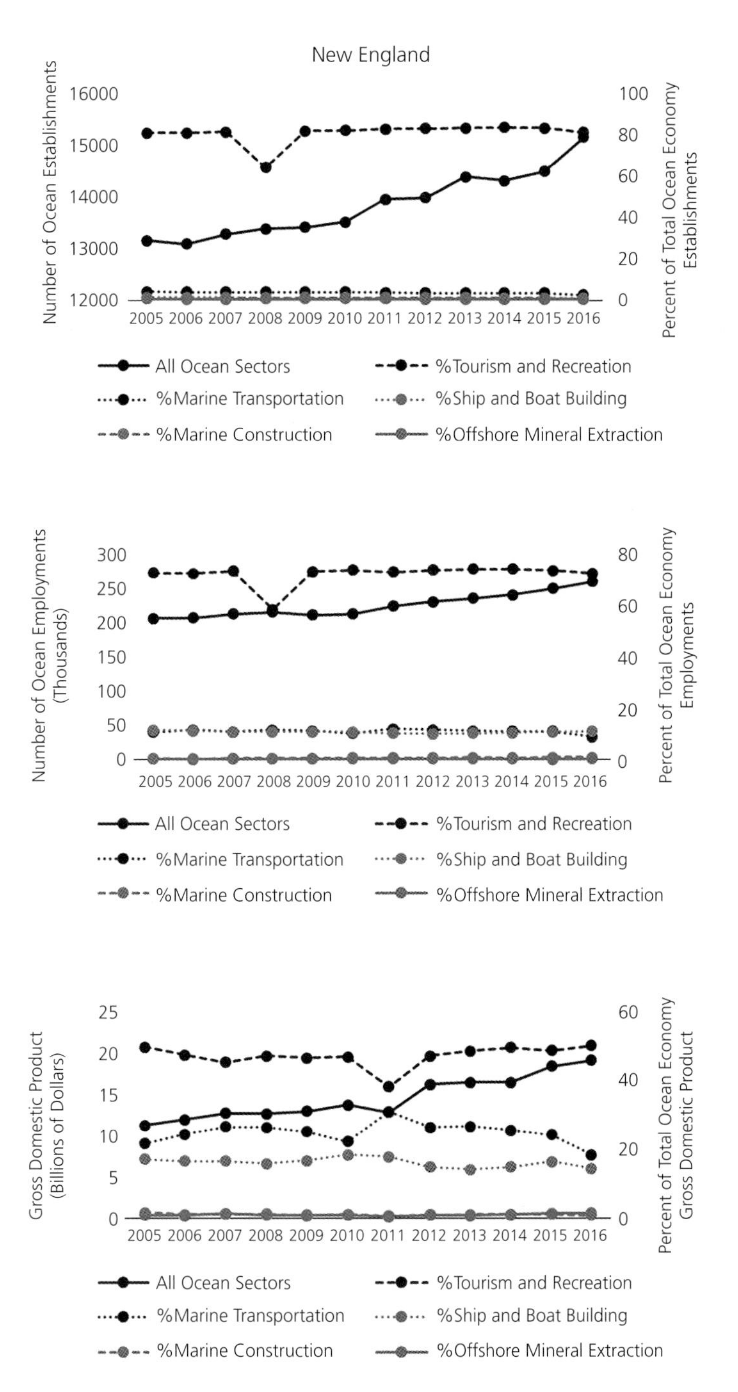

Figure 3.20 Number and percent contribution of multisector establishments, employments, and Gross Domestic Product (GDP) and their percent contribution to total multisector oceanic economy in the New England region (years 2005–2016).

Data derived from the National Ocean Economics Program.

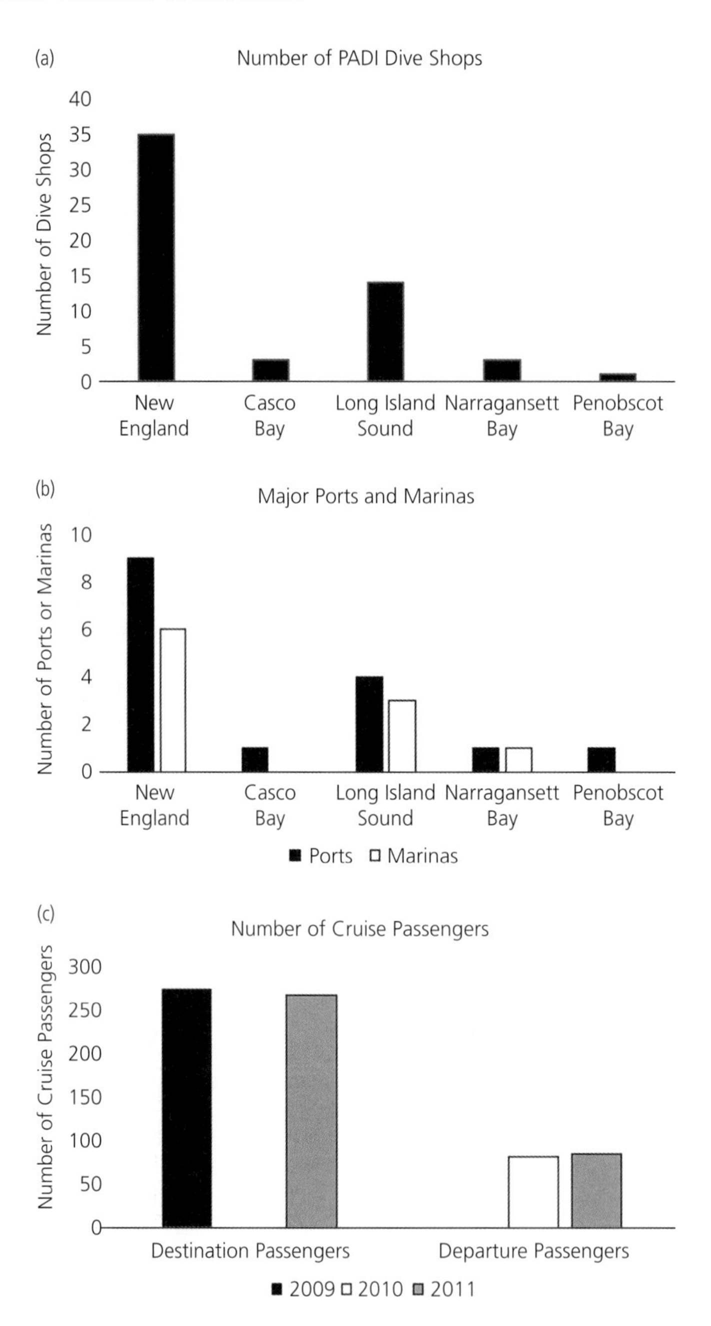

Figure 3.21 Tourism indicators for the New England region and identified subregions: (a) Number of Professional Association of Diving Instructors (PADI) dive shops. Data derived from http://apps.padi.com/scuba-diving/dive-shop-locator/. (b) Number of major ports and marinas. (c) Number of cruise destination and departure port passengers (years 2009–2011; values in thousands).

subregions, greater values have been observed over time for Long Island Sound (average: 90.9 taxa year^{-1} ± 3.1, SE), with values currently near their historically highest at 136. Lower, relatively consistent values have been observed for Narragansett Bay (average: 69.0 taxa year^{-1} ± 1.6, SE) and Casco and Penobscot Bays (average: 51.2 taxa year^{-1} ± 1.5, SE).

Total surveyed fish and invertebrate biomass for the U.S. Northeast (Mid-Atlantic-New England) region has remained relatively steady over time (Fig. 3.24),

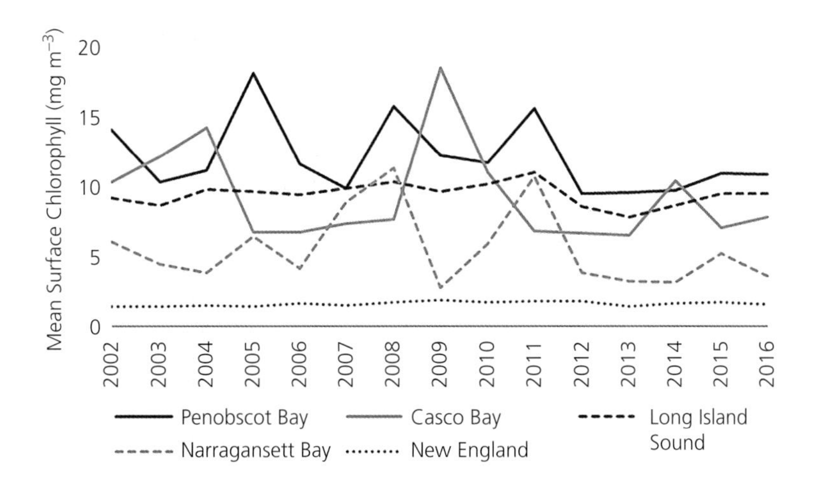

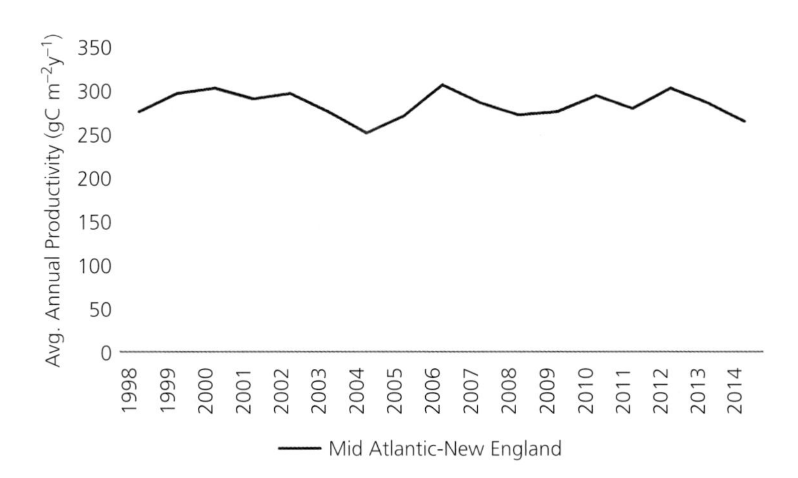

Figure 3.22 Mean surface chlorophyll (mg m^{-3}; top panel) for the New England region and identified subregions and average annual productivity (grams carbon m^{-2} year^{-1}; bottom panel) for the Mid-Atlantic and New England regions over time.

Data derived from NASA Ocean Color Web (https://oceancolor.gsfc.nasa.gov/) and productivity calculated using the Vertically Generalized Production Model—VGPM.

with gradual increases observed from the 1960s to 1990s at 6–13 million metric tons per year, and more prominent increases occurring during the 2000s at up to 25 million metric tons per year. In 2013 however, biomass values decreased to similar levels observed during the mid to late 2000s. Comparing across regions, U.S. northeast biomass values are fourth-highest on average and have been observed to be increasing over time as associated with more recent actions toward sustainable fisheries management, ongoing rebuilding plans, and increasing restrictions on fishing pressure throughout the region. Exploitation rates (i.e., landings/biomass ratio) in the U.S. Northeast are second-highest in the nation, and have ranged from 0.02 to 0.1 over time, with values fluctuating during the 1970s and having decreased as of the 1980s onward (Fig. 3.25). Values for total landings/productivity (up to 0.0006) are highest nationally and have remained relatively constant, while total biomass/productivity values (up to 0.0025) have gradually increased over the past two decades.

3.4.5.3. Targeted and Non-Targeted Resources

Overall commercial and recreational fisheries landings are fourth-highest nationally in the New England region (Fig. 3.26) and exhibited declines from peak values in the 1960s to around 200–300 thousand metric tons since the 1970s. Decreases in the proportion of recreational catch were observed during the 1980s, with recreational fisheries continuing to contribute

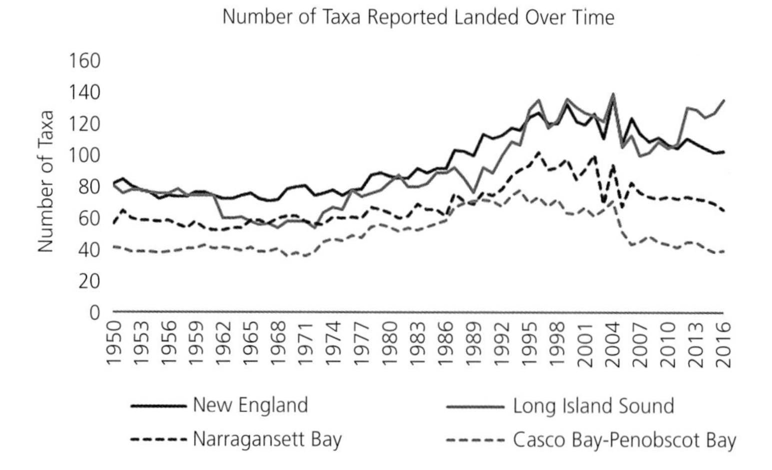

Figure 3.23 Number of taxa reported captured by commercial and recreational fisheries in the New England region and identified subregions. Data derived from NOAA National Marine Fisheries Service commercial and recreational fisheries statistics.

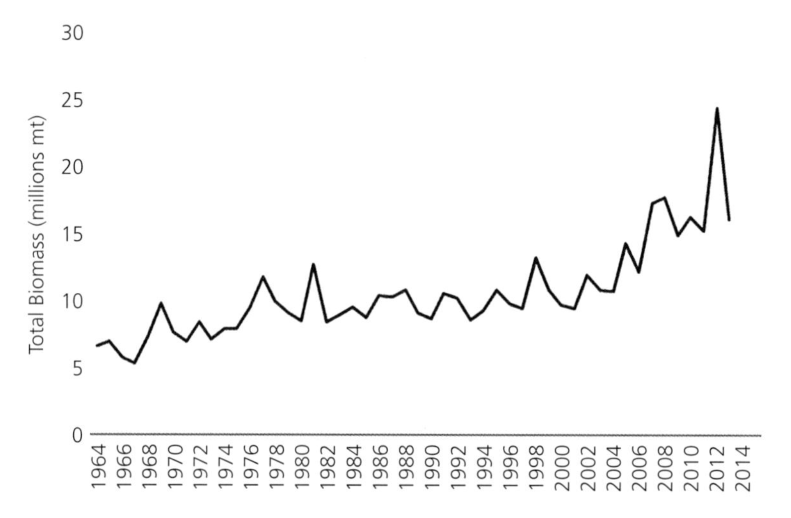

Figure 3.24 Total combined surveyed fish and invertebrate biomass (metric tons; mt) for the Mid-Atlantic and New England regions, over time.

5–10% of total landings. Highest commercial landings were observed for Long Island Sound during the 1950s to mid-1960s, after which values plummeted to approximately 20% of their 1962 historical peak (~105,000 metric tons). Similar trends were likewise observed for recreational fisheries landings (New York-Connecticut waters) by weight and number. Although some moderate increases were observed during the 1990s, values have remained consistently lower over time at a similar level to commercial landings observed over time in Casco and Penobscot Bays. However, these three regions have seen increases during more recent years in terms of recreational fisheries landings (both weight and number) at comparable values. Among subregions, Narragansett Bay commercial landings have rebounded over time following early declines in the 1950s and 1960s, allowing this area to emerge as the leading subregion in terms of commercial landings (average: 43,824.3 metric tons ± 1458, SE), although this subregion is much less pronounced in terms of recreational landings. When standardized per square kilometer (Fig. 3.27), landings for New England and its subregions follow very similar patterns as observed for total landings values, with this region containing the highest concentration of fisheries landings in the nation (up to 3.5 metric tons km^{-2}). Concentrated

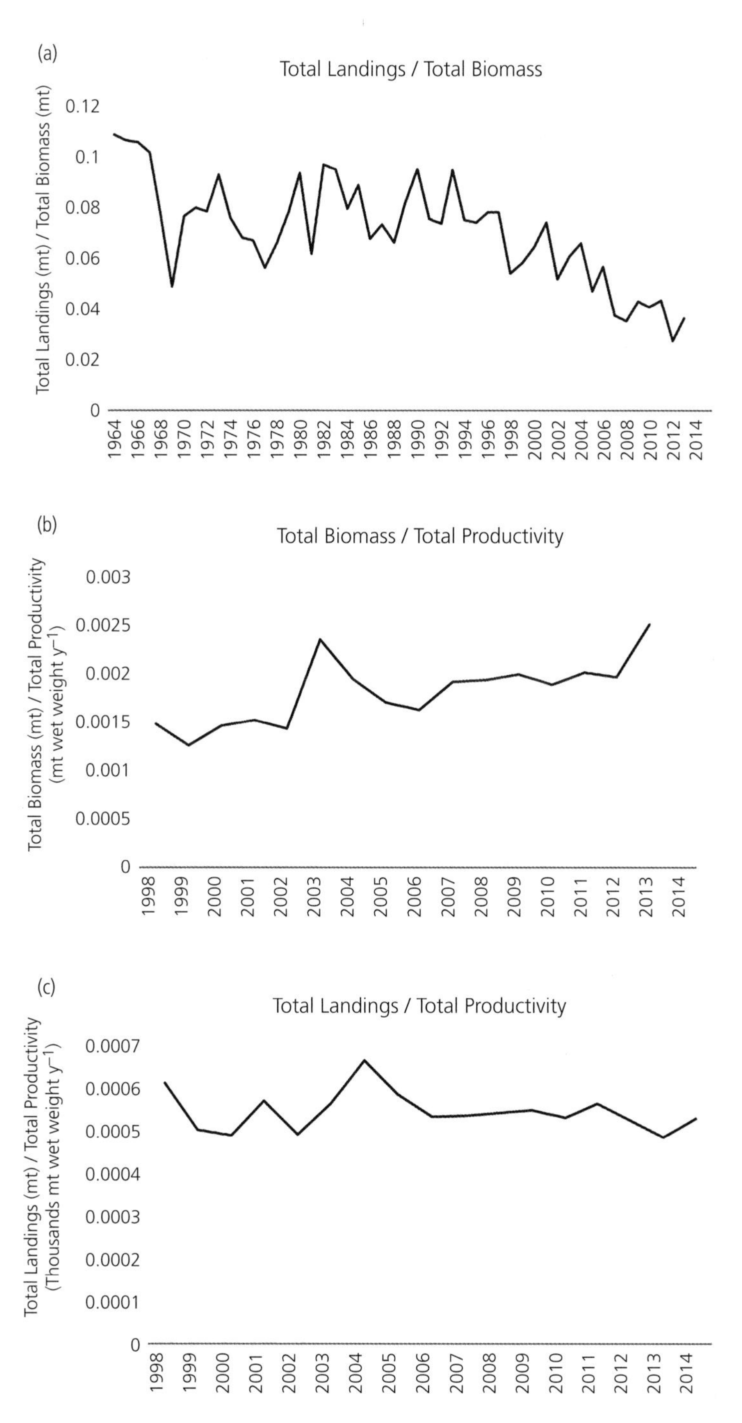

Figure 3.25 Ratios of (a) total commercial and recreational landings (metric tons) to total biomass (metric tons; i.e. exploitation index); (b) total biomass (metric tons) to total productivity (metric tons wet weight year^{-1}); (c) total commercial and recreational landings (metric tons) to total productivity (metric tons wet weight yr^{-1}) for the U.S. Northeast (Mid-Atlantic and New England) region.

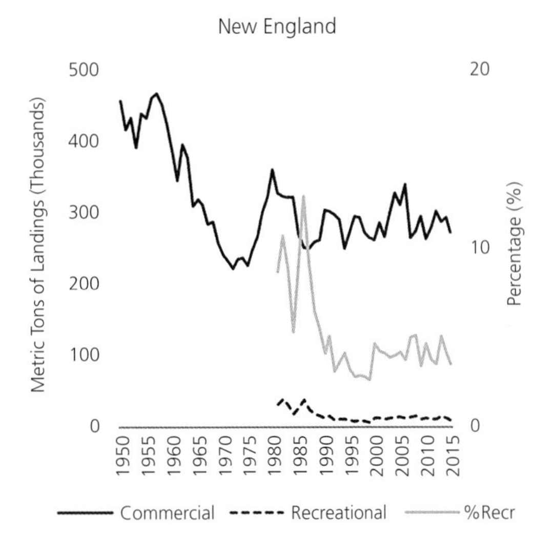

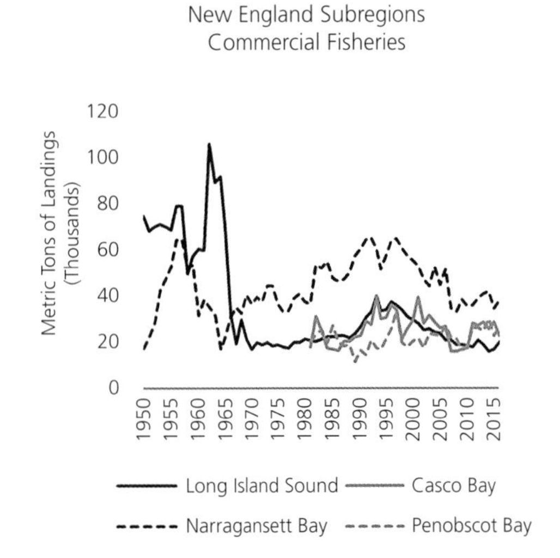

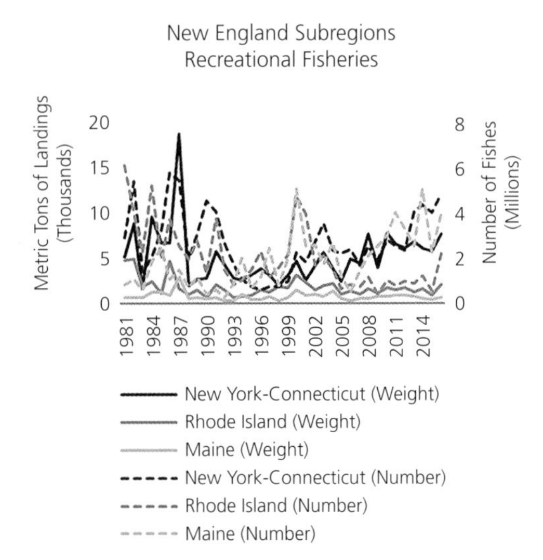

Figure 3.26 Total commercial and recreational landings and percent contribution of recreational landings (%Recr) over time for the New England region and identified subregions (commercial: 1950–2016; recreational: 1981–2016).

Data derived from NOAA National Marine Fisheries Service commercial and recreational fisheries statistics.

fishing landings (up to 124 metric tons km^{-2}) have been historically reported from waters in, surrounding and even offshore of Narragansett Bay over time. Per unit area, fishing landings in Casco and Penobscot Bays have remained relatively constant, while concentrated landings in Long Island Sound have been historically much lower when compared to other subregions.

Bycatch continues to persist throughout the New England region (Fig. 3.28), with values increasing over time for bony fishes, sharks, and marine mammals. In terms of weight, bycatch is most pronounced for bony fishes and invertebrates, while bycatch for marine mammal and seabirds has been most dominant by number. Bycatch for the New England region is among the highest regions in the nation, with the highest

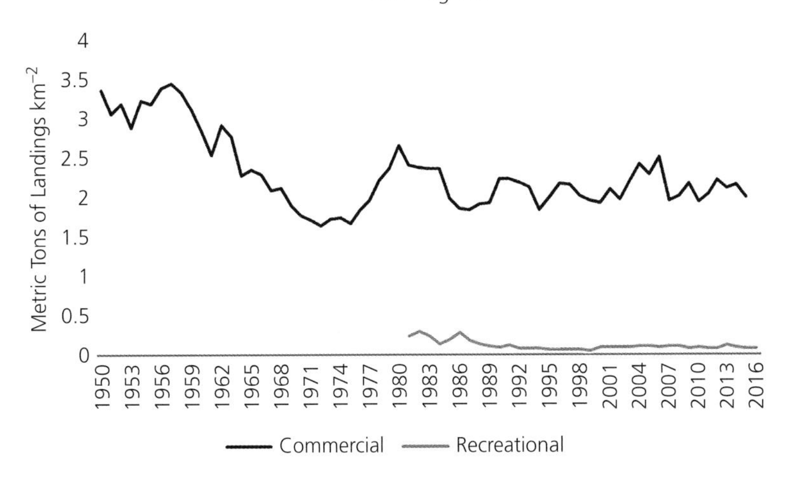

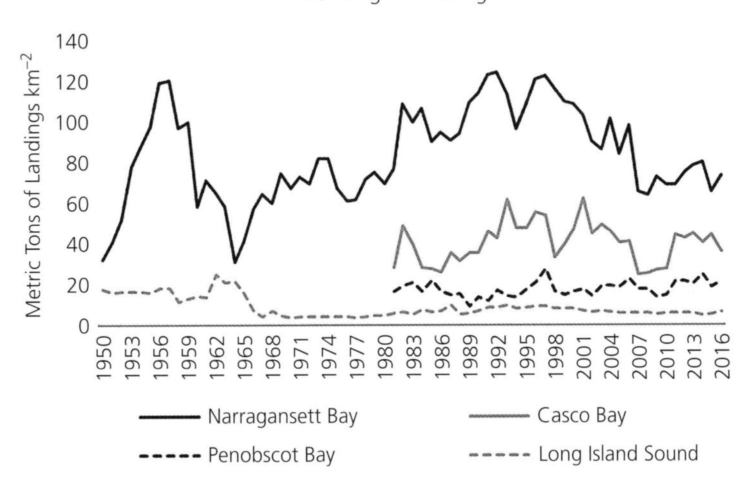

Figure 3.27 Total commercial and recreational landings (metric tons) per square kilometer over time (1950–2016) for the New England region and identified subregions.

Data derived from NOAA National Marine Fisheries Service commercial and recreational fisheries statistics.

bycatch values observed for sharks and marine mammals, and mid-level values for invertebrates (third-highest) and bony fishes (fourth-highest). However, New England has the comparatively lowest bycatch for seabirds in the country.

3.5 Synthesis

The New England socioecological system is an environment with low to moderate numbers of managed species that is responding to the consequences of overfishing, habitat loss, coastal development, and nutrient loading. More recent stressors including regional warming, climate-related species shifts, and proliferation of invasive species continue to affect this system, and alter its composition, dynamics, and LMR production (Silliman & Bertness 2004, Drinkwater 2005, Pederson et al. 2005, Harley et al. 2006, Fogarty et al. 2007, Frumhoff et al. 2007, Mathieson et al. 2008a, Carman et al. 2009, Horton & McKenzie 2009, Nye et al. 2009, Valentine et al. 2009, Lucey & Nye 2010, Griffen et al. 2012, Pershing et al. 2015, Kleisner et al.

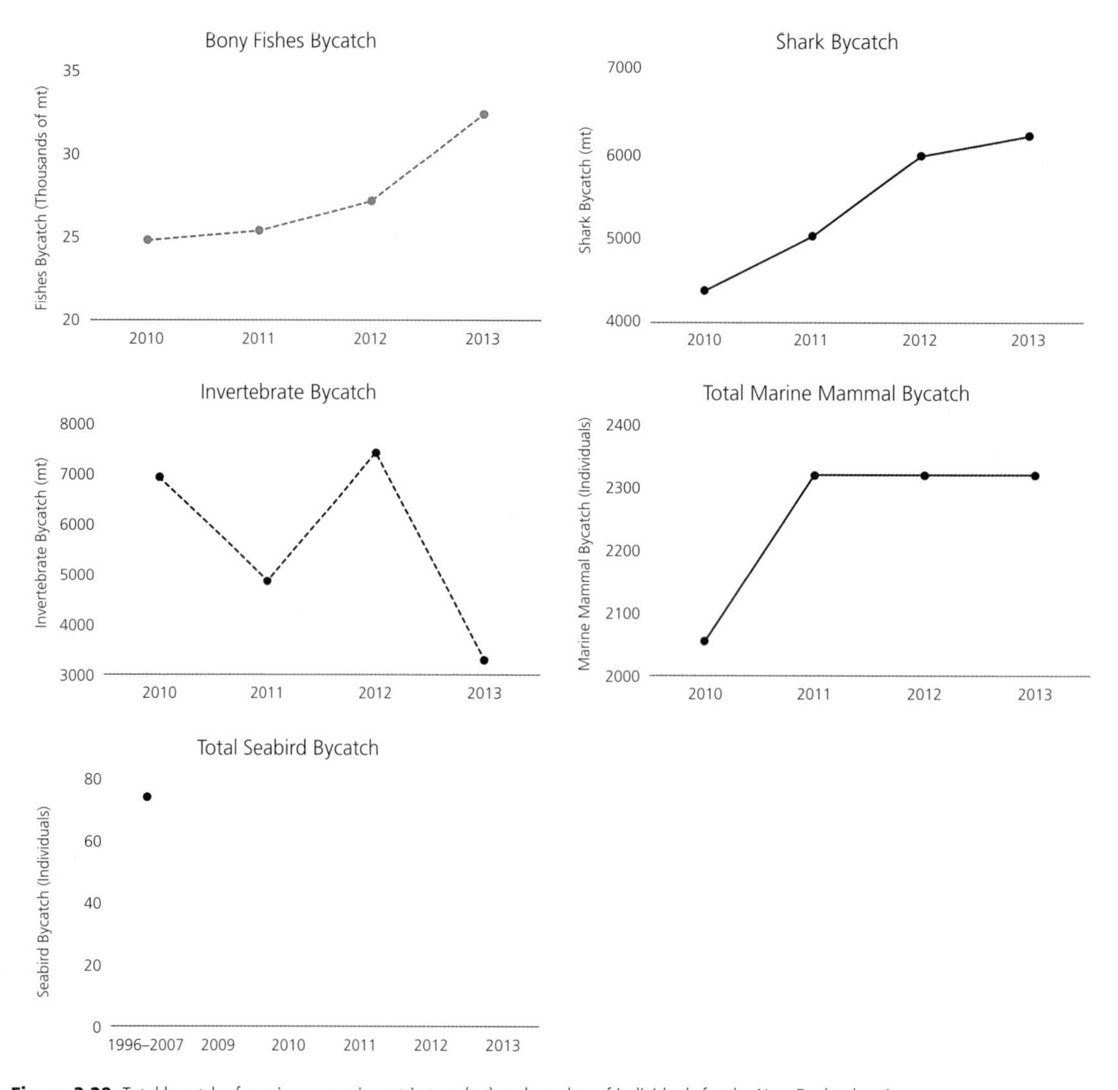

Figure 3.28 Total bycatch of species groups in metric tons (mt) and number of individuals for the New England region.

Data derived from NOAA National Marine Fisheries Service U.S. National Bycatch Report Database System. Note: New England seabird bycatch is only reported for the1996–2007 time period as a combined estimate.

2016, Klein et al. 2017, Meyer-Gutbrod 2017). Despite intense overharvesting, fisheries revenue in this system continues to increase, but landings since the 1980s remain consistently lower than those observed during peak values of the 1950s. Landings in Long Island Sound have especially decreased over time, while those for Casco and Narragansett Bay increased in the 1980s-1990s and have declined in more recent decades. Similar trends have been observed in their fisheries revenues over time. This region contains the highest number and proportion of stocks that are overfished or experiencing overfishing, but also contains low numbers of unassessed stocks and lower numbers of strategic marine mammal stocks (Link & Marshak 2019). These trends illustrate that despite the New England system being a data-rich region, or having made progress in reducing human impacts on marine mammals, these advancements have not necessarily led to improved rebuilding of fisheries stocks. Currently, this system continues to show the effects of sequential depletion, for which longer time periods and more comprehensive management approaches remain necessary.

With high human population density, increasing tourism, and other ocean uses, pressures from fishing, recreational, energy, and industrial sectors continue to affect New England. Management of these pressures is required to ensure environmentally sound practices and to account for their effects on New England's LMR economy. While this region has above-average socioeconomic status regarding its LMRs (Link & Marshak 2019), it currently remains mid-range in its contributions toward national other ocean uses, while contributing only 5% of the U.S. total ocean economy. Given increasing seafood demand and intensifying trade-offs in this region with a growing human population, the continued sustainability of this marine economy is of heightened concern. As certain New England commercial fisheries (e.g., lobster, cod) have been predicted to stall, decline, or further shift northward in response to regional warming or their ongoing overfished status (Pinsky & Fogarty 2012, Pinsky et al. 2013, Meng et al. 2016, Klein et al. 2017, Morley et al. 2017, Rheuban et al. 2017, Frisk et al. 2018, Bell et al. 2018, Morley et al. 2018, Tanaka et al. 2019), these factors require urgent consideration in management. In addition, increasing abundance of Mid-Atlantic species moving into this system has led to their interactions with prominent New England fisheries species, which may cause changes to offshore assemblages and affect future fishing allocations (Wahle et al. 2013, McMahan & Grabowski 2015, McMahan 2017, Morley et al. 2018, Dubik et al. 2019). With the effects of ongoing interrelated stressors in a system that has been significantly impacted by past human activities, EBFM strategies for New England must consider these historic and emerging factors. There has been comprehensive progress toward a broader understanding of this ecosystem and in synthesizing advanced ecosystem-level metrics, sophisticated models, and quantitative frameworks for estimating system thresholds (Brodziak & Link 2002, Brodziak et al. 2004, Bakun et al. 2009, Link 2010, Link et al. 2011a, Jin et al. 2012, 2013, NEFSC 2012, Large et al. 2015a, b, Tam et al. 2017, Gaichas et al. 2019b, NMFS 2019). Although partial progress has been made, management actions that can set system-wide and control rules as related to the effects of ongoing and emerging stressors on basal ecosystem productivity, fisheries and protected species biomass, and marine socioeconomics are needed for this system.

The New England ecosystem is well-known for aspects of its LMR-based socioeconomic status. This region is currently second-highest in terms of its total fisheries revenue and fourth-highest for total landings nationally, while recent system exploitation rates have been below suggested sustainable thresholds of 0.1 (Samhouri et al. 2010, Large et al. 2013, 2015a, b, Tam et al. 2017, Link & Marshak 2019). This region is also second-highest in terms of the proportional contributions of its LMR establishments, employments, and GDP toward its total marine economy, emphasizing continued importance toward the economic prosperity and cultural identity of this socioecological system. These trends are also fostered by the second-highest (in the nation; tied with the Mid-Atlantic) basal ecosystem productivity, which has also facilitated the fourth-highest LMR-based GDP among regions. Aspects of New England LMR governance are also noteworthy, especially regarding its concentrated Congressional and management attention on lower numbers of managed species, strategic use of fisheries closures, lower numbers of NEPA EIS actions, and cooperative management among the NEFMC and ASMFC (and MAFMC). However, this region is also subject to higher litigation as seen by the highest number of fisheries-related lawsuits in the country, contains a less-balanced FMC that is dominated by commercial representatives, and leads in terms of amendments for its FMPs, suggesting higher instability regarding its management actions. Priorities for NEFMC EBFM actions include accounting for trophic interactions among protected and fisheries species, and applying climate vulnerability assessment results into management strategies (NEFSC 2012, Hare et al. 2016, Gaichas et al. 2019b, NMFS 2019). Progress toward the creation of an FEP for Georges Bank is underway, which has been complemented by the priorities of a recently released EBFM implementation plan for the U.S. Northeast region (NMFS 2019) and focuses on implementing the NOAA Fisheries EBFM Policy and Roadmap (NMFS 2016a, b). Specifically, these priorities are focused on the following:

- Developing engagement strategies to facilitate the participation of partners and stakeholders in the EBFM process.
- Conducting science and providing ESRs to understand the New England ecosystem.
- Identifying ecosystem-level, cumulative risk (across LMRs, habitats, ecosystem functions, and associated fisheries communities) and vulnerability to human and natural pressures in New England.
- Identifying the individual and cumulative pressures that pose the most risk to vulnerable resources and dependent communities.
- Analyzing trade-offs for optimizing benefits from all fisheries within each ecosystem or jurisdiction, taking into account ecosystem-specific

policy goals and objectives, cognizant that ecosystems are composed of interconnected components.
- Developing Management Strategy Evaluation capabilities to better conduct ecosystem-level analyses to provide ecosystem-wide management advice.
- Developing and monitoring Ecosystem-Level Reference Points for New England.
- Incorporating ecosystem considerations into appropriate LMR assessments, control rules, and management decisions.
- Providing integrated advice for other management considerations, particularly applied across multiple species within an ecosystem.
- Evaluate ecosystem-level measures of resilience and community well-being.

The continuation of ESRs for the northeast region (Link et al. 2002, NEFSC 2009, 2012), which have now morphed into a State of the Ecosystem Report for New England (Gaichas et al. 2019b), have advanced this effort. These focused efforts have allowed for more thorough integration of ecosystem-level information, with the potential for its application in developing system-wide reference points and more advanced multispecies approaches to management. These efforts also facilitate the application of information for ecosystem-level planning and risk assessment to bolster management actions. Recurring release of ESRs that include the identification, monitoring, and comprehensive assessment of ecosystem-level indicators enhance the development of holistic management practices that account for the physical, biological, and socioeconomic components of this ecosystem and provide comprehensive information needed to support management as it shifts toward more holistic practices (Brodziak & Link 2002, Brodziak et al. 2004, Bakun et al. 2009, Gamble & Link 2009, Link 2010, Link et al. 2011a, Jin et al. 2012, NEFSC 2012, Large et al. 2015a, b, Tam et al. 2017, Gaichas et al. 2019b, NMFS 2019). Ongoing efforts, including the application of multispecies and aggregate production models that are concentrated on Georges Bank and the Gulf of Maine are proving useful for more informative assessments. These models work toward a broader understanding of New England regions and their responses to ongoing stressors (Link et al. 2010a, b, Fulton et al. 2011, Townsend et al. 2014, 2017, 2018). All of these advancements are enhancing the potential for the successful application of ecosystem approaches to management in New England.

Despite many of these large-scale efforts toward greater scientific understanding of the New England ecosystem, challenges remain toward effectively implementing formalized EBFM management actions and enacting ecosystem-level control rules. Namely, this region currently lacks a completed FEP, and though there is a solid analytical basis, only partial progress has occurred toward considering system catch limits for this region (Brodziak & Link 2002, Brodziak et al. 2004, Bakun et al. 2009, Link 2010, Link et al. 2011a, Jin et al. 2012, 2013, NEFSC 2012, Gaichas et al. 2019b, Link & Marshak 2019, NMFS 2019). While efforts to implement EBFM for this region are underway, actions remain necessary to incorporate broader ecosystem considerations into management, particularly in recognition of system-level shifts, exceeded thresholds, and ongoing and predicted consequences from regional impacts of climate change. With the remaining consequences of overharvesting on the New England ecosystem, in addition to the effects of coastal development and regional warming, the ability to curb system overexploitation in this changing environment through comprehensive management actions remains a requirement. These actions must also consider the cumulative effects of these stressors on both biological and human communities. Management actions that explicitly consider the consequences of altered ecosystem functioning as a result of human activities are especially warranted. Given the advanced understanding of many of these relationships in New England systems, a next critical step for this region is to directly apply this ecosystem-level information into expanded system-level reference points for consideration as sustainable thresholds to facilitate improved management of this system.

With a rising human population density that is already the third-highest in the nation, ongoing stressors affecting this system are likely to increase from human activities with concurrent competition for resources within and among its marine sectors. In particular, conflicts among fisheries, tourism, marine transportation, and emerging wind energy development are predicted to increase, while Mid-Atlantic species are also likely to become more abundant in this region with regional warming and associated shifts in distribution (Hall-Arber et al. 2001, Ihde et al. 2011, Portman et al. 2011, Bailey et al. 2014, Samoteskul et al. 2014, McMahan & Grabowski 2015, McMahan 2017, Smythe 2017, Wyatt et al. 2017). These range shifts may also foster increased resource conflicts among commercial and recreational entities in New England. With enhanced understanding of New England ecosystem functioning, including predicted and observed multi-level responses to climate change, natural forcing, and human activities on its dynamics and marine socioeconomics, these concerns should also be reflected in strategies that account for their effects on other ocean sectors, in addition to their relationships with LMRs and fisheries. As for other regions, systematic management strategies that consider spatially overlapping ocean uses, resource partitioning, multiple species, and the socioeconomic effects of changing environments are warranted in this system. With expanding wind energy development, tourism, and marine transportation (Martin & Hall-Arber 2008, Anthony et al. 2009, Link 2010, Holland et al. 2012, Bailey et al. 2014, Samoteskul et al. 2014, Guida et al. 2017, Wyatt et al. 2017, Gaichas et al. 2019a, b, Link & Marshak 2019), additional nonfishing effects are anticipated, which warrant continued spatial management considerations of the New England EEZ.

Broader engagement strategies for ecosystem-level management remain needed to enhance cooperation by stakeholders. Increasing awareness of climate change and environmental threats to the sustainability of commercially important species have focused community attention on the future of LMRs and their marine economies (da Silva & Kitts 2006, Hartley & Robertson 2008, Hartley & Robertson 2006, 2008, Hartley 2010, Holland et al. 2013). As a result, greater interest in EBM, and an understanding of the interconnectivity of these stressors, ecological interactions among multiple species, fisheries productivity, and human activities through the lens of a socioecological system is emerging in this region (Brodziak & Link 2002, Brodziak et al. 2004, Bakun et al. 2009, Link 2010, Link et al. 2011a, Jin et al. 2012, 2013, NEFSC 2012, Samhouri et al. 2013, Gaichas et al. 2019b, NMFS 2019). Continuing to build on this interest, together with ongoing initiatives and scientific advancements regarding the functioning of the New England ecosystem, allows for continued advances in EBM strategies for this region. More developed engagement strategies are especially needed given the enhanced conflicts that are likely to develop among fisheries and sectors as a result of shifting species or development of offshore wind energy. Efforts that work to address these concerns, and build upon progress by the Northeast Regional Planning Body (NRPB 2016) toward ocean planning, regarding the development of ecosystem-based frameworks are greatly needed. While governance frameworks in this region are focused and targeted toward a limited few managed species, they require re-evaluation toward more systematic approaches, reduced lawsuits, and enhanced cooperative stakeholder involvement.

3.5.1 Progress Toward EBFM

This ecosystem is excelling in the socioeconomic status of its LMRs, and is relatively productive, as related to the determinants of successful LMR management (Fig. 3.29; Link & Marshak 2019).

New England fisheries management continues evolving into approaches that address cumulative issues affecting its fisheries and LMRs. Through a more systematic and prioritized consideration of all fisheries, pressures, risks, and outcomes, holistic EBFM practices are being developed and implemented across the U.S., including this ecosystem (NMFS 2016a, b,

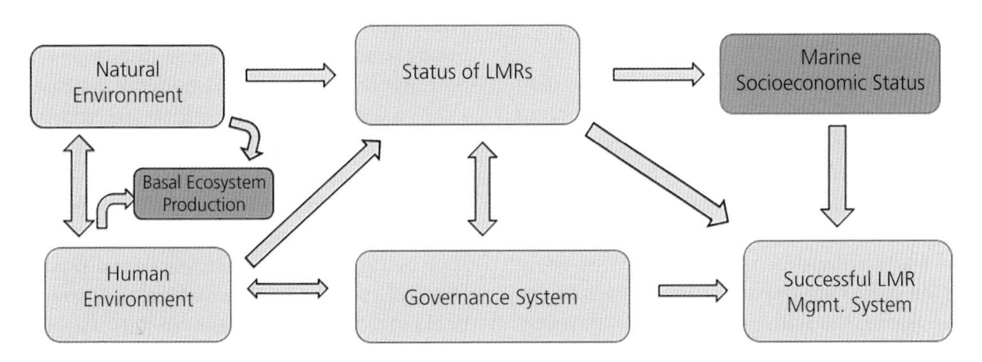

Figure 3.29 Schematic of the determinants and interconnectivity of successful living marine resource (LMR) systems management criteria as modified from Link and Marshak (2019). Those highlighted in blue indicate noteworthy criteria for the New England ecosystem.

Levin et al. 2018, Marshall et al. 2018, Link & Marshak 2019). In accordance with the NOAA Fisheries EBFM Policy and Roadmap (NMFS 2016a, b), an EBFM implementation plan (NMFS 2019) for the New England region has been developed through efforts overseen by the NOAA NEFSC, NOAA Fisheries Greater Atlantic Regional Fisheries Office (GARFO), NEFMC, and the ASMFC. This plan specifies approaches to conduct EBFM and to address NOAA Fisheries Roadmap priorities and gaps. Progress for New England and its major subregions in terms of EBFM Roadmap Guiding Principles and Goals is shown in Table 12.9 in Chapter 12.

Overall, a moderate degree of progress has been made at the regional and subregional levels in implementing ecosystem-level planning, advancing knowledge of ecosystem principles, and examining system trade-offs. Similarly, moderate progress has been observed in assessing risks and vulnerabilities to ecosystems, such as through ongoing investigations into climate vulnerability and species prioritizations for stock and habitat assessments. Pilot initiatives toward incorporating habitat and ecosystem-based considerations into management have also been undertaken (Brodziak & Link 2002, Link 2002, Link et al. 2002, Stevenson et al. 2004, Link et al. 2005, Link et al. 2011a, Tyrrell et al. 2011, Gaichas et al. 2012, Essington et al. 2016, Gaichas et al. 2016, 2017, 2019, NMFS 2019) to complement multispecies federal FMPs for the region's offshore groundfish species and the high number of single-species FMPs under the jurisdiction of the ASMFC. While much information has been obtained and applied regarding ecosystem-level calculations, syntheses, and models, only partial progress has been observed in using these system-wide emergent properties in management actions. However, compared to other regions, New England still leads in terms of much of its progress toward these EBFM guiding principles and goals.

Progress on the recommendations of the Ecosystem Principles Advisory Panel (EPAP) for the development of FEPs (EPAP 1999, Wilkinson & Abrams 2015) is shown in Table 12.10 (in Chapter 12). While there is no current formalized FEP for the New England ecosystem, a substantial degree of progress has been observed toward characterizing, monitoring, and assessing aspects of its ecosystem dynamics in support of this ecosystem approach, with several ESRs partially fulfilling that role. An FEP is under development by the NEFMC EBFM Plan Development Team for Georges Bank (NEFMC 2019). These advancements are also complemented by ASMFC efforts, including the development of ecosystem reference points for management actions and discussions of ecosystem-level data needs for more effective EBFM implementation. Efforts are underway to adopt ecosystem reference points in the near future, with recent adoption for Atlantic Menhaden (ASMFC 2018, 2020a).

3.5.2 Conclusions

Continuing with a business-as-usual approach or a scaled-down ecosystem approach when managing the resources of this ecologically significant region, without considering their effects on the whole ecosystem, limits efforts toward sustainable practices. The improved sustainability of this region is dependent on management approaches that consider the effects of natural stressors and human activities on interacting sectors, system productivity, and at multiple trophic levels, while accounting for socioeconomic factors. System-wide considerations of these relationships are needed to refine and maintain this system within needed sustainable thresholds. EBFM provides opportunities that focusing on one stock or species complex or one fishery would not. Simultaneous harvest of interacting species complexes and fisheries, which account for their trophic interdependence and contributions toward system productivity are essential for optimal management. Overlapping fishing effort for these species, in addition to the gear-related effects of their harvest on essential habitats and subsequent ecological consequences to multispecies complexes is considered in these approaches. The system-wide consequences from increasingly abundant range-expanding and invasive species are likely to be magnified in this region, for which system-level management frameworks are necessary when accounting for their ecological effects. With many ecologically important species occurring throughout this system, these approaches allow for portfolio and multilevel considerations that can provide for a more effective, responsive, and precautionary management strategy (Edwards et al. 2004, Kellner et al. 2011, Link 2018).

As the New England ecosystem continues to be subject to multiple pressures, this more comprehensive approach is essential for its continued prosperity. The effects of overexploitation continue to overshadow New England fisheries and their habitats. Given the comparatively advanced knowledge of this system and its socioecological interconnectivity, there is great potential to move forward in monitoring and managing the effects of regional warming and emergent human activities using biological and socioeconomic metrics that are related to defined system-wide thresholds. It is incumbent on managers, stakeholders, and scientists to cooperatively ensure that future fisheries management strategies work toward maintaining the New England ecosystem within sustainable limits, especially in consideration of rebuilding its depleted fisheries populations. As compounding effects of climate change, sea level rise, increasing storm intensity, habitat loss, eutrophication, and the consequences of historical overfishing continue to affect this system management throughout this socioecological system can be more effectively addressed using an ecosystem approach. Applying comprehensive management strategies in this system that consider its inherent productivities and well-known integrated ecosystem properties, and work toward enhancing its LMR status and socioeconomics can build upon many of the advanced frameworks that are in development for this region. These efforts provide a pathway toward ensuring the continuation of New England's maritime culture, its nationally important LMRs and ocean-based economies, and for preserving the legacy of this historically iconic fisheries ecosystem.

3.6 References

Acheson JM. 1980. Attitudes towards limited entry among finfishermen in northern New England. *Fisheries*. 5(6): 20–5.

Acheson JM. 2006. Lobster and groundfish management in the Gulf of Maine: a rational choice perspective. *Human Organization* 65(3):240.

Acheson JM. 2011. Coming up empty: management failure of the New England groundfishery. *Maritime Studies* 10(1):57–86.

Acheson JM, Gardner R. 2011. Modeling disaster: the failure of the management of the New England groundfish industry. *North American Journal of Fisheries Management* 31(6):1005–18.

Acheson J, Gardner R. 2014. Fishing failure and success in the Gulf of Maine: lobster and groundfish management. *Maritime Studies* 13(1):8.

Adamowicz SC, Roman CT. 2005. New England salt marsh pools: a quantitative analysis of geomorphic and geographic features. *Wetlands* 25(2):279.

Adams JT. 1923. *Revolutionary New England, 1691–1776*. New York, NY: Atlantic Monthly Press.

Alexander KE, Leavenworth WB, Cournane J, Cooper AB, Claesson S, Brennan S, Smith G, et al. 2009. Gulf of Maine cod in 1861: historical analysis of fishery logbooks, with ecosystem implications. *Fish and Fisheries* 10(4):428–49.

Ames EP. 2004. Atlantic cod stock structure in the Gulf of Maine. *Fisheries* 29(1):10–28.

Ames, EP, Lichter J. 2013. Gadids and alewives: structure within complexity in the gulf of Maine. *Fisheries Research* 141:70–78.

Andrews M, Rossi D. 1986. The economic impact of commercial fisheries and marine-related activities: a critical review of northeastern input-output studies. *Coastal Management* 13(3–4):335–67.

Anthony A, Atwood J, August PV, Byron C, Cobb S, Foster C, Fry C, et al. 2009. Coastal lagoons and climate change: ecological and social ramifications in the US Atlantic and Gulf coast ecosystems. *Ecology and Society* 14(1):8.

Anthony VC. 1990. The New England groundfish fishery after 10 years under the magnuson fishery conservation and management act. *North American Journal of Fisheries Management* 10(2):175–84.

Anthony VC. 1993. The state of groundfish resources off the northeastern United States. *Fisheries* 18(3):12–7.

ASMFC (Atlantic States Marine Fisheries Commission). 2018. *2018 Review of the Atlantic States Marine Fisheries Commission Fishery Management Plan and State Compliance for Atlantic Menhaden (Brevoortia tyrannus)*. Arlington, VA: ASMFC (p. 17).

ASMFC (Atlantic States Marine Fisheries Commission). 2020a. *ASMFC Atlantic Menhaden Board Adopts Ecological Reference Points*. Arlington, VA: ASMFC (p. 2).

ASMFC (Atlantic States Marine Fisheries Commission). 2020b. *ASMFC Stock Status Overview*. Arlington, VA: ASMFC (p. 43).

Auster PJ. 2007. Linking deep-water corals and fish populations. *Bulletin of Marine Science* 81(3):93–9.

Auster PJ, Malatesta RJ, Langton RW, Watting L, Valentine PC, Donaldson CL, Langton EW, Shepard AN, Babb WG. 1996. The impacts of mobile fishing gear on seafloor habitats in the Gulf of Maine. Northwest Atlantic): implications for conservation of fish populations. *Reviews in Fisheries Science* 4(2):185–202.

Auster PJ, Malatesta RJ, LaRosa SC. 1995. Patterns of microhabitat utilization by mobile megafauna on the southern New England (USA) continental shelf and slope. *Marine Ecology Progress Series* 127:77–85.

Auster PJ, Moore J, Heinonen KB, Watling L. 2005. "A habitat classification scheme for seamount landscapes: assessing the functional role of deep-water corals as fish habitat." In: *Cold-Water Corals and Ecosystems*. Edited by A Freiwald, JM Roberts. Berlin, Germany: Springer (pp. 761–9).

Azarovitz TR. 1981. A brief historical review of the Woods Hole Laboratory trawl survey time series. Canadian Special Publication of Fisheries and Aquatic Sciences 58:62–67.

Bailey H, Brookes KL, Thompson PM. 2014. Assessing environmental impacts of offshore wind farms: lessons learned and recommendations for the future. *Aquatic Biosystems* 10(1):8.

Backus RH. 1987. *Georges Bank*. Cambridge, MA: MIT Press (p. 593).

Bakun A, Babcock EA, Santora C. 2009. Regulating a complex adaptive system via its wasp-waist: grappling with ecosystem-based management of the New England herring fishery. *ICES Journal of Marine Science* 66(8):1768–75.

Bassett T. 1987. Documenting recreation and tourism in New England. *The American Archivist* 50(4):550–69.

Beland KF, Jordan RM, Meister AL. 1982. Water depth and velocity preferences of spawning Atlantic salmon in Maine rivers. *North American Journal of Fisheries Management*. 2(1):11–3.

Bell EL. 2009. Cultural resources on the New England coast and continental shelf: research, regulatory, and ethical considerations from a Massachusetts perspective. *Coastal Management* 37(1):17–53.

Bell RJ, Richardson DE, Hare JA, Lynch PD, Fratantoni PS. 2015. Disentangling the effects of climate, abundance, and size on the distribution of marine fish: an example based on four stocks from the Northeast US shelf. *ICES Journal of Marine Science* 72(5):1311–22.

Bell RJ, Wood A, Hare J, Richardson D, Manderson J, Miller T. 2018. Rebuilding in the face of climate change. *Canadian Journal of Fisheries and Aquatic Sciences* 75:1405–1414.

Bentley MG. 2011. "The global spread of the Chinese mitten crab Eriocheir sinensis." In: *In the Wrong Place—Alien Marine Crustaceans: Distribution, Biology and Impacts*. Edited by BS Galil, PF Clark, JT Carlton. Dordrecht, Netherlands: Springer (pp. 107–127).

Bigelow HB, Schroeder WC. 1953. *Fishes of the Gulf of Maine*. Washington, DC: US Government Printing Office.

Bolster WJ. 2008. Putting the ocean in Atlantic history: maritime communities and marine ecology in the Northwest Atlantic, 1500–1800. *The American Historical Review*. 113(1):19–47.

Bourque BJ, Johnson BJ, Steneck RS. 2008. Possible prehistoric fishing effects on coastal marine food webs in the Gulf of Maine. *Human Impacts on Ancient Marine Ecosystems* 29:165–85.

Brodziak J, Cadrin SX, Legault CM, Murawski SA. 2008a. Goals and strategies for rebuilding New England groundfish stocks. *Fisheries Research* 94(3):355–66.

Brodziak J, Link J. 2002. Ecosystem-based fishery management: what is it and how can we do it? *Bulletin of Marine Science* 70(2):589–611.

Brodziak JK, Mace PM, Overholtz WJ, Rago PJ. 2004. Ecosystem trade-offs in managing New England fisheries. *Bulletin of Marine Science* 74(3):529–48.

Brodziak J, Traver ML, Col LA. 2008b. The nascent recovery of the Georges Bank haddock stock. *Fisheries Research* 94(2):123–32.

Bromberg KD, Burtness MD. 2005. Reconstructing New England salt marsh losses using historical maps. *Estuaries* 28(6):823–32.

Brooks DA. 1985. Vernal circulation in the Gulf of Maine. *Journal of Geophysical Research: Oceans* 90(C3):4687–706.

Brooks DA. 1987. The influence of warm-core rings on slope water entering the Gulf of Maine. *Journal of Geophysical Research: Oceans* 92(C8):8183–96.

Brousseau DJ, Kriksciun K, Baglivo JA. 2003. Fiddler crab burrow usage by the Asian crab, Hemigrapsus sanguineus, in a Long Island Sound salt marsh. *Northeastern Naturalist* 10(4):415–21.

Brown BE, Brennan JA, Grosslein MD, Heyerdahl EG, Hennemuth RC. 1976. The effect of fishing on the marine finfish biomass in the Northwest Atlantic from the Gulf of Maine to Cape Hatteras. *ICNAF Research Bulletin* 12:49–68.

Brown OB, Cornillon PC, Emmerson SR, Carle HM. 1986. Gulf Stream warm rings: a statistical study of their behavior. Deep sea research part A. *Oceanographic Research Papers* 33(11–12):1459–73.

Buchsbaum R, Pederson J, Robinson WE. 2005. *The Decline of Fisheries Resources in New England: Evaluating the Impact of Overfishing, Contamination, and Habitat Degradation*. Cambridge, MA: MIT Press.

Bullard SG, Carman MR. 2009. Current trends in invasive ascidian research. *Invasive Species: Detection, Impact and Control* 2009:57–79.

Busch WD, Lary SJ, Castilione CM, McDonald RP. 1998. *Distribution and Availability of Atlantic Coast Freshwater Habitats for American Eel (Anguilla Rostrata)*. Amherst, NY: US Fish and Wildlife Service (p. 982).

Cargnelli LM. 1999a. *Essential Fish Habitat Source Document. Atlantic Surfclam, Spisula Solidissima, Life History and Habitat Characteristics. NOAA Technical Memorandum NMFS-NE-142*. Washington, DC: NOAA.

Cargnelli LM. 1999b. *Essential Fish Habitat Source Document. Ocean Quahog, Arctica Islandica, Life History and Habitat Characteristics. NOAA Technical Memorandum NMFS-NE-148*. Washington, DC: NOAA.

Carman MR, Grunden DW. 2010. First occurrence of the invasive tunicate Didemnum vexillum in eelgrass habitat. *Aquatic Invasions* 5(1):23–9.

Carman MR, Hoagland KE, Green-Beach E, Grunden DW. 2009. Tunicate faunas of two North Atlantic-New England islands: Martha's Vineyard, Massachusetts and Block Island, Rhode Island. *Aquatic Invasions* 4(1):65–70.

Carmichael RH, Shriver AC, Valiela I. 2004. Changes in shell and soft tissue growth, tissue composition, and survival of quahogs, Mercenaria mercenaria, and softshell clams, Mya arenaria, in response to eutrophic-driven changes in food supply and habitat. *Journal of Experimental Marine Biology and Ecology* 313(1):75–104.

Carp BL. 2010. *Defiance of the Patriots: The Boston Tea Party and the Making of America*. New Haven, CT: Yale University Press.

Carretta JV, Forney KA, Oleson EM, Weller DW, Lang AR, Baker J, Muto MM, et al. 2019. *U.S. Pacific Marine Mammal Stock Assessments: 2018. U.S. Department of Commerce, NOAA Technical Memorandum NMFSSWFSC-617*. Washington, DC: NOAA.

Chen C, Huang H, Beardsley RC, Xu Q, Limeburner R, Cowles GW, Sun Y, Qi J, Lin H. 2011. Tidal dynamics in the Gulf of Maine and New England Shelf: an application of FVCOM. *Journal of Geophysical Research: Oceans* 116(C12):12010.

Christensen V, Walters C. 2004. Ecopath with Ecosim: methods, capabilities and limitations. *Ecological Modeling* 172:109–39.

Christensen V, Walters CJ, Pauly D. 2005. *Ecopath with Ecosim: A User's Guide*. Vancouver, Canada: Fisheries Centre of University of British Columbia (p. 154).

Clapham PJ, Link JS. 2006. *Whales, Whaling and Ecosystems in the North Atlantic Ocean. Whales, Whaling and Ocean Ecosystems*. Berkeley, CA: University of California Press (pp. 314–23).

Clark SH, Brown BE. 1977. Changes in biomass of finfishes and squids from the Gulf of Maine to Cape Hatteras, 1963–74, as determined from research vessel survey data. *Fishery Bulletin* 75(1):1–21.

Colburn LL, Clay PM, Olson J, da Silva PP, Smith SL, Westwood A, Ekstrom J. 2010. *Community Profiles for Northeast US Marine Fisheries*. Woods Hole, MA: US Department of Commerce.

Collette BB. 1986. Resilience of the fish assemblage in New-England tidepools. *Fishery Bulletin* 84(1):200–4.

Collie JS, Escanero GA, Valentine PC. 1997. Effects of bottom fishing on the benthic megafauna of Georges Bank. *Marine Ecology Progress Series* 155:159–7.

Craddock JE, Polloni PT, Hayward B, Wenzel F. 2009. Food habits of Atlantic white-sided dolphins (*Lagenorhynchus acutus*) off the coast of New England. *Fishery Bulletin* 107(3):384–94.

Crawford S, Whelan G, Infante DM, Blackhart K, Daniel WM, Fuller PL, Birdsong T, et al. 2016. *Through a Fish's Eye: The Status of Fish Habitats in the United States 2015*. National Fish Habitat Partnership. http://assessment.fishhabitat.org/.

Curti KL, Collie JS, Legault CM, Link JS. 2013. Evaluating the performance of a multispecies, statistical catch-at-age model. *Canadian Journal of Fisheries and Aquatic Sciences* 70:470–84.

Curtis TH, Sosebee KA. 2015. Landings composition of the Northeast US skate, Rajidae, wing fishery and the effectiveness of prohibited species regulations. *Marine Fisheries Review* 77(4):1–8.

da Silva PP, Kitts A. 2006. Collaborative fisheries management in the Northeast US: emerging initiatives and future directions. *Marine Policy* 30(6):832–41.

Dahl TE. 2000. *Status and Trends of Wetlands in the Conterminous United States 1986 to 1997*. Onalaska, WI: US Fish and Wildlife Service.

Daley BA, Scavia D. 2008. *An Integrated Assessment of the Continued Spread and Potential Impacts of the Colonial Ascidian, Didemnum sp. A, in U.S. Waters. NOAA Technical Memorandum NOS NCCOS 78*. Washington, DC: NOAA (p. 61).

Davis RC, Short FT, Burdick DM. 1998. Quantifying the effects of Green Crab damage to Eelgrass transplants. *Restoration Ecology* 6:297–302.

DeAlteris JT, Skrobe LG, Castro KM. 2000. Effects of mobile bottom fishing gear on biodiversity and habitat in offshore New England waters. *Northeastern Naturalist* (4):379–95.

Deegan LA, Buchsbaum RO. 2005. "The effect of habitat loss and degradation on fisheries." In: *The Decline of Fisheries Resources in New England*. Edited by R Buchsbaum, J Pederson, WE Robinson. Amherst, MA: University of Massachusetts Press (pp. 67–96).

Deegan LA, Johnson DS, Warren RS, Peterson BJ, Fleeger JW, Fagherazzi S, Wollheim WM. 2012. Coastal eutrophication as a driver of salt marsh loss. *Nature* 490(7420):388.

Deegan LA, Wright A, Ayvazian SG, Finn JT, Golden H, Merson RR, Harrison J. 2002. Nitrogen loading alters seagrass ecosystem structure and support of higher trophic levels. *Aquatic Conservation: Marine and Freshwater Ecosystems* 12(2):193–212.

Demos J. 1965. Notes on life in Plymouth Colony. *The William and Mary Quarterly* 22(2):264–86.

DePiper GS, Gaichas SK, Lucey SM, Pinto da Silva P, Anderson MR, Breeze H, Bundy A, et al. 2017. Operationalizing integrated ecosystem assessments within a multidisciplinary team: lessons learned from a worked example. *ICES Journal of Marine Science* 74(8):2076–86.

Deroba JJ, Gaichas SK, Lee MY, Feeney RG, Boelke D, Irwin BJ. 2019. The dream and the reality: meeting decision-making time frames while incorporating ecosystem and economic models into management strategy evaluation. *Canadian Journal of Fisheries and Aquatic Sciences* 76(7):1112–33.

Dijkstra J, Harris LG, Westerman E. 2007. Distribution and long-term temporal patterns of four invasive colonial ascidians in the Gulf of Maine. *Journal of Experimental Marine Biology and Ecology* 342(1):61–8.

Drinkwater KF. 2005. The response of Atlantic cod (Gadus morhua) to future climate change. *ICES Journal of Marine Science* 62:1327–37.

Drinkwater KF, Belgrano A, Borja A, Conversi A, Edwards M, Greene CH, Ottersen G, Pershing AJ, Walker H. 2003. The response of marine ecosystems to climate variability associated with the North Atlantic Oscillation. *Geophysical Monograph-American Geophysical Union* 134:211–34.

Drinkwater KF, Mountain DB, Herman A. 1998. Recent changes in the hydrography of the Scotian Shelf and Gulf of Maine—a return to conditions of the 1960s? *NAFO Scientific Council Research Document* 1998(98):16.

Driscoll CT, Whitall D, Aber J, Boyer E, Castro M, Cronan C, Goodale CL, et al. 2003. Nitrogen pollution in the northeastern United States: sources, effects, and management options. *BioScience* 53(4):357–74.

Dubik BA, Clark EC, Young T, Zigler SB, Provost MM, Pinsky ML, Martin KS. 2019. Governing fisheries in the face of change: social responses to long-term geographic shifts in a US fishery. *Marine Policy* 99:243–51.

Dunton KJ, Jordaan A, McKown KA, Conover DO, Frisk MG. 2010. Abundance and distribution of Atlantic sturgeon (Acipenser oxyrinchus) within the Northwest Atlantic Ocean, determined from five fishery-independent surveys. *Fishery Bulletin* 108(4):450.

Edwards SF, Link JS, Rountree BP. 2004. Portfolio management of wild fish stocks. *Ecological Economics* 49(3):317–29.

Epifianio CE. 2013. Invasion biology of the Asian shore crab *Hemigrapsus sanguineus*: a review. *Journal of Experimental Marine Biology and Ecology* 441:33–49.

Essington TE, Levin PS, Marshall KN, Koehn L, Anderson LG, Bundy A, Carothers C, et al. 2016. *Building Effective Fishery Ecosystem Plans: A Report from the Lenfest Fishery Ecosystem Task Force.* Washington, DC: Lenfest Ocean Program.

Fahay MP, Berrien PL, Johnson DL, Morse WW. 1999. *Essential Fish Habitat Source Document. Atlantic Cod, Gadus Morhua, Life History and Habitat Characteristics. NOAA Technical Memorandum NMFS-NE-124.* Washington, DC: NOAA (p. 41).

Fay G, Link JS, Hare JA. 2017. Assessing the effects of ocean acidification in the Northeast US using an end-to-end marine ecosystem model. *Ecological Modelling* 347: 1–10.

Fay G, DePiper GS, Steinback S, Gamble R, Link JS. 2019. Economic and ecosystem effects of fishing on the Northeast US shelf. *Frontiers in Marine Science* 6:133.

Ferguson L, Srinivasan M, Oleson E, Hayes S, Brown SK, Angliss R, Carretta J, et al. 2017. *Proceedings of the First National Protected Species Assessment Workshop. U.S. Dept. of Commer., NOAA. NOAA Technical Memorandum NMFS-F/SPO-172.* Washington, DC: NOAA (p. 92).

Flanagan PH, Jensen OP, Morley JW, Pinsky ML. 2019. Response of marine communities to local temperature changes. *Ecography* 42(1):214–24.

Flanagan SM, Nielsen MG, Robinson KW, Coles JF. 1999. *Water-Quality Assessment of the New England Coastal Basins in Maine, Massachusetts, New Hampshire, and Rhode Island: Environmental Settings and Implications for Water Quality and Aquatic Biota.* Washington, DC: US Department of the Interior, US Geological Survey.

Fogarty M, Incze L, Hayhoe K, Mountain D, Manning J. 2008. Potential climate change impacts on Atlantic cod (Gadus morhua) off the Northeastern United States. *Mitigation and Adaptation Strategies for Global Change* 13:453–66.

Fogarty M, Incze L, Wahle R, Mountain D, Robinson A, Pershing A, Hayhoe K, Richards A, Manning J. 2007. Potential climate change impacts on marine resources of the Northeastern United States. *Mitigation and Adaptation Strategies for Global Change* 13(5–6):453–66.

Fogarty MJ, Murawski SA. 1998. Large-scale disturbance and the structure of marine systems: fishery impacts on Georges Bank. *Ecological Applications* 8(sp1):S6–22.

Fogarty MJ, Overholtz WJ, Link JS. 2012. Aggregate surplus production models for demersal fishery resources of the Gulf of Maine. *Marine Ecology Progress Series* 459:247–58.

Foster DR. 1995. *Land-Use History and Four Hundred Years of Vegetation Change in New England. Principles, Patterns and Processes of Land Use Change: Some Legacies of the Columbian Encounter.* New York, NY: Wiley (pp. 253–319).

Frank KT, Perry RI, Drinkwater KF. 1990. Predicted response of northwest Atlantic invertebrate and fish stocks to CO_2-induced climate change. *Transactions of the American Fisheries Society* 119(2):353–65.

Franks PJ, Chen C. 2001. A 3-D prognostic numerical model study of the Georges Bank ecosystem. Part II: biological–physical model. *Deep Sea Research Part II: Topical Studies in Oceanography* 48(1–3):457–82.

Fratantoni PS, Holzwarth-Davis T, Melrose DC, Taylor MH. 2019. *Description of Oceanographic Conditions on the Northeast US Continental Shelf During 2016. US Department of Commerce, Northeast Fisheries Science Center Reference Document 19-07.* Washington, DC: NOAA (p. 39).

Fratantoni PS, Holzworth-Davis T, Taylor MH. 2015. *Description of Oceanographic Conditions on the Northeast US Continental Shelf during 2014. US Department of Commerce, Northeast Fisheries Science Center Reference Document 15–21.* Washington, DC: NOAA (p. 41).

Friedland KD, Langan JA, Large SI, Selden RL, Link JS, Watson RA, Collie JS. 2020. Changes in higher trophic level productivity, diversity and niche space in a rapidly warming continental shelf ecosystem. *Science of the Total Environment* 704(20):135270.

Friedland KD, McManus MC, Morse RE, Link JS. 2019. Event scale and persistent drivers of fish and invertebrate distributions on the US Northeast Shelf. *ICES Journal of Marine Science* 76:1316–34.

Frisk MG, Dolan TE, McElroy AE, Zacharias JP, Xu H, Hice LA. 2018. Assessing the drivers of the collapse of Winter Flounder: implications for management and recovery. *Journal of Sea Research* 141:1–3.

Frumhoff PC, McCarthy JJ, Melillo JM, Moser SC, Wuebbles DJ. 2007. "Confronting climate change in the U.S. Northeast: science, impacts, and solutions." In: *Synthesis Report of the Northeast Climate Impacts Assessment (NECIA).* Cambridge, MA: Union of Concerned Scientists.

Fullard KJ, Early G, Heide-Jørgensen MP, Bloch D, Rosing-Asvid A, Amos W. 2000. Population structure of long-finned pilot whales in the North Atlantic: a correlation with sea surface temperature? *Molecular Ecology* 9(7):949–58.

Fulton EA, Smith ADM, Smith DC. 2007. *Alternative Management Strategies for Southeast Australian Commonwealth Fisheries: Stage 2: Quantitative Management Strategy Evaluation*. Canberra, Australia: Australian Fisheries Management Authority Report (p. 378).

Fulton EA, Link JS, Kaplan IC, Savina-Rolland M, Johnson P, Ainsworth C, Horne P, et al. 2011. Lessons in modelling and management of marine ecosystems: the Atlantis experience. *Fish and Fisheries* 12(2):171–88.

Gaichas SK, Bundy A, Miller TJ, Moksness E, Stergiou KI. 2012a. What drives marine fisheries production? *Marine Ecology Progress Series* 459:159–63.

Gaichas SK, Fogarty M, Fay G, Gamble R, Lucey S, Smith L. 2017. Combining stock, multispecies, and ecosystem level fishery objectives within an operational management procedure: simulations to start the conversation. *ICES Journal of Marine Science* 74(2):552–65.

Gaichas S, Gamble R, Fogarty M, Benoît H, Essington T, Fu C, Koen-Alonso M, Link J. 2012b. Assembly rules for aggregate-species production models: simulations in support of management strategy evaluation. *Marine Ecology Progress Series* 459:275–92.

Gaichas S, Hardison S, Large S, Lucey S. 2019a. *State of the Ecosystem 2019: Mid Atlantic*. Washington, DC: NOAA (p. 28).

Gaichas S, Hardison S, Large S, Lucey S. 2019b. *State of the Ecosystem 2019: New England*. New England Fishery Management Council (p. 31). https://noaa-edab.github.io/presentations/20190417_NEFMC_Lucey.html#1

Gaichas SK, Seagraves RJ, Coakley JM, DePiper GS, Guida VG, Hare JA, Rago PJ, Wilberg MJ. 2016. A framework for incorporating species, fleet, habitat, and climate interactions into fishery management. *Frontiers in Marine Science* 3:105.

Gamble RJ, Link JS. 2009. Analyzing the tradeoffs among ecological and fishing effects on an example fish community: a multispecies fisheries production model. *Ecological Modelling* 220:2570–82.

Gamble RJ, Link JS. 2012. Using an aggregate production fisheries simulation model with ecological interactions to explore effects of fishing and climate on an example fish community. *Marine Ecology Progress Series* 459:259–74.

Garfield III N, Evans DL. 1987. Shelf water entrainment by Gulf Stream warm-core rings. *Journal of Geophysical Research: Oceans* 92(C12):13003–12.

Garrison L, Link J. 2004. *An Expanded Multispecies Virtual Population Analysis Approach (MSVPA-X) to Evaluate Predator–Prey Interactions in Exploited Fish Ecosystems*. Arlington, VA: Atlantic States Marine Fisheries Commission (p. 90).

Garrison LP. 2000. Spatial and dietary overlap in the Georges Bank groundfish community. *Canadian Journal of Fisheries and Aquatic Sciences* 57(8):1679–91.

Garrison LP, Link JS. 2000a. Dietary guild structure of the fish community in the Northeast United States continental shelf ecosystem. *Marine Ecology Progress Series* 202:231–40.

Garrison LP, Link JS. 2000b. Fishing effects on spatial distribution and trophic guild structure of the fish community in the Georges Bank region. *ICES Journal of Marine Science* 57(3):723–30.

Garrison LP, Link JS, Kilduff DP, Cieri MD, Muffley B, Vaughan DS, Sharov A, Mahmoudi B, Latour RJ. 2010. An expansion of the MSVPA approach for quantifying predator–prey interactions in exploited fish communities. *ICES Journal of Marine Science* 67(5):856–70.

Gawarkiewicz GG, Todd RE, Plueddemann AJ, Andres M, Manning JP. 2012. Direct interaction between the Gulf Stream and the shelfbreak south of New England. *Scientific Reports*. 2:553.

Gedan KB, Altieri AH, Bertness MD. 2011. Uncertain future of New England salt marshes. *Marine Ecology Progress Series* 434:229–37.

Gedan KB, Bertness MD. 2009. Experimental warming causes rapid loss of plant diversity in New England salt marshes. *Ecology Letters* 12(8):842–8.

Gedan KB, Bertness MD. 2010. How will warming affect the salt marsh foundation species Spartina patens and its ecological role? *Oecologia* 164(2):479–87.

Gedan KB, Silliman BR, Bertness MD. 2009. Centuries of human-driven change in salt marsh ecosystems. *Annual Review of Marine Science* 1:117–41.

Gilbert JR, Guldager N. 1998. *Status of Harbor and Gray Seal Populations in Northern New England*. Orono, ME: University of Maine, Department of Wildlife Ecology.

GMCME (Gulf of Maine Council on the Marine Environment). 2010a. *Marine Invasive Species: State of the Gulf of Maine Report*. Gulf of Maine Council on the Marine Environment (p. 21). http://www.gulfofmaine.org/state-of-the-gulf/docs/marine-invasive-species.pdf

GMCME (Gulf of Maine Council on the Marine Environment). 2010b. *State of the Gulf of Maine Report*. http://www.gulfofmaine.org/state-of-the-gulf/.

Gornitz V. 1990. Vulnerability of the East Coast, USA to future sea level rise. *Journal of Coastal Research* 1:201–37.

Greene CH, Meyer-Gutbrod E, Monger BC, McGarry LP, Pershing AJ, Belkin IM, Fratantoni PS, et al. 2013. Remote climate forcing of decadal-scale regime shifts in Northwest Atlantic shelf ecosystems. *Limnology and Oceanography* 58(3):803–16.

Grieve BD, Hare JA, Saba VS. 2017. Projecting the effects of climate change on Calanus finmarchicus distribution within the US Northeast Continental Shelf. *Scientific Reports* 7(1):6264.

Griffen BD, Altman I, Bess BM, Hurley J, Penfield A. 2012. The role of foraging in the success of invasive Asian shore crabs in New England. *Biological Invasions* 14(12):2545–58.

Grosholz ED, Ruiz GM. 1996. Predicting the impact of introduced marine species: lessons from the multiple invasions of the European green crab Carcinus maenas. *Biological Conservation* 78(1–2):59–66.

Guida V, Drohan A, Welch H, McHenry J, Johnson D, Kentner V, Brink J, Timmons D, Estela-Gomez E. 2017. *Habitat Mapping and Assessment of Northeast Wind Energy Areas. OCS Study BOEM 2017-088.* Sterling, VA: US Department of the Interior, Bureau of Ocean Energy Management (p. 312).

Hall-Arber et al. 2001. *New England's Fishing Communities.* Cambridge, MA: Massachusetts Institute of Technology, Sea Grant College Program.

Hameed S, Wolfe CL, Chi L. 2018. Impact of the Atlantic meridional mode on Gulf Stream North Wall position. *Journal of Climate* 31(21):8875–94.

Hamilton PK, Kraus SD. 2019. Frequent encounters with the seafloor increase right whales' risk of entanglement in fishing groundlines. *Endangered Species Research* 39:235–46.

Hamilton PK, Mayo CA. 1990. Population characteristics of right whales (Eubalaena glacialis) observed in Cape Cod and Massachusetts Bays, 1978–1986. *Reports of the International Whaling Commission, Special Issue* 12:203–8.

Hapke CJ, Himmelstoss EA, Kratzmann MG, List JH, Thieler ER. 2010. National Assessment of Shoreline Change: Historical Shoreline Change Along the New England and Mid-Atlantic Coasts Open-File Report 2010-1118. US Geological Survey (p. 57). https://pubs.usgs.gov/of/2010/1118/pdf/ofr2010-1118_report_508_rev042312.pdf

Hare JA, Able KW. 2007. Mechanistic links between climate and fisheries along the east coast of the United States: explaining population outbursts of Atlantic croaker (Micropogonias undulatus). *Fisheries Oceanography* 16(1):31–45.

Hare JA, Alexander MA, Fogarty MJ, Williams EH, Scott JD. 2010. Forecasting the dynamics of a coastal fishery species using a coupled climate–population model. *Ecological Applications* 20(2):452–64.

Hare JA, Kocik JF, Link JS. 2019. Atlantic salmon recovery informing and informed by ecosystem-based fisheries management. *Fisheries* 44(9):403–11.

Hare JA, Morrison WE, Nelson MW, Stachura MM, Teeters EJ, Griffis RB, Alexander MA, et al. 2016. A vulnerability assessment of fish and invertebrates to climate change on the Northeast US Continental Shelf. *PloS One* 11(2):e0146756.

Harley CD, Randall Hughes A, Hultgren KM, Miner BG, Sorte CJ, Thornber CS, Rodriguez LF, Tomanek L, Williams SL. 2006. The impacts of climate change in coastal marine systems. *Ecology Letters* 9(2):228–41.

Hart DR, Rago PJ. 2006. Long-term dynamics of US Atlantic sea scallop *Placopecten magellanicus* populations. *North American Journal of Fisheries Management* 26(2):490–501.

Hartley TW. 2010. Fishery management as a governance network: examples from the Gulf of Maine and the potential for communication network analysis research in fisheries. *Marine Policy* 34(5):1060–7.

Hartley TW, Robertson RA. 2006. Emergence of multi-stakeholder-driven cooperative research in the Northwest Atlantic: the case of the Northeast Consortium. *Marine Policy* 30(5):580–92.

Hartley TW, Robertson RA. 2008. Cooperative research program goals in New England: perceptions of active commercial fishermen. *Fisheries* 33(11):551–9.

Hatch JM, Wiley D, Murray KT, Welch L. 2016. Integrating satellite-tagged seabird and fishery-dependent data: a case study of great shearwaters (Puffinus gravis) and the US new England sink gillnet fishery. *Conservation Letters* 9(1):43–50.

Hayes SA, Josephson E, Maze-Foley K, Rosel PE. 2019. *US Atlantic and Gulf of Mexico Marine Mammal Stock Assessments—2018. NOAA Tech Memo NMFS-NE-258.* Washington, DC: NOAA (p. 298).

Healey TH. 2000. Ludwig's Ratchet and the collapse of New England groundfish stocks. *Coastal Management* 28(3):187–213.

Heck KL, Able KW, Roman CT, Fahay MP. 1995. Composition, abundance, biomass, and production of macrofauna in a New England estuary: comparisons among eelgrass meadows and other nursery habitats. *Estuaries* 18(2):379–89.

Heinonen KB, Auster PJ. 2012. Prey selection in crustacean-eating fishes following the invasion of the Asian shore crab *Hemigrapsus sanguineus* in a marine temperate community. *Journal of Experimental Marine Biology and Ecology* 413:177–83.

Hetland RD, Signell RP. 2005. Modeling coastal current transport in the Gulf of Maine. *Deep Sea Research Part II: Topical Studies in Oceanography* 52(19–21):2430–49.

Hiddink JG, Jennings S, Sciberras M, Szostek CL, Hughes KM, Ellis N, Rijnsdorp AD, et al. 2017. Global analysis of depletion and recovery of seabed biota after bottom trawling disturbance. *Proceedings of the National Academy of Sciences* 114(31):8301–6.

Hilborn R. 2007. Defining success in fisheries and conflicts in objectives. *Marine Policy* 31(2):153–8.

Hoagland P, Jin D, Thunberg E, Steinback S. 2005. Economic activity associated with the Northeast Shelf large marine ecosystem: application of an input-output approach. *Elsevier* 13:157–179.

Hobbs NV, Lazo-Wasem E, Faasse M, Cordell JR, Chapman JW, Smith CS, Prezant R, Shell R, Carlton JT. 2015. Going global: the introduction of the Asian isopod Ianiropsis serricaudis Gurjanova (Crustacea: Peracarida) to North America and Europe. *Aquatic Invasions* 10(2):177–87.

Hoffmann EE, Powell EN, Klinck JM, Munroe DM, Mann R, Haidvogel DB, Narváez DA, Zhang X, Kuykendall KM.

2018. An overview of factors affecting distribution of the Atlantic surfclam (Spisula solidissima), a continental shelf biomass dominant, during a period of climate change. *Journal of Shellfish Research* 37(4):821–32.

Holland DS, Sutinen JG. 1999. An empirical model of fleet dynamics in New England trawl fisheries. *Canadian Journal of Fisheries and Aquatic Sciences* 56(2):253–64.

Holland DS. 2000. A bioeconomic model of marine sanctuaries on Georges Bank. *Canadian Journal of Fisheries and Aquatic Sciences* 57(6):1307–19.

Holland DS, Silva PP, Wiersma JB. 2010. *A Survey of Social Capital and Attitudes Toward Management in the New England Groundfish Fishery. Northeast Fisheries Science Center Reference Document 10-12*. Washington, DC: NOAA.

Holland DS, Kitts AW, Da Silva PP, Wiersma J. 2013. Social capital and the success of harvest cooperatives in the New England groundfish fishery. *Marine Resource Economics* 28(2):133–53.

Horton S, McKenzie K. 2009. *Identifying Coastal Habitats at Risk from Climate Change Impacts in the Gulf of Maine*. Climate Change Network. http://www.gulfofmaine.org/council/publications/Identifying%20Coastal%20Habitats%20at%20Risk_CCN_HortonMcKenzie_2009.pdf

Hourigan TF, Etnoyer PJ, Cairns SD. 2017. *Introduction to the State of Deep-Sea Coral and Sponge Ecosystems of the United States. NOAA Technical Memorandum NMFS-OHC-4*. Washington, DC: NOAA (p. 38).

Howell P. 2012. The status of the southern New England lobster stock. *Journal of Shellfish Research* 31(2):573–80.

Howland J, Gallager S, Singh H, Girard A, Abrams L, Griner C, Taylor R, Vine N. 2006. Development of a Towed Survey System for Deployment by the Fishing Industry. IEEE Oceans (p. 5). https://sph.med.unsw.edu.au/node/191280517

Howle LE, Kraus SD, Werner TB, Nowacek DP. 2019. Simulation of the entanglement of a North Atlantic right whale (Eubalaena glacialis) with fixed fishing gear. *Marine Mammal Science* 35(3):760–78.

Hughes JE, Deegan LA, Wyda JC, Weaver MJ, Wright A. 2002. The effects of eelgrass habitat loss on estuarine fish communities of southern New England. *Estuaries* 25(2):235–49.

Hutchings JA. 2005. Life history consequences of overexploitation to population recovery in Northwest Atlantic cod (Gadus morhua). *Canadian Journal of Fisheries and Aquatic Sciences* 62(4):824–32.

Hutchings JA, Baum JK. 2005. Measuring marine fish biodiversity: temporal changes in abundance, life history and demography. *Philosophical Transactions of the Royal Society B: Biological Sciences* 360(1454):315–38.

ICES. 2016. *Final Report of the Working Group on the Northwest Atlantic Regional Sea (WGNARS)*. ICES CM 2016/SSGIEA:03. Falmouth, MA: ICES SSGIEA Committee (p. 42).

Ihde TF, Wilberg MJ, Loewensteiner DA, Secor DH, Miller TJ. 2011. The increasing importance of marine recreational fishing in the US: challenges for management. *Fisheries Research* 108(2–3):268–76.

Jackson JB, Kirby MX, Berger WH, Bjorndal KA, Botsford LW, Bourque BJ, Bradbury RH, et al. 2001. Historical overfishing and the recent collapse of coastal ecosystems. *Science* 293(5530):629–37.

Jensen AC. 1967. *Brief History of the New England Offshore Fisheries*. Fishery Leaflet 594. Washington, DC: Department of the Interior Fish and Wildlife Service Bureau of Commercial Fisheries (p. 14).

Jin D, DePiper G, Hoagland P. 2016. Applying portfolio management to implement ecosystem-based fishery management (EBFM). *North American Journal of Fisheries Management* 36(3):652–69.

Jin D, Hoagland P, Dalton TM. 2003. Linking economic and ecological models for a marine ecosystem. *Ecological Economics* 46(3):367–85.

Jin D, Hoagland P, Dalton TM, Thunberg EM. 2012. Development of an integrated economic and ecological framework for ecosystem-based fisheries management in New England. *Progress in Oceanography* 102:93–101.

Jin D, Hoagland P, Wikgren B. 2013. An empirical analysis of the economic value of ocean space associated with commercial fishing. *Marine Policy* 42:74–84.

Jin D, Thunberg E, Kite-Powell H, Blake K. 2002. Total factor productivity change in the New England groundfish fishery: 1964–1993. *Journal of Environmental Economics and Management* 44(3):540–56.

Johnson A, Salvador G, Kenney J, Robbins J, Kraus S, Landry S, Clapham P. 2005. Fishing gear involved in entanglements of right and humpback whales. *Marine Mammal Science* 21(4):635–45.

Johnson CL, Runge JA, Curtis KA, Durbin EG, Hare JA, Incze LS, Link JS, et al. 2011. Biodiversity and ecosystem function in the Gulf of Maine: pattern and role of zooplankton and pelagic nekton. *PLoS One* 6(1):e16491.

Johnson MR, Boelke C, Chiarella LA, Colosi PD, Greene KE, Lellis-Dibble KA, Ludemann H, et al. 2008. *Impacts to Marine Fisheries Habitat from Nonfishing Activities in the Northeastern United States*. Washington, DC: Department of Commerce.

Jordaan A, Chen Y, Townsend DW, Sherman S. 2010. Identification of ecological structure and species relationships along an oceanographic gradient in the Gulf of Maine using multivariate analysis with bootstrapping. *Canadian Journal of Fisheries and Aquatic Sciences* 67(4):701–19.

Kane J. 2007. Zooplankton abundance trends on Georges Bank, 1977–2004. *ICES Journal of Marine Science* 64(5):909–19.

Karlson RH, Osman RW. 2012. Species composition and geographic distribution of invertebrates in fouling communities along the east coast of the USA: a regional perspective. *Marine Ecology Progress Series* 458:255–68.

Karp MA, Blackhart K, Lynch PD, Deroba J, Hanselman D, Gertseva V, Teo S, et al. 2019. *Proceedings of the 13th National Stock Assessment Workshop: Model Complexity, Model Stability, and Ensemble Modeling. U.S. Department of Commerce, NOAA. NOAA Technical Memoranda NMFS-F/SPO-189.* Washington, DC: NOAA (p. 49).

Kellner JB, Sanchirico JN, Hastings A, Mumby PJ. 2011. Optimizing for multiple species and multiple values: tradeoffs inherent in ecosystem-based fisheries management. *Conservation Letters* 4(1):21–30.

Kenney RD, Scott GP, Thompson TJ, Winn HE. 1997. Estimates of prey consumption and trophic impacts of cetaceans in the USA northeast continental shelf ecosystem. *Journal of Northwest Atlantic Fishery Science* 22:155–71.

Keser M, Swenarton JT, Vozarik JM, Foertch JF. 2003. Decline in eelgrass (Zostera marina L.) in Long Island Sound near Millstone Point, Connecticut (USA) unrelated to thermal input. *Journal of Sea Research* 49(1):11–26.

Klein ES, Smith SL, Kritzer JP. 2017. Effects of climate change on four New England groundfish species. *Reviews in Fish Biology and Fisheries* 27(2):317–38.

Kleisner KM, Fogarty MJ, McGee S, Barnett A, Fratantoni P, Greene J, Hare JA, et al. 2016. The effects of sub-regional climate velocity on the distribution and spatial extent of marine species assemblages. *PloS One* 11(2):e0149220.

Kleisner KM, Fogarty MJ, McGee S, Hare JA, Moret S, Perretti CT, Saba VS. 2017. Marine species distribution shifts on the US Northeast Continental Shelf under continued ocean warming. *Progress in Oceanography* 153:24–36.

Knowlton AR, Hamilton PK, Marx MK, Pettis HM, Kraus SD. 2012. Monitoring North Atlantic right whale Eubalaena glacialis entanglement rates: a 30 year retrospective. *Marine Ecology Progress Series* 466:293–302.

Knowlton AR, Kraus SD. 2001. Mortality and serious injury of northern right whales (Eubalaena glacialis) in the western North Atlantic Ocean. *Journal of Cetacean Research and Management* 2:193–208.

Kordas RL, Harley CD, O'Connor MI. 2011. Community ecology in a warming world: the influence of temperature on interspecific interactions in marine systems. *Journal of Experimental Marine Biology and Ecology* 400(1–2):218–26.

Kraus SD, Brown MW, Caswell H, Clark CW, Fujiwara M, Hamilton PK, Kenney RD, et al. 2005. North Atlantic right whales in crisis. *Science* 309(5734):561–2.

Krumhansl KA, Okamoto DK, Rassweiler A, Novak M, Bolton JJ, Cavanaugh KC, Connell SD, et al. 2016. Global patterns of kelp forest change over the past half-century. *Proceedings of the National Academy of Sciences* 113(48):13785–90.

Kurlansky M. 2011. *Cod: A Biography of the Fish that Changed the World.* London, UK: Vintage Books.

Lake TRT, Ravana KR, Saunders R. 2012. Evaluating changes in diadromous species distributions and habitat accessibility following the Penobscot River restoration project. *Marine and Coastal Fisheries* 4:284–93.

Large SI, Fay G, Friedland KD, Link JS. 2013. Defining trends and thresholds in responses of ecological indicators to fishing and environmental pressures. *ICES Journal of Marine Science* 70:755–67.

Large SI, Fay G, Friedland KD, Link JS. 2015a. Critical points in ecosystem responses to fishing and environmental pressures. *Marine Ecology Progress Series* 521:1–7.

Large SI, Fay G, Friedland KD, Link JS. 2015b. Quantifying patterns of change in marine ecosystem response to multiple pressures. *PLoS One* 10(3):e0119922.

Lazzari MA, Sherman S, Kanwit JK. 2003. Nursery use of shallow habitats by epibenthic fishes in Maine nearshore waters. *Estuarine, Coastal and Shelf Science* 56(1):73–84.

Ledesma ME, O'Connor NJ. 2001. Habitat and diet of the non-native crab Hemigrapsus sanguineus in southeastern New England. *Northeastern Naturalist* 8:63–78.

Lengyel NL, Collie JS, Valentine PG. 2009. The invasive colonial ascidian Didemnum vexillum on Georges Bank. Ecological effects and genetic identification. *Aquatic Invasion* 4:143–52.

Leonard GH. 2000. Latitudinal variation in species interactions: a test in the New England rocky intertidal zone. *Ecology* 81(4):1015–30.

Lindsay JA, Savage NB. 1978. Northern New England's threatened soft-shell clam populations. *Environmental Management* 2(5):443–52.

Link JS, Brodziak JK, Edwards SF, Overholtz WJ, Mountain D, Jossi JW, Smith TD, Fogarty MJ. 2001. *Ecosystem Status in the Northeast United States Continental Shelf Ecosystem: Integration, Synthesis, Trends and Meaning of Ecosystem Metrics.* ICES CM. 2001:T10. ICES. https://www.ices.dk/sites/pub/CM%20Doccuments/2001/T/T1001.pdf

Link JS. 2002. What does ecosystem-based fisheries management mean. *Fisheries* 27(4):18–21.

Link JS, Brodziak JK, Dow DD, Edwards SF, Fabrizio MC, Fogarty MJ, Hart DR, et al. 2002a. Status of the Northeast US Continental Shelf Ecosystem.

Link JS, Brodziak JK, Edwards SF, Overholtz WJ, Mountain D, Jossi JW, Smith TD, Fogarty MJ. 2002b. Marine ecosystem assessment in a fisheries management context. Canadian *Journal of Fisheries and Aquatic Sciences* 59(9):1429–40.

Link J, Almeida F, Valentine P, Auster P, Reid R, Vitaliano J. 2005. The effects of area closures on Georges Bank. *American Fisheries Society Symposium* 41:345–68.

Link J, Overholtz W, O'Reilly J, Green J, Dow D, Palka D, Legault C, et al. 2005. *An Overview of EMAX: The Northeast U.S. Continental Shelf Ecological Network. ICES CM 2005/L:02.* ICES (p. 28). https://www.ices.dk/sites/pub/CM%20Doccuments/2005/L/L0205.pdf

Link JS, Griswold CA, Methratta ET, Gunnard J. 2006. *Documentation for the Energy Modeling and Analysis eXercise*

(EMAX). US Department of Commerce, Northeast Fisheries Science Center Reference Document 06-15. Washington, DC: NOAA (p. 166).

Link JS, Marshak AR. 2019. Characterizing and comparing marine fisheries ecosystems in the United States: determinants of success in moving toward ecosystem-based fisheries management. *Reviews in Fish Biology and Fisheries* 29(1):23–70.

Link J, O'Reilly J, Fogarty M, Dow D, Vitaliano J, Legault C, Overholtz W, et al. 2008a. Energy flow on Georges Bank revisited: the energy modeling and analysis eXercise (EMAX) in historical context. *Journal of Northwest Atlantic Fishery Science* 39:83–101.

Link J, Overholtz W, O'Reilly J, Green J, Dow D, Palka D, Legault C, et al. 2008b. The Northeast US continental shelf Energy Modeling and Analysis exercise (EMAX): Ecological network model development and basic ecosystem metrics. *Journal of Marine Systems* 74(1–2):453–74.

Link JS, Yemane D, Shannon LJ, Coll M, Shin YJ, Hill L, Borges MD. 2009. Relating marine ecosystem indicators to fishing and environmental drivers: an elucidation of contrasting responses. *ICES Journal of Marine Science* 67(4):787–95.

Link J, Col L, Guida V, Dow D, O'Reilly J, Green J, Overholtz W, et al. 2009a. Response of balanced network models to large-scale perturbation: implications for evaluating the role of small pelagics in the Gulf of Maine. *Ecological Modelling* 220(3):351–69.

Link JS, Bogstad B, Sparholt H, Lilly GR. 2009b. Trophic role of Atlantic cod in the ecosystem. *Fish and Fisheries* 10:58–87.

Link J. 2010. *Ecosystem-Based Fisheries Management: Confronting Tradeoffs*. Cambridge, UK: Cambridge University Press.

Link JS. 2018. System-level optimal yield: increased value, less risk, improved stability, and better fisheries. *Canadian Journal of Fisheries and Aquatic Sciences* 75(1):1–6.

Link JS, Fulton EA, Gamble RJ. 2010a. The northeast US application of Atlantis: a full system model exploring marine ecosystem dynamics in a living marine resource management context. *Progress in Oceanography* 87(1–4):214–34.

Link JS, Ihde TF, Townsend HM, Osgood KE, Schirripa MJ, Kobayashi DR, Gaichas SK, et al. 2010b. *Report of the 2nd National Ecosystem Modeling Workshop (NEMoW II): Bridging the Credibility Gap Dealing with Uncertainty in Ecosystem Models. NOAA Technical Memorandum NMFS-F/SPO-102*. Washington, DC: NOAA.

Link JS, Bundy A, Overholtz WJ, Shackell N, Manderson J, Duplisea D, Hare J, Koen-Alonso M, Friedland KD. 2011a. Ecosystem-based fisheries management in the Northwest Atlantic. *Fish and Fisheries* 12(2):152–70.

Link JS, Gamble RJ, Fogarty MJ. 2011b. *An Overview of the NEFSC's Ecosystem Modeling Enterprise for the Northeast US Shelf Large Marine Ecosystem: Towards Ecosystem-based Fisheries Management. NEFSC CRD 11–23*. Washington, DC: NOAA (p. 89).

Link JS, Bell RJ, Auster PJ, Smith BE, Overholtz WJ, Methrattra ET, Pranovi F, Stockhausen WT. 2012. *Food Web and Community Dynamics of the Northeast U.S. Large Marine Ecosystem. US Department of Commerce, Northeast Fisheries Science Center Reference Documents 12–15*. Washington, DC: NOAA (p. 96).

Link JS, Tolman HL, Bayler E, Holt C, Brown CW, Burke PB, Carman JC, et al. 2017. *High-Level NOAA Unified Modeling Overview*. Washington, DC: NOAA.

Lord JP, Dalvano BE. 2015. Differential response of the American lobster Homarus americanus to the invasive Asian shore crab Hemigrapsus sanguineus and green crab Carcinus maenas. *Journal of Shellfish Research* 34(3):1091–7.

Lough RG. 2004. *Essential Fish Habitat Source Document. Atlantic Cod, Gadus Morhua, Life History and Habitat Characteristics. Second Edition. NMFS-NE-190*. Washington, DC: NOAA (p. 94).

Lubchenco J. 1980. Algal zonation in the New England rocky intertidal community: an experimental analysis. *Ecology* 61(2):333–44.

Lucey SM, Nye JA. 2010. Shifting species assemblages in the northeast US continental shelf large marine ecosystem. *Marine Ecology Progress Series* 415:23–33.

Lynch PD, Methot RD, Link JS. 2018. *Implementing a Next Generation Stock Assessment Enterprise. An Update to the NOAA Fisheries Stock Assessment Improvement Plan. NOAA Tech. Memo. NMFS-F/ SPO-183*. Washington, DC: NOAA (p. 127).

Mathieson AC, Dawes CJ, Harris LG, Hehre EJ. 2003. Expansion of the Asiatic green alga Codium fragile subsp. tomentosoides in the Gulf of Maine. *Rhodora* 1:1–53.

Mathieson AC, Dawes CJ, Pederson J, Gladych RA, Carlton JT. 2008a. The Asian red seaweed *Grateloupia turuturu* (Rhodophyta) invades the Gulf of Maine. *Biological Invasions* 10(7):985–8.

Mathieson AC, Pederson JR, Neefus CD, Dawes CJ, Bray TL. 2008b. Multiple assessments of introduced seaweeds in the Northwest Atlantic. *ICES Journal of Marine Science* 65(5):730–41.

Mattila DK, Guinee LN, Mayo CA. 1987. Humpback whale songs on a North Atlantic feeding ground. *Journal of Mammalogy* 68(4):880–3.

Maxwell SM, Hazen EL, Lewison RL, Dunn DC, Bailey H, Bograd SJ, Briscoe DK, et al. 2015. Dynamic ocean management: defining and conceptualizing real-time management of the ocean. *Marine Policy* 58:42–50.

McDermott JJ. 1991. A breeding population of the Western Pacific crab Hemigrapsus sanguineus (Crustacea: Decapoda: Grapsidae) established on the Atlantic coast of North America. *Biological Bulletin* 181:195–8.

McDermott SP, Bransome NC, Sutton SE, Smith BE, Link JS, Miller TJ. 2015. Quantifying Alosine prey in the diets of marine piscivores in the Gulf of Maine. *Journal of Fish Biology* 86:1811–29.

McElroy WD, O'Brien LO, Blaylock J, Martin MH, Rago, PJ, Hoey JJ, Sheremet VA. 2019. *Design, Implementation, and Results of a Cooperative Research Gulf of Maine Longline Survey, 2014–2017.* NOAA Technical Memorandum NMFS-NE-249. Washington, DC: NOAA (p. 155).

McFarland R. 1911. *A History of the New England Fisheries.* New York, NY: D. Appleton & Company.

McMahan MD. 2017. *Ecological and Socioeconomic Implications of a Northern Range Expansion of Black Sea Bass, Centropristis striata.* Boston, MA: Northeastern University.

McMahan MD, Grabowski JH. 2015. Ecological implications of a northern range expansion of Black Sea Bass, Centropristis striata. *Diving for Science* 28:11.

Meng KC, Oremus KL, Gaines SD. 2016. New England cod collapse and the climate. *PloS One* 11(7):e0158487.

Menge BA. 1976. Organization of the New England rocky intertidal community: role of predation, competition, and environmental heterogeneity. *Ecological Monographs* 46(4):355–93.

Menge BA. 1978. Predation intensity in a rocky intertidal community. *Oecologia* 34(1):17–35.

Mercer JM, Whitlatch RB, Osman RW. 2009. Potential effects of the invasive colonial ascidian (Didemnum vexillum Kott, 2002) on pebble-cobble bottom habitats in Long Island Sound, USA. *Aquatic Invasions* 4(1):133–42.

Meyer-Gutbrod EL. 2017. *Impacts of Climate-Associated Changes in Prey Availability on North Atlantic Right Whale Population Dynamics.* Ithaca, NY: Cornell University.

Moore MJ. 2019. How we can all stop killing whales: a proposal to avoid whale entanglement in fishing gear. *ICES Journal of Marine Science* 76(4):781–6.

Morley JW, Batt RD, Pinsky ML. 2017. Marine assemblages respond rapidly to winter climate variability. *Global Change Biology* 23(7):2590–601.

Morley JW, Selden RL, Latour RJ, Frölicher TL, Seagraves RJ, Pinsky ML. 2018. Projecting shifts in thermal habitat for 686 species on the North American continental shelf. *PloS One* 13(5):e0196127.

Mountain DG. 2012. Labrador slope water entering the Gulf of Maine—response to the North Atlantic Oscillation. *Continental Shelf Research* 47:150–5.

Mountain DG, Kane J. 2010. Major changes in the Georges Bank ecosystem, 1980s to the 1990s. *Marine Ecology Progress Series* 398:81–91.

Moustahfid H, Link JS, Overholtz WJ, Tyrrell MC. 2009a. The advantage of explicitly incorporating predation mortality into age-structured stock assessment models: an application for Atlantic mackerel. *ICES Journal of Marine Science* 66(3):445–54.

Moustahfid H, Tyrrell MC, Link JS. 2009b. Accounting explicitly for predation mortality in surplus production models: an application to longfin inshore squid. *North American Journal of Fisheries Management* 29(6):1555–66.

Murawski SA. 1991. Can we manage our multispecies fisheries? *Fisheries* 16(5):5–13.

Murawski SA. 1993. Climate change and marine fish distributions: forecasting from historical analogy. *Transactions of the American Fisheries Society* 122(5):647–58.

Murawski SA. 2005. *The New England Groundfish Resource: A History of Population Change in Relation to Harvesting. The Decline of Fisheries Resources in New England.* Cambridge, MA: MIT Press (p. 11).

Murawski SA, Brown R, Lai HL, Rago PJ, Hendrickson L. 2000. Large-scale closed areas as a fishery-management tool in temperate marine systems: the Georges Bank experience. *Bulletin of Marine Science* 66(3):775–98.

Muto MM, Helker VT, Angliss RP, Boveng PL, Breiwick JM, Cameron MF, Clapham PJ, et al. 2019. *Alaska Marine Mammal Stock Assessments, 2018. NOAA Tech. Memo. NMFS-AFSC-393.* Washington, DC: NOAA (p. 390).

NAS (The National Academy of Sciences, Engineering, and Medicine). 2017. *Review of the Marine Recreational Information Program.* Washington, DC: The National Academies Press.

Neckles HA. 2015. Loss of eelgrass in Casco Bay, Maine, linked to green crab disturbance. *Northeastern Naturalist* 22(3):478–501.

NEFC (Northeast Fisheries Center). 1988. An Evaluation of the Bottom Trawl Survey Program of the Northeast Fisheries Center. NOAA Tech. Memo. NMFS-F/NEC – 52. Washington, DC: NOAA (p. 83).

NEFMC (New England Fishery Management Council). 2019. *Ecosystem-Based Fishery Management Committee.* https://www.nefmc.org/committees/ecosystem-based-fisheries-management

NEFSC (Northeast Fisheries Science Center). 2009. *Ecosystem Status Report for the Northeast U.S. Continental Shelf Large Marine Ecosystem. By the Ecosystem Assessment Program.* Northeast Fisheries Science Center Reference Document 09-11. Washington, DC: U.S. Department of Commerce (p. 34).

NEFSC (Northeast Fisheries Science Center). 2012. *Ecosystem Status Report for the Northeast Shelf Large Marine Ecosystem—2011. NOAA-NMFS, Northeast Fisheries Science Center Reference Document 12-07.* Washington, DC: NOAA (p. 33).

NEFSC (Northeast Fisheries Science Center). 2018b. *Ecosystem-Based Fishery Management Strategy Georges Bank Prototype Study Summary Document.* Woods Hole, MA: NEFSC Ecosystem Dynamics and Assessment Branch (p. 109).

NERACOOS (Northeastern Regional Association of Coastal Ocean Observing Systems). 2016. *Annual Impact Report 2016.* (p. 6).

Nettleton JC, Mathieson AC, Thornber C, Neefus CD, Yarish C. 2013. Introduction of Gracilaria vermiculophylla (Rhodophyta, Gracilariales) to New England, USA: estimated arrival times and current distribution. *Rhodora* 115(961):28–41.

Newell ME. 1998. *From Dependency to Independence: Economic Revolution in Colonial New England.* Ithaca, NY: Cornell University Press.

Niering WA, Scott-Warren R. 1980. Vegetation patterns and processes in New England salt marshes. *Bioscience* 30(5):301–7.

NMFS (National Marine Fisheries Service). 2017a. *National Marine Fisheries Service—2nd quarter 2017 update.* http://www.nmfs.noaa.gov/sfa/fisheries_eco/status_of_fisheries/archive/2017/second/q2-2017-stock-status-table.pdf.

NMFS (National Marine Fisheries Service). 2017b. *National Observer Program FY 2013 Annual Report. NOAA Tech. Memo. NMFS F/SPO-178.* Washington, DC: NOAA (p. 34).

NMFS (National Marine Fisheries Service). 2018. *Fisheries Economics of the United States, 2016. NOAA Technical Memo NMFS-F/SPO-187.* Washington, DC: NOAA (p. 243).

NMFS (National Marine Fisheries Service). 2019. *Northeast Regional Implementation Plan of the NOAA Fisheries Ecosystem-Based Fishery Management Roadmap.* Washington, DC: NOAA (p. 25).

NMFS (National Marine Fisheries Service). 2020a. *National Marine Fisheries Service—2nd Quarter 2020 Update.* Washington, DC: NOAA (p. 51).

NMFS (National Marine Fisheries Service). 2020b. *Northeast Stock Assessment Documents.* https://www.fisheries.noaa.gov/new-england-mid-atlantic/northeast-stock-assessment-documents

NROC (Northeast Ocean Data Portal). 2020. *Northeast Ocean Data.* https://www.northeastoceandata.org

NRPB (Northeast Regional Planning Body). 2016. *Northeast Ocean Plan.* (p. 206). https://neoceanplanning.org/wp-content/uploads/2018/01/Northeast-Ocean-Plan_Full.pdf

Nutters HM, da Silva PP. 2012. Fishery stakeholder engagement and marine spatial planning: lessons from the Rhode Island Ocean SAMP and the Massachusetts Ocean Management Plan. *Ocean & Coastal Management* 67:9–18.

Nye JA, Gamble RJ, Link JS. 2013. The relative impact of warming and removing top predators on the Northeast US large marine biotic community. *Ecological Modelling* 264:157–68.

Nye JA, Link JS, Hare JA, Overholtz WJ. 2009. Changing spatial distribution of fish stocks in relation to climate and population size on the Northeast United States continental shelf. *Marine Ecology Progress Series* 393:111–29.

Orciari RD, Leonard GH, Mysling DJ, Schluntz EC. 1994. Survival, growth, and smolt production of Atlantic salmon stocked as fry in a southern New England stream. *North American Journal of Fisheries Management* 14(3):588–606.

Orphanides C. 2019. *Estimates of Cetacean and Pinniped Bycatch in the 2016 New England Sink and Mid-Atlantic Gillnet Fisheries. Northeast Fisheries Science Center Reference Document 19-04.* Washington, DC: NOAA.

Orson RA, Warren RS, Niering WA. 1998. Interpreting sea level rise and rates of vertical marsh accretion in a southern New England tidal salt marsh. *Estuarine, Coastal and Shelf Science* (4):419–29.

Osman RW, Whitlatch RB. 2007. Variation in the ability of Didemnum sp. to invade established communities. *Journal of Experimental Marine Biology and Ecology* 342(1):40–53.

Overholtz WJ. 2002. The Gulf of Maine–Georges Bank Atlantic herring (Clupea harengus): spatial pattern analysis of the collapse and recovery of a large marine fish complex. *Fisheries Research* 57(3):237–54.

Overholtz WJ, Edwards SF, Brodziak JK. 1995. Effort control in the New England groundfish fishery: a bioeconomic perspective. *Canadian Journal of Fisheries and Aquatic Sciences* 52(9):1944–57.

Overholtz WJ, Jacobson LD, Link JS. 2008. An ecosystem approach for assessment advice and biological reference points for the Gulf of Maine–Georges Bank Atlantic herring complex. *North American Journal of Fisheries Management* 28(1):247–57.

Overholtz WJ, Link JS. 2007. Consumption impacts by marine mammals, fish, and seabirds on the Gulf of Maine–Georges Bank Atlantic herring (Clupea harengus) complex during the years 1977–2002. *ICES Journal of Marine Science* 64(1):83–96.

Overholtz WJ, Link JS, Suslowicz LE. 2000. Consumption of important pelagic fish and squid by predatory fish in the northeastern USA shelf ecosystem with some fishery comparisons. *ICES Journal of Marine Science* 57(4):1147–59.

Packer DB, Cargnelli LM, Griesbach SJ, Shumway SE. 1999. *Essential Fish Habitat Source Document: Sea Scallop, Placopecten Magellanicus, Life History and Habitat Characteristics. US Department of Commerce, National Oceanic and Atmospheric Administration. NMFS-NE-134.* Washington, DC: NOAA (p. 21).

Packer DB, Zetlin CA, Vitaliano JJ. 2003. *Essential Fish Habitat Source Document. Little Skate, Leucoraja Erinacea, Life History and Habitat Characteristics. NOAA Technical Memorandum NMFS-NE-175.* Washington, DC: NOAA.

Parrish DL, Behnke RJ, Gephard SR, McCormick SD, Reeves GH. 1998. Why aren't there more Atlantic salmon (Salmo salar)? *Canadian Journal of Fisheries and Aquatic Sciences* 55(S1):281–7.

Pauly D, Christensen V, Walters C. 2000. Ecopath, Ecosim and Ecospace as tools for evaluating ecosystem impact of fisheries. *ICES Journal of Marine Science* 57:697–706.

Payne PM, Selzer LA. 1989. The distribution, abundance and selected prey of the harbor seal, Phoca vitulina concolor, in southern New England. *Marine Mammal Science* 5(2):173–92.

Pederson J, Bullock R, Carlton J, Dijkstra J, Dobroski N, Dyrynda P, Fisher R, et al. 2005. *Marine Invaders in the Northeast: Rapid Assessment Survey of Non-Native and Native Marine Species of Floating Dock Communities, August 2003.* Cambridge, MA: MIT Sea Grant College Program.

Perretti CT, Fogarty MJ, Friedland KD, Hare JA, Lucey SM, McBride RS, Miller TJ, et al. 2017. Regime shifts in fish recruitment on the Northeast US Continental Shelf. *Marine Ecology Progress Series* 574:1–1.

Pershing AJ, Alexander MA, Hernandez CM, Kerr LA, Le Bris A, Mills KE, Nye JA, et al. 2015. Slow adaptation in the face of rapid warming leads to collapse of the Gulf of Maine cod fishery. *Science* 350(6262):809–12.

Peterson BJ, Fournier AM, Furman BT, Carroll JM. 2014. Hemigrapsus sanguineus in Long Island salt marshes: experimental evaluation of the interactions between an invasive crab and resident ecosystem engineers. *PeerJ* 2:e472.

Petraitis PS. 1987. Factors organizing rocky intertidal communities of New England: herbivory and predation in sheltered bays. *Journal of Experimental Marine Biology and Ecology* 109(2):117–36.

Petrie B, Drinkwater K. 1993. Temperature and salinity variability on the Scotian Shelf and in the Gulf of Maine 1945–1990. *Journal of Geophysical Research: Oceans* 98(C11):20079–89.

Pettigrew NR, Churchill JH, Janzen CD, Mangum LJ, Signell RP, Thomas AC, Townsend DW, Wallinga JP, Xue H. 2005. The kinematic and hydrographic structure of the Gulf of Maine Coastal Current. *Deep Sea Research Part II: Topical Studies in Oceanography* 52(19–21):2369–91.

Pinsky ML, Fogarty M. 2012. Lagged social-ecological responses to climate and range shifts in fisheries. *Climatic Change* 115(3–4):883–91.

Pinsky ML, Worm B, Fogarty MJ, Sarmiento JL, Levin SA. 2013. Marine taxa track local climate velocities. *Science* 341(6151):1239–42.

Pollnac RB, Colburn LL, Seara T, Weng C, Yentes K. 2011. Job satisfaction, well-being and change in Southern New England fishing communities. Washington, DC: NOAA. http://www.nefsc.noaa. gov/read/socialsci/pdf/publications/Job-Satisfaction_ FinalRpt_REVISED. pdf

Pollnac RB, Poggie Jr JJ. 1988. The structure of job satisfaction among New England fishermen and its application to fisheries management policy. *American Anthropologist* 90(4):888-901.

Polovina JJ. 1984. Model of a coral reef ecosystem I: the Ecopath model and its application to French Frigate Shoals. *Coral Reefs* 3:1–10.

Portman ME, Jin D, Thunberg E. 2011. The connection between fisheries resources and spatial land use change: the case of two New England fish ports. *Land Use Policy* 28(3):523–33.

Pranovi F, Link JS. 2009. Ecosystem exploitation and trophodynamic indicators: a comparison between the Northern Adriatic Sea and Southern New England. *Progress in Oceanography* 81(1–4):149–64.

Raphael R. 2011. *The First American Revolution: Before Lexington and Concord*. New York, NY: The New Press.

Raposa KB, Roman CT. 2001. Seasonal habitat-use patterns of nekton in a tide-restricted and unrestricted New England salt marsh. *Wetlands* 21(4):451–61.

Read AJ, Brownstein CR. 2003. Considering other consumers: fisheries, predators, and Atlantic herring in the Gulf of Maine. *Conservation Ecology* 7(1):2.

Read AJ, Drinker P, Northridge S. 2006. Bycatch of marine mammals in US and global fisheries. *Conservation Biology* 20(1):163–9.

Reeves RR, McClellan K, Werner TB. 2013. Marine mammal bycatch in gillnet and other entangling net fisheries, 1990 to 2011. *Endangered Species Research* 20(1):71–97.

Reid RN, Almeida FP, Zetlin CA. 1999. *Essential Fish Habitat Source Document: Fishery-Independent Surveys, Data Sources, and Methods. NOAA Technical Memorandum NMFS-NE-122*. Washington, DC: NOAA.

Repetto R. 2001. A natural experiment in fisheries management. *Marine Policy* 25(4):251–64.

Rheuban JE, Kavanaugh MT, Doney SC. 2017. Implications of future Northwest Atlantic bottom temperatures on the American lobster. Homarus americanus fishery. *Journal of Geophysical Research: Oceans* 122(12):9387–98.

Richardson DE, Hare JA, Overholtz WJ, Johnson DL. 2010. Development of long-term larval indices for Atlantic herring. Clupea harengus on the northeast US continental shelf. *ICES Journal of Marine Science* 67(4):617–27.

Robadue Jr DD, Tippie VK. 1980. Public involvement in offshore oil development: lessons from New England. *Coastal Management* 7(2–4):237–70.

Robertson RA, Caporossi G. 2003. *New England Recreational Fishers? Attitudes Toward Marine Protected Areas: A Preliminary Investigation*. USDA Forest Service. https://citeseerx.ist.psu.edu/viewdoc/download?doi=10.1.1.543.8802&rep=rep1&type=pdf

Robinson BS, Jacobson GL, Yates MG, Spiess AE, Cowie ER. 2009. Atlantic salmon, archaeology and climate change in New England. *Journal of Archaeological Science* 36(10):2184–91.

Roland A, Bolster WJ, Keyssar A. 2008. *The Way of the Ship: America's Maritime History Reenvisoned, 1600–2000*. New York, NY: John Wiley & Sons.

Roman CT, Jaworski N, Short FT, Findlay S, Warren RS. 2000. Estuaries of the northeastern United States: habitat and land use signatures. *Estuaries* 23(6):743–64.

Roman CT, Peck JA, Allen JR, King JW, Appleby PG. 1997. Accretion of a New England (USA) salt marsh in response to inlet migration, storms, and sea-level rise. *Estuarine, Coastal and Shelf Science* 45(6):717–27.

Rosenberg AA, Swasey JH, Bowman M. 2006. Rebuilding US fisheries: progress and problems. *Frontiers in Ecology and the Environment* 4(6):303–8.

Ross MR, Biagi RC. 1991. *Recreational Fisheries of Coastal New England*. Amherst, MA: University of Massachusetts Press.

Rothschild BJ, Keiley EF, Jiao Y. 2013. Failure to eliminate overfishing and attain optimum yield in the New England groundfish fishery. *ICES Journal of Marine Science* 71(2):226–33.

Royce WF, Buller RJ, Premetz ED. 1959. *Decline of the Yellowtail Flounder (Limanda Ferruginea) Off New England*. Washington, DC: US Government Printing Office.

Runge JA, Kovach AI, Churchill JH, Kerr LA, Morrison JR, Beardsley RC, Berlinsky DL, et al. 2010. Understanding climate impacts on recruitment and spatial dynamics of Atlantic cod in the Gulf of Maine: integration of observations and modeling. *Progress in Oceanography* 87(1–4):251–63.

Russell BA, Knowlton AR, Zoodsma B. 2001. *Recommended Measures to Reduce Ship Strikes of North Atlantic Right Whales. Report Submitted to the National Marine Fisheries Service in Partial Fulfillment of NMFS Contract 40EMF9000223*. Silver Spring, MD: National Marine Fisheries Service, Office of Protected Resources (p. 23).

Ryan JP, Yoder JA, Townsend DW. 2001. Influence of a Gulf Stream warm-core ring on water mass and chlorophyll distributions along the southern flank of Georges Bank. *Deep Sea Research Part II: Topical Studies in Oceanography* 48(1–3):159–78.

Safina C, Rosenberg AA, Myers RA, Quinn II TJ, Collie JS. 2005. US ocean fish recovery: staying the course. *Science* 309(5735):707–8.

Salz RJ, Loomis DK. 2004. Saltwater anglers' attitudes towards marine protected areas. *Fisheries* 29(6):10–7.

Samhouri JF, Haupt AJ, Levin PS, Link JS, Shuford R. 2013. Lessons learned from developing integrated ecosystem assessments to inform marine ecosystem-based management in the USA. *ICES Journal of Marine Science* 71(5):1205–15.

Samoteskul K, Firestone J, Corbett J, Callahan J. 2014. Changing vessel routes could significantly reduce the cost of future offshore wind projects. *Journal of Environmental Management* 141:146–54.

Saunders R, Hachey MA, Fay CW. 2006. Maine's diadromous fish community: past, present, and implications for Atlantic salmon recovery. *Fisheries* 31(11):537–47.

Saunders RL. 1981. Atlantic salmon (Salmo salar) stocks and management implications in the Canadian Atlantic provinces and New England, USA. *Canadian Journal of Fisheries and Aquatic Sciences* 38(12):1612–25.

Savoy T, Pacileo D. 2003. Movements and important habitats of subadult Atlantic sturgeon in Connecticut waters. *Transactions of the American Fisheries Society* 132(1):1–8.

Serchuk FM, Wigley SE. 1992. Assessment and management of the Georges Bank cod fishery: an historical review and evaluation. *Journal of Northwest Atlantic Fishery Science* 13:25–52.

Shackell NL, Bundy A, Nye JA, Link JS. 2012. Common large-scale responses to climate and fishing across Northwest Atlantic ecosystems. *ICES Journal of Marine Science* 69(2):151–62.

Shalev E. 2009. A perfect republic: the mosaic constitution in revolutionary New England, 1775–1788. *The New England Quarterly* 82(2):235–63.

Shelley P. 2012. Have the managers finally gotten it right: federal groundfish management in New England. *Roger Williams University Law Review* 17:21.

Shelley P, Atkinson J, Dorsey E, Brooks P. 1995. The New England fisheries crisis: what have we learned. *Tulane Environmental Law Journal* 9:221.

Sherman K, Jaworski NJ, Smayda T. 1996. *The Northeast Shelf Ecosystem: Assessment, Sustainability and Management*. Cambridge, MA: Blackwell Science (p. 564).

Sherman K, Kane J, Murawski S, Overholtz W, Solow A. 2002. 6 The US Northeast shelf large marine ecosystem: zooplankton trends in fish biomass recovery. *Large Marine Ecosystems, Elsevier* 10:195–215.

Shepherd GR, Terceiro M. 1994. *The Summer Flounder, Scup, and Black Sea Bass Fishery of the Middle Atlantic Bight and Southern New England Waters*. NOAA Technical Report NMFS 122. Seattle, WA: Department of Commerce.

Shoemaker N. 2014. *Living with Whales: Documents and Oral Histories of Native New England Whaling History*. Amherst, MA: University of Massachusetts Press.

Sibunka JD, Silverman MJ. 1989. *MARMAP Surveys of the Continental Shelf from Cape Hatteras, North Carolina, to Cape Sable, Nova Scotia 1984–1987*. Atlas No. 3. Summary of Operations. Woods Hole, MA: National Marine Fisheries Service.

Silliman BR, Bertness MD. 2004. Shoreline development drives invasion of Phragmites australis and the loss of plant diversity on New England salt marshes. *Conservation Biology* 18(5):1424–34.

Simkanin C, Fofonoff PW, Larson K, Lambert G, Dijkstra JA, Ruiz GM. 2016. Spatial and temporal dynamics of ascidian invasions in the continental United States and Alaska. *Marine Biology* 163(7):163.

Sissenwine MP. 1974. Variability in recruitment and equilibrium catch of the Southern New England yellowtail flounder fishery. *ICES Journal of Marine Science* 36(1):15–26.

Sissenwine MP, Cohen EB. 1991. "Resource productivity and fisheries management of the Northeast Shelf ecosystem." In: *Food Chains, Yields, Models, And Management Of Large Marine Ecosoystems.* Edited by K Sherman, LM Alexander, BD Gold. Abington, UK: Routledge (p. 28).

Smith BE, Ligenza TJ, Almeida FP, Link JS. 2007. The trophic ecology of Atlantic cod: insights from tri-monthly, localized scales of sampling. *Journal of Fish Biology* 71:749–62.

Smith L, Gamble R, Gaichas S, Link J. 2015. Simulations to evaluate management trade-offs among marine mammal consumption needs, commercial fishing fleets and finfish biomass. *Marine Ecology Progress Series* 523:215–32.

Smith PC, Pettigrew NR, Yeats P, Townsend DW, Han G. 2012. Regime shift in the Gulf of Maine. *American Fisheries Society Symposium* 79:185–203.

Smythe TC. 2017. Marine spatial planning as a tool for regional ocean governance? An analysis of the New England ocean planning network. *Ocean & Coastal Management* 135:11–24.

Smith TD. 2002. The Woods Hole bottom-trawl resource survey: development of fisheries-independent multispecies monitoring. *ICES Marine Science Symposia* 215:474–82.

Sorte CJ, Williams SL, Carlton JT. 2010. Marine range shifts and species introductions: comparative spread rates and community impacts. *Global Ecology and Biogeography* 19(3):303–16.

Sparks-McConkey PJ, Watling L. 2001. Effects on the ecological integrity of a soft-bottom habitat from a trawling disturbance. *Hydrobiologia* 456:73–8.

Steele JH, Collie JS, Bisagni JJ, Gifford DJ, Fogarty MJ, Link JS, Sullivan BK, Sieracki ME, Beet AR, Mountain DG, Durbin EG. 2007. Balancing end-to-end budgets of the Georges Bank ecosystem. *Progress in Oceanography* 74(4):423–48.

Steneck RS. 1997. *Fisheries-Induced Biological Changes to the Structure and Function of the Gulf of Maine Ecosystem. Proceedings of the Gulf of Maine Ecosystem Dynamics, Scientific Symposium and Workshop, RARGOM Report 1997*. Hanover, NH: Regional Association for Research on the Gulf of Maine (pp. 91–1).

Steneck RS, Graham MH, Bourque BJ, Corbett D, Erlandson JM, Estes JA, Tegner MJ. 2002. Kelp forest ecosystems: biodiversity, stability, resilience and future. *Environmental Conservation* 29(4):436–59.

Steneck RS, Wahle RA. 2013. American lobster dynamics in a brave new ocean. *Canadian Journal of Fisheries and Aquatic Sciences* 70(11):1612–24.

Stevenson D, Chiarella L, Stephan D, Reid R, Wilhelm K, McCarthy J, Pentony M. 2004. *Characterization of the Fishing Practices and Marine Benthic Ecosystems of the Northeast US Shelf, and an Evaluation of the Potential Effects of Fishing on Essential Fish Habitat. NOAA Tech Memo NMFS NE 181*. Washington, DC: NOAA (p. 179).

Stone HH, Gavaris S, Legault CM, Neilson JD, Cadrin SX. 2004. Collapse and recovery of the yellowtail flounder (Limanda ferruginea) fishery on Georges Bank. *Journal of Sea Research* 51(3–4):261–70.

Sutcliffe Jr WH, Loucks RH, Drinkwater KF. 1976. Coastal circulation and physical oceanography of the Scotian Shelf and the Gulf of Maine. *Journal of the Fisheries Board of Canada* 33(1):98–115.

Sutton RT, Hodson DLR. 2007. Climate response to basin-scale warming and cooling of the North Atlantic ocean. *Journal of Climate* 20:891–907.

Tableau A, Collie JS, Bell RJ, Minto C. 2018. Decadal changes in the productivity of New England fish populations. *Canadian Journal of Fisheries and Aquatic Sciences* 21(999):1–3.

Tam JC, Link JS, Large SI, Andrews K, Friedland KD, Gove J, Hazen E, et al. 2017. Comparing apples to oranges: common trends and thresholds in anthropogenic and environmental pressures across multiple marine ecosystems. *Frontiers in Marine Science* 4:282.

Tanaka KR, Cao J, Shank BV, Truesdell SB, Mazur MD, Xu L, Chen Y. 2019. A model-based approach to incorporate environmental variability into assessment of a commercial fishery: a case study with the American lobster fishery in the Gulf of Maine and Georges Bank. *ICES Journal of Marine Science*. 76(4), 884–96.

Taylor R, Vine N, York A, Lerner S, Hart D, Howland J, Prasad L, Mayer L, Gallager S. 2008. *Evolution of a Benthic Imaging System from a Towed Camera to an Automated Habitat Characterization System*. New York, NY: IEEE (p. 1–7).

Thayer PE, Stahlnecker JF. 2006. *Non-Native Invasive Marine Species in Maine, A Report to the Maine State Legislature, Marine Resources Committee and Natural Resources Committee*. Publication 244. Portland, ME: Casco Bay Estuary Partnership.

Thrush SF, Dayton PK. 2002. Disturbance to marine benthic habitats by trawling and dredging: implications for marine biodiversity. *Annual Review of Ecology and Systematics* 33(1):449–73.

Tourtellot AB. 2000. Lexington and Concord: The Beginning of the War of the American Revolution. New York, NY: WW Norton & Company.

Townsend DW. 1991. Influences of oceanographic processes on the biological productivity of the Gulf of Maine. *Reviews in Aquatic Sciences* 5(3):211–30.

Townsend HK, Aydin K, Holsman C, Harvey I, Kaplan E, Hazen P, Woodworth-Jefcoats M. et al. 2017. *Report of the 4th National Ecosystem Modeling Workshop (NEMoW 4): Using Ecosystem Models to Evaluate Inevitable Trade-offs. NOAA Technical Memo NMFS-F/SPO-173*. Washington, DC: NOAA (p. 77).

Townsend H, Harvey CJ, deReynier Y, Davis D, Zador S, Gaichas S, Weijerman M, et al. 2019. Progress on implementing ecosystem-based fisheries management in the US Through the use of ecosystem models and analysis. *Frontiers in Marine Science* 6:641.

Townsend HM, Harvey CJ, Aydin KY, Gamble RJ, Gruss A, Levin PS, Link JS, et al. 2014. *Report of the 3rd National Ecosystem Modeling Workshop (NEMoW 3): Mingling Models for Marine Resource Management, Multiple Model Inference. NOAA Technical Memorandum NMFS-F/SPO-173*. Washington, DC: NOAA.

Townsend HM, Link JS, Osgood KE, Gedamke T, Watters GM, Polovina JJ, Levin PS, Cyr EC, Aydin KY. 2008. *Report of the National Ecosystem Modeling Workshop (NEMoW)*. Washington, DC: NOAA.

Tsou TS, Collie JS. 2001. Estimating predation mortality in the Georges Bank fish community. *Canadian Journal of Fisheries and Aquatic Sciences* 58(5):908–22.

Tyrrell MC, Link JS, Moustahfid H. 2011. The importance of including predation in fish population models: implications for biological reference points. *Fisheries Research* 108(1):1–8.

Tyrrell MC, Link JS, Moustahfid H, Overholtz WJ. 2008. Evaluating the effect of predation mortality on forage species population dynamics in the Northeast US continental shelf ecosystem using multispecies virtual population analysis. *ICES Journal of Marine Science* 65(9):1689–700.

Valentine PC, Carman MR, Dijkstra J, Blackwood DS. 2009. Larval recruitment of the invasive colonial ascidian Didemnum vexillum, seasonal water temperatures in New

England coastal and offshore waters, and implications for spread of the species. *Aquatic Invasions* 4(1):153–68.

Vickers D. 1994. *Farmers and Fishermen: Two Centuries of Work in Essex County, Massachusetts, 1630–1850*. Chapel Hill, NC: UNC Press Books.

Wahle RA, Brown C, Hovel K. 2013. The geography and body-size dependence of top-down forcing in New England's lobster-groundfish interaction. *Bulletin of Marine Science* 89(1):189–212.

Walsh HJ, Richardson DE, Marancik KE, Hare JA. 2015. Long-term changes in the distributions of larval and adult fish in the northeast US shelf ecosystem. *PloS one* 10(9):e0137382.

Ward-Geiger LI, Silber GK, Baumstark RD, Pulfer TL. 2005. Characterization of ship traffic in right whale critical habitat. *Coastal Management* 33(3):263–78.

Warren RS, Niering WA. 1993. Vegetation change on a northeast tidal marsh: interaction of sea-level rise and marsh accretion. *Ecology* 74(1):96–103.

Watling L, Auster PJ. 2005. "Distribution of deep-water Alcyonacea off the Northeast Coast of the United States." In: *Cold-Water Corals and Ecosystems*. Edited by A Freiwald, JM Roberts. Berlin, Germany: Springer (pp. 279–96).

Watling L, Findlay RH, Mayer LM, Schick DF. 2001. Impact of a scallop drag on the sediment chemistry, microbiota, and faunal assemblages of a shallow subtidal marine benthic community. *Journal of Sea Research* 46:309–24.

Watling L, Norse EA. 1998. Disturbance of the seafloor by mobile fishing gear: a comparison to forest clear cutting. *Conservation Biology* 12:1180–97.

Watson EB, Wigand C, Davey EW, Andrews HM, Bishop J, Raposa KB. 2017. Wetland loss patterns and inundation-productivity relationships prognosticate widespread salt marsh loss for southern New England. *Estuaries and Coasts* 40(3):662–81.

Weigle S. 2007. *Non-Shipping Pathways for Marine Invasive Species in Maine*. Portland, ME: Casco Bay Estuary Partnership.

Weigold ME, Pillsbury E. 2014. "Long Island sound: a socio-economic perspective." In: *Long Island Sound*. Edited by JS Latimer, MA Tedesco, RL Swanson, C Yarish, PE Stacey, C Garza. New York, NY: Springer (pp. 1–46).

Weinrich MT, Belt CR, Morin D. 2001. Behavior and ecology of the Atlantic white-sided dolphin (*Lagenorhynchus acutus*) in coastal New England waters. *Marine Mammal Science* 17(2):231–48.

Wells CD, Pappal AL, Cao Y, Carlton JT. 2013. *Report on the 2013 Rapid Assessment Survey of Marine Species at New England Bays and Harbors*. Boston, MA: Massachussetts Office of Coastal Management.

Wigley RL. 1968. Benthic invertebrates of the New England fishing banks. *Underwater Naturalist* 5(1):8–13.

Wigley SE, McBride HM, McHugh NJ. 2003. *Length–Weight Relationships for 74 Fish Species Collected During NEFSC Research Vessel Bottom Trawl Surveys, 1992–99. NOAA Technical Memorandum NMFS-NE-171*. Washington, DC: NOAA (p. 26).

Willis TV, Wilson KA, Johnson BJ. 2017. Diets and stable isotope derived food web structure of fishes from the inshore Gulf of Maine. *Estuaries and Coasts* 40(3):889–904.

Wyatt KH, Griffin R, Guerry AD, Ruckelshaus M, Fogarty M, Arkema KK. 2017. Habitat risk assessment for regional ocean planning in the US Northeast and Mid-Atlantic. *PloS One* 12(12):e0188776.

Xue H, Incze L, Xu D, Wolff N, Pettigrew N. 2008. Connectivity of lobster populations in the coastal Gulf of Maine. Part I: circulation and larval transport potential. *Ecological Modelling* 210(1–2):193–211.

Zemeckis DR, Martins D, Kerr LA, Cadrin SX. 2014. Stock identification of Atlantic cod (Gadus morhua) in US waters: an interdisciplinary approach. *ICES Journal of Marine Science* 71(6):1490–506.

Zollett EA, Rosenberg AA. 2005. *A Review of Cetacean Bycatch in Trawl Fisheries*. Woods Hole, MA: Northeast Fisheries Science Center.

CHAPTER 4

The U.S. Mid-Atlantic Region

➲ The Region in Brief

- While containing lower numbers of managed taxa among the eight regional U.S. marine ecosystems, this region has relatively well-managed state and federal fisheries that are important both nationally and along the U.S. Atlantic coast, including Atlantic menhaden, blue crab, eastern oyster, Black sea bass, Summer flounder, and Striped bass.
- Human population and population density are nationally highest in this region, contributing significant pressures to marine habitats and living marine resources (LMR) while also maintaining a pronounced coastal and oceanic economy.
- The U.S. Mid-Atlantic has very high socioeconomic status compared to other U.S. regions, the third-highest LMRs-based economy in the U.S, and nationally the highest establishments and employments for all ocean uses and for the tourism and recreation sector.
- The Mid-Atlantic is tied with New England for the second-highest basal productivity in the U.S., contributing toward higher system biomass, landings, and LMR-dependent jobs and revenue.
- Highly productive habitats (salt marsh, seagrass, oyster reef) in large estuarine bays and offshore rocky reefs support economically and ecologically important species throughout their life histories.
- There are relatively few depleted marine mammal stocks and threatened/endangered species populations in this region.
- Within this region and its subregions, average sea surface temperatures have increased at >0.74°C since the mid-twentieth century.
- Other ocean uses and coastal tourism are pronounced in the Mid-Atlantic. This region also leads the nation in the development of offshore wind energy areas.
- Collaborative fisheries management among state and federal entities has been focused upon commercially and recreationally important species for which overfishing has been minimized, stock status is well known, and rebuilding programs and management enterprises have been mostly successful.
- Among subregions, the Chesapeake and Delaware Bays contribute most heavily toward fisheries landings, LMR revenue, and LMR employments, while human population stressors surrounding these areas are some of the highest nationally.
- Overall, EBFM progress has been made at the regional and subregional level in terms of implementing ecosystem-level planning, advancing knowledge of ecosystem principles, and in assessing risks and vulnerabilities to ecosystems through ongoing investigations into climate vulnerability and species prioritizations for stock and habitat assessments.
- This ecosystem is excelling in the areas of LMR and socioeconomic status, the quality of its governance system, and is relatively productive, as related to the determinants of successful LMR management.

Ecosystem-Based Fisheries Management: Progress, Importance, and Impacts in the United States. Jason S. Link and Anthony R. Marshak, Oxford University Press. © U.S. Department of Commerce, U.S. Government 2021. DOI: 10.1093/oso/9780192843463.003.0004

4.1 Introduction

The U.S. Mid-Atlantic fisheries ecosystem evokes thoughts of iconic coastal and estuarine-associated fisheries species such as blue crab (*Callinectes sapidus*), Eastern oyster (*Crassostrea virginica*), Black sea bass (*Centropristis striata*), Summer flounder (*Paralichthys dentatus*), and Striped bass (*Morone saxatilis*). Scenes of crab houses, watermen fishing for shellfish, oyster dredges, lighthouses, schooners, and shipwrecks come to mind when thinking of the maritime culture of this historically important region, which is nicely accompanied by a sampling of local tastes that include raw or fried oysters, crab cakes, and seafood dusted in Old Bay seasoning. The Mid-Atlantic ecosystem also invokes the sounds and history of Appalachian bluegrass and mountain music, American folk standards, traditional gospel and jazz roots, Bruce Springsteen tunes, and Chesapeake hymns, work songs, and sea shanties, with the occasional tugboat horn heard in the distance. Many literary depictions of this marine setting are rooted and focused in its major coastal cities and bays, including James Michener's *Chesapeake*, works and poems by Edgar Allen Poe (Baltimore, MD; Philadelphia, PA) and Walt Whitman (Atlantic City, NJ), and F. Scott Fitzgerald's *The Great Gatsby* (Long Island, NY). More recent novels that portray the Mid-Atlantic coastal heritage include Chip Cheek's *Cape May* (Cape May, NJ), while North Carolina swamps and marshes also serve as the setting for Delia Owens' 2018 bestseller, *Where the Crawdads Sing*. In addition, William Warner's Pulitzer Prize-winning *Beautiful Swimmers* documents the history of blue crabs and watermen within Chesapeake Bay, the nation's largest estuary. Many of these cultural components are closely tied to the natural environments comprising this major marine ecosystem, particularly its shorelines and beaches near which iconic and extensive boardwalks are found, and the living marine resources (LMRs) that are associated with this region. They contribute heavily toward its significant tourism, marine transportation, and emerging marine energy-based economy. Additionally, they serve as a major foundation for the livelihood of its populated coastal communities.

There are many studies that characterize the Mid-Atlantic ecosystem (Sherman et al. 1996, Link et al. 2002a, 2002b, Sherman et al. 2002, Stevenson et al. 2004, Link et al. 2008, Link et al. 2011, NEFSC 2012, Shackell et al. 2012, Wyatt et al. 2017, Gaichas et al. 2018b, 2019, Link & Marshak 2019, Nichols et al. 2019), including its notably large estuaries (Newell 1988, Monaco & Ulanowicz 1997, Paul et al. 1998, Llansó et al. 2002, Jung & Houde 2003, Kiddon et al. 2003, Kemp et al. 2005, Ihde et al. 2016, Ihde & Townsend 2017), its important and major fisheries (e.g., blue crab, oysters, estuarine fishes, Black seabass, bluefish, tunas; de Sylva & Davis 1963, Musick & Mercer 1977, Epifanio et al. 1984, Stanley & Sellers 1986, Hill et al. 1989, Lipcius & Van Engel 1990, Eklund & Targett 1991, Ford 1997, Able & Fahay 1998, Munch & Conover 2000, Steimle & Zetlin 2000, Martino & Able 2003, Kirby 2004, Rudershausen et al. 2010, Capossela et al. 2013, Schulte 2017), protected species (e.g., whales, dolphins, seals, sea turtles, sturgeons; Selzer & Payne 1988, Mead & Potter 1990, Shoop & Kenney 1992, Smith & Clugston 1997, Pierre 1999, Knowlton et al. 2002, Weinrich & Clapham 2002, Gannon & Waples 2004, Stein et al. 2004, Belden & Orphanides 2007, Firestone et al. 2008, Ampela 2009, Smolowitz et al. 2015), and their contributions toward the marine socioeconomics of this region, especially in how humans use, enjoy, and interact with this system (Gentner & Lowther 2002, Colburn et al. 2010, Kaufman 2011, Tegen et al. 2015, Martin et al. 2018). This marine region, its resources, and its goods and services have played a major role in American History, as illustrated by its major settings during the Revolutionary War and War of 1812, its historical strategic location for many military installations such as Yorktown, Fort McHenry, Camp Meade, Norfolk, and Quantico, and as a Port of Embarkation at Newport News, VA during World War I (Danysh 1979, Steffen 1984, Marston 2003, Borneman 2004, Stewart 2004, Halpern 2012, Cogliano 2016).

What follows is a brief background for this region, describing all the major considerations of this marine ecosystem. Then there is a short synopsis of how all the salient data for those factors are collected, analyzed, and used for LMR management. We then present an evaluation of key facets of ecosystem-based fisheries management (EBFM) for this region.

4.2 Background and Context

4.2.1 Habitat and Geography

The Mid-Atlantic ecosystem is part of the Northeast Shelf Large Marine Ecosystem (Sherman et al. 1996). It makes up the second-largest component of the United States Atlantic Exclusive Economic Zone (EEZ). It connects the South Atlantic through to the New England ecosystem via meandering of the Gulf Stream north of Cape Hatteras (Fig. 4.1). The region also serves as a major focus area and transitional point for migratory species, including increasingly abundant tropically

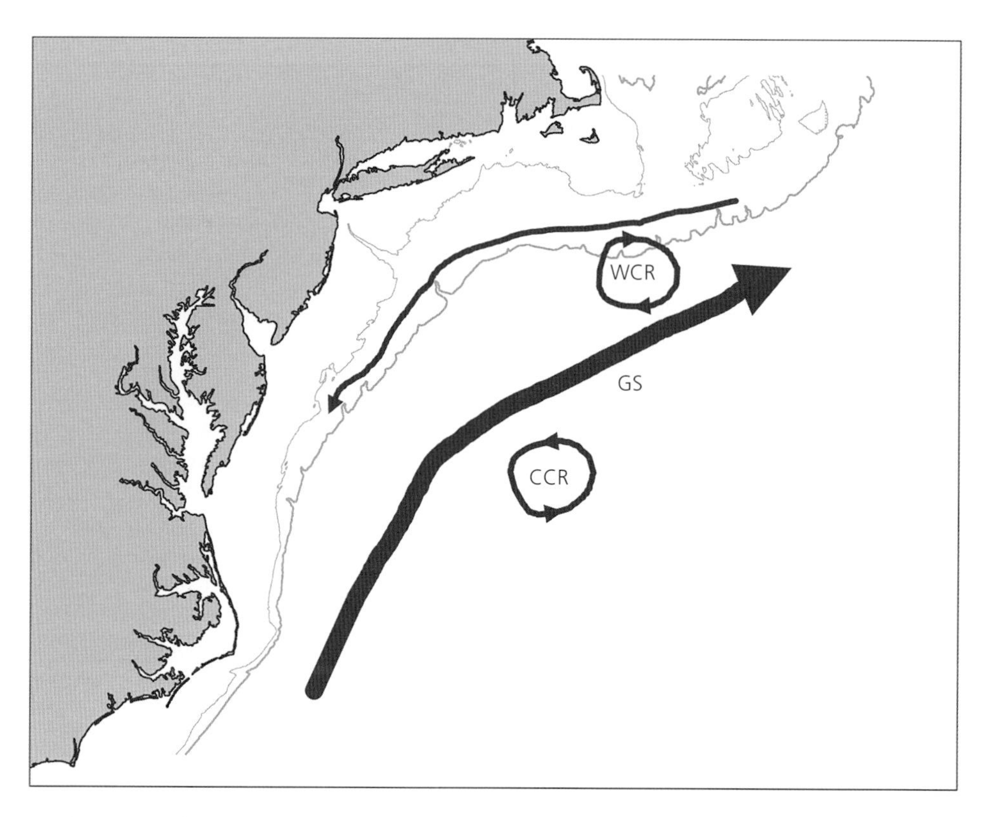

Figure 4.1 Major currents and circulation patterns throughout and encompassing the Mid-Atlantic region and subregions. GS = Gulf Stream; WCR = warm core ring; CCR = cold core ring.

associated flora and fauna, which may foster changes to the overall composition of this temperate ecosystem (Hare & Able 2007, Nye et al. 2009, Karlson & Osman 2012, Pinsky & Fogarty 2012, Ribeiro et al. 2015, Kleisner et al. 2016, Morley et al. 2017, 2018).

Significant concentrations of human population, fisheries landings, and economic production are found throughout the Mid-Atlantic region, particularly in two key subregions: the **Chesapeake Bay**—District of Columbia, Maryland, and Virginia; and **Delaware Bay**—Delaware, New Jersey, and Pennsylvania (Fig. 4.2). There are other subregional features in this region, but we emphasize these two herein as exemplary, smaller-scale ecosystems.

The main features of the Mid-Atlantic ecosystem include large expanses of coastal salt marshes, seagrasses, and previously quite extensive oyster beds, offshore rocky reefs, low to mid-range numbers of managed species, and an oceanography driven by the continuation of the Gulf Stream (Morgan & Bishop 1977, Orth et al. 1984, Stevenson et al. 1988, Myers & Drinkwater 1989, Rountree & Able 1992, Furbish & Albano 1994, Roman et al. 2000, Steimle & Zetlin 2000, Angradi et al. 2001, Koch & Orth 2003, Nordlie 2003, Kirby 2004, Orth et al. 2006, Ezer et al. 2013, Link & Marshak 2019). Many taxa in this region are associated with nearshore oyster reefs and vegetated habitats, in addition to offshore rocky reefs, sandy shoals, and hardbottoms, which support many ecologically and commercially important species (Orth et al. 1984, Stanley & Sellers 1986, Steimle & Zetlin 2000, Beck et al. 2001, Deegan et al. 2002, Minello et al. 2003, Koch & Orth 2003, Nordlie 2003, Peterson et al. 2003, Rodney & Paynter 2006, Erbland & Ozbay 2008, Kritzer et al. 2016).

4.2.2 Climate and Other Major Stressors

The Mid-Atlantic ecosystem is subject to many stressors, which include habitat loss, coastal development and associated increases in nutrient loading and pollution, climate-related regional warming, sea level rise, species range shifts, hurricanes, differential overfishing, and recent invasions from exotic species (Joseph 1972, Chambers et al. 1999, Najjar et al. 2000, Rogers &

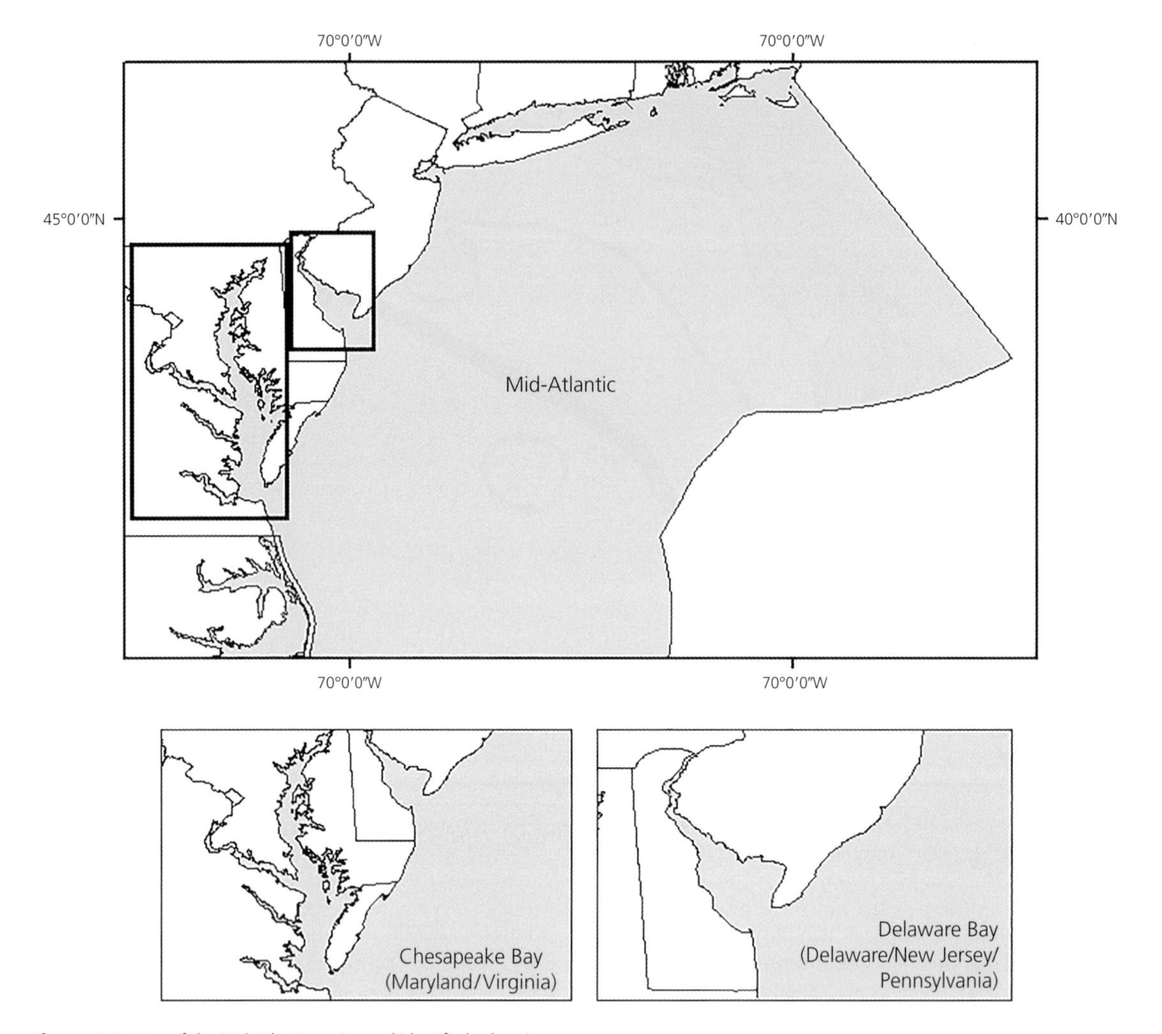

Figure 4.2 Map of the Mid-Atlantic region and identified subregions.

McCarty 2000, Neff et al. 2000, Roman et al. 2000, Kennish 2001, Koch & Orth 2003, Kirby 2004, Kemp et al. 2005, Montane & Austin 2005, Orth et al. 2006, Grabowski & Peterson 2007, Alber et al. 2008, Stedman & Dahl 2008, Anthony et al. 2009, Najjar et al. 2010, Carruthers et al. 2013, Ezer et al. 2013, Patrick et al. 2014). Major degradation of Mid-Atlantic coastal wetlands, including dominant cordgrass (*Spartina alterniflora*) salt marsh, has occurred over the past century, with highest losses along the U.S. Atlantic coast reported for this region (Stevenson et al. 1985, Wray et al. 1995, Najjar et al. 2000, Kennish 2001, Wilson et al. 2007, Alber et al. 2008, Stedman & Dahl 2008, Smith et al. 2017, Jordan et al. 2018). These are compounded by natural and human-related stressors and upstream land-based activities such as agriculture, associated pollution, non-point nutrient loading, and inland development that have significantly affected Mid-Atlantic wetlands (Paul et al. 1998, 2002, Jantz et al. 2005, Fisher et al. 2006, Ozbay et al. 2017, Jordan et al. 2018, Prosser et al. 2018). Reductions of oyster reefs and eelgrass (*Zostera marina*) beds have also occurred over time as a result of overharvesting, disease, and eutrophication, though degrees of recovery have been observed through protection and extensive restoration efforts (Rothschild et al. 1994, Koch & Orth 2003, Kemp et al. 2005, Orth et al. 2006, Patrick et al. 2014, Lefcheck et al. 2017, 2018, Patrick et al. 2018). Large investments toward restoring Chesapeake oysters have been ongoing, with focused interests on enhancing spawning stock biomass (Mann et al. 2014, Lipcius et al. 2015, Hornick & Plough 2019), while efforts on the lower eastern shore of Virginia have restored up to 10,000 acres of seagrass, constituting the world's largest successful restoration project (Lefcheck et al. 2018, Orth et al. 2019). However, ongoing stressors continue to affect this system, including the overall quality of its habitats and dependent LMRs (Najjar et al. 2000, Najjar et al. 2010, Wyatt et al. 2017, Jordan et al. 2018, Patrick et al. 2018). Compounding habitat loss and post-glacial land subsidence in this region is increasing sea level throughout coastal environments and prominent bays, with average values observed at 3.5 mm year^{-1} in Chesapeake Bay and forecasted at up to 1.6 m in this subregion over the next century (Zervas 2001, Najjar et al. 2010, Beckett et al. 2016, Shaw et al. 2016, Lee et al. 2017, Boesch et al. 2018). In particular, Hampton Roads, Virginia is suggested to be the second-most at risk community from sea level rise after New Orleans, Louisiana (Sadler et al. 2017, Stafford & Abramowitz 2017). Region-wide values for the Mid-Atlantic region are also expected to increase by up to 0.6 m (Najjar et al. 2000), with above average observations in sea level rise (up to 2 mm y^{-1}) for this region related to dynamics of the North Atlantic Oscillation (NAO) and Atlantic Meridional Overturning Circulation (AMOC), and differential influences of the Gulf Stream (Ezer et al. 2013, Kopp 2013, Valle-Levinson et al. 2017). Shoreline armoring in response to sea level rise is also causing loss and damage of nearshore habitats while also inhibiting inland migration of salt marshes (Wilson et al. 2007, Patrick et al. 2014, 2016).

Climate-related changes to this system are from an increasing sea surface temperature (SST), which has facilitated range expansions and increasingly abundant lower latitude species including South Atlantic Tarpon (*Megalops atlanticus*) and penaeid shrimps (*Farfantapeneus* spp.), with thermally stressed species emigrating toward the northern limits of this system into New England habitats (Hare et al. 2010a, b, Jones et al. 2010, Najjar et al. 2010, Pinsky et al. 2013, Morley et al. 2017, 2018). The Mid-Atlantic is a hurricane prone region, particularly given frequent hurricane landfalls in coastal North Carolina and subsequent passing storm paths throughout Mid-Atlantic watersheds (Smith 1999, Paerl et al. 2000, Goldenberg et al. 2001, Paerl et al. 2006, Link & Marshak 2019). Over time, increasing frequency and intensity of storms, and consequential effects to local marine ecosystems and coastal communities have occurred as a result of significant storms that include Donna (1964), Hazel (1964), Agnes (1972), Bertha (1996), Fran (1996), Floyd (1999), Isabel (2003), Irene (2011), and Sandy (2012; Bailey et al. 1975, Montane & Austin 2005, Mallin & Corbett 2006, Kemp & Horton 2013). During the past few decades, Hurricanes Floyd, Isabel, Irene, and Sandy especially impacted marine habitats and assemblages of targeted and protected species in this region. These storms caused major effects to local fisheries, with damage and loss observed to marine economies (Paerl et al. 2000, Peierls et al. 2003, Houde et al. 2005, Montane & Austin 2005, Miller et al. 2006, Paerl et al. 2006, Hughes et al. 2009, Munroe et al. 2013, Clay et al. 2016, Seara et al. 2016, Warner et al. 2017). Furthermore, the increasing effects of ocean acidification and coastal hypoxia continue to be of concern in this region, particularly in the Chesapeake Bay (Stierhoff et al. 2006, Costantini et al. 2008, Shen et al. 2008, Cai et al. 2017).

Addressing the cumulative impacts of these stressors, particularly in terms of how they interact with the LMRs of this ecosystem, their fisheries, and with its coastal communities poses significant challenges in this highly populated region that is also heavily dependent on coastal tourism, marine transportation, and marine construction (Joseph 1972, Najjar et al. 2000, Rogers & McCarty 2000, Roman et al. 2000, Wu et al. 2002, Daily 2003, Koch & Orth 2003, Kirby 2004, Kemp et al. 2005, Anthony et al. 2009, Najjar et al. 2010, Carruthers et al. 2013, Colburn et al. 2016).

4.2.3 Exploitation

Overfishing has affected the Mid-Atlantic system, particularly its commercially and recreationally important estuarine fishes, sharks, pelagic species, and invertebrates (Smith 1985, Richards & Rago 1999, Steimle & Zetlin 2000, Kirby 2004, Grout 2006, Grubbs & Musick 2007, Hare & Able 2007, Mann & Powell 2007, Babcock 2008, Sartwell 2009, Hale et al. 2016, Gaichas et al. 2019). However, with low proportions of federally managed stocks currently experiencing overfishing or having overfished status, and comparatively low proportions of unassessed stocks (Link & Marshak 2019), progress toward recovery of this key fisheries ecosystem is also underway. Recreational fishing occurs throughout several major portions of the Mid-Atlantic system (i.e., Chesapeake, Delaware Bays; New Jersey Shoreline; Southern Long Island Sound), with prominent fishing communities targeting nearshore and reef-associated species (Richards & Deuel 1987, Steimle & Zetlin 2000, Kerr et al. 2009, Gaichas et al. 2019, Link & Marshak 2019). Recreational fishing for sharks has occurred off New York since the mid-twentieth century (Babcock 2008). However, shifts in the targeted geographies and landings of regional fisheries have been observed, with commercial and recreational fishers increasing trip times and expanding into areas farther north and offshore as a result of changing species distributions (Kerr et al. 2009, Nye et al. 2009, McCay 2012, Pinsky & Fogarty 2012, Pinsky et al. 2013, Young et al. 2018, Dubik et al. 2019).

Landings from the southern limits of this system are becoming increasingly composed of lower latitude-associated species (Kerr et al. 2009, Najjar et al. 2010, Morley et al. 2018, Young et al. 2018, Dubik et al. 2019). Catches landed in Mid-Atlantic ports are also not necessarily those captured from their associated state or nearby federal waters, as fishers from North Carolina and Virginia continue to target preferred species (e.g., Summer flounder, Atlantic croaker—*Micropogonias undulatus*) and harvest in areas off New Jersey and New York in response to those taxa shifting their distributions (Young et al. 2018, Dubik et al. 2019). Given static proportional allocations of state-level fisheries quotas (e.g., Summer flounder) that have not been updated to reflect species distribution shifts, increased frequency of captures in the waters off northern states being landed in higher-allocation southern states also has been observed (Young et al. 2018, Dubik et al. 2019). Additionally, sportfishing continues to be a major source of tourism and revenue in this region, given the abundance of large migratory pelagic fishes throughout offshore waters and northern expansion of South Atlantic recreational fishery species (McConnell & Strand 1994, Gentner & Lowther 2002, Kauffman 2011).

While consequences of past overfishing and compounding environmental stressors still affect this system, particularly on oysters and the community composition of offshore and estuarine-associated pelagic and demersal species (Colvocoresses & Musick 1984, Rothschild et al. 1994, Paul et al. 1998, Steimle & Zetlin 2000, Uphoff 2003, Rodney & Paynter 2006, MacKenzie 2007, Jacques 2017), additional stock rebuilding of previously depleted species has occurred, with the U.S. Mid-Atlantic emerging as a leading example of sustainable fisheries management (Richards & Rago 1999, Grout 2006, Link & Marshak 2019). As a result of historical overexploitation, there have been consequences to species life histories, depletions of higher trophic levels, degradation of essential habitats, and shifts in biomass among species complexes in this region (Stevens et al. 2000, Coleman et al. 2004, Stevenson et al. 2004, Link et al. 2008, Limburg & Waldman 2009, Gaichas et al. 2019). Improvements toward developing more accurate assessments of targeted and bycaught species, such as Atlantic butterfish (*Peprilus triacanthus*; Manderson et al. 2011, Kohut et al. 2012, Adams et al. 2014, Adams 2017), which have accounted for environmental influences on stock abundance and distribution, are also enhancing management of this system. Efforts toward greater ecological understanding of species that are overfished or experiencing overfishing (e.g., Atlantic mackerel *Scomber scombrus*), and the overall importance of forage fish, continue as Mid-Atlantic Fishery Management Council (MAFMC) priorities (MAFMC 2017).

4.2.4 Invasive Species

The emerging prevalence of invasive species, particularly European green crab (*Carcinus maenas*), Asian shore crab (*Hemigrapsus sanguineus*), Blue catfish (*Ictalurus furcatus*), Asiatic clam (*Corbicula fluminea*), and common reed (*Phragmites australis*), has also raised concerns regarding ecological stability in the Mid-Atlantic (Diaz 1974, Phelps 1994, McDermott 1998, Rice & Rooth 2000, Schloesser et al. 2011, Epifiano 2013, Matheson et al. 2016). European green crabs have invaded salt marshes in North Carolina and northward throughout the Mid-Atlantic and New England coasts, with effects on eelgrass beds (i.e., declining biomass and damage to rhizomes) and anticipated effects on native fish community structure and abundance (Matheson et al. 2016). However, predatory control by blue crabs also appears to limit some of their propagation and associated ecological consequences throughout the Mid-Atlantic (de Rivera et al. 2005). Asian shore crabs have expanded throughout the U.S. Atlantic coast and are found within rocky intertidal communities, where they have been observed preying upon and outcompeting both resident and exotic crab species, including blue crabs and European green crabs (Lohrer & Whitlach 2002, MacDonald et al. 2007, Epifanio 2013). Following introduction, Blue catfish have become increasingly dominant in the Chesapeake Bay system (Schloesser et al. 2011, Schmitt et al. 2017, Fabrizio et al. 2018), growing to large sizes, with the consequences from their predation affecting local community composition and prey populations (including American shad *Alosa sapidissima*, River herring *Alosa* spp., Atlantic menhaden *Brevoortia tyrannus*, and blue crabs) in this system (Schloesser et al. 2011, Schmitt et al. 2017, 2019). Additional system-wide changes, including resurgence of some species of submerged aquatic vegetation (SAV) in portions of the Chesapeake watershed, have been observed with expanding populations of Asiatic clam, as a result of their associated filtration rates (Phelps 1994, Crisman 2008). However, their ability to locally deplete phytoplankton in upper reaches of the Chesapeake has also raised concerns regarding their competitive effects on other species of bivalves, including oysters (Newell 1988, Strayer 1999). Extensive expansions of the competitively superior common reed throughout Chesapeake and Mid-Atlantic salt marsh systems has also occurred, causing alterations in the dominance of foundational cordgrass and changes to macrofaunal assemblages, particularly larval and juvenile estuarine prey fishes (Chambers et al. 1999, Able & Hagan 2000, Rice & Rooth 2000). Concerns regarding the economic effects of invasive hydrozoans that block power plant intakes and whose potent stings affect coastal communities have become more commonplace in the Mid-Atlantic (Bayha & Graham 2014, Gaynor et al. 2016, Restaino et al. 2018). Cumulatively, all of these natural and human-induced stressors have imparted significant consequences and shifts to the Mid-Atlantic ecosystem and its associated LMRs.

4.2.5 Ecosystem-Based Management (EBM) and Multisector Considerations

In addition to its iconic species, the Mid-Atlantic ecosystem supports multiple human uses, including military interests, which result in inevitable trade-offs between objectives for LMR management and other societal needs (Martin & Hall-Arber 2008, Link et al. 2011, NEFSC 2012, NUWC 2012, Samhouri et al. 2013, Samoteskul et al. 2014, Gaichas et al. 2018b. 2019). Beyond the trade-offs among regional commercial and recreational fisheries, multiple coastal and ocean uses that support the Mid-Atlantic marine economy co-occur within areas providing habitat to many managed species that support these local fisheries, and are foundational for marine tourism, marine transportation, and emerging wind energy sites (Martin & Hall-Arber 2008, Anthony et al. 2009, BOEM 2012, Holland et al. 2012, Bailey et al. 2014, Samoteskul et al. 2014, BOEM 2016, 2017, Guida et al. 2017, Carr-Harris & Lang 2019; Link & Marshak 2019). These sectors contribute heavily toward the region's marine economy, while their overlapping and interrelated activities with fisheries emphasize the need to consider multisector ocean uses holistically within this ecosystem. In Mid-Atlantic offshore habitats, the effects of wind energy siting on essential fish habitat (EFH) and accessibility to major fishing grounds are of concern to fishermen, while the attraction, production, and mortality potential of these large structures toward fishery populations are also receiving attention (Bailey et al. 2014, Samoteskul et al. 2014, Wyatt et al. 2017, Cruz-Marrero et al. 2019). Limited past regulation of ship ballast water releases resulted in the introduction of invasive species into the Mid-Atlantic region, with higher prevalence of exotic bivalves, crustaceans, and pathogens attributed to this source (MacDermott 1998, Mann & Harding 2000, Fofonoff et al. 2009, Karlson & Osman 2012, LaPointe et al. 2016). Sand mining has also emerged as a significant stressor to this region, causing serial depletion of nearshore habitats and expanding into offshore regions (Middleton 2014, MRB 2018).

While conflicts among commercial and recreational fisheries occur in the Mid-Atlantic, their magnitude and effects have been relatively more muted than in other regions due in part to less contentious fisheries (Dawson & Wilkins 1981, Wilson & McCay 1998, Sutinen & Johnston 2003). However, conversations regarding management decisions on Summer flounder, Striped bass, and Atlantic menhaden fisheries have been quite antagonistic (Lavelle 2014, Dance 2019, Vogelsong 2020). While not as apparent as in other regions, low-moderate risk of conflicts among some Mid-Atlantic fisheries and protected species have been identified, which are annually monitored (NEFSC 2012, Gaichas et al. 2018a). While successful management measures controlling harbor porpoise bycatch have been implemented (Gaichas et al. 2019), bycatch associated with longlining and other pelagic gear types is still of concern, with marine mammal bycatch (particularly incidental captures of dolphins in net and bottom trawl fisheries) in this region being second-highest nationally among the eight regional U.S. marine ecosystems (Rossman 2007, McClellan et al. 2011, Murray & Orphanides 2013, Link & Marshak 2019). Management approaches considering the effects of coastal and tourism-related development on LMR habitat quality, together with socioecological impacts of increasing aquaculture for hard clams (*Mercenaria mercenaria*) and oysters and marine energy siting in this region remain needed (Leslie & McLeod 2007, Ehler & Douvere 2010, Samhouri et al. 2013, Olsen et al. 2014, Smythe & McCann 2019). Exploring, identifying, and resolving these differences in priorities and goals for the Mid-Atlantic is the essence of EBFM, and the emerging basis of its EAFM and EBFM implementation plans (MAFMC 2019a, NMFS 2019b). With its targeted taxa, multiple habitats, trade-offs among marine sectors, recovering fisheries and protected species populations, and ongoing human-related pressures, continued efforts toward strengthening and executing ecosystem approaches to fisheries management in the Mid-Atlantic remain a priority.

Given these multiple considerations, trade-offs, and dynamics, it remains appropriate to continue employing

I Can't Believe It's Not Butter: Butterfish, Thermal Habitat, Frontal Boundaries, Distributions and Changing Catchabilities (q)

—(c.f. Chapter 3)

Did you ever try to catch fireflies as a kid? But as you did so, you kept catching moths instead. To the point that you just stopped? Frustrating, wasn't it? That's essentially what happens with fisheries bycatch, albeit with slightly bigger stakes.

The butterfish, *Peprilus triacanthus*, is a small, pelagic, forage fish often found schooling with squid along the east coast (Murawski & Waring 1979, Lange & Waring 1992, Manderson et al. 2011, Kohut et al. 2012, 2015). It supports a fishery unto itself, but because it is caught with squid, is primarily a bycatch species caught as part of the squid fishery. Because the population size of this species had limited information, the estimates were thought to be quite small (<5,000 metric tons; Manderson et al. 2011, Adams et al. 2015). Thus, a limit on butterfish catch shut down the entire squid fishery.

Butterfish, squids, and generally most pelagic fishes, are known to have distinct thermal preferences. They often follow frontal boundaries of water discontinuities to maintain these preferred thermal conditions. In other words, when one sees a convergence of two water masses, often one can find an agglomeration of fish in such locales. The location of butterfish is highly correlated to these locations with clear differences in temperature.

So what if we could predict where these different temperatures were located, could we then predict butterfish? And if so, could we then get that information to the fishermen to avoid butterfish bycatch? Yep, and that's what a team of ecologists, oceanographers, and fishermen did along the Atlantic coast (Manderson et al. 2011, Kohut et al. 2012, Adams et al. 2015, Kohut et al. 2015).

But there is more. What if we could use those predictions to inform surveys of these fishes, ground-truth those predictions with catches from the fishing fleet, and provide that additional information to the modeling teams that estimate how many butterfish are in the ocean? This occurred, and an important parameter that assumes how "catchable" these fish are in the surveys was drastically changed. And from that, past estimates of butterfish were recalculated to be approximately 10× (Adams et al. 2015) more abundant than was previously thought. And because of that, the limits on how much butterfish could be caught were increased. Which means that more squid could be caught with a much lower risk of being closed because of butterfish bycatch.

It was a win-win situation (except the modelers who complained that it made the math more complicated, but we always complain about something in the models...).

This is an excellent example of including relatively basic information in the process of setting the limits for how much fish or squid we can catch, and by doing so providing more accurate information and the opportunity for more catch. It is a great example of using ecosystem considerations to better inform fisheries management. It also demonstrates the value of acknowledging trade-offs in multiple fishing objectives and regulations, and conducting research to better inform those trade-offs.

References

Adams CF, Miller TJ, Manderson JP, Richardson DE, Smith BE. 2015. *Butterfish 2014 Stock Assessment. Northeast Fisheries Science Center Reference Documents 15-06*. Washington, DC: NOAA (p. 110).

Kohut J, Palamara L, Bochenek E, Jensen O, Manderson J, Oliver M, Gray S, Roebuck C. 2012. *Using Ocean Observing Systems and Local Ecological Knowledge to Nowcast Butterfish Bycatch Events in the Mid-Atlantic Bight Longfin Squid Fishery*. New York, NY: IEEE (pp. 1–6).

Kohut J, Palamara L, Curchitser E, Manderson J, DiDomenico G. 2015. *Cooperative Development of Dynamic Habitat Models Informed by the Integrated Ocean Observing System (IOOS) Informs Fisheries Management Decision Making in the Coastal Ocean. 2015-MTS/IEEE*. New York, NY: IEEE (pp. 1–8).

Lange AM, Waring GT. 1992. Fishery interactions between long-finned squid (Loligo pealei) and butterfish (Peprilus triacanthus) off the Northeast USA. *Journal of Northwest Atlantic Fishery Science* 12:49–62.

Manderson J, Palamara L, Kohut J, Oliver MJ. 2011. Ocean observatory data are useful for regional habitat modeling of species with different vertical habitat preferences. *Marine Ecology Progress Series* 438:1–7.

Murawski SA, Waring GT. 1979. A population assessment of butterfish, Peprilus triacanthus, in the northwestern Atlantic Ocean. *Transactions of the American Fisheries Society* 108(5):427–39.

an ecosystem-based approach that systematically incorporates abiotic, biotic, and human factors into the Mid-Atlantic fisheries management frameworks. Efforts toward more comprehensive approaches, including the development of a regional EAFM strategy by the MAFMC and pilot initiatives to incorporate habitat and ecosystem-based considerations into state and federal management efforts have been undertaken (Sanchirico et al. 2008, Link et al. 2011, MAFMC 2017, 2019a, 2019b). They complement multispecies federal fishery management plans (FMP) for the region's coastal, estuarine, and offshore pelagic and demersal fisheries species and the high number of single-species FMPs under the jurisdiction of the Atlantic States Marine Fisheries Commission (ASMFC). The MAFMC has developed and supported EAFM Guidance Documents, NOAA Northeast Fisheries Science Center (NEFSC) annual ecosystem-level risk assessments and state of the ecosystem reports, and efforts to incorporate broader considerations of forage fish, protected resources, and habitat factors into management strategies (MAFMC 2017, 2019a, 2019b). Advancements by ASMFC toward more holistic management of nearshore fisheries resources include the development of ecosystem reference points for management actions (including for Atlantic Menhaden) and accounting for ecosystem-level data needs for more effective implementation of EBFM (ASMFC 2012, 2020a; Buchheister et al. 2016, 2017). Efforts to merge these components into a broader EBFM framework, apply many ecosystem-level metrics, and incorporate broader ecosystem considerations into management actions remain needed (Link et al. 2011, Gaichas et al. 2018b, 2019, MAFMC 2019a, NMFS 2019b).

4.3 Informational and Analytical Considerations for this Region

4.3.1 Observation Systems and Data Sources

Considerable data exist in the Mid-Atlantic region to support EBFM. In moving toward a more holistic LMR management approach, data regarding the physical, biological, and socioeconomic components of the Mid-Atlantic ecosystem are available through regional observation systems and multiple monitoring programs spanning several decades. The Mid-Atlantic Coastal Ocean Observing System (MARACOOS) provides oceanographic and atmospheric data from multiple sensor stations throughout coastal and offshore locations (MARACOOS 2016). The system is a composite of aggregated data from ~15 different federal and regional sources, with the majority of observations from coastal sensors maintained by National Oceanic and Atmospheric Administration's (NOAA) National Weather Service (NWS) and National Ocean Service (NOS), the National Estuarine Research Reserve (NERR) system, Chesapeake Bay regional sensors, and sensors monitored by the Maryland Department of Natural Resources. Since 2008, physical data reported at daily to annual scales have been available at the MARACOOS portal <https://maracoos.org/> with monitored atmospheric variables including biological (chlorophyll, animal tracking and abundance), chemical (pH, carbon dioxide, dissolved oxygen), and physical parameters (current speed and direction, wind speed and direction, wave height, period, and direction, air and water temperature, salinity, air pressure, and sea level). This information is also complemented by underwater gliders, high-frequency radar, satellite data, and other sensors on offshore buoys. These data are being applied toward forecasting regional sea-level rise, enhancing preparations for coastal flooding, and predicting regional atmospheric and oceanographic patterns. MARACOOS priorities include providing information for safe maritime practices, improved water quality, and in facilitating environmentally sound ocean uses (MARACOOS 2016). The Mid-Atlantic Regional Council on the Ocean (MARCO) supported Mid-Atlantic Ocean Data Portal also serves as a data repository, with information from federal, state, and independent platforms (including MARACOOS) used to create maps in advancing regional planning efforts (MARCO 2015).

Several oceanographic and atmospheric data sources have been incorporated toward a broader understanding of the Mid-Atlantic ecosystem, particularly its overall status (NEFSC 2009, 2012, MAFMC 2015, Gaichas et al. 2018a, b, 2019, MAFMC 2019a), focal components, and their applicability toward regional ecosystem status reports (ESRs) and Integrated Ecosystem Assessments (IEAs; Link et al. 2002b, Levin et al. 2009, NEFSC 2006, 2012, Samhouri et al. 2013, Gaichas et al. 2018a, 2019). Data regarding the Atlantic multidecadal oscillation (AMO) and NAO indices and their influences on satellite-derived SST have been applied toward a broader understanding of climate forcing in the Mid-Atlantic (Friedland & Hare 2007, NEFSC 2012, Nye et al. 2014). Trends in freshwater input and physical factors such as position of the Gulf Stream, and wind and thermohaline dynamics, have been derived from NWS datasets and oceanographic data taken aboard NOAA northeast regional surveys (Politis et al. 2014, MARACOOS 2016). This information has been applied toward characterizing major water masses throughout the Mid-Atlantic, including changes in their stratification (NEFSC 2006, 2012, Fratantoni et al. 2015). The U.S. Environmental Protection Agency (EPA) continues to collect and maintain data on Chesapeake Bay water quality during its multidecadal program (CBP 2009). All of this physical information has been incorporated into ESRs for this region and the broader Northeastern U.S, including dynamics of the Gulf Stream, Labrador Current, and riverine input to this system (NEFSC 2006, 2012, Gaichas et al. 2018a, 2019).

Information regarding habitat factors and aspects of primary and secondary production are monitored via satellite information and ongoing monitoring programs throughout the Mid-Atlantic accounting for water quality, chlorophyll-a concentrations, phytoplankton biomass, and estimated coverage and risk assessments of nearshore vegetated habitats (EPA 2012, MAFMC 2012, NEFSC 2012, Crawford et al. 2015, MAFMC 2015 Gaichas et al. 2018a, b, 2019, MAFMC 2019a). These data also account for phytoplankton blooms. Zooplankton abundance and biomass are monitored in surveys (i.e., Marine Resources Monitoring Assessment and Prediction Program—MARMAP, Northeast Shelf Ecosystem Monitoring Program—ECOMON) from the NEFSC, which are performed six times per year and began in the 1970s (Sibunka & Silverman 1989, Kane 2007, Richardson et al. 2010, Politis et al. 2014).

Comprehensive databases for secondary through upper trophic-level species include fishery-independent information collected from seasonal trawl surveys. These surveys account for biomass and abundance of benthic invertebrates, particularly crustaceans, oysters, and sea scallops, while other sessile benthic species have been examined and their habitats later protected through recently conducted remotely operated vehicle (ROV) surveys of deep corals (Azarovitz 1981, NEFC 1988, Smith, 2002, NEFSC 2006, 2012, 2013a, b, NMFS 2016a, Hourigan et al. 2017, Gaichas et al. 2018a, NEFSC 2018, Gaichas et al. 2019). They are also complemented by nearshore trawl surveys within Chesapeake and Delaware Bay undertaken by state agencies and academic institutions, such as the Virginia Institute of Marine Science (VIMS) conducted nearshore Northeast Area Monitoring and Assessment Program (NEAMAP; Bonzek et al. 2016, 2017). These surveys provide fishery-independent information regarding the biomass and abundance of commercially important finfish species and composition of offshore assemblages. Data are applied toward broader understanding of ecosystem functioning and informing assessments of Mid-Atlantic stock status. These surveys are also complemented by NOAA Fisheries fishery-dependent monitoring programs for commercial and recreational harvest, landings, revenue, and effort, including surveys such as the Marine Recreational Information Program (MRIP), and fisheries observation data on commercial landings, catches, and bycatch gathered by the National Observer Program (NAS 2017, NMFS 2017b). Factors including trends in average size, spatial distributions, aggregate biomass of pelagic and demersal species groupings and their ratios, predator–prey relationships, and mean trophic level, condition indices, and thermal preferences represent many of the ecosystem-level metrics that have been derived from this fishery-dependent and independent survey information (Azarovitz 1981, NEFC 1988, Smith, 2002, NEFSC 2012, MAFMC 2017, Gaichas et al. 2018a, b, Lynch et al. 2018, Gaichas et al. 2019, MAFMC 2019a). Social and economic information for fisheries and other uses, particularly regarding revenue and employment, are also collected by NOAA Fisheries, the National Ocean Economics Program (NOEP), and NOAA's Office for Coastal Management Economics: National Ocean Watch (ENOW) program (NMFS 2018a, OCM 2019). Monitoring programs exist for evaluating the status, recovery, and bycatch of protected species that include sea turtles and cetaceans, and several monitoring programs for marine birds are overseen by NOAA (as collected during ECOMON and protected species surveys) and the U.S. Department of Interior (Bureau of Environmental Management Environmental Studies Program, United States Geological Survey Breeding Bird Survey; O'Connell et al. 2009, NMFS 2018b). In the Mid-Atlantic region, all of these aforementioned data sources have been applied toward integrative ecosystem measurements that examine cumulative changes in these metrics over time and their interdependent relationships (NEFSC 2012, MAFMC 2017, Gaichas et al. 2018a, b, 2019, MAFMC 2019a).

4.3.2 Models and Assessments

To support management of LMRs and foster EBFM approaches, assessments of Mid-Atlantic economically important and protected species are conducted by NOAA Fisheries in collaboration with the MAFMC; Ferguson et al. 2017, Lynch et al. 2018). Stock assessment models are used to generate estimates of stock status with a multitude of data sources (e.g., landings, catch per unit effort, life history characteristics, survey biomass, etc.), and assessments for most Mid-Atlantic species are conducted through the Stock Assessment Workshop/Stock Assessment Review Committee (SAW/SARC) process (Lynch et al. 2018). Other species such as Atlantic menhaden are assessed through the Southeast Data Assessment and Review (SEDAR) process (SEDAR 2020). The ASMFC conducts stock assessments for those species under its jurisdiction (Kilduff et al. 2009), with individual states also assessing stock status for species such as blue crab and oyster in Mid-Atlantic bays (Miller et al. 2005, Wong 2009, Miller et al. 2011). NOAA Fisheries also examines the effects of commercial and recreational fisheries on Mid-Atlantic marine economics through the application of the NMFS Commercial Fishing Industry Input/Output and Recreational Economic Impact Models (NMFS 2018a). Data from NOAA Fisheries marine mammal stock abundance surveys are used to conduct population assessments as published in regional marine mammal stock assessment reports and peer-reviewed by regional scientific review groups (SRGs; Carretta et al. 2019, Hayes et al. 2019, Muto et al. 2019). The FWS prepares stock assessment reports for nearshore and inland threatened and protected species under its jurisdiction. Information collected on Mid-Atlantic protected species is also applied toward five-year status reviews of those listed under the Endangered Species Act (ESA).

Available data and species assessments have also been applied toward broader ecosystem models for the Mid-Atlantic, while efforts to incorporate IEA products (e.g., ESRs) into single-species stock assessments and LMR management decisions are also ongoing (Townsend et al. 2008, Link et al. 2010b, Townsend et al. 2014, Buchheister et al. 2016, 2017, Townsend et al. 2017). Multispecies and functional group production and catch at-age models, which incorporate trophic information, have been developed, while emerging economic modules also examine aspects of revenue, profitability, and fisheries portfolios (Townsend et al. 2014, 2017). A length-based multispecies model (Hydra) examines the effects of the environment on growth, maturity, fecundity, and recruitment (Gaichas et al. 2017), while components of dietary relationships and the influences of multiple fishing fleets are also being incorporated. Ecosystem modeling efforts throughout the Mid-Atlantic region are closely associated with the Northeast IEA program, which employs models that range from the single-species to ecosystem level. These include qualitative conceptual models of the Mid-Atlantic Bight that are based on submodels regarding aspects of fisheries ecology, climate, and socioeconomic components (DePiper et al. 2017). Aggregate group production models are also being used to account for trade-offs between management objectives related to fisheries and marine mammals, including the effects of human and fishery interactions and trophic relationships on their biomass (Smith et al. 2015). The effects of decreasing forage fish biomass on predatory fishes throughout the Mid-Atlantic have also been accounted for in larger models, including Ecopath with Ecosim (EwE) and Atlantis-based platforms for the Northeastern EEZ, which also examined influences of climate change and fisheries-related metrics on species functional groups (Nye et al. 2013, Buchheister et al. 2016, 2017, Fay et al. 2017, Townsend et al. 2017, McNamee 2018). A multispecies production assessment model also examines species interactions that account for predation, functional responses, and competition and how they vary spatially (Townsend et al. 2014, 2017). These efforts all build upon past foundational Multispecies Virtual Population Analysis (MSVPA) and its expansion (MSVPA-X) that were focused on Atlantic forage fishes (e.g., Atlantic menhaden, Atlantic herring *Clupea harengus*, Atlantic mackerel) and their predators (Tyrrell et al. 2008, Garrison et al. 2010). Furthermore, work by NOAA Fisheries and the ASMFC has examined Atlantic menhaden as a case species for establishing ecosystem-level reference points (Buchheister et al. 2016, 2017, ASMFC 2020a). Several models have been considered for enhancing assessments for this species and other interacting species, including single and multispecies surplus production models, and multispecies catch at age models (Townsend et al. 2017). These include proposals to develop guidance for managing Mid-Atlantic forage species at a proposed threshold of higher than 40% of their virgin biomass (Pikitch et al. 2012, Houde et al. 2014, Buchheister et al. 2017).

Efforts are also underway in the Chesapeake subregion to examine the effects of nutrient loading and its remediation on the ecosystem, including a U.S. Environmental Protection Agency (EPA) developed Chesapeake Eutrophication Model (CEM), and habitat-based models (e.g., Chesapeake Bay Fisheries Ecosystem Model, CBFEM) accounting for multispecies production and consumption as related to water quality, habitat overlap, and watershed dynamics. This information has also been synthesized with components from the EPA Chesapeake Watershed Model (CSM) toward the development of the Chesapeake Atlantis Model to account for nutrient loading, habitat dynamics, and other stressors on ecosystem functioning and multispecies interactions (Townsend et al. 2014, Ihde et al. 2016, Ihde & Townsend 2016, 2017). Information from these models is also being applied toward a broader socioeconomic understanding of this system, including efforts to estimate the commercial and recreational value of Chesapeake fisheries and the consequences of ecosystem-level effects to their revenues (Townsend et al. 2017). Delaware Bay specific models have also been developed. These include an EwE model examining the effects of marsh restoration actions on system productivity, which found an increase in total ecosystem biomass of 47.7 t km^{-2} year^{-1} (~3%; Frisk et al. 2011, Vasslides et al. 2017). Consumption rates of prominent Delaware Bay estuarine species also have been calculated via EwE modeling (Frisk et al. 2006). There are other models that inform management discussions in this region and its subregions (Polsky et al. 2000, Jones et al. 2001a, b, Koneff & Royle 2004, Hare et al. 2010a, 2010b, Link et al. 2010a), but most are not directly incorporated in or reviewed by the SAW/SARC process as noted above. Continued application of this type of information has been identified as a key factor toward making stock assessments more ecologically robust, which is essential for successful implementation of EBFM (Lynch et al. 2018, Karp et al. 2019; Link & Marshak 2019; Marshall et al. 2019).

4.4 Evaluation of Major Facets of EBFM for this Region

To elucidate how well the Mid-Atlantic ecosystem is doing in terms of LMR management and progress toward EBFM, here we characterize the status and trends of socioeconomic, governance, environmental, and ecological criteria pertinent for employing an EBFM framework for the Mid-Atlantic, and examine their inter-relationships as applied toward its successful implementation. We also quantify those factors that emerge as key considerations under an ecosystem-based approach for the region and for its two major subregions of interest when available. Ecosystem indicators for the Mid-Atlantic related to the (1) human environment and socioeconomic status, (2) governance system and LMR status, (3) natural environmental forcing and features, and (4) systems ecology, exploitation, and major fisheries of this system are presented and synthesized next. We explore all these facets to evaluate the strengths and weaknesses along the pathway:

$$PP \rightarrow B_{\text{targeted, protected spp, ecosystem}} \leftrightarrow L_{\uparrow\text{targeted spp, }\downarrow\text{bycatch}} \rightarrow \text{jobs, economic revenue.}$$

Where PP is primary production, B is biomass of either targeted or protected species or the entire ecosystem, L is landings of targeted or bycaught species, all leading to the other socioeconomic factors. This operates in the context of an ecological and human system, with governance feedbacks at several of the steps (i.e., between biomass and landings, jobs, and economic revenue), implying that fundamental ecosystem features can determine the socioeconomic value of fisheries in a region, as modulated by human interventions.

4.4.1 Socioeconomic Criteria

4.4.1.1 Social and Regional Demographics

In the U.S. Mid-Atlantic region, high coastal human population values and population densities (Fig. 4.3) have persisted from 1970 to present with values currently around 33 million residents and ~295 people km^{-2}, respectively. This region makes up 10.6% of the United States total population (year 2010), while density values have risen at ~1.2 people km^{-2} year^{-1}. Human population is heavily concentrated in the Chesapeake Bay and Delaware Bay subregions, given their proximities to the cities of Fairfax, Norfolk, and Virginia Beach (VA), Baltimore (MD), Washington (DC), Philadelphia (PA), Camden (NJ), New Castle and Wilmington (DE), and other populated coastal counties (i.e., Fairfax, Virginia Beach—VA; Anne Arundel, Baltimore, Prince George's—MD; Philadelphia, Delaware—PA; Burlington, Camden, Cape May, Cumberland, Gloucester—NJ; Kent, New Castle, Sussex—DE counties). The proximity of many of these coastal areas to New York City and Long Island makes the Mid-Atlantic ecosystem subject to considerable human population stressors. Population estimates have increased (~80.5 thousand people year^{-1}) near the Chesapeake Bay over the past 40 years, with currently up to 9.5 million residents and 184 people km^{-2} in surrounding areas, while values have remained steady near Delaware Bay (up to ~4.5 million residents). Human population density surrounding Delaware Bay, however, is one of the highest nationally at nearly 300 people km^{-2}. The Chesapeake Bay subregion encompasses 40 coastal counties in Virginia and 17 in Maryland, while the Delaware Bay subregion contains three aforementioned coastal counties in Delaware, six in New Jersey (those mentioned above and Salem County), and two in Pennsylvania.

4.4.1.2 Socioeconomic Status and Regional Fisheries Economics

High numbers of LMR establishments and employments, and high LMR-associated Gross Domestic Product (GDP) values (~$1.3 billion USD), are found throughout the Mid-Atlantic region (Fig. 4.4). Third-highest nationally, the Mid-Atlantic contains significant numbers of LMR establishments, employments,

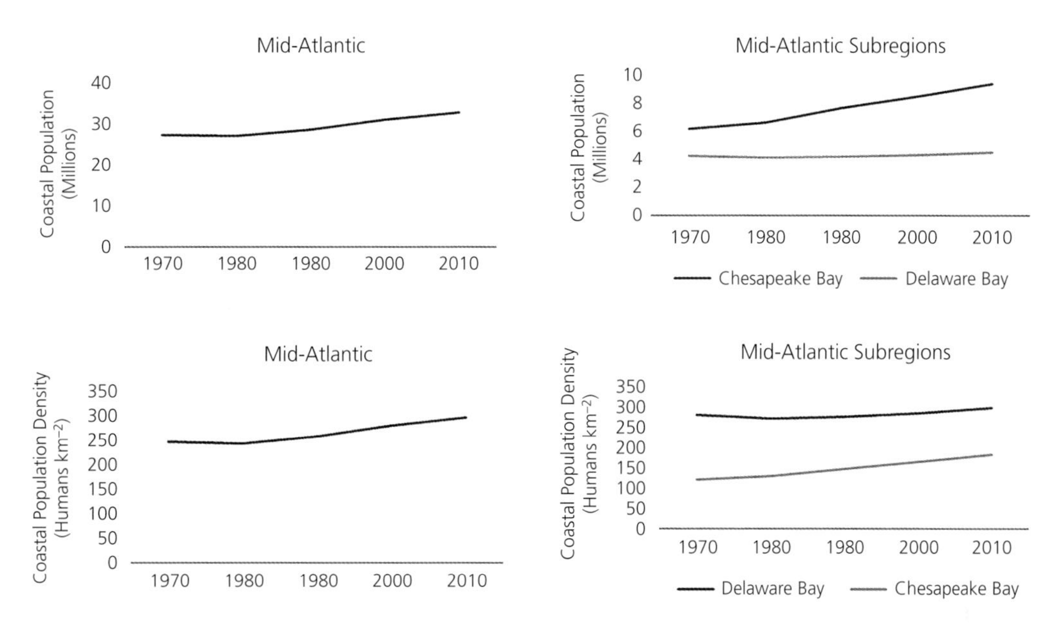

Figure 4.3 Decadal trends (1970–2010) in coastal human population (top) and population density (bottom; humans km^{-2}) for the Mid-Atlantic region and identified subregions.

Data derived from U.S. censuses and taken from https://coast.noaa.gov/digitalcoast/data/demographictrends.html.

and GDP value, which especially contribute toward the national ocean economy. Although declining since 2005, their trend has remained steady in recent years with LMR establishments continuing to comprise 3–4% of regional ocean economy establishments and LMR employments contributing ~1–2% of regional ocean employments over time. LMR-associated GDP has slowly increased over time, and contributes ~2–3% toward the regional total ocean economy. Values have remained relatively steady throughout subregions, with the highest numbers of LMR establishments and employments, and GDP value, historically occurring within Chesapeake Bay (135.3 ± 5.1 standard error, SE, establishments; 893.1 ± 45.2 SE employments; \$864.7 ± 8 million SE, USD), and more recently (2015) emerging as highest in Delaware Bay. Overall, values between subregions have been relatively similar, especially during recent years, with LMR establishments in the Chesapeake and Delaware Bays comprising 2–3% of the total ocean economy and LMR employments ~1% of a given subregional ocean economy. The LMR-associated GDP has comprised ~1–2% of their total ocean economies over time.

Since 1990, and relative to other regions, high numbers of permitted fishing vessels have operated in the Mid-Atlantic (Fig. 4.5), with values expanding up to ~25,000 during the mid- to late 2000s. In more recent years, decreases in numbers have been observed with values currently similar to those during 1999–2000 at ~15,000 vessels. Total revenue (year 2017 USD) of landed commercial fishery catches (Fig. 4.6) has increased over time as the fifth-highest in the nation, peaking at \$552 million in 2011. Among subregions, commercial catch revenues have similarly increased, with highest values (~\$300 million) observed in the Chesapeake Bay as of 2016, and peak values for Delaware Bay (~\$228 million) occurring during 2011. While trends in the recent state of the ecosystem report for the Mid-Atlantic (Gaichas et al. 2019) have shown declines in total revenue for MAFMC-managed species, the values presented here reflect commercial fisheries values reported by NOAA Fisheries for both federally and state-managed species throughout the entire Mid-Atlantic shelf.

Ratios of commercial fisheries revenue to total biomass (up to \$140/metric ton; Fig. 4.7) and jobs to biomass (up to 1.9 jobs/thousand metric tons) are second and third-highest nationally, respectively, in the Mid-Atlantic (tied with New England) where revenue/biomass values have increased and jobs/biomass has decreased over time. Ratios of the same variables noted above relative to primary production (metric tons wet weight y^{-1}) are, respectively, first- and second-highest nationally. Jobs/productivity ratios have ranged over the past decade from 13.9 to 15.0 jobs/million metric tons wet weight y^{-1}, while ratios of total

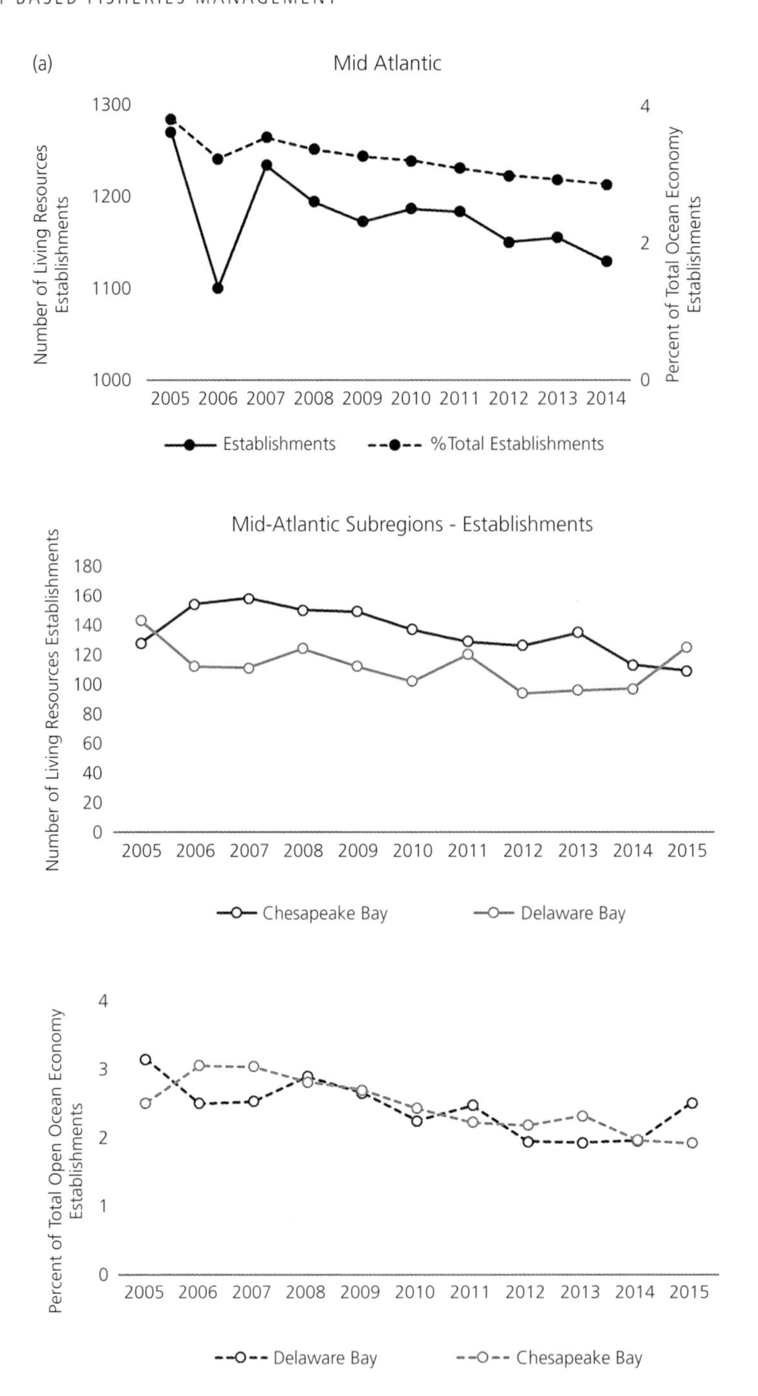

Figure 4.4a Number of living marine resources establishments and their percent contribution to total multisector oceanic economy establishments in the Mid-Atlantic region and identified subregions (years 2005–2015).

Data derived from the National Ocean Economics Program.

Figure 4.4b Number of living marine resources employments and their percent contribution to total multisector oceanic economy employments in the Mid-Atlantic region and identified subregions (years 2005–2015).

Data derived from the National Ocean Economics Program.

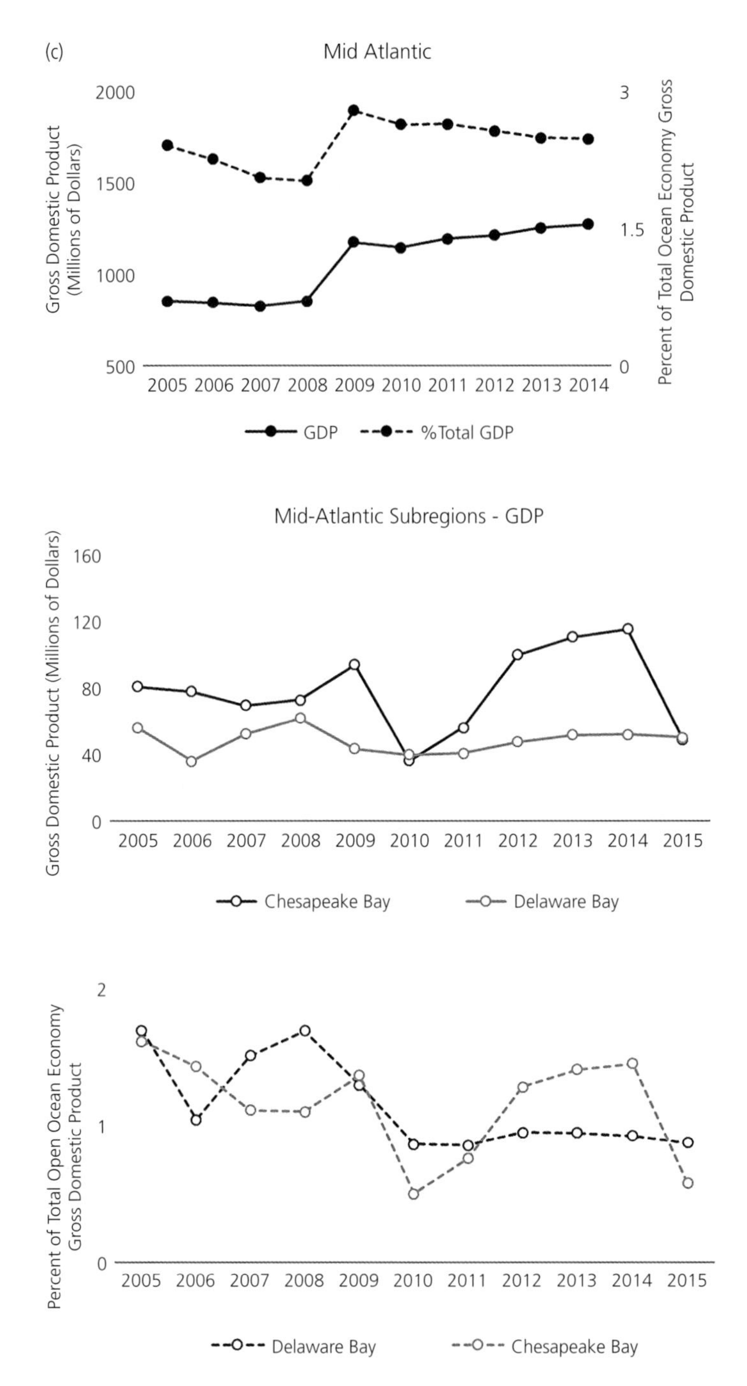

Figure 4.4c Gross domestic product value (GDP; USD) from living marine resources revenue and percent contribution to total multisector oceanic economy GDP for the Mid-Atlantic region and identified subregions (years 2005–2015).

Data from the National Ocean Economics Program.

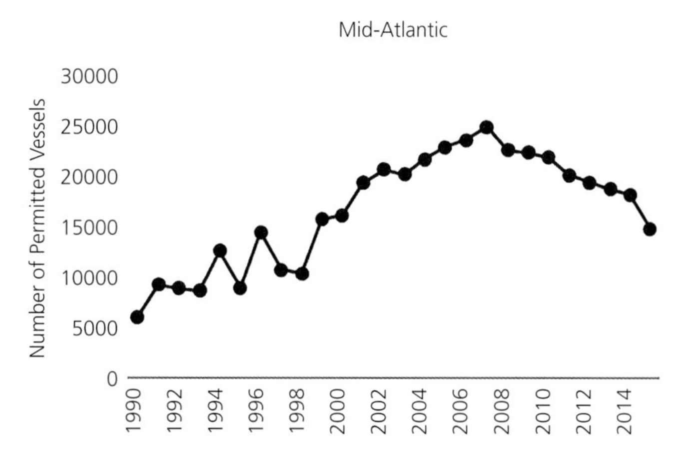

Figure 4.5 Total number of permitted vessels for the Mid-Atlantic region over time (years 1990–2015).

Data derived from NOAA National Marine Fisheries Service Council Reports to Congress.

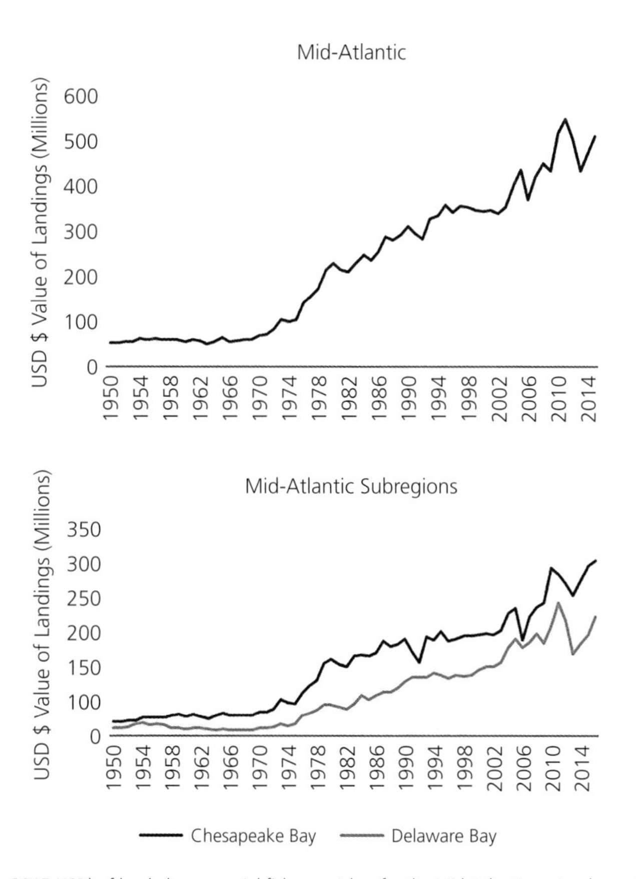

Figure 4.6 Total revenue (year 2017 USD) of landed commercial fishery catches for the Mid-Atlantic region (years 1950–2015) and identified subregions (years 1950–2016), over time.

Data derived from NOAA National Marine Fisheries Service commercial fisheries statistics.

Figure 4.7 Ratios of (a) total living marine resources employments to total biomass (thousands of metric tons); (b) total commercial landings revenue (USD) to total biomass (metric tons); (c) total living marine resources employments to total productivity (metric tons wet weight year^{-1}); (d) total commercial landings revenue (USD) to total productivity (metric tons wet weight year^{-1}) for the Mid-Atlantic region.

commercial landings revenue to productivity have increased steadily over time to ~\$1.5/metric ton wet weight y^{-1}.

4.4.2 Governance Criteria

4.4.2.1 Human Representative Context

Twenty-two federal and state agencies and organizations operate within the Mid-Atlantic and are directly responsible for fisheries and environmental management (Table 4.1). Fisheries resources throughout state waters (0–3 nautical miles) are managed by individual state agencies and the ASMFC, while those in federal waters are managed by both NOAA Fisheries (Greater Atlantic Region) and the MAFMC. The Chesapeake Bay falls under the jurisdictions of the Virginia Marine Resources Commission, Maryland Department of Natural Resources, Washington, D.C. Department of Energy and Environment—Fisheries and Wildlife Division, and the Potomac River Fisheries Commission. While not a management body, the interjurisdictional Chesapeake Bay Commission also has the power to advocate and influence management decisions regarding the Bay. The Chesapeake Bay Program also consists of state, federal, academic, and local watershed organizations to support restoration and minimize pollution and overenrichment of the Bay. Marine resources in Delaware Bay are under the jurisdictions of the Delaware Department of Natural Resources and Environmental Control (DNREC) Division of Fish and Wildlife, New Jersey Department of Environmental Protection (NJDEP) Division of Fish and Wildlife (Marine Fisheries Council), and Pennsylvania Fish and Boat Commission.

Nationwide, the Mid-Atlantic contains the highest number of Congressional representatives, which are concentrated within seven states (Fig. 4.8). Similar numbers of representatives are found in both the Chesapeake and Delaware Bays, although when standardized per mile of shoreline these values are much higher in Delaware Bay. The composition of the MAFMC has been split relatively evenly among members from the commercial and recreational fishing sector and other representatives, although expanding recreational membership has been observed starting in the early 2000s (Fig. 4.9a), with oscillating numbers of representatives from other interests. With 21 voting members, the MAFMC contains the highest number of council members nationally, while regional representa-

Table 4.1 Federal and state agencies and organizations responsible for fisheries and environmental consulting and management in the Mid-Atlantic region and subregions

Region	Subregion	Federal Agencies with Marine Resource Interests	Council	State Agencies and Cooperatives	Multijurisdictional
Mid-Atlantic		U.S. Department of Commerce **DoC** (NOAA Fisheries Northeast Fisheries Science Center **NEFSC**, Greater Atlantic Regional Fisheries Office **GARFO**, NOAA Office of National Marine Sanctuaries **NMS**, NOAA Chesapeake Bay Office **CBO**); U.S. Department of Interior **DoI** (U.S. Fish and Wildlife Service **FWS**, U.S. National Park Service **NPS**, U.S. Bureau of Ocean Energy Management **BOEM**, U.S. Geological Survey **USGS**); U.S. Environmental Protection Agency **EPA**	Mid-Atlantic Fishery Management Council **MAFMC**	Atlantic States Marine Fisheries Commission **ASMFC**; N Carolina Department of Environmental Quality—Division of Marine Fisheries; Virginia Marine Resources Commission; Maryland Department of Natural Resources; Delaware Department of Natural Resources and Environmental Control **DNREC**—Division of Fish and Wildlife; New Jersey Department of Environmental Protection **NJDEP**—Division of Fish and Wildlife (Marine Fisheries Council); Pennsylvania Fish and Boat Commission; New York Department of Environmental Conservation—Division of Fish and Wildlife	Chesapeake Bay Commission; International Commission for the Conservation of Atlantic Tunas **ICCAT**; Northwest Atlantic Fisheries Organization **NAFO**
	Chesapeake Bay	**DoC** (**NEFSC**, **GARFO**, **CBO**); **DoI** (**FWS**, **BOEM**, **NPS**, **USGS**); **EPA**	**MAFMC**	**ASMFC**; Virginia Marine Resources Commission; Maryland Department of Natural Resources; Washington, DC Fisheries and Wildlife Division; Potomac River Fisheries Commission	Chesapeake Bay Commission; **NAFO**
	Delaware Bay	**DoC** (**NEFSC**, **GARFO**); **DoI** (**FWS**, **BOEM**, **USGS**); **EPA**	**MAFMC**	**ASMFC; DNREC**; **NJDEP**; Pennsylvania Fish and Boat Commission	**NAFO**

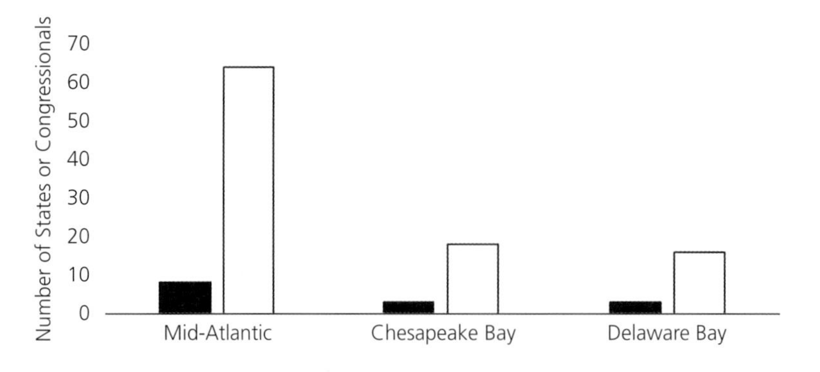

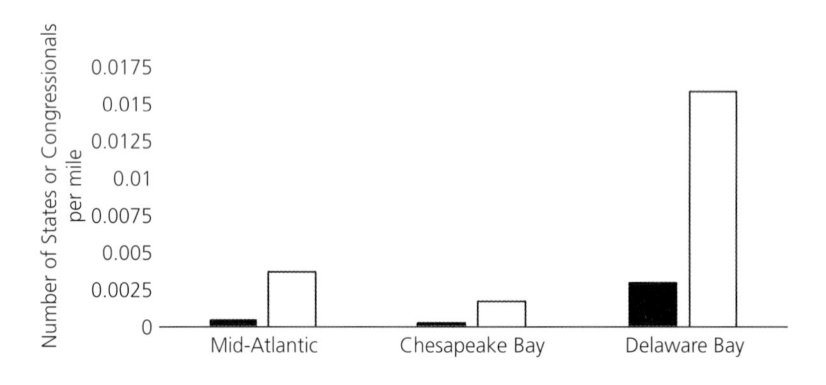

Figure 4.8 Number of states and U.S. Congressional representatives (top panel) and as standardized by miles of shoreline (bottom panel) for the Mid-Atlantic region and its identified subregions.

tives from the Mid-Atlantic also serve on the Atlantic Highly Migratory Species Advisory Panel. The Atlantic marine mammal scientific review group (SRG) is responsible for advising federal agencies on the status of marine mammal stocks covered under the Marine Mammal Protection Act (MMPA), for regions including the Mid-Atlantic. Its composition and membership have remained steady over time (Fig. 4.9b), with individuals from the academic and private institutional sectors making up the majority of its representatives.

4.4.2.2 Fishery and Systematic Context

Five current Mid-Atlantic-specific FMPs are managed by NOAA Fisheries and the MAFMC for Atlantic mackerel, squid, and butterfish; Atlantic surf clam and ocean quahog; Summer flounder, Scup, and Black sea bass; tilefish; and Atlantic bluefish (Table 4.2). Two more FMPs are shared among the MAFMC and New England Fishery Management Council (NEFMC) for Monkfish and Spiny dogfish. In total, these FMPs have been subject to 113 modifications (i.e., amendments and framework/regulatory amendments). The ASMFC manages 26 FMPs for state fisheries resources, including bony fishes (20 species), coastal sharks (22 species), horseshoe crab, and Jonah crab. In total, Atlantic statewide FMPs have been subject to 230 modifications over time. Currently, no explicitly named fishery ecosystem plan (FEP) exists for the Mid-Atlantic region; while one does exist for Chesapeake Bay, it is not currently used for any regulatory purposes. However, the MAFMC has developed an ecosystem approach to fisheries management guidance document toward its implementation that has been viewed as a functional FEP.

Two national seashores (Assateague Island and Fire Island), seven coastal national memorials or historic sites, one NOAA Habitat Focus Area (Choptank/Delaware River Complex), and four NERRs are found throughout the Mid-Atlantic region (Table 4.3). Four

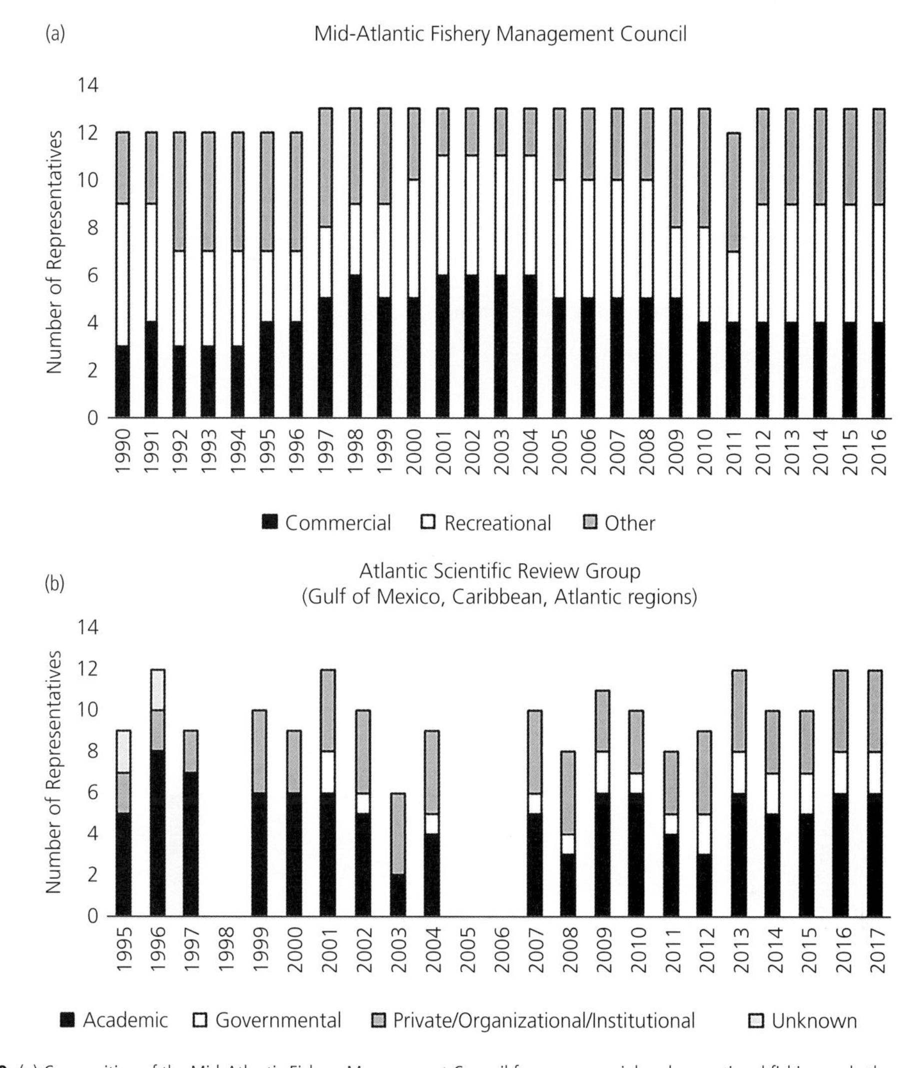

Figure 4.9 (a) Composition of the Mid-Atlantic Fishery Management Council from commercial and recreational fishing and other sectors and the (b) Atlantic marine mammal scientific review group (data unavailable 1998, 2005–2006) over time.

Data derived from NOAA National Marine Fisheries Service Council Reports to Congress and Scientific Review Group reports.

offshore canyons are identified as Habitat Areas of Particular Concern (HAPCs) in the Mid-Atlantic: Lydonia, Norfolk, Oceanographer, and Viatch. Few named fishing closure zones occur in the Mid-Atlantic, while fishing is prohibited in 51 specific areas and restricted in 159 of them (Fig. 4.10). Few of these closures occur in Mid-Atlantic subregions, of which the Chesapeake Bay contains the highest number with 13 prohibited and 77 restricted areas. Very little area of the Mid-Atlantic contains named closures or permanent fishing prohibitions (~0.6% of the EEZ), while ~289,000 km^2 is subject to certain fishing restrictions. Among subregions, the largest restricted area occurs in the Chesapeake Bay (~11,100 km^2), while ~4.7% of Delaware Bay is permanently prohibited from fishing especially due to the presence of the Cape May National Wildlife Refuge and Salem River Wildlife Management Area.

4.4.2.3 Organizational Context

Relative to the total commercial value of its managed fisheries, the MAFMC budget is mid-ranged (fourth-highest) when compared to other regions in the U.S. (Fig. 4.11), and makes up a small proportion of its overall operating costs. These relative values also increased in the early and mid-2010s and have remained relatively

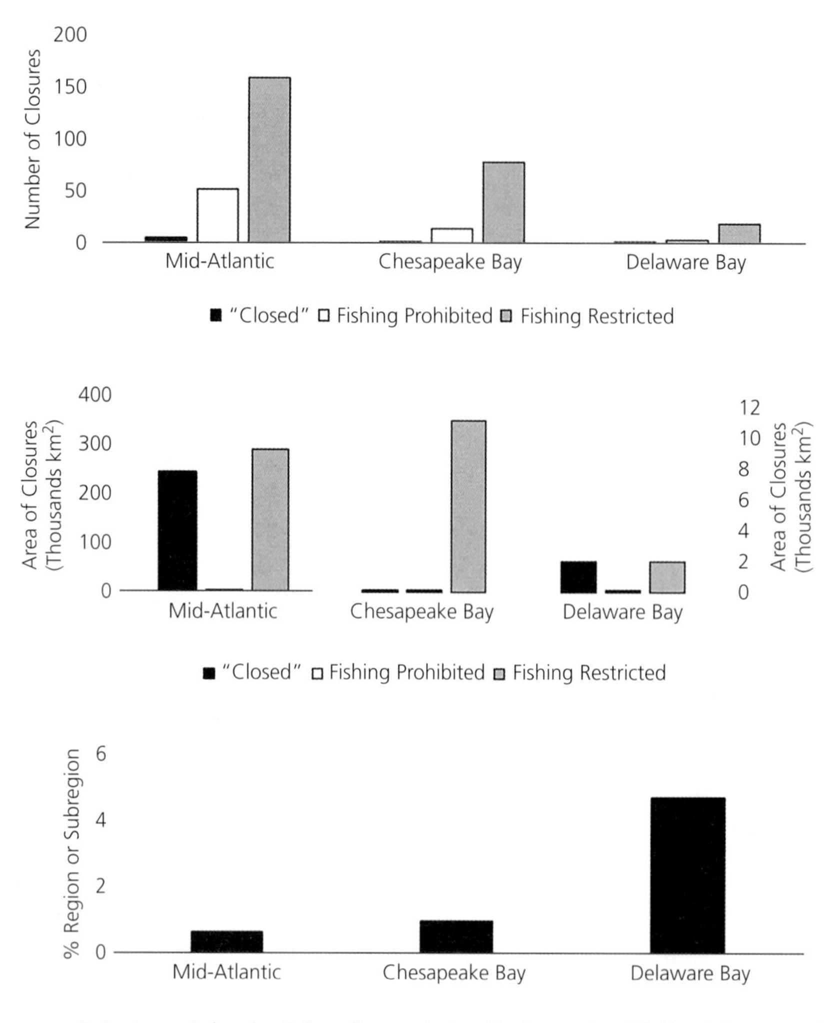

Figure 4.10 Number (top panel) and area (middle panel; thousands of km²) of named fishing closures, and prohibited or restricted fishing areas of the Mid-Atlantic region and identified subregions, including percent region or subregion (bottom panel) where commercial and/or recreational fishing is permanently prohibited. Values for Chesapeake Bay and Delaware Bay areas (middle panel) plotted on the secondary (right) axis.

Data derived from NOAA Marine Protected Areas inventory.

stable in more recent years. Cumulative National Environmental Policy Act Environmental Impact Statement (NEPA-EIS) actions from 1987 to 2016 in the Mid-Atlantic have been relatively low compared to other regions, with 48 observed over the past ~30 years (Fig. 4.12). Relatively few actions have occurred within the Chesapeake and Delaware Bays, with a cumulative peak value of 26 actions within the Chesapeake Bay subregion. Only three domestic fisheries-related lawsuits have occurred in the Mid-Atlantic region since 2010.

4.4.2.4 Status of Living Marine Resources (Targeted and Protected Species)

Nationally, the Mid-Atlantic contains the lowest number of managed species. Of the 68 federally managed species in the region, 13 are targeted fishery species covered under FMPs, while only 9 species (6 bony fishes and 3 skate species) are prohibited from harvest. Forty-six species of marine mammals, sea turtles, and fishes are also federally protected under the ESA or MMPA (Fig. 4.13, Tables 4.2, 4.4–4.5). An overlapping 46 species are also managed under the state jurisdiction of the ASMFC, of

Table 4.2 List of major managed fishery species in the Mid-Atlantic region, under state and federal jurisdictions

Council/Agency	Fishery Management Plan (FMP)	FMP Modifications	Major Species/ Species Group	Scientific Name(s) or Family Name(s)
Mid-Atlantic Fishery Management Council **MAFMC**	Atlantic mackerel, squid, and butterfish (4 species)	19 merged FMP amendments, 9 merged FMP frameworks	All Species	
		2 amendments	Atlantic mackerel	*Scomber scombrus*
		1 amendment	Squid	*Ilex* spp., *Doryteuthis* spp.
			American butterfish	*Peprilus triacanthus*
	Atlantic surf clam and ocean quahog (2 species)	2 amendments	Atlantic surfclams	*Spisula solidissima*
		18 amendments	Ocean quahogs	*Arctica islandica*
	Summer flounder, scup, and Black Sea bass (3 species)	19 amendments, 9 frameworks, 1 regulatory amendment	Summer flounder Scup Black sea bass	*Paralichthys dentatus* *Stenotomus chrysops* *Centropristis striata*
	Tilefish	1 framework, 4 amendments	Golden tilefish, Blueline tilefish	*Lopholatilus chamaelonticeps*, *Caulolatilus microps*
	Atlantic bluefish	1 framework, 5 amendments	Atlantic bluefish	*Pomatomus saltatrix*
MAFMC/ New England Fishery Management Council **NEFMC**	Monkfish	7 amendments, 9 frameworks	Monkfish	*Lophius americanus*
	Spiny dogfish	2 frameworks, 4 amendments	Spiny dogfish	*Squalus acanthias*
Atlantic States Marine Fisheries Commission **ASMFC**	American eel	4 addendum	American eel	*Anguilla rostrata*
	American lobster (co-managed with NOAA Fisheries; New England)	3 amendments, 24 addendum, 2 technical addendum	American lobster	*Homarus americanus*
	Atlantic croaker	1 amendment, 2 addendum	Atlantic croaker	*Micropogonias undulatus*
	Atlantic herring	3 amendments, 7 addendum, 2 technical addendum	Atlantic herring	*Clupea harengus*

Continued

Table 4.2 Continued

Council/Agency	Fishery Management Plan (FMP)	FMP Modifications	Major Species/ Species Group	Scientific Name(s) or Family Name(s)
Atlantic States Marine Fisheries Commission **ASMFC**	Atlantic menhaden	2 amendments, 1 r + revision, 1 supplement, 6 addendum, 2 technical addendum	Atlantic menhaden	*Brevoortia tyrannus*
	Atlantic striped bass	6 amendments, 1 supplement, 9 addendum	Atlantic striped bass	*Morone saxatilis*
	Atlantic sturgeon	1 amendment, 1 technical addendum, 4 addendum	Atlantic sturgeon	*Acipenser oxyrhynchus*
	Black drum		Black drum	*Pogonias cromis*
	Black sea bass	2 amendments, 27 addendum	Black sea bass	*Centropristis striata*
	Bluefish	1 amendment, 1 addendum	Bluefish	*Pomatomus saltatrix*
	Coastal sharks (22 species)	4 addendum	Aggregated large coastal	*Carcharhinus* spp., *Galeocerdo cuvier, Ginglymostoma cirratum, Negaprion brevirostris*
	Cobia		Cobia	*Rachycentron canadum*
		4 addendum	Hammerhead	*Sphyrna* spp.
		7 addendum	Pelagic	*Alopias vulpinus, Carcharhinus longimanus, Isurus oxyrinchus, Lamna nasus, Prionace glauca*
			Blacknose	*Carcharhinus acronotus*
			Non-blacknose small coastal	*Carcharhinus isodon, Rhizoprionodon terraenovae, Sphyrna tiburo*
			Smoothhound	*Mustelus* spp.
	Horseshoe crab		Horseshoe crab	*Limulus polyphemus*
	Jonah crab	2 addendum	Jonah crab	*Cancer borealis*
	Northern shrimp (New England)	2 amendments, 1 addendum	Northern shrimp	*Pandalus borealis*

Council/Agency	Fishery Management Plan (FMP)	FMP Modifications	Major Species/ Species Group	Scientific Name(s) or Family Name(s)
Atlantic States Marine Fisheries Commission **ASMFC**	Red drum (Atlantic coast)	2 amendments, 1 addendum	Red drum	*Sciaenops ocellatus*
	Scup	3 amendments, 29 addendum	Scup	*Stenotomus chrysops*
	Shad and river herring (3 species)	2 addendum, 3 amendments, 1 technical addendum	American shad and river herring	*Alosa aestivalis, A. pseudoharengus, A. sapidissima*
	Spanish mackerel	1 amendment, 1 addendum	Spanish mackerel	*Scomberomorus maculatus*
	Spiny dogfish	5 addendum	Spiny dogfish	*Squalus acanthias*
	Spot	1 addendum	Spot	*Leiostomus xanthurus*
	Spotted seatrout	1 amendment	Spotted seatrout	*Cynoscion nebulosus*
	Summer flounder	2 amendments, 28 addendum	Summer flounder	*Paralichthys dentatus*
	Tautog	6 addendum	Tautog	*Tautoga onitis*
	Weakfish	4 amendments, 1 technical addendum, 4 addendum	Weakfish	*Cynoscion regalis*
	Winter flounder	1 amendment, 5 addendum	Winter flounder	*Pseudopleuronectes americanus*

Table 4.3 List of major bays and islands, NOAA habitat focus areas (HFAs), NOAA national estuarine research reserves (NERRs), NOAA national marine monuments and sanctuaries (NMS), coastal national parks, national seashores, and number of habitats of particular concern (HAPCs) in the Mid-Atlantic region and subregions

Region	Subregion	Major Bays	Major Islands	HFAs	NERRs	Monuments and NMS	Coastal National Parks	National Seashores	HAPCs
Mid-Atlantic		Chesapeake Bay, Delaware Bay, Block Island Sound, Long Island Sound, Lower Bay, Raritan Bay, Sandy Hook Bay, Great South Bay, Barnegat Bay, Albermarie Sound	Outer Banks, Smith Island, Bloodsworth Island, Tangier, Long Island, Block Island	Choptank River Complex, Maryland/ Delaware HFA	Jacques Cousteau NERR, Hudson River NERR, Delaware NERR, Chesapeake Bay (Maryland/ Virginia) NERRs		Fort Raleigh National Historic Site, Wright Brothers National Memorial, Fort Monroe National Monument, George Washington Birthplace National Monument, Gloria Dei National Historic Site, Great Egg Harbor River, Gateway National Recreation Area	Assateague Island National Seashore, Fire Island National Seashore	5
	Chesapeake Bay		Kent Island, Smith Island, Bloodsworth Island, Tangier Island	Choptank River Complex, Maryland/ Delaware HFA	Chesapeake Bay (Maryland/ Virginia) NERRs		George Washington Birthplace National Monument		
	Delaware Bay		Pea patch island		Delaware NERR		Gloria Dei National Historic Site		

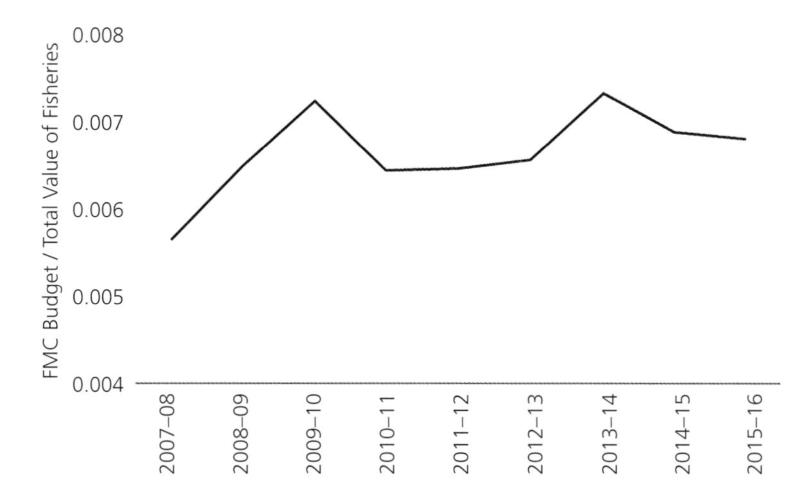

Figure 4.11 Total annual budget of the Mid-Atlantic Fishery Management Council as compared to the total value of its marine fisheries. All values in USD.

Data from NOAA National Marine Fisheries Service Office of Sustainable Fisheries and NOAA National Marine Fisheries commercial landings database.

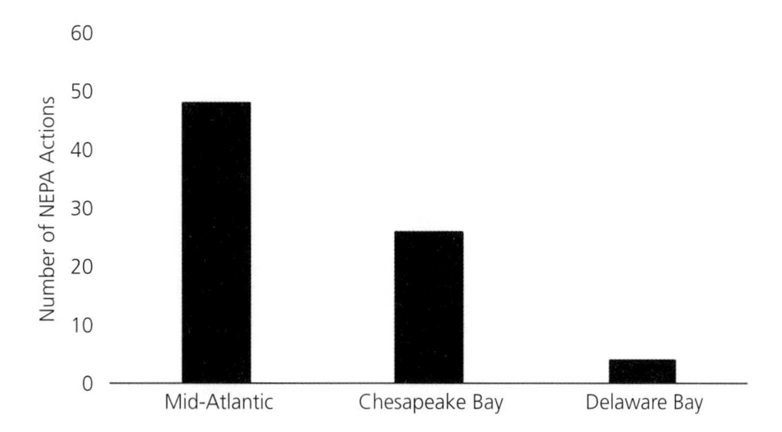

Figure 4.12 Number of National Environmental Policy Act Environmental Impact Statement (NEPA-EIS) actions from 1987 to 2016 for the Mid-Atlantic region and identified subregions.

Data derived from U.S. Environmental Protection Agency EIS database.

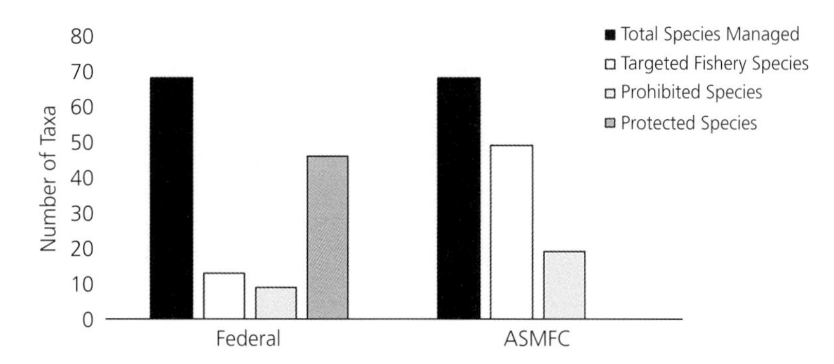

Figure 4.13 Number of managed taxa (species or families) for the Mid-Atlantic region, including federal and state (Atlantic States Marine Fisheries Commission, ASMFC) jurisdictions.

Table 4.4 List of prohibited fishery species in the Mid-Atlantic region, including federal and state jurisdictions

Council/Authority	Region	Prohibited Species	Scientific Name
Mid-Atlantic Fishery Management Council **MAFMC**; Greater Atlantic Regional Fisheries Office **GARFO**	Mid-Atlantic (9 species)	Atlantic salmon Federal Red drum Ocean pout Windowpane flounder Atlantic wolffish Barndoor skate Thorny skate Smooth skate Federal Striped bass	*Salmo salar* *Sciaenops ocellatus* *Zoarces americanus* *Scophthalmus aquosus* *Anarhichas lupus* *Dipturus laevis* *Amblyraja radiata* *Malacoraja senta* *Morone saxatilis*
Atlantic States Marine Fisheries Commission **ASMFC**	Atlantic state waters (19 species)	Coastal sharks	*Alopias superciliosus*, *Carcharhinus* spp., *Carcharias taurus*, *Carcharodon carcharias*, *Hexachus* spp., *Isurus paucus*, *Notorynchus cepedianus*, *Odontaspis noronhai*, *Rhincodon typus*, *Rhizoprionodon porosus*, *Squatina demeril*

Table 4.5 List of protected species under the Endangered Species Act (ESA) or Marine Mammal Protection Act (MMPA) in the Mid-Atlantic region

Region	Managing Agency	Protected Species Group	Common Name	Scientific Name
Mid-Atlantic (46 species)	Mid-Atlantic Fishery Management Council **MAFMC**; Greater Atlantic Regional Fisheries Office **GARFO**	Whales	Blainville's beaked whale Blue whale Cuvier's beaked whale Dwarf sperm whale False killer whale Fin whale Gervais' beaked whale Humpback whale Killer whale (Orca) Long-finned pilot whale Melon-headed whale Minke whale North Atlantic right whale Northern bottlenose whale Pygmy killer whale Pygmy sperm whale Sei whale Short-finned pilot whale Sowerby's beaked whale Sperm whale True beaked whale	*Mesoplodon densirostris* *Balaenoptera musculus* *Ziphius cavirostris* *Kogia sima* *Pseudorca crassidens* *Balaenoptera physalus* *Mesoplodon europaeus* *Megaptera novaeangliae* *Orcinus orca* *Globicephala melas* *Peponocephala electra* *Balaenoptera acutorostrata* *Eubalaena glacialis* *Hyperoodon ampullatus* *Feresa attenuata* *Kogia breviceps* *Balaenoptera borealis* *Globicephala macrorhynchus* *Mesoplodon bidens* *Physeter macrocephalus* *Mesoplodon mirus*
		Dolphins	Atlantic spotted dolphin Atlantic white-sided dolphin Common bottlenose dolphin	*Stenella frontalis* *Lagenorhynchus acutus* *Tursiops truncatus*

Region	Managing Agency	Protected Species Group	Common Name	Scientific Name
Mid-Atlantic (46 species)	Mid-Atlantic Fishery Management Council **MAFMC**; Greater Atlantic Regional Fisheries Office **GARFO**	Dolphins	Clymene dolphin	*Stenella clymene*
			Common dolphin	*Delphinus delphis*
			Fraser's dolphin	*Lagenodelphis hosei*
			Harbor porpoise	*Phocoena phocoena*
			Pantropical spotted dolphin	*Stenella attenuata*
			Risso's dolphin	*Grampus griseus*
			Rough-toothed dolphin	*Steno bredanensis*
			Spinner dolphin	*Stenella longirostris*
			Striped dolphin	*Stenella coeruleoalba*
			White-beaked dolphin	*Lagenorhynchus albirostris*
		Sea Turtles	Green sea turtle	*Chelonia mydas*
			Hawksbill sea turtle	*Eretmochelys imbricata*
			Kemp's Ridley sea turtle	*Lepidochelys kempii*
			Leatherback sea turtle	*Dermochelys coriacea*
			Loggerhead sea turtle	*Caretta*
		Fishes	Atlantic salmon	*Salmo salar*
			Atlantic sturgeon	*Acipenser oxyrinchus*
			Shortnose sturgeon	*Acipenser brevirostrum*
		Seals	Gray seal	*Halichoerus grypus*
			Harbor seal	*Phoca vitulina*
			Harp seal	*Phoca groenlandica*
			Hooded seal	*Crystophora cristata*

which 19 sharks have fishing prohibitions. There are 16 forage species and species groups (e.g., anchovies—*Engraulidae*, halfbeaks—*Hemiramphidae*, Atlantic chub mackerel—*Scomber colias*, copepods, krill, amphipods) that have been designated as ecosystem component species in all MAFMC FMPs (MAFMC 2017).

As of mid-2017, 14 stocks were federally managed in the Mid-Atlantic with each listed as a NOAA Fish Stock Sustainability Index (FSSI) stock (Fig. 4.14; NMFS 2017a, 2020).[1] Of all stocks, only one was identified as experiencing overfishing at this time (**Summer flounder**—*Paralichthys dentatus*) and none were overfished.[2] Only two stocks had unknown overfishing and overfished status (**Atlantic mackerel**—*Scomber scombrus*; **northern shortfin squid**—*Illex illecebrosus*).[3] Thus, 14.3% of all Mid-Atlantic stocks were unclassified as to whether they were overfished or experiencing overfishing, making it one of the lowest among regions in terms of the proportion of federally managed fishery stocks with unknown status.[4] For ASMFC-managed stocks, however, only seven of its Mid-Atlantic stocks are considered sustainable or rebuilt as of 2020, while five are depleted (**American eel**—*Anguilla rostrata*; **American shad**—*Alosa sapidissima*; **Atlantic sturgeon**—*Acipenser oxyrhynchus*; **River herring**—*Alosa* spp.; **Weakfish**—*Cynoscion regalis*), five are overfished (**Bluefish**—*Pomatomus saltatrix*; **Striped bass**—*Morone saxatilis*; **Tautog**—*Tautoga onitis* New Jersey/New York Bight and Delaware/Maryland/Virginia stocks; **Winter flounder**—*Pseudopleuronectes americanus*), and two are currently undergoing overfishing (**Striped bass**;

[1] As of June 2020, 16 stocks are federally managed in the Mid-Atlantic, with 14 listed as FSSI stocks. Blueline tilefish-Mid-Atlantic Coast was added to the MAFMC Tilefish FMP in 2017, and Atlantic chub mackerel—*Scomber colias*, Mid-Atlantic Coast was added to the MAFMC mackerel, squid, butterfish FMP in 2019. Both are listed as non-FSSI stocks.

[2] As of June 2020, one Mid-Atlantic stock is experiencing overfishing (Atlantic mackerel—Gulf of Maine/Cape Hatteras) and two are overfished (Bluefish—Atlantic coast; Atlantic mackerel—Gulf of Maine/Cape Hatteras).

[3] Three Mid-Atlantic stocks have unknown overfishing (Atlantic chub mackerel; Blueline tilefish; longfin inshore squid—*Doryteuthis pealeii*, Georges Bank/Cape Hatteras) or overfished status (Atlantic chub mackerel; Blueline tilefish; northern shortfin squid, northwestern Atlantic Coast) as of June 2020.

[4] In June 2020, 18.8% of all Mid-Atlantic stocks had unknown overfishing or overfished status.

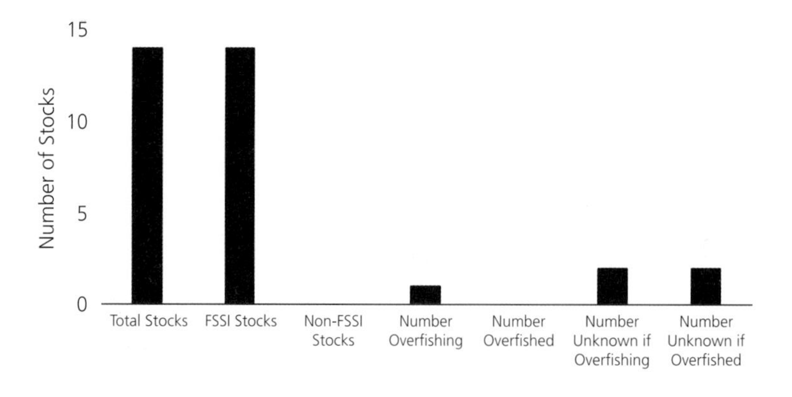

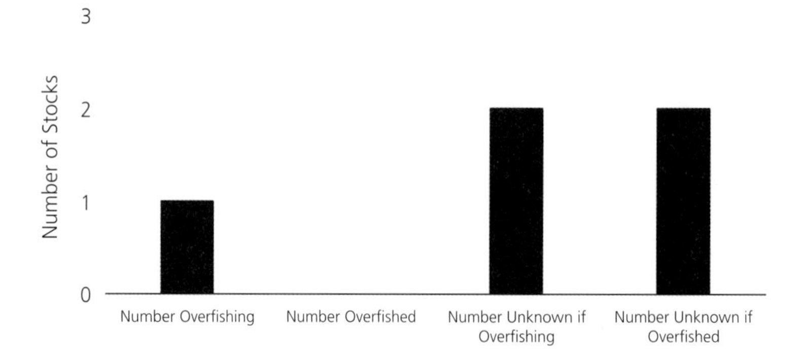

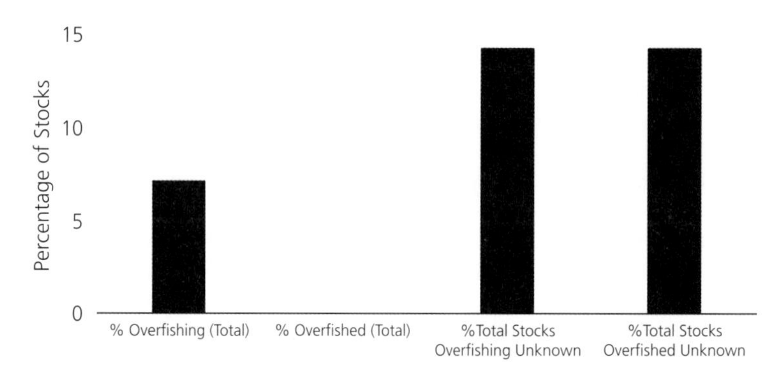

Figure 4.14 For the Mid-Atlantic region as of June 2017 (a) Total number of managed Fish Stock Sustainability Index (FSSI) stocks and non-FSSI stocks, and breakdown of stocks experiencing overfishing, classified as overfished, and of unknown status. (b) Number of stocks experiencing overfishing, classified as overfished, and of unknown status. (c) Percent of stocks experiencing overfishing, classified as overfished, and of unknown status.

Data from NOAA National Marine Fisheries Service. Note: stocks may refer to a species, family, or complex.

Tautog New Jersey/New York Bight stock; ASMFC 2020b). The protected species of the Mid-Atlantic are divided into 39 independent marine mammal stocks, and 15 ESA-listed distinct population segments, of which 80% are endangered (Fig. 4.15). Lower numbers of marine mammal stocks are found in the Mid-Atlantic than other regions (12% of total marine mammal stocks), with one-third listed as strategic marine mammal stocks. There are 28% of marine mammal stocks with unknown population size occurring in the

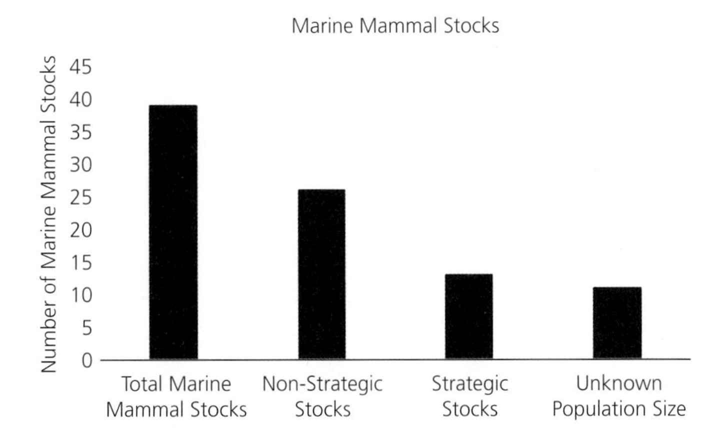

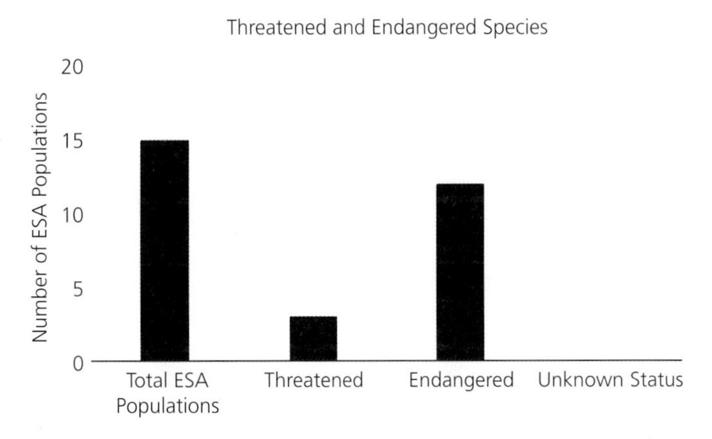

Figure 4.15 Number and status of federally protected species (marine mammal stocks, top panel; distinct population segments of species listed under the endangered species act (ESA), bottom panel) occurring in the Mid-Atlantic region.

Mid-Atlantic, which is tied with New England for fifth-highest in the nation.

4.4.3 Environmental Forcing and Major Features

4.4.3.1 Oceanographic and Climatological Context

Circulation patterns throughout the Mid-Atlantic EEZ are driven primarily by the Gulf Stream (Fig. 4.1). Connected to and originating from the Gulf of Mexico Loop Current, the Gulf Stream is the major western boundary current of the United States and spans its entire Atlantic coast. Seasonal southern counterclockwise "cold-core" and northern clockwise "warm-core" rings also develop due to the meandering of the Gulf Stream, which break off the main current and provide additional means of propagule transport along coastal regions and throughout the Atlantic EEZ. Due to high riverine input in both the Chesapeake and Delaware Bay subregions, a north to south circulation flow is observed. The oceanography of this region is also influenced by the seasonal presence of the "cold pool" in the Mid-Atlantic Bight, which extends between Cape Hatteras and southern Georges Bank off New England. This band of cold deep water exists between spring and fall, is surrounded by along-shelf and cross-shelf warmer waters, and is separated from surface waters by a seasonal thermocline (Lentz 2017), making the Mid-Atlantic Bight strongly stratified (Schofield et al. 2008). The "cold pool" and convergence of the Cape Hatteras Front (Savidge & Austin 2007) also serve as southernmost habitat for more northerly species, with their duration significantly influencing recruitment patterns (Miller et al. 2016) in addition to

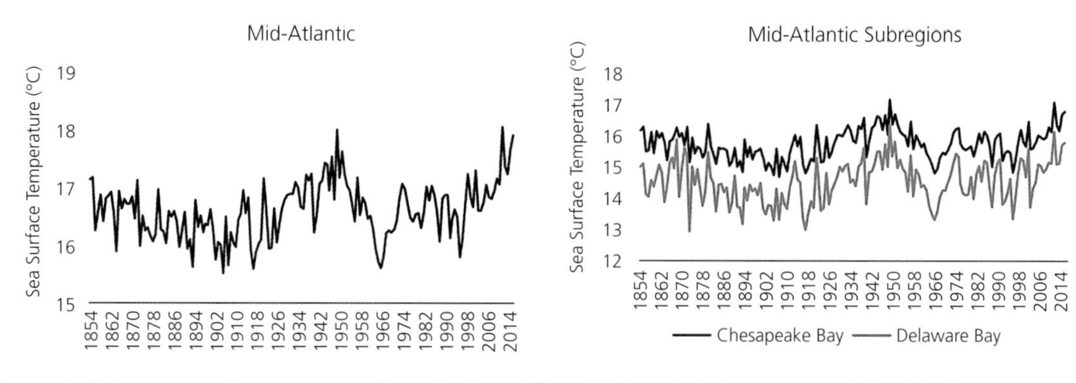

Figure 4.16 Average sea surface temperature (°C) over time (years 1854–2016) for the Mid-Atlantic region and identified subregions.

Data derived from the NOAA Extended Reconstructed Sea Surface Temperature dataset (https://www.ncdc.noaa.gov/).

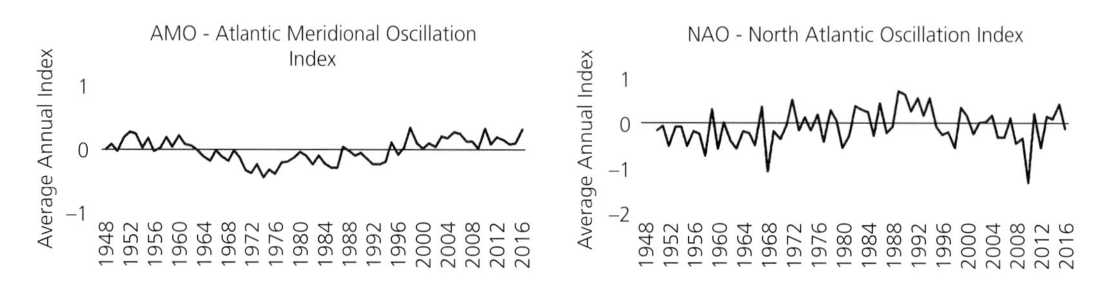

Figure 4.17 Climate forcing indices for the Mid-Atlantic region over time (years 1948–2016).

Data derived from NOAA Earth System Research Laboratory data framework (https://www.esrl.noaa.gov/).

the timing of fish migrations, plankton blooms, and storm intensities.

Clear interannual and multidecadal patterns in average SST have been observed for the Mid-Atlantic region (Fig. 4.16), with temperatures averaging 16.6°C ± 0.03 (SE) over time (1854–2016), and having increased by 0.74°C since the mid-twentieth century. Similar trends have been observed within key subregions, with average temperatures over the 162-year time period higher in the Chesapeake Bay at 15.8°C ± 0.03 (SE) and slightly lower in Delaware Bay (14.6°C ± 0.03, SE). Coincident with these temperature observations are Atlantic basin-scale climate oscillations (Fig. 4.17), including the AMO and NAO, which exhibit decadal cycles that can influence the environmental conditions and ecologies of several U.S. regions, including the Mid-Atlantic.

4.4.3.2 Notable Physical Features and Destabilizing Events and Phenomena

With two prominently identified subregions, the Mid-Atlantic contains 10 major bays (Table 4.3), primarily off the coasts of North Carolina, Virginia, Maryland, New Jersey, and New York, and 9 major islands concentrated off the coasts of North Carolina and New York, and within the Chesapeake Bay. Proportionally, the Mid-Atlantic region makes up 2% of the U.S. EEZ (~289,000 km^2), 12% of the national shoreline (~19,800 km), while containing the deepest average depth (~1800 m) for U.S. Atlantic coast regions and maximum depths of ~4.4 km (Fig. 4.18). Both Chesapeake Bay and Delaware Bay comprise ~4.5% of the Mid-Atlantic EEZ, with the Chesapeake being the larger of the two areas (~11,100 km^2) with ~10,400 km of shoreline and deeper average depths (8.8 m). Delaware Bay reaches a deeper maximum depth (41 m), but is much smaller (~2000 km^2; ~1000 km shoreline) and shallower on average (5.2 m).

In the Mid-Atlantic region, 255 hurricanes have been reported since 1850 with an average of 15 hurricanes per decade and peak numbers (~24 per decade) recorded in the 1880s, 1950s, and 2000s (Fig. 4.19). Of U.S. Atlantic coast regions, the Mid-Atlantic has been most prone to hurricanes and their continued trajectories following landfall. While at lower intensities,

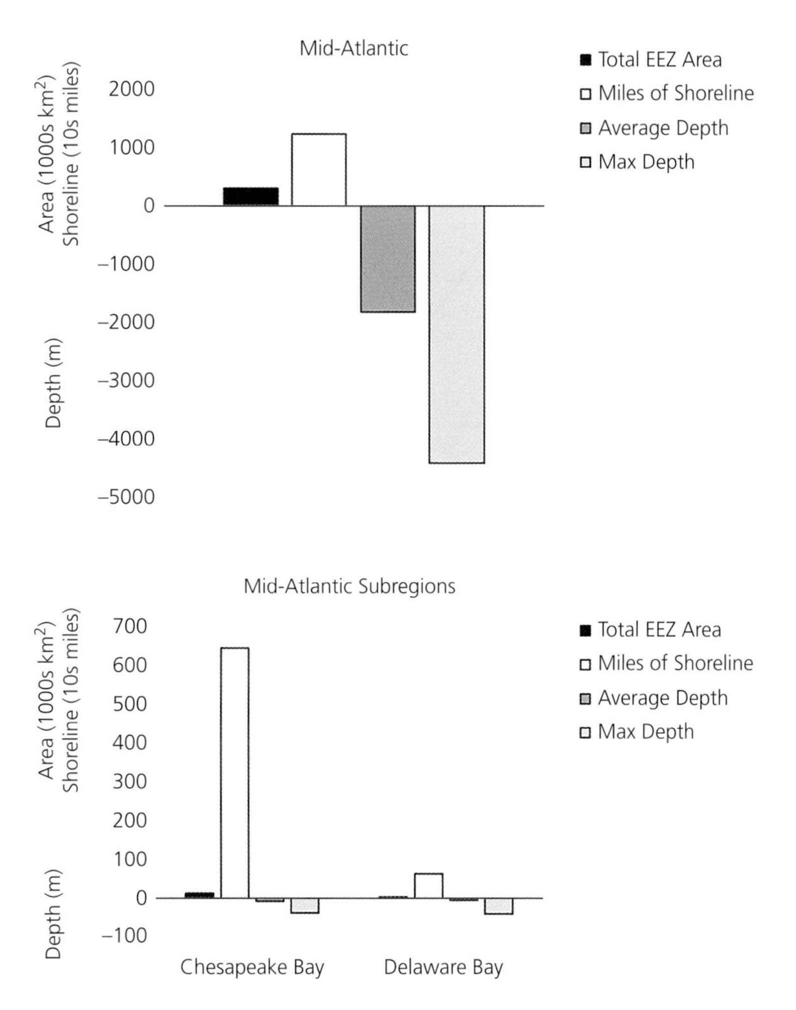

Figure 4.18 Total shelf area, miles of shoreline, and average and maximum depth of marine regions and subregions of the Mid-Atlantic. EEZ = Exclusive Economic Zone

these peak trends have also been observed for the Chesapeake Bay (up to 8 hurricanes/decade), and more recently in Delaware Bay (up to 4 hurricanes during the 2000s).

4.4.4 Major Pressures and Exogenous Factors

4.4.4.1 Other Ocean Use Context

Ocean uses in the Mid-Atlantic are currently most associated with the tourism/recreation, marine transportation, and ship and boatbuilding sectors (Fig. 4.20), with offshore wind energy development becoming a greater issue for this region. Among sectors, trends for Mid-Atlantic Ocean establishments and employments have increased over the past few decades, with tourism establishments and employments comprising 86–89% of total Mid-Atlantic Ocean establishments and 73–76% of total Mid-Atlantic Ocean employments. Highest values for total ocean metrics occurred in 2016 at ~41,000 establishments and ~790,000 employments (nationally highest). The GDP from tourism (including recreational fisheries) is highest among sectors, contributing 53–60% toward total Mid-Atlantic Ocean GDP. Over the past few decades, especially during the 2010s, ocean GDP has also increased, with most recent estimated values at ~$57.2 billion. Overall, the Mid-Atlantic contributes toward 15% of the U.S. total ocean GDP, which is third-highest nationally. Establishments and employments in the Mid-Atlantic marine transportation sector contribute 5–6% toward total ocean establishments and comprise 14–18% of total ocean employments. The Mid-Atlantic

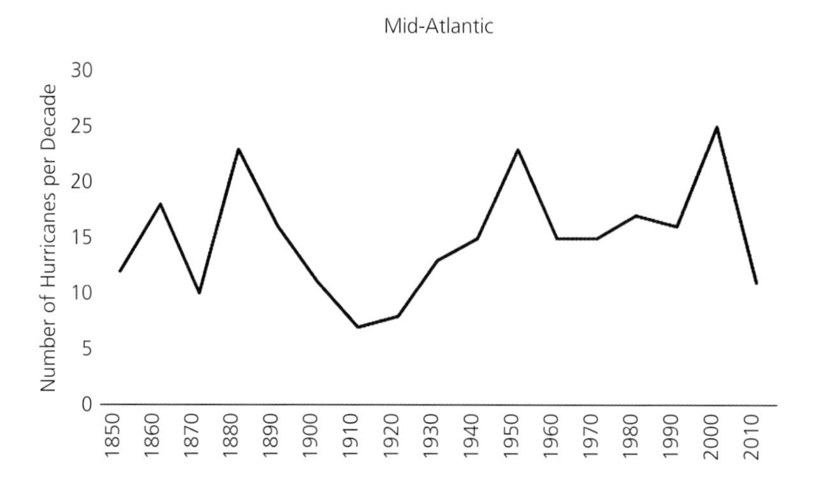

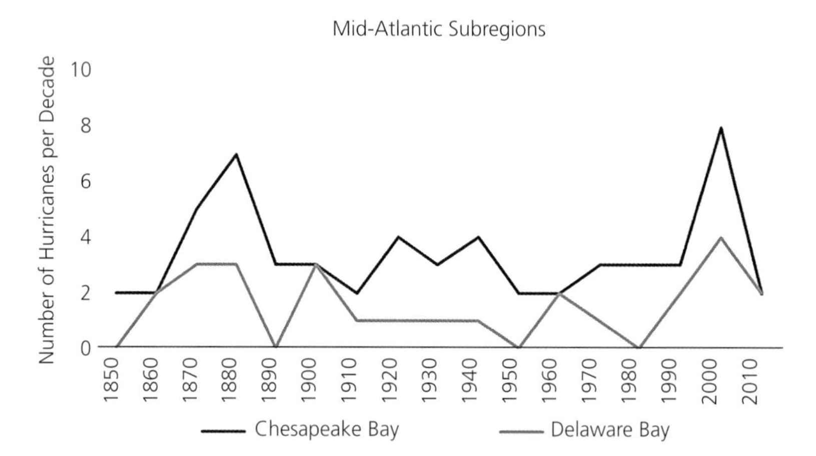

Figure 4.19 Total number of hurricanes recorded per decade for the Mid-Atlantic region (top panel) and identified subregions (bottom panel) since year 1850.

Data derived from NOAA Office for Coastal Management (https://coast.noaa.gov/hurricanes/).

contributes 31% toward U.S. tourism establishments and ~28% of national tourism employments and GDP, all of which are highest in the nation. It is also the second-highest nationally for marine construction establishments (24% of U.S. marine construction establishments) and marine transportation establishments and employments (26–28% of U.S. marine transportation establishments and employments). This region is second-highest for ship and boatbuilding GDP (26% of U.S. ship and boatbuilding GDP). Overall, the Mid-Atlantic is high-ranking in terms of its marine socioeconomic status, especially as associated with other ocean uses and its LMR economy.

The degree of tourism, as measured by number of dive shops (fifth-highest), number of major ports and marinas, and number of cruise ship departure passengers is mid-range nationally for the Mid-Atlantic (Fig. 4.21), with substantial numbers of dive shops and moderate numbers of major ports and marinas, which are subject to considerable maritime commerce. Relatively fewer dive shops, ports, or marinas are found throughout the key subregions, with most concentrated near the Chesapeake Bay (24 dive shops, 4 major ports, and 11 major marinas), while 8 major ports and 1 major marina are found along Delaware Bay. Cruise departure passengers are prominent in the

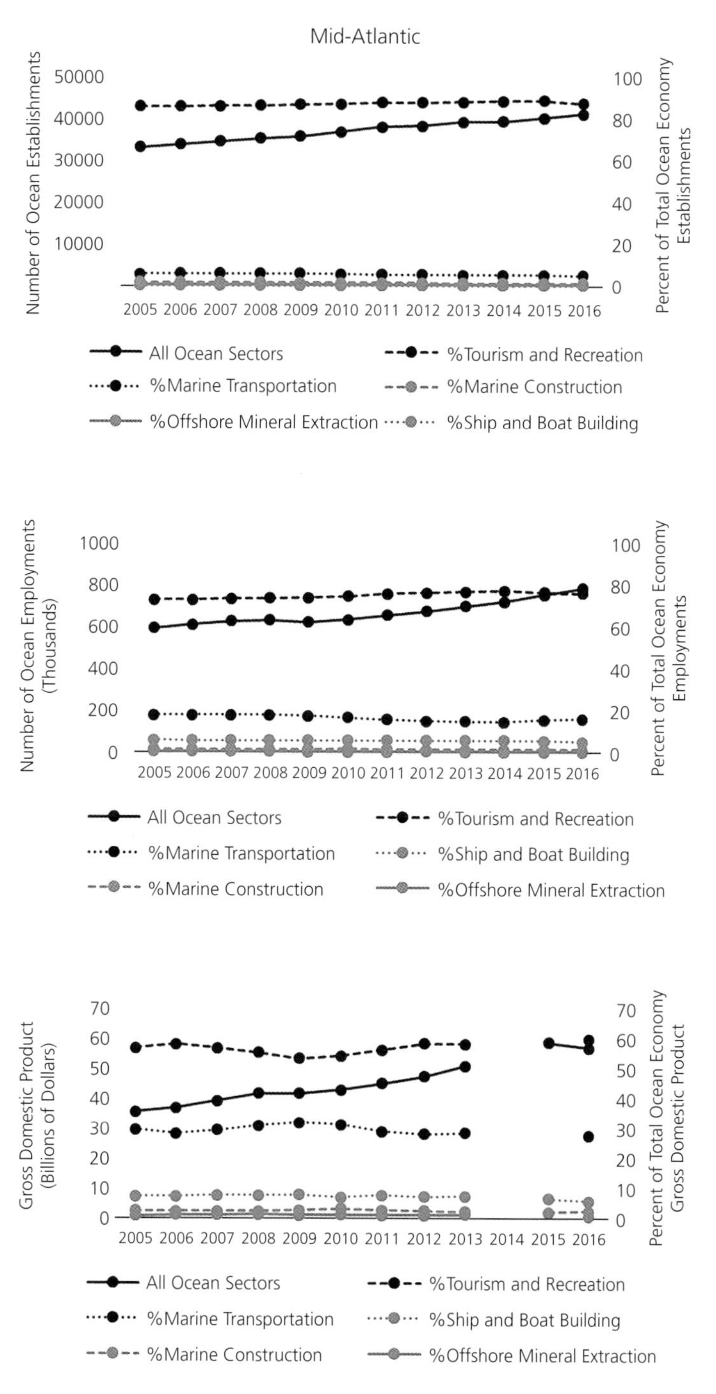

Figure 4.20 Number and percent contribution of multisector establishments, employments, and Gross Domestic Product (GDP) and their percent contribution to total multisector oceanic economy in the Mid-Atlantic region (years 2005–2016).

Data derived from the National Ocean Economics Program.

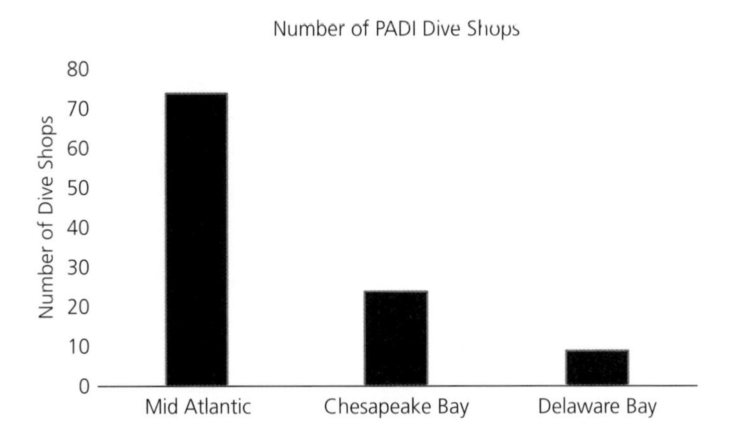

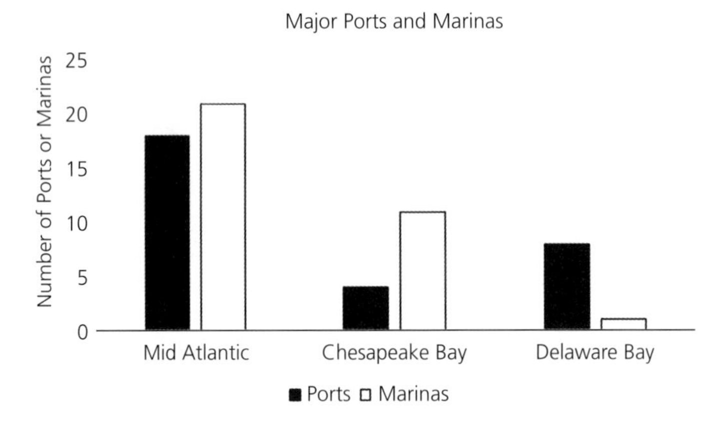

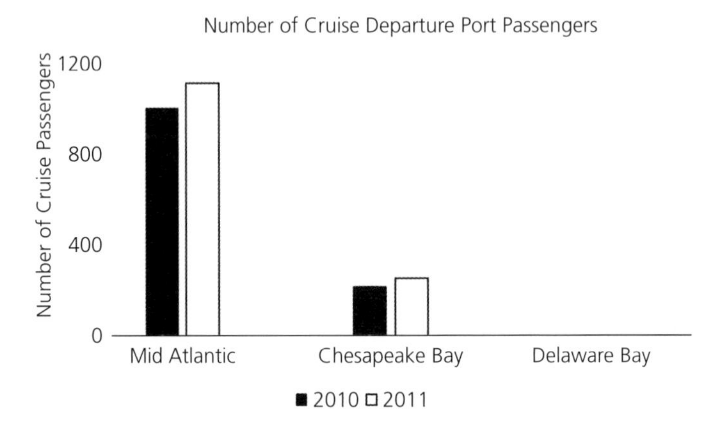

Figure 4.21 Tourism indicators for the Mid-Atlantic region and identified subregions: (a) Total number of Professional Association of Diving Instructors (PADI) dive shops. Data derived from http://apps.padi.com/scuba-diving/dive-shop-locator/. (b) Number of major ports and marinas. (c) Number of cruise departure port passengers (years 2010 and 2011; values in thousands).

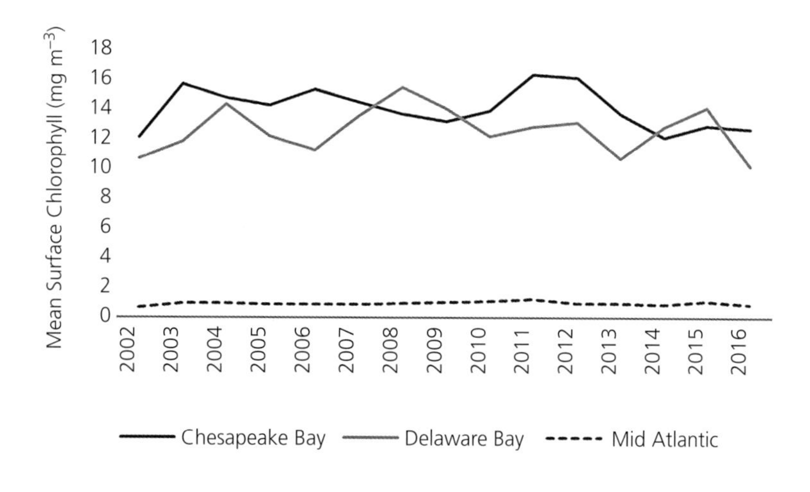

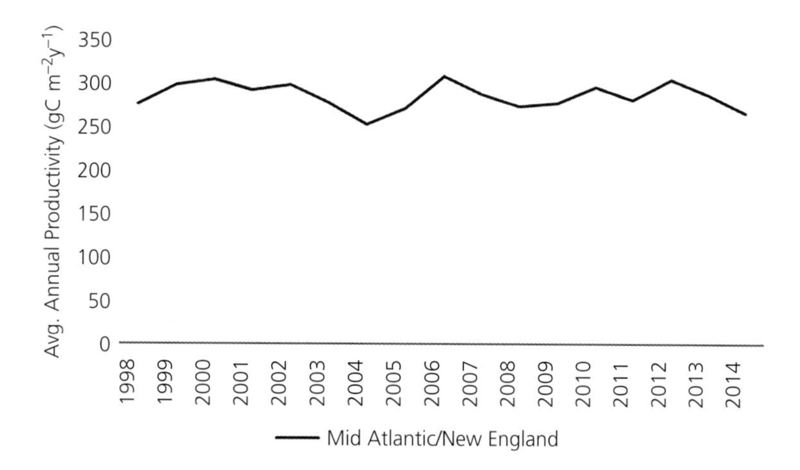

Figure 4.22 Mean surface chlorophyll (mg m^{-3}; top panel) for the Mid-Atlantic region and identified subregions and average annual productivity (grams carbon m^{-2} $year^{-1}$; bottom panel) for the Mid-Atlantic and New England regions over time.

Data derived from NASA Ocean Color Web (https://oceancolor.gsfc.nasa.gov/) and productivity calculated using the Vertically Generalized Production Model—VGPM.

Mid-Atlantic region (fourth-highest nationally), with 200–250 thousand departure passengers observed in the Chesapeake Bay region annually. While no active offshore oil rigs are found throughout the U.S. Atlantic continental shelf, the Mid-Atlantic region has emerged as the leading region for offshore wind energy production with nine identified U.S. Bureau of Ocean Energy Management (BOEM) sites. Currently, construction and Operational Plans (COPs) are set to be approved for two Mid-Atlantic wind energy sites (François 2018).

4.4.5 Systems Ecology and Fisheries

4.4.5.1 Basal Ecosystem Production

Since 2002, mean surface chlorophyll-a values throughout the Mid-Atlantic EEZ have been relatively low at ~1.00 ± 0.03 (SE) mg m^{-3} (Fig. 4.22). Comparatively, nearshore subregions subject to higher freshwater input and nutrient loading have averaged 14.0 ± 0.4 (SE) mg m^{-3} in the Chesapeake Bay and 12.6 ± 0.4 (SE) mg m^{-3} in Delaware Bay, with higher chlorophyll-a values generally observed for Delaware Bay from 2014 to 2015. Since 2002, average annual productivity within

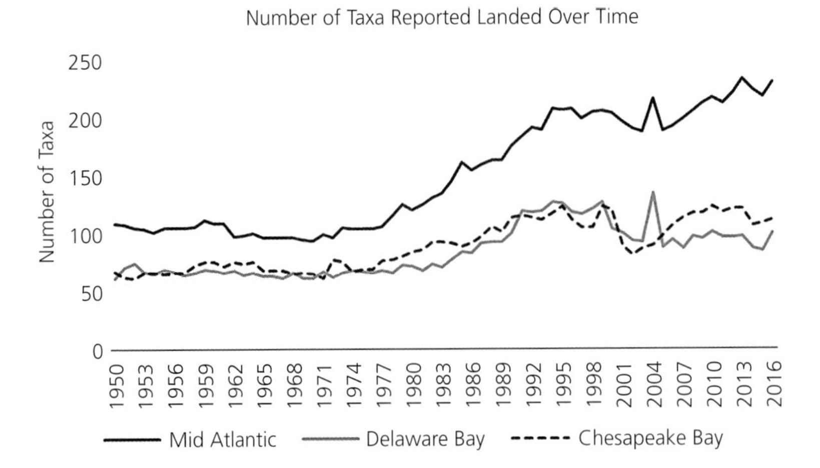

Figure 4.23 Number of taxa reported captured by commercial and recreational fisheries in the Mid-Atlantic region and identified subregions. Data derived from NOAA National Marine Fisheries Service commercial and recreational fisheries statistics.

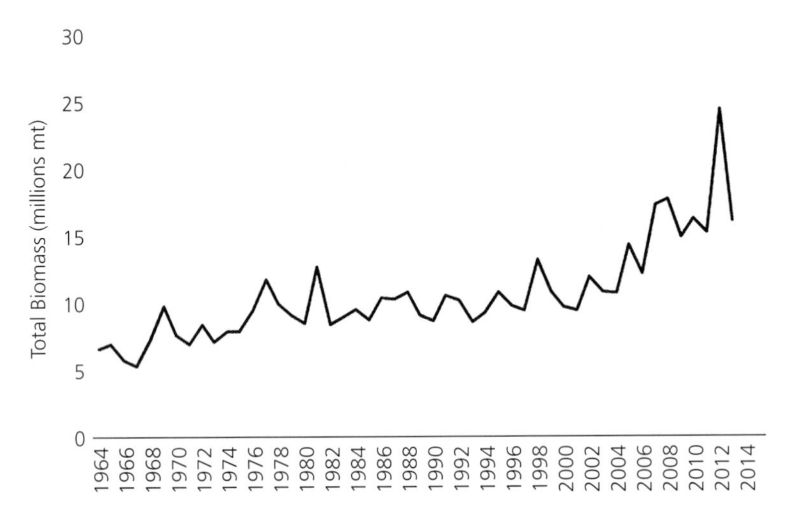

Figure 4.24 Total combined surveyed fish and invertebrate biomass (metric tons; mt) for the Mid-Atlantic and New England regions, over time.

the U.S. Northeastern (Mid-Atlantic-New England) EEZ has remained relatively steady between 250–300 g C m^{-2} $year^{-1}$ (second-highest nationally).

4.4.5.2 System Exploitation

The highest numbers of reported landed taxa in the nation have been captured in the Mid-Atlantic region over time, with values continuing to increase since the 1980s and peaking at 235 taxa in 2013 (Fig. 4.23). Among subregions, similar values have been observed over time for both the Chesapeake and Delaware Bays (average: 85–90 taxa $year^{-1}$), with values currently slightly higher in the Chesapeake subregion since 2005 (average: ~114 taxa landed) as compared to Delaware Bay (average: ~95 taxa landed).

Total NEFSC surveyed fish and invertebrate biomass for the U.S. Northeast (Mid-Atlantic-New England) region has remained relatively steady over time (Fig. 4.24), with gradual increases observed from the 1960s to 1990s from 6 to 13 million metric tons per year, and more prominent increases occurring during the 2000s up to 25 million metric tons per year. We acknowledge the composition of this surveyed biomass may have changed, but the overall magnitude

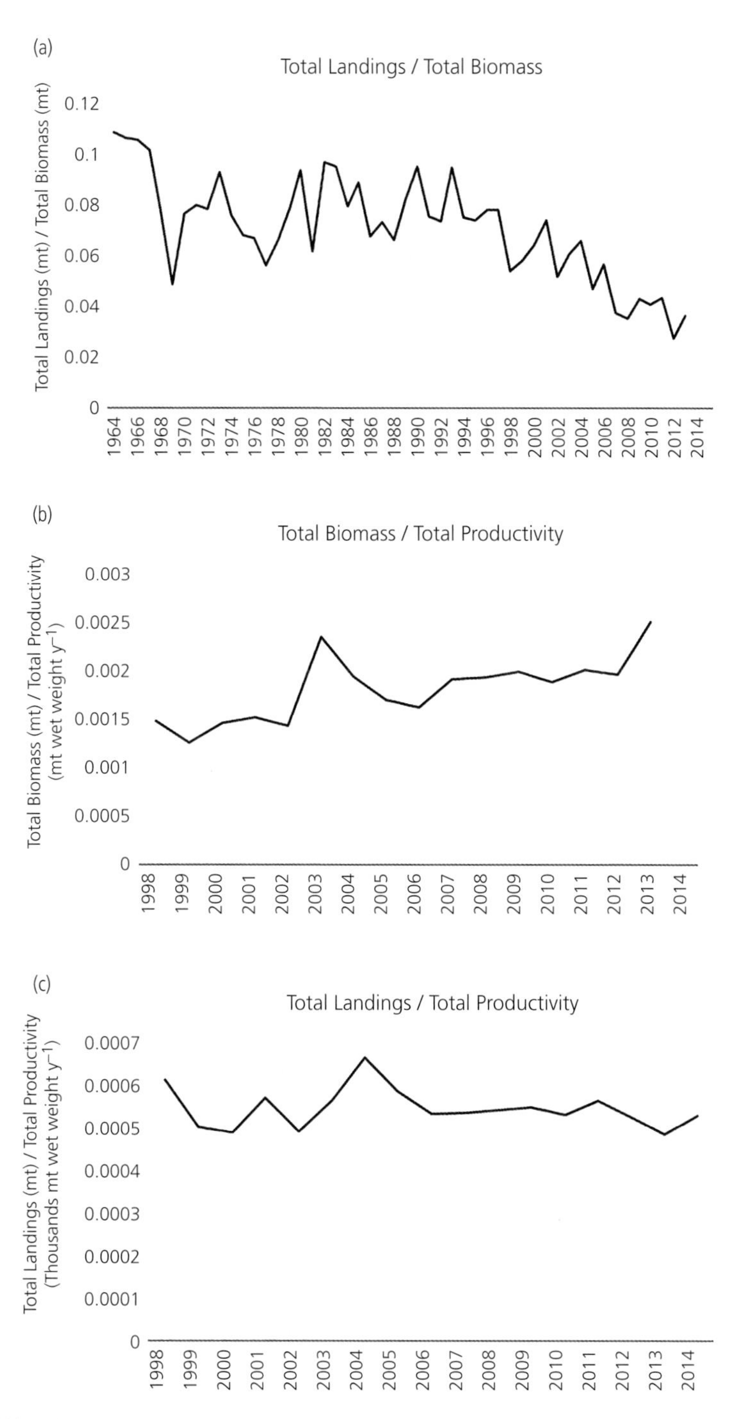

Figure 4.25 Ratios of (a) total reported commercial and recreational landings (metric tons) to total biomass (metric tons, i.e., exploitation index); (b) total biomass (metric tons) to total productivity (metric tons wet weight year^{-1}); (c) total reported commercial and recreational landings (metric tons) to total productivity (metric tons wet weight yr^{-1}) for the Mid-Atlantic region.

has remained relatively consistent for the ecosystem (NEFSC 2006, 2012, Gaichas et al. 2019). During 2013 however, values decreased to ~16 million metric tons. U.S. northeast biomass values are fourth-highest on average in the nation and have been observed increasing over time. This trend has been partially associated with successful fisheries management in the Mid-Atlantic Bight survey regions, ongoing rebuilding plans, and increasing restrictions on fishing pressure throughout the region. Exploitation rates (i.e., landings/biomass ratio) in the U.S. northeast are second-highest nationally, and have ranged from 0.02 to 0.1 over time, with values fluctuating and then decreasing from the 1980s onward (Fig. 4.25). Values for total landings/productivity (up to 0.0006) on average are highest in the nation and have remained relatively constant, while total biomass/productivity values (up to 0.0025) have gradually increased over the past two decades.

4.4.5.3 Targeted and Non-Targeted Resources

Commercial and recreational fisheries landings are fourth-highest in the nation for the Mid-Atlantic region (Fig. 4.26). They have exhibited declines from previous peak values in the 1960s to remain around 200–400 thousand metric tons since the 1970s. Decreases in the proportion of recreational catch were observed during the 1980s, with recreational fisheries

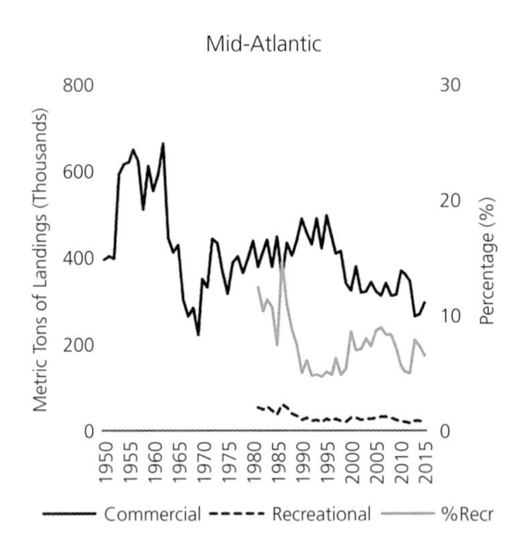

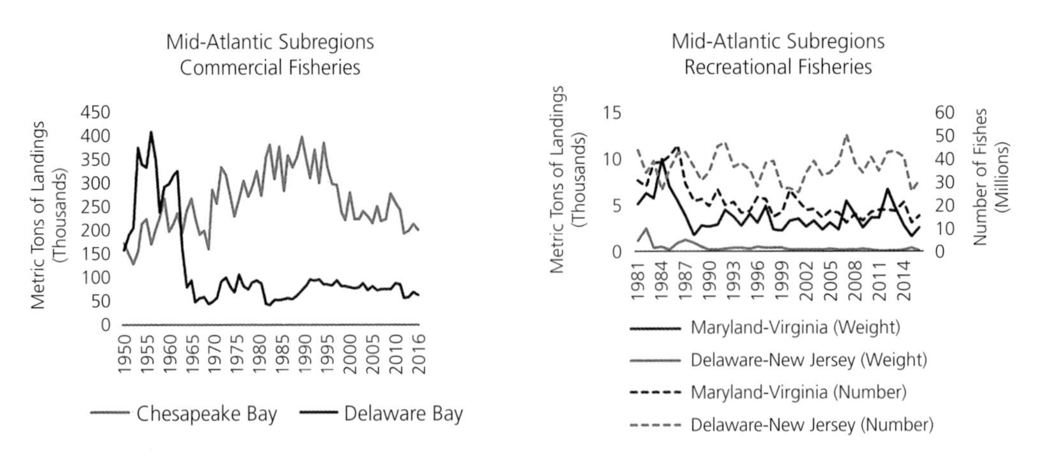

Figure 4.26 Total reported commercial and recreational landings and percent contribution of recreational landings (%Recr) over time for the Mid-Atlantic region and identified subregions (commercial: 1950–2016; recreational: 1981–2016).

Data derived from NOAA National Marine Fisheries Service commercial and recreational fisheries statistics.

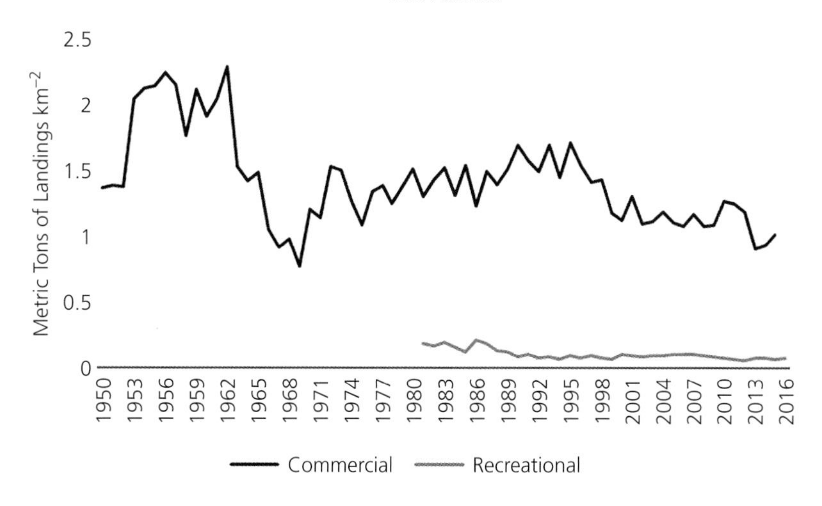

Mid-Atlantic Subregions

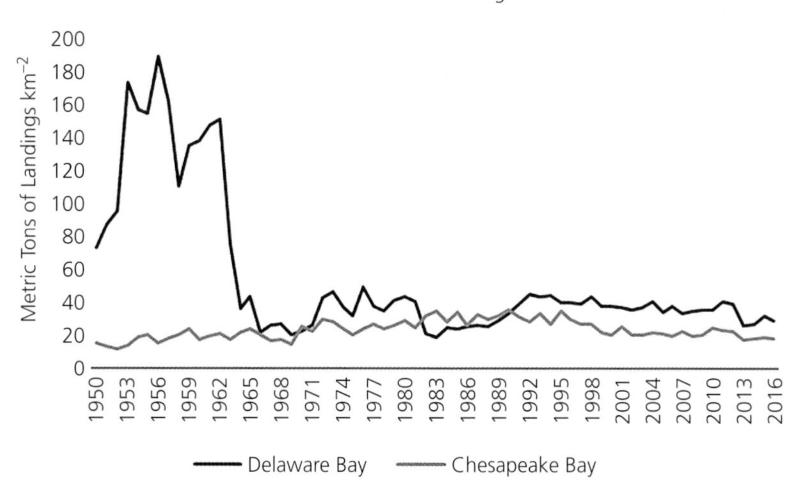

Figure 4.27 Total reported commercial and recreational landings (metric tons) per square kilometer over time (1950–2016) for the Mid-Atlantic region and identified subregions.

Data derived from NOAA National Marine Fisheries Service commercial and recreational fisheries statistics.

continuing to contribute 5–10% of total landings. Among subregions, commercial landings were initially higher for the Delaware Bay region during the 1950s–1960s and followed by significant declines. Landings remained relatively steady in the Chesapeake Bay region until the 1990s during which decreases occurred and values returned to those observed in the 1950s. Recreational fisheries catches (reported by weight) have been lower in Delaware Bay than in Chesapeake Bay, although reported numbers of fishes landed over time are higher than those landed in Chesapeake Bay. Pronounced decreases in recreational landings (by weight) were observed in Chesapeake Bay during the 1980s, associated with a moratorium on striped bass during this period (Secor et al. 1995), with values having remained around 2000–5000 metric tons in recent decades. When standardized per square kilometer (Fig. 4.27), Mid-Atlantic landings

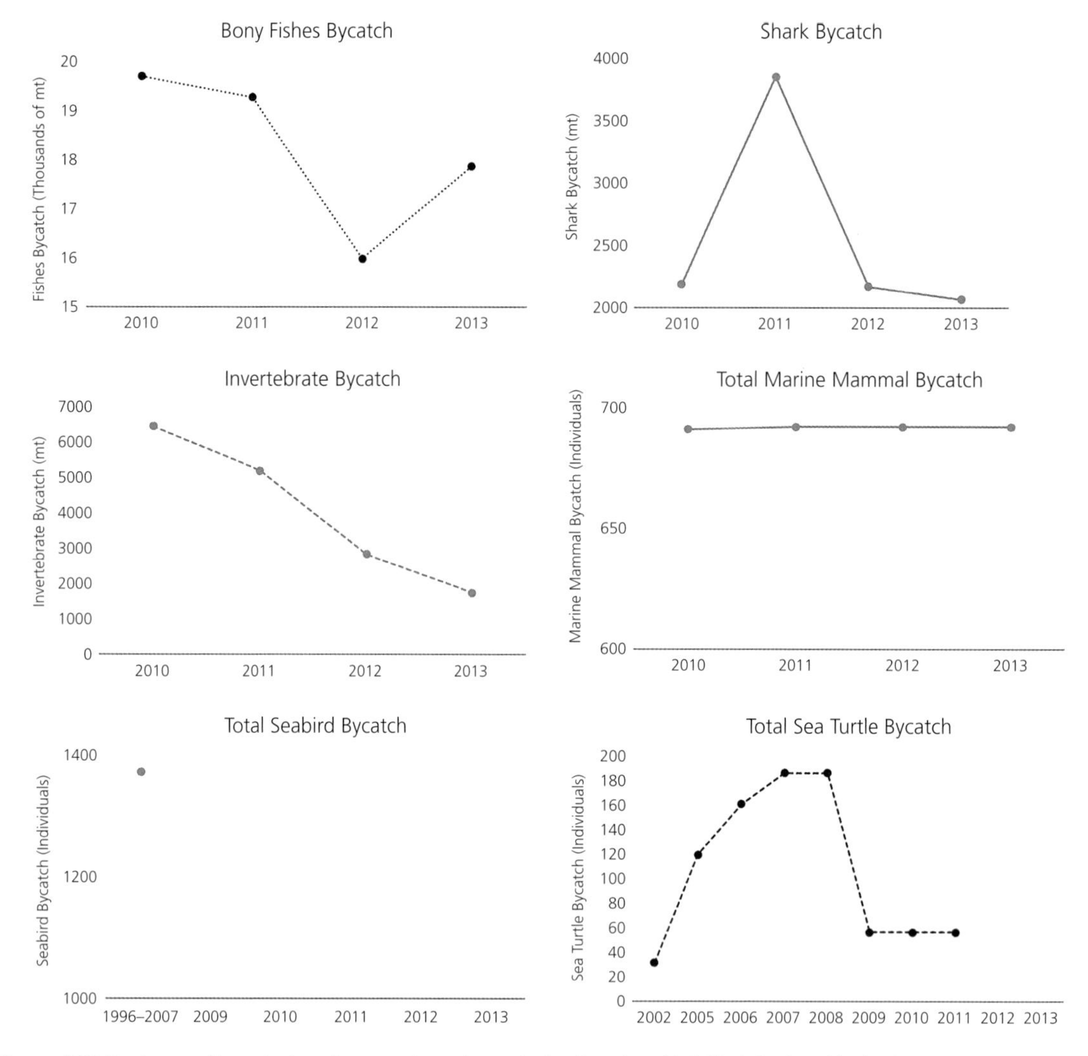

Figure 4.28 Total reported bycatch of species groups in metric tons (mt) and number of individuals for the Mid-Atlantic region.

Data derived from NOAA National Marine Fisheries Service U.S. National Bycatch Report Database System. Note: Mid-Atlantic seabird bycatch is only reported for the 1996–2007 time period as a combined estimate.

follow very similar patterns to total landings values. This region contains the second-highest concentration of fisheries landings over time in the nation. Intense fishing pressure (up to 200 metric tons km^{-2}) was observed in Delaware Bay, leading to pronounced decreases in fisheries landings during the mid-1960s. Per unit area, fishing landings in the Chesapeake Bay have remained relatively constant, at similar values to current Delaware Bay areal landings.

Although values have decreased over time, bycatch continues to persist throughout the Mid-Atlantic region (Fig. 4.28). By weight, bycatch is most pronounced for bony fishes, invertebrates, and sharks, whereas marine mammal and seabird bycatch has been most dominant for those groups reported by number. Bycatch for the Mid-Atlantic region is comparatively mid-range overall in the nation, although marine mammal bycatch is second-highest nationally and has remained relatively constant. Notably, decreases in bycatch for bony fishes, sharks, invertebrates, seabirds, and sea turtles have been observed over time.

4.5 Synthesis

The Mid-Atlantic is an environment with moderate numbers of managed species, and one that is subject to stressors that include habitat loss, coastal development, nutrient loading, climate-related species range shifts, hurricanes, other ocean uses, and proliferation of invasive species. The consequences of past overfishing continue to affect state and some federally managed LMRs in this region (Smith 1985, Packer & Hoff 1999, Richards & Rago 1999, Steimle & Zetlin 2000, Kirby 2004, Grout 2006, Mann & Powell 2007, Babcock 2008, Sartwell 2009, Hale et al. 2016, Terceiro 2016, Gaichas et al. 2019, NMFS 2019a) in concert with some shifting fishing effort in response to range expansions of historically targeted species toward northern limits of this system (Nye et al. 2009, Lucey & Nye 2010, Pinsky & Fogarty 2012, Nye et al. 2013, Ribeiro et al. 2015, Hare et al. 2016, Kleisner et al. 2016, Morley et al. 2017, 2018, Gaichas et al. 2019). Despite early overharvesting in the 1960s and 1970s, total commercial fisheries landings and revenue in this system have been maintained at what are likely sustainable levels, especially for federally managed species. However, catches in Chesapeake and Delaware Bays continue to remain much lower than in previous decades, while several state-managed stocks remain overfished or depleted. Stressors in these estuarine systems particularly associated with climate change, eutrophication, habitat loss, and partial displacements of native species by invasive taxa continue to affect LMR production in this region (Joseph 1972, Chambers et al. 1999, Najjar et al. 2000, Rogers & McCarty 2000, Neff et al. 2000, Roman et al. 2000, Kennish 2001, Koch & Orth 2003, Kirby 2004, Kemp et al. 2005, Montane & Austin 2005, Orth et al. 2006, Grabowski & Peterson 2007, Alber et al. 2008, Steadman & Dahl 2008, Anthony et al. 2009, Najjar et al. 2010, Carruthers et al. 2013, Ezer et al. 2013, Patrick et al. 2014). The Mid-Atlantic contains the highest human population density in the country, with increasing tourism and other ocean uses also occurring throughout this region. While contributing toward its nationally significant tourism, marine transportation, and LMR-based economies, pressures from fishing and other ocean uses also continue to affect this system and its subregions. Shifts in the distribution and concentration of commercial and recreational fishing effort in this system continue to occur as a result of regional warming, with continued seafood demand and tradeoffs among marine sectors are predicted to intensify with the increasing human population. With these consequential and interrelated marine sectors and stressors, enhanced ecosystem-based strategies for the Mid-Atlantic and its large prominent semi-enclosed subregions enable consideration of influential and emerging factors on its coupled socioecological systems. Much progress has been made in creating advanced ecosystem-level measurements, with consideration of these components in sophisticated models and quantitative frameworks. However, management actions that set system-wide limits and control rules as related to the effects of these ongoing and emerging stressors on basal ecosystem productivities, fisheries biomass, and marine socioeconomics are still needed for this system. Although current calculations suggest that fishing, revenue, and employments are aligned with biomass and productivity limits, continued stressors and increasing human pressures are likely to affect their thresholds and warrant greater consideration on their system-wide effects.

The Mid-Atlantic ecosystem is well known for aspects of its LMRs and socioeconomic status, particularly its lower numbers and proportions of fishery and protected species stocks that are overfished, experiencing overfishing, depleted, or unassessed. It contains the fourth-highest fisheries landings in the nation, lower total bycatch among regions, and system exploitation below suggested sustainable thresholds of 0.1 (Link & Marshak 2019, Samhouri et al. 2010, Large et al. 2013, 2015, Tam et al. 2017). This region also leads the nation in terms of its tourism-based economy (i.e., establishments, employments, and GDP) and in total ocean establishments and employments, while it is also third-highest in terms of economic contributions of its LMRs. These trends are fostered by having the second-highest basal ecosystem productivity in the nation, which also facilitates high LMR revenue and employments as related to its total production and biomass. Governance frameworks are also relatively advanced in this region, especially regarding the multispecies nature of its FMPs, lower numbers of managed species, and cooperative management among the MAFMC and ASMFC. Ongoing consideration of ecosystem-based factors in FMPs and Council initiatives include efforts toward accounting for the importance of forage fish and trophic dynamics in regard to maintaining sustainable fisheries populations, continued focus on benthic and pelagic EFH toward fisheries productivity and refining overfishing limits (MAFMC 2016, Gaichas et al. 2018b, 2019, MAFMC 2019a), applying climate vulnerability assessments, and developing associated management strategies. Guidelines for developing EAFM frameworks that account for habitat conservation, protected species, deep sea corals, ecosystem risk assessments, and continued ecosystem-level evaluations have been developed or are underway (Gaichas et al. 2018b, MAFMC 2012, 2015, NMFS 2016a, MAFMC 2017, 2019a, b). These efforts are also complemented by the priorities of a recently completed EBFM implementation plan for the U.S. Northeast region (NMFS 2019b), including the Mid-Atlantic Bight, focused on the following:

- Developing engagement strategies to facilitate the participation of partners and stakeholders in the EBFM process.
- Conducting science and providing ESRs to understand the Mid-Atlantic Bight ecosystem.
- Identifying ecosystem-level, cumulative risk (across LMRs, habitats, ecosystem functions, and associated fisheries communities) and vulnerability to human and natural pressures in the Mid-Atlantic.
- Identifying the individual and cumulative pressures that pose the most risk to vulnerable resources and dependent communities.
- Analyzing trade-offs for optimizing benefits from all fisheries within each ecosystem or jurisdiction, taking into account ecosystem-specific policy goals and objectives, cognizant that ecosystems are composed of interconnected components.
- Developing management strategy evaluation capabilities to better conduct ecosystem-level analyses to provide ecosystem-wide management advice.
- Developing and monitoring ecosystem-level reference points for the Mid-Atlantic Bight.
- Incorporating ecosystem considerations into appropriate LMR assessments, control rules, and management decisions.
- Providing integrated advice for other management considerations, particularly applied across multiple species within an ecosystem.
- Evaluate ecosystem-level measures of resilience and community well-being.

These considerations allow for enhanced characterization of Mid-Atlantic ecosystem dynamics, increasingly robust LMR assessments, and more thorough integration of ecosystem-level information and data streams toward the development of system-wide reference points and multispecies approaches to management. Implementation is also underway to apply available and synthesized information regarding ecosystem properties for ecosystem-level planning, risk assessments, and management actions. These efforts have been bolstered by recurring development of ESRs and identification, monitoring, and assessment of ecosystem-level indicators, and the development of frameworks for conducting risk assessments that account for the physical, biological, and socioeconomic components of this ecosystem (NEFSC 2018, Gaichas et al. 2018a, b, 2019, MAFMC 2019a). Recently, these investigations have been directly conducted for the Mid-Atlantic Bight and have provided enhanced regionally specific ecosystem assessments of its socio-ecological system (Gaichas et al. 2018b, 2019). Many modeling frameworks have allowed for greater understanding of ecosystem functioning in this region and can help with advancing EBFM. These frameworks allow for examinations of ecosystem responses to ongoing stressors and management actions, the findings of which are enhancing ecosystem approaches to management for the Mid-Atlantic.

While the Mid-Atlantic is leading in many aspects of its LMR and ecosystem-centric efforts, challenges remain toward effectively implementing additional facets of EBFM, particularly enacting ecosystem-level control rules. Currently, this region has adopted EAFM for its federally managed fisheries, with initial steps being taken toward EAFM in state waters, and its guidance document serving as a functional FEP. An FEP also exists for the Chesapeake Bay subregion (Miller et al. 2001, NOAA & CBFEAP 2006). Actions toward setting system and trophic-level catch limits have also been undertaken in this subregion (Essington et al. 2016, Buchheister et al. 2017), creating the potential for more systematic management practices. In particular, the Chesapeake Bay FEP contains information on Atlantic menhaden, food web interactions, habitat, socioeconomic factors, and biological reference points (NOAA & CBFEAP 2006, Essington et al. 2016). All of this information is being applied toward more systematic management, including sustaining the Atlantic menhaden fishery and its population as forage for predatory species (ASMFC 2012). The ASMFC is developing ecosystem-based reference points for Atlantic menhaden in consideration of its role as an important prey species for the Chesapeake Bay system (ASMFC 2020a). At present, this information has been applied toward calculating more accurate mortality rates and reference points for setting revised catch limits (Essington et al. 2016). While these advancements are noteworthy, additional efforts are needed to incorporate broader ecosystem considerations into management actions, particularly in recognition of system-level shifts, multiple stressors, broader species interactions, commercial and recreational fishing patterns, and exceeded thresholds. Given the advanced understanding of these relationships in the Mid-Atlantic, the next step for this region is to apply this information into expanded system-level reference points that reflect sustainable thresholds for LMRs, ecosystem services, and human activities.

As a heavily concentrated and increasing human population contributes to growing stressors affecting this system, competition for resources within and among its marine sectors is likely to occur, with anticipated conflicts among commercial and recreational fisheries, tourism, marine transportation, and emerging wind energy sectors. There is abundant information reflecting an advanced understanding of the

Where's Waldo? Shifting Fish Distributions

—(c.f., Cod and Gulf of Maine, Chapter 3; All of Atlantic, winners and losers among the fish community, Chapters 3–5; Central tropical Pacific HMS and Turtles, Chapter 10; West Coast hakes, other species, Chapter 8; GoMEX deep zones, Chapter 7; Eastern Bering Sea groundfish, Chapter 9)

Once when I was a kid I (JL) decided, against all logic, to trade my Pete Rose baseball card for the entire card set of the early 1970s Oakland A's team. Never Johnny Bench, but Pete Rose was possible, especially for such a (perceived) great return. The hindsight value of such a trade aside, I thought it was the culmination of finally developing excellent negotiating skills against one of my childhood neighbors and sometime nemesis. I recall with great pain, even to this day, that when I went to grab Pete's card from my shoebox containing my treasury of cards, it wasn't there. I couldn't find it. I searched

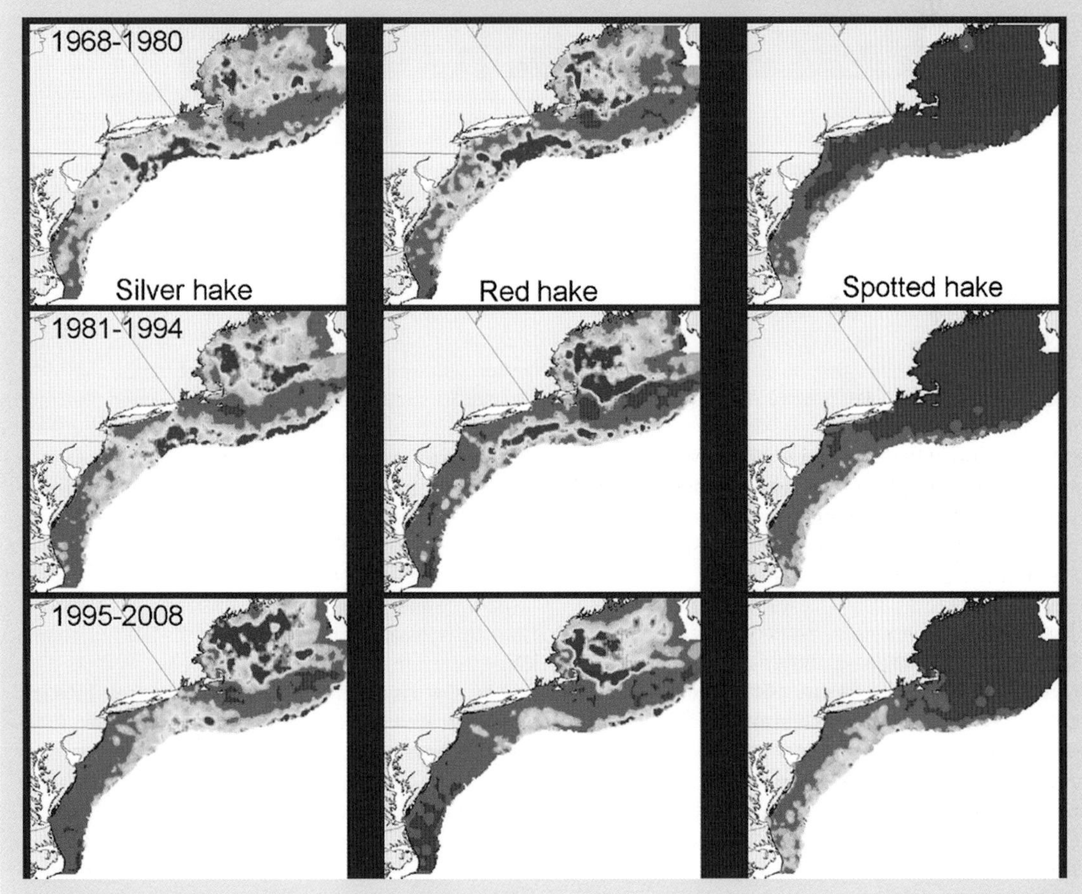

From Nye et al. 2009.

continued

Where's Waldo? Shifting Fish Distributions: *Continued*

frantically. I thought I had left it out near the shoebox, if not on top of it. After 45 minutes, my nemesis/neighbor called the deal off. I had no capital to work with, and essentially, I was out of commission for trading that day, and even quite a while thereafter. More than that, I just couldn't find Pete's card. Turns out he was no longer there, as while cleaning my room my visiting and helpful grandmother "moved him" along; as in I painfully suspect he was thrown out. What became clear was that I lost opportunities to conduct business because something wasn't where it was usually located, not where it was supposed to be.

It's hard to work with something if it ain't there anymore.

Marine fish species distributions are changing in response to climate change. East coast. West coast. Gulf coast. Pacific. Atlantic. Doesn't matter, we see these shifts in North America, and really all around the world. The majority of fish species are either moving to deeper basins or poleward (Perry et al. 2005, Nye et al. 2009, Cheung et al. 2010, Fodrie et al. 2010, Lucey & Nye 2010, Pinsky & Fogarty 2012, Pinsky et al. 2013, Whitfield et al. 2014, Cheung et al. 2015, Hare et al. 2016, Kleisner et al. 2017, Marshak & Heck 2017, Morley et al. 2017, 2018) in response to drastically changing water temperatures. Doesn't matter the kind of fish either, small or large, economically valuable or not, a broad range of species across a wide taxonomic basis are all exhibiting these changes.

Which means that the fisheries pursuing these fish are changing. Some fleets are traveling further and further. But that only works to a point, and then you run into jurisdictional issues. As in you can't fish in someone else's country or zone or management area or whatever. For example, the Atlantic stock of blueline tilefish (*Caulolatilus microps*) largely supported a fishery in the South Atlantic region. But as that species has moved north, the Mid-Atlantic jurisdiction has had to become involved (MAFMC 2017, SEDAR 2017, Schmidtke 2017, Gaichas et al. 2018). This resulted in a lot of discussions in an "emergency" rule-making context (MAFMC 2017). Thus, entire fisheries built around a particular fish or two are facing some significant challenges. Cod in New England, hake in the Pacific northwest, even tropical tuna fisheries are all also needing to alter where they catch their fish, and some are starting to realize that they may not be able to access those fish much longer.

Which means that those trying to manage these fisheries are also challenged. How do you manage something if it's not there anymore? It's hard to manage something that's no longer in your jurisdiction. As these fish stocks shift their distributions, it is equally challenging to disentangle the effects of such movement along with fishing pressure, change in productivity, etc. that also influence fish stock dynamics.

Yet just as key species are moving out of ecosystems, others are moving in. And therein lies opportunity. It may mean that eating norms need to adapt, that fisheries markets need to be developed, that fishers need to shift what they target to account for the new taxa. But therein lies the opportunity to compensate for those taxa that are exiting.

The implications are simple. In any given ecosystem, some species will leave and others will enter. Focusing solely on one stock at a time can result in a negative perspective. Adopting a broader, systematic focus results in a recognition that sure, the markets and targeted stocks will [need to] adjust, but that there will continue to be fish available for harvest.

References

Cheung WW, Brodeur RD, Okey TA, Pauly D. 2015. Projecting future changes in distributions of pelagic fish species of Northeast Pacific shelf seas. *Progress in Oceanography* 130:19–31.

Cheung WW, Lam VW, Sarmiento JL, Kearney K, Watson RE, Zeller D, Pauly D. 2010. Large-scale redistribution of maximum fisheries catch potential in the global ocean under climate change. *Global Change Biology* 16(1):24–35.

Gaichas SK, DePiper GS, Seagraves RJ, Muffley BW, Sabo MG, Colburn LL, Loftus AJ. 2018. Implementing ecosystem approaches to fishery management: risk assessment in the US Mid-Atlantic. *Frontiers in Marine Science* 5:442.

Fodrie FJ, Heck Jr KL, Powers SP, Graham WM, Robinson KL. 2010. Climate-related, decadal-scale assemblage changes of seagrass-associated fishes in the northern Gulf of Mexico. *Global Change Biology* 16(1):48–59.

Hare JA, Morrison WE, Nelson MW, Stachura MM, Teeters EJ, Griffis RB, Alexander MA, et al. 2016. A vulnerability assessment of fish and invertebrates to climate change on the Northeast US Continental Shelf. *PloS One* 11(2):e0146756.

Kleisner KM, Fogarty MJ, McGee S, Hare JA, Moret S, Perretti CT, Saba VS. 2017. Marine species distribution shifts on the US Northeast Continental Shelf under continued ocean warming. *Progress in Oceanography* 153:24–36.

Lucey SM, Nye JA. 2010. Shifting species assemblages in the northeast US continental shelf large marine ecosystem. *Marine Ecology Progress Series* 415:23–33.

MAFMC (Mid-Atlantic Fishery Management Council). 2017. *Amendment 6 to the Tilefish Fishery Management Plan—Measures to Manage Blueline Tilefish—Includes Environmental Assessment and Initial Regulatory Flexibility Analysis.* Washington, DC: NOAA (p. 150).

Marshak AR, Heck Jr KL. 2017. Interactions between range-expanding tropical fishes and the northern Gulf of Mexico red snapper Lutjanus campechanus. *Journal of Fish Biology* 91(4):1139–65.

Morley JW, Batt RD, Pinsky ML. 2017. Marine assemblages respond rapidly to winter climate variability. *Global Change Biology* 23(7):2590–601.

Morley JW, Selden RL, Latour RJ, Frölicher TL, Seagraves RJ, Pinsky ML. 2018. Projecting shifts in thermal habitat for 686 species on the North American continental shelf. *PloS One* 13(5).

Nye JA, Link JS, Hare JA, Overholtz WJ. 2009. Changing spatial distribution of fish stocks in relation to climate and population size on the Northeast United States continental shelf. *Marine Ecology Progress Series* 393:111–29.

Perry AL, Low PJ, Ellis JR, Reynolds JD. 2005. Climate change and distribution shifts in marine fishes. *Science* 308(5730): 1912–5.

Pinsky ML, Fogarty M. 2012. Lagged social-ecological responses to climate and range shifts in fisheries. *Climatic Change* 115(3–4):883–91.

Pinsky ML, Worm B, Fogarty MJ, Sarmiento JL, Levin SA. 2013. Marine taxa track local climate velocities. *Science* 341(6151):1239–42.

Schmidtke MA. 2017. *Life History and Management Methods for Blueline Tilefish (Caulolatilus microps) from the United States Mid-Atlantic Region.* Dissertation. Norfolk, VA: Old Dominion University.

SEDAR (Southeast Data, Assessment, and Review). 2017. *SEDAR 50—South Atlantic Blueline Tilefish Assessment Report.* North Charleston, SC: SEDAR.

Whitfield PE, Muñoz RC, Buckel CA, Degan BP, Freshwater DW, Hare JA. 2014. Native fish community structure and Indo-Pacific lionfish Pterois volitans densities along a depth-temperature gradient in Onslow Bay, North Carolina, USA. *Marine Ecology Progress Series* 509:241–54.

Mid-Atlantic ecosystem, including the predicted and observed system-wide effects of climate change, natural forcing, and human activities on its dynamics and marine socioeconomics. Therefore, it is imperative that this information serves toward developing comprehensive management strategies that account for these effects on other ocean sectors and their trade-offs with fisheries. Comprehensive and systematic management strategies that consider spatially overlapping ocean uses, multiple species, and the socioeconomic effects of changing environments and resource partitioning are especially warranted in this region. These especially include concerns at both the northern and southern ends of the Mid-Atlantic in response to shifting fishing patterns, species distributions, wind energy development, and continued marine transportation. Greater assessment of these concerns through available data portals, planning bodies, and interagency agreements will allow for broader ecosystem-based approaches to be realized in this region. With expanding tourism, marine transportation, energy development, and mineral extraction of this system (Martin & Hall-Arber 2008, Anthony et al. 2009, Link et al. 2010a, 2010b; Holland et al. 2012, Bailey et al. 2014, Samoteskul et al. 2014, Wyatt et al. 2017, Gaichas et al. 2018b, 2019, Link & Marshak 2019), additional non-fishing effects on the Mid-Atlantic ecosystem are anticipated, which warrant continued focus toward habitat conservation, and continued spatial management considerations of its EEZ.

Comprehensive information is being collected regarding the Mid-Atlantic ecosystem, while policy drivers, planning bodies, and ocean data portals have been developed to advance ecosystem-based approaches. LMR and socioeconomic status for this region is currently among some of the highest nationally. However, emerging issues require more robust application of ecosystem information toward broader management strategies to sustain current sustainable practices. Increased conflicts among fishing interests (particularly recreational fishers) are likely to develop as a result of shifting species, lags in responses by their fisheries and management actions, and limited progress toward management actions that directly account for these factors. Increasing trade-offs among fisheries and other ocean uses require their more direct consideration in management practices. Organizations such as the Northeast Regional Planning Body, Mid-Atlantic Regional Planning Body, and MARCO that address these multisector concerns remain in place, while well-developed, existing governance frameworks in this region provide great opportunity to advance EBFM. Initial progress is underway through foundational management actions that consider certain species interactions, habitat conservation actions, fishing effects on multiple LMRs, and human dimensions. Continuing to apply recently synthesized multifactor information toward creating broader reference points will advance current ecosystem approaches to management in this region. Through this approach, the continued sustainability of its comparatively well-managed fisheries will be even better ensured.

4.5.1 Progress Toward Ecosystem-Based Fisheries Management (EBFM)

This ecosystem is excelling in the areas of LMR and socioeconomic status, the quality of its governance system, and is relatively productive, as related to the determinants of successful LMR management (Fig. 4.29; Link & Marshak 2019).

The development and implementation of EBFM practices is occurring across all U.S. regions to allow

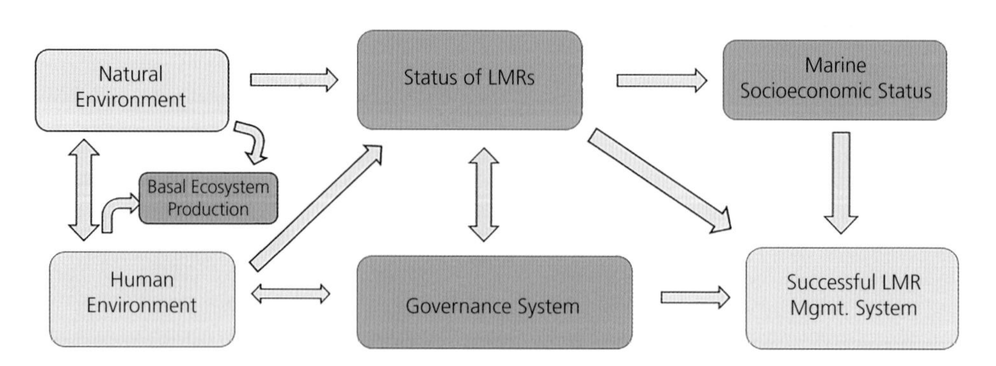

Figure 4.29 Schematic of the determinants and interconnectivity of successful Living Marine Resource (LMR) systems management criteria as modified from Link and Marshak (2019). Those highlighted in blue indicate noteworthy criteria for the Mid-Atlantic ecosystem.

for more systematic LMR management, with prioritized consideration of all fisheries, pressures, risks, and outcomes (NMFS 2016b, c, Levin et al. 2018, Marshall et al. 2018, Link & Marshak 2019). Fisheries management for the Mid-Atlantic ecosystem continues to evolve toward holistic approaches. Over the past years, the refinement of local management strategies has allowed for advancements in EBFM implementation. In accordance with the NOAA Fisheries EBFM Policy and Roadmap (NMFS 2016b, c), an EBFM implementation plan (NMFS 2019b) for the Mid-Atlantic has been developed through efforts overseen by the NEFSC, NOAA Fisheries Greater Atlantic Regional Fisheries Office (GARFO), and MAFMC, in addition to ASMFC efforts to develop ecosystem-level reference points for Atlantic menhaden (Buchheister et al. 2016, 2017, ASMFC 2020a). This regional implementation plan specifies approaches to be taken toward conducting EBFM and in addressing NOAA Fisheries Roadmap priorities and gaps. Progress for the Mid-Atlantic and its major subregions in terms of EBFM Roadmap Guiding Principles and Goals is shown in Table 12.9. Overall, progress has been made at the regional and subregional level in terms of implementing ecosystem-level planning, advancing knowledge of ecosystem principles, and in assessing risks and vulnerabilities to ecosystems through ongoing investigations into climate vulnerability and species prioritizations for stock and habitat assessments. While information has been obtained and models developed, only partial progress has been observed toward applying ecosystem-level emergent properties or reference points into management frameworks.

Progress toward recommendations of the Ecosystem Principles Advisory Panel (EPAP) for the development of FEPs (EPAP 1999, Wilkinson & Abrams 2015) is shown in Table 12.10. An EAFM guidance plan has been developed for the Mid-Atlantic, which is considered by NOAA and the MAFMC as a functional FEP. A subregional FEP has also been developed for Chesapeake Bay. Additionally, much progress has been observed in characterizing, monitoring, and assessing aspects of its ecosystem dynamics in support of this ecosystem approach and its implementation.

4.5.2 Conclusions

Fisheries management in the Mid-Atlantic represents some of the most progressive efforts toward holistic strategies and concentrated rebuilding plans, and serves as a leading example for well-managed species and cooperation with engaged stakeholders (Wilson & McCay 1998, da Silva & Kitts 2006, Hartley & Robertson 2006, Kitts et al. 2007, Olsen et al. 2014, Link & Marshak 2019). Greater reception toward robust ecosystem-based management frameworks, and an understanding of the interconnectivity of regional stressors—together with ecological interactions among multiple species, fisheries productivity, and human activities through the lens of a socioecological system (SES)—has emerged in this region (Link et al. 2011, NEFSC 2012, Biedron & Knuth 2014, 2016, Gaichas et al. 2018a, b, 2019, MAFMC 2019a, NMFS 2019b). Fully implementing an ecosystem-based approach in the Mid-Atlantic will result in many advancements toward its overall management, especially by accounting for the cumulative effects of ongoing and increasing human and natural stressors and mitigating their impacts. These are of concern especially for continued LMR productivity, harvest, and well-being of biological communities in nearshore vegetated and offshore pelagic and hard-bottom habitats. The expansion of offshore wind energy may result in closures of shelf regions to harvest, which while alleviating some of these concerns will also lead to additional shifts in fishing effort and species vulnerabilities. The tourism industry associated with this region is highly dependent on ensuring a sustainable seafood supply, ecological integrity of coastal and offshore habitats to support diving, other marine recreational activities, and the increasingly prominent recreational fisheries, which all rely upon healthy ecosystem functioning. All of these considerations are prioritized by adopting an EBFM framework.

Continuing with a business-as-usual approach or a scaled-down ecosystem approach toward managing the resources of this ecologically significant region, without considering cumulative effects on the ecosystem as a whole, limits effort toward sustainable practices. It also ignores the significance of these factors toward nationally significant LMR production and marine socioeconomic growth in this region. The continued sustainability of this region is dependent on management that concurrently considers interacting sectors, multiple trophic levels, ongoing stressors, and governance and socioeconomic feedbacks. System-wide considerations of the inherent relationships between productivity, biomass, landings, and socioeconomics are needed to refine and maintain this system at sustainable thresholds, while also accounting for the influences from natural and human activities. EBFM provides insights and opportunities that focusing on one stock or species complex or one fishery would not. Optimal management allows for simultaneously controlled harvest of interacting species complexes and fisheries that also serve as foundational habitat or prey for commercially important species. Overlapping and/or differential fishing effort for these species and the effects of different harvest methodologies among fisheries is also considered in these frameworks, while also accounting for ecological consequences to multispecies complexes. With moderate species richness in this system, this approach facilitates portfolio and multilevel considerations that provide for a more effective management strategy (Edwards et al. 2004, Kellner et al. 2011, Link 2018), which is warranted in the Mid-Atlantic. As compounding effects of sea-level rise, increasing storm intensity, habitat loss, eutrophication, and the consequences of historical overfishing continue to affect this region, their management throughout this socioecological system is more effectively addressed using an ecosystem approach. EBFM in the Mid-Atlantic provides a pathway toward ensuring the future sustainability of its nationally important LMRs and marine economy, and in continuing its maritime cultural heritage.

4.6 References

Able KW, Fahay MP. 1998. *The First Year in the Life of Estuarine Fishes in the Middle Atlantic Bight.* New Brunswick, NJ: Rutgers University Press.

Able KW, Hagan SM. 2000. Effects of common reed (Phragmites australis) invasion on marsh surface macrofauna: response of fishes and decapod crustaceans. *Estuaries* 23(5):633–46.

Adams CF. 2017. Age-specific differences in the seasonal spatial distribution of butterfish (Peprilus triacanthus). *ICES Journal of Marine Science* 74(1):170–9.

Adams CF, Miller TJ, Manderson JP, Richardson DE, Smith BE. 2014. *Butterfish 2014 Stock Assessment. Northeast Fisheries Science Center Reference Document 15-06.* Washington, DC: NOAA.

Alber M, Swenson EM, Adamowicz SC, Mendelssohn IA. 2008. Salt marsh dieback: an overview of recent events in the US. *Estuarine, Coastal and Shelf Science* 80(1):1–1.

Ampela K. 2009. *The Diet and Foraging Ecology of Gray Seals (Halichoerusgrypus) in United States Waters.* New York, NY: City University of New York.

Angradi TR, Hagan SM, Able KW. 2001. Vegetation type and the intertidal macroinvertebrate fauna of a brackish marsh: Phragmites vs. Spartina. *Wetlands* 21(1):75–92.

Anthony A, Atwood J, August PV, Byron C, Cobb S, Foster C, Fry C, et al. 2009. Coastal lagoons and climate change: ecological and social ramifications in the US Atlantic and Gulf coast ecosystems. *Ecology and Society* 14(1):8.

ASMFC (Atlantic States Marine Fisheries Commission). 2012. *Amendment 2 to the Interstate Fishery Management Plan for Atlantic Menhaden.* Washington, DC: NOAA (p. 102).

ASMFC (Atlantic States Marine Fisheries Commission). 2020a. *ASMFC Atlantic Menhaden Board Adopts Ecological Reference Points.* Washington, DC: NOAA (p. 2).

ASMFC (Atlantic States Marine Fisheries Commission). 2020b. *ASMFC Stock Status Overview.* (p. 43). http://www.asmfc.org/files/pub/ASMFC_StockStatus_Jan2020.pdf

Azarovitz TR. 1981. A brief historical review of the Woods Hole Laboratory trawl survey time series. Canadian Special Publications of Fishery and Aquatic Science 58:62–7.

Babcock EA. 2008. Recreational fishing for pelagic sharks worldwide. *Sharks of the Open Ocean: Biology, Fisheries and Conservation* 29:193–204.

Bailey H, Brookes KL, Thompson PM. 2014. Assessing environmental impacts of offshore wind farms: lessons learned and recommendations for the future. *Aquatic Biosystems* 10(1):8.

Bailey JF, Patterson JL, Paulhus JL. 1975. *Hurricane Agnes Rainfall and Floods, June–July 1972.* Washington, DC: US Government Printing Office.

Bayha KM, Graham WM. 2014. "Nonindigenous marine jellyfish: invasiveness, invasibility, and impacts." In: *Jellyfish Blooms.* Edited by KA Pitt, CH Lucas. Dordrecht, Netherlands: Springer (pp. 45–77).

Beck MW, Heck KL, Able KW, Childers DL, Eggleston DB, Gillanders BM, Halpern B, et al. 2001. The identification, conservation, and management of estuarine and marine nurseries for fish and invertebrates. *Bioscience* 51(8):633–41.

Beckett LH, Baldwin AH, Kearney MS. 2016. Tidal marshes across a Chesapeake Bay subestuary are not keeping up with sea-level rise. *PloS One* 11(7):e0159753.

Belden DL, Orphanides C. 2007. *Estimates of Cetacean and Pinniped Bycatch in the 2006 Northeast Sink Gillnet and Mid-Atlantic Coastal Gillnet Fisheries. Northeast Fisheries Science Center Reference Document 14-02.* Washington, DC: NOAA.

Biedron IS, Knuth BA. 2014. *Ecosystem-Based Fisheries Management: Perspectives of Fishery Management Councils and Stakeholders in the New England and Mid-Atlantic regions. Human Dimensions Research Unit Publication Series 14-11.* Ithaca, NY: Cornell University (p. 32).

Biedron IS, Knuth BA. 2016. Toward shared understandings of ecosystem-based fisheries management among fishery management councils and stakeholders in the US Mid-Atlantic and New England regions. *Marine Policy* 70:40–8.

BOEM (Bureau of Ocean Energy Management). 2012. *Commercial Wind Lease Issuance and Site Assessment Activities on the Atlantic Outer Continental Shelf Offshore New Jersey, Delaware, Maryland, and Virginia. Final Environmental Assessment. OCS EIS/EA.* BOEM 2012-003. Herndon, VA: U.S. Department of the Interior (p. 341).

BOEM (Bureau of Ocean Energy Management). 2016. *BOEM Wind Planning Areas (Polygon Shapefile).* Herndon, VA: Washington, DC: U.S. Department of the Interior.

Boesch DF, Boicourt WC, Cullather RI, Ezer T, Galloway Jr GE, Johnson ZP, Kilbourne KH, et al. 2018. *Sea-level Rise: Projections for Maryland 2018.* Cambridge, MD: University of Maryland Center for Environmental Science (p. 27)

Bonzek CF, Gartland J, Gauthier DJ, Latour RJ. 2016. *Northeast Area Monitoring and Assessment Program (NEAMAP) Data collection and Analysis in Support of Single and Multispecies Stock Assessments in the Mid-Atlantic: Northeast Area Monitoring and Assessment Program Near Shore Trawl Survey Annual Report.* Gloucester Point, VA: Virginia Institute of Marine Science, College of William and Mary.

Bonzek CF, Gartland J, Gauthier DJ, Latour RJ. 2017. *Northeast Area Monitoring and Assessment Program (NEAMAP) Data collection and Analysis in Support of Single and Multispecies Stock Assessments in the Mid-Atlantic: Northeast Area Monitoring and Assessment Program Near Shore Trawl Survey.* Gloucester Point, VA: Virginia Institute of Marine Science, College of William and Mary.

Borneman WR. 2004. *1812: The War that Forged a Nation.* New York, NY: Harper Collins Publishers.

Buchheister A, Miller TJ, Houde ED. 2017. Evaluating ecosystem-based reference points for Atlantic Menhaden. *Marine and Coastal Fisheries* 9(1):457–78.

Buchheister A, Miller TJ, Houde ED, Secor DH, Latour RJ. 2016. Spatial and temporal dynamics of Atlantic menhaden (Brevoortia tyrannus) recruitment in the Northwest Atlantic ocean. *ICES Journal of Marine Science* 73(4):1147–59.

Cai WJ, Huang WJ, Luther GW, Pierrot D, Li M, Testa J, Xue M, et al. 2017. Redox reactions and weak buffering capacity lead to acidification in the Chesapeake Bay. *Nature Communications* 8(1):1–2.

Capossela KM, Fabrizio MC, Brill RW. 2013. Migratory and within-estuary behaviors of adult Summer Flounder (Paralichthys dentatus) in a lagoon system of the southern mid-Atlantic Bight. *Fishery Bulletin* 111(2):189–201.

Carr-Harris A, Lang C. 2019. Sustainability and tourism: a case study of the United States' first offshore wind farm. *Resource and Energy Economics* 57(C):51–67.

Carretta JV, Forney KA, Oleson EM, Weller DW, Lang AR, Baker J, Muto MM, et al. 2019. *U.S. Pacific Marine Mammal Stock Assessments: 2018. NOAA Technical Memorandum NMFSSWFSC-617.* Washington, DC: NOAA.

Carruthers TJ, Beckert K, Schupp CA, Saxby T, Kumer JP, Thomas J, Sturgis B, et al. 2013. Improving management of a mid-Atlantic coastal barrier island through assessment of habitat condition. *Estuarine, Coastal and Shelf Science* 116:74–86.

CBP (Chesapeake Bay Program). 2009. *Bay Barometer A Health and Restoration Assessment of the Chesapeake Bay and Watershed in 2008. CBP/TRS 293-09. EPA-903-R-09-001.* Washington, DC: United States Environmental Protection Agency (p. 35).

Chambers RM, Meyerson LA, Saltonstall K. 1999. Expansion of Phragmites australis into tidal wetlands of North America. *Aquatic Botany* 64(3–4):261–73.

Clay PM, Colburn LL, Seara T. 2016. Social bonds and recovery: an analysis of Hurricane Sandy in the first year after landfall. *Marine Policy* 74:334–40.

Cogliano FD. 2016. *Revolutionary America, 1763–1815: A Political History.* Abingdon, UK: Routledge.

Colburn LL, Clay PM, Olson J, da Silva PP, Smith SL, Westwood A, Ekstrom J. 2010. *Community Profiles for Northeast US Marine Fisheries.* Woods Hole, MA: US Department of Commerce.

Colburn LL, Jepson M, Weng C, Seara T, Weiss J, Hare JA. 2016. Indicators of climate change and social vulnerability in fishing dependent communities along the Eastern and Gulf Coasts of the United States. *Marine Policy* 74:323–33.

Coleman FC, Figueira WF, Ueland JS, Crowder LB. 2004. The impact of United States recreational fisheries on marine fish populations. *Science* 305(5692):1958–60.

Colvocoresses JA, Musick J. 1984. Species associations and community composition of middle Atlantic Bight continental-shelf demersal fishes. *Fishery Bulletin* 82(2):295.

Costantini M, Ludsin SA, Mason DM, Zhang X, Boicourt WC, Brandt SB. 2008. Effect of hypoxia on habitat quality of striped bass (Morone saxatilis) in Chesapeake Bay. *Canadian Journal of Fisheries and Aquatic Sciences* 65(5): 989–1002.

Crawford S, Whelan G, Infante DM, Blackhart K, Daniel WM, Fuller PL, Birdsong T, et al. 2016. *Through a Fish's Eye: The Status of Fish Habitats in the United States 2015.* Washington, DC: National Fish Habitat Partnership.

Crisman TL. 2008. "An ecological perspective on management of the Chesapeake Bay." In: *Large-Scale Ecosystem Restoration: Five Case Studies from the United States.* Edited by M Doyle, C Drew. Washington, DC: Island Press (p. 204).

Cruz-Marrero W, Cullen DW, Gay NR, Stevens BG. 2019. Characterizing the benthic community in Maryland's offshore wind energy areas using a towed camera sled: Developing a method to reduce the effort of image analysis and community description. *PloS One* 14(5):e0215966.

da Silva PP, Kitts A. 2006. Collaborative fisheries management in the Northeast US: Emerging initiatives and future directions. Marine Policy 30(6):832–41.

Daily G. 2003. "What are ecosystem services." In: *Global environmental challenges for the twenty-first century: Resources, Consumption and Sustainable Solutions.* Edited by DE Lorey. Lanham, MD: Rowman & Littlefield Publishers (pp. 227–31).

Dance S. 2019. *There's Going to be No Fish to Fight Over at All: The Chesapeake Bay's Rockfish Population is Falling. The Baltimore Sun.* https://www.baltimoresun.com/news/

environment/bs-md-striped-bass-decline-20190424-story.html

Danysh R. 1979. "Military history in the Department of Defense." In: *A Guide to the Study and Use of Military History.* Edited by JE Jessup Jr, RW Coakley. Washington, DC: Center of Military History United States Army (p. 401).

Dawson CP, Wilkins BT. 1981. Motivations of New York and Virginia marine boat anglers and their preferences for potential fishing constraints. *North American Journal of Fisheries Management* 1(2):151–8.

de Sylva DP, Davis WP. 1963. White marlin, Tetrapturus albidus, in the Middle Atlantic Bight, with observations on the hydrography of the fishing grounds. *Copeia* 30:81–99.

DePiper GS, Gaichas SK, Lucey SM, Pinto da Silva P, Anderson MR, Breeze H, Bundy A, et al. 2017. Operationalizing integrated ecosystem assessments within a multidisciplinary team: lessons learned from a worked example. *ICES Journal of Marine Science* 74(8):2076–86.

De Rivera CE, Ruiz GM, Hines AH, Jivoff P. 2005. Biotic resistance to invasion: native predator limits abundance and distribution of an introduced crab. *Ecology* 86(12):3364–76.

Deegan LA, Hughes JE, Rountree RA. 2002. "Salt marsh ecosystem support of marine transient species." In: *Concepts and Controversies in Tidal Marsh Ecology.* Edited by MP Weinstein, DA Kreeger. Dordrecht, Netherlands: Springer (pp. 333–65).

Diaz RJ. 1974. Asiatic clam, Corbicula manilensis (Philippi), in the tidal James River, Virginia. *Chesapeake Science* 15(2):118–20.

Dubik BA, Clark EC, Young T, Zigler SB, Provost MM, Pinsky ML, Martin KS. 2019. Governing fisheries in the face of change: social responses to long-term geographic shifts in a US fishery. *Marine Policy* 99:243–51.

Edwards SF, Link JS, Rountree BP. 2004. Portfolio management of wild fish stocks. *Ecological Economics* 49(3):317–29.

Ehler C, Douvere F. 2010. An international perspective on marine spatial planning initiatives. *Environments: A Journal of Interdisciplinary Studies* 37(3):9–20.

Eklund AM, Targett TE. 1991. Seasonality of fish catch rates and species composition from the hard bottom trap fishery in the Middle Atlantic Bight (US east coast). *Fisheries Research* 12(1):1–22.

EPA (U.S. Environmental Protection Agency). 2012. *National Coastal Condition Report IV, EPA-842-R-10-003.* Washington, DC: United States Environmental Protection Agency Office of Research.

EPAP (Ecosystem Principles Advisory Panel). 1999. *Ecosystem-Based Fishery Management Report to Congress.* Silver Spring, MD: National Marine Fisheries Service (p. 62).

Epifanio CE. 2013. Invasion biology of the Asian shore crab Hemigrapsus sanguineus: a review. *Journal of Experimental Marine Biology and Ecology* 441:33–49.

Epifanio CE, Valenti CC, Pembroke AE. 1984. Dispersal and recruitment of blue crab larvae in Delaware Bay, USA. *Estuarine, Coastal and Shelf Science* 18(1):1–2.

Erbland PJ, Ozbay G. 2008. A comparison of the macrofaunal communities inhabiting a Crassostrea virginica oyster reef and oyster aquaculture gear in Indian River Bay, Delaware. *Journal of Shellfish Research* 27(4):757–69.

Essington TE, Levin PS, Marshall KN, Koehn L, Anderson LG, Bundy A, Carothers C, et al. 2016. *Building Effective Fishery Ecosystem Plans: A Report from the Lenfest Fishery Ecosystem Task Force.* Washington, DC: Lenfest Ocean Program.

Ezer T, Atkinson LP, Corlett WB, Blanco JL. 2013. Gulf Stream's induced sea level rise and variability along the US mid-Atlantic coast. *Journal of Geophysical Research: Oceans* 118(2):685–97.

Fabrizio MC, Tuckey TD, Latour RJ, White GC, Norris AJ. 2018. Tidal habitats support large numbers of invasive blue catfish in a Chesapeake Bay subestuary. *Estuaries and Coasts* 41(3):827–40.

Fay G, Link JS, Hare JA. 2017. Assessing the effects of ocean acidification in the Northeast US using an end-to-end marine ecosystem model. *Ecological Modelling* 347:1–10.

Ferguson L, Srinivasan M, Oleson E, Hayes S, Brown SK, Angliss R, Carretta J, et al. 2017. *Proceedings of the First National Protected Species Assessment Workshop. NOAA Technical Memorandum NMFS-F/SPO-172.* Washington, DC: NOAA (p. 92).

Firestone J, Lyons SB, Wang C, Corbett JJ. 2008. Statistical modeling of North Atlantic right whale migration along the mid-Atlantic region of the eastern seaboard of the United States. *Biological Conservation* 141(1):221–32.

Fisher TR, Benitez JA, Lee KY, Sutton AJ. 2006. History of land cover change and biogeochemical impacts in the Choptank River basin in the mid-Atlantic region of the US. *International Journal of Remote Sensing* 27(17):3683–703.

Fofonoff PW, Ruiz GM, Hines AH, Steves BD, Carlton JT. 2009. "Four centuries of biological invasions in tidal waters of the Chesapeake Bay region." In: *Biological Invasions in Marine Ecosystems.* Edited by G Rilov, JA Crooks. Berlin, Germany: Springer (pp. 479–506).

Ford SE. 1997. History and present status of molluscan shellfisheries from Barnegat Bay to Delaware Bay. *The History, Present Condition, and Future of the Molluscan Fisheries of North and Central America and Europe* 1:119–40.

François D. 2018. *Atlantic Coast Renewable Energy Status Update. BOEM's Offshore Wind and Maritime Industry Knowledge Exchange.* https://www.boem.gov/sites/default/files/renewable-energy-program/BOEM-Atlantic-Coast-Renewable-Energy-Leases.pdf

Fratantoni PS, Holzwarth T, Bascuñán C, Taylor MH. 2015. *Description of the 2010 Oceanographic Conditions on the Northeast US Continental Shelf. Northeast Fisheries Science Center Reference Document 15–21.* Washington, DC: NOAA (p. 41).

Friedland KD, Hare JA. 2007. Long-term trends and regime shifts in sea surface temperature on the continental shelf of the northeast United States. *Continental Shelf Research* 27(18):2313–28.

Frisk MG, Miller TJ, Latour RJ, Martell S. 2006. *An Ecosystem Model of Delaware Bay: Final Report.* Baltimore, MD: University of Maryland Center for Environmental Science.

Frisk MG, Miller TJ, Latour RJ, Martell SJ. 2011. Assessing biomass gains from marsh restoration in Delaware Bay using Ecopath with Ecosim. *Ecological Modelling* 222(1):190–200.

Furbish CE, Albano M. 1994. Selective herbivory and plant community structure in a mid-Atlantic salt marsh. *Ecology* 75(4):1015–22.

Gaichas S, DePiper G, Seagraves R, Colburn L, Loftus A, Sabo M, Muffley B. 2018a. *Mid-Atlantic EAFM Risk Assessment Documentation and Results.* (p. 36). https://static1.squarespace.com/static/511cdc7fe4b00307a2628ac6/t/5b7efd0c03ce64cbcbc2dea7/1535048976510/SOE_MAB_RiskAssess_08_18.pdf

Gaichas SK, DePiper GS, Seagraves RJ, Muffley BW, Sabo M, Colburn LL, Loftus AL. 2018b. Implementing ecosystem approaches to fishery management: risk assessment in the US Mid-Atlantic. *Frontiers in Marine Science* 5.

Gaichas SK, Fogarty M, Fay G, Gamble R, Lucey S, Smith L. 2017. Combining stock, multispecies, and ecosystem level fishery objectives within an operational management procedure: simulations to start the conversation. *ICES Journal of Marine Science* 74: 552–65.

Gaichas S, Hardison S, Large S, Lucey S. 2019. *State of the Ecosystem 2019: Mid-Atlantic.* Washington, DC: NOAA (p. 28).

Gannon DP, Waples DM. 2004. Diets of coastal bottlenose dolphins from the US mid-Atlantic coast differ by habitat. *Marine Mammal Science* 20(3):527–45.

Garrison LP, Link JS, Kilduff DP, Cieri MD, Muffley B, Vaughan DS, Sharov A, Mahmoudi B, Latour RJ. 2010. An expansion of the MSVPA approach for quantifying predator–prey interactions in exploited fish communities. *ICES Journal of Marine Science* 67(5):856–70.

Gaynor JJ, Bologna PA, Restaino DJ, Barry C. 2016. First occurrence of the invasive hydrozoan Gonionemus vertens A. Agassiz, 1862 (Cnidaria: Hydrozoa) in New Jersey, USA. *BioInvasions Records* 5(4):233–7.

Gentner B, Lowther A. 2002. Evaluating marine sport fisheries in the USA. *Recreational Fisheries: Ecological, Economic and Social Evaluation* 11:186–206.

Goldenberg SB, Landsea CW, Mestas-Nuñez AM, Gray WM. 2001. The recent increase in Atlantic hurricane activity: causes and implications. *Science* 293(5529):474–9.

Grabowski JH, Peterson CH. 2007. Restoring oyster reefs to recover ecosystem services. *Ecosystem Engineers: Plants to Protists* 4:281–98.

Grout DE. 2006. Interactions between striped bass (Morone saxatilis) rebuilding programmes and the conservation of Atlantic salmon (Salmo salar) and other anadromous fish species in the USA. *ICES Journal of Marine Science* 63(7):1346–52.

Grubbs RD, Musick JA. 2007. *Spatial Delineation of Summer Nursery Areas for Juvenile Sandbar Sharks in Chesapeake Bay, Virginia. American Fisheries Society Symposium 50.* Bethesda, MD: American Fisheries Society (p. 63).

Guida, V., A. Drohan, H. Welch, J. McHenry, D. Johnson, V. Kentner, J. Brink, D. Timmons, E. Estela-Gomez. 2017. *Habitat Mapping and Assessment of Northeast Wind Energy Areas. Sterling, VA. OCS Study BOEM 2017-088.* Washington, DC: US Department of the Interior, Bureau of Ocean Energy Management (p. 312).

Hale EA, Park IA, Fisher MT, Wong RA, Stangl MJ, Clark JH. 2016. Abundance estimate for and habitat use by early juvenile Atlantic sturgeon within the Delaware River Estuary. *Transactions of the American Fisheries Society* 145(6):1193–201.

Halpern P. 2012. *A Naval History of World War I.* Annapolis, MD: Naval Institute Press.

Hare JA, Able KW. 2007. Mechanistic links between climate and fisheries along the east coast of the United States: explaining population outbursts of Atlantic croaker (Micropogonias undulatus). *Fisheries Oceanography* 16:31–45.

Hare JA, Alexander MA, Fogarty MJ, Williams EH, Scott JD. 2010a. Forecasting the dynamics of a coastal fishery species using a coupled climate–population model. *Ecological Applications* 20(2):452–64.

Hare MP, Weinberg J, Peterfalvy O, Davidson M. 2010b. The "southern" surfclam (Spisula Solidissima Similis) found north of its reported range: a commercially harvested population in Long Island Sound, New York. *Journal of Shellfish Research* 29(4):799–807.

Hare JA, Morrison WE, Nelson MW, Stachura MM, Teeters EJ, Griffis RB, Alexander MA, et al. 2016. A vulnerability assessment of fish and invertebrates to climate change on the Northeast US Continental Shelf. *PloS One* 11(2):e0146756.

Hartley TW, Robertson RA. 2006. Emergence of multi-stakeholder-driven cooperative research in the Northwest Atlantic: the case of the Northeast Consortium. *Marine Policy* 30(5):580–92.

Hayes SA, Josephson E, Maze-Foley K, Rosel PE. 2019. *US Atlantic and Gulf of Mexico Marine Mammal Stock Assessments-2018. NOAA Tech Memo NMFS-NE-258.* Washington, DC: NOAA (p. 298).

Hill J, Fowler DL, Van Den Avyle MJ. 1989. *Species Profiles. Life Histories and Environmental Requirements of Coastal Fishes and Invertebrates (Mid-Atlantic). Blue Crab.* Athens, GA: Georgia Cooperative Fishery and Wildlife Research Unit Athens.

Holland D, Sanchirico J, Johnston R, Jogleka D. 2012. *Economic Analysis for Ecosystem-Based Management:*

Applications to Marine and Coastal Environments. Abingdon, UK: Routledge.

Hornick KM, Plough LV. 2019. Tracking genetic diversity in a large-scale oyster restoration program: effects of hatchery propagation and initial characterization of diversity on restored vs. wild reefs. *Heredity* 123(2):92–105.

Houde E, Gaichas S, Seagraves R. 2014. *Managing Forage Fishes in the Mid-Atlantic Region*. Dover, DE: Mid-Atlantic Fishery Management Council (p. 41).

Houde ED, Bichy J, Jung S. 2005. Effects of hurricane Isabel on fish populations and communities in Chesapeake Bay. *Hurricane Isabel in Perspective: Proceedings of a Conference CRC Publication* 5:193–9.

Hourigan TF, Etnoyer PJ, Cairns SD. 2017. *The State of Deep-Sea Coral and Sponge Ecosystems of the United States. NOAA Technical Memorandum NMFS-OHC-4*. Silver Spring, MD: NOAA (p. 467).

Hughes C, Richardson CA, Luckenbach M, Seed R. 2009. Difficulties in separating hurricane induced effects from natural benthic succession: Hurricane Isabel, a case study from Eastern Virginia, USA. *Estuarine, Coastal and Shelf Science* 85(3):377–86.

Ihde TF, Kaplan IC, Fulton EA, Gray IA, Hasan M, Bruce D, Slacum W, Townsend HM. 2016. *Design and Parameterization of the Chesapeake Bay Atlantis Model: A Spatially Explicit End-To-End Ecosystem Model. NOAA Technical Memorandum NMFS-F/SPO-166*. Washington, DC: NOAA.

Ihde TF, Townsend HM. 2017. Accounting for multiple stressors influencing living marine resources in a complex estuarine ecosystem using an Atlantis model. *Ecological Modelling* 365:1–9.

Jacques PJ. 2017. The origins of coastal ecological decline and the great Atlantic oyster collapse. *Political Geography* 60:154–64.

Jantz P, Goetz S, Jantz C. 2005. Urbanization and the loss of resource lands in the Chesapeake Bay watershed. *Environmental Management* 36(6):808–25.

Jones KB, Neale AC, Nash MS, Van Remortel RD, Wickham JD, Riitters KH, O'Neill RV. 2001a. Predicting nutrient and sediment loadings to streams from landscape metrics: a multiple watershed study from the United States Mid-Atlantic Region. *Landscape Ecology* 16(4):301–12.

Jones KB, Neale AC, Wade TG, Wickham JD, Cross CL, Edmonds CM, Loveland TR, et al. 2001b. The consequences of landscape change on ecological resources: an assessment of the United States Mid-Atlantic region, 1973–1993. *Ecosystem Health* 7(4):229–42.

Jones SJ, Lima FP, Wethey DS. 2010. Rising environmental temperatures and biogeography: poleward range contraction of the blue mussel, Mytilus edulis L., in the western Atlantic. *Journal of Biogeography* 37(12):2243–59.

Jordan TE, Weller DE, Pelc CE. 2018. Effects of local watershed land use on water quality in mid-Atlantic coastal bays and subestuaries of the Chesapeake Bay. *Estuaries and Coasts* 41(1):38–53.

Joseph EB. 1972. The status of the sciaenid stocks of the middle Atlantic coast. Chesapeake Science. 1972 Jun 1;13(2):87–100.

Jung S, Houde ED. 2003. Spatial and temporal variabilities of pelagic fish community structure and distribution in Chesapeake Bay, USA. *Estuarine, Coastal and Shelf Science* 58(2):335–51.

Kane J. 2007. Zooplankton abundance trends on Georges Bank, 1977–2004. *ICES Journal of Marine Science* 64(5): 909–19.

Karlson RH, Osman RW. 2012. Species composition and geographic distribution of invertebrates in fouling communities along the east coast of the USA: a regional perspective. *Marine Ecology Progress Series* 458:255–68.

Karp MA, Blackhart K, Lynch PD, Deroba J, Hanselman D, Gertseva V, Teo S, Townsend H, Williams E, Yau A (eds). 2019. *Proceedings of the 13th National Stock Assessment Workshop: Model Complexity, Model Stability, and Ensemble Modeling*. U.S. Dept. of Commerce, NOAA. NOAA Technical Memoranda NMFS-F/SPO-189 (p. 49).

Kauffman GJ. 2011. *Socioeconomic Value of the Delaware River Basin in Delaware, New Jersey, New York, and Pennsylvania: The Delaware River Basin, an economic engine for over 400 years*. Newark, NJ: University of Delaware.

Kellner JB, Sanchirico JN, Hastings A, Mumby PJ. 2011. Optimizing for multiple species and multiple values: tradeoffs inherent in ecosystem-based fisheries management. *Conservation Letters* 4(1):21–30.

Kemp AC, Horton BP. 2013. Contribution of relative sea-level rise to historical hurricane flooding in New York City. *Journal of Quaternary Science* 28(6):537–41.

Kemp WM, Boynton WR, Adolf JE, Boesch DF, Boicourt WC, Brush G, Cornwell JC, et al. 2005. Eutrophication of Chesapeake Bay: historical trends and ecological interactions. *Marine Ecology Progress Series* 303:1–29.

Kennish MJ. 2001. Coastal salt marsh systems in the US: a review of anthropogenic impacts. *Journal of Coastal Research* 1:731–48.

Kerr LA, Connelly WJ, Martino EJ, Peer AC, Woodland RJ, Secor DH. 2009. Climate change in the US Atlantic affecting recreational fisheries. *Reviews in Fisheries Science* 17(2):267–89.

Kiddon JA, Paul JF, Buffum HW, Strobel CS, Hale SS, Cobb D, Brown BS. 2003. Ecological condition of US mid-Atlantic estuaries, 1997–1998. *Marine Pollution Bulletin* 46(10):1224–44.

Kilduff P, Carmichael J, Latour R. 2009. *Guide to Fisheries Science and Stock Assessments. National Oceanic and Atmospheric Administration Grant No. NA05NMF4741025*. Washington, DC: Atlantic States Marine Fisheries Commission.

Kirby MX. 2004. Fishing down the coast: historical expansion and collapse of oyster fisheries along continental margins. *Proceedings of the National Academy of Sciences* 101(35):13096–9.

Kitts A, da Silva PP, Rountree B. 2007. The evolution of collaborative management in the Northeast USA tilefish fishery. *Marine Policy* 31(2):192–200.

Kleisner KM, Fogarty MJ, McGee S, Barnett A, Fratantoni P, Greene J, Hare JA, et al. 2016. The effects of sub-regional climate velocity on the distribution and spatial extent of marine species assemblages. *PLoS One* 11(2):e0149220.

Knowlton AR, Ring JB, Russell B, Aquarium NE. 2002. *Right Whale Sightings and Survey Effort in the Mid Atlantic Region: Migratory Corridor, Time Frame, and Proximity to Port Entrances*. Silver Spring, MD: NMFS Ship Strike Working Group.

Koch EW, Orth RJ. 2003. "The Mid-Atlantic coast of the United States." In: *World Atlas of Seagrasses*. Edited by EP Green, FT Short. Berkely, CA: University of California Press (pp. 216–23).

Kohut J, Palamara L, Bochenek E, Jensen O, Manderson J, Oliver M, Gray S, Roebuck C. 2012. *Using Ocean Observing Systems and Local Ecological Knowledge to Nowcast Butterfish Bycatch Events in the Mid-Atlantic Bight Longfin Squid Fishery*. New York, NY: IEEE (pp. 1–6).

Koneff MD, Royle JA. 2004. Modeling wetland change along the United States Atlantic coast. *Ecological Modelling* 177(1–2):41–59.

Kopp RE. 2013. Does the mid-Atlantic United States sea level acceleration hot spot reflect ocean dynamic variability? *Geophysical Research Letters* 40(15):3981–5.

Kritzer JP, DeLucia MB, Greene E, Shumway C, Topolski MF, Thomas-Blate J, Chiarella LA, Davy KB, Smith K. 2016. The importance of benthic habitats for coastal fisheries. *BioScience* 66(4):274–84.

Lapointe NW, Fuller PL, Neilson M, Murphy BR, Angermeier PL. 2016. Pathways of fish invasions in the Mid-Atlantic region of the United States. *Management of Biological Invasions* 7(3):221–33.

Large SI, Fay G, Friedland KD, Link JS. 2013. Defining trends and thresholds in responses of ecological indicators to fishing and environmental pressures. *ICES Journal of Marine Science* 70:755–67.

Large SI, Fay G, Friedland KD, Link JS. 2015. Critical points in ecosystem responses to fishing and environmental pressures. *Marine Ecology Progress Series* 521:1–17.

Lavelle M. 2014. *Uncivil War Breaks Out Over Fluke as Habitat Shifts North*. The Daily Climate. https://www.scientificamerican.com/article/uncivil-war-breaks-out-over-fluke-as-habitat-shifts-north/

Lee SB, Li M, Zhang F. 2017. Impact of sea level rise on tidal range in Chesapeake and Delaware Bays. *Journal of Geophysical Research: Oceans* 122(5):3917–38.

Lefcheck JS, Marion SR, Orth RJ. 2017. Restored eelgrass (Zostera marina L.) as a refuge for epifaunal biodiversity in mid-Western Atlantic coastal bays. *Estuaries and Coasts* 40(1):200–12.

Lefcheck JS, Orth RJ, Dennison WC, Wilcox DJ, Murphy RR, Keisman J, Gurbisz C, et al. 2018. Long-term nutrient reductions lead to the unprecedented recovery of a temperate coastal region. *Proceedings of the National Academy of Sciences* 115(14):3658–62.

Lentz SJ. 2017. Seasonal warming of the Middle Atlantic Bight Cold Pool. *Journal of Geophysical Research: Oceans* 122(2):941–54.

Leslie HM, McLeod KL. 2007. Confronting the challenges of implementing marine ecosystem-based management. *Frontiers in Ecology and the Environment* 5(10):540–8.

Levin PS, Essington TE, Marshall KN, Koehn LE, Anderson LG, Bundy A, Carothers C, et al. 2018. Building effective fishery ecosystem plans. *Marine Policy* 92:48–57.

Levin PS, Fogarty MJ, Murawski SA, Fluharty D. 2009. Integrated ecosystem assessments: developing the scientific basis for ecosystem-based management of the ocean. *PLoS Biology* 7(1):e1000014.

Limburg KE, Waldman JR. 2009. Dramatic declines in North Atlantic diadromous fishes. *BioScience* 59(11):955–65.

Link JS. 2018. System-level optimal yield: increased value, less risk, improved stability, and better fisheries. *Canadian Journal of Fisheries and Aquatic Sciences* 75(1):1–6.

Link J, Overholtz W, O'Reilly J, Green J, Dow D, Palka D, Legault C, et al. 2008. The Northeast US continental shelf energy modeling and analysis exercise (EMAX): ecological network model development and basic ecosystem metrics. *Journal of Marine Systems* 74(1–2):453–74.

Link JS, Brodziak JK, Edwards SF, Overholtz WJ, Mountain D, Jossi JW, Smith TD, Fogarty MJ. 2002a. Marine ecosystem assessment in a fisheries management context. *Canadian Journal of Fisheries and Aquatic Sciences* 59(9):1429–40.

Link JS, Brodziak JKT. 2002b. *Status of the Northeast U.S. Continental Shelf Ecosystem: A Report of the Northeast Fisheries Science Center's Ecosystem Status Working Group. Northeast Fishery Science Center Reference Document 02-11*. Woods Hole, MA: National Marine Fisheries Service (p. 245).

Link JS, Bundy A, Overholtz WJ, Shackell N, Manderson J, Duplisea D, Hare J, Koen-Alonso M, Friedland KD. 2011. Ecosystem-based fisheries management in the Northwest Atlantic. *Fish and Fisheries* 12(2):152–70.

Link JS, Fulton EA, Gamble RJ. 2010a. The northeast US application of Atlantis: a full system model exploring marine ecosystem dynamics in a living marine resource management context. *Progress in Oceanography* 87(1–4):214–34.

Link JS, Ihde TF, Townsend HM, Osgood KE, Schirripa MJ, Kobayashi DR, Gaichas SK, et al. 2010b. *Report of the 2nd National Ecosystem Modeling Workshop (NEMoW II): Bridging the Credibility Gap Dealing with Uncertainty in Ecosystem Models. NOAA Technical Memorandum NMFS-F/SPO-102*. Washington, DC: NOAA.

Link JS, Marshak AR. 2019. Characterizing and comparing marine fisheries ecosystems in the United States: determinants of success in moving toward ecosystem-based fisheries management. *Fish Biology and Fisheries* 29(1): 23–70.

Lipcius RN, Burke RP, McCulloch DN, Schreiber SJ, Schulte DM, Seitz RD, Shen J. 2015. Overcoming restoration paradigms: value of the historical record and metapopulation dynamics in native oyster restoration. *Frontiers in Marine Science* 2:65.

Lipcius RN, Van Engel WA. 1990. Blue crab population dynamics in Chesapeake Bay: variation in abundance (York River, 1972–1988) and stock-recruit functions. *Bulletin of Marine Science* 46(1):180–94.

Llansó RJ, Scott LC, Hyland JL, Dauer DM, Russell DE, Kutz FW. 2002. An estuarine benthic index of biotic integrity for the mid-Atlantic region of the United States. II. Index development. *Estuaries* 25(6):1231–42.

Lohrer AM, Whitlatch RB. 2002. Interactions among aliens: apparent replacement of one exotic species by another. *Ecology* 83:719–732.

Lucey SM, Nye JA. 2010. Shifting species assemblages in the northeast US continental shelf large marine ecosystem. *Marine Ecology Progress Series* 415:23–33.

Lynch PD, Methot RD, Link JS. 2018. *Implementing a Next Generation Stock Assessment Enterprise. An Update to the NOAA Fisheries Stock Assessment Improvement Plan. NOAA Technical Memo NMFS-F/SPO-183.* Washington, DC: NOAA (p. 127).

MacDonald JA, Roudez R, Glover T, Weis JS. 2007. The invasive green crab and Japanese shore crab: behavioral interactions with a native crab species, the blue crab. *Biological Invasions* 9:837–48.

Mackenzie CL. 2007. Causes underlying the historical decline in eastern oyster (Crassostrea virginica Gmelin, 1791) landings. *Journal of Shellfish Research* 26(4):927–39.

Matheson K, McKenzie CH, Gregory RS, Robichaud DA, Bradbury IR, Snelgrove PV, Rose GA. 2016. Linking eelgrass decline and impacts on associated fish communities to European green crab Carcinus maenas invasion. *Marine Ecology Progress Series* 548:31–45.

McDermott JJ. 1998. The western Pacific brachyuran (Hemigrapsus sanguineus: Grapsidae), in its new habitat along the Atlantic coast of the United States: geographic distribution and ecology. *ICES Journal of Marine Science* 55(2):289–98.

MAFMC (Mid-Atlantic Fishery Management Council). 2012. *Council Policy on Impacts of Fishing Activities on Fish Habitat.* (p. 4).

MAFMC (Mid-Atlantic Fishery Management Council). 2015. *Policies on Non-Fishing Activities and Projects that Impact Fish Habitat.* (p. 19).

MAFMC (Mid-Atlantic Fishery Management Council). 2017. *Unmanaged Forage Omnibus Amendment.* (p. 223).

MAFMC (Mid-Atlantic Fishery Management Council). 2019a. *Ecosystem Approach to Fisheries Management Guidance Document.* (p. 38).

MAFMC (Mid-Atlantic Fishery Management Council). 2019b. *Northeast Regional Marine Fish Habitat Assessment Workplan July 1, 2019–June 31, 2022.* (p. 15).

Mallin MA, Corbett CA. 2006. How hurricane attributes determine the extent of environmental effects: multiple hurricanes and different coastal systems. *Estuaries and Coasts* 29(6):1046–61.

Manderson J, Palamara L, Kohut J, Oliver MJ. 2011. Ocean observatory data are useful for regional habitat modeling of species with different vertical habitat preferences. *Marine Ecology Progress Series* 438:1–7.

Mann R, Harding JM. 2000. Invasion of the North American Atlantic coast by a large predatory Asian mollusc. *Biological Invasions* 2(1):7–22.

Mann R, Powell EN. 2007. Why oyster restoration goals in the Chesapeake Bay are not and probably cannot be achieved. *Journal of Shellfish Research* 26(4):905–18.

Mann R, Southworth M, Carnegie RB, Crockett RK. 2014. Temporal variation in fecundity and spawning in the eastern oyster, Crassostrea virginica, in the Piankatank River, Virginia. *Journal of Shellfish Research* 33(1):167–76.

MARACOOS (Mid-Atlantic Regional Association Coastal Ocean Observing System). 2016. *MARACOOS Semi-Annual Report: 12/1/2015–5/31/2016. NOAA Award Number NA11NOS0120038 (June 2011–May 2016).* Washington, DC: NOAA (p. 10).

MARCO (Mid-Atlantic Regional Council on the Ocean). 2015. *Mid-Atlantic Ocean Data Portal.* Washington, DC: NOAA.

Marshall KN, Koehn LE, Levin PS, Essington TE, Jensen OP. 2019. Inclusion of ecosystem information in US fish stock assessments suggests progress toward ecosystem-based fisheries management. *ICES Journal of Marine Science* 76(1):1–9.

Marshall KN, Levin PS, Essington TE, Koehn LE, Anderson LG, Bundy A, Carothers C, et al. 2018. Ecosystem-based fisheries management for social–ecological systems: renewing the focus in the United States with next generation fishery ecosystem plans. *Conservation Letters* 11(1):e12367.

Marston D. 2003. *The American Revolution 1774–1783.* Abingdon, UK: Routledge.

Martin J, Young J, Kauffman GJ, Homsey A. 2018. *Economic Value of the Maryland Coastal Bays Watershed.* Berlin, MD: Maryland Coastal Bays Program.

Martin KS, Hall-Arber M. 2008. The missing layer: geo-technologies, communities, and implications for marine spatial planning. *Marine Policy* 32(5):779–86.

Martino EJ, Able KW. 2003. Fish assemblages across the marine to low salinity transition zone of a temperate estuary. *Estuarine, Coastal and Shelf Science* 56(5–6):969–87.

McCay, BJ. 2012. Anthropology: shifts in fishing grounds. *Nature Climate Change* 2:840–1.

McClellan CM, Read AJ, Cluse WM, Godfrey MH. 2011. Conservation in a complex management environment: the by-catch of sea turtles in North Carolina's commercial fisheries. *Marine Policy* 35(2):241–8.

McConnell KE, Strand IE. 1994. *The Economic Value of Mid and South Atlantic Sportfishing*. Baltimore, MD: University of Maryland.

McDermott JJ. 1998. The western Pacific brachyuran (Hemigrapsus sanguineus: Grapsidae), in its new habitat along the Atlantic coast of the United States: geographic distribution and ecology. *ICES Journal of Marine Science* 55(2):289–98.

McNamee JE. 2018. *A Multispecies Statistical Catch-At-Age (MSSCAA) Model for a Mid-Atlantic Species Complex*. Doctoral Dissertation. Kingston, RI: University of Rhode Island (p. 288).

Mead IG, Potter CW. 1990. "Natural history of bottlenose dolphins along the Central Atlantic coast." In: *The Bottlenose Dolphin*. Edited by S Leatherwood, R Reeves. Cambridge, MA: Academic Press (p. 165).

Middleton L. 2014. *Nourishing Our Coastal Beaches: Adaptive Strategy or Fiscal Sand Trap?* https://www.chesapeakequarterly.net/sealevel/main15/

Miller WD, Harding Jr LW, Adolf JE. 2006. Hurricane Isabel generated an unusual fall bloom in Chesapeake Bay. *Geophysical Research Letters* 33(6).

Miller, T. 2001. *Chesapeake Bay Fisheries Ecosystem Plan: Patterns in Total Removals*. Solomons, MD: Chesapeake Biological Laboratory.

Minello TJ, Able KW, Weinstein MP, Hays CG. 2003. Salt marshes as nurseries for nekton: testing hypotheses on density, growth and survival through meta-analysis. *Marine Ecology Progress Series* 246:39–59.

Miller TJ, Hare JA, Alade LA. 2016. A state-space approach to incorporating environmental effects on recruitment in an age-structured assessment model with an application to southern New England yellowtail flounder. *Canadian Journal of Fisheries and Aquatic Sciences* 73(8):1261–70.

Miller TJ, Martell SJ, Bunnell DB, Davis G, Fegley L, Sharov A, Bonzek C, et al. 2005. *Stock Assessment of Blue Crab in Chesapeake Bay. Technical Report Series Number TS-487–05*. Cambridge, MD: University of Maryland Center for Environmental Science.

Miller TJ, Wilberg MJ, Colton AR, Davis GR, Sharov A, Lipcius RN, Ralph GM, Johnson EG, Kaufman AG. 2011. *Stock Assessment of Blue Crab in Chesapeake Bay 2011: Final Assessment Report*. Cambridge, MD: University of Maryland Center for Environmental Science.

Monaco ME, Ulanowicz RE. 1997. Comparative ecosystem trophic structure of three US mid-Atlantic estuaries. *Marine Ecology Progress Series* 161:239–54.

Montane MM, Austin HM. 2005. *Effects of Hurricanes on Atlantic Croaker (Micropogonias undulatus) Recruitment to Chesapeake Bay*. Edgewater, MD: Chesapeake Research Consortium (pp. 185–92).

Morgan CW, Bishop JM. 1977. An example of Gulf Stream eddy-induced water exchange in the Mid-Atlantic Bight. *Journal of Physical Oceanography* 7(3):472–9.

Morley JW, Batt RD, Pinsky ML. 2017. Marine assemblages respond rapidly to winter climate variability. *Global Change Biology* 23(7):2590–601.

Morley JW, Selden RL, Latour RJ, Frölicher TL, Seagraves RJ, Pinsky ML. 2018. Projecting shifts in thermal habitat for 686 species on the North American continental shelf. *PLoS One* 13(5):e0196127.

MRB (Mid-Atlantic Regional Planning Body). 2018. *Mid-Atlantic Regional Ocean Assessment*. https://roa.midatlanticocean.org/

Munch SB, Conover DO. 2000. Recruitment dynamics of bluefish (Pomatomus saltatrix) from Cape Hatteras to Cape Cod, 1973–1995. *ICES Journal of Marine Science* 57(2):393–402.

Munroe D, Tabatabai A, Burt I, Bushek D, Powell EN, Wilkin J. 2013. Oyster mortality in Delaware Bay: impacts and recovery from Hurricane Irene and Tropical Storm Lee. *Estuarine, Coastal and Shelf Science* 135:209–19.

Murray KT, Orphanides CD. 2013. Estimating the risk of loggerhead turtle Caretta caretta bycatch in the US mid-Atlantic using fishery-independent and-dependent data. *Marine Ecology Progress Series* 477:259–70.

Musick JA, Mercer LP. 1977. Seasonal distribution of black sea bass, Centropristis striata, in the Mid-Atlantic Bight with comments on the ecology and fisheries of the species. *Transactions of the American Fisheries Society* 106(1):12–25.

Muto MM, Helker VT, Angliss RP, Boveng PL, Breiwick JM, Cameron MF, Clapham PJ, et al. 2019. *Alaska Marine Mammal Stock Assessments, 2018. NOAA Technical Memo NMFS-AFSC-393*. Washington, DC: NOAA (p. 390).

Myers RA, Drinkwater K. 1989. The influence of Gulf Stream warm core rings on recruitment of fish in the northwest Atlantic. *Journal of Marine Research* 47(3):635–56.

Najjar RG, Pyke CR, Adams MB, Breitburg D, Hershner C, Kemp M, Howarth R, et al. 2010. Potential climate-change impacts on the Chesapeake Bay. *Estuarine, Coastal and Shelf Science* 86(1):1–20.

Najjar RG, Walker HA, Anderson PJ, Barron EJ, Bord RJ, Gibson JR, Kennedy VS, et al. 2000. The potential impacts of climate change on the mid-Atlantic coastal region. *Climate Research* 14(3):219–33.

NAS (The National Academy of Sciences, Engineering, and Medicine). 2017. *Review of the Marine Recreational Information Program*. Washington, DC: The National Academies Press.

Neff R, Chang H, Knight CG, Najjar RG, Yarnal B, Walker HA. 2000. Impact of climate variation and change on Mid-Atlantic Region hydrology and water resources. *Climate Research* 14(3):207–18.

NEFC (Northeast Fisheries Center). 1988. *An Evaluation of the Bottom Trawl Survey Program of the Northeast Fisheries Center. NOAA Technical Memo NMFS-F/NEC—52.* Washington, DC: NOAA (p. 83).

NEFSC (Northeast Fisheries Science Center). 2006. *Cruise Results NOAA Ship Albatross IV Cruise No. AL 06-02 (Parts I-II) Winter Bottom Trawl Survey*. Washington, DC: NOAA (p. 6).

NEFSC (Northeast Fisheries Science Center). 2012. *Ecosystem Status Report for the Northeast Shelf Large Marine Ecosystem—2011. NOAA-NMFS. Northeast Fisheries Science Center Reference Document 12-07*. Washington, DC: NOAA (p. 33).

NEFSC (Northeast Fisheries Science Center). 2013a. *Cruise Results NOAA Ship Henry B. Bigelow (R-225) Cruise No. HB 12-06 (Parts I–IV) Fall Bottom Trawl Survey*. Washington, DC: NOAA (p. 9).

NEFSC (Northeast Fisheries Science Center). 2013b. *Cruise Results NOAA Ship Henry B. Bigelow (R-225) Cruise No. HB 12-01 (Parts I–IV) Spring Bottom Trawl Survey*. Washington, DC: NOAA (p. 9).

NEFSC (Northeast Fisheries Science Center). 2018. *Cruise Results UNOLS R/V Hugh R. Sharp Cruise No. S1 18-01 (Parts I –III) Sea Scallop Survey*. Washington, DC: NOAA (p. 9).

Newell RI. 1988. Ecological changes in Chesapeake Bay: are they the result of overharvesting the American oyster, Crassostrea virginica. *Understanding the Estuary: Advances in Chesapeake Bay Research* 129:536–46.

Nichols CR, Zarillo G, D'Elia CF. 2019. "Mid-Atlantic Bight and Chesapeake Bay." In: *Tomorrow's Coasts: Complex and Impermanent.* Edited by LD Wright, CR Nichols. Cham, Switzerland: Springer (pp. 241–59).

NMFS (National Marine Fisheries Service). 2016a. *Fisheries of the Northeastern United States; Atlantic Mackerel, Squid, and Butterfish Fisheries; Amendment 16. National Oceanic and Atmospheric Administration 50 CFR Part 648 [Docket No.: 160706587–6999–02] RIN 0648–BG21.* Washington, DC: NOAA (p. 9).

NMFS (National Marine Fisheries Service). 2016b. *Ecosystem-Based Fisheries Management Policy of the National Marine Fisheries Service, National Oceanic and Atmospheric Administration*. Washington, DC: NOAA.

NMFS (National Marine Fisheries Service). 2016c. *NOAA Fisheries Ecosystem-Based Fisheries Management Roadmap.* Washington, DC: NOAA.

NMFS (National Marine Fisheries Service). 2017a. *National Marine Fisheries Service—2nd Quarter 2017 Update.* Washington, DC: NOAA (p. 53).

NMFS (National Marine Fisheries Service). 2017b. *National Observer Program FY 2013 Annual Report. NOAA Technical Memo NMFS F/SPO-178.* Washington, DC: NOAA (p. 34).

NMFS (National Marine Fisheries Service). 2018a. *Fisheries Economics of the United States, 2016. NOAA Technical Memo NMFS-F/SPO-187*. Washington, DC: NOAA (p. 243).

NMFS (National Marine Fisheries Service). 2018b. *National Seabird Program 2018 Annual Report.* Washington, DC: NOAA (p. 16).

NMFS (National Marine Fisheries Service). 2019a. *Atlantic Bluefish Operational Assessment for 2019.* Washington, DC: NOAA (p. 12).

NMFS (National Marine Fisheries Service). 2019b. *Northeast Regional Implementation Plan of the NOAA Fisheries Ecosystem-Based Fishery Management Roadmap.* Washington, DC: NOAA (p. 25).

NMFS (National Marine Fisheries Service). 2020. *National Marine Fisheries Service—2nd Quarter 2020 Update.* Washington, DC: NOAA (p. 51).

NOAA (National Oceanic and Atmospheric Administration) and CBFEAP (Chesapeake Bay Office—Chesapeake Bay Fisheries Ecosystem Advisory Panel). 2006. *Fisheries Ecosystem Planning for Chesapeake Bay. Trends in Fisheries Science and Management 3.* Bethesda, MD: American Fisheries Society.

Nordlie FG. 2003. Fish communities of estuarine salt marshes of eastern North America, and comparisons with temperate estuaries of other continents. *Reviews in Fish Biology and Fisheries* 13(3):281–325.

NUWC (US Naval Undersea Warfare Center Division). 2012. *Determination of Acoustic Effects on Marine Mammals and Sea Turtles for the Atlantic Fleet Training and Testing Environmental Impact Statement/Overseas Environmental Impact Statement. Marine Species Modeling Team. NUWC-NPT Technical Report 12,071.* Newport, RI: NAVSEA (p. 88).

Nye JA, Baker MR, Bell R, Kenny A, Kilbourne KH, Friedland KD, Martino E, et al. 2014. Ecosystem effects of the Atlantic multidecadal oscillation. *Journal of Marine Systems* 133:103–16.

Nye JA, Gamble RJ, Link JS. 2013. The relative impact of warming and removing top predators on the Northeast US large marine biotic community. *Ecological Modelling* 264:157–68.

Nye JA, Link JS, Hare JA, Overholtz WJ. 2009. Changing spatial distribution of fish stocks in relation to climate and population size on the Northeast United States continental shelf. *Marine Ecology Progress Series* 393:111–29.

O'Connell AF, Gardner B, Gilbert AT, Laurent K. 2009. *Compendium of Avian Occurrence Information for the Continental Shelf Waters Along the Atlantic Coast of the United States, Final Report (Database Section—Seabirds). OCS Study BOEM 2012-076.* Beltsville, MD: USGS Patuxent Wildlife Research Center.

OCM (Office for Coastal Management). 2019. *NOAA Report on the U.S. Ocean and Great Lakes Economy.* Charleston, SC: NOAA Office of Coastal Management.

Olsen E, Fluharty D, Hoel AH, Hostens K, Maes F, Pecceu E. 2014. Integration at the round table: marine spatial planning in multi-stakeholder settings. *PLoS One* 9(10):e109964.

Orth RJ, Dennison WC, Gurbisz C, Hannam M, Keisman J, Landry JB, Lefcheck JS, et al. 2019. Long-term annual aerial surveys of submersed aquatic vegetation (SAV) support science, management, and restoration. *Estuaries and Coasts* 4:1–6.

Orth RJ, Heck KL, van Montfrans J. 1984. Faunal communities in seagrass beds: a review of the influence of plant structure and prey characteristics on predator-prey relationships. *Estuaries* 7(4):339–50.

Orth RJ, Luckenbach ML, Marion SR, Moore KA, Wilcox DJ. 2006. Seagrass recovery in the Delmarva coastal bays, USA. *Aquatic Botany* 84(1):26–36.

Ozbay G, Fan C, Yang Z. 2017. "Relationship between land use and water quality and its assessment using hyperspectral remote sensing in Mid-Atlantic estuaries." In: *Water Quality*. Edited by H Tutu. London, UK: IntechOpen.

Packer DB, Hoff T. 1999. "Life history, habitat parameters, and essential habitat of Mid-Atlantic summer flounder." In: *Fish Habitat: Essential Fish Habitat and Rehabilitation. Symposium 22.* Edited by LR Benaka. Bethesda, MD: American Fisheries Society (pp. 76–92).

Paerl HW, Bales JD, Ausley LW, Buzzelli CP, Crowder LB, Eby LA, Go M, et al. 2000. Hurricanes' hydrological, ecological effects linger in major US estuary. *Eos, Transactions American Geophysical Union* 81(40):457–62.

Paerl HW, Valdes LM, Joyner AR, Peierls BL, Piehler MF, Riggs SR, Christian RR, et al. 2006. Ecological response to hurricane events in the Pamlico Sound system, North Carolina, and implications for assessment and management in a regime of increased frequency. *Estuaries and Coasts* 29(6):1033–45.

Patrick CJ, Weller DE, Li X, Ryder M. 2014. Effects of shoreline alteration and other stressors on submerged aquatic vegetation in subestuaries of Chesapeake Bay and the mid-Atlantic coastal bays. *Estuaries and Coasts* 37(6):1516–31.

Patrick CJ, Weller DE, Orth RJ, Wilcox DJ, Hannam MP. 2018. Land use and salinity drive changes in SAV abundance and community composition. *Estuaries and Coasts* 41(1):85–100.

Patrick CJ, Weller DE, Ryder M. 2016. The relationship between shoreline armoring and adjacent submerged aquatic vegetation in Chesapeake Bay and nearby Atlantic coastal bays. *Estuaries and Coasts* 39(1):158–70.

Paul JF, Comeleo RL, Copeland J. 2002. Landscape metrics and estuarine sediment contamination in the Mid-Atlantic and southern New England regions. *Journal of Environmental Quality* 31(3):836–45.

Paul JF, Strobel CJ, Melzian BD, Kiddon JA, Latimer JS, Campbell DE, Cobb DJ. 1998. State of the estuaries in the Mid-Atlantic region of the United States. *Environmental Monitoring and Assessment* 51(1–2):269–84.

Peierls BL, Christian RR, Paerl HW. 2003. Water quality and phytoplankton as indicators of hurricane impacts on a large estuarine ecosystem. *Estuaries* 26(5):1329–43.

Peterson CH, Grabowski JH, Powers SP. 2003. Estimated enhancement of fish production resulting from restoring oyster reef habitat: quantitative valuation. *Marine Ecology Progress Series* 264:249–64.

Phelps HL. 1994. The Asiatic clam (Corbicula fluminea) invasion and system-level ecological change in the Potomac River estuary near Washington, DC. *Estuaries* 17(3):614–21.

Pierre RS. 1999. Restoration of Atlantic sturgeon in the northeastern USA with special emphasis on culture and restocking. *Journal of Applied Ichthyology* 15(4-5):180–2.

Pikitch E, Boersma PD, Boyd I, Conover D, Cury P, Essington T, Heppell S, et al. 2012. *Little Fish, Big Impact: Managing a Crucial Link in Ocean Food Webs.* Washington, DC: Lenfest Ocean Program (p. 108).

Pinsky ML, Fogarty M. 2012. Lagged social-ecological responses to climate and range shifts in fisheries. *Climatic Change* 115(3–4):883–91.

Pinsky ML, Worm B, Fogarty MJ, Sarmiento JL, Levin SA. 2013. Marine taxa track local climate velocities. *Science* 341(6151):1239–42.

Politis PJ, Galbraith JK, Kostovick P, Brown RW. 2014. *Northeast Fisheries Science Center Bottom Trawl Survey Protocols for the NOAA Ship Henry B. Bigelow. Northeast Fisheries Science Center Reference Document 14-06.* Washington, DC: NOAA (p. 138).

Polsky C, Allard J, Currit N, Crane R, Yarnal B. 2000. The Mid-Atlantic Region and its climate: past, present, and future. *Climate Research* 14(3):161–73.

Prosser DJ, Jordan TE, Nagel JL, Seitz RD, Weller DE, Whigham DF. 2018. Impacts of coastal land use and shoreline armoring on estuarine ecosystems: an introduction to a special issue. *Estuaries and Coasts* 41(1):2–18.

Restaino DJ, Bologna PA, Gaynor JJ, Buchanan GA, Bilinski JJ. 2018. Who's lurking in your lagoon? First occurrence of the invasive hydrozoan Moerisia sp. (Cnidaria: Hydrozoa) in New Jersey, USA. *BioInvasions Record* 7(3).

Ribeiro F, Hale E, Hilton EJ, Clardy TR, Deary AL, Targett TE, Olney JE. 2015. Composition and temporal patterns of larval fish communities in Chesapeake and Delaware Bays, USA. *Marine Ecology Progress Series* 527:167–80.

Rice D, Rooth J. 2000. Colonization and expansion of Phragmites australis in upper Chesapeake Bay tidal marshes. *Wetlands* 20(2):280.

Richards RA, Deuel DG. 1987. Atlantic striped bass: stock status and the recreational fishery. *Marine Fisheries Review* 49(2):58–66.

Richards RA, Rago PJ. 1999. A case history of effective fishery management: Chesapeake Bay striped bass. *North American Journal of Fisheries Management* 19(2):356–75.

Richardson DE, Hare JA, Overholtz WJ, Johnson DL. 2010. Development of long-term larval indices for Atlantic herring (Clupea harengus) on the northeast US continental shelf. *ICES Journal of Marine Science* 67(4):617–27.

Rodney WS, Paynter KT. 2006. Comparisons of macrofaunal assemblages on restored and non-restored oyster reefs in mesohaline regions of Chesapeake Bay in Maryland.

Journal of Experimental Marine Biology and Ecology 335(1):39–51.

Rogers CE, McCarty JP. 2000. Climate change and ecosystems of the Mid-Atlantic Region. *Climate Research* 14(3):235–44.

Roman CT, Jaworski N, Short FT, Findlay S, Warren RS. 2000. Estuaries of the Northeastern United States: habitat and land use signatures. *Estuaries* 23(6):743–64.

Rountree RA, Able KW. 1992. Fauna of polyhaline subtidal marsh creeks in southern New Jersey: composition, abundance and biomass. *Estuaries* 15(2):171–85.

Rossman MC. 2007. *Allocating Observer Sea Days to Bottom Trawl and Gillnet Fisheries in the Northeast and Mid-Atlantic Regions to Monitor and Estimate Incidental Bycatch of Marine Mammals. Northeast Fisheries Science Center Reference Document 07-19*. Washington, DC: NOAA.

Rothschild BJ, Ault JS, Goulletquer P, Héral M. 1994. Decline of the Chesapeake Bay oyster population: a century of habitat destruction and overfishing. *Marine Ecology Progress Series* 11:29–39.

Rudershausen PJ, Buckel JA, Edwards J, Gannon DP, Butler CM, Averett TW. 2010. Feeding ecology of blue marlins, dolphinfish, yellowfin tuna, and wahoos from the North Atlantic Ocean and comparisons with other oceans. *Transactions of the American Fisheries Society* 139(5):1335–59.

Sadler JM, Haselden N, Mellon K, Hackel A, Son V, Mayfield J, Blase A, Goodall JL. 2017. Impact of sea-level rise on roadway flooding in the Hampton Roads region, Virginia. *Journal of Infrastructure Systems* 23(4):05017006.

Samhouri JF, Haupt AJ, Levin PS, Link JS, Shuford R. 2013. Lessons learned from developing integrated ecosystem assessments to inform marine ecosystem-based management in the USA. *ICES Journal of Marine Science* 71(5):1205–15.

Samhouri JF, Levin PS, Ainsworth CH. 2010. Identifying thresholds for ecosystem-based management. *PLoS One* 5(1):e8907.

Samoteskul K, Firestone J, Corbett J, Callahan J. 2014. Changing vessel routes could significantly reduce the cost of future offshore wind projects. *Journal of Environmental Management* 141:146–54.

Sanchirico JN, Smith MD, Lipton DW. 2008. An empirical approach to ecosystem-based fishery management. *Ecological Economics* 64(3):586–96.

Sartwell T. 2009. *What Can Be Done to Save the East Coast Blue Crab Fishery. Master's Thesis.* Durham, NC: Duke University (p. 46).

Savidge DK, Austin JA. 2007. The Hatteras Front: August 2004 velocity and density structure. *Journal of Geophysical Research: Oceans* 112(C7).

Seara T, Clay PM, Colburn LL. 2016. Perceived adaptive capacity and natural disasters: a fisheries case study. *Global Environmental Change* 38:49–57.

Selzer LA, Payne PM. 1988. The distribution of white-sided (Lagenorhynchus acutus) and common dolphins (Delphinus delphis) vs. environmental features of the continental shelf of the northeastern United States. *Marine Mammal Science* 4(2):141–53.

Schmitt JD, Hallerman EM, Bunch A, Moran Z, Emmel JA, Orth DJ. 2017. Predation and prey selectivity by nonnative catfish on migrating alosines in an Atlantic slope estuary. *Marine and Coastal Fisheries* 9(1):108–25.

Schmitt JD, Peoples BK, Castello L, Orth DJ. 2019. Feeding ecology of generalist consumers: a case study of invasive blue catfish Ictalurus furcatus in Chesapeake Bay, Virginia, USA. *Environmental Biology of Fishes* 102(3): 443–65.

Schloesser RW, Fabrizio MC, Latour RJ, Garman GC, Greenlee B, Groves M, Gartland J. 2011. Ecological role of Blue Catfish in Chesapeake Bay communities and implications for management. In: *Conservation, Ecology, and Management of Catfish: The Second International Symposium*. Edited by PH Michaletz, VH Travnichek. Bethesda, MD: American Fisheries Society, Symposium 77 (pp. 369–82).

Schofield O, Chant R, Cahill B, Castelao R, Gong D, Kahl A, Kohut J, et al. 2008. The decadal view of the Mid-Atlantic Bight from the COOLroom: is our coastal system changing? *Oceanography* 21(4):108–17.

Schulte DM. 2017. History of the Virginia oyster fishery, Chesapeake Bay, USA. *Frontiers in Marine Science* 4:127.

Secor DH, Trice TM, Hornick HT. 1995. Validation of otolith-based ageing and a comparison of otolith and scale-based ageing in mark-recaptured Chesapeake Bay striped bass, Morone saxatilis. *Fishery Bulletin* 93(1):186–90.

SEDAR (Southeast Data Assessment and Review). 2020. *SEDAR 69. Benchmark Stock Assessment Report. Atlantic Menhaden*. North Charleston, SC: SEDAR (p. 481).

Shackell NL, Bundy A, Nye JA, Link JS. 2012. Common large-scale responses to climate and fishing across Northwest Atlantic ecosystems. *ICES Journal of Marine Science* 69(2):151–62.

Shaw T, García-Artola A, Engelhart S, Kemp A, Cahill N, Nikitina D, Corbett R, et al. 2016. *Sea-Level in the US Mid-Atlantic Coast During the Common Era. EPSC2016-8725.* Vienna, Austria: EGU General Assembly Conference Abstracts 2016.

Sherman K, Jaworski NJ, Smayda T. 1996. *The Northeast Shelf Ecosystem: Assessment, Sustainability and Management.* Cambridge, MA: Blackwell Science (p. 564).

Sherman K, Kane J, Murawski S, Overholtz W, Solow A. 2002. 6 The US Northeast shelf large marine ecosystem: zooplankton trends in fish biomass recovery. *Large Marine Ecosystems, Elsevier* 10:195–215.

Shen J, Wang T, Herman J, Mason P, Arnold GL. 2008. Hypoxia in a coastal embayment of the Chesapeake Bay: a model diagnostic study of oxygen dynamics. *Estuaries and Coasts* 31(4):652–63.

Shoop CR, Kenney RD. 1992. Seasonal distributions and abundances of loggerhead and leatherback sea turtles in waters of the northeastern United States. *Herpetological Monographs* 1:43–67.

Sibunka JD, Silverman MJ. 1989. *MARMAP Surveys of the Continental Shelf from Cape Hatteras, North Carolina, to Cape Sable, Nova Scotia 1984–1987. Atlas No. 3. Summary of Operations*. Woods Hole, MA: National Marine Fisheries Service.

Smith E. 1999. Atlantic and east coast hurricanes 1900–98: a frequency and intensity study for the twenty-first century. *Bulletin of the American Meteorological Society* 80(12):2717–20.

Smith L, Gamble R, Gaichas S, Link J. 2015. Simulations to evaluate management tradeoffs among marine mammal consumption needs, commercial fishing fleets and finfish biomass. *Marine Ecology* 523:215–32.

Smith SL, Cunniff SE, Peyronnin NS, Kritzer JP. 2017. Prioritizing coastal ecosystem stressors in the Northeast United States under increasing climate change. *Environmental Science & Policy* 78:49–57.

Smith, T.D. 2002. The Woods Hole bottom-trawl resource survey: development of fisheries-independent multi-species monitoring. *ICES Marine Science Symposia* 215:474–482.

Smith TI. 1985. The fishery, biology, and management of Atlantic sturgeon, Acipenser oxyrhynchus, in North America. *Environmental Biology of Fishes* 14(1):61–72.

Smith TI, Clugston JP. 1997. Status and management of Atlantic sturgeon, Acipenser oxyrinchus, in North America. *Environmental Biology of Fishes* 48(1–4):335–46.

Smolowitz RJ, Patel SH, Haas HL, Miller SA. 2015. Using a remotely operated vehicle (ROV) to observe loggerhead sea turtle (Caretta caretta) behavior on foraging grounds off the Mid-Atlantic United States. *Journal of Experimental Marine Biology and Ecology* 471:84–91.

Smythe TC, McCann J. 2019. Achieving integration in marine governance through marine spatial planning: findings from practice in the United States. *Ocean & Coastal Management* 167:197–207.

Stafford S, Abramowitz J. 2017. An analysis of methods for identifying social vulnerability to climate change and sea level rise: a case study of Hampton Roads, Virginia. *Natural Hazards* 85(2):1089–117.

Stanley JG, Sellers MA. 1986. Species Profiles. Life Histories and Environmental Requirements of Coastal Fishes and Invertebrates (Mid-Atlantic). American Oyster. *US Fish and Wildlife Service Biological Report* 82(11.65):25.

Stevenson JC, Ward LG, Kearney MS. 1988. Sediment transport and trapping in marsh systems: implications of tidal flux studies. *Marine Geology* 80(1–2):37–59.

Strayer DL. 1999. Effects of alien species on freshwater mollusks in North America. *Journal of the North American Benthological Society* 18(1):74–98.

Stedman S, Dahl TE. 2008. *Status and Trends of Wetlands in the Coastal Watersheds of the Eastern United States 1998 to 2004*. National Oceanic and Atmospheric Administration, National Marine Fisheries Service and U.S. Department of the Interior, Fish and Wildlife Service (p. 32).

Steffen CG. 1984. *The Mechanics of Baltimore: Workers and Politics in the Age of Revolution, 1763–1812*. Champaign, IL: University of Illinois Press.

Steimle FW, Zetlin C. 2000. Reef habitats in the Mid-Atlantic bight: abundance, distribution, associated biological communities, and fishery resource use. *Marine Fisheries Review* 62(2):24–42.

Stein AB, Friedland KD, Sutherland M. 2004. Atlantic sturgeon marine distribution and habitat use along the northeastern coast of the United States. *Transactions of the American Fisheries Society* 133(3):527–37.

Stevens JD, Bonfil R, Dulvy NK, Walker PA. 2000. The effects of fishing on sharks, rays, and chimaeras (chondrichthyans), and the implications for marine ecosystems. ICES Journal of Marine Science 57(3):476–94.

Stevenson D, Chiarella L, Stephan D, Reid R, Wilhelm K, McCarthy J, Pentony M. 2004. *Characterization of the Fishing Practices and Marine Benthic Ecosystems of the Northeast US Shelf, and an Evaluation of the Potential Effects of Fishing on Essential Fish Habitat. NOAA Technical Memorandum NMFS-NE-181*. Washington, DC: NOAA (p. 179).

Stevenson JC, Kearney MS, Pendleton EC. 1985. Sedimentation and erosion in a Chesapeake Bay brackish marsh system. *Marine Geology* 67(3–4):213–35.

Stewart RW. 2004. *American Military History: The United States Army in a Global Era, 1917–2003*. Washington, DC: Government Printing Office.

Stierhoff KL, Targett TE, Miller K. 2006. Ecophysiological responses of juvenile summer and winter flounder to hypoxia: experimental and modeling analyses of effects on estuarine nursery quality. *Marine Ecology Progress Series* 325:255–66.

Sutinen JG, Johnston RJ. 2003. Angling management organizations: integrating the recreational sector into fishery management. *Marine Policy* 27(6):471–87.

Tam JC, Link JS, Large SI, Andrews K, Friedland KD, Gove J, Hazen E, et al. 2017. Comparing apples to oranges: common trends and thresholds in anthropogenic and environmental pressures across multiple marine ecosystems. *Frontiers in Marine Science* 4:282.

Tegen S, Keyser D, Flores-Espino F, Miles J, Zammit D, Loomis D. 2015. *Offshore Wind Jobs and Economic Development Impacts in the United States: Four Regional Scenarios*. Golden, CO: National Renewable Energy Laboratory.

Terceiro M. 2016. *Stock Assessment of Summer Flounder for 2016. Northeast Fishery Science Center Reference Document 16–15*. Woods Hole, MA National Marine Fisheries Service (p. 117).

Townsend HM, Harvey CJ, Aydin KY, Gamble RJ, Gruss A, Levin PS, Link JS, et al. 2014. *Report of the 3rd National Ecosystem Modeling Workshop (NEMoW 3): Mingling Models for Marine Resource Management, Multiple Model Inference. NOAA Technical Memorandum NMFS-F/SPO-149.* Washington, DC: NOAA.

Townsend HM, Link JS, Osgood KE, Gedamke T, Watters GM, Polovina JJ, Levin PS, Cyr EC, Aydin KY. 2008. *Report of the National Ecosystem Modeling Workshop (NEMoW). NOAA Technical Memorandum NMFS-F/SPO-87.* Washington, DC: NOAA.

Townsend HK, Holsman AK, Harvey C, Kaplan I, Hazen E, Woodworth-Jefcoats P, et al. 2017. *Report of the 4th National Ecosystem Modeling Workshop (NEMoW 4): Using Ecosystem Models to Evaluate Inevitable Trade-offs. NOAA Technical Memorandum NMFS-F/SPO-173.* Washington, DC: NOAA (p. 77).

Tyrrell MC, Link JS, Moustahfid H, Overholtz WJ. 2008. Evaluating the effect of predation mortality on forage species population dynamics in the Northeast US continental shelf ecosystem using multispecies virtual population analysis. *ICES Journal of Marine Science* 65(9):1689–700.

Uphoff Jr JH. 2003. Predator–prey analysis of striped bass and Atlantic menhaden in upper Chesapeake Bay. *Fisheries Management and Ecology* 10(5):313–22.

Valle-Levinson A, Dutton A, Martin JB. 2017. Spatial and temporal variability of sea level rise hot spots over the eastern United States. *Geophysical Research Letters* 44(15):7876–82.

Vasslides JM, de Mutsert K, Christensen V, Townsend H. 2017. Using the Ecopath with Ecosim modeling approach to understand the effects of watershed-based management actions in coastal ecosystems. *Coastal Management* 45(1):44–55.

Vogelsong S. 2020. Menhaden regulations have plagued lawmakers for decades. why were they able to find a fix this year? *The Virginia Mercury.* https://www.virginiamercury.com/2020/02/17/menhaden-have-plagued-lawmakers-for-decades-why-were-they-able-to-find-a-fix-this-year/

Warner JC, Schwab WC, List JH, Safak I, Liste M, Baldwin W. 2017. Inner-shelf ocean dynamics and seafloor morphologic changes during Hurricane Sandy. *Continental Shelf Research* 138:1–8.

Weinrich MT, Clapham PJ. 2002. Population identity of humpback whales (Megaptera novaeangliae) in the waters of the US mid-Atlantic states. *Journal of Cetacean Research and Management* 4(2):135–41.

Wilkinson EB, Abrams K. 2015. *Benchmarking the 1999 EPAP Recommendations with Existing Fishery Ecosystem Plans. NOAA Technical Memorandum NMFS-OSF-5.* Washington, DC: NOAA (p. 22).

Wilson DC, McCay BJ. 1998. How the participants talk about "participation" in Mid-Atlantic fisheries management. *Ocean & Coastal Management* 41(1):41–61.

Wilson MD, Watts BD, Brinker DF. 2007. Status review of Chesapeake Bay marsh lands and breeding marsh birds. *Waterbirds* 30(sp1):122–38.

Wong RA. 2009. *2009 Assessment of the Delaware Bay Blue Crab (Callinectes sapidus) Stock. Delaware Division of Fish and Wildlife.* Dover, DE: Department of Natural Resources & Environmental Control (p. 109).

Wu SY, Yarnal B, Fisher A. 2002. Vulnerability of coastal communities to sea-level rise: a case study of Cape May County, New Jersey, USA. *Climate Research* 22(3): 255–70.

Wray RD, Leatherman SP, Nicholls RJ. 1995. Historic and future land loss for upland and marsh islands in the Chesapeake Bay, Maryland, US. *Journal of Coastal Research* 11:1195–203.

Wyatt KH, Griffin R, Guerry AD, Ruckelshaus M, Fogarty M, Arkema KK. 2017. Habitat risk assessment for regional ocean planning in the US Northeast and Mid-Atlantic. *PloS One* 12(12):e0188776.

Young T, Fuller EC, Provost MM, Coleman KE, St. Martin K, McCay BJ, Pinsky ML. 2018. Adaptation strategies of coastal fishing communities as species shift poleward. ICES Journal of Marine Science 76(1):93–103.

Zervas, C. 2001. *Sea Level Variations of the United States, 1854–1999, NOAA Technical Report NOS CO-OPS 36.* Silver Spring, MD: National Ocean Service (p. 66).

CHAPTER 5

The U.S. South Atlantic Region

➲ The Region in Brief

- The South Atlantic contains the third-highest number of managed taxa of the eight regional U.S. marine ecosystems, including commercially and recreationally important reef fishes (snappers and groupers), penaeid shrimps, coastal migratory pelagic fishes (Cobia, mackerels, Dolphin/Wahoo), and coral reef resources.
- This region contains the second-highest percentage of stocks that are overfished (10%) or experiencing overfishing (14%) in the U.S., and a high percentage of stocks of unknown status.
- Human population and population density are high in this region, with increasing values over time, contributing significant pressures to marine habitats and maintaining coastal and ocean economies that are highly dependent on tourism.
- The South Atlantic has a comparatively low to mid-range socioeconomic status relative to other U.S. regions, and the fifth-highest living marine resource (LMR)-based economy in the U.S.
- This system is driven primarily by circulation of the Gulf Stream, which is the major boundary current for the U.S. Atlantic coast.
- Warming and species distribution shifts within this region have been less pronounced than in other regions, with average sea surface temperatures remaining relatively consistent and having increased by >0.65°C since the mid-twentieth century, particularly over the past decade.
- Although the South Atlantic has the fifth-highest basal productivity in the U.S., among regions it contains lower overall fisheries landings, LMR-dependent jobs, and revenue.
- Offshore coral reef and hard-bottom habitats, in addition to nearshore estuarine oyster reefs, salt marshes, and submerged aquatic vegetation, support high numbers of economically and ecologically important species throughout their life histories.
- High numbers of marine mammal stocks (of which approximately half are strategic stocks) and mid-level numbers of threatened/endangered species are found in this region.
- The South Atlantic is tied nationally for second-highest in the development of offshore wind energy areas.
- Fisheries management among state and federal entities has been focused upon rebuilding plans for historically overfished commercially and recreationally important reef species, with low litigation and one of the most consistently balanced fishery management councils.
- Among subregions, Cape Hatteras-Pamlico Sound contributes more heavily toward fisheries landings, LMR revenue, and LMR employments, while recreational fishing and other ocean uses are more prominent in the Florida Keys.
- Overall, EBFM progress has been made in terms of implementing ecosystem-level planning, advancing knowledge of ecosystem principles, and in assessing risks and vulnerabilities to ecosystems through ongoing investigations into climate vulnerability and species prioritizations for stock and habitat assessments.
- As related to the determinants of successful LMR management, this ecosystem is above average in terms of its higher-ranked human environment while also being low to mid-ranked for total primary production and socioeconomic status.

Ecosystem-Based Fisheries Management: Progress, Importance, and Impacts in the United States. Jason S. Link and Anthony R. Marshak, Oxford University Press. © U.S. Department of Commerce, U.S. Government 2021. DOI: 10.1093/oso/9780192843463.003.0005

5.1 Introduction

Although the U.S. South Atlantic is not especially known by any single iconic fisheries species, it brings to mind significant pelagic sport fisheries that include billfishes and tunas, Dolphinfish (*Coryphaena hippurus*), Cobia (*Rachycentron canadum*), and Wahoo (*Acanthocybium solandri*), major reef fisheries largely composed of the economically important snapper-grouper complex, and popular invertebrate fisheries such as Blue crab (*Callinectes sapidus*), penaeid shrimp, Spiny lobster (*Panulirus* spp.) and Florida stone crab (*Menippe mercenaria*). Other important nearshore species include commercially and recreationally important Red drum (*Sciaenops ocellatus*) and Spotted seatrout (*Cynoscion nebulosus*). Additionally, regional images of expansive sandy beaches, cruise ships, shrimping and charter fishing boats, salt marshes, and diving on shipwrecks complement the local tastes from its marine environments, such as a nice grouper or hogfish sandwich, Calabash shrimp, or those enjoyed at a traditional Carolina oyster roast. The South Atlantic ecosystem invokes many regional sounds such as Appalachian bluegrass, rhythm and blues, folk and country music, influences from Latin American and Caribbean music (e.g., salsa, merengue, and reggae), especially in South Florida, in addition to revving motors from Daytona stock cars further north. Classic tales capturing its culture and set in its major coastal cities include works by *Gone with the Wind* author, Alexandra Ripley, the opera *Porgy and Bess* (Charleston, SC), *Midnight in the Garden of Good and Evil* (Savannah, GA), and novels by Elmore Leonard such as *Rum Punch* and *Get Shorty* (Miami, FL). The Florida Keys also invoke strong ties to Ernest Hemingway, where he lived and roamed while writing several popular works. At its other limit, the North Carolina Outer Banks are considered the "Graveyard of the Atlantic," given the high numbers of shipwrecks that have occurred in treacherous waters off this area (Stick 1989). Many of these cultural overtones are related to the marine environments that support its beach and ocean-based culture and the living marine resources (LMRs) associated with this region. They contribute heavily toward its significant tourism and marine transportation-based marine economy, while also serving as a major cultural resource for native and visiting members of its coastal communities.

There are many works that characterize the South Atlantic ecosystem (Atkinson et al. 1985, Yoder 1991, SAFMC 1998, Dame et al. 2000, Sedberry 2001, Sedberry et al. 2001, Carpenter 2002, Joye et al. 2006, Marancik & Hare 2007, Schaefer & Alber 2007, SAFMC 2009, Fautin et al. 2010, Conley et al. 2017, SAFMC 2018, Link & Marshak 2019, Shertzer et al. 2019), its important and major fisheries (e.g., large and coastal pelagic fishes, snappers, groupers, reef fishes, Spiny lobster; Labisky et al. 1980, Schmied & Burgess 1987, McGowan & Richards 1989, Wenner & Sedberry 1989, Sedberry et al. 1994, Ault et al. 1998, Manooch et al. 1998, Coleman et al. 2000, Brown-Peterson et al. 2001, Harper & Muller 2001, Gentner & Lowther 2002, McGovern et al. 2002, Coleman et al. 2004, Kendall et al. 2008, Shertzer & Williams 2008, Flaherty et al. 2014, Shertzer et al. 2019), its protected species (e.g., dolphins, whales, sea turtles, manatees, sawfishes; Caldwell & Golley 1965, Irvine & Campbell 1978, Schmidly 1981, Hersh 1987, Epperly et al. 1995, Wells et al. 1999, Deutsch et al. 2003, Craig & Reynolds 2004, Keller et al. 2006, Carlson et al. 2007, Shamblin et al. 2011, Garrison et al. 2012, Norton et al. 2012), and their contributions toward the marine socioeconomics of this region, especially in how humans use, enjoy, and interact with this system (Buchanan 1973, Milon et al. 1983, Roehl et al. 1993, Ditton & Stoll 2000, Leeworthy & Wiley 2001, Bohnsack et al. 2002, Jepson et al. 2002, Klein et al. 2004, Johns et al. 2014). This marine region, its resources, and its goods and services have played a major role in American history (Shepherd & Williamson 1972, Shepherd & Walton 1976, Cecelski 2001), as seen by its iconic setting during the Revolutionary and American Civil Wars, Sherman's March to the Sea, the Reconstruction era, and as a major site (North Carolina Shipbuilding Company, Wilmington, NC) for the U.S. Government Emergency Shipbuilding Program during World War II (Randall 1968, Coleman & Klein 1976, Edgar 1998, Browning 2002, Seeb 2013).

What follows is a brief background for this region, describing all the major considerations of this marine ecosystem. Then there is a short synopsis of how all the salient data for those factors are collected, analyzed, and used for LMR management. We then present an evaluation of key facets of ecosystem-based fisheries management (EBFM) for this region.

5.2 Background and Context

5.2.1 Habitat and Geography

The South Atlantic ecosystem comprises the Southeast U.S. Continental Shelf Large Marine Ecosystem (LME;

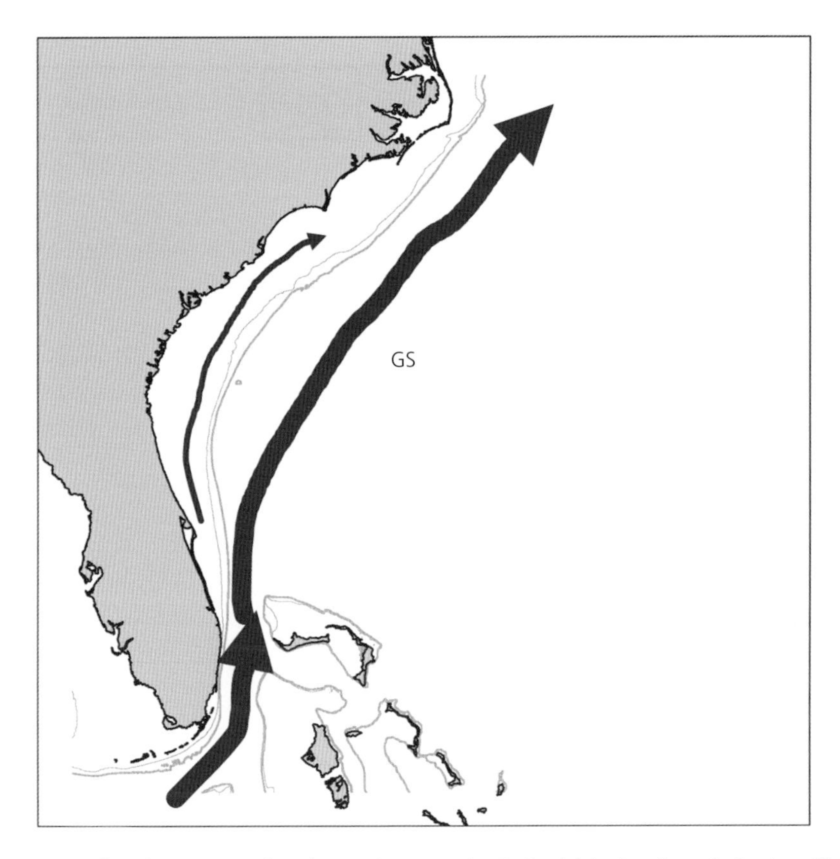

Figure 5.1 Major currents and circulation patterns throughout and encompassing the South Atlantic region and subregions. GS = Gulf Stream.

Sherman 1991). It makes up the second-largest component of the southeastern region of the United States Exclusive Economic Zone (EEZ). This ecosystem has connectivity from the Gulf of Mexico and tropical Lucayan (i.e., Bahamian) Archipelago, and continuing toward the northeastern U.S. Atlantic via the Gulf Stream (Fig. 5.1). Connectivity with the Cuban Archipelago and Greater Antilles occurs through the Antilles Current, providing a major source for recruitment of corals and other tropically associated species throughout South Florida.

Significant fisheries landings and economic production are found throughout the South Atlantic region, particularly in or near two key subregions: **Cape Hatteras-Pamlico Sound**—North Carolina; and the **Florida Keys**—Florida (Fig. 5.2). There are other subregional features in this region, but we emphasize these two herein as exemplary, smaller-scale ecosystems.

The main features of the South Atlantic ecosystem include large expanses of coastal salt marsh, offshore coral and rocky reefs, and an oceanography driven by the Gulf Stream (Teal & Teal 1969, Miller & Richards 1980, Atkinson et al. 1983, Hefner & Brown 1984, Wiegert & Freeman 1990, Bell 1997, Epifanio & Garvine 2001, Reed 2002, Collier et al. 2008, Stedman & Dahl 2008, Więski et al. 2010, Walker 2012, Angelini et al. 2015, Lee 2015, Bacheler & Smart 2016, NCDEQ 2016a, b, Bacheler et al. 2019, Geraldi et al. 2019). The South Atlantic is a highly transitional ecosystem with subtropical conditions at its southern end and conditions similar to many temperate ecosystems at its northernmost limits (SAFMC 2009). It also contains one of the largest lagoonal estuaries in the U.S. (Pamlico Sound) with connectivity to the extensive Albemarle Sound (Bales & Nelson 1988, Paerl et al. 2001). Many taxa in this region are associated

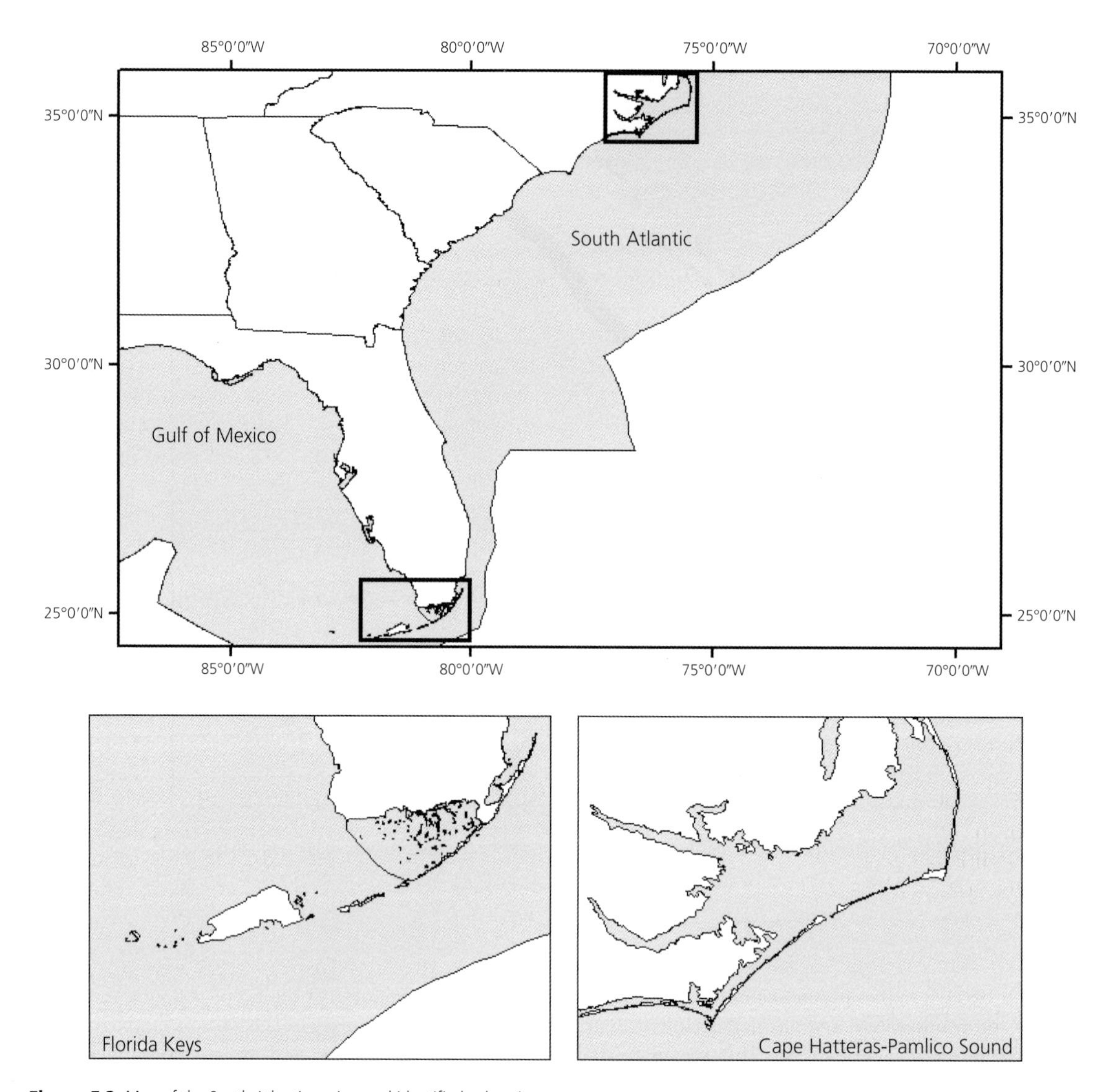

Figure 5.2 Map of the South Atlantic region and identified subregions.

with nearshore oyster reefs and vegetated habitats, in addition to offshore rocky hard-bottom and coralline substrates, which support ecologically and commercially important species (Weinstein et al. 1980, Rogers et al. 1984, Bell 1997, Koenig et al. 2000, Fourqurean et al. 2001, Hovel et al. 2002, Lehnert & Allen 2002, Peterson et al. 2003, Kendall et al. 2008, Shertzer & Williams 2008, Bacheler & Smart 2016, NCDEQ 2016a, b, Bacheler et al. 2019, Geraldi et al. 2019, Shertzer et al. 2019).

5.2.2 Climate and Other Major Stressors

The South Atlantic ecosystem is subject to many stressors, which include habitat loss, climate-associated factors such as sea-level rise and species range shifts, hurricanes, historical overfishing, and ongoing consequences from invasive species (Mager & Ruebsamen 1988, Gornitz et al. 1994, Van Dolah et al. 1999, Collins et al. 2000, Morris et al. 2002, Whitfield et al. 2002, Stedman & Dahl 2008, Bettinger et al. 2009, Engelhart et al. 2009, Barlow & Reichard 2010, Najjar et al. 2010,

Sorte et al. 2010, Muller & Stone 2013, Whitfield et al. 2014, Ballew et al. 2016, Hauer et al. 2016, NCDEQ 2016a, b; Geraldi et al. 2019). Reduction of South Atlantic coastal wetlands, including prominent *Spartina alterniflora* salt marshes, has been observed during the past century (Kennish 2001, Sharitz 2003, Alber et al. 2008, Stedman & Dahl 2008, Napton et al. 2010, NCDEQ 2016a, b). While losses are not as pronounced as in the Gulf of Mexico and Mid-Atlantic regions, significant natural stressors and upstream land-based activities in the southeastern United States continue to contribute to degraded nearshore water quality and negative wetland trends in the South Atlantic (Holland et al. 2004, Stedman & Dahl 2008, Napton et al. 2010). Loss of oyster reef and submerged aquatic vegetation, including seagrasses, has also occurred (Kirby 2004, Coen et al. 2007, Grabowski & Peterson 2007, Beck et al. 2011, NCDEQ 2016a, b), while poor water quality off South Florida has severely affected its coral reef systems (Lapointe & Clark 1992, Lapointe et al. 2004, Wagner et al. 2010). Efforts to protect and restore coastal habitats in the South Atlantic have resulted in progress toward their restoration, although ongoing stressors continue to affect the degree of habitat recovery observed for this region (Mager & Ruebsamen 1988, Van Dolah et al. 1999, Coen & Luckenbach 2000, Peterson et al. 2003, Grabowski et al. 2005, Grabowski & Peterson 2007, Paling et al. 2009, NCDEQ 2016a, 2016b; Bersoza-Hernández et al. 2018). Increased use of "living shorelines," such as restored oyster reefs, has also emerged as an environmentally friendly alternative for shoreline protection in place of coastal bulkheads or other man-made structures (Gittman et al. 2016, Smith et al. 2018).

Mostly associated with wetland losses, increasing sea level and upland saltwater intrusion have been observed in this region, with values having increased by 3–4 mm/year in South Florida since the 1990s, accelerating to rates of >20 mm/year throughout areas south of Cape Hatteras during the 2010s (Gornitz et al. 1994, Valle-Levinson et al. 2017). Increasing ocean acidification has been documented in the South Atlantic as driven by both increasing atmospheric carbon dioxide and terrestrial inputs to nearshore coastal areas (Reimer et al. 2017). This increasing vulnerability of the South Atlantic region has also been associated with regional climate forcing in response to weakening of the Atlantic Meridional Oscillation (AMO) and changes in the North Atlantic Oscillation (NAO; Marshall et al. 2001, McCarthy et al. 2015, Valle-Levinson et al. 2017). Compounded with climate-associated sea-level rise has been the increased prevalence of tropically associated species such as Spotted drum (*Equetus punctatus*) and the Rock beauty (*Holacanthus tricolor*) in the upper regions of this system (Whitfield et al. 2014). These observations include the encroachment of black mangroves (*Avicennia germinans*) into northern Florida and Georgia salt marsh habitat (Osland et al. 2013, Riley et al. 2014, Coldren et al. 2016), and northward expansion of South Atlantic resident species (e.g., Spotted sea trout *Cynoscion nebulosus*, Tarpon *Megalops atlanticus*, and Blueline tilefish *Caulolatilus microps*) into the Mid-Atlantic region (Najjar et al. 2010, MAFMC 2017, Peterson et al. 2017). It is worth noting, however, that despite some of these observations, limited evidence of overall distributional shifts for the South Atlantic is currently found in fishery-independent surveys (Morley et al. 2017). Greater thermal stability has been observed in the South Atlantic than other systems (Morley et al. 2017, Link & Marshak 2019), with fewer predicted changes in species distributions as compared to other regions (Morley et al. 2017, 2018) and decreasing species range sizes over time (Batt et al. 2017).

This region has also been prone to intermittent hurricane activity, with increased frequency and intensity of higher category storms such as Emily (1983), Andrew (1992), Hugo (1989), Fran (1996), Charley (2004), Irma (2017), and Florence (2018) occurring over the past decades, causing significant effects to its local marine ecosystems and coastal communities (Mallin et al. 1999, 2000, Blake et al. 2007, Keim et al. 2007, Smith et al. 2009, Muller & Stone 2013, Lindner & Neuhauser 2018, Paerl et al. 2019). Hurricanes Andrew, Hugo, and Irma especially affected marine habitats and assemblages of targeted and protected species in this region, with major damages and revenue loss to marine economies (Gardner et al. 1992, Milton et al. 1994, Tilmant et al. 1994, Blake et al. 2007, Xie et al. 2008, Smith et al. 2009). In addition, three sequential hurricanes in 1999 (Dennis, Floyd, and Irene) and continued large-scale flooding events from Hurricanes Matthew (2016) and Florence have also caused large-scale ecosystem and economic effects in this region (Paerl et al. 2001, 2019).

5.2.3 Exploitation

Overfishing has also affected the South Atlantic system, especially its commercially and recreationally important reef fishes, sharks, pelagic species, and invertebrates (Vaughan et al. 1995, Ault et al. 1998, Huntsman et al. 1999, Coleman et al. 2004, Zatcoff et al. 2004, Ault et al. 2005, Rudershausen et al. 2008, Figueira & Coleman 2010, NMFS 2013, Bacheler & Ballenger

2018). A significant proportion (up to 38%) of fishery species of concern that have been historically listed as overfished or experiencing overfishing are also taken by recreational fisheries in the South Atlantic, which is the third-highest volume of recreational fishing in the nation (Coleman et al. 2004, Shertzer et al. 2019). More recently it has been calculated that since 1981, 71% of landings for the preponderance of South Atlantic species (i.e., those in the coastal migratory, Dolphin/ Wahoo, and snapper-grouper Fishery Management Plans) have come from recreational captures (NMFS 2018b, Shertzer et al. 2019). Estuarine fisheries also contribute high economic value to this region, with many harvested species spending a portion of their life histories in estuarine habitats (Dame et al. 2000, Lellis-Dibble et al. 2008, NMFS 2018b). Limited commercial, recreational, and fishery-independent data have inhibited assessments for many South Atlantic species, with a large fraction either unassessed or in unknown status (Link & Marshak 2019). But, rebuilding plans remain in effect for several species with known overfishing or overfished status (Berkson & Thorson 2014, Carruthers et al. 2014, Newman et al. 2015). Overall, the consequences of overfishing have been observed in terms of shifts in the community composition of offshore pelagic and reef fishes, with consequences to species life histories (particularly for the snapper-grouper complex), depletions of higher trophic levels, and potential cascading effects (Ault et al. 1998, Beerkircher et al. 2002, Coleman & Williams 2002, Coleman et al. 2004, Zatcoff et al. 2004, Ault et al. 2005, Myers et al. 2007, Rudershausen et al. 2008, Loh et al. 2015). Overfishing of Florida stone crab, Spiny lobster, and sponges has also occurred, with continued biological effects observed for these species (Davis 1977, Pollock 1993, Eggleston et al. 2003, Andrews et al. 2005, Muller et al. 2006, McClenachan 2008). Stricter management of finfish and invertebrate fisheries has been implemented in more recent years with some degree of resilience and improvement occurring for some of these populations (Matthews 1995, Hunt 2000, Muller et al. 2006, Koenig et al. 2010). However, the Queen conch (*Aliger gigas*) population of the Florida Keys remains depleted, with rebuilding efforts having resulting in limited population growth, and the fishery currently remaining closed (Stoner & Glazer 1998, Delgado et al. 2004, Lirman et al. 2019).

In general, fisheries management in the South Atlantic is less contentious than in other regions, with surveys reporting the highest national angler satisfaction with management processes and outcomes (Brinson & Wallmo 2017). However, challenges have persisted in gathering accurate fishery-dependent data for effective population estimates and in advancing techniques beyond catch-only methods for more robust assessment methods (Bacheler et al. 2013b, Berkson & Thorson 2014, Carruthers et al. 2014, Newman et al. 2015, Shertzer et al. 2020). The multispecies nature of this system warrants more advanced efforts to evaluate fishery status, particularly for data-limited reef fishes whose fishery-independent based assessments remain constrained (Bacheler et al. 2013a, b, Berkson & Thorson 2014, Carruthers et al. 2014, Newman et al. 2015). Efforts to more accurately characterize this region's recreational fisheries have emerged over the past couple of decades (Coleman et al. 2004, Figueira & Coleman 2010, Gallagher et al. 2017, Farmer & Froeschke 2015, Potts & Burton 2017, Shertzer et al. 2019), though progress has been hindered by limited survey coverage and irregular reporting of landings (Backstrom & Woodward 2017, Abbot et al. 2018). With major contributions from recreational fishing to reef and coastal pelagic fisheries in this region, moderate climatic vulnerability to fishing-dependent communities, and ongoing closures to regional commercial fisheries, potential conflicts among fishing interests are anticipated to increase (Colburn et al. 2016). Focus on the effects of bycatch toward these conflicting interests has remained a high priority, given the vulnerabilities of multiple targeted sportfishes and protected species to trawling and longline gears associated with this region's major fisheries (Trent et al. 1997, Diamond et al. 1999, Watson et al. 1999, Carlson & Lee 2000, Beerkircher et al. 2002, Epperly et al. 2002, Zollett 2009, Finkbeiner et al. 2011). Bycatch and discards of other fishes is also a concern as many South Atlantic species co-occur and are subject to varying annual quotas (Beerkircher et al. 2002, Stephen & Harris 2010, Stephen et al. 2011). High discard mortality is often observed for those species whose catch limits have been reached as fishing continues for other targeted species found in close proximity (Stephen & Harris 2010, Stephen et al. 2011).

5.2.4 Invasive Species

The emerging prevalence of invasive species, particularly Red lionfish (*Pterois* spp.), European green crab (*Carcinus maenas*), and Asian tiger shrimp (*Penaeus monodon*), has become a concern in the South Atlantic. Since the 2000s, lionfish have been observed throughout the South Atlantic, raising concerns about their effects on native species and local ecosystems (Whitfield et al. 2002, Hare & Whitfield 2003, Whitfield et al. 2014, Bryan et al. 2018, Sancho et al. 2018). In this and similar systems, lionfish have been found to deplete prey fishes (Muñoz et al. 2011, Green et al. 2012, Layman et al. 2014, Ballew et al. 2016), possess partial competitive advantages over similar mesopredatory species (Albins 2013, Raymond et al. 2015, Marshak et al. 2018), and are subject to limited predatory control due to historical overexploitation of large fishes (Huntsman et al. 1994, McClenachan 2009, Bacheler et al. 2019) and their protective spines (Albins & Hixon 2008, Morris & Akins 2009, Morris & Whitfield 2009). This species has become increasingly abundant within nearshore estuarine and offshore systems spanning from South Florida to the Carolinas, emerging as a dominant fish species in coral and hard-bottom habitats (Jud et al. 2011, Whitfield et al. 2002, Whitfield et al. 2014). European green crabs have also invaded seagrasses and salt marshes in the Carolinas, with destructive effects on eelgrass (*Zostera marina*) shoots and rhizomes having occurred while crabs burrow and dig for prey, and with observed consequences to fish communities due to diminished eelgrass habitats (Matheson et al. 1999). Invasive Asian tiger shrimp have also been shown to prey on commercially important penaeid shrimps, blue crab, and bivalves, with potential effects on South Atlantic ecosystems (Fuller et al. 2014, Hill et al. 2017).

In concert, these stressors are causing major consequences to the South Atlantic ecosystem, and also affecting the status of its LMRs.

5.2.5 Ecosystem-Based Management (EBM) and Multisector Considerations

The South Atlantic ecosystem is known not only for its cultural images, species, and stressors, but also the trade-offs that arise across LMRs (and usage thereof) and between LMR management and other ocean uses (Lindeman et al. 2000, Islam & Tanaka 2004, Anthony et al. 2009, Barbier et al. 2011, Marshall et al. 2014). In addition to trade-offs among South Atlantic commercial and recreational fisheries, multiple coastal and ocean uses supporting marine economies co-occur along a mosaic of areas providing habitat to many managed species, supporting these local fisheries, and serving as the foundation for robust marine tourism and marine transportation industries (Milon et al. 1983, Roehl et al. 1993, Bell 1997, Lindeman et al. 2000, Peterson et al. 2000, Islam & Tanaka 2004, Bin et al. 2005, Deale et al. 2008, Anthony et al. 2009, Barbier et al. 2011, Colgan 2013, Marshall et al. 2014). These sectors contribute heavily toward the region's marine economy, while their overlapping and interrelated activities emphasize the connectivity among this regional fisheries ecosystem and its other ocean uses. In South Florida, for example, impacts from marine tourism include anchor damage to coral reefs, boat propeller scarring in seagrasses, recreational overexploitation of marine habitats, pollution and marine debris, and the effects of sunscreens on coral viability (Frankovich & Zieman 2005, Ault et al. 2009, Collado-Vides et al. 2011, Wood 2018, Lirman et al. 2019). The release of ship ballast waters has resulted in the introduction of several exotic and invasive species into the South Atlantic region, with the increasing prevalence of Asian Tiger Shrimp and Green mussel (*Perna viridis*) attributed to this source (Power et al. 2004, Fuller et al. 2014). Conflicting interests among commercial and recreational fisheries and their interactions with resident and migratory protected species populations, including efforts to reduce associated bycatch, are also significant in this system (Diamond et al. 1999, Watson et al. 1999, Epperly et al. 2002, Epperly 2003, VanderKooy & Muller 2003 Scott-Denton et al. 2012, Chavez & Southward-Williard 2017). The effects of military sonar testing on marine mammal populations have also been an area of concern for this region (NUWC 2012). Furthermore, local diving revenues are heavily dependent on shipwrecks and decommissioned sunk ships from local marine transportation industries (Lindquist & Pietrafesa 1989, Spirek & Harris 2003, Voss et al. 2013). More recent siting efforts for offshore wind energy turbines in the South Atlantic have been undertaken in consideration of these ecotourism-related sectors and access to recreational fishing grounds (Voss et al. 2013). Holistic management approaches are necessary to account for overlapping recreational practices, coastal and tourism-related development, interconnectivity among fishing interests, and their effects on LMR habitat quality. Greater cumulative consideration of these factors will work to minimize resource conflicts and account for emerging priorities in this region, such as aquaculture siting (Islam & Tanaka 2004, Anthony et al. 2009, Brumbaugh & Coen 2009, Barbier et al. 2011, Marshall et al. 2014). Exploring, identifying, and resolving these differences in priorities and goals for the South Atlantic is the essence of EBFM, and the emerging basis of its fishery ecosystem plan (FEP; SAFMC 2009, 2018).

5.3 Informational and Analytical Considerations for this Region

5.3.1 Observation Systems and Data Sources

In moving toward a more holistic LMR management approach, data regarding the physical, biological, and socioeconomic components of the South Atlantic ecosystem are available through regional observation systems and multiple monitoring programs. The Southeast Coastal Ocean Observing Regional Association (SECOORA) provides oceanographic and atmospheric data from ~1500 sensor stations throughout coastal and offshore locations of the South Atlantic and portions of the eastern Gulf of Mexico (SECOORA 2016). The system is a composite of aggregated data from ~25 different federal and regional sources, with the majority of observations from coastal and inland sensors maintained by the United States Geological Survey (USGS) National Water Information System (NWIS), NOAA's National Weather Service (NWS), National Data Buoy Center (NDBC), the National Estuarine Research Reserve (NERR) System, and the World Meteorological Organization. SECOORA also incorporates sensor stations and data components from the Gulf of Mexico Coastal Ocean Observing System (GCOOS). Housing information collected since 1994, the SECOORA portal[1] contains oceanographic and atmospheric data at hourly to annual scales for a suite of 76 variables such as air temperature, winds, pH, chlorophyll, currents, oxygen, sea level, and thermohaline indicators. Similar to GCOOS, these data are being applied toward enhancing marine safety, minimizing the effects from marine weather and coastal hazards, and monitoring the South Atlantic ecosystem in response to human pressures and climate change (SECOORA 2016).

Much of this oceanographic and atmospheric information has been incorporated into its regional FEPs for greater understanding of the South Atlantic ecosystem and in identifying climatological considerations for EBFM (Blanton et al. 2003, SAFMC 2009, 2018). This information is being applied toward the development of an ecosystem status report for the region (NMFS 2019, Craig et al., in prep) as a component of a broader Integrated Ecosystem Assessment (IEA; Levin et al. 2009). Data have been applied toward characterizing the climatic influences of the NAO, El Niño Southern Oscillation (ENSO), and Arctic Oscillation (AO) on this system, with complementary analyses of the International Comprehensive Ocean-Atmosphere Data Set (ICOADS; Blanton et al. 2003, ICOADS 2009) documenting the dynamics in the wind-driven climatology of this region. These datasets have been applied to characterizing hydrographic variability of this system, especially in the South Atlantic Bight and Charleston Gyre, as related to stratification of the Gulf Stream (Blanton et al. 2003, SAFMC 2018). Nutrient loading is monitored through the USGS National Water Quality Assessment (NWQA) program, which quantifies and assesses total nitrogen and phosphorous yields throughout southeastern United States stream reaches (SAFMC 2018). Basal ecosystem production data for the South Atlantic (i.e., chlorophyll concentration and primary production) are available from NASA Ocean Color Web Data Sea-viewing Wide Field-of-view Sensor (SeaWiFS) and Moderate Resolution Imaging Spectroradiometer (MODIS)-Aqua satellite imagery. Harmful algal bloom concentrations in the South Atlantic are monitored by NOAA's National Centers for Coastal Ocean Science (NCCOS) using MODIS satellite data.

Information characterizing habitat factors and aspects of benthic production are available through disparate data sources regarding the extent, status, and condition of South Atlantic coral reefs, oyster reefs, and vegetated nearshore habitats. However, comprehensive habitat information at the scale of many LMR population distributions has remained limited (Marshak & Brown 2017, Peters et al. 2018). Most studies examining trends in habitat quality and delineation have been conducted at relatively localized scales, and have shown declines in cover and degradation in quality over time (Hefner & Brown 1984, Fonseca et al. 1994, Roman et al. 2000, Kirby 2004, Beck et al. 2011, Lirman et al. 2019). These have been particularly focused on restoring oyster reefs and vegetated habitats (Fonseca et al. 1994, Coen & Luckenbach 2000, Coen et al. 2007, Brumbaugh & Coen 2009, Paling et al. 2009, Beck et al. 2011, Peterson et al. 2003) in response to declining seagrass and salt marsh extent and quality (Fourqurean et al. 2001, Hovel et al. 2002, Stedman & Dahl 2008, Carlson et al. 2010, Osland et al. 2013). Declines in coral cover over time have also been observed within the Florida Keys as a result of poor water quality, more frequent bleaching events, the ecological consequences from overfishing, and proliferations of coral disease (Lapointe & Clarke 1992, Porter et al. 1999, 2001, Miller et al. 2002, Downs et al. 2005, Soto et al. 2011). More recently, outbreaks of stony coral tissue loss disease throughout this subregion have also led to increasing coral mortality on Florida Keys reefs (Walton et al. 2018, Precht 2019, Sharp et al. 2019). As observed for the northern Gulf of Mexico (nGOM), areal information from the NOAA Coastal Change Analysis Program (C-CAP), together with tidal gage data from the NOAA Center for Operational Oceanographic Products and Services (CO-OPS) and stations operated by the National Water Level Observation Network (NWLON),

[1] https://secoora.org

have also been used to elucidate the extent, quality, and risks posed to estuarine watersheds and emergent salt marsh habitat, including effects from changes in land use (NFHP 2010, Crawford et al. 2015) and sea-level rise.

Comprehensive databases for secondary through upper trophic level South Atlantic species include some fishery-independent information collected from the Southeast Area Monitoring and Assessment Program (SEAMAP) zooplankton and ichthyoplankton surveys (1982–present), which mostly sample the Gulf of Mexico, including the Florida Keys. The South Atlantic shelf has been sampled much less frequently over time, with limited surveys currently occurring in this region; SEAMAP South Atlantic Coastal Trawl Survey, and the Southeast Reef Fish (trap-video) survey (Conn 2011, Ballenger 2015, Ballenger & Smart 2015, Carmichael et al. 2015, Ballenger & Smart 2016, Conley et al. 2017, SEAMAP 2017, Bacheler & Ballenger 2018). While no large-scale surveys presently exist for South Atlantic zooplankton and ichthyoplankton, there are localized long-term surveys in this region such as the Beaufort Bridgenet Ichthyoplankton Sampling Program for Beaufort Inlet, North Carolina, and the North Inlet-Winyah Bay Ichthyoplankton Survey in South Carolina (SEAMAP 2016). The South Carolina Department of Natural Resources conducts the Marine Resources Monitoring, Assessment, and Prediction (MARMAP) short and long bottom longline surveys to assess groundfish, reef fish, and coastal pelagic fishes for the region (Broome et al. 2011, Conn 2011, Ballenger & Smart 2016). These data have been applied for examining the magnitude of regional species shifts in the South Atlantic (Morley et al. 2017, 2018), and are being applied to investigating aspects of mean trophic level, biomass, species richness, diversity, and other components in emergent ecosystem models and indicator assessments (Townsend et al. 2014, 2017). Recent efforts have also included expanding sampling information through the use of multibeam mapping and improving the resolution of surveys to allow for more comprehensive biomass estimates (BOEM 2016, Marshak et al. 2018). These data inform South Atlantic stock status and abundance, especially for its snapper-grouper complex, and are complemented by NOAA Fisheries fishery-dependent monitoring programs for commercial and recreational harvest, landings, revenue, and effort, including surveys such as the Marine Recreational Information Program (MRIP) and the Southeast Region Headboat Survey (Fitzpatrick et al. 2017). Socioeconomic information for fisheries and other ocean uses is also collected by NOAA Fisheries, the National Ocean Economics Program (NOEP), and NOAA's Office for Coastal Management Economics: National Ocean Watch (ENOW) Program. Other NOAA monitoring programs exist for protected species, including sea turtles and cetaceans, and several monitoring programs for resident and migratory marine birds, such as the USGS Breeding Bird Survey, are overseen by NOAA, the U.S. Department of Interior, and state agencies.

5.3.2 Models and Assessments

Assessments of economically important and protected species in the South Atlantic are conducted by NOAA Fisheries in collaboration with the South Atlantic Fishery Management Council (SAFMC), the U.S. Fish and Wildlife Service (FWS), and the Gulf of Mexico Fishery Management Council (GMFMC) for managed coral reef, coastal pelagic species, and spiny lobster (FWS 2014, Ferguson et al. 2017, Lynch et al. 2018). The Atlantic States Marine Fisheries Commission (ASMFC) conducts stock assessments for those species under its jurisdiction (Kilduff et al. 2009). Stock assessment models are used to generate estimates of stock status with a multitude of data sources (e.g., landings, catch per unit effort, life history characteristics, aspects of survey biomass, etc.), and assessment reviews for South Atlantic species are conducted through the Southeast Data, Assessment, and Review (SEDAR) process (Lynch et al. 2018). However, this region continues to remain data-limited for many species, with a significant percentage (~48%) of its stocks remaining unassessed (Berlson & Thorson 2014, Carruthers et al. 2014, Newman et al. 2015, Link & Marshak 2019). NOAA Fisheries also examines the effects of commercial and recreational fisheries on South Atlantic marine socioeconomics through the application of the NMFS Commercial Fishing Industry Input/Output and Recreational Economic Impact Models (NMFS 2018a). Data from NOAA Fisheries marine mammal stock abundance surveys are used to conduct population assessments as published in regional marine mammal stock assessment reports and peer-reviewed by regional scientific review groups (SRGs; Carretta et al. 2019, Hayes et al. 2019, Muto et al. 2019). The FWS prepares stock assessment reports for South Atlantic manatees under its jurisdiction. Each of the four South Atlantic states also has a monitoring program for sea turtles (Day et al. 2005, Hawkes et al. 2005, Norton 2005, Witherington et al. 2009). Information collected on South Atlantic protected species is also applied to five-year status reviews of those listed under the Endangered Species Act (ESA; Bettridge et al. 2015, Seminoff et al. 2015, Valdivia et al. 2019).

Available data and species assessments have been used in broader ecosystem models for the South Atlantic (Okey & Pugliese 2001, Okey et al. 2014, Townsend et al. 2017), and are also being applied to assessments in the development of a comprehensive ESR. This information is also working to enhance single-species stock assessments and LMR management decisions (Townsend et al. 2008, Link et al. 2010, Townsend et al. 2014, 2017). While most ecosystem modeling conducted in the southeast region has occurred in the nGOM (Chapter 7), multispecies production models have also been developed for the South Atlantic (Townsend et al. 2017). In particular, work by NOAA Fisheries and the AMSFC identified Atlantic menhaden (*Brevoortia tyrannus*) as a case species for establishing ecosystem reference points, given their broad spatial range and overlap with multiple predators and environments. Several models were applied, which include the single-species Beaufort Assessment Model and the Northwest Atlantic Coastal Shelf Model of Intermediate Complexity (NWACS-MICE), in addition to multispecies surplus production models, multispecies catch at age models, and Ecopath with Ecosim (EwE) models (Buchheister et al. 2017, Townsend et al. 2017, ASMFC 2020a) for enhancing assessments of Atlantic menhaden and other interacting species. In August 2020, the ASMFC approved the use of ecosystem reference points in the management of Atlantic Menhaden (ASMFC 2020a).

There are a few other models that have informed the management discussions in this region, but none that are being directly applied in the SEDAR process like those noted above. They are, however, being applied toward refining ecosystem approaches to management and being incorporated into SAFMC efforts to understand South Atlantic food webs and their connectivity, and climate and environmental variability on fisheries (SAFMC 2018). An EwE model has been developed and applied toward the South Atlantic FEP (South Atlantic Bight Ecopath Model; Okey et al. 2014), while the utility of an object-oriented simulator of marine ecosystems (OSMOSE) has been promoted by the SAFMC to examine the effects of size-based predation rules in the context of trophic interactions (Shin & Cury 2004, Shin et al. 2004, SAFMC 2016, 2018). The Spatially Referenced Regression on Watershed Attributes (SPARROW) model has been used to examine relationships among nutrient loading (i.e., nitrogen and phosphorous additions) in the context of its FEP. The influences of sedimentation through application of the HydroTrend 3.0 model (McCarney-Castle et al. 2010) have also been examined toward understanding anthropogenic impacts on land use and their consequences for local productivities. Furthermore, several models of the South Florida ecosystem have been developed to bolster ecosystem approaches to resource management in consideration of climate change and human-related stressors (Acosta et al. 1998, Gentile et al. 2001, Ault et al. 2003, Ogden et al. 2005, Rudnick et al. 2005, Hyun & He 2010, Kelble et al. 2013, Nuttle & Fletcher 2013, Xue et al. 2015).

5.4 Evaluation of Major Facets of EBFM for this Region

To elucidate how well the South Atlantic ecosystem is doing in terms of LMR management and progress toward EBFM, we characterize the status and trends of socioeconomic, governance, environmental, and ecological criteria pertinent for employing an EBFM framework for this region, and examine their interrelationships as applied toward its successful implementation. We also quantify those factors that emerge as key considerations under an ecosystem-based approach for the region, with emphasis for the status and trends in Cape Hatteras-Pamlico Sound and the Florida Keys when available. Ecosystem indicators for the South Atlantic related to the (1) human environment and socioeconomic status, (2) governance system and LMR status, (3) natural environmental forcing and features, and (4) systems ecology, exploitation, and major fisheries of this system are presented and synthesized next. We explore all these facets to evaluate the strengths and weaknesses along the pathway:

$$PP \rightarrow B_{\text{targeted, protected spp, ecosystem}} \leftrightarrow L_{\uparrow\text{targeted spp, }\downarrow\text{bycatch}} \rightarrow \text{jobs, economic revenue.}$$

Where PP is primary production, B is biomass of either targeted or protected species or the entire ecosystem, L is landings of targeted or bycaught species, all leading to the other socioeconomic factors. This operates in the context of an ecological and human system, with governance feedbacks at several of the steps (i.e., between biomass and landings, jobs, and economic revenue), implying that fundamental ecosystem features can determine the socioeconomic value of fisheries in a region, as modulated by human interventions.

The Whole is Greater Than The Sum of the Parts: The Importance of Evaluating Emergent, Common Properties of Ecosystems with Respect to Pressures and Change

—(see also every other chapter)

Have you ever stared at one of those mosaic walls? The individual tiles are intriguing, but you just can't make sense of them, there's no real overarching story, no real pattern beyond what's on the individual tiles, just some interesting art. Then you step back, and you maybe see a bit of a pattern. Then you step back even further and the overall picture just explodes at you, becoming so obvious. You needed to step back, look at the bigger picture, and see the wall as a whole and not as individual tiles. Once you do so, the pattern becomes clear as a bell, you just couldn't see it because you were focused on the smaller scale. It's like the change in perspective made everything so much clearer to you.

That's a lot like emergent properties from marine ecosystems. So often we focus on individual fish populations, or fisheries, or individual species, or even groups of species. The dynamics thereof are indeed intriguing. But if we step back and look at all these components of the ecosystem *as a system* major patterns start to emerge. There is much exploration of complex systems as applied to marine ecosystems (Jorgensen 1997, Levin 1998, Jorgensen & Muller 2000, Wu & Marceau 2002, Levin 2003, Levin & Lubchenco 2008, Levin et al. 2013, Fuller et al. 2017, Hagstrom & Levin 2017). Admittedly, much of it is amazing and not always useful or practical, but some of it is.

Think of it this way in a fisheries ecosystem context. Fish grow. When fish are caught, they die. When they die more than they grow, that's bad. This is the level of explanation we have to communicate to many interested parties about fisheries science and how we manage fish. You and I get that there is a bit more to it than that, but trying to communicate complex concepts in accessible terms is a challenge. Basically, the rate of removal needs to be less than or equal to the rate of replenishment. Now imagine trying to communicate at that level about emergent ecosystem properties based on second derivatives of cumulative trophodynamic functioning. Just a wee bit harder. But doable.

Several works have begun to show that there are fundamental, repeatable, widely observed emergent properties of marine ecosystems. For instance, removing more than 1–3, and certainly more than 3–5 t km^{-2} can lead to overall, ecosystem overfishing (Conti & Scardi 2010, Bundy et al. 2012, Tam et al. 2017, Link & Marshak 2019, Link & Watson 2019). Or that comparing cumulative biomass to trophic level results in an S-shaped curve (Pranovi & Link 2009, Pranovi et al. 2012, 2014, 2019, Link et al. 2015, Libralato et al. 2019). Or cumulative biomass relative to cumulative production almost always results in a hockey stick-shaped curve (Pranovi et al. 2012, 2014, 2019, Link et al. 2015, Libralato et al. 2019). Or that there is only so much primary production that can support a certain amount of fish caught in an ecosystem (Ryther 1969, Pauly & Christensen 1995, Chassot et al. 2007, 2010, Friedland et al. 2012, Watson et al. 2014, Fogarty et al. 2016, Stock et al. 2017, Link & Watson 2019), and that this primary production required has limits (Pomeroy 1991, Nielsen & Richardson 1996, Falkowski et al. 1998, Chassot et al. 2010, Watson et al. 2015), and removing more than 1‰ causes problems (Pauly et al. 2005, Libralato et al. 2008, Large et al. 2015, Watson et al. 2015, Tam et al. 2017, Link & Marshak 2019, Link & Watson 2019). Or that marine food webs contain a lot of information, and as they mature the degree of flexibility or resilience based on all this information/energy flow allows for resilience until it passes the point where it gets more brittle and then less resilient (Odum 1969, DeAngelis et al. 1989, Pérez-España & Arreguín-Sánchez 2001, Link 2002, Link et al. 2005, Vasseur & Fox 2007, Griffin et al. 2009, Yaragina & Dolgov 2009, Vallina & Le Quere 2011, Stäbler et al. 2018). And so on.

The salient point of all these emergent examples is not that they exist or are fascinating depictions of marine ecosystems at a higher level of hierarchical organization, nor that these patterns emerge almost universally in all marine ecosystems, although that is indeed amazing. The important point is that many of these can now identify thresholds upon which management can be based. And these tipping points, thresholds, etc. can be used as reference levels to denote when mitigating action is required as well as when recovery *of the full system* has been obtained. Additionally, given the nature of these emergent properties and their higher order features in the hierarchy of components of ecosystems, they tend to detect major perturbations to the entire system more rapidly, and hence can act as early warning signals. Having decision criteria that effects *all* components of an ecosystem seems both prudent and efficient, perhaps a way to head off more ramifications from continued perturbations. Just like trying to piece together the picture tile-by-tile, too often we neglect these emergent patterns by trying to understand the marine ecosystem on a species-by-species basis, piecing together patterns that a more emergent view would readily and more accurately detect.

The challenge is that most scientists, and by extension resource managers, have been trained to think with a reductionist view. There is a place and purpose for that. Yet a holistic view, as seen by these emergent features, also

continued

The Whole is Greater Than The Sum of the Parts: *Continued*

has value. Ignoring the reality of these emergent features does not make them go away, does not inviolate their presence, and negates the advantages they can provide. Perhaps this is an area for growth in the ongoing challenge to operationalize EBFM.

WHOLE = $\sum$ parts + ?

References

Bundy A, Bohaboy EC, Hjermann DO, Mueter FJ, Fu C, Link JS. 2012. Common patterns, common drivers: comparative analysis of aggregate surplus production across ecosystems. *Marine Ecology Progress Series* 459:203–18.

Chassot E, Bonhommeau S, Dulvy NK, Mélin F, Watson R, Gascuel D, Le Pape O. 2010. Global marine primary production constrains fisheries catches. *Ecology Letters* 13(4):495–505.

Chassot E, Mélin F, Le Pape O, Gascuel D. 2007. Bottom-up control regulates fisheries production at the scale of eco-regions in European seas. *Marine Ecology Progress Series* 343:45–55.

Conti L, Scardi M. 2010. Fisheries yield and primary productivity in large marine ecosystems. *Marine Ecology Progress Series* 410:233–44.

DeAngelis DL, Mulholland PJ, Palumbo AV, Steinman AD, Huston MA, Elwood JW. 1989. Nutrient dynamics and food-web stability. *Annual Review of Ecology and Systematics* 20(1):71–95.

Falkowski PG, Barber RT, Smetacek V. 1998. Biogeochemical controls and feedbacks on ocean primary production. *Science* 281(5374):200–6.

Fogarty MJ, Rosenberg AA, Cooper AB, Dickey-Collas M, Fulton EA, Gutiérrez NL, Hyde KJ, et al. 2016. Fishery production potential of large marine ecosystems: a prototype analysis. *Environmental Development* 17:211–9.

Friedland KD, Stock C, Drinkwater KF, Link JS, Leaf RT, Shank BV, Rose JM, Pilskaln CH, Fogarty MJ. 2012. Pathways between primary production and fisheries yields of large marine ecosystems. *PloS One* 7(1).

Fuller EC, Samhouri JF, Stoll JS, Levin SA, Watson JR. 2017. Characterizing fisheries connectivity in marine social–ecological systems. *ICES Journal of Marine Science* 74(8):2087–96.

Griffin JN, O'Gorman EJ, Emmerson MC, Jenkins SR, Klein AM, Loreau M, Symstad A. 2009. Biodiversity and the stability of ecosystem functioning. *Biodiversity, Ecosystem Functioning, and Human Wellbeing—An Ecological and Economic Perspective* 30:78–93.

Hagstrom GI, Levin SA. 2017. Marine ecosystems as complex adaptive systems: emergent patterns, critical transitions, and public goods. *Ecosystems* 20(3):458–76.

Jorgensen SE. 1997. *Integration of Ecosystem Theories: A Pattern.* Boston, MA: Kluwer.

Jorgensen SE, Muller F. 2000. *Handbook of Ecosystem Theories and Management.* Boca Raton, FL: Lewis Publishers.

Large SI, Fay G, Friedland KD, Link JS. 2015. Critical points in ecosystem responses to fishing and environmental pressures. *Marine Ecology Progress Series* 521:1–7.

Levin S, Xepapadeas T, Crépin AS, Norberg J, De Zeeuw A, Folke C, Hughes T, et al. 2013. Social-ecological systems as complex adaptive systems: modeling and policy implications. *Environment and Development Economics* 18(2):111–32.

Levin SA. 1998. Ecosystems and the biosphere as complex adaptive systems. *Ecosystems* 1:431–6.

Levin SA. 2003. Complex adaptive systems: exploring the known, the unknown, and the unknowable. *Bulletin of the American Mathematical Society* 40:3–19.

Levin SA, Lubchenco J. 2008. Resilience, robustness, and marine ecosystem based management. *BioScience* 58:27–32.

Libralato S, Pranovi F, Zucchetta M, Monti MA, Link JS. 2019. Global thresholds in properties emerging from cumulative curves of marine ecosystems. *Ecological Indicators* 103:554–62.

Link J. 2002. Does food web theory work for marine ecosystems? *Marine Ecology Progress Series* 230:1–9.

Link JS, Marshak AR. 2019. Characterizing and comparing marine fisheries ecosystems in the United States: determinants of success in moving toward ecosystem-based fisheries management. *Reviews in Fish Biology and Fisheries* 29(1):23–70.

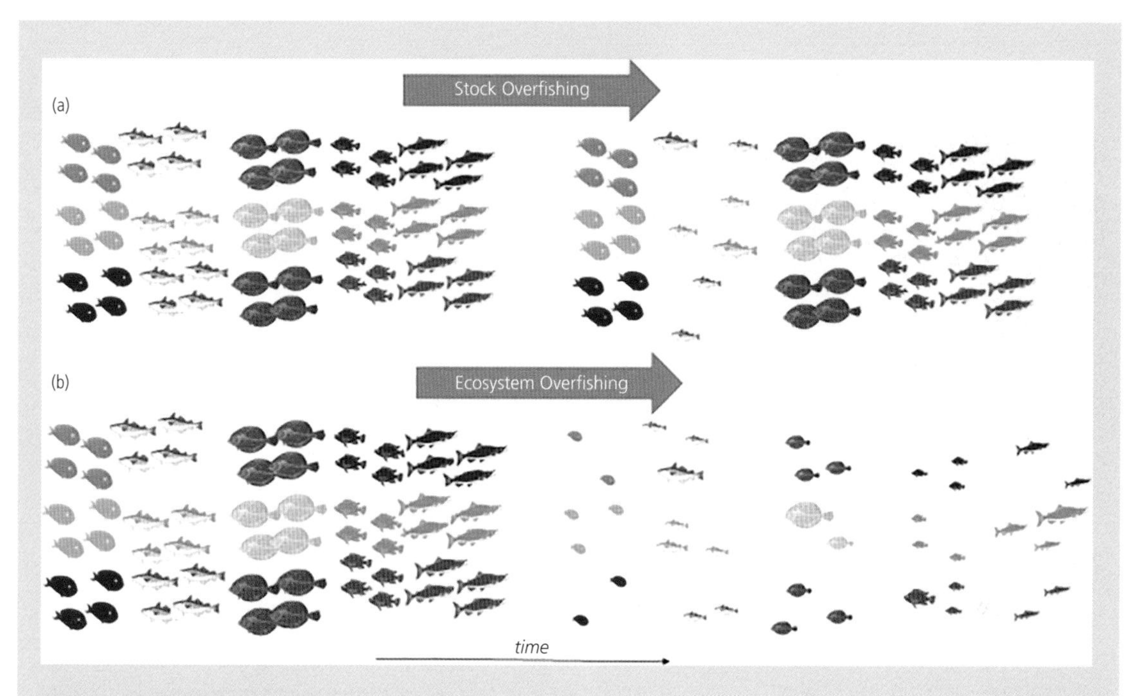

From Link and Watson 2019.

Link JS, Pranovi F, Libralato S, Coll M, Christensen V, Solidoro C, Fulton EA. 2015. Emergent properties delineate marine ecosystem perturbation and recovery. *Trends in Ecology & Evolution* 30(11):649–61.

Link JS, Stockhausen WT, Methratta ET. 2005. Food-web theory in marine ecosystems. Aquatic food webs: an ecosystem approach. *Oxford University Press* 7:98–114.

Link JS, Watson RA. 2019. Global ecosystem overfishing: clear delineation within real limits to production. *Science Advances* 5(6):eaav0474.

Nielsen E, Richardson K. 1996. Can changes in the fisheries yield in the Kattegat (1950–1992) be linked to changes in primary production? *ICES Journal of Marine Science* 53(6):988–94.

Odum EP. 1969. The strategy of ecosystem development. *Science* 164:262–70.

Pauly D, Alder J, Bakun A, Heileman S, Kock KH, Mace P, Perrin W, et al. 2005. "Marine fish eries systems." In: *Ecosystems and Human Well-being: Current States and Trends*, Volume 1. Edited by Hassan R, Scholes R, Ash N. Millennium Ecosystem Assessment. Washington, DC: Island Press (pp. 477–511).

Pauly D, Christensen V. 1995. Primary production required to sustain global fisheries. *Nature* 374(6519):255–7.

Pérez-España H, Arreguín-Sánchez F. 2001. An inverse relationship between stability and maturity in models of aquatic ecosystems. *Ecological Modelling* 145(2–3):189–96.

Pomeroy LR. 1991. "Relationships of primary and secondary production in lakes and marine ecosystems." In: *Comparative Analyses of Ecosystems.* Edited by J Cole, G Lovett, S Findlay. New York, NY: Springer (pp. 97–119).

Pranovi F, Libralato S, Zucchetta M, Anelli Monti M, Link JS. 2019. Cumulative biomass curves describe past and present conditions of large marine ecosystems. *Global Change Biology* 26(2):786–97.

Pranovi F, Libralato S, Zucchetta M, Link J. 2014. Biomass accumulation across trophic levels: analysis of landings for the Mediterranean Sea. *Marine Ecology Progress Series* 512:201–16.

Pranovi F, Link J, Fu C, Cook AM, Liu H, Gaichas S, Friedland KD, Utne KR, Benoît HP. 2012. Trophic-level determinants of biomass accumulation in marine ecosystems. *Marine Ecology Progress Series* 459:185–201.

Pranovi F, Link JS. 2009. Ecosystem exploitation and trophodynamic indicators: a comparison between the Northern Adriatic Sea and Southern New England. *Progress in Oceanography* 81(1–4):149–64.

Ryther JH. 1969. Photosynthesis and fish production in the sea. *Science* 166(3901):72–6.

Stäbler M, Kempf A, Temming A. 2018. Assessing the structure and functioning of the southern North Sea ecosystem with a food-web model. *Ocean & Coastal Management* 165:280–97.

Stock CA, John JG, Rykaczewski RR, Asch RG, Cheung WW, Dunne JP, Friedland KD, et al. 2017. Reconciling fisheries catch and ocean productivity. *Proceedings of the National Academy of Sciences* 114(8):E1441–9.

continued

The Whole is Greater Than The Sum of the Parts: *Continued*

Tam JC, Link JS, Large SI, Andrews K, Friedland KD, Gove J, Hazen E, et al. 2017. Comparing apples to oranges: common trends and thresholds in anthropogenic and environmental pressures across multiple marine ecosystems. *Frontiers in Marine Science* 4:282.

Vallina SM, Le Quéré C. 2011. Stability of complex food webs: resilience, resistance and the average interaction strength. *Journal of Theoretical Biology* 272(1):160–73.

Vasseur DA, Fox JW. 2007. Environmental fluctuations can stabilize food web dynamics by increasing synchrony. *Ecology Letters* 10(11):1066–74.

Watson RA, Nowara GB, Hartmann K, Green BS, Tracey SR, Carter CG. 2015. Marine foods sourced from farther as their use of global ocean primary production increases. *Nature Communications* 6:7365.

Watson R, Zeller D, Pauly D. 2014. Primary productivity demands of global fishing fleets. *Fish and Fisheries* 15(2):231–41.

Wu J, Marceau D. 2002. Modeling complex ecological systems: an introduction. *Ecology Modelling* 153:1–6.

Yaragina NA, Dolgov AV. 2009. Ecosystem structure and resilience—a comparison between the Norwegian and the Barents Sea. *Deep Sea Research Part II: Topical Studies in Oceanography* 56(21–22):2141–53.

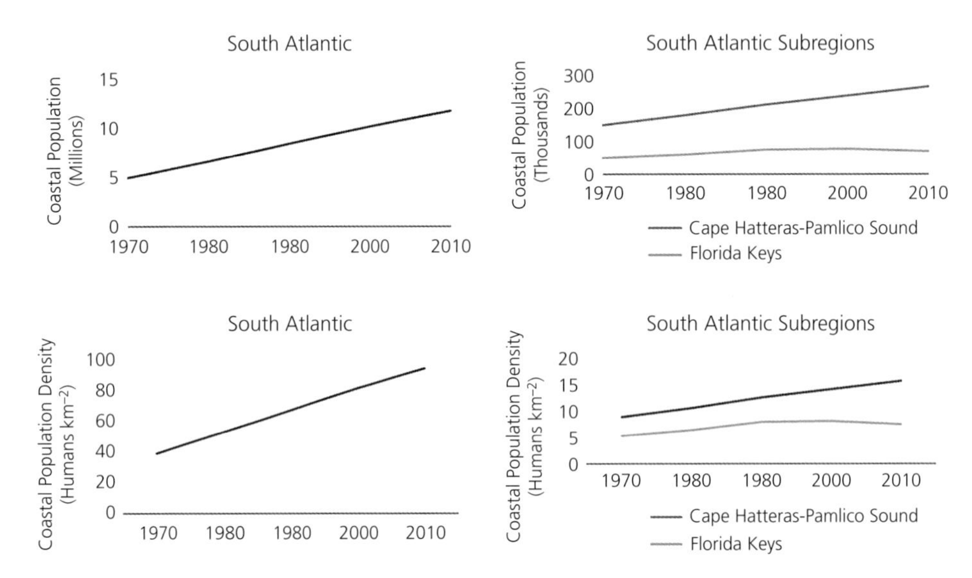

Figure 5.3 Decadal trends (1970–2010) in coastal human population (top) and population density (bottom; humans km^{-2}) for the South Atlantic region and identified subregions.

Data derived from U.S. censuses and taken from https://coast.noaa.gov/digitalcoast/data/demographictrends.html.

5.4.1 Socioeconomic Criteria

5.4.1.1 Social and Regional Demographics

In the U.S. South Atlantic region, coastal human population values and population densities (Fig. 5.3) have consistently increased from 1970 to present with values currently around 11.7 million residents and ~94 people km^{-2}, respectively. This region makes up 3.8% of the United States total population (year 2010), while density values have risen at ~1.4 people km^{-2} year^{-1}. Comparatively among the two subregions, human population is more heavily concentrated in areas surrounding Cape Hatteras-Pamlico Sound, given its proximity to the North Carolina Outer Banks and more populated coastal counties (i.e., Beaufort, Carteret, Craven, and Dare, NC). In this subregion, human population has increased (~3000 people year^{-1}) over the past 40 years, with currently up to ~271 thousand residents and 15.8 people km^{-2} in surrounding areas. Values have remained relatively steady in the much less populated Florida Keys (up to ~74 thousand residents in Munroe County, FL), with year 2010 human population density at 7.6 people km^{-2}. However, coastal population metrics have also increased considerably over the past decades (~83,000 people year^{-1}) in adjacent South Florida (i.e., Palm Beach, Broward, and Miami-Dade counties) with up to 5.6 million residents and ~350 people km^{-2} as of 2010. These trends also suggest likely related increased visitation to the Florida Keys and associated human impacts. The Cape Hatteras-Pamlico Sound subregion encompasses two additional, less populated North Carolina coastal counties (i.e., Hyde, Pamlico), while the Florida Keys comprise a majority of Florida's Munroe County.

5.4.1.2 Socioeconomic Status and Regional Fisheries Economics

Low to mid-range numbers of LMR establishments and employments, and LMR-associated gross domestic product (GDP) values (up to $381 million USD), are found throughout the South Atlantic region (Fig. 5.4). Ranking fifth-highest (out of eight regional ecosystems) nationally, the South Atlantic contains moderate numbers of LMR establishments, employments, and GDP value. Although the number of LMR establishments and employments declined during the 2000s, increases in more recent years have been observed. LMR establishments continue to comprise 2–4% of regional ocean economy establishments, while LMR employments have contributed ~1% of regional ocean employments over time. LMR-associated GDP has increased by ~100 million since 2011, and has contributed ~0.5–2.0% toward the regional total ocean economy. Values have remained relatively steady throughout subregions, with the highest numbers of LMR establishments and employments, and GDP value historically occurring within the Cape Hatteras-Pamlico Sound area (48 ± 0.8 standard error, SE, establishments; 303.9 ± 17.4 SE employments; $33.7 ± 1.8 million SE, USD). Overall, values between subregions have been relatively similar, but numbers of LMR establishments have been decreasing in Cape Hatteras-Pamlico Sound and increasing in the Florida Keys during recent years to nearly equivalent levels, with LMR establishments contributing up to 5% of the Florida Keys total ocean economy. LMR employments contribute 1–2% to the total ocean economies of Cape Hatteras-Pamlico Sound and the Florida Keys, while greater contributions of LMR-associated GDP (6–8%) have been observed in the Cape Hatteras-Pamlico Sound subregion over time.

During the 1990s, low reported numbers of commercially permitted fishing vessels operated in the South Atlantic (Fig. 5.5), with values expanding upward to ~10,000 in the early to mid-2000s. While values have fluctuated, increases in recent years have been observed with values peaking in 2012 at ~16,100 vessels. Total revenue (year 2017 USD) of landed commercial fishery catches (Fig. 5.6) has increased over time as the sixth-highest in the nation, peaking at $243 million in 1995 and remaining around $185 million in more recent years. Among subregions, commercial catch revenues have been similar over time with peak values occurring in 1982 (Cape Hatteras-Pamlico Sound: $78.5 million; Florida Keys: $71.2 million). Over time decreases in revenue have been observed for both subregions, with current landings revenue (year 2016) higher in the Florida Keys ($27 million) than in Cape Hatteras-Pamlico Sound ($12.7 million). Ratios of LMR jobs and commercial fisheries revenue to primary production (Fig. 5.7) are fifth and sixth-highest nationally in the South Atlantic, respectively. Revenue/productivity values have remained relatively stable over the past decade with some interannual fluctuation (range: $0.13 to $0.22 metric ton^{-1}), while jobs/productivity have marginally increased over time (up to 3.9 jobs thousand metric tons^{-1}).

5.4.2 Governance Criteria

5.4.2.1 Human Representative Context

Twenty federal and state agencies and organizations operate within the South Atlantic and are directly responsible for fisheries and environmental management (Table 5.1). Fisheries resources throughout state waters (0–3 nautical miles) are managed by individual state agencies and the ASMFC, while those in federal waters are managed by both NOAA Fisheries (Southeast Region) and the SAFMC. The Cape Hatteras-Pamlico Sound subregion falls under the jurisdiction of the North Carolina Department of Environmental Quality—Division of Marine Fisheries, while marine resources of the Florida Keys are managed as a NOAA National Marine Sanctuary (NMS) and by the Florida Fish and Wildlife Conservation Commission (FWC).

The South Atlantic contains lower numbers of Congressional representatives (fifth-highest nationally), which are concentrated within four states (Fig. 5.8). Low representation is found in both the Cape Hatteras-Pamlico Sound and Florida Keys subregions. When

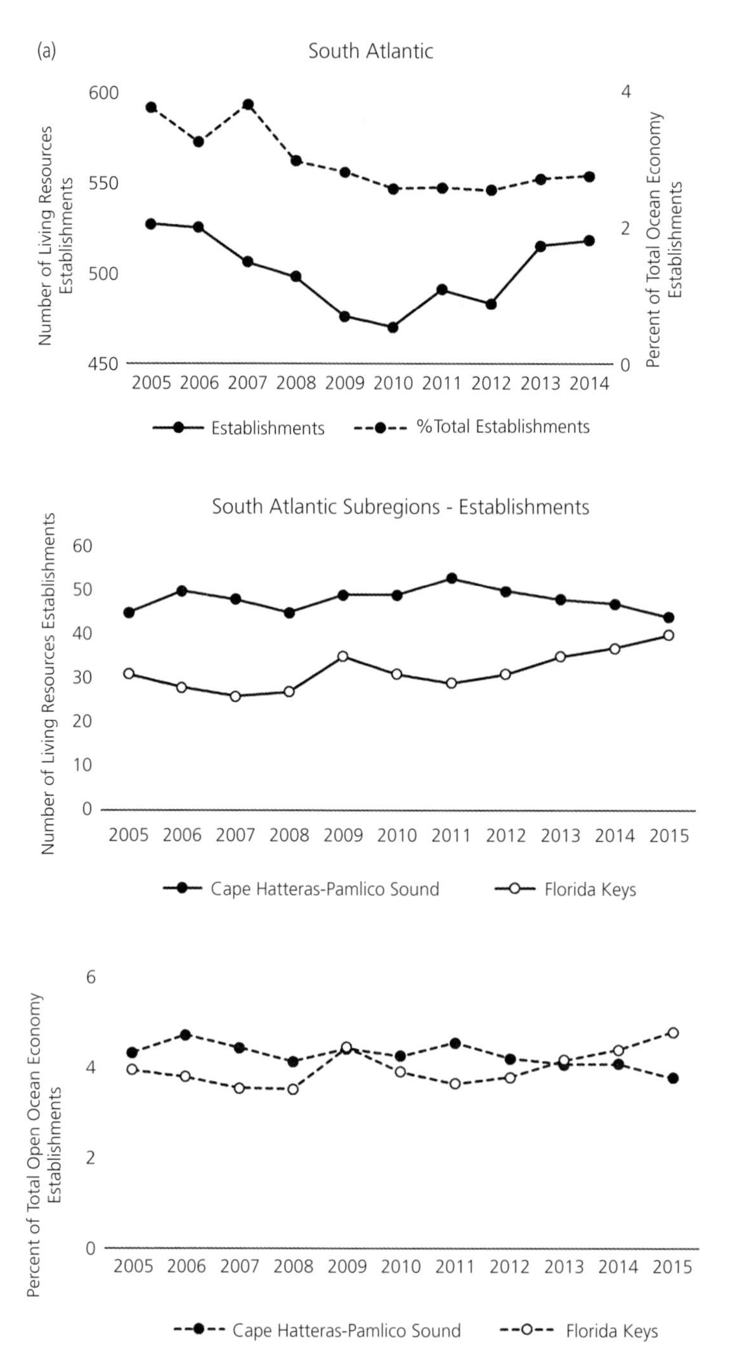

Figure 5.4a Number of living marine resources establishments and their percent contribution to total multisector oceanic economy establishments in the South Atlantic region and identified subregions (years 2005–2015).

Data derived from the National Ocean Economics Program.

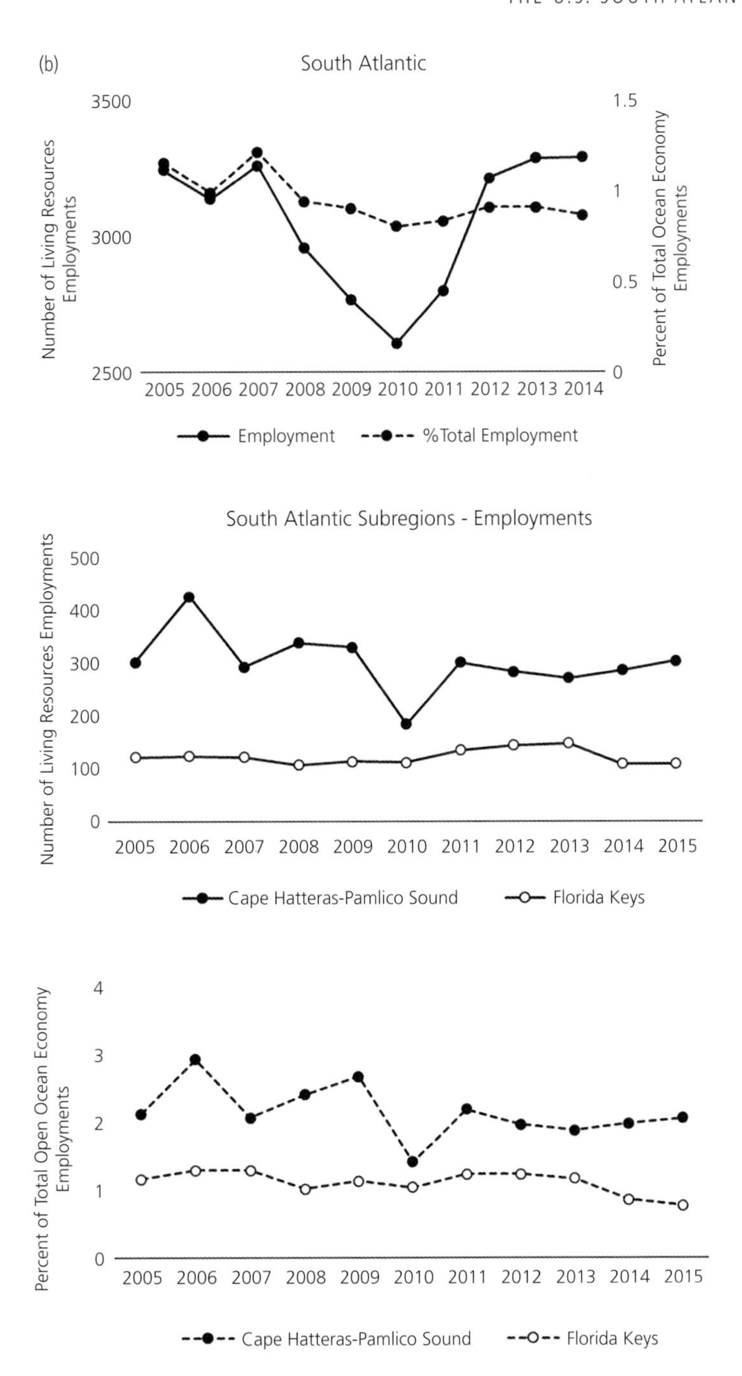

Figure 5.4b Number of living marine resources employments and their percent contribution to total multisector oceanic economy employments in the South Atlantic region and identified subregions (years 2005–2015).

Data derived from the National Ocean Economics Program.

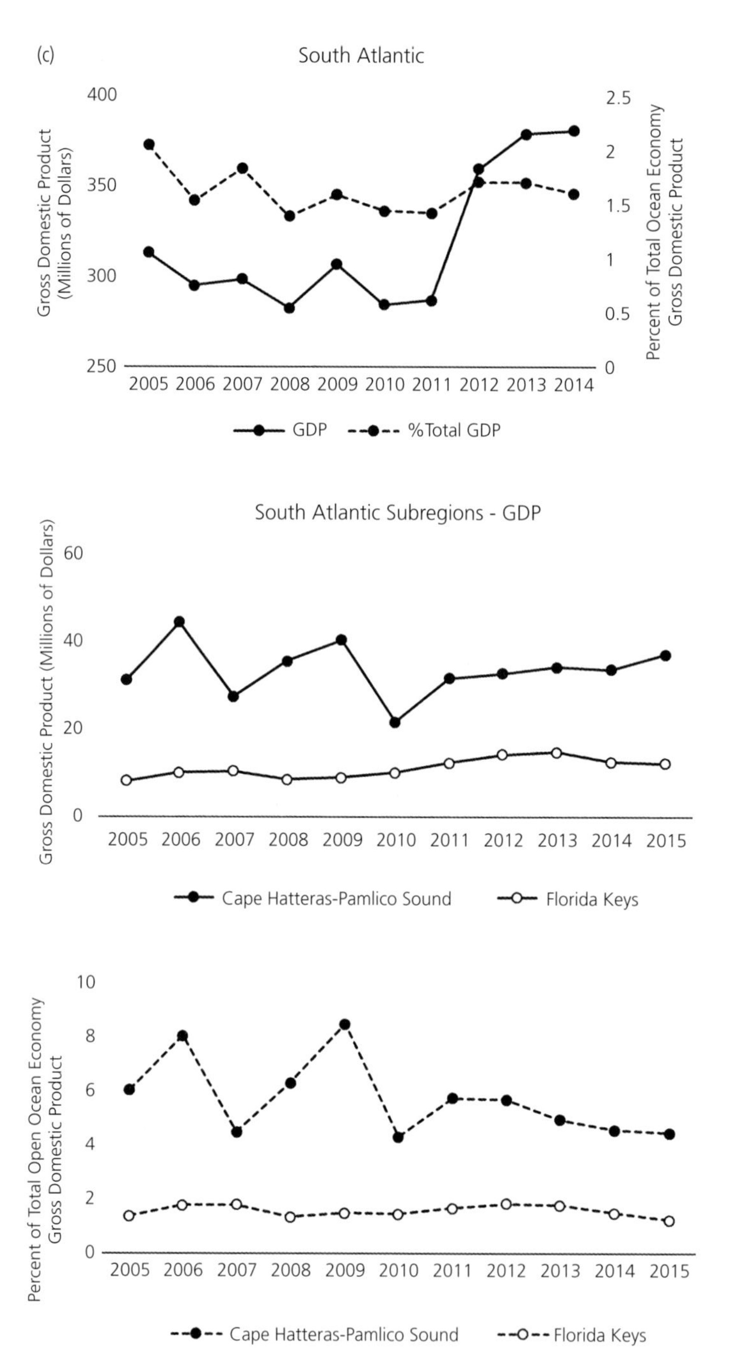

Figure 5.4c Gross domestic product value (GDP; USD) from living marine resources revenue and percent contribution to total multisector oceanic economy GDP for the South Atlantic region and identified subregions (years 2005–2015).

Data from the National Ocean Economics Program.

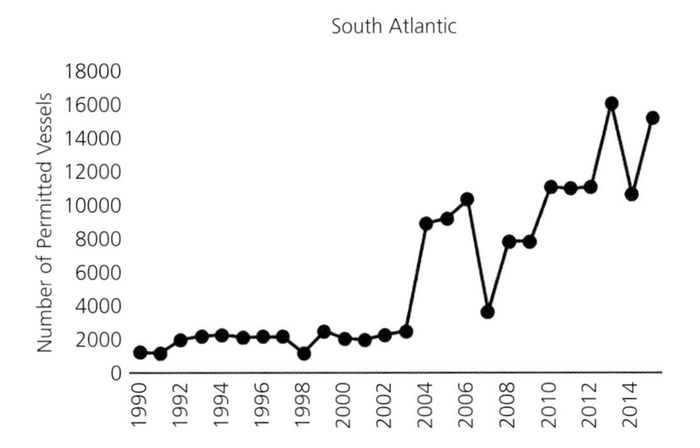

Figure 5.5 Total number of permitted vessels for the South Atlantic region over time (years 1990–2015).
Data derived from NOAA National Marine Fisheries Service Council Reports to Congress.

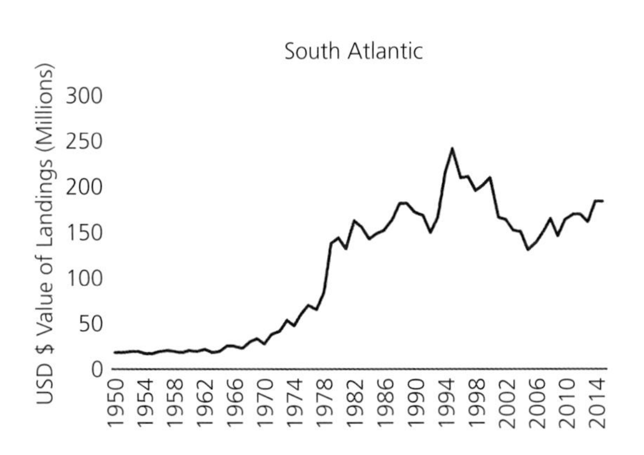

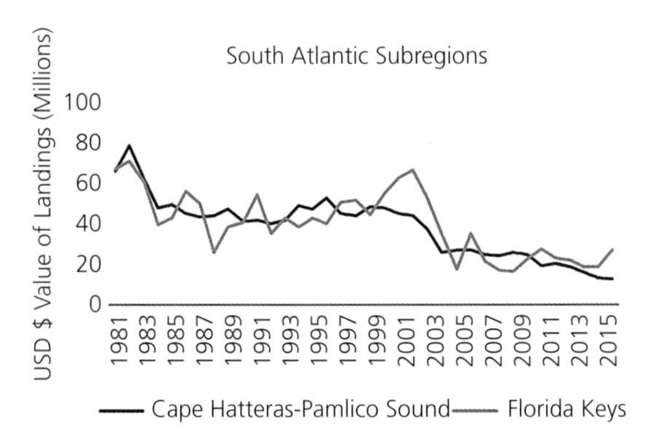

Figure 5.6 Total revenue (year 2017 USD) of landed commercial fishery catches for the South Atlantic region (years 1950–2015) and identified subregions (years 1981–2016), over time.

Data derived from NOAA National Marine Fisheries Service commercial fisheries statistics.

standardized per mile of shoreline, Congressional representation for the South Atlantic is fourth-highest out of eight regions nationally (mid-range). Composition of the SAFMC has been split relatively evenly among members from the commercial and recreational fishing sectors, with minimal representation from other organizations (Fig. 5.9a). This composition has remained steady over time, representing one of the more stably balanced fishery management councils (FMCs) nationally. With 13 total voting members, including one Federal and two State government representatives, and eight appointed members from fishing or other sectors, the SAFMC contains a medium number of council members nationally (tied for fifth-highest with the North Pacific FMC). Regional representatives from the South Atlantic also serve on the Atlantic Highly Migratory Species Advisory Panel. The Atlantic marine mammal SRG is responsible for advising federal agencies on the status of marine mammal stocks covered under the Marine Mammal Protection Act (MMPA), for regions including the South Atlantic. Its composition and membership have remained steady over time (Fig. 5.9b), with individuals from the academic and private institutional sectors making up the majority of its representatives.

5.4.2.2 Fishery and Systematic Context

Five current South Atlantic-specific fishery management plans (FMPs) are managed by NOAA Fisheries and the SAFMC for golden crab; the snapper/grouper fishery; the South Atlantic shrimp fishery; *Sargassum*

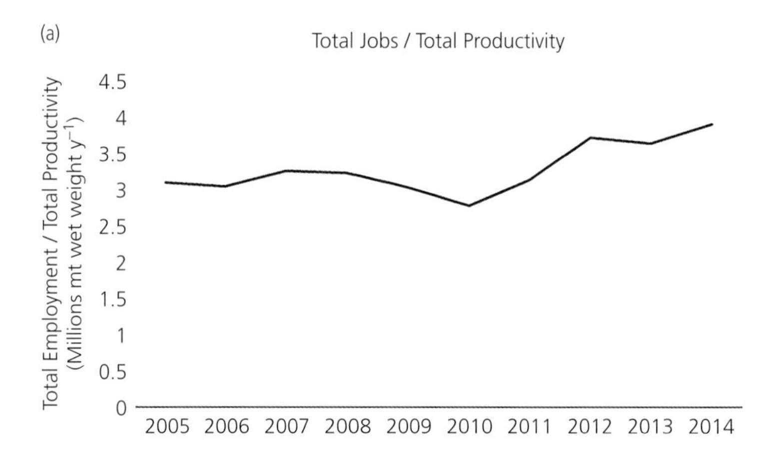

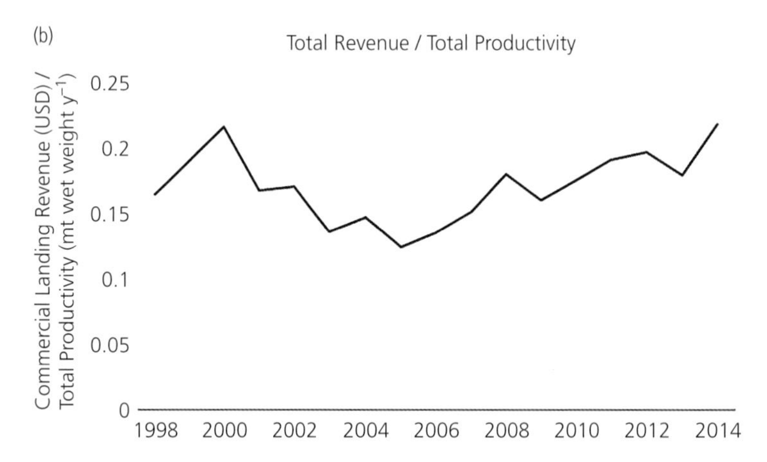

Figure 5.7 Ratios of (a) total living marine resources employments to total productivity (metric tons wet weight year^{-1}); (b) total commercial landings revenue (USD) to total productivity (metric tons wet weight year^{-1}) for the South Atlantic region.

species and its associated biological community; and Atlantic Dolphin/Wahoo (Table 5.2). Three more FMPs are shared between the SAFMC and the GMFMC for coral reef species; coastal migratory pelagic resources (Cobia, King mackerel, and Spanish mackerel); and Spiny lobster. In total, these FMPs have been subject to 137 modifications (i.e., amendments and framework/regulatory amendments). The ASMFC manages 26 FMPs for state fisheries resources, including bony fishes (20 species), coastal sharks (22 species), Horseshoe crab, and Jonah crab. In total, Atlantic state-wide FMPs have been subject to 230 modifications over time. An FEP exists for the South Atlantic region, having evolved from the SAFMC essential habitat plan. This FEP contains comprehensive information regarding South Atlantic fisheries, the habitat and biology of its managed species, social and economic management impacts, and the ecological consequences of conservation and management (SAFMC 2009).

Four national seashores (Canaveral, Cape Hatteras, Cape Lookout, Cumberland Island), eight coastal national parks, monuments/memorials, or historic sites, one NOAA habitat focus area (Biscayne Bay), and five NERRs) are found throughout the South Atlantic region (Table 5.3). Eleven inshore and offshore regions are identified as habitat areas of particular concern (HAPCs) in the South Atlantic. These include identified coral reef and hard-bottom habitats, five specific deep sea coral areas, and snapper-grouper habitats (including for tilefish) from the Florida Keys to off North Carolina; North Carolina Penaeid shrimp inshore habitats, including in Pamlico Sound; Spiny lobster habitats

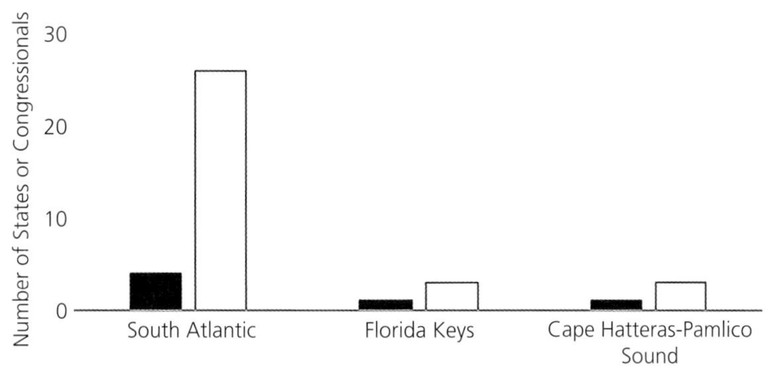

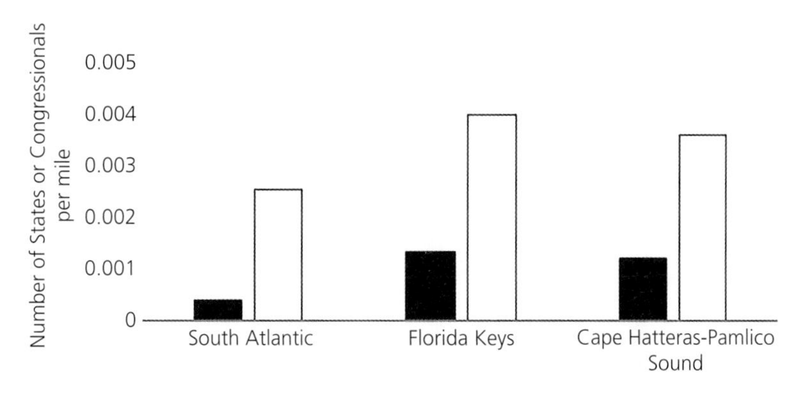

Figure 5.8 Number of states and U.S. Congressional representatives (top panel) and as standardized by miles of shoreline (bottom panel) for the South-Atlantic region and its identified subregions.

Table 5.1 Federal and State agencies and organizations responsible for fisheries and environmental consulting and management in the South Atlantic region and subregions

Region	Subregion	Federal Agencies with Marine Resource Interests	Council	State Agencies and Cooperatives	Multi-Jurisdictional
South Atlantic		U.S. Department of Commerce **DoC** (NOAA Fisheries Southeast Fisheries Science Center **SEFSC**, Southeast Regional Fisheries Office **SERO**, NOAA Office of National Marine Sanctuaries **NMS**) U.S. Department of Interior **DoI** (U.S. Fish and Wildlife Service **FWS**, U.S. National Park Service **NPS**, U.S. Bureau of Ocean Energy Management **BOEM**, U.S. Geological Survey **USGS**) U.S. Environmental Protection Agency **EPA**	South Atlantic Fishery Management Council **SAFMC**	Atlantic States Marine Fisheries Commission **ASMFC**; Florida Fish and Wildlife Conservation Commission **FWC**; Georgia Department of Natural Resources; North Carolina Department of Environmental Quality—Division of Marine Fisheries; South Carolina Department of Natural Resources	Coral Reef Task Force **CRTF**; International Commission for the Conservation of Atlantic Tunas **ICCAT**; International Society for Reef Studies **ISRS**; National Coral Reef Institute **NCRI**

Continued

Table 5.1 Continued

Region	Subregion	Federal Agencies with Marine Resource Interests	Council	State Agencies and Cooperatives	Multi-Jurisdictional
	Cape Hatteras-Pamlico Sound	**DoC (SEFSC, SERO); DoI (FWS, BOEM, NPS, USGS); EPA**	**SAFMC**	**ASMFC**; N Carolina Department of Environmental Quality—Division of Marine Fisheries	**ICCAT**
	Florida Keys	**DoC (SEFSC, SERO, NMS); DoI (FWS, BOEM, USGS); EPA**	**SAFMC**	**ASMFC;** Florida Fish and Wildlife Conservation Commission **FWC**	**CRTF, ICCAT, ISRS, NCRI**

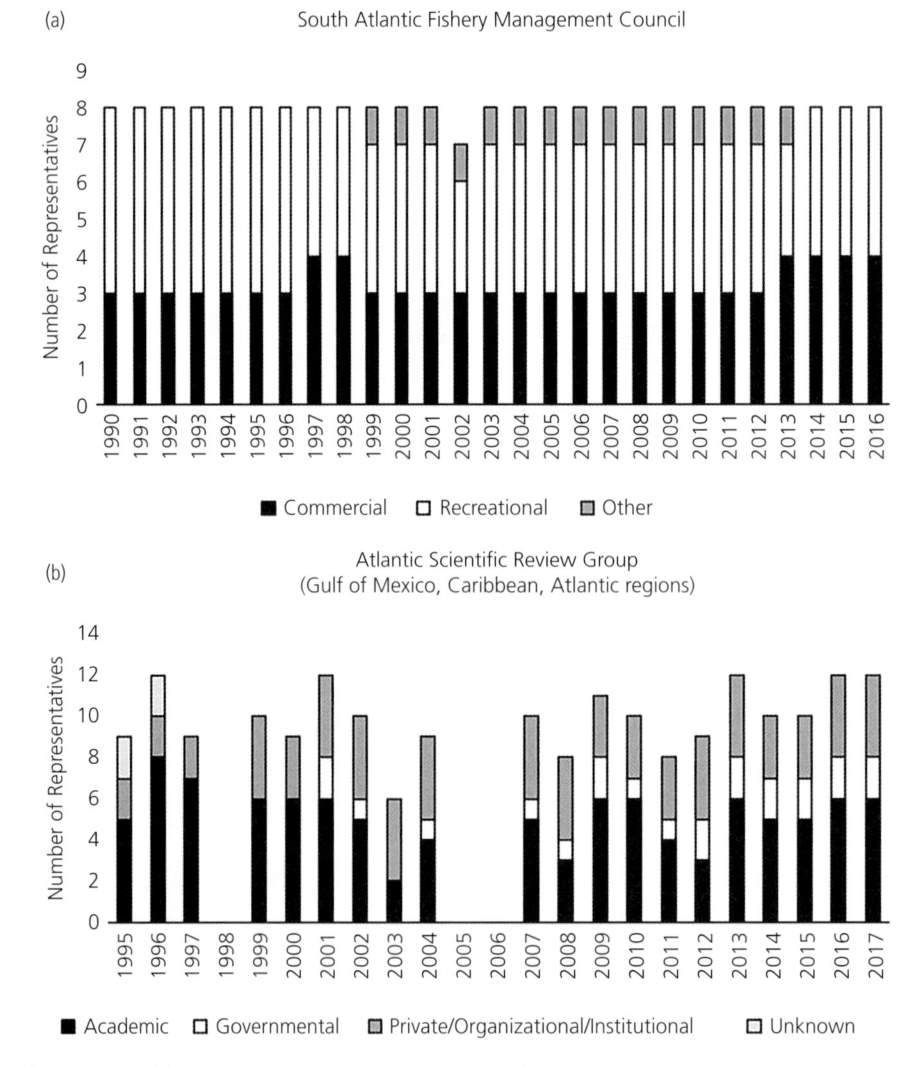

Figure 5.9 (a) Composition of the South Atlantic Fishery Management Council from commercial and recreational fishing and other sectors and the (b) Atlantic marine mammal scientific review group (data unavailable 1998, 2005–2006) over time.

Data derived from NOAA National Marine Fisheries Service Council Reports to Congress and scientific review group reports.

Table 5.2 List of major managed fishery species in the South Atlantic region, under state and federal jurisdictions

Council/Agency	Fishery Management Plan (FMP)	FMP Modifications	Major Species/ Species Group	Scientific Name(s) or Family Name(s)
Gulf of Mexico Fishery Management Council **GMFMC**/ South Atlantic Fishery Management Council **SAFMC**	Coral, coral reefs, and live/ hard-bottom habitat (142 species)	12 amendments	All species	Octocorallina, Hydrozoans, Scleractinians, Antipatharians, Live Rock
	Coastal migratory pelagic resources (3 species)	26 amendments, 3 framework amendments, 19 regulatory amendments	King mackerel Cobia	*Scomberomorus cavalla* *Rachycentron canadum*
			Spanish mackerel	*Scomberomorus maculatus*
	Spiny lobster	11 amendments, 2 regulatory amendments	Spiny lobster	*Panulirus argus*
SAFMC	Golden crab	5 amendments	Golden crab	*Chaceon fenneri*
	Snapper/grouper fishery (55 species)	43 amendments	Groupers/ seabasses	Family Serranidae: *Cephalopholis* spp., *Centropristis* spp., *Epinephelus* spp., *Hypothodus* spp., *Mycteroperca* spp.
			Wreckfish	*Polyprion americanus*
			Snappers	Family Lutjanidae: *Etilis* spp., *Lutjanus* spp., *Ocyurus* sp., *Rhomboplites* sp.
			Other reef fishes	Species from families Balistidae (Triggerfishes), Carangidae (Jacks), Ephippidae (Spadefishes), Haemulidae (Grunts), Labridae (Wrasses), Malacanthidae (Tilefishes), Sparidae (Porgies)
	S. Atlantic shrimp fishery (5 species)	9 amendments	All shrimp	*Farfantepenaeus* spp., *Litopenaeus* spp., *Pleoticus* spp., *Sicyonia* spp.
	Sargassum (2 species)	1 amendment	All species	*Sargassum* spp.
	Atlantic dolphin/wahoo (2 species)	6 amendments	Atlantic Dolphin and Wahoo	*Coryphaena hippurus*, *Acanthocybium solandri*

Continued

Table 5.2 Continued

Council/Agency	Fishery Management Plan (FMP)	FMP Modifications	Major Species/ Species Group	Scientific Name(s) or Family Name(s)
Atlantic States Marine Fisheries Commission **ASMFC**	American eel	4 addendum	American eel	*Anguilla rostrata*
	American lobster (co-managed with NOAA Fisheries; New England)	3 amendments, 24 addendum, 2 technical addendum	American lobster	*Homarus americanus*
	Atlantic croaker	1 amendment, 2 addendum	Atlantic croaker	*Micropogonias undulatus*
	Atlantic herring	3 amendments, 7 addendum, 2 technical addendum	Atlantic herring	*Clupea harengus*
	Atlantic menhaden	2 amendments, 1 revision, 1 supplement, 6 addendum, 2 technical addendum	Atlantic menhaden	*Brevoortia tyrannus*
	Atlantic striped bass	6 amendments, 1 supplement, 9 addendum	Atlantic striped bass	*Morone saxatilis*
	Atlantic sturgeon	1 amendment, 1 technical addendum, 4 addendum	Atlantic sturgeon	*Acipenser oxyrhynchus*
	Black drum		Black drum	*Pogonias cromis*
	Black sea bass	2 amendments, 27 addendum	Black sea bass	*Centropristis striata*
	Bluefish	1 amendment, 1 addendum	Bluefish	*Pomatomus saltatrix*
	Coastal sharks (22 species)	4 addendum	Aggregated large coastal	*Carcharhinus* spp., *Galeocerdo cuvier, Ginglymostoma cirratum, Negaprion brevirostris*
			Hammerhead	*Sphyrna* spp.
			Pelagic	*Alopias vulpinus, Carcharhinus longimanus, Isurus oxyrinchus, Lamna nasus, Prionace glauca*

Council/Agency	Fishery Management Plan (FMP)	FMP Modifications	Major Species/ Species Group	Scientific Name(s) or Family Name(s)
Atlantic States Marine Fisheries Commission **ASMFC**			Blacknose	*Carcharhinus acronotus*
			Non-Blacknose Small Coastal	*Carcharhinus isodon*, *Rhizoprionodon terraenovae*, *Sphyrna tiburo*
			Smoothhound	*Mustelus* spp.
	Cobia		Cobia	*Rachycentron canadum*
	Horseshoe crab	7 addendum	Horseshoe crab	*Limulus polyphemus*
	Jonah crab	2 addendum	Jonah crab	*Cancer borealis*
	Northern shrimp	2 amendments, 1 addendum	Northern shrimp	*Pandalus borealis*
	Red drum (Atlantic Coast)	2 amendments, 1 addendum	Atlantic Coast red drum	*Sciaenops ocellatus*
	Scup	3 amendments, 29 addendum	Scup	*Stenotomus chrysops*
	Shad and river herring (3 species)	2 addendum, 3 amendments, 1 technical addendum	American shad and river herring	*Alosa aestivalis*, A. pseudoharengus, A. sapidissima
	Spanish mackerel	1 amendment, 1 addendum	Spanish mackerel	*Scomberomorus maculatus*
	Spiny dogfish	5 addendum	Spiny dogfish	*Squalus acanthias*
	Spot	1 addendum	Spot	*Leiostomus xanthurus*
	Spotted seatrout	1 amendment	Spotted seatrout	*Cynoscion nebulosus*
	Summer flounder	2 amendments, 28 addendum	Summer flounder	*Paralichthys dentatus*
	Tautog	6 addendum	Tautog	*Tautoga onitis*
	Weakfish	4 amendments, 1 technical addendum, 4 addendum	Weakfish	*Cynoscion regalis*
	Winter flounder	1 amendment, 5 addendum	Winter flounder	*Pseudopleuronectes americanus*

Table 5.3 List of major bays and islands, NOAA habitat focus areas (HFAs), NOAA national estuarine research reserves (NERRs), NOAA national marine monuments and sanctuaries (NMS), coastal national parks, national seashores, and number of habitats of particular concern (HAPCs) in the South Atlantic region and subregions

Region	Subregion	Major Bays	Major Islands	HFAs	NERRs	Monuments and NMS	Coastal National Parks	National Seashores	HAPCs
South Atlantic		Biscayne Bay, Calbiogue Sound, Pamlico Sound, Wassaw Sound	Bald Head Island, Cumberland Island, Emerald Island, Florida Keys, Hilton Head Island, Hunting Island, Jekyll Island, Jupiter Island, Key Biscayne, Merritt Island, North Carolina Outer Banks, Parris Island, Pawleys Island, Pritchards Island, Sapelo Island, Skidaway Island, Tybee Island, Virginia Key	Biscayne Bay, Florida HFA	ACE Basin NERR, Guana Tolomato Matanzas NERR, North Carolina NERR, North Inlet-Winyah Bay NERR, Sapelo Island NERR	Florida Keys NMS, Gray's Reef NMS, Monitor NMS	Biscayne National Park, Castillo de San Marcos National Monument, Fort Caroline National Memorial, Fort Frederica National Monument, Fort Motanzas National Monument, Fort Pulaski National Monument, Fort Sumter National Monument, Timucuan Ecological & Historic Preserve	Canaveral National Seashore, Cape Hatteras National Seashore, Cape Lookout National Seashore, Cumberland Island National Seashore	11
	Cape Hatteras-Pamlico Sound		North Carolina Outer Banks, Roanoke Island		North Carolina NERR	Monitor NMS		Cape Hatteras National Seashore	2
	Florida Keys		Florida Keys			Florida Keys NMS	Biscayne National Park		4

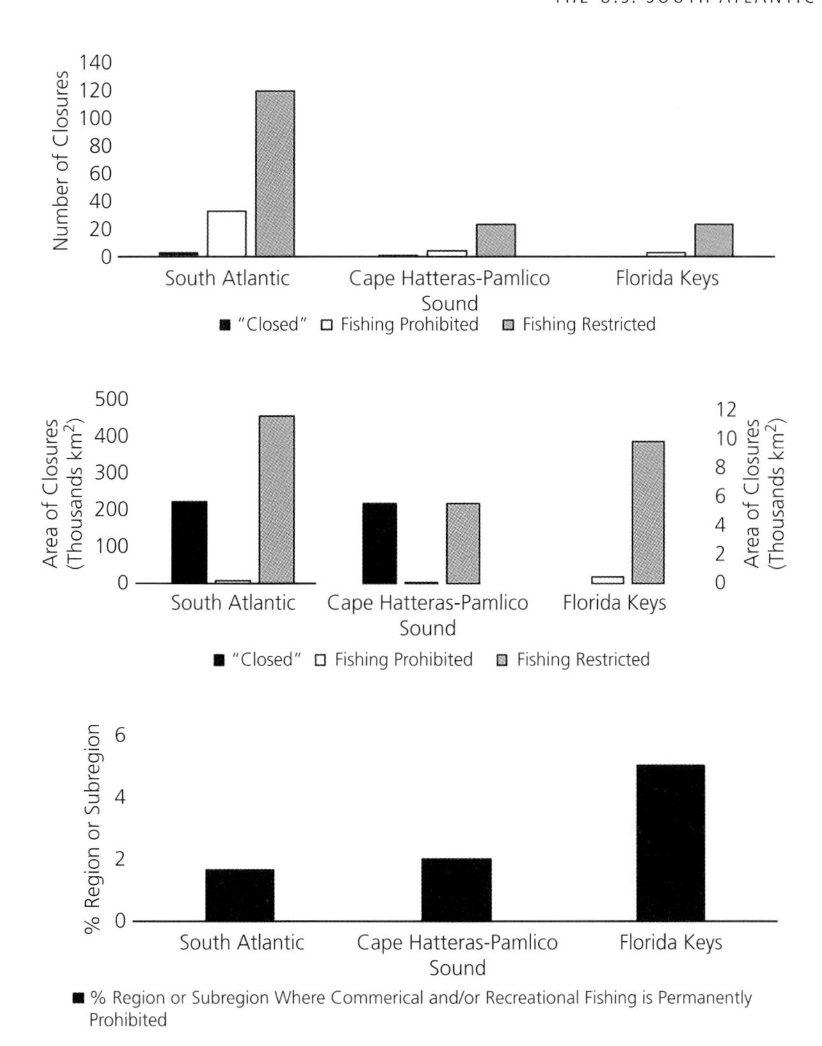

Figure 5.10 Number (top panel) and area (middle panel; km^2) of named fishing closures, and prohibited or restricted fishing areas of the South Atlantic region and identified subregions, including percent region or subregion (bottom panel) where commercial and/or recreational fishing is permanently prohibited. Values for Cape Hatteras-Pamlico Sound and Florida Keys (top and middle panels) plotted on the secondary (right) axis.

Data derived from NOAA Marine Protected Areas inventory.

surrounding South Florida and the Florida Keys; and pelagic offshore Dolphin and Wahoo habitats throughout the South Atlantic and Florida Keys. Only three named fishing closure zones occur in the South Atlantic, while fishing is permanently prohibited in 33 specific areas and restricted in 120 of them (Fig. 5.10). Very few of these closures occur in South Atlantic subregions, of which the Florida Keys and Cape Hatteras-Pamlico Sound areas 3 and 4 are prohibited fishing areas, respectively, and each contain or comprise 23 restricted areas. A small component of the South Atlantic contains named closures or permanent fishing prohibitions (~1.7% of the EEZ), while ~450,000 km^2 is subject to some fishing restrictions. Among subregions, the largest restricted area occurs in the Florida Keys NMS (~10,000 km^2), while ~2% of Cape Hatteras-Pamlico Sound is permanently prohibited from fishing due to the presence of four surrounding national wildlife refuges (NWRs), the largest of which is the Swanquarter NWR.

5.4.2.3 Organizational Context

Relative to the total commercial value of its managed fisheries, the SAFMC budget is third-highest when compared to other regions (Fig. 5.11), and makes up

Figure 5.11 Total annual budget of the South Atlantic Fishery Management Council as compared to the total value of its marine fisheries. All values in USD.

Data from NOAA National Marine Fisheries Service Office of Sustainable Fisheries and NOAA National Marine Fisheries commercial landings database.

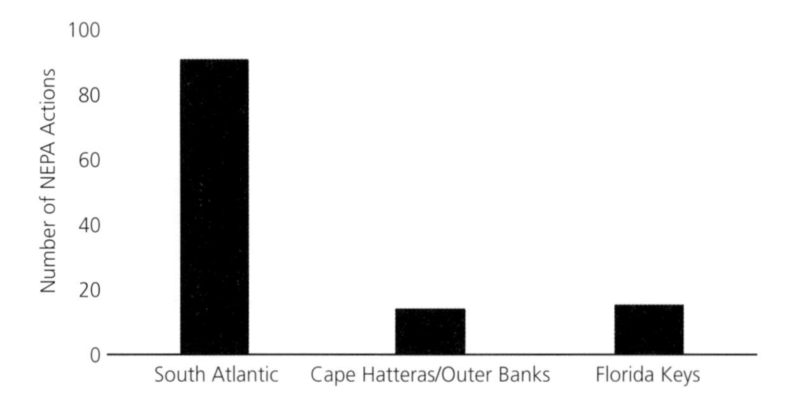

Figure 5.12 Number of National Environmental Policy Act Environmental Impact Statement (NEPA-EIS) actions from 1987 to 2016 for the South Atlantic region and identified subregions.

Data derived from U.S. Environmental Protection Agency EIS database.

a small proportion of its overall costs. These relative values increased in the late 2000s and have remained relatively stable in the 2010s. Cumulative National Environmental Policy Act Environmental Impact Statement (NEPA-EIS) actions from 1987 to 2016 in the South Atlantic have been relatively low compared to other regions, although highest among regions along the U.S. Atlantic coast, with 91 observed over the past ~30 years (Fig. 5.12). Relatively few actions have occurred within the Cape Hatteras-Pamlico Sound and the Florida Keys subregions, with cumulative values of 14 and 15, respectively. Only five fisheries-related lawsuits have occurred in the South Atlantic region since 2010, making it one of the least litigious among all U.S. regions.

5.4.2.4 Status of Living Marine Resources (Targeted and Protected Species)

Nationally, the South Atlantic contains the third-highest number of managed species. Of the 268 federally managed species in the region, 69 are targeted fishery species covered under FMPs, while 149 species (5 bony fishes, 144 coral reef species) are prohibited from harvest. Red snapper (*Lutjanus campechanus*) is subject to limited harvest. Fifty species of marine mammals, sea turtles, corals, seagrasses, and fishes are also federally protected under the ESA or MMPA (Fig. 5.13, Tables 5.2, 5.4–5). An overlapping 68 species are also managed under the state jurisdiction of the ASMFC, of which 19 sharks have fishing prohibitions. The Wreckfish

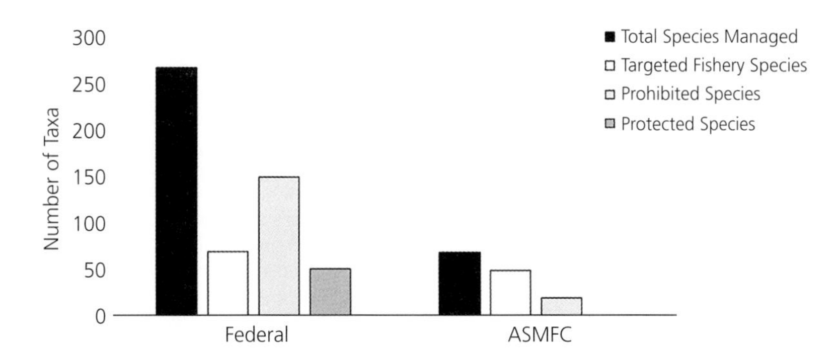

Figure 5.13 Number of managed taxa (species or families) for the South Atlantic region, including federal and state (Atlantic States Marine Fisheries Commission—ASMFC) jurisdictions.

Table 5.4 List of prohibited fishery species in the South Atlantic region, including federal and state jurisdictions

Council/Authority	Region	Prohibited Species	Scientific Name
South Atlantic Fishery Management Council **SAFMC**; Southeast Regional Office **SERO**	South Atlantic (5 species fishes, 2 species sea fans, 142 hydrozoan, stony coral, and black coral taxa)	Goliath grouper	*Epinephelus itajara*
		Speckled hind	*Epinephelus drummondhayi*
		Warsaw grouper	*Epinephelus nigritus*
		Nassau grouper	*Epinephelus striatus*
		Red snapper (limited harvest)	*Lutjanus campechanus*
		Federal Red drum	*Sciaenops ocellatus*
		Sea fans	*Gorgonia* spp.
		Hydrozoans, hard corals, and black corals	Families Acroporidae, Agaricidae, Anthemiphylliidae, Astrocoeniidae, Cayophyllidae, Dendrophyllidae, Faviidae, Flabellidae, Fungiidae, Guyniidae, Meandrinidae, Milleporidae, Mussidae, Oculinidae, Pocillopridae, Poritidae, Rhizangiidae, Siderastreidae, Orders Antipatharia, Stylasterina
		Live rock	Any coral reef resource attached to a hard substrate, including dead coral or rock (excluding individual mollusk shells)
Atlantic States Marine Fisheries Commission **ASMFC**	Atlantic state waters (19 species)	Coastal sharks	*Alopias superciliosus, Carcharhinus* spp., *Carcharias taurus, Carcharodon carcharias, Hexachus* spp., *Isurus paucus, Notorynchus cepedianus, Odontaspis noronhai, Rhincodon typus, Rhizoprionodon porosus, Squatina demeril*

(*Polyprion americanus*) commercial fishery managed under the snapper-grouper FMP is subject to an Individual Transferable Quota (ITQ) program with limited entry for recreational fishers (Yandle & Crosson 2015, NMFS 2020b). Seasonal commercial closures also occur during its spawning season (mid-January through mid-April), while its recreational fishery is only open during July and August of each year (NMFS 2020b).

As of mid-2017, 50 stocks were federally managed in the South Atlantic with 29 listed as a NOAA Fish Stock

Table 5.5 List of protected species under the Endangered Species Act (ESA) or Marine Mammal Protection Act (MMPA) in the South Atlantic region

Region	Managing Agency	Protected Species Group	Common Name	Scientific Name
South Atlantic (50 species)	South Atlantic Fishery Management Council **SAFMC**; Southeast Regional Office **SERO**	Whales	Blainsville's beaked whale	*Mesoplodon densirostris*
			Blue whale	*Balaenoptera musculus*
			Cuvier's beaked whale	*Ziphius cavirostris*
			Dwarf sperm whale	*Kogia sima*
			False killer whale	*Pseudorca crassidens*
			Fin whale	*Balaenoptera physalus*
			Gervais' beaked whale	*Mesoplodon europaeus*
			Humpback whale	*Megaptera novaeangliae*
			Killer whale (Orca)	*Orcinus orca*
			Long-finned pilot whale	*Globicephala melas*
			Melon-headed whale	*Peponocephala electra*
			Minke whale	*Balaenoptera acutorostrata*
			North Atlantic right whale	*Eubalaena glacialis*
			Pygmy killer whale	*Feresa attenuata*
			Pygmy sperm whale	*Kogia breviceps*
			Sei whale	*Balaenoptera borealis*
			Short-finned pilot whale	*Globicephala macrorhyn-chus*
			Sowerby's beaked whale	*Mesoplodon bidens*
			Sperm whale	*Physeter macrocephalus*
			True's beaked whale	*Mesoplodon mirus*
		Dolphins	Atlantic spotted dolphin	*Stenella frontalis*
			Common bottlenose dolphin	*Tursiops truncatus truncatus*
			Clymene dolphin	*Stenella clymene*
			Common dolphin	*Delphinus delphis*
			Fraser's dolphin	*Lagenodelphis hosei*
			Harbor porpoise	*Phocoena phocoena*
			Pantropical spotted dolphin	*Stenella attenuata*
			Risso's dolphin	*Grampus griseus*
			Rough-toothed dolphin	*Steno bredanensis*
			Spinner dolphin	*Stenella longirostris*
			Striped dolphin	*Stenella coeruleoalba*
		Sea turtles	Green sea turtle	*Chelonia mydas*
			Hawksbill sea turtle	*Eretmochelys imbricata*
			Kemp's Ridley sea turtle	*Lepidochelys kempii*
			Leatherback sea turtle	*Dermochelys coriacea*
			Loggerhead sea turtle	*Caretta caretta*
		Fishes	Atlantic sturgeon	*Acipenser oxyrinchus oxyrinchus*
			Nassau grouper	*Epinephelus striatus*
			Scalloped hammerhead shark	*Sphyrna lewini*
			Shortnose sturgeon	*Acipenser brevirostrum*
			Smalltooth sawfish	*Pristis pectinata*
		Corals	Rough cactus coral	*Mycetophyllia ferox*
			Pillar coral	*Dendrogyra cylindrus*
			Lobed star coral	*Orbicella annularis*
			Mountainous star coral	*Orbicella faveolata*
			Boulder star coral	*Orbicella franksi*
			Staghorn coral	*Acropora cervicornis*
			Elkhorn coral	*Acropora palmata*
		Seagrasses	Johnson's seagrass	*Halophila johnsonii*
	U.S. Fish and Wildlife Service **FWS**	Other species	West Indian manatee	*Trichechus manatus*

Sustainability Index (FSSI) stock (Fig. 5.14; NMFS 2017, 2020a).[2] Of all stocks, seven were identified as experiencing overfishing (**Hogfish**—*Lachnolaimus maximus*, southeast Florida stock; **Red snapper**, southern Atlantic coast **Blueline tilefish**—*Caulolatilus microps*; **Speckled hind**—*Epinephelus drummondhayi*; **Warsaw grouper**—*Hyporthodus nigritus*; **Golden tilefish**—*Lopholatilus chamaeleonticeps*, southern Atlantic coast and **Red grouper**—*E. morio*, southern Atlantic coast), and five were overfished (**Hogfish**, southeast Florida stock; **Red snapper**, southern Atlantic coast; **Red porgy**—*Pagrus pagrus*; **Snowy grouper**—*E. niveatus*; and **Red grouper**, southern Atlantic coast).[3] There are 11 stocks that remain unclassified as to whether they are experiencing overfishing (22.5%; second-highest nationally) and 24 have unknown overfished status (49%; fourth-highest nationally), making the South Atlantic mid-range nationally in terms of the proportion of fishery stocks with unknown status.[4] For ASMFC managed stocks, six of its South Atlantic stocks are considered sustainable or rebuilt as of 2020, while five are depleted (**American eel**—*Anguilla rostrata*; **American shad**—*Alosa sapidissima*; **Atlantic sturgeon**—*Acipenser oxyrhynchus oxyrhynchus*; **River herring**—*Alosa* spp.; **Weakfish**—*Cynoscion regalis*), two are overfished (**Bluefish**—*Pomatomus saltatrix*; **Striped bass**—*Morone saxatilis*), and one is currently undergoing overfishing (**Striped bass**; ASMFC 2020b). Among the protected species of the South Atlantic are 46 independent marine mammal stocks, and 25 ESA-listed distinct population segments, of which 52% are endangered (Fig. 5.15). Nationwide, the South Atlantic has mid-range numbers of marine mammal stocks (14.4% of total marine mammal stocks), with one-third listed as strategic marine mammal stocks. There are 33% of South Atlantic marine mammal stocks with unknown population size, which is third-highest among all U.S. regions (Link & Marshak 2019).

[2] As of June 2020, 49 stocks are federally managed in the South Atlantic, with 25 listed as FSSI stocks. In 2019, cessation of joint federal management of the Cobia-Southern Atlantic Coast stock by the SAFMC and GMFMC occurred, which is now singularly managed by the ASMFC. In addition, Hogfish—southeast Florida; Scamp-southern Atlantic Coast; Speckled hind-southern Atlantic Coast; and Warsaw grouper—southern Atlantic Coast are no longer listed as FSSI stocks.

[3] As of June 2020, six South Atlantic stocks are experiencing overfishing (Red snapper—southern Atlantic Coast; Speckled hind; Warsaw grouper; Red porgy; Red grouper; and Greater amberjack—*Seriola dumerili*) and five are overfished (Hogfish-southeast Florida; Red snapper-southern Atlantic Coast; Red porgy; Snowy grouper; and Red grouper).

[4] These values remain the same as of June 2020.

5.4.3 Environmental Forcing and Major Features

5.4.3.1 Oceanographic and Climatological Context

Circulation patterns throughout the South Atlantic EEZ are driven primarily by the Gulf Stream (Fig. 5.1). Connected to and originating from the Caribbean Current, Gulf of Mexico Loop Current and the Antilles current, the Gulf Stream is the major current along the United States Atlantic Coast (Fuglister 1951, Murphy et al. 1999). The flow of warmer waters emanating from these currents continues up toward Cape Hatteras, where they then move further offshore (Murphy et al. 1999). As a result, Cape Hatteras, and nearby Pamlico Sound, serves as a major biogeographic break along the U.S. Atlantic coast (Hayden et al. 1984, Fautin et al. 2010, Bangley et al. 2018). Southern counterclockwise "cold-core" rings also break off past Cape Hatteras where the Gulf Stream is no longer constrained by the continental shelf. While "cold-core" rings do drift southward, they always stay offshore of the Gulf Stream. In addition, meanderings and filaments of the Gulf Stream along the southeast shelf provide additional means of propagule transport along coastal regions and throughout the Atlantic EEZ. Due to high riverine input (i.e., Neuse and Pamlico Rivers) into the southwestern portion of Pamlico Sound, a mean eastward flow is observed emanating from the western extent of Pamlico Sound, which converges with a mean north-south flow originating from its connection with Albermarle Sound (Mid-Atlantic region). This convergence produces a southward flow moving into the more saline offshore waters comprising the Gulf Stream (Jia & Li 2012). Additionally, the continuity of the Gulf of Mexico Loop Current and Gulf Stream around South Florida influence mean north-south circulation dynamics in the Florida Keys. These circulation dynamics are also influenced by meandering portions of the South Equatorial current and Antillean current, creating additional sinks along South Florida for propagules originating from the Bahamas and the Caribbean Greater Antilles.

Clear interannual and multidecadal patterns in average sea surface temperature (SST) have been observed for the South Atlantic region (Fig. 5.16), with temperatures averaging 25.0°C ± 0.02 (SE) over

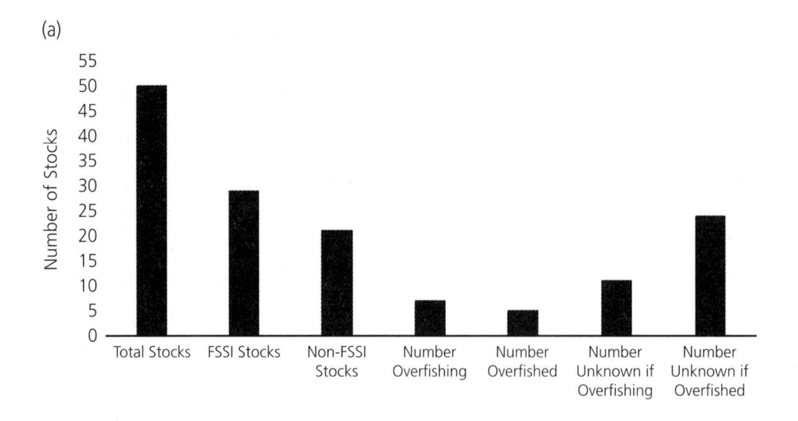

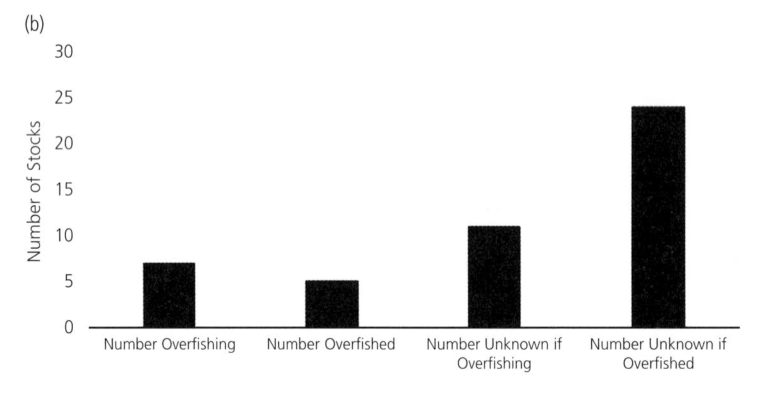

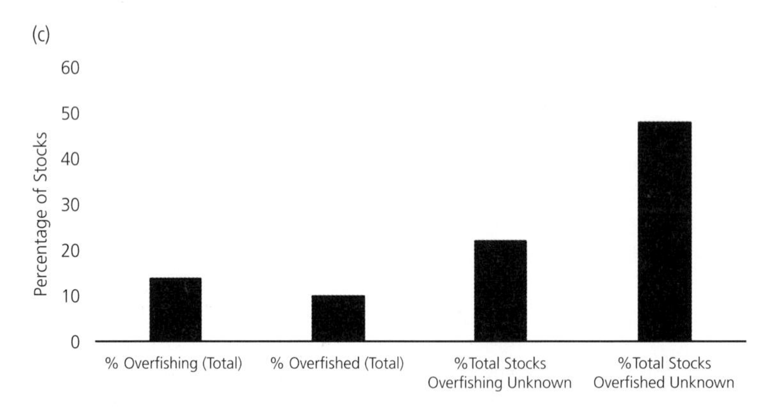

Figure 5.14 For the South Atlantic region as of June 2017 (a) Total number of managed Fish Stock Sustainability Index (FSSI) stocks and non-FSSI stocks, and breakdown of stocks experiencing overfishing, classified as overfished, and of unknown status. (b) Number of stocks experiencing overfishing, classified as overfished, and of unknown status. (c) Percent of stocks experiencing overfishing, classified as overfished, and of unknown status.

Data from NOAA National Marine Fisheries Service. Note: stocks may refer to a species, family, or complex.

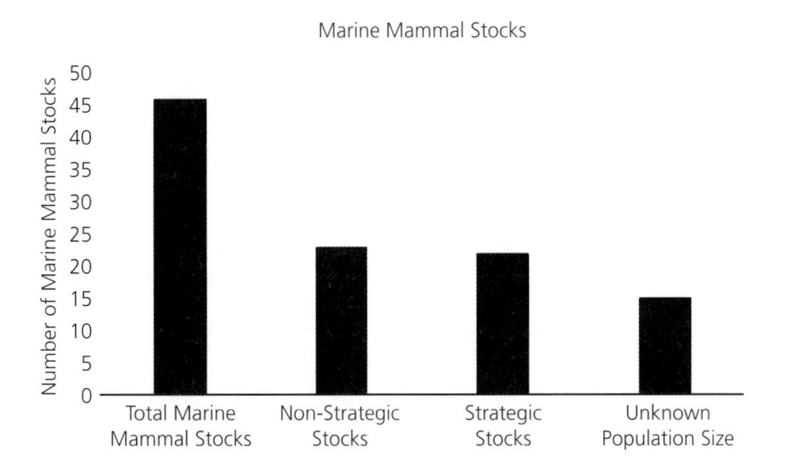

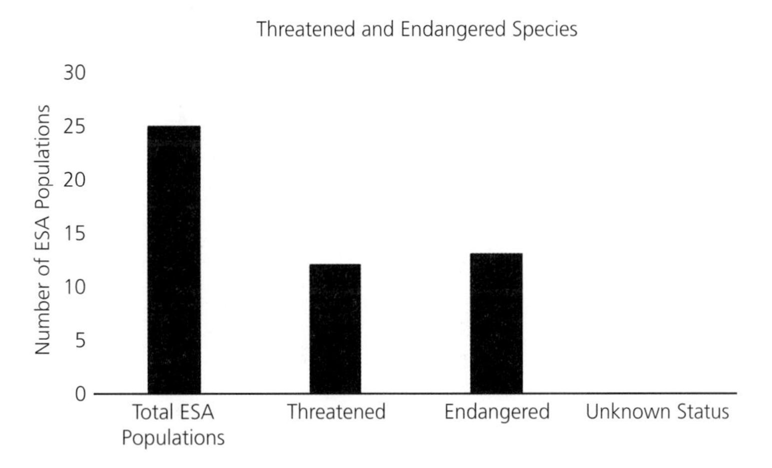

Figure 5.15 Number and status of federally protected species (marine mammal stocks, top panel; distinct population segments of species listed under the Endangered Species Act—ESA, bottom panel) occurring in the South Atlantic region.

time (1854–2016), and having increased by 0.65°C since the mid-twentieth century. These increases have particularly occurred over the past decade since 2013, with otherwise consistent temperatures observed. Similar trends have been observed within key subregions, with average temperatures over the 162-year time period higher in the subtropical Florida Keys at 27.2°C ± 0.02 (SE) and lower in the Cape Hatteras-Pamlico Sound area (23.9°C ± 0.03, SE), which comprises the most northern inshore extent of the South Atlantic. Temperatures in the Florida Keys have increased by 0.83°C since 1950, particularly during the past decade, while increases have been less pronounced in Cape Hatteras-Pamlico Sound (0.34°C since 1950). Coincident with these temperature observations are Atlantic basin-scale climate oscillations (Fig. 5.17), including the AMO and NAO, which exhibit decadal cycles that can influence the environmental conditions and ecologies of several U.S. regions, including the South Atlantic. Trends in NAO and AMO fluctuations appear to be closely associated with those observed for SST over time (Marshall et al. 2001, McCarthy et al. 2015, Valle-Levinson et al. 2017).

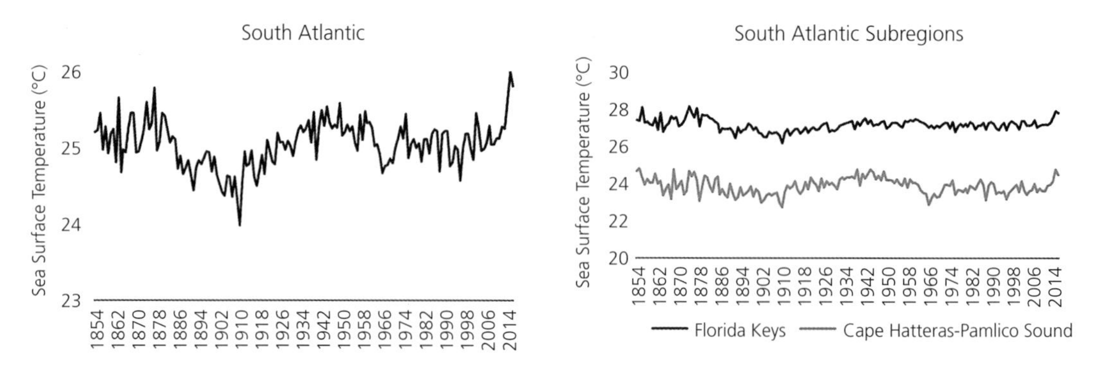

Figure 5.16 Average sea surface temperature (°C) over time (years 1854–2016) for the South Atlantic region and identified subregions.
Data derived from the NOAA Extended Reconstructed Sea Surface Temperature dataset (https://www.ncdc.noaa.gov/).

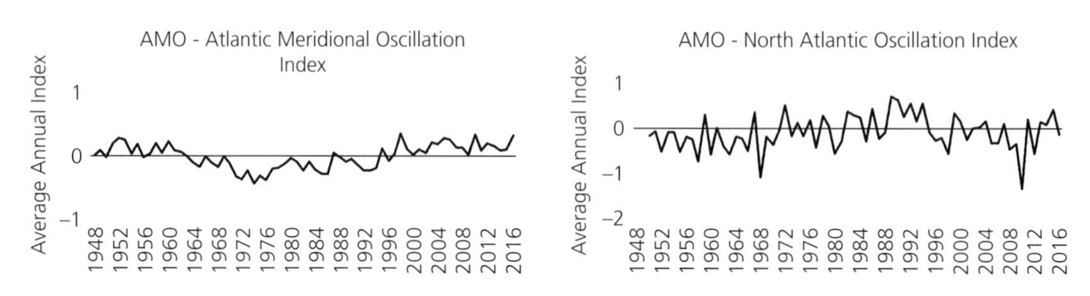

Figure 5.17 Climate forcing indices for the South Atlantic region over time (years 1948–2016).
Data derived from NOAA Earth System Research Laboratory data framework (https://www.esrl.noaa.gov/).

5.4.3.2 Notable Physical Features and Destabilizing Events and Phenomena

With two prominently identified subregions at its northern and southern limits, the South Atlantic contains four major bays (Table 5.3), one each off the coasts of Florida, Georgia, South Carolina, and North Carolina, and 18 major island groupings concentrated throughout inshore areas of all four states. Proportionally, the South Atlantic region makes up 3.1% of the U.S. EEZ (~455,000 km²), 10% of the national shoreline (~16,500 km), while containing a shallower average depth (~1600 m) among regions and maximum depths of ~5.3 km (Fig. 5.18), which comprise the deepest waters off the U.S. Atlantic Coast. Together, the Florida Keys and Cape Hatteras-Pamlico Sound comprise ~3.6% of the South Atlantic EEZ, with the Florida Keys having much greater area (~10,000 km²) and shoreline extent (~2700 km), and deeper average (19.7 m) and maximum depths (236 m). Cape Hatteras-Pamlico Sound is approximately half the size of the Florida Keys (~5500 km² area; 1335 km shoreline), with a much shallower average (3.4 m) and maximum depths (7 m). The width of the South Atlantic shelf varies latitudinally where it is broader off Georgia and South Carolina, while narrower off South Florida and the Cape Hatteras-Pamlico Sound subregion.

In the South Atlantic region, 229 hurricanes have been reported since records were kept in the mid-1800s, with an average of 13.5 hurricanes per decade occurring since 1850 and peak numbers (~25 per decade) recorded in the 1950s and 2000s (Fig. 5.19). Of U.S. Atlantic coast regions, the South Atlantic has been the second-most prone to hurricanes. While at lower intensities, these peak trends have also been observed for Cape Hatteras-Pamlico Sound (up to 11 hurricanes/decade during the 2000s; 96 total since 1850), while somewhat less frequent in the Florida Keys (up to 9 hurricanes during the 1960s; 57 total since 1850).

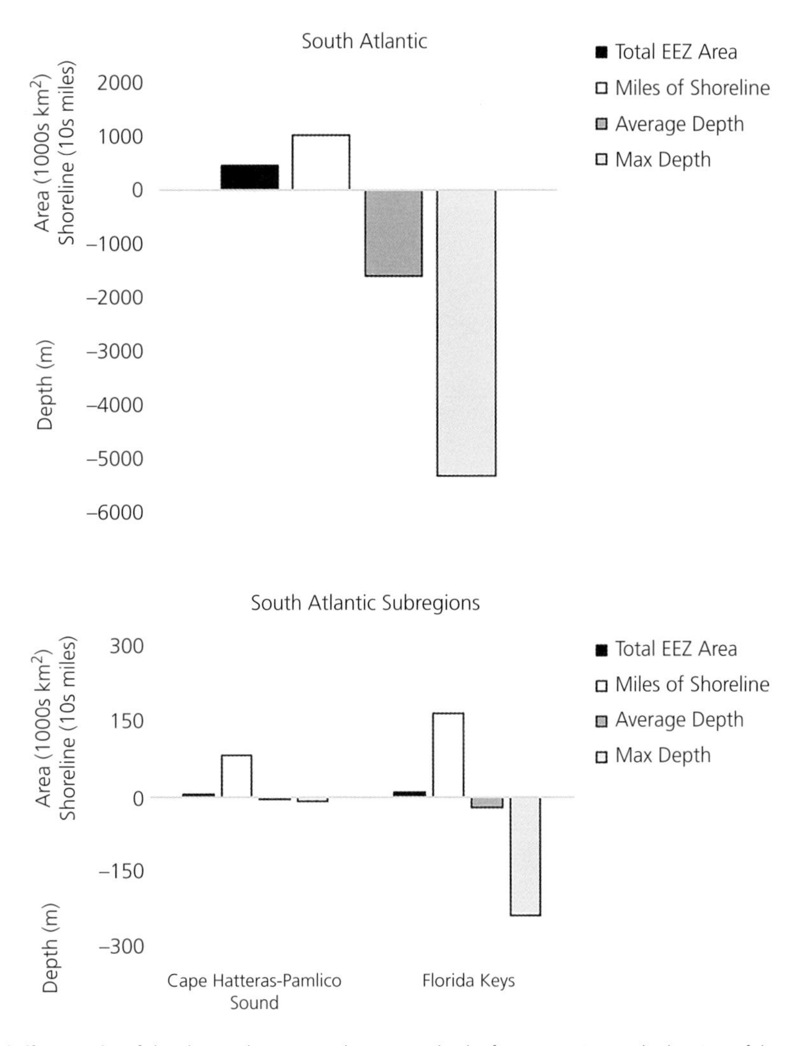

Figure 5.18 Total shelf area, miles of shoreline, and average and maximum depth of marine regions and subregions of the South Atlantic.

5.4.4 Major Pressures and Exogenous Factors

5.4.4.1 Other Ocean Use Context

Ocean uses in the South Atlantic are most associated with the tourism/recreation, marine transportation, and ship and boatbuilding sectors (Fig. 5.20). In total and among sectors, trends for South Atlantic ocean establishments and employments have remained relatively consistent over the past decades, with tourism establishments and employments comprising 80–86% of total South Atlantic ocean establishments and 79–85% of total South Atlantic ocean employments. The GDP from tourism/recreation and marine transportation is highest among sectors, with tourism contributing 56–69% toward total South Atlantic ocean GDP, and contributions from marine transportation ranging from 12 to 24%. During the past decades, South Atlantic ocean GDP has remained steady, with the most recent estimated values at ~$17.6 billion. Overall, the South Atlantic contributes toward 7% of the U.S. total ocean GDP, which is in the lower tier nationally. Establishments and employments in the South Atlantic marine transportation sector contribute 6–7% toward total ocean establishments and comprise 10–12% of total ocean employments. Nationally, the South Atlantic contributes

Figure 5.19 Total number of hurricanes recorded per decade for the South Atlantic region (top panel) and identified subregions (bottom panel) since year 1850.

Data derived from NOAA Office for Coastal Management (https://coast.noaa.gov/hurricanes/).

toward 18–19% each of U.S. marine construction and ship and boatbuilding establishments. This region contributes 11–12% of national marine construction employments and GDP, and 14–15% of national tourism/recreation employments and GDP. It is third-highest nationally for ship and boat building establishments (18.5% of U.S. ship and boat building establishments) and for tourism/recreation GDP (15.2% of tourism/recreation GDP).

The degree of tourism in the South Atlantic, as measured by number of dive shops, number of major ports and marinas, and number of cruise ship departure passengers is among some of the highest nationally (Fig. 5.21), with the highest numbers of dive shops nationally, third-highest numbers of marinas, and 18 major ports (also third-highest nationally). There are 35 dive shops throughout the Florida Keys, with overlapping proximity to two major Florida Ports (Port Everglades, Port of Miami), and three major marinas (Haulover Beach Park Marina, Loggerhead Club and Marina, Miami Beach Marina; all in Miami). Tourism and recreation is much less concentrated in the Cape Hatteras-Pamlico Sound area, with only 3 dive shops, and proximity to two major ports (Morehead City, Wilmington), and no major marinas.

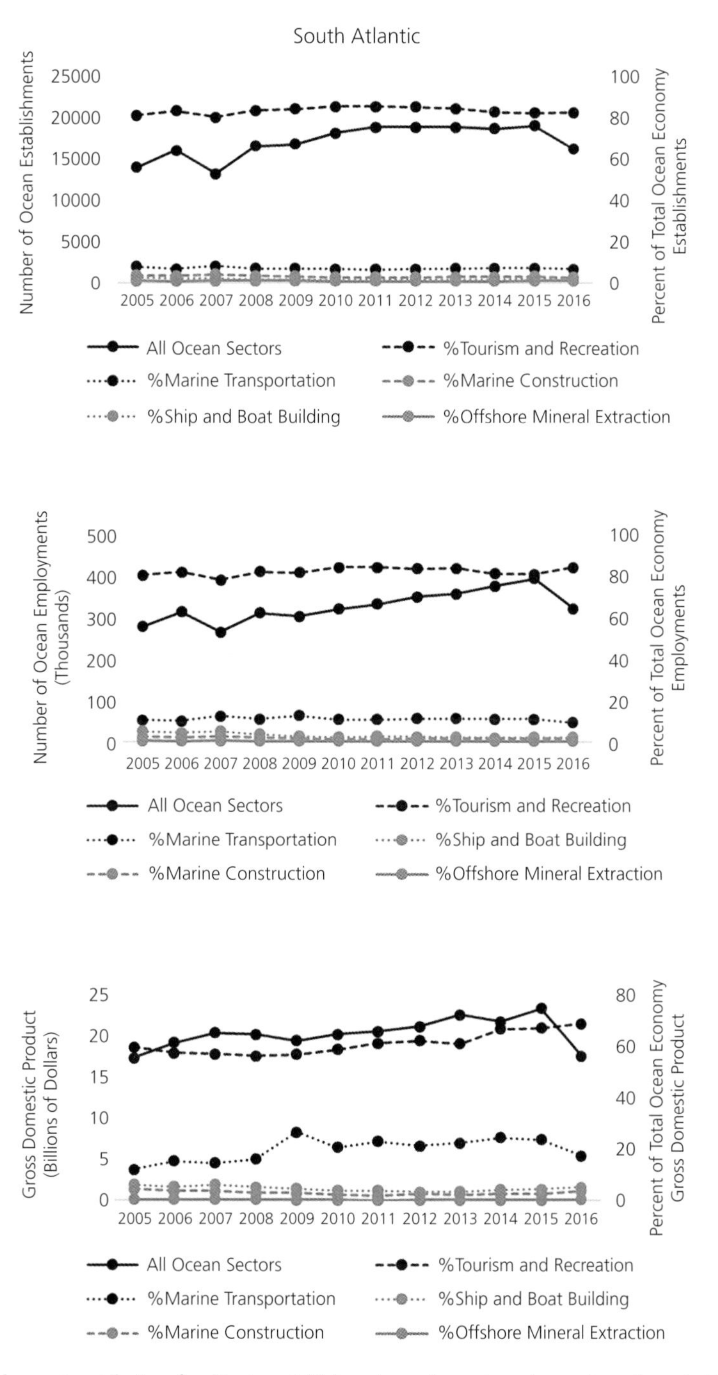

Figure 5.20 Number and percent contribution of multisector establishments, employments, and gross domestic product (GDP) and their percent contribution to total multisector oceanic economy in the South Atlantic region (years 2005–2016).

Data derived from the National Ocean Economics Program.

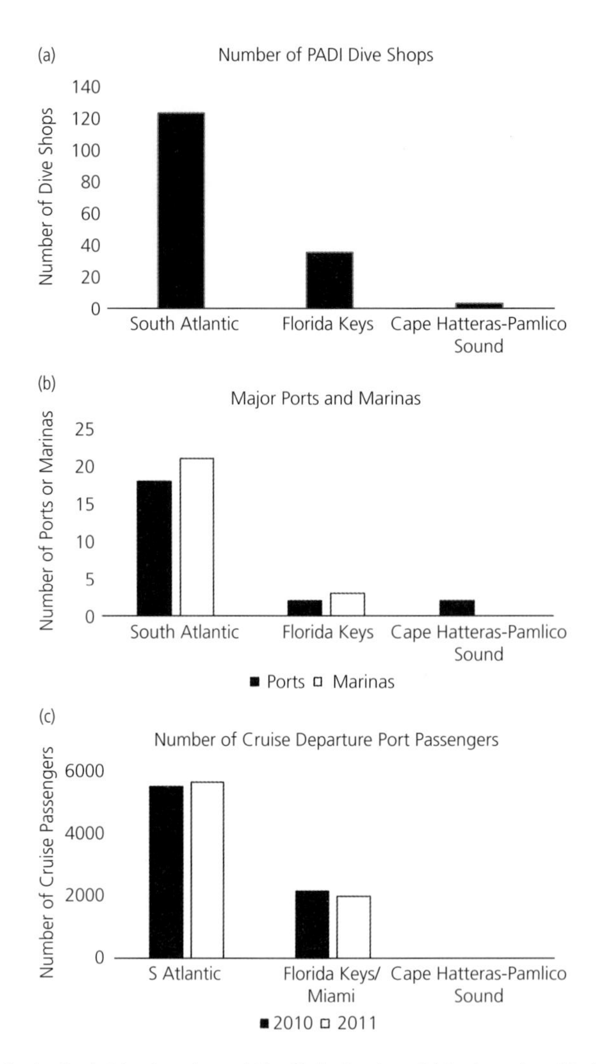

Figure 5.21 Tourism indicators for the South Atlantic region and identified subregions: (a) Total number of Professional Association of Diving Instructors (PADI) dive shops. Data derived from http://apps.padi.com/scuba-diving/dive-shop-locator/. (b) Total number of major ports and marinas. (c) Total number of cruise departure port passengers (years 2010–2011; values in thousands).

Cruise departure passengers are most prominent in the South Atlantic region (highest nationally), with 5.4–5.6 million departure passengers observed throughout the area, with most concentrated in the Miami, Fort Lauderdale, and Port Canaveral regions of Florida. While no active offshore oil rigs are found throughout the U.S. Atlantic continental shelf, the South Atlantic has emerged as a leading area for offshore wind energy production with six established U.S. Bureau of Ocean Energy Management (BOEM) wind energy sites (tied with New England for second-highest nationally).

5.4.5 Systems Ecology and Fisheries

5.4.5.1 Basal Ecosystem Production

Since 2002, mean surface chlorophyll values throughout the South Atlantic EEZ have been very low at ~0.5 ± 0.01 (SE) mg m^{-3} (Fig. 5.22), as associated with its clearer, subtropical waters. Low, relatively constant chlorophyll values are observed throughout the Florida Keys (1.7 ± 0.04, SE), while values are much higher in the Cape Hatteras-Pamlico Sound area (10.2 ± 0.5, SE), given its higher riverine input and nutrient loading. Over the past two decades, chlorophyll values have

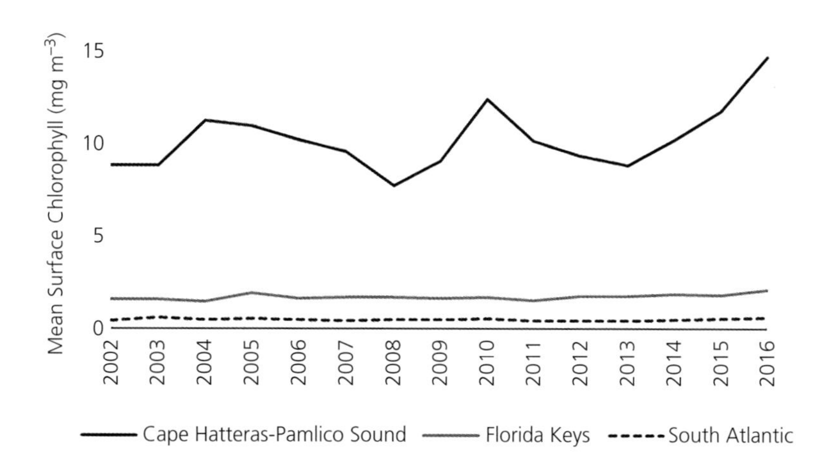

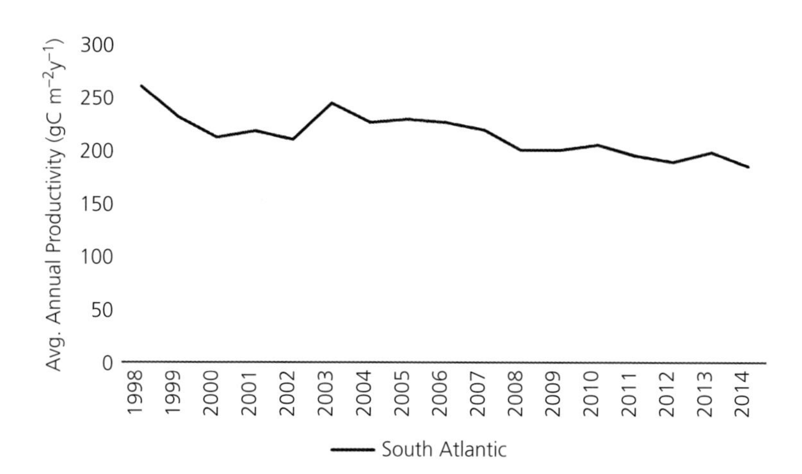

Figure 5.22 Mean surface chlorophyll (mg m^{-3}; top panel) for the South Atlantic region and identified subregions and average annual productivity (grams Carbon m^{-2} $year^{-1}$; bottom panel) for the South Atlantic region over time.

Data derived from NASA Ocean Color Web (https://oceancolor.gsfc.nasa.gov/) and productivity calculated using the Vertically Generalized Production Model—VGPM.

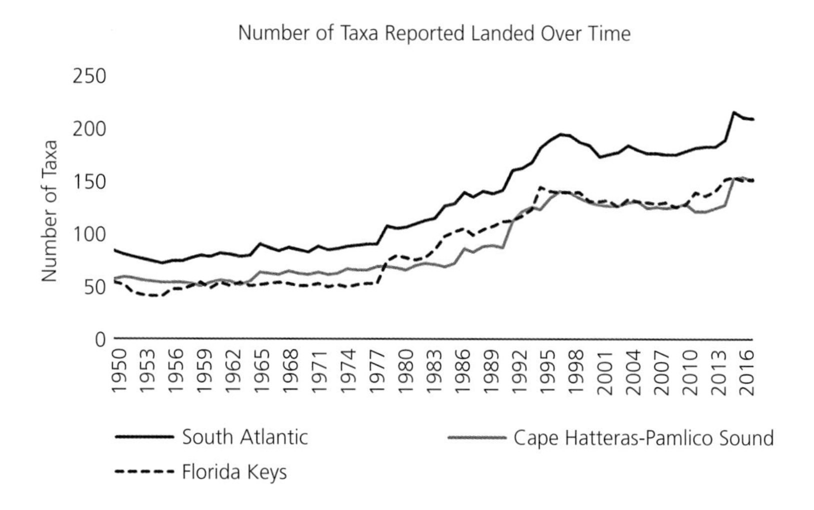

Figure 5.23 Number of taxa reported captured by commercial and recreational fisheries in the South Atlantic region and identified subregions.

Data derived from NOAA National Marine Fisheries Service commercial and recreational fisheries statistics.

oscillated in the Cape Hatteras-Pamlico Sound area, with increases observed over the past four years (up to 14.7 mg m^{-3} $year^{-1}$). Average annual productivity within the South Atlantic EEZ has remained relatively steady between 200–250 g C m^{-2} $year^{-1}$ (fifth-highest nationally) over the same time period.

5.4.5.2 System Exploitation

Nationally, the second-highest numbers of landed taxa have been captured in the South Atlantic region over time, with values continuing to increase since the late 1970s and peaking at 218 reported taxa in 2014 (Fig. 5.23). Among subregions, similar values have been observed over time for both the Florida Keys and Cape Hatteras-Pamlico Sound (average: 91–92 taxa $year^{-1}$), with increases paralleling those for the South Atlantic. Values currently remain nearly equivalent in both regions with 153 taxa reported in 2016 for the Florida Keys and 152 reported for Cape Hatteras-Pamlico Sound. Surveyed nearshore fish and invertebrate biomass for the South Atlantic region has remained relatively constant over time (Fig. 5.24), averaging 148.8 ± 0.7 (SE) metric tons $year^{-1}$, with increases observed from 1997 onward and peaking in 2011 at 249.1 metric tons. In addition to successful efforts to limit overfishing and to rebuild depleted stocks, incrementally increasing biomass has also been associated with higher abundance of range-shifting, tropically associated species becoming established in South Atlantic inshore habitats (Morley et al. 2018). Values for total landings/productivity in the South Atlantic (up to 0.0001; Fig. 5.25) decreased during the mid-2000s and have remained relatively constant in more recent years. Overall, landings/productivity ratios are sixth-highest nationally.

5.4.5.3 Targeted and Non-Targeted Resources

Total commercial and recreational landings are sixth-highest nationally in the South Atlantic region (Fig. 5.26), which experienced declines from previous peak values in the 1980s to remain around 50 thousand metric tons since the mid-2000s. Subsequently major increases in the proportion of recreational catch have occurred, with peak contributions (33.5%) observed during 2007 and currently remaining at ~23%. While Shertzer et al. (2019) found overwhelming contributions from recreational fisheries (up to 85% of total landings) for coastal pelagic and reef fishes, the findings shown here account for total finfish and invertebrate landings

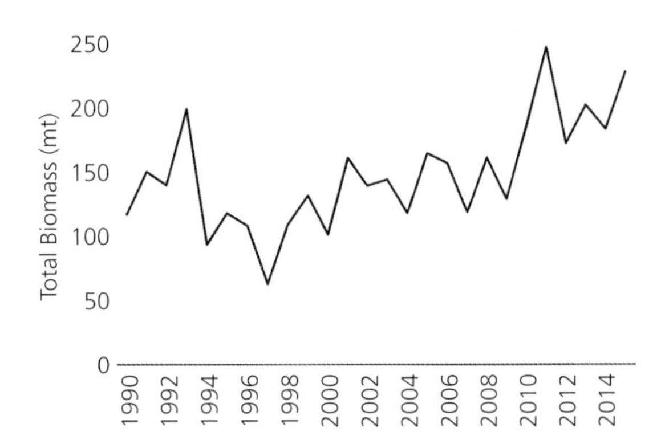

Figure 5.24 Total surveyed nearshore (3–10 nautical miles offshore) fish and invertebrate biomass (metric tons; mt) for the South Atlantic region, over time.

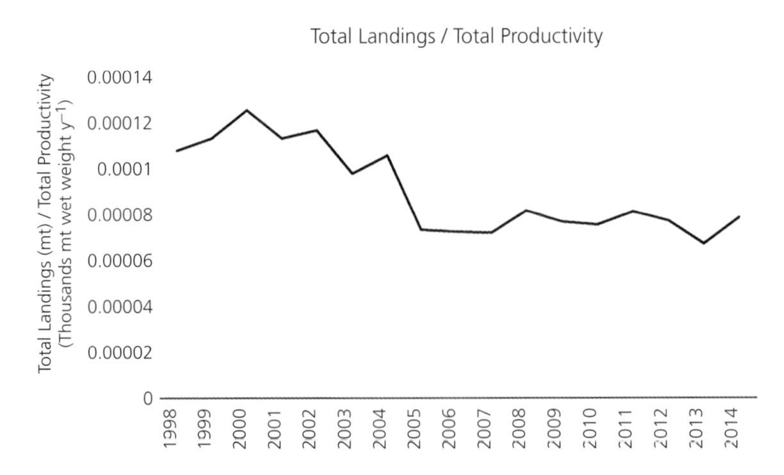

Figure 5.25 Ratios of total reported commercial and recreational landings (metric tons) to total productivity (metric tons wet weight yr^{-1}) for the South Atlantic region.

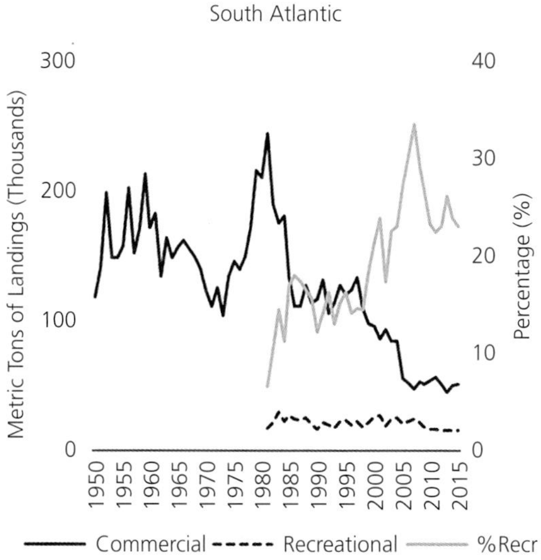

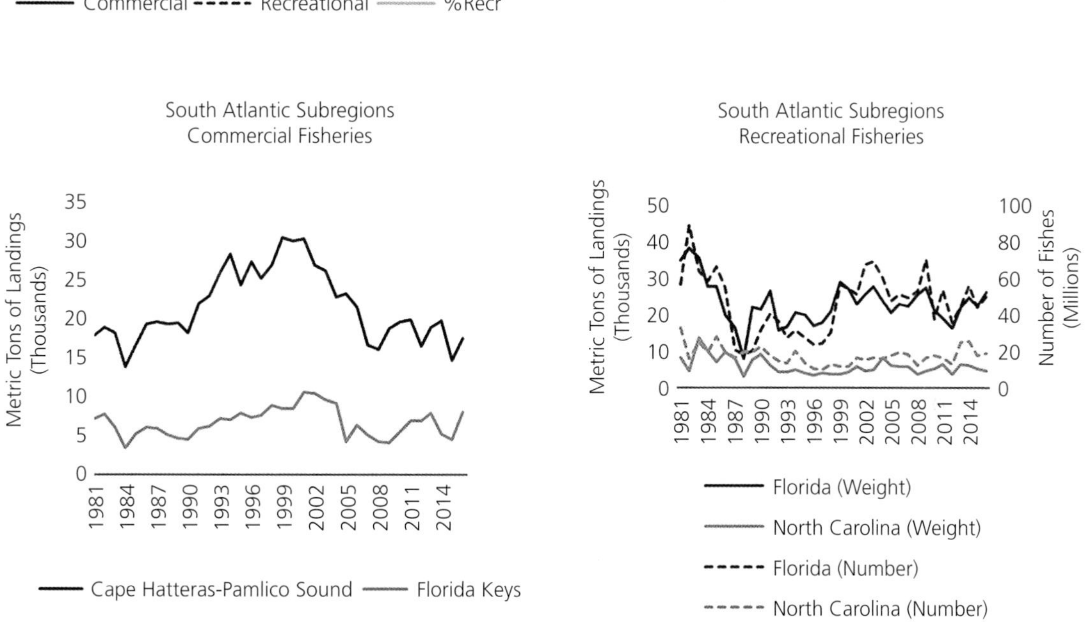

Figure 5.26 Total reported commercial and recreational landings and percent contribution of recreational landings (%Recr) over time for the South Atlantic region (1950–2016) and identified subregions (1981–2016).

Data derived from NOAA National Marine Fisheries Service commercial and recreational fisheries statistics.

that are dominated by commercially important species including Blue crab, whelks, penaeid shrimps, and Atlantic menhaden (NMFS 2018b). However, increases in recreational fishing effort from ~45 million to ~75 million annual trips have also been observed since 1981 (NMFS 2018b). These are also compounded by increases in recreational discards and releases from ~70 million to ~200 million individuals during the same time period, which are now accounted for in many regional stock assessments (NMFS 2018b, Atkinson et al. 2019, Courtney 2019).

Increases in Cape Hatteras-Pamlico Sound landings were observed during the 1990s (up to 30.6 thousand metric tons in 1999), which then declined in the 2000s. As of 2007, landings have remained around 16–17 thousand metric tons. Substantially lower values (~6700 ± 311 metric tons year^{-1}) have been landed in the Florida Keys over time. Similarly, while Florida

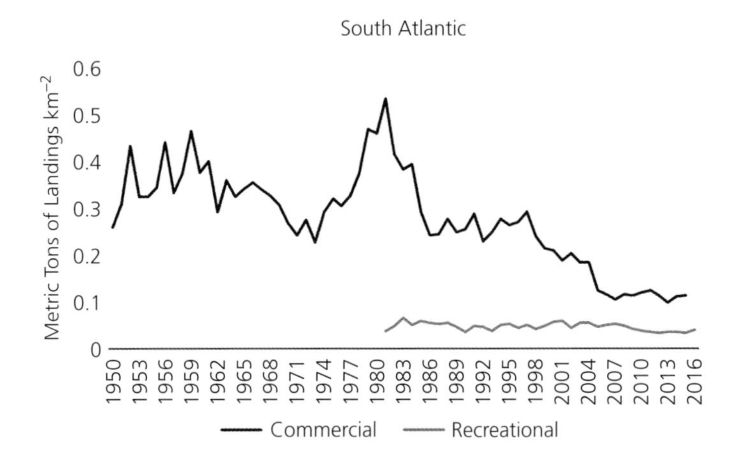

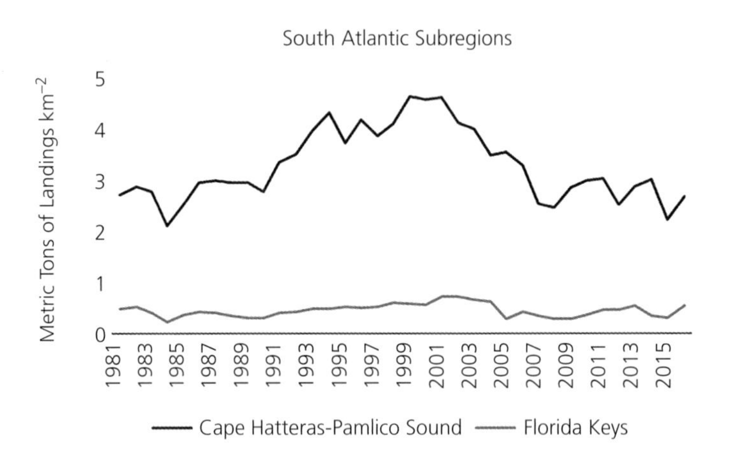

Figure 5.27 Total reported commercial and recreational landings (metric tons) per square kilometer over time (1950–2016) for the South Atlantic region and identified subregions (commercial landings; 1981–2016).

Data derived from NOAA National Marine Fisheries Service commercial and recreational fisheries statistics.

Keys commercial landings increased during the 1990s, lower values have been observed in more recent years following a peak of 10.8 thousand metric tons in 2001. Recreational landings by weight and number have been persistently higher in the Florida Keys and Florida waters, and follow similar patterns. Decreases in Florida recreational landings were observed during the 1980s to nearly equivalent values of those found in Cape Hatteras-Pamlico Sound and North Carolina (~8 thousand metric tons; 20 million fishes). Only a marginal rebound was observed in later years, with Florida recreational landings continuing to persist around 20–30 thousand metric tons (50 million fishes). Recreational landings for Cape Hatteras-Pamlico Sound and North Carolina decreased during the 1980s from peak values of ~13.5 thousand metric tons (~32.5 million fishes) in the early 1980s, and have remained around 5 thousand metric tons (~16 million fishes) from 1990 onward.

When standardized per square kilometer (Fig. 5.27), South Atlantic total landings follow very similar patterns in terms of declines and overall trend, with this region also containing the fifth-highest concentration of fisheries landings over time nationally. Concentrated fishing pressure (up to ~4 metric tons km^{-2}) has been historically observed in and around Cape Hatteras-Pamlico Sound, while lower values have been observed in the Florida Keys (up to 0.7 metric tons km^{-2}).

Although values have decreased over time for most taxa, bycatch continues to persist throughout the South Atlantic region (Fig. 5.28). By number, bycatch is most pronounced for invertebrates, bony fishes, sea turtles,

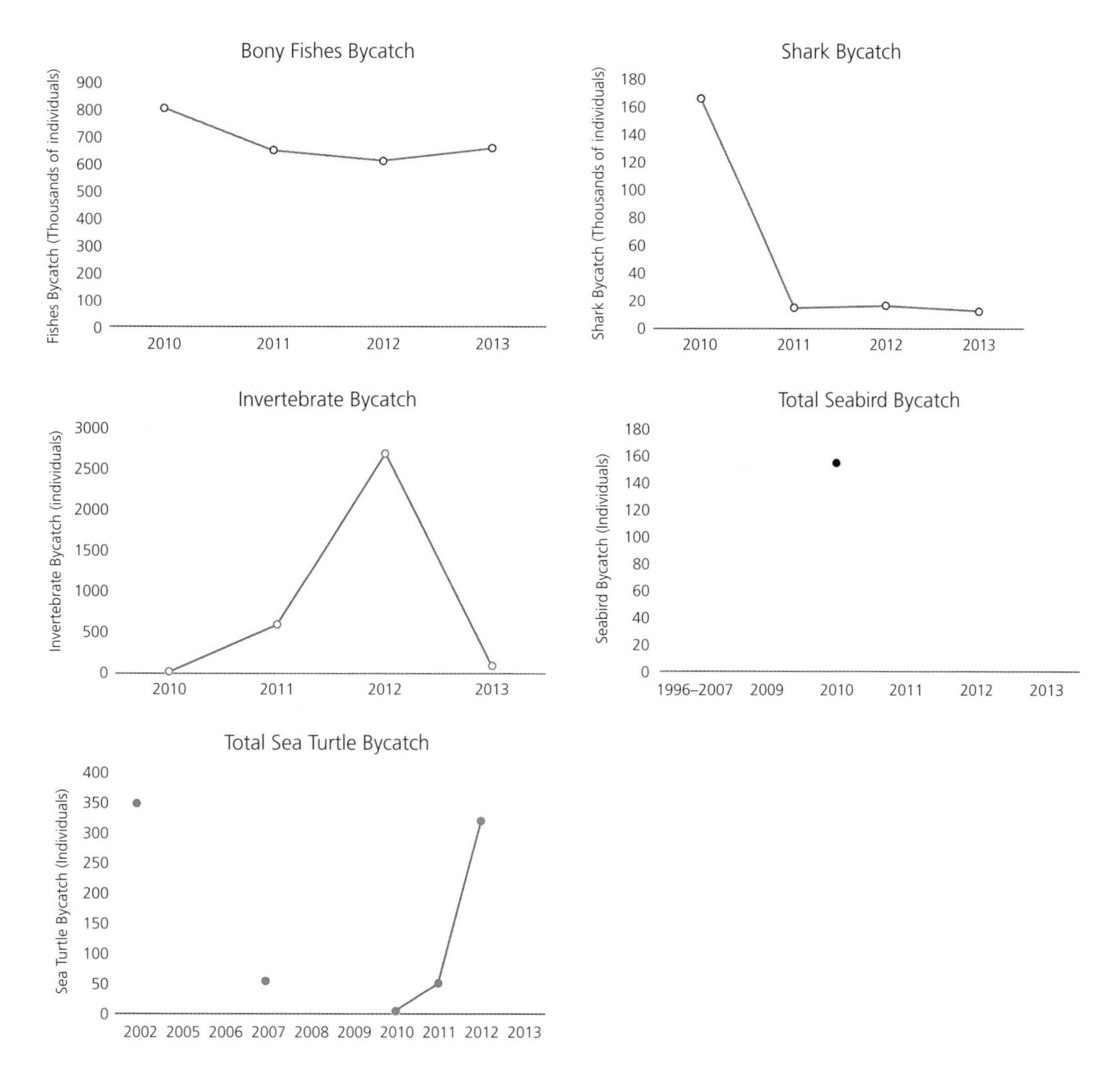

Figure 5.28 Total reported bycatch of species groups in metric tons (mt) and number of individuals for the South Atlantic region.
Data derived from NOAA National Marine Fisheries Service U.S. National Bycatch Report Database System.

and sharks, while unreported for marine mammals and minimally reported for seabirds. Overall, reported bycatch for the South Atlantic region is second-highest nationally by number, with most prominent decreases observed over time for sharks and bony fishes. Increases in sea turtle bycatch have been observed since 2010 at up to 322 individuals, at a level similar to values observed in 2002.

5.5 Synthesis

The South Atlantic is a species rich environment subject to several major stressors that include habitat loss, sea-level rise, ocean acidification, and intermittent high category hurricanes with increasing frequency over the past decades. The consequences of overfishing continue to affect LMRs in this region, particularly as a result of higher recreational fishing effort for reef fishes and coastal pelagic species (Shertzer et al. 2019). Overharvesting and increased reliance on imported seafood (NMFS 2018a, T. Kellison pers. comm.), together with the consequences of these other stressors and economic factors (NMFS 2018a, K. Craig pers. comm.), has led to declining fisheries revenue in the Florida Keys and Cape Hatteras-Pamlico Sound over time, in addition to major declines in commercial fisheries landings throughout the entire South Atlantic region. These effects are also compounded by the increased prevalence of invasive species, limited data resulting in unknown status of multiple species, and significant reduction of essential salt marshes and seagrasses, which all potentially affect LMR production (Parker & Dixon 1998, Shertzer et al. 2009, Sorte et al. 2010, Osland et al. 2013, Whitfield et al. 2014). Increasing residential and tourism-related human population densities throughout this region (sixth-highest nationally; Link & Marshak 2019), and especially concentrated in the Cape Hatteras-Pamlico Sound subregion, impart continued fishing and recreational pressures on this system while also contributing major revenue toward local and regional marine economies. Commercial fishing effort remains concentrated in the northern locations of this region, while recreational effort is more prevalent throughout its southern reaches, especially off Florida, with tourism communities contributing substantially toward seafood consumption and recreational harvest. Given the prevalence of these highly interrelated marine sectors, and the effects of ongoing human and natural stressors in this system, EBFM strategies for the South Atlantic and its prominent subregions must consider these factors on its coupled socioecological systems. In particular, attention should be directed toward their consequences to LMR biomass, biodiversity, and marine socioeconomics in a changing environment.

The South Atlantic ecosystem is well known for aspects of its human environment, particularly its tourism-based industry, where it contains the highest numbers of dive shops and cruise departure passengers nationally (Link & Marshak 2019), in addition to numerous ports and marinas. It also contains the fifth-highest basal productivity among all U.S. regions, which fosters ongoing tourism and contributes toward its LMR-based cultural history. Despite certain data limitations, management in this region has been comparatively progressive, especially through the development and update of a comprehensive FEP (SAFMC 2009, 2018), and ongoing consideration of ecosystem-based factors in FMPs and council-based initiatives. These include efforts toward incorporating forage fish concerns, climate variability and fisheries, habitat conservation, and trophic interactions into management practices. This is especially seen via implementation of approaches to address these ideas in the updated FEP (SAFMC 2018). Progress regarding these considerations is reflected in the multispecies focus of three FMPs (i.e., coastal migratory pelagics, coral, snapper-grouper), a separate FMP and ecosystem-based amendments focused on *Sargassum* habitat, and overall essential habitat identification, conservation, and enhancement. Emerging effects of aquaculture and other ocean uses on habitat, and the consequences of non-native and invasive species are also priorities in the continued implementation of ecosystem-based approaches to management (SAFMC 2009, 2016, 2018, NMFS 2019). These efforts are complemented by the priorities of a recently completed South Atlantic EBFM Implementation Plan (NMFS 2019), which include:

- Maintaining or improving ecosystem structure and function; economic, social, and cultural benefits from resources; and biological, economic, and cultural diversity.
- Identifying and consolidating available climate change and EBM guidance and incorporating this information into management products and analyses.
- Adding ecosystem considerations to the stock assessment process and developing ecosystem considerations chapters to be added to SAFMC FMP amendments.
- Developing an ESR for the South Atlantic.
- Conducting South Atlantic climate vulnerability analyses for targeted and protected species, and community vulnerability assessments.
- Increasing capacity and support for ecosystem modeling efforts and continued development of multispecies production modeling.
- Developing methods and best practices to evaluate, conduct, and track ecosystem goods and services.
- Developing best practices and quantitative measures for evaluating trade-offs between overall ecosystem health, and community resilience and well-being.

Together, these considerations are allowing for enhanced characterization of South Atlantic ecosystem dynamics, more robust assessments, and comprehensive application of information streams with the goal of creating system-wide reference points and multispecies approaches to management, ultimately to improve the status of the ecosystem and its LMRs as the major outcome. These efforts provide utility in formulating a strategic approach for implementing ecosystem-level planning, conducting ecosystem-level risk assessments, and applying ecosystem properties directly into management frameworks. Expanding upon ecosystem-level information for the South Atlantic, addressing data gaps, and applying them toward broader system-wide assessments will continue to advance holistic management practices for this region. Efforts to more

Who Wants a Jelly Donut? Explosion of Gelatinous Zooplankton

The monitoring of these understudied species has diminished, but they cause known and suspected impacts to entire ecosystems and just about all major trust species
—(see also every other chapter)

Imagine if there were a surefire way to tell whether an ecosystem was experiencing significant perturbation, but no one wanted to bother to monitor that item because it was difficult to measure or not viewed as economically important or even a moderate priority for sampling? What would you think of such short-sighted planning? A bit alarming, huh?

Yet that is what we tend to do with jellyfish.

Globally, whenever major blooms of jellyfish occur it is indicative of eutrophication, overfishing, warmed waters, or related types of perturbation (Lynam et al. 2006, Attrill et al. 2007, Pauly et al. 2009, Lynam et al. 2011, Brotz et al. 2012, Purcell 2012, Henschke 2019). There are some projections that these organisms may be all we will be left with for targeted fishing in the (near) future (Pauly et al. 2003, 2009, Pauly 2009, Richardson et al. 2009, Roux et al. 2013). Certainly there are natural occurrences of jellyfish, but the trend has been for jellyfish blooms to be of larger and larger magnitudes and more frequently occurring (Mills 2001, Link & Ford 2006, Gibbons & Richardson 2008, Richardson et al. 2009, Brotz et al. 2012, Condon et al. 2013, Brodeur et al. 2016).

The impacts to economically and ecologically important taxa from jellyfish are well documented (Purcell & Sturdevant 2001, Purcell et al. 2001, Graham et al. 2003, Pauly et al. 2009, Nastav et al. 2013, Quiñones et al. 2013, Palmieri et al. 2014, Robinson et al. 2014, Brodeur et al. 2016, Schnedler-Meyer et al. 2016). Many ecosystem modeling and synthesis efforts show that jellyfish are known to drive the dynamics of energy flows for entire marine ecosystems (Link et al. 2008a, b, Pauly et al. 2009, Jiang et al. 2010, Link et al. 2010, Ruzicka et al. 2012, Decker et al. 2014, Robinson et al. 2015).

There are even emerging markets for jellyfish (Kingsford et al. 2000, Omori & Nakano 2001, Kitamura & Omori 2010, López-Martinez & Alvarez-Tello 2013, Brotz & Pauly 2017, Brotz et al. 2017), with some fisheries becoming established (Nishikawa et al. 2008, Kitamura & Omori 2010, López-Martinez & Alvarez-Tello 2013, Brotz et al. 2017), including for cannonball jellyfish (*Stomolophus meleagris*) off the coast of Georgia (West-Page 2015). We suppose that if you salt and marinade (or dehydrate and salt and marinate) rubbery goo enough, even that can taste good. But the degree to which the global market for jellies will grow remains to be seen.

The sampling of gelatinous zooplankton certainly can pose some challenges, but they are not insurmountable. There are a range of emerging technologies and clever use of old sampling methods (e.g., fish stomachs) to track jellyfish (Purcell & Sturdevant 2001, Link & Ford 2006, Brodeur et al. 2008, Purcell 2009, D'Ambra et al. 2018, Hays et al. 2018). The cost-to-benefit ratios are likely quite low (Purcell et al. 2007, Gibbons & Richardson 2013, Palmieri et al. 2015, Brodeur et al. 2016).

Yet these organisms are not routinely monitored even though they are important features of marine ecosystems. Perhaps it is because they are hard to see and sample. But as a class of organism that may be quite indicative of the future state of marine ecosystems, it would wise to start considering them, even qualitatively (i.e., presence-absence) on extant surveys. There is actually a piece of legislation in the U.S. requiring monitoring of these taxa (Jellyfish Control Act of 1966; 16 U.S.C. §§1201–1205), but it has not been funded or enforced for over four decades (Purcell et al. 2001, Condon et al. 2013, Brodeur et al. 2016). Maybe revisiting that Act would reinvigorate the discussion about how to consider gelatinous zooplankton.

As ecosystem-based fisheries management considers the full range of factors influencing and influenced by fisheries, a reconsideration of jellyfish is in order. Omitting them

continued

Who Wants a Jelly Donut? *Continued*

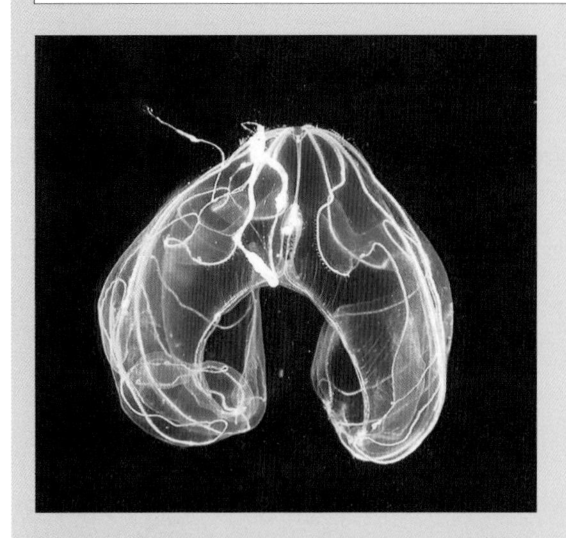

misses major interactions, trade-offs, and impacts to many living marine resources in most marine ecosystems. Ignoring jellyfish may cause us to miss major signals to which we would do well to pay attention.

References

Attrill MJ, Wright J, Edwards M. 2007. Climate-related increases in jellyfish frequency suggest a more gelatinous future for the North Sea. *Limnology and Oceanography* 52(1):480–5.

Brodeur RD, Link JS, Smith BE, Ford MD, Kobayashi DR, Jones TT. 2016. Ecological and economic consequences of ignoring jellyfish: a plea for increased monitoring of ecosystems. *Fisheries* 41(11):630–7.

Brodeur RD, Suchman CL, Reese DC, Miller TW, Daly EA. 2008. Spatial overlap and trophic interactions between pelagic fish and large jellyfish in the northern California Current. *Marine Biology* 154(4):649–59.

Brotz L, Cheung WW, Kleisner K, Pakhomov E, Pauly D. 2012. "Increasing jellyfish populations: trends in large marine ecosystems." In: *Jellyfish Blooms IV.* Edited by J Purcell, H Mianzan, JR Frost. Dordrecht, Netherlands: Springer (pp. 3–20).

Brotz L, Pauly D. 2017. "Studying jellyfish fisheries: toward accurate national catch reports and appropriate methods for stock assessments." In: *Jellyfish: Ecology, Distribution Patterns and Human Interactions.* Edited by GL Mariottini. New York, NY: Nova Publishers (pp. 313–29).

Condon RH, Duarte CM, Pitt KA, Robinson KL, Lucas CH, Sutherland KR, Mianzan HW, et al. 2013. Recurrent jellyfish blooms are a consequence of global oscillations. *Proceedings of the National Academy of Sciences* 110(3):1000–5.

D'Ambra ID, Graham WM, Carmichael RH, Hernandez Jr FJ. 2018. Dietary overlap between jellyfish and forage fish in the northern Gulf of Mexico. *Marine Ecology Progress Series* 587:31–40.

Decker MB, Cieciel K, Zavolokin A, Lauth R, Brodeur RD, Coyle KO. 2014. "Population fluctuations of jellyfish in the Bering Sea and their ecological role in this productive shelf ecosystem." In: *Jellyfish Blooms.* Edited by KA Pitt, CH Lucas. Dordrecht, Netherlands: Springer (pp. 153–83).

Gibbons MJ, Richardson AJ. 2008. "Patterns of jellyfish abundance in the North Atlantic." In: *Jellyfish Blooms: Causes, Consequences, and Recent Advances.* Edited by KA Pitt, JE Purcell. Dordrecht, Netherlands: Springer (pp. 51–65).

Gibbons MJ, Richardson AJ. 2013. Beyond the jellyfish joyride and global oscillations: advancing jellyfish research. *Journal of Plankton Research* 35(5):929–38.

Graham WM, Martin DL, Felder DL, Asper VL, Perry HM. 2003. "Ecological and economic implications of a tropical jellyfish invader in the Gulf of Mexico." In: *Marine Bioinvasions: Patterns, Processes and Perspectives.* Edited by J Pederson. Dordrecht, Netherlands: Springer (pp. 53–69).

Hays GC, Doyle TK, Houghton JD. 2018. A paradigm shift in the trophic importance of jellyfish? *Trends in Ecology & Evolution* 33(11): 874–84.

Henschke N. 2019. "Jellyfishes in a changing ocean." In: *Predicting Future Oceans.* Edited by AM Cisneros-Montemayor, WWL Cheung, Y Ota. Amsterdam, Netherlands: Elsevier (pp. 137–148).

Jiang H, Cheng HQ, Xu HG, Arreguín-Sánchez F, Quesne WL. 2010. Impact of large jellyfish bloom on energy balance of middle and upper ecosystem in East China Sea. *Marine Environmental Science* 29(1):91–5.

Kingsford MJ, Pitt KA, Gillanders BM. 2000. Management of jellyfish fisheries, with special reference to the order Rhizostomeae. *Oceanography and Marine Biology: An Annual Review* 38:85–156.

Kitamura M, Omori M. 2010. Synopsis of edible jellyfishes collected from Southeast Asia, with notes on jellyfish fisheries. *Plankton and Benthos Research* 5(3):106–18.

Link J, O'Reilly J, Fogarty M, Dow D, Vitaliano J, Legault C, Overholtz W, et al. 2008a. Energy flow on Georges Bank revisited: the energy modeling and analysis exercise (EMAX) in historical context. *Journal of Northwest Atlantic Fishery Science* 39:83–101.

Link JS, Ford MD. 2006. Widespread and persistent increase of Ctenophora in the continental shelf ecosystem off NE USA. *Marine Ecology Progress Series* 320:153–9.

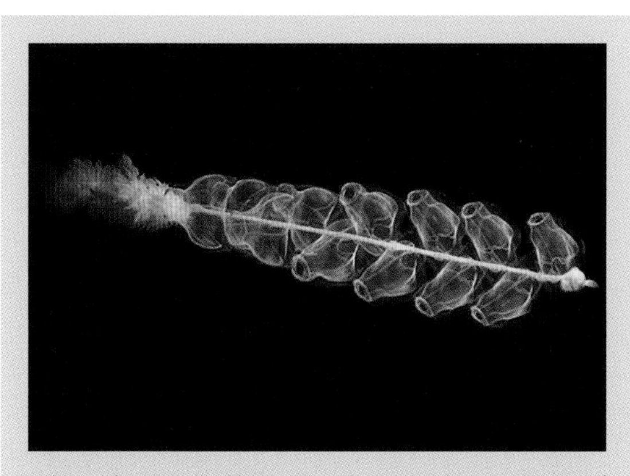

Link JS, Fulton EA, Gamble RJ. 2010. The northeast US application of Atlantis: a full system model exploring marine ecosystem dynamics in a living marine resource management context. *Progress in Oceanography* 87(1–4):214–34.

Link J, Overholtz W, O'Reilly J, Green J, Dow D, Palka D, Legault C, et al. 2008b. The Northeast US continental shelf energy modeling and analysis exercise (EMAX): ecological network model development and basic ecosystem metrics. *Journal of Marine Systems* 74(1–2):453–74.

López-Martinez J, Álvarez-Tello J. 2013. The jellyfish fishery in Mexico. *Agricultural Sciences* 4(6A):57–61.

Lynam CP, Gibbons MJ, Axelsen BE, Sparks CA, Coetzee J, Heywood BG, Brierley AS. 2006. Jellyfish overtake fish in a heavily fished ecosystem. *Current Biology* 16(13):R492–3.

Lynam CP, Lilley MK, Bastian T, Doyle TK, Beggs SE, Hays GC. 2011. Have jellyfish in the Irish Sea benefited from climate change and overfishing? *Global Change Biology* 17(2):767–82.

Mills CE. 2001. Jellyfish blooms: are populations increasing globally in response to changing ocean conditions? *Hydrobiologia* 451(1–3):55–68.

Nastav B, Malej M, Malej Jr A, Malej A. 2013. Is it possible to determine the economic impact of jellyfish outbreaks on fisheries? A case study–Slovenia. *Mediterranean Marine Science* 14(1):214–23.

Nishikawa J, Thu NT, Ha TM. 2008. Jellyfish fisheries in northern Vietnam. *Plankton and Benthos Research* 3(4):227–34.

Omori M, Nakano E. 2001. Jellyfish fisheries in southeast Asia. *Hydrobiologia* 451(1–3):19–26.

Palmieri MG, Barausse A, Luisetti T, Turner K. 2014. Jellyfish blooms in the Northern Adriatic Sea: fishermen's perceptions and economic impacts on fisheries. *Fisheries Research* 155:51–8.

Palmieri MG, Schaafsma M, Luisetti T, Barausse A, Harwood A, Sen A, Turner RK. 2015. "Jellyfish blooms and their impacts on welfare benefits: recreation in the UK and fisheries in Italy." In: *Coastal Zones Ecosystem Services*. Edited by RK Turner, M Schaafsma. Cham, Switzerland: Springer (pp. 219–40).

Pauly D. 2009. Aquacalypse now the end of fish. *New Republic* 240(18):24–7.

Pauly D, Alder J, Bennett E, Christensen V, Tyedmers P, Watson R. 2003. The future for fisheries. *Science* 302(5649): 1359–61.

Pauly D, Graham W, Libralato S, Morissette L, Palomares MD. 2009. "Jellyfish in ecosystems, online databases, and ecosystem models." In: *Jellyfish Blooms: Causes, Consequences, and Recent Advances*. Edited by KA Pitt, JE Purcell. Dordrecht, Netherlands: Springer (pp. 67–85).

Purcell JE. 2009. Extension of methods for jellyfish and ctenophore trophic ecology to large-scale research. *Hydrobiologia* 616(1):23–50.

Purcell JE. 2012. Jellyfish and ctenophore blooms coincide with human proliferations and environmental perturbations. *Annual Review of Marine Science* 4:209–35.

Purcell JE, Graham WM, Dumont HJ. 2001a. *Jellyfish Blooms: Ecological and Societal Importance*. Dordrecht, Netherlands: Kluwer Academic.

Purcell JE, Sturdevant MV. 2001b. Prey selection and dietary overlap among zooplanktivorous jellyfish and juvenile fishes in Prince William Sound, Alaska. *Marine Ecology Progress Series* 210:67–83.

Purcell JE, Uye SI, Lo WT. 2007. Anthropogenic causes of jellyfish blooms and their direct consequences for humans: a review. *Marine Ecology Progress Series* 350:153–74.

Quiñones J, Monroy A, Acha EM, Mianzan H. 2013. Jellyfish bycatch diminishes profit in an anchovy fishery off Peru. *Fisheries Research* 139:47–50.

Richardson AJ, Bakun A, Hays GC, Gibbons MJ. 2009. The jellyfish joyride: causes, consequences and management responses to a more gelatinous future. *Trends in Ecology & Evolution* 24(6):312–22.

Robinson KL, Ruzicka JJ, Decker MB, Brodeur RD, Hernandez FJ, Quiñones J, Acha EM, et al. 2014. Jellyfish, forage fish, and the world's major fisheries. *Oceanography* 27(4):104–15.

Robinson KL, Ruzicka JJ, Hernandez FJ, Graham WM, Decker MB, Brodeur RD, Sutor M. 2015. Evaluating energy flows through jellyfish and gulf menhaden (Brevoortia patronus) and the effects of fishing on the northern Gulf of Mexico ecosystem. *ICES Journal of Marine Science* 72(8):2301–12.

Roux JP, van der Lingen CD, Gibbons MJ, Moroff NE, Shannon LJ, Smith AD, Cury PM. 2013. Jellyfication of marine ecosystems as a likely

continued

Who Wants a Jelly Donut? *Continued*

consequence of overfishing small pelagic fishes: lessons from the Benguela. *Bulletin of Marine Science* 89(1):249–84.

Ruzicka JJ, Brodeur RD, Emmett RL, Steele JH, Zamon JE, Morgan CA, Thomas AC, Wainwright TC. 2012. Interannual variability in the Northern California Current food web structure: changes in energy flow pathways and the role of forage fish, euphausiids, and jellyfish. *Progress in Oceanography* 102:19–41.

Schnedler-Meyer NA, Mariani P, Kiørboe T. 2016. The global susceptibility of coastal forage fish to competition by large jellyfish. *Proceedings of the Royal Society B: Biological Sciences* 283(1842):20161931.

West-Page J. 2015. Characterization of bycatch in the cannonball jellyfish fishery in the coastal waters off Georgia. *Marine and Coastal Fisheries* 7(1):190–9.

accurately address marine socioeconomic considerations and to address their associated data gaps are explicitly mentioned in South Atlantic FMC EBFM priorities, and in those of its updated FEP. These efforts are also bolstered by prioritized development of more robust modeling frameworks and the application of EwE models for broader understanding of the South Atlantic ecosystem and its responses to management actions. While data limitations continue to hinder progress for some of these priorities, efforts to integrate available data toward assessing the status of this socio-ecological system are ongoing, and additional ecosystem indicators are being identified, calculated, and tracked.

While the South Atlantic is advancing in terms of its LMR management priorities and ecosystem efforts, some challenges remain to effectively implement formalized EBFM planning. Limited information regarding the status and biomass of fishery stocks and protected species in this region, and data gaps for many environmental factors have constrained EBFM implementation and prevented the application of ecosystem-level properties into management actions. While frameworks are being developed, data limitations need to be addressed in order to incorporate broader ecosystem considerations into management, especially in recognition of system-level shifts and thresholds. In light of concentrated overharvesting of this system, ongoing habitat loss, nutrient loading, excessive sedimentation (Pollock 1993, Ault et al. 1998, Eggleston et al. 2003, Ault et al. 2009, Smith et al. 2009, McCarney-Castle et al. 2010, Patrick et al. 2010, Collado-Vides et al. 2011, Lirman et al. 2019), and increasing recreational pressure (Shertzer et al. 2019), the ability to mitigate system-level overexploitation and its consequences to both natural and human ecosystem components through comprehensive management actions remains urgently needed. System-wide consideration of natural stressors and the effects of human activities on top-down and bottom-up forcing that have resulted in changes to basal and secondary production and biomass still need to be explicitly incorporated into management strategies. Continued effort in addressing data limitations and directly considering relationships among fisheries harvest, ocean use, and ecosystem properties is essential to advance EBFM via development of system-wide thresholds and reference points. These are especially warranted in determining South Atlantic exploitation potential as related to inherent primary production, biomass, and overall sustainable resource use. With suggested shifts in some fisheries species distributions and loss of species richness (Morley et al. 2017, 2018), increasing hurricane potential, and expanding exploitation of the southeastern shelf all affecting South Atlantic LMRs, there is an urgent need for more cumulative ecosystem-level management actions.

As human population continues to increase in the South Atlantic along with increases in associated stressors affecting this system, competition for resources among interrelated marine sectors is likely to occur, with continued conflicts among commercial and recreational fisheries, tourism, and marine transportation. As greater understanding of climate vulnerabilities of both natural and human communities remains a high priority in this region, including the socioeconomic effects of regional warming, shifting fisheries, and sea-level rise, these concerns also need to account for effects on other ocean sectors, including tourism and marine transportation. More comprehensive and systematic management strategies that consider spatially overlapping ocean uses, multiple species, and the socioeconomic effects of changing environment and resource partitioning are especially warranted in this region. Given the overwhelming importance of tourism in the South Atlantic and lower proportional contributions of LMRs toward its marine economy, emerging trade-offs among fishing and other marine sectors are essential components in regional management initiatives. With increasing tourism and expanding ocean uses, increases in non-fishing effects on this ecosystem are anticipated, which warrant continued focus toward habitat conservation and continued spatial management considerations of the South Atlantic EEZ. Ongoing investments in LMR and environmental surveys, more comprehensive fishery-independent and commercial and recreational fishery-dependent data collection, improved data archiving and accessibility, and communication among scientific, managerial, and stakeholder entities are all needed to resolve these issues. Although fisheries management is not perceived as being as contentious in this region as in other areas of the southeastern U.S. (e.g., Gulf of Mexico), conflicts among commercial and recreational interests are likely to continue, while strict rebuilding plans for newly assessed species and species complexes (particularly reef fishes, which have slow maturity rates) may also emerge with improved assessment potential. Efforts that work to address these concerns in the context of emerging regional ecosystem-based frameworks remain needed.

Although leading nationally in aspects of its tourism (i.e., dive shops, cruise departure passengers), the South Atlantic is ranked low in terms of the status of its marine socioeconomics. This ranking is especially due to low overall fisheries value and low LMR-based employments and revenue as related to basal production. While containing the third-highest number of managed species nationally, this region also contains a high proportion of strategic marine mammal stocks and the second-lowest national status of its LMRs (Link & Marshak 2019). This ranking is due mainly to low and declining fisheries landings, high bycatch, poorly assessed biomass and fisheries status, high numbers of prohibited species, and ongoing effects of overexploitation. Nationally, the South Atlantic is second-highest in the percentage of stocks that are overfished (10%) and undergoing overfishing (12%), and third-highest for unassessed stocks (Link & Marshak 2019), with remaining uncertainty regarding the status of many key economic species in this multispecies fishery. Governance in this region, however, is ranked as one of the highest nationally (Link & Marshak 2019), with balanced approaches to species-level and ecosystem approaches in management plans, an FMC with more equivalent representation among commercial and recreational fishing representatives, fewer lawsuits and NEPA-EIS actions, and higher Congressional representation as related to shoreline and fishery value. Thus, the governance system seems very well positioned to explore potential enhancements to the LMR economic considerations.

5.5.1 Progress Toward Ecosystem-Based Fisheries Management (EBFM)

As related to the determinants of successful LMR management (Fig. 5.29; Link & Marshak 2019), this ecosystem is above average in terms of its higher-ranked human environment while also being low- to mid-ranked for total primary production and socioeconomic status.

Holistic EBFM practices are being developed and implemented across all U.S. regions in consideration of a more systematic and prioritized consideration of all fisheries, pressures, risks, and outcomes (NMFS 2016a, b, Levin et al. 2018, Marshall et al. 2018, Link & Marshak 2019). Fisheries management for the South Atlantic continues evolving into approaches that address cumulative issues affecting the LMRs and fisheries of this ecosystem. Over the past few years, efforts to implement EBFM in this region have arisen through refinement of local management strategies (NMFS 2019). In accordance with the NOAA Fisheries EBFM Policy and Roadmap (NMFS 2016a, b), an EBFM implementation plan (NMFS 2019) for the South Atlantic has been developed through efforts overseen by the NOAA Southeast Fisheries Science Center (SEFSC), NOAA Fisheries Southeast Regional Office (SERO), and the SAFMC. This plan specifies

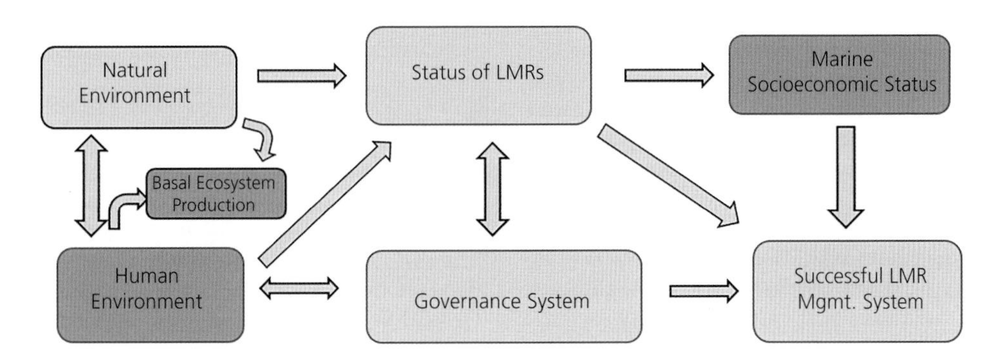

Figure 5.29 Schematic of the determinants and interconnectivity of successful Living Marine Resource (LMR) systems management criteria as modified from Link and Marshak (2019). Those highlighted in blue indicate noteworthy criteria for the South Atlantic ecosystem.

approaches to be taken toward conducting EBFM and in addressing NOAA Fisheries Roadmap priorities and gaps. Progress for the South Atlantic and its major subregions in terms of EBFM Roadmap Guiding Principles and Goals is shown in Table 12.9.

As observed for other regions, progress has been made in terms of implementing ecosystem-level planning, advancing knowledge of ecosystem principles, and in assessing risks and vulnerabilities to ecosystems through ongoing investigations into climate vulnerability and species prioritizations for stock and habitat assessments. Although the South Atlantic is progressing toward EBFM, little overall progress has been observed in applying ecosystem-level emergent properties into management frameworks. Progress toward recommendations of the Ecosystem Principles Advisory Panel (EPAP) for the development of FEPs (EPAP 1999, Wilkinson & Abrams 2015) is shown in Table 12.10. A comprehensive FEP for the South Atlantic has been in place (SAFMC 2018). Progress has been observed in characterizing and monitoring aspects of ecosystem dynamics in support of an ecosystem approach to management, particularly as related to the North Carolina Coastal Habitat Protection Plan (NCDEQ 2016a, b) and its applications for enhancing the South Atlantic FEP (SAFMC 2018).

5.5.2 Conclusions

Fully implementing an ecosystem-based approach in the South Atlantic will result in many advancements in the management of LMRs, especially in accounting for the cumulative effects of ongoing and increasing human and natural stressors and mitigating their impacts. These are especially of concern for continued sustainable LMR harvest and well-being of biological communities in vegetated and reef habitats. The major tourism industry associated with this region is highly dependent on ensuring a sustainable seafood supply and ecological integrity of coastal and offshore habitats to support diving, beach culture, sport fisheries, and other marine recreational activities. Ecosystem-based approaches allow for these inter-relationships to be prioritized and mutually considered. Efforts by management entities to monitor, consider, and incorporate the natural, biological, and socioeconomic factors that affect the degree to which marine resources may be sustainably extracted and harvested, especially in a changing, underassessed, and potentially overexploited environment warrant continued expansions. Continuing with a business-as-usual approach for exploiting and managing the resources of this region without considering their effects on the ecosystem as a whole limits effectiveness toward sustainable practices. Managing fisheries and their species in a vacuum, independent of interacting sectors, historic and ongoing stressors, or governance feedbacks ignores the significance of these factors toward LMR production and marine socioeconomic growth. System-wide considerations of the inherent relationships between productivity, biomass, landings, and socioeconomics are more explicitly needed to refine and maintain this system at sustainable thresholds, while also accounting for the influences from natural and human activities and other ocean uses. This framework provides opportunities that focusing on one stock or species complex or one fishery would not. EBFM allows for simultaneous controlled harvest of interacting species complexes and fisheries as found in reef systems that include important shellfish species (i.e., Spiny lobster, Stone crab) that serve as prey for commercially important reef fishes, and account for trophic interactions among other important species, including forage fish and apex predators (i.e., sharks, groupers, coastal pelagic species). Overlapping and/or differential fishing effort for these species, and the effects of different harvest methodologies among fisheries, are also considered in these frameworks, while also accounting for ecological consequences to multispecies complexes. With very high species richness in this system, this portfolio approach allows for multiple foci on functional groups and trophic levels and enables more effective management strategies (Edwards et al. 2004, Kellner et al. 2011, Link 2018) that are especially warranted in the South Atlantic. Ultimately, continuing to implement comprehensive management approaches in the South Atlantic will facilitate sustainable continuation of its LMR-dependent tourism-based economy and cultural marine heritage. As increasing pressures and ocean uses continue to impact fisheries, coastal communities, and the natural components of this system, ecosystem-based approaches are essential for systematically managing and mitigating their effects.

5.6 References

Abbott JK, Lloyd-Smith P, Willard D, Adamowicz W. 2018. Status-quo management of marine recreational fisheries undermines angler welfare. *Proceedings of the National Academy of Sciences* 115(36):8948–53.

Acosta A, Dunmire T, Venier J. 1998. A Preliminary Trophic Model of the Fish Communities of Florida Bay. Proceedings of 1998 Florida Bay Science Conference. Miami, Fl: University of Miami (p. 58).

Alber M, Swenson EM, Adamowicz SC, Mendelssohn IA. 2008. Salt marsh dieback: an overview of recent events in the US. *Estuarine, Coastal and Shelf Science* 80(1):1–11.

Albins MA. 2013. Effects of invasive Pacific red lionfish Pterois volitans versus a native predator on Bahamian coral-reef fish communities. *Biological Invasions* 15(1): 29–43.

Albins MA, Hixon MA. 2008. Invasive Indo-Pacific lionfish Pterois volitans reduce recruitment of Atlantic coral-reef fishes. *Marine Ecology Progress Series* 367:233–8.

Andrews K, Nall L, Jeffrey C, Pittman S, et al. 2005. *The State of Coral Reef Ecosystems of Florida. The State of Coral Reef Ecosystems of the United States and Pacific Freely Associated States.* Fort Lauderdale, FL: Nova Southeastern University NSUWorks (pp. 150–201).

Angelini C, van der Heide T, Griffin JN, Morton JP, Derksen-Hooijberg M, Lamers LP, Smolders AJ, Silliman BR. 2015. Foundation species' overlap enhances biodiversity and multifunctionality from the patch to landscape scale in southeastern United States salt marshes. *Proceedings of the Royal Society B: Biological Sciences* 282(1811):20150421.

Anthony A, Atwood J, August PV, Byron C, Cobb S, Foster C, Fry C, et al. 2009. Coastal lagoons and climate change: ecological and social ramifications in the US Atlantic and Gulf coast ecosystems. *Ecology and Society* 14(1):8.

ASMFC (Atlantic States Marine Fisheries Commission). 2020a. *ASMFC Atlantic Menhaden Board Adopts Ecological Reference Points.* Arlington,VA: ASMFC Offices (p. 2).

ASMFC (Atlantic States Marine Fisheries Commission) 2020b. *ASMFC Stock Status Overview.* Arlington,VA: ASMFC Offices (p. 43).

Atkinson LP, Lee TN, Blanton JO, Chandler WS. 1983. Climatology of the southeastern United States continental shelf waters. *Journal of Geophysical Research: Oceans* 88(C8):4705–18.

Atkinson LP, Menzel DW, Bush KA.1985. *Oceanography of the Southeastern U.S. Continental Shelf.* Washington DC: American Geophysical Union (p. 156).

Atkinson SF, McCarthy KJ, Shideler AC. 2019. *Length Distribution and Release Discard Mortality for Southeastern Yellowtail Snapper. SEDAR64-DW-15.* North Charleston, SC: SEDAR (p. 6).

Ault JS, Bohnsack JA, Meester GA. 1998. A retrospective (1979–1996) multispecies assessment of coral reef fish stocks in the Florida Keys. *Fishery Bulletin* 96(3):395–414.

Ault JS, Bohnsack JA, Smith SG, Luo J. 2005. Towards sustainable multispecies fisheries in the Florida, USA, coral reef ecosystem. *Bulletin of Marine Science* 76(2):595–622.

Ault JS, Luo J, Wang JD. 2003. "A spatial ecosystem model to assess spotted seatrout population risks from exploitation and environmental changes." In: *Biology of Spotted Seatrout.* Edited by SA Bortone. Boca Raton, FL: CRC Press (pp. 267–96)

Ault JS, Smith SG, Tilmant JT. 2009. Are the coral reef finfish fisheries of south Florida sustainable? *Proceedings International Coral Reef Symposium* 11:989–93.

Bacheler NM, Ballenger JC. 2018. Decadal-scale decline of scamp (Mycteroperca phenax) abundance along the southeast United States Atlantic coast. *Fisheries Research* 204:74–87.

Bacheler NM, Bartolino V, Reichert MJ. 2013a. Influence of soak time and fish accumulation on catches of reef fishes in a multispecies trap survey. *Fishery Bulletin* 111(3):218–32.

Bacheler NM, Schobernd ZH, Gregalis KC, Schobernd CM, Teer BZ, Gillum Z, Glasgow DM, et al. 2019. Patterns in fish biodiversity associated with temperate reefs on the Southeastern US continental shelf. *Marine Biodiversity* 49(5):2411–28.

Bacheler NM, Schobernd CM, Schobernd ZH, Mitchell WA, Berrane DJ, Kellison GT, Reichert MJ. 2013b. Comparison of trap and underwater video gears for indexing reef fish presence and abundance in the southeast United States. *Fisheries Research* 143:81–8.

Bacheler NM, Smart TI. 2016. Multi-decadal decline in reef fish abundance and species richness in the southeast USA assessed by standardized trap catches. *Marine Biology* 163(2):26.

Backstrom J, Woodward RT. 2017. *Using Qualitative Site Characteristics Data in Marine Recreational Fishing Models: A New Site Aggregation Approach.* Annual Meeting, July 30–August 1. Chicago, IL: Agricultural and Applied Economics Association.

Bales JD, Nelson TM. 1988. *Bibliography of Hydrologic and Water-quality Investigations Conducted in Or Near the Albemarle-Pamlico Sounds Region, North Carolina.* Washington, DC: Department of the Interior, US Geological Survey.

Ballenger JC. 2015. *SERFS Chevron Trap Red Snapper Index of Abundance: An Investigation of the Utility of Historical (1990–2009) Chevron Trap Catch Data. SEDAR41-DW51 MARMAP/SEAMAP-SA Reef Fish Survey Technical Report #2015-009.* North Charleston, SC: SEDAR (p. 39).

Ballenger JC, Smart TI. 2015. *Gray Triggerfish Fishery-Independent Index of Abundance in US South Atlantic Waters Based on a Chevron Trap Survey (1990–2014). SEDAR41-DW52 MARMAP/SEAMAP-SA Reef Fish Survey Technical Report # 2015-010.* North Charleston, SC: SEDAR (p. 45).

Ballenger JC, Smart TI. 2016. *Modifications to the Southeast Reef Fish Survey Chevron Trap Age Compositions for Red Snapper and Gray Triggerfish. SEDAR 41-RW07 MARMAP/SEAMAP-SA Reef Fish Survey Technical Report # 2016-005.* North Charleston, SC: SEDAR (p. 13).

Ballew NG, Bacheler NM, Kellison GT, Schueller AM. 2016. Invasive lionfish reduce native fish abundance on a regional scale. *Scientific Reports* 6:32169.

Bangley CW, Paramore L, Dedman S, Rulifson RA. 2018. Delineation and mapping of coastal shark habitat within a shallow lagoonal estuary. *PloS One* 13(4).

Barbier EB, Hacker SD, Kennedy C, Koch EW, Stier AC, Silliman BR. 2011. The value of estuarine and coastal ecosystem services. *Ecological Monographs* 81(2):169–93.

Barlow PM, Reichard EG. 2010. Saltwater intrusion in coastal regions of North America. *Hydrogeology Journal* 18(1):247–60.

Batt RD, Morley JW, Selden RL, Tingley MW, Pinsky ML. 2017. Gradual changes in range size accompany long-term trends in species richness. *Ecology Letters* 20(9):1148–57.

Beck MW, Brumbaugh RD, Airoldi L, Carranza A, Coen LD, Crawford C, Defeo O, et al. 2011. Oyster reefs at risk and recommendations for conservation, restoration, and management. *Bioscience* 61(2):107–16.

Beerkircher LR, Cortes E, Shivji M. 2002. Characteristics of shark bycatch observed on pelagic longlines off the southeastern United States, 1992–2000. *Marine Fisheries Review* 64(4):40–9.

Bell FW. 1997. The economic valuation of saltwater marsh supporting marine recreational fishing in the southeastern United States. *Ecological Economics* 21(3):243–54.

Berkson J, Thorson JT. 2014. The determination of data-poor catch limits in the United States: is there a better way? *ICES Journal of Marine Science* 72(1):237–42.

Bersoza-Hernández A, Brumbaugh RD, Frederick P, Grizzle R, Luckenbach MW, Peterson CH, Angelini C. 2018. Restoring the eastern oyster: how much progress has been made in 53 years? *Frontiers in Ecology and the Environment* 16(8):463–71.

Bettinger P, Merry K, Hepinstall J. 2009. Average tropical cyclone intensity along the Georgia, Alabama, Mississippi, and north Florida coasts. *Southeastern Geographer* 49(1):50–66.

Bettridge SOM, Baker SC, Barlow J, Clapham P, Ford MJ, Gouveia D, et al. 2015. *Status Review of the Humpback Whale (Megaptera Novaeangliae) Under the Endangered Species Act. National Oceanic and Atmospheric Administration, National Marine Fisheries Service Report NOAA-TM-NMFS-SWFSC-540*. Washington, DC: NOAA.

Bin O, Landry CE, Ellis CL, Vogelsong H. 2005. Some consumer surplus estimates for North Carolina beaches. *Marine Resource Economics* 20(2):145–61.

Blake ES, Rappaport EN, Landsea CW. 2007. *The Deadliest, Costliest, and Most Intense United States Tropical Cyclones from 1851 to 2006 (and Other Frequently Requested Hurricane Facts). NOAA Technical Memorandum NWS TPC-5*. Miami, FL: NOAA/National Weather Service, National Centers for Environmental Prediction, National Hurricane Center.

Blanton BO, Aretxabaleta A, Werner FE, Seim HE. 2003. Monthly climatology of the continental shelf waters of the South Atlantic Bight. *Journal of Geophysical Research: Oceans* 108(C8).

BOEM (Bureau of Ocean Energy Management). 2016. *Benthic Habitat Mapping and Assessment in the Wilmington-East Wind Energy Call Area. OCS Study BOEM 2016-003 and NOAA Technical Memorandum 196*. Sterling, VA: Department of the Interior.

Bohnsack BL, Ditton RB, Stoll JR, Chen RJ, Novak R, Smutko LS. 2002. The economic impacts of the recreational bluefin tuna fishery in Hatteras, North Carolina. *North American Journal of Fisheries Management* 22(1):165–76.

Brinson AA, Wallmo K. 2017. Determinants of saltwater anglers' satisfaction with fisheries management: regional perspectives in the United States. *North American Journal of Fisheries Management* 37(1):225–34.

Broome M, Claar D, Hamman E, Matthews T, Salazar M, Shugart-Schmidt K, Tillman A, Vincent M, Berkson J. 2011. *Exploratory Assessment of Four Stocks in the U.S. South Atlantic: Bank Sea Bass (Centropristis Ocyurus), Graytriggerfish (Balistes Capriscus), Sandperch (Diplectrum Formosum), and Tomtate (Haemulon Aurolineatum). NOAA Technical Memorandum NMFS 5 SEFSC 5617*. Washington, DC: NOAA (p. 133).

Brown-Peterson NJ, Overstreet RM, Lotz JM, Franks JS, Burns KM. 2001. Reproductive biology of cobia, Rachycentron canadum, from coastal waters of the southern United States. *Fishery Bulletin* 99(1):15.

Browning RM. 2002. *Success is All That Was Expected: The South Atlantic Blockading Squadron During the Civil War*. Dulles, VA: Potomac Books Incorporated.

Brumbaugh RD, Coen LD. 2009. Contemporary approaches for small-scale oyster reef restoration to address substrate versus recruitment limitation: a review and comments relevant for the Olympia oyster, *Ostrea lurida* Carpenter 1864. *Journal of Shellfish Research* 28(1):147–62.

Bryan DR, Blondeau J, Siana A, Ault JS. 2018. Regional differences in an established population of invasive Indo-Pacific lionfish (*Pterois volitans* and *P. miles*) in south Florida. *PeerJ* 6:e5700.

Buchanan CC. 1973. Effects of an artificial habitat on the marine sport fishery and economy of Murrells Inlet, South Carolina. *Marine Fisheries Review* 35(9):15–22.

Buchheister A, Miller TJ, Houde ED. 2017. Evaluating ecosystem-based reference points for Atlantic menhaden. *Marine and Coastal Fisheries* 9(1):457–78.

Caldwell DK, Golley FB. 1965. Marine mammals from the coast of Georgia to Cape Hatteras. *Journal of the Elisha Mitchell Scientific Society* 1:24–32.

Carmichael J, Duval MJ, Reichert MJM, Bacheler N, Kellison T. 2015. *Workshop to Determine Optimal Approaches for Surveying The Deep-Water Species Complex Off the Southeastern U.S. Atlantic Coast 7–9 April. NOAA Technical Memorandum NMFS-SEFSC-685*. Beaufort, NC: NOAA Beaufort Laboratory (p. 23).

Carlson JK, Lee DW. 2000. *The Directed Shark Drift Gillnet Fishery: Catch and Bycatch 1998–1999. NMFS/SEFC Sustainable Fisheries Division Contribution No. SFD-99/00-87*. Silver Spring, MD: National Marine Fisheries Service.

Carlson JK, Osborne J, Schmidt TW. 2007. Monitoring the recovery of smalltooth sawfish, Pristis pectinata, using standardized relative indices of abundance. *Biological Conservation* 136(2):195–202.

Carlson PR, Yarbro LA, Kaufman KA, Mattson RA. 2010. Vulnerability and resilience of seagrasses to hurricane and runoff impacts along Florida's west coast. *Hydrobiologia* 649(1):39–53.

Carpenter KE. 2002. *The Living Marine Resources of the Western Central Atlantic*. Volumes 1–3. Rome: FAO (p. 2127).

Carretta JV, Forney KA, Oleson EM, Weller DW, Lang AR, Baker J, Muto MM, et al. 2019. *U.S. Pacific Marine Mammal Stock Assessments: 2018. NOAA Technical Memorandum NMFSSWFSC-617*. Washington, DC: NOAA.

Carruthers TR, Punt AE, Walters CJ, MacCall A, McAllister MK, Dick EJ, Cope J. 2014. Evaluating methods for setting catch limits in data-limited fisheries. *Fisheries Research* 153:48–68.

Cecelski DS. 2001. *The Waterman's Song: Slavery and Freedom in Maritime North Carolina*. Chapel Hill, NC: University of North Carolina Press.

Chavez S, Southward-Williard A. 2017. The effects of bycatch reduction devices on diamondback terrapin and blue crab catch in the North Carolina commercial crab fishery. *Fisheries Research* 186:94–101.

Coen LD, Brumbaugh RD, Bushek D, Grizzle R, Luckenbach MW, Posey MH, Powers SP, Tolley SG. 2007. Ecosystem services related to oyster restoration. *Marine Ecology Progress Series* 341:303–7.

Coen LD, Luckenbach MW. 2000. Developing success criteria and goals for evaluating oyster reef restoration: ecological function or resource exploitation? *Ecological Engineering* 15(3–4):323–43.

Colburn LL, Jepson M, Weng C, Seara T, Weiss J, Hare JA. 2016. Indicators of climate change and social vulnerability in fishing dependent communities along the Eastern and Gulf Coasts of the United States. *Marine Policy* 74:323–33.

Coldren GA, Barreto CR, Wykoff DD, Morrissey EM, Langley JA, Feller IC, Chapman SK. 2016. Chronic warming stimulates growth of marsh grasses more than mangroves in a coastal wetland ecotone. *Ecology* 97(11):3167–75.

Coleman FC, Figueira WF, Ueland JS, Crowder LB. 2004. The impact of United States recreational fisheries on marine fish populations. *Science* 305(5692):1958–60.

Coleman FC, Koenig CC, Huntsman GR, Musick JA, Eklund AM, McGovern JC, Sedberry GR, Chapman RW, Grimes CB. 2000. Long-lived reef fishes: the grouper-snapper complex. *Fisheries* 25(3):14–21.

Coleman FC, Williams SL. 2002. Overexploiting marine ecosystem engineers: potential consequences for biodiversity. *Trends in Ecology & Evolution* 17(1):40–4.

Coleman K, Klein MM. 1976. *Colonial Georgia: A History*. New York, NY: Scribner.

Colgan CS. 2013. The ocean economy of the United States: Measurement, distribution, & trends. *Ocean & Coastal Management* 71:334–43.

Collado-Vides L, Mazzei V, Thyberg T, Lirman D. 2011. Spatio-temporal patterns and nutrient status of macroalgae in a strongly managed region of Biscayne Bay, Florida, USA. *Botanica Marina* 54:377–90.

Collier C, Ruzicka R, Banks K, Barbieri L, Beal J, Bingham D, Bohnsack JA, et al. 2008. *The State of Coral Reef Ecosystems of Southeast Florida. The State of Coral Reef Ecosystems of the United States and Pacific Freely Associated States: 2008*. Fort Lauderdale, FL: Nova Southeastern University NSUWorks.

Collins MR, Rogers SG, Smith TI, Moser ML. 2000. Primary factors affecting sturgeon populations in the southeastern United States: fishing mortality and degradation of essential habitats. *Bulletin of Marine Science* 66(3):917–28.

Conley MF, Anderson MG, Steinberg N, Barnett A. 2017. *The South Atlantic Bight Marine Assessment: Species, Habitats, and Ecosystems*. Boston, MA: The Nature Conservancy, Eastern Conservation Science.

Conn PB. 2011. *An Evaluation and Power Analysis of Fishery Independent Reef Fish Sampling in the Gulf of Mexico and U.S. South Atlantic. NOAA Technical Memorandum NMFS-SEFSC-610*. Washington, DC: NOAA (p. 38).

Courtney D. 2019. *Updated Post-Release Live-Discard Mortality Rate and Range of Uncertainty Developed for Blacktip Sharks Captured in Hook and Line Recreational Fisheries for use in the SEDAR 29-Update. SEDAR65- RD04*. North Charleston, SC: SEDAR (p. 7).

Craig BA, Reynolds III JE. 2004. Determination of manatee population trends along the Atlantic coast of Florida using a Bayesian approach with temperature-adjusted aerial survey data. *Marine Mammal Science* 20(3):386–400.

Crawford S, Whelan G, Infante DM, Blackhart K, Daniel WM, Fuller PL, Birdsong T, et al. 2016. *Through a Fish's Eye: The Status of Fish Habitats in the United States 2015*. Washington, DC: National Fish Habitat Partnership.

Dame R, Alber M, Allen D, Mallin M, Montague C, Lewitus A, Chalmers A, et al. 2000. Estuaries of the south Atlantic coast of North America: their geographical signatures. *Estuaries* 23(6):793–819.

Davis GE. 1977. Effects of recreational harvest on a spiny lobster, Panulirus argus, population. *Bulletin of Marine Science* 27(2):223–36.

Day RD, Christopher SJ, Becker PR, Whitaker DW. 2005. Monitoring mercury in the loggerhead sea turtle, Caretta caretta. *Environmental Science & Technology* 39(2):437–46.

Deale C, Norman WC, Jodice LW. 2008. Marketing locally harvested shrimp to South Carolina coastal visitors: the development of a culinary tourism supply chain. *Journal of Culinary Science & Technology* 6(1):5–23.

Delgado GA, Bartels CT, Glazer RA, Brown-Peterson NJ, McCarthy KJ. 2004. Translocation as a strategy to rehabilitate the queen conch (Strombus gigas) population in the Florida Keys. *Fishery Bulletin* 102(2):278–88.

Deutsch CJ, Reid JP, Bonde RK, Easton DE, Kochman HI, O'Shea TJ. 2003. Seasonal movements, migratory behavior, and site fidelity of West Indian manatees along the Atlantic coast of the United States. *Wildlife Monographs* 1:1–77.

Diamond SL, Crowder LB, Cowell LG. 1999. Catch and bycatch: the qualitative effects of fisheries on population vital rates of Atlantic croaker. *Transactions of the American Fisheries Society* 128(6):1085–105.

Ditton RB, Stoll JR. 2000. A socio-economic review of recreational billfish fisheries. *Proceedings of the Gulf and Caribbean Fisheries Institute* 51:666–81.

Downs CA, Fauth JE, Robinson CE, Curry R, Lanzendorf B, Halas JC, Halas J, Woodley CM. 2005. Cellular diagnostics and coral health: declining coral health in the Florida Keys. *Marine Pollution Bulletin* 51(5–7):558–69.

Edgar, W. 1998. *South Carolina: A History.* Columbia, SC: University of South Carolina Press.

Edwards SF, Link JS, Rountree BP. 2004. Portfolio management of wild fish stocks. *Ecological Economics* 49(3):317–29.

Eggleston DB, Johnson EG, Kellison GT, Nadeau DA. 2003. Intense removal and non-saturating functional responses by recreational divers on spiny lobster Panulirus argus. *Marine Ecology Progress Series* 257:197–207.

Engelhart SE, Horton BP, Douglas BC, Peltier WR, Törnqvist TE. 2009. Spatial variability of late Holocene and 20th century sea-level rise along the Atlantic coast of the United States. *Geology* 37(12):1115–8.

EPAP (Ecosystem Principles Advisory Panel). 1999. *Ecosystem-Based Fishery Management Report to Congress.* Silver Spring, MD: National Marine Fisheries Service (p. 62).

Epifanio CE, Garvine RW. 2001. Larval transport on the Atlantic continental shelf of North America: a review. *Estuarine, Coastal and Shelf Science* 52(1):51–77.

Epperly SP. 2003. "Fisheries-related mortality and turtle excluder devices (TEDs)." In: *The Biology of Sea Turtles. Volume II.* Edited by PL Lutz, JA Musick, J Wyneken. Boca Raton, FL: CRC Press (pp. 339–354).

Epperly S, Avens L, Garrison L, Henwood T, Hoggard W, Mitchell J, Nance J, et al. 2002. *Analysis of Sea Turtle Bycatch in the Commercial Shrimp Fisheries of Southeast USA Waters and the Gulf of Mexico. NOAA Technical Memorandum NMFS-SEFSC-490.* Miami, FL: NOAA (p. 88).

Epperly SP, Braun J, Veishlow A. 1995. Sea turtles in North Carolina waters. *Conservation Biology* 9(2):384–94.

Farmer NA, Froeschke JT. 2015. Forecasting for recreational fisheries management: what's the catch? *North American Journal of Fisheries Management* 35(4):720–35.

Fautin D, Dalton P, Incze LS, Leong JA, Pautzke C, Rosenberg A, Sandifer P, et al. 2010. An overview of marine biodiversity in United States waters. *PLoS One* 5(8):e11914.

Ferguson L, Srinivasan M, Oleson E, Hayes S, Brown SK, Angliss R et al. (eds.). 2017. *Proceedings of the First National Protected Species Assessment Workshop.* U.S. Dept. of Commer. NOAA. NOAA Technical Memorandum NMFS-F/SPO-172 (p. 92).

Figueira WF, Coleman FC. 2010. Comparing landings of United States recreational fishery sectors. *Bulletin of Marine Science* 86(3):499–514.

Finkbeiner EM, Wallace BP, Moore JE, Lewison RL, Crowder LB, Read AJ. 2011. Cumulative estimates of sea turtle bycatch and mortality in USA fisheries between 1990 and 2007. *Biological Conservation* 144(11):2719–27.

Fitzpatrick EE, Williams EH, Shertzer KW, Siegfried KI, Craig JK, Cheshire RT, Kellison GT, Fitzpatrick KE, Brennan K. 2017. The NMFS southeast region headboat survey: history, methodology, and data integrity. *Marine Fisheries Review* 79(1):1–28.

Flaherty KE, Switzer TS, Winner BL, Keenan SF. 2014. Regional correspondence in habitat occupancy by Gray Snapper (Lutjanus griseus) in estuaries of the southeastern United States. *Estuaries and Coasts* 37(1):206–28.

Fonseca MS, Kenworthy WJ, Courtney FX, Hall MO. 1994. Seagrass planting in the southeastern United States: methods for accelerating habitat development. *Restoration Ecology* 2(3):198–212.

Fourqurean JW, Willsie A, Rose CD, Rutten LM. 2001. Spatial and temporal pattern in seagrass community composition and productivity in south Florida. *Marine Biology* 138(2):341–54.

Frankovich TA, Zieman JC. 2005. Periphyton light transmission relationships in Florida bay and the Florida keys, USA. *Aquatic Botany* 83:14–30.

Fuller PL, Knott DM, Kingsley-Smith PR, Morris JA, Buckel CA, Hunter ME, Hartman LD. 2014. Invasion of Asian tiger shrimp, Penaeus monodon Fabricius, 1798, in the western North Atlantic and Gulf of Mexico. *Aquatic Invasions* 9(1):59–70.

Fuglister FC. 1951. Annual variations in current speeds in the Gulf Stream system. *Journal of Marine Research* 10(1):119–27.

FWS (U.S. Fish and Wildlife Service). 2014. *West Indian Manatee (Trichechus manatus) Stock Assessment Report* (p. 17).

Gallagher AJ, Hammerschlag N, Danylchuk AJ, Cooke SJ. 2017. Shark recreational fisheries: status, challenges, and research needs. *Ambio* 46(4):385–98.

Gardner LR, Michener WK, Williams TM, Blood ER, Kjerve B, Smock LA, Lipscomb DJ, Gresham C. 1992. Disturbance effects of Hurricane Hugo on a pristine coastal landscape: North Inlet, South Carolina, USA. *Netherlands Journal of Sea Research* 30:249–63.

Garrison CA, Baumstark R, Ward-Geiger LI, Hines E. 2012. Application of a habitat model to define calving habitat of the North Atlantic right whale in the southeastern United States. *Endangered Species Research* 18(1):73–87.

Gentile JH, Harwell MA, Cropper Jr W, Harwell CC, DeAngelis D, Davis S, Ogden JC, Lirman D. 2001. Ecological conceptual models: a framework and case study on ecosystem management for South Florida sustainability. *Science of the Total Environment* 274(1–3):231–53.

Gentner B, Lowther A. 2002. Evaluating marine sport fisheries in the USA. Recreational fisheries: ecological, economic and social evaluation. *Blackwell Scientific Publications* 11:186–206.

Geraldi NR, Kellison GT, Bacheler NM. 2019. Climate indices, water temperature, and fishing predict broad scale variation in fishes on temperate reefs. *Frontiers in Marine Science* 6:30.

Gittman RK, Peterson CH, Currin CA, Joel Fodrie F, Piehler MF, Bruno JF. 2016. Living shorelines can enhance the nursery role of threatened estuarine habitats. *Ecological Applications* 26(1):249–63.

Gornitz VM, Daniels RC, White TW, Birdwell KR. 1994. The development of a coastal risk assessment database: vulnerability to sea-level rise in the US Southeast. *Journal of Coastal Research* 1:327–38.

Grabowski JH, Hughes AR, Kimbro DL, Dolan MA. 2005. How habitat setting influences restored oyster reef communities. *Ecology* 86(7):1926–35.

Grabowski JH, Peterson CH. 2007. "Restoring oyster reefs to recover ecosystem services." In: *Ecosystem Engineers: Concepts, Theory and Applications*. Edited by K Cuddington, JE Byers, WG Wilson, A Hastings. Amsterdam, Netherlands: Elsevier-Academic Press (pp. 281–98).

Green SJ, Akins JL, Maljković A, Côté IM. 2012. Invasive lionfish drive Atlantic coral reef fish declines. *PloS One* 7(3):e32596.

Hare JA, Whitfield PE. 2003. *An Integrated Assessment of the Introduction of Lionfish (Pterois Volitans/Miles Complex) to the Western Atlantic Ocean. NOAA Technical Memorandum NOS NCCOS 2*. Washington, DC: NOAA (p. 21).

Harper D, Muller R. 2001. "14 Spiny lobster fisheries of the United States of America." In: *Report on the FAO/DANIDA/CFRAMP/WECAFC Regional Workshops on the Assessment of the Caribbean Spiny Lobster (Panulirus argus)*. FAO Fisheries Report No. 619. Edited by S Venema. Rome, Italy: FAO (p. 258).

Hauer ME, Evans JM, Mishra DR. 2016. Millions projected to be at risk from sea-level rise in the continental United States. *Nature Climate Change* 6(7):691.

Hawkes LA, Broderick AC, Godfrey MH, Godley BJ. 2005. Status of nesting loggerhead turtles Caretta caretta at Bald Head Island (North Carolina, USA) after 24 years of intensive monitoring and conservation. *Oryx* 39(1):65–72.

Hayden BP, Ray GC, Dolan R. 1984. Classification of coastal and marine environments. *Environmental Conservation* 11(3):199–207.

Hayes SA, Josephson E, Maze-Foley K, Rosel PE. 2019. *US Atlantic and Gulf of Mexico Marine Mammal Stock Assessments-2018. NOAA Tech Memo NMFS-NE-258*. Washington, DC: NOAA (p. 298).

Hefner JM, Brown JD. 1984. Wetland trends in the southeastern United States. Wetlands 4(1):1–11.

Hersh SL. 1987. *Characterization and Differentiation of Bottlenose Dolphin Populations (Genus Tursiops) in the Southeastern US based on Mortality Patterns and Morphometrics*. PhD dissertation. Miami, FL: University of Miami (p. 213).

Hill JM, Caretti ON, Heck Jr KL. 2017. Recently established Asian tiger shrimp *Penaeus monodon* Fabricius, 1798 consume juvenile blue crabs Callinectes sapidus Rathbun, 1896 and polychaetes in a laboratory diet-choice experiment. *BioInvasions Record* 6(3):233–8.

Holland AF, Sanger DM, Gawle CP, Lerberg SB, Santiago MS, Riekerk GH, Zimmerman LE, Scott GI. 2004. Linkages between tidal creek ecosystems and the landscape and demographic attributes of their watersheds. *Journal of Experimental Marine Biology and Ecology* 298(2):151–78.

Hovel KA, Fonseca MS, Myer DL, Kenworthy WJ, Whitfield PE. 2002. Effects of seagrass landscape structure, structural complexity and hydrodynamic regime on macrofaunal densities in North Carolina seagrass beds. *Marine Ecology Progress Series* 243:11–24.

Hunt JH. 2000. "Status of the fishery for Panulirus argus in Florida." In: *Spiny Lobsters: Fisheries and Culture*. Edited by BF Phillips, J Kittaka. Oxford, UK: Fishing News Books (pp. 321–33).

Huntsman GR, Potts J, Mays RW. 1994. A preliminary assessment of the populations of seven species of grouper (Serranidae, Epinephelinae) in the western Atlantic Ocean from Cape Hatteras, North Carolina to the Dry Tortugas, Florida. *Proceedings of the Gulf and Caribbean Fisheries Institute* 43:193–213.

Huntsman GR, Potts J, Mays RW, Vaughan D. 1999. Groupers (Serranidae, Epinephelinae): endangered apex predators of reef communities. *American Fisheries Society Symposium* 23:217–31.

Hyun KH, He R. 2010. Coastal upwelling in the South Atlantic Bight: a revisit of the 2003 cold event using long term observations and model hindcast solutions. *Journal of Marine Systems* 83(1–2):1–3.

ICOADS (International Comprehensive Ocean-Atmosphere Data Set). 2009. *DOC/NOAA/NESDIS/NCDC. National Climatic Data Center, NESDIS, NOAA, U.S.* Department of Commerce. https://data.nodc.noaa.gov/cgi-bin/iso?id=gov.noaa.ncdc:C00606

Irvine AB, Campbell HW. 1978. Aerial census of the West Indian manatee, Trichechus manatus, in the southeastern United States. *Journal of Mammalogy* 59(3):613–7.

Islam MS, Tanaka M. 2004. Impacts of pollution on coastal and marine ecosystems including coastal and marine fisheries and approach for management: a review and synthesis. *Marine Pollution Bulletin* 48(7–8):624–49.

Jepson M, Kitner K, Pitchon A, Perry WW, Stoffle B. 2002. *Potential Fishing Communities in the Carolinas, Georgia and*

Florida: An Effort in Baseline Profiling and Mapping. Washington, DC: South Atlantic Fishery Management Council, National Ocean and Atmospheric Administration, National Marine Fisheries Service.

Jia P, Li M. 2012. Circulation dynamics and salt balance in a lagoonal estuary. *Journal of Geophysical Research: Oceans* 117:C01003.

Johns G, Lee DJ, Leeworthy VB, Boyer J, Nuttle W. 2014. Developing economic indices to assess the human dimensions of the South Florida coastal marine ecosystem services. *Ecological Indicators* 44:69–80.

Joye SB, Bronk DA, Koopmans DJ, Moore WS. 2006. "Evaluating the potential importance of groundwater-derived carbon, nitrogen, and phosphorus inputs to South Carolina and Georgia coastal ecosystems." In: *Changing Land Use Patterns in the Coastal Zone*. Edited by GS Kleppel, MR DeVoe, MV Rawson. New York, NY: Springer (pp. 139–78).

Jud ZR, Layman CA, Lee JA, Arrington DA. 2011. Recent invasion of a Florida (USA) estuarine system by lionfish Pterois volitans/P. miles. *Aquatic Biology* 13(1):21–6.

Keim BD, Muller RA, Stone GW. 2007. Spatiotemporal patterns and return periods of tropical storm and hurricane strikes from Texas to Maine. *Journal of Climate* 20(14):3498–509.

Kelble CR, Loomis DK, Lovelace S, Nuttle WK, Ortner PB, Fletcher P, Cook GS, Lorenz JJ, Boyer JN. 2013. The EBM-DPSER conceptual model: integrating ecosystem services into the DPSIR framework. *PloS One* 8(8):e70766.

Keller CA, Ward-Geiger LI, Brooks WB, Slay CK, Taylor CR, Zoodsma BJ. 2006. North Atlantic right whale distribution in relation to sea-surface temperature in the southeastern United States calving grounds. *Marine Mammal Science* 22(2):426–45.

Kellner JB, Sanchirico JN, Hastings A, Mumby PJ. 2011. Optimizing for multiple species and multiple values: tradeoffs inherent in ecosystem-based fisheries management. *Conservation Letters* 4(1):21–30.

Kendall MS, Bauer LJ, Jeffrey CF. 2008. Influence of benthic features and fishing pressure on size and distribution of three exploited reef fishes from the southeastern United States. *Transactions of the American Fisheries Society* 137(4):1134–46.

Kennish MJ. 2001. Coastal salt marsh systems in the US: a review of anthropogenic impacts. *Journal of Coastal Research* 1:731–48.

Kilduff P, Carmichael J, Latour R. 2009. *Guide to Fisheries Science and Stock Assessments. Atlantic States Marine Fisheries Commission, National Oceanic and Atmospheric Administration Grant No. NA05NMF4741025*. Washington, DC: NOAA.

Kirby MX. 2004. Fishing down the coast: historical expansion and collapse of oyster fisheries along continental margins. *Proceedings of the National Academy of Sciences* 101(35):13096–9.

Koenig CC, Coleman FC, Grimes CB, Fitzhugh GR, Scanlon KM, Gledhill CT, Grace M. 2000. Protection of fish spawning habitat for the conservation of warm-temperate reef-fish fisheries of shelf-edge reefs of Florida. *Bulletin of Marine Science* 66(3):593–616.

Koenig CC, Coleman FC, Kingon KC. 2010. Recovery of the goliath grouper (Epinephelus itajara) population of the Southeastern U.S. *Proceedings of the Gulf and Caribbean Fisheries Institute* 62:219–23).

Klein YL, Osleeb JP, Viola MR. Tourism-generated earnings in the coastal zone: a regional analysis. *Journal of Coastal Research* 20.4(204):1080–8.

Labisky RF, Gregory Jr DR, Conti JA. 1980. Florida's spiny lobster fishery: an historical perspective. *Fisheries* 5(4):28–37.

Lapointe BE, Barile PJ, Matzie WR. 2004. Anthropogenic nutrient enrichment of seagrass and coral reef communities in the Lower Florida Keys: discrimination of local versus regional nitrogen sources. *Journal of Experimental Marine Biology and Ecology* 308(1):23–58.

Lapointe BE, Clark MW. 1992. Nutrient inputs from the watershed and coastal eutrophication in the Florida Keys. *Estuaries* 15(4):465–76.

Layman CA, Jud ZR, Nichols P. 2014. Lionfish alter benthic invertebrate assemblages in patch habitats of a subtropical estuary. *Marine Biology* 161(9):2179–82.

Lee DS. 2015. *Gulf Stream Chronicles: A Naturalist Explores Life in an Ocean River*. Chapel Hill, NC: UNC Press Books.

Leeworthy VR, Wiley PC. 2001. *Current Participation Patterns in Marine Recreation*. Washington, DC: National Oceanic and Atmospheric Administration, National Ocean Service, Special Projects (p. 53).

Lehnert RL, Allen DM. 2002. Nekton use of subtidal oyster shell habitat in a southeastern US estuary. *Estuaries* 25(5):1015–24.

Lellis-Dibble KA, McGlynn KE, Bigford TE. 2008. *Estuarine Fish and Shellfish Species in U.S. Commercial and Recreational Fisheries: Economic Value as an Incentive to Protect and Restore Estuarine Habitat. NOAA Technical Memo NMFSF/SPO-90*. Washington, DC: NOAA (p. 94).

Levin PS, Essington TE, Marshall KN, Koehn LE, Anderson LG, Bundy A, Carothers C, et al. 2018. Building effective fishery ecosystem plans. *Marine Policy* 92:48–57.

Levin PS, Fogarty MJ, Murawski SA, Fluharty D. 2009. Integrated ecosystem assessments: developing the scientific basis for ecosystem-based management of the ocean. *PLoS Biology* 7(1):e1000014.

Lindeman KC, Pugliese R, Waugh GT, Ault JS. 2000. Developmental patterns within a multispecies reef fishery: management applications for essential fish habitats and protected areas. *Bulletin of Marine Science* 66(3):929–56.

Lindner BL, Neuhauser A. 2018. Climatology and variability of tropical cyclones affecting Charleston, South Carolina. *Journal of Coastal Research* 34(5):1052–64.

Lindquist DG, Pietrafesa LJ. 1989. Current vortices and fish aggregations: the current field and associated fishes around a tugboat wreck in Onslow Bay, North Carolina. *Bulletin of Marine Science* 44(2):533–44.

Link JS. 2018. System-level optimal yield: increased value, less risk, improved stability, and better fisheries. *Canadian Journal of Fisheries and Aquatic Sciences* 75(1):1–6.

Link JS, Ihde TF, Townsend HM, Osgood KE, Schirripa MJ, Kobayashi DR, Gaichas SK, et al. 2010. *Report of the 2nd National Ecosystem Modeling Workshop (NEMoW II): Bridging the Credibility Gap Dealing with Uncertainty in Ecosystem Models. NOAA Technical Memorandum NMFS-F/SPO-102*. Washington, DC: NOAA.

Link JS, Marshak AR. 2019. Characterizing and comparing marine fisheries ecosystems in the United States: determinants of success in moving toward ecosystem-based fisheries management. *Reviews in Fish Biology and Fisheries* 29(1):23–70.

Lirman D, Ault JS, Fourqurean JW, Lorenz JJ. 2019. "The coastal marine ecosystem of South Florida, United States." In: *World Seas: An Environmental Evaluation*. Edited by C Sheppard. London, UK: Academic Press (pp. 427–44).

Loh TL, McMurray SE, Henkel TP, Vicente J, Pawlik JR. 2015. Indirect effects of overfishing on Caribbean reefs: sponges overgrow reef-building corals. *PeerJ* 3:e901.

Lynch PD, Methot RD, Link JS. 2018. *Implementing a Next Generation Stock Assessment Enterprise. An Update to the NOAA Fisheries Stock Assessment Improvement Plan. NOAA Technical Memo NMFS-F/ SPO-183*. Washington, DC: NOAA (p. 127).

MAFMC (Mid-Atlantic Fishery Management Council). 2017. *Amendment 6 to the Tilefish Fishery Management Plan. Measures to Manage Blueline Tilefish. Includes Environmental Assessment and Initial Regulatory Flexibility Analysis*. Washington, DC: Mid-Atlantic Fishery Management Council, NOAA Fisheries (p. 150).

Mager Jr AN, Ruebsamen R. 1988. National Marine Fisheries Service habitat conservation efforts in the coastal southeastern United States for 1987. *Marine Fisheries Review* 50(3):43–50.

Mallin MA, Posey MH, Shank GC, McIver MR, Ensign SH, Alphin TD. 1999. Hurricane effects on water quality and benthos in the Cape Fear watershed: natural and anthropogenic impacts. *Ecological Applications* 9(1):350–62.

Mallin MA, Williams KE, Esham EC, Lowe RP. 2000. Effect of human development on bacteriological water quality in coastal watersheds. *Ecological Applications* 10(4):1047–56.

Manooch III CS, Potts JC, Burton ML, Vaughan DS. 1998. *Population Assessment of the Vermilion Snapper, Rhomboplites Aurorubens, from the Southeastern United States. NOAA Technical Memorandum NMFS-SEFSC-411*. Washington, DC: NOAA (p. 63).

Marancik KE, Hare JA. 2007. Large scale patterns in fish trophodynamics of estuarine and shelf habitats of the southeast United States. *Bulletin of Marine Science* 80(1):67–91.

Marshak AR, Brown SK. 2017. Habitat science is an essential element of ecosystem-based fisheries management. *Fisheries* 42(6):300.

Marshak AR, Heck Jr KL, Jud ZR. 2018. Ecological interactions between Gulf of Mexico snappers (Teleostei: Lutjanidae) and invasive red lionfish (Pterois volitans). *PloS One* 13(11):e0206749.

Marshak AR, Kracker L, Peters R. 2018. *Report from the Using Acoustic Multibeam Echosounder (ME70) Technologies for Habitat Mapping Workshop. NOAA Technical Memo NMFS-F/SPO-191*. Washington, DC: NOAA (p. 36).

Marshall FE, Banks K, Cook GS. 2014. Ecosystem indicators for Southeast Florida beaches. *Ecological Indicators* 44:81–91.

Marshall J, Johnson H, Goodman J. 2001. A study of the interaction of the North Atlantic Oscillation with ocean circulation. *Journal of Climate* 14:1399–421.

Marshall KN, Levin PS, Essington TE, Koehn LE, Anderson LG, Bundy A, Carothers C, et al. 2018. Ecosystem-based fisheries management for social–ecological systems: renewing the focus in the United States with next generation fishery ecosystem plans. *Conservation Letters* 11(1):e12367.

Matheson RE, Camp DK, Sogard SM, Bjorgo KA. 1999. Changes in seagrass-associated fish and crustacean communities on Florida Bay mud banks: the effects of recent ecosystem changes? *Estuaries* 22(2):534.

Matthews T. 1995. Fishing effort reduction in the Florida spiny lobster fishery. *Proceedings of the Gulf and Caribbean Fisheries Institute* 48:111–21.

McCarney-Castle K, Voulgaris G, Kettner AJ. 2010. Analysis of fluvial suspended sediment load contribution through anthropocene history to the South Atlantic Bight? *The Journal of Geology* 118(4):399–416.

McCarthy GD, Haigh ID, Hirschi JJM, Grist JP, Smeed DA. 2015. Ocean impact on decadal Atlantic climate variability revealed by sea-level observations. *Nature* 521:508–10.

McClenachan L. 2009. Documenting loss of large trophy fish from the Florida Keys with historical photographs. *Conservation Biology* 23(3):636–43.

McClenachan L. 2008. "Social conflict, overfishing and disease in the Florida sponge fishery, 1849–1939." In: *Oceans Past: Management Insights from the History of Marine Animal Populations*. Edited by D Starky. Abingdon, UK: Routledge (pp. 25–46).

McGovern JC, Collins MR, Pashuk O, Meister HS. 2002. Temporal and spatial differences in life history parameters of black sea bass in the southeastern United States. *North American Journal of Fisheries Management* 22(4):1151–63.

McGowan MF, Richards WJ. 1989. Bluefin tuna, Thunnus thynnus, larvae in the Gulf Stream off the Southeastern United States: satellite and shipboard observations of their environment. *Fishery Bulletin* 87(3):615–31.

Miller GC, Richards WC. 1980. Reef fish habitat, faunal assemblages, and factors determining distributions in the

South Atlantic Bight. *Proceeding of the Gulf and Caribbean Fisheries Institute* 32:114–30.

Miller M, Bourque A, Bohnsack J. 2002. An analysis of the loss of acroporid corals at Looe Key, Florida, USA: 1983–2000. *Coral Reefs* 21(2):179–82.

Milon JW, Mulkey WD, Riddle PH, Wilkowske GH. 1983. The economic impact of marine recreational boating on the Florida economy. *Florida Sea Grant College* 54:26.

Milton SL, Leone-Kabler S, Schulman AA, Lutz PL. 1994. Effects of Hurricane Andrew on the sea turtle nesting beaches of South Florida. *Bulletin of Marine Science* 54(3):974–81.

Morley JW, Batt RD, Pinsky ML. 2017. Marine assemblages respond rapidly to winter climate variability. *Global Change Biology* 23(7):2590–601.

Morley JW, Selden RL, Latour RJ, Frölicher TL, Seagraves RJ, Pinsky ML. 2018. Projecting shifts in thermal habitat for 686 species on the North American continental shelf. *PloS One* 13(5):e0196127.

Morris Jr JA, Akins JL. 2009. Feeding ecology of invasive lionfish (Pterois volitans) in the Bahamian archipelago. *Environmental Biology of Fishes* 86:389–98.

Morris Jr JA, Whitfield PE. 2009. *Biology, Ecology, Control and Management of the Invasive Indo-Pacific Lionfish: An Updated Integrated Assessment. NOAA Technical Memorandum NOS NCCOS 99*. Washington, DC: NOAA (p. 57).

Morris JT, Sundareshwar PV, Nietch CT, Kjerfve B, Cahoon DR. 2002. Responses of coastal wetlands to rising sea level. *Ecology* 83(10):2869–77.

Muller RA, Stone GW. 2013. A climatology of tropical storm and hurricane strikes to enhance vulnerability prediction for the southeast US coast. *Journal of Coastal Research* 17(4).

Muller RG, Bert TM, Gerhart SD. 2006. *The 2006 Stock Assessment Update for the Stone Crab, Menippe spp., Fishery in Florida. IHR 2006-011*. Tallahassee, FL: Florida Fish and Wildlife Conservation Commission (p. 43).

Muñoz RC, Currin CA, Whitfield PE. 2011. Diet of invasive lionfish on hard bottom reefs of the Southeast USA: insights from stomach contents and stable isotopes. *Marine Ecology Progress Series* 432:181–193.

Murphy SJ, Hurlburt HE, O'Brien JJ. 1999. The connectivity of eddy variability in the Caribbean Sea, the Gulf of Mexico, and the Atlantic Ocean. *Journal of Geophysical Research: Oceans* 104(C1):1431–53.

Muto MM, Helker VT, Angliss RP, Boveng PL, Breiwick JM, Cameron MF, Clapham PJ, et al. 2019. *Alaska Marine Mammal Stock Assessments, 2018. NOAA Technical Memo NMFS-AFSC-393*. Washington, DC: NOAA (p. 390).

Myers RA, Baum JK, Shepherd TD, Powers SP, Peterson CH. 2007. Cascading effects of the loss of apex predatory sharks from a coastal ocean. *Science* 315(5820):1846–50.

Najjar RG, Pyke CR, Adams MB, Breitburg D, Hershner C, Kemp M, Howarth R, et al. 2010. Potential climate-change impacts on the Chesapeake Bay. *Estuarine, Coastal and Shelf Science* 86(1):1–20.

Napton DE, Auch RF, Headley R, Taylor JL. 2010. Land changes and their driving forces in the Southeastern United States. *Regional Environmental Change* 10(1):37–53.

NCDEQ (North Carolina Department of Environmental Quality). 2016a. *North Carolina Coastal Habitat Protection Plan*. Raleigh, NC: NCDEQ (p. 32).

NCDEQ (North Carolina Department of Environmental Quality). 2016b. *North Carolina Coastal Habitat Protection Plan Source Document*. Raleigh, NC: NCDEQ (p. 475).

Newman D, Berkson J, Suatoni L. 2015. Current methods for setting catch limits for data-limited fish stocks in the United States. *Fisheries Research* 164:86–93.

NFHP (National Fish Habitat Partnership). 2010. *Through a Fish's Eye: The Status of Fish Habitats in the United States*. Washington, DC: National Fish Habitat Board, Association of Fish and Wildlife Agencies (p. 68).

NMFS (National Marine Fisheries Service). 2013. *Final Amendment 5a to the 2006 Consolidated Atlantic Highly Migratory Species Fishery Management Plan*. Silver Spring, MD: Highly Migratory Species Management Division (p. 410).

NMFS (National Marine Fisheries Service). 2016a. *Ecosystem-Based Fisheries Management Policy of the National Marine Fisheries Service*. Washington, DC: NOAA.

NMFS (National Marine Fisheries Service). 2016b. *NOAA Fisheries Ecosystem-Based Fisheries Management Roadmap*. Washington, DC: NOAA.

NMFS (National Marine Fisheries Service). 2017. *National Marine Fisheries Service—2nd Quarter 2017 Update*. Washington, DC: NOAA (p. 53).

NMFS (National Marine Fisheries Service). 2018a. *Fisheries Economics of the United States, 2016. NOAA Technical Memo. NMFS-F/SPO-187*. Washington, DC: NOAA (p. 243).

NMFS (National Marine Fisheries Service). 2018b. *Fisheries of the United States, NOAA Current Fishery Statistics No. 2017*. Washington, DC: NOAA.

NMFS (National Marine Fisheries Service). 2019. *Ecosystem-Based Fisheries Management Plan for the South Atlantic, May 2019*. Washington, DC: NOAA (p. 16).

NMFS (National Marine Fisheries Service). 2020a. National Marine Fisheries Service—2nd Quarter 2020 Update. Washington, DC: NOAA (p. 51).

NMFS (National Marine Fisheries Service). 2020b. *Species Directory: Wreckfish*. Washington, DC: NOAA.

Norton SL, Wiley TR, Carlson JK, Frick AL, Poulakis GR, Simpfendorfer CA. 2012. Designating critical habitat for juvenile endangered smalltooth sawfish in the United States. *Marine and Coastal Fisheries* 4(1):473–80.

Norton TM. 2005. Sea turtle conservation in Georgia and an overview of the Georgia sea turtle center on Jekyll Island, Georgia. *Georgia Journal of Science* 63(4):208.

Nuttle WK, Fletcher PJ. 2013. *Integrated Conceptual Ecosystem Model Development for the Florida Keys/Dry Tortugas Coastal Marine Ecosystem. NOAA Technical Memorandum, OAR-AOML-101 & NOS-NCCOS-161*. Miami, FL: NOAA, National Ocean Service (p. 91).

NUWC (Naval Undersea Warfare Center). 2012. *Determination of Acoustic Effects on Marine Mammals and Sea Turtles for the Atlantic Fleet Training and Testing Environmental Impact Statement/Overseas Environmental Impact Statement. Marine Species Modeling Team. NUWC-NPT Technical Report 12,071*. Newport, RI. US Naval Undersea Warfare Center Division (p. 88).

Ogden JC, Davis SM, Barnes TK, Jacobs KJ, Gentile JH. 2005. Total system conceptual ecological model. *Wetlands* 25(4):955–79.

Okey TA, Cisneros-Montemayor AM, Puliese R, Sumaila UR. 2014. *Exploring the Trophodynamic Signatures of Forage Species in the U.S. South Atlantic Bight Ecosystem to Maximize System-Wide Values. Fisheries Centre Working Paper #2014-14*. Vancouver, Canada: Fisheries Centre, The University of British Columbia (p. 80).

Okey TA, Pugliese R. 2001. A preliminary Ecopath model of the Atlantic continental shelf adjacent to the southeastern United States. Fisheries impacts on North Atlantic ecosystems: models and analyses. *Fisheries Centre Research Reports* 9(4):167–81.

Osland MJ, Enwright N, Day RH, Doyle TW. 2013. Winter climate change and coastal wetland foundation species: salt marshes vs. mangrove forests in the southeastern United States. *Global Change Biology* 19(5):1482–94.

Paerl HW, Bales JD, Ausley LW, Buzzelli CP, Crowder LB, Eby LA, Fear JM, et al. 2001. Ecosystem impacts of three sequential hurricanes (Dennis, Floyd, and Irene) on the United States' largest lagoonal estuary, Pamlico Sound, NC. *Proceedings of the National Academy of Sciences* 98(10):5655–60.

Paerl HW, Hall NS, Hounshell AG, Luettich RA, Rossignol KL, Osburn CL, Bales J. 2019. Recent increase in catastrophic tropical cyclone flooding in coastal North Carolina, USA: long-term observations suggest a regime shift. *Scientific Reports* 9(1):1–9.

Paling EI, Fonseca M, van Katwijk MM, van Keulen M. 2009. "Seagrass restoration." In: *Coastal Wetlands: An Integrated Ecosystems Approach*. Edited by GME Perillo, E Wolanski, DR Cahoon, M Brinson. Amsterdam, Netherlands: Elsevier (pp. 687–713).

Parker Jr RO, Dixon RL. 1998. Changes in a North Carolina reef fish community after 15 years of intense fishing—global warming implications. *Transactions of the American Fisheries Society* 127(6):908–20.

Patrick WS, Spencer P, Link J, Cope J, Field J, Kobayashi D, Lawson P, et al. 2010. Using productivity and susceptibility indices to assess the vulnerability of United States fish stocks to overfishing. *Fishery Bulletin* 108(3):305–22.

Peters R, Marshak AR, Brady MM, Brown SK, Osgood K, Greene C, Guida V, et al. 2018. *Habitat Science is a Fundamental Element in an Ecosystem-Based Fisheries Management Framework: An Update to the Marine Fisheries Habitat Assessment Improvement Plan. NOAA Technical Memorandum NMFS-F/SPO-181*. Washington, DC: NOAA (p. 29).

Peterson CH, Grabowski JH, Powers SP. 2003. Estimated enhancement of fish production resulting from restoring oyster reef habitat: quantitative valuation. *Marine Ecology Progress Series* 264:249–64.

Peterson CH, Summerson HC, Thomson E, Lenihan HS, Grabowski J, Manning L, Micheli F, Johnson G. 2000. Synthesis of linkages between benthic and fish communities as a key to protecting essential fish habitat. *Bulletin of Marine Science* 66(3):759–74.

Peterson J, Griffis R, Zador SG, Sigler MF, Joyce JE, Hunsicker M, Bograd S, et al. 2017. Climate change impacts on fisheries and aquaculture of the United States. *Climate Change Impacts on Fisheries and Aquaculture: A Global Analysis* 1:159–218.

Pollock DE. 1993. Recruitment overfishing and resilience in spiny lobster populations. *ICES Journal of Marine Science* 50(1):9–14.

Porter JW, Dustan P, Jaap WC, Patterson KL, Kosmynin V, Meier OW, Patterson ME, Parsons M. 2001. "Patterns of spread of coral disease in the Florida Keys." In: *The Ecology and Etiology of Newly Emerging Marine Diseases*. Edited by JW Porter. Dordrecht, Netherlands: Springer (pp. 1–24).

Porter JW, Lewis SK, Porter KG. 1999. The effect of multiple stressors on the Florida Keys coral reef ecosystem: a landscape hypothesis and a physiological test. *Limnology and Oceanography* 44(3part2):941–9.

Potts JC, Burton ML. 2017. Preliminary observations on the age and growth of dog snapper (Lutjanus jocu) and mahogany snapper (Lutjanus mahogoni) from the Southeastern US. *PeerJ* 5:e3167.

Power AJ, Walker RL, Payne K, Hurley D. 2004. First occurrence of the nonindigenous green mussel, Perna viridis (Linnaeus, 1758) in coastal Georgia, United States. *Journal of Shellfish Research* 23(3):741–5.

Precht WF. 2019. Failure to respond to a coral disease outbreak: potential costs and consequences. *PeerJ Preprints* 7:e27860v2.

Randall DP. 1968. Wilmington, North Carolina: the historical development of a port city. *Annals of the Association of American Geographers* 58(3):441–51.

Raymond WW, Albins MA, Pusack TJ. 2015. Competitive interactions for shelter between invasive Pacific red lionfish and native Nassau grouper. *Environmental Biology of Fishes* 98(1):57–65.

Reed JK. 2002. Deep-water Oculina coral reefs of Florida: biology, impacts, and management. *Hydrobiologia* 471(1–3):43–55.

Reimer JJ, Cai WJ, Xue L, Vargas R, Noakes S, Hu X, Signorini SR, et al. 2017. Time series pCO2 at a coastal mooring: Internal consistency, seasonal cycles, and interannual variability. *Continental Shelf Research* 145:95–108.

Riley ME, Johnston CA, Feller IC, Griffen BD. 2014. Range expansion of Aratus pisonii (mangrove tree crab) into novel vegetative habitats. *Southeastern Naturalist* 13(4): N43–8.

Roehl WS, Ditton RB, Holland SM, Perdue RR. 1993. Developing new tourism products: sport fishing in the south-east United States. *Tourism Management* 14(4): 279–88.

Rogers SG, Targett TE, Van Sant SB. 1984. Fish-nursery use in Georgia salt-marsh estuaries: the influence of springtime freshwater conditions. *Transactions of the American Fisheries Society* 113(5):595–606.

Roman CT, Jaworski N, Short FT, Findlay S, Warren RS. 2000. Estuaries of the northeastern United States: habitat and land use signatures. *Estuaries* 23(6):743–64.

Rudershausen PJ, Williams EH, Buckel JA, Potts JC, Manooch III CS. 2008. Comparison of reef fish catch per unit effort and total mortality between the 1970s and 2005–2006 in Onslow Bay, North Carolina. *Transactions of the American Fisheries Society* 137(5):1389–405.

Rudnick DT, Ortner PB, Browder JA, Davis SM. 2005. A conceptual ecological model of Florida Bay. *Wetlands* 25(4):870–83.

SAFMC (South Atlantic Fishery Management Council). 1998. *Final Habitat Plan for the South Atlantic Region: Essential Fish Habitat Requirements for Fishery Management Plans of the South Atlantic Fishery Management Council.* Charleston, SC: South Atlantic Fishery Management Council (p. 458).

SAFMC (South Atlantic Fishery Management Council). 2009. *Fishery Ecosystem Plan of the South Atlantic Region.* Charleston, SC: South Atlantic Fishery Management Council (p. 689).

SAFMC (South Atlantic Fishery Management Council). 2016. *Policy Considerations for South Atlantic Food Webs and Connectivity and Essential Fish Habitats.* Charleston, SC: South Atlantic Fishery Management Council.

SAFMC (South Atlantic Fishery Management Council) 2018. *SAFMC Fishery Ecosystem Plan II Implementation Plan.* Charleston, SC: South Atlantic Fishery Management Council (p. 83).

Sancho G, Kingsley-Smith PR, Morris JA, Toline CA, McDonough V, Doty SM. 2018. Invasive Lionfish (Pterois volitans/miles) feeding ecology in Biscayne National Park, Florida, USA. *Biological Invasions* 20(9):2343–61.

Schaefer SC, Alber M. 2007. Temperature controls a latitudinal gradient in the proportion of watershed nitrogen exported to coastal ecosystems. *Biogeochemistry* 85(3):333–46.

Schmidly DJ. 1981. *Marine Mammals of the Southeastern United States Coast and the Gulf of Mexico. FWS/OBS-SO/41.* Washington, DC: Fish and Wildlife Service, Office of Biological Services.

Schmied RL, Burgess EE. 1987. Marine recreational fisheries in the southeastern United States: an overview. *Marine Fisheries Review* 49(2):1–7.

Scott-Denton E, Cryer PF, Duffy MR, Gocke JP, Harrelson MR, Kinsella DL, Nance JM, et al. 2012. Characterization of the US Gulf of Mexico and South Atlantic penaeid and rock shrimp fisheries based on observer data. *Marine Fisheries Review* 74(4):1–27.

SEAMAP (Southeast Area Monitoring and Assessment Program) 2016. *2016–2020 Management Plan. Collection, Management, and Dissemination of Fishery-independent Data from the Waters of the Southeastern United States. No. NA16NMF4350112.* Washington, DC: NOAA (p. 114).

SEAMAP (Southeast Area Monitoring and Assessment Program). 2017. *Annual Report of the Southeast Area Monitoring and Assessment Program (Seamap) October 1, 2015 – September 30, 2016. No. 261.* Washington, DC: NOAA (p. 21).

SECOORA (Southeast Coastal Ocean Observing Regional Association) 2016. *Southeast Coastal Ocean Observing Regional Association Strategic Plan 2016–2020.* Silver Spring, MD: Integrated Ocean Observing System.

Sedberry GR. 2001. *Island in the Stream: Oceanography and Fisheries of the Charleston Bump. Symposium 25.* Bethesda, MD: American Fisheries Society.

Sedberry GR, McGovern JC, Pashuk OL. 2001. *The Charleston Bump: An Island of Essential Fish Habitat in the Gulf Stream. Symposium 2001.* Bethesda, MD: American Fisheries Society (pp. 3–24).

Sedberry GR, Ulrich GF, Applegate AJ. 1994. Development and status of the fishery for wreckfish (Polyprion americanus) in the southeastern United States. *Proceedings of the Gulf and Caribbean Fisheries Institute* 43:168–92.

Seeb SK. 2013. "Cape Fear's forgotten fleet: the Eagles Island ship graveyard, Wilmington, North Carolina." In: *The Archaeology of Watercraft Abandonment.* Edited by N Richards, SK Seeb. New York, NY: Springer (pp. 215–38).

Seminoff JA, Allen CD, Balazs GH, Dutton PH, Eguchi T, Haas HL, Hargrove SA, et al. 2015. *Status Review of the Green Turtle (Chelonia mydas) Under the U.S. Endangered Species Act. NOAA Technical Memorandum NOAANMFS-SWFSC-539.* Washington, DC: NOAA (p. 571).

Shamblin BM, Dodd MG, Bagley DA, Ehrhart LM, Tucker AD, Johnson C, Carthy RR, et al. 2011. Genetic structure of the southeastern United States loggerhead turtle nesting aggregation: evidence of additional structure within the peninsular Florida recovery unit. *Marine Biology* 158(3):571–87.

Sharitz RR. 2003. Carolina bay wetlands: unique habitats of the southeastern United States. *Wetlands* 23(3):550–62.

Sharp W, Maxwell K, Hunt J. 2019. *Investigating the Ongoing Coral Disease Outbreak in the Florida Keys: Evaluating its Small-Scale Epidemiology and Mitigation Techniques.* Tallahassee, FL: Florida Fish and Wildlife Conservation Commission (p. 34).

Shepherd JF, Walton GM. 1976. Economic change after the American Revolution: pre-and post-war comparisons of maritime shipping and trade. *Explorations in Economic History* 13(4):397–422.

Shepherd JF, Williamson SH. 1972. The coastal trade of the British North American colonies, 1768–1772. *The Journal of Economic History* 32(4):783–810.

Sherman K. 1991. The large marine ecosystem concept: research and management strategy for living marine resources. *Ecological Applications* 1(4):349–60.

Shertzer KW, Bacheler NM, Pine III WE, Runde BJ, Buckel JA, Rudershausen PJ, MacMahan JH. 2020. Estimating population abundance at a site in the open ocean: combining information from conventional and telemetry tags with application to gray triggerfish (Balistes capriscus). *Canadian Journal of Fisheries and Aquatic Sciences* 77: 34–43.

Shertzer KW, Williams EW. 2008. Fish assemblages and indicator species: reef fishes off the southeastern United States. *Fishery Bulletin* 106(3):257–69.

Shertzer KW, Williams EH, Taylor JC. 2009. Spatial structure and temporal patterns in a large marine ecosystem: Exploited reef fishes of the southeast United States. *Fisheries Research* 100(2):126–33.

Shertzer KW, Williams EH, Craig JK, Fitzpatrick EE, Klibansky N, Siegfried KI. 2019. Recreational sector is the dominant source of fishing mortality for oceanic fishes in the Southeast United States Atlantic Ocean. *Fisheries Management and Ecology* 26(6):621–9.

Shin YJ, Cury P. 2004. Using an individual-based model of fish assemblages to study the response of size spectra to changes in fishing. *Canadian Journal of Fisheries and Aquatic Sciences* 61(3):414–31.

Shin Y, Shannon L, Cury P. 2004. Simulations of fishing effects on the southern Benguela fish community using an individual-based model: learning from a comparison with ECOSIM. *African Journal of Marine Science* 26:95–114.

Smith CS, Puckett B, Gittman RK, Peterson CH. 2018. Living shorelines enhanced the resilience of saltmarshes to Hurricane Matthew. 2016. *Ecological Applications* 28(4):871–7.

Smith TJ, Anderson GH, Balentine K, Tiling G, Ward GA, Whelan KR. 2009. Cumulative impacts of hurricanes on Florida mangrove ecosystems: sediment deposition, storm surges and vegetation. *Wetlands* 29(1):24.

Sorte CJ, Williams SL, Carlton JT. 2010. Marine range shifts and species introductions: comparative spread rates and community impacts. *Global Ecology and Biogeography* 19(3):303–16.

Soto IM, Muller Karger FE, Hallock P, Hu C. 2011. Sea surface temperature variability in the Florida Keys and its relationship to coral cover. *Journal of Marine Biology* 2011:1–10.

Spirek J, Harris L. 2003. "Maritime heritage on display: underwater examples from South Carolina." In: *Submerged Cultural Resource Management.* Edited by JD Spirek, DA Scott-Ireton. Boston, MA: Springer (pp. 165–175).

Stedman S, Dahl TE. 2008. *Status and Trends of Wetlands in the coastal Watersheds of the Eastern United States 1998 to 2004.* Washington, DC: Fish and Wildlife Service.

Stephen JA, Harris PJ. 2010. Commercial catch composition with discard and immediate release mortality proportions off the southeastern coast of the United States. *Fisheries Research* 103(1–3):18–24.

Stephen JA, Harris PJ, Reichert MJ. 2011. Comparison of life history parameters for landed and discarded fish captured off the southeastern United States. *Fishery Bulletin* 109(3):292–304.

Stick D. 1989. *Graveyard of the Atlantic: Shipwrecks of the North Carolina Coast.* Chapel Hill, NC: University of North Carolina Press.

Stoner AW, Glazer RA. 1998. Variation in natural mortality: implications for queen conch stock enhancement. *Bulletin of Marine Science* 62(2):427–42.

Teal J, Teal M. 1969. *Life and Death of the Salt Marsh.* New York, NY: Ballantine Books (p. 279).

Tilmant JT, Curry RW, Jones R, Szmant A, Zieman JC, Flora M, Robblee MB, et al. 1994. Hurricane Andrew's effects on marine resources: the small underwater impact contrasts sharply with the destruction in mangrove and upland-forest communities. *BioScience* 44(4):230–7.

Townsend HM, Harvey CJ, Aydin KY, Gamble RJ, Gruss A, Levin PS, Link JS, et al. 2014. *Report of the 3rd National Ecosystem Modeling Workshop (NEMoW 3): Mingling Models for Marine Resource Management, Multiple Model Inference. NOAA Technical Memorandum NMFS-F/SPO-173.* Washington, DC: NOAA.

Townsend HM, Link JS, Osgood KE, Gedamke T, Watters GM, Polovina JJ, Levin PS, Cyr EC, Aydin KY. 2008. *Report of the National Ecosystem Modeling Workshop (NEMoW).* Washington, DC: NOAA.

Townsend HK, Aydin K, Holsman C, Harvey I, Kaplan E, Hazen P, Woodworth-Jefcoats M. et al. 2017. *Report of the 4th National Ecosystem Modeling Workshop (NEMoW 4): Using Ecosystem Models to Evaluate Inevitable Trade-offs. NOAA Technical Memo NMFS-F/SPO-173.* Washington, DC: NOAA (p. 77).

Trent L, Parshley DE, Carlson JK. 1997. Catch and bycatch in the shark drift gillnet fishery off Georgia and East Florida. *Marine Fisheries Review* 59(1):19–28.

Valle-Levinson A, Dutton A, Martin JB. 2017. Spatial and temporal variability of sea level rise hot spots over the eastern United States. *Geophysical Research Letters* 44(15):7876–82.

Valdivia A, Wolf S, Suckling K. 2019. Marine mammals and sea turtles listed under the US Endangered Species Act are recovering. *PloS One* 14(1):e0210164.

Van Dolah RF, Hyland JL, Holland AF, Rosen JS, Snoots TR. 1999. A benthic index of biological integrity for assessing habitat quality in estuaries of the southeastern USA. *Marine Environmental Research* 48(4–5):269–83.

VanderKooy SJ, Muller RG. 2003. Management of spotted seatrout and fishery participants in the US. *Biology of the Spotted Seatrout* 2003:227–45.

Vaughan DS, Collins MR, Schmidt DJ. 1995. Population characteristics of the black sea bass Centropristis striata from the southeastern US. *Bulletin of Marine Science* 56(1):250–67.

Voss CM, Peterson CH, Fegley SR. 2013. *Fishing, Diving, and Ecotourism Stakeholder Uses and Habitat Information for North Carolina Wind Energy Call Areas. OCS Study BOEM 2013-210.* Herndon, VA: Bureau of Ocean Energy Management, Office of Renewable Energy Programs (p. 23).

Wagner DE, Kramer P, Van Woesik R. 2010. Species composition, habitat, and water quality influence coral bleaching in southern Florida. *Marine Ecology Progress Series* 408:65–78.

Walker BK. 2012. Spatial analyses of benthic habitats to define coral reef ecosystem regions and potential biogeographic boundaries along a latitudinal gradient. *PloS One* 7(1):e30466.

Walton CJ, Hayes NK, Gilliam DS. 2018. Impacts of a regional, multi-year, multi-species coral disease outbreak in Southeast Florida. *Frontiers in Marine Science* 5:323.

Watson J, Foster D, Nichols S, Shah A, Scoll-oenlon E, Nanc J. 1999. The development of bycatch reduction technology in the southeastern United States shrimp fishery. *Marine Technology Society Journal* 33(2):51–6.

Weinstein MP, Weiss SL, Walters MF. 1980. Multiple determinants of community structure in shallow marsh habitats, Cape Fear River Estuary, North Carolina, USA. *Marine Biology* 58(3):227–43.

Wells RS, Rhinehart HL, Cunningham P, Whaley J, Baran M, Koberna C, Costa DP. 1999. Long distance offshore movements of bottlenose dolphins. *Marine Mammal Science* 15(4):1098–114.

Wenner CA, Sedberry GR. 1989. *Species Composition, Distribution, and Relative Abundance of Fishes in the Coastal Habitat Off the Southeastern United States. 1989. NOAA Technical Report NMFS 79.* Washington, DC: NOAA (p. 49).

Whitfield PE, Gardner T, Vives SP, Gilligan MR, Courtenay Jr WR, Ray GC, Hare JA. 2002. Biological invasion of the Indo-Pacific lionfish Pterois volitans along the Atlantic coast of North America. *Marine Ecology Progress Series* 235:289–97.

Whitfield PE, Muñoz RC, Buckel CA, Degan BP, Freshwater DW, Hare JA. 2014. Native fish community structure and Indo-Pacific lionfish Pterois volitans densities along a depth-temperature gradient in Onslow Bay, North Carolina, USA. *Marine Ecology Progress Series* 509:241–54.

Wiegert RG, Freeman BJ. 1990. Tidal salt marshes of the southeast Atlantic coast: a community profile. Georgia University, Athens, GA. *U.S. Fish and Wildlife Service Biology Report* 85(7.29):1–70.

Więski K, Guo H, Craft CB, Pennings SC. 2010. Ecosystem functions of tidal fresh, brackish, and salt marshes on the Georgia coast. *Estuaries and Coasts* 33(1):161–9.

Wilkinson, EB, Abrams K. 2015. *Benchmarking the 1999 EPAP Recommendations with Existing Fishery Ecosystem Plans. NOAA Technical Memorandum NMFS-OSF-5.* Washington, DC: NOAA (p. 22).

Witherington B, Kubilis P, Brost B, Meylan A. 2009. Decreasing annual nest counts in a globally important loggerhead sea turtle population. *Ecological Applications* 19(1):30–54.

Wood E. 2018. *Impacts of Sunscreens on Coral Reefs.* Stockholm, Sweden: ICRI, Government Offices of Sweden, Ministry of the Environment and Energy.

Xie L, Liu H, Peng M. 2008. The effect of wave–current interactions on the storm surge and inundation in Charleston Harbor during Hurricane Hugo 1989. *Ocean Modelling* 20(3):252–69.

Xue Z, Zambon J, Yao Z, Liu Y, He R. 2015. An integrated ocean circulation, wave, atmosphere, and marine ecosystem prediction system for the South Atlantic Bight and Gulf of Mexico. *Journal of Operational Oceanography* 8(1):80–91.

Yandle T, Crosson S. 2015. Whatever happened to the wreckfish fishery? An evaluation of the oldest finfish ITQ program in the United States. *Marine Resource Economics* 30(2):193–217.

Yoder JA. 1991. "Warm-temperate food chains of the southeast shelf ecosystem." In: *Food Chains, Yields, Models and Management of Large Marine Ecosystems.* Edited by K Sherman, LM Alexander, BD Gold. Boulder, CO: Westview Press Inc (pp. 49–66).

Zatcoff MS, Ball AO, Sedberry GR. 2004. Population genetic analysis of red grouper, Epinephelus morio, and scamp, Mycteroperca phenax, from the southeastern US Atlantic and Gulf of Mexico. *Marine Biology* 144(4):769–77.

Zollett EA. 2009. Bycatch of protected species and other species of concern in US east coast commercial fisheries. *Endangered Species Research* 9(1):49–59.

CHAPTER 6

The U.S. Caribbean Region

➲ The Region in Brief

- While containing the highest number of managed taxa among the eight regional U.S. marine ecosystems, including over 200 distinctly managed coral reef species, this region has been challenged by historical exploitation of its important fisheries, particularly Caribbean spiny lobster, Queen conch, and its snapper-grouper complex.
- Historical exploitation, habitat degradation, and limited shelf area have put constraints on recent fisheries landings and living marine resource (LMR)-associated revenue.
- The U.S. Caribbean is primarily driven by circulation of the Caribbean and North Equatorial currents, with additional oceanographic and productivity influences associated with Orinoco river outflow.
- This region has a relatively deep Exclusive Economic Zone (EEZ).
- Although there is limited congressional representation in this region and low absolute human population numbers, human population density is the second-highest in the nation, contributing significant pressures to coral reef and vegetated marine habitats and LMRs.
- The U.S. Caribbean has lower LMR-associated marine socioeconomic status than other U.S. regions, with the majority of its marine economy related to a billion-dollar tourism and multi-million dollar marine transportation industry.
- The U.S. Caribbean contains the sixth-highest basal productivity in the U.S., contributing toward significant coral reef system biomass.
- Within this region, average sea surface temperatures have been increased >1.5°C (second-highest among regions) since the mid-twentieth century.
- Fringing coastal and offshore regions, the mangrove-seagrass-coral reef continuum supports economically and ecologically important species throughout their life histories. However, fishing exploitation and historic habitat loss have potentially created recruitment bottlenecks for several important fisheries species.
- Compared to other regions, lower numbers of depleted marine mammal stocks and threatened or endangered species are found in the U.S. Caribbean.
- Efforts toward stock rebuilding, the implementation of seasonal and permanent closures to reef fish and Queen conch spawning aggregations, partial improvements to enforcement of fisheries legislation, improved communication and outreach with stakeholders, and the development of more comprehensive fisheries survey and management approaches have been implemented toward enhancing the U.S. Caribbean fisheries ecosystem.
- Overall, ecosystem-based fisheries management (EBFM) progress has included implementing ecosystem-level planning, advancing knowledge of ecosystem principles, and in assessing risks and vulnerabilities to ecosystems through ongoing investigations into climate vulnerability and species prioritizations for stock and habitat assessments.
- This ecosystem is excelling in marine tourism and in aspects of its governance frameworks as related to the determinants of successful LMR management.

Ecosystem-Based Fisheries Management: Progress, Importance, and Impacts in the United States. Jason S. Link and Anthony R. Marshak,
Oxford University Press. © U.S. Department of Commerce, U.S. Government 2021. DOI: 10.1093/oso/9780192843463.003.0006

6.1 Introduction

The U.S. Caribbean ecosystem brings to mind iconic fisheries species such as Caribbean spiny lobster "*langosta*" (*Panulirus argus*), Queen conch "*carrucho*" (*Aliger gigas*), Nassau grouper "*mero cherna*" (*Epinephelus striatus*), Yellowtail snapper "*colirubia*" (*Ocyurus chrysurus*), and Dolphinfish "*dorado*" (*Coryphaena hippurus*). Images of sandy beaches with nesting sea turtles, snorkeling and diving on fringing coral reefs, fish and lobster traps, bioluminescent bays, and sport fishing in open waters emerge as a background. Eating some lobster empanadillas, conch mofongo, whelks, octopus salad, fried whole snapper "*chillo entero*," or "pot fish" are popular ways to enjoy the cultural seafood of this region. Thoughts of the U.S. Caribbean ecosystem invoke the sounds of salsa, soca, reggae, and *reguetón* music, steel drum bands, skiff boat ("*yola*") motors, and tree frogs ("*coqui*"), in addition to stories relayed in Hunter S. Thompson's *The Rum Diary*, Herman Wouk's *Don't Stop the Carnival*, and the social atmosphere captured in *The House on the Lagoon* by Rosario Ferré. Many of these cultural aspects are rooted in the natural environments that comprise the U.S. Caribbean ecosystem, especially its beaches, coral reef-associated habitats, and dependent living marine resources (LMRs). These serve as the foundation for its significant tourism-based marine economy and as a major source for the livelihoods of its coastal communities.

Many studies have characterized components of the U.S. Caribbean ecosystem (Odum et al. 1959, Lugo & Snedaker 1974, Lewis 1977, Rogers 1979, Lugo 1980, McGeehe 1994, Lirman 1999, Edmunds 2000, Hernandez-Delgado et al. 2000, Christensen et al. 2003, Nemeth et al. 2003, Edmunds 2004, Kendall et al. 2004, 2005, Armstong et al. 2006, Harborne et al. 2006, Aguilar-Perera & Appeldoorn 2007, Pittman et al. 2007a, Aguilar-Perera & Appeldoorn 2008, Ballantine et al. 2008, Garcias-Sais et al. 2008, Pittman et al. 2008, Rothenberger et al. 2008, Nemeth & Appeldoorn 2009, Appeldoorn et al. 2009, Bauer & Kendall 2010, Garcia-Sais 2010, Pittman et al. 2010, Rogers et al. 2008, Sherman et al. 2010, Smith et al. 2010, Whitall et al. 2011, Bejarano & Appeldoorn 2013, Friedlander et al. 2013, Pittman et al. 2013, Bejarano et al. 2014, Appeldoorn et al. 2016, Smith et al. 2016a, Appeldoorn 2018), its important and iconic fisheries species (e.g., lobster, conch, reef fishes, and pelagic sport fishes; Mattox 1952, Feliciano 1958, Erdman 1962, Randall 1964, Olsen & LaPlace 1979, Castillo-Barahona 1981, Munro 1983, Manooch et al. 1987, Appeldoorn 1988a, 1991, Acosta & Appeldoorn 1992, Appeldoorn 1992, Sadovy & Figuerola 1992, Friedlander 1995, Rooker 1995, Matos-Caraballo 1997, Mateo et al. 1998, Matos-Caraballo 2001, 2002, Nemeth 2005, Thiele 2005, Marshak et al. 2006, Matos-Caraballo et al. 2006, Nemeth et al. 2006, Monaco et al. 2007, Ojeda-Serrano et al. 2007, Ault et al. 2008, Marshak & Appeldoorn 2008, Matos-Caraballo 2009, Schärer-Umpierre et al. 2014, Rowell et al. 2015, Baker et al. 2016), its protected species (e.g. sea turtles, cetaceans, manatees, and corals; Powell et al. 1981, Eckert et al. 1986, Collazo et al. 1992, Mignucci-Giannoni 1998, Lefebvre et al. 1999, Van Dam & Diez 1998, Cardona-Maldonado et al. 1999, Roden & Mullin 2000, Diez & Van Dam 2002, Swartz et al. 2002, Sanders et al. 2005, Van Dam et al. 2008, Schärer et al. 2008, Bruckner & Hill 2009, Miller et al. 2009, Alves-Stanley et al. 2010, Wirt et al. 2015, Rodriguez-Ferrer et al. 2019), and the sociocultural context (Valdés-Pizzini 1990, Fiske 1992, Uysal et al. 1994, Pendleton 2002, Kojis & Quinn 2006, Griffith et al. 2007, Agar et al. 2008, Tonioli & Agar 2009, Valdés-Pizzini et al. 2010, Kojis & Quinn 2011, Matos-Caraballo & Agar 2011a, b, Valdés-Pizzini & Schärer-Umpierre 2014, Kojis et al. 2017) in which humans use, enjoy, and interact with this marine ecosystem. This part of the ocean, its resources, and its goods and services have played an important part in American and Caribbean history (López 1980, Dookhan 1994, Caban 2018), as seen by its major setting for slave revolts on St. Croix and the eventual abolition of slavery by Denmark in the 1840s, its significance during the Spanish–American War, the economic importance of the Port of San Juan, and the enhancement of American-Scandinavian relations when acquiring the U.S. Virgin Islands (USVI) through the 1916 Treaty of the Danish West Indies (Hall 1992, Bastian 2001, Caban 2018, Dietz 2018). In addition, major *Arawak*, *Taino*, and *Carib* indigenous communities inhabited Puerto Rico and the USVI prior to their colonization, whose cultures still influence U.S. Caribbean fishing practices (Wilson 1993, Hoogland & Hoffman 1999, Carlson & Keegan 2004).

What follows is a brief background for this region, describing all the major considerations of this marine ecosystem. Then there is a short synopsis of how all the salient data for those factors are collected, analyzed, and used for LMR management. We then present an evaluation of key facets of ecosystem-based fisheries management (EBFM) for this region.

6.2 Background and Context

6.2.1 Habitat and Geography

The U.S. Caribbean makes up 12.3% of the Caribbean large marine ecosystem. It comprises a portion of the southeastern region of the United States Exclusive Economic Zone (EEZ). This region is composed of two key archipelago subregions (Fig. 6.1), including: **Puerto Rico** (including Culebra, Desecheo, Mona, Monito, and Vieques Islands) and the **U.S. Virgin Islands** (USVI), which consists of St. Croix and the sub-archipelagos of St. Thomas and St. John. A disputed territory, Navassa Island, is found off the west coast of Haiti, but is not specifically included in the U.S. Caribbean EEZ. Admittedly there are many Caribbean overseas territories

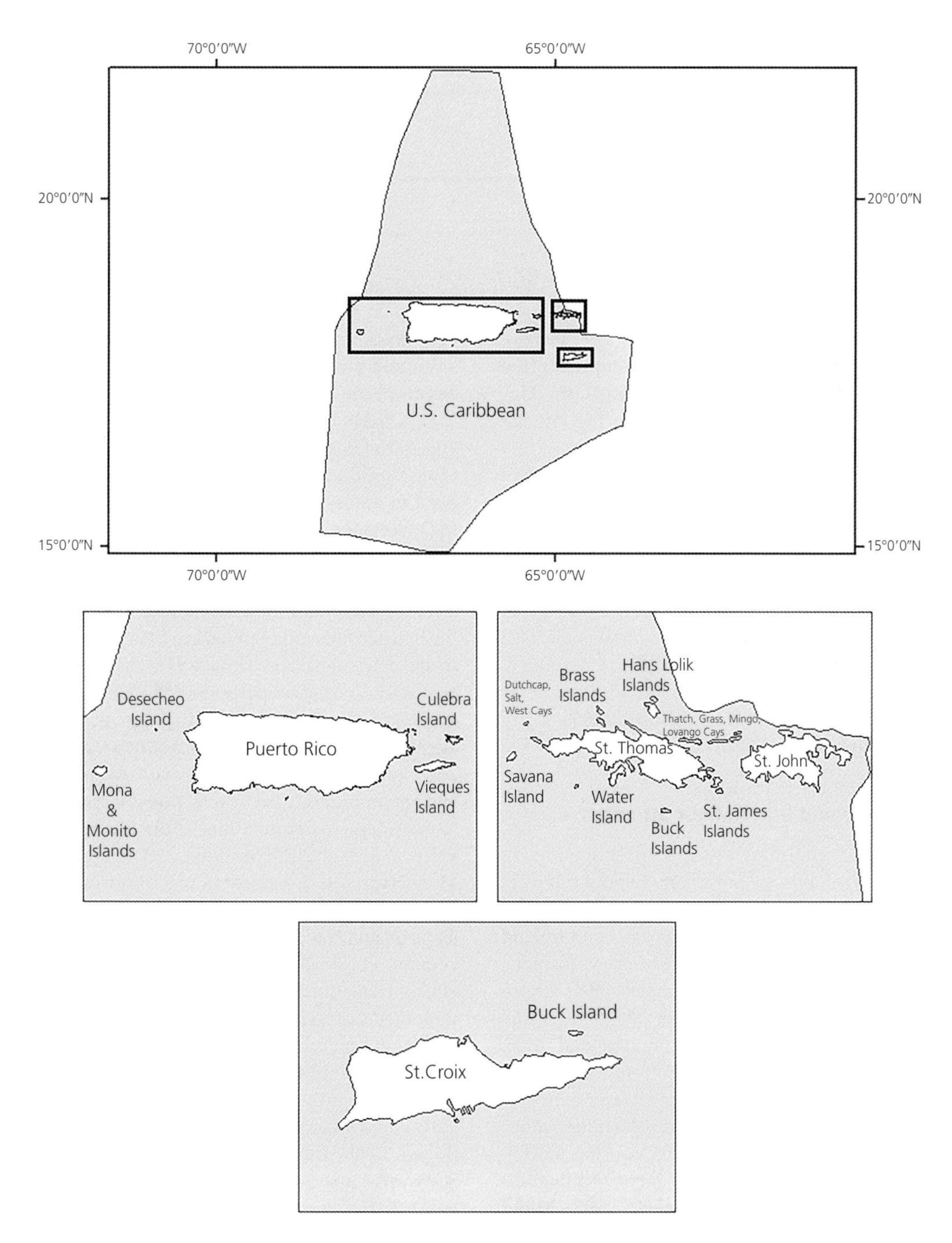

Figure 6.1 Map of the U.S. Caribbean region.

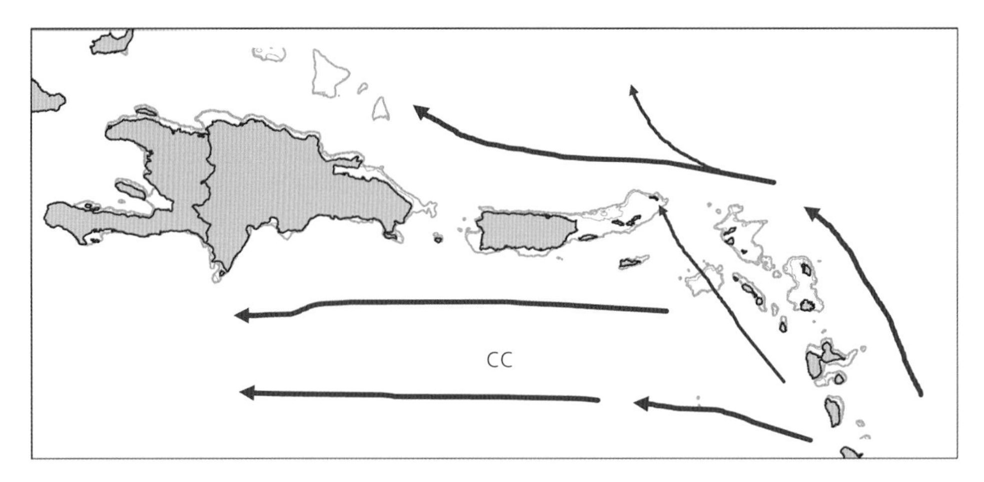

Figure 6.2 Major currents and circulation patterns throughout and encompassing the U.S. Caribbean region. CC = Caribbean Current.

(e.g., French, Netherlands Antilles, British West Indies), but these are the main archipelagos comprising the U.S. portion of its Caribbean EEZ. As such, the U.S. Caribbean has extensive connectivity to the Greater Caribbean.

The main features of the U.S. Caribbean ecosystem include the importance and connectivity of its mangrove-seagrass-coral reef continuum, the influences of outflow from the Amazon and Orinoco rivers, and the Caribbean and North Equatorial Currents (Fig. 6.2; Lessios et al. 1984, Roberts 1997, Jordan & Winter 2000, Morrell & Corredor 2001, Christensen et al. 2003, Kendall et al. 2004, Armstrong et al. 2006, Chérubin & Richardson 2007, Monaco et al. 2007, Pittman et al. 2007b, 2008, Aguilar-Perera & Appeldoorn 2008, Clark et al. 2009, Pittman et al. 2010, López et al. 2013).

6.2.2 Climate and Other Major Stressors

Compounded with fishing pressure (described next), there are above-average stressors facing the U.S. Caribbean ecosystem. These especially include regional warming, hurricanes, and coastal and upland development (Rogers et al. 1982, Yoshioka & Yoshioka 1987, Rogers 1990, Scatena & Larsen 1991, Lugo-Fernández et al. 1994, Winter et al. 1998, Lugo et al. 2000, Pandolfi et al. 2005, Sammarco et al. 2006, Donner et al. 2007, Ryan et al. 2008, Smith et al. 2008, Wilkinson & Souter 2008). Increasing water temperature has led to more frequent coral bleaching events, including a major occurrence in 2005, causing mortality to reef-building organisms and increased degradation of reef habitats (Winter et al. 1998, Miller et al. 2006, Whelan et al. 2007, Miller et al. 2009, Eakin et al. 2010, Smith et al. 2013, 2016b). Increasing prevalence and susceptibility of bleached corals to disease has also been observed with rising temperatures in the U.S. Caribbean (Calnan et al. 2008, Bruckner & Hill 2009, Cróquer & Weil 2009, Miller et al. 2009, Weil et al. 2009, Brandt et al. 2013). More recently, emerging coral mortality has also been occurring as a result of the recently introduced stony coral disease (Martin 2019). These factors continue to affect reef habitat integrity, which also has been severely affected from high mortality and limited recovery of the herbivorous long-spined sea urchin (*Diadema antillarum*) since the 1980s (Lessios et al. 1984, Ruiz-Ramos et al. 2011, Rodriguez-Barreras et al. 2018). Phenological effects on temperature-driven spawning activities of aggregating species, including groupers, Queen conch, and highly migratory species, have also been observed and are predicted to occur with potential consequences to their depleted populations (Appeldoorn et al. 2011, Erisman & Asch 2014, Muhling et al. 2015, Oxenford & Monnereau 2017). Concerns regarding prevalence of Ciguatera and its effects on local fishes and their fisheries also remain an issue (Azziz-Baumgartner et al. 2012, Loeffler et al. 2018). Although infrequent, overall effects from hurricanes are observed in this region, with major consequences to essential habitats, coastal and inland populations, fishing communities, and territorial economies having occurred in recent decades from Hurricanes Hugo (1989), Georges (1998), Irma (2017), and Maria (2017; Edmunds & Whitman 1991, Rogers 1990, Hubbard et al. 1991, Rogers et al. 1991, Scatena & Larsen 1991, Rodriguez et al. 1994, Larsen & Webb 2009, Lugo 2018, NOAA 2018a, b, Barreto-Orta et al. 2019, Cox et al. 2019).

These impacts are exacerbated by one of the highest human population densities in the U.S., having led to historical overexploitation of coral reef communities (Garrison et al. 2004, Stallings 2009, Ward-Paige et al. 2010, Pestle 2013), and increasing coastal and upland development on highly erodible lands resulting in nutrient and sediment runoff (Rogers 1990, Torres 2001, Ryan et al. 2008, Larsen & Webb 2009). Much of this developed area once housed fringing mangroves that served to stabilize shorelines, baffle terrigenous sediments, and house juveniles of many commercially and ecologically important species (Ellison & Farnsworth 1996, Rogers & Beets 2001, Martinuzzi et al. 2009). In addition to those stressors above, increased sedimentation and nutrient loading, together with higher phytoplankton concentrations and algal overgrowth, have also contributed to the loss of integrity in nearshore coral reef and seagrass habitats (Loya 1976, Acevedo et al. 1989, Rogers 1990, Rogers & Beets 2001, Warne et al. 2005, Rogers & Miller 2006, Ballantine et al. 2008, Larsen & Webb 2009, Bejarano & Appeldoorn 2013). Continued sea-level rise and increasing surge potential with storm intensity also threaten the viability of coastal communities in response to changes to climate forcing (Cambers 1997, Michener et al. 1997, Lugo 2000, Hopkinson et al. 2008, López-Marrero & Yarnal 2010, Moore et al. 2010). Acidification of U.S. Caribbean waters also threatens coral reefs and calcified organisms, including harvested shellfish, macroinvertebrates and their dependent economies (Hoegh-Guldberg et al. 2007, Gledhill et al. 2008, Kleypas & Yates 2009, Cooley et al. 2012, Melendez & Salisbury 2017). Global and local stressors have also combined to result in seven species of hard corals, which are essential fish habitat for many U.S. Caribbean-managed species, being listed as threatened under the U.S. Endangered Species Act (ESA; NMFS 2014).

6.2.3 Exploitation

Concentrated commercial, recreational, and artisanal fishing in the U.S. Caribbean, coupled with continued degradation of the coral reef ecosystem upon which fishery resources depend, has led to differential depletion of commercially important fishes (i.e., snappers, Nassau and Goliath—*Epinephelus itajara*—groupers, parrotfishes) and Queen conch in shallow coral reef habitats (Appeldoorn 1991, 1992, Sadovy & Figuerola 1992, Matos-Caraballo 1997, Garrison et al. 1998, Beets & Friedlander 1999, Sadovy & Eklund 1999, Hernandez-Delgado & Sabat 2000, CFMC 2005, Nemeth 2005, Valdés-Pizzini 2007, Ault et al. 2008, Garcia-Sais et al. 2008, Marshak & Appeldoorn, 2008, Ballantine et al. 2008, Stallings 2009, Ward-Paige et al. 2010, Baker et al. 2016, Kadison et al. 2017). For several species that have been assessed, overfished statuses have been identified with some progress toward stock rebuilding from fishing closures (Appeldoorn 1991, Appeldoorn et al. 1992, Beets & Friedlander 1992, CFMC 2005, Nemeth 2005, Ault et al. 2008, Valle-Esquivel et al. 2011, Baker et al. 2016). For example, permanent closure of the Hind Bank Marine Conservation District in the USVI has led to increases in abundance and size of targeted groupers (e.g., Red hind—*Epinephelus guttatus*) off St. Thomas (Beets & Friedlander 1992, Nemeth 2005, Nemeth et al. 2006, Kadison et al. 2010) as compared to more marginal effects of seasonal closures for these species off Puerto Rico (Marshak & Appeldoorn 2008, Tonioli & Agar 2009) or within smaller closures off St. Croix (Nemeth et al. 2006a). More recently, other stocks such as Yellowtail snapper and spiny lobster have been shown to be relatively healthy (SEDAR 2005, O'Hop et al. 2012, SEDAR 2019), while no overfishing is currently documented for the majority (n = 8/10) of those species complexes that have been assessed (NMFS 2017, 2020). Increases in the abundance of Queen conch, including mature adults, following protective management initiatives have also been observed over time (Mateo et al. 1998, Marshak et al. 2006, Appeldoorn 2008, Baker et al. 2016). While some increases in U.S. Caribbean fishery populations have been observed, progress in the recovery of other stocks, including groupers and queen conch, has continued to lag due to population effects of past growth and recruitment overfishing, loss of spawning aggregations, and potential Allee effects (Appeldoorn 1988b, Sadovy 1999, Marshak et al. 2006, Nemeth et al. 2006b, Appeldoorn 2008, Marshak & Appeldoorn 2008, Kadison et al. 2010, Baker et al. 2016, Doerr & Hill 2018). Degradation and deforestation of essential nursery mangroves and seagrasses habitats also continue to affect biological communities (Hernandez-Delagado & Sabat 2000, Rogers & Beets 2001, Christensen et al. 2003, Halpern 2004, Watson & Munro 2004, Sheaves 2005, Pittman et al. 2007a, Garcia-Sais 2008, Bejarano & Appeldoorn 2013, Doerr & Hill 2018).

Data limitations, together with historical fishing exploitation, differing perceptions by stakeholders regarding the status of local fisheries, and mixed stakeholder opinion regarding the causes of fishery population declines (e.g., pollution, habitat degradation) have led to a sometimes contentious fisheries management environment in the U.S. Caribbean (Garcia-Quijano 2009, Tonioli & Agar 2009, Matos-Caraballo & Agar 2011, Valdés-Pizzini et al. 2012).

While monitoring programs have been in effect for this region, cooperative participation in reporting fishing trips and catches, in allowing sampling of harvest, and in sharing information regarding fishing sites by the fishing community has oscillated (particularly in Puerto Rico) in response to enacted legislation that has been perceived as overly restrictive or non-transparent (Tonioli & Agar 2011, Matos-Caraballo et al. 2011, Valdés-Pizzini et al. 2012). Influence from fishing organizations and their representatives in political and regulatory fora, including Caribbean Fishery Management Council (CFMC) meetings, remains highly significant, and responses to past proposed sanctuary and marine reserve designations have been met with opposition. Given the historical importance of fishing and seafood to many coastal communities, its demand by tourists, and the cultural identity and pride associated with fishing U.S. Caribbean LMRs, sensitivity to perceived infringements on individual access to marine resources is of great concern when drafting regional management approaches. Additionally, given concentrated efforts and territoriality among fishers and sectors, there is concern among fishers that restrictions on one's yield would lead to gains for other fishers or entities, and to displacements or economic loss over time to particular individuals or the fishing community as a whole (Griffith et al. 2007, Craig 2008, Garcia-Quijano 2009, Tonioli & Agar 2009, Valdés-Pizzini et al. 2010, Matos-Caraballo et al. 2011, Valdés-Pizzini et al. 2012). The USVI Department of Planning and Natural Resources, Division of Fish and Wildlife (DFW) currently has strong relationships with the fishing communities of St. Thomas and St. John (T. Gerard pers. comm.), while some management actions remain politically sensitive in St. Croix (Agar et al. 2019). Improved enforcement of environmental laws also remains needed for the U.S. Caribbean, particularly greater federal and territorial capacity for local enforcement, improved regulation of unlicensed fisheries sales, and efforts to limit higher exploitation of LMRs.

In more recent years, however, there has been greater progress toward less strained relationships among fishers and managers, with improved communication and outreach with stakeholders, active participation by fishermen's associations in the regulatory process, and increasing compliance with reporting requirements and fishing regulations (W. Arnold, T. Gerard pers. comms.). These efforts are leading to a more knowledgeable fishing fleet. An ecosystem approach may provide benefits toward greater consideration of these socioeconomic factors, in constructing a more robust managerial framework, and in continuing to communicate the long-term utility of certain management strategies to stakeholders. An approach that accounts for the natural, societal, environmental, and ecological factors affecting fisheries populations and their dependent economies, beyond traditional single-stock or single-sector methods, allows for more effective regional management of this system. Management through the lens of a coupled socioecological system, where the marine economy is connected to and dependent on its underlying fisheries ecology and productivity will lead to a more complete approach toward U.S. Caribbean fisheries management. Through greater consideration of environmental influences (i.e., system productivity, multispecies factors, human and natural stressors), fishing practices, habitat delineations, sectoral trade-offs, regional governance frameworks, stakeholder perceptions, and the socioeconomic effects and considerations of natural and human environments, a more robust approach to management of the U.S. Caribbean marine ecosystem is emerging.

6.2.4 Invasive Species

In addition to the direct effects noted above on vegetated and coralline habitats, increasing prevalence of marine invasive species, particularly Red lionfish (*Pterois* spp.) and the invasive seagrass *Halophila stipulacea* (Olinger et al. 2017, Ruiz et al. 2017), has raised concerns regarding their effects on native species, trophic interactions, and coral reef ecosystems at large (Albins & Hixon 2008, 2013, Albins 2013). Lionfish were first reported around the U.S. Caribbean during 2008, as likely related to dispersal from populations that were introduced off Florida (Schofield 2009, Toledo-Hernandez et al. 2014). In similar systems, depletion of prey (Green et al. 2012, Layman et al. 2014), partial competitive advantages of lionfish (Albins 2013, Raymond et al. 2015, Marshak et al. 2018), and suggested limited predatory control due to historical extirpation of larger groupers have been observed (Sadovy 1999, Bruno et al. 2013, Hackerott et al. 2013, Mumby et al. 2013, Valdiva et al. 2014). Lionfish predation on a variety of reef fish families and species, including damselfishes, surgeonfishes, gobies, cardinalfishes, and parrotfishes, has been documented in southwest Puerto Rico (Harms-Tuohy et al. 2016). Although culling efforts continue, and are suggested to be effective in certain inshore locations (Côté et al. 2014, Green et al. 2014), variable habitat-dependent outcomes and high densities of year-round spawning lionfish occurring in deeper mesophotic areas have brought into question the overall effectiveness of these removals (Andradi-Brown et al. 2017a, b, Smith et al. 2017,

Harms-Tuohy et al. 2018). Given their high prey consumption of cryptic and post-settlement juvenile reef species, their effects may exacerbate those from fishing pressure and habitat degradation for certain species (Albins & Hixon 2013, Rocha et al. 2015). Overall, their consequence to commercially important species or any major ecosystem effects of their invasion have yet to be demonstrated in this region, and may be less extensive in more speciose tropical environments or at larger scales (Albins 2015, Chagaris et al. 2017, Côté & Smith 2018). However, as a lionfish fishery has developed throughout the U.S. and Greater Caribbean, and ciguatoxins have been found in lionfish, concerns remain regarding the susceptibility of consumers to *Ciguatera* fish poisoning (Robertson et al. 2014, Hardison et al. 2018).

Additionally, the continued spread of the Indo-Pacific seagrass *H. stipulacea* into eastern Puerto Rico (Ruiz et al. 2017) and the USVI has led to altered fish community composition, while also associated with reduced fish diversity in St. Thomas nearshore seagrasses (Olinger et al. 2017). Rapid proliferation in St. John bays has been observed, with densities of up to 600 shoots/m^2 occurring over a 5-year period (Willette et al. 2020). This species is highly resilient to hurricanes and other disturbances, with fast regeneration and asexual spread through fragmentation, which is leading to a transformative effect on U.S. Caribbean seagrass communities (Hernandez-Delgado et al. 2020, James et al. 2020, Willette et al. 2020). In addition, it is a less preferred species for grazers, including sea turtles, whose herbivory on native seagrasses is likely facilitating its invasion (Whitman et al. 2019). Despite these effects, investigations also suggest that *H. stipulacea* could potentially replace some services of submerged aquatic vegetation in areas where seagrass cover has been reduced or lost (Viana et al. 2019, James et al. 2020).

6.2.5 Ecosystem-Based Management (EBM) and Multisector Considerations

The U.S. Caribbean ecosystem is known not only for these iconic species, cultures, and images, but also for the trade-offs that arise across LMRs (and usage thereof), and between LMR management and other ocean uses (Johnson 2002, Burke et al. 2004, Craig 2008, Ogden 2010, Carriger et al. 2013). In the U.S. Caribbean, multiple coastal and ocean uses supporting marine economies co-occur along a mosaic of areas that provide habitat to many managed species, local fisheries, and a robust marine tourism industry (Johnson 2002, Burke et al. 2004, Craig 2008, Ogden 2010, Carriger et al. 2013). Overlapping expansive fishing effort, ongoing coastal development, recreational boating, and ecotourism throughout inshore and offshore regions where LMRs migrate and co-occur emphasizes the connectivity of this regional fisheries ecosystem and its other uses. For example, other stressors to coral reef habitats occur through marine pollution, anchor damage, boat propeller scarring in seagrasses, recreational overuse, and increasing tourism. These all compound the effects of fishing in these systems and compete with their use in a given area. Sociopolitical issues and infrastructural challenges have also impeded environmental progress and stalled larger-scale marine conservation efforts (Pittman et al. 2017). Management approaches need to account for resource competition and interconnectivity among fishing interests, coastal and tourism-related development and LMR habitat quality, and spatial efforts and practices among diving, fishing, and recreational interests. Exploring, identifying, and resolving these differences in priorities and goals is the essence of EBFM (Craig 2008, Link 2010, Link & Browman 2014, Patrick & Link 2015, Link & Marshak 2019). With diverse taxa among multiple habitats, concentrated trade-offs among tourism and fishing, historical effects of fishing pressure, and high human population pressures, the U.S. Caribbean is a priority area for EBFM execution.

In light of these multiple considerations, trade-offs, and dynamics, there is utility in employing an ecosystem-based approach to systematically incorporate abiotic, biotic, and human factors into the current U.S. Caribbean fisheries management framework. While efforts toward more comprehensive approaches, including island-wide frameworks and the development of a U.S. Caribbean Fishery Ecosystem Plan (FEP) have emerged, comprehensive fisheries management has generally remained restricted due to limited resources, inadequate enforcement of regulations, and focused interests on a few fishery management plans (FMPs). Fisheries management in the region remains challenging given the significant influence of the fishing community in lobbying for legislation, government responses that have favored expanding fishing effort into previously unexploited deep-water regions, and the underassessed nature of local fisheries populations (Valdés-Pizzini 1990, Appeldoorn et al. 2005, Appeldoorn 2008, Valdés-Pizzini et al. 2012, Friedlander et al. 2014, Newman et al. 2015, Sagarese et al. 2018). The multispecies nature of this historically exploited system, emerging fishing practices targeting multiple trophic levels, non-comprehensive monitoring programs, and only recently improved cooperation among

Lionfish, Tigers, and Bears, Oh My: Invasion of Lionfish and Food Web Effects

—(c.f. Chapters 4, 5, and 7)

"In the jungle, the coral jungle, the lionfish will bite." — Richard Appeldoorn

Red lionfish (*Pterois spp.*) have invaded the Caribbean, Gulf of Mexico, and South Atlantic. Since lionfish are not native to Atlantic waters, they have very few predators in these ecosystems. Their spines have a venom which tends to further discourage predation (and also makes for a painful event if poked into a human, one that usually requires medical attention). They feed on fish and small crustaceans, including the young of important commercial fish species such as snapper and grouper (Albins & Hixon 2008, Morris & Akins 2009, Green et al. 2012, Dahl & Patterson 2014). They are basically a not very welcome invasive species (really, only a few ever are) that are beginning to notably restructure food webs in southeast U.S. marine ecosystems.

The thing about food webs is that they are never a linear proposition, nor are they simple pairs of interacting species. A lot of indirect effects and unintended consequences happen when changes occur to food webs. The complexity of food webs, especially marine ones, is well known (Link 2002), and any dynamics thereto is akin to pushing on a bowl of jello. There's lots of wiggle, but difficulty in predicting precise, species-pairing responses. Certainly, if extant fish that are well established in an ecosystem wax and wane in abundance, there is functional redundancy (Naeem 1998, Rosenfeld 2002, Ricotta et al. 2016) to compensate. Climate change is facilitating even more introductions of species into these Atlantic ecosystems via species distribution shifts, and there are always populations waxing and waning there as well. But when one throws in an invasive species that other organisms have not had much exposure to, and hence an inability to adapt to, it is not impossible for entire food webs to be restructured. Inserting a new species into an ecosystem that the other species have not had any familiarity with can be catastrophic. For example, in several places new invasive predators have reduced ecosystem biomass and functionally, if not entirely wiped out commercially important stocks (Zaitsev 1992, Walton et al. 2002, Pinnegar et al. 2014, Doherty et al. 2015, 2016).

Whether it is happening to that degree in the Atlantic due to lionfish remains to be seen, but the experiment has begun. Recent observations suggest that some of the worst-case scenarios have come to pass in some locations, and may be less founded than initially predicted in others (Côté & Smith 2018). Density-dependent effects and the emergence of ulcerative disease appear to exert some controls on non-native lionfish populations (Benkwitt 2013, Dahl et al. 2019, Harris et al. 2020). But, the ecosystem effects of lionfish continue to be documented, with greater effects currently observed in the northern limits of their invaded range (Côté & Smith 2018). Calls to mitigate or remove the invasive species always follow not the invasion or detection thereof, but recognition of the consequences of an invasion on endemic species that result in economic impacts. Once an invasive is established, there's not much one can do to eradicate it. Can we establish a lionfish fishery to lower the abundance and impact of this species? Sure. Will it be effective? Perhaps, but not likely at the scale and scope of the invasion (Green et al. 2014, Andradi-Brown et al. 2017, Harms-Tuhoy et al. 2018).

The best one can aim for in such invasive situations is to foster a robust, healthy food web that can absorb such perturbations and still maintain core structure and functioning. The best way to do that is to manage a system of living marine resources as a system, not as disconnected parts or pairs of species.

References

Albins MA, Hixon MA. 2008. Invasive Indo-Pacific lionfish Pterois volitans reduce recruitment of Atlantic coral-reef fishes. *Marine Ecology Progress Series* 367:233–8.

Andradi-Brown DA, Vermeij MJ, Slattery M, Lesser M, Bejarano I, Appeldoorn R, Goodbody-Gringley G, et al. 2017. Large-scale invasion of western Atlantic mesophotic reefs by lionfish potentially undermines culling-based management. *Biological Invasions* 19(3):939–54.

Benkwitt CE. 2013. Density-dependent growth in invasive lionfish (Pterois volitans). *PLoS One* 8(6).

Côté IM, Smith NS. 2018. The lionfish Pterois sp. invasion: has the worst-case scenario come to pass? *Journal of Fish Biology* 92(3):660–89.

Dahl KA, Edwards MA, Patterson III WF. 2019. Density-dependent condition and growth of invasive lionfish in the northern Gulf of Mexico. *Marine Ecology Progress Series* 623:145–59.

Dahl KA, Patterson III WF. 2014. Habitat-specific density and diet of rapidly expanding invasive red lionfish, Pterois volitans, populations in the northern Gulf of Mexico. *PloS One* 9(8).
Doherty TS, Dickman CR, Nimmo DG, Ritchie EG. 2015. Multiple threats, or multiplying the threats? Interactions between invasive predators and other ecological disturbances. *Biological Conservation* 190:60–8.
Doherty TS, Glen AS, Nimmo DG, Ritchie EG, Dickman CR. 2016. Invasive predators and global biodiversity loss. *Proceedings of the National Academy of Sciences* 113(40):11261–5.
Green SJ, Dulvy NK, Brooks AM, Akins JL, Cooper AB, Miller S, Côté IM. 2014. Linking removal targets to the ecological effects of invaders: a predictive model and field test. *Ecological Applications* 24(6): 1311–22.
Green SJ, Akins JL, Maljković A, Côté IM. 2012. Invasive lionfish drive Atlantic coral reef fish declines. *PLoS One* 7(3).
Harms-Tuohy CA, Appeldoorn RS, Craig MT. 2018. The effectiveness of small-scale lionfish removals as a management strategy: effort, impacts and the response of native prey and piscivores. *Management of Biological Invasions* 9(2):149.
Harris HE, Fogg AQ, Allen MS, Ahrens RN, Patterson WF. 2020. Precipitous declines in northern Gulf of Mexico invasive Lionfish populations following the emergence of an ulcerative skin disease. *Scientific Reports* 10(1):1–7.
Link J. 2002. Does food web theory work for marine ecosystems? *Marine Ecology Progress Series* 230:1–9.
Morris JA, Akins JL. 2009. Feeding ecology of invasive lionfish (Pterois volitans) in the Bahamian archipelago. *Environmental Biology of Fishes* 86(3):389.
Naeem S. 1998. Species redundancy and ecosystem reliability. *Conservation Biology* 12(1):39–45.
Pinnegar JK, Tomczak MT, Link JS. 2014. How to determine the likely indirect food-web consequences of a newly introduced non-native species: a worked example. *Ecological Modelling* 272:379–87.
Ricotta C, de Bello F, Moretti M, Caccianiga M, Cerabolini BE, Pavoine S. 2016. Measuring the functional redundancy of biological communities: a quantitative guide. *Methods in Ecology and Evolution* 7(11):1386–95.
Rosenfeld JS. 2002. Functional redundancy in ecology and conservation. *Oikos* 98(1):156–62.
Walton WC, MacKinnon C, Rodriguez LF, Proctor C, Ruiz GM. 2002. Effect of an invasive crab upon a marine fishery: green crab, Carcinus maenas, predation upon a venerid clam, Katelysia scalarina, in Tasmania (Australia). *Journal of Experimental Marine Biology and Ecology* 272(2):171–89.
Zaitsev YP. 1992. Recent changes in the trophic structure of the Black Sea. *Fisheries Oceanography* 1(2):180–9.

the fishing community and management entities have created unique management challenges for the region (Garrison et al. 1998, Salas et al. 2007, Appeldoorn 2008, Marshak et al. 2008, Appeldoorn et al. 2009, Garcia-Quijano 2009, Tonioli & Agar 2009, Matos-Caraballo & Agar 2011).

6.3 Informational and Analytical Considerations for this Region

6.3.1 Observation Systems and Data Sources

Although the U.S. Caribbean is considered data-limited overall, a regional observation system and several monitoring programs are available to inform aspects of the physical, biological, and socioeconomic components of this ecosystem. This information helps to facilitate EBM for the U.S. Caribbean. The Caribbean Coastal Ocean Observing System (CARICOOS) provides oceanographic and atmospheric information from a network of data buoys, coastal meteorological stations, vessels, instruments, and radars throughout Puerto Rico and the USVI (CARICOOS 2016). The system is composed of aggregated information from several ocean data buoys, including a directional wave buoy and ocean acidification monitoring station off the west coast of Puerto Rico. In addition, 5 high-frequency radar stations, 16 coastal weather stations, drifters, and seagliders provide oceanographic and atmospheric data over the U.S. Caribbean coast and EEZ. Thermohaline information, and data for winds, waves, tides, and currents are readily available through this platform, and have been applied toward reporting and forecasting on water quality, storm surge, beach quality, population connectivity patterns, and other coastal hazards (CARICOOS 2016). These efforts are undertaken together with support from NOAA's National Weather Service, the University of Puerto Rico (UPR), the CFMC, the Puerto Rico Department of Natural and Environmental Resources (PR-DNER), and the Caribbean Regional Association. This information has also been complemented by the development of enhanced wave forecast systems for the U.S. Caribbean (Anselmi-Molina et al. 2012).

Information on the status and trends of coral reef and pelagic ecosystems and their dependent fishery species has been collected through monitoring efforts. Since 2000, comprehensive assessments of U.S. Caribbean coral reef habitats and their associated flora and fauna have been carried out by the NOAA National Centers for Coastal Ocean Science (NCCOS; e.g., Kendall et al. 2001, Christensen et al. 2003, Kendall et al. 2004, 2005, Monaco et al. 2007, Pittman et al. 2007a, b, Garcia-Sais et al. 2008, Pittman et al. 2008, Rothenberger et al. 2008, Clark et al. 2009, Monaco et al. 2009, Bauer & Kendall 2010, Pittman et al. 2010, Whitall et al. 2011, Clark et al. 2012, Friedlander et al. 2013, Pittman et al. 2013, 2017). These investigations have provided information regarding large-scale benthic habitat characterizations, biogeographic and spatial ecology of associated species, and have been applied toward greater understanding of U.S. Caribbean coral reef ecosystems. In 2013, ongoing NCCOS monitoring in the region became part of NOAA's National Coral Reef Monitoring Program (NCRMP) as the primary method to monitor status and trends of biological, climatic, and socioeconomic indicators for U.S. coral reef ecosystems, sampling biennially across the insular shelves of Puerto Rico, Vieques, Culebra, St. Croix, St. Thomas, and St. John (CRCP 2014). Complementary efforts have also included the NOAA-funded and UPR executed Coral Reef Ecosystem and Deep Coral Reef Ecosystem Studies (CRES, Deep CRES), and investigations supported by the Caribbean Coral Reef Institute (CCRI; e.g., Ault et al. 2008, Ballantine et al. 2008, Garcia-Sais et al. 2008, Appeldoorn et al. 2009, Cróquer & Weil 2009, Guénette & Hill 2009, Nemeth & Appeldoorn 2009, Hinderstein et al. 2010, Sherman et al. 2010, Bejarano et al. 2014, Valdés-Pizzini & Schärer-Umpierre 2014, Appeldoorn et al. 2016). The Comprehensive U.S. Caribbean Coral Reef Monitoring Program (C-CCREMP) has worked toward amassing information and addressing data gaps for Puerto Rico and USVI coral reefs and for improved monitoring of threats that include land-based sources of pollution and climate change (CRCP 2007, Pittman et al. 2012). Furthermore, the Territorial Coral Reef Monitoring Program (TCRMP) has been funded by the NOAA Coral Reef Conservation Program (CRCP) and implemented by the University of the Virgin Islands (UVI) since 2001. This program monitors coral reef status and benthic structure, including more recently in mesophotic reefs (Smith et al. 2015). These programs have also been complemented by efforts of the NOAA Integrated Coral Observing Network, which includes an offshore coral monitoring station in the La Parguera Natural Reserve (PR) to inform about coral bleaching and acidification potential. Investigations into improving the efficacy of coral restoration programs, developing enhanced watershed management strategies, and providing data on the effects of recreational and fishing pressure are being carried out through NOAA and PR-DNER-supported programs within the Northeast Marine Corridor and Culebra Island Habitat Focus Area (HFA; NOAA 2016). Application of this information toward developing emerging integrated ecosystem assessments (IEAs; Levin et al. 2009, 2013) for the U.S. Caribbean is also underway (NMFS 2019).

These coral reef-focused studies and programs enhance ongoing fishery-independent and dependent monitoring efforts led by PR-DNER, USVI DFW, and NOAA Fisheries (Matos-Caraballo 1997, 2001, 2002, Valle-Esquivel & Diaz 2003, Matos-Caraballo 2009, NMFS 2018a, b). Information regarding abundance, size, and reproductive status of Puerto Rico and USVI reef fishes has been collected through the Caribbean Southeast Area Monitoring and Assessment Program (SEAMAP-C) since 1988 (Cummings et al. 2007, Marshak & Appeldoorn 2008). This program has conducted stock abundance surveys for Queen conch along the eastern and western Puerto Rico shelves and investigated recruitment indices for Caribbean spiny lobster (Cummings et al. 2007, Marshak & Appeldoorn 2008). Fishery-independent investigations into grouper aggregations have been carried out by UPR, UVI, PR-DNER, Puerto Rico Sea Grant, CFMC (Garcia-Sais et al. 2012), and CCRI, with trends documented over time and following management efforts to limit overfishing on remaining aggregations (Sadovy & Figuerola 1992, Shapiro et al. 1993, Sadovy et al. 1994, Beets & Friedlander 1999, Nemeth 2005, Nemeth et al. 2006a, b, Nemeth et al. 2007, Ojeda-Serrano et al. 2007, Marshak & Appeldoorn 2008, Kadison et al. 2010, Schärer et al. 2012, 2014, Rowell et al. 2015). These types of fishery-independent, mainly visual census-based surveys have also been expanded in an effort to improve assessments of fishery and general LMRs following guidance from the Southeast Fisheries Science Center (SEFSC) regarding appropriate sample size (Cass-Calay et al. 2016, W. Arnold pers. comm.). Assessments of landed fishery species, fishing effort, and regular biological port samplings of commercial (Puerto Rico, USVI) and recreational (Puerto Rico) catches have been carried out by PR-DNER and USVI DFW throughout municipalities. In the USVI, voluntary reporting of recreationally landed fish began in 2017, with 97% of the reports submitted by anglers on St. Thomas and St. John (T. Gerard pers. comm.). USVI DFW has also begun collecting this information from charter fishing vessels, private-use recreational anglers, and divers, which remain unpublished (T. Gerard pers. comm.). The U.S. Fish and Wildlife Service (FWS) also supports the PR-DNER Fisheries Laboratory in their collection of age, growth, and reproduction information. These efforts are also incorporated into NOAA Fisheries fishery-dependent monitoring programs for commercial (Puerto Rico and USVI; Matos-Caraballo 1997, 2001, 2002, Valle-Esquivel & Diaz 2003, Matos-Caraballo 2009, NMFS 2018a, b) and recreational harvest (Puerto Rico; NMFS 2018b). Other examinations of USVI recreational fisheries have been carried out by NOAA NCCOS (Goedeke et al. 2016) and Kojis & Tobias (2016). Although limited, socioeconomic information for fisheries and other ocean uses are collected by NOAA Fisheries, Puerto Rico Sea Grant, PR-DNER, USVI DFW (Valdés-Pizzini 1990, 2007, Tonioli & Agar 2009, Valdés-Pizzini et al. 2010, Kojis & Quinn 2011, Matos-Caraballo & Agar 2011a, b, Valdés-Pizzini et al. 2012, Brander & van Beukering 2013, Valdés-Pizzini & Schärer-Umpierre 2014, Crosson & Hibbert 2017, Fleming et al. 2017, Kojis et al. 2017), and the National Ocean Economics Watch (ENOW) Program (NOAA 2020). The SEFSC is supporting a commercial data validation effort that will also overlap with recreational data sources, given the integrated nature of these fisheries (W. Arnold pers. comm.). Monitoring programs by NOAA, the Department of Interior, and territorial agencies also exist for protected species, including sea turtles, cetaceans, manatees, and marine birds (Swartz et al. 2002, FWS 2007, Bjorkland 2011, Waring et al. 2013, FWS 2017).

6.3.2 Models and Assessments

While limited in number, assessments of economically important and protected species in the U.S. Caribbean are conducted by NOAA Fisheries in collaboration with the CFMC, FWS, PR-DNER, and USVI DFW (FWS 2007, Ferguson et al. 2017, FWS 2017, Lynch et al. 2018, Collazo et al. 2019, Srinivasan et al. 2019). Stock assessment models to estimate stock status are dependent on data collected through academic, state, and federal monitoring programs regarding landings, catch per unit effort, life history characteristics, composition (size or age), and stock abundance. Assessments are conducted through the Southeast Data, Assessment, and Review Process (SEDAR) with input from the CFMC Scientific and Statistical Committee (Lynch et al. 2018, Karp et al. 2019). However, this region is very data-limited with only a small fraction of its stocks (i.e., species or family) currently assessed (Appeldoorn et al. 1992, 2005, Appeldoorn 2008, Ault et al. 2008, Friedlander et al. 2014, Newman et al. 2015, Sagarese et al. 2018) and potentially subject to current or historical overfished status (Baker & Appeldoorn 2016, Cummings et al. 2016, SEDAR 2016, Sagarese et al. 2018, Harford et al. 2019). Efforts to enhance assessment potential and create frameworks for data-limited species, including length-based assessments from surveys (Ault et al. 2008, SEDAR 2016), are ongoing (Cummings et al. 2016, SEDAR 2016, Sagarese et al. 2018, Harford et al. 2019), with recent applications toward estimating overfishing limits (Sagarese et al. 2015). Enhanced monitoring, outreach, refined stock evaluation frameworks, and the incorporation of socioeconomic information are improving data accuracies and assessment potential in the U.S. and broader Caribbean (Ault et al. 2008, Narozanski et al. 2013, Cummings et al. 2015, Gill et al. 2017). Recently, the first fully successful stock assessment for this region was undertaken (for Caribbean spiny lobster), which will allow for its assessment among U.S. Caribbean islands as related to a newly established acceptable biological catch rule (SEDAR 2019). These continued and collective efforts strengthen successful LMR management capacity to allow for broader, more systematic approaches at the ecosystem level (Appeldoorn 2008, Ban et al. 2009).

While NOAA Fisheries examines the socioeconomics of U.S. commercial and recreational fishing communities, these assessments generally have not included the U.S. Caribbean (NMFS 2018a). Some assessments of the Puerto Rican and USVI fishing communities have been carried out by PR-DNER, USVI DFW, Puerto Rico Sea Grant, the CFMC, the NOAA SEFSC, and academic researchers (Valdés-Pizzini 1990, 2007, Tonioli & Agar 2009, Valdés-Pizzini et al. 2010, 2012, Valdés-Pizzini & Schärer-Umpierre 2014, NOAA 2020). However, this information has mostly remained limited. Additionally, data from PR-DNER and NOAA Fisheries marine mammal stock abundance surveys are used to conduct population assessments as published in regional marine mammal stock assessment reports and peer-reviewed by regional Scientific Review Groups (SRGs; Carretta et al. 2019, Hayes et al. 2019, Muto et al. 2019). The FWS prepares stock assessment reports for U.S. Caribbean manatees under its jurisdiction, while NOAA, territorial agencies, and academic institutes and organizations also monitor listed coral species. Information collected on U.S. Caribbean protected species is also applied toward 5-year status reviews of those listed under the ESA.

Despite data limitations, partial biological assessments and ecosystem modeling efforts have been undertaken for the U.S. Caribbean ecosystem. Several investigations using Ecopath with Ecosim (EwE) have examined trophic interactions in coral reefs off Puerto Rico and the USVI (Opitz 1993, 1996, Searles 2004, Townsend et al. 2008, Smikle et al. 2010), including within the southwest Puerto Rico (La Parguera) coral reef ecosystem (Guénette & Hill 2009). Investigations with EwE modeling have also provided information regarding the ecological effects of recovering green sea turtles (*Chelonia mydas*) on Puerto Rico and USVI seagrasses (Wabnitz 2010). Investigations into the shallow water seascape of the U.S. Caribbean has led to the development of predictive models of species diversity, distribution, and abundance in southwest Puerto Rico and USVI seascapes (Pittman et al. 2007a, b, 2009, Pittman & Brown, 2011). Complementary examination of the La Parguera socioecological system has been carried out using a Drivers-Pressures-State-Impacts-Response (DPSIR) model to describe the ecosystem and incorporate human factors that affect ecological processes (Valdés-Pizzini & Schärer-Umpierre 2014). These efforts are allowing for the development of more refined approaches to ecosystem modeling, assessment, and investigation into U.S. Caribbean socioecological systems.

6.4 Evaluation of Major Facets of EBFM for this Region

To elucidate how well the U.S. Caribbean ecosystem is doing in terms of LMR management and progress toward EBFM, here we characterize the status and trends of socioeconomic, governance, environmental, and ecological criteria pertinent toward employing an EBFM framework for the U.S. Caribbean, and examine their inter-relationships as applied toward its successful implementation. We also quantify those factors that emerge as key considerations under an ecosystem-based approach for the region, with evaluation of trends in Puerto Rico and the USVI when available. Ecosystem indicators for the U.S. Caribbean related to the (1) human environment and socioeconomic status, (2) governance system and LMR status, (3) natural environmental forcing and features, and (4) systems ecology, exploitation, and major fisheries of this system are presented and synthesized next. We explore all these facets to evaluate the strengths and weaknesses along the pathway:

$$PP \rightarrow B_{\text{targeted, protected spp, ecosystem}} \leftrightarrow L_{\uparrow\text{targeted spp, }\downarrow\text{bycatch}} \rightarrow \text{jobs, economic revenue.}$$

Where PP is primary production, B is biomass of either targeted or protected species or the entire ecosystem, L is landings of targeted or bycaught species, all leading to the other socioeconomic factors. This operates in the context of an ecological and human system, with governance feedbacks at several of the steps (i.e., between biomass and landings, jobs, and economic revenue), implying that fundamental ecosystem features can determine the socioeconomic value of fisheries in a region, as modulated by human interventions.

6.4.1 Socioeconomic Criteria

6.4.1.1 Social and Regional Demographics

In the U.S. Caribbean, human population values and some of the highest population densities (Fig. 6.3) have persisted from 1970 to present, with values currently around 2.6 million residents and ~221 people km^{-2}, respectively, in coastal municipalities. Although these coastal regions only make up 0.9% of the United States total population (2010 values), the U.S. Caribbean is the second-highest in coastal population density nationally, with density values increasing at a comparatively slower rate of ~1.2 people km^{-2} year^{-1}. Most recently (year 2010), total population has been estimated at ~3.7 million in Puerto Rico and ~106,000 in the USVI. As of 2010, human population is especially concentrated in the coastal regions of Puerto Rico (~252 people km^{-2}), with the majority of the population

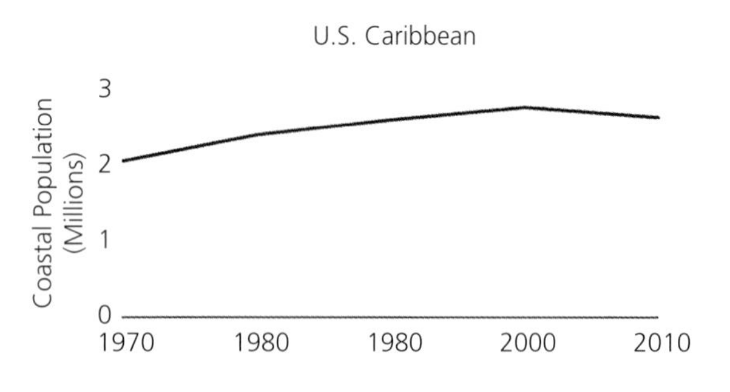

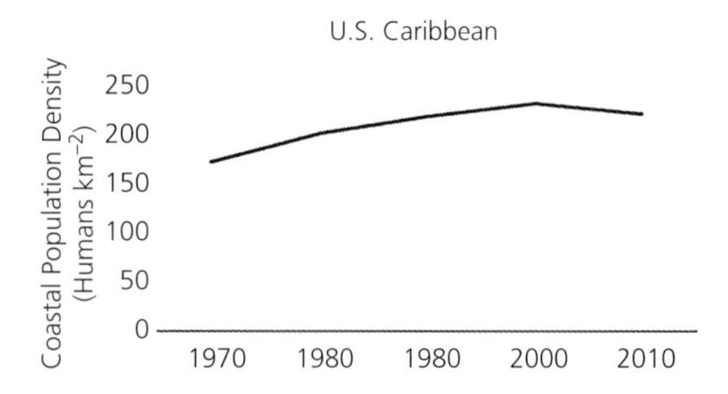

Figure 6.3 Decadal trends (1970–2010) in coastal human population (top) and population density (bottom; humans km^{-2}) for the U.S. Caribbean region.

Data derived from U.S. censuses and taken from https://coast.noaa.gov/digitalcoast/data/demographictrends.html.

found within the northeast (~780,000 residents in San Juan, Bayamon, and Carolina) and south-central (~166,000 residents in Ponce) municipalities. Much lower population values and densities (56 people km^{-2}) are observed in the USVI, with most residents found in St. Thomas (~52,000) and St. Croix (~51,000), and significantly less living in St. John (~4000). Over the past 40 years, population values have increased for St. Thomas and St. Croix by ~450–550 individuals per $year^{-1}$, and for Puerto Rico by ~13,000 individuals per $year^{-1}$, while coastal population densities for Puerto Rico and the USVI have increased by 1.34 and 0.57 people km^{-2} per $year^{-1}$, respectively. Following Hurricanes Irma and Maria in 2017, >100,000 residents also moved from the U.S. Caribbean to the continental U.S (Melendez & Vargas-Ramos 2017), potentially affecting the accuracy of these population estimations.

6.4.1.2 Socioeconomic Status and Regional Fisheries Economics

Limited marine socioeconomic data are available for the U.S. Caribbean, which are detailed in work by Clements et al. (2016) that examined the contributions of LMR-associated establishments, employments, and gross domestic product (GDP) to the U.S. Caribbean total ocean economy during years 2012 and 2014 (Table 6.1). Overall, LMRs in the U.S. Caribbean contribute 1–3% of total ocean establishments and employments, and only 0.2% of total ocean GDP, making it one of the lowest national contributors toward the total U.S. ocean economy. Most of the LMR-associated economy is concentrated in Puerto Rico, with 1000+ commercial fishermen, 32+ seafood markets, 15 seafood wholesale businesses, 44 fishing cooperatives, and ~3000 employees associated with these LMR establishments (Clements et al. 2016). Comparatively, the USVI LMR-associated economy is composed of 225–250+ commercial fishermen, 9+ identified seafood markets (3 government run, the remainder informal operations), and 1 aquaculture operation (Clements et al. 2016, Kojis et al. 2017).

Irregular reporting of permitted fishing vessels has occurred since 1990, with ~3000–4000 vessels reported per year during the 1990s, and discontinued reporting to NOAA Fisheries as of 1998 (Fig. 6.4). Although data have been reported to PR-DNER and USVI Division of Fish and Wildlife, ongoing challenges in collecting accurate information regarding regional fishing effort and vessel activity have been prominent throughout Puerto Rico, as influenced by tenuous relationships among local government representatives, fishery managers, and the fishing community (Matos-Caraballo & Agar 2011). Given notable declines in fisheries landings throughout the region during the 1980s–1990s (Matos-Caraballo 1997, 2001, 2002, 2009), and observed declines for fishing vessels in other regions, it is likely that reported U.S. vessel numbers represent a lower subset of past numbers and that current values are lower than those reported here (Matos-Caraballo & Agar 2011). In the USVI, however, reports of commercial fishing activity have been more consistent regarding fishing trips, effort, and vessel registration numbers (T. Gerard, pers. comm.). Total revenue (year 2017 USD) of landed commercial fishery catches (Fig. 6.5) has fluctuated and increased over time, peaking at

Table 6.1 The number of living marine resources establishments and employments, gross domestic product value (GDP; in USD) from living marine resources revenue, and their percent contribution to total multisector oceanic economy values in the U.S. Caribbean (years 2012, 2014)

Years 2012, 2014 U.S. Caribbean	
Establishments	124
Employments	4454
GDP	$11,305,331
%Total Establishments	1.5
%Total Employment	2.7
%Total GDP	0.2

Data derived from the National Ocean Economics Program and Clements et al. 2016.

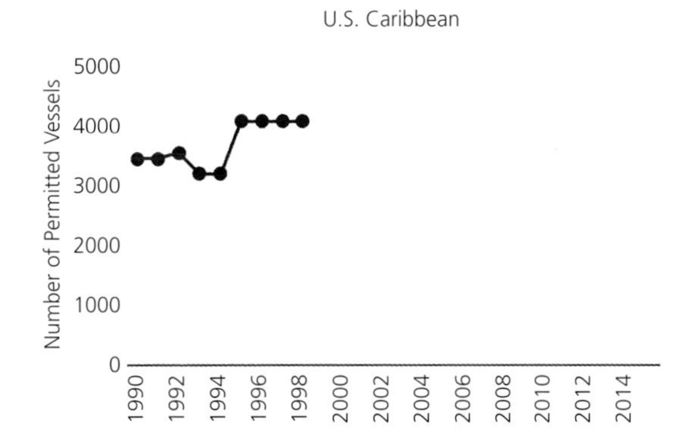

Figure 6.4 Total number of permitted vessels for the U.S. Caribbean region over time (years 1990–2015).

Data derived from NOAA National Marine Fisheries Service Council Reports to Congress. No reporting of U.S. Caribbean vessels to this source has continued post-1998.

$15 million in 2003 during the past 20 years, but with values remaining as the lowest nationally among regions. From 2003 to 2014, the majority (65.5% ± 2.1, SE) of reported fisheries revenue for the U.S. Caribbean was associated with higher valued reported landings occurring in the USVI (NMFS 2018a, b). Since 2014, Puerto Rico again contributes the majority of fisheries revenue for the region (56.4% ± 2.8, SE; NMFS 2018).

While ratios of commercial fisheries revenue to total primary production (metric tons wet weight y^{-1}; Fig. 6.6) increased in the U.S. Caribbean during the early 2000s and have remained relatively stable in years since, their values continue to be the lowest on average nationally. Data for total LMR employments to total primary production are only available for 2012 (~0.3 employments/million metric tons wet weight $year^{-1}$), which are comparatively lowest nationally among all U.S. regions.

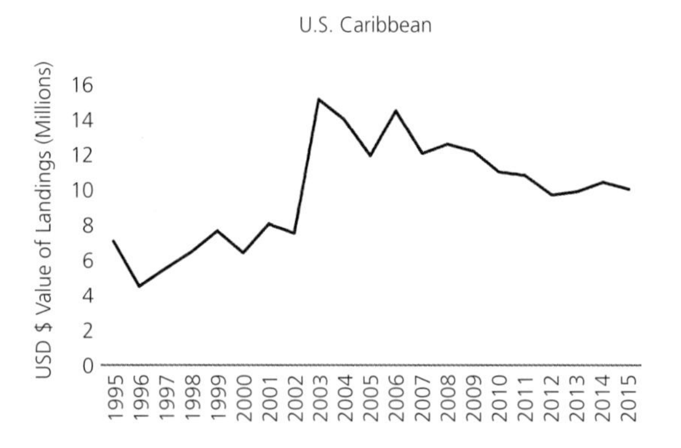

Figure 6.5 Total revenue (year 2017 USD) of landed commercial fishery catches for the U.S. Caribbean region (years 1995–2015) over time.

Data derived from NOAA National Marine Fisheries Service commercial fisheries statistics.

6.4.2 Governance Criteria

6.4.2.1 Human Representative Context

Nineteen federal and state agencies and organizations operate within the U.S. Caribbean and are directly responsible for fisheries and environmental management (Table 6.2). Fisheries resources throughout territorial waters (0–9 nautical miles Puerto Rico; 0–3 nautical miles USVI) are managed by individual U.S. territorial agencies, while those in federal waters are managed by both NOAA Fisheries (Southeast Region) and the CFMC (PR-DNER 2010, DPNR-DFW 2016). Puerto Rico territorial waters fall under the jurisdiction of PR-DNER, while USVI waters are under the jurisdictions of the USVI DFW.

Nationwide, the U.S. Caribbean contains low numbers of congressional representatives, with two non-voting representatives in the U.S. House of Representatives (i.e., Resident Commissioner of Puerto Rico and USVI Delegate), which are concentrated within two territories (Fig. 6.7). There is no congressional representation for Navassa Island. When standardized per mile of shoreline, representation is comparatively similar to that in other southeast regional locations (i.e., northern Gulf of Mexico and South Atlantic). Over time, composition of the CFMC has been generally composed of 1–4 members from the commercial fishing sector, 1–2 members from the recreational fishing sector, and 1–2 other representatives, and has remained relatively constant (Fig. 6.8a). While containing seven total members, including one federal and two territorial government representatives, the CFMC contains the lowest total number nationally of appointed council members from fishing or other sectors at only four. In addition, regional representatives from the U.S. Caribbean also serve on the Atlantic Highly Migratory Species Advisory Panel. The Atlantic marine mammal

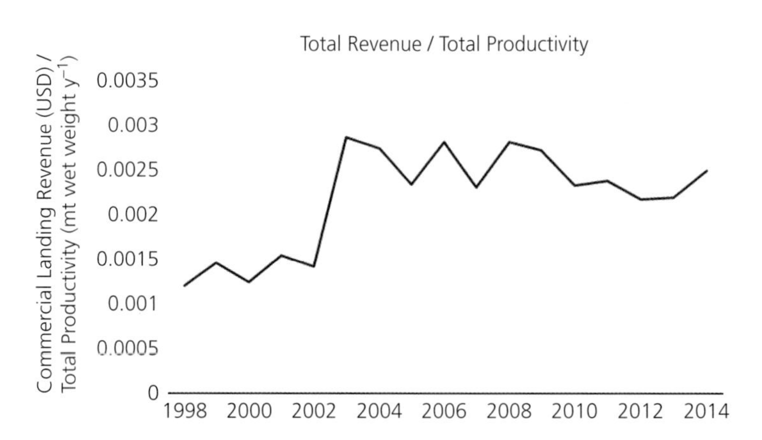

Figure 6.6 Ratios of total commercial landings revenue (USD) to total productivity (metric tons wet weight $year^{-1}$) for the U.S. Caribbean region.

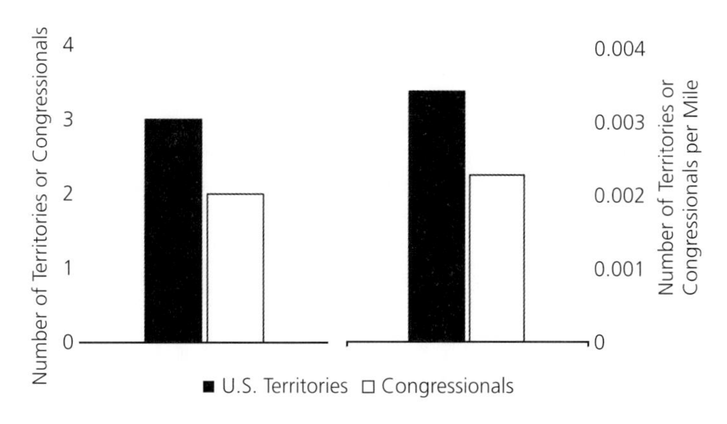

Figure 6.7 Number of territories and U.S. Congressional representatives (primary axis, left) and as standardized by miles of shoreline (secondary axis, right) for the U.S. Caribbean region.

Table 6.2 Federal and state agencies and organizations responsible for fisheries and environmental consulting and/or management in the U.S. Caribbean region

Region	Federal Agencies with Marine Resource Interests	Council	State Agencies and Cooperatives	Multijurisdictional
U.S. Caribbean	U.S. Department of Commerce **DoC** (NOAA Fisheries Southeast Fisheries Science Center **SEFSC**, Southeast Regional Office **SERO**); U.S. Department of Interior **DoI** (U.S. Fish and Wildlife Service **FWS**, U.S. Bureau of Ocean Energy Management **BOEM**, U.S. National Park Service **NPS**, U.S. Geological Survey **USGS**); U.S. Environmental Protection Agency **EPA**	Caribbean Fishery Management Council **CFMC**	Puerto Rico Department of Natural and Environmental Resources **PR-DNER**; U.S. Virgin Islands Department of Planning and Natural Resources, Division of Fish and Wildlife **USVI DFW**	Allied Marine Laboratories of the Caribbean **AMLC**; Caribbean Coral Reef Institute **CCRI**; Coral Reef Task Force **CRTF**; Gulf and Caribbean Fisheries Institute **GCFI**; Western Central Atlantic Fisheries Commission **WECAFC**; International Commission for the Conservation of Atlantic Tunas **ICCAT**; International Society for Reef Studies **ISRS**

SRG is responsible for advising federal agencies on the status of marine mammal stocks covered under the Marine Mammal Protection Act (MMPA), for regions including the U.S. Caribbean. Its composition and membership have remained steady over time (Fig. 6.8b), with individuals from the academic and private institutional sectors making up the majority of its representatives.

6.4.2.2 Fishery and Systematic Context

Currently, four U.S. Caribbean-specific FMPs are managed by NOAA Fisheries and the CFMC for Caribbean spiny lobster, reef fishes, coral reef resources, and Queen conch (Table 6.3). In total, these FMPs have been subject to 32 modifications (i.e., amendments and framework/regulatory amendments). Broader approaches to fisheries management are underway with an FEP being developed for the U.S. Caribbean region, in addition to the enactment of three recently approved island-specific FMPs (CFMC 2019a, b, c, NMFS 2019). When implemented, these island-based FMPs will replace the four U.S. Caribbean-wide FMPs that are currently in place.

One marine national monument (Buck Island), three coastal national historic sites, one NOAA HFA (Northeast Marine Corridor and Culebra Island), and one National Estuarine Research Reserve (NERR; Jobos Bay) are found throughout the U.S. Caribbean region (Table 6.4). Thirty-nine named coastal and offshore areas are identified as habitat areas of particular concern (HAPCs) in the U.S. Caribbean, with 23 occurring off Puerto Rico, 12 off St. Croix, and 4 off St. Thomas. Several of these locations also coincide with seasonally or permanently protected reef fish spawning aggregation sites, particularly for commercially important Red

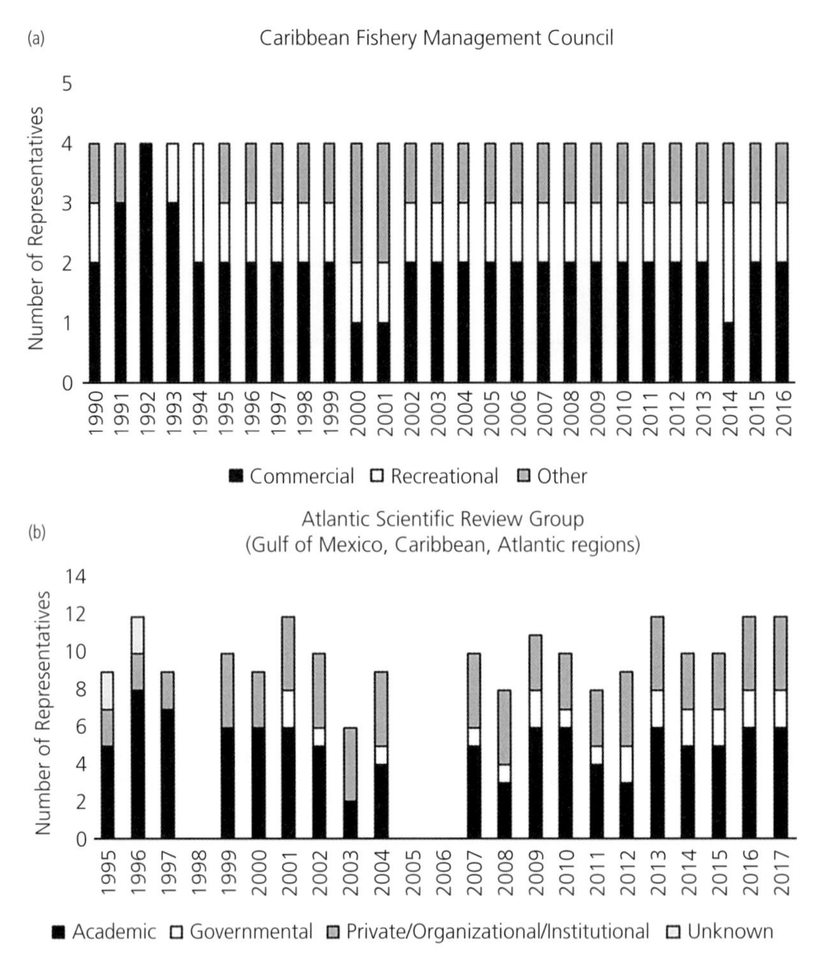

Figure 6.8 (a) Composition of the Caribbean Fishery Management Council from commercial and recreational fishing and other sectors, and the (b) Atlantic marine mammal Scientific Review Group over time (data unavailable 1998, 2005–2006).

Data derived from NOAA National Marine Fisheries Service Council Reports to Congress and Scientific Review Group reports.

hind grouper (*E. guttatus*) and Mutton snapper (*Lutjanus analis*). An additional seven unnamed locations also exist. Fishing closure zones in the U.S. Caribbean are not specifically named as such, although fishing is prohibited in 12 specific areas and restricted in 23 of them (Fig. 6.9), several of which are also identified HAPCs. Very little area of the U.S. Caribbean contains permanent fishing prohibitions (~0.7% of the EEZ), while ~2700 km^2 is subject to some fishing restrictions. The majority of these prohibited and restricted fishing areas are found off Puerto Rico (n = 21), with the degree of their success toward rebuilding reef fish populations remaining mixed (Nemeth 2005, Marshak & Appeldoorn 2008).

6.4.2.3 Organizational Context

Relative to the total commercial value of its managed fisheries, the CFMC budget is ranked highest when compared to other regions (Fig. 6.10), and is equal to 10–20% of its overall costs (i.e., fisheries value). These relative values have also increased during the early and mid-2010s and have remained relatively stable in more recent years. Cumulative National Environmental Policy Act Environmental Impact Statement (NEPA EIS) actions from 1987 to 2016 in the U.S. Caribbean have been relatively low (sixth-highest) compared to other regions with 72 observed over the past 20 years. Only one fisheries-related lawsuit (related to parrotfish)

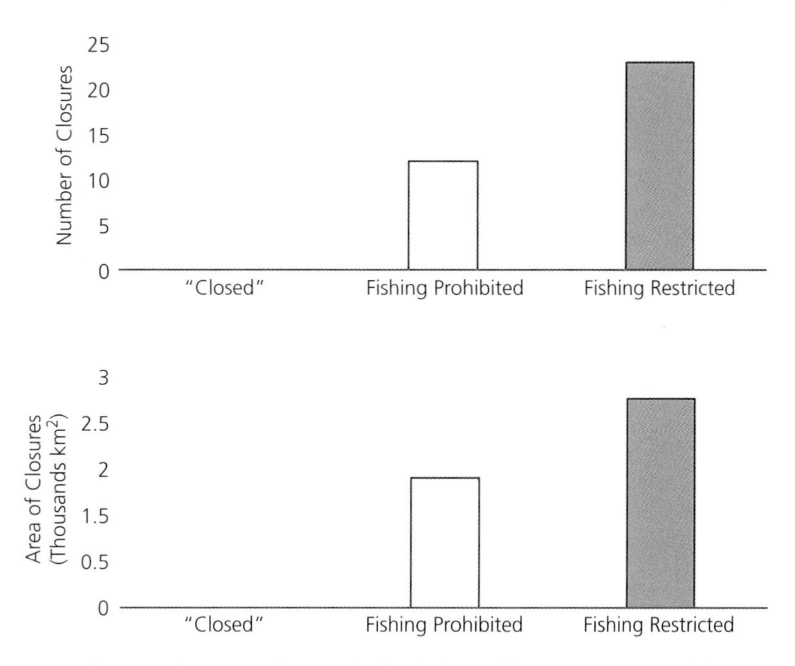

Figure 6.9 Number (top panel) and area (bottom panel; thousands of km²) of named fishing closures, and prohibited or restricted fishing areas of the U.S. Caribbean region.

Data derived from NOAA Marine Protected Areas inventory.

Table 6.3 List of major managed fishery species in the U.S. Caribbean region, under state and federal jurisdictions as reflected in the current four fishery management plans (FMPs) under the Caribbean Fishery Management Council (CFMC)

Council/Agency	Fishery Management Plan (FMP)	FMP Modifications	Major Species/ Species Group	Scientific Name(s) or Family Name(s)
Caribbean Fishery Management Council **CFMC**	Spiny lobster	7 amendments	Caribbean spiny lobster	*Panulirus argus*
	Reef fish (81 species; additionally 58 species of reef fish aquarium trade finfish)	8 plan amendments, 6 regulatory amendments	Groupers	Family Serranidae: *Cephalopholis* spp., *Epinephelus* spp., *Mycteroperca* spp.
			Snappers	Family Lutjanidae: *Apsilus* sp., *Etilis* sp., *Lutjanus* spp., *Pristipomoides* spp., *Ocyurus* sp., *Rhomboplites* sp.
			Other reef fishes	Species from Families Acanthuridae, Balistidae, Carangidae, Chaetodontidae, Haemulidae, Holocentridae, Labridae, Malacanthidae, Monacanthidae, Mullidae, Ostraciidae, Pomacanthidae, Scaridae, Sparidae
	Coral reef resources (157 species)	5 amendments	All species	Hydrozoans, Octocorallina, Scleractinians, Antipatharia, Porifera, Anemones, Zoanthids, False Corals, Annelida, Mollusca, Crustaceans, Echinoderms, Bryozoa, Algae
	Queen conch	4 amendment, 2 regulatory amendment	Queen conch	*Aliger gigas*; 12 other species were included pre-2012 (*Astrea* spp., *Cassis* spp., *Charonia* spp., *Cittarium* spp., *Fasciolaria* spp., *Lobatus* spp., *Strombus* spp., *Vasum* spp.)

Table 6.4 List of major bays and islands, NOAA habitat focus areas (HFAs), NOAA National Estuarine Research Reserves (NERRs), NOAA national marine monuments and sanctuaries (NMS), coastal national parks, national seashores, and number of habitats of particular concern (HAPCs) in the U.S. Caribbean region and subregions

Region	Major Bays	Major Islands	HFAs	NERRs	Monuments and NMS	Coastal National Parks	National Sea/ Lakeshores	HAPCs
U.S. Caribbean	**Puerto Rico -** Aguadialla Bay, Boqueron Bay, Guayanilla Bay, Mayaguez Bay, Jobos Bay, Mosquito Bay, Anasco Bay, San Juan Bay, Ensenada Honda, Rincon Bay, Flamenco Bay, Marejada Bay, Almodovar Bay, Puerto Del Manglar, Salina del Sur Bay; **United States Virgin Islands**—Long Bay, Magens Bay, Ensomhed Bay, Turquoise Bay, Coral Bay, Francis Bay, Maho Bay, Cinnamon Bay, Hawksnest Bay, Salt River Bay, Great Pond Bay	**Puerto Rico** (including Mona, Monito, Desecheo, Culebra, Vieques) **United States Virgin Islands** (St. Croix, including Buck Island; St. John, including Congo Cay, Grass Cay, Lovango Cay, Mingo Cay; St. Thomas, including Buck Island, Capella Island, Great Saint James, Hans Lollik Island, Inner Brass Island, Little Saint James, Outer Brass Island, Thatch Cay, Savana Island, Water Island)	Northeast Marine Corridor and Culebra Island, Puerto Rico HFA	Jobos Bay NERR	Buck Island Reef National Monument	San Juan National Historic Site, Virgin Islands National Park, Hassel Island Historic District		45

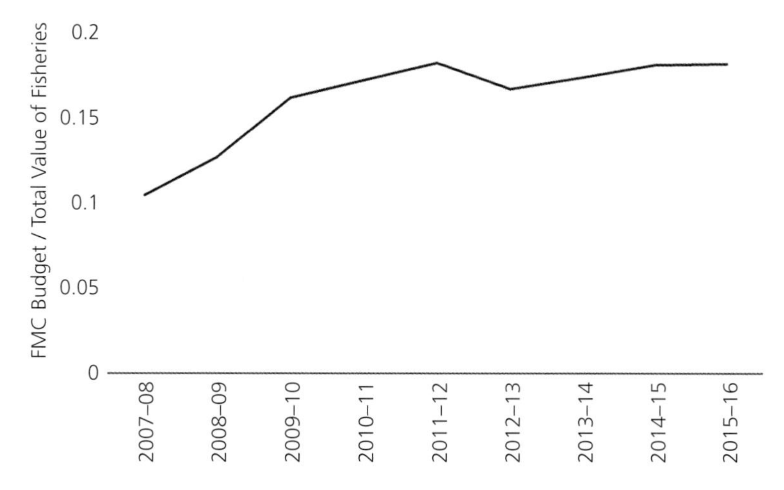

Figure 6.10 Total annual budget of the U.S. Caribbean Fishery Management Council as compared to the total value of its marine fisheries. All values in USD.

Data from NOAA National Marine Fisheries Service Office of Sustainable Fisheries and NOAA National Marine Fisheries commercial landings database.

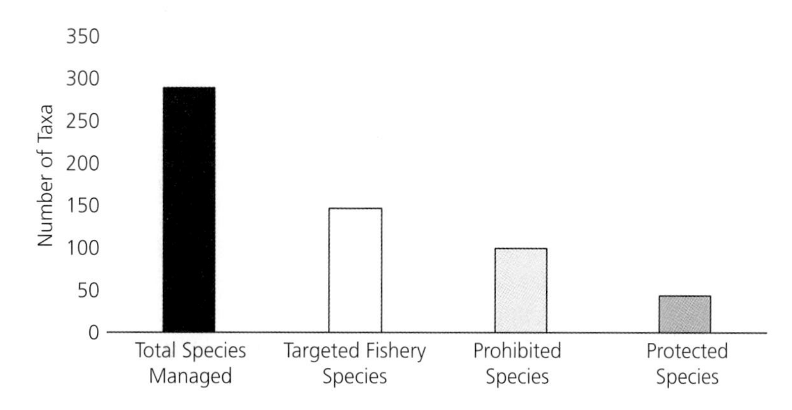

Figure 6.11 Number of managed taxa (species or families) for the U.S. Caribbean region, including federal and state jurisdictions.

has occurred in the U.S. Caribbean region since 2010. There is also ongoing Queen conch litigation.

6.4.2.4 Status of Living Marine Resources (Targeted and Protected Species)

Nationally, the U.S. Caribbean contains the highest number of managed species. Of the 288 federally managed species in the region, 146 are targeted fishery species covered under FMPs, with an additional 99 species (5 fishes, 44 soft coral and gorgonian taxa, 50 hydrozoan, stony coral, and black coral taxa) prohibited from harvest. Forty-five species of marine mammals, sea turtles, and fishes that occur in the region are also federally protected under the ESA or MMPA (Fig. 6.11, Tables 6.3, 6.5–6.6).

As of June 2017, 42 stocks (i.e., species or complexes) are federally managed in the U.S. Caribbean with 10 listed as NOAA Fish Stock Sustainability Index (FSSI) stocks (Fig. 6.12).[1] Of all stocks, three were identified as experiencing overfishing (**Puerto Rico triggerfishes**

[1] As of June 2020, only three stocks (Caribbean spiny lobster—Puerto Rico; Caribbean spiny lobster—St. Croix; Caribbean spiny lobster—St. Thomas/St. John) are listed as FSSI stocks. All other stocks (n = 39) are classified as non-FSSI stocks.

Table 6.5 List of currently prohibited fishery species in the U.S. Caribbean region, including federal and state jurisdictions

Council/Authority	Region	Prohibited Species	Scientific Name
Caribbean Fishery Management Council **CFMC**; NOAA Fisheries Southeast Regional Office **SERO** Puerto Rico Department of Natural and Environmental Resources **PR-DNER**; U.S. Virgin Islands Division of Fish and Wildlife **DFW**; U.S. Virgin Islands Department of Planning and Natural Resources	Caribbean (5 species fishes, 44 soft coral and gorgonian taxa, 50 hydrozoan, stony coral, and black coral taxa)	Hydrozoans, hard corals, and black corals	Families Acroporidae, Agaricidae, Astrocoeniidae, Cayophyllidae, Faviidae, Meandrinidae, Milleporidae, Mussidae, Oculinidae, Pocillopridae, Poritidae, Rhizangiidae, Siderastreidae, Stylasteridae, Order Antipatharia
		Live rock	Any coral reef resource attached to a hard substrate, including dead coral or rock (excluding individual mollusk shells)
		Soft corals and gorgonians	Families Anthothelidae, Briaridae, Clavulariidae, Ellisellidae, Gorgoniidae, Plexauridae
		Goliath grouper	*Epinephelus itajara*
		Nassau grouper	*Epinephelus striatus*
		Midnight parrotfish	*Scarus coelestinus*
		Blue parrotfish	*Scarus coeruleus*
		Rainbow parrotfish	*Scarus guacamaia*

and filefishes complex; **Puerto Rico Caribbean spiny lobster**—*Panulirus argus*; and **Puerto Rico wrasses complex**) and as overfished (**Goliath grouper**—*Epinephelus itajara*; **Nassau grouper**—*Epinephelus striatus*; and **Queen conch**—*Aliger gigas*).[2] Several stocks have unknown overfishing status (n = 5; 11.9% of total stocks), while overfished status for the majority of stocks is unknown (n = 38; 90.5% of total stocks), making the U.S. Caribbean one of the regions with the highest proportion of data-poor stocks with unknown stock status nationally.[3] Protected species are divided into 29 independent marine mammal stocks, and 20 ESA-listed distinct population segments, of which 45% are endangered (Fig. 6.13). Nationwide, lower numbers of marine mammal stocks are found in the U.S. Caribbean (9% of total marine mammal stocks), with 41% listed as strategic marine mammal stocks. An equivalent percentage of marine mammal stocks with unknown population size occurs in the U.S. Caribbean, which is the second-highest among regions.

[2] As of June 2020, no Caribbean stocks are undergoing overfishing, while the same three stocks are overfished (Goliath grouper; Nassau grouper; Queen conch).

[3] These values remain the same as of June 2020.

6.4.3 Environmental Forcing and Major Features

6.4.3.1 Oceanographic and Climatological Context

Circulation patterns throughout and surrounding the U.S. Caribbean EEZ are driven primarily by the Caribbean Current (Fig. 6.2). Connected to and originating from the North and South Equatorial Currents, the Caribbean Current passes to the south of the U.S. Caribbean, while its northern portions are influenced by the North Equatorial Current. A northward flow is observed bringing warmer equatorial waters throughout the region and entering into the Gulf of

Table 6.6 List of protected species under the Endangered Species Act (ESA) or Marine Mammal Protection Act (MMPA) in the U.S. Caribbean region

Region	Managing Agency	Protected Species Group	Common Name	Scientific Name
U.S. Caribbean (45 species)	Caribbean Fishery Management Council **CFMC**; Southeast Regional Office **SERO**	Whales	Blainville's beaked whale	*Mesoplodon densirostris*
			Blue whale	*Balaenoptera musculus*
			Bryde's whale	*Balaenoptera edeni*
			Cuvier's beaked whale	*Ziphius cavirostris*
			Dwarf sperm whale	*Kogia sima*
			False killer whale	*Pseudorca crassidens*
			Fin whale	*Balaenoptera physalus*
			Gervais' beaked whale	*Mesoplodon europaeus*
			Humpback whale	*Megaptera novaeangliae*
			Killer whale (Orca)	*Orcinus orca*
			Melon-headed whale	*Peponocephala electra*
			Minke whale	*Balaenoptera acutorostrata*
			North Atlantic right whale	*Eubalaena glacialis*
			Pygmy killer whale	*Feresa attenuata*
			Pygmy sperm whale	*Kogia breviceps*
			Sei whale	*Balaenoptera borealis*
			Short-finned pilot whale	*Globicephala macrorhynchus*
			Sowerby's beaked whale	*Mesoplodon bidens*
			Sperm whale	*Physeter macrocephalus*
		Dolphins	Atlantic spotted dolphin	*Stenella frontalis*
			Common Bottlenose Dolphin	*Tursiops truncatus truncatus*
			Clymene dolphin	*Stenella clymene*
			Fraser's dolphin	*Lagenodelphis hosei*
			Pantropical spotted dolphin	*Stenella attenuata*
			Risso's dolphin	*Grampus griseus*
			Rough-toothed dolphin	*Steno bredanensis*
			Spinner dolphin	*Stenella longirostris*
			Striped dolphin	*Stenella coeruleoalba*
		Sea turtles	Green sea turtle	*Chelonia mydas*
			Hawksbill sea turtle	*Eretmochelys imbricata*
			Kemp's Ridley sea turtle	*Lepidochelys kempii*
			Leatherback sea turtle	*Dermochelys coriacea*
			Loggerhead sea turtle	*Caretta caretta*
		Fishes	Nassau grouper	*Epinephelus striatus*
			Scalloped hammerhead shark	*Sphyrna lewini*
			Giant oceanic manta ray	*Manta birostris*
			Oceanic whitetip shark (suggested, not confirmed)	*Carcharhinus longimanus*
		Corals	Rough cactus coral	*Mycetophyllia ferox*
			Pillar coral	*Dendrogyra cylindrus*
			Lobed star coral	*Orbicella annularis*
			Mountainous star coral	*Orbicella faveolata*
			Boulder star coral	*Orbicella franksi*
			Staghorn coral	*Acropora cervicornis*
			Elkhorn coral	*Acropora palmata*
	U.S. Fish and Wildlife Service **FWS**	Other species	West Indian manatee	*Trichechus manatus*

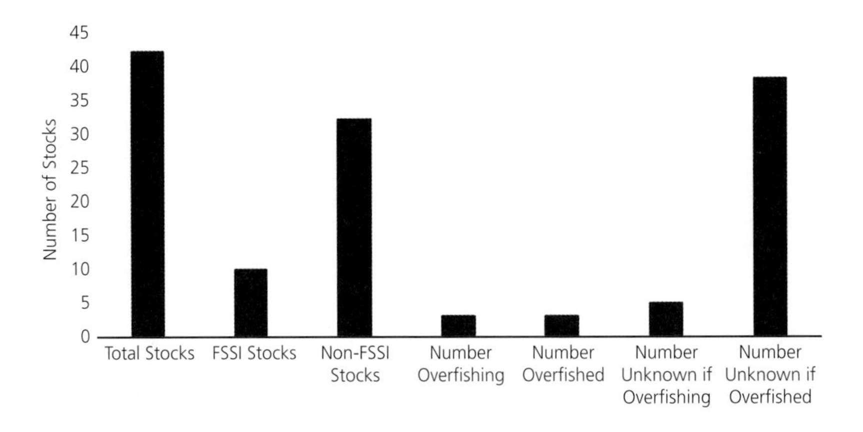

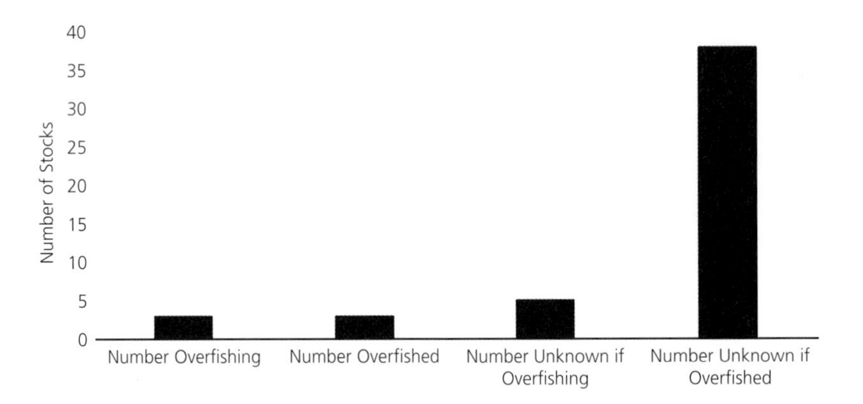

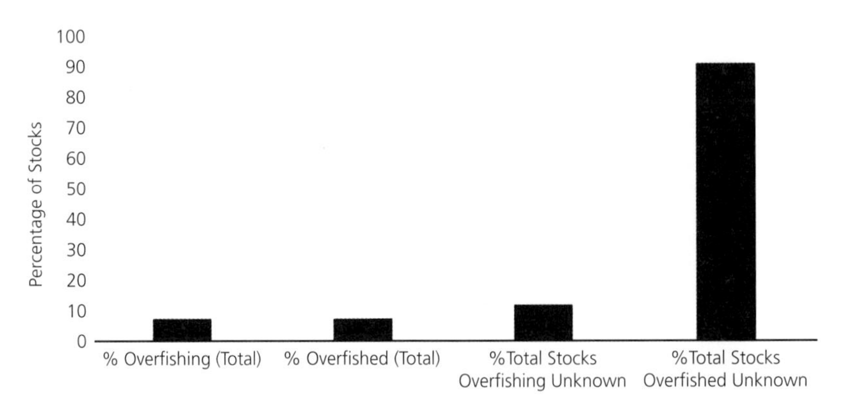

Figure 6.12 For the U.S. Caribbean region as of June 2017 (a) Total number of managed Fish Stock Sustainability Index (FSSI) stocks and non-FSSI stocks, and breakdown of stocks experiencing overfishing, classified as overfished, and of unknown status. (b) The number of stocks experiencing overfishing, classified as overfished, and of unknown status. (c) Percent of stocks experiencing overfishing, classified as overfished, and of unknown status.

Data from NOAA National Marine Fisheries Service. Note: stocks may refer to a species, family, or complex.

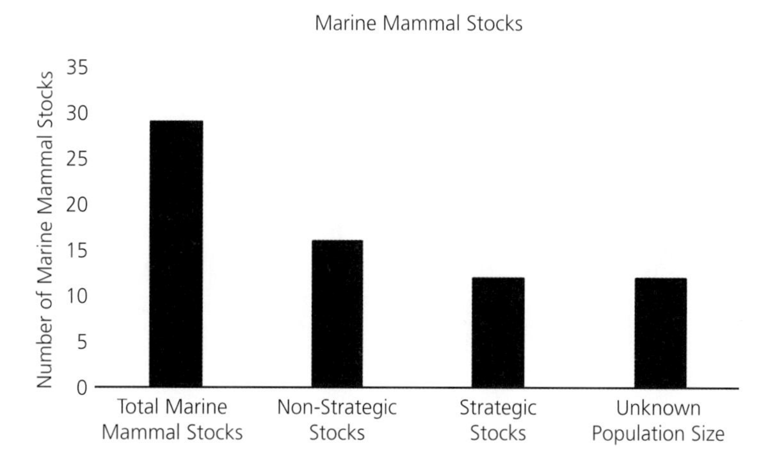

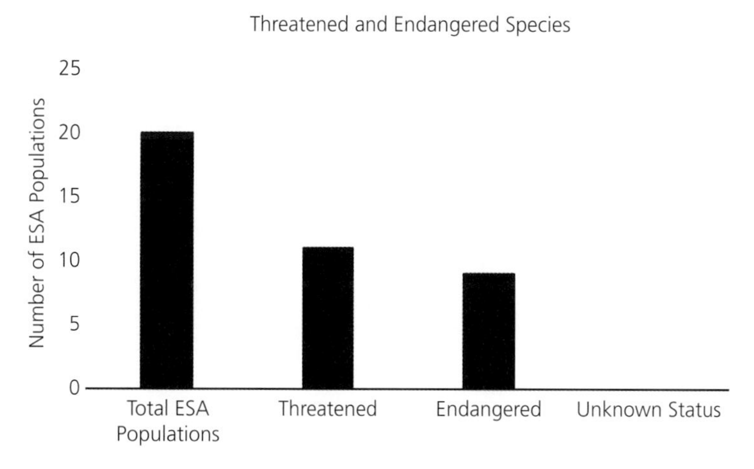

Figure 6.13 Number and status of federally protected species (marine mammal stocks, top panel; distinct population segments of species listed under the Endangered Species Act (ESA), bottom panel) occurring in the U.S. Caribbean region.

Mexico as the formation of the Gulf Loop Current, and provides a means of high connectivity between these regions. The North Equatorial Current feeds into the Gulf Stream along the U.S. Atlantic EEZ. Periodic freshwater input from the Orinoco and Amazon rivers influence nutrient concentrations and system productivity throughout the Caribbean Sea, influencing the U.S. Caribbean region (Morell & Corredor 2001, Cherubin & Richardson 2007, Lopez et al. 2013).

Clear interannual and multidecadal patterns in average sea surface temperature (SST) have been observed for the U.S. Caribbean (Fig. 6.14), with temperatures averaging 27.2°C ± 0.03 (SE) over time (1854–2016) and having increased by 1.5°C since the mid-twentieth century. Among regions, this increase is second-highest in the nation. To place these temperature trends in broader, basin-scale, and climatological context, the Atlantic Meridional Oscillation (AMO) and North Atlantic Oscillation (NAO) exhibit decadal cycles that can influence the environmental conditions and ecologies of several U.S. Atlantic regions, including the U.S. Caribbean. These both show periodic shifts in broad-scale climatological conditions, with a slight trend in the AMO over the past 50 years (Fig. 6.15).

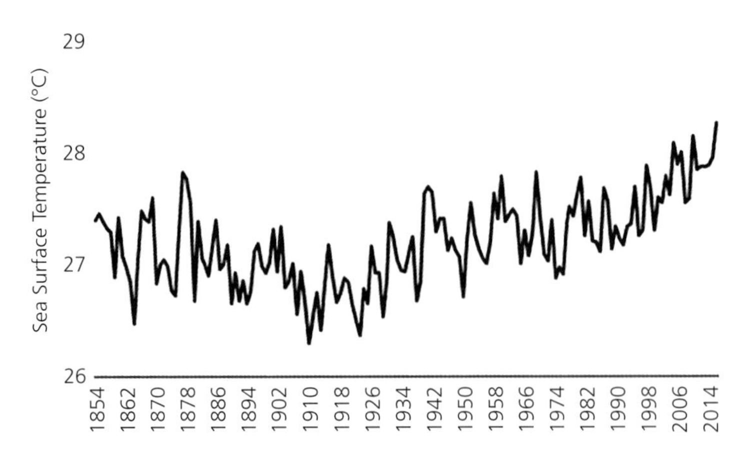

Figure 6.14 Average sea surface temperature (°C) over time (years 1854–2016) for the U.S. Caribbean region.

Data derived from the NOAA Extended Reconstructed Sea Surface Temperature dataset (https://www.ncdc.noaa.gov/).

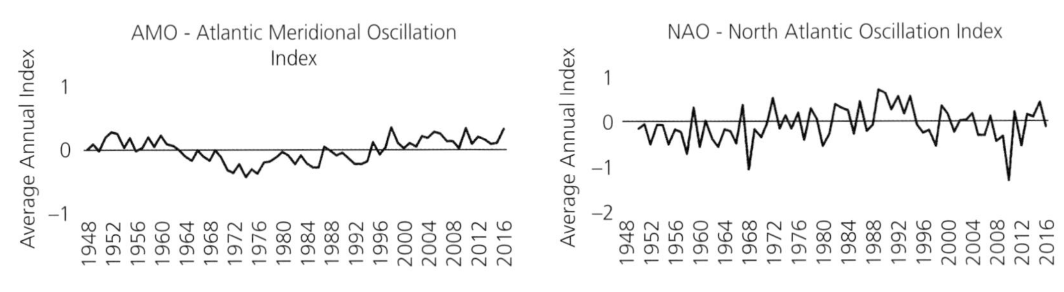

Figure 6.15 Climate forcing indices for the U.S. Caribbean region over time (years 1948–2016).

Data derived from NOAA Earth System Research Laboratory data framework (https://www.esrl.noaa.gov/).

6.4.3.2 Notable Physical Features and Destabilizing Events and Phenomena

With two prominently identified territories, the U.S. Caribbean contains 26 major bays (Table 6.4), primarily off the coast of Puerto Rico, with 6 major islands throughout the Puerto Rico archipelago and 18 major islands throughout the USVI, mostly surrounding St. Thomas. Proportionally, the U.S. Caribbean region makes up 1.5% of the U.S. EEZ (~238,000 km^2) and 1% of the national shoreline (~1400 km). It contains the deepest average depth (~4300 m) for the western Atlantic region and maximum depths of ~8.5 km (Fig. 6.16) due to the presence of the Puerto Rico Trench off the north coasts of Puerto Rico, St. Thomas, and St. John.

In the U.S. Caribbean region, 213 hurricanes have been reported since 1850, with an average of 12.5 hurricanes occurring per decade and peak numbers (17–18 per decade) recorded in the 1930s, 1950s, and 2000s (Fig. 6.17). Of all regions, the U.S. Caribbean has been the fifth-most prone to hurricanes.

6.4.4 Major Pressures and Exogenous Factors

6.4.4.1 Other Ocean Use Context

Data regarding the economics of other ocean uses in the U.S. Caribbean (Table 6.7) are limited to cumulative information available from years 2012 and 2014 as compiled by Clements et al. (2016). Ocean uses in the U.S. Caribbean are most associated with the tourism/recreation and marine transportation sectors. The tourism and recreation sector comprises 95% of total U.S. Caribbean ocean establishments and 87% of total ocean employments. GDP from tourism and recreation contributes toward the bulk (97% of its total revenue) of the U.S. Caribbean ocean economy at \$5.7 billion (Puerto Rico ~\$4.3 billion year^{-1}, USVI \$1.4 billion year^{-1}).

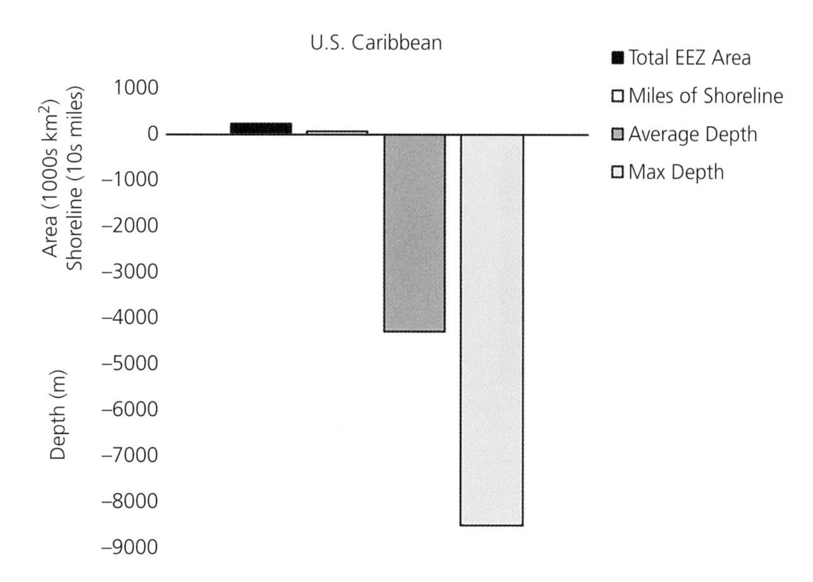

Figure 6.16 Total shelf area, miles of shoreline, and average and maximum depth of marine regions and subregions of the U.S. Caribbean.

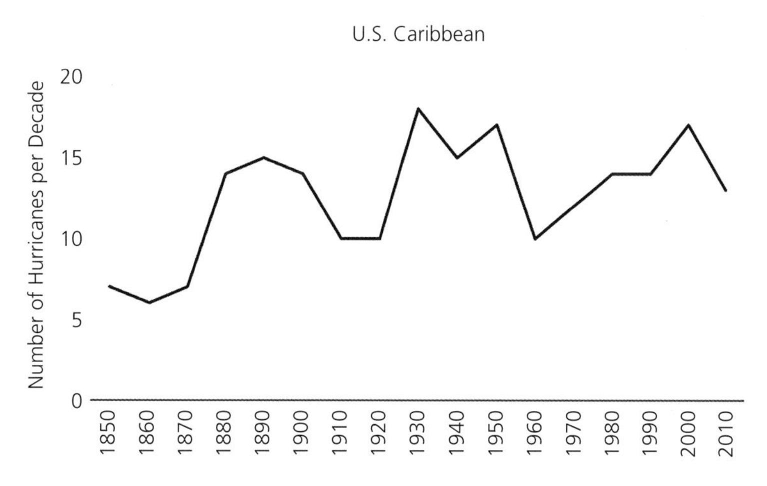

Figure 6.17 Total number of hurricanes recorded per decade for the U.S. Caribbean since year 1850.
Data derived from NOAA Office for Coastal Management (https://coast.noaa.gov/hurricanes/).

Marine transportation contributes much less toward total ocean establishments and employments (2% of total ocean establishments, 7.4% of total ocean employments), and comprises 2.3% of the U.S. Caribbean ocean economy at ~$136 million (Puerto Rico $103 million year^{-1}, USVI $33.4 million year^{-1}). Additionally, marine construction contributes 2.5% of total ocean employments and an annual revenue of $19.8 million (0.3% of the total marine economy), while offshore mineral extraction and ship and boatbuilding comprise much smaller proportions of the regional marine economy.

The degree of tourism in the U.S. Caribbean, as measured by number of dive shops, number of major ports and marinas, and number of cruise ship passengers, is one of the highest throughout U.S. marine regions (Fig. 6.18). While there are mid-level numbers of dive shops nationally (sixth-highest), and relatively low numbers of major ports (eighth-highest) and marinas (sixth-highest), much regional tourism and revenue has been associated with cruise destinations and departures (highest nationally), with ~500,000 departure passengers and ~3.5 million destination passengers observed annually. No active offshore oil rigs or

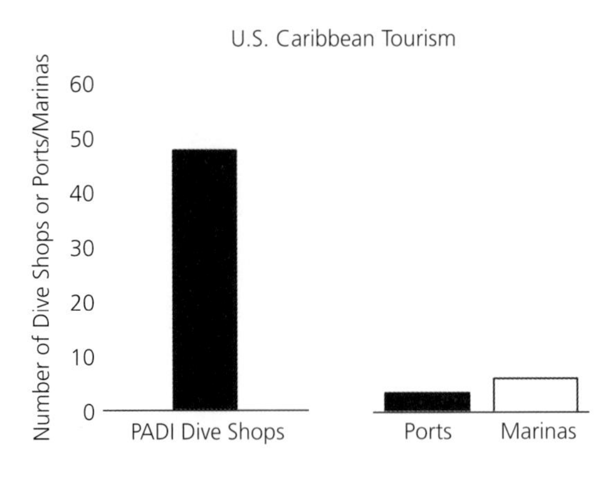

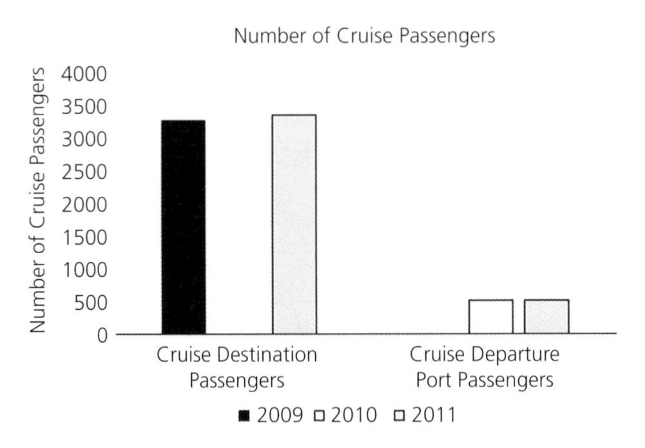

Figure 6.18 Tourism indicators for the U.S. Caribbean region and identified subregions. Total number of Professional Association of Diving Instructors (PADI) dive shops (data derived from http://apps.padi.com/scuba-diving/dive-shop-locator/) and total number of major ports and marinas (top panel). The number of cruise destination and departure port passengers is shown in the bottom panel. (years 2010 and 2011; values in thousands).

Table 6.7 Number and percent contribution of multisector establishments, employments, and gross domestic product (GDP) and their percent contribution to total multisector oceanic economy in the U.S. Caribbean (years 2012, 2014)

Sector	Establishments	Employment	Revenue	% Total Establishments	%Total Employments	%Total Revenue
Marine construction	24	4086	$19,816,175	0.3	2.5	0.3
Marine transportation	166	12,378	$136,477,049	2.0	7.4	2.3
Offshore mineral Resources	50	17	$334,582	0.6	0.0	0.0
Ship and boat building	54	188	No Data	0.6	0.1	No data
Tourism and recreation	7899	145,335	$5,706,247,065	95.0	87.3	97.1
Total	**8317**	**166,458**	**$5,874,180,202**			

Data derived from the National Ocean Economics Program and Clements et al. (2016)

offshore wind energy production sites are found throughout the U.S. Caribbean EEZ.

6.4.5 Systems Ecology and Fisheries

6.4.5.1 Basal Ecosystem Production

Since 1998, mean surface chlorophyll values throughout the U.S. Caribbean have been very low at ~0.12 ± 0.004 (SE) mg m^{-3} (Fig. 6.19), reflecting the oligotrophic nature of the region. Comparatively, average annual productivity throughout this region is sixth-highest nationally, with seasonal influences from South American rivers and contributions of Saharan dust in providing sources of bioavailable iron toward stimulating phytoplankton production (Griffin & Kellogg 2004, Lopez et al. 2013). Productivity values have remained relatively steady between 160 and 210 g C m^{-2} $year^{-1}$, with marginal decreases observed in more recent years. While oligotrophic, the production associated with this system has also led to the development of significant coral reef-associated biomass (Opitz 1993).

6.4.5.2 System Exploitation

Nationally, the second-lowest numbers of landed taxa have been reported in the U.S. Caribbean region over time, with 41 taxa having been consistently reported since 1995. Given the high diversity associated with the region, underreporting and inaccurate characterization of landed species by weight and number remain significant challenges for the U.S. Caribbean (Matos-Caraballo & Agar 2011, SEDAR 2016). Although fishery-independent surveys of U.S. Caribbean reef fish and

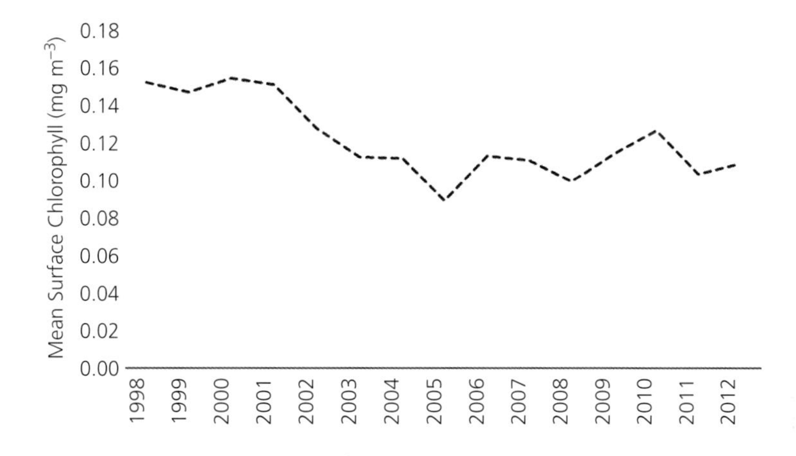

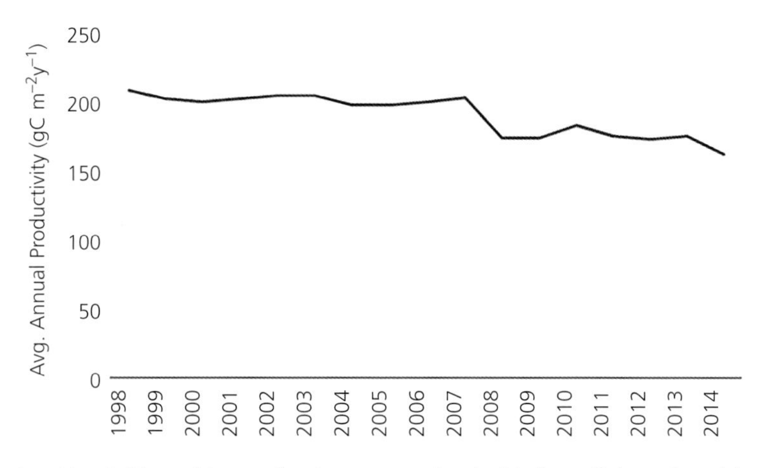

Figure 6.19 Mean surface chlorophyll (mg m^{-3}; top panel) and average annual productivity (grams Carbon m^{-2} $year^{-1}$; bottom panel) for the U.S. Caribbean region over time.

Data derived from NASA Ocean Color Web (https://oceancolor.gsfc.nasa.gov/) and productivity calculated using the Vertically Generalized Production Model, VGPM.

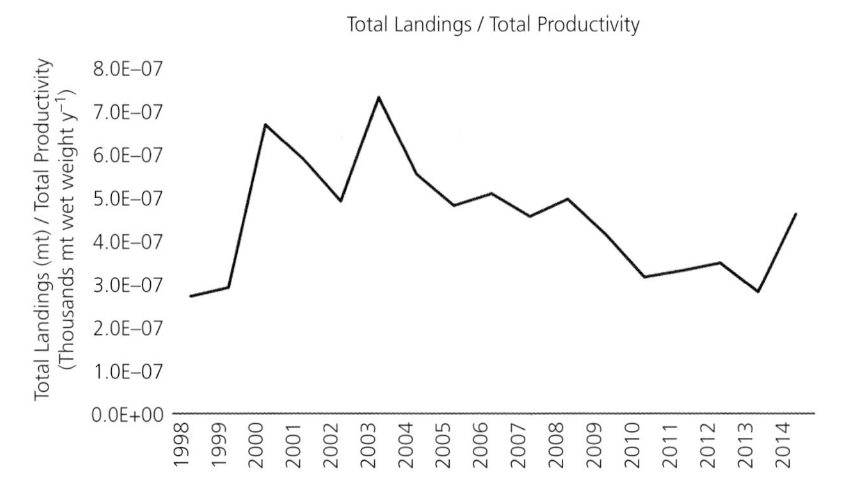

Figure 6.20 Ratios of total reported commercial and recreational landings (metric tons) to total productivity (metric tons wet weight yr^{-1}) for the U.S. Caribbean region. Recreational landings are only reported in Puerto Rico, which together with U.S. Caribbean commercial landings are of uncertain accuracy (Sagarese pers. comm.).

invertebrate biomass have occurred through the Caribbean Southeast Area Monitoring and Assessment Program (SEAMAP-C), with additional studies by academics, PR-DNER, and USVI territorial agencies, survey methodology constraints have precluded comprehensive biomass estimates or use in stock assessments (Cass-Calay et al. 2016). Stock abundance survey data for reef fishes, Queen conch, and Caribbean spiny lobster still have not been accepted for use toward estimating stock biomass given limitations in current fishery-independent sampling designs and data availability (Cass-Calay et al. 2016). As a result, the calculation of ratios examining exploitation rates (i.e., fisheries landings to biomass) and other socioecological relationships for this region remain quite limited to those related to system productivity, and also limit the application of ecosystem-level thresholds for this region. Values for total landings/productivity are lowest nationally, on average (up to $7.0E^{-7}$) and have been observed declining over the past decade (Fig. 6.20).

6.4.5.3 Targeted and Non-Targeted Resources

On average, reported commercial and recreational fisheries landings are around 1–2 thousand mt and have generally remained closer to 1000 mt since the past decade (Fig. 6.21). These values are less than previous peak values in the 1970s of over 3000 mt per year (Matos-Caraballo 1997, 2001, 2002, 2009) and are lowest nationally (Link & Marshak 2019). Additional decreases since the 2000s have been observed. Increases in the proportion of reported recreational catch have occurred since the 2010s, with recreational catches now contributing up to 60% of total reported landings for the region at similar values to 1990s reported commercial landings. Until 2017, reporting of recreational landings only occurred in Puerto Rico, where on average a slight majority of commercial catches have been landed over time (51% ± 2.8, SE). It is worth noting that there are still considerable uncertainties regarding total reported landings estimates. Factors including confusion and misreporting of species, lack of enforcement, sampling issues, poor accounting for fisheries expansions, and limited recreational fishing data have hindered accurate accounting for U.S. Caribbean fishing activity, especially in Puerto Rico (T. Gerard, S. Sagarese pers. comms.). Reported data for the USVI commercial fishery is assumed to be relatively more complete, however (T. Gerard pers. comm.). Therefore, while these data contain the most comprehensive information currently available, these patterns are not necessarily representative of total U.S. Caribbean fishing trends.

When standardized per square kilometer (Fig. 6.22), U.S. Caribbean landings over the past two decades follow very similar patterns as observed for non-standardized total landings values, with this region containing the second-lowest concentration of fisheries landings over time nationally. Historically, however, concentrated fishing pressure as a result of fishing subsidies was observed throughout Puerto Rico and USVI reef systems during the 1970s (Appeldoorn 1991, Kimmel & Appeldoorn 1992, Appeldoorn et al. 1995, Beets & Friedlander 1998, Sadovy 1999), leading to

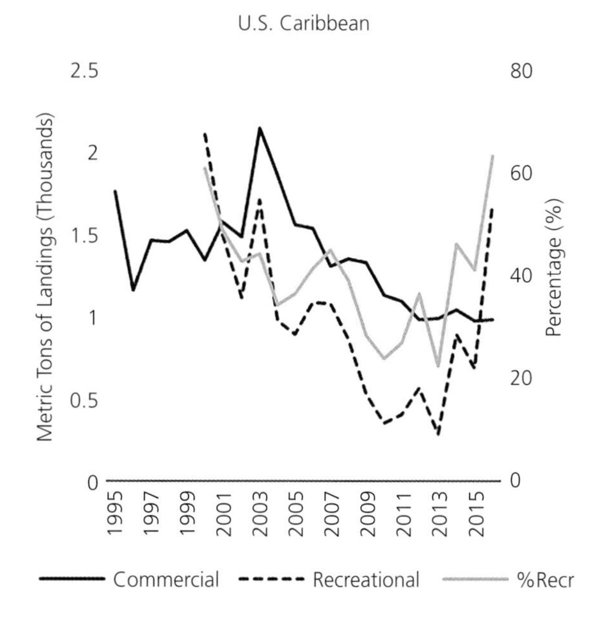

Figure 6.21 Total reported commercial and recreational landings and percent contribution of recreational landings (%Recr) over time for the U.S. Caribbean region (commerical: 1995–2016; recreational: 2000–2016).

Data derived from NOAA National Marine Fisheries Service commercial and recreational fisheries statistics. Recreational landings are only reported in Puerto Rico, which together with U.S. Caribbean commercial landings are of uncertain accuracy (Sagarese pers. comm.).

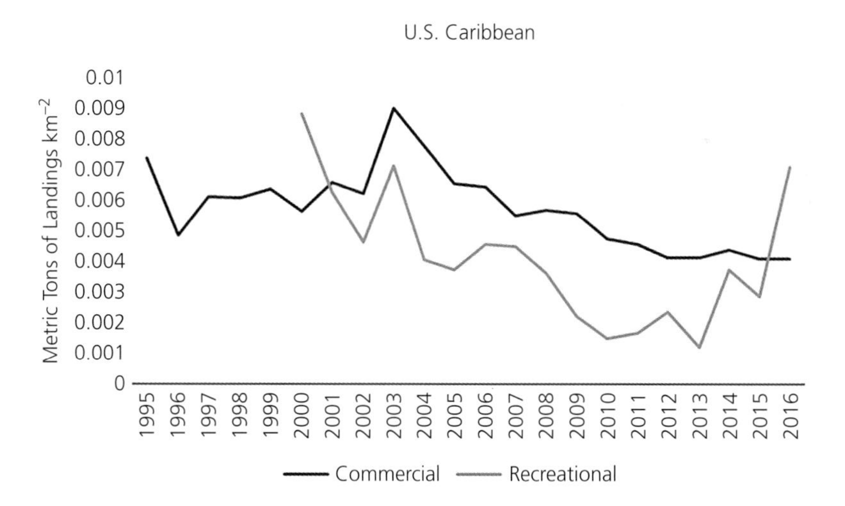

Figure 6.22 Total reported commercial and recreational landings (metric tons) per square kilometer over time for the U.S. Caribbean region. (commercial: 1995–2016; recreational: 2000–2016)

Data derived from NOAA National Marine Fisheries Service commercial and recreational fisheries statistics. Recreational landings are only reported in Puerto Rico, which together with U.S. Caribbean commercial landings are of uncertain accuracy (Sagarese pers. comm.).

high exploitation and decreases in fisheries landings during the 1980s and 1990s (Matos-Caraballo 1997, 2001, 2002, 2009), and significant depletion of inshore fish and invertebrate species over time (Kimmel & Appeldoorn 1992, Appeldoorn et al. 1995, Appeldoorn 2008, Garcia-Sais 2010, Bejarano et al. 2014). While bycatch of fishes, sharks, marine mammals, seabirds, and sea turtles remain a concern throughout the U.S. Caribbean (Garrison et al. 1998, Rogers & Beets 2001, Marshak et al. 2008, Agar et al. 2008, Matos-Caraballo

& Agar 2011, Clark et al. 2012), little reported data are available regarding trends in bycatch over time. Some efforts have been undertaken toward reducing bycatch in less selective gears. In 2017, the Virgin Islands Fish Trap Reduction Program was signed by the USVI Governor and enacted, which limited participation in the trap fishery to those participants since 2011 and required that all fishers reduce their traps by 10–20% (Arnold 2019, T. Gerard pers. comm.). A gill and trammel net ban and buyback plan has also been enacted, with mixed reception observed in St. Croix (Agar et al. 2019). Over the past decade, efforts toward stock rebuilding, the implementation of seasonal and permanent closures to reef fish and Queen conch spawning aggregations, partial improvements to enforcement of fisheries legislation, and the development of more comprehensive fisheries survey and management approaches are assisting in enhancing the health of the U.S. Caribbean fisheries ecosystem.

6.5 Synthesis

The U.S. Caribbean has been affected by above average natural and human stressors that include the nationally second-highest rate of SST increase over the past 70 years (Link & Marshak 2019), increasing frequency and intensity of hurricanes, high coastal development, and concentrated historical fishing pressure in nearshore reef systems. These factors have led to consequential effects on this marine ecosystem, especially regarding the status of its LMRs and dependent economies (Rogers & Beets 2001, Rogers & Miller 2006, Appeldoorn 2008, Ault et al. 2008, Wilkinson & Souter 2008, Larsen & Webb 2009, Eakin et al. 2010, Fanning et al. 2011, Lugo 2018, Cox et al. 2019). Among U.S. marine ecosystems, comparatively lower-ranked LMR status and marine socioeconomics are currently observed in the U.S. Caribbean (Clements et al. 2016, Link & Marshak 2019), with emerging and ongoing natural stressors exacerbating the effects of human impacts to this region (Lugo 2000, Lugo et al. 2000, Smith et al. 2008, Weil et al. 2009, Eakin et al. 2010, Moore et al. 2010, Appeldoorn et al. 2011, Appeldoorn et al. 2016, Melendez & Salisbury 2017, Oxenford & Monnereau 2017, Harms-Tuohy et al. 2018). Associated with the second-highest human population density among U.S. regions, the environmental effects of significantly concentrated fisheries exploitation during the 1970s and 1980s along with increased coastal development have altered top-down and bottom-up forces in this system, with direct effects on LMR production (Rogers & Beets 2001, Christensen et al. 2003, Warne et al. 2005, Rogers & Miller 2006, Ballantine et al. 2008, Bejarano & Appeldoorn 2013, Appeldoorn et al. 2016). These are compounded by the effects of climate change, regional warming, species invasions, and habitat degradation that continue to affect the U.S. Caribbean. Although monitoring of marine socioeconomics remains limited, a billion-dollar tourism industry in this region contributes toward ongoing seafood demand, recreational access to marine resources, sportfishing, and complementary human pressures. Emerging EBFM strategies for the U.S. Caribbean must consider the cumulative effects of these natural and human stressors on its coupled socioecological system, particularly their influences on basal ecosystem productivity, fishery demand, sustainable fisheries landings and biomass, and their ongoing ecological consequences in a changing environment.

There are components of the U.S. Caribbean fisheries ecosystem that are performing relatively well, particularly given recent ecosystem-based advances in aspects of its governance. Recent efforts by the fishing community and other stakeholders to participate in conservation and regulatory actions are leading to increasing compliance with reporting requirements and fishing regulations (W. Arnold, T. Gerard pers. comms.). Although variable, this system has the sixth-highest primary productivity among all U.S. regions, which contributes toward its coral reef-associated biomass (Hernandez-Delgado et al. 2000, Christensen et al. 2003, Kendall et al. 2004, 2005, Aguilar-Perera & Appeldoorn 2007, Pittman et al. 2007a, Ballantine et al. 2008, Garcia-Sais et al. 2008, Bejarano et al. 2014, Appeldoorn 2018, Link & Marshak 2019). The emerging development of comprehensive island-wide FMPs for Puerto Rico, St. Thomas/St. John, and St. Croix, plus a U.S. Caribbean-wide FEP are underway (CFMC 2019a, b, c, NMFS 2019). These focus attention away from single-species approaches to those of stock complexes by island, and aim to provide improved consideration of commercial and recreational fishing sectors, their markets, and fisheries socioeconomics (CFMC 2019a, b, c). These efforts are complemented by advances in the understanding of ecosystem functioning for this system and initiatives toward ecosystem approaches to management (CFMC 2019a, b, c, NMFS 2019). Specifically, these include the development of outlines and strategies toward developing an FEP, compiling available ecosystem information and assessing data gaps, undertaking environmental assessments, and refining approaches in ecosystem modeling for this region to support ecosystem-based approaches (CFMC 2019a, b, c). They also build upon the multispecies focuses of previous reef fish and corals FMPs, enhance information regarding essential fish habitat (EFH) of managed species, account for other natural stressors (i.e., hurricanes), and detail anticipated effects of management actions on U.S. Caribbean physical, biological, and social environments (CFMC 2019a, b, c). These efforts are further complemented by the completion of an EBFM implementation plan for the U.S. Caribbean (NMFS 2019), the priorities of which include:

- Developing and maintaining core data and information streams.
- Developing an EBFM guidance document that identifies factors impacting managed species.
- Developing community-level species vulnerability and catch diversity measures, and social vulnerability assessments to examine the social/community effects of management changes.
- Developing and enhancing stock assessment methods, particularly for data-poor species and modeling ecosystems at species complex levels.
- Working toward incorporating information into national and the newly proposed regional IEAs.

These efforts advance current progress toward enhanced characterization and monitoring of U.S. Caribbean ecosystem dynamics and address data limitations that have hindered ecosystem-based approaches to management of this system (such as implementing system-wide reference points). They also contribute to a strategic approach for finalizing an FEP that provides guidance for EBFM of this system and identifies quantitative information regarding indicators needed for ecosystem-level assessments. Marine socioeconomic considerations are reflected in island-wide FMPs and in EBFM implementation priorities, especially in examining socioeconomic effects of coastal development and land management on fishing communities, quantifying the effects of management practices on community health and socioeconomic well-being, and in expanding community vulnerability assessments. Although fishery-dependent data limitations, patchy monitoring, and only partial data integration have severely limited the ability to assess this ecosystem and its LMRs at present, these priorities are working toward providing foundational information to support needed investigations into ecosystem status and broader management strategies.

Although ranked low overall regarding the status of its marine socioeconomics, the U.S. Caribbean leads nationally in terms of aspects of its marine tourism, particularly cruise ship destinations, which contribute heavily to its local economy. While containing the highest number of managed fisheries species nationally and comparatively lower numbers of depleted marine mammal stocks or endangered species, the U.S. Caribbean is ranked lowest among regions in terms of the status of its LMRs (Link & Marshak 2019), with lowest total landings, indicators of high system exploitation (Rogers & Beets 2001, Rogers & Miller 2006, Appeldoorn 2008, Ault et al. 2008, Bejarano & Appeldoorn 2013, Bejarano et al. 2014), and poorly assessed stocks, which all continue to exhibit major consequences from historical overfishing of certain species. For the U.S. Caribbean, high uncertainty remains regarding the status of many key economic species in this multispecies fishery, with past resistance by stakeholders to cooperate with strict fishing restrictions. Overall governance in this region is ranked fifth-highest among regions (Link & Marshak 2019), with higher congressional representation related to its shoreline and value of fisheries (albeit with non-voting members), comparatively higher use of spatial protections and HAPCs, low numbers of lawsuits, and few NEPA EIS activities. However, many of these factors are directly related to this region also possessing the lowest national fisheries value among all regions (Link & Marshak 2019).

Given limited information availability and compounding natural and human effects on this system, many challenges remain in enacting ecosystem approaches to management for the U.S. Caribbean. Efforts to mitigate fishing effects, assess populations, and to rebuild overfished stocks have remained constrained by limited monitoring of fishery populations and complementary environmental indicators, as well as restricted resources for environmental enforcement. The U.S. Caribbean has the highest proportion of unassessed stocks in terms of overfished status throughout the U.S. (Link & Marshak 2019). While investigations into regional ecosystem dynamics and ecological assessments have occurred, they have generally been restricted to small portions of the U.S. Caribbean EEZ. Efforts toward resolving these data gaps and creating ecosystem-level management frameworks that incorporate broader ecosystem considerations remain needed, especially in consideration of system-level shifts and thresholds that may be potentially exceeded. Given the historical overharvesting of this system, compounded with the effects of habitat loss and degradation, regional mortalities of keystone species, and recreational pressure, there remains an urgent need to examine and mitigate system overexploitation and address these concerns in more robust management strategies. Direct consideration of relationships between fisheries harvest, other ocean use, and ecosystem properties is necessary for a sustainable approach to management in the U.S. Caribbean.

As the effects of a highly concentrated human population continue to exacerbate current stressors to this system, intensified resource competition among marine sectors is likely to occur, with anticipated conflicts among fishing interests, tourism, and marine transportation. Emerging priorities for the U.S. Caribbean include greater understanding of the socioeconomic effects of management practices on fishing communities (including changes in fishing effort) along with consideration of other ocean sectors to allow for more comprehensive management strategies. This information can be included in a systematic framework that accounts for overlapping ocean uses, multiple species, and conflicting factors. Because LMRs contribute a low proportion toward the marine economy of the U.S. Caribbean compared to its tourism industry, the trade-offs emerging among tourism, fisheries, and other sectors need to be considered in a fisheries management context. While the U.S. Caribbean leads the nation in terms of the number of HAPCs and protected areas within its EEZ, very little of its waters remain permanently closed to fishing, with more strategic considerations of habitat and spatial management necessary to combat the effects of human and natural stressors. As noted earlier, what is especially needed for this region is investment toward enforcement of environmental legislation, improved LMR surveying, more comprehensive fishery-independent and commercial and recreational fishery-dependent data collection (including for bycatch), improved data archiving and accessibility, and enhanced communication with managers and stakeholders. All of these efforts will be instrumental in resolving many of the current hurdles facing this ecosystem. Therefore, efforts that address these concerns and build upon current management priorities for the U.S. Caribbean are especially needed.

6.5.1 Progress toward Ecosystem-Based Fisheries Management (EBFM)

This ecosystem is excelling in the marine tourism component of its socioeconomic status and in aspects of its governance frameworks as related to the determinants of successful LMR management (Fig. 6.23; Link & Marshak 2019).

Fisheries management for the U.S. Caribbean is broadening from a traditional single-species approach into one addressing cumulative issues affecting the LMRs and fisheries of this ecosystem. Through a more systematic and prioritized consideration of all fisheries, pressures, risks, and outcomes, holistic EBFM practices are being developed and implemented across U.S. regions, including this ecosystem (NMFS 2016a, b, Levin et al. 2018, Marshall et al. 2018, Link & Marshak 2019, NMFS 2019). Over the past few years, efforts to implement EBFM in the U.S. Caribbean have arisen through the refinement of local management strategies and the development of several ecosystem working groups (CFMC 2019a, b, c, NMFS 2019). In accordance with the NOAA Fisheries EBFM Policy and Roadmap (NMFS 2016a, b), an EBFM implementation plan (NMFS 2019) for the U.S. Caribbean has been developed through efforts overseen by the SEFSC, NOAA Fisheries Southeast Regional Office (SERO), and CFMC. That plan specifies approaches to be taken toward conducting EBFM and in addressing NOAA Fisheries Roadmap priorities and gaps. Progress for the U.S. Caribbean and its major subregions in terms of EBFM Roadmap Guiding Principles and Goals is shown in Table 12.9 (Chapter 12).

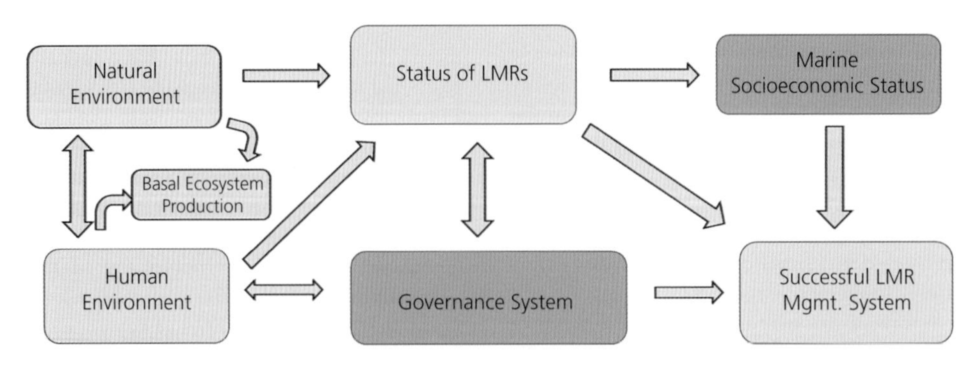

Figure 6.23 Schematic of the determinants and interconnectivity of successful living marine resource (LMR) systems management criteria as modified from Link and Marshak (2019). Those highlighted in blue indicate noteworthy criteria regarding the U.S. Caribbean ecosystem.

Overall, progress has been made in terms of implementing ecosystem-level planning, advancing knowledge of ecosystem principles, and in assessing risks and vulnerabilities to ecosystems through ongoing investigations into climate vulnerability and species prioritizations for stock and habitat assessments. However, little overall progress has been observed toward applying ecosystem-level emergent properties into management frameworks.

Progress toward recommendations of the Ecosystem Principles Advisory Panel (EPAP) for the development of FEPs (EPAP 1999, Wilkinson & Abrams 2015) is shown in Table 12.10 (Chapter 12). Island-based FMPs have been approved for the U.S. Caribbean, and a CFMC EBFM Technical Advisory Committee has been established to support the development of an FEP, shifting focus away from previous management approaches toward more comprehensive, system-wide considerations (CFMC 2019a, b, c, NMFS 2019). Progress has been observed in characterizing and monitoring aspects of ecosystem dynamics in support of an ecosystem-based approach to management.

6.5.2 Conclusions

Applying an ecosystem-based approach for the U.S. Caribbean allows for many advancements in its overall management, especially accounting for the cumulative effects of ongoing and increasing human and natural stressors, as well as mitigating their impacts. These are particularly germane for LMR productivity, harvest, and well-being of coral reef-associated biological communities. The local billion-dollar tourism industry associated with this region is highly dependent on ensuring a sustainable seafood supply, ecological integrity of coastal and offshore habitats to support diving and other marine recreational activities, and well-managed sport fisheries, all of which need to be prioritized in an EBFM context. Management entities must continue to monitor, consider, and incorporate the natural, biological, and socioeconomic factors that affect the degree to which marine resources can be sustainably extracted and harvested, especially in a changing, underassessed, and potentially overexploited environment. Continuing with a business-as-usual approach toward exploiting and managing the resources of this ecologically significant region without considering their effects on the ecosystem as a whole, or the influences of interacting sectors, historic and ongoing stressors, and socioeconomic feedbacks, limits effectiveness toward sustainable practices and ignores the significance of these factors for ensuring LMR productivity and marine socioeconomic (i.e., blue economy) growth.

System-wide considerations of the inherent relationships between productivity, biomass, landings, and socioeconomics are needed to refine and maintain this system toward needed sustainable thresholds, while also accounting for the influences from natural and human activities and other ocean uses. EBFM provides opportunities that focusing on one stock or species complex or one fishery would not. Optimal management allows for simultaneous harvest of interacting species complexes and fisheries found in coral reef systems, including shellfish fishery species that serve as prey for commercially important reef fishes, and accounts for other coral reef species and their trophic interactions. Overlapping and at times differential fishing effort for these species and the effects of different harvest methodologies among fisheries (e.g., effects of lobster traps on coral reef habitats and reef fish communities; effects of spearfishing on spawning aggregations of multiple reef fish species) also needs to be considered in these frameworks. With very high species richness in this system, this approach that allows for portfolio and multilevel considerations provides for a more effective and lucrative management strategy (Edwards et al. 2004, Kellner et al. 2011, Link 2018). As compounding effects of coastal development, habitat loss, eutrophication, and fishing pressure continue to affect this region, their consequences throughout this socioecological system will be more effectively addressed using an ecosystem approach. Applying comprehensive management strategies that consider EBM actions provides a pathway toward ensuring future sustainability for the U.S. Caribbean and its significant LMR-based culture.

6.6 References

Acevedo R, Morelock J, Olivieri RA. 1989. Modification of coral reef zonation by terrigenous sediment stress. *Palaios* 1:92–100.

Acosta A, Appeldoorn RS. 1992. Estimation of growth, mortality and yield per recruit for Lutjanus synagris (Linnaeus) in Puerto Rico. *Bulletin of Marine Science* 50(2):282–91.

Agar JJ, Fleming CS, Tonioli F. 2019. The net buyback and ban in St. Croix, US Virgin Islands. *Ocean & Coastal Management* 167:262–70.

Agar JJ, Waters JR, Valdés-Pizzini M, Shivlani M, Murray T, Kirkley JE, Suman D. 2008. US Caribbean fish trap fishery socioeconomic study. *Bulletin of Marine Science* 82(3): 315–31.

Aguilar-Perera A, Appeldoorn RS. 2007. Variation in juvenile fish density along the mangrove–seagrass–coral reef continuum in SW Puerto Rico. *Marine Ecology Progress Series* 348:139–48.

Aguilar-Perera A, Appeldoorn RS. 2008. Spatial distribution of marine fishes along a cross-shelf gradient containing a continuum of mangrove–seagrass–coral reefs off southwestern Puerto Rico. *Estuarine, Coastal and Shelf Science* 76(2):378–94.

Albins MA. 2013. Effects of invasive Pacific red lionfish Pterois volitans versus a native predator on Bahamian coral-reef fish communities. *Biological Invasions*. 15(1):29–43.

Albins MA. 2015. Invasive Pacific lionfish Pterois volitans reduce abundance and species richness of native Bahamian coral-reef fishes. *Marine Ecology Progress Series* 522:231–43.

Albins MA, Hixon MA. 2008. Invasive Indo-Pacific lionfish Pterois volitans reduce recruitment of Atlantic coral-reef fishes. *Marine Ecology Progress Series* 367:233–8.

Albins MA, Hixon MA. 2013. Worst case scenario: potential long-term effects of invasive predatory lionfish (Pterois volitans) on Atlantic and Caribbean coral-reef communities. *Environmental Biology of Fishes* 96(10–11):1151–7.

Alves-Stanley CD, Worthy GA, Bonde RK. 2010. Feeding preferences of West Indian manatees in Florida, Belize, and Puerto Rico as indicated by stable isotope analysis. *Marine Ecology Progress Series* 402:255–67.

Andradi-Brown DA, Grey R, Hendrix A, Hitchner D, Hunt CL, Gress E, Madej K, et al. 2017a. Depth-dependent effects of culling—do mesophotic lionfish populations undermine current management? *Royal Society Open Science* 4(5):170027.

Andradi-Brown DA, Vermeij MJ, Slattery M, Lesser M, Bejarano I, Appeldoorn R, Goodbody-Gringley G, et al. 2017b. Large-scale invasion of western Atlantic mesophotic reefs by lionfish potentially undermines culling-based management. *Biological Invasions* 19(3):939–54.

Anselmi-Molina CM, Canals M, Morell J, Gonzalez J, Capella J, Mercado A. 2012. Development of an operational nearshore wave forecast system for Puerto Rico and the US Virgin Islands. *Journal of Coastal Research* 28(5):1049–56.

Appeldoorn RS. 1988a. Age determination, growth, mortality and age of first reproduction in adult queen conch, Strombus gigas L., off Puerto Rico. *Fisheries Research* 6(4):363–78.

Appeldoorn RS. 1988b. Fishing pressure and reproductive potential in strombid conchs: is there a critical stock density for reproduction? *Memoria de la Sociedad de Ciencias Naturales La Salle* 48:275–88.

Appeldoorn RS. 1991. History and recent status of the Puerto Rican conch fishery. *Proceedings of the Gulf and Caribbean Fisheries Institute* 40:267–82.

Appeldoorn RS. 1992. Preliminary calculations of sustainable yield for queen conch (Strombus gigas) in Puerto Rico and the US Virgin Islands. *Proceedings of the Gulf and Caribbean Fisheries Institute* 41:95–105.

Appeldoorn RS. 2008. Transforming reef fisheries management: application of an ecosystem-based approach in the USA Caribbean. *Environmental Conservation* 35(3): 232–41.

Appeldoorn RS. 2018. Connectivity is everything. *Gulf and Caribbean Research* 29(1):ii–xix.

Appeldoorn R, Ballantine D, Bejarano I, Carlo M, Nemeth M, Otero E, Pagan F, et al. 2016. Mesophotic coral ecosystems under anthropogenic stress: a case study at Ponce, Puerto Rico. *Coral Reefs* 35(1):63–75.

Appeldoorn R, Beets J, Bohnsack J, Bolden S, Matos D, Meyers S, Rosario A, Sadovy Y, Tobias W. 1992. *Shallow Water Reef Fish Stock Assessment for the U.S. Caribbean. NOAA Technical Memorandum NMFS-SEFSC 304*. Washington, DC: NOAA (p. 70).

Appeldoorn RS, Gonzalez EC, Glazer R, Prada M. 2011. "Applying an ecosystem-based management to Queen Conch (Strombus gigas) fisheries in the Caribbean." In: *Towards Marine Ecosystem-Based Management in the Wider Caribbean*. Edited by L Fanning, R Mahon, P McConney. Amsterdam, Netherlands: Amsterdam University Press.

Appeldoorn RS, Kimmel JJ, Meyers ST, Sadovy YV, Valdés-Pizzini M. 2005. Fisheries management policy in Puerto Rico: a progress report. *Proceedings of the Gulf and Caribbean Fisheries Institute* 47:111–22.

Appeldoorn RS, Yoshioka PM, Ballantine DL. 2009. Coral reef ecosystem studies: integrating science and management in the Caribbean. *Caribbean Journal of Science* 45(2–3): 134–8.

Armstrong RA, Singh H, Torres J, Nemeth RS, Can A, Roman C, Eustice R, Riggs L, Garcia-Moliner G. 2006. Characterizing the deep insular shelf coral reef habitat of the Hind Bank marine conservation district (US Virgin Islands) using the Seabed autonomous underwater vehicle. *Continental Shelf Research* 26(2):194–205.

Arnold W. 2019. *Managing Trap Fisheries in the U.S. Virgin Islands: Review of Pertinent State and Federal Regulations. A Presentation to the Caribbean Fishery Management Council. 168th Caribbean Fishery Management Council Regular Meeting December 10–11, 2019 Ponce, Puerto Rico*. Washington, DC: NOAA.

Ault JS, Smith SG, Luo J, Monaco ME, Appeldoorn RS. 2008. Length-based assessment of sustainability benchmarks for coral reef fishes in Puerto Rico. *Environmental Conservation* 35(3):221–31.

Azziz-Baumgartner E, Luber G, Conklin L, Tosteson TR, Granade HR, Dickey RW, Backer LC. 2012. Assessing the incidence of ciguatera fish poisoning with two surveys conducted in Culebra, Puerto Rico, during 2005 and 2006. *Environmental Health Perspectives* 120(4):526–9.

Baker N, Appeldoorn RS, Torres-Saavedra PA. 2016. Fishery-independent surveys of the queen conch stock in western Puerto Rico, with an assessment of historical trends and management effectiveness. *Marine and Coastal Fisheries* 8(1):567–79.

Ballantine DL, Appeldoorn RS, Yoshioka P, Weil E, Armstrong R, Garcia JR, Otero E, et al. 2008. "Biology and ecology of Puerto Rican coral reefs." In: *Coral Reefs of the USA*. Edited by BM Riegl, RE Dodge, RS Appledoorn. Dordrecht, Netherlands: Springer (pp. 375–406).

Ban NC, Hansen GJ, Jones M, Vincent AC. 2009. Systematic marine conservation planning in data-poor regions: socio-economic data is essential. *Marine Policy* 33:794–800.

Barreto-Orta M, Méndez-Tejeda R, Rodríguez E, Cabrera N, Díaz E, Pérez K. 2019. State of the beaches in Puerto Rico after Hurricane Maria (2017). *Shore & Beach* 87(1):16.

Bastian J. 2001. A question of custody: the colonial archives of the United States Virgin Islands. *The American Archivist* 64(1):96–114.

Bauer LJ, Kendall MS. 2010. *An Ecological Characterization of the Marine Resources of Vieques, Puerto Rico Part II: Field Studies of Habitats, Nutrients, Contaminants, Fish, and Benthic Communities. NOAA Technical Memorandum NOS NCCOS 110*. Silver Spring, MD: NOAA (p. 174).

Beets J, Friedlander AL. 1992. Stock analysis and management strategies for Red Hind, Epinephelus guttatus in the US Virgin Islands. *Proceedings of the Gulf and Caribbean Fisheries Institute* 42:66–85.

Beets J, Friedlander A. 1999. Evaluation of a conservation strategy: a spawning aggregation closure for red hind, Epinephelus guttatus, in the US Virgin Islands. *Environmental Biology of Fishes* 55(1–2):91–8.

Bejarano I, Appeldoorn RS. 2013. Seawater turbidity and fish communities on coral reefs of Puerto Rico. *Marine Ecology Progress Series* 474:217–26.

Bejarano I, Appeldoorn RS, Nemeth M. 2014. Fishes associated with mesophotic coral ecosystems in La Parguera, Puerto Rico. *Coral Reefs* 33(2):313–28.

Bjorkland RH. 2011. *An Assessment of Sea Turtle, Marine Mammal and Seabird Bycatch in the Wider Caribbean Region. Doctoral Dissertation*. Durham, NC: Duke University.

Brander L, van Beukering P. 2013. *NOAA Coral Reef Conservation Program. The Total Economic Value of U.S. Coral Reefs: A Review of the Literature*. Silver Spring, MD: NOAA.

Brandt ME, Smith TB, Correa AMS, Vega-Thurber R. 2013. Disturbance driven colony fragmentation as a driver of a coral disease outbreak. *PLoS One* 8:e57164.

Bruckner AW, Hill RL. 2009. Ten years of change to coral communities off Mona and Desecheo Islands, Puerto Rico, from disease and bleaching. *Diseases of Aquatic Organisms* 87(1–2):19–31.

Bruno JF, Valdivia A, Hackerott S, Cox CE, Green S, Côté I, Akins L, Layman C, Precht W. 2013. Testing the grouper biocontrol hypothesis: a response to Mumby et al. *PeerJ PrePrints* 2013:e139v1.

Burke LM, Maidens J, Spalding M, Kramer P, Green E. 2004. *Reefs at Risk in the Caribbean*. Washington, DC: World Resources Institute.

Cabán PA. 2018. *Constructing a Colonial People: Puerto Rico and the United States, 1898–1932*. Abingdon, UK: Routledge.

Calnan J, Smith T, Nemeth R, Kadison E, Blondeau J. 2008. Coral disease prevalence and host susceptibility on mid-depth and deep reefs in the US Virgin Islands. *Revista Biologia Tropical* 56(Supplement 1):223–4.

Cambers G. 1997. Beach changes in the Eastern Caribbean Islands: hurricane impacts and implications for climate change. *Journal of Coastal Research* 1:29–47.

Cardona-Maldonado MA, Mignucci-Giannoni AA. 1999. Pygmy and dwarf sperm whales in Puerto Rico and the Virgin Islands, with a review of Kogia in the Caribbean. *Caribbean Journal of Science* 35(1–2):29–37.

CARICOOS (Caribbean Coastal Ocean Observing System). 2016. *CARICOOS Strategic Operational Plan 2016–2021*. Washington, DC: IOOS (p. 34).

Carlson LA, Keegan WF. 2004. "Resource depletion in the prehistoric northern West Indies." In: *Voyages of Discovery: The Archaeology of Islands*. Edited by SM Fitzpatrick. Santa Barbara, CA: Praeger (pp. 85–107).

Carretta JV, Karin. A. Forney, Erin M. Oleson, David W. Weller, Aimee R. Lang, Jason Baker, Marcia M. Muto, et al. 2019. *U.S. Pacific Marine Mammal Stock Assessments: 2018. NOAA Technical Memorandum NMFSSWFSC-617*. Washington, DC: NOAA.

Carriger JF, Fisher WS, Stockton Jr TB, Sturm PE. 2013. Advancing the Guánica Bay (Puerto Rico) watershed management plan. *Coastal Management* 41(1):19–38.

Cass-Calay SL, Arnold WS, Bryan MD, Schull J. 2016. *Report of the US Caribbean Fishery-Independent Survey Workshop. NOAA Technical Memorandum NMFS-SEFSC-688*. Washington, DC: NOAA (p. 128).

Castillo-Barahona F. 1981. Composition of the commercial catch of spiny lobster in Puerto Rico and the U.S. Virgin Islands. *Proceedings of the Gulf and Caribbean Fisheries Institute* 33:267–70.

CFMC (Caribbean Fishery Management Council). 2005. *Comprehensive Amendment to the Fishery Management Plans (FMPs) of the U.S. Caribbean to Address Required Provisions of the Magnuson-Stevens Fishery Conservation and Management Act*. San Juan, PR: Caribbean Fishery Management Council (p. 624).

CFMC (Caribbean Fishery Management Council). 2019a. *Comprehensive Fishery Management Plan for the Puerto Rico Exclusive Economic Zone, Including Draft Environmental Assessment. Prepared for the Caribbean Fishery Management Council 165th Meeting. Briefing Book Version (Draft)*. San Juan, PR: Caribbean Fishery Management Council (p. 607).

CFMC (Caribbean Fishery Management Council). 2019b. *Comprehensive Fishery Management Plan for the St. Croix Exclusive Economic Zone, Including Draft Environmental Assessment. Prepared for the Caribbean Fishery Management Council 165th Meeting. Briefing Book Version (Draft).* San Juan, PR: Caribbean Fishery Management Council (p. 487).

CFMC (Caribbean Fishery Management Council). 2019c. *Comprehensive Fishery Management Plan for the St. Thomas/St. John Exclusive Economic Zone, Including Draft Environmental Assessment. Prepared for the Caribbean Fishery Management Council 165th Meeting. Briefing Book Version (Draft).* San Juan, PR: Caribbean Fishery Management Council (p. 477).

Chagaris D, Binion-Rock S, Bogdanoff A, Dahl K, Granneman J, Harris H, Mohan J, et al. 2017. An ecosystem-based approach to evaluating impacts and management of invasive lionfish. *Fisheries* 42(8):421–31.

Chérubin LM, Richardson PL. 2007. Caribbean current variability and the influence of the Amazon and Orinoco freshwater plumes. *Deep Sea Research Part I: Oceanographic Research Papers* 54(9):1451–73.

Christensen JD, Jeffrey CF, Caldow C, Monaco ME, Kendall MS, Appeldoorn RS. 2003. Cross-shelf habitat utilization patterns of reef fishes in southwestern Puerto Rico. *Gulf and Caribbean Research* 14(2):9–27.

Clark R, Pittman SJ, Battista TA, Caldow C. 2012. *Survey and Impact Assessment of Derelict Fish Traps in St. Thomas and St. John, U.S. Virgin Islands. NOAA Technical Memorandum NOS NCCOS 147.* Silver Spring, MD: NOAA (p. 51).

Clark RD, Pittman S, Caldow C, Christensen J, Roque B, Appeldoorn RS, Monaco ME. 2009. Nocturnal fish movement and trophic flow across habitat boundaries in a coral reef ecosystem (SW Puerto Rico). *Caribbean Journal of Science* 45(2–3):282–304.

Clements J, Feliciano V, Almodovar-Caraballo BI, Colgan C. 2016. *Describing the Ocean Economies of the U.S. Virgin Islands and Puerto Rico.* Washington, DC: NOAA Office of Coastal Management.

Colin PL. 1995. Surface currents in Exuma Sound, Bahamas and adjacent areas with reference to potential larval transport. *Bulletin of Marine Science* 56(1):48–57.

Colin PL. 2003. Larvae retention: genes or oceanography? *Science* 300(5626):1657–9.

Collazo JA, Boulon Jr R, Tallevast TL. 1992. Abundance and growth patterns of Chelonia mydas in Culebra, Puerto Rico. *Journal of Herpetology* 1:293–300.

Collazo JA, Krachey MJ, Pollock KH, Pérez-Aguilo FJ, Zegarra JP, Mignucci-Giannoni AA. 2019. Population estimates of Antillean manatees in Puerto Rico: an analytical framework for aerial surveys using multi-pass removal sampling. *Journal of Mammalogy* 100(4):1340–9.

Cooley SR, Lucey N, Kite-Powell H, Doney SC. 2012. Nutrition and income from molluscs today imply vulnerability to ocean acidification tomorrow. *Fish and Fisheries* 13(2):182–215.

Côté IM, Akins L, Underwood E, Curtis-Quick J, Green SJ. 2014. Setting the record straight on invasive lionfish control: culling works. *PeerJ PrePrints* 2:e398v1.

Côté IM, Smith NS. 2018. The lionfish Pterois sp. invasion: has the worst-case scenario come to pass? *Journal of Fish Biology* 92(3):660–89.

Cox D, Arikawa T, Barbosa A, Guannel G, Inazu D, Kennedy A, Li Y, et al. Hurricanes Irma and Maria post-event survey in US Virgin Islands. *Coastal Engineering Journal* 5:1–4.

Craig RK. 2008. Coral reefs, fishing, and tourism: tensions in US Ocean Law Policy Reform. *Stanford Environmental Law Journal* 27:3.

CRCP (Coral Reef Conservation Program). 2007. *Comprehensive U.S. Caribbean Coral Reef Ecosystem Monitoring Project (C-CCREMP): Report on the Results of the C-CCREMP FY2006 Workshops.* Silver Spring, MD: NOAA (p. 12).

CRCP (Coral Reef Conservation Program). 2014. *National Coral Reef Monitoring Plan.* Silver Spring, MD, NOAA (p. 39).

Crosson S, Hibbert L. 2017. Integrating commercial fisheries registration, education, and social science in the U.S. Virgin Islands. *North American Journal of Fisheries Management* 37(2): 349–52.

Cróquer A, Weil E. 2009. Changes in Caribbean coral disease prevalence after the 2005 bleaching event. *Diseases of Aquatic Organisms* 87(1–2):33–43.

Cummings NJ, Karnauskas M, Harford W, Michaels WL, Acosta A. 2015. *Report of a GCFI Workshop: Strategies for Improving Fishery-Dependent Data for Use in Data-Limited Stock Assessments in the Wider Caribbean Region. Gulf and Caribbean Fisheries Institute Conference, Christ Church, Barbados, November 3–7, 2014.* NOAA Technical Memorandum NMFS-SEFSC-681. Washington, DC: NOAA (p. 25).

Cummings N, Sagarese S, Huynh QC. 2016. *An Alternative Approach to Setting Annual Catch Limits for Data-Limited Fisheries: Use of the DLM Tool and Mean Length Estimator for Six US Caribbean Stocks. SEDAR46-RW-03.* North Charleston, SC: SEDAR.

Cummings N, Trumble R, Wakeford R. 2007. *Expansion of the SEAMAP_C Fishery-Independent Sampling Program Overview Document. CFMC/NMFS/NOAA.* Washington, DC: NOAA.

Diez CE, Van Dam RP. 2002. Habitat effect on hawksbill turtle growth rates on feeding grounds at Mona and Monito Islands, Puerto Rico. *Marine Ecology Progress Series* 234:301–9.

Dietz JL. 2018. *Economic History of Puerto Rico: Institutional Change and Capitalist Development.* Princeton, NJ: Princeton University Press.

Doerr JC, Hill RL. 2018. Spatial distribution, density, and habitat associations of queen conch Strombus gigas in St. Croix, US Virgin Islands. *Marine Ecology Progress Series* 594:119–33.

Donner S, Knutson T, Oppenheimer M. 2007. Model-based assessment of the role of human-induced climate change in the 2005 Caribbean coral bleaching event. *Proceedings of the National Academy of Science* 104(13):5483–8.

Dookhan I. 1994. *A History of the Virgin Islands of the United States.* Kingston, Jamaica: Canoe Press.

DPNR-DFW (Department of Planning and Natural Resources, Division of Fish and Wildlife). 2016. *United States Virgin Islands Commercial and Recreational Fishers' Information Handbook.* St. Thomas, VI: DPNR (p. 52).

Eakin CM, Morgan JA, Heron SF, Smith TB, Liu G, Alvarez-Filip L, Baca B, et al. 2010. Caribbean corals in crisis: record thermal stress, bleaching, and mortality in 2005. *PloS One* 5(11):e13969.

Eckert SA, Nellis DW, Eckert KL, Kooyman GL. 1986. Diving patterns of two leatherback sea turtles (Dermochelys coriacea) during internesting intervals at Sandy Point, St. Croix, US Virgin Islands. *Herpetologica* 1:381–8.

Edmunds PJ, Witman JD. 1991. Effect of Hurricane Hugo on the primary framework of a reef along the south shore of St. John, US Virgin Islands. *Marine Ecology Progress Series* 78(2):201–4.

Edmunds PJ. 2000. Patterns in the distribution of juvenile corals and coral reef community structure in St. John, US Virgin Islands. *Marine Ecology Progress Series* 202:113–24.

Edmunds PJ. 2004. Juvenile coral population dynamics track rising seawater temperature on a Caribbean reef. *Marine Ecology Progress Series* 269:111–9.

Edwards SF, Link JS, Rountree BP. 2004. Portfolio management of wild fish stocks. *Ecological Economics* 49(3):317–29.

Ellison AM, Farnsworth EJ. 1996. Anthropogenic disturbance of Caribbean mangrove ecosystems: past impacts, present trends, and future predictions. *Biotropica* 1:549–65.

EPAP (Ecosystem Principles Advisory Panel). 1999. *Ecosystem-Based Fishery Management Report to Congress. National Marine Fisheries Service.* Silver Spring, MD: NOAA (p. 62).

Erdman DS. 1962. The sport fishery for blue marlin off Puerto Rico. *Transactions of the American Fisheries Society* 91(2):225–7.

Erisman BE, Asch R. 2015. Spatio-temporal interactions between fish spawning aggregations, fisheries, and climate change. *Proceedings of the Gulf and Caribbean Fisheries Institute* 67:230–1.

Fanning L, Mahon R, McConney P, Verhart L. 2011. *Towards Marine Ecosystem-Based Management in the Wider Caribbean.* Amsterdam, Netherlands: Amsterdam University Press.

Feliciano C. 1958. The lobster fishery of Puerto Rico. *Proceedings of the Gulf and Caribbean Fisheries Institute* 10:147–56.

Ferguson L, Srinivasan M, Oleson E, Hayes S, Brown SK, Angliss R, Carretta J, et al. 2017. *Proceedings of the First National Protected Species Assessment Workshop. NOAA Technical Memorandum NMFS-F/SPO-172.* Washington, DC: NOAA (p. 92).

Fiske SJ. 1992. Sociocultural aspects of establishing marine protected areas. *Ocean & Coastal Management* 17(1):25–46.

Fleming CS, Armentrout A, Crosson S. 2017. *Economic Survey Results for United States Virgin Islands Commercial Fisheries. NOAA Technical Memorandum NMFS-SEFSC-718.* Washington, DC: NOAA (p. 33).

Friedlander A. 1995. The recreational fishery for blue marlin, Makaira nigricans (Pisces: Istiophoridae), in the US Virgin Islands. *Fisheries Research* 22(3–4):163–73.

Friedlander AM, Jeffrey CF, Hile SD, Pittman SJ, Monaco ME, Caldow C. 2013. *Coral Reef Ecosystems of St. John, US Virgin Islands: Spatial and Temporal Patterns in Fish and Benthic Communities (2001–2009). NOAA Technical Memorandum NOS NCCOS 152.* Silver Spring, MD: NOAA.

Friedlander AM, Nowlis JO, Koike HA, Kittinger JN, McClenachan L, Gedan KB, Blight LK. 2014. Improving fisheries assessments using historical data. *Marine Historical Ecology in Conservation: Applying the Past to Manage for the Future* 24:91.

FWS (United States Fish and Wildlife Service). 2007. *5-Year Review—West Indian Manatee (Trichechus Manatus): Summary and Evaluation.* Jacksonville, FL: Fish and Wildlife Service Ecological Services.

FWS (United States Fish and Wildlife Service). 2017. Endangered and threatened wildlife and plants. Reclassification of the West Indian manatee from endangered to threatened. *Federal Register* 82(64):16668–704.

Garcia-Sais JR. 2010. Reef habitats and associated sessile-benthic and fish assemblages across a euphotic–mesophotic depth gradient in Isla Desecheo, Puerto Rico. *Coral Reefs* 29(2):277–88.

García-Sais J, Appeldoorn R, Battista T, Bauer L, Bruckner A, Caldow C, Carrubba L, et al. 2008. *The State of Coral Reef Ecosystems of Puerto Rico. The State of Coral Reef Ecosystems of the United States and Pacific Freely Associated States. NOAA Technical Memorandum NOS NCCOS 78.* Silver Spring, MD: NOAA (pp. 75–116).

García-Sais JR, Sabater-Clavell J, Esteves R, Carlo M. 2012. *Fishery-Independent Survey of Commercially Exploited Fish and Shellfish Populations from Mesophotic Reefs Within the Puerto Rican Exclusive Economic Zone.* San Juan, Puerto Rico: Caribbean Fisheries Management Council.

García-Quijano CG. 2007. Fishers' knowledge of marine species assemblages: bridging between scientific and local ecological knowledge in Southeastern Puerto Rico. *American Anthropologist* 109(3):529–36.

García-Quijano CG. 2009. Managing complexity: ecological knowledge and success in Puerto Rican small-scale fisheries. *Human Organization* 1:1–7.

Garrison V, Rogers C, Beets J, Friedlander A. 2004. The habitats exploited and the species trapped in a Caribbean island trap fishery. *Environmental Biology of Fishes* 71(3):247–60.

Garrison VH, Rogers CS, Beets J. 1998. Of reef fishes, overfishing and in situ observations of fish traps in St. John, US Virgin Islands. *Revista de Biologia Tropical* 46(5):41–59.

Gill DA, Oxenford HA, Turner RA, Schuhmann PW. 2017. Making the most of data-poor fisheries: low cost mapping of small island fisheries to inform policy. *Marine Policy* 101:198–207.

Gledhill DK, Wanninkhof R, Millero FJ, Eakin M. 2008. Ocean acidification of the greater Caribbean region 1996–2006. *Journal of Geophysical Research: Oceans* 113(C10).

Goedeke TL, Orthmeyer A, Edwards P, Dillard MK, Gorstein M, Jeffrey CFG. 2016. Characterizing Participation in Non-Commercial Fishing and Other Shore-based Recreational Activities on St. Croix, U.S. Virgin Islands. NOAA Technical Memorandum NOS NCCOS 209. Silver Spring, MD: NOAA (p. 93).

Green SJ, Akins JL, Maljković A, Côté IM. 2012. Invasive lionfish drive Atlantic coral reef fish declines. *PloS One* 7(3):e32596.

Green SJ, Dulvy NK, Brooks AM, Akins JL, Cooper AB, Miller S, Côté IM. 2014. Linking removal targets to the ecological effects of invaders: a predictive model and field test. *Ecological Applications* 24(6):1311–22.

Griffin DW, Kellogg CA. 2004. Dust storms and their impact on ocean and human health: dust in Earth's atmosphere. *EcoHealth* 1(3):284–95.

Griffith DM, Pizzini V, García Quijano CG. 2007. *Entangled Communities: Socioeconomic Profiles of Fishers, their Communities, and their Responses to Marine Protective Measures in Puerto Rico. NOAA Series on U.S. Caribbean Fishing Communities. NOAA Technical Memorandum NMFS-SEFSC-556*. Washington, DC: NOAA (p. 524).

Guénette S, Hill RL. 2009. A trophic model of the coral reef ecosystem of La Parguera, Puerto Rico: synthesizing fisheries and ecological data. *Caribbean Journal of Science* 45(2–3):317–37.

Hackerott S, Valdivia A, Green SJ, Côté IM, Cox CE, Akins L, Layman CA, Precht WF, Bruno JF. 2013. Native predators do not influence invasion success of Pacific lionfish on Caribbean reefs. *PLoS One* 8(7):e68259.

Hall NA. 1992. *Slave Society in the Danish West Indies: St. Thomas, St. John, and St. Croix*. Aarhus, Denmark: Aarhus Universitetsforlag.

Halpern BS. 2004. Are mangroves a limiting resource for two coral reef fishes? *Marine Ecology Progress Series* 272:93–8.

Harborne AR, Mumby PJ, Micheli F, Perry CT, Dahlgren CP, Holmes KE, Brumbaugh DR. 2006. The functional value of Caribbean coral reef, seagrass and mangrove habitats to ecosystem processes. *Advances in Marine Biology* 50:57–189.

Hardison DR, Holland WC, Darius HT, Chinain M, Tester PA, Shea D, Bogdanoff AK, et al. 2018. Investigation of ciguatoxins in invasive lionfish from the greater Caribbean region: implications for fishery development. *Plos One* 13(6):e0198358.

Harford WJ, Sagarese SR, Karnauskas M. 2019. Coping with information gaps in stock productivity for rebuilding and achieving maximum sustainable yield for grouper–snapper fisheries. *Fish and Fisheries* 20(2):303–21.

Harms-Tuohy CA, Appeldoorn RS, Craig MT. 2018. The effectiveness of small-scale lionfish removals as a management strategy: effort, impacts and the response of native prey and piscivores. *Management of Biological Invasions* 9:149–62.

Harms-Tuohy CA, Schizas NV, Appeldoorn RS. 2016. Use of DNA metabarcoding for stomach content analysis in the invasive lionfish Pterois volitans in Puerto Rico. *Marine Ecology Progress Series* 558:181–91.

Hayes SA, Josephson E, Maze-Foley K, Rosel PE. 2019. *US Atlantic and Gulf of Mexico Marine Mammal Stock Assessments-2018. NOAA Tech Memo NMFS-NE-258*. Washington, DC: NOAA. (p. 298).

Hernandez-Delgado EA, Alicea-Rodriguez L, Toledo CG, Sabat AM. 2000. Baseline characterization of coral reefs and fish communities within the proposed Culebra Island marine fishery reserve, Puerto Rico. *Proceedings of the Gulf and Caribbean Fisheries Institute* 51:537–55.

Hernández-Delgado EA, Sabat AM. 2000. Ecological status of essential fish habitats through an anthropogenic environmental stress gradient in Puerto Rican coral reefs. *Proceedings of the Gulf and Caribbean Fisheries Institute* 51:457–70.

Hernández-Delgado EA, Toledo-Hernández C, Ruíz-Díaz CP, Gómez-Andújar N, Medina-Muñiz JL, Canals-Silander MF, Suleimán-Ramos SE. 2020. Hurricane impacts and the resilience of the invasive sea vine, Halophila stipulacea: a case study from Puerto Rico. *Estuaries and Coasts* 8:1–21.

Hinderstein LM, Marr JC, Martinez FA, Dowgiallo MJ, Puglise KA, Pyle RL, Zawada DG, Appeldoorn R. 2010. Theme section on Mesophotic coral ecosystems: characterization, ecology, and management. *Coral Reefs* 29:247–51.

Hoegh-Guldberg O, Mumby PJ, Hooten AJ, Steneck RS, Greenfield P, Gomez E, Harvell CD, et al. 2007. Coral reefs under rapid climate change and ocean acidification. *Science* 318(5857):1737–42.

Hoogland ML, Hofman CL. 1999. Expansion of the Taino cacicazgos towards the Lesser Antilles. *Journal de la Société des Américanistes* 1:93–113.

Hopkinson CS, Lugo AE, Alber M, Covich AP, Van Bloem SJ. 2008. Forecasting effects of sea-level rise and windstorms on coastal and inland ecosystems. *Frontiers in Ecology and the Environment* 6(5):255–63.

Hubbard DK, Parsons KM, Bythell JC, Walker ND. 1991. The effects of Hurricane Hugo on the reefs and associated environments of St. Croix, US Virgin Islands—a preliminary assessment. *Journal of Coastal Research* 1:33–48.

James RK, Christianen MJ, van Katwijk MM, de Smit JC, Bakker ES, Herman PM, Bouma TJ. 2020. Seagrass coastal protection services reduced by invasive species expansion and megaherbivore grazing. *Journal of Ecology* 108(5):2025–37.

Johnson D. 2002. Environmentally sustainable cruise tourism: a reality check. *Marine Policy* 26(4):261–70.

Jordan RW, Winter A. 2000. Assemblages of coccolithophorids and other living microplankton off the coast of Puerto Rico during January–May 1995. *Marine Micropaleontology* 39(1–4):113–30.

Kadison E, Brandt M, Nemeth R, Martens J, Blondeau J, Smith TB. 2017. Abundance of commercially important reef fish indicates different levels of over-exploitation across shelves of the U.S. Virgin Islands. *PLoS One* 12:e0180063.

Kadison E, Nemeth RS, Blondeau J, Smith T, Calnan J. 2010. Nassau grouper (Epinephelus striates) in St. Thomas, US Virgin Islands, with evidence for a spawning aggregation site recovery. *Proceedings of the Gulf and Caribbean Fisheries Institute* 62:273–9.

Karp MA, Blackhart K, Lynch PD, Deroba J, Hanselman D, Gertseva V, Teo S, Townsend H, Williams E, Yau A. 2019. *Proceedings of the 13th National Stock Assessment Workshop: Model Complexity, Model Stability, and Ensemble Modeling. NOAA Technical Memoranda NMFS-F/SPO-189*. Washington, DC: NOAA (p. 49).

Kellner JB, Sanchirico JN, Hastings A, Mumby PJ. 2011. Optimizing for multiple species and multiple values: tradeoffs inherent in ecosystem-based fisheries management. *Conservation Letters* 4(1):21–30.

Kendall MS, Buja KR, Christensen JD, Kruer CR, Monaco ME. 2004. The seascape approach to coral ecosystem mapping: an integral component of understanding the habitat utilization patterns of reef fish. *Bulletin of Marine Science* 75(2):225–37.

Kendall MS, Monaco ME, Buja KR, Christensen JD, Kruer CR, Finkbeiner M, Warner RA. 2001. *Methods Used to Map the Benthic Habitats of Puerto Rico and the U.S. Virgin Islands.* Silver Spring, MD: NOAA.

Kendall MS, Takata LT, Jensen O, Hillis-Starr Z, Monaco ME. 2005. *An Ecological Characterization of Salt River Bay National Historical Park and Ecological Preserve, U.S. Virgin Islands. NOAA Technical Memorandum NOS NCCOS 14.* Washington, DC: NOAA (p. 116).

Kimmel JJ, Appeldoorn RS. 1992. A critical review of fisheries and fisheries management policy in Puerto Rico. *Proceedings of the Gulf and Caribbean Fisheries Institute* 41:349–60.

Kleypas JA, Yates KK. 2009. Coral reefs and ocean acidification. *Oceanography* 22(4):108–17.

Kojis BL, Quinn NJ. 2006. A census of US Virgin Islands commercial fishers at the start of the 21st century. *Proceedings of the 10th International Coral Reef Symposium* 1326:1334.

Kojis BL, Quinn NJ. 2011. *Census of Marine Commercial Fishers of the US Virgin Islands in 2010.* Washington, DC: NOAA National Marine Fisheries Service (p. 125).

Kojis B, Quinn N, Agar J. 2017. *Census of Licensed Fishers of the U.S. Virgin Islands. 2016. NOAA Technical Memorandum NMFS-SEFSC-715.* Washington, DC: NOAA (p. 160).

Kojis BL, Tobias WJ. 2016. *Survey of Boat-Based Recreational Fishers in the US Virgin Islands.* Honolulu, Hawaii: Proceedings of the 13th International Coral Reef Symposium (pp. 170–83).

Larsen MC, Webb RM. 2009. Potential effects of runoff, fluvial sediment, and nutrient discharges on the coral reefs of Puerto Rico. *Journal of Coastal Research* 251:189–208.

Layman CA, Jud ZR, Nichols P. 2014. Lionfish alter benthic invertebrate assemblages in patch habitats of a subtropical estuary. *Marine Biology* 161(9):2179–82.

Lefebvre LW, Reid JP, Kenworthy WJ, Powell JA. 1999. Characterizing manatee habitat use and seagrass grazing in Florida and Puerto Rico: implications for conservation and management. *Pacific Conservation Biology* 5(4):289–98.

Lessios HA, Robertson DR, Cubit JD. 1984. Spread of diadema mass mortality through the Caribbean. *Science* 226(4672):335–7.

Levin PS, Essington TE, Marshall KN, Koehn LE, Anderson LG, Bundy A, Carothers C, et al. 2018. Building effective fishery ecosystem plans. *Marine Policy* 92:48–57.

Levin PS, Fogarty MJ, Murawski SA, Fluharty D. 2009. Integrated ecosystem assessments: developing the scientific basis for ecosystem-based management of the ocean. *PLoS Biology* 7(1):e1000014.

Levin PS, Kelble CR, Shuford RL, Ainsworth C, deReynier Y, Dunsmore R, Fogarty MJ, et al. 2013. Guidance for implementation of integrated ecosystem assessments: a US perspective. *ICES Journal of Marine Science* 71(5):1198–204.

Lewis JB. 1977. Processes of organic production on coral reefs. *Biological Reviews* 52:305–47.

Link J. 2010. *Ecosystem-Based Fisheries Management: Confronting Tradeoffs.* Cambridge, UK: Cambridge University Press.

Link JS. 2018. System-level optimal yield: increased value, less risk, improved stability, and better fisheries. *Canadian Journal of Fisheries and Aquatic Sciences* 75(1):1–6.

Link JS, Browman HI. 2014. Integrating what? Levels of marine ecosystem-based assessment and management. *ICES Journal of Marine Science* 71(5):1170–3.

Link JS, Marshak AR. 2019. Characterizing and comparing marine fisheries ecosystems in the United States: determinants of success in moving toward ecosystem-based fisheries management. *Reviews in Fish Biology and Fisheries* 29(1):23–70.

Lirman D. 1999. Reef fish communities associated with Acropora palmata: relationships to benthic attributes. *Bulletin of Marine Science* 65:235–52.

Loeffler CR, Robertson A, Flores Quintana HA, Silander MC, Smith TB, Olsen D. 2018. Ciguatoxin prevalence in 4 commercial fish species along an oceanic exposure gradient in the US Virgin Islands. *Environmental Toxicology and Chemistry* 37(7):1852–63.

López A. 1980. *The Puerto Ricans: Their History, Culture, and Society*. Cambridge, MA: Schenkman Publishing Company.

López-Marrero T, Yarnal B. 2010. Putting adaptive capacity into the context of people's lives: a case study of two flood-prone communities in Puerto Rico. *Natural Hazards* 52(2):277–97.

López R, López JM, Morell J, Corredor JE, Castillo CE. 2013. Influence of the Orinoco River on the primary production of eastern Caribbean surface waters. *Journal of Geophysical Research: Oceans* 118(9):4617–32.

Lugo AE. 1980. Mangrove ecosystems: successional or steady state? *Biotropica* 1:65–72.

Lugo AE. 2000. Effects and outcomes of Caribbean hurricanes in a climate change scenario. *Science of the Total Environment* 262(3):243–51.

Lugo AE. 2002. Conserving Latin American and Caribbean mangroves: issues and challenges. *Madera y Bosques* 8(Es1):5–25.

Lugo AE. 2018. *Social-Ecological-Technological Effects of Hurricane María on Puerto Rico: Planning for Resilience under Extreme Events*. Berlin, Germany: Springer.

Lugo AE, Snedaker SC. 1974. The ecology of mangroves. *Annual Review of* Ecology and Systematics 5:39–64.

Lugo AE, Rogers CS, Nixon SW. 2000. Hurricanes, coral reefs and rainforests: resistance, ruin and recovery in the Caribbean. *AMBIO: A Journal of the Human Environment* 29(2):106–15.

Lugo-Fernández A, Hernández-Avila ML, Roberts HH. 1994. Wave-energy distribution and hurricane effects on Margarita Reef, southwestern Puerto Rico. *Coral Reefs* 13(1):21–32.

Lynch PD, Methot RD, Link JS. 2018. *Implementing a Next Generation Stock Assessment Enterprise. An Update to the NOAA Fisheries Stock Assessment Improvement Plan. NOAA Technical Memo NMFS-F/ SPO-183*. Washington, DC: NOAA (p. 127).

Manooch III CS, Drennon CL. 1987. Age and growth of yellowtail snapper and queen triggerfish collected from the US Virgin Islands and Puerto Rico. *Fisheries Research* 6(1):53–68.

Marshak AR, Appeldoorn RS. 2008. Evaluation of seasonal closures of red hind, Epinephelus guttatus, spawning aggregations to fishing of the west coast of Puerto Rico using fishery-dependent and independent time series data. *Proceedings of the Gulf and Caribbean Fisheries Institute* 60:566–72.

Marshak AR, Appeldoorn RS, Jimenez N. 2006. Utilization of GIS mapping in the measurement of the spatial distribution of queen conch (Strombus gigas) in Puerto Rico. *Proceedings Gulf and Caribbean Fisheries Institute* 57:31–48.

Marshak AR, Heck Jr KL, Jud ZR. 2018. Ecological interactions between Gulf of Mexico snappers (Teleostei: Lutjanidae) and invasive red lionfish (Pterois volitans). *PloS One* 13(11):e0206749.

Marshak AR, Hill RL, Sheridan PE, Scharer MT, Appeldoorn RS. 2008. In-situ observations of Antillean fish trap contents in southwest Puerto Rico: relating catch to habitat and damage potential. *Proceedings of the Gulf and Caribbean Fisheries Institute* 60:447–553.

Marshall KN, Levin PS, Essington TE, Koehn LE, Anderson LG, Bundy A, Carothers C, et al. 2018. Ecosystem-based fisheries management for social–ecological systems: renewing the focus in the United States with next generation fishery ecosystem plans. *Conservation Letters* 11(1):e12367.

Martin CA. 2019. A mysterious coral disease is ravaging Caribbean reefs. *Science News* 9.

Martinuzzi S, Gould WA, Lugo AE, Medina E. 2009. Conversion and recovery of Puerto Rican mangroves: 200 years of change. *Forest Ecology and Management* 257(1):75–84.

Mateo I, Appeldoorn RS, Rolke W. 1998. Spatial variations in stock abundance of queen conch, Strombus gigas, (Gastropoda; Strombidae) in the west and east coast of Puerto Rico. *Proceedings of the Gulf and Caribbean Fisheries Institute* 50:32–48.

Matos-Caraballo D. 1997. Status of the groupers in Puerto Rico, 1970–1995. *Proceedings of the Gulf and Caribbean Fisheries Institute* 49:340–53.

Matos-Caraballo D. 2001. Overview of the spiny lobster, Panulirus argus, commercial fishery in Puerto Rico during 1992–1998. *Proceedings of the Gulf and Caribbean Fisheries Institute* 52:194–203.

Matos-Caraballo D. 2002. Portrait of the commercial fishery of the Red Hind, Epinephelus guttatus, in Puerto-Rico during 1992–1999. *Proceedings of the Gulf and Caribbean Fisheries Institute* 53:446–59.

Matos-Caraballo DA. 2009. Lessons learned from the Puerto Rico's commercial fishery, 1988–2008. *Proceedings of the Gulf and Caribbean Fisheries Institute* 61:123–8.

Matos-Caraballo D, Agar JJ. 2011. Census of active commercial fishermen in Puerto Rico: 2008. *Marine Fisheries Review* 73(1):13–27.

Matos-Caraballo DA, Agar JU. 2011b. Comprehensive census of the marine commercial fishery of Puerto Rico, 2008. *Proceedings of the Gulf and Caribbean Fisheries Institute* 63: 99–112.

Matos-Caraballo D, Posada JM, Luckhurst BE. 2006. Fishery-dependent evaluation of a spawning aggregation of tiger grouper (Mycteroperca tigris) at Vieques Island, Puerto Rico. *Bulletin of Marine Science* 79(1):1–6.

Mattox NT. 1952. A preliminary report on the biology and economics of the spiny lobster in Puerto Rico. *Proceedings of the Gulf and Caribbean Fisheries Institute* 4:69–70.

McGehee MA. 1994. Correspondence between assemblages of coral reef fishes and gradients of water motion, depth, and substrate size off Puerto Rico. *Marine Ecology Progress Series* 105:243–55.

Melendez M, Salisbury J. 2017. Impacts of ocean acidification in the coastal and marine environments of Caribbean small island developing states (SIDS). *Commonwealth Marine Economies Programme: Caribbean Marine Climate Change Report Card: Science Review* 2017:31–9.

Melendez E, Vargas-Ramos C. 2017. *Estimates of Post-Hurricane Maria Exodus from Puerto Rico (Centro RB2017-01)*. New York, NY: City University of New York.

Michener WK, Blood ER, Bildstein KL, Brinson MM, Gardner LR. 1997. Climate change, hurricanes and tropical storms, and rising sea level in coastal wetlands. *Ecological Applications* 7(3):770–801.

Miller J, Muller E, Rogers C, Waara R, Atkinson A, Whelan KR, Patterson M, Witcher B. 2009. Coral disease following massive bleaching in 2005 causes 60% decline in coral cover on reefs in the US Virgin Islands. *Coral Reefs* 28(4):925.

Mignucci-Giannoni AA. 1998. Zoogeography of cetaceans off Puerto Rico and the Virgin Islands. *Caribbean Journal of Science* 34(3–4):173–90.

Miller J, Muller E, Rogers C, Waara R, Atkinson A, Whelan KR, Patterson M, Witcher B. 2009. Coral disease following massive bleaching in 2005 causes 60% decline in coral cover on reefs in the US Virgin Islands. *Coral Reefs* 28(4):925.

Miller J, Waara R, Muller E, Rogers C. 2006. Coral bleaching and disease combine to cause extensive mortality on reefs in US Virgin Islands. *Coral Reefs* 25(3):418.

Monaco ME, Friedlander AM, Caldow C, Christensen JD, Rogers C, Beets J, Miller J, Boulon R. 2007. Characterising reef fish populations and habitats within and outside the US Virgin Islands Coral Reef National Monument: a lesson in marine protected area design. *Fisheries Management and Ecology* 14(1):33–40.

Monaco ME, Friedlander AM, Caldow C, Hile SD, Menza C, Boulon RH. 2009. Long-term monitoring of habitats and reef fish found inside and outside the US Virgin Islands Coral Reef National Monument: a comparative assessment. *Caribbean Journal of Science* 45(2–3):338–48.

Moore WR, Harewood L, Grosvenor T. 2010. The supply side effects of climate change on tourism. *MPRA Paper* 21469.

Morell JM, Corredor JE. 2001. Photomineralization of fluorescent dissolved organic matter in the Orinoco River plume: estimation of ammonium release. *Journal of Geophysical Research: Oceans* 106(C8):16807–13.

Muhling BA, Liu Y, Lee SK, Lamkin JT, Malca E, Llopiz J, Ingram Jr GW, et al. 2015. Past, ongoing and future research on climate change impacts on tuna and billfishes in the Western Atlantic. *ICCAT Collective Volume of Scientific Papers* 71:1716–2.

Mumby PJ, Brumbaugh DR, Harborne AR, Roff G. 2013. On the relationship between native grouper and invasive lionfish in the Caribbean. *PeerJ PrePrints* 2:e45v1.

Munro JL. 1983. *Caribbean Coral Reef Fishery Resources (Vol. 7)*. Makati City, Philippines: ICLARM Studies and Reviews (pp. 1–276).

Murphy SJ, Hurlburt HE, O'Brien JJ. 1999. The connectivity of eddy variability in the Caribbean Sea, the Gulf of Mexico, and the Atlantic Ocean. *Journal of Geophysical Research: Oceans* 104(C1):1431–53.

Muto MM, Helker VT, Angliss RP, Boveng PL, Breiwick JM, Cameron MF, Clapham PJ, et al. 2019. *Alaska Marine Mammal Stock Assessments, 2018. NOAA Technical Memo NMFS-AFSC-393*. Washington, DC: NOAA (p. 390).

Narozanski A, Box S, Stoyle G. 2013. Developing management tools to enhance efficiency of marine protected areas management in Honduras. *Proceedings of the Gulf and Caribbean Fisheries Institute* 62:221–6.

Nemeth M, Appeldoorn R. 2009. The distribution of herbivorous coral reef fishes within fore-reef habitats: the role of depth, light and rugosity. *Caribbean Journal of Science* 45(2–3):247–54.

Nemeth RS. 2005. Population characteristics of a recovering US Virgin Islands red hind spawning aggregation following protection. *Marine Ecology Progress Series* 286:81–97.

Nemeth RS, Blondeau J, Herzlieb S, Kadison E. 2007. Spatial and temporal patterns of movement and migration at spawning aggregations of red hind, Epinephelus guttatus, in the US Virgin Islands. *Environmental Biology of Fishes* 78(4):365–81.

Nemeth RS, Herzlieb S, Blondeau J. 2006a. Comparison of two seasonal closures for protecting red hind spawning aggregations in the US Virgin Islands. *Proceedings of the 10th International Coral Reef Symposium* 4:1306–13.

Nemeth RS, Kadison EL, Herzlieb ST, Blondeau JE, Whiteman EA. 2006b. Status of a Yellowfin (Mycteroperca venenosa) grouper spawning aggregation in the US Virgin Islands with notes on other species. *Proceedings of the Gulf and Caribbean Fisheries Institute* 57:543–58.

Nemeth RS, Whaylen LD, Pattengill-Semmens CV. 2003. A rapid assessment of coral reefs in the Virgin Islands (Part 2: Fishes). *Atoll Research Bulletin* 496:566–89.

Newman D, Berkson J, Suatoni L. 2015. Current methods for setting catch limits for data-limited fish stocks in the United States. *Fisheries Research* 164:86–93.

NMFS (National Marine Fisheries Service). 2014. *Endangered and Threatened Wildlife and Plants: Final Listing Determinations on Proposal To List 66 Reef-Building Coral Species and To Reclassify Elkhorn and Staghorn Corals. Federal Register 79, No. 175*. Washington, DC: NOAA.

NMFS (National Marine Fisheries Service). 2016a. *Ecosystem-Based Fisheries Management Policy of the National Marine Fisheries Service.* Washington, DC: NOAA.

NMFS (National Marine Fisheries Service). 2016b. *NOAA Fisheries Ecosystem-Based Fisheries Management Roadmap.* Washington, DC: NOAA.

NMFS (National Marine Fisheries Service). 2017. *National Marine Fisheries Service – 2nd Quarter 2017 Update.* Washington, DC: NOAA (p. 53).

NMFS (National Marine Fisheries Service). 2018a. *Fisheries Economics of the United States, 2016. NOAA Technical Memo NMFS-F/SPO-187.* Washington, DC: NOAA (p. 243).

NMFS (National Marine Fisheries Service). 2018b. *Fisheries of the United States, 2017. NOAA Current Fishery Statistics No. 2017.* Washington, DC: NOAA.

NMFS (NOAA Fisheries). 2019. *Ecosystem-Based Fishery Management in the U.S. Caribbean Region: Roadmap Implementation Plan.* Washington, DC: NOAA (p. 16).

NMFS (National Marine Fisheries Service). 2020. *National Marine Fisheries Service – 2nd Quarter 2020 Update.* Washington, DC: NOAA (p. 51).

NOAA (National Oceanic and Atmospheric Administration). 2016. *An Implementation Framework for NOAA's Habitat Blueprint Focus Area in the Caribbean – The Northeast Marine Corridor and Culebra Island, Puerto Rico.* Washington, DC: NOAA (p. 42).

NOAA (National Oceanic and Atmospheric Administration). 2018. *Hurricanes Irma and Maria Damage Assessment: Provisional Results for the U.S. Virgin Islands Commercial and For-Hire Fisheries. 60-Day Interim Report.* Washington, DC: NOAA (p. 19).

NOAA (National Oceanic and Atmospheric Administration). 2018. *Hurricanes Irma and Maria Damage Assessment: Provisional Results for the Puerto Rican Commercial and For-Hire Fisheries. 60-Day Interim Report.* Washington, DC: NOAA (p. 25).

NOAA (National Oceanic and Atmospheric Administration). 2020. *NOAA Report on the U.S. Marine Economy.* Charleston, SC: NOAA Office for Coastal Management (p. 23).

O'Hop J, Murphy M, Chagaris D. 2012. *The 2012 Stock Assessment Report for Yellowtail Snapper in the South Atlantic and Gulf of Mexico.* Fish and Wildlife Conservation Commission. St. Petersburg, FL: Fish and Wildlife Research Institute (p. 342).

Odum HT, Burkholder P, Rivero J. 1959. Measurements of productivity of turtle grass flats, reefs, and the Bahia Fosforescente of southern Puerto Rico. *Publications of the Institute of Marine Science* 15:9–170.

Ogden JC. 2010. Marine spatial planning (MSP): a first step to ecosystem-based management (EBM) in the Wider Caribbean. *Revista de Biologıa Tropical* 58(3):71–9.

Ojeda-Serrano ED, Appeldoorn RS, Ruiz-Valentin ID. 2007. Reef fish spawning aggregations of the Puerto Rican shelf. *Proceedings of the Gulf and Caribbean Fisheries Institute* 59:401–8.

Olinger LK, Heidmann SL, Durdall AN, Howe C, Ramseyer T, Thomas SG, Lasseigne DN, et al. 2017. Altered juvenile fish communities associated with invasive Halophila stipulacea seagrass habitats in the US Virgin Islands. *PLoS One* 12(11).

Olsen DA, LaPlace JA. 1979. A study of a Virgin Islands grouper fishery based on a breeding aggregation. *Proceedings of the Gulf and Caribbean Fisheries Institute* 31:130–44.

Opitz S. 1993. A quantitative model of the trophic interactions in a Caribbean coral reef ecosystem. Trophic models of aquatic ecosystems. *ICLARM Conference Proceedings* 26:259–68.

Opitz S. 1996. *Trophic Interactions in Caribbean Coral Reefs. ICLARM Tech Report 43.* Makati City, Philippines: ICLARM (p. 341).

Oxenford HA, Monnereau I. 2017. Impacts of climate change on fish and shellfish in the coastal and marine environments of Caribbean small island developing states (SIDS). *Caribbean Marine Climate Change Report Card: Science Review* 2017:83–114.

Pandolfi JM, Jackson JB, Baron N, Bradbury RH, Guzman HM, Hughes TP, Kappel CV, et al. 2005. Are US coral reefs on the slippery slope to slime? *Science* 307:1725–26.

Patrick WS, Link JS. 2015. Myths that continue to impede progress in ecosystem-based fisheries management. *Fisheries* 40(4):155–60.

Pendleton LH. 2002. *A Preliminary Study of the Value of the Coastal Tourism in Rincón, Puerto Rico.* San Clemente, CA: Environmental Defense, Surfer's Environmental Alliance & The Surfrider Foundation.

Pestle WJ. 2013. Fishing down a prehistoric Caribbean marine food web: Isotopic evidence from Punta Candelero, Puerto Rico. *The Journal of Island and Coastal Archaeology* 8(2):228–54.

Pittman SJ, Brown KA. 2011. Multi-scale approach for predicting fish species distributions across coral reef seascapes. *PloS One* 6(5):e20583.

Pittman SJ, Caldow C, Hile SD, Monaco ME. 2007a. Using seascape types to explain the spatial patterns of fish in the mangroves of SW Puerto Rico. *Marine Ecology Progress Series* 348:273–84.

Pittman SJ, Christensen JD, Caldow C, Menza C, Monaco ME. 2007b. Predictive mapping of fish species richness across shallow-water seascapes in the Caribbean. *Ecological Modelling* 204(1–2):9–21.

Pittman SJ, Dorfman DS, Hile SD, Jeffrey CFG, Edwards MA, Caldow C. 2013. *Land-Sea Characterization of the St. Croix East End Marine Park, U.S. Virgin Islands. NOAA Technical Memorandum NOS NCCOS 170.* Silver Spring, MD: NOAA (p. 119).

Pittman SJ, Hile SD, Jeffrey CFG, Caldow C, Kendall MS, Monaco ME, Hillis-Starr Z. 2008. *Fish Assemblages and Benthic Habitats of Buck Island Reef National Monument (St. Croix, U.S. Virgin Islands) and the Surrounding Seascape:*

A Characterization of Spatial and Temporal Patterns. NOAA Technical Memorandum NOS NCCOS 71. Silver Spring, MD: NOAA (p. 96).

Pittman SJ, Hile SD, Jeffrey CF, Clark R, Woody K, Herlach BD, Caldow C, Monaco ME, Appeldoorn R. 2010. *Coral Reef Ecosystems of Reserva Natural La Parguera (Puerto Rico): Spatial and Temporal Patterns in Fish and Benthic Communities (2001–2007). NOAA Technical Memorandum NOS NCCOS 107*. Silver Spring, MD: NOAA (p. 202).

Pittman SJ, Hitt S, Renchen GF, Jeffrey CFG. 2012. *Synthesis of Marine Ecosystem Monitoring Activities for the United States Virgin Islands: 1990–2009. NOAA Technical Memorandum NOS NCCOS 148*. Silver Spring, MD: NOAA (p. 55).

Pittman SJ, Jeffrey CFG, Menza C, Kågesten G, Orthmeyer A, Dorfman DS, Mateos-Molina D, et al. 2017. *Mapping Ecological Priori*θ*es and Human Impacts to Support Land-Sea Management of Puerto Rico's Northeast Marine Corridor. NOAA Technical Memorandum NOS NCCOS 218*. Silver Spring, MD: NOAA (p. 71).

Powell JA, Belitsky DW, Rathbun GB. 1981. Status of the West Indian manatee (Trichechus manatus) in Puerto Rico. *Journal of Mammalogy* 62(3):642–6.

PR-DNER (Puerto Rico Department of Natural and Environmental Resources). 2010. *Reglamento de Pesca de Puerto Rico*. San Juan, Puerto Rico: Departamento de Recursos Naturales y Ambientales (p. 99).

Randall JE. 1964. Contributions to the biology of the queen conch, Strombus gigas. *Bulletin of Marine Science* 14(2):246–95.

Raymond WW, Albins MA, Pusack TJ. 2015. Competitive interactions for shelter between invasive Pacific red lionfish and native Nassau grouper. *Environmental Biology of Fishes* 98(1):57–65.

Roberts CM. 1997. Connectivity and management of Caribbean coral reefs. *Science* 278(5342):1454–7.

Robertson A, Garcia AC, Quintana HA, Smith TB, Castillo II BF, Reale-Munroe K, Gulli JA, et al. 2014. Invasive lionfish (Pterois volitans): a potential human health threat for ciguatera fish poisoning in tropical waters. *Marine Drugs* 12(1):88–97.

Rocha LA, Rocha CR, Baldwin CC, Weigt LA, McField M. 2015. Invasive lionfish preying on critically endangered reef fish. *Coral Reefs* 34(3):803–6.

Roden CL, Mullin KD. 2000. Sightings of cetaceans in the northern Caribbean Sea and adjacent waters, winter 1995. *Caribbean Journal of Science*. 36(3–4):280–8.

Rodriguez RW, Webb RM, Bush DM. 1994. Another look at the impact of Hurricane Hugo on the shelf and coastal resources of Puerto Rico, USA. *Journal of Coastal Research* 1:278–96.

Rodríguez-Barreras R, Montanez-Acuna A, Otano-Cruz A, Ling SD. 2018. Apparent stability of a low-density Diadema antillarum regime for Puerto Rican coral reefs. *ICES Journal of Marine Science* 75(6):2193–201.

Rodriguez-Ferrer G, Reyes R, Hammerman NM, García-Hernández JE. 2019. Cetacean sightings in Puerto Rican waters: including the first underwater photographic documentation of a minke whale (Balaenoptera acutorostrata). *Latin American Journal of Aquatic Mammals* 13(1–2):26–36.

Rogers CS. 1979. The productivity of San Cristobal Reef, Puerto Rico. *Limnology and Oceanography* 24:342–9.

Rogers CS. 1990. Responses of coral reefs and reef organisms to sedimentation. *Marine Ecology Progress Series* 62(1):185–202.

Rogers CS, Beets J. 2001. Degradation of marine ecosystems and decline of fishery resources in marine protected areas in the US Virgin Islands. *Environmental Conservation* 28(4):312–22.

Rogers CS, McLain LN, Tobias CR. 1991. Effects of Hurricane Hugo(1989) on a coral reef in St. John, USVI. *Marine Ecology Progress Series* 78(2):189–99.

Rogers CS, Miller J. 2006. Permanent phase shifts' or reversible declines in coral cover? Lack of recovery of two coral reefs in St. John, US Virgin Islands. *Marine Ecology Progress Series* 306:103–14.

Rogers CS, Miller J, Muller EM, Edmunds P, Nemeth RS, Beets JP, Friedlander AM, et al. 2008. *Ecology of Coral Reefs in the US Virgin Islands. Coral Reefs of the USA*. Dordrecht, Netherlands: Springer (pp. 303–73).

Rogers CS, Suchanek TH, Pecora FA. 1982. Effects of hurricanes David and Frederic (1979) on shallow Acropora palmata reef communities: St. Croix, US Virgin Islands. *Bulletin of Marine Science* 32(2):532–48.

Rooker JR. 1995. Feeding ecology of the schoolmaster snapper, Lutjanus apodus (Walbaum), from southwestern Puerto Rico. *Bulletin of Marine Science* 56(3):881–94.

Rothenberger P, Blondeau J, Cox C, Curtis S, Fisher WS, Garrison V, Hillis-Starr Z, et al. 2008. The state of coral reef ecosystems of the US Virgin Islands. *The State of Coral Reef Ecosystems of the United States and Pacific Freely Associated States* 73:29–73.

Rowell TJ, Nemeth RS, Schärer MT, Appeldoorn RS. 2015. Fish sound production and acoustic telemetry reveal behaviors and spatial patterns associated with spawning aggregations of two Caribbean groupers. *Marine Ecology Progress Series* 518:239–54.

Ruiz H, Ballantine DL, Sabater J. 2017. Continued Spread of the seagrass Halophila stipulacea in the Caribbean: documentation in Puerto Rico and the British Virgin Islands. *Gulf and Caribbean Research* 28(1):SC5–7.

Ruiz-Ramos DV, Hernández-Delgado EA, Schizas NV. 2011. Population status of the long-spined urchin Diadema antillarum in Puerto Rico 20 years after a mass mortality event. *Bulletin of Marine Science* 87(1):113–27.

Ryan KE, Walsh JP, Corbett DR, Winter A. 2008. A record of recent change in terrestrial sedimentation in a coral-reef environment, La Parguera, Puerto Rico: a response to coastal development? *Marine Pollution Bulletin* 56(6): 1177–83.

Sadovy Y. 1999. The case of the disappearing grouper: Epinephelus striatus, the Nassau grouper, in the Caribbean and western Atlantic. *Proceedings of the Gulf and Caribbean Fisheries Institute* 45:5–22.

Sadovy Y, Colin PL, Domeier ML. 1994. Aggregation and spawning in the tiger grouper, Mycteroperca tigris (Pisces: Serranidae). *Copeia* 1994(2):511–16.

Sadovy Y, Eklund AM. 1999. *Synopsis of Biological Data On the Nassau Grouper, Epinephelus Striatus (Bloch, 1792), and the Jewfish, E. Itajara (Lichtenstein, 1822). NOAA Technical Reports NMFS 146*. Washington, DC: NOAA (p. 65).

Sadovy Y, Figuerola M. 1992. The status of the red hind fishery in Puerto Rico and St. Thomas as determined by yield-per-recruit analysis. *Proceedings of the Gulf and Caribbean Fisheries Institute* 42:23–38.

Sagarese SR, Rios AB, Cass-Calay SL, Cummings NJ, Bryan MD, Stevens MH, Harford WJ, McCarthy KJ, Matter VM. 2018. Working Towards a Framework for Stock Evaluations in Data-Limited Fisheries. *North American Journal of Fisheries Management* 38(3):507–37.

Sagarese SR, Walter JF, Isely JJ, Bryan MD, Cummings N. 2015. *A Comparison of Datarich Versus Data-Limited Methods in Estimating Overfishing Limits. SEDAR46-DW-01.* North Charleston, SC: SEDAR (p. 28).

Sammarco PW, Winter A, Stewart JC. 2006. Coefficient of variation of sea surface temperature (SST) as an indicator of coral bleaching. *Marine Biology* 149(6):1337–44.

Salas S, Chuenpagdee R, Seijo JC, Charles A. 2007. Challenges in the assessment and management of small-scale fisheries in Latin America and the Caribbean. *Fisheries Research* 87(1):5–16.

Sanders IM, Barrios-Santiago JC, Appeldoorn RS. 2005. Distribution and relative abundance of humpback whales off western Puerto Rico during 1995–1997. *Caribbean Journal of Science* 41:101–7.

Scatena FN, Larsen MC. 1991. Physical aspects of hurricane Hugo in Puerto Rico. *Biotropica* 1:317–23.

Schärer M, Nemeth M, Valdivia A, Miller M, Williams D, Diez C. 2008. Elkhorn coral distribution and condition throughout the Puerto Rican Archipelago. *Proceedings of the 11th International Coral Reef Symposium* 11:815–819.

Schärer MT, Nemeth MI, Mann D, Locascio J, Appeldoorn RS, Rowell TJ. 2012. Sound production and reproductive behavior of yellowfin grouper, Mycteroperca venenosa (Serranidae) at a spawning aggregation. *Copeia* 2012(1): 135–44.

Schärer MT, Rowell TJ, Nemeth MI, Appeldoorn RS. 2012. Sound production associated with reproductive behavior of Nassau grouper Epinephelus striatus at spawning aggregations. *Endangered Species Research* 19(1):29–38.

Schärer-Umpierre MT, Mateos-Molina D, Appeldoorn R, Bejarano I, Hernández-Delgado EA, Nemeth RS, Nemeth MI, Valdés-Pizzini M, Smith TB. 2014. Marine managed areas and associated fisheries in the US Caribbean. *Advances in Marine Biology* 69:129–52.

Schofield PJ. 2009. Geographic extent and chronology of the invasion of non-native lionfish (Pterois volitans [Linnaeus 1758] and P. miles [Bennett 1828]) in the Western North Atlantic and Caribbean Sea. *Aquatic Invasions* 4(3):473–9.

SEDAR (Southeast Data, Assessment, and Review). 2005. *Stock Assessment Report of SEDAR 8 Caribbean Yellowtail Snapper.* Charleston, SC: SEDAR (p. 179).

SEDAR (Southeast Data, Assessment, and Review). 2019. *SEDAR 57 Stock Assessment Report U.S. Caribbean Spiny Lobster.* Charleston, SC: SEDAR (p. 232).

Searles TA. 2004. *Ecological Modeling of Coral Reef Ecosystems in Puerto Rico using ECOPATH with Ecosim. Paper read at NOAA Undergraduate Summer Research Symposium, August 2004.* Washington, DC: NOAA.

Shapiro DY, Sadovy Y, McGehee MA. 1993. Size, composition, and spatial structure of the annual spawning aggregation of the red hind, Epinephelus guttatus (Pisces: Serranidae). *Copeia* 3:399–406.

Sherman C, Nemeth M, Ruíz H, Bejarano I, Appeldoorn R, Pagán F, Schärer M, Weil E. 2010. Geomorphology and benthic cover of mesophotic coral ecosystems of the upper insular slope of southwest Puerto Rico. *Coral Reefs* 29(2):347–60.

Smikle SG, Christensen V, Aiken KA. 2010. A review of Caribbean ecosytems and fishery resources using ECOPATH models. *Études Caribéennes* 1(15).

Smith NS, Green SJ, Akins JL, Miller S, Côté IM. 2017. Density-dependent colonization and natural disturbance limit the effectiveness of invasive lionfish culling efforts. *Biological Invasions* 19(8):2385–99.

Smith TB, Nemeth RS, Blondeau J, Calnan JM, Kadison E, Herzlieb S. 2008. Assessing coral reef health across onshore to offshore stress gradients in the US Virgin Islands. *Marine Pollution Bulletin* 56(12):1983–91.

Smith TB, Blondeau J, Nemeth RS, Pittman SJ, Calnan JM, Kadison E, Gass J. 2010. Benthic structure and cryptic mortality in a Caribbean mesophotic coral reef bank system, the Hind Bank Marine Conservation District, US Virgin Islands. *Coral Reefs* 29(2):289–308.

SEDAR (Southeast Data, Assessment, and Review). 2016. *SEDAR 46. Stock Assessment Report Caribbean Data-Limited Species.* North Charleston, SC: SEDAR (p. 373).

Smith TB, Brandt ME, Calnan JM, Nemeth RS, Blondeau J, Kadison E, Taylor M, Rothenberger JP. 2013. Convergent mortality responses of Caribbean coral species to seawater warming. *Ecosphere* 4:art87.

Smith TB, Brandtneris VW, Canals M, Brandt ME, Martens J, Brewer RS, Kadison E, et al. 2016a. Potential structuring forces on a shelf edge upper mesophotic coral ecosystem in the US Virgin Islands. *Frontiers in Marine Science* 3:115.

Smith TB, Ennis RS, Kadison E, Weinstein DW, Jossart J, Gyory J, Henderson L. 2015. *The United States Virgin Islands Territorial Coral Reef Monitoring Program. Year 15 Annual Report.* St. Croix, Virgin Islands: University of the Virgin Islands (p. 288).

Smith TB, Gyory J, Brandt ME, Miller WJ, Jossart J, Nemeth RS. 2016b. Caribbean mesophotic coral ecosystems are unlikely climate change refugia. *Global Change Biology* 22:2756–65.

Srinivasan M, Brown SK, Markowitz E, Soldevilla M, Patterson E, Forney K, Murray K, et al. 2019. *Proceedings of the 2nd National Protected Species Assessment Workshop. NOAA Technical Memo NMFS-F/SPO-198*. Washington, DC: NOAA (p. 46).

Stallings C. 2009. Fishery-independent data reveal negative effect of human population density on Caribbean predatory fish communities. *PLoS One* 4(5):e5333.

Swartz SL, Martinez A, Stamates J, Burks C, Mignucci-Giannoni AA. 2002. *Acoustic and Visual Survey of Cetaceans in the Waters of Puerto Rico and the Virgin Islands: February–March 2001. NOAA Technical Memo NMFS-SEFSC 463*. Washington, DC: NOAA (p. 62).

Theile S. 2005. Status of the queen conch Strombus gigas stocks, management and trade in the Caribbean: a CITES review. *Proceedings of the Gulf and Caribbean Fisheries Institute* 56:675–94.

Toledo-Hernández C, Vélez-Zuazo X, Ruiz-Diaz CP, Patricio AR, Mege P, Navarro M, Sabat AM, Betancur-R R, Papa R. 2014. Population ecology and genetics of the invasive lionfish in Puerto Rico. *Aquatic Invasions* 9(2).

Tonioli FC, Agar JJ. 2009. Extending the Bajo de Sico, Puerto Rico, seasonal closure: an examination of small-scale fishermen's perceptions of possible socio-economic impacts on fishing practices, families, and community. *Marine Fisheries Review* 71(2):15–23.

Tonioli, FC, Agar JJ. 2011. *Synopsis of Puerto Rican Commercial Fisheries. NOAA Technical Memorandum NMFS-SEFSC-622* (p. 69).

Torres JL. 2001. Impacts of sedimentation on the growth rates of Montastraea annularis in southwest Puerto Rico. *Bulletin of Marine Science* 69(2):631–7.

Townsend HM, Link JS, Osgood KE, Gedamke T, Watters GM, Polovina JJ, Levin PS, Cyr EC, Aydin KY. 2008. *Report of the National Ecosystem Modeling Workshop (NEMoW). NOAA Technical Memorandum NMFS-F/SPO-102*. Washington, DC: NOAA.

Uysal M, Jurowski C, Noe FP, McDonald CD. 1994. Environmental attitude by trip and visitor characteristics: US Virgin Islands National Park. *Tourism Management* 15(4):284–94.

Valdés-Pizzini M. 1990. Fishermen associations in Puerto Rico: praxis and discourse in the politics of fishing. *Human Organization* 49(2):164.

Valdés-Pizzini M. 2007. Reflections of the way life used to be: anthropology, history and the decline of the fish stocks in Puerto Rico. *Proceedings of the Gulf and Caribbean Fisheries Institute* 59:37–48.

Valdés-Pizzini M, Agar JJ, Kitner K, García-Quijano C, Tust M, Forrestal F. 2010. *Cruzan Fisheries: A Rapid Assessment of the Historical, Social, Cultural and Economic Processes that Shaped Coastal Communities' Dependence and Engagement in Fishing in the Island of St. Croix, U.S. Virgin Islands. NOAA Technical Memorandum NMFS-SEFSC-597*. Washington, DC: NOAA (p. 144).

Valdés-Pizzini M, García-Quijano CG, Schärer-Umpierre MT. 2012. Connecting humans and ecosystems in tropical fisheries: social sciences and the ecosystem-based fisheries management in Puerto Rico and the Caribbean. *Caribbean Studies* 1:95–128.

Valdés-Pizzini M, Schärer-Umpierre M. 2014. *People, Habitats, Species, and Governance: An Assessment of the Social-Ecological System of La Parguera, Puerto Rico. Interdisciplinary Center for Coastal Studies*. Mayagüez, Puerto Rico: University of Puerto Rico.

Valdivia A, Bruno JF, Cox CE, Hackerott S, Green SJ. 2014. Re-examining the relationship between invasive lionfish and native grouper in the Caribbean. *PeerJ* 2:e348.

Valle-Esquivel M, Díaz G. 2003. *Preliminary Estimation of Reported Landings, Expansion Factors and Expanded Landings for the Commercial Fisheries of the United States Virgin Islands. Sustainable Fisheries Division Contribution SFD-2003-0027*. Silver Spring, MD: NOAA NMFS.

Valle-Esquivel M, Shivlani M, Matos-Caraballo M, Die D. 2011. *Coastal Fisheries of Puerto Rico. FAO Fisheries and Aquaculture Technical Paper. No. 544*. Rome, Italy: FAO (pp. 285–313).

Van Dam RP, Diez CE. 1998. Home range of immature hawksbill turtles (Eretmochelys imbricata (Linnaeus)) at two Caribbean islands. *Journal of Experimental Marine Biology and Ecology* 220(1):15–24.

Van Dam RP, Diez CE, Balazs GH, Colón LA, McMillan WO, Schroeder B. 2008. Sex-specific migration patterns of hawksbill turtles breeding at Mona Island, Puerto Rico. *Endangered Species Research* 4(1–2):85–94.

Viana IG, Siriwardane-de Zoysa R, Willette DA, Gillis LG. 2019. Exploring how non-native seagrass species could provide essential ecosystems services: a perspective on the highly invasive seagrass Halophila stipulacea in the Caribbean Sea. *Biological Invasions* 21(5):1461–72.

Wabnitz CCC. 2010. Sea turtle conservation and ecosystem-based management with a focus on green turtles (Chelonia mydas) and seagrass beds (Doctoral dissertation, University of British Columbia).

Ward-Paige CA, Mora C, Lotze HK, Pattengill-Semmens C, McClenachan L, Arias-Castro E, Myers RA. 2010. Large-scale absence of sharks on reefs in the greater-Caribbean: a footprint of human pressures. *PloS One* 5(8):e11968.

Waring GT, Josephson E, Maze-Foley K, Rosel PE. 2015. *US Atlantic and Gulf of Mexico Marine Mammal Stock Assessments—2014. NOAA Technical Memorandum NMFS-NE-231*. Washington, DC: NOAA (p. 361).

Warne AG, Webb RM, Larsen MC. 2005. *Water, Sediment, and Nutrient Discharge Characteristics of Rivers in Puerto Rico, and Their Potential Influence On Coral Reefs*. Washington, DC: Department of the Interior, US Geological Survey.

Watson M, Munro JL. 2004. Settlement and recruitment of coral reef fishes in moderately exploited and overexploited Caribbean ecosystems: implications for marine protected areas. *Fisheries Research* 69(3):415–25.

Weil E, Croquer A, Urreiztieta I. 2009. Temporal variability and impact of coral diseases and bleaching in La Parguera, Puerto Rico from 2003–2007. *Caribbean Journal of Science* 45(2–3):221–47.

Whelan KRT, Miller J, Sanchez O, Patterson M. 2007. Impact of the 2005 coral bleaching event on Porites porites and Colpophyllia natans at Tektite Reef, US Virgin Islands. *Coral Reefs* 26:689–93.

Whitall DR, Costa BM, Bauer LJ, Dieppa A, Hile SD. 2011. *A Baseline Assessment of the Ecological Resources of Jobos Bay, Puerto Rico. NOAA Technical Memorandum NOS NCCOS 133*. Silver Spring, MD: NOAA (p.188).

Whitman ER, Heithaus MR, Barcia LG, Brito DN, Rinaldi C, Kiszka JJ. 2019. Effect of seagrass nutrient content and relative abundance on the foraging behavior of green turtles in the face of a marine plant invasion. *Marine Ecology Progress Series* 628:171–82.

Wilkinson CR, Souter D. 2008. *Status of Caribbean Coral Reefs After Bleaching and Hurricanes in 2005*. Townsville, Australia: Global Coral Reef Monitoring Network, and Reef and Rainforest Research Centre (p. 152).

Wilkinson, EB, Abrams K. 2015. *Benchmarking the 1999 EPAP Recommendations with existing Fishery Ecosystem Plans. NOAA Technical Memorandum NMFS-OSF-5*. Washington, DC: NOAA (p. 22).

Willette DA, Chiquillo KL, Cross C, Fong P, Kelley T, Toline CA, Zweng R, Muthukrishnan R. 2020. Growth and recovery after small-scale disturbance of a rapidly-expanding invasive seagrass in St. John, US Virgin Islands. *Journal of Experimental Marine Biology and Ecology* 523:151265.

Wilson SM. 1993. The cultural mosaic of the indigenous Caribbean. *Proceedings of the British Academy* 81:37–66.

Winter A, Appeldoorn RS, Bruckner A, Williams Jr EH, Goenaga C. 1998. Sea surface temperatures and coral reef bleaching off La Parguera, Puerto Rico (northeastern Caribbean Sea). *Coral Reefs* 17(4):377–82.

Wirt KE, Hallock P, Palandro D, Lunz KS. 2015. Potential habitat of Acropora spp. on reefs of Florida, Puerto Rico, and the US Virgin Islands. *Global Ecology and Conservation* 3:242–55.

Yoshioka PM, Yoshioka BB. 1987. Variable effects of hurricane David on the shallow water gorgonians of Puerto Rico. *Bulletin of Marine Science* 40(1):132–44.

CHAPTER 7

The Northern Gulf of Mexico

➲ The Region in Brief

- The northern Gulf of Mexico (nGOM) contains one of the nation's largest marine economies (among the eight U.S. regional marine ecosystems), which is dependent on offshore mineral extractions, tourism, marine transportation, living marine resources (LMRs), and other ocean uses.
- This system is driven primarily by circulation of the Loop Current, which originates from the Caribbean current at the Yucatán Channel, and flows into the Florida Current at the Florida Straits.
- The nGOM contains the fourth-highest basal productivity in the U.S., contributing to some of the highest levels of system biomass, landings, and LMR-dependent jobs and revenue in the nation.
- Within this region, average sea surface temperature has increased by 0.7°C since the mid-twentieth century.
- The region contains high numbers of marine species comprising commercially and recreationally important invertebrates (e.g., penaeid shrimp, Blue crab, Eastern oyster) and finfish (e.g., Red snapper, groupers, Red drum, pelagic sportfishes) fisheries, which contribute heavily to national landings and seafood supply.
- Productive vegetated inshore habitats (salt marsh, seagrass, mangrove), oyster beds, and offshore artificial, rocky, and coral reef habitats (e.g., Flower Gardens National Marine Sanctuary) support numerous economically and ecologically important species throughout their life histories.
- This region is subject to natural stressors that include the highest frequency of hurricanes in the U.S. Atlantic basin, recurrent harmful algal blooms, and a large concentrated area of hypoxic bottom water.
- Effects of climate change in the nGOM are being observed with further tropicalization of the biota in this region, shifts of species into deeper habitats, sea-level rise, and effects of ocean acidification on shellfish and corals.
- Trade-offs among interconnected commercial and recreational fisheries, protected species, and other ocean economies are concentrated in this diverse fisheries ecosystem.
- Much fisheries management has been focused upon strict rebuilding programs for historically overfished reef fishes, including Red snapper, groupers, and Gray triggerfish. However, data limitations and a significant percentage of unassessed stocks (~45%) continue to affect LMR management in this region.
- Among subregions, the Mississippi River Delta and Mobile Bay contribute most toward Gulf of Mexico fisheries landings, LMR revenue, and LMR employments, while human stressors are also high surrounding Galveston Bay and the Mississippi River Delta.
- Overall, ecosystem-based fisheries management (EBFM) progress has been made in terms of implementing ecosystem-level planning, advancing knowledge of ecosystem principles, and in assessing risks and vulnerabilities to ecosystems through ongoing investigations into climate vulnerability and species prioritizations for stock and habitat assessments.
- This ecosystem is excelling in the areas of socioeconomic status, human environment, and is relatively productive, as related to the determinants of successful LMR management.

Ecosystem-Based Fisheries Management: Progress, Importance, and Impacts in the United States. Jason S. Link and Anthony R. Marshak, Oxford University Press. © U.S. Department of Commerce, U.S. Government 2021. DOI: 10.1093/oso/9780192843463.003.0007

7.1 Introduction

When one thinks of the northern Gulf of Mexico (nGOM) ecosystem, iconic fisheries species like Brown (*Farfantepenaeus aztecus*) and White shrimp (*Litopenaeus setiferus*), Blue crab (*Callinectes sapidus*), and Red snapper (*Lutjanus campechanus*) immediately come to mind. Images of shrimp boats, oil rigs, Gulf coast beaches, charter fishing, and oyster houses emerge, and eating a nice bowl of seafood gumbo, some blackened redfish, or a "fully dressed" shrimp or oyster po'boy conjure up the seafood famous from this region. Thoughts of the nGOM ecosystem invoke the sounds of New Orleans jazz, blues, and zydeco, Mardi Gras celebrations and Blue Angels airshows, in addition to stories by Kate Chopin, John Grisham, and Anne Rice, or musings from Mark Twain throughout portions of the Mississippi River Delta. Popular songs, including those about Galveston, Biloxi, and Mobile, remind us that this region has been historically known as "The Playground of the South." These cultural aspects are rooted in the natural environments that comprise the nGOM ecosystem, especially its diverse marine habitats and dependent living marine resources (LMRs), which are the foundation for its significant marine economy and the well-being of its coastal communities.

There are many works characterizing the nGOM ecosystem (Deegan et al. 1986, Conner et al. 1989, Rezak et al. 1990, Kumpf et al. 1999, Chesney & Baltz 2001, Yáñez-Arancibia & Day 2004, Vidal & Pauly 2004, Cowan et al. 2008, Felder & Camp 2009, Heileman & Rabalais 2009, Engle 2011, Karnauskas et al. 2013, Yáñez-Arancibia et al. 2013, Yoskowitz et al. 2013, Ainsworth et al. 2015, Karnauskas et al. 2015, De Mutsert et al. 2016, Karnauskas et al. 2017a, b, Ward & Tunnell 2017, Kilborn et al. 2018), the important and iconic fisheries species (e.g., shrimps, oysters, Blue crab, drums, reef fishes, mackerels; Butler 1954, Hildebrand 1954, Houde 1971, Houde et al. 1979, GMFMC 1981a, b, Perry et al. 1982, GMFMC & SAFMC 1983, Johnson et al. 1994, Goodyear 1995, Coleman et al. 1996, Gallaway et al. 1999, Zimmerman et al. 2000, Heck et al. 2001, DeVries et al. 2002, Zimmerman et al. 2002, Nance 2004, Caillouet et al. 2008, Roth et al. 2008, Wells et al. 2008b, Beck et al. 2011, Scott-Denton et al. 2011, Hart 2012, VanderKooy 2012, Schrandt & Powers 2015, Farmer et al. 2016, Chen 2017), protected species therein (e.g., sea turtles, dolphins, manatees, whales, sharks, and corals; Bright et al. 1984, Powell & Rathbun 1984, Plotkin et al. 1993, Blaylock & Hoggard 1994, Renaud & Carpenter 1994, Würsig et al. 2000, Cortés et al. 2002, Fertl et al. 2005, Carlson et al. 2007, Tyson et al. 2011, Chen 2017), and the sociocultural context (Adams et al. 2004, Jepson 2007, Jacob et al. 2013, Fleming et al. 2014, Colburn et al. 2016) within which humans use, enjoy, and interact with this marine ecosystem. This part of the ocean, its resources, and its goods and services have had an important part in American history (Gramling & Hagelman 2005, Barbier 2013, Batker et al. 2014), as seen in the 1803 Louisiana Purchase, its strategic role during the War of 1812, U.S. Civil War, and World War II, and major increases in contributions toward national seafood and energy production during the twentieth century (Flick & Martin 1990, Cotham 1998, Adams et al. 2004, Austin et al. 2004, Buker et al. 2004, Kukla 2004, Gruesz 2006, Hearn 2010).

What follows is a brief background for this region, describing all the major considerations of this marine ecosystem. Then there is a short synopsis of how all the salient data for those factors are collected, analyzed, and used for LMR management. We then present an evaluation of key facets of ecosystem-based fisheries management (EBFM) for this region.

7.2 Background and Context

7.2.1 Habitat and Geography

The nGOM ecosystem encompasses a major portion of the southeastern United States Exclusive Economic Zone (EEZ) and consists of the coastal and offshore waters surrounding the states of Texas, Louisiana, Mississippi, Alabama, and western Florida. This region forms the northern component of the Gulf of Mexico Large Marine Ecosystem and is bounded to the south by the Gulf waters of Mexico, including the Gulf of Campeche and Cuba (Kumpf et al. 1999). Extensive connectivity occurs among these northern and southern areas and the Caribbean Sea through the Loop Current, which also connects with the Florida Current in the Florida Straits (Fig. 7.1).

Human population, fisheries landings, and economic production are concentrated in five key subregions from west to east: **Matagorda** and **Galveston Bays**—Texas; **The Mississippi River Delta**—Louisiana and Mississippi; **Mobile Bay**—Alabama, and **Tampa Bay**—Florida (Fig. 7.2). Additionally, major economically important areas of the Florida Gulf coast include Apalachicola Bay and Apalachee Bay. There are other subregional features in this region, but we emphasize these five herein as exemplary, smaller-scale ecosystems. Throughout the nGOM, these inshore subregions contain diverse vegetated (salt marsh, seagrass, and more recently climate-associated range expansions of tropical

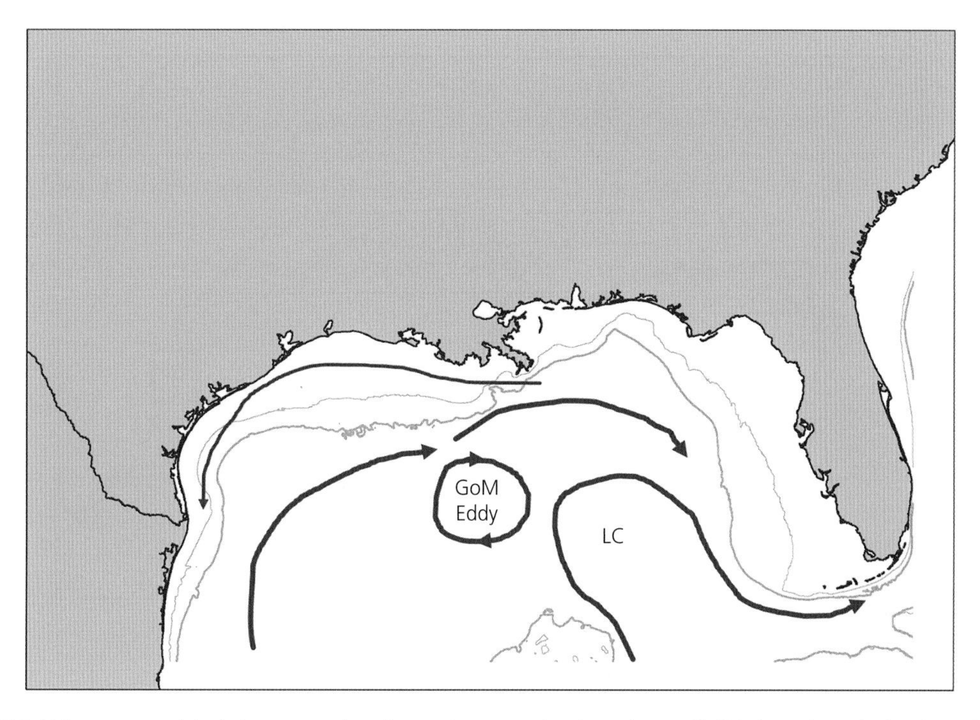

Figure 7.1 Major currents and circulation patterns throughout and encompassing the northern Gulf of Mexico region and subregions. LC = Loop Current (aka Gulf Loop Current); GoM Eddy = Gulf of Mexico eddy.

mangroves) and non-vegetated (oyster reef) habitats that support the early and adult life stages of many commercially and recreationally important fishes (e.g., Red drum—*Sciaenops ocellatus*, Spotted sea trout—*Cynoscion nebulosus*, some snappers and groupers).

The main features of the nGOM ecosystem include productive wetlands and offshore environments, with circulation and climate forcing driven by the Loop Current and Atlantic Multidecadal Oscillation (AMO) (Rezak et al. 1990, Cowan et al. 2008, Engle 2011, Liu et al. 2012, Karnauskas et al. 2013, 2015, 2017). Diverse natural environments within this region range from nearshore soft sediments and vegetated areas to offshore rocky and sandy substrates, including natural and artificial reefs, which provide habitats to support a multitude of ecologically and economically important species (Weinstein & Heck 1979, Dennis & Bight 1988, Minello & Rozas 2002, Zimmerman et al. 2002, Patterson et al. 2005, Wells et al. 2008a, Fodrie et al. 2010, Ajemian et al. 2015, Karnauskas et al. 2017a, Minello et al. 2017). Shallow and deep coral reefs are found within the Gulf of Mexico, including the Flower Gardens National Marine Sanctuary in the northwestern portion, which support high numbers of protected and managed taxa (Moore & Bullis 1960, Gittings et al. 1992, Rooker et al. 1997, Aronson et al. 2005, Reed et al. 2006, Hickerson et al. 2008, Semmler et al. 2016).

7.2.2 Climate and Other Major Stressors

The nGOM is subject to many stressors, which include hypoxia, eutrophication, harmful algal blooms (HABs), overfishing, climate change, and biological invasions. The human population along the northern Gulf of Mexico coast has expanded rapidly, causing many stressors to increase (Stumpf et al. 2003, Walsh et al. 2006, Steidinger 2009, Villareal & Magaña 2016, Walker et al. 2018). The most well-known example of these pressures is an expansive hypoxic area off Louisiana driven by high nutrient loading from Mississippi River discharge, commonly referred to as "the dead zone" (Rabalais et al. 2001, 2002, Scavia et al. 2003, Bianchi 2007, Bianchi et al. 2010). Nutrient loading from increasing coastal development, together with local circulation patterns, is suspected to fuel HABs as well, such as those observed during past and recent red tide (*Karenia brevis*) events throughout southwest Florida and the nGOM (Hu et al. 2006, Walsh et al. 2006, Vargo et al. 2008, Weisberg et al. 2019). Eutrophication is also persistent throughout nGOM estuaries due to coastal

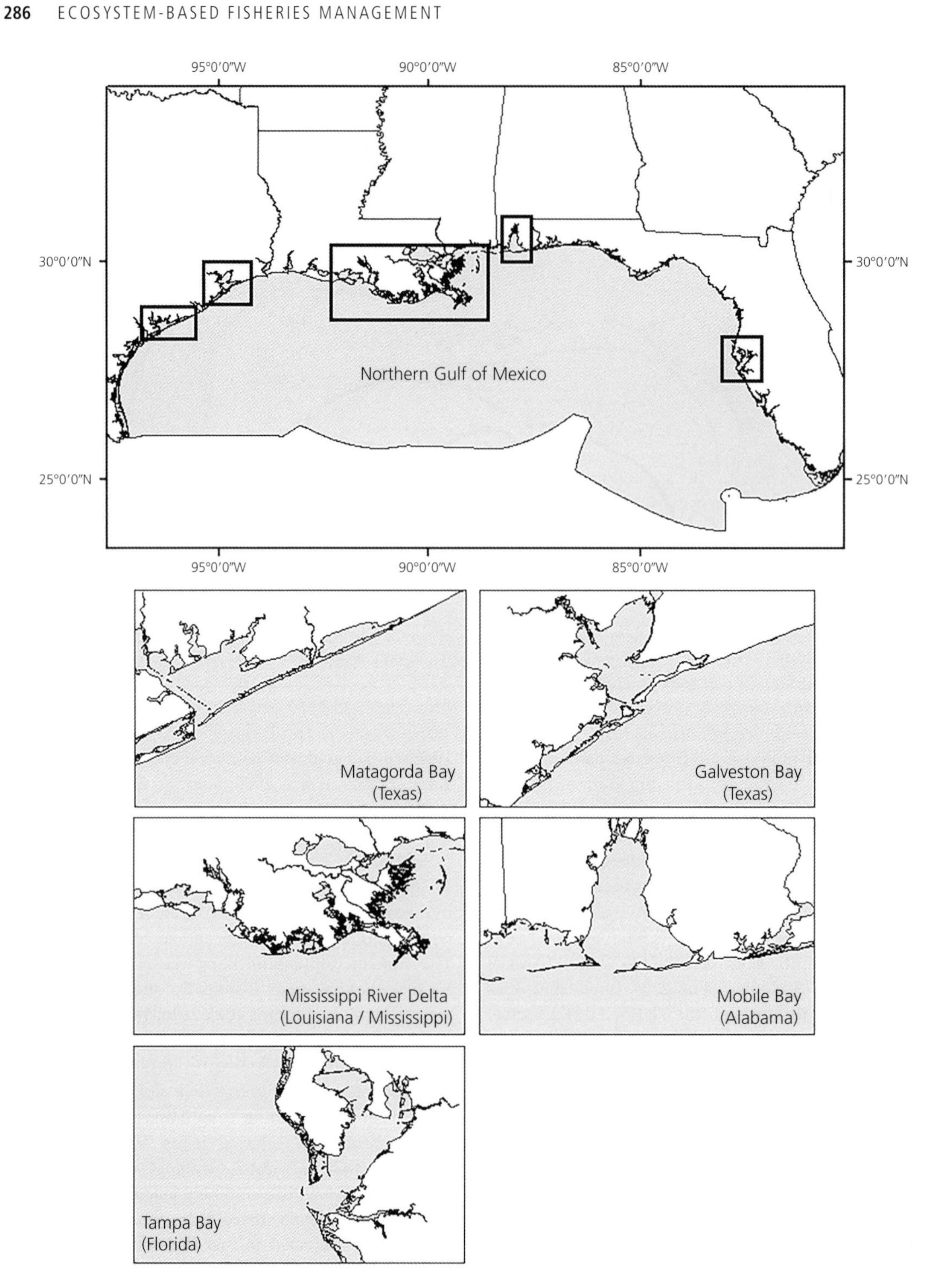

Figure 7.2 Map of the northern Gulf of Mexico region and identified subregions.

nutrient loading, with upstream runoff of agricultural fertilizers throughout the entire Mississippi River drainage system serving as a major contributor to these nutrient loads (Justić et al. 2005, Scavia & Bricker 2006, Rabalais et al. 2007, Laurent et al. 2018, Price et al. 2018). These inputs lead to large phytoplankton blooms in the northwest Gulf, their eventual senescence, and oxygen depletions via bacterial breakdown of this organic material in deeper areas where low mixing occurs (Rabalais et al. 1999, Mitsch et al. 2001, Rabalais et al. 2002, Bianchi et al. 2010). High category hurricanes including Katrina (2005), Rita (2005), Harvey (2017), and Michael (2018) have also significantly affected marine economies, fishing communities, and essential habitats (Buck 2005, Colgan & Adkins 2006, Williams 2010, Lucas & Carter 2013, Solís et al. 2013, Du et al. 2019). The effects of freshwater diversions on marine and estuarine LMRs, their fisheries, and their essential habitats are being observed in this region (Lane et al. 2004, Das et al. 2012, de Mutsert et al. 2017, White et al. 2018, 2019, Tupitza & Glaspie 2020). Consequences from these events are compounded by increasing human population density, which has fostered high seafood and marine energy demand from this system, resulting in growing human-related stressors on the Gulf marine environment over the past century (Twilley et al. 2001, Bricker et al. 2008, Burkett 2008).

The cumulative effects of other human and biological stressors continue to be observed in the nGOM. For instance, the 2010 Deepwater Horizon Oil Spill (DWH) introduced 210 million gallons of crude oil into the nGOM system, differentially affecting pelagic, deep reef, inshore oyster, and vegetated habitats, their dependent species, and multiple fisheries (Fodrie & Heck 2011, Mendelssohn et al. 2012, Silliman et al. 2012, Sumaila et al. 2012, White et al. 2012, Incardona et al. 2014, Beyer et al. 2016). There remains continued potential for oil toxicity (Fodrie & Heck 2011, Barron 2012, Beyer et al. 2016, DeMare et al. 2018).

The tropicalization of nGOM taxa is occurring as higher temperatures lead to increased abundance of range-shifting, tropically associated species (Precht & Aronson 2004, Zimmer et al. 2006, Fodrie et al. 2010, Heck et al. 2015, Marshak & Heck 2017, Scheffel et al. 2018, Macy et al. 2019, Marshak & Heck 2019) and biological communities migrate into deeper habitats in response to regional warming (Pinsky et al. 2013, Kleisner et al. 2016, Morley et al. 2018). Sea-level rise and ocean acidification also have the potential to affect emergent wetlands, shallow and deep-water corals, harvested shellfish, and dependent coastal communities (Burkett 2008, Jankowski et al. 2017, Feely et al. 2018). Together with other potential and ongoing threats to essential habitats, and their loss (Bauman & Turner 1990, Chesney et al. 2000, Milazzo 2012, Zu Ermgassen et al. 2012, Karnauskas et al. 2013, EPA 2015), these factors are significantly affecting the nGOM marine ecosystem, especially the sustainability and composition of its LMRs.

7.2.3 Exploitation

Heavily concentrated commercial and recreational fisheries harvest has led to overfishing and overfished status for many species over time, the result of which is a large number of rebuilding plans (Gracia 1996, Chesney et al. 2000, Caillouet et al. 2008, Cowan et al. 2011a, b, Milazzo 2012, Karanauskas et al. 2013, Link & Marshak 2019). This past overharvesting has raised concerns regarding future recruitment and sustainability for some fisheries populations, such as those from the snapper-grouper complex, the effects of which may be exacerbated by other human-related stressors.

Much of the current management emphasis is on overfished reef species such as groupers, Red snapper, and Gray triggerfish (*Balistes capriscus*; GMFMC 2018a, 2019). Fisheries management in the region remains challenging given the significant importance of the recreational sector. This is especially of concern regarding reef-associated and pelagic finfishes, for which data remain limited and recreational access is perceived as currently over-regulated (Gillig et al. 2000, Adams et al. 2004, Coleman et al. 2004, Johnston et al. 2007, Cowan 2011, Doerpinghaus et al. 2014, Abbott 2015, Arlinghaus et al. 2016). The importance of benthic invertebrates (shrimps, crabs, oysters) and forage fish such as Gulf menhaden (*Brevoortia patronus*) that comprise the bulk of commercial catches, as contrasted with many finfish species that are important for the recreational sector, creates unique management challenges and trade-offs for the region (Coleman et al. 2004, Cooke & Cowx 2006, Vaughan et al. 2007, De Mutsert et al. 2008, Abbott 2015, Sagarese et al. 2016). These challenges could be more effectively addressed through an ecosystem-based approach.

Data limitations regarding stock status and recreational effort, together with historical overfishing, past perceptions of stock rebuilding programs by stakeholders as restrictive, and conflicts between commercial and recreational interests have led to a contentious fisheries management environment in the nGOM (Cooke & Cowx 2006, Johnston et al. 2007, Abbott 2015). Historically, stakeholders and policymakers in this region have been cautious regarding government

Gag Me with a Red Tide: The Contribution of HABs to Mortality Estimates of Gag Grouper

What if there was a population of fish that started dying? And you knew it wasn't because of fishing. And you knew it wasn't because of some kind of predator. And you knew it wasn't because of some kind of competitor. And it was actually more than one fish population, i.e., many of them. And you knew this because you could see (and smell) literally boatloads of fish kills washing ashore and streaking for miles along slicks in the ocean. And there wasn't a darn thing in your stock assessment models or management protocols that could predict or account for it. What would you do?

It would probably seem wise to figure out what was going on. Fishery detective work (*sensu* Hilborn & Mangel 1997) if you will. And to somehow include what you found in your modeling and management advice for the population.

Let's take a look at Gag grouper (*Mycteroperca microlepis*) in the Gulf of Mexico. The assessments for that stock weren't always exhibiting the most robust model diagnostics. It posed all the usual challenges typically facing managing an important recreational and commercial fishery. There were problems in fitting, and some of the dynamics of the stock were just not being captured.

Then there were several fish kills of Gag grouper. After further exploration, such fish kills seemed to coincide with red tide (a type of harmful algal bloom, or HAB) events. To make a long, really cool detective story short, when analysts included HABs in the assessment models, the model behavior improved and the estimates actually resulted in slightly better model fits, which led to better management decisions, which led to better stock status (Walter et al. 2013, Grüss et al. 2014, Sagarese et al. 2014, 2016, Grüss et al. 2016). That is, from that better model fitting, fishery managers in the region realized that this additional source of mortality would be prudent to consider when setting management measures. There are the usual caveats to this situation, chief of which is we still have to monitor the situation to ensure the mechanisms are understood and the functional relationships don't break down. And to this day people still argue about whether to include this ancillary information and whether it is robust enough, whatever that means. But the point is, by taking into account facets of the ocean environment (i.e., adopting an ecosystem approach) for this taxa helped our understanding and may have led to a better evaluation of the stock available for fishing.

The important point from this? It allowed adjustments to how much Gag grouper could be caught, removed suspicions from various sectors of the fleet against each other, and in low HAB situations actually afforded the opportunity for increased catches.

This approach is now being operationalized for several other species in the Gulf (including commercially important Red grouper *Epinephelus morio*) with respect to HABs. It represents an important example of including whatever information one can, doing a bit of detective work, and allowing the models to be flexible enough to consider other, atypical factors that can influence fish stocks.

Sorting out the causes of HABs is an entirely different story, but this situation also highlights the importance, even for offshore fishes, of managing land use practices in the watershed that can magnify algal blooms and affect large swaths of marine ecosystems.

References

Grüss A, Sagarese S, Schirripa MJ, Tetzlaff JC, Bryan M, Walter III J, Chagaris D, et al. 2014. Incorporating Integrated Ecosystem Assessment Products into Stock Assessments in the Gulf of Mexico:

A First Experience with Gag Grouper (Mycteroperca microlepis). ICES CM 2014/3064 C:08. Copenhagen, Denmark: International Council for the Exploration of the Sea.

Grüss A, Schirripa MJ, Chagaris D, Velez L, Shin YJ, Verley P, Oliveros-Ramos R, Ainsworth CH. 2016. Estimating natural mortality rates and simulating fishing scenarios for Gulf of Mexico red grouper (Epinephelus morio) using the ecosystem model OSMOSE-WFS. Journal of Marine Systems 154:264–79.

Hilborn R, Mangel M. 1997. The Ecological Detective: Confronting Models with Data. Princeton, NJ: Princeton University Press.

Sagarese SR, Bryan MD, Walter JF, Schirripa M, Grüss A, Karnauskas M. 2015. Incorporating Ecosystem Considerations Within the Stock Synthesis Integrated Assessment Model for Gulf of Mexico Red Grouper (Epinephelus Morio). SEDAR42-RW-01. North Charleston, SC: SEDAR.

Sagarese SR, Grüss A, Karnauskas M, Walter III JF. 2014. Ontogenetic Spatial Distributions of Red Grouper (Epinephelus Morio) Within the Northeastern Gulf of Mexico and Spatiotemporal Overlap with Red Tide Events. SEDAR42-DW-04. North Charleston, SC: SEDAR.

Walter J, Christman MC, Landsberg JH, Linton B, Steidinger K, Stumpf R, Tustison J. 2013. Satellite Derived Indices of Red Tide Severity for Input for Gulf of Mexico Gag Grouper Stock Assessment. SEDAR33-DW08. North Charleston, SC: SEDAR (p. 43).

regulation of nGOM resources and coastal economies, and at times have been reluctant to support federal management approaches (Tokotch et al. 2012, Alhale 2017). Favored approaches have included movements toward state-specific resource management following successful rebuilding efforts, as evinced by variable durations and extensions of the Red snapper fishing season among state jurisdictions (Alhale 2017). Influence from shrimping and recreational fishing organizations and representatives in political and regulatory fora remains significant. Given the historical importance of fishing and seafood to this region and cultural pride and identity associated with nGOM marine resources, sensitivity toward policies perceived to be infringing on livelihoods and individual access to marine resources is quite high in this region (Tokotch et al. 2012, Smith et al. 2018).

7.2.4 Invasive Species

Increasing prevalence of invasive species including Red lionfish (*Pterois* spp.), Asian tiger shrimp (*Penaeus monodon*), Nile tilapia (*Oreochromis niloticus*), Eurasian watermilfoil (*Myriophyllum spicatum*), the Amazonian apple snail (*Pomacea maculata*), and exotic jellyfish and corals has been observed in the nGOM over the past decades (Showalter 2003). Following their successful invasion of the South Atlantic and Caribbean, lionfish populations were first documented in the nGOM during 2010 (Schofield 2010), and have proliferated throughout the entire region (Schofield 2010, Fogg et al. 2013, Scyphers et al. 2014, Chagaris et al. 2017). Individuals have been found in multiple habitats, including inshore seagrasses (Byron et al. 2014), and primarily occur in high densities within offshore artificial and natural reefs (Dahl & Patterson 2014, Dahl et al. 2016). Studies have demonstrated their effects on nGOM native fishes and invertebrates at localized scales, including habitat displacements of reef fishes (Harrison 2015, Dahl et al. 2016, Marshak et al. 2018) and high consumption leading to localized depletions of juvenile fishes and small cryptic species (Dahl & Patterson 2014, Dahl et al. 2016, 2017). Investigations also suggest that their invasion has altered trophic interactions in the nGOM foodweb, with some anticipated effects on broader ecosystem functioning (Chagaris et al. 2017, 2020). Nearshore culling efforts continue throughout the nGOM in an attempt to control the invasion, but the persistence of deeper spawning populations continues to call into question the long-term effectiveness of these efforts (Dahl et al. 2016, Chagaris et al. 2017, Harris et al. 2019). More recently observed density-dependent effects on their populations suggest that intra- and interspecific competition could differentially influence their long-term abundance in some nGOM habitats (Dahl et al. 2019, Harris et al. 2019).

As a result of presumed introductions through aquaculture escapements, Asian tiger shrimp and Nile Tilapia have become more commonplace within the nGOM (Fuller et al. 2014, Lowe et al. 2012, Hill et al. 2017, Petatán-Ramírez et al. 2020). Aggressive predation by Asian tiger shrimp on nGOM crustaceans (including juvenile Blue crabs) and other invertebrates has been documented, causing concerns about their effects on commercial shrimp, crab, and bivalve populations (Fuller et al. 2014, Hill et al. 2017). Nile tilapia have been observed aggressively competing with estuarine nGOM fishes, leading to habitat displacements of native species, including interfering with native fish nests, and potentially increasing susceptibility of displaced species to larger predators (Peterson et al. 2004, Martin et al. 2010, Lowe et al. 2012). Invasive plants, including Eurasian watermilfoil, have been hypothesized to outcompete nGOM native flora, raising concerns about overall nekton habitat quality (Boylen et al. 1999). However, studies following its nGOM invasion have indicated that watermilfoil may also potentially replace some ecosystem services, including serving as suitable nursery habitat for multiple estuarine fishes under favorable hydrological conditions (Martin & Valentine 2011, Kauffman et al. 2018, Alford & Rozas 2019). In addition, physical and environmental factors in nGOM estuaries may serve to enhance its spread in low-salinity environments and limit expansion into more downstream habitats (Martin & Valentine 2012, Kauffman et al. 2018). Concerns about Amazonian apple snails in similar habitats have arisen, given the susceptibility of native low-salinity marsh plants to their herbivory (Martin et al. 2012, Martin & Valentine 2014, Low & Anderson 2017). More recent invasions by non-indigenous Indo-Pacific damselfish (*Neopomacentrus cyanomos*) have also been observed in offshore nGOM habitats, with their ecological effects on offshore species currently unknown (Bennett et al. 2019). Furthermore, exotic corals, such as Indo-Pacific *Tubastraea micranthus* and *T. coccinea*, and jellyfish species (*Phyllorhiza punctata*) continue to become more prevalent in offshore nGOM waters, with concerns regarding their effects on native fauna (Graham et al. 2003, Johnson et al. 2005, Barord et al. 2007, Sammarco et al. 2010, Precht et al. 2014, Sammarco et al. 2014).

7.2.5 Ecosystem-Based Management (EBM) and Multisector Considerations

The nGOM ecosystem is known for not only its iconic species and images, but also the trade-offs that arise across LMRs (and usage thereof) and between LMR management and other ocean uses (Twilley et al. 2001, Adams et al. 2004, Burkett 2008, Yáñez-Arancibia et al. 2013, Fleming et al. 2014). In the nGOM, diverse coastal and ocean uses supporting regional and national marine economies are prevalent in this ecosystem, particularly offshore oil and gas production, tourism and recreation, and marine transportation sectors (Adams et al. 2004, Fleming et al. 2014, NOAA 2018). The ecological effects of fishing effort and marine economic development throughout inshore and offshore environments where LMRs migrate emphasizes the connectivity of this regional fisheries ecosystem. For example, major shrimp fisheries occur within unstructured offshore habitats, and are focused on species that are highly estuarine-dependent during their early life histories. However, their associated bycatch of coastal and offshore species has fluctuated over time (Goodyear 1995, Wells et al. 2008b), including during observed decreases of shrimping effort in the 2000s (Gallaway et al. 2017). Changes in bycatch susceptibility are partially related to altered species concentrations in response to the development of nearshore oil and gas platforms and artificial reefs (Wells et al. 2008b, Jaxion-Harm & Szedlmayer 2015). Future sustainability of these fisheries depends on consideration of interconnected inshore and offshore human activities in addition to cross-shelf species distributions and ecosystem processes. Not only do these cumulative effects and trade-offs result in interactions among fisheries and other ocean uses, but also in conflicts and interconnectivity among commercial and recreational fishing interests, national agriculture, LMR habitat quality, fisheries harvest, protected species conservation (e.g., trawls requiring turtle excluder devices (TEDs) to minimize bycatch), as well as offshore energy production and its influences on both fishing effort and LMR dynamics. Exploring, identifying, and resolving these differences in priorities and goals is the essence of EBFM (Link 2010, Link & Browman 2014, Patrick & Link 2015, Link & Marshak 2019).

7.3 Informational and Analytical Considerations for this Region

7.3.1 Observation Systems and Data Sources

In moving toward a more holistic managerial approach, data regarding aspects of the physical, biological, and socioeconomic components of the nGOM ecosystem are available through regional observation systems and multiple monitoring programs. The Gulf of Mexico Coastal Ocean Observing System (GCOOS) provides oceanographic and atmospheric data from ~400 sensor stations throughout coastal and offshore locations (GCOOS 2016). The system consists of data from ~20 different federal and regional sources, with the majority of observations from coastal sensors and offshore industrial Acoustic Doppler Current Profilers (ADCPs) maintained by NOAA's National Ocean Service (NOS) and the National Estuarine Research Reserve (NERR) System. Since 2008, physical data reported at daily to annual scales have been available at the GCOOS portal[1] with monitored atmospheric variables including air pressure, air temperature, relative humidity, winds, and oceanographic information including chlorophyll, currents, dissolved oxygen, thermohaline information, turbidity, water level, and wave features. This information is being applied toward ensuring safer marine operations, minimizing human effects from coastal hazards and natural stressors (e.g., HABs, hypoxia, and hurricanes), and in monitoring long-term changes to the nGOM ecosystem (GCOOS 2016).

Several oceanographic and atmospheric data sources have been used for broader understanding of the nGOM ecosystem, particularly regarding its focal components and cumulative functioning (Karanuskas et al. 2013, 2017). They have also been incorporated in Integrated Ecosystem Assessments (IEAs; Levin et al. 2009) and regional Ecosystem Status Reports (ESRs; Karnauskas et al. 2013, 2017). Data regarding the AMO and Atlantic Warm Pool (AWP) indices, and shifts in location of the Loop Current and geostrophic transport across the Yucatán Channel and Florida (AOML 2012, ESRL 2013), have been incorporated into ESRs for the region. Physical pressures including hurricane activity, accounting for both storm frequency and intensity, also have been reported (Wang et al. 2011, Karnauskas et al. 2013). Data related to factors that influence the size and extent of the Gulf of Mexico hypoxic zone are available through several monitoring programs that track nutrient input, precipitation (runoff), stream flow, and agricultural activities throughout the Mississippi River watershed (Karnauskas et al. 2013). These specifically include the U.S. Department of Agriculture (USDA) fertilizer consumption (use and price) database, information from the USDA Conservation Effects Assessment Project, and merged satellite data with U.S. Geological Survey (USGS) river gage and total basin load data as reported from water quality sampling to inform on precipitation and stream flow. The spatial extent of the hypoxic zone is monitored by the Louisiana Universities Marine Consortium (LUMCON) and through bottom water hypoxia sampling as a component of the fishery-independent Southeast Area Monitoring and Assessment Program (SEAMAP) trawl and hydrographic survey.

Chemical contaminants, HAB-associated toxins, and water pollutant uptake into the nGOM food web are monitored through the U.S. Food and Drug Administration (FDA) Gulf Coast Seafood Laboratory and NOAA National Status and Trends Mussel Watch Program (MWP), the latter of which accounts for metal contamination (particularly cadmium and mercury) in coastal sediments and bivalves. The Bureau of Ocean Energy Management (BOEM) and Baker Hughes Rig Count database records the number of oil platforms created and removed from the nGOM, while regional oil spills are accounted for by the Bureau of Safety and Environmental Enforcement (BSEE). Other contaminants, in terms of total suspended solids, are captured by only a few permanent measurement stations located near the coast, with the most robust dataset occurring for the Mississippi River drainage at Tarbert Landing and collected by U.S. Army Corps of Engineers gauges. Since 2000, the Environmental Protection Agency and state monitoring programs of Texas, Mississippi, Alabama, and Florida have also accounted for *Enterococcus* bacteria counts at recreational beaches. USGS and the U.S. Environmental Protection Agency (EPA) also track annual nutrient loads throughout the Mississippi/Atchafalaya River Basin at ~40 stations, from which long-term trends have been observed for more than 30 years (Karnauskas et al. 2013, EPA 2017). These efforts are complemented by state-specific coastal monitoring programs, including for nutrients, which collectively contribute to the Cooperative Hypoxia Assessment and Monitoring Program (CHAMP; Meckley et al. 2017).

Information characterizing habitat factors and aspects of benthic production are available but generally incomplete. Comprehensive habitat information at the scale of many LMRs has remained limited (Marshak & Brown 2017, Peters et al. 2018). Most studies examining trends in habitat quality and habitat delineation have been conducted at sub-regional scales, and have shown declines in cover and quality over time (Cowan et al. 2008, Peterson & Lowe 2009, Steffen et al. 2010, Walter et al. 2013, Gittman et al. 2019). A recent synthesis of nGOM-wide information for oyster reefs over time relied upon data from surveys conducted at differing

[1] http://gcoos.org

It's Getting Kinda Stuffy in Here: Hypoxia, Limiting Habitat, and Declining Health of Fish Stocks and Fisheries, or the Trade-offs of Dealing with Conflicting Objectives

—(c.f. Chapters 8, 4)

We often forget that water contains dissolved oxygen in it. And the vast majority of sea creatures need oxygen to live. You know, basic metabolism stuff that you forgot from Introductory Biology.

When you remove oxygen from the water, critters that can move do. Critters that can't die.

That's what we're seeing in many places, but particularly in the Gulf of Mexico.

Nutrient rich waters flow down the Mississippi River from the nation's bread-basket, carrying a very large load of not only silt, but organic materials. Over time, this has built up in the Gulf of Mexico. The result has been a hypoxic zone the size of New Jersey (Rabalais et al. 2002, Dodds 2006, Turner et al. 2008). This "dead zone" has impacted fishes, invertebrates, and other critters that drive the commercial and recreational fisheries in the Gulf region (Malakoff 1998, Rabalais et al. 2002, Dodds 2006, Turner et al. 2008, Craig 2012). Vessels have a harder time finding living creatures, they have to travel farther, they drag up dead crabs, there is lower abundance of fish around, and so forth. As the situation has continued, the size of the dead zone has expanded.

This is an important example of the need for ecosystem-based management, well beyond just the fishery sector. There are many parties interested in how the Mississippi River is used and managed, beyond just fisheries.

This situation particularly represents another trade-off between fisheries and big agriculture (c.f. salmon and water textbox in Chapter 8). Both produce food for the nation. Both are vitally important. Both are huge parts of regional economies. Both even interact to a degree. But the downstream effects of one are disproportionately impactful on the other.

At one level, the solution is simple—stop the nutrient loading. Really it is basic non-point source solutions and mitigations that merit consideration, just scaled up to a continental

From Link et al. 2019

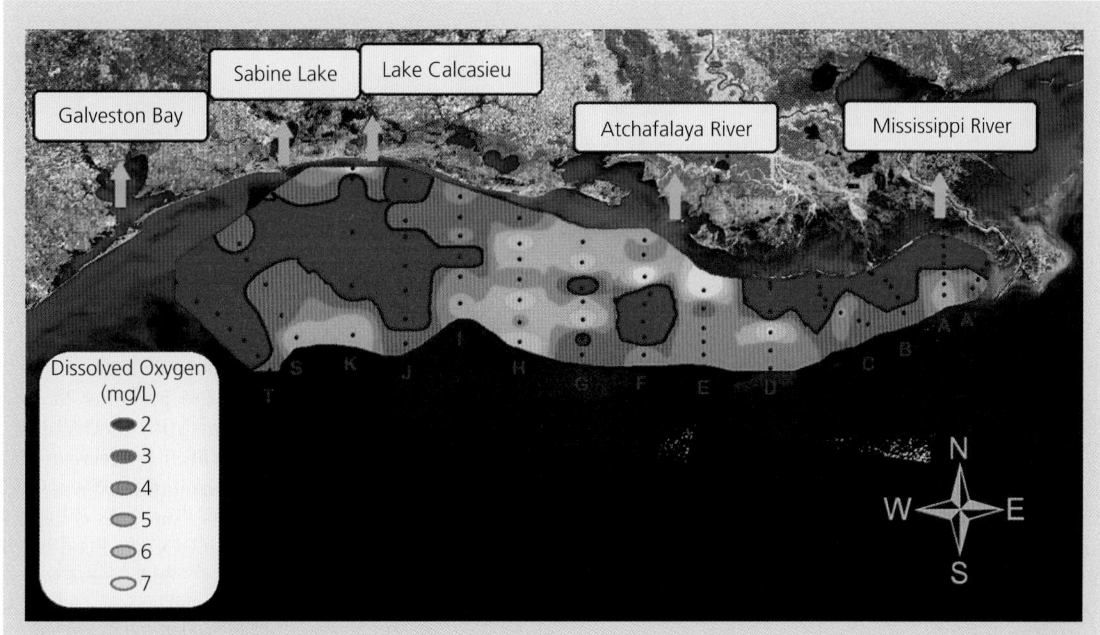

From Rabalais et al. 2002.

scope. Yet at another level, the practice of doing so has so many ramifications as to be a non-starter relative to those many diverse, varied, and powerful interests. Given the costs involved and potential economic effects at larger scales (Ribaudo et al. 2016, Ribaudo & Shortle 2019), often these kinds of trade-offs are easier to deal with in a more local or regional level. However, increased awareness of how agricultural runoff affects the Gulf of Mexico has led to proposed actions such as purchasing nitrogen reduction credits to limit discharges (Ribaudo et al. 2005). Additionally, more cost-effective approaches would include those where farmers are subsidized not just for switching to pollution-reducing practices, but for actual improvements in water quality (i.e., pay for performance), and which allocate subsidies to high-priority locations (Shortle et al. 2012, Ribaudo 2016, Ribaudo & Shortle 2019). The magnitude of the issues involved here really does represent the need for some type of national policy discussion, with broader consideration of best management practices for agriculture beyond individual state boundaries (Fernholz et al. 2018).

Awareness of the problem is a start. Coordinating and seeking win-win solutions to this situation need to happen. The key will be to establish a venue for such discussions and then have the agreement by all parties involved to abide by any joint decisions made. A national discussion of policies and priorities would ultimately be beneficial, but that may be hard to establish given the scope and diversity of interests involved.

References

Craig JK. 2012. Aggregation on the edge: effects of hypoxia avoidance on the spatial distribution of brown shrimp and demersal fishes in the Northern Gulf of Mexico. *Marine Ecology Progress Series* 445:75–95.

Dodds WK. 2006. Nutrients and the "dead zone": the link between nutrient ratios and dissolved oxygen in the northern Gulf of Mexico. *Frontiers in Ecology and the Environment* 4(4):211–17.

Fernholz K, Feeney W, Groot H, Pepke E, Henderson C, Erickson G. 2018. *Water Quality Best Management Practices in US Midwestern Agricultural Landscapes: What Can be Learned from the Experience of the Forest Sector.* Minneapolis, MN: Dovetail Partners, Inc (p. 15).

Link JS, Kohler B, Griffis R, Brady P, Ito S, Garçon V, Hollowed A, et al. 2019. A graphical novel from the 4th International Symposium on the Effects of Climate Change on the World's Oceans. *ICES Journal of Marine Science* 76:1221–43.

Malakoff D. 1998. Death by suffocation in the Gulf of Mexico. *Science* 281(5374):190–2.

Rabalais NN, Turner RE, Wiseman Jr WJ. 2002. Gulf of Mexico hypoxia, aka "The Dead Zone". *Annual Review of Ecology and Systematics* 33(1):235–63.

Ribaudo M. 2017. Choices: conservation programs can accomplish more with less by improving cost-effectiveness. *Agriculture & Applied Economics Association* 32(4).

Ribaudo MO, Heimlich R, Peters M. 2005. Nitrogen sources and Gulf hypoxia: potential for environmental credit trading. *Ecological Economics* 52(2):159–68.

continued

It's Getting Kinda Stuffy in Here: *Continued*

Ribaudo M, Marshall E, Aillery M, Malcolm S. 2016. *Reducing the Dead Zone in the Gulf of Mexico: Assessing the Costs to Agriculture. Annual Meeting, July 31–August 2.* Boston, MA: Agricultural and Applied Economics Association.

Ribaudo M, Shortle J. 2019. Reflections on 40 Years of Applied Economics Research on Agriculture and Water Quality. Agricultural and Resource Economics Review 48(3):519–30.

Shortle JS, Ribaudo M, Horan RD, Blandford D. 2012. Reforming agricultural nonpoint pollution policy in an increasingly budget-constrained environment. Environmental Science & Technology 46(3):1316–25.

Turner RE, Rabalais NN, Justic D. 2008. Gulf of Mexico hypoxia: alternate states and a legacy. Environmental Science & Technology 42(7):2323–7.

resolutions quantifying overall declines in the extent, density, and biomass of oysters within ~30 sites throughout the nGOM (Zu Ermgassen et al. 2012). Monitoring data for seagrass coverage are available across 15 nGOM sites expanding from Laguna Madre, TX to Florida Bay, FL, which inform the declining status and quality of nGOM seagrasses (EPA 2004). Areal information from the NOAA Coastal Change Analysis Program (C-CAP) together with tidal gauge data from the NOAA Center for Operational Oceanographic Products and Services (CO-OPS) and stations operated by the National Water Level Observation Network (NWLON) have also been used to characterize the extent, quality, and risks posed to estuarine watersheds and emergent salt marsh habitats (NFHP 2010, Karnauskas et al. 2013, Crawford et al. 2015, Karanuskas et al. 2017). Basal ecosystem production data for the nGOM (i.e., chlorophyll concentration and primary production) are available from NASA Ocean Color Web Data Sea-viewing Wide Field-of-view Sensor (SeaWiFS) and Moderate Resolution Imaging Spectroradiometer (MODIS)-Aqua satellite imagery. Additionally, the Florida Fish and Wildlife Conservation Commission (FWC) and NOAA National Centers for Coastal Ocean Science (NCCOS) maintain large databases of HAB and red tide events, particularly along the southwest Florida shelf.

Comprehensive databases for nGOM secondary through upper trophic level species include fishery-independent information collected from SEAMAP zooplankton and ichthyoplankton surveys, and SEAMAP trap-video, trawl, and midwater surveys for penaeid shrimp and pelagic and demersal fishes (Conn 2011, Smith et al. 2011, Vaughan et al. 2011, Karnauskas et al. 2013, Hart 2016, SEAMAP 2017, Switzer & Rester 2019). Several ecosystem-level metrics, including mean trophic level, species richness, diversity, and pelagic to demersal biomass have been derived from these datasets (Karnauskas et al. 2013). These monitoring programs assist in informing nGOM stock status and abundance, and are complemented by NOAA Fisheries fishery-dependent monitoring programs for commercial and recreational harvest, landings, revenue, and effort; this includes surveys such as the previous Marine Recreational Information Program (MRIP) and recently adopted Fishing Effort Survey (FES) for recreational fisheries (NMFS 2018b). Socioeconomic information for fisheries and other ocean uses are also collected by NOAA Fisheries, the National Ocean Economics Program (NOEP), and NOAA's Office for Coastal Management Economics: National Ocean Watch (ENOW) Program. NOAA monitoring programs also exist for protected species that include sea turtles and cetaceans, and several monitoring programs for marine birds, such as the USGS Breeding Bird Survey, are overseen by NOAA, the U.S. Department of Interior, and state agencies. The BOEM Gulf of Mexico Marine Assessment Program for Protected Species (GoMMAPPS) has also been in effect since 2017, and is currently the largest protected species monitoring program in the nGOM (BOEM 2017). The Texas A&M University Gulf of Mexico Species Interactions (GoMexSI) database also provides information on multiple species interactions in the nGOM, including species diets and trophic interactions (Simons et al. 2013).

7.3.2 Models and Assessments

Assessments of economically important and protected species in the nGOM are conducted by NOAA Fisheries in collaboration with the Gulf of Mexico (GMFMC) and South Atlantic Fishery Management Councils (SAFMC), the U.S. Fish and Wildlife Service (FWS), the Gulf States Marine Fisheries Commission (GSMFC), and the various state agencies throughout the region (FWS 2014, Ferguson et al. 2017, Lynch et al. 2018). Stock assessment models are used to generate estimates of stock status and incorporate a number of sources of information about the stock and the fishery (e.g., landings, catch per unit effort, life history characteristics, survey biomass, etc.). Assessments for nGOM fishery species are conducted through the Southeast Data, Assessment, and Review (SEDAR) process (Lynch et al. 2018). However, this region continues to remain data-limited for many species with a significant percentage (~45%) of its stocks remaining unassessed (Newman et al. 2015, Link & Marshak 2019). NOAA Fisheries also examines the effects of commercial and recreational fisheries on nGOM marine socioeconomics through application of the NMFS Commercial Fishing Industry Input/Output and Recreational Economic Impact Models (NMFS 2018a). Data from NOAA Fisheries marine mammal stock abundance surveys are used to conduct population assessments as published in regional marine mammal stock assessment reports and peer-reviewed by regional scientific review groups (SRGs; Carretta et al. 2019, Hayes et al. 2019, Muto et al. 2019). The FWS prepares stock assessment reports for nGOM manatees under its jurisdiction (FWS 2014). Information collected on nGOM protected species is also applied toward 5-year status reviews of those listed under the Endangered Species Act (ESA; Bettridge et al. 2015, Rosel et al. 2016, Valdiva et al. 2019).

Available data and species assessments have been applied toward broader ecosystem models for the nGOM, while efforts to incorporate IEA products into single-species stock assessments and LMR management decisions are also ongoing (Townsend et al. 2008, Link et al. 2010, Townsend et al. 2014, 2017). In addition, information developed in the Gulf of Mexico IEA program has been synthesized (Kelble et al. 2013, Karnauskas et al. 2013, 2017) to inform EBM for the region. Simulation models for the nGOM include those for both single-species and ecosystems (Carey et al. 2013, Gray et al. 2013, Townsend et al. 2014, 2017). Additionally, multimodel approaches such as those using a reference set of operating models, using inter-model comparisons, or using ensemble modeling approaches continue to be preliminary or targeted toward reef fishes or invertebrate fishery species (Townsend et al. 2014, 2017). Most simulation models used for the region include Ecopath with Ecosim (EwE) and Object-Oriented Simulator of Marine Ecosystems (OSMOSE) models for the West Florida Shelf, EwE models for the U.S. Gulf coast ('GOM Shark EwE'), Galveston Bay, the northern Gulf of Mexico ('nGOM Ecopath'), and a Gulf of Mexico-wide Atlantis model ('Atlantis-GOM'; Townsend et al. 2014, Ainsworth et al. 2015, Townsend et al. 2017). EwE simulation models for the West Florida Shelf are applied toward understanding the reef fish community ('WFS reef fish EwE'; Chagaris 2013, Chagaris & Mamoudi 2013) and red tide impacts ('WFS Red tide EwE'), while 'OSMOSE-WFS' describes trophic interactions throughout the West Florida Shelf (Grüss et al. 2013a, b). All three of these models have been used to inform SEDAR historical estimates of natural mortality for Red grouper (*Epinephelus morio*) and to examine and document the increasing effects of recent red tide events on this parameter (Sagarese et al. 2015, Grüss et al. 2016).

Complementary efforts for Gulf of Mexico reef fishes are also in progress through the application of a biophysical model (the Connectivity Modeling System (CMS)) toward understanding environmental effects on Red snapper, Gag grouper (*Mycteroperca microlepis*), and Red grouper populations (Townsend et al. 2017). The 'GOM Shark EwE' model has been used to assess top-down ecosystem impacts of alternative fishing policies for large coastal sharks, particularly on interactions among higher and lower trophic level organisms (Townsend et al. 2014, 2017). The 'Atlantis-GOM' model has been used to examine the ecosystem-level effects of stressors including hypoxia and DWH, and in testing the effects of harvest control rules in the nGOM (Townsend et al. 2014, Ainsworth et al. 2015, Townsend et al. 2017). The South Florida Water Management District has also developed habitat suitability and regional simulation models, with application toward evaluating restoration projects, enhancing understanding of ecosystem functioning, and examining the effects of climate change on Everglades systems (Ogden et al. 2005, SFWMD 2006, Obeysekera et al. 2015, Kearney et al. 2015). There are other models that have informed management discussions in this region (Browder 1993, DeAngelis et al. 1998, Mazzotti et al. 2006, Teh et al. 2008, Walters et al. 2008, Evans & Scavia 2010, Drexler & Ainsworth 2013, De Mutsert et al. 2016, Geers et al. 2016, Chagaris et al. 2017, Grüss et al. 2017, O'Farrell et al. 2017 and references therein), including recently developed hydrodynamic and biogeochemical models (de Rada et al. 2009, Fennel et al. 2011, Zheng & Weisberg 2012, Justić & Wang 2014, Grüss et al. 2017, Gomez et al. 2019a, b) and those examining the ecosystem effects of river diversions (CPRA 2012, de Mutsert et al. 2012, 2016, Dynamic Solutions 2016, de Mutsert et al. 2017). However, few of these have been directly used in the SEDAR process like those noted above.

7.4 Evaluation of Major Facets of EBFM for this Region

To elucidate how well the nGOM ecosystem is doing in terms of LMR management and progress toward EBFM, here we characterize the status and trends of socioeconomic, governance, environmental, and ecological criteria pertinent toward employing an EBFM framework for the nGOM, and examine their inter-relationships as applied toward its successful implementation. We also quantify those factors that emerge as key considerations under an ecosystem-based approach for the region and for five major subregions of interest. Ecosystem indicators for the nGOM related to the (1) human environment and socioeconomic status, (2) governance system and LMR status, (3) natural environmental forcing and features, and (4) systems ecology, exploitation, and major fisheries of this system are presented and synthesized next. We explore all these facets to evaluate the strengths and weaknesses along the pathway:

$$PP \rightarrow B_{\text{targeted, protected spp, ecosystem}} \leftrightarrow L_{\uparrow\text{targeted spp, }\downarrow\text{bycatch}} \rightarrow \text{jobs, economic revenue.}$$

Where PP is primary production, B is biomass of either targeted or protected species or the entire ecosystem, L is landings of targeted or bycaught species, all leading to the other socioeconomic factors. This operates in the context of an ecological and human system, with governance feedbacks at several of the steps (i.e., between biomass and landings, jobs, and economic revenue), implying that fundamental ecosystem features can determine the socioeconomic value of fisheries in a region, as modulated by human interventions.

7.4.1 Socioeconomic Criteria

7.4.1.1 Social and Regional Demographics

In the nGOM, substantial (i.e., doubling) increases in coastal human population and population density (Fig. 7.3) have been observed from 1970 to present with values currently around 15 million residents and ~80 people km^{-2}, respectively. These values account for permanent residents, while seasonal inhabitants and tourists also contribute heavily to this region, with ~16.4 million travelers visiting the Mississippi and Alabama Gulf coasts alone in 2013 (Guo et al. 2017). This region makes up 5% of the United States total population, while density values have risen at ~1.1 people km^{-2} year^{-1}. Human population is most heavily concentrated in the Galveston Bay subregion, given its proximity to the city of Houston and other populated Texas coastal counties (i.e., Brazoria, Chambers, Galveston, and Harris counties). Population estimates have also doubled near Galveston Bay over the past 40 years, with currently up to 4.7 million residents and 357 people km^{-2}. Similarly increasing trends have been observed for the Tampa Bay subregion (i.e., Hernando, Hillsborough, Manatee, Pasco, Pinellas, and Sarasota counties, FL) with up to 3.5 million residents and 272 people km^{-2}. Lower, steadier population values have been observed over this timeframe throughout the

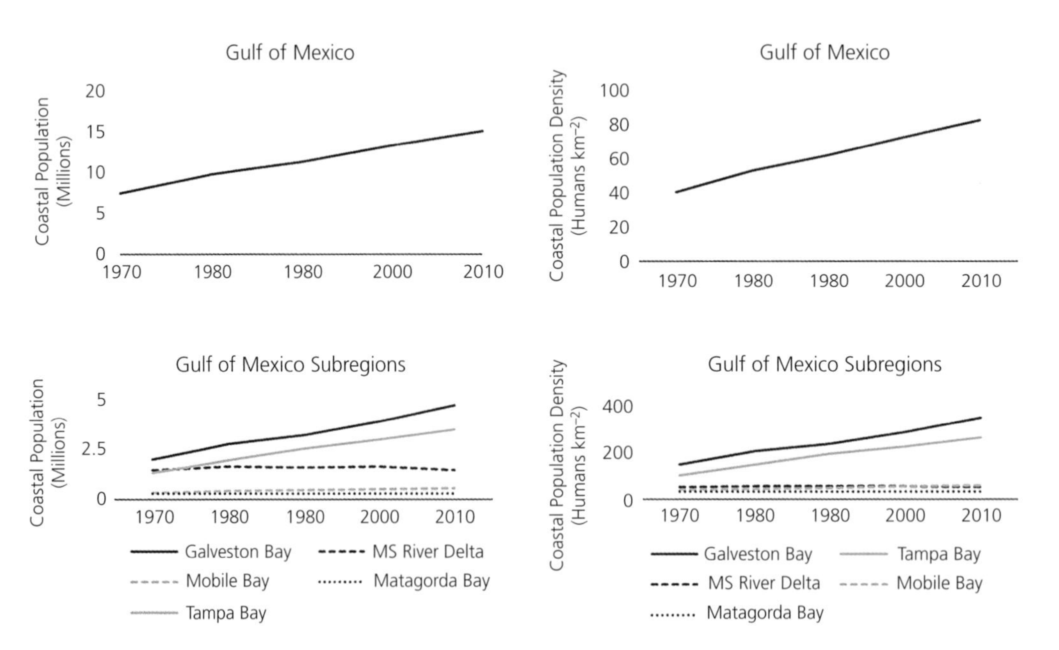

Figure 7.3 Decadal trends (1970–2010) in coastal human population (top) and population density (bottom; humans km^{-2}) for the northern Gulf of Mexico region and identified subregions.

Data derived from U.S. censuses and taken from https://coast.noaa.gov/digitalcoast/data/demographictrends.html.

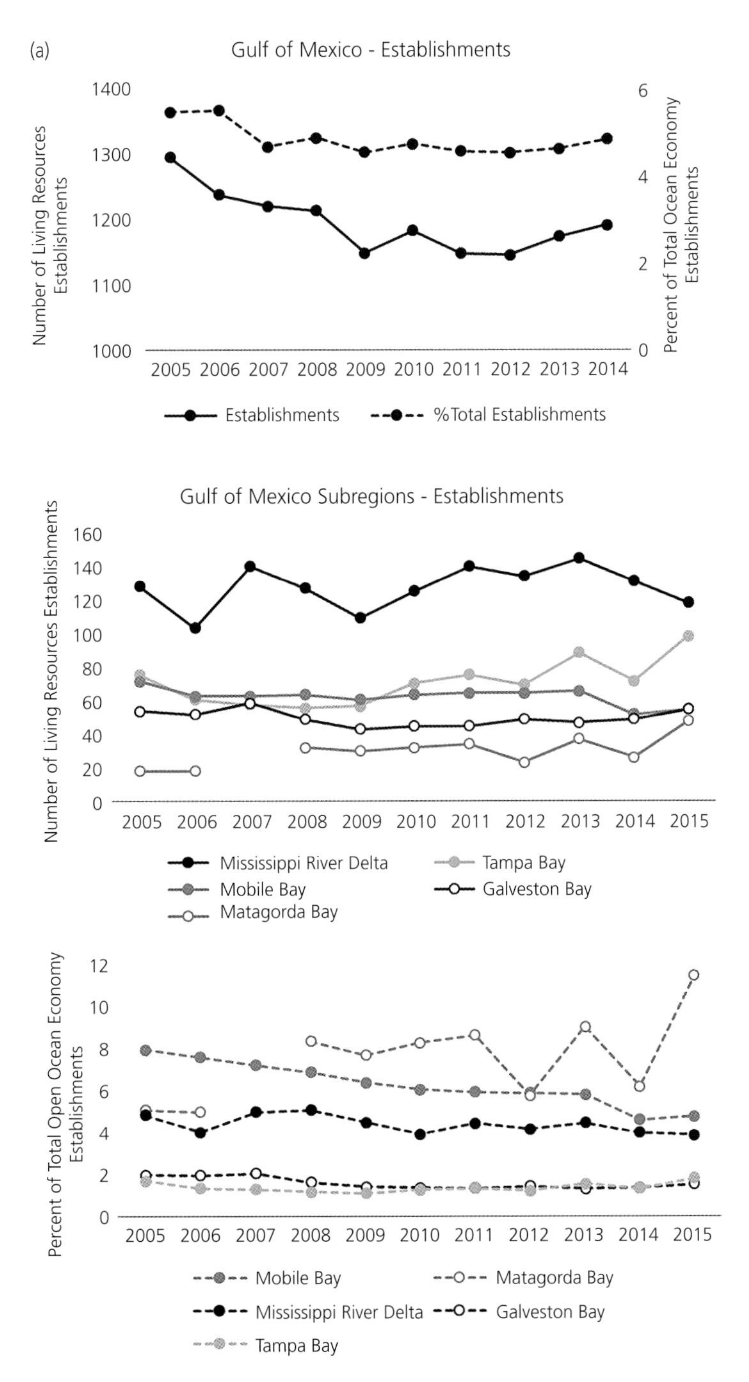

Figure 7.4a Number of living marine resources establishments and their percent contribution to total multisector oceanic economy establishments in the northern Gulf of Mexico region and identified subregions (years 2005–2015).

Data derived from the National Ocean Economics Program.

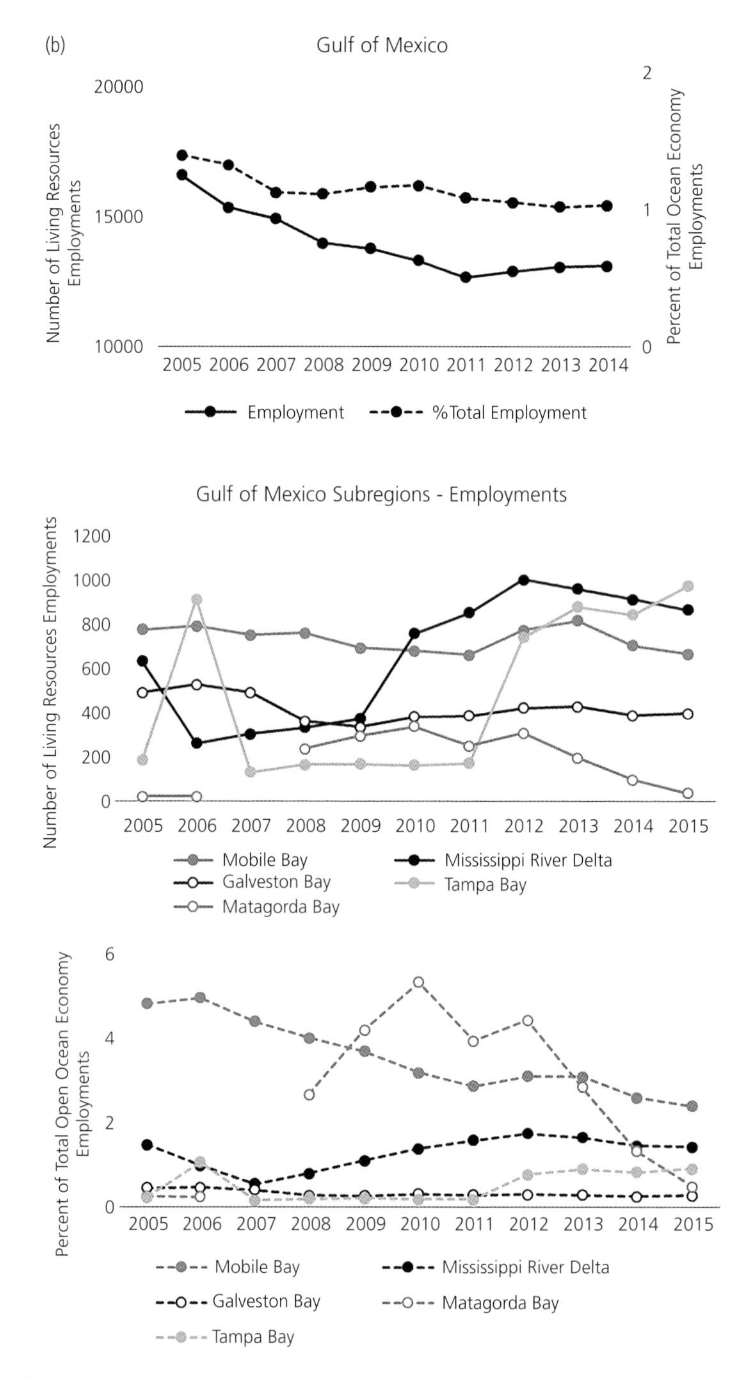

Figure 7.4b Number of living marine resources employments and their percent contribution to total multisector oceanic economy employments in the northern Gulf of Mexico region and identified subregions (years 2005–2015).

Data derived from the National Ocean Economics Program.

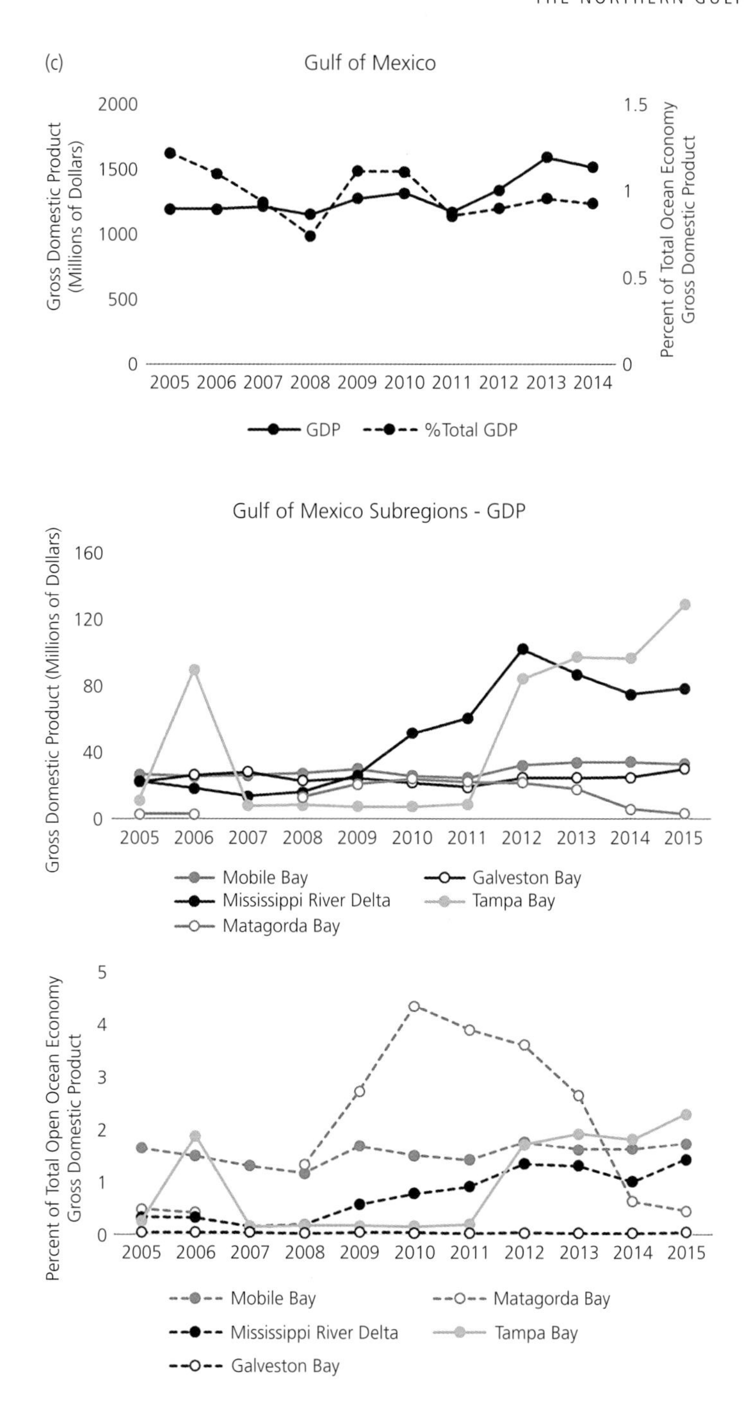

Figure 7.4c Gross domestic product value (GDP; USD) from living marine resources revenue and percent contribution to total multisector oceanic economy GDP for the northern Gulf of Mexico region and identified subregions (years 2005–2015).

Data from the National Ocean Economics Program.

Mississippi River Delta (MRD; ~1.5 million residents), which is composed of nine Louisiana Parishes (i.e., Iberia, Jefferson, LaFourche, Orleans, Plaquemines, St. Bernard, St. Mary, Terrebonne, and Vermillion) and two Mississippi counties (i.e., Hancock, Harrison). Consistent values have also been observed over time for the Matagorda Bay (Calhoun, Jackson, Matagorda counties, TX; ~300,000 residents) and Mobile Bay (Broward, Mobile counties, AL; 400,000–600,000 residents) subregions. Population densities in these latter three subregions have also remained at lower values over time ranging from 32 to 62 people km^{-2}, with moderate increases (~0.6 people km^{-2} year^{-1}) observed in the areas surrounding Mobile Bay.

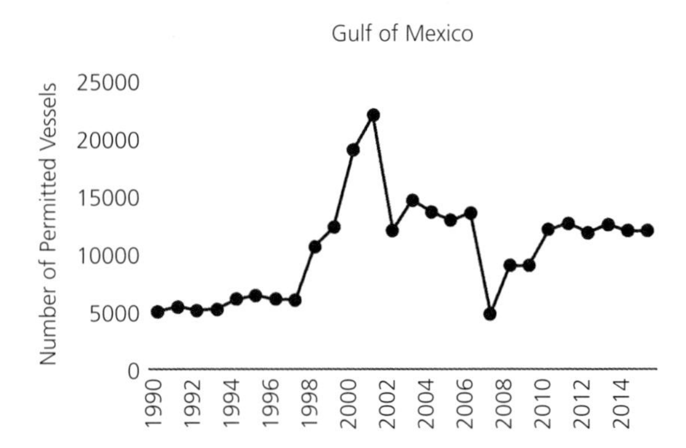

Figure 7.5 Total number of permitted vessels for the northern Gulf of Mexico region over time (years 1990–2015).

Data derived from NOAA National Marine Fisheries Service Council Reports to Congress.

7.4.1.2 Socioeconomic Status and Regional Fisheries Economics

High numbers of LMR establishments and employments, and high LMR-associated Gross Domestic Product (GDP) values (~$1.5 billion USD), are found throughout the nGOM (Fig. 7.4). Second only to the Pacific region (Link & Marshak 2019), the nGOM contains high numbers of LMR establishments, employments, and GDP, which notably add to the national ocean economy and for which this region is excelling. Although numbers have slightly declined since 2005, their trend has leveled-off in recent years with LMR establishments comprising a steady 5% of regional ocean economy establishments and LMR employments contributing ~2% of regional ocean employments over time. LMR-associated GDP has remained steady over time, and contributes ~1% to the regional total ocean economy.

Values have remained relatively steady throughout subregions, with the highest numbers of LMR establishments and employments currently occurring in Tampa Bay and the MRD. While lower numbers of LMR establishments are found in the other three subregions, those in Matagorda Bay and Mobile Bay comprise more (6–12%) of the total ocean economy establishments in their areas, with recent increases occurring for Matagorda Bay and slight decreases over time observed for Mobile Bay establishments and employments. Similar decreases over the past decade have been observed for LMR employments in Galveston Bay, and have been more pronounced in Matagorda Bay, while substantial increases have occurred in the MRD and more recently in Tampa Bay. Although the MRD subregion was greatly affected by Hurricane Katrina in 2005 and DWH in 2010, trends in the number and percent contribution of LMR employments for the MRD do not appear to have been strongly impacted. However, greater decreases in the percent contributions of LMR employments toward total ocean employments have occurred more recently in Matagorda Bay after brief expansion, and in Mobile Bay, while they have remained consistently low in Galveston Bay. LMR-associated GDP has also expanded

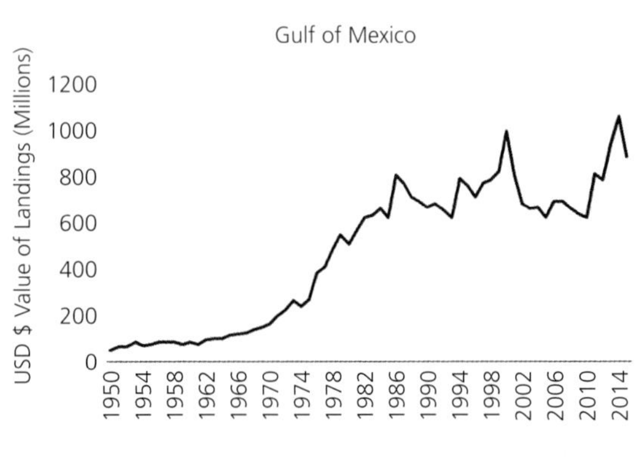

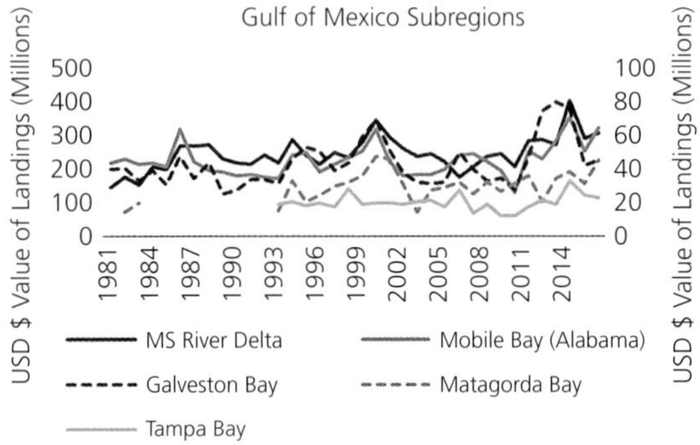

Figure 7.6 Total revenue (year 2017 USD) of landed commercial fishery catches for the northern Gulf of Mexico region (years 1950–2015) and identified subregions (years 1981–2016), over time. Values for the Mississippi (MS) River Delta subregion presented on the primary (left) axis. Values for all other subregions plotted on the secondary (right) axis.

Data derived from NOAA National Marine Fisheries Service commercial fisheries statistics.

Figure 7.7 Over time, ratios of (a) total living marine resources employments to total biomass (thousands of metric tons); (b) total commercial landings revenue (USD) to total biomass (metric tons); (c) total living marine resources employments to total productivity (metric tons wet weight year^{-1}); (d) total commercial landings revenue (USD) to total productivity (metric tons wet weight year^{-1}) for the northern Gulf of Mexico region.

in the MRD and Tampa Bay, recently peaking at up to 100 million USD for the MRD in 2012 and up to 148 million USD for Tampa Bay in 2015, while remaining lower in other subregions at 2–32 million USD over time.

Since 1990, high numbers of commercial and for-hire permitted (federal and Florida state-level reported) fishing vessels have operated in the Gulf of Mexico (Fig. 7.5), with values expanding up to ~22,000 during the mid to late 1990s. Associated with losses from Hurricane Katrina, substantial decreases in permitted vessels occurred in 2006, with only 4,800 remaining in service, while numbers in recent years have rebounded to around 12,000 vessels. Total revenue (year 2017 USD) of landed commercial fishery catches (Fig. 7.6) has increased over time to the third-highest in the nation (out of eight marine regions), peaking at $1.1 billion in 2014. Among subregions, commercial catch revenue has remained relatively steady, with highest values (~$200 to 300 million) observed in the MRD until the mid-2000s, when revenue declined and later rebounded. Commercial catches from Mobile Bay and Galveston Bay also contribute a similar range of revenues. However, their values declined during the early to mid-2000s. Matagorda Bay has historically contributed much lower commercial revenue to the nGOM at approximately one-third to one-half that of the other subregions. Similarly, values for Tampa Bay have also remained quite low at ~$20–30 million per year.

Ratios of jobs to biomass (up to 2.6 jobs/thousand metric tons; Fig. 7.7) and commercial fisheries revenue as compared to total biomass (up to $180/metric ton) are highest nationally in the nGOM, with values remaining relatively steady over time. Ratios of the same variables noted above relative to pelagic primary production (metric tons wet weight y^{-1}) are fourth-highest nationally. Jobs/productivity ratios have ranged over the past decade from 8.5 to 10.1 jobs/million metric tons wet weight y^{-1}, while ratios of total commercial landings revenue to productivity have remained stable and slightly increased over time to ~$0.8/metric ton wet weight y^{-1}.

7.4.2 Governance Criteria

7.4.2.1 Human Representative Context

Twenty-six major federal and state agencies and organizations operate within the nGOM and are directly responsible for fisheries and environmental management (Table 7.1). Fisheries resources throughout state waters (up to 9 nautical miles for Texas, Louisiana, Mississippi, Alabama, and the west coast of Florida for certain federally managed reef fishes) are managed by individual

Table 7.1 Federal and State agencies and organizations responsible for fisheries and environmental consulting and management in the Northern Gulf of Mexico region and subregions

Region	Subregion	Federal Agencies with Living Marine Resource Interests	Council	State Agencies and Cooperatives	Multijurisdictional
Gulf of Mexico		U.S. Department of Commerce **DoC** (NOAA Fisheries Southeast Fisheries Science Center **SEFSC**, Southeast Regional Office **SERO**, **NMS**); U.S. Department of Interior **DoI** (U.S. Fish and Wildlife Service **FWS**, Bureau of Ocean Energy Management **BOEM**, National Parks Service **NPS,** U.S. Geological Survey **USGS,** U.S. Department of Defense—U.S. Army Corps of Engineers **DoD—USACE**) U.S. Environmental Protection Agency **EPA**	Gulf of Mexico Fishery Management Council **GMFMC**	Atlantic States Marine Fisheries Commission **ASMFC**; Gulf States Marine Fisheries Commission **GSMFC**; Texas Parks and Wildlife Department **TPWD**; Louisiana Department of Natural Resources **LDNR**; Louisiana Department of Wildlife and Fisheries **LDWF**; Louisiana Coastal Protection and Restoration Authority **CPRA**; Mississippi Department of Marine Resources **MDMR**; Alabama Department of Conservation and Natural Resources **ADCNR**; Florida Fish and Wildlife Conservation Commission **FWC;** Florida Department of Environmental Protection; Gulf of Mexico Alliance	Coral Reef Task Force **CRTF**; International Commission for the Conservation of Atlantic Tunas **ICCAT**; International Society for Reef Studies **ISRS**; National Coral Reef Institute **NCRI**; United States-Mexico Fisheries Cooperation Program **FCP**
	Galveston Bay	**DoC** (**SEFSC**, **SERO**, Flower Gardens **NMS** Based in Galveston); **DoI** (**FWS**, **BOEM**, **USGS**); **EPA**	**GMFMC**	**GSMFC, TPWD**	
	Matagorda Bay	**DoC** (**SEFSC**, **SERO**); **DoI** (**FWS**, **BOEM**, **USGS**); **EPA**	**GMFMC**	**GSMFC, TPWD**	
	Mississippi River Delta	**DoC** (**SEFSC**, **SERO**); **DoI** (**FWS**, **BOEM**, **NPS**, **USGS**); **EPA**	**GMFMC**	**GSMFC, LDWF, CPRA, MDMR**	
	Mobile Bay	**DoC** (**SEFSC**, **SERO**); **DoI** (**FWS**, **BOEM**, **USGS**); **EPA**	**GMFMC**	**GSMFC, ADCNR**	
	Tampa Bay	**DoC** (**SEFSC**, **SERO**); **DoI** (**FWS**, **BOEM**, **USGS**); **EPA**	**GMFMC**	**GSMFC, FWC**	

state agencies and the GSMFC, while those in federal waters are managed by both NOAA Fisheries (Southeast Region) and the GGMFMC. Both Galveston and Matagorda Bays fall under the jurisdiction of the Texas Parks and Wildlife Department (TPWD), while portions of the MRD are subject to either the Louisiana Department of Wildlife and Fisheries (LDWF), the Louisiana Department of Natural Resources (LDNR) Office of Coastal Management, or the Mississippi Department of Marine Resources (MDMR). Additionally, marine resources in Mobile Bay are managed by the Alabama Department of Conservation and Natural Resources (ADCNR), while those in Tampa Bay are managed by the Florida FWC.

Nationally, the nGOM contains the third-highest number of congressional representatives, which are concentrated within five states (Fig. 7.8). Composition of the GMFMC has been split relatively evenly among the commercial and recreational fishing sector and other representatives, although greater recreational membership was observed in the early 2000s (Fig. 7.9a). The GMFMC also contains one of the highest numbers of appointed council members nationally, while regional representatives from the Gulf of Mexico also serve on the Atlantic Highly Migratory Species Advisory Panel. The Atlantic marine mammal Scientific Review Group (SRG) is responsible for advising federal agencies on the status of marine mammal stocks covered under the Marine Mammal Protection Act (MMPA), for regions including the nGOM. Its composition and membership have remained steady over time (Fig. 7.9b), with individuals from the academic and private institutional sectors making up the majority of its representatives.

7.4.2.2 Fishery and Systematic Context

Three current nGOM-specific fishery management plans (FMPs) are managed by NOAA Fisheries and the GMFMC for federal Red drum, 49 species of reef fishes, and 4 species of shrimp (Table 7.2). There is also an FMP for regulating offshore aquaculture. Three more

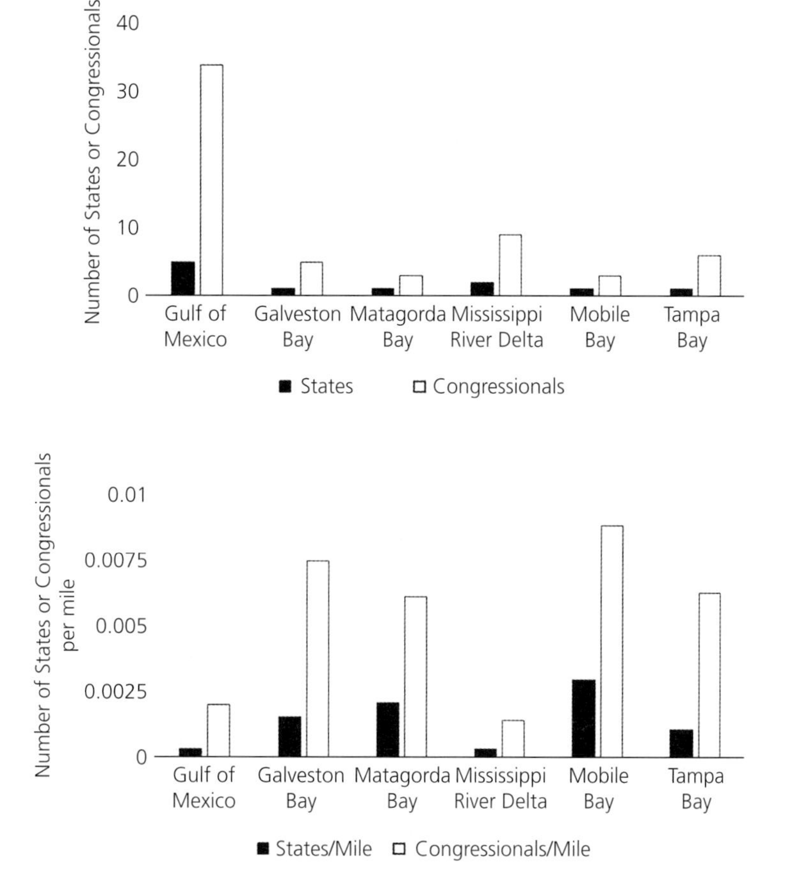

Figure 7.8 Number of states and U.S. Congressional representatives (top panel) and as standardized by miles of shoreline (bottom panel) for the northern Gulf of Mexico region and its identified subregions.

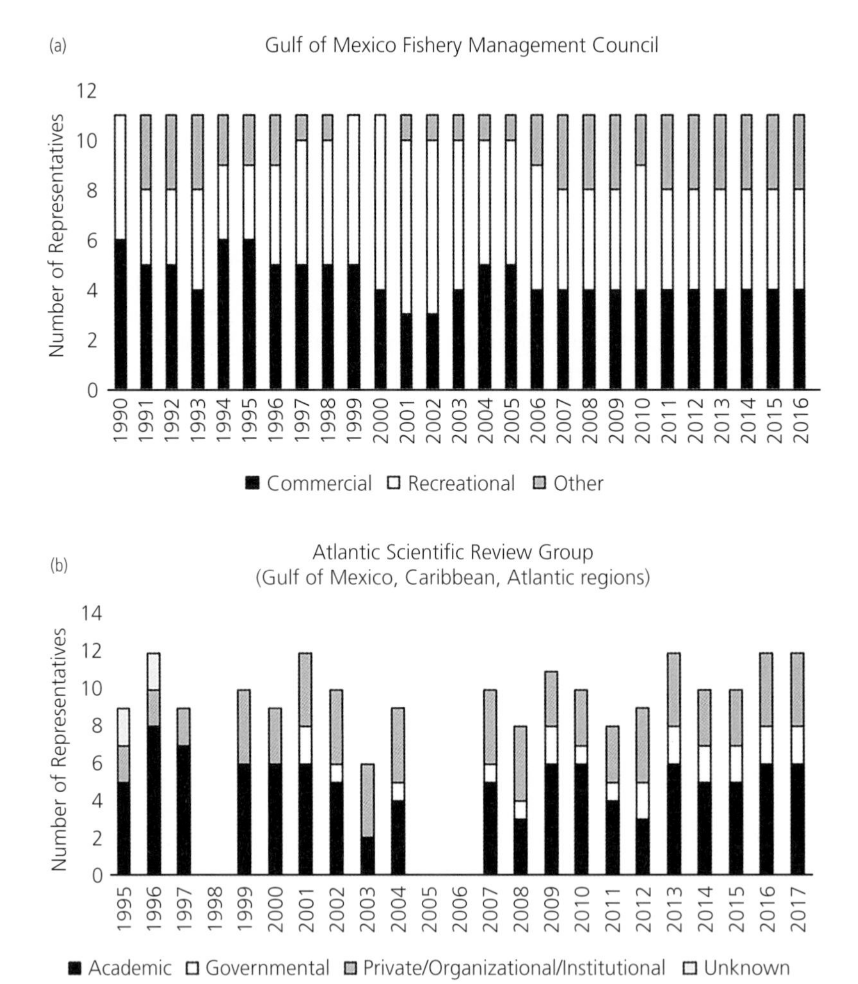

Figure 7.9 (a) Composition of the Gulf of Mexico Fishery Management Council and the (b) Atlantic Marine Mammal Scientific Review Group (data unavailable 1998, 2005–2006) over time.

Data derived from NOAA National Marine Fisheries Service Council Reports to Congress and Scientific Review Group reports.

FMPs are shared among the GMFMC and SAFMC for 142 species of coral reef taxa, 3 species of coastal migratory pelagic resources, and Spiny lobster. In total, these FMPs have been subject to 138 modifications (i.e., amendments and framework/regulatory amendments). The GSMFC manages 11 FMPs for state fisheries resources, including shrimp (8 species), flounders (12 fishery species, with particular emphasis on Southern *Paralichthys lethostigma* and Gulf flounder *P. albigutta*; VanderKooy 2015), oyster, Blue crab, and inshore fishes. None of these state-wide FMPs have been subject to modification. No fishery ecosystem plan (FEP) exists for the nGOM region, but one is currently under development.

Two national seashores (Padre Island and Gulf Islands), four coastal national parks, and five NERRs are found throughout the nGOM (Table 7.3).

Additionally, the Flower Gardens National Marine Sanctuary (NMS) is located in the offshore waters southeast of Galveston Bay, and contains coral reef habitats. Thirteen reef and bank Habitat Areas of Particular Concern (HAPCs) are found in offshore areas of the northwestern Gulf of Mexico, while four additional HAPCs (Madison-Swanson Marine Reserve, Florida Middle Grounds, Pulley Ridge, and the Dry Tortugas) are found off north, central, and southwest Florida. Few named fishing closure zones occur in the nGOM, while fishing is prohibited in 35 specific areas and restricted in 93 of them (Fig. 7.10).

Table 7.2 List of major managed fishery species in the Gulf of Mexico, under state and federal jurisdictions

Council/Agency	Fishery Management Plan (FMP)	FMP Modifications	Major Species/Species Group	Scientific Name(s) or Family Name(s)
Gulf of Mexico Fishery Management Council **GMFMC**	Stone crab, Gulf Mexico (*Now managed by Florida as of 2011*)	8 amendments	Stone crab	*Menippe mercenaria*
	Red drum, Gulf Mexico	3 amendments	Red drum	*Sciaenops ocellatus*
	Reef fish fishery, Gulf Mexico (49 species)	45 amendments	Groupers/seabasses (18 species)	Family Serranidae—*Centropristis* spp., *Epinephelus* spp., *Mycteroperca* spp.
			Red snapper	*Lutjanus campechanus*
			Other snappers (14 species)	Family Lutjanidae—*Etilis spp., Lutjanus spp., Ocyurus* sp., *Pristipomoides* spp., *Rhomboplites* sp.
			Other reef fishes included in the fishery (16 species)	Species from families Balistidae (Triggerfishes), Carangidae (Jacks), Haemulidae (Grunts), Labridae (Wrasses), Malacanthidae (Tilefishes), Serranidae (Sand perches), Sparidae (Porgies)
	Shrimp fishery, Gulf Mexico (4 species)	17 amendments	All shrimp	*Farfantepenaeus* spp., *Litopenaeus* spp., *Pleoticus* spp.
GMFMC/ South Atlantic Fishery Management Council **SAFMC**	Coral, coral reefs and live/hard bottom habitat, Gulf Mexico (142 species)	12 amendments	All species	Octocorallina, Hydrozoans, Scleractinians, Antipatharians, Live Rock
	Coastal migratory pelagic resources (3 species)	26 amendments, 3 framework amendments, 19 regulatory amendments	King mackerel Cobia *(now managed by the Atlantic States Marine Fisheries Commission as of 2019)* Spanish mackerel	*Scomberomorus cavalla* *Rachycentron canadum* *Scomberomorus maculatus*
	Spiny lobster	11 amendments, 2 regulatory amendments	Spiny lobster	*Panulirus argus*

Continued

Table 7.2 Continued

Council/Agency	Fishery Management Plan (FMP)	FMP Modifications	Major Species/Species Group	Scientific Name(s) or Family Name(s)
Gulf States Marine Fisheries Commission **GSMFC**	Gulf Shrimp (8 species)	None	All shrimp	*Farfantepenaeus* spp., *Litopenaeus* spp., *Pleoticus* spp., *Sicyonia* spp., *Trachypenaeus* spp., *Xiphopenaeus* spp.
	Flounder (27 species)		All flounder	*Paralichthys albigutta*, *P. lethostigma* and other species of Family Bothidae
	Spotted seatrout		Spotted seatrout	*Cynoscion nebulosus*
	Striped bass		Striped bass	*Morone saxatilis*
	Oyster		Eastern oyster	*Crassostrea virginica*
	Gulf sturgeon		Gulf sturgeon	*Acipenser oxyrinchus desotoi*
	Blue crab		Blue crab	*Callinectes sapidus*
	Striped mullet		Striped mullet	*Mugil cephalus*
	Spanish mackerel		Spanish mackerel	*Scomberomorus maculatus*
	Gulf menhaden		Gulf menhaden	*Brevoortia patronus*
	Black drum		Black drum	*Pogonias cromis*

Table 7.3 List of major bays and islands, NOAA Habitat Focus Areas (HFAs), NOAA National Estuarine Research Reserves (NERRs), NOAA National Marine Monuments and Sanctuaries (NMS), Coastal National Parks, National Seashores, and number of Habitats of Particular Concern (HAPCs) in the Northern Gulf of Mexico region and subregions

Region	Subregion	Major Bays	Major Islands	HFAs	NERRs	Monuments and NMS	Coastal National Parks	National Seashores	HAPCs
Gulf of Mexico		Corpus Christi Bay, Aransas Bay, Matagorda Bay, Galveston Bay, Vermillion Bay, Bay St. Louis, Mississippi River Delta, Grand Bay, Mobile Bay, Pensacola Bay, Perdido Bay, Apalachicola Bay, Choctawhatchee Bay, West Bay, St. Andrew Bay, Apalachee Bay, Tampa Bay, Sarasota Bay, Estero Bay, Rookery Bay, Gullivan Bay, Whitewater Bay	Padre Island, San Jose Island, Matagorda Island, Galveston Island, Marsh Island, Chandeleur Islands, Cat Island, Shipp Island, Horn Island, Petit Bois Island, Dauphin Island, Ono Island, Perdido Key, Santa Rosa Island, St. George Island, Dog Island, Anclote Key, Honeymoon Island, Caladesi Island, Sanibel Island, Marco Island		Rookery Bay NERR, Apalachicola Bay NERR, Weeks Bay NERR, Grand Bay NERR, Mission-Aransas NERR	Flower Gardens NMS	Jean Lafitte National Historical Park and Preserve, De Soto National Memorial, Everglades National Park, Dry Tortugas National Park	Gulf Islands National Seashore, Padre Island National Seashore	17
	Galveston Bay		Galveston Island						
	Matagorda Bay		Matagorda Island						
	Mississippi River Delta		Chandeleur Islands				Jean Lafitte National Historical Park and Preserve	Gulf Islands National Seashore	
	Mobile Bay		Dauphin Island, Gaillard Island		Weeks Bay NERR				
	Tampa Bay						De Soto National Memorial		

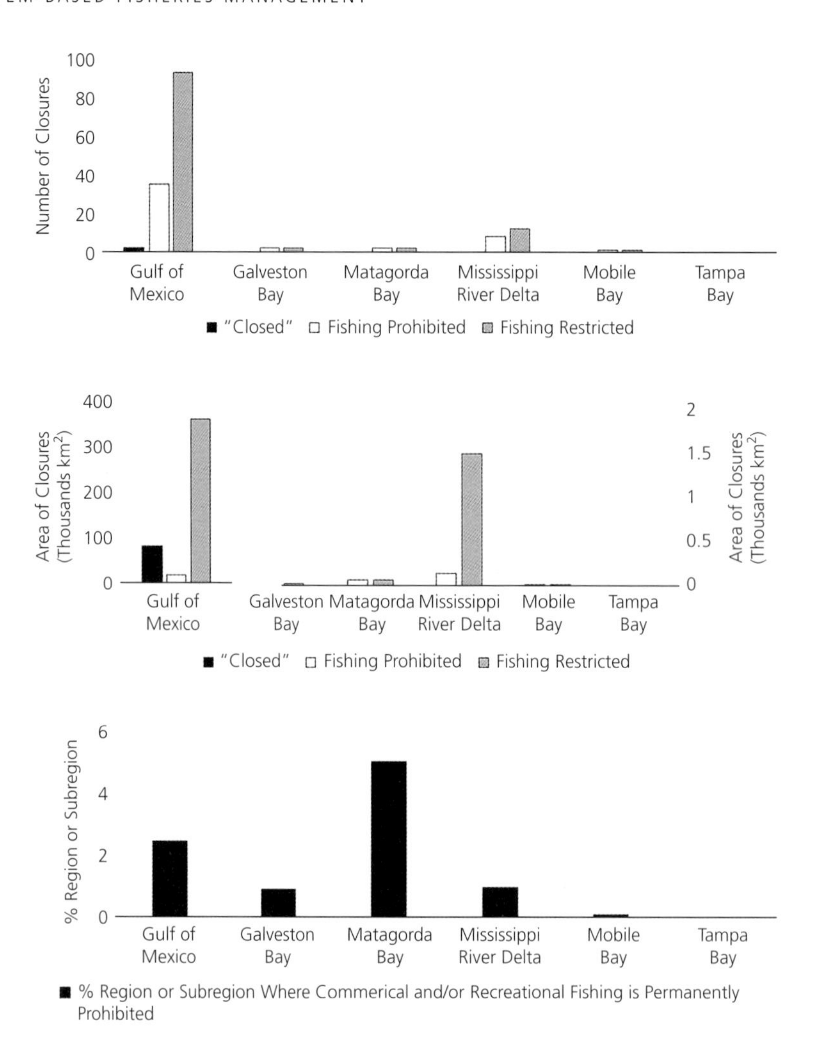

Figure 7.10 Number (top panel) and area (middle panel; thousands of km^2) of named fishing closures, and prohibited or restricted fishing areas of the northern Gulf of Mexico region and identified subregions, including percent region or subregion (bottom panel) where commercial and/or recreational fishing is permanently prohibited.

Data derived from NOAA Marine Protected Areas inventory.

Few of these closures occur in the five key nGOM subregions, of which the MRD contains the highest number with 8 prohibited and 12 restricted areas. Very little area of the nGOM is set aside as named closures or permanent fishing prohibitions (~2% of the EEZ), while ~360,000 km^2 is subject to some fishing restrictions. Among subregions, the largest restricted area occurs in the MRD (~1500 km^2), while ~5% of Matagorda Bay is permanently prohibited from fishing due to the presence of the Aransas National Wildlife Refuge. Only 0.09% of Mobile Bay is permanently closed to fishing (within the Bon Secour National Wildlife Refuge), while no closed, prohibited, or restricted areas occur within Tampa Bay.

7.4.2.3 Organizational Context

Relative to the total commercial value of its managed fisheries, the GMFMC budget is mid-ranged when compared to other regions (Fig. 7.11; Link & Marshak 2019), and makes up a small proportion of its overall costs. These relative values also peaked in the early 2010s and have recently begun trending upward due to fluctuating and increasing fishery value. Cumulative National Environmental Policy Act Environmental Impact Statement (NEPA-EIS) actions from 1987 to 2016 have been highest in the Gulf of Mexico with an average of ~10 per year for the past ~30 years (Fig. 7.12). Relatively few actions have occurred within each of the five key nGOM subregions, with a peak-cumulative

Figure 7.11 Total annual budget of the Gulf of Mexico Fishery Management Council (FMC) as compared to the total value of its marine fisheries. All values in USD.

Data from NOAA National Marine Fisheries Service Office of Sustainable Fisheries and NOAA National Marine Fisheries commercial landings database.

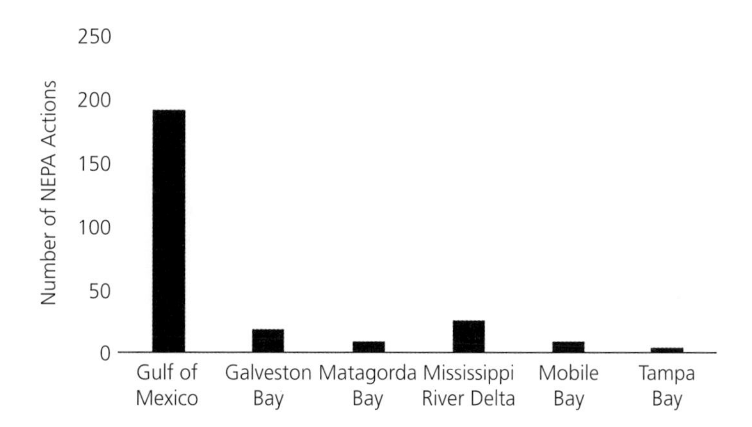

Figure 7.12 Number of National Environmental Policy Act Environmental Impact Statement (NEPA-EIS) actions from 1987 to 2016 for the northern Gulf of Mexico region and identified subregions.

Data derived from U.S. Environmental Protection Agency EIS database.

value of only 25 actions for the MRD. Nine fisheries-related lawsuits have occurred in the nGOM region since 2010.

7.4.2.4 Status of Living Marine Resources (Targeted and Protected Species)

Nationally, the nGOM contains the fourth-highest number of managed species (Link & Marshak 2019). Of the 249 federally managed species in the region, 58 are targeted fishery species covered under FMPs, while an additional 146 species (mostly coral reef taxa) are prohibited from harvest. Forty-five species of marine mammals, sea turtles, fishes, and corals are also federally protected under the ESA or MMPA (Fig. 7.13, Tables 7.2, 7.4–5). An overlapping 120 managed species are also managed under the state jurisdiction of the GSMFC, of which 76 have fishing prohibitions. Throughout all regions, the nGOM contains the highest number of prohibited species.

As of mid-2017, 38 stocks were federally managed in the nGOM, with 23 of them listed as NOAA Fish Stock Sustainability Index (FSSI) stocks (Fig. 7.14; NMFS 2017, 2020).[2] Of all stocks, only one was identified as

[2] As of June 2020, 37 stocks are federally managed in the Gulf of Mexico. In 2019, cessation of joint federal management of the Cobia-Southern Atlantic Coast stock by the GMFMC and SAFMC occurred, which is now singularly managed by the Atlantic States Marine Fisheries Commission (ASMFC).

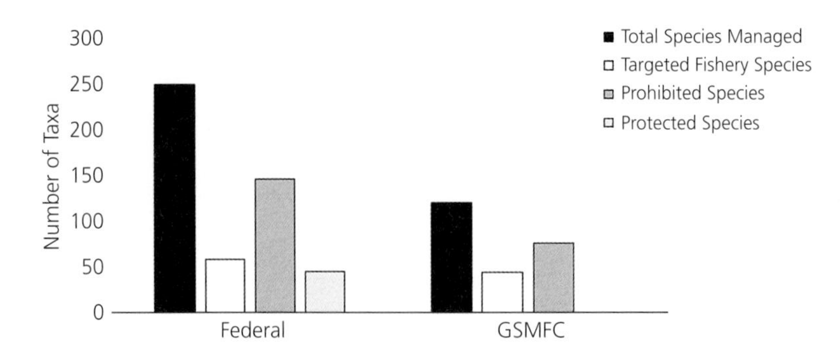

Figure 7.13 Number of managed taxa (species or families) for the northern Gulf of Mexico region, including federal and state (Gulf States Marine Fisheries Commission, GSMFC) jurisdictions.

experiencing overfishing (**Greater amberjack**—*Seriola dumerili*) and three were listed as overfished (**Greater amberjack**, **Gray triggerfish**—*Balistes capriscus*, and **Red snapper**—*Lutjanus campechanus*).[3] Additionally, only one stock continues to have unknown overfishing status (**Gulf of Mexico shallow-water grouper complex**). However, 44.7% of all nGOM stocks were unclassified as to whether they were overfished, making it the fifth-highest among regions in terms of the proportion of fishery stocks with unknown status.[4] The protected species of the nGOM are divided into 64 independent marine mammal stocks (including 36 common bottlenose dolphin—*Tursiops truncatus* stocks), and 24 ESA-listed distinct population segments, of which half are endangered (Fig. 7.15). Nationally, marine mammal stocks are most abundant in the nGOM (20% of total marine mammal stocks), with the highest national percentage of strategic marine mammal stocks (57.8%; including 31 common bottlenose dolphin strategic stocks). The highest national percentage of marine mammal stocks with unknown population size occurs in the nGOM (53.1%).

7.4.3 Environmental Forcing and Major Features

7.4.3.1 Oceanographic and Climatological Context

Circulation patterns throughout the nGOM EEZ are driven primarily by the Gulf Loop Current (Fig. 7.1). Originating from the Caribbean current, which enters the Gulf of Mexico between the Mexican Yucatán Peninsula and western Cuba and is responsible for connectivity between the Gulf of Mexico and U.S. Caribbean regions. The Gulf Loop Current moves northward to flow into the Florida Current at the Florida Straits and joins the Gulf Stream. The Loop Current is dynamic, varying from a winding loop that stretches northward onto the shelf in the nGOM to shedding this loop as a large, anticyclonic eddy. Both alongshore counterclockwise and offshore clockwise Coriolis-driven circulation patterns occur throughout the Gulf of Mexico, allowing turbid waters containing sediments and nutrients from the Mississippi and Mobile/Tensaw Rivers to be carried westward and warmer waters to be carried northward. Seasonal clockwise "cold-core" and counterclockwise "warm-core" eddies also develop in the western Gulf of Mexico, as the result of these differential water masses converging with those emanating from the Gulf Loop Current as it passes eastward toward the Gulf Stream. Together, these components allow for the transport of warm water masses and propagules in both easterly and westerly directions throughout the region (Teo et al. 2007, Johnson et al. 2009, Rooker et al. 2012, Johnson et al. 2013). The dynamics of the Loop Current are among the most important physical features in the Gulf of Mexico.

Clear interannual and multidecadal patterns in average sea surface temperature (SST) have been observed for the nGOM (Fig. 7.16), with temperatures averaging 25.6°C ± 0.03 (standard error, SE) over time (1854–2016). Temperatures in the nGOM have continued to increase as of the early 1900s, including by 0.7°C since the mid-twentieth century. Similar trends have been observed within key subregions, with

[3] As of June 2020, two Gulf of Mexico stocks are experiencing overfishing (Greater amberjack; Gray triggerfish) and only one is overfished (Greater amberjack).

[4] As of June 2020, 43.2% of Gulf of Mexico stocks have unknown overfished status.

Table 7.4 List of prohibited fishery species in the Gulf of Mexico, including federal and state jurisdictions

Council/Authority	Region	Prohibited Species	Scientific Name
Gulf of Mexico Fishery Management Council **GMFMC**; Southeast Regional Office **SERO**	Gulf of Mexico (2 species fishes, 2 species sea fans, 142 hydrozoan, stony coral, and black coral taxa)	Red drum Goliath grouper	*Sciaenops ocellatus* *Epinephelus itajara*
		Sea fans	*Gorgonia* spp.
		Hydrozoans, hard corals, and black corals	Families Acroporidae, Agaricidae, Anthemiphylliidae, Astrocoeniidae, Cayophyllidae, Dendrophyllidae, Faviidae, Flabellidae, Fungiidae, Guyniidae, Meandrinidae, Milleporidae, Mussidae, Oculinidae, Pocillopridae, Poritidae, Rhizangiidae, Siderastreidae, Orders Antipatharia, Stylasterina
		Live rock	Any coral reef resource attached to a hard substrate, including dead coral or rock (excluding individual mollusk shells)
Gulf States Marine Fisheries Commission **GSMFC**	Gulf of Mexico state waters (76 taxa)	Bonefish	*Albula vulpes*
		Snook	*Centropomus undecimalis*
		Stony, hard, black, fire corals (17 taxa)	Families Acroporidae, Agariciidae, Astrocoeniidae, Balanophyllidae, Caryophyllidae, Coralliidae, Dendrophylliidae, Faviidae, Fungiidae, Merulinidae, Mussidae, Oculinidae, Pocilloporidae, Poritidae, Siderastreadae, *Antipathes* spp., *Millepora* spp.
		Goliath grouper Nassau grouper Speckled hind Warsaw grouper Striped marlin Red drum Blue marlin Roundscale spearfish White marlin	*Epinephelus itajara* *Epinephelus striatus* *Epinephelus drummondhayi* *Epinephelus nigritus* *Kajikia audax* *Sciaenops ocellatus* *Makaira nigricans* *Tetrapturus georgii* *Kajikia albida*
		Sharks (25 species)	*Alopias superciliosus*, *Carcharhinus* spp., *Carcharias taurus*, *Carcharodon carcharias*, *Galeocerdo cuvier*, *Hexachus* spp., *Isurus paucus*, *Notorynchus cepedianus*, *Odontaspis noronhai*, *Rhincodon typus*, *Rhizoprionodon porosus*, *Sphyrna* spp., *Squatina demeril*
		Sailfish Gulf sturgeon Atlantic sturgeon Atlantic tarpon Sawfishes (2 species) Spiny dogfish Manta rays (2 species) Devil ray Spotted eagle ray Longbill spearfish Mediterranean spearfish	*Istiophorus albicans* *Acipenser oxyrinchus desotoi* *Acipenser oxyrinchus* *Megalops atlanticus* *Pristis pectinata*, *P. pristis* *Squalus acanthias* *Manta alfredi*, *M. birostris* *Manta hypostoma* *Aetobatus narinari* *Tetrapturus pfluegeri* *Tetrapturus belone*

Continued

Table 7.4 Continued

Council/Authority	Region	Prohibited Species	Scientific Name
Gulf States Marine Fisheries Commission **GSMFC**	Gulf of Mexico state waters (76 taxa)	Longspine urchin	*Diadema antillarum*
		Live rock	
		Queen conch	*Aliger gigas*
		Calico scallop	*Argopecten gibbus*
		Mitten crab	*Eriocheir sinensis*
		Bay scallop	*Argopecten irradians*
		Seafans (2 species)	*Gorgonia* spp.
		Bahama starfish	*Oreaster reticulatus*
		Horseshoe crab	*Limulus polyphemus*

average temperatures over the 162-year period ranging from 24.2 to 26.7°C, being highest in Tampa Bay and Galveston and Matagorda Bays. Coincident with these temperature observations are Atlantic basin-scale climate oscillations (Fig. 7.17), including the AMO and North Atlantic Oscillation (NAO), which exhibit decadal cycles that can influence the environmental conditions and ecologies of the nGOM.

7.4.3.2 Notable Physical Features and Destabilizing Events and Phenomena

With five prominently identified subregions, the northern Gulf of Mexico contains 19 major bays (Table 7.3), the majority of which occur along Florida, and 21 major barrier islands concentrated off the coasts of Texas, the MRD, the Gulf Islands National Seashore, and western Florida. Proportionally, the nGOM makes up ~5% of the U.S. EEZ (~685,000 km^2), 17% of the national shoreline (~27,600 km), and contains the shallowest maximum depth (~3500 m) and second-shallowest average depth (~1000 m) as compared to other U.S. regions (Fig. 7.18; Link & Marshak 2019). The five key subregions contribute very small proportions to the nGOM EEZ. Among them, the MRD comprises the largest area (~17,600 km^2), the longest extent of shoreline (~10,500 km), and reaches the deepest depths among those subregions (86 m). However, on average, the subregion is only 3.1 m deep. Galveston, Matagorda, and Mobile Bays are similarly sized (1078.6–1456.4 km^2, 545.7–1073.9 km of shoreline), each also having shallow average (1.9–2.0 meters) and maximum (5–11 m) depths. Tampa Bay is somewhat smaller (926.1 km^2, 955.7 km of shoreline) than the other subregions, but contains comparatively moderate average (3.7 m) and maximum (25 m) depths.

In the nGOM, 635 hurricanes have been reported since 1850 with an average of 37 hurricanes per decade having occurred and peak numbers (~60 per decade) recorded in the 1970s and 2000s (Fig. 7.19). While at lower intensities, these peak trends have also been observed for the MRD (up to 15 hurricanes/decade), Tampa Bay, and Galveston Bay (up to 6 hurricanes/decade each), which also experienced significant storms during the 1940s. Over time, 117 hurricanes have been reported for the MRD, followed by 59 in Tampa Bay, 49 in Mobile Bay, and 39 and 31 for Galveston and Matagorda Bays, respectively. During 2005, Hurricane Katrina caused substantial damage to the MRD, especially affecting fisheries' landings, value, exploitation rates, and marine socioeconomics. Having been documented since the mid-1980s, several peaks in the extent of Gulf of Mexico bottom water hypoxia events were observed in the mid-2000s, with a record-setting event also having occurred in 2017 (~22,700 km^2; LUMCON 2019); summer hypoxic conditions ($\leq$2 mg L^{-1}) have repeatedly encompassed over 20,000 km^2 of bottom surface area (Fig. 7.20).

7.4.4 Major Pressures and Exogenous Factors

7.4.4.1 Other Ocean Use Context

Ocean uses in the nGOM are most associated with offshore mineral extraction, tourism/recreation, and marine transportation sectors (Fig. 7.21). In total and

Table 7.5 List of protected species under the Endangered Species Act (ESA) or Marine Mammal Protection Act (MMPA) in the Gulf of Mexico region

Region	Managing Agency	Protected Species Group	Common Name	Scientific Name
Gulf of Mexico (45 species)	Gulf of Mexico Fishery Management Council **GMFMC**; Southeast Regional Office **SERO**	Whales	Blainville's beaked whale	*Mesoplodon densirostris*
			Blue whale	*Balaenoptera musculus*
			Bryde's whale	*Balaenoptera edeni*
			Cuvier's beaked whale	*Ziphius cavirostris*
			Dwarf sperm whale	*Kogia sima*
			False killer whale	*Pseudorca crassidens*
			Fin whale	*Balaenoptera physalus*
			Gervais' beaked whale	*Mesoplodon europaeus*
			Humpback whale	*Megaptera novaeangliae*
			Killer whale (Orca)	*Orcinus orca*
			Melon-headed whale	*Peponocephala electra*
			Minke whale	*Balaenoptera acutorostrata*
			North Atlantic right whale	*Eubalaena glacialis*
			Pygmy killer whale	*Feresa attenuata*
			Pygmy sperm whale	*Kogia breviceps*
			Sei whale	*Balaenoptera borealis*
			Short-finned pilot whale	*Globicephala macrorhynchus*
			Sowerby's beaked whale	*Mesoplodon bidens*
			Sperm whale	*Physeter macrocephalus*
		Dolphins	Atlantic spotted dolphin	Stenella frontalis
			Common bottlenose dolphin	*Tursiops truncatus*
			Clymene dolphin	*Stenella clymene*
			Fraser's dolphin	*Lagenodelphis hosei*
			Pantropical spotted dolphin	*Stenella attenuata*
			Risso's dolphin	*Grampus griseus*
			Rough-toothed dolphin	*Steno bredanensis*
			Spinner dolphin	*Stenella longirostris*
			Striped dolphin	*Stenella coeruleoalba*
		Sea turtles	Green sea turtle	*Chelonia mydas*
			Hawksbill sea turtle	*Eretmochelys imbricata*
			Kemp's Ridley sea turtle	*Lepidochelys kempii*
			Leatherback sea turtle	*Dermochelys coriacea*
			Loggerhead sea turtle	*Caretta caretta*
		Fishes	Gulf sturgeon	*Acipenser oxyrinchus desotoi*
			Nassau grouper	*Epinephelus striatus*
			Largetooth sawfish	*Pristis microdon*
			Smalltooth sawfish	*Pristis pectinata*
		Corals	Rough cactus coral	*Mycetophyllia ferox*
			Pillar coral	*Dendrogyra cylindrus*
			Lobed star coral	*Orbicella annularis*
			Mountainous star coral	*Orbicella faveolata*
			Boulder star coral	*Orbicella franksi*
U.S. Fish and Wildlife Service **FWS**			Staghorn coral	*Acropora cervicornis*
			Elkhorn coral	*Acropora palmata*
		Other species	West Indian manatee	*Trichechus manatus*

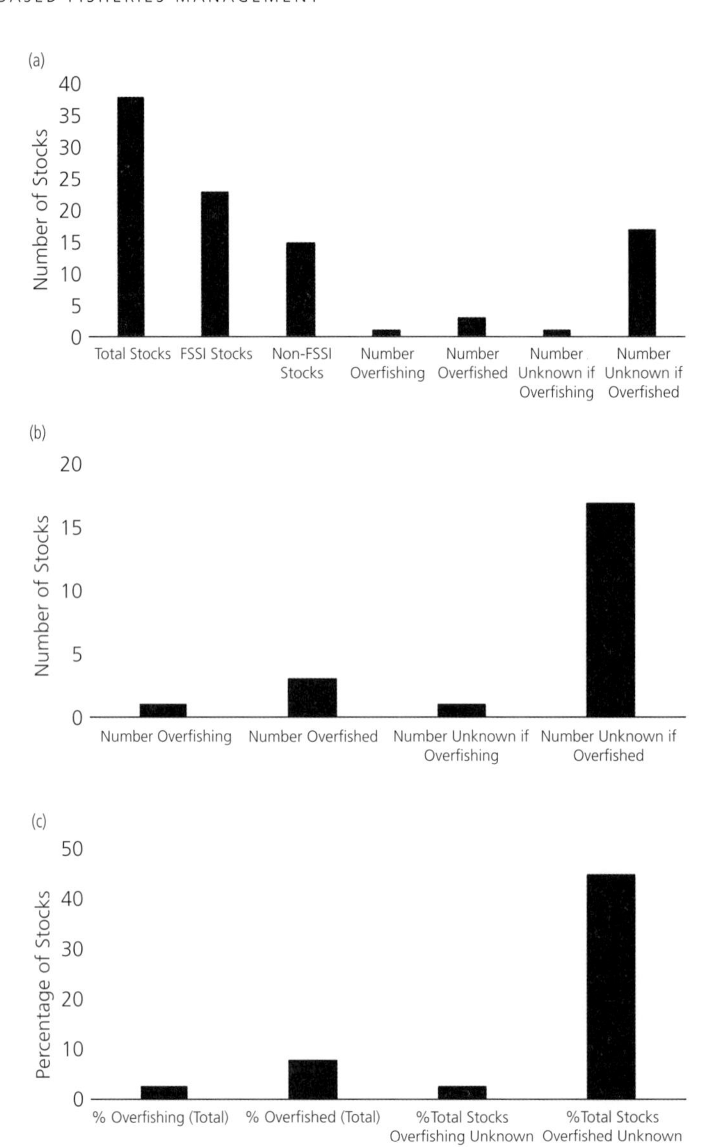

Figure 7.14 For the northern Gulf of Mexico region as of June 2017 (a) Total number of managed Fish Stock Sustainability Index (FSSI) stocks and non-FSSI stocks, and breakdown of stocks experiencing overfishing, classified as overfished, and of unknown status. (b) Number of stocks experiencing overfishing, classified as overfished, and of unknown status. (c) Percent of stocks experiencing overfishing, classified as overfished, and of unknown status.

Data from NOAA National Marine Fisheries Service. Note: stocks may refer to a species, family, or complex.

among sectors, trends for nGOM ocean establishments and employments have remained consistent over the past decades, with tourism establishments and employments comprising 65–70% of total nGOM ocean establishments and 45–60% of total nGOM ocean employments. Additionally, GDP from offshore mineral extraction is highest among sectors, contributing 60–80% toward total nGOM ocean GDP. During the past decades, ocean GDP has oscillated, with marked increases occurring during the late 2000s and mid-2010s, and most recent estimated values at ~$110 billion. Overall, the nGOM contributes toward 46% of the

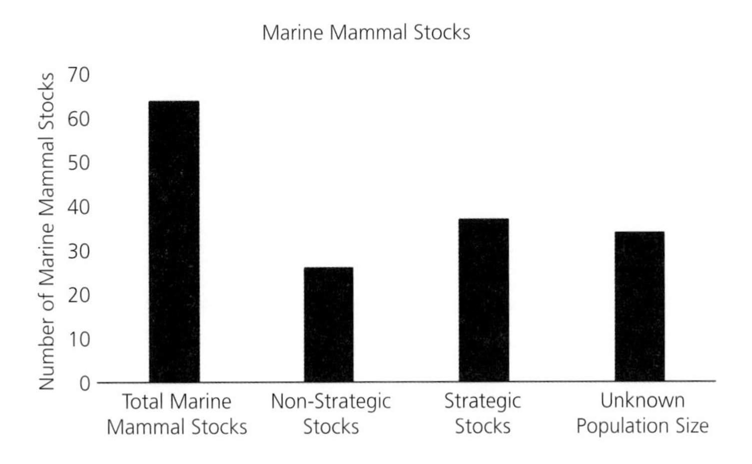

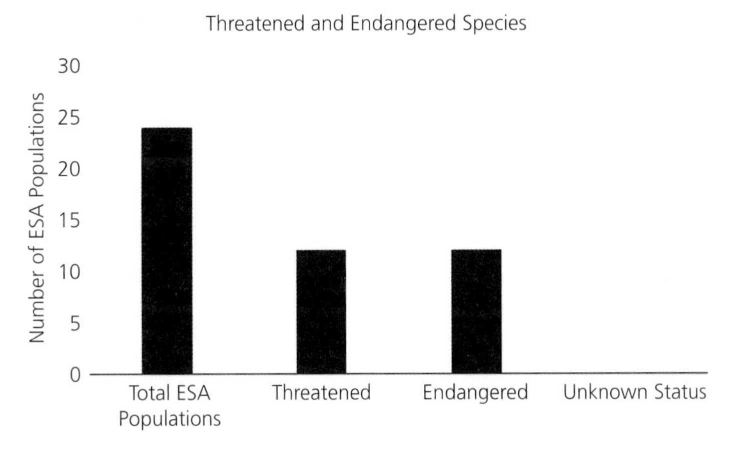

Figure 7.15 Number and status of federally protected species (marine mammal stocks, top panel; distinct population segments of species listed under the Endangered Species Act (ESA), bottom panel) occurring in the northern Gulf of Mexico.

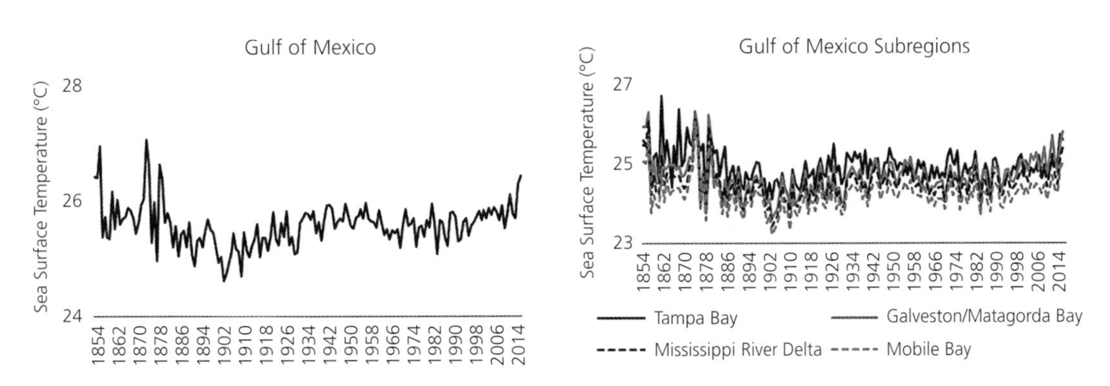

Figure 7.16 Average sea surface temperature (°C) over time (years 1854–2016) for the northern Gulf of Mexico region and identified subregions. Values are combined for both Galveston and Matagorda Bays.

Data derived from the NOAA Extended Reconstructed Sea Surface Temperature dataset (https://www.ncdc.noaa.gov/).

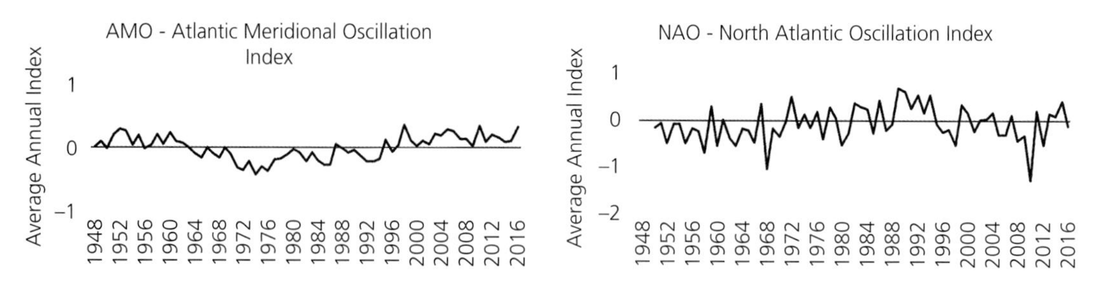

Figure 7.17 Climate forcing indices for the northern Gulf of Mexico region over time (years 1948–2016).

Data derived from NOAA Earth System Research Laboratory data framework (https://www.esrl.noaa.gov/).

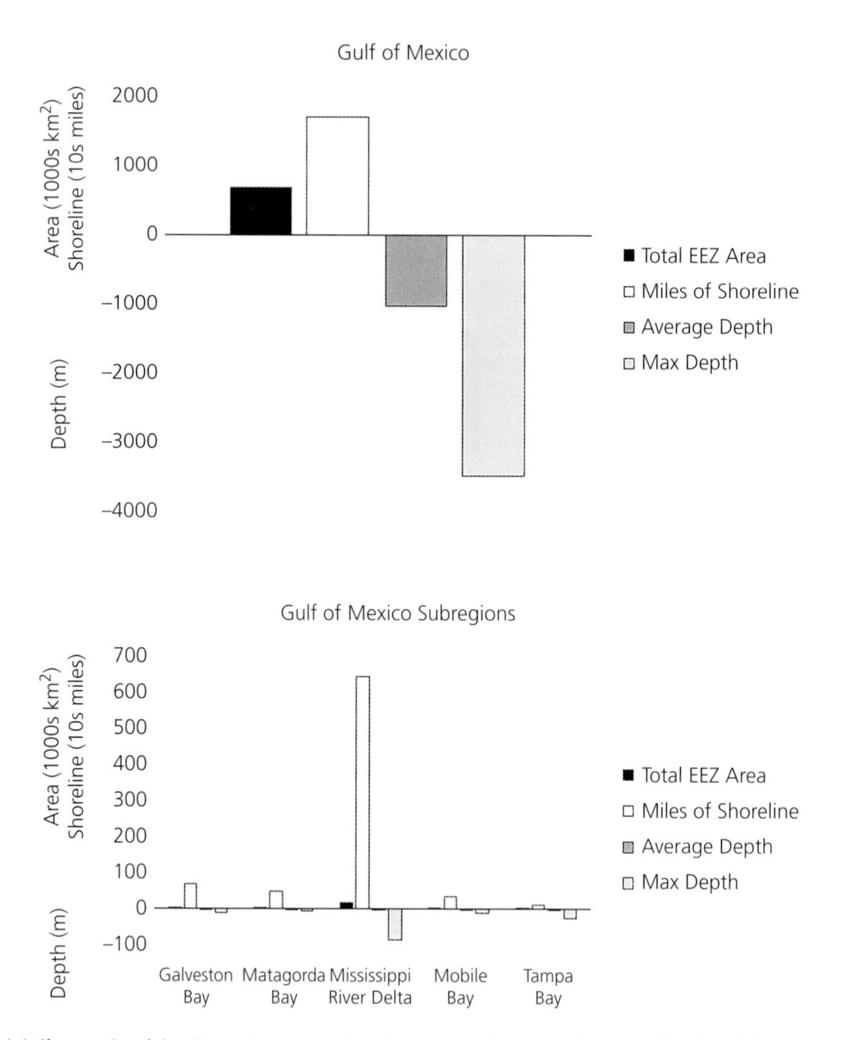

Figure 7.18 Total shelf area, miles of shoreline, and average and maximum depth of marine regions and subregions of the northern Gulf of Mexico.

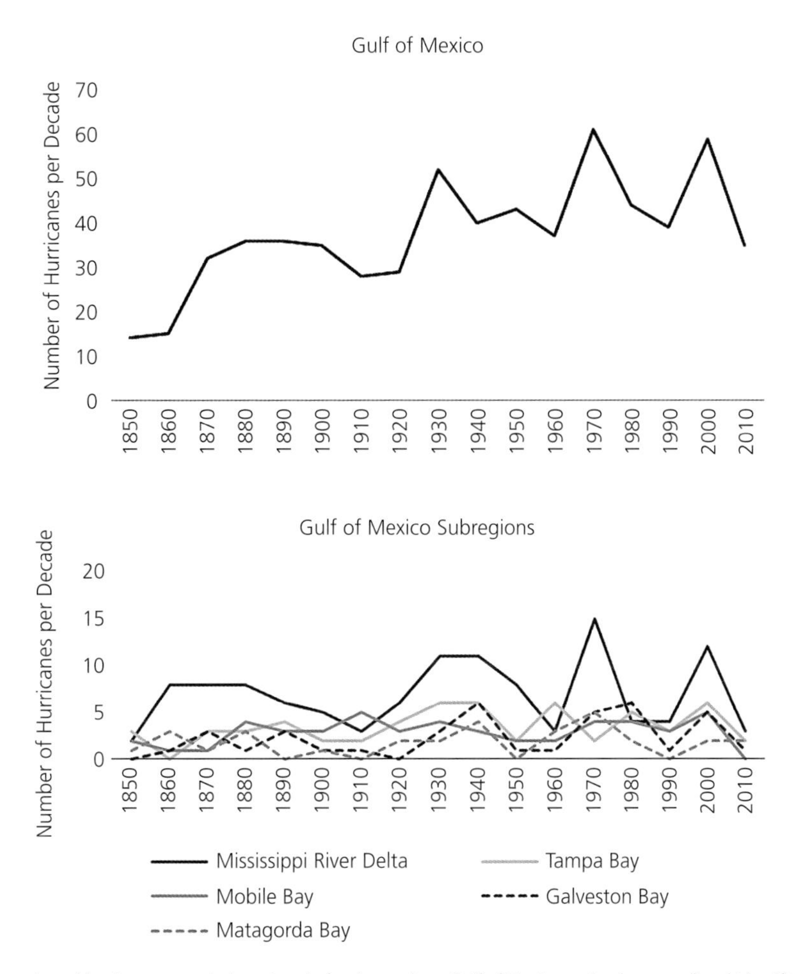

Figure 7.19 Total number of hurricanes recorded per decade for the northern Gulf of Mexico region (top panel) and identified subregions (bottom panel) since year 1850.

Data derived from NOAA Office for Coastal Management (https://coast.noaa.gov/hurricanes/).

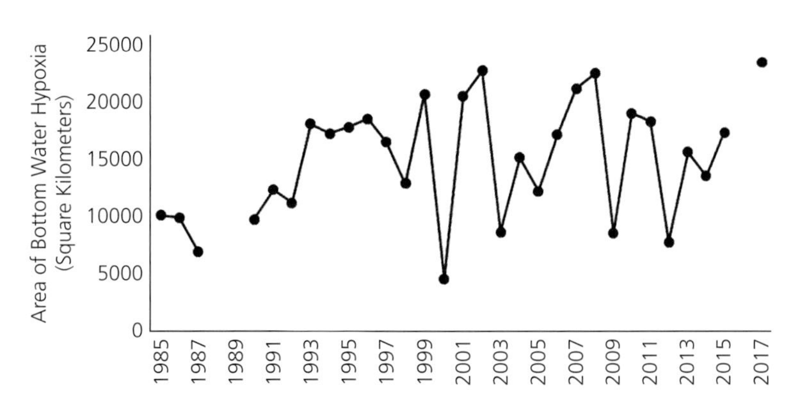

Figure 7.20 Area (square kilometers) of the mid-summer bottom water hypoxia (≤2 mg L^{-1} of oxygen) event in the northern Gulf of Mexico over time.

Data from Louisiana Universities Marine Consortium (LUMCON) Gulf of Mexico Hypoxia Database (https://gulfhypoxia.net/research/shelfwide-cruises/). Data unavailable for 1988–1989, 2016.

U.S. total ocean GDP, which is the highest in the nation. Establishments and employments in the nGOM offshore mineral extraction sector contribute 11–15% toward total ocean establishments and comprise 15–25% of total ocean employments. Nationally, the nGOM contributes toward 72% of U.S. offshore mineral extraction establishments and ~82% of national offshore mineral extraction employments and GDP. It is also the highest in the nation for ship and boat building establishments and employments (36% of U.S. ship and boat building establishments and 30% of U.S. ship and boat building employments), and for marine construction establishments, employments, and GDP (28% of U.S. marine construction establishments, 40% of U.S. marine construction employments, and 35% of U.S. marine construction GDP). Overall, the nGOM is excelling in terms of maintaining its high marine socioeconomic status, especially as associated with other ocean uses and its LMR-associated economy. Nationally, offshore oil production, as indexed by the average count of offshore oil rigs per region, is most pronounced in the nGOM (Fig. 7.22). However, values have varied over time with declines having occurred in Baker Hughes rig counts since 2000. Most oil production is concentrated offshore of the MRD, following similar trends as those for the region, with much lower values and steadier trends observed for Galveston, Matagorda, and Mobile Bays, and very little production off western Florida.

The degree of tourism, as measured by number of dive shops (fourth-highest), number of major ports (highest) and marinas, and number of cruise ship departure passengers (third-highest) is among the highest nationally in the nGOM (Fig. 7.23; Link & Marshak 2019). Few dive shops, ports, or marinas are found throughout the key subregions, with many concentrated in Florida (Tampa Bay with 7 dive shops, 2 major ports, and 3 major marinas), while major ports and marinas are found along the MRD. Cruise departure passengers are more equally prominent throughout four of the five subregions, although highest values are found for departures from near Galveston and Tampa Bays.

7.4.5 Systems Ecology and Fisheries

7.4.5.1 Basal Ecosystem Production

Since 2002, mean surface chlorophyll values throughout the nGOM EEZ have been relatively low at ~1.05 mg m^{-3} (Fig. 7.24). Nearshore subregions subject to higher freshwater input and nutrient loading average 15.6 mg m^{-3} in Tampa Bay, 11.2 mg m^{-3} in the MRD and Mobile Bay, and 11.1 mg m^{-3} in Galveston and Matagorda Bays. Higher values were generally observed for the MRD and Mobile Bay from 2008 to 2016. Average annual productivity within the nGOM EEZ has remained relatively steady between 200 and 250 gC m^{-2} year^{-1} (fourth-highest nationally), with trends slightly decreasing over time (1998–2014). High benthic production as observed throughout coastal wetlands, submerged aquatic vegetation, and even deep seep communities is also found in the Gulf of Mexico, with localized estimates in Alabama and Florida estuaries ranging from ~400 to ~800 gC m^{-2} year^{-1} (Caffrey et al. 2014). Overall, the inherent productivity of this region has facilitated its strong marine economy and historic seafood production.

7.4.5.2 System Exploitation

Nationally, the third-highest numbers of landed taxa have been captured in the nGOM over time (Link & Marshak 2019), with values continuing to increase since the 1980s and recently peaking at 229 reported taxa (Fig. 7.25). Among subregions, the highest values have been observed over time for Tampa Bay (ranging from 126 to 225 taxa from 1991–present) and the MRD (ranging from 94 to 156 taxa from 1999–present), while similarly lower values have remained steady for Galveston, Matagorda, and Mobile Bays (~60 taxa on average from 1999 to present).

Total surveyed fish and invertebrate biomass for the nGOM has remained relatively steady, ranging from 4 to 8 million metric tons per year (Fig. 7.26). Comparing among regions nationally, nGOM biomass values are fifth-highest and are estimated to be substantially lower than in years prior to the 1990s due to intensive overfishing by both commercial and recreational fisheries. Exploitation rates (i.e., landings/biomass ratio) in the nGOM are highest in the nation, and have ranged from 0.1 to 0.15 over time, with values decreasing in 2005 (as associated with Hurricane Katrina) and leveling off until 2010 (Fig. 7.27). Overall, these values have been near or above sustainable exploitation limits as defined in the literature (Samhouri et al. 2010, Large et al. 2013, 2015, Tam et al. 2017). Values for total biomass/productivity (up to 0.007) and total landings/productivity (up to 0.0006) are third-highest nationally, and have remained relatively constant over time. Landings/productivity values have likewise been

Figure 7.21 Number and percent contribution of multisector establishments, employments, and Gross Domestic Product (GDP) and their percent contribution to the total multisector oceanic economy of the northern Gulf of Mexico region (years 2005–2016).

Data derived from the National Ocean Economics Program.

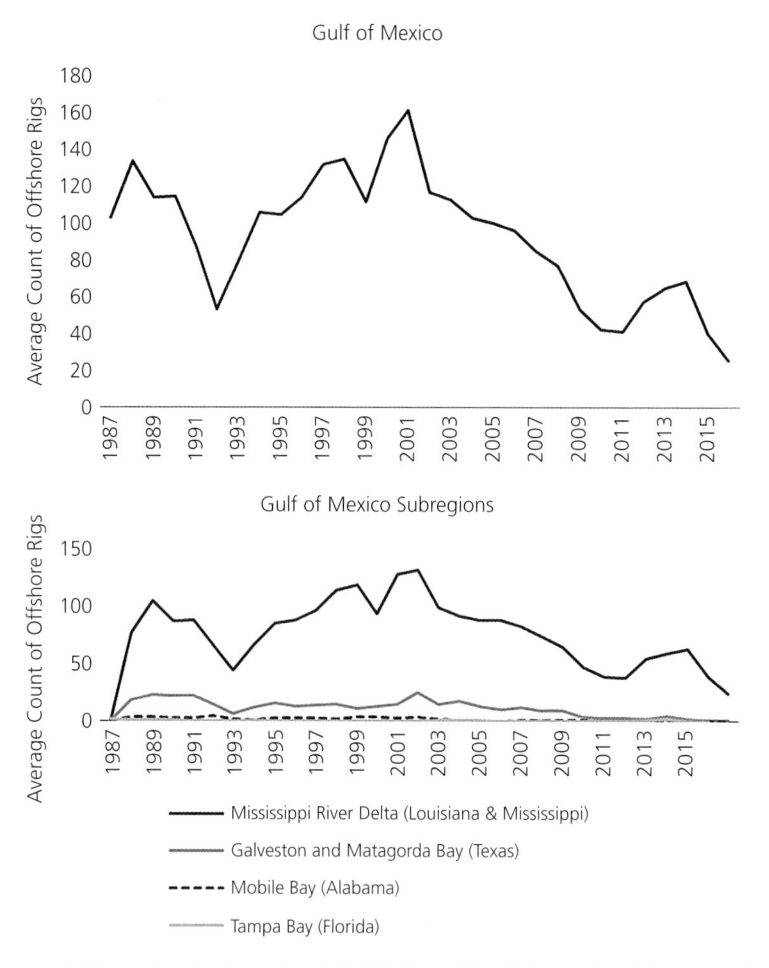

Figure 7.22 Average count of offshore oil rigs for the northern Gulf of Mexico and identified subregions. Values are combined for both Galveston and Matagorda Bays.

Data derived from Baker Hughes Rotary Rig Counts.

above those stipulated in other studies as being sustainable for a given system (Link & Marshak 2019).

7.4.6 Targeted and Non-Targeted Resources

Total nGOM commercial and recreational fisheries landings (Fig. 7.28) have increased since the 1950s. A doubling of commercially landed catch occurred from 1950 to 1985, peaking at 1.2 million metric tons and related to increases in Gulf Menhaden and shrimp catches off Louisiana, Mississippi, and Texas, with substantial decreases during the 1990s–2000s to 700 thousand metric tons. Although recreational fisheries for reef fishes and many other target species are of great importance in the nGOM (Coleman et al. 2004), cumulatively recreational fisheries for the region have remained steady albeit modest, contributing 7% of total reported landings and are most prominent off Tampa Bay. These overall proportional contributions are due to the high abundance of commercially important Gulf Menhaden, penaeid shrimps, and Blue crab in total landings (NMFS 2018b). Among key subregions, commercial fisheries landings are highest in the MRD (Louisiana-Mississippi), Mobile Bay (Alabama), and Galveston Bay (Texas), while substantially lower in Matagorda Bay (Texas) and Tampa Bay (Florida).

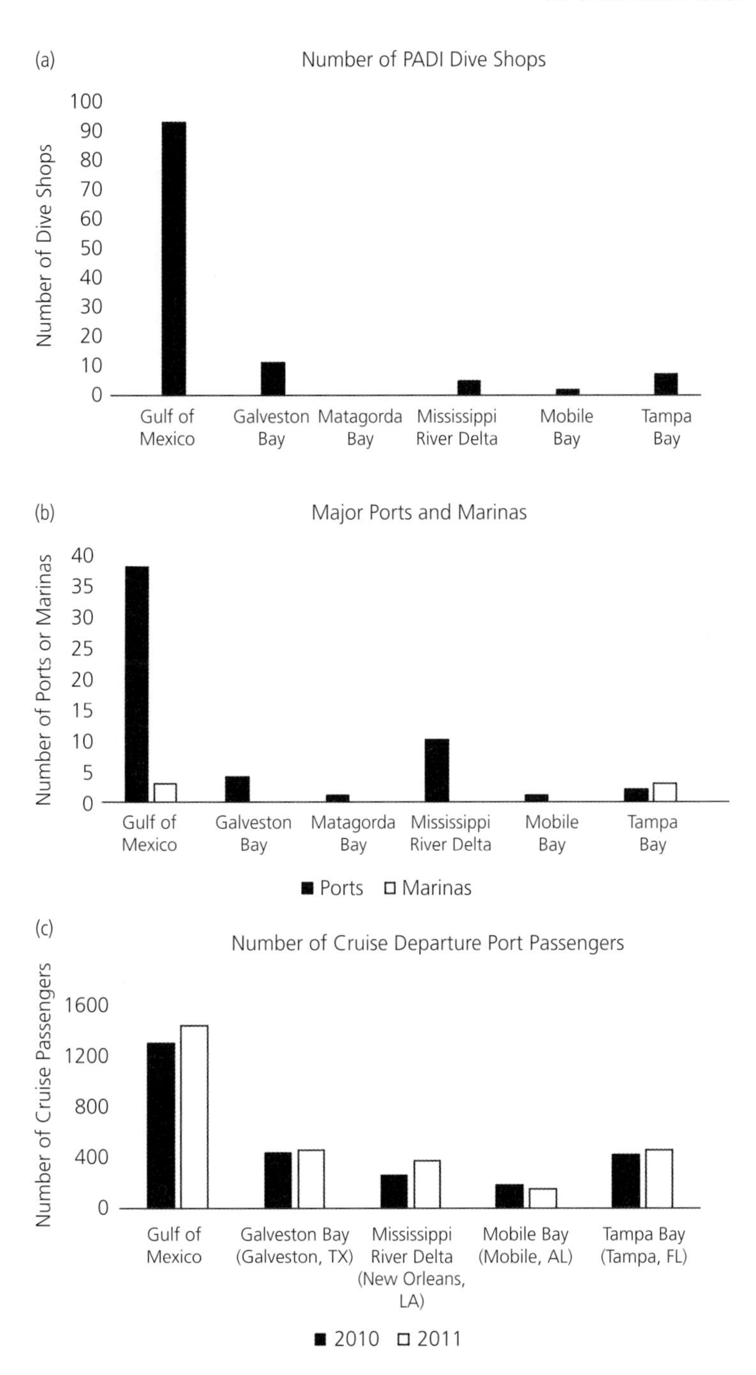

Figure 7.23 Tourism indicators for the northern Gulf of Mexico region and identified subregions: (a) Number of Professional Association of Diving Instructors (PADI) dive shops. Data derived from http://apps.padi.com/scuba-diving/dive-shop-locator/. (b) Number of major ports and marinas. (c) Number of cruise departure port passengers (years 2010 and 2011).

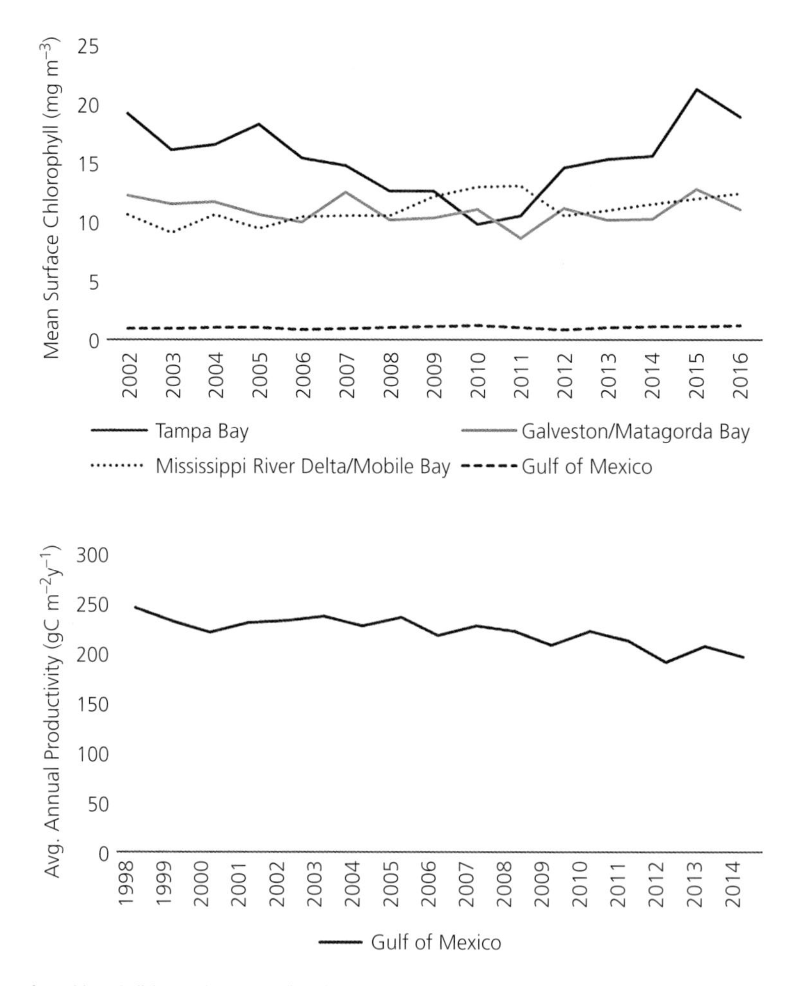

Figure 7.24 Mean surface chlorophyll (mg m^{-3}; top panel) and average annual productivity (grams carbon m^{-2} $year^{-1}$; bottom panel) for the northern Gulf of Mexico region and identified subregions, over time. Values are depicted for both Galveston and Matagorda Bays (combined), the Mississippi River Delta and Mobile Bay (combined), and Tampa Bay.

Data derived from NASA Ocean Color Web (https://oceancolor.gsfc.nasa.gov/) and productivity calculated using the Vertically Generalized Production Model (VGPM).

Recreational fisheries landings are highest for both weight and number in the Tampa Bay and MRD subregions. When standardized per square kilometer (Fig. 7.29), nGOM landings follow very similar patterns as observed for total landings values, with the nGOM emerging nationally as having the third-highest concentration of fisheries landings over time. Tampa Bay also emerges as an important location with highest concentration of fishing effort surrounding this smallest subregion of interest. These values have historically been near or above the 1-ton km^{-2} threshold for a given system (Bundy et al. 2012, Tam et al. 2017).

Although values have decreased over time, bycatch continues to persist throughout the nGOM (Fig. 7.30). By weight, bycatch is most pronounced for bony fishes, invertebrates and sharks, while shark and bony fish bycatch is most dominant by number. Bycatch for invertebrates and bony fishes is nationally highest in the nGOM, with lower values observed for marine mammal bycatch. Additionally, seabird bycatch has been minimally observed or reported for the nGOM, while sea turtles have been most vulnerable in the Gulf of Mexico as compared to anywhere in the nation. As a result

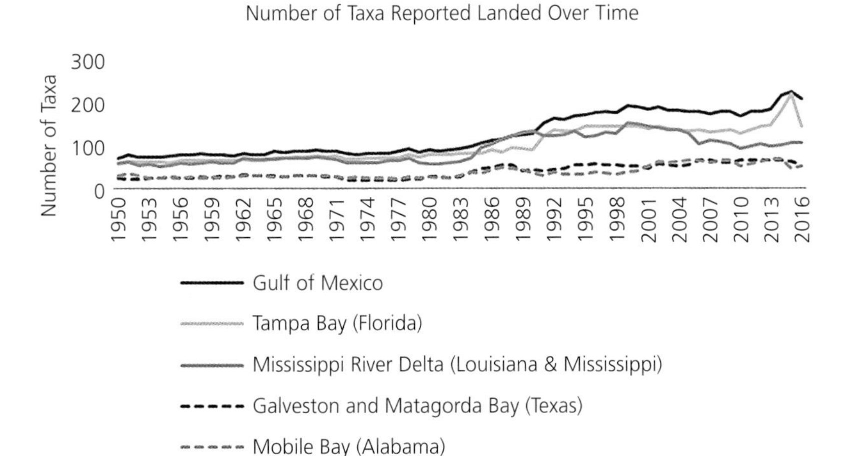

Figure 7.25 Number of taxa reported captured by commercial and recreational fisheries in the northern Gulf of Mexico region and identified subregions. Values are combined for both Galveston and Matagorda Bays.

Data derived from NOAA National Marine Fisheries Service commercial and recreational fisheries statistics.

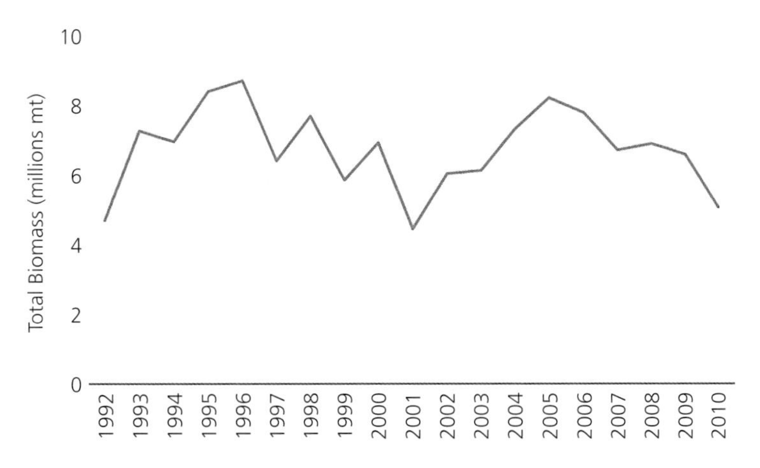

Figure 7.26 Total surveyed fish and invertebrate biomass (metric tons; mt) for the northern Gulf of Mexico region, over time.

of regulations that now require TEDs to be installed on trawling nets, substantial decreases in sea turtle bycatch have occurred over time (Finkbeiner et al. 2011).

7.5 Synthesis

The nGOM provides critical social and economic benefits to the region and the nation. However, exploitation of these resources and an increasing human population makes the nGOM an area subject to significant natural and human stressors, including the highest number of hurricanes in the U.S. Atlantic region (including increased frequency of higher category storms in recent years), large expanses of hypoxic bottom water, overfishing, and major oil spills like the 2010 DWH event. Together, these factors have affected the status of LMRs and their dependent economies, with recent hurricanes and DWH affecting nGOM vessels, shrimping effort, and aspects of fisheries value (Buck 2005, Colgan & Adkins 2006, Williams 2010, Lucas & Carter 2013, Solís et al. 2013, Du et al. 2019). DWH impacts have been most visible in the MRD. The impacts from hurricanes have been strongly pronounced in the Galveston Bay, Matagorda Bay, and MRD subregions, with lingering and cumulative

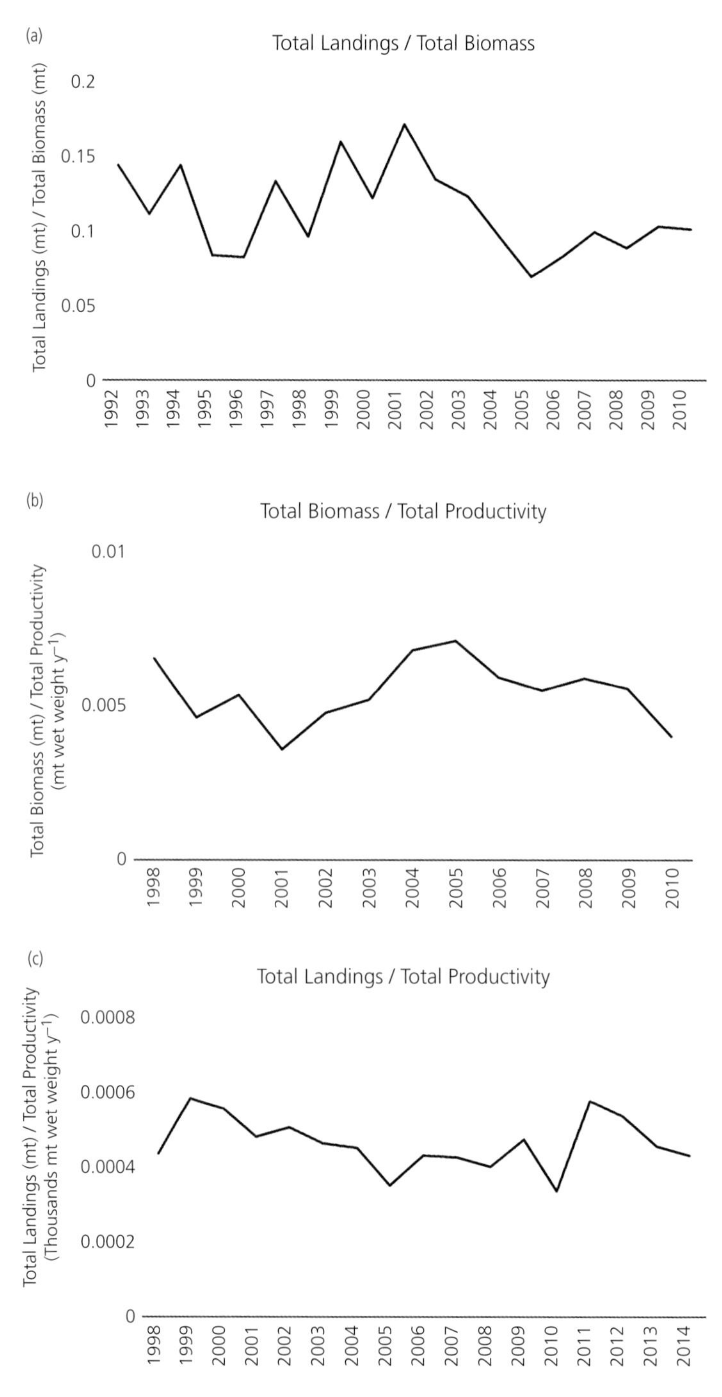

Figure 7.27 Ratios of (a) total commercial and recreational landings (metric tons) to total biomass (metric tons; i.e., exploitation index); (b) total biomass (metric tons) to total productivity (metric tons wet weight year^{-1}); (c) total commercial and recreational landings (metric tons) to total productivity (metric tons wet weight yr^{-1}) for the northern Gulf of Mexico region.

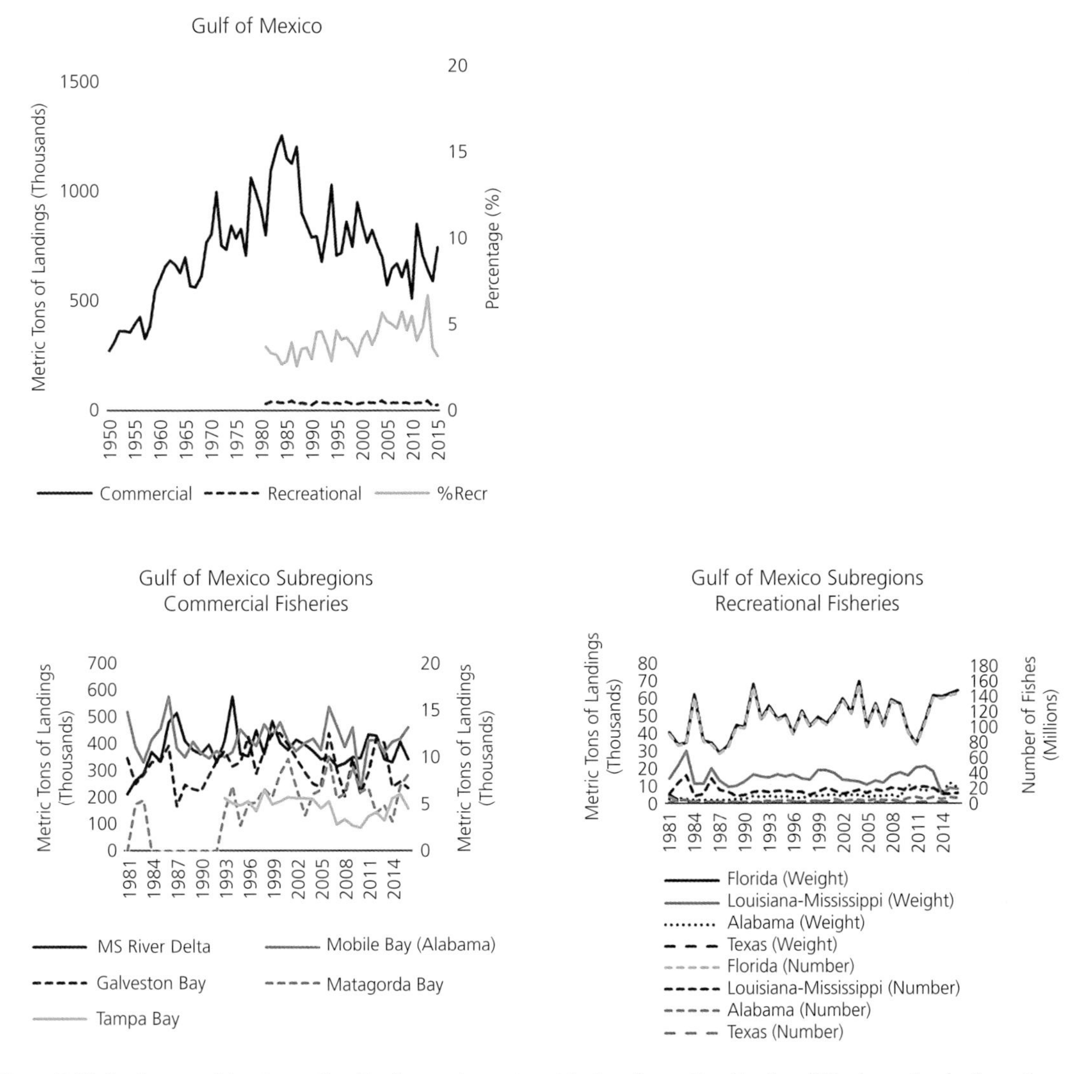

Figure 7.28 Total commercial and recreational landings and percent contribution of recreational landings (%Recr) over time for the northern Gulf of Mexico region (1950–2016) and identified subregions (1981–2016). Landings for Mississippi (MS) River Delta commercial fisheries plotted on the primary (left) axis. Commercial fisheries landings for all other subregions plotted on the secondary (right) axis.

Data derived from NOAA National Marine Fisheries Service commercial and recreational fisheries statistics.

effects still observable, which have notably accelerated marsh loss, coastal erosion, and impacts to human health (Norris et al. 2010, Lichtveld et al. 2016, Lin et al. 2016, Rangoonwala et al. 2016, Khanna et al. 2017, Dietz et al. 2018, Deis et al. 2019, Du & Park 2019). These effects are compounded by climatological oscillations and increasing temperatures over time, which have begun to foster tropicalization of the region, potentially affecting LMR production (Precht & Aronson 2004, Zimmer et al. 2006, Fodrie et al. 2010, Heck et al. 2015, Karanuskas et al. 2015, Marshak & Heck 2017, Scheffel et al. 2018, Macy et al. 2019). Concerns about ongoing vulnerability to sea-level rise also continue to affect this region, especially MRD wetlands and components of Mobile Bay (Burkett 2008, Passeri et al. 2016, Jankowski et al. 2017). Higher human populations found in Galveston Bay and the MRD are associated with significant fisheries landings, ports, marinas, oil and gas extraction, and exploitation pressure in these regions. Tourism pressures throughout these

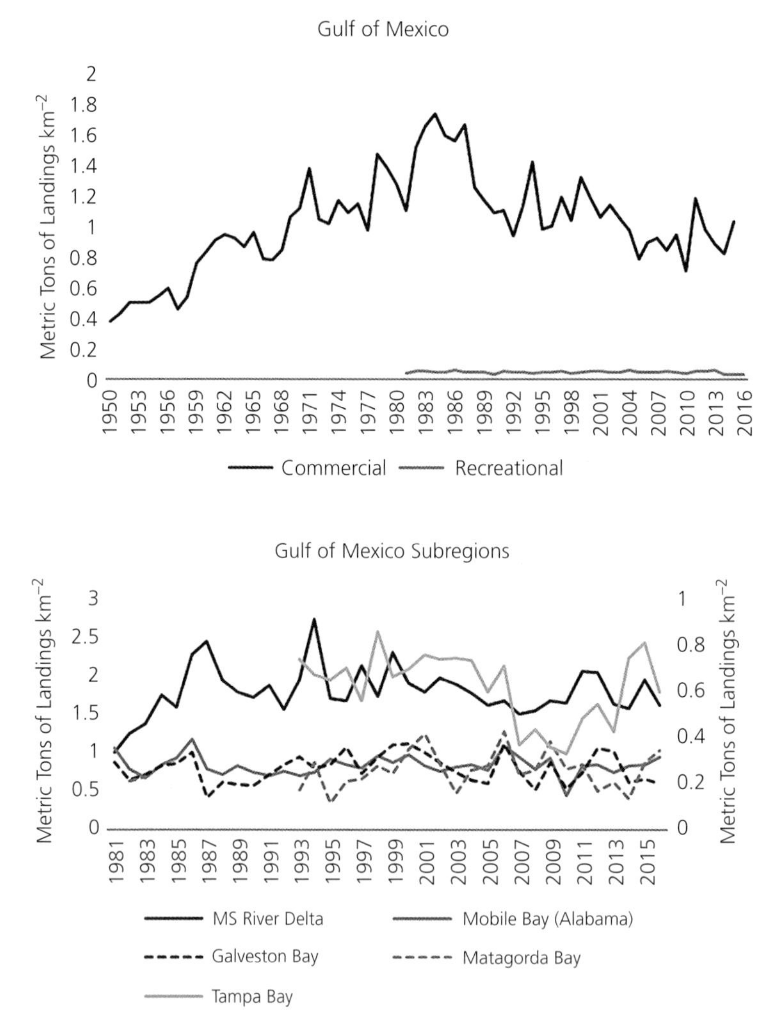

Figure 7.29 Total commercial and recreational landings (metric tons) per square kilometer over time (1950–2016) for the northern Gulf of Mexico region and commercial landings per square kilometer (1981–2016) for identified subregions. Values for Mississippi (MS) River Delta commercial fisheries are plotted on the primary (left) axis. Values for all other subregions are plotted on the secondary (right) axis.

Data derived from NOAA National Marine Fisheries Service commercial and recreational fisheries statistics.

subregions, including the Tampa Bay area, are especially prominent, with recreational and sport fishing also contributing to regional and subregional marine economies. Their landings are particularly highest and significantly concentrated in the Tampa Bay, MRD, and Mobile Bay subregions. Considering the effects of these cumulative natural and human stressors on coupled, regional socioecological systems, particularly their effects upon basal ecosystem productivities and the biomass and diversity of LMRs, is an essential component for EBFM strategies for the nGOM and its subregions.

The nGOM ecosystem is excelling in terms of its marine socioeconomic status related to its inherent basal productivity and importance of LMRs, oil and gas production, and tourism. Recent management priorities of curbing overfishing and ensuring sustainability of LMRs has led to significant decreases in the number of species experiencing overfishing (Karnauskas et al. 2013, 2017) and has led to removal of overfished status for several key fishery species over time (Karnauskas et al. 2013, 2017a). Although no current FEP exists for the region, advances toward ecosystem-level planning and enhanced knowledge of

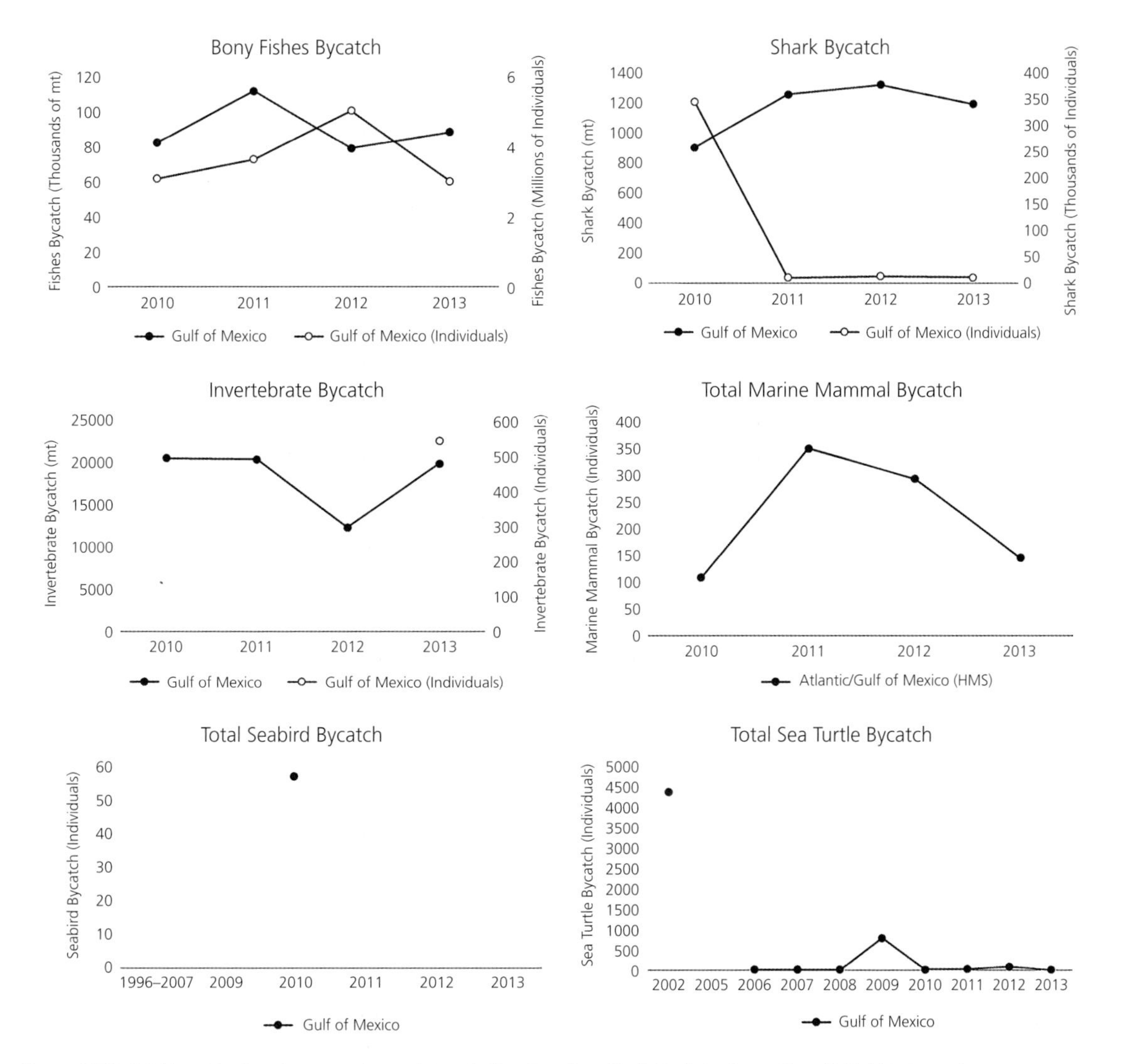

Figure 7.30 Total bycatch of species groups in metric tons (mt) and number of individuals for the northern Gulf of Mexico region.

Data derived from NOAA National Marine Fisheries Service U.S. National Bycatch Report Database System. For marine mammal bycatch, data from Gulf of Mexico and Atlantic highly migratory species (HMS) fisheries are presented.

ecosystem functioning have occurred over the past decade (Karnauskas et al. 2013, 2017, GMFMC 2018c, Link & Marshak 2019, NMFS 2019, Dell'Apa et al. 2020), with ongoing integration of this information toward EBFM implementation and more holistic management strategies underway. Partial progress regarding these efforts is currently reflected in specific FMPs and their amendments, which include multispecies foci (i.e., coastal pelagics, coral, reef fish, shrimp), considerations of trophic interactions (reef fish, Stone crab), environmental and catastrophic factors such as red tide (coastal pelagics, reef fish), gear conflicts and interactions with other fisheries and protected species (reef fish, shrimp, Spiny lobster, Stone crab), spatial management and habitat considerations (coral, reef fish, shrimp, Spiny lobster), climate influences (coastal pelagics), and socioeconomic components (reef fish; GMFMC 2018c).

Ongoing priorities toward implementation of EBFM in the nGOM, as specified in the NOAA Fisheries EBFM Road Map Regional Implementation Plan (NMFS 2019) include:

- Incorporating environmental and habitat considerations into stock assessments and advancing assessments for data-limited species.
- Advancing tracking of ecosystem trends, including improved basic monitoring information regarding fisheries and protected species, open ocean dynamics, and incorporating emerging and acoustic technologies toward these purposes.
- Examining ecosystem responses to climate change.
- Using ESRs, species profiles, local ecological knowledge, risk assessments, and biophysical modeling for enhanced decision-making.
- Improved recognition of multispecies interactions, appropriate ecosystem-level spatial scales and connectivity, and of human dimensions into management.

These efforts build upon current progress toward characterizing and monitoring ecosystem dynamics and advancing knowledge of the nGOM ecosystem. They also move toward implementing ecosystem-level planning, conducting ecosystem-level risk assessments, and applying ecosystem properties into management frameworks. Initial efforts to develop a FEP are also underway (GMFMC 2018c, NMFS 2019). This region contains several federal FMPs that manage multispecies complexes. However, the majority of state and federal FMPs are focused on relatively few species per plan. Marine socioeconomic considerations are reflected in management plans and strategies, especially given the significant trade-offs among sectors observed in this system. Movement toward these more comprehensive approaches is also facilitated by the development of more robust modeling frameworks to allow for a broader understanding of the ecosystem effects of management strategies and targeted fishing policies. While still data-limited in many respects, current nGOM abiotic and biological survey data has been incorporated into assessing the region via two ESRs that have relied on a socioecological approach to its characterization (Karnauskas et al. 2013, 2017). Major highlights have included a greater understanding of AMO influence on this ecosystem, assessment of ecosystem pressures, habitats, and productivity of lower and higher trophic levels, and documentation of the influences of natural pressures on the biological and human components of this system.

In addition to these areas where the nGOM is excelling and advancing in terms of LMR management and ecosystem-centric efforts, there are remaining challenges toward implementing EBFM and integrating available information into tactical and strategic ecosystem approaches to management. The main one is that FEPs for this region and its subregions have not yet been developed. However, initiatives toward ecosystem-based frameworks include council-supported modeling workshops, the establishment of an Ecosystem Science and Statistical Committee (SSC), and finalization of a regional EBFM implementation plan assisting toward its development (GMFMC 2018b, NMFS 2019). This region has remained hindered in aspects of its EBFM progress as a result of ongoing data limitations, including incomplete monitoring of broader ecosystem indicators. Efforts toward creating ecosystem-level management actions that incorporate broader ecosystem considerations are needed, especially applying ecosystem-level emergent properties in this context. These remain particularly warranted given the historical overharvesting of this system in terms of landings and overfishing, and its ongoing system-wide overexploitation as related to biomass and production. System-wide consideration of bottom-up environmental factors (e.g., habitat, eutrophication) that influence ecosystem functioning, productivity, and species composition, besides top-down effects of human impacts and overharvesting, still need to be more widely and directly incorporated into management strategies. For most effective EBFM approaches, direct consideration of relationships between fisheries harvest, ocean use, and ecosystem properties would be factored into system-wide thresholds and reference points for management. Cumulative consideration of these components is needed to prevent continued overexploitation of the nGOM and allow for the long-term sustainability of its LMRs and marine economy.

With emerging trade-offs among marine sectors, increasing human population size, numerous stressors, and contentious interactions between concentrated commercial and recreational fisheries, there is an urgent need for comprehensive management strategies in the nGOM. Those explicitly considering the social and ecological consequences of management interventions, especially as related to overlapping ocean uses, multiple species, and conflicting factors, need to be incorporated in a more systematic manner. This is particularly evident not just for the nGOM, but also in the Matagorda and Mobile Bay subregions, where greater economic contributions from LMRs occur and trade-offs with other marine sectors would be reflected in more holistic approaches to management. Additional consideration of habitat and spatial management is needed for the more strategic use of protected and closed areas in this region for fishing and other marine resource activities. Other considerations needed for more effective LMR management in the nGOM include improved balance in terms of FMC attention toward multiple species beyond those that are iconic and controversial, and the consequences of fishing activities on their interactions. Consideration of the proportional effects of commercial and recreational fishing on various trophic levels, species complexes, and in relation to basal and fisheries productivity is also warranted. Enhanced monitoring of recreational fisheries is also necessary. Overall investments toward improved LMR surveying, fisheries data collection, accessibility, and communication of information with managers and councils will be instrumental in resolving many of these hurdles. As fisheries management remains highly contentious in the region, in part due to past responses to overfishing seen in differing state and federal approaches, the need to account for commercial and recreational interests, skepticism regarding management actions, and the prioritization of more comprehensive assessments to address stock status is severely needed.

In terms of human population and population density, this region is currently mid-ranked in the U.S., with significant tourism and residential communities contributing heavily to its marine economy. This region is leading the nation in the fields of oil and gas production, offshore mineral extraction, marine construction, ship and boatbuilding, and in its proportional contribution toward total U.S. marine GDP. These successes are also related to the significant primary productivity of this system (fourth-highest nationally), which fosters the second-highest national fisheries landings and third-highest national fisheries value for this region. However, this region is also unique in terms of having significant natural and human stressors, which have affected productivity across the Gulf and among its subregions, resulting in significant overexploitation of this system. Declines in benthic habitats and ongoing effects of DWH continue to raise concerns regarding their effects on basal ecosystem production. Among regions, the nGOM is ranked fifth-highest in the nation regarding the overall status of its LMRs (Link & Marshak 2019), with particularly concentrated landings, high system exploitation, under-assessed stocks, and past overfishing. Additionally, it is the highest in the nation regarding the number of prohibited species. Nationally, the nGOM is fifth-highest in the percentage of stocks for which overfished status is unknown (44.7%), while uncertainty regarding overfished status of key economic species and resistance by stakeholders to accept strict fishing restrictions continues. These concerns are also reflected in the nGOM containing the fourth-highest number of LMR-related lawsuits nationally, especially related to fishing restrictions. Hence, there are challenges facing the governance of LMRs in the nGOM, and EBFM provides a means to address these myriad challenges.

7.5.1 Progress Toward Ecosystem-Based Fisheries Management (EBFM)

This ecosystem is excelling in the areas of socioeconomic status, human environment, and is relatively productive, as related to the determinants of successful LMR management (Fig. 7.31; Link & Marshak 2019).

Through a systematic and prioritized consideration of fisheries, pressures, risks, and outcomes, holistic EBFM practices are being developed and implemented across U.S. regions (NMFS 2016a, b, Levin et al. 2018, Marshall et al. 2018, Link & Marshak 2019). Approaches for nGOM fisheries management are evolving into frameworks that can address cumulative issues affecting the LMRs and fisheries of this ecosystem. Over the past few years, efforts to implement EBFM in this region have arisen through refinement of federal and local management approaches. In accordance with the NOAA Fisheries EBFM Policy and Roadmap (NMFS 2016a, b), an EBFM implementation plan (NMFS 2019) for the nGOM has been developed through efforts overseen by the NOAA Southeast Fisheries Science Center (SEFSC), NOAA Fisheries Southeast Regional Office (SERO), GMFMC, and the GSMFC. This plan specifies approaches to be taken to conduct EBFM and to address NOAA Fisheries Roadmap priorities and gaps. Progress for the nGOM and its major subregions in terms of EBFM Roadmap Guiding Principles and Goals is shown in Table 12.9. Overall, progress has been made in terms of implementing ecosystem-level planning, advancing knowledge of ecosystem principles, and in assessing risks and vulnerabilities to ecosystems through ongoing investigations into climate vulnerability and species prioritizations for stock and habitat assessments. However, little overall progress has been observed toward applying ecosystem-level emergent properties into management frameworks.

Progress toward recommendations of the Ecosystem Principles Advisory Panel (EPAP) for the development of FEPs (EPAP 1999, Wilkinson & Abrams 2015) is shown in Table 12.10. While there is no current FEP for the nGOM, a GMFMC technical committee has been formed to assist with its development, and an outline for the document (including management, biological/

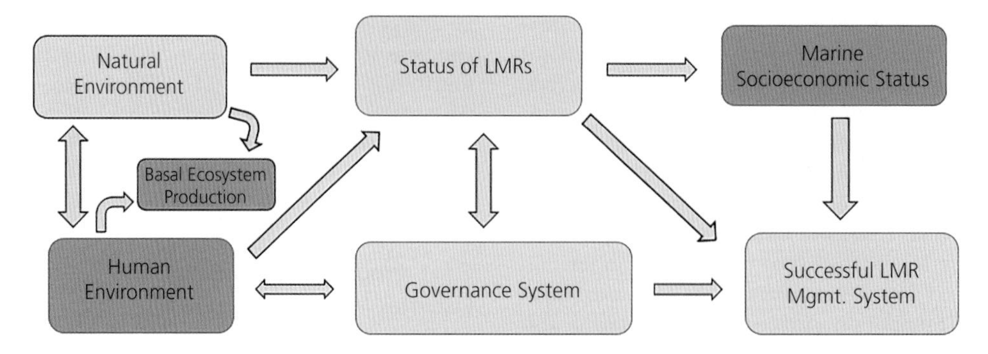

Figure 7.31 Schematic of the determinants and interconnectivity of successful living marine resource (LMR) systems management criteria as modified from Link and Marshak (2019). Those highlighted in blue indicate criteria for which the northern Gulf of Mexico ecosystem is excelling.

ecological, and socioeconomic sections) has been recently proposed and approved by this committee (GMFMC 2020). Under the EPAP recommendations, most progress for the nGOM has occurred through characterizing, monitoring, and assessing aspects of its ecosystem dynamics in support of a developing ecosystem approach to management.

7.5.2 Conclusions

With increasing stressors and human population, and a national marine economy heavily dependent on nGOM resources, applying an ecosystem-based approach to this system will result in major improvements and advancements toward its overall management. Namely, this includes finalizing an FEP, accounting for data limitations, and more comprehensively accounting for trade-offs among fisheries sectors, among targeted and protected species, and between fisheries and other ocean uses. Major feedbacks have been observed in terms of emerging natural and human stressors greatly affecting the productivity, continued harvest, and biological resilience of LMRs in this region. A marine economy that includes the nation's second-highest LMR-derived GDP and substantial national seafood production depends on sustainable harvests in consideration of ecosystem productivity. Consideration of the natural, biological, and socioeconomic factors that affect the degree to which marine resources are sustainably extracted and harvested, especially in a changing environment, is necessary for successful LMR management. Continuing with a business-as-usual approach to resource management and exploitation without considering the ecosystem as a whole, the influences of other interacting sectors, or governance and socioeconomic feedbacks limits sustainable practices and detrimentally ignores the significance of these factors toward LMR production and marine socioeconomic growth. A system-wide consideration of the relationships between productivity, biomass, landings, and socioeconomics is needed to maintain the system at needed thresholds, while also accounting for additional influences from natural and human activities. More comprehensive consideration of these factors is needed to understand their collective effects on the ecosystem and to alleviate their cumulative consequences. EBFM provides opportunities that focusing on one stock at a time would not. For instance, optimal multispecies management allows for simultaneous harvest of species complexes such as reef fishes, while accounting for trophic interactions through concurrent management of predatory species (i.e., groupers, sharks, coastal pelagic predators) and differential fishing effort by commercial and recreational fisheries on these complexes, predators, and other prey species. Given the high species richness found in this system, such an approach with portfolio considerations provides for a more effective and sustainable management strategy (Edwards et al. 2004, Kellner et al. 2011, Link 2018), particularly in the culturally rich nGOM.

7.6 References

Abbott JK. 2015. Fighting over a red herring: The role of economics in recreational-commercial allocation disputes. *Marine Resource Economics* 30(1):1–20.

Adams CM, Hernandez E, Cato JC. 2004. The economic significance of the Gulf of Mexico related to population, income, employment, minerals, fisheries and shipping. *Ocean & Coastal Management* 47(11–12):565–80.

Ainsworth CH, Schirripa MJ, Morzaria-Luna HN. 2015. *An Atlantis Ecosystem Model for the Gulf of Mexico Supporting Integrated Ecosystem Assessment. NOAA Technical Memorandum NMFS-SEFSC-676*. Washington, DC: NOAA (p. 149).

Ajemian MJ, Wetz JJ, Shipley-Lozano B, Shively JD, Stunz GW. 2015. An analysis of artificial reef fish community structure along the northwestern Gulf of Mexico shelf: potential impacts of "Rigs-to-Reefs" programs. *PLoS One* 10(5).

Alford SB, Rozas LP. 2019. Effects of Nonnative Eurasian watermilfoil, Myriophyllum spicatum, on nekton habitat quality in a Louisiana Oligohaline estuary. *Estuaries and Coasts* 42(3):613–28.

Alhale SA. 2017. *Who Should Manage Red Snapper (Lutjanus campechanus) in the Gulf of Mexico? A Study of the Social Dynamics of the Red Snapper Fishery*. St. Petersburg, FL: University of South Florida.

AOML (Atlantic Oceanographic and Meteorological Laboratory). 2012. *NOAA/AOML Regional Satellite Products.* http://www.aoml.noaa.gov/phod/regsatprod/index.php

Arlinghaus R, Cooke SJ, Sutton SG, Danylchuk AJ, Potts W, Freire KDMF, Alós J, et al. 2016. Recommendations for the future of recreational fisheries to prepare the social-ecological system to cope with change. *Fisheries Management and Ecology* 23:177–186.

Aronson RB, Precht WF, Murdoch TJ, Robbart ML. 2005. Long-term persistence of coral assemblages on the Flower Garden Banks, northwestern Gulf of Mexico: implications for science and management. *Gulf of Mexico Science* 23(1):84–94.

Austin D, Carriker B, McGuire T, Pratt J, Priest T, Pulsipher AG. 2004. *History of the Offshore Oil and Gas Industry in Southern Louisiana: Interim Report; Volume I: Papers on the Evolving Offshore Industry. Minerals Management Service, Gulf of Mexico OCS Region, New Orleans, LA. OCS Study MMS 2004-049*. Silver Spring, MD: NOAA (p. 98).

Barbier EB. 2013. Valuing ecosystem services for coastal wetland protection and restoration: Progress and challenges. *Resources* 2(3):213–30.

Barron MG. 2012. Ecological impacts of the Deepwater Horizon oil spill: implications for immunotoxicity. *Toxicologic Pathology* 40(2):315–20.

Barord GJ, Graham WM, Bahya KM. 2007. First report of the invasive medusa, Phyllorhiza punctata von Lendenfeld (1884), in Galveston Bay, Texas. *Gulf of Mexico Science* 25(2):8.

Batker D, de la Torre I, Costanza R, Day JW, Swedeen P, Boumans R, Bagstad K. 2014. "The threats to the value of ecosystem goods and services of the Mississippi delta." In: *Perspectives on the Restoration of the Mississippi Delta.* Edited by JW Day, GP Kemp, AM Freeman, DP Muth. Dordrecht, Netherlands: Springer (pp. 155–73).

Baumann RH, Turner RE. 1990. Direct impacts of outer continental shelf activities on wetland loss in the central Gulf of Mexico. *Environmental Geology and Water Sciences* 15(3):189–s98.

Beck MW, Brumbaugh RD, Airoldi L, Carranza A, Coen LD, Crawford C, Defeo O, et al. 2011. Oyster reefs at risk and recommendations for conservation, restoration, and management. *Bioscience* 61(2):107–16.

Bennett CT, Robertson A, Patterson WF. 2019. First record of the non-indigenous Indo-Pacific damselfish, Neopomacentrus cyanomos (Bleeker, 1856) in the northern Gulf of Mexico. *BioInvasions Records* 8(1):154–66.

Bettridge SOM, Baker SC, Barlow J, Clapham P, Ford MJ, Gouveia D, et al. 2015. *Status Review of the Humpback Whale (Megaptera novaeangliae) Under the Endangered Species Act. Report NOAA-TM-NMFS-SWFSC-540*. Washington, DC: NOAA.

Beyer J, Trannum HC, Bakke T, Hodson PV, Collier TK. 2016. Environmental effects of the Deepwater Horizon oil spill: a review. *Marine Pollution Bulletin* 110(1):28–51.

Bianchi TS. 2007. *Biogeochemistry of Estuaries*. Oxford, UK: Oxford University Press (p. 706).

Bianchi TS, DiMarco SF, Cowan Jr JH, Hetland RD, Chapman P, Day JW, Allison MA. 2010. The science of hypoxia in the Northern Gulf of Mexico: a review. *Science of the Total Environment* 408(7):1471–84.

Blaylock RA, Hoggard W. 1994. *Preliminary Estimates of Bottlenosed Dolphin Abundance in Southern US Atlantic and Gulf of Mexico Continental Shelf Waters*. NOAA Technical Memorandum NMFS-SEFSC-356. Washington, DC: NOAA.

BOEM (US Bureau of Ocean Energy Management). 2017. *BOEM Joins Federal Partners to Launch Intensive Marine Assessment Program for Protected Species in the Gulf of Mexico.* Washington, DC: Department of the Interior.

Boylen CW, Eichler LW, Madsen JD. 1999. Loss of native aquatic plant species in a community dominated by Eurasian watermilfoil. *Hydrobiologia* 415:207–11.

Bricker SB, Longstaff B, Dennison W, Jones A, Boicourt K, Wicks C, Woerner J. 2008. Effects of nutrient enrichment in the nation's estuaries: a decade of change. *Harmful Algae* 8(1):21–32.

Bright TJ, Kraemer GP, Minnery GA, Viada ST. 1984. Hermatypes of the Flower Garden Banks, northwestern Gulf of Mexico: a comparison to other western Atlantic reefs. *Bulletin of Marine Science* 34(3):461–76.

Browder JA. 1993. A pilot model of the Gulf of Mexico continental shelf. Trophic models of aquatic ecosystems. *ICLARM Conference Proceedings* 26(390):279–84.

Buck EH. 2005. *Hurricanes Katrina and Rita: Fishing and Aquaculture Industries—Damage and Recovery.* Washington, DC: Congressional Research Service, Library of Congress.

Buker GE. 2004. *Blockaders, Refugees, and Contrabands: Civil War on Florida's Gulf Coast, 1861–1865*. Tuscaloosa, AL: University of Alabama Press.

Bundy A, Bohaboy EC, Hjermann D O, Mueter FJ, Fu C, Link JS. 2012. Common patterns, common drivers: comparative analysis of aggregate surplus production across ecosystems. *Marine Ecology Progress Series* 459:203–18.

Burkett V. 2008. *The Northern Gulf of Mexico Coast: Human Development Patterns, Declining Ecosystems, and Escalating Vulnerability to Storms and Sea Level Rise*. Sterling, VA: Earthscan Publications.

Butler PA. 1954. Summary of our knowledge of the oyster in the Gulf of Mexico. *Fishery Bulletin* 55(89):479–89.

Byron D, Heck Jr KL, Kennedy MA. 2014. Presence of juvenile lionfish in a northern Gulf of Mexico nursery habitat. *Gulf of Mexico Science* 32(1-2):75–7.

Caffrey JM, Murrell MC, Amacker KS, Harper JW, Phipps S, Woodrey MS. 2014. Seasonal and inter-annual patterns in primary production, respiration, and net ecosystem metabolism in three estuaries in the northeast Gulf of Mexico. *Estuaries and Coasts* 37(1):222–41.

Caillouet Jr CW, Hart RA, Nance JM. 2008. Growth overfishing in the brown shrimp fishery of Texas, Louisiana, and adjoining Gulf of Mexico EEZ. *Fisheries Research* 92(2–3):289–302.

Carey MP, Levin PS, Townsend H, Minello TJ, Sutton GR, Francis TB, Harvey CJ, et al. 2013. Characterizing coastal foodwebs with qualitative links to bridge the gap between the theory and the practice of ecosystem-based management. *ICES Journal of Marine Science* 71(3):713–24.

Carlson JK, Osborne J, Schmidt TW. 2007. Monitoring the recovery of smalltooth sawfish, Pristis pectinata, using standardized relative indices of abundance. *Biological Conservation* 136(2):195–202.

Carretta JV, Forney KA, Oleson EM, Weller DW, Lang AR, Baker J, Muto MM, et al. 2019. *U.S. Pacific Marine Mammal Stock Assessments: 2018. NOAA Technical Memorandum NMFSSWFSC-617*. Washington, DC: NOAA.

Chagaris D, Binion-Rock S, Bogdanoff A, Dahl K, Granneman J, Harris H, Mohan J, et al. 2017. An ecosystem-based approach to evaluating impacts and management of invasive lionfish. *Fisheries* 42(8):421–31.

Chagaris D, Mahmoudi B. 2013. *Natural Mortality of Gag Grouper from 1950 to 2009 Generated by an Ecosim Model. SEDAR 33-DW07*. North Charleston, S.C: SEDAR.

Chagaris DD. 2013. *Ecosystem-Based Evaluation of Fishery Policies and Tradeoffs on the West Florida Shelf*. PhD Thesis. Gainesville, FL: University of Florida.

Chagaris DD, Patterson III WF, Allen MS. 2020. Relative effects of multiple stressors on reef food webs in the northern Gulf of Mexico revealed via ecosystem modeling. *Frontiers in Marine Science* 7:513.

Chen Y. 2017. "Fish resources of the Gulf of Mexico." In: *Habitats and Biota of the Gulf of Mexico: Before the Deepwater Horizon Oil Spill*. Edited by CH Ward. New York, NY: Springer (pp. 869–1038).

Chesney EJ, Baltz DM. 2001. The effects of hypoxia on the northern Gulf of Mexico coastal ecosystem: a fisheries perspective. *Coastal and Estuarine Studies* 9:321–54.

Chesney EJ, Baltz DM, Thomas RG. 2000. Louisiana estuarine and coastal fisheries and habitats: perspectives from a fish's eye view. *Ecological Applications* 10(2):350–66.

Colburn LL, Jepson M, Weng C, Seara T, Weiss J, Hare JA. 2016. Indicators of climate change and social vulnerability in fishing dependent communities along the Eastern and Gulf Coasts of the United States. *Marine Policy* 74:323–33.

Coleman FC, Figueira WF, Ueland JS, Crowder LB. 2004. The impact of United States recreational fisheries on marine fish populations. *Science* 305:1958–60.

Coleman FC, Koenig CC, Collins LA. 1996. Reproductive styles of shallow-water groupers (Pisces: Serranidae) in the eastern Gulf of Mexico and the consequences of fishing spawning aggregations. *Environmental Biology of Fishes* 47(2):129–41.

Colgan CS, Adkins J. 2006. Hurricane damage to the ocean economy in the US gulf region in 2005. *Monthly Labor Review* 8:76–8.

Conn, Paul B. 2011. *An Evaluation and Power Analysis of Fishery Independent Reef Fish Sampling in the Gulf of Mexico and U.S. South Atlantic. NOAA Technical Memorandum NMFS-SEFSC-610*. Washington, DC: NOAA (p. 38).

Conner WH, Day JW, Baumann RH, Randall JM. 1989. Influence of hurricanes on coastal ecosystems along the northern Gulf of Mexico. *Wetlands Ecology and Management* 1(1):45–56.

Cooke SJ, Cowx IG. 2006. Contrasting recreational and commercial fishing: searching for common issues to promote unified conservation of fisheries resources and aquatic environments. *Biological Conservation* 128(1):93–108.

Cortés E, Brooks L, Scott G. 2002. *Stock Assessment of Large Coastal Sharks in the U.S. Atlantic and Gulf of Mexico. National Marine Fisheries Service, Sustainable Fisheries Division contribution SFD-02/03-177*. Panama City, FL: Southeast Fisheries Science Center, (p. 222).

Cotham ET. 1998. *Battle on the Bay: The Civil War Struggle for Galveston*. Austin, TX: University of Texas Press.

Cowan Jr JH. 2011a. Red snapper in the Gulf of Mexico and US South Atlantic: data, doubt, and debate. *Fisheries* 36(7):319–31.

Cowan JH, Grimes CB, Patterson WF, Walters CJ, Jones AC, Lindberg WJ, Sheehy DJ, et al. 2011b. Red snapper management in the Gulf of Mexico: science-or faith-based? *Reviews in Fish Biology and Fisheries* 21(2):187–204.

Cowan Jr JH, Grimes CB, Shaw RF. 2008. Life history, history, hysteresis, and habitat changes in Louisiana's coastal ecosystem. *Bulletin of Marine Science* 83(1):197–215.

CPRA (Coastal Protection and Restoration Authority of Louisiana). 2012. *Louisiana's Comprehensive Master Plan for a Sustainable Coast*. Baton Rouge, LA: CPRA.

Crawford S, Whelan G, Infante DM, Blackhart K, Daniel WM, Fuller PL, Birdsong T, et al. 2016. *Through a Fish's Eye: The Status of Fish Habitats in the United States 2015*. Washington, DC: National Fish Habitat Partnership.

Dahl KA, Edwards MA, Patterson III WF. 2019. Density-dependent condition and growth of invasive lionfish in the northern Gulf of Mexico. *Marine Ecology Progress Series* 623:145–59.

Dahl KA, Patterson III WF. 2014. Habitat-specific density and diet of rapidly expanding invasive red lionfish, Pterois volitans, populations in the northern Gulf of Mexico. *PloS One* 9(8):e105852.

Dahl KA, Patterson WF, Robertson A, Ortmann AC. 2017. DNA barcoding significantly improves resolution of invasive lionfish diet in the Northern Gulf of Mexico. *Biological Invasions* 19(6):1917–33.

Dahl KA, Patterson III WF, Snyder RA. 2016. Experimental assessment of lionfish removals to mitigate reef fish community shifts on northern Gulf of Mexico artificial reefs. *Marine Ecology Progress Series* 558:207–21.

Das A, Justic D, Inoue M, Hoda A, Huang H, Park D. 2012. Impacts of Mississippi River diversions on salinity gradients in a deltaic Louisiana estuary: ecological and management implications. *Estuarine, Coastal and Shelf Science* 111:17–26.

De Mutsert K, Cowan JH, Essington TE, Hilborn R. 2008. Reanalyses of Gulf of Mexico fisheries data: landings can be misleading in assessments of fisheries and fisheries ecosystems. *Proceedings of the National Academy of Sciences* 105(7):2740–4.

De Mutsert K, Cowan Jr JH, Walters CJ. 2012. Using Ecopath with Ecosim to explore nekton community response to freshwater diversion into a Louisiana estuary. *Marine and Coastal Fisheries: Dynamics, Management, and Ecosystem Science* 4:104–16.

De Mutsert K, Lewis K, Buszowki J, Steenbeek J, Millroy S. 2016. *Delta Management Fish and Shellfish Ecosystem Model: Ecopath with Ecosim Plus Ecospace (EwE) Model Description.* Baton Rouge, LA: Coastal Protection and Restoration Authority of Louisiana.

De Mutsert K, Lewis K, Milroy S, Buszowski J, Steenbeek J. 2017. Using ecosystem modeling to evaluate trade-offs in coastal management: effects of large-scale river diversions on fish and fisheries. *Ecological Modelling* 360:14–26.

De Mutsert K, Steenbeek J, Lewis K, Buszowski J, Cowan Jr JH, Christensen V. 2016. Exploring effects of hypoxia on fish and fisheries in the northern Gulf of Mexico using a dynamic spatially explicit ecosystem model. *Ecological Modelling* 331:142–50.

de Rada S, Arnone R, Anderson S. 2009. *Bio-Physical Ocean Modeling in the Gulf of Mexico. Pages 1–7 in OCEANS 2009, MTS/IEEE Biloxi, Marine Technology for Our Future: Global and Local Challenges (Conference Proceedings).* Biloxi, MS: Institute of Electrical and Electronics Engineers.

DeAngelis DL, Gross LJ, Huston MA, Wolff WF, Fleming DM, Comiskey EJ, Sylvester SM. 1998. Landscape modeling for Everglades ecosystem restoration. *Ecosystems* 1(1):64–75.

Deegan LA, Day Jr JW, Gosselink JG, Yáñez-Arancibia A, Chavez GS, Sanchez-Gil P. 1986. "Relationships among physical characteristics, vegetation distribution and fisheries yield in Gulf of Mexico estuaries." In: *Estuarine Variability.* Edited by DA Wolfe. Cambridge, MA: Academic Press (pp. 83–100).

Deis DR, Mendelssohn IA, Fleeger JW, Bourgoin SM, Lin Q. 2019. Legacy effects of Hurricane Katrina influenced marsh shoreline erosion following the Deepwater Horizon oil spill. *Science of The Total Environment* 672:456–67.

Dell'Apa A, Kilborn JP, Harford WJ. 2020. Advancing ecosystem management strategies for the Gulf of Mexico's fisheries resources: implications for the development of a fishery ecosystem plan. *Bulletin of Marine Science* 96(3):1–24.

Dennis GD, Bright TJ. 1988. Reef fish assemblages on hard banks in the northwestern Gulf of Mexico. *Bulletin of Marine Science* 43(2):280–307.

DeVries DA, Grimes CB, Prager MH. 2002. Using otolith shape analysis to distinguish eastern Gulf of Mexico and Atlantic Ocean stocks of king mackerel. *Fisheries Research* 57(1):51–62.

Dietz M, Liu KB, Bianchette T. 2018. Hurricanes as a Major Driver of coastal erosion in the Mississippi River Delta: a multi-decadal analysis of shoreline retreat rates at Bay Champagne, Louisiana (USA). *Water* 10(10):1480.

Doerpinghaus J, Hentrich K, Troup M, Stavrinaky A, Anderson S. 2014. An assessment of sector separation on the Gulf of Mexico recreational red snapper fishery. *Marine Policy* 50:309–17.

Drexler M, Ainsworth CH. 2013. Generalized additive models used to predict species abundance in the Gulf of Mexico: an ecosystem modeling tool. *PloS One* 8(5):e64458.

Du J, Park K. 2019. Estuarine salinity recovery from an extreme precipitation event: Hurricane Harvey in Galveston Bay. *Science of the Total Environment* 670:1049–59.

Du J, Park K, Dellapenna TM, Clay JM. 2019. Dramatic hydrodynamic and sedimentary responses in Galveston Bay and adjacent inner shelf to Hurricane Harvey. *Science of the Total Environment* 653:554–64.

Dynamic Solutions. 2016. *Development of the CASM for Evaluation of Fish Community Impacts for the Mississippi River Delta Management Study: Model Setup, Calibration and Validation for Existing Conditions. Contract 2503-13-42.* Baton Rouge, LA: Coastal Protection and Restoration Authority of Louisiana.

EPA (Environmental Protection Agency). 2017. *Mississippi River/Gulf of Mexico Watershed Nutrient Task Force 2017 Report to Congress. Second Biennial Report.* Washington, DC: EPA (p. 121).

ESRL (Earth System Research Laboratory). 2013. *Climate Indices: Monthly Atmospheric and Ocean Time Series.* Washington, DC: NOAA.

EPAP (Ecosystem Principles Advisory Panel). 1999. *Ecosystem-Based Fishery Management Report to Congress.*

Silver Spring, MD: National Marine Fisheries Service (p. 62).

Edwards SF, Link JS, Rountree BP. 2004. Portfolio management of wild fish stocks. *Ecological Economics* 49(3):317–29.

Engle VD. 2011. Estimating the provision of ecosystem services by Gulf of Mexico coastal wetlands. *Wetlands* 31(1):179–93.

Evans MA, Scavia D. 2010. Forecasting hypoxia in the Chesapeake Bay and Gulf of Mexico: Model accuracy, precision, and sensitivity to ecosystem change. *Environmental Research Letters* 6(1):015001.

Farmer NA, Malinowski RP, McGovern MF, Rubec PJ. 2016. Stock complexes for fisheries management in the Gulf of Mexico. *Marine and Coastal Fisheries* 8(1):177–201.

Feely RA, Okazaki RR, Cai WJ, Bednaršek N, Alin SR, Byrne RH, Fassbender A. 2018. The combined effects of acidification and hypoxia on pH and aragonite saturation in the coastal waters of the California current ecosystem and the northern Gulf of Mexico. *Continental Shelf Research* 152:50–60.

Felder DL, Camp DK. 2009. *Gulf of Mexico Origin, Waters, and Biota: Biodiversity.* College Station, TX: Texas A&M University Press.

Fennel K, Hetland R, Feng Y, DiMarco S. 2011. A coupled physical-biological model of the northern Gulf of Mexico shelf: model description, validation and analysis of phytoplankton variability. *Biogeosciences* 8:1881–99.

Ferguson L, Srinivasan M, Oleson E, Hayes S, Brown SK, Angliss R, Carretta J. et al. 2017. *Proceedings of the First National Protected Species Assessment Workshop. NOAA Technical Memorandum NMFS-F/SPO-172.* Washington, DC: NOAA (p. 92).

Fertl D, Schiro AJ, Regan GT, Beck CA, Adimey N, Price-May L, Amos A, Worthy GA, Crossland R. 2005. Manatee occurrence in the northern Gulf of Mexico, west of Florida. *Gulf and Caribbean Research* 17(1):69–94.

Finkbeiner EM, Wallace BP, Moore JE, Lewison RL, Crowder LB, Read AJ. 2011. Cumulative estimates of sea turtle bycatch and mortality in USA fisheries between 1990 and 2007. *Biological Conservation* 144(11):2719–27.

Fleming CS, Tonioli FC, Agar JJ. 2014. *A Review of Principal Coastal Economic Sectors Within the Southeast United States and the US Caribbean. NOAA Technical Memorandum NMFS-SEFSC-669.* Washington, DC: NOAA (p. 44).

Flick GJ, Martin RE. 1990. *The Seafood Industry.* Berlin, Germany: Springer Science & Business Media.

Fodrie FJ, Heck Jr KL. 2011. Response of coastal fishes to the Gulf of Mexico oil disaster. *PLoS One* 6(7):e21609.

Fodrie FJ, Heck Jr KL, Powers SP, Graham WM, Robinson KL. 2010. Climate-related, decadal-scale assemblage changes of seagrass-associated fishes in the northern Gulf of Mexico. *Global Change Biology* 16(1):48–59.

Fogg AQ, Hoffmayer ER, Driggers III WB, Campbell MD, Pellegrin GJ, Stein W. 2013. Distribution and length frequency of invasive lionfish (Pterois sp.) in the northern Gulf of Mexico. *Gulf and Caribbean Research* 25(1):111–5.

Fuller PL, Knott DM, Kingsley-Smith PR, Morris JA, Buckel CA, Hunter ME, Hartman LD. 2014, Invasion of Asian tiger shrimp, Penaeus monodon Fabricius, 1798, in the western north Atlantic and Gulf of Mexico. *Aquatic Invasions* 9(1).

FWS (Fish and Wildlife Service). 2014. *West Indian Manatee (Trichechus manatus) Stock Assessment Report.* Boquerón, Puerto Rico: U.S. Fish and Wildlife Service, Caribbean Ecological Services Field Office (p. 17).

Gallaway BJ, Cole JG, Meyer R, Roscigno P. 1999. Delineation of essential habitat for juvenile red snapper in the northwestern Gulf of Mexico. *Transactions of the American Fisheries Society* 128(4):713–26.

Gallaway BJ, Gazey WJ, Cole JG. An updated description of the benefits and consequences of Red Snapper shrimp trawl bycatch management actions in the Gulf of Mexico. *North American Journal of Fisheries Management* 37(2):414–9.

GCOOS (Gulf of Mexico Coastal Ocean Observing System). 2016. *Strategic Plan 2017–2022.* College Station, TX: Department of Oceanography (p. 45).

Gillig D, Ozuna Jr T, Griffin WL. 2000. The value of the Gulf of Mexico recreational red snapper fishery. *Marine Resource Economics* 15(2):127–39.

Gittman RK, Baillie CJ, Arkema KK, Bennett RO, Benoit J, Blitch S, Brun J, et al. 2019. Voluntary restoration: mitigation's silent partner in the quest to reverse coastal wetland loss in the USA. *Frontiers in Marine Science* 6:511.

Gittings SR, Boland GS, Deslarzes KJ, Combs CL, Holland BS, Bright TJ. 1992. Mass spawning and reproductive viability of reef corals at the East Flower Garden Bank, northwest Gulf of Mexico. *Bulletin of Marine Science* 51(3):420–8.

GMFMC (Gulf of Mexico Fishery Management Council). 1981a. *Environmental Impact Statement and Fishery Management Plan for the Reef Fish Resources of the Gulf of Mexico.* Tampa, FL: Gulf of Mexico Fishery Management Council.

GMFMC (Gulf of Mexico Fishery Management Council). 1981b. *Fishery Management Plan for the Shrimp Fishery of the Gulf of Mexico, United States Waters.* Tampa, FL: Gulf of Mexico Fishery Management Council.

GMFMC (Gulf of Mexico Fishery Management Council). 2018a. *Gulf Fishery News. Spring 2018. Volume 40, Number 2.* Tampa, FL: Gulf of Mexico Fishery Management Council.

GMFMC (Gulf of Mexico Fishery Management Council). 2018b. *Gulf of Mexico Fishery Management Council Ecosystem Committee Minutes.* Tampa, FL: Gulf of Mexico Fishery Management Council (p. 18).

GMFMC (Gulf of Mexico Fishery Management Council). 2018c. *Ecosystem Approaches to Fishery Management in the Gulf of Mexico. White Paper. Tab E, No. 8(a).* Tampa, FL: Gulf of Mexico Fishery Management Council (p. 8).

GMFMC (Gulf of Mexico Fishery Management Council). 2019. *Gulf Fishery News. Winter 2019. Volume 41, Number 1.* Tampa, FL: Gulf of Mexico Fishery Management Council.

GMFMC (Gulf of Mexico Fishery Management Council). 2020. *Ecosystem Technical Committee Meeting Summary. March 2, 2020.* Tampa, FL: Gulf of Mexico Fishery Management Council (p. 5).

GMFMC (Gulf of Mexico Fishery Management Council), SAFMC (South Atlantic Fishery Management Council). 1983. *Fishery Management Plan Final Environmental Impact Statement Regulatory Impact Review Final Regulations for the Coastal Migratory Pelagic Resources (Mackerels).* Tampa, FL: Gulf of Mexico Fishery Management Council.

Gomez FA, Lee SK, Hernandez FJ, Chiaverano LM, Muller-Karger FE, Liu Y, Lamkin JT. 2019a. ENSO-induced co-variability of Salinity, Plankton Biomass and Coastal Currents in the Northern Gulf of Mexico. *Scientific Reports* 9(1):1–10.

Gomez FA, Wanninkhof R, Barbero L, Lee SK, Hernandez Jr FJ. 2019b. Seasonal patterns of surface inorganic carbon system variables in the Gulf of Mexico inferred from a regional high-resolution ocean-biogeochemical model. *Biogeosciences Discussions* 17(6):1685–700.

Goodyear CP. 1995. *Red Snapper in US waters of the Gulf of Mexico. Contribution: MIA-95/96-05.* Miami, FL: National Marine Fisheries Service (p. 171).

Gracia A. 1996. White shrimp (Penaeus setiferus) recruitment overfishing. *Marine and Freshwater Research* 47(1):59–65.

Graham WM, Martin DL, Felder DL, Asper VL, Perry HM. 2003. "Ecological and economic implications of a tropical jellyfish invader in the Gulf of Mexico." In: *Marine Bioinvasions: Patterns, Processes and Perspectives.* Edited by J Pederson. Dordrecht, Netherlands: Springer, (pp. 53–69).

Gramling R, Hagelman R. 2005. A working coast: people in the Louisiana wetlands. *Journal of Coastal Research* SI(44):112–33.

Gray A, Ainsworth C, Chagaris D, Mahmoudi B. 2013. *Red Tide Mortality On Gag Grouper 1980–2009. SEDAR 33-AW21.* North Charleston, SC: SEDAR.

Gruesz KS. 2006. The Gulf of Mexico System and the "Latinness" of New Orleans. *American Literary History* 18(3):468–95.

Grüss A, Schirripa MJ, Chagaris D, Velez L, Shin YJ, Verley P, Oliveros-Ramos R, Ainsworth CH. 2016. Estimating natural mortality rates and simulating fishing scenarios for Gulf of Mexico red grouper (Epinephelus morio) using the ecosystem model OSMOSE-WFS. *Journal of Marine Systems* 154:264–79.

Grüss A, Rose KA, Simons J, Ainsworth CH, Babcock EA, Chagaris DD, De Mutsert K, et al. 2017. Recommendations on the use of ecosystem modeling for informing ecosystem-based fisheries management and restoration outcomes in the Gulf of Mexico. *Marine and Coastal Fisheries* 9(1):281–95.

Grüss A, Schirripa MJ, Chagaris D, Drexler MD, Simons J, Verley P, Shin, YJ, et al. 2013a. *Natural Mortality Rates and Diet Patterns of Gag Grouper (Mycteroperca Microlepis) in the West Florida Shelf Ecosystem in the 2000s: Insights from the Individual-Based, Multi-Species Model OSMOSE-WFS. SEDAR33-AW24.* North Charleston, SC: SEDAR.

Grüss A, Schirripa MJ, Chagaris D, Drexler MD, Simons J, Verley P, Shin YJ, et al. 2013b. *Evaluation of Natural Mortality Rates and Diet Composition for Gag (Mycteroperca Microlepis) in the West Florida Shelf Ecosystem Using the Individual-Based, Multi-Species Model OSMOSE. SEDAR33-DW11.* North Charleston, S.C: SEDAR.

Guo Z, Robinson D, Hite D. 2017. Economic impact of Mississippi and Alabama Gulf Coast tourism on the regional economy. *Ocean & Coastal Management* 145:52–61.

Harris HE, Patterson III WF, Ahrens RN, Allen MS. 2019. Detection and removal efficiency of invasive lionfish in the northern Gulf of Mexico. *Fisheries Research* 213:22–32.

Hart RA. 2012. *Stock Assessment of White Shrimp (Litopenaeus Setiferus) in the U.S. Gulf of Mexico for 2011. NOAA Technical Memorandum NMFS-SEFSC-637.* Washington, DC: NOAA (p. 36).

Hart RA. 2016. *Stock Assessment Update for Pink Shrimp (Farfantepenaeus Duorarum) in the U.S. Gulf of Mexico for 2015.* Washington, DC: NOAA (p. 16).

Hayes SA, Josephson E, Maze-Foley K, Rosel PE. 2019. *US Atlantic and Gulf of Mexico Marine Mammal Stock Assessments-2018. NOAA Tech Memo NMFS-NE-258.* Washington, DC: NOAA (p. 298).

Hearn CG. 2010. *Mobile Bay and the Mobile Campaign: The Last Great Battles of the Civil War.* Jefferson, NC: McFarland.

Heck Jr KL, Coen LD, Morgan SG. 2001. Pre-and post-settlement factors as determinants of juvenile blue crab Callinectes sapidus abundance: results from the north-central Gulf of Mexico. *Marine Ecology Progress Series* 222:163–76.

Heck Jr KL, Fodrie FJ, Madsen S, Baillie CJ, Byron DA. 2015. Seagrass consumption by native and a tropically associated fish species: potential impacts of the tropicalization of the northern Gulf of Mexico. *Marine Ecology Progress Series* 520:165–73.

Heileman S, Rabalais N. 2009. *The UNEP Large Marine Ecosystem Report. A Perspective on the Changing Condition in LMEs of the World's Regional Seas. UNEP Regional Seas Report and Studies No. 182.* Nairobi, Kenya: United Nations Environment Programme.

Hickerson EL, Schmahl GP, Robbart M, Precht WF, Caldow C. 2008. "State of coral reef ecosystems of the Flower Garden Banks, Stetson Bank, and Other Banks in the Northwestern Gulf of Mexico." In: *The State of Coral Reef Ecosystems of the United States and Pacific Freely Associated States: 2008. NOAA Technical Memorandum NOS NCCOS 73. NOAA/NCCOS Center for Coastal Monitoring and Assessment's Biogeography Team.* Edited by JE Waddell, AM Clarke. Silver Spring, MD: NOAA (pp. 189–217).

Hildebrand HH. 1954. A study of the fauna of the brown shrimp (Penaeus aztecus Ives) grounds in the western Gulf of Mexico. *Publications of the Institute of Marine Science* 3(2):225–366.

Hill JM, Caretti ON, Heck Jr KL. 2017. Recently established Asian tiger shrimp Penaeus monodon Fabricius, 1798 consume juvenile blue crabs Callinectes sapidus Rathbun, 1896 and polychaetes in a laboratory diet-choice experiment. *BioInvasions Record* 6(3).

Houde ED. 1971. *Survey of the Literature Relating to Sport and Commercial Fishes of South Florida*. University of Miami, FL: Rosenstiel School of Marine and Atmospheric Sciences.

Houde ED, Leak JC, Dowd CE, Berkeley SE, Richards WJ. 1979. *Ichthyoplankton Abundance and Diversity in the Eastern Gulf of Mexico*. Miami, FL: University of Miami (p. 546).

Hu C, Muller-Karger FE, Swarzenski PW. 2006. Hurricanes, submarine groundwater discharge, and Florida's red tides. *Geophysical Research Letters* 33:L11601.

Incardona JP, Gardner LD, Linbo TL, Brown TL, Esbaugh AJ, Mager EM, Stieglitz JD, et al. 2014. Deepwater horizon crude oil impacts the developing hearts of large predatory pelagic fish. *Proceedings of the National Academy of Sciences* 111(15):E1510–8.

Jacob S, Weeks P, Blount B, Jepson M. 2013. Development and evaluation of social indicators of vulnerability and resiliency for fishing communities in the Gulf of Mexico. *Marine Policy* 37:86–95.

Jepson M. 2007. Social indicators and measurements of vulnerability for Gulf Coast fishing communities. *Napa Bulletin* 28(1):57–68.

Jankowski KL, Törnqvist TE, Fernandes AM. 2017. Vulnerability of Louisiana's coastal wetlands to present-day rates of relative sea-level rise. *Nature Communications* 8:14792.

Jaxion-Harm J, Szedlmayer ST. 2015. Depth and artificial reef type effects on size and distribution of red snapper in the Northern Gulf of Mexico. *North American Journal of Fisheries Management* 35(1):86–96.

Johnson AG, Fable Jr WA, Grimes CB, Trent L, Vasconcelos Perez J. 1994. Evidence for distinct stocks of king mackerel, Scomberomorus cavalla, in the Gulf of Mexico. *Fishery Bulletin* 92(1):91–101.

Johnson DR, Perry HM, Graham WM. 2005. Using nowcast model currents to explore transport of non-indigenous jellyfish into the Gulf of Mexico. *Marine Ecology Progress Series* 305:139–46.

Johnson DR, Perry HM, Lyczkowski-Shultz J. 2013. Connections between Campeche Bank and red snapper populations in the Gulf of Mexico via modeled larval transport. *Transactions of the American Fisheries Society* 142(1):50–8.

Johnson DR, Perry HM, Lyczkowski-Shultz J, Hanisko D. 2009. Red snapper larval transport in the northern Gulf of Mexico. *Transactions of the American Fisheries Society* 138(3):458–70.

Johnston RJ, Holland DS, Maharaj V, Campson TW. 2007. Fish harvest tags: An alternative management approach for recreational fisheries in the US Gulf of Mexico. *Marine Policy* 31(4):505–16.

Justić D, Wang L. 2014. Assessing temporal and spatial variability of hypoxia over the inner Louisiana–upper Texas shelf: application of an unstructured-grid three-dimensional coupled hydrodynamic-water quality model. *Continental Shelf Research* 72:163–79.

Justić D, Rabalais NN, Turner RE. 2005. Coupling between climate variability and coastal eutrophication: evidence and outlook for the northern Gulf of Mexico. *Journal of Sea Research* 54(1):25–35.

Karnauskas M, Kelble CR, Regan S, Quenée C, Allee R, Jepson M, Freitag A, et al. 2017a *Ecosystem Status Report Update for the Gulf of Mexico. NOAA Technical Memorandum NMFS-SEFSC 706*. Washington, DC: NOAA (p. 51).

Karnauskas M, Schirripa MJ, Craig JK, Cook GS, Kelble CR, Agar JJ, Black BA, et al. 2015. Evidence of climate-driven ecosystem reorganization in the Gulf of Mexico. *Global Change Biology* 21(7):2554–68.

Karnauskas M, Schirripa MJ, Kelble CR, Cook GS, Craig JK. 2013. *Ecosystem Status Report for the Gulf of Mexico. NOAA Technical Memorandum NMFS-SEFSC 653*. Washington, DC: NOAA (p. 52).

Karnauskas M, Walter III JF, Campbell MD, Pollack AG, Drymon JM, Powers S. 2017b. Red snapper distribution on natural habitats and artificial structures in the northern Gulf of Mexico. *Marine and Coastal Fisheries* 9(1):50–67.

Kauffman TC, Martin CW, Valentine JF. 2018. Hydrological alteration exacerbates the negative impacts of invasive Eurasian milfoil Myriophyllum spicatum by creating hypoxic conditions in a northern Gulf of Mexico estuary. *Marine Ecology Progress Series* 592:97–108.

Kearney KA, Butler M, Glazer R, Kelble CR, Serafy JE, Stabenau E. 2015. Quantifying Florida bay habitat suitability for fishes and invertebrates under climate change scenarios. *Environmental Management* 55:836–56.

Kelble CR, Loomis DK, Lovelace S, Nuttle WK, Ortner PB, Fletcher P, Cook GS, Lorenz JJ, Boyer JN. 2013. The EBM-DPSER conceptual model: integrating ecosystem services into the DPSIR framework. *PLoS One* 8(8):e70766.

Kellner JB, Sanchirico JN, Hastings A, Mumby PJ. 2011. Optimizing for multiple species and multiple values: tradeoffs inherent in ecosystem-based fisheries management. *Conservation Letters* 4(1):21–30.

Khanna S, Santos M, Koltunov A, Shapiro K, Lay M, Ustin S. 2017. Marsh loss due to cumulative impacts of Hurricane Isaac and the Deepwater Horizon oil spill in Louisiana. *Remote Sensing* 9(2):169.

Kilborn JP, Drexler M, Jones DL. 2018. Fluctuating fishing intensities and climate dynamics reorganize the Gulf of Mexico's fisheries resources. *Ecosphere* 9(11):e02487.

Kleisner KM, Fogarty MJ, McGee S, Barnett A, Fratantoni P, Greene J, et al. 2016. The effects of sub-regional climate velocity on the distribution and spatial extent of marine species assemblages. *PLoS One* 11(2):e0149220.

Kukla J. 2004. *A Wilderness So Immense: The Louisiana Purchase and the Destiny of America*. New York, NY: Anchor.

Kumpf H, Steidinger K, Sherman K. 1999. The Gulf of Mexico large marine ecosystem: assessment, sustainability, and management. *Marine Ecology Progress Series* 190:271–87.

Lane RR, Day JW, Justic D, Reyes E, Marx B, Day JN, Hyfield E. 2004. Changes in stoichiometric Si, N and P ratios of Mississippi River water diverted through coastal wetlands to the Gulf of Mexico. *Estuarine, Coastal and Shelf Science* 60(1):1–10.

Large SI, Fay G, Friedland KD, Link JS. 2013. Defining trends and thresholds in responses of ecological indicators to fishing and environmental pressures. *ICES Journal of Marine Science* 70:755–67.

Large SI, Fay G, Friedland KD, Link JS. 2015. Critical points in ecosystem responses to fishing and environmental pressures. *Marine Ecology Progress Series* 521:1–17.

Laurent A, Fennel K, Ko DS, Lehrter J. 2018. Climate change projected to exacerbate impacts of coastal eutrophication in the northern Gulf of Mexico. *Journal of Geophysical Research: Oceans* 123(5):3408–26.

Levin PS, Essington TE, Marshall KN, Koehn LE, Anderson LG, Bundy A, Carothers C, et al. 2018. Building effective fishery ecosystem plans. *Marine Policy* 92:48–57.

Levin PS, Fogarty MJ, Murawski SA, Fluharty D. 2009. Integrated ecosystem assessments: developing the scientific basis for ecosystem-based management of the ocean. *PLoS Biology* 7(1):e1000014.

Lichtveld M, Sherchan S, Gam KB, Kwok RK, Mundorf C, Shankar A, Soares L. 2016. The Deepwater Horizon oil spill through the lens of human health and the ecosystem. *Current Environmental Health Reports* 3(4):370–8.

Lin Q, Mendelssohn IA, Graham SA, Hou A, Fleeger JW, Deis DR. 2016. Response of salt marshes to oiling from the Deepwater Horizon spill: implications for plant growth, soil surface-erosion, and shoreline stability. *Science of the Total Environment* 557:369–77.

Link J. 2010. *Ecosystem-Based Fisheries Management: Confronting Tradeoffs*. Cambridge, UK: Cambridge University Press.

Link JS. 2018. System-level optimal yield: increased value, less risk, improved stability, and better fisheries. *Canadian Journal of Fisheries and Aquatic Sciences* 75(1):1–6.

Link JS, Browman HI. 2014. Integrating what? Levels of marine ecosystem-based assessment and management. *ICES Journal of Marine Science* 71(5):1170–3.

Link JS, Ihde TF, Townsend HM, Osgood KE, Schirripa MJ, Kobayashi DR, Gaichas SK, et al. 2010. *Report of the 2nd National Ecosystem Modeling Workshop (NEMoW II): Bridging the Credibility Gap Dealing with Uncertainty in Ecosystem Models. NOAA Technical Memorandum NMFS-F/SPO-102*. Washington, DC: NOAA.

Link JS, Marshak AR. 2019. Characterizing and comparing marine fisheries ecosystems in the United States: determinants of success in moving toward ecosystem-based fisheries management. *Reviews in Fish Biology and Fisheries* 29(1):23–70.

Low L, Anderson CJ. 2017. The threat of a nonnative, invasive apple snail to oligohaline marshes along the Northern Gulf of Mexico. *Journal of Coastal Research* 33(6):1376–82.

Lowe MR, Wu W, Peterson MS, Brown-Peterson NJ, Slack WT, Schofield PJ. 2012. Survival, growth and reproduction of non-native Nile tilapia II: fundamental niche projections and invasion potential in the northern Gulf of Mexico. *PLoS One* 7(7):e41580.

Liu Y, Lee SK, Muhling BA, Lamkin JT, Enfield DB. 2012. Significant reduction of the Loop Current in the 21st century and its impact on the Gulf of Mexico. *Journal of Geophysical Research: Oceans* 117(C5).

Lucas KL, Carter GA. 2013. Change in distribution and composition of vegetated habitats on Horn Island, Mississippi, northern Gulf of Mexico, in the initial five years following Hurricane Katrina. *Geomorphology* 199:129–37.

LUMCON (Louisiana Universities Marine Consortium). 2019. *Shelfwide Cruises*. https://gulfhypoxia.net/research/shelfwide-cruises/

Lynch PD, Methot RD, Link JS. 2018. *Implementing a Next Generation Stock Assessment Enterprise. An Update to the NOAA Fisheries Stock Assessment Improvement Plan. NOAA Technical Memo NMFS-F/SPO-183*. Washington, DC: NOAA (p. 127).

Macy A, Sharma S, Sparks E, Goff J, Heck KL, Johnson MW, Harper P, Cebrian J. 2019. Tropicalization of the barrier islands of the northern Gulf of Mexico: a comparison of herbivory and decomposition rates between smooth cordgrass (Spartina alterniflora) and black mangrove (Avicennia germinans). *PloS One* 14(1):e0210144.

Marshak AR, Brown SK. 2017. Habitat science is an essential element of ecosystem-based fisheries management. *Fisheries* 42(6):300.

Marshak AR, Heck Jr KL. 2017. Interactions between range-expanding tropical fishes and the northern Gulf of Mexico red snapper Lutjanus campechanus. *Journal of Fish Biology* 91(4):1139–65.

Marshak AR, Heck Jr KL, Jud ZR. 2018. Ecological interactions between Gulf of Mexico snappers (Teleostei: Lutjanidae) and invasive red lionfish (Pterois volitans). *PloS One* 13(11):e0206749.

Marshall KN, Levin PS, Essington TE, Koehn LE, Anderson LG, Bundy A, Carothers C, et al. 2018. Ecosystem-based fisheries management for social–ecological systems: renewing the focus in the United States with next generation fishery ecosystem plans. *Conservation Letters* 11(1):e12367.

Martin CW, Bahya KM, Valentine JF. 2012. Establishment of the invasive island apple snail Pomacea insularum (Gastropoda: Ampullaridae) and eradication efforts in Mobile, Alabama, USA. *Gulf of Mexico Science* 30(1):5.

Martin CW, Valentine JF. 2011. Impacts of a habitat-forming exotic species on estuarine structure and function: an experimental assessment of Eurasian milfoil. *Estuaries and Coasts* 34(2):364–72.

Martin CW, Valentine JF. 2012. Eurasian milfoil invasion in estuaries: physical disturbance can reduce the proliferation of an aquatic nuisance species. *Marine Ecology Progress Series* 449:109–19.

Martin CW, Valentine JF. 2014. Tolerance of embryos and hatchlings of the invasive apple snail Pomacea maculata to estuarine conditions. *Aquatic Ecology* 48(3):321–6.

Martin CW, Valentine MM, Valentine JF. 2010. Competitive interactions between invasive Nile tilapia and native fish: the potential for altered trophic exchange and modification of food webs. *PLoS One* 5(12):e14395.

Mazzotti FJ, Pearlstine LG, Barnes T, Volety A, Chartier K, Weinstein A, DeAngleis D. 2006. *Stressor Response Models for the Blue Crab, Callinectes, sapidus. JEM Technical Report.* Fort Lauderdale, FL: University of Florida, Florida Lauderdale Research and Education Center (p. 12).

Meckley TD, Ashby S, DiMarco SD, Giordano SD, Greene RM, Hilmer DM, Howden SF, et al. 2017. *Building a Cooperative Monitoring Program for Gulf of Mexico Hypoxia and Interrelated Issues. Proceedings Paper from the 6th Annual NOAA/NGI Hypoxia Research Coordination Workshop: Establishing a Cooperative Hypoxic Zone Monitoring Program, 12–13 September 2016.* Starkville, MS: Mississippi State University Science and Technology Center (p. 36).

Mendelssohn IA, Andersen GL, Baltz DM, Caffey RH, Carman KR, Fleeger JW, Joye SB, et al. 2012. Oil impacts on coastal wetlands: implications for the Mississippi River Delta ecosystem after the Deepwater Horizon oil spill. *BioScience* 62(6):562–74.

Milazzo MJ. 2012. Progress and problems in US marine fisheries rebuilding plans. *Reviews in Fish Biology and Fisheries* 22(1):273–96.

Minello TJ, Caldwell PA, Rozas LP. 2017. *Fishery Habitat in Estuaries of the US Gulf of Mexico: a Comparative Assessment of Gulf Estuarine Systems (Cages). NOAA Technical Memorandum NMFS-SEFSC-702.* Washington, DC: NOAA.

Minello TJ, Rozas LP. 2002. Nekton in gulf coast wetlands: fine-scale distributions, landscape patterns, and restoration implications. *Ecological Applications* 12(2):441–55.

Mitsch W, Day J, Gilliam J, Groffman P, Hey D, Randall G, Wang N. 2001. Reducing nitrogen loading to the Gulf of Mexico from the Mississippi River basin: strategies to counter a persistent problem. *Bioscience* 51(5):373–88.

Moore DR, Bullis Jr HR. 1960. A deep-water coral reef in the Gulf of Mexico. *Bulletin of Marine Science* 10(1):125–8.

Morley JW, Selden RL, Latour RJ, Frölicher TL, Seagraves RJ, Pinsky ML. 2018. Projecting shifts in thermal habitat for 686 species on the North American continental shelf. *PLoS One* 13(5):e0196127.

Muto MM, Helker VT, Angliss RP, Boveng PL, Breiwick JM, Cameron MF, Clapham PJ, et al. 2019. *Alaska Marine Mammal Stock Assessments, 2018. NOAA Technical Memo NMFS-AFSC-393.* Washington, DC: NOAA (p. 390).

Nance JM. 2004. *Estimation of Effort in the Offshore Shrimp Trawl Fishery of the Gulf of Mexico. Working Paper to the Red Snapper Stock Assessment Data Workshop, April, 2004. Document No. SEDAR7-DW-24.* Tampa, FL: Gulf of Mexico Fishery Management Council.

Newman D, Berkson J, Suatoni L. 2015. Current methods for setting catch limits for data-limited fish stocks in the United States. *Fisheries Research* 164:86–93.

NFHP (National Fish Habitat Partnership). 2010. *Through a Fish's Eye: The Status of Fish Habitats in the United States.* Washington, DC: Association of Fish and Wildlife Agencies (p. 68).

NMFS (National Marine Fisheries Service). 2016a. *Ecosystem-based fisheries management policy of the National Marine Fisheries Service, National Oceanic and Atmospheric Administration.* Washington, DC: NOAA.

NMFS (National Marine Fisheries Service). 2016b. *NOAA Fisheries Ecosystem-Based Fisheries Management Roadmap.* Washington, DC: NOAA.

NMFS (National Marine Fisheries Service). 2017. *National Marine Fisheries Service – 2nd Quarter 2017 Update.* Washington, DC: NOAA (p. 53).

NMFS (National Marine Fisheries Service). 2018a. *Fisheries Economics of the United States, 2016. NOAA Technical Memo NMFS-F/SPO-187.* Washington, DC: NOAA (p. 243).

NMFS (National Marine Fisheries Service). 2018b. *National Seabird Program 2018 Annual Report.* Washington, DC: NOAA (p. 16).

NMFS (National Marine Fisheries Service). 2019. *Gulf of Mexico Ecosystem-Based Fishery Management Road Map Implementation Plan.* Washington, DC: NOAA.

NMFS (National Marine Fisheries Service). 2020. *National Marine Fisheries Service – 2nd Quarter 2020 Update.* Washington, DC: NOAA (p. 51).

NOAA (National Oceanic and Atmospheric Administration). 2018. *NOAA Report on the Ocean and Great Lakes Economy of the United States.* Charleston, SC: NOAA Office for Coastal Management.

Norris FH, Sherrieb K, Galea S. 2010. Prevalence and consequences of disaster-related illness and injury from Hurricane Ike. *Rehabilitation Psychology* 55(3):221.

Obeysekera J, Barnes J, Nungesser M. 2015. Climate sensitivity runs and regional hydrologic modeling for predicting the response of the greater Florida Everglades ecosystem to climate change. *Environmental Management* 55(4):749–62.

O'Farrell H, Grüss A, Sagarese SR, Babcock EA, Rose KA. 2017. Ecosystem modeling in the Gulf of Mexico: current status and future needs to address ecosystem-based fisheries management and restoration activities. *Reviews in Fish Biology and Fisheries* 27(3):587–614.

Ogden JC, Davis SM, Jacobs KJ, Barnes T, Fling HE. 2005. The use of conceptual ecological models to guide ecosystem restoration in South Florida. *Wetlands* 25(4):795–809.

Passeri DL, Hagen SC, Plant NG, Bilskie MV, Medeiros SC, Alizad K. 2016. Tidal hydrodynamics under future sea level rise and coastal morphology in the Northern Gulf of Mexico. *Earth's Future* 4(5):159–76.

Patrick WS, Link JS. 2015. Myths that continue to impede progress in ecosystem-based fisheries management. *Fisheries* 40(4):155–60.

Patterson WF, Wilson CA, Bentley SJ, Cowan JH. 2005. Delineating Juvenile Red Snapper Habitat on the Northern Gulf of Mexico Continental Shelf. *American Fisheries Society Symposium* 41:277–88.

Peters R, Marshak AR, Brady MM, Brown SK, Osgood K, Greene C, Guida V, et al. 2018. *Habitat Science is a Fundamental Element in an Ecosystem-Based Fisheries Management Framework: An Update to the Marine Fisheries Habitat Assessment Improvement Plan. NOAA Technical Memorandum NMFS-F/SPO-181*. Washington, DC: NOAA (p. 29).

Perry HM, Adkins G, Condrey R, Hammerschmidt PC, Heath S, Herring JR, Moss C, Perkins G, Steele P. 1982. *A Profile of the Blue Crab Fishery of the Gulf of Mexico*. Ocean Springs, MS: Gulf States Marine Fisheries Commission.

Petatán-Ramírez D, Hernández L, Becerril-García EE, Berúmen-Solórzano P, Auliz-Ortiz D, Reyes-Bonilla H. 2020. Potential distribution of the tiger shrimp Penaeus monodon (Decapoda: Penaeidae), an invasive species in the Atlantic Ocean. *Revista de Biología Tropical* 68(1):156–66.

Peterson MS, Lowe MR. 2009. Implications of cumulative impacts to estuarine and marine habitat quality for fish and invertebrate resources. *Reviews in Fisheries Science* 17(4):505–23.

Peterson MS, Slack WT, Brown-Peterson NJ, McDonald JL. 2004. Reproduction in nonnative environments: establishment of Nile tilapia, Oreochromis niloticus, in coastal Mississippi watersheds. *Copeia* 2004(4):842–9.

Pinsky ML, Worm B, Fogarty MJ, Sarmiento JL, Levin SA. 2013. Marine taxa track local climate velocities. *Science* 341(6151):1239–42.

Plotkin PT, Wicksten MK, Amos AF. 1993. Feeding ecology of the loggerhead sea turtle Caretta caretta in the Northwestern Gulf of Mexico. *Marine Biology* 115(1):1–5.

Powell JA, Rathbun GB. 1984. Distribution and abundance of manatees along the northern coast of the Gulf of Mexico. *Gulf of Mexico Science* 7(1):1.

Precht WF, Aronson RB. 2004. Climate flickers and range shifts of reef corals. *Frontiers in Ecology and the Environment* 2(6):307–14.

Precht WF, Hickerson EL, Schmahl GP, Aronson RB. 2014. The invasive coral Tubastraea coccinea (Lesson, 1829): implications for natural habitats in the Gulf of Mexico and the Florida Keys. *Gulf of Mexico Science* 32(1):5.

Price AM, Baustian MM, Turner RE, Rabalais NN, Chmura GL. 2018. Dinoflagellate cysts track eutrophication in the Northern Gulf of Mexico. *Estuaries and Coasts* 41(5):1322–36.

Rabalais NN, Turner RE, Gupta BK, Platon E, Parsons ML. 2007. Sediments tell the history of eutrophication and hypoxia in the northern Gulf of Mexico. *Ecological Applications* 17(sp5):S129–43.

Rabalais NN, Turner RE, Justic D, Dortch Q, Wiseman WJ, Sen Gupta BK. 1999. *Characterization of Hypoxia: Topic 1 Report for the Integrated Assessment on Hypoxia in the Gulf of Mexico. NOAA Coastal Ocean Program Decision Analysis Series, Vol. 15*. Silver Spring, MD: NOAA Coastal Ocean Program (p. 167).

Rabalais NN, Turner RE, Wiseman WJ. 2001. Hypoxia in the Gulf of Mexico. *Journal of Environmental Quality* 30(2):320–9.

Rabalais NN, Turner RE, Wiseman Jr WJ. 2002. Gulf of Mexico hypoxia, aka "The Dead Zone". *Annual Review of Ecology and Systematics* 33(1):235–63.

Rangoonwala A, Jones CE, Ramsey E. 2016. Wetland shoreline recession in the Mississippi River Delta from petroleum oiling and cyclonic storms. *Geophysical Research Letters* 43(22):11–652.

Reed JK, Weaver DC, Pomponi SA. 2006. Habitat and fauna of deep-water Lophelia pertusa coral reefs off the southeastern US: Blake Plateau, Straits of Florida, and Gulf of Mexico. *Bulletin of Marine Science* 78(2):343–75.

Renaud ML, Carpenter JA. 1994. Movements and submergence patterns of loggerhead turtles (Caretta caretta) in the Gulf of Mexico determined through satellite telemetry. *Bulletin of Marine Science* 55(1):1–5.

Rezak R, Gittings SR, Bright TJ. 1990. Biotic assemblages and ecological controls on reefs and banks of the northwest Gulf of Mexico. *American Zoologist* 30(1):23–35.

Rooker JR, Dokken QR, Pattengill CV, Holt GJ. 1997. Fish assemblages on artificial and natural reefs in the Flower Garden Banks National Marine Sanctuary, USA. *Coral Reefs* 16(2):83–92.

Rooker JR, Simms JR, Wells RD, Holt SA, Holt GJ, Graves JE, Furey NB. 2012. Distribution and habitat associations of billfish and swordfish larvae across mesoscale features in the Gulf of Mexico. *PloS One* 7(4):e34180.

Rosel PE, Corkeron PJ, Engleby L, Epperson DM, Mullin K, Soldevilla MS, Taylor BL. 2016. *Status Review of Bryde's Whales (Balaenoptera Edeni) in the Gulf of Mexico Under the Endangered Species Act. NOAA Technical Memorandum NMFS-SEFSC-692*. Washington, DC: NOAA.

Roth BM, Rose KA, Rozas LP, Minello TJ. 2008. Relative influence of habitat fragmentation and inundation on brown shrimp Farfantepenaeus aztecus production in northern Gulf of Mexico salt marshes. *Marine Ecology Progress Series* 359:185–202.

Sagarese SR, Bryan MD, Walter JF, Schirripa M, Grüss A, Karnauskas M. 2015. *Incorporating Ecosystem Considerations Within the Stock Synthesis Integrated Assessment Model for Gulf of Mexico Red Grouper (Epinephelus Morio). SEDAR42-RW-01*. North Charleston, SC: SEDAR (p. 27).

Sagarese SR, Nuttall MA, Geers TM, Lauretta MV, Walter III JF, Serafy JE. 2016. Quantifying the trophic importance of Gulf menhaden within the northern Gulf of Mexico ecosystem. *Marine and Coastal Fisheries* 8(1):23–45.

Samhouri JF, Levin PS, Ainsworth CH. 2010. Identifying thresholds for ecosystem-based management. *PLoS One* 5(1):e8907.

Sammarco PW, Porter SA, Cairns SD. 2010. A new coral species introduced into the Atlantic Ocean Tubastraea micranthus (Ehrenberg 1834) (Cnidaria, Anthozoa, Scleractinia): an invasive threat? *Aquatic Invasions* 5(2):131–40.

Sammarco PW, Porter SA, Sinclair J, Genazzio M. 2014. Population expansion of a new invasive coral species, Tubastraea micranthus, in the northern Gulf of Mexico. *Marine Ecology Progress Series* 495:161–73.

Scavia D, Bricker SB. 2006. "Coastal eutrophication assessment in the United States." In: *Nitrogen Cycling in the Americas: Natural and Anthropogenic Influences and Controls.* Edited by LA Martinelli, RW Howarth. Dordrecht, Netherlands: Springer (pp. 187–208).

Scavia D, Rabalais NN, Turner RE, Justić D, Wiseman Jr WJ. 2003. Predicting the response of Gulf of Mexico hypoxia to variations in Mississippi River nitrogen load. *Limnology and Oceanography* 48(3):951–6.

Scheffel WA, Heck KL, Johnson MW. 2018. Tropicalization of the northern Gulf of Mexico: impacts of salt marsh transition to black mangrove dominance on faunal communities. *Estuaries and Coasts* 41(4):1193–205.

Schrandt MN, Powers SP, Mareska JF. 2015. Habitat use and fishery dynamics of a heavily exploited coastal migrant, Spanish mackerel. *North American Journal of Fisheries Management* 35(2):352–63.

Schofield PJ. 2010. Update on geographic spread of invasive lionfishes (*Pterois volitans* [Linnaeus, 1758] and P. miles [Bennett, 1828]) in the Western North Atlantic Ocean, Caribbean Sea and Gulf of Mexico. *Aquatic Invasions* 5:S117–22.

Scott-Denton E, Cryer PF, Gocke JP, Harrelson MR, Kinsella DL, Pulver JR, Smith RC, Williams JA. 2011. Descriptions of the US Gulf of Mexico reef fish bottom longline and vertical line fisheries based on observer data. *Marine Fisheries Review* 73(2):1–26.

Scyphers SB, Powers SP, Akins JL, Drymon JM, Martin CW, Schobernd ZH, Schofield PJ, Shipp RL, Switzer TS. 2015. The role of citizens in detecting and responding to a rapid marine invasion. *Conservation Letters* 8(4):242–50.

SEAMAP (Southeast Area Monitoring and Assessment Program). 2017. *Annual Report of the Southeast Area Monitoring and Assessment Program (SEAMAP) October 1, 2015–September 30, 2016. No. 261*. Ocean Springs, MS: Gulf States Marine Fisheries Commission (p. 21).

Semmler RF, Hoot WC, Reaka ML. 2017. Are mesophotic coral ecosystems distinct communities and can they serve as refugia for shallow reefs? *Coral Reefs* 36(2):433–44.

SFWMD (South Florida Water Management District). 2006. *Regional Simulation Model (RSM).* West Palm Beach, FL: South Florida Water Management District.

Showalter S. 2003. *Aquatic Nuisance Species in the Gulf of Mexico: A Guide for Future Action by the Gulf of Mexico Regional Panel and the Gulf States.* Washington, DC: NOAA (p. 22).

Simons JD, Yuan M, Carollo C, Vega-Cendejas M, Shirley T, Palomares MLD, Roopnarine P, et al. 2013. Building a fisheries trophic interaction database for management and modeling research in the Gulf of Mexico large marine ecosystem. *Bulletin of Marine Science* 89:135–60.

Smith JN, Snyder SM, Berkson J, Murphy BR, McMullin SL. 2018. Fisheries management of red snapper in the Gulf of Mexico: a case study. *A Collection of Case Studies* (casestudies):72–84.

Smith SG, Ault JS, Bohnsack JA, Harper DE, Luo JG, McClellan DB. 2011. Multispecies survey design for assessing reef-fish stocks, spatially explicit management performance, and ecosystem condition. *Fisheries Research* 109(1):25–41.

Solís D, Perruso L, del Corral J, Stoffle B, Letson D. 2013. Measuring the initial economic effects of hurricanes on commercial fish production: the US Gulf of Mexico grouper (Serranidae) fishery. *Natural Hazards* 66(2):271–89.

Steffen M, Estes MG, Al-Hamdan M. 2010. *Using Remote Sensing Data to Evaluate Habitat loss in the Mobile, Galveston, and Tampa Bay Watersheds.* Lawrence, KS: Coastal Education and Research Foundation (p. 24).

Steidinger KA. 2009. Historical perspective on Karenia brevis red tide research in the Gulf of Mexico. *Harmful Algae* 8(4):549–61.

Stumpf RP, Culver ME, Tester PA, Tomlinson M, Kirkpatrick GJ, Pederson BA, Truby E, Ransibrahmanakul V, Soracco M. 2003. Monitoring Karenia brevis blooms in the Gulf of Mexico using satellite ocean color imagery and other data. *Harmful Algae* 2(2):147–60.

Sumaila UR, Cisneros-Montemayor AM, Dyck A, Huang L, Cheung W, Jacquet J, Kleisner K, et al. 2012. Impact of the Deepwater Horizon well blowout on the economics of US Gulf fisheries. *Canadian Journal of Fisheries and Aquatic Sciences* 69(3):499–510.

Switzer T, Rester JK. 2019. *Annual report to the SEAMAP technical coordinating committee. October 1, 2018–September 30, 2019*. Ocean Springs, MS: Gulf States Marine Fisheries Commission (p. 14).

Tam JC, Link JS, Large SI, Andrews K, Friedland KD, Gove J, Hazen E, et al. 2017. Comparing apples to oranges: common trends and thresholds in anthropogenic and environmental pressures across multiple marine ecosystems. *Frontiers in Marine Science* 4:282.

Teh SY, DeAngelis DL, Sternberg LD, Miralles-Wilhelm FR, Smith TJ, Koh HL. 2008. A simulation model for projecting changes in salinity concentrations and species dominance in the coastal margin habitats of the Everglades. *Ecological Modelling* 213(2):245–56.

Teo SL, Boustany AM, Block BA. 2007. Oceanographic preferences of Atlantic bluefin tuna, Thunnus thynnus, on their Gulf of Mexico breeding grounds. *Marine Biology*. 152(5):1105–19.

Tokotch BN, Meindl CF, Hoare A, Jepson ME. 2012. Stakeholder perceptions of the northern Gulf of Mexico grouper and tilefish individual fishing quota program. *Marine Policy* 36(1):34–41.

Townsend HM, Harvey CJ, Aydin KY, Gamble RJ, Gruss A, Levin PS, Link JS, et al. 2014. *Report of the 3rd National Ecosystem Modeling Workshop (NEMoW 3): Mingling Models for Marine Resource Management, Multiple Model Inference. NOAA Technical Memorandum NMFS-F/SPO-149*. Washington, DC: NOAA.

Townsend HM, Link JS, Osgood KE, Gedamke T, Watters GM, Polovina JJ, Levin PS, Cyr EC, Aydin KY. 2008. *Report of the National Ecosystem Modeling Workshop (NEMoW). NOAA Technical Memorandum NMFS-F/SPO-87*. Washington, DC: NOAA.

Townsend HK, Holsman AK, Harvey C, Kaplan I, Hazen E, Woodworth-Jefcoats P, et al. 2017. *Report of the 4th National Ecosystem Modeling Workshop (NEMoW 4): Using Ecosystem Models to Evaluate Inevitable Trade-offs. NOAA Technical Memorandum NMFS-F/SPO-173*. Washington, DC: NOAA (p. 77).

Tupitza JC, Glaspie CN. 2020. Restored freshwater flow and estuarine benthic communities in the northern Gulf of Mexico: research trends and future needs. *PeerJ* 8:e8587.

Tyson RB, Nowacek SM, Nowacek DP. 2011. Community structure and abundance of bottlenose dolphins Tursiops truncatus in coastal waters of the northeast Gulf of Mexico. *Marine Ecology Progress Series* 438:253–65.

Twilley RR, Barron E, Gholz HL, Harwell MA, Miller RL, Reed DJ, Rose JB, et al. 2001. *Confronting Climate Change in the Gulf Coast Region: Prospects for Sustaining Our Ecological Heritage*. Cambridge, MA: Union of Concerned Scientists and Washington DC: Ecological Society of America.

Vargo GA, Heil CA, Fanning KA, Dixon LK, Neely MB, Lester K, Ault D, Murasko S, Havens J, Walsh J, Bell S. 2008. Nutrient availability in support of Karenia brevis blooms on the central West Florida Shelf: what keeps Karenia blooming?. *Continental Shelf Research* 28(1):73–98.

Valdivia A, Wolf S, Suckling K. 2019. Marine mammals and sea turtles listed under the US Endangered Species Act are recovering. *PloS One* 14(1):e0210164.

VanderKooy S. 2012. *The Oyster Fishery of the Gulf of Mexico, United States: A Regional Management Plan—2012 Revision*. Ocean Springs, MS: Gulf States Marine Fisheries Commission.

VanderKooy SJ. 2015. *Management Profile for the Gulf and Southern Flounder Fishery in the Gulf of Mexico*. Ocean Springs, MS: GSMFC Publication 247 (p. 200).

Vaughan D, Schueller A, Smith J, VanderKooy S. 2011. *SEDAR 27: Gulf Menhaden Stock Assessment Report*. North Charleston, SC: Southeast Data, Assessment, and Review.

Vaughan DS, Shertzer KW, Smith JW. 2007. Gulf Menhaden (Brevoortia patronus) in the U.S. Gulf of Mexico: fishery characteristics and biological reference points for management. *Fisheries Research* 83:263–75.

Vidal L, Pauly D. 2004. Integration of subsystems models as a tool toward describing feeding interactions and fisheries impacts in a large marine ecosystem, the Gulf of Mexico. *Ocean & Coastal Management* 47(11–12):709–25.

Villareal TA, Magaña HA. 2016. *A Red Tide Monitoring Program for Texas Coastal Waters*. Port Aransas, TX: Marine Science Institute.

Walker JS, Shaver DJ, Stacy BA, Flewelling LJ, Broadwater MH, Wang Z. 2018. Brevetoxin exposure in sea turtles in south Texas (USA) during Karenia brevis red tide. *Diseases of Aquatic Organisms* 127(2):145–50.

Walsh JJ, Jolliff JK, Darrow BP, Lenes JM, Milroy SP, Remsen A, Dieterle DA, et al. 2006. Red tides in the Gulf of Mexico: where, when, and why? *Journal of Geophysical Research: Oceans* 111(C11).

Walter ST, Carloss MR, Hess TJ, Leberg PL. 2013. Hurricane, habitat degradation, and land loss effects on Brown Pelican nesting colonies. *Journal of Coastal Research* 29(6a):187–95.

Walters C, Martell SJ, Christensen V, Mahmoudi B. 2008. An Ecosim model for exploring Gulf of Mexico ecosystem management options: implications of including multistanza life-history models for policy predictions. *Bulletin of Marine Science* 83(1):251–71.

Wang C, Liu H, Lee SK, Atlas R. 2011. Impact of the Atlantic warm pool on United States landfalling hurricanes. *Geophysical Research Letters* 38(19):L19702.

Ward CH, Tunnell JW. 2017. "Habitats and biota of the Gulf of Mexico: an overview." In: *Habitats and Biota of the Gulf of Mexico: Before the Deepwater Horizon Oil Spill*. Edited by CH Ward. New York, NY: Springer (pp. 1–54).

Wells RD, Cowan Jr JH, Patterson III WF. 2008a. Habitat use and the effect of shrimp trawling on fish and invertebrate communities over the northern Gulf of Mexico continental shelf. *ICES Journal of Marine Science* 65(9):1610–9.

Wells RD, Cowan JH, Patterson WF, Walters CJ. 2008b. Effect of trawling on juvenile red snapper (Lutjanus campechanus) habitat selection and life history parameters. *Canadian Journal of Fisheries and Aquatic Sciences* 65(11):2399–411.

Weinstein MP, Heck KL. 1979. Ichthyofauna of seagrass meadows along the Caribbean coast of Panama and in the

Gulf of Mexico: composition, structure and community ecology. *Marine Biology* 50(2):97–107.

Weisberg RH, Liu Y, Lembke C, Hu C, Hubbard K, Garrett M. 2019. The Coastal Ocean Circulation Influence on the 2018 West Florida Shelf K. brevis Red Tide Bloom. *Journal of Geophysical Research: Oceans* 124(4):2501–12.

White ED, Messina F, Moss L, Meselhe E. 2018. Salinity and marine mammal dynamics in Barataria Basin: historic patterns and modeled diversion scenarios. *Water* 10(8):1015.

White HK, Hsing PY, Cho W, Shank TM, Cordes EE, Quattrini AM, Nelson RK, et al. 2012. Impact of the Deepwater Horizon oil spill on a deep-water coral community in the Gulf of Mexico. *Proceedings of the National Academy of Sciences* 109(50):20303–8.

White JR, DeLaune RD, Justic D, Day JW, Pahl J, Lane RR, Boynton WR, Twilley RR. 2019. Consequences of Mississippi River diversions on nutrient dynamics of coastal wetland soils and estuarine sediments: a review. *Estuarine, Coastal and Shelf Science* 224:209–16.

Wilkinson, EB, Abrams K. 2015. *Benchmarking the 1999 EPAP Recommendations with Existing Fishery Ecosystem Plans. NOAA Technical Memorandum NMFS-OSF-5*. Washington, DC: NOAA (p. 22).

Williams VJ. 2010. Identifying the economic effects of salt water intrusion after Hurricane Katrina. *Journal of Sustainable Development* 3(1):29.

Würsig BG, Jefferson TA, Schmidly DJ, Foster L, Foster L, Foster L, Artiste EU. 2000. *The Marine Mammals of the Gulf of Mexico.* College Station, TX: Texas A & M University Press.

Yáñez-Arancibia A, Day JW. 2004. The Gulf of Mexico: towards an integration of coastal management with large marine ecosystem management. *Ocean & Coastal Management* 47(11–12):537–63.

Yáñez-Arancibia A, Day JW, Reyes E. 2013. Understanding the coastal ecosystem-based management approach in the Gulf of Mexico. *Journal of Coastal Research* 63(sp1):244–62.

Yoskowitz D, Leon C, Gibeaut J, Lupher B, Lopez M, Santos C, Sutton G, McKinney L. 2013. *Gulf 360: State of the Gulf of Mexico.* Corpus Christi, TX: Texas A&M University, Harte Research Institute for Gulf of Mexico Studies.

Zheng L, Weisberg RH. 2012. Modeling the west Florida coastal ocean by downscaling from the deep ocean, across the continental shelf and into the estuaries. *Ocean Modelling* 48:10–29.

Zimmer B, Precht W, Hickerson E, Sinclair J. 2006. Discovery of Acropora palmata at the flower garden banks national marine sanctuary, northwestern Gulf of Mexico. *Coral Reefs* 25(2):192.

Zimmerman RJ, Minello TJ, Rozas LP. 2002. "Salt marsh linkages to productivity of penaeid shrimps and blue crabs in the northern Gulf of Mexico." In: *Concepts and Controversies in Tidal Marsh Ecology.* Edited by MP Weinstein, DA Kreeger. Dordrecht, Netherlands: Springer (pp. 293–314).

Zimmerman RJ, Minnillo T, Jand Rozas LP. 2000. "Salt marsh linkages to productivity of Penaeid shrimps and blue crabs in the Northern Gulf of Mexico." In: *Concepts and Controversies in Marsh Ecology.* Edited by MP Weinsteinand, DA Kreeger. Amsterdam, Netherlands: Kluwer Academic (pp. 293–314).

Zu Ermgassen PS, Spalding MD, Blake B, Coen LD, Dumbauld B, Geiger S, Grabowski JH, et al. 2012. Historical ecology with real numbers: past and present extent and biomass of an imperiled estuarine habitat. *Proceedings of the Royal Society B: Biological Sciences* 279(1742):3393–400.

CHAPTER 8

The U.S. Pacific Region

➲ The Region in Brief

- The Pacific region currently contains the third-highest commercial fisheries landings and fourth-highest fisheries value among all U.S. marine ecosystems, with pronounced variability in landings having occurred over time.
- The Pacific contains the sixth-highest number of managed taxa in the nation, including commercially and recreationally important salmon, Pacific sardine and other coastal pelagic species, Pacific groundfish (e.g., rockfishes, flatfishes, halibut, Pacific hake, Pacific cod, Sablefish, Lingcod), cephalopods, Dungeness crab, and highly migratory fishes.
- This region contains the highest number of fish stocks, a very low number and percentage of which are overfished or experiencing overfishing, and a mid-range percentage of stocks of unknown overfished status. Additionally, the highest numbers of prohibited and restricted fishing areas in the U.S. are found in the Pacific region.
- Human population and population density are second and fourth-highest respectively in this region, with values having slowly increased over time. Human activities in this region include fishing, energy exploration, coastal development, and coastal tourism, which continue to exert heavy pressures on marine habitats.
- The Pacific has a comparatively high-range socioeconomic status among all U.S. regions, with the largest living marine resources (LMR)-based economy in the U.S. Proportional contributions of LMRs to the Pacific total ocean economy are third-highest nationally among regions.
- Although the Pacific currently contains the third-highest overall fisheries landings and LMR-dependent jobs and revenue, on average total basal ecosystem productivity is seventh-highest nationally despite seasonal upwellings of deeper nutrient-rich waters occurring along its narrow continental shelf.
- Within this region and its subregions, average sea surface temperatures have increased >1.4°C since the mid-twentieth century.
- Large expanses of offshore rocky reefs, kelp canopies, estuarine salt marshes, and seagrasses support economically and ecologically important species throughout their life histories.
- High numbers of marine mammal stocks (of which approximately one-third are strategic stocks under the Marine Mammal Protection Act (MMPA)) and the nationally highest numbers of threatened/endangered species are found in this region.
- The Pacific is third-highest nationally in offshore oil production, while also containing the highest number of wind energy areas for Pacific Ocean regions.
- Fisheries management among state, federal, and tribal entities has been focused upon rebuilding plans for historically overfished or prohibited (commercially and recreationally important) species, and has generally maintained good stock status for the majority of its managed species. Representation on the Pacific Fishery Management Council (PFMC) has oscillated over time with shifting dominance among recreational and commercial sectors.

continued

Ecosystem-Based Fisheries Management: Progress, Importance, and Impacts in the United States. Jason S. Link and Anthony R. Marshak, Oxford University Press. © U.S. Department of Commerce, U.S. Government 2021. DOI: 10.1093/oso/9780192843463.003.0008

- Among subregions, the Channel Islands, Columbia River Estuary, and Puget Sound-Salish Sea contribute more heavily toward fisheries landings, LMR economies, and tourism, while shipping and tourism are pronounced in San Francisco Bay.
- Overall, ecosystem-based fisheries management (EBFM) progress has been made at the regional level, and to a certain degree within subregions, in terms of implementing ecosystem-level planning, advancing knowledge of ecosystem principles, and in assessing risks and vulnerabilities to ecosystems through ongoing investigations into climate vulnerability and species prioritizations for stock and habitat assessments.
- This ecosystem is excelling in the areas of its natural and human environments, including comparatively limited natural stressors, LMR and socioeconomic status, and the quality of its governance system as related to the determinants of successful LMR management.

8.1 Introduction

The California Current (Pacific) fisheries ecosystem is strongly associated with historically important fisheries species, including Pacific sardine (*Sardinops sagax caerulea*), various species of Pacific salmon (*Oncorhynchus* spp.), Pacific halibut (*Hippoglossus stenolepis*), large tunas (*Thunnus* spp.), over 90 species of groundfish that are dominated by commercially important rockfishes (*Sebastes* spp., *Sebastolobus* spp., *Scorpaena* spp.), and Pacific hake (*Merluccius productus*; also see Chapter 11 for international fisheries). Images of beaches with surfers, sailboats, trawling vessels, ferries, lighthouses, pinnipeds, whale watching tours, iconic rocky shorelines, and salmon fishing span the entire U.S. west coast throughout which plates of raw oysters (*Crassostrea gigas*), Dungeness crab (*Metacarcinus magister*), salmon, and calamari adorn numerous tables. Sounds of the Pacific region include crashing waves, blasts from the foghorns of its numerous ships and abundant marine traffic, barks of seals and sea lions, splashes of breaching whales, seabird calls, and a plethora of music styles (e.g., folk, surf, psychedelic, Latin, alternative, indigenous, hip-hop). The U.S. Pacific brings to mind tales like *Cannery Row* and the iconic travels of Steinbeck and Ricketts, and many other works set in coastal California or the Pacific Northwest. These may be found along the shelves of stores like Portland's Powell's Books, along with titles by Jack Kerouac, Ken Kesey, Tom Robbins, Pam Muñoz Ryan, and Amy Tan. This region is renowned for other tales and legends, with countless studios and theme parks covering the landscape, from Hollywood to Disney. Legends handed down by the many indigenous tribes and nations of northern California and the Pacific Northwest illustrate the connectivity of diverse cultures to this marine ecosystem, its upstream watersheds, and abundant living marine resources (LMRs), upon which foundational livelihoods and a nationally dominant maritime economy were built. Major tourism and marine transportation sectors (both domestic and international) are grounded in the continued sustainable management of the Pacific marine ecosystem and its ongoing productivity.

There are many works that characterize the Pacific ecosystem (Chelton et al. 1982, Hayward et al. 1994, 1995, 1996, Francis et al. 1998, McGowan et al. 1998, Bograd et al. 2000, Bograd & Lynn 2001, Schwing et al. 2002, Venrick et al. 2003, Goericke et al. 2004, Di Lorenzo et al. 2005, Field & Francis 2006, Lees et al. 2006, Goericke et al. 2007, Di Lorenzo et al. 2008, McClatchie et al. 2008, Checkley & Barth 2009, McClatchie et al. 2009, Bjorkstedt et al. 2010, 2011, King et al. 2011, Bjorkstedt et al. 2012, CCIEA 2012, Wells et al. 2013, CCIEA 2014, Peterson et al. 2014, Leising et al. 2015, CCIEA 2015, 2016, Harvey et al. 2016, Koehn et al. 2016, Levin et al. 2016, CCIEA 2017, Samhouri et al. 2017, Wells et al. 2017, CCIEA 2018, Lindegren et al. 2018, CCIEA 2019, Link & Marshak 2019, Thompson et al. 2019), its important and major fisheries (e.g., sardines, groundfishes, highly migratory species, Dungeness crab, shrimp, squid; Wild 1980, Sund et al. 1981, Bedford & Hagerman 1983, Johnson et al. 1986, Norton 1987, Brown 1988, Vojkovich 1998, Parker et al. 2000, Reeb et al. 2000, Botsford 2001, Botsford & Lawrence 2002, Emmett et al. 2005, Zeidberg et al. 2006, Burge et al. 2007, Field et al. 2007, Weng et al. 2007, Yoklavich et al. 2007, Juan-Jorda et al. 2009, Bradburn et al. 2011, Carlisle et al. 2012, Wedding & Yoklavich 2015), its protected species (e.g., salmon, seals, sea lions, whales, dolphins, sea turtles, seabirds; Antonelis & Fiscus 1980, DeLong & Antonelis 1991, Brown et al. 1994, Mantua et al. 1997, Defran & Weller 1999, Mate et al. 1999, Sydeman & Allen 1999, Gresh et al. 2000, Barlow & Forney 2007, Pitcher et al. 2007, Barlow et al. 2008, Calambokidis et al. 2009, Sydeman et al. 2009, 2012, Benson et al. 2011, Bailey et al. 2012, Ulanski 2016), and their contributions toward the marine socioeconomics of this region, especially in how humans use, enjoy, and interact with this system, including its significant importance to tribal communities (NRC 1996, Leet 2001, Bailey et al. 2003, Stercho 2006, Norman 2007, Teck et al. 2010, Warlick et al. 2018). This marine region, its resources, and its goods and services have all significantly contributed to American history and U.S. maritime expansions, as illustrated through the acquisition of the Pacific Northwest in the 1803 Louisiana Purchase and subsequent exploration by Lewis and Clark (Gibson 1992, Schwantes 1996, Pomeroy 2003), the 1848 Treaty of Guadelupe Hidalgo and subsequent ratification of California to the Union in 1850, and significant expansion of the U.S. west coast marine economy (Fitzgerald 1986, McEvoy 1990, Walker 2001). In addition, efforts in this region fostered U.S.-Canadian relationships through the formulation of the Oregon Treaty and bilateral agreements for transboundary resources that followed (Sage 1946, Jensen 1985, Munro et al. 1998), and factors such as the 1974 Boldt Decision were instrumental for reserving Pacific tribal treaty rights to fisheries resources and ensuring tribal voice in environmental decision making (Knutson 1987, Brown 1994).

What follows is a brief background for this region, describing all the major considerations of this marine ecosystem. Then there is a short synopsis of how all the salient data for those factors are collected, analyzed, and used for LMR management. We then present an evaluation of key facets of ecosystem-based fisheries management (EBFM) for this region.

8.2 Background and Context

8.2.1 Habitat and Geography

For the purposes here, we term the Pacific region as not implying the entirety of the Pacific Ocean, but rather Pacific waters in the U.S. Exclusive Economic Zone (EEZ) that occur on the U.S. West Coast contiguous with coastal California, Oregon, and Washington, and including riverine systems in those states and Idaho. This region is also referred to as the U.S. Pacific, West Coast, Cal-Cofi (after a notable, longstanding scientific effort; noted next), or the California Current, and is synonymous with the California Current Large Marine Ecosystem (LME). The U.S. Pacific region makes up the third-largest portion of the United States EEZ, with connectivity between Mexican and Canadian waters, with the latter continuing into the Gulf of Alaska ecosystem (Fig. 8.1).

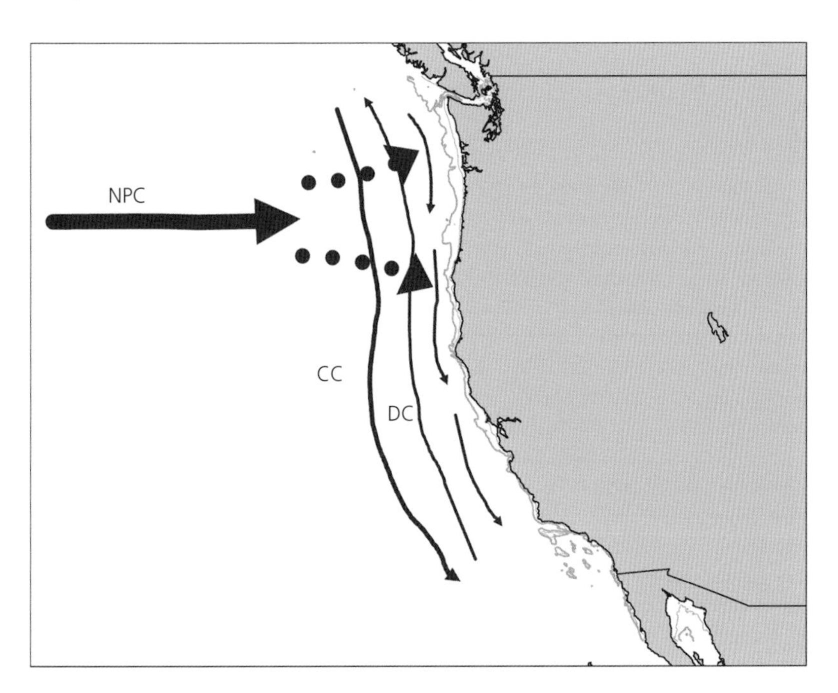

Figure 8.1 Major currents and circulation patterns throughout and encompassing the Pacific region (California Current) and subregions. CC = California Current; DC = Davis Current (seasonal); NPC = North Pacific current. Dashed line for the NPC indicates where it can shift position relative to continental intercept, influencing upwelling and flows of the other major currents.

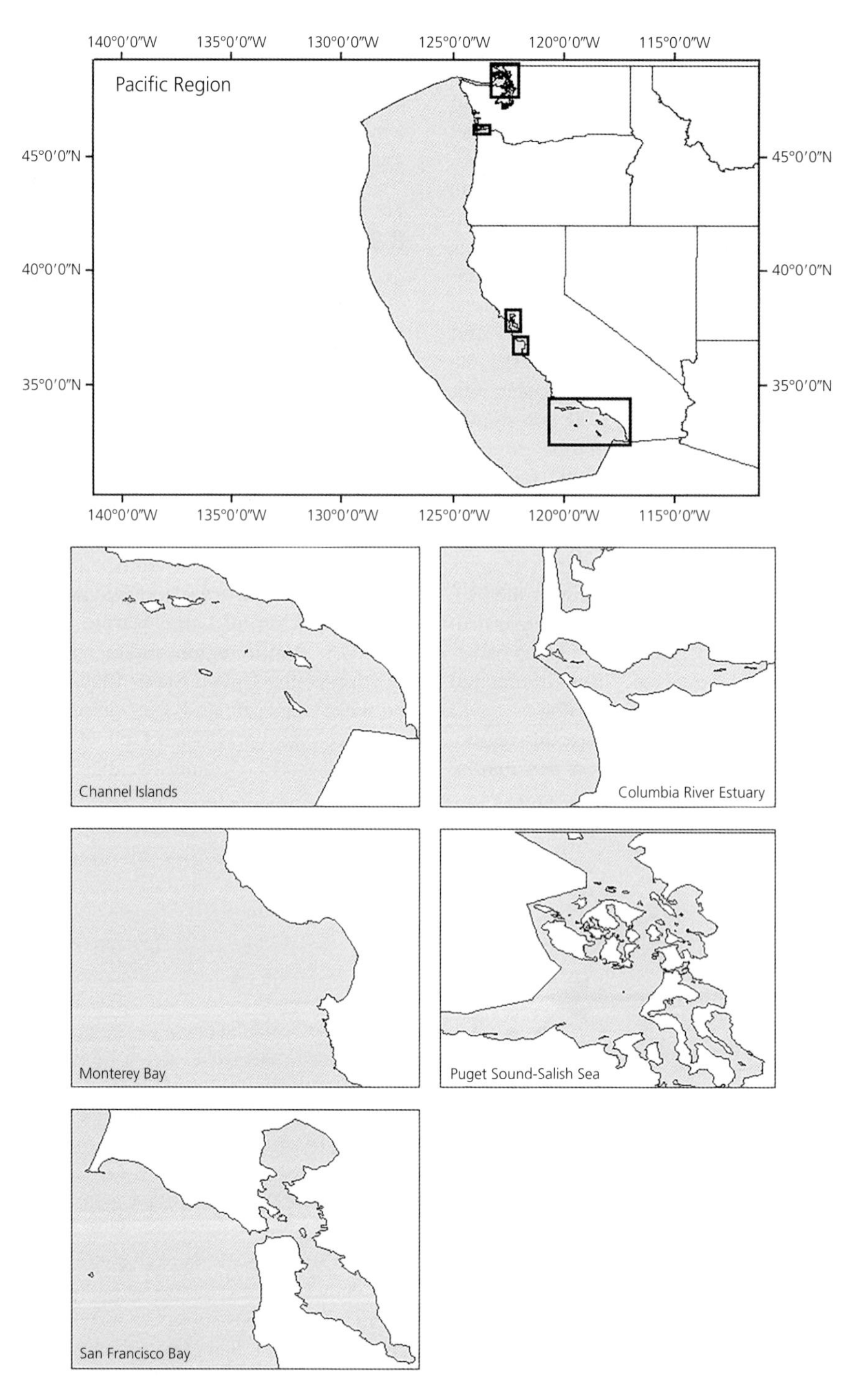

Figure 8.2 Map of the Pacific region and identified subregions.

With high human population and comparatively moderate to high concentrations of population density occuring in this region, significant fisheries landings and economic production are found throughout the U.S. Pacific, particularly in five subregions of interest: **Channel Islands**, **Monterey Bay**, **San Francisco Bay** (with connection to the California Delta)—California; **Columbia River Estuary**—Oregon, Washington; **Puget Sound-Salish Sea**—Washington (Fig. 8.2). There are other subregional features in this region, but we emphasize these five herein as exemplary, relatively smaller-scale ecosystems.

The main features of the Pacific ecosystem include expansive rocky intertidal zonation, *Macrocystis* and *Nereocystis* kelp forests, eelgrass (*Zostera marina*) beds, offshore rocky reefs and deep coral systems, seamounts and undersea canyons, as well as pelagic habitats that host abundant forage fish populations and highly migratory species (Dayton 1971, Connell 1972, Paine 1974, Tegner & Dayton 2000, Steneck et al. 2002, Ward et al. 2003, Emmett et al. 2005, Blanchette et al. 2006, Brodeur et al. 2006, Kaldy & Lee 2007, Whitmire & Clarke 2007, Blanchette et al. 2008, Reiss et al. 2008, Tallis 2009, Tallis et al. 2009, Menge & Menge 2013, Schroeder et al. 2013, Hemery & Henkel 2015, Lam et al. 2015). The California Current drives this significant coastal boundary system along its narrow shelf, which is dependent on seasonally recurrent upwellings of cold, nutrient-rich deeper waters that support its productive environment and is susceptible to influence by eastern tropical Pacific circulation and equatorial currents (Bruland et al. 2001, Carr & Kearns 2003, Marchesiello et al. 2003, Marchesiello & Estrade 2009, Schroeder et al. 2013, Kämpf & Chapman 2016).

8.2.2 Climate and Other Major Stressors

This region has experienced significant regional warming, with third-highest in the nation increases of sea surface temperature (SST) over the past half-century (Link & Marshak 2019). Noticeable changes in species composition have occurred as a result of range-shifting species from lower latitudes becoming more abundant (Roemmich & McGowan 1995, Holbrook et al. 1997, Mendelssohn et al. 2003, Di Lorenzo et al. 2005, Zeidberg & Robison 2007, Kudela et al. 2008, Hsieh et al. 2009, Hazen et al. 2013, Stewart et al. 2014, Asch 2015). Many of these species are entering Canadian and Alaskan waters, potentially spurring interest in shifts in fisheries allocations throughout the California Current system (Stewart et al. 2014, Jurgens et al. 2018, Morley et al. 2018, Pinsky et al. 2018, Selden et al. 2020). Concerns regarding sea-level rise, harmful algal blooms (HABs) associated with regional warming, the associated effects of ocean acidification, and ecosystem-level shifts in response to climate forcing from basin-scale processes such as the El Niño-Southern Oscillation (ENSO) and Pacific Decadal Oscillation (PDO) are also very prominent in this region (Tegner & Dayton 1987, Mantua et al. 1997, McGowan et al. 1998, Brinton & Townsend 2003, Chavez et al. 2003, McGowan et al. 2003, Edwards 2004, Hauri et al. 2009, Hsieh et al. 2009, Board 2012, McCabe et al. 2016, Hoover et al. 2017, Marshall et al. 2017, Parker & Ollier 2017, Ryan et al. 2017, Thorne et al. 2018, CCIEA 2019, Santora et al. 2020).

More recently during 2014–16, an unanticipated major marine heatwave event ("the Blob") significantly affected this system, causing very warm water temperatures, weaker alongshore currents, and increased nearshore abundance of species usually associated with southern and offshore waters (Leising et al. 2015, McClatchie et al. 2016, Wells et al. 2017, CCIEA 2018). In addition, dominance of lipid-poor zooplankton (i.e., subtropical or offshore copepods and pyrosomes; Broduer et al. 2018) led to poor feeding conditions for small pelagic fishes. Warmer waters also led to an unprecedented, economically disruptive HAB, shifts in sardine spawning, delayed opening of the Dungeness crab fishery, and reduced overall abundance of forage fishes (McCabe et al. 2016, Ryan et al. 2017, Santora et al. 2020). Additionally, increased recruitment of northern and southern California rockfishes, reduced seabird reproduction, and high California sea lion pup mortality were observed (McClatchie et al. 2016, C. Harvey, pers. comm.). This warming event also had major effects on northern California salmon populations, and led to several federal fishery disasters along the Pacific coast (Bond et al. 2015, Jacox et al. 2016, Daly et al. 2017, Richerson & Holland 2017, Herbold et al. 2018). In addition, other studies have shown an overall strong influence of ENSO and North Pacific Gyre Oscillation (NPGO) events on Chinook (*O. tshawytscha*) and Coho (*O. kisutch*) salmon survival (Kilduff et al. 2015, Daly et al. 2017, 2019).

Together with climate-associated stressors, the Pacific ecosystem is subject to additional stressors that also affect other U.S. regions, which include habitat loss, increasing coastal and offshore development, nutrient loading and pollution, historical overfishing, and proliferations of invasive species (Tiner 1984, Francis 1986, Ayres et al. 1999, Steneck et al. 2002, Ruesink et al. 2005, Levin et al. 2006, Halpern et al. 2009, Hughes et al. 2009, Holt & Punt 2009, Tepolt et al. 2009,

Teck et al. 2010, Cope et al. 2011, Gleason et al. 2011, Lockwood & Somero 2011, Miller et al. 2011, Cloern & Jassby 2012, Howard et al. 2014, Andrews et al. 2015). Nearshore habitat quality has also been greatly reduced as a consequence of high coastal population density and development, as well as upstream land-based activities that include hydropower, legacy dams and riverine fish passage barriers, agriculture, industry, and their associated pollution (Simenstad et al. 1992, Gibson et al. 2002, Halpern et al. 2009, Hughes et al. 2009, Teck et al. 2010, Howard et al. 2014, Brophy et al. 2019). Heavy nutrient concentrations in this system and its estuaries have led to heightened phytoplankton blooms and macroalgal growth (Anderson et al. 2008, Lewitus et al. 2012). Related to these factors, decreasing salt marsh and eelgrass cover has been observed over time, raising concerns about the continued sustainability of this stressed system (Kentula & Dewitt 2003, Anderson et al. 2008, Gleason et al. 2011, Sherman & DeBruyckere 2018). Reductions in the coverage of offshore kelp forests throughout the Pacific region have also been observed as a result of ongoing stressors, including continued regional warming, climate forcing, and trophic cascades leading to increased urchin grazing (Tegner & Dayton 1987, Steneck et al. 2002, Edwards 2004, Edwards & Estes 2006, Cavanaugh et al. 2011, Pfister et al. 2018, Bell et al. 2020). Additionally, offshore habitats have been subjected to the consequences of trawling activities, and given a narrow shelf, concentrations of natural and human-derived nutrients have led to hypoxic conditions in these areas, which have constrained fisheries productivity (Grantham et al. 2004, Bellman et al. 2005, Bograd et al. 2008, Halpern et al. 2009, Hilborn et al. 2012, Kaplan et al. 2013, Howard et al. 2014, Klinger et al. 2017, Chan et al. 2019). Within these stressed coastal and offshore habitats, shifts in community composition have been observed as a result of these natural and human-derived stressors (Grantham et al. 2004, Keller et al. 2010, Rykaczewski & Dunne 2010, Klinger et al. 2017). The ecological consequences from climate-related species range expansions and human activities are becoming increasingly observed in the U.S. Pacific, with predicted socioeconomic effects (Holbrook et al. 1997, Field et al. 2007, Hsieh et al. 2009, Sumaila et al. 2011, Hazen et al. 2013, Stewart et al. 2014, Asch 2015, Morley et al. 2018, Jurgens et al. 2018). For example, range expansions and intermittently increasing abundance (i.e., periodic "outbreaks") of Humboldt squid (*Dosidicus gigas*) have been associated with predatory release from declining populations of heavily harvested tunas and billfishes, which may also have potential consequences to economically important Pacific groundfishes and other fisheries species, particularly Pacific hake (Field et al. 2007, Zeidberg & Robison 2007, Stewart et al. 2012, Miller et al. 2013, Stewart et al. 2014). Shifts in the distributions of Red rock crab (*Cancer productus*), Shiner perch (*Cymatogaster aggregata*), Northern anchovy (*Engraulis mordax*), White croaker (*Genyonemus lineatus*), North Pacific bigeye octopus (*Octopus californicus*), English sole (*Parophrys vetulus*), Greenspotted rockfish (*Sebastes chlorostictus*), and other commercially and ecologically important species toward more northern portions of this system have been observed (Brodeur et al. 2006, Hazen et al. 2013, Jurgens et al. 2018, Morley et al. 2018). Pelagic colonial tunicates (*Pyrosoma atlanticum*) have also increased in abundance as associated with the warm blob (Brodeur et al. 2017, Sutherland et al. 2018). Given the high number of commercially important shellfish species in this region, abundant deep corals supporting many demersal fish species and macroinvertebrates, and the importance of calcareous zooplankton as a major prey source for dominant forage fishes, the effects of ocean acidification are also a major concern in the U.S. Pacific (Hauri et al. 2009, Feeley et al. 2012, Barton et al. 2015, Haigh et al. 2015, Feely et al. 2016, Marshall et al. 2017).

Although hurricanes and typhoons are relatively infrequent in this region, some consequential effects and those of more common extreme weather events (e.g., windstorms, snowstorms) are still observed in this system, with impacts to coastal communities (Chenoweth et al. 2004, Corbosiero et al. 2009, Warner et al. 2012). Recurring droughts throughout California estuaries and upstream environments have also had major consequences for salmon populations, plankton productivity, system functioning, and Pacific socioeconomics (Bogard & Lynn 2001, MacDonald 2007, Harou et al. 2009, Griffin & Anchukaitis 2015, Swain 2015). These issues are also reflected in significant litigation and management concerns regarding water management, especially in a changing climate (MacDonald 2007, Harou et al. 2009, Swain 2015). As observed throughout other U.S. regions, concerns about increasing sea-level rise and its effects on wetland environments, coastal communities, and fisheries infrastructure remain, particularly given the severe vulnerability of west coast marshes (Thorne et al. 2018). Increasing temperatures have also given rise to recurring *Pseudo-nitszchia* blooms, which have resulted in fisheries closures, caused human health concerns, led to deaths of marine mammals and seabirds, and contributed to whale entanglements

(Santora et al. 2020). All of these factors continue to influence LMRs and their fisheries in the California Current ecosystem.

8.2.3 Exploitation

The effects of historical overfishing, compounded by the influences of climatic oscillations on planktonic communities, forage fish, and upper trophic levels, have been observed in this region (Beamish & Bouillon 1993, Hare & Francis 1995, Mantua et al. 1997, Francis et al. 1998, McFarlane et al. 2002, Chavez et al. 2003, Holt & Punt 2009, Ainley & Hyrenbach 2010, Punt 2011, Keller et al. 2012, Asch 2015). Once the nation's largest fishery, the collapse of the Pacific sardine population during the 1950s is a famous example illustrating the consequences of overexploitation and the necessity of accounting for the environmental factors that drive fisheries production when regulating harvest (Sharp & McLain 1993, Hare & Francis 1995, McFarlane et al. 2002, Chavez et al. 2003, Norton & Mason 2005, Herrick et al. 2007, Zwolinksi & Demer 2012, Ishimura et al. 2013, Tommasi et al. 2017). Shifts in ENSO and PDO climate forcing significantly reduced the intensity of upwelling phenomena in this region, depressing nutrient availability and limiting sardine population growth. These factors led to shifts in the abundance of other forage fishes (i.e., anchovies), which outcompeted Pacific sardine and thus limited the ability of its population to continue supporting its historically extraordinary landings (up to 700,000 metric tons; Lluch-Belda et al. 1989, Chavez et al. 2003, MacCall 2011, Lindegren et al. 2013). Catches fell to an average of 24,000 tons in later years, with oscillating increases in abundance, moratoria intermittently enacted during the 1960s onward, and the population currently closed to harvest since 2015 (Radovich 1982, Wolf 1992, Lindegren et al. 2013, PFMC 2019). Drastic declines in Pacific salmon populations have also been observed throughout this region as a result of historical overharvesting, loss of habitats, constraints to upstream spawning areas from hydropower development and other fish passage barriers, and effects from climate forcing (Nehlsen et al. 1991, Beamish & Bouillon 1993, Frissell 1993, Brown et al. 1994, Gresh et al. 2000, Ruckelshaus et al. 2002a, b, Ruckelshaus & Heppel 2004, Waples et al. 2008, Mueter et al. 2002, Moore et al. 2011, Katz et al. 2013, Thayer et al. 2014, Crozier et al. 2019). Habitat and water quality degradation and historical overharvesting of shellfish and Pacific groundfish populations have led to threatened and endangered listings of several past fishery species (e.g., Yelloweye and Boccacio rockfish, Green sturgeon, Gulf grouper), while depleted populations of marine mammals, sea turtles, and other protected species persist as a result of human-related stressors (Frissell 1993, Mason 2004, Barlow & Forney 2007, Benson et al. 2007, Oreskes & Finley 2007, Halpern et al. 2009, Carretta et al. 2013, Cooley et al. 2017, Richards 2017, Link & Marshak 2019). Despite these examples, Pacific fisheries have become some of the most well managed within the U.S., with nationally significant catches (8% of total fisheries landings) still observed for this region, and low proportions of stocks that are overfished, experiencing overfishing, or unassessed (Yoklavich 1998, Parker et al. 2000, Holt & Punt 2009, Punt 2011, Hilborn & Ovando 2014, Link & Marshak 2019, NMFS 2019). However, past and currently increasing tourism-related fishing effort has led to significant pressures on resident populations, particularly for several overfished highly migratory species (Leet 2001, Coleman et al. 2004, Bellquist et al. 2016).

Since the early 2000s, significant rebuilding of multispecies groundfish populations has occurred in the California Current ecosystem (Holt & Punt 2009, Punt 2011, Heery & Cope 2014, CCIEA 2018, Kauer et al. 2018, CCIEA 2019). During this period, harvest limits for many rebuilding and co-occurring species were strictly lowered to allow for stock rebuilding, as also reflected in commercial landings for those species that were available under more conservative levels. Of these, flatfishes (i.e., Dover sole *Solea solea*, Petrale sole *Eopsetta jordani*), Sablefish (*Anoplopoma fimbria*), and Arrrowtooth flounder (*Atheresthes stomias*) were more abundant from 2000 onward (Cope & Haltuch 2012, Keller et al. 2012, NMFS 2020a). In more recent years, high levels of rockfish recruits have been observed, which are thought to be reflective of enhanced adult spawning populations and positive environmental conditions (Thorson et al. 2013, Thompson et al. 2017, Schroeder et al. 2019, Thorson et al. 2019, Markel & Shurin 2020). Juveniles and young-of-the-year of these species serve as nutrient-rich prey for many higher trophic level species, including Chinook salmon and multiple seabirds (Mills et al. 2007, Wells et al. 2017). The broader ecosystem and socioeconomic impacts of these stock rebuilding efforts will continue to be observed, along with changes to their interactions with other important species in this system (Harvey et al. 2008, Samhouri et al. 2017, Thompson et al. 2017, Warlick et al. 2018). Maintaining rockfish stocks at rebuilt levels, preventing repeated overharvesting, and considering environmental factors for ensuring their sustainability all remain major priorities for managers in this region (Marshall et al. 2017, Kauer et al. 2018, Warlick et al. 2018, Y. DeReynier, pers. comm.).

Fisheries management in the Pacific represents highly progressive efforts toward holistic strategies, and in many ways serves as a leading example for sustainably managed fisheries through cooperation with engaged stakeholders, including tribal interests (Field & Francis 2006, Field et al. 2007, Holt & Punt 2009, Kaplan & Levin 2009, Lester et al. 2010, Punt 2011, Kaplan & Leonard 2012, Kaplan et al. 2012, PFMC 2013, Moffitt et al. 2016, NMFS 2019). As also observed for the North Pacific, emerging holistic management approaches have been enabled by comprehensive multidecadal data regarding coastal and offshore fisheries, essential habitats, climate forcing, and similar ecological and environmental information (Field & Francis 2006, Lester et al. 2010, PFMC 2013, CCIEA 2019, NMFS 2019). While additional ecological data for nearshore and offshore habitats remains needed for more robust environmental assessments, efforts toward addressing these data gaps and incorporating this information into broader ecosystem-based frameworks have been undertaken (Greene et al. 2013, Lederhouse & Link 2016, Beechie et al. 2017, Marshak & Brown 2017, Brophy et al. 2019). Accounting for spatial and environmental factors in stock and protected species surveys and assessments, and foundational understanding of system-wide processes, has become increasingly common in the Pacific region (Parker et al. 2000, Mueter et al. 2002, Reiss et al. 2008, Zwolinski et al. 2011, Cope & Haltuch 2012, Forney et al. 2012, Wedding & Yoklavich 2015, Tolimieri et al. 2018, Crozier et al. 2019, Haltuch et al. 2019, 2020). Initial efforts to apply this information into broader management strategies have also been undertaken, especially considering the effects of fishing and other ocean uses on protected and targeted species, while also accounting for inter-relationships among multiple protected and fisheries populations and trophic levels (Yoklavich 1998, Parker et al. 2000, Field & Francis 2006, Field et al. 2007, Holt & Punt 2009, Kaplan & Levin 2009, Lester et al. 2010, Punt 2011, Kaplan & Leonard 2012, Kaplan et al. 2012, PFMC 2013, Moffitt et al. 2016, NMFS 2019). Continued attention on the vulnerabilities of essential habitats and their species to changes in climate forcing and regional warming warrants more advanced efforts to evaluate ecosystem status and develop management strategies accounting for and mitigating the effects of concurrent natural and human stressors. These priorities are explicitly stated in the fishery ecosystem plan (FEP) for the California Current, including the need for increased biophysical and socioeconomic information on these interactions, and creating system-level buffers in consideration of their uncertainties and emerging effects (PFMC 2013). In particular, these buffers can also account for the effects of climate forcing on forage fishes, their impacts on and trade-offs with higher trophic levels, and when setting catch limits and harvest control rules for multiple species (PFMC 2013, NOAA 2016). In addition, anticipated shifts in commercial and recreational fishing effort, together with fisheries conflicts with protected species and robustly developed marine sectors suggest that spatial conflicts in the Pacific are very likely to increase with regional warming, becoming heavily concentrated along its narrow shelf. Increases in species vulnerability to commercial fishing gears, offshore energy planning, and increasing marine transportation throughout the Pacific basin are likely to increase, with expanded spatial management of essential and critical habitats needing to consider these factors (Simmonds & Elliot 2009, Alter et al. 2010, Bailey et al. 2014, Hayes et al. 2019, Santona et al. 2020).

Despite the progress made for groundfish species, west coast salmon populations continue to remain low since their crash during the mid to latter decades of the twentieth century (Lackey et al. 2006, Thayer et al. 2014, Anderson et al. 2020). Degradation, loss, and alteration of terrestrial and riverine habitats continue to constrain salmon populations, while intermittent stock declines have continued despite focused restocking efforts (Thayer et al. 2014, Anderson et al. 2020). Continued environmental stressors including regional warming, climate forcing, and sea-level rise are anticipated to significantly affect salmon stocks over the next decades (Wainwright et al. 2013, Cline et al. 2019, Crozier et al. 2019). Ongoing attention to Pacific salmon fisheries and endangered populations, their critical habitats, and environmental effects on their production all continue to drive a large proportion of LMR management efforts in this region (Crozier et al. 2019, Knudsen & McDonald 2020, Sattherwaite et al. 2020, Wells et al. 2020).

In addition to focused efforts on addressing human and environmental influences on the California Current ecosystem and its LMRs, interactions among marine mammals, fisheries populations, and fishing fleets are also of significant concern (Hilborn et al. 2012, Chasco et al. 2017a, b, Koehn et al. 2017, Surma et al. 2018). Examples such as trophic effects on sea otters and their top-down influences on urchins, kelp beds, and species that depend on them (Estes et al. 1998, Shelton et al. 2018, Gregr et al. 2020), interactions between sea lions, salmon populations, and other fisheries, or environmental influences leading to greater whale entanglements in fishing gears have become greater priorities in this system (Hilborn et al. 2012, Adams et al. 2016, Chasco et al. 2017a, b, Koehn et al. 2017, Hamilton & Baker 2019, Lebon & Kelly 2019, Santora et al. 2020). As whale and pinniped populations increase and their protections are enforced, concerns regarding greater marine mammal predation on commercially important salmon and coastal pelagic species continue to be expressed by stakeholders (Koehn et al. 2017, C. Harvey pers. comm.). In addition, efforts to continue minimizing bycatch on these species remain a high priority (Carretta et al. 2017, Jannot et al. 2018, Hamilton & Baker 2019, Wainwright et al. 2019, Savoca et al. 2020).

8.2.4 Invasive Species

With increasing temperatures and large-scale human activities throughout the Pacific, the proliferation of invasive species in this region, and their ecological effects on native flora and fauna, has become increasingly of concern (Grosholz & Ruiz 1995, Jamieson et al. 1998, Grosholz et al. 2000, Boyd et al. 2002, Waldek et al. 2003, Wonham & Carlton 2005, Daley & Scavia 2008, Simkanin et al. 2009, Lockwood & Somero 2011, Ashelby et al. 2013, Grosholz et al. 2015, Yamada et al. 2015, Strong & Ayres 2016, Zabin et al. 2018). Significant changes to fouling communities have occurred throughout rocky intertidal habitats and human-made structures as a result of species being introduced through ship traffic, ballast water exchange, and via commercial oyster transplantations (Boyd et al. 2002, de Rivera et al. 2005, Foss 2008). Of particular concern is the increasing proliferation of European green crabs (*Carcinus maenas*) throughout nearshore environments, which has affected native bivalves, crustaceans, and eelgrass beds in this region (Cohen et al. 1995, Grozholz & Ruiz 1995, Jamieson et al. 1998, Jamieson et al. 2002, Jensen et al. 2002, Tepolt et al. 2009, See & Fiest 2010, Yamada et al. 2015, 2017). Similar impacts have been observed within coastal habitats along the U.S. east coast (Davis et al. 1998, Jensen et al. 2002, Neckles 2015, Matheson et al. 2016). Introduced rodents and small mammals have caused increased predation on Pacific shorebirds (Knowlton et al. 2007, Jones et al. 2008). Past introductions of Japanese (Pacific) oyster (*C. gigas*) and other non-native bivalves have led to both the development of extensive fisheries and significant changes to coastal shellfish communities over the past century (Chew 1984, Lavoie 2005, Ruesink et al. 2005). Introduction of Pacific oyster-associated species such as the Estuarine mud snail (*Batillaria attramentaria*) has led to their rapid spread and associated displacement of native mollusks, including California horn snails (*Cerithideopsis californica*; Byers 1999). All of these proliferations of invasive and exotic species have led to significant changes in the Pacific, particularly in coastal estuaries and freshwater systems supporting salmon fisheries (Carlton et al. 1990, Ruiz et al. 1997, Grosholz 2000, Grosholz et al. 2000, Sorte et al. 2010, Ainouche & Gray 2016), with notable consequences to commercially and ecologically important species (Grosholz & Ruiz 1996, Ruiz et al. 1997, Lord 2017).

8.2.5 Ecosystem-Based Management (EBM) and Multisector Considerations

The California Current ecosystem is known for the trade-offs that arise within LMR management and between that and other ocean uses (Airamé et al. 2003, Halpern et al. 2009, Lester et al. 2010, Kaplan et al. 2013, Lester et al. 2013, Chasco et al. 2017a, b). Greater importance of recreational fisheries in this region, among other Pacific basin regions, has been reflected in historically high contributions of highly migratory species to landings and sportfisheries (Bedford & Hagerman 1983, Leet 2001, Coleman et al. 2004, Norman 2007, Bellquist et al. 2016). With increasing tourism, higher demand for seafood and recreational fishing opportunities has occurred in this region. Shifts in prominent commercially and recreationally important fisheries species, commercial effort, and climate-related closures have also altered fishing strategies, landings compositions, and parsing of fishing territories (Norman 2007, Williams & Blood 2011, Cavole et al. 2016, Takada 2017). Concurrent with decreasing recreational landings has been the increasing importance of commercial representation in management (Cavole et al. 2016, Takada 2017, Link & Marshak 2019). As shifts in the distributions and abundance of California Current sportfishes continue, together with rebuilding of offshore commercial populations, these conflicts may increase with potential economic losses (Cheung et al. 2015), especially in those areas where commercial and recreational fishing interests overlap. This potential outcome is related to species becoming more abundant in other subregions, including Canadian and Alaskan waters, with increasing abundance of tropical and subtropical species (Cheung et al. 2015). These increasingly probable fisheries trade-offs also compound those among fisheries and highly developed other ocean uses in this region, which support the Pacific

Dam It Jim, I'm Just a Doctor: The Role of Habitat, Water Quality, Water Quantity, and Water Access

—(c.f., river herrings in Chapter 3; other salmon in Chapter 9)

This might be going out on a limb, but for fish to live they usually need water.

Now in a world with large oceans, that's usually not a concern. But for some fish that need to swim up rivers, it can be.

Salmon might have a few of these concerns.

Most people know that salmon live in the ocean and then amazingly find their natal river system, homing in on it to swim upstream and spawn. This life cycle has provided an amazing bounty for millenia, establishing major economies and cultures in coastal regions.

People also use rivers for many other things. For generating electricity. For providing water to irrigate land. For navigation. For, as the saying goes, a solution via dilution of some pollution. For transport of goods. For recreation. And so on. There are multiple uses of rivers and streams.

Often those other uses result in fundamental, physical alterations to the river. Via engineered efforts like diversions, channels, and dams.

Take an example place, say like the Delta region of central California, and bring those various uses into consideration. What we see is that due to the large demand for irrigation to support agriculture, a lowering of the water level results in notable impacts to water flow for some rivers and tributaries (Fisher et al. 1991, Yoshiyama et al. 1998, Kimmerer 2008, Newman & Brandes 2010). Similar demands to provide safe drinking water to the human population also impact the quantity of water. That alone—quantity and quality—provides some potential contrast and conflict. But then contrast those demands with the needs of salmon, and they have been having less and less water quantity. And we're not even considering water quality (low nutrients, high oxygen, etc.) or habitat (physical structure, flow impediments, thermal conditions, etc.) considerations for these fish.

From a legal perspective, there are U.S. Department of Agriculture (USDA)-oriented policies and legislation to grow food for the country. But there is also the Water Quality Act to provide suitable standards for drinking water. And there is the Endangered Species Act which posits that there needs to be adequate habitat for salmon.

There are many possible solutions to this situation. There are also as many challenges. What this situation does describe is highlighting the conflict among legislation. The conflict among user-groups. The competing objectives for the same commodity—water. And it is precisely because of these trade-offs among the different priorities that an ecosystem-based approach is needed. A more integrated, coordinated way to manage the water to accommodate these various interests may in fact show that the workable solutions can be beneficial to all parties *simultaneously*.

References

Fisher AC, Hanemann WM, Keeler AG. 1991. Integrating fishery and water resource management: a biological model of a California salmon fishery. *Journal of Environmental Economics and Management* 20(3):234–61.

Kimmerer WJ. 2008. Losses of Sacramento River Chinook salmon and delta smelt to entrainment in water diversions in the Sacramento–San Joaquin Delta. *San Francisco Estuary and Watershed Science* 6(2).

Newman KB, Brandes PL. 2010. Hierarchical modeling of juvenile Chinook salmon survival as a function of Sacramento–San Joaquin Delta water exports. *North American Journal of Fisheries Management* 30(1):157–69.

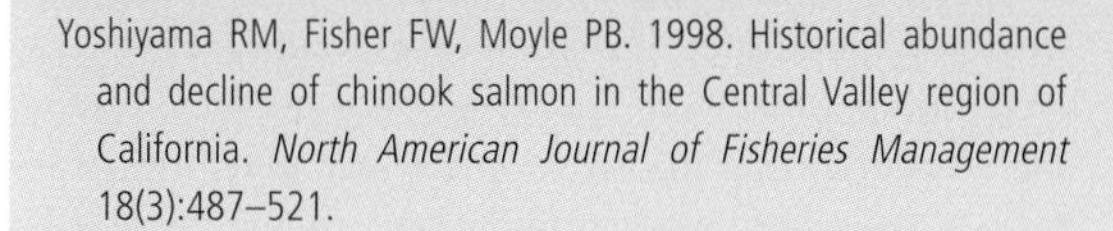
Yoshiyama RM, Fisher FW, Moyle PB. 1998. Historical abundance and decline of chinook salmon in the Central Valley region of California. *North American Journal of Fisheries Management* 18(3):487–521.

c.f.—https://californiawaterblog.com/2018/03/08/is-ecosystem-based-management-legal-for-the-sacramento-san-joaquin-delta/
c.f.—https://www.latimes.com/business/hiltzik/la-fi-hiltzik-caltrump-salmon-20180802-story.html

marine economy and co-occur within areas providing habitat to many managed species. Sectors including tourism, marine transportation, ship and boatbuilding, marine construction, and emerging marine energy interests are especially notable in the Pacific, while the total economic contributions of LMRs from this region are also highest in the nation (Smith 1983, Miller 1987, Boehlert et al. 2008, Halpern et al. 2009, Simkanin 2009, Boehlert & Gill 2010, Teck et al. 2010, Salcido 2011, Sherman et al. 2013, Halpern et al. 2014, Jensen et al. 2015, Link & Marshak 2019). The effects of these activities on protected and targeted species continue to be documented (Baraff & Loughlin 2000, Read et al. 2006, Boehlert et al. 2008, Halpern et al. 2009, Boehlert & Gill 2010, Teck et al. 2010, Maxwell et al. 2013, Redfern et al. 2013, Carretta et al. 2014, Halpern et al. 2014, Hanson et al. 2019). Upstream land use, the pace of urban development, energy exploration, and water use by multiple sectors (including agriculture) have significant trade-offs with salmon populations and their fisheries (Bisbal & McConnaha 1998, Thompson et al. 2011, Hanak & Lund 2012, Herbold et al. 2018). Additional concerns regarding the effects of predation and abundance on salmon and other economically important fisheries populations, and how water quality, drought, and changing environmental conditions may alter these relationships also remain of concern (Best & St-Pierre 1986, Fresh 1997, DeMaster et al. 2001, Wright et al. 2007, Weise & Harvey 2008, Hazen et al. 2013, Chasco et al. 2017a, b, Herbold et al. 2018). The interconnectivity of this region to Mexican, Canadian, and North and Western Pacific commerce, tourism, and shared fisheries resources demonstrates its broader importance to and dependence upon other regional ecosystems and their associated economies (Alper 1996, Oh 2009, Song et al. 2017). However, the expanding influence of U.S. Pacific-based marine sectors into these regions has also led to enhanced human pressures throughout much of the Pacific basin (Boehlert et al. 2008, Halpern et al. 2009, Boehlert & Gill 2010, Teck et al. 2010, Halpern et al. 2014).

First-hand evidence of the consequences of historical overfishing, together with intermittent recovery of many iconic fisheries in this region and collective efforts toward more sustainable practices, have facilitated broader support for more holistic management strategies and fostered increased cooperation from stakeholders toward their enactment (PFMC 2013, NMFS 2019). These include observing the consequences of continued environmental cycling on major pelagic fishery species (anchovies/sardines), and witnessing their broader effects on Pacific fisheries, food webs, and marine economies (Lluch-Belda et al. 1989, Chavez et al. 2003, MacCall 2011, Powell & Xu 2011, Lindegren et al. 2013, Fleming et al. 2016). Increasing awareness of environmental threats to the sustainability of habitats and commercially important species, together with an environmentally engaged public and conservation-minded elected officials, have focused community attention on the future of foundational Pacific LMRs and their marine economies (Arkema et al. 2006, Field & Francis 2006, Lester et al. 2010, Kittinger et al. 2014, Lynham et al. 2017, Norman et al.

2018). These concerns have been especially magnified in light of many climate-associated effects, including regional warming, on this highly responsive system, and concerns regarding continued sustainable practices in light of increasing human population, coastal development, and emerging other ocean uses. In more recent years, extreme marine heatwaves, intermittent climatic oscillations (e.g., ENSO, NPGO), and their continued combined effects on fisheries production have caused stakeholders to be increasingly affected by these environmental stressors (Leising et al. 2015, McClatchie et al. 2016, Wells et al. 2017, CCIEA 2018). Greater reception toward robust ecosystem-based managerial frameworks, and an understanding of the interconnectivity of environmental stressors, ecological interactions among multiple species, fisheries productivity, and human activities through the lens of a socioecological system has emerged in this region (DeReynier 2012, Samhouri et al. 2013, PFMC 2013, Levin et al. 2016, 2018, Dawson & Levin 2019, NMFS 2019). Continuing to build on advancements through cumulative consideration of ecosystem-level factors and the effects of human and natural stressors, sectoral trade-offs, and shifting fishing practices and ocean uses allows for continued advances in EBM strategies for the U.S. Pacific.

8.3 Informational and Analytical Considerations for this Region

8.3.1 Observation Systems and Data Sources

Facilitating an EBM approach for the Pacific region, data regarding the physical, biological, and socioeconomic components of its ecosystem are available through regional observation systems and multiple monitoring programs spanning several decades. Three ocean observing systems: the Southern California Coastal Ocean Observing System (SCCOOS), the Central and Northern California Coastal Ocean Observing System (CeNCOOS), and the Northwest Association of Networked Ocean Observing Systems (NANOOS) provide oceanographic and atmospheric data from multiple sensor stations throughout coastal and offshore locations of the U.S. Pacific region (CeNCOOS 2016, NANOOS 2016, SCCOOS 2016). Each of these systems represents a composite of aggregated data from federal, state, academic, private, industrial, and tribal sources, including automated and manual shore stations, gliders, radar, buoys, in-situ water sampling data, ocean acidification sensors, and meteorological stations. Containing sensors that have been collecting physical data at daily to annual scales as early as the 1910s–1920s in southern California, and over several decades in more northern regions, these portals provide information for multiple physical variables, including atmospheric (e.g., air temperature, pressure, precipitation, humidity, winds) and oceanographic (e.g., water temperature, salinity, oxygen, pH, waves, carbon dioxide, aragonite, photosynthetically available radiation, etc.) parameters. Data collections provide information on HABs and hypoxia, periodic and systematic regional warming events, climatological shifts, and long-term patterns, for predicting storm surge and other severe weather events, and for producing regional oceanographic and climatological forecasts and models for the Pacific and its subcomponents. Additional priorities for these observing systems include providing information for safe maritime practices, assessing, and predicting coastal hazards, improving water quality, and forecasting climatological effects on marine ecosystems (CeNCOOS 2016, SCCOOS 2016, NANOOS 2017).

Information from multiple oceanographic and atmospheric data sources has led to enhanced understanding of the California Current ecosystem, as reflected in its California Current Integrated Ecoystem Assessments (CCIEAs) and annual ecosystem status reports (ESRs) that have been produced since 2012 (IEAs; Levin et al. 2009, Samhouri et al. 2013, CCIEA 2019). Applied climatological data for this region include satellite-derived SST interpolated from NOAA's Advanced Very High-Resolution Radiometer (AVHRR) datasets and hydrographic data obtained through the Newport Hydrographic (NH) and California Cooperative Oceanic Fisheries Investigations (CalCOFI) databases. Information regarding Pacific climate indices (e.g., ENSO, PDO, NPGO) is also maintained at the NOAA Earth Systems Research Laboratory (ESRL) and applied toward examining their effects on system productivity and LMRs. Information regarding upwelling intensity is accounted for through the Bakun Upwelling Index and the more recent incorporation of other ocean models (Jacox et al. 2018) that estimate vertical transport (Cumulative Upwelling Transport Index, CUTI) and nitrate flux (Biologically Effective Upwelling Index, BEUTI), for which data are available from the NOAA Southwest Fisheries Science Center (SWFSC) Pacific Fisheries Environmental Laboratory (PFEL). Water column dissolved oxygen data regarding naturally recurring hypoxic events are also obtained from NH and CalCOFI line sources, while properties related to ocean acidification (e.g., aragonite saturation state with depth) are monitored by universities, shellfish industries, and NOAA laboratories, including the Northwest Fisheries Science Center (NWFSC; CCIEA 2017, 2019).

Habitat factors and aspects of primary and secondary production are monitored via satellite information and ongoing monitoring programs throughout Pacific subregions that account for chlorophyll-a concentrations, phytoplankton biomass, and their estimated coverage. Hydrological data such as freshwater and snow-water condition and physical parameters are collected by the California Department of Water Resources and the Natural Resources Conservation Service, with application toward monitoring estuarine and upstream habitats that support multiple marine species including important salmon populations. These also include data applied toward conducting habitat risk assessments, examining recovery of nearshore vegetated habitats, and examining the effects of bottom fishing gears on the seafloor, including the amount of area disturbed by bottom tending fisheries gear and status and effects to associated structural epifauna (Halpern et al. 2009, Halpern et al. 2014, Greene et al. 2015, CCIEA 2019). In addition, recent habitat restoration efforts in support of anadromous species that include dam removals, the restoration of wetlands, oyster populations, submerged aquatic vegetation, or mitigating the effects of lost or degraded coastal and upstream habitats have emerged as major habitat-related priorities for this region (Dumbauld et al. 2011, Thom et al. 2011, Muething 2018).

Comprehensive databases for secondary through upper trophic level species include fishery-independent information collected from several sources. Zooplankton, jellyfish, and ichthyoplankton abundance and biomass are seasonally monitored through NOAA NWFSC and SWFSC trawl surveys dating back to the early 1970s. Extensive monitoring of forage fish populations through these programs and in connection with CalCOFI has also continued within this region (Bograd et al. 2003, CCIEA 2019, Thompson et al. 2019). Offshore NWFSC (benthic) and SWFSC (midwater) seasonal trawl surveys also measure biomass, distribution, abundance, and condition of groundfish and benthic invertebrates (especially crustaceans) throughout their life histories, while shelf and deep corals, benthic communities, and other sessile epifauna are also examined using remotely operated vehicles (ROVs), video, submersible, and acoustic technologies (Bograd et al. 2003, CCIEA 2019, Thompson et al. 2019). The notable Newport Hydorgraphic Line (Peterson et al. 2002, 2014, Peterson & Keister 2003, Huyer et al. 2007, Peterson 2009) has routinely sampled oceanographic measurements and zooplankton along the same transect since the late 1990s. Upstream and nearshore surveys of Pacific salmon populations are conducted through cooperative efforts by NOAA Fisheries, the California, Oregon, and Washington Departments of Fish & Wildlife and the Idaho Department of Fish & Game (Weitkamp et al. 1995, Hard et al. 1996, Johnson et al. 1997, NWFSC 2015). There are also acoustic surveys for Pacific hake and small pelagic species (Stewart & Hamel 2010, Zwolinski et al. 2012, Chu et al. 2016, Stierhoff et al. 2018, 2020). These surveys provide fishery-independent information regarding abundance and biomass of commercially important fisheries species and composition of offshore assemblages (Ambrose et al. 2006, Baltz et al. 2006, Watson et al. 2007, Keller et al. 2017). Data are collected for multiple trophic levels, which are applied toward informing assessments of West Coast region fisheries, stock status and broader Pacific ecosystem functioning. These surveys are also complemented by NOAA Fisheries fishery-dependent monitoring programs for commercial harvest, landings, revenue, and effort, including fisheries observation data on commercial landings, catches, and bycatch gathered by the National Observer Program (NMFS 2017b). Additional surveys undertaken by the Pacific States Marine Fisheries Commission (PSMFC) and state agencies account for recreational fishing efforts throughout Pacific waters (PSMFC 2018). Factors including trends in average size, spatial distributions, aggregate biomass of pelagic and demersal species groupings and their ratios, predator–prey relationships mean trophic level, condition indices, multispecies mortality estimates, recruitment predictions, community assessments, and thermal preferences represent many of the ecosystem-level metrics that have been derived from this fishery-dependent and independent survey information (CCIEA 2019).

Socioeconomic information for fisheries and other uses, particularly regarding revenue and employment, are also collected by NOAA Fisheries, the National Ocean Economics Program (NOEP), and NOAA's Office for Coastal Management Economics: National Ocean Watch (ENOW) program. Ongoing coastal community vulnerability assessments have also relied on datasets regarding fishery dependence and social indicators as derived from the U.S. Census Bureau and its American Community Survey (ACS), in addition to data from the PSMFC Pacific Fisheries Information (PacFIN) database. Other monitoring programs exist for evaluating the status, recovery, and bycatch of protected species that include pinnipeds and cetaceans, while several monitoring programs for extensive seabird populations in this region are overseen by NOAA, the U.S. Department of Interior, and state agencies (Nevins et al. 2011, Heinemann et al. 2016, Ballance et al. 2017). In the Pacific region, all of these data sources have been applied toward integrative ecosystem measurements that examine cumulative changes in these metrics over time and their interdependent relationships (Levin et al. 2009, Samhouri et al. 2013, 2017, CCIEA 2019).

8.3.2 Models and Assessments

Assessments of commercially important and protected species in the Pacific are conducted by NOAA Fisheries in collaboration with the Pacific Fishery Management Council (PFMC), the U.S. Fish and Wildlife Service (FWS), the California, Oregon, and Washington Departments of Fish & Wildlife, and the Idaho Department of Fish & Game (Ferguson et al. 2017, Lynch et al. 2018, Srinivasan et al. 2019). Both the PSMFC and PFMC support studies and assessments of Pacific species, their habitats, and the application of emerging technologies (PSMFC 2018), including a temperature-based harvest control rule for Pacific sardine (Punt et al. 2016). Stock assessment models are used to generate estimates of stock status with multiple data sources (e.g., landings, catch per unit effort, life history characteristics, fish condition, survey biomass, etc.), and evaluation of assessments for Pacific species are conducted through the Stock Assessment Review (STAR) process (Lynch et al. 2018, Karp et al. 2019). More recently, relationships between oceanographic drivers and recruitment for Sablefish and Petrale sole have been determined, as these factors continue to receive greater attention in this region (Tolimieri et al. 2018, Haltuch et al. 2020). NOAA Fisheries also examines the effects of commercial and recreational fisheries on west coast marine socioeconomics through the application of the NMFS Commercial Fishing Industry Input/Output and Recreational Economic Impact Models (NMFS 2018). In addition, socioeconomic vulnerability assessments have been undertaken for the region, including the Community Social Vulnerability Index (CSVI), to examine the degree of community dependence on commercial and recreational fisheries in a given county. Data from NOAA Fisheries marine mammal stock abundance surveys are used to conduct population assessments as published in regional marine mammal stock assessment reports and peer-reviewed by scientific review groups (SRGs; Carretta et al. 2019, Hayes et al. 2019, Muto et al. 2019). The FWS prepares stock assessment reports for nearshore and inland threatened and protected species under its jurisdiction, such as the sea otter (*Enhydra lutris*). Information collected on protected species of the Pacific region is also applied toward 5-year status reviews of those listed under the Endangered Species Act (ESA).

Data have also been applied toward broader ecosystem models for the Pacific region and its notable subregions, while efforts to incorporate IEA products (e.g., ESRs) into single and multispecies stock assessments and LMR management decisions are also ongoing (Townsend et al. 2008, 2014, 2017). Major efforts include examining trade-offs between natural, social, and economic systems with qualitative network models (Reum et al. 2015, Harvey et al. 2016), developing models and accounting for trade-offs between recovering predators and protected prey species such as killer whales (*Orcinus orca*), pinnipeds, and salmon populations (Kaplan & Levin 2009, Kaplan et al. 2012, Marshall et al. 2016, Chasco et al. 2017a, b, Kaplan & Ward 2017, Wells et al. 2020), and investigating the effects of forage fish harvest (i.e., Pacific sardine and Northern anchovy) on predator populations using multimodeling frameworks (i.e., Atlantis, Models of Intermediate Complexity (MICE), and Ecopath frameworks; Kaplan et al. 2012, Shelton et al. 2014, Punt et al. 2016, Koehn et al. 2017, Kaplan et al. 2019). Efforts to examine marine mammal and sea turtle distributions relative to anthropogenic stressors rely on additive mixed models, boosted regression trees, and Bayesian approaches to understand species-habitat relationships. Other modeling applications include examining spatial trade-offs between wave energy development and fisheries, and climate-associated effects of ocean acidification on fisheries species, including predators dependent on calcareous prey, with Regional Ocean Modeling System (ROMS) and Atlantis ecosystem models (Brand et al. 2007, Horne et al. 2010, Kaplan et al. 2014, 2017, Townsend et al. 2017, Kaplan et al. 2019). These approaches have also examined the effects of environmental forcing, anthropogenic impacts, and groundfish and pelagic fisheries on Pacific food webs and protected species (Kaplan et al. 2014, 2019). Other applications evaluating the risks of climate change and ocean acidification, strengthening species assessments, and conducting management strategy evaluations have been considered (Kaplan et al. 2014, PFMC 2014, Kaplan et al. 2017, 2019). Complementary approaches using both Atlantis and Ecopath models have also examined the role of Pacific sardine in the California

Flip-Flopping, Fair-Weather Fish: Sardines, Anchovies, and Thermal Considerations—Thermal Dependencies, Oscillations, El Niños, and Scripps Pier

Have you ever seen a camera zoomed out on the stands at a tennis match? It's a bit humorous to watch the heads turn back-and-forth, back-and-forth, back-and-forth with each volley. It gets really interesting if there is one of those big lobs, and the time to return a volley is extended. Or even more comical is when there are rapid-fire returns, with the crowd's heads whipping back and forth.

That's kinda like watching coastal small pelagics in the California Current.

It used to be that when anchovy (*Engraulis mordax*) populations were up, sardine (*Sardinops sagax caerulea*) populations were down in the California Current Ecosystem. And vice versa (Chavez et al. 2003, Lindegren et al. 2013). These alternating regimes were thought to be a combination of both population-dependent and independent factors. They were often driven by large-scale ocean-atmospheric features, as tracked by the Pacific Decadal Oscillation (PDO) and experienced by individual fish as thermal conditions, concurrent with the level of fishing these stocks were experiencing. Some competition between the two species has been suspected, or at least biomass dominance related to climate forcing setting up differential conditions between them (Lluch-Belda et al. 1991, MacCall 2011, Lindegren et al. 2013). It was to the point that thermal conditions (often the temperature at the Scripps Institute of Oceanography Pier) were included in (at least some) models of sardine populations to help provide advice to manage these stocks. These forage fish exhibit short life cycles and are quite responsive to environmental conditions, thus making them a key indicator of environmental conditions and a key indicator (as forage) for much of the food web dynamics.

These asynchronous population dynamics were pretty predictable and the alternating populations were routinely occurring. Except when they were not. What used to be asynchronous dynamics have now become synchronous. And what used to be clearly linked to temperature and cycles of El Niño have now become disrupted. It's unclear if that is due to shifts in the periodicity of the PDO, if upwellings and currents have shifted, if the recent "warm blob" phenomena have impacted the ecosystem, or if the measures of fishing

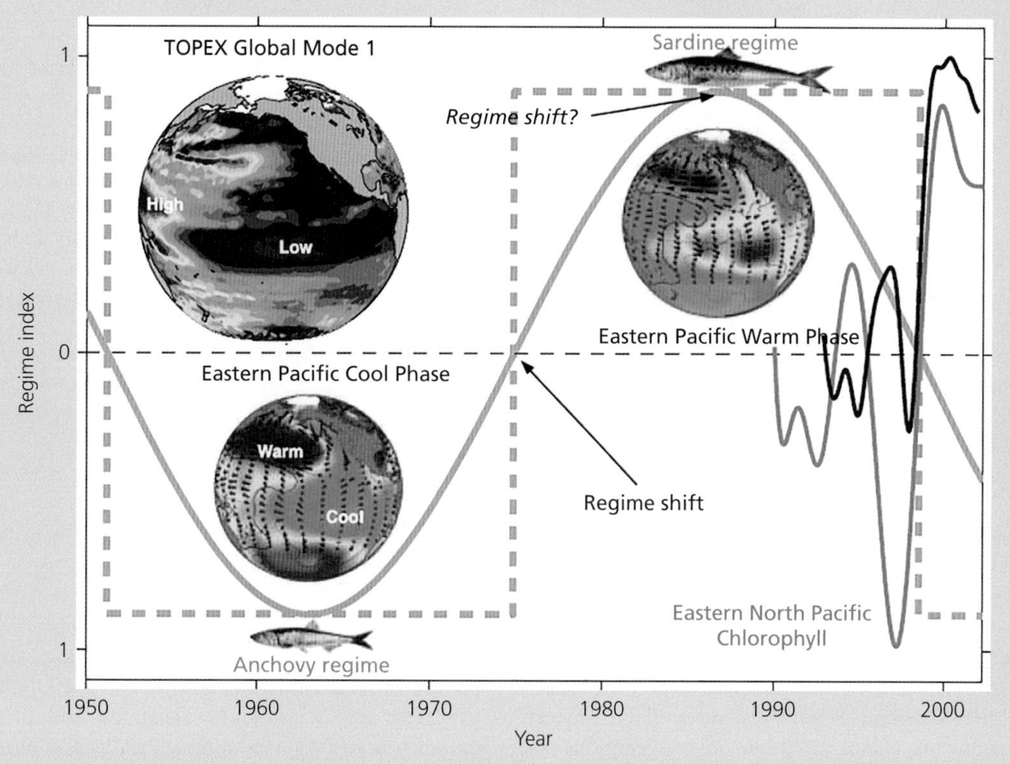

From Chavez et al. 2003.

continued

Flip-Flopping, Fair-Weather Fish: *Continued*

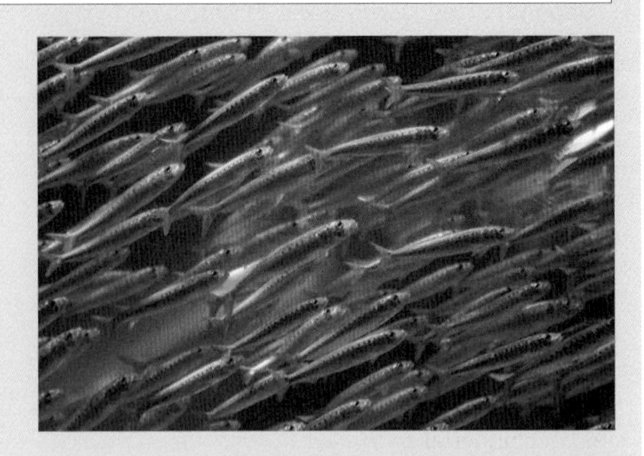

were somehow missed. Regardless of the causal mechanism, two things are clear. One is that climate-induced shifts to the ocean are impacting these stocks, likely altering the periodicity of their cycles. And two is that many other components of this ecosystem are dependent on these populations of forage fish, and the disruptions to the cycles of these forage fish have large ramifications for the rest of the food web (Miller et al. 2010, Kaplan et al. 2013, Koehn et al. 2016, Kaplan et al. 2017).

What this situation highlights is the importance of ongoing monitoring of multiple aspects of an ecosystem, of ongoing consideration of oceanographic conditions, and recognizing that what happens in one stock, or even two, is not independent and has sweeping ramifications for the entire food web.

References

Chavez FP, Ryan J, Lluch-Cota SE, Ñiquen M. 2003. From anchovies to sardines and back: multidecadal change in the Pacific Ocean. *Science* 299(5604):217–21.

Kaplan IC, Brown CJ, Fulton EA, Gray IA, Field JC, Smith AD. 2013. Impacts of depleting forage species in the California Current. *Environmental Conservation* 40(4):380–93.

Kaplan IC, Francis TB, Punt AE, Koehn LE, Curchitser E, Hurtado-Ferro F, Johnson KF, et al. 2019. A multi-model approach to understanding the role of Pacific sardine in the California Current food web. *Marine Ecology Progress Series* 617:307–21.

Koehn LE, Essington TE, Marshall KN, Kaplan IC, Sydeman WJ, Szoboszlai AI, Thayer JA. 2016. Developing a high taxonomic resolution food web model to assess the functional role of forage fish in the California Current ecosystem. *Ecological Modelling* 335:87–100.

Lindegren M, Checkley DM, Rouyer T, MacCall AD, Stenseth NC. 2013. Climate, fishing, and fluctuations of sardine and anchovy in the California Current. *Proceedings of the National Academy of Sciences* 110(33):13672–7.

Lluch-Belda DA, Lluch-Cota DB, Hernandez-Vazquez SE, Salinas-Zavala CA, Schwartzlose RA. 1991. Sardine and anchovy spawning as related to temperature and upwell in the California current system. *CalCOFI Reports* 32:105–11.

MacCall AD. 2011. "The sardine-anchovy puzzle." In: *Shifting Baselines*. Edited by JBC Jackson, K Alexander, E Sala. Washington, DC: Island Press (pp. 47–57).

Miller TW, Brodeur RD, Rau G, Omori K. 2010. Prey dominance shapes trophic structure of the northern California Current pelagic food web: evidence from stable isotopes and diet analysis. *Marine Ecology Progress Series* 420:15–26.

Current food web (Kaplan et al. 2017, 2019). The PFMC Ecosystem Workgroup, Ecosystem Advisory Panel, and Scientific and Statistical Committee have supported these efforts and continued application of Atlantis modeling approaches in providing information to enhance fisheries management (PFMC 2014, Kaplan & Marshall 2016). Continued efforts toward examining the effects of future ocean conditions on groundfish recruitment dynamics are also underway using these modeling frameworks (Townsend et al. 2017).

More specific modeling efforts also include the development of a seasonal ocean prediction system for the Pacific Northwest (J-SCOPE), based on the influences of climate forcing toward producing a higher-resolution ROMS model, and predicting specific oceanic properties including nutrients, planktonic distributions and community dynamics, and physical parameters such as wind forcing (Siedlecki et al. 2016). The model has also been applied toward forecasting the distributions of sardines, Pacific hake, and Dungeness crab, and incorporated into broader IEA frameworks (Kaplan et al. 2016, Townsend et al. 2018, Norton et al. 2020). Complementing these efforts are the use of coupled biophysical ROMS-based models examining spatiotemporal dynamics of krill and seabirds, and applying them toward forecasting future abundance. Similar frameworks have also been used toward predicting habitats for Green sturgeon (*Acipenser medirostris*) together with maximum entropy (MaxEnt) models. Effects of climate change and PDO on Pacific salmon populations have been examined using downscaled temperature and streamflow projections from Global Circulation Models (GCMs) as based on predicted emission scenarios, including responses to potential management strategies (Townsend et al. 2017). Additionally, dynamic modeling investigations have been undertaken to examine the effects of climate change on highly mobile species (Mannocci et al. 2017, Abrahms et al. 2019), predict species distributions and fisheries bycatch susceptibility (Hazen et al. 2017, 2018), and to enhance dynamic ocean management (Maxwell et al. 2015, Welch et al. 2019). Multiscale modeling frameworks have also been used to investigate marine mammal habitats as related to biophysical oceanographic influences (Scales et al. 2017a, b, Abrahms et al. 2019, Woodman et al. 2019, Dodson et al. 2020).

Other physical-biological and Bayesian modeling frameworks have examined physiological responses of salmon populations to changing water temperatures and stream flows, including examining growth and survival potential related to oceanic properties, and at finer scales with mass-balance models (Stock et al. 2011, Fiechter et al. 2015, Jeffrey et al. 2017, Litzow et al. 2020). The Salmon Ecosystem Simulation and Management Evaluation (SESAME) project examines the effects of habitat variability throughout salmon life stages based on specific habitat variables throughout a river-estuary-coastal ocean continuum. This project has been applied toward the creation of a Dynamic Energy Budget (DEB) model examining growth from eggs to mature adults throughout this landscape (Townsend et al. 2014, MacWilliams et al. 2016). Coupled ecological-economic models such as Economic Impact Analysis for Planning (IMPLAN) have been used to examine economic responses of groundfish and salmon harvest under different management scenarios for dams and catch shares (Thomson 2012, Gislason et al. 2017). There are other models that inform management discussions in this region and subregion (Wonham & Carlton 2005, Field et al. 2006, Herrick et al. 2007, Lester et al. 2010, Rykaczewski & Dunne 2010, Forney et al. 2012, Kaplan et al. 2013, Bakun et al. 2015, Harvey et al. 2016, Koehn et al. 2016, Marshall et al. 2017, Parker & Ollier 2017), but most are not directly incorporated into the stock assessment review process.

8.4 Evaluation of Major Facets of EBFM for this Region

To elucidate how well the Pacific ecosystem is doing in terms of LMR management and progress toward EBFM, here we characterize the status and trends of socioeconomic, governance, environmental, and ecological criteria pertinent toward employing an EBFM framework for this region, and examine their inter-relationships as applied toward its successful implementation. We also quantify those factors that emerge as key considerations under an ecosystem-based approach for the region and for five major subregions of interest. Ecosystem indicators for the Pacific region related to the (1) human environment and socioeconomic status, (2) governance system and LMR status, (3) natural environmental forcing and features, and (4) systems ecology, exploitation, and major fisheries of this system are presented and synthesized next. We explore all these facets to evaluate the strengths and weaknesses along the pathway:

$$PP \rightarrow B_{\text{targeted, protected spp, ecosystem}} \leftrightarrow L_{\uparrow\text{targeted spp, }\downarrow\text{bycatch}} \rightarrow \text{jobs, economic revenue.}$$

Where PP is primary production, B is biomass of either targeted or protected species or the entire ecosystem, L is landings of targeted or bycaught species, all leading to the other socioeconomic factors. This operates in the context of an ecological and human system, with governance feedbacks at several of the steps (i.e., between biomass and landings, jobs, and economic revenue), implying that fundamental ecosystem features can determine the socioeconomic value of fisheries in a region, as modulated by human interventions.

8.4.1 Socioeconomic Criteria

8.4.1.1 Social and Regional Demographics

In the U.S. Pacific region, coastal human population values and population densities (Fig. 8.3) have increased from 1970 to present with values currently around 32.8 million residents and 151.6 people km^{-2}, respectively. This region is second-highest nationally in human population, comprising 10.6% of United States residents (year 2010), while density values have risen at ~1.5 people km^{-2} year^{-1}.

Among subregions, human population is most heavily concentrated in areas surrounding southern

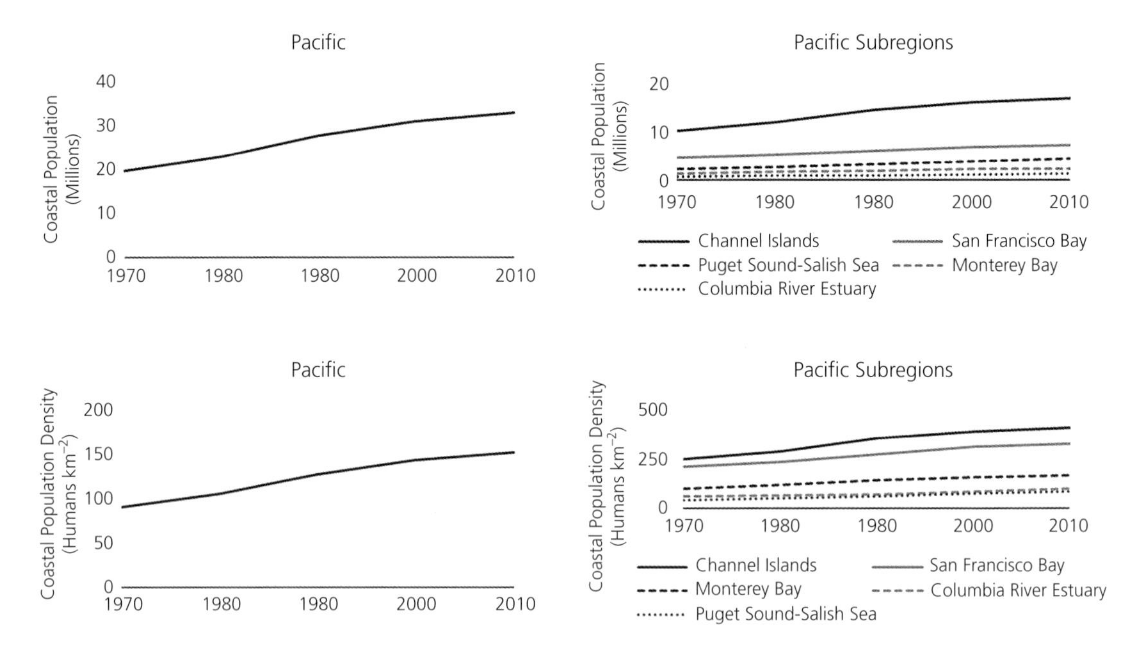

Figure 8.3 Decadal trends (1970–2010) in coastal human population (top) and population density (bottom; humans km^{-2}) for the Pacific region and identified subregions.

Data derived from U.S. censuses and taken from https://coast.noaa.gov/digitalcoast/data/demographictrends.html.

California and near the Channel Islands subregion, where the cities and counties of Los Angeles, San Diego, and Orange are found. In this area that includes Santa Barbara and Ventura Counties, CA, human population has increased over the past 40 years (~178,000 people year^{-1}), with currently up to ~17.2 million residents and ~408.7 people km^{-2} in surrounding areas. Human population is also high in the San Francisco Bay area, with inhabitants concentrated in cities including San Francisco, San Jose, Oakland, Berkeley, and Palo Alto, in addition to ten populated coastal counties (i.e., Alameda, Contra Costa, Marin, Napa, San Francisco, San Mateo, Santa Clara, Santa Cruz, Solano, and Sonoma, CA). Over the past 40 years, human population surrounding San Francisco Bay has increased by ~66,500 people year^{-1} with currently up to ~7.4 million residents and ~325.2 people km^{-2}.

Comparatively lower human populations are found surrounding the Puget Sound-Salish Sea, Monterey Bay, and then the Columbia River Estuary subregions. Population in the Puget Sound-Salish Sea area has increased (~57,200 people year^{-1}) to currently ~4.6 million residents and ~83.1 people km^{-2}, with most individuals living near the City of Seattle and in King, Pierce, and Snohomish Counties, WA. The area includes ten other coastal counties (i.e., Clallam, Grays Harbor, Island, Jefferson, Kitsap, Mason, San Juan, Skagit, Thurston, and Whatcom, WA). Population in areas surrounding Monterey Bay has increased (~25,500 people year^{-1}) to currently ~2.5 million residents and ~167.1 people km^{-2}, with most individuals living in Santa Clara County, CA. Significant populations in this subregion are also found in Monterey and Santa Cruz Counties, CA. Human population has increased in areas surrounding the Columbia River Estuary over the past 40 years (~13,600 people year^{-1}) to ~1.4 million residents and ~97.3 people km^{-2}. Most individuals reside near the City of Portland and in Multnomah County, OR, with additional concentrations in Clark and Cowlitz Counties, WA. Four other less populated coastal counties surround the subregion (i.e., Clatsop, Columbia Counties, OR and Pacific, Wahkiakum Counties, WA) with current populations ranging from ~4,000 (Wahkiakum, WA) to 49,000 residents (Columbia, OR).

8.4.1.2 Socioeconomic Status and Regional Fisheries Economics

Although marginal decreases have been observed in recent years, high numbers of LMR establishments and employments and LMR-associated gross domestic product (GDP) values (~1.85 billion USD) are found throughout the Pacific region (Fig. 8.4). Highest in the

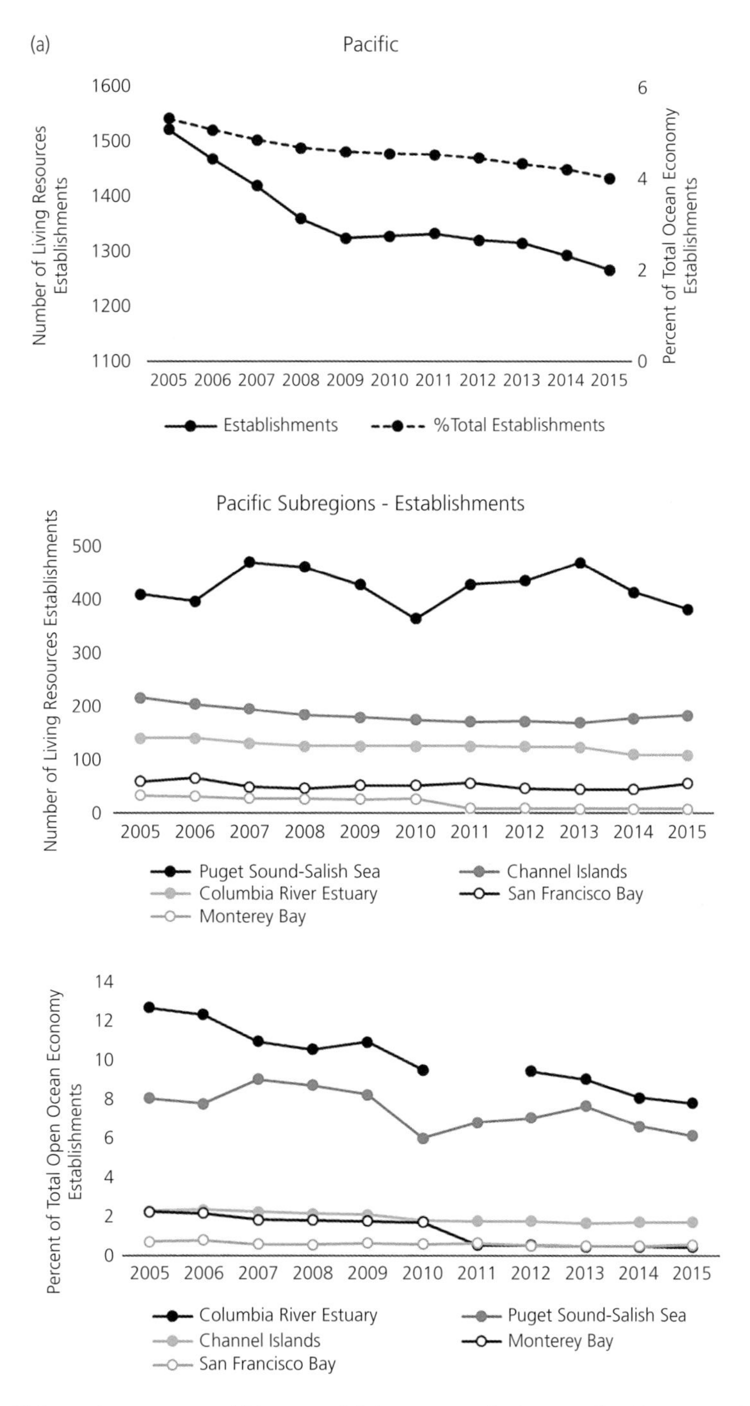

Figure 8.4a Number of living marine resources establishments and their percent contribution to total multisector oceanic economy establishments in the Pacific region and identified subregions (years 2005–2015).

Data derived from the National Ocean Economics Program.

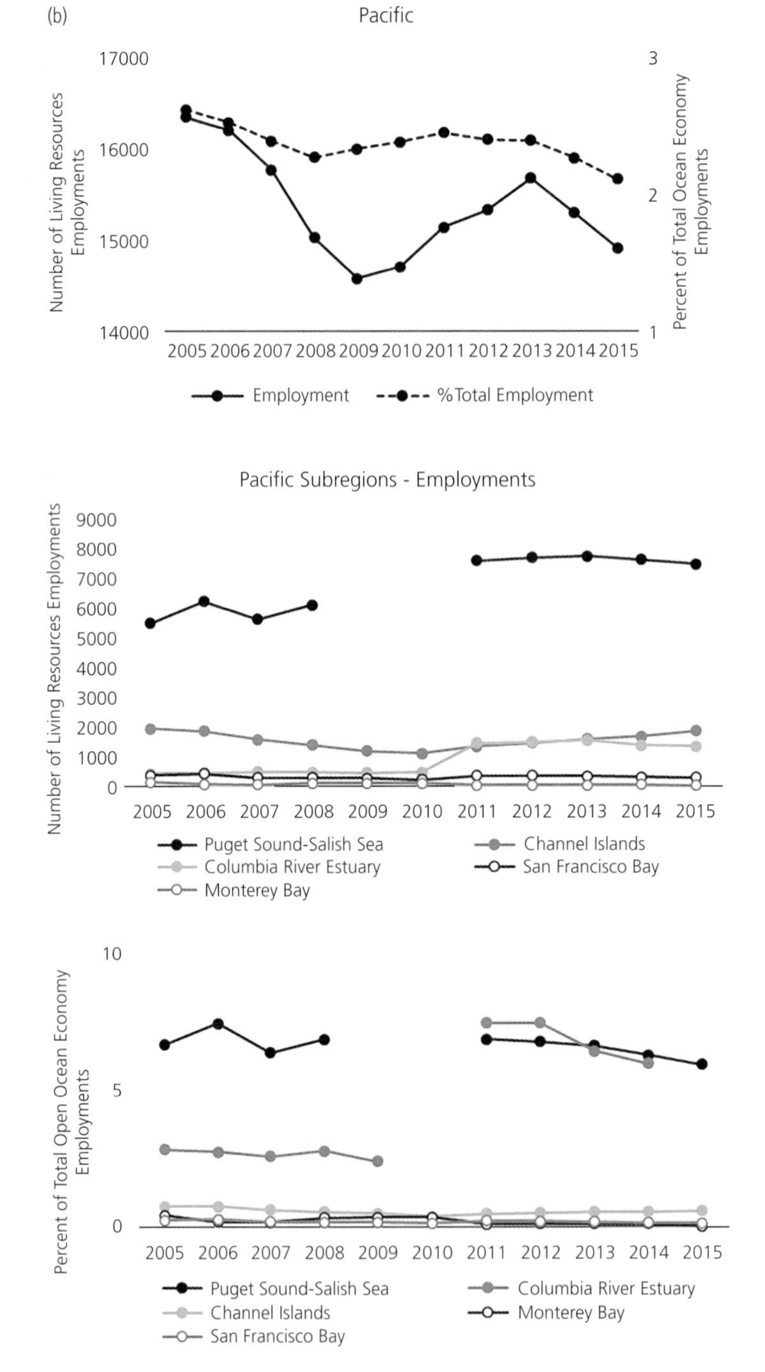

Figure 8.4b Number of living marine resources employments and their percent contribution to total multisector oceanic economy employments in the Pacific region and identified subregions (years 2005–2015).

Data from the National Ocean Economics Program.

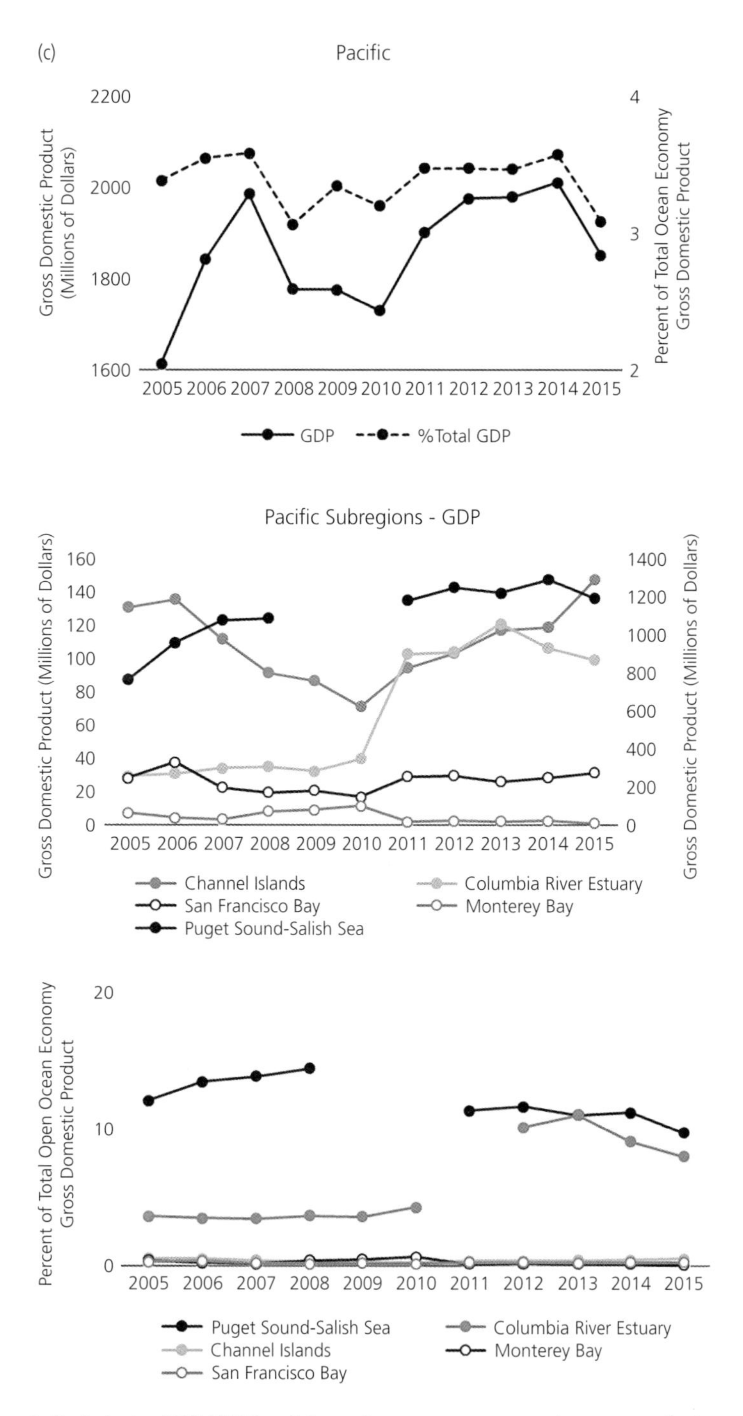

Figure 8.4c Gross Domestic Product value (GDP; USD) from living marine resources revenue and percent contribution to total multisector oceanic economy GDP for the Pacific region and identified subregions (years 2005–2015). GDP values for Puget Sound-Salish Sea are included on the secondary axis (middle panel).

Data derived from the National Ocean Economics Program.

nation, the Pacific contains >1000 LMR establishments, >10,000 employments, and significant GDP value that have increased over the past decade. The percent contribution of the Pacific LMR-associated economy toward its regional ocean economy is third-highest nationally, with LMR establishments comprising up to 5% of regional ocean economy establishments and LMR employments contributing 2–3% of regional ocean employments over time. LMR-associated GDP has fluctuated over the past decade with an overall increase of ~\$238.7 million since 2005, and has contributed up to 3.6% toward the regional total ocean economy in recent years.

In nearly all subregions, LMR economics have remained relatively stable or increased over time, with the Puget Sound-Salish Sea and Columbia River Estuary subregions contributing the highest percentages toward their regional ocean economies. The highest numbers of LMR establishments and employments and GDP value have historically occurred in the Puget Sound-Salish Sea subregion (424.3 ± 10.5 standard error, SE, establishments; 6886.9 ± 321.5 SE employments; \$1.1 ± 0.06 billion SE, USD), including increases in LMR employments and LMR-associated GDP over the past decade. Although decreasing over time, in Puget Sound-Salish Sea LMRs have historically contributed up to 9% of total ocean establishments, 7.5% of total ocean employments, and 14.4% toward total ocean GDP, making it highest overall among subregions for the latter two values.

Comparatively high numbers of LMR establishments, employments, and GDP value also occur in the Channel Islands (184 ± 4.5 SE establishments; 1548.7 ± 83.3 SE employments; \$110.2 ± 7.0 million SE, USD), with LMRs contributing 1–2% toward its total ocean economy, and in the Columbia River Estuary (125 ± 3.1 SE establishments; 907.5 ± 156.4 SE employments; \$66.8 ± 11.7 million SE, USD). However, LMRs have a much higher contribution in the Columbia River subregion LMRs contributing up to 12.6% of total ocean establishments (highest overall), 7.5% of total ocean employments, and 11% toward total ocean GDP, with increased percent GDP contribution occurring post-2010. Lower contributions of LMRs toward subregional total ocean economies are found in Monterey Bay and San Francisco Bay, where values have remained steady over the past decade.

During the early 1990s, decreases in the number of reported permitted commercial fishing vessels operating in the Pacific (Fig. 8.5) occurred, particularly for commercial salmon troll permits. From the mid-1990s to early 2000s, values remained similar until increases

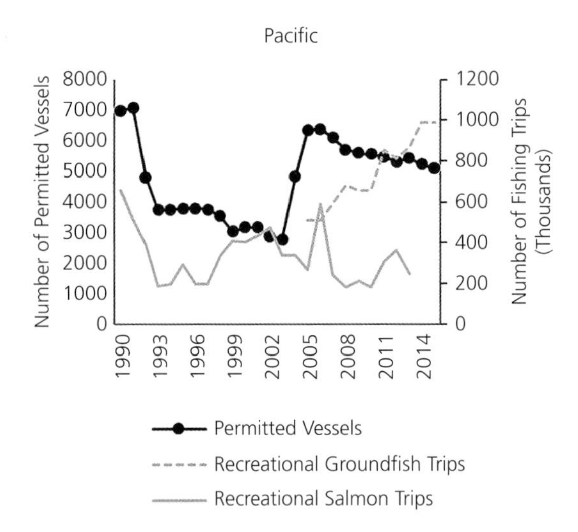

Figure 8.5 Total number of permitted vessels, and recreational salmon and groundfish fishing trips, for the Pacific region over time (years 1990–2015).

Data derived from NOAA National Marine Fisheries Service Council Reports to Congress.

were observed from 2004 to 2006 (up to ~6400 vessels), during which increased reporting of permitted vessels for highly migratory species (HMS) captures occurred. In recent years, reported values have decreased to ~5000 vessels. These values are seventh-highest nationally. Reported recreational fishing trips for salmon also decreased during the early 1990s, and have fluctuated since, with higher numbers observed in the 2000s and peaking in 2006 at ~590,000 trips. More recently, reported values are similar to those observed in the mid to late 1990s at ~240,000 trips. Trends for recreational groundfish trips have increased since 2005 from ~500,000 to nearly 1 million trips per year.

The Pacific is fourth-highest in the nation in terms of total revenue (year 2017 USD) of its landed commercial fishery catches (Fig. 8.6). Increases in commercial landings value have occurred over time for this region, particularly since the late 1990s, with peak values observed at \$814.8 million in 2011. Among most subregions, commercial catch revenues have remained relatively steady since 1981. However, a major decrease in revenue was observed in the Channel Islands during the early to mid-1980s from \$204 million to \$50.2 million. Although fluctuating, values in the following years have slightly increased and have averaged ~\$61.3 ± 2.8 SE since 1986. Overall, increased revenue for the Puget Sound-Salish Sea subregion has been observed over time. In the 2000s, pronounced increases occurred with values peaking at \$143.9 million in 2013,

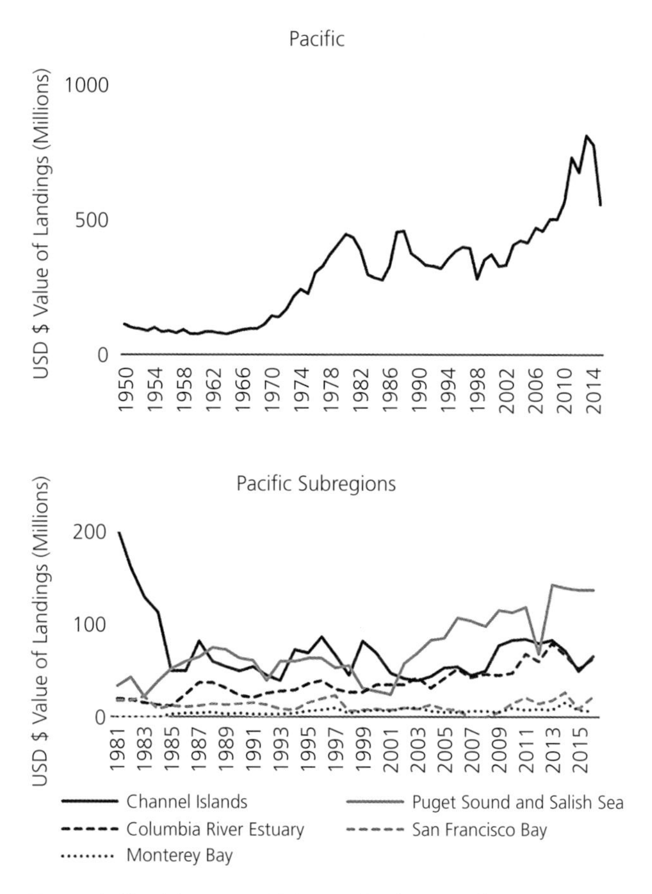

Figure 8.6 Total revenue (year 2017 USD) of landed commercial fishery catches for the Pacific region (years 1950–2015) and identified subregions (years 1981–2016), over time.

Data derived from NOAA National Marine Fisheries Service commercial fisheries statistics.

making Puget Sound-Salish Sea commercial catch revenue currently highest among subregions. Similarly, increased revenue has been observed over time for the Columbia River Estuary, with highest values also occurring in 2013 at $80.2 million. Lower reported commercial catch revenues are found in the San Francisco Bay (range: $5.6 to $26.7 million) and Monterey Bay ($2.7 to $16.2 million) subregions, with relatively stable trends over time despite interannual fluctuations for both regions during the 1990s and 2000s.

Ratios of LMR jobs and commercial fisheries revenue to biomass and primary production (Fig. 8.7) are at mid-high rankings nationally (first to fourth-highest) in the Pacific. Although slightly decreasing over time, the Pacific is highest in the nation in terms of the number of jobs relative to primary production (up to 17.4 jobs million metric tons^{-1}). Additionally, the Pacific is third-highest nationally in jobs/total surveyed biomass (up to 2.1 jobs thousand metric tons^{-1}) and revenue/productivity (up to $0.85 metric ton^{-1}), with a near-tripling in value having occurred for the latter over the past two decades. Overall, the Pacific has been fifth-highest nationally in terms of fisheries revenue/biomass since the 1980s, although values have increased (up to $85.6 metric ton^{-1}) over the past two decades making the region currently fourth-highest for this indicator nationally.

8.4.2 Governance Criteria

8.4.2.1 Human Representative Context

Twenty-five federal, state, tribal, and international agencies and organizations operate within the Pacific

Figure 8.7 Over time, ratios of (a) total living marine resources employments to total biomass (thousands of metric tons); (b) total commercial landings revenue (USD) to total biomass (metric tons); (c) total living marine resources employments to total productivity (metric tons wet weight year^{-1}); (d) total commercial landings revenue (USD) to total productivity (metric tons wet weight year^{-1}) for the Pacific region.

and are directly responsible for fisheries, protected species, and environmental management (Table 8.1). Fisheries resources throughout state waters (0–3 nautical miles) are managed by individual state agencies, while those in federal waters are managed by both NOAA Fisheries (West Coast Region) and the PFMC. While California, Oregon, Idaho, and Washington are members of the PSMFC, this organization does not directly manage species in the Pacific under any fishery management plans (FMPs) and instead serves as a coordinating body for fisheries management and conservation issues in this region. Focused PSMFC programs include data and monitoring efforts for anadromous fish and their habitats, preventing aquatic invasive species, controlling Northern pikeminnow (*Ptychocheilus oregonensis*) predation on salmon populations, and assessing Pacific recreational fishing effort (Recreational Fisheries Information Network, RecFIN). Pacific fishery and environmental resources also fall under several international and multi-jurisdictional agreements, including those of the Inter-American Tropical Tuna Commission (IATTC), International Pacific Halibut Commission (IPHC), Pacific Salmon Commission (PSC), and Pacific Whiting Treaty (PWT). There is also engagement in the Convention for the Conservation of Antarctic Marine Living Resources (CCAMLR) (see Chapter 11 for more details on these international fisheries contexts).

The Channel Islands and Monterey Bay subregions contain separate NOAA National Marine Sanctuaries (NMS) and include management by the California Department of Fish & Wildlife – Fish and Game Commission, which also manages the San Francisco Bay subregion. The Puget Sound-Salish Sea subregion falls under the jurisdiction of the Washington Department of Fish & Wildlife, with several aquatic reserves also managed by the Washington Department of Natural Resources. Salmon resources and groundfish off the State of Washington are subject to treaty agreements with suites of federally recognized fishing tribes. The Columbia River Estuary subregion is managed by the Washington Department of Fish & Wildlife, Oregon Department of Fish & Wildlife Marine Fisheries, and Idaho Department of Fish & Game throughout its watershed. The fishery resources of this subregion are co-managed by federally recognized fishing tribes, particularly through the Columbia River InterTribal Fish Commission and the Northwest Indian Fisheries Commission with regards to exclusive tribal groundfish and salmon rights.

Table 8.1 Federal and State agencies and organizations responsible for fisheries and environmental consulting and management in the Pacific region and subregions

Region	Subregions	Federal Agencies with Marine Resource Interests	Council	State Agencies and Cooperatives	Multi-Jurisdictional
Pacific		U.S. Department of Commerce **DoC** NOAA Fisheries Northwest Fisheries Science Center **NWFSC**, Southwest Fisheries Science Center **SWFSC**, West Coast Regional Office **WCRO**, NOAA Office of National Marine Sanctuaries **NMS**; U.S. Department of Interior **DoI** U.S. Fish and Wildlife Service **FWS**, National Park Service **NPS**, Bureau of Ocean Energy Management **BOEM**, U.S. Geological Survey **USGS**; U.S. Environmental Protection Agency **EPA**	Pacific Fishery Management Council **PFMC**	Pacific States Marine Fisheries Commission **PSMFC**; California Department of Fish & Wildlife—Fish & Game Commission; Idaho Department of Fish & Game; Oregon Department of Fish & Wildlife Marine Fisheries; Washington Department of Fish & Wildlife	Columbia River InterTribal Fish Commission; Convention for the Conservation of Antarctic Marine Living Resources **CCAMLR**; Inter-American Tropical Tuna Commission **IATTC**; International Pacific Halibut Commission **IPHC**; International Scientific Committee for Tuna and Tuna-Like Species in the Northern Pacific Ocean **ISCTTS**; NW Indian Fisheries Commission; Pacific Salmon Commission **PSC**; Pacific Whiting Treaty **PWT**
	Channel Islands Monterey Bay	**DoC (NMS, SWFSC, WCRO)**; **DoI (FWS, BOEM, NPS, USGS)**; **EPA**	**PFMC**	**PSMFC**; California Department of Fish & Wildlife—Fish & Game Commission	**IATTC**; **IPHC**; **ISCTTS**; **PSC**; **PWT**
	Columbia River Estuary	**DoC (NWFSC, WCRO)**; **DoI (FWS, BOEM, NPS, USGS)**; **EPA**	**PFMC**	**PSMFC**; Idaho Department of Fish & Game; Oregon Department of Fish & Wildlife Marine Fisheries; Washington Department of Fish & Wildlife	Columbia River InterTribal Fish Commission; **IPHC**; NW Indian Fisheries Commission; **PSC**; **PWT**
	Puget Sound-Salish Sea	**DoC (NWFSC, WCRO)**; **DoI (FWS, BOEM, NPS, USGS)**; **EPA**	**PFMC**	**PSMFC**; Washington Department of Fish & Wildlife Washington Department of Natural Resources	**IPHC**; NW Indian Fisheries Commission; **PSC**; **PWT**
	San Francisco Bay	**DoC (NMS, SWFSC, WCRO)**; **DoI (FWS, BOEM, NPS, USGS)**; **EPA**	**PFMC**	**PSMFC**; California Department of Fish & Wildlife—Fish & Game Commission	**IPHC**; **ISCTTS**; **PSC**; **PWT**

Nationally, the Pacific region contains the second-highest number of Congressional representatives (n = 43, including Idaho), which are concentrated within four states (Fig. 8.8). Until the mid-2000s, composition of the PFMC has been dominated by members of the recreational fishing sector, with additional representation from other sectors (Fig. 8.9a). From 2005 onward, a significant change in FMC composition occurred, resulting in dominance by the commercial fishing sector and very low representation from recreational and other sectors. These trends have continued to present. While containing 14 total voting members, the PFMC contains a lower number

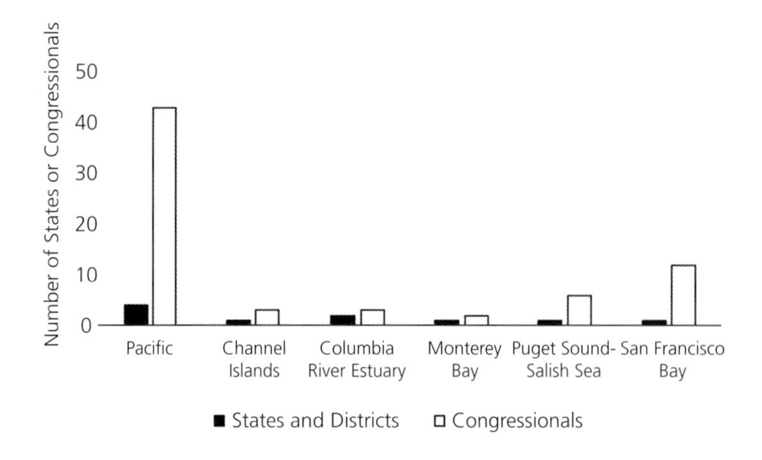

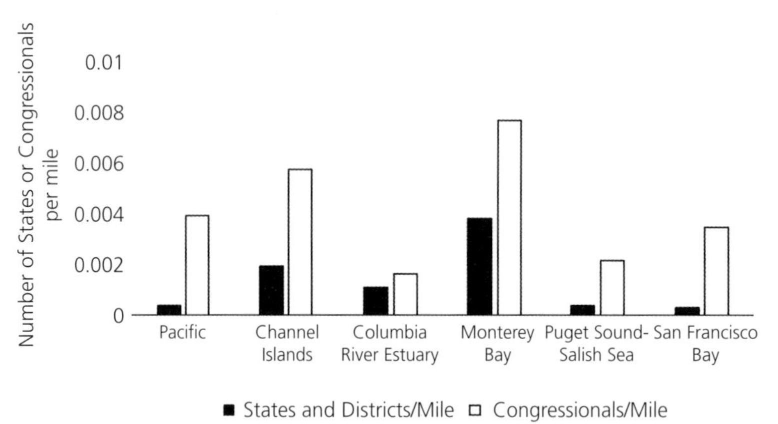

Figure 8.8 Number of states and U.S. Congressional representatives (top panel) and as standardized by miles of shoreline (bottom panel) for the Pacific region and its identified subregions.

of appointed council members from fishing or other sectors at of only seven (seventh-highest nationally).

The Pacific marine mammal SRG is responsible for advising federal agencies on the status of marine mammal stocks covered under the Marine Mammal Protection Act (MMPA) for the Pacific and Western Pacific regions. Its composition and membership have remained relatively steady over time (Fig. 8.9b), with individuals from the governmental and private institutional sectors making up the majority of its representatives, with periodic representation from academics and the tribal community.

8.4.2.2 Fishery and Systematic Context

In the Pacific region, FMPs (n = 4) are managed by NOAA Fisheries and the PFMC for ocean salmon, Pacific groundfish, coastal pelagic species, and HMS (Table 8.2). In total, these FMPs have been subject to 65 amendments, which is third-lowest for the nation. In addition, U.S., and Canadian Pacific halibut resources are managed through the IPHC. Similarly, the PWT also accounts for shared Pacific hake resources between the U.S. and Canada, while the PSC manages transboundary salmon stocks (see Chapter 11 on RFMOs). Tribal management, particularly for salmon, was considerably affected by the 1974 Boldt decision, where select Native American tribes currently retain title to 50% of Puget Sound and western Washington salmon resources (Knutson 1987, Brown 1994). A FEP has existed for the Pacific Coast since 2013, and has been subject to one modification. This FEP contains comprehensive information regarding Pacific coastal and offshore fisheries, the habitat and biology of its managed species, and the socioeconomic

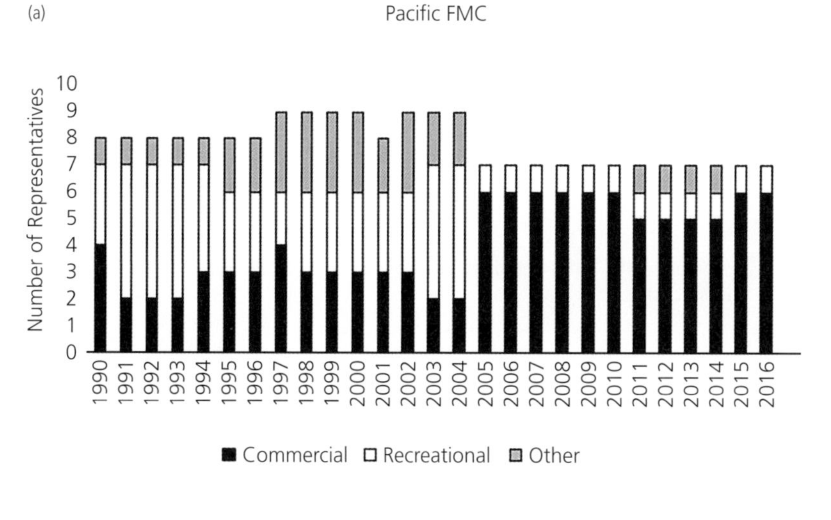

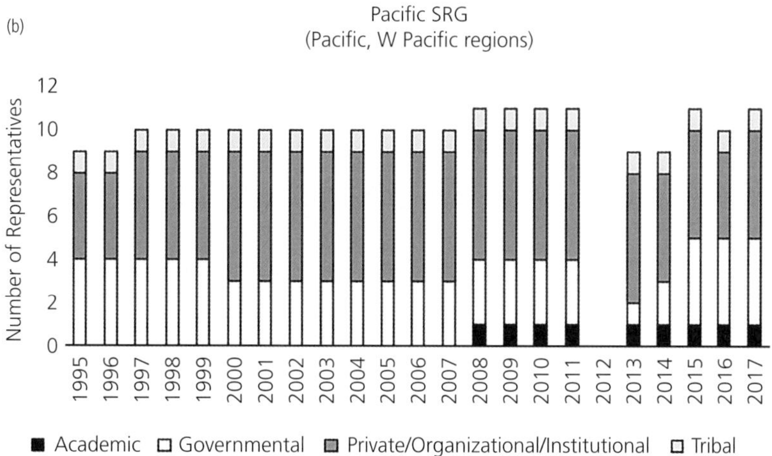

Figure 8.9 (a) Composition of the Pacific Fishery Management Council (FMC) and the (b) Pacific marine mammal scientific review group (SRG) over time.

Data derived from NOAA National Marine Fisheries Service Council Reports to Congress and scientific review group reports. SRG composition was unreported for 2012.

and ecological impacts of conservation and fisheries management policies.

Thirteen coastal national parks, monuments/memorials, or historic sites, one NOAA Habitat Focus Area (Russian River), five National Estuarine Research Reserves (NERRs), and five NMS are found throughout the Pacific region (Table 8.3), with the majority occurring in the San Francisco Bay subregion. Thirty-four expansive inshore and offshore regions are also identified as Habitat Areas of Particular Concern (HAPCs) in the Pacific. These include five generic alongshore habitat classifications (estuaries, kelp canopy, rocky reefs, seagrass, Washington State waters 0–3 nautical miles) and 29 distinct offshore banks, canyons, and seamounts throughout the Pacific EEZ, mostly concentrated off California and the Channel Islands.

There are 17 named fishing closure zones in the Pacific, while as of 2016, fishing was permanently prohibited in 139 specific areas and restricted in 178 of them (Fig. 8.10). These latter two numbers are highest in the nation, while the number of closures is second-highest after New England. While numerically high, the area of closures (8.1 km^2) and prohibited areas (1522.7 km^2) is comparatively modest in the Pacific region. Locations with fishing restrictions make up the majority of protected areas (~361,800 km^2), with the Channel Islands being the subregion with the greatest amount of restricted area (~21,000 km^2). Similar

Table 8.2 List of major managed fishery species in the Pacific region, under state, federal, and international jurisdictions

Council/Agency	FMP/FEP	FMP Modifications	Major Species/ Species Group	Scientific Name(s) or Family Name(s)
Pacific Fishery Management Council **PFMC**	Ocean salmon (4 species)	19 amendments	Chinook salmon Coho salmon Pink salmon Sockeye salmon	*Oncorhynchus tshawytscha* *Oncorhynchus kisutch* *Oncorhynchus gorbuscha* *Oncorhynchus nerka*
	Pacific groundfish (87 species)	28 amendments	Elasmobranchs	Big skate *Raja binoculata*, Leopard shark *Triakis semifasciata*, Longnose skate *Raja rhina*, Pacific Spiny dogfish *Squalus suckley*
			Flatfish	Species from Families Pleuronectidae and Paralichthyidae
			Rockfish	Species from Families Sebastidae and/or Scorpaenidae
			Pacific whiting Pacific cod Sablefish Lingcod	*Merluccius productus* *Gadus macrocephalus* *Anoplopoma fimbria* *Ophiodon elongatus*
			Other roundfish	Cabezon *Scorpaenichthys marmoratus*, Kelp greenling *Hexagrammos decagrammus*
	Coastal pelagic species (13 species)	15 amendments	Pacific sardine Pacific mackerel Jack mackerel Market squid Northern anchovy	*Sardinops sagax* *Scomber japonicus* *Trachurus symmetricus* *Doryteuthis opalescens* *Engraulis mordax*
			Krill or euphausiids	*Euphausia pacifica*, *Thysanoessa spinifera* and *Euphausia* spp., *Nematocelis* spp., *Nyctiphanes* spp., *Thysanoessa* spp.
	Highly migratory species (13 species)	3 amendments	North Pacific albacore	*Thunnus alalunga*
			Other tunas	Yellowfin tuna *Thunnus albacares*, Bigeye tuna *Thunnus obesus*, Skipjack tuna *Katsuwonus pelamis*, Pacific bluefin tuna *Thunnus orientalis*
			Billfish/swordfish	Striped marlin *Tetrapturus audax*, Swordfish *Xiphias gladius*
			Sharks	Common thresher shark *Alopias vulpinus*, Pelagic thresher shark *Alopias pelagicus*, Bigeye thresher shark *Alopias superciliosus*, Shortfin mako shark *Isurus oxyrinchus*, Blue shark *Prionace glauca*
			Dolphinfish	*Coryphaena hippurus*
International Pacific Halibut Commission **IPHC**—includes North Pacific and Pacific regions	Pacific halibut (1 species)	92 Annual regulations	Pacific halibut	*Hippoglossus stenolepis*

Table 8.3 List of major bays and islands, NOAA habitat focus areas (HFAs), NOAA National Estuarine Research Reserves (NERRs), NOAA National Marine Monuments and Sanctuaries (NMS), Coastal National Parks, National Seashores, and number of habitats of particular concern (HAPCs) in the Pacific region and subregions

Region	Subregion	Major Bays	Major Islands	HFAs	NERRs	Monuments and NMS	Coastal National Parks	National Sea/ Lakeshores	HAPCs
Pacific		Arcata Bay, Baker Bay, Columbia River, Monterey Bay, Nestucca Bay, Netarts Bay, North Bay, Puget Sound, San Francisco Bay, Siletz Bay, Tillamook Bay, Willapa Bay, Yaquina Bay	Channel Islands, San Juan Islands	Russian River, California HFA	Padilla Bay NERR, South Slough NERR, San Francisco NERR, Elkhorn Slough NERR, Tijuana River NERR	Cordell Bank NMS, Channel Islands NMS, Gulf of the Farallones NMS, Monterey Bay NMS, Olympic Coast NMS	Ebey's Landing National Historical Reserve, Olympic National Park, Lewis and Clark National Historical Park, Redwood National Park, Golden Gate National Recreation Area, Fort Point National Historic Site, Presidio of San Francisco, Alcatraz Island, Rosie the Riveter WWII Home Front National Historical Park, Port Chicago Naval Magazine National Memorial, Channel Islands National Park, Santa Monica Mountains National Recreation Area, Cabrillo National Monument	Point Reyes National Seashore	34
Pacific	**Channel Islands**		Anacapa Island, San Clemente Island, San Miguel Island, San Nicolas Island, Santa Barbara Island, Santa Catalina Island, Santa Cruz Island, Santa Rosa Island, Sutil Island		Tijuana River NERR	Channel Islands NMS	Cabrillo National Monument, Channel Islands National Park		13

Continued

Table 8.3 Continued

Region	Subregion	Major Bays	Major Islands	HFAs	NERRs	Monuments and NMS	Coastal National Parks	National Sea/ Lakeshores	HAPCs
Pacific	**Columbia River Estuary**		Brush Island, Crims Island, Horseshoe Island, Goose Island, Government Island, Grassy Island, Hayden Island, Marsh Island, Sand Island, Tronson Island, Wallace Island, Welch Island, Woody Island				Fort Vancouver National Historical Site, Lewis and Clark National Historical Park, McLoughlin House (Fort Vancouver NHS)		1
	Monterey Bay				Elkhorn Slough NERR	Monterey Bay NMS			3
	Puget Sound—Salish Sea		Anderson Island, Bainbridge Island, Blake Island, Fidalgo Island, Harstine Island, Marrowstone Island, Maury Island McNeil Island, San Juan Islands, Vashon Island, Whidbey Island		Padilla Bay NERR		Klondike Gold Rush National Historical Park—Seattle Unit, San Juan Island National Historical Park		3
	San Francisco Bay	Grizzly Bay, Gulf of the Farallones, San Pablo Bay, Suisun Bay	Alcatraz, Angel Island, Bair Island, Bethel Island, Brooks Island, Browns Island, Farallon Islands, Ryer Island, Winter Island		San Francisco NERR	Cordell Bank NMS, Gulf of the Farallones NMS	Alcatraz Island, Golden Gate National Recreational Area, Muir Woods National Monument, Port Chicago Naval Magazine National Monument, Presidio of San Francisco, Rosie the Riveter World War II Homefront National Historical Park	Point Reyes National Seashore	2

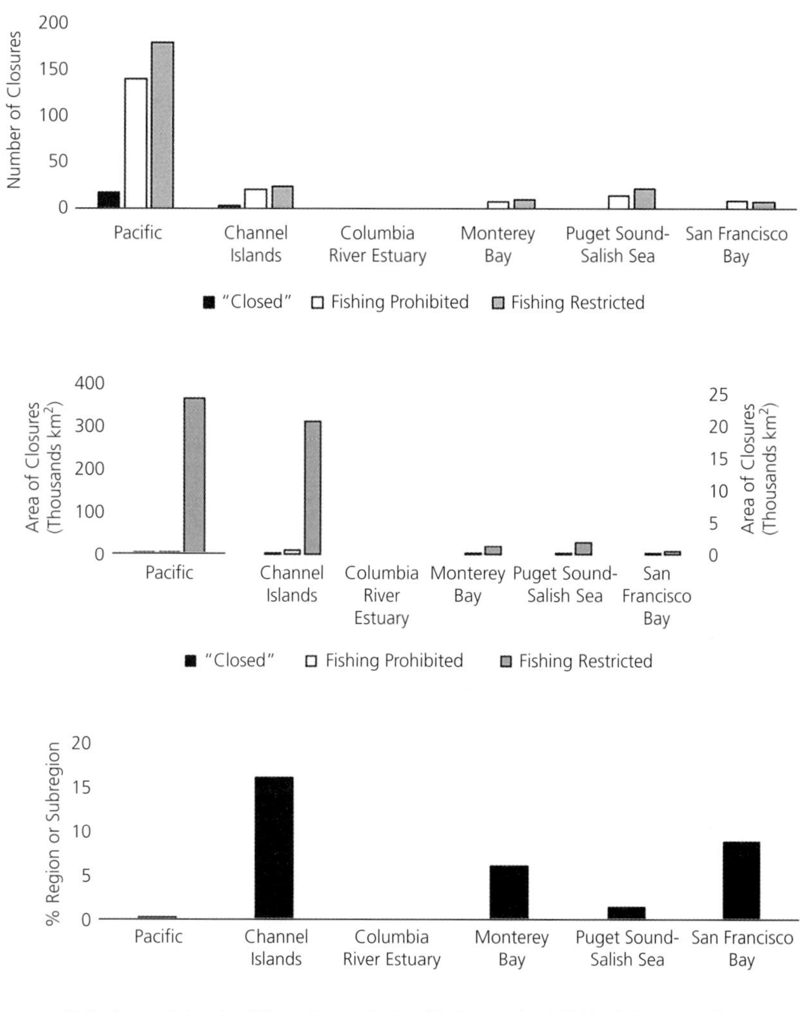

Figure 8.10 Number (top panel) and area (middle panel; km^2) of named fishing closures, and prohibited or restricted fishing areas of the Pacific region and identified subregions, including percent region or subregion (bottom panel) where commercial and/or recreational fishing is permanently prohibited. Values for Pacific subregions (middle panel) plotted on the secondary (right) axis.

Data derived from NOAA Marine Protected Areas inventory.

numbers of protected and restricted sites are found in the Channel Islands (20 protected, 24 restricted), Puget Sound-Salish Sea (14 protected, 21 restricted), Monterey Bay (7 protected, 10 restricted) and San Francisco Bay (8 protected, 7 restricted), with most of these sites likewise occurring over relatively limited areas. Approximately 0.2% of the Pacific EEZ contains permanent fishing prohibitions, with the largest concentration of prohibited areas occurring in the Channel Islands (528.7 km^2). In this subregion, 15.9% of its area is permanently protected from fishing with the largest fishing protection site (106.4 km^2) occurring at Richardson Rock (San Miguel Island). Fishing prohibitions occur throughout 6% of Monterey Bay and 8.7% of San Francisco Bay, with Año Nuevo State Marine Conservation Area (Monterey Bay; 28.8 km^2) and San Pablo Bay National Wildlife Refuge (San Francisco Bay; 91 km^2) being the largest prohibited sites in these subregions. Only 1.3% of Puget Sound has permanent fishing prohibitions, while no closures or prohibited and restricted sites occur in the Columbia River Estuary.

8.4.2.3 Organizational Context

Relative to the total commercial value of its fisheries, the PFMC budget is fifth-highest in the country when compared to other regions (Fig. 8.11), and makes up a small proportion of the value of its managed resources. These relative values decreased slightly in the early 2010s and have increased in more recent years, with current values similar to those in 2009–10. Cumulative National Environmental Policy Act Environmental Impact Statement (NEPA-EIS) actions from 1987–2016 in the Pacific are third-highest nationally, with 234 observed over the past 30 years (Fig. 8.12) and the majority occurring in the Columbia River Estuary (n = 46), San Francisco Bay (n = 46), and Puget Sound-Salish Sea (n = 36) subregions. Additionally, 13 fisheries-related lawsuits have occurred in the Pacific region since 2010, making it the third-most litigious among regions.

8.4.2.4 Status of Living Marine Resources (Targeted and Protected Species)

Nationally, the Pacific contains the sixth-highest number of managed species. Of the 210 federally managed species in the region, 118 are targeted fishery species covered under FMPs, while 11 species (Pacific halibut—prohibited in the west coast trawling fishery and only accessible during times and conditions specified by the IPHC, five species of Pacific salmon, Pacific sardine, Dungeness crab—occasionally prohibited, and three species of sharks) are prohibited from harvest. Fifty species of marine mammals, sea turtles, fishes, and invertebrates are also federally protected under the ESA or MMPA (Fig. 8.13, Tables 8.2, 8.4–5).

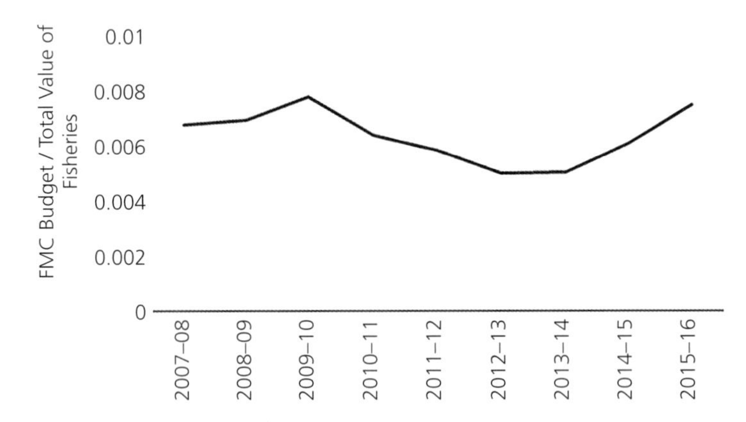

Figure 8.11 Total annual budget of the Pacific Fishery Management Council (FMC) as compared to the total value of its marine fisheries. All values in USD.

Data from NOAA National Marine Fisheries Service Office of Sustainable Fisheries and NOAA National Marine Fisheries commercial landings database.

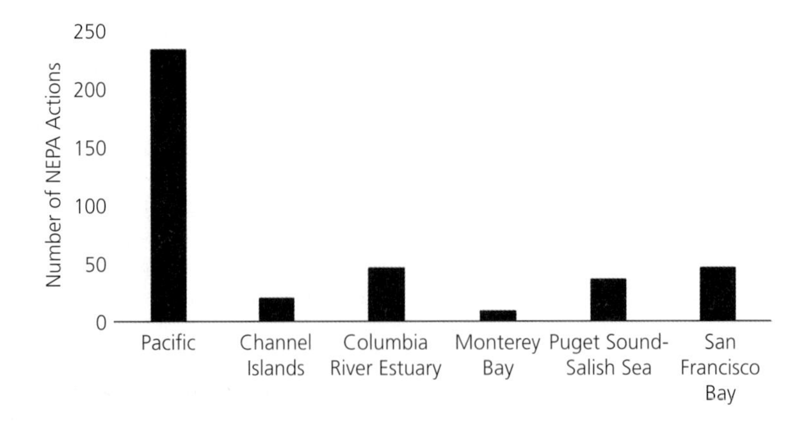

Figure 8.12 Number of National Environmental Policy Act Environmental Impact Statement (NEPA-EIS) actions from 1987 to 2016 for the Pacific region and identified subregions.

Data derived from U.S. Environmental Protection Agency EIS database.

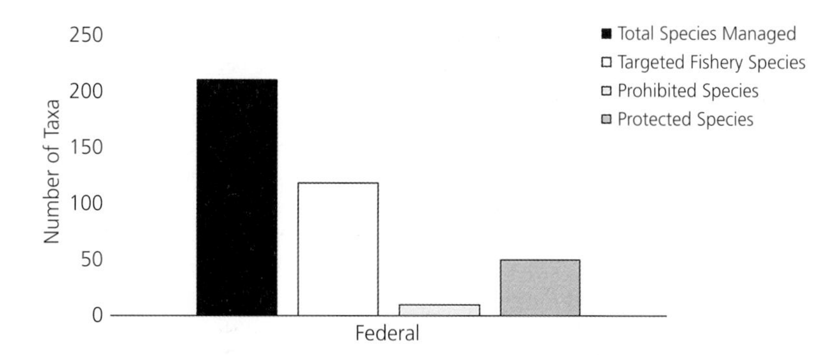

Figure 8.13 Number of federally managed taxa (species or families) for the Pacific region.

Table 8.4 List of prohibited fishery species in the Pacific region, including federal and state jurisdictions

Council/Authority	Region	Prohibited Species	Scientific Name
International Pacific Halibut Commission **IPHC**; Pacific Fishery Management Council **PFMC**; NOAA Fisheries West Coast Regional Office **WCRO**	Pacific (11 species)	Pacific halibut	*Hippoglossus stenolepis*
		Pacific salmon (5 species)	*Oncorhynchus* spp.
		Dungeness crab	*Metacarcinus magister*
		Pacific sardine	*Sardinops sagax caerulea*
		Sharks (3 species)	Great white shark *Carcharodon carcharias*, Basking shark *Cetorhinus maximus*, Mega mouth shark *Megachasma pelagio*

As of mid-2017, 144 fishery stocks, including stock complexes, (highest nationally) were federally managed in the Pacific with 43 listed as a NOAA Fish Stock Sustainability Index (FSSI) stock (Fig. 8.14; NMFS 2017a, 2020b).[1] Among all these stocks at that time, three were experiencing overfishing (**Coho salmon, Puget Sound: Hood Canal stock**—*Oncorhynchus kisutch*; **Pacific bluefin tuna**—*Thunnus orientalus*; and **Swordfish**—*Xiphias gladius*) and three were overfished (**Pacific bluefin tuna**—*Thunnus orientalus*; **Pacific ocean perch**—*Sebastes alutus*; and **Yelloweye rockfish**—*Sebastes ruberrimus*).[2] There were 32 stocks that remained unclassified as to whether they were experiencing overfishing (22.2%; third-highest nationally) and 22 had unknown overfished status (15.3%; third-lowest nationally), making the Pacific mid-range nationally in terms of the proportion of fishery stocks with unknown status.[3]

There are 45 independent marine mammal stocks, and 54 ESA-listed distinct population segments

[1] As of June 2020, 151 stocks or stock complexes (highest nationally) are federally managed in the U.S. Pacific, with 43 still listed as FSSI stocks. Between 2017 and 2020, three PFMC-managed stocks were added (Big skate—*Beringraja binoculata*; Cabezon—*Scorpaenichthys marmoratus*, Oregon; Black rockfish—*Sebastes melanops*, Washington) and three stocks were split into regional components. China Rockfish—*Sebastes nebulosus* is now managed as three separate stocks (China Rockfish–Central Pacific Coast, Northern Pacific, and Southern Pacific Coast stocks). Blue Rockfish—*Sebastes mystinus* is now managed as two stock complexes (California blue and Deacon rockfish complex; Oregon blue and Deacon rockfish complex). Two Lingcod—*Ophiodon elongatus* stocks (Lingcod, Northern Pacific Coast and Southern Pacific Coast) are now managed by the PFMC.

[2] As of June 2020, four Pacific stocks are experiencing overfishing (Pacific bluefin tuna; Striped marlin—*Kajikia audax*, western and central North Pacific; Swordfish; Yellowfin tuna—*Thunnus albacares*, eastern Pacific), and eight are overfished (Pacific sardine—*Sardinops sagax caerulea*; Chinook salmon—*Oncorhynchus tshawytscha*, California Central Valley: Sacramento River fall; Chinook salmon, Northern California Coast: Klamath River fall; Coho salmon—*O. kisutch*, Puget Sound: Snohomish; Coho salmon, Washington coast: Queets; Coho salmon, Washington Coast: Strait of Juan de Fuca; Pacific bluefin tuna; Striped marlin, western and central North Pacific).

[3] In June 2020, 33 stocks (21.9%) remained unclassified regarding their overfishing status and 21 stocks (13.9%) had unknown overfished status.

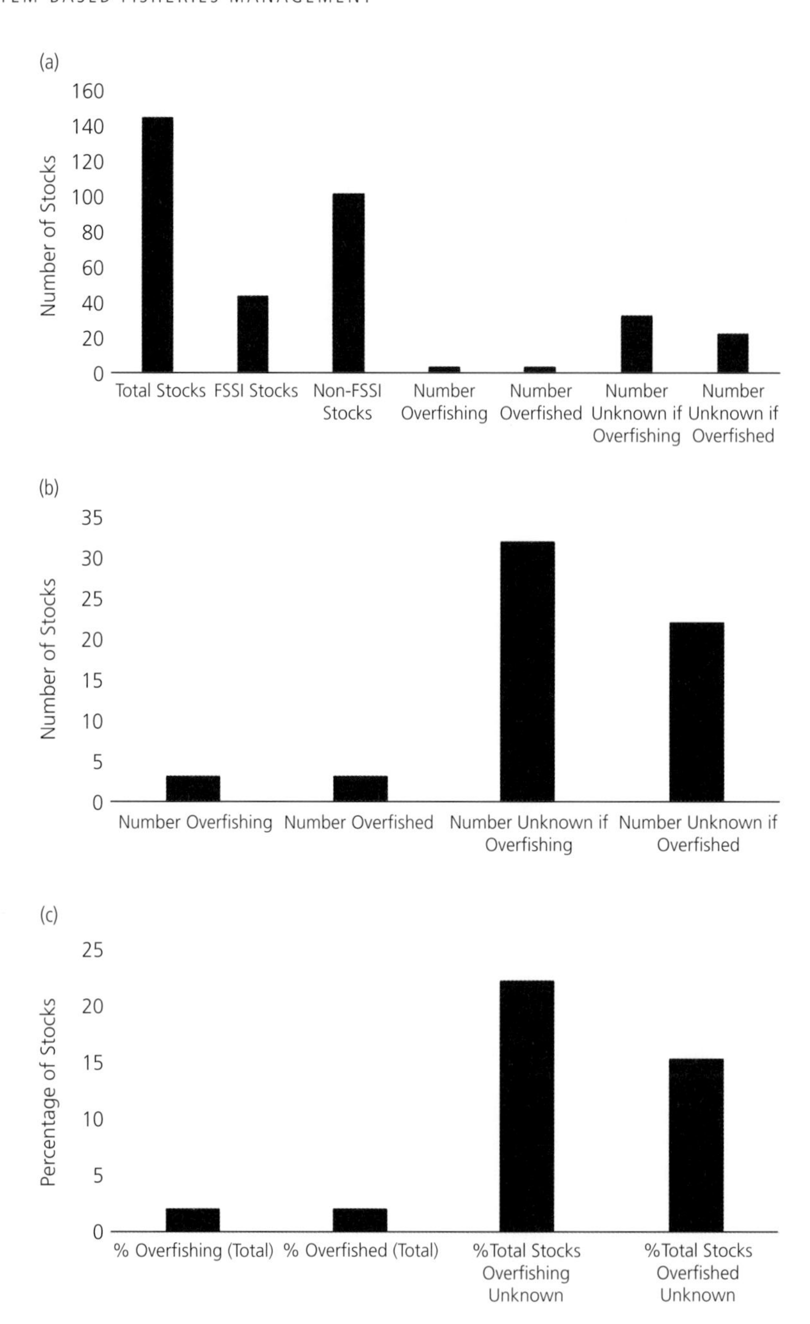

Figure 8.14 For the Pacific region as of June 2017 (a) Total number of managed Fish Stock Sustainability Index (FSSI) stocks and non-FSSI stocks, and breakdown of stocks experiencing overfishing, classified as overfished, and of unknown status. (b) Number of stocks experiencing overfishing, classified as overfished, and of unknown status. (c) Percent of stocks experiencing overfishing, classified as overfished, and of unknown status.

Note: stocks may refer to a species, family, or complex.

Table 8.5 List of protected species under the Endangered Species Act (ESA) or Marine Mammal Protection Act (MMPA) in the Pacific region

Region	Managing Agency	Protected Species Group	Common Name	Scientific Name
Pacific—California Current (50 species)	Pacific Fishery Management Council **PFMC**; West Coast Regional Office **WCRO**	Baleen whales	Blue whale	*Balaenoptera musculus*
			Fin whale	*Balaenoptera physalus*
			Gray whale	*Eschrichtius robustus*
			Humpback whale	*Megaptera novaeangliae*
			Northern minke whale	*Balaenoptera acutorostrata*
			North Pacific right whale	*Eubalaena japonica*
			Sei whale	*Balaenoptera borealis*
		Toothed whales	Baird's beaked whale	*Berardius bairdii*
			Cuvier's beaked whale	*Ziphius cavirostris*
			Dwarf sperm whale	*Kogia sima*
			Hubb's beaked whale	*Mesoplodon carlhubbsi*
			Killer whale (Orca)	*Orcinus orca*
			Pygmy sperm whale	*Kogia breviceps*
			Short-finned pilot whale	*Globicephala macrorhynchus*
			Sperm whale	*Physeter macrocephalus*
			Stejneger's beaked whale	*Mesoplodon stejnegeri*
		Dolphins/Porpoises	Common bottlenose dolphin	*Tursiops truncatus*
			Dall's porpoise	*Phocoenoides dalli*
			Harbor porpoise	*Phocoena phocoena*
			Long-beaked common dolphin	*Delphinus capensis*
			Northern right whale dolphin	*Lissodelphis borealis*
			Pacific white-sided dolphin	*Lagenorhynchus obliquidens*
			Risso's dolphin	*Grampus griseus*
			Short beaked common dolphin	*Delphinus delphis*
			Striped dolphin	*Stenella coeruleoalba*
		Pinnipeds	Harbor seal	*Phoca vitulina*
			Northern elephant seal	*Mirounga ngustirostris*
			Northern fur seal	*Callorhinus ursinus*
			California sea lion	*Zalophus californianus*
			Steller sea lion	*Eumetopias jubatus*
			Guadalupe fur seal	*Arctocephalus townsendi*
		Sea turtles	Leatherback sea turtle	*Dermochelys coriacea*
			Green sea turtle	*Chelonia mydas*
			Olive Ridley sea turtle	*Lepidochelys olivacea*
			Loggerhead sea turtle	*Caretta caretta*
		Fishes	Puget sound yelloweye Rockfish	*Sebastes ruberrimus*
			Puget sound bocaccio rockfish	*Sebastes paucispinis*
			Eulachon	*Thaleichthys pacificus*
			Green sturgeon	*Acipenser medirostris*
			Gulf grouper	*Mycteroperca jordani*
			Pacific salmon and steelhead	*Oncorhynchus* spp. (n = 6)
			Scalloped hammerhead shark	*Sphyrna lewini*
		Invertebrates	White abalone	*Haliotis sorenseni*
			Black abalone	*Haliotis cracherodii*
	U.S. Fish and Wildlife Service **FWS**	Other species	Sea otter	*Enhydra lutris*

(highest nationally), of which ~41% are endangered and three have unknown status (Fig. 8.15). Nationally, mid-range numbers of marine mammal stocks are found in the Pacific (14.1% of total USA marine mammal stocks), with 28.9% listed as strategic marine mammal stocks. There are 11.1% of Pacific marine mammal stocks that have unknown population size, which is nationally lowest among regions.

8.4.3 Environmental Forcing and Major Features

8.4.3.1 Oceanographic and Climatological Context

Circulation patterns throughout the Pacific EEZ are driven primarily by the California Current (Checkley & Barth 2009; Fig. 8.1). The California Current derives from the North Pacific Current as it splits near the North American landmass. It forms the lower southward alongshore current of the U.S. Pacific coast, contributing to the North Pacific Gyre. As the California Current continues to move southward, it forms the eastern portion of the North Equatorial current and the southern portion of the North Pacific Gyre, within which much of the Eastern Pacific region is found (Cummins & Freeland 2007, Sydeman et al. 2011). Additionally, southward propagule transport occurs along inshore portions of the Pacific Coast via the Coastal Jet (Checkley & Barth 2009). Within the Channel Islands, the California Current also meanders to form the Southern California Eddy, which enhances connectivity along regions of the southern California coast. This eddy serves as the origin for the subsurface California Undercurrent through which northward transport, local propagule retention, and connectivity

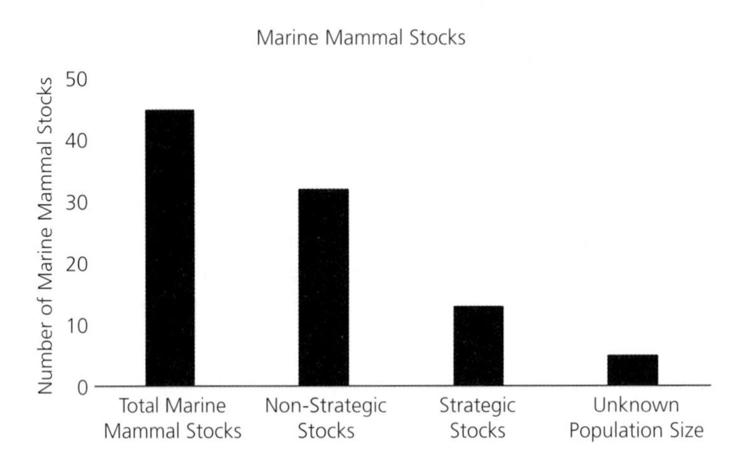

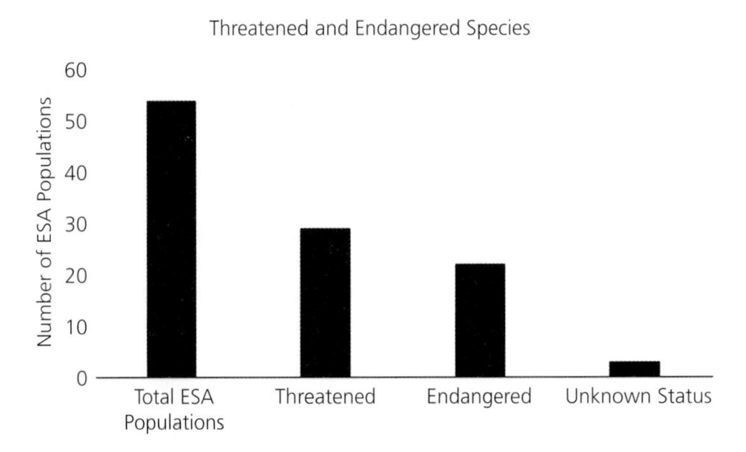

Figure 8.15 Number and status of federally protected species (marine mammal stocks, top panel; distinct population segments of species listed under the Endangered Species Act (ESA), bottom panel) occurring in the Pacific region.

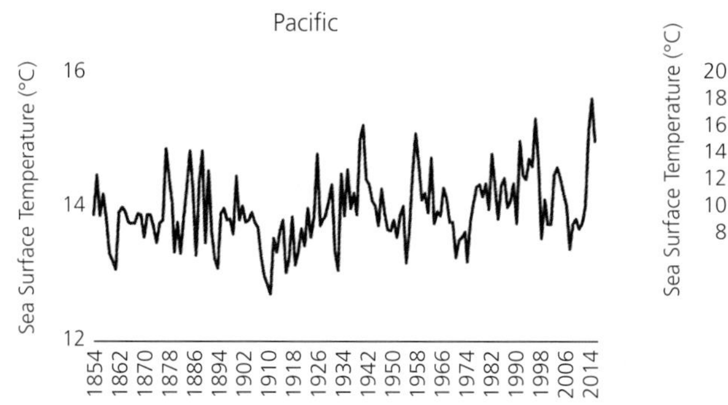

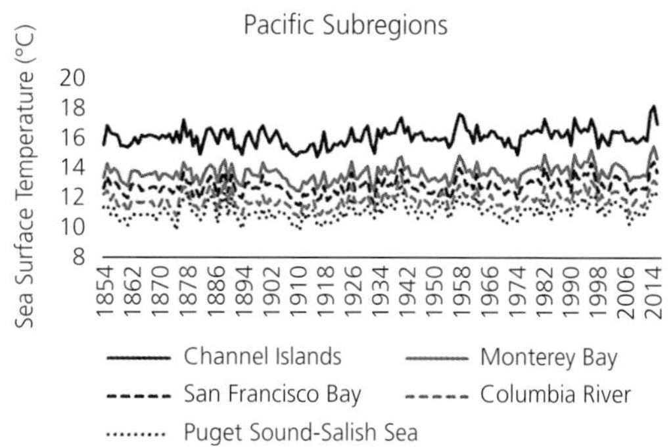

Figure 8.16 Average sea surface temperature (°C) over time (years 1854–2016) for the Pacific region and identified subregions.
Data derived from the NOAA Extended Reconstructed Sea Surface Temperature dataset (https://www.ncdc.noaa.gov/).

occur. As northern winds regularly move surface waters offshore along its relatively narrow continental shelf, the Pacific coast region is also subject to regular upwelling intensities of cold, nutrient-rich water.

Clear interannual and multidecadal patterns in average SST have been observed for the Pacific region (Fig. 8.16), with temperatures averaging 13.9°C ± 0.04 (SE) over time (1854–2016), and having increased by 1.34°C since the mid-twentieth century (Link & Marshak 2019). Similar trends have been observed within key subregions, with average temperatures over the 162-year period highest in the lower latitude Channel Islands at 16.1°C ± 0.04 (SE) and Monterey Bay (13.6°C ± 0.04 SE). SST is lowest in northernmost Puget Sound-Salish Sea at 11.1°C ± 0.04 (SE), while temperatures in the Columbia River Estuary (11.8°C ± 0.04 SE) and San Francisco Bay (12.8°C ± 0.04 SE) are only slightly higher. Among subregions, temperatures have all increased >1°C since the mid-twentieth century, most notably in Puget Sound-Salish Sea (1.59°C), the Columbia River Estuary (1.56°C), and San Francisco Bay (1.44°C). Increases in temperature for Monterey Bay (1.42°C) and the Channel Islands (1.34°C) have been slightly less pronounced. Coincident with these temperature observations are Pacific basin-scale climate oscillations (Fig. 8.17), including ENSO, Multivatiate El Niño Index (MEI), Northern Oscillation Index (NOI), NPGO, and PDO, which exhibit cycles that can influence the environmental conditions and ecologies of several U.S. regions, including the Pacific.

8.4.3.2 Notable Physical Features and Destabilizing Events and Phenomena

With five prominently identified subregions, the Pacific contains 14 major bays and estuaries (Table 8.3) off the coasts of California, Oregon, and Washington, and three major island groupings (Channel Islands, Farallon Islands, and San Juan Islands) concentrated throughout inshore and offshore areas of California and Washington. Proportionally, the Pacific makes up ~6% of the U.S. EEZ (~811,000 km^2), and 10.5% of the national shoreline (~17,620 km), while containing an average depth of ~2.9 km and maximum depths of ~4.9 km (Fig. 8.18), making it the shallowest of regions throughout the U.S. Pacific basin EEZ. Cumulatively, the five subregions make up only ~1.5% of the Pacific EEZ, with Puget Sound-Salish Sea (~4900 km^2) and the Channel Islands (~3300 km^2) having the greatest areas and San Francisco Bay (~5600 km) and Puget Sound-Salish Sea (~4500 km) having the longest shorelines. Among subregions, the deepest average depths occur off the Channel Islands (181.9 m), while maximum depths are greatest (1.4 km) in the canyons off Monterey Bay. The Columbia River Estuary is the smallest (616.7 km^2) and shallowest (average depth: 0.5 m) among subregions, although it contains ~3000 km of shoreline. Additionally, this region is also known for its narrow continental shelf, significant freshwater input from coastal mountain ranges and major rivers, and many coastal estuaries.

In the Pacific region, only five hurricanes have been reported over time off southern California, with three touching the Channel Islands during the 1970s. Additional hurricanes were observed in the Pacific during 1965 (Hurricane Claudia) and more recently in 1997 (Hurricane Ignacio). Major winter storms have also occurred within this region (Warner et al. 2012), while portions are also vulnerable to tsunamis (Barrick 1979, Dengler et al. 2008). The Pacific is also subject to natural hypoxic conditions with depth (Connolly et al. 2010) that are asso-

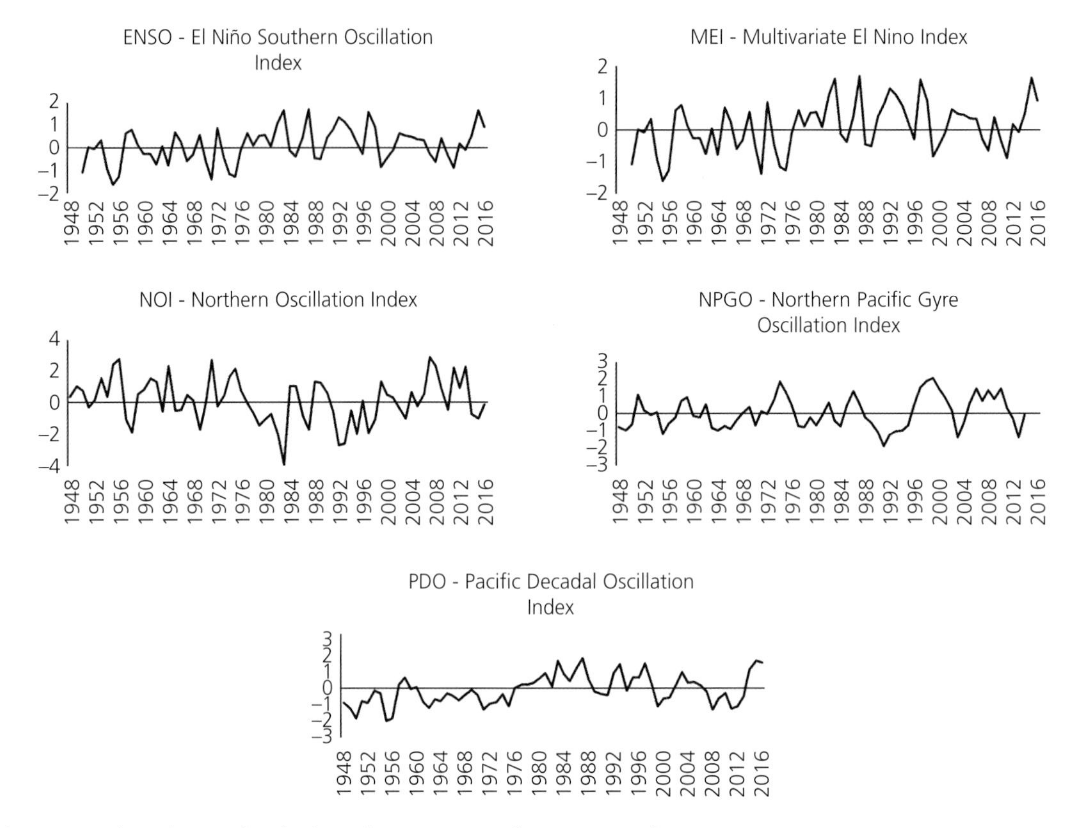

Figure 8.17 Climate forcing indices for the Pacific region over time (years 1948–2016).

Data derived from NOAA Earth System Research Laboratory data framework (https://www.esrl.noaa.gov/).

ciated with dynamics in nutrient-rich water upwelling events and climate forcing intensity (Kahru et al. 2009).

8.4.4 Major Pressures and Exogenous Factors

8.4.4.1 Other Ocean Use Context

Ocean uses in the Pacific are mostly associated with the tourism/recreation and marine transportation sectors (Fig. 8.19). Among sectors, trends for Pacific Ocean establishments and employments have shown significant growth over the past several decades, with consistent dominance of tourism establishments and employments comprising 82–84% of total Pacific Ocean establishments and 62–74% of total Pacific Ocean employments. Increases in total ocean establishments and employments were most prominent during the 2010s, with highest values occurring in 2016 at ~33,000 establishments and ~733,000 employments (nationally second-highest). The GDP from tourism is highest among sectors, contributing 35–47% toward total Pacific Ocean GDP. Over the past few decades, ocean GDP has more steadily increased, with most recent estimated values at ~$62.1 billion. Overall, the Pacific contributes 19% of the U.S. total ocean GDP, which is second-highest in the nation. Establishments and employments in the Pacific marine transportation sector contribute 7–8% toward its total ocean establishments and comprise 17–22% of its total ocean employments. The Pacific contributes 24% of U.S. tourism establishments and ~25–26% of national tourism employments and GDP, all of which are second-highest in the nation. It is highest nationally for marine transportation establishments and employments (26.7% of U.S. marine transportation establishments and 32.8% of employments), as well as marine transportation and ship and boatbuilding GDP (35.9% of U.S. marine transportation GDP and 27.4% of U.S. ship and boatbuilding GDP). In terms of its marine establishments, the Pacific is also the second-highest

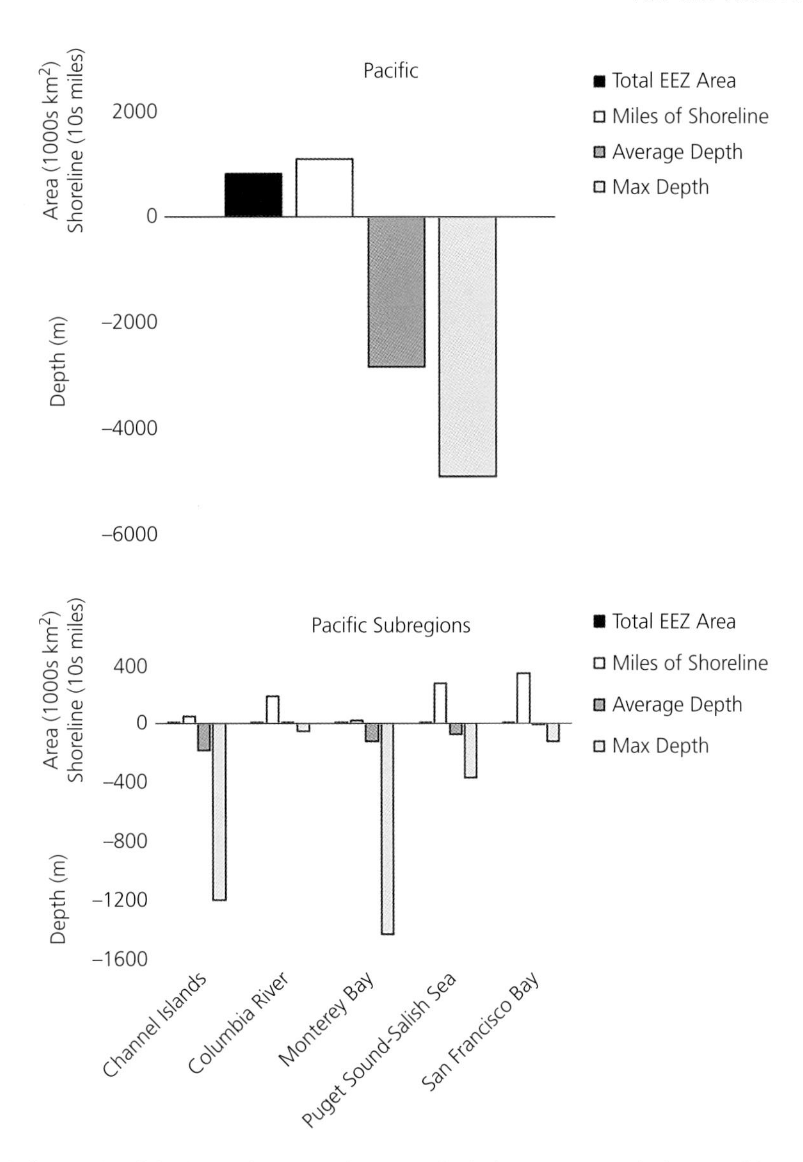

Figure 8.18 Total shelf area, miles of shoreline, and average and maximum depth of marine region and subregions of the Pacific.

nationally for its offshore mineral extraction (including sand/gravel mining and oil and natural gas production) and ship and boat building sectors (11.6% of U.S. offshore mineral extraction establishments and 19.9% of U.S. ship and boatbuilding establishments). It is also second-highest in the nation for marine construction employments and GDP (21.7% of U.S. marine construction employments and 26.3% of GDP). Overall, the Pacific is doing quite well in terms of its marine socioeconomic status, especially as associated with other ocean uses and its LMR-related economy.

The degree of tourism in the Pacific, as measured by number of dive shops, number of major ports and marinas, and number of cruise ship departure passengers is among the mid to high-range nationally (Fig. 8.20), with the second-highest numbers of dive shops and ports, and highest numbers of marinas (mostly concentrated near the Channel Islands and Puget Sound-Salish Sea). Dive shops are most abundant near the Channel Islands (n = 54), while six and seven ports are found in San Francisco Bay and Puget Sound-Salish Sea, respectively. Cruise destination pas-

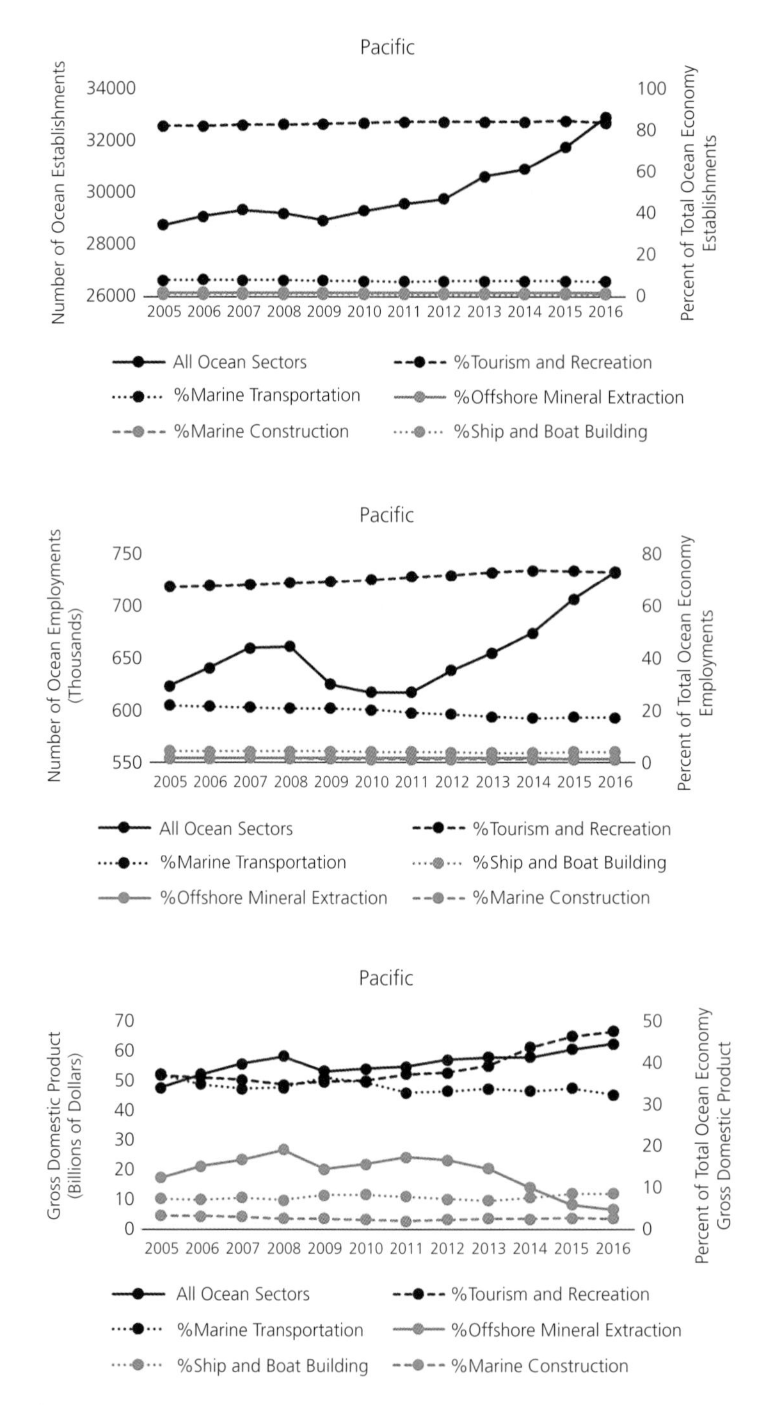

Figure 8.19 Number and percent contribution of multisector establishments, employments, and Gross Domestic Product (GDP) and their percent contribution to total multisector oceanic economy in the Pacific region (years 2005–2016).

Data derived from the National Ocean Economics Program.

sengers in the Pacific are fifth-highest in the nation with ~50,000 per year. However, departure passengers are second-highest nationally, with the majority leaving from populated ports near the Channel Islands (i.e., Los Angeles and San Diego) at rates of 800,000 to 1 million per year. Cruise traffic is also pronounced in the Puget Sound-Salish Sea subregion, and to a lesser extent in San Francisco Bay.

The Pacific is third-highest nationally in terms of offshore oil production and the number of oil rigs over time, although values have decreased since the late 1980s (Fig. 8.21). Peak values ($n = 7$) occurred in 1988, with numbers ranging from 0 to 3 observed in more recent years. Additionally, four proposed U.S. Bureau of Ocean Energy Management (BOEM) wind energy sites are found in this region, which is highest among Pacific Ocean regions and fourth-highest nationally.

8.4.5 Systems Ecology and Fisheries

8.4.5.1 Basal Ecosystem Production

Since 2002, mean surface chlorophyll values (Fig. 8.22) throughout the U.S. Pacific EEZ have generally been low (range: 0.19–1.1 mg m^{-3}). Among subregions, the highest values are found in the Columbia River Estuary (12.3 mg $m^{-3} \pm 0.5$, SE), Puget Sound-Salish Sea (8.3 mg $m^{-3} \pm 0.3$, SE), and Monterey Bay (8.2 mg $m^{-3} \pm 0.6$, SE). Values have slightly increased over time in Puget Sound-Salish Sea, and decreased during the 2000s and 2010s in Monterey Bay. Low, stable values have been observed in the Channel Islands (range: 0.80–3.3 mg m^{-3}) over time. Average annual productivity within the Pacific EEZ has remained relatively steady with some interannual fluctuation between ~115–132 g C m^{-2} $year^{-1}$ (seventh-highest nationally) over the same time period, with current values similar to those observed during the late 1990s. While the region is subject to intense seasonal productivities as associated with upwelling events, these are not necessarily captured when productivity values are annually averaged for the entire Pacific EEZ.

8.4.5.2 System Exploitation

Nationally, the fourth-highest numbers of reported taxa have been landed in the Pacific region, with values having increased in the late 1970s and having steadily increased up to 157 taxa (Fig. 8.23). Within subregions, trends follow similar patterns with highest numbers reported for the Columbia River Estuary (up to 157 taxa) and California subregions (up to 131 taxa) in 2016. Overall lower values have been observed in Puget Sound-Salish Sea, with up to 78 taxa reported in recent years. Combined surveyed nearshore fish and invertebrate biomass for the Pacific region has remained relatively constant over time (Fig. 8.24); values have averaged ~10.4 million ± 87,600 (SE) metric tons $year^{-1}$, with fluctuations of 8–16 million metric tons $year^{-1}$ occurring from the late 1980s through the early 2000s. More recent values have remained around an average of 9.7 million metric tons up to 2012.

Values for total landings/biomass decreased in the early 1980s, following which values remained steady through the early 2000s (Fig. 8.25). A return to similar higher values of the early 1980s occurred in 2004 and continued with peak exploitation during the late 2000s (up to 0.07 in 2008 and 2010). Since 2010, decreases in exploitation have been observed with current values at 0.05. Values for total biomass/productivity have decreased over time from a peak value in 1998 (0.015) to a minimum of 0.007 in 2008. In more recent years, slight rebounds have been observed with a value of 0.011 in 2012. Total landings/productivity values have remained relatively steady (0.0004–0.0006) over time, with increasing values observed in more recent years since 2009.

8.4.5.3 Targeted and Non-Targeted Resources

Total commercial and recreational landings in the Pacific are currently the third-highest in the nation (Fig. 8.26). Pacific landings experienced a major decline in the early 1950s, following which values remained around 450,000 metric tons through the 1950s–1960s. During the 1970s, increases were observed with peak values occurring in 1976 (667,000 metric tons) followed by another decline in the early 1980s to ~350,000 metric tons. In following years, values have regularly oscillated between ~360,000 and ~600,000 metric tons, although current values are currently lower at ~340,000 metric tons following another significant decline in reported landings in 2015. Increases in recreational catches since the mid-2000s have been observed, with values currently contributing ~6% of total catches at 20.8 thousand metric tons. However, in terms of numbers of fish, recreational landings have consistently decreased since the 1980s from peak values of ~37.4 million fishes to current values of ~7.9 million fishes.

Since 1981, high commercial and recreational catch contributions have occurred in southern California and in areas surrounding the Channel Islands.

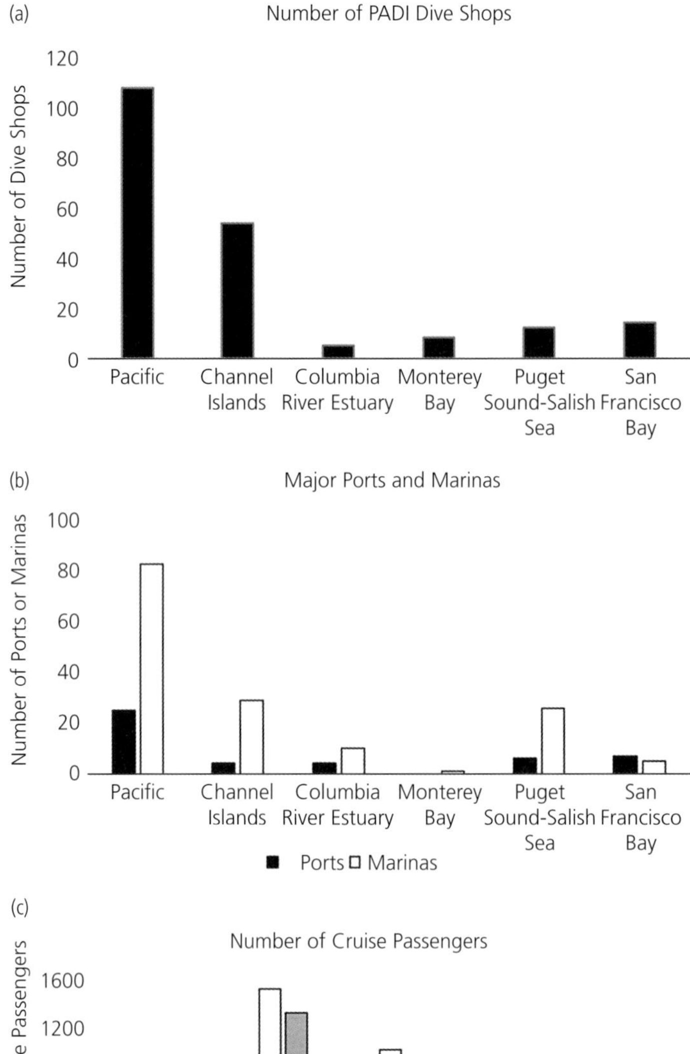

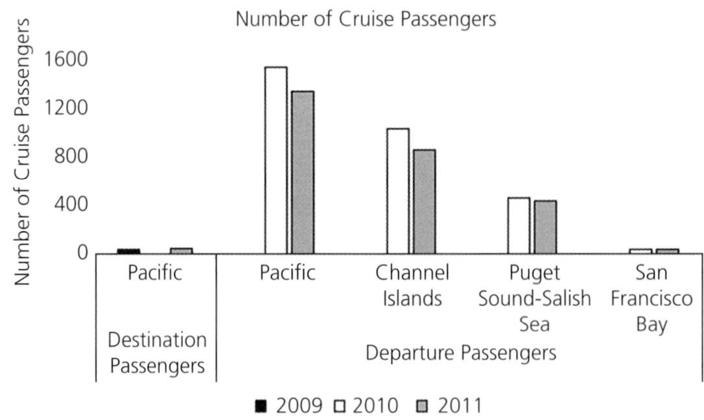

Figure 8.20 Tourism indicators for the Pacific region and identified subregions: (a) Number of Professional Association of Diving Instructors (PADI) dive shops. Data derived from http://apps.padi.com/scuba-diving/dive-shop-locator/. (b) Number of major ports and marinas. (c) Number of cruise destination (years 2009, 2011) and cruise departure port passengers (years 2010–2011); values in thousands.

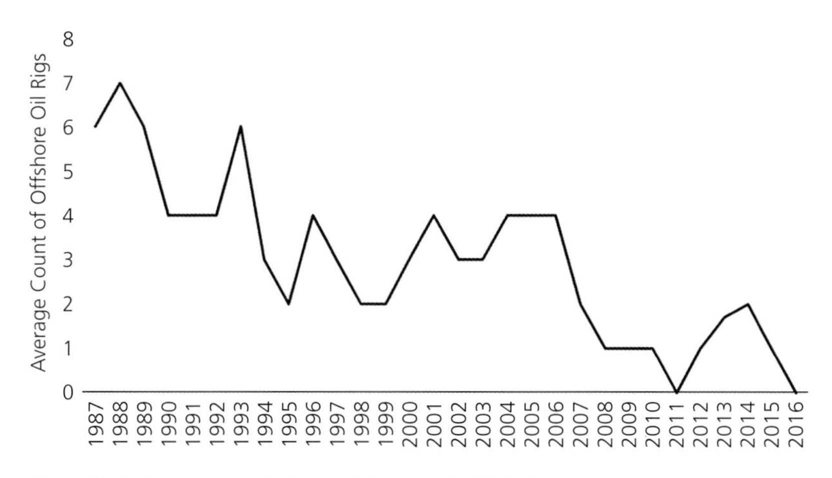

Figure 8.21 Average count of offshore oil rigs for the Pacific Region.

Data derived from Baker Hughes Rotary Rig Counts.

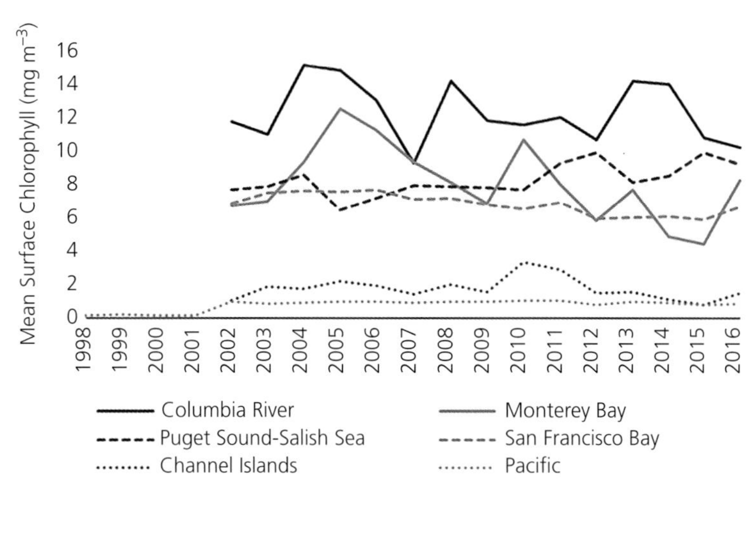

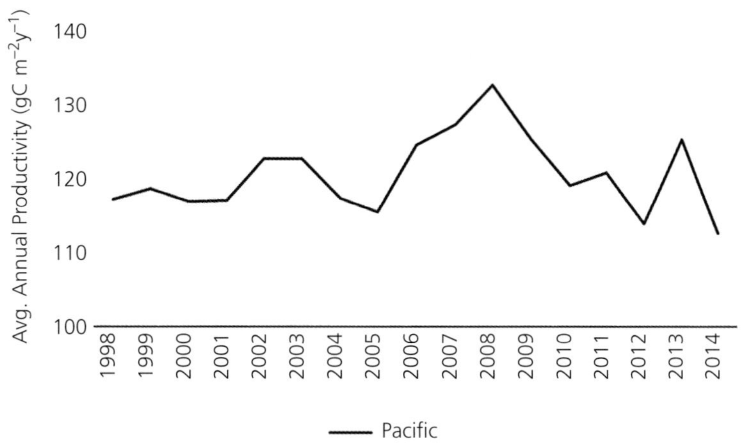

Figure 8.22 Mean surface chlorophyll (mg m^{-3}; top panel) for the Pacific region and identified subregions and average annual productivity (grams carbon m^{-2} year^{-1}; bottom panel) for the Pacific (California Current) region over time.

Data derived from NASA Ocean Color Web (https://oceancolor.gsfc.nasa.gov/) and productivity calculated using the Vertically Generalized Production Model (VGPM).

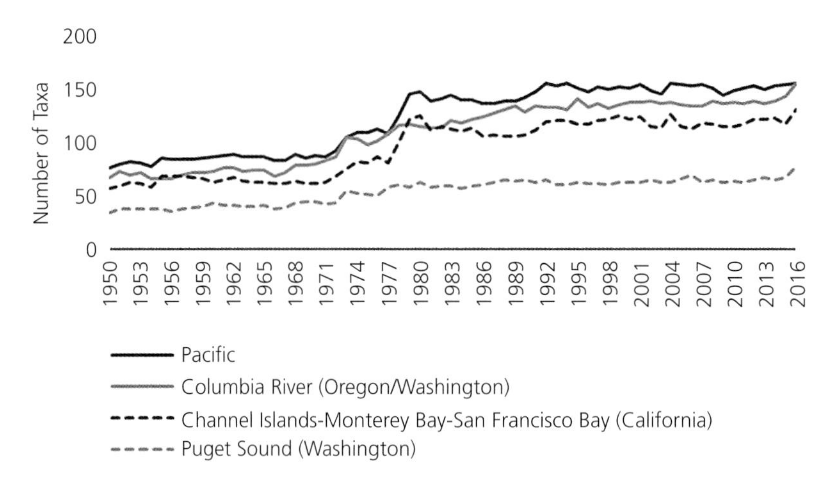

Figure 8.23 Number of taxa reported captured by commercial and recreational fisheries in the Pacific region and identified subregions. Data derived from NOAA National Marine Fisheries Service commercial and recreational fisheries statistics.

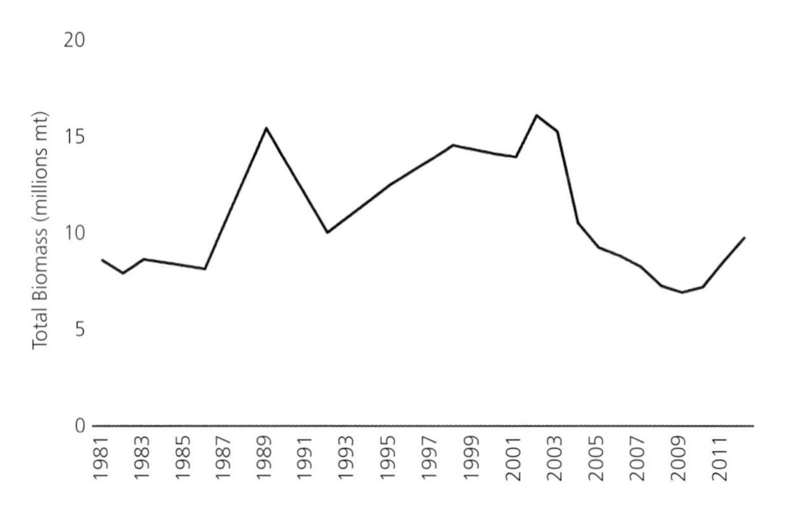

Figure 8.24 Total combined surveyed fish and invertebrate biomass (metric tons; mt) for the Pacific region, over time.

Although Channel Islands landings declined in the 1980s and have oscillated in more recent years, they have averaged ~115,000 metric tons over time, with current values much lower at 37,000 metric tons following declines in the 2010s. Commercial landings have fluctuated in the Puget Sound-Salish Sea subregion with peak values observed during the 1990s at ~200,000 metric tons, particularly in the Port of Seattle, following which values averaged around 21,000 metric tons. From the 1980s to mid-2000s, increases in commercial landings occurred in the Columbia River Estuary with values peaking in 2006 at ~93,000 metric tons. In more recent years, marginal declines have been observed with current landings at 48.5 thousand metric tons. Lower values have persisted for Monterey Bay (average: 13.8 thousand tons)

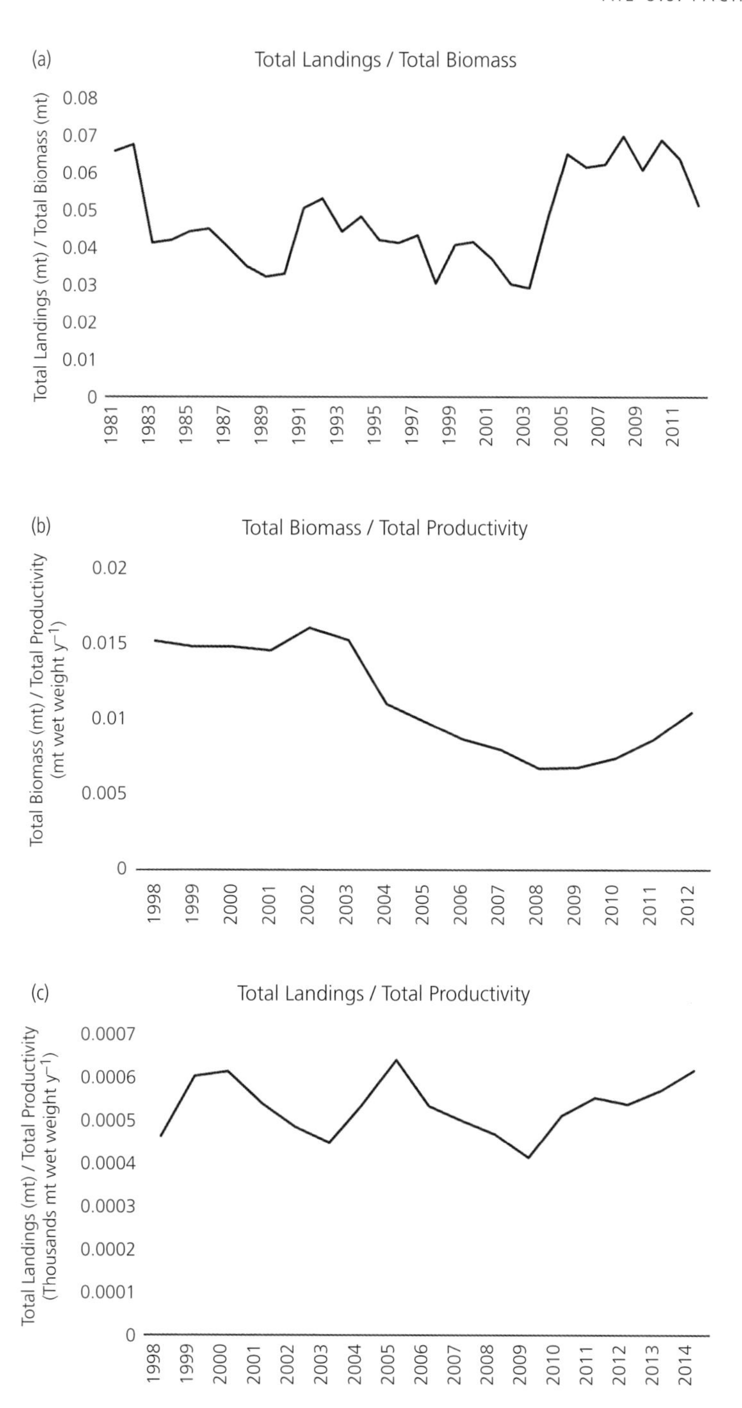

Figure 8.25 Ratios of (a) total commercial and recreational landings (metric tons) to total biomass (metric tons; i.e., exploitation index); (b) total biomass (metric tons) to total productivity (metric tons wet weight year^{-1}); (c) total commercial and recreational landings (metric tons) to total productivity (metric tons wet weight year^{-1}) for the Pacific region.

Figure 8.26 Total commercial and recreational landings and percent contribution of recreational landings (%Recr) over time for the Pacific region (1950–2016) and identified subregions (commerical and recreational: 1981–2016; artisanal: 1976–2016).

Data derived from NOAA National Marine Fisheries Service (NMFS) commercial and recreational fisheries statistics, as well as recreational and artisanal data from the Pacific States Marine Fisheries Commission (PSMFC).

and San Francisco Bay (average: ~7700 metric tons) over time, with the highest values occurring for Monterey Bay during the 1990s and 2000s, and an increase in landings (up to 28.2 thousand metric tons in 2014) also occurring in more recent years.

Since the 1980s, decreases in recreational catches have been observed throughout the Pacific coast, particularly in southern (from 16 million to 3.9 million fishes) and northern California (from 13 million to 3.1 million fishes). Values in the Pacific Northwest have historically been lower than in California (average: 1.1 million fishes), but also experienced declines in more recent years. However, reported values of artisanal catches dramatically increased in the early 2000s (up to 25 million fishes) and have remained between 16–20 million fishes in years since. When

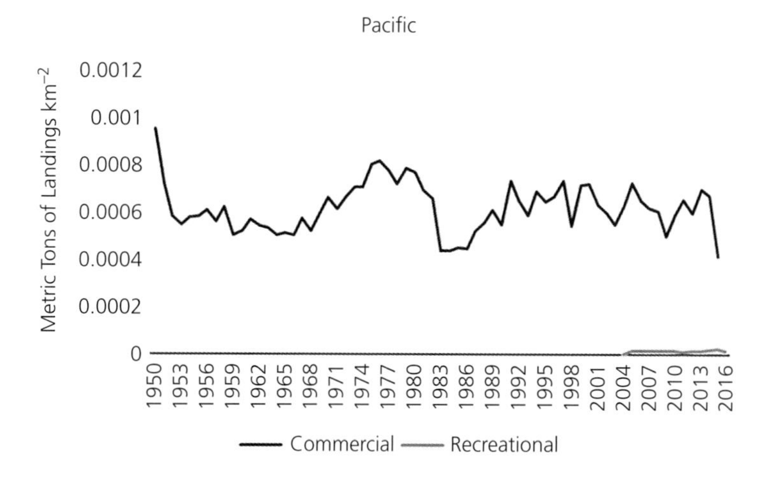

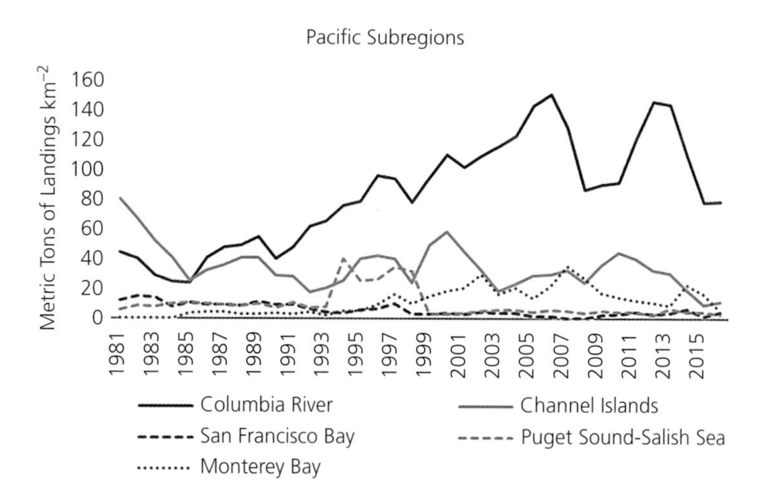

Figure 8.27 Total commercial landings (metric tons) per square kilometer over time (1950–2016) for the Pacific region and identified subregions (1981–2016).

Data derived from NOAA National Marine Fisheries Service commercial and recreational fisheries statistics.

standardized per square kilometer (Fig. 8.27), Pacific landings also follow similar trends observed for total landings values. This region contains the fourth-highest concentration of fisheries landings over time in the nation. The highest concentrated fishing pressure (up to ~150.8 metric tons km^{-2}) has been historically observed in and around the Columbia River Estuary, with values increasing over time, while lower values have been observed in the Channel Islands (up to 79.8 metric tons km^{-2}), Puget Sound-Salish Sea (up to 39.9 metric tons km^{-2}), Monterey Bay (up to 35.2 metric tons km^{-2}), and San Francisco Bay (up to 14.5 metric tons km^{-2}).

Although values have decreased over time for certain taxa, bycatch continues to persist throughout the Pacific region (Fig. 8.28). By weight, bycatch is most pronounced for bony fishes and invertebrates, and is moderate for sharks, marine mammals, and seabirds. On average, bycatch in this region is fifth-highest by weight and second-highest by number in the nation, albeit with decreases observed over time for bony fishes, sharks, and seabirds. For other taxa (i.e., marine mammals, sea turtles, invertebrates), reported bycatch values have been more consistent or minimal (i.e., sea turtles), with number of bycaught invertebrates increasing since 2010.

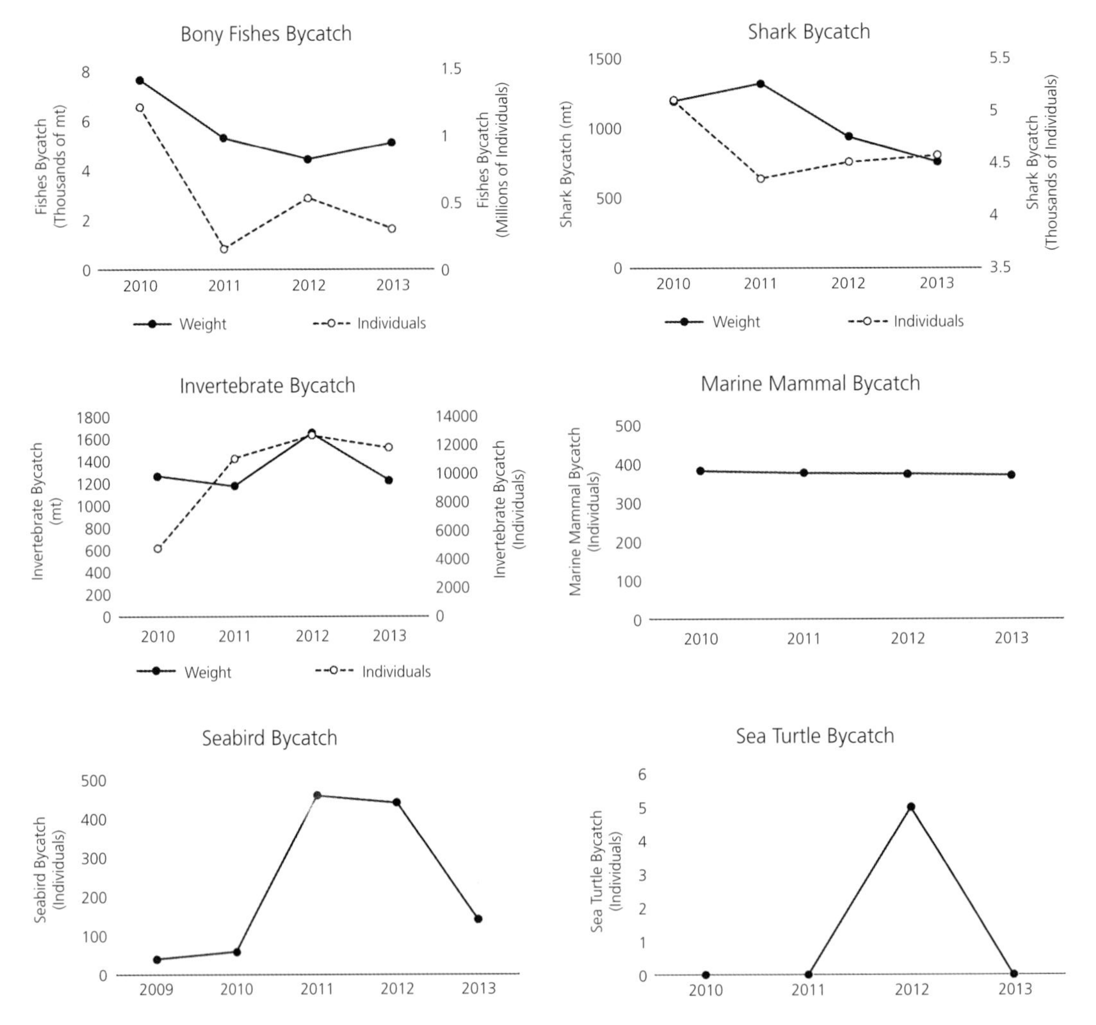

Figure 8.28 Total bycatch of species groups in metric tons (mt) and number of individuals for the Pacific region.
Data derived from NOAA National Marine Fisheries Service U.S. National Bycatch Report Database System.

8.5 Synthesis

The U.S. Pacific (California Current) ecosystem emerges as an environment housing moderately high numbers of managed species, with biota and marine communities that are responding to the consequences of historical overexploitation of its fisheries resources, habitat loss, increasing coastal development, nutrient loading, HABs, ocean acidification, climate forcing, marine heatwaves, and other ocean uses. Additionally, more recent stressors including substantial regional warming and climate-related species shifts continue to affect this system, and alter its composition, dynamics, and LMR production (Holbrook et al. 1997, Brodeur et al. 2006, Field et al. 2007, Hsieh et al. 2009, Sumaila et al. 2011, Ashelby et al. 2013, Hazen et al. 2013, Stewart et al. 2014, Asch 2015, Grosholz et al. 2015, Yamada et al. 2015, Strong & Ayres 2016, Jurgens et al. 2018, Morley et al. 2018, Zabin et al. 2018, Link & Marshak 2019). Despite historical overfishing, oscillating values of various stocks, and collapse of several key fisheries stocks, together with the effects of regional warming and climate forcing on LMR populations, many species have been rebuilt in recent years, and fisheries landings and revenue in this system remain the third- and fourth-highest in the nation, respectively. Although lower landings have been observed in Puget Sound, Monterey Bay, and San Francisco Bay, their trends have remained relatively consistent over the past few decades. Declines in landings and value for the Channel Islands continue to occur, although steady increases have been observed in the Columbia River subregion, as related to regular catches of salmonids and increased offshore groundfish landings to the Astoria area (NMFS 2020a). Fisheries value has also increased in Puget Sound. This region contains relatively low numbers and proportions of stocks that are overfished or experiencing overfishing, with low to moderate numbers of unassessed fisheries stocks, and low numbers of strategic marine mammal stocks (Link & Marshak 2019). With increasing human population density (especially near the Channel Islands and San Francisco Bay) and some of the higher tourism and other ocean uses levels in the nation, associated pressures from fishing, recreational, and industrial sectors continue to affect the U.S. Pacific. They also contribute to its nationally significant LMR-, tourism-, and marine transportation-based economy. This region is one of the highest in the nation in terms of LMR socioeconomic status (Link & Marshak 2019), while also comprising ~19% of the U.S. total ocean economy. As regional warming and oscillations in climate forcing continue to affect this ecosystem and its LMRs, continued sustainable exploitation and marine economic status of the Pacific region is important.

The Pacific ecosystem is well known for its high LMR-based socioeconomic status and the significant contribution of LMRs to its subregional marine economies. This is especially related to the overall sustainable management of its fisheries, habitats, and protected resources over more recent decades. In addition, recent system exploitation is below suggested sustainable thresholds of 0.1 and is ranked third-lowest in the nation (Samhouri et al. 2010, Large et al. 2013, 2015a, b, Tam et al. 2017, Link & Marshak 2019). This region is also one of the highest ranked nationally in terms of the total contributions of its LMR-based employments and GDP toward its total marine economy. These trends emphasize the continued importance of LMRs toward the economic prosperity and cultural identity of this socioecological system. Significant production, in addition to high LMR and marine socioeconomic status, occurs in the U.S. Pacific ecosystem despite having a comparatively lower average primary productivity as a result of intermittent nutrient upwellings (Bruland et al. 2001, Kudela et al. 2008, Menge & Menge 2013, Kämpf & Chapman 2016, Link & Marshak 2019). This production supports a valuable fisheries biomass that is dominated by forage fish and pelagic species, which contributes heavily to secondary and tertiary trophic levels that are concentrated along a narrow continental shelf (Chavez et al. 2003, Field & Francis 2006, Kaplan & Levin 2009, Lester et al. 2010, Kaplan et al. 2012, 2013, PFMC 2013, Koehn et al. 2016, Levin et al. 2016, Dawson & Levin 2019, PFMC 2019). Governance frameworks are also advanced in this region, especially regarding the multispecies nature of its few, stable FMPs, concentrated management attention on moderate numbers of managed species, strategic and abundant use of fisheries closures, and cooperative management among the PFMC and state agencies. However, this region is also subject to higher litigation potential in terms of the third-highest number of fisheries-related lawsuits, low congressional attention as related to its fisheries value and shoreline, and low AREAL proportion of permanently protected EEZ despite high numbers of marine protected areas.

Priorities for PFMC EBFM actions include protecting unfished and unmanaged forage fish species, accounting for the effects of fisheries harvest policies on co-interacting species, and efforts toward minimizing bycatch and the effects of human activities on essential fish habitat (EFH). Additional priorities include examining the multispecies and socioeconomic effects of regional warming and climate forcing shifts. Other considerations include applying climate vulnerability assessment outputs into management strategies, and cross-FMP actions for bycatch, catch monitoring, and protecting critical habitats and EFH (PFMC 2013, 2017, NMFS 2019). These priorities are also included in a recently released EBFM Implementation Plan for the West Coast region (NMFS 2019), which focuses on implementing the NOAA Fisheries EBFM Policy and Roadmap (NMFS 2016a, b). Specifically, these priorities are focused on the following:

- Continued implementation and improvement of the Pacific FEP.
- Ongoing development and enhancement of annual ESRs for the PFMC.
- Conduct gap analyses of California Current ecosystem-level science and monitoring efforts, including benthic habitat and trophic information and social science capacity.
- Identify overarching pressures for west coast species, including conducting climate vulnerability assessments.
- Conduct risk assessments (e.g., habitat, cumulative impacts, fishing community vulnerability), and apply this information toward prioritizing ecosystem/socioeconomic resilience.

- Advance ecosystem modeling and ecosystem forecasting efforts, and apply toward management strategy evaluations.
- Continue developing productivity-based harvest control rules using Atlantis and other modeling frameworks.
- Refine projections of biophysical parameters to much finer scales.
- Examine climate-driven future scenarios for west coast hydrology and water supply and their effects on protected salmon/sturgeon populations and coastal socioeconomics.
- Analyze regional trade-offs between habitat management, salmon population recovery, and other water use sectors.
- Conduct cumulative impact analyses linking environmental conditions, climate, target species distributions, fisheries management, and socioeconomics in coastal pelagic and HMS fisheries.

As observed for other temperate systems, advanced understanding of the Pacific ecosystem is leading to refined stock assessments, more sophisticated ecosystem models, and dynamic ocean management approaches that consider climate and environmental pressures (Ruzicka et al. 2014, 2016, Hazen et al. 2017, Hazen et al. 2018, Abrahms et al. 2019, Townsend et al. 2019). Application of this information remains needed for developing system-wide reference points and enhancing multispecies approaches to management. These advancements facilitate continued ecosystem-level planning and risk assessment to bolster current management actions and expand upon their applications. Additionally, the recurring release of ESRs has provided comprehensive information for PFMC ecosystem indicator initiatives and in supporting pilot studies that examine environmental effects on fisheries productivity (Samhouri et al. 2013, Levin et al. 2016, Samhouri et al. 2017, PFMC 2017, CCIEA 2019). Continued application of ecosystem models is allowing for more informative and localized assessments that work toward a broader understanding of Pacific subregions and their responses to ongoing stressors, especially regional warming and climate forcing (Field et al. 2006, Herrick et al. 2007, Lester et al. 2010, Kaplan et al. 2012, 2013, Harvey et al. 2016, Koehn et al. 2016, Levin et al. 2016, Marshall et al. 2017, Townsend et al. 2018, CCIEA 2019). Investigating and testing ecosystem harvest control rules in the Pacific region with tools such as management strategy evaluations will help to realize their needed applications. All of these advancements are enhancing emergent ecosystem approaches to management in the Pacific, and providing a framework for its potential expanded application throughout the California Current LME.

There has been considerable progress toward a broader understanding of the U.S. Pacific ecosystem, especially in examining ecosystem-level effects of climate forcing, developing sophisticated biophysical models, and in creating quantitative frameworks for estimating system thresholds (Field & Francis 2006, Field et al. 2007, Holt & Punt 2009, Kaplan & Levin 2009, Lester et al. 2010, Punt 2011, Kaplan & Leonard 2012, Kaplan et al. 2012, PFMC 2013, Moffitt et al. 2016, Tam et al. 2017, CCIEA 2019, NMFS 2019). However, applying this information in setting regional and subregional system-wide control rules that consider species complexes and trophic levels remains to be fully implemented. These actions are especially warranted given past and emerging stressors affecting basal ecosystem productivity, fisheries and protected species biomass, marine socioeconomics, and differential community and system-level responses among subregions to these pressures (Tegner & Dayton 1987, Mantua et al. 1997, McGowan et al. 1998, Ayres et al. 1999, Steneck et al. 2002, Brinton & Townsend 2003, Chavez et al. 2003, McGowan et al. 2003, Halpern et al. 2009, Hughes et al. 2009, Holt & Punt 2009, Tepolt et al. 2009, Teck et al. 2010, Cope et al. 2011, Gleason et al. 2011, Lockwood & Somero 2011, Miller et al. 2011, Cloern & Jassby 2012, Howard et al. 2014, CCIEA 2019). Given the significant contributions of forage fish toward ecosystem functioning in this system, and concurrent effects of climate forcing and regional warming on basal productivity and their planktonic prey, management practices must continue to account for these bottom-up effects on upper trophic levels in consideration of setting harvest limits (Chavez et al. 2003, Field & Francis 2006, Kaplan & Levin 2009, Lester et al. 2010, Kaplan et al. 2012, 2013, PFMC 2013, Koehn et al. 2016, Levin et al. 2016, Dawson & Levin 2019, PFMC 2019). These have begun with prohibitions on newly directed fisheries for unmanaged forage species until thorough scientific review (PFMC 2013, NOAA 2016). But additional considerations of forage fish harvest and its implications for setting groundfish and salmon catch limits, in addition to trade-offs with predatory mammals and seabirds, are also needed. Proliferation of range-expanding species, including periodically observed Humboldt Squid during the 2000s, and their subsequent predation and ecological effects on California Current species, are also important to consider when setting harvest limits. Predatory release of this species by overfishing is also thought to have contributed to aspects of these expansions (Field et al. 2007, Zeidberg & Robison 2007, Stewart et al. 2012, Miller et al. 2013, Stewart et al. 2014), and its system-level effects must also be factored into future management actions.

With the fourth-highest national population density, which continues to increase, there is very high potential for future resource competition among prominent marine sectors in this region. As natural and human-related stressors continue to increase, it is imperative that management measures such as closures and cumulative consideration of stressors on baseline production be used to manage concurrent human activities and avoid repeating past unsustainable exploitation of the Pacific ecosystem, which requires being more precise and adaptive (e.g., Hazen et al. 2018). Consideration of system-level interconnected effects from cumulative marine sectors, and conflicts among fisheries, tourism, marine transportation, and marine energy development, remain necessary for comprehensive cooperative management strategies, especially in light of species and system-level shifts in response to regional warming and climate forcing. Shifting species compositions and climate-related effects on the condition of economically important species and forage fish prey resources are also likely to affect the viability of commercial fisheries landings and revenue. These are anticipated to be especially significant in the Channel Islands, where fisheries landings and revenue have declined, while revenue may be further enhanced in more northern portions of this system with converging species complexes and targeted recreational fisheries. Alternatively, continued natural and human effects in critical salmon habitats and loss of species out of this system are also likely to negatively affect long-term landings and revenue in these more northern areas, including Puget Sound, the Columbia River, and the California delta connected to San Francisco Bay. Increasingly abundant sportfishes and recreational fishery species from lower latitudes may enhance conflicts among fishing sectors. With expanding marine sectors, especially related to marine energy and mineral extraction, into historical commercial fishing grounds (Boehlert et al. 2008, Boehlert & Gill 2010, Salcido 2011, Sherman et al. 2013), systematic management strategies must account for these interacting sectors and their effects on LMRs, their fisheries, and their essential habitats. As in other U.S. regions, management strategies that consider spatially overlapping ocean uses, resource partitioning, multiple species, and the socioeconomic effects of changing environments are especially warranted in this system. Predictive models that account for these cumulative effects will also enhance these multisector management frameworks.

8.5.1 Progress Toward Ecosystem-Based Fisheries Management (EBFM)

This ecosystem is excelling in the areas of its natural and human environments, including (comparatively) limited natural stressors, LMR and socioeconomic status, and the quality of its governance system as related to the determinants of successful LMR management (Fig. 8.29; Link & Marshak 2019).

Fisheries management in the Pacific continues advancing into approaches that address the cumulative issues affecting its LMRs and fisheries. Through a more systematic and prioritized consideration of all

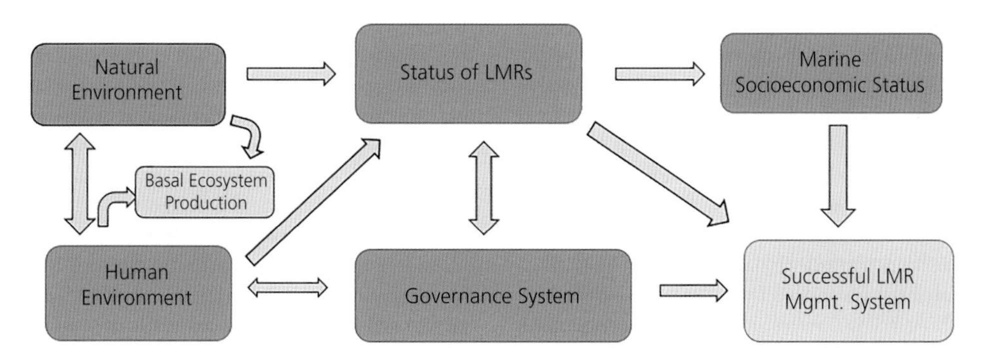

Figure 8.29 Schematic of the determinants and interconnectivity of successful living marine resource (LMR) systems management criteria as modified from Link and Marshak (2019). Those highlighted in blue indicate criteria for which the Pacific ecosystem is excelling.

fisheries, pressures, risks, and outcomes, holistic EBFM practices are being developed and implemented across the U.S., including this ecosystem (NMFS 2016a, b, Levin et al. 2018, Marshall et al. 2018, Link & Marshak 2019). In accordance with the NOAA Fisheries EBFM Policy and Roadmap (NMFS 2016a, b), an EBFM implementation plan (NMFS 2019) for the Pacific region has been developed through efforts overseen by the NOAA Fisheries West Coast Regional Office (WCRO), NWFSC and SWFSC, in addition to collaborative efforts with the PFMC and PSMFC. This plan specifies approaches to be taken toward conducting EBFM and in addressing NOAA Fisheries Roadmap priorities and gaps. Progress for the Pacific and its major subregions in terms of EBFM Roadmap Guiding Principles and Goals is shown in Table 12.9.

Overall, progress has been made at the regional level, and to a certain degree within subregions, in terms of implementing ecosystem-level planning, advancing knowledge of ecosystem principles, and in assessing risks and vulnerabilities to ecosystems through ongoing investigations into climate vulnerability and species prioritizations for stock and habitat assessments. While information has been obtained and calculations and models developed, and some progress has been made toward incorporating ecosystem information in LMR management, limited progress has been made on using ecosystem-level emergent properties in management frameworks or exploring system trade-offs within the Pacific region.

Progress toward recommendations from the Ecosystem Principles Advisory Panel (EPAP) for the development of FEPs (EPAP 1999, Wilkinson & Abrams 2015) is shown in Table 12.10. An FEP has been in place for the Pacific region since 2013 (PFMC 2013), with a substantial degree of progress observed toward characterizing, monitoring, and assessing aspects of its ecosystem dynamics. Specific EBFM-related advancements have occurred through three major initiatives undertaken by the PFMC, which are included in the Pacific Coast FEP (PFMC 2013). The first initiative was the enactment of a comprehensive ban on targeting unfished forage species, which spanned all four regional FMPs. The second was a review of the indicators included in annual ESRs, which were also formally established under the FEP, to better align ESRs with Council needs. Finally, the third initiative, which is an exploration of the climate-readiness of coastal fishing communities and PFMC management plans, is still underway and currently through its scenario planning stage. These initial steps toward EBFM by the PFMC have been built into its FEP to serve as a vehicle for addressing emerging issues. Other ecosystem-based approaches by the PFMC include a complete overhaul of its groundfish FMP to reduce overfishing and stock bycatch (PFMC 2019), controls over the drift gillnet fishery for swordfish and other HMS to prevent protected species bycatch (Carretta et al. 2017), and enhanced identification of EFH (PFMC 2014, 2018a, b, NMFS 2019). Efforts toward providing greater ecosystem-level data and more in-depth ecosystem assessments of Pacific subregions remain warranted. However, efforts to apply IEA approaches in the Channel Islands, Monterey Bay, and Olympic Coast NMS condition reports are currently underway (Brown et al. 2019), while ecosystem-based advancements have also been observed for Puget Sound (Kershner et al. 2011, Samhouri et al. 2011). The implementation of EBFM and development of regional FEPs have considered interconnected system-wide factors for this fisheries ecosystem, including basal productivity, LMR biomass, fisheries landings, and LMR-dependent employments, revenue, and socioeconomics.

Engagement strategies for ecosystem-level management have been successful in Pacific region, with a long history in the California Current Integrated Ecosystem Assessment (CCIEA) work (CCIEA 2012, 2014, 2016, 2019), given a cooperative relationship among scientists, managers, and stakeholders; these have been further developed in the implementation of this region's FEP and EBFM plans (NMFS 2019). With this foundation, enhanced application of ecosystem-level management in addressing ongoing and emerging concerns for the sustainable future of this socioecological system is possible. Although not without limitations, governance frameworks in this area remain very advanced, focused, and targeted toward species complexes, while also emerging toward enhanced cooperative management actions. Building on these current frameworks, applying the breadth of ecosystem-level knowledge toward addressing emerging issues, resolving constraints, and accounting for the environmental and human effects directly in management actions will strengthen opportunities for advancing ecosystem-based sustainable practices in this region. Building on the knowledge of human dimensions in the Pacific will allow this region to continue developing comprehensive management strategies toward preserving coastal economies. Advances made in coupled understanding of these components also serve as a model for other regions in their progress toward ecosystem-level management.

8.5.2 Conclusions

In light of these factors, and building upon advanced ecosystem-focused integrated models, assessments, and research priorities for this region, there is an opportunity to continue applying information on the Pacific ecosystem in an approach that adopts the determinants of successful LMR management (Link & Marshak 2019) to resolve many of the issues affecting this region and its subregions. This approach directly enables management actions that consider differential primary and secondary productivities throughout this system. It also complements regional priorities for EBFM implementation by accounting for the influences of human and natural stressors on fisheries biomass, landings, and socioeconomics. Continued system-wide considerations of the inherent relationships between productivity, biomass, landings, and socioeconomics are needed to maintain this system within needed sustainable thresholds, especially in light of multiple stressors. EBFM provides opportunities that focusing on one stock or species complex, one fishery, or the effects of only the fishing sector would not. Simultaneous controlled harvest of interacting species complexes and fisheries, which account for their trophic interdependence and contributions toward system productivity are essential for optimal management. Overlapping fishing effort for these species, in addition to the gear-related effects of their harvest on essential habitats and subsequent ecological consequences to multispecies complexes is partially considered in current Pacific frameworks, with potential for their expansion. In addition, cross-FMP initiatives are enhancing broader management approaches (NOAA 2016, PFMC 2017), while IEA approaches to NMS condition reports are allowing for more comprehensive assessments in subregions (Brown et al. 2019). Continued application of portfolio and multilevel management approaches (Edwards et al. 2004, Kellner et al. 2011, Link 2018) provides for more effective, responsive, and precautionary management strategies that consider the interconnectivity of this ecosystem and its cumulative response to ongoing pressures.

Broadening ongoing holistic approaches to management for the Pacific ecosystem allows for continuation of its nationally significant cultural marine heritage, LMR-dependent economy, and an ecologically sound approach to resource use. Enhancing EBFM in this region advances current fisheries management strategies toward a system-wide focus that accounts for the interdependence of fisheries and protected species complexes, habitat quality, climate forcing, and bottom-up forcing related to productivity thresholds. Factors including more sustainable land use, sectoral trade-offs, and cumulative effects of human activities on ecosystem dynamics will strengthen conservation and sustainable management efforts. Thorough integration of ecosystem-level factors remains needed toward setting limits on resource use and harvest under mutual consideration of the interacting components of this system in response to multiple stressors, and for ensuring its continued sustainability. Applying available, comprehensive information toward the development of system-level reference points and thresholds with respect to basal productivity and biomass, with which dependent LMR status and economies are interconnected, is especially possible in this relatively data-rich and highly productive system where advanced management approaches occur. With advanced knowledge of this system and emerging knowledge regarding its socioecological responses to multiple stressors, there is potential to move forward in monitoring and managing the effects of regional warming and emergent human activities using biological and socioeconomic metrics that are related to defined system-wide thresholds. As compounding effects of climate change, habitat loss, eutrophication, fishing pressure, and other ocean uses continue to affect this region, their management throughout this socioecological system can be more effectively addressed by building upon current ecosystem approaches. Applying and expanding comprehensive management strategies in this system and its subregions that consider inherent productivities and ecosystem properties will work toward maintaining its LMR and socioeconomic status in light of ongoing pressures.

8.6 References

Abrahms B, Welch H, Brodie S, Jacox MG, Becker EA, Bograd SJ, Irvine LM, et al. 2019. Dynamic ensemble models to predict distributions and anthropogenic risk exposure for highly mobile species. *Diversity and Distributions* 26(8):1182–93.

Adams J, Kaplan IC, Chasco B, Marshall KN, Acevedo-Gutiérrez A, Ward EJ. 2016. A century of Chinook salmon consumption by marine mammal predators in the Northeast Pacific Ocean. *Ecological Informatics* 34:44–51.

Ainley DG, Hyrenbach KD. 2010. Top-down and bottom-up factors affecting seabird population trends in the California current system (1985–2006). *Progress in Oceanography* 84(3–4):242–54.

Ainouche M, Gray A. 2016. Invasive Spartina: lessons and challenges. *Biological Invasions* 18(8):2119–22.

Airamé S, Dugan JE, Lafferty KD, Leslie H, McArdle DA, Warner RR. 2003. Applying ecological criteria to marine reserve design: a case study from the California Channel Islands. *Ecological Applications* 13(sp1):170–84.

Alper DK. 1996. The idea of Cascadia: Emergent transborder regionalisms in the pacific northwest-western Canada. *Journal of Borderlands Studies* 11(2):1–22.

Alter SE, Simmonds MP, Brandon JR. 2010. Forecasting the consequences of climate-driven shifts in human behavior on cetaceans. *Marine Policy* 34(5):943–54.

Ambrose DA, Charter RL, Manion SM. 2006. *Ichtyoplankton and Station Data for Surface (Manta) and Oblique (Bongo) Plankton Tows for California Cooperative Oceanic Fisheries Investigations survey cruises in 2005. NOAA Technical Memorandum NOAA-TM-NMFS-SWFSC-392*. Washington, DC: NOAA (p. 125).

Anderson DM, Burkholder JM, Cochlan WP, Glibert PM, Gobler CJ, Heil CA, Kudela RM, et al. 2008. Harmful algal blooms and eutrophication: examining linkages from selected coastal regions of the United States. *Harmful Algae* 8(1):39–53.

Anderson JJ. 2020. "Decadal climate cycles and declining Columbia River salmon." In: *Sustainable Fisheries Management: Pacific Salmon.* Edited by EE Knudsen, CR Steward, DD MacDonald, JE Williams, DW Reiser. New York, NY: CRC Press (pp. 467–84).

Andrews K, Williams G, Samhouri J, Marshall K, Gertseva V, Levin P. 2015. The legacy of a crowded ocean: indicators, status, and trends of anthropogenic pressures in the California Current ecosystem. *Environmental Conservation* 42(2):139–51.

Antonelis GA, Fiscus CH. 2980. The pinnipeds of the California Current. *CalCOFI Report* 21:68–78.

Arkema KK, Abramson SC, Dewsbury BM. 2006. Marine ecosystem-based management: from characterization to implementation. *Frontiers in Ecology and the Environment* 4(10):525–32.

Asch RG. 2015. Climate change and decadal shifts in the phenology of larval fishes in the California Current ecosystem. *Proceedings of the National Academy of Sciences* 112(30):E4065–74.

Ashelby CW, De Grave S, Johnson ML. 2013. The global invader Palaemon macrodactylus (Decapoda, Palaemonidae): an interrogation of records and a synthesis of data. *Crustaceana* 86(5):594–624.

Ayres DR, Garcia-Rossi D, Davis HG, Strong DR. 1999. Extent and degree of hybridization between exotic (Spartina alterniflora) and native (S. foliosa) cordgrass (Poaceae) in California, USA determined by random amplified polymorphic DNA (RAPDs). *Molecular Ecology* 8(7):1179–86.

Bailey B, Jennings T, Jennings A. 2003. *Estuary Management in the Pacific Northwest: An Overview of Programs and Activities in Washington, Oregon, and Northern California.* Corvallis, OR: Oregon Sea Grant.

Bailey H, Benson SR, Shillinger GL, Bograd SJ, Dutton PH, Eckert SA, Morreale SJ, et al. 2012. Identification of distinct movement patterns in Pacific leatherback turtle populations influenced by ocean conditions. *Ecological Applications* 22(3):735–47.

Bailey H, Brookes KL, Thompson PM. 2014. Assessing environmental impacts of offshore wind farms: lessons learned and recommendations for the future. *Aquatic Biosystems* 10:8.

Bakun A, Black BA, Bograd SJ, Garcia-Reyes M, Miller AJ, Rykaczewski RR, Sydeman WJ. 2015. Anticipated effects of climate change on coastal upwelling ecosystems. *Current Climate Change Reports* 1(2):85–93.

Ballance L, Srinivasan M, Henry A, Angliss R, Barre L, Barlow J, Bengtson J, et al. 2017. *A Strategic Plan for Conducting Large Geographic Scale, Ship-Based Surveys to Support the U.S. Marine Mammal Protection and Endangered Species Acts. NOAA Technical Memorandum NMFS-F/SPO-169.* Washington, DC: NOAA (p. 20).

Baltz KA, Sakuma KM, Ralston S. 2006. *The Physical Oceanography Off the Central California Coast During May–June 2001: A Summary of CTD and Other Hydrographic Data from Young of the Year Juvenile Rockfish Surveys. NOAA Technical Memorandum NOAA-TM-NMFS-SWFSC-395.* Washington, DC: NOAA (p. 83).

Baraff LS, Loughlin TR. 2000. Trends and potential interactions between pinnipeds and fisheries of New England and the US West Coast. *Marine Fisheries Review* 62(4):1–39.

Barlow J, Forney KA. 2007. Abundance and population density of cetaceans in the California Current ecosystem. *Fishery Bulletin* 105(4):509–26.

Barlow J, Kahru M, Mitchell BG. 2008. Cetacean biomass, prey consumption, and primary production requirements in the California Current ecosystem. *Marine Ecology Progress Series* 371:285–95.

Barrick DE. 1979. A coastal radar system for tsunami warning. *Remote Sensing of Environment* 8(4):353–8.

Barton A, Waldbusser GG, Feely RA, Weisberg SB, Newton JA, Hales B, Cudd S, et al. 2015. Impacts of coastal acidification on the Pacific Northwest shellfish industry and adaptation strategies implemented in response. *Oceanography* 28(2):146–59.

Beamish RJ, Bouillon DR. 1993. Pacific salmon production trends in relation to climate. *Canadian Journal of Fisheries and Aquatic Sciences* 50(5):1002–16.

Bedford DW, Hagerman FB. 1983. The billfish fishery resource of the California Current. *California Cooperative Oceanic Fisheries Investigations Report* 24:70–8.

Beechie TJ, Stefankiv O, Timpane-Padgham BL, Hall J, Pess GR, Rowse ML, Liermann MC, Fresh KL, Ford MD. 2017. *Monitoring Salmon Habitat Status and Trends in Puget Sound: Development of Sample Designs, Monitoring Metrics, and Sampling Protocols for Large River, Floodplain, Delta, and Nearshore Environments.* Washington, DC: NOAA https://

www.fisheries.noaa.gov/resource/map/salmon-habitat-status-and-trend-monitoring-program-data

Bell TW, Allen JG, Cavanaugh KC, Siegel DA. 2020. Three decades of variability in California's giant kelp forests from the Landsat satellites. *Remote Sensing of Environment* 238:110811.

Bellman MA, Heppell SA, Goldfinger C. 2005. Evaluation of a US west coast groundfish habitat conservation regulation via analysis of spatial and temporal patterns of trawl fishing effort. *Canadian Journal of Fisheries and Aquatic Sciences* 62(12):2886–900.

Bellquist LF, Graham JB, Barker A, Ho J, Semmens BX. 2016. Long-term dynamics in "trophy" sizes of pelagic and coastal pelagic fishes among California recreational fisheries (1966–2013). *Transactions of the American Fisheries Society* 145(5):977–89.

Benson SR, Eguchi T, Foley DG, Forney KA, Bailey H, Hitipeuw C, Samber BP, et al. 2011. Large-scale movements and high-use areas of western Pacific leatherback turtles, Dermochelys coriacea. *Ecosphere* 2(7):1–27.

Benson SR, Forney KA, Harvey JT, Carretta JV, Dutton PH. 2007. Abundance, distribution, and habitat of leatherback turtles (Dermochelys coriacea) off California, 1990– 2003. *Fishery Bulletin* 105(3):337–47.

Bernal PA. 1981. A review of the low-frequency response of the pelagic ecosystem. *California Cooperative Oceanic Fisheries Investigations Report* 22:49–62.

Best EA, St-Pierre G. 1986. *Pacific Halibut as Predator and Prey. Technical Report No. 21.* Seattle, WA: International Pacific Halibut Commission.

Bisbal GA, McConnaha WE. 1998. Consideration of ocean conditions in the management of salmon. *Canadian Journal of Fisheries and Aquatic Sciences* 55(9):2178–86.

Bjorkstedt EP, Bograd SJ, Sydeman WJ, Thompson SA, Goericke R, Durazo R, Warzybok P, et al. 2012. State of the California Current 2011–2012: ecosystems respond to local forcing as La Niña wavers and wanes. *Reports of California Cooperative Oceanic Fisheries Investigations* 53:41–76.

Bjorkstedt EP, Goericke R, McClatchie S, Weber E, Watson W, Lo N, Peterson B, et al. 2011. State of the California Current 2010–2011: regionally variable responses to a strong (but fleeting?) La Nina. *California Cooperative Oceanic Fisheries Investigations Reports* 52:36–68.

Bjorkstedt EP, Goericke R, McClatchie S, Weber E, Watson W, Lo N, Peterson B, et al. 2010. State of the California Current 2009–2010: regional variation persists through transition from La Niña to El Niño (and back?). *California Cooperative Oceanic Fisheries Investigations Reports* 51:39–69.

Blanchette CA, Broitman BR, Gaines SD. 2006. Intertidal community structure and oceanographic patterns around Santa Cruz Island, CA, USA. *Marine Biology* 149(3):689–701.

Blanchette CA, Melissa Miner C, Raimondi PT, Lohse D, Heady KE, Broitman BR. 2008. Biogeographical patterns of rocky intertidal communities along the Pacific coast of North America. *Journal of Biogeography* 35(9):1593–607.

Boehlert GW, Gill AB. 2010. Environmental and ecological effects of ocean renewable energy development: a current synthesis. *Oceanography* 23(2):68–81.

Boehlert GW, McMurray GR, Tortorici CE. 2008. *Ecological Effects of Wave Energy Development in the Pacific Northwest. NOAA Technical Memorandum NMFS-F/SPO-92.* Washington, DC: NOAA.

Bograd SJ, Castro CG, Di Lorenzo E, Palacios DM, Bailey H, Gilly W, Chavez FP. 2008. Oxygen declines and the shoaling of the hypoxic boundary in the California Current. *Geophysical Research Letters* 35(12).

Bograd SJ, DiGiacomo PM, Durazo RE, Hayward TL, Hyrenbach KD, Lynn RJ, Mantyla AW, et al. 2000. The state of the California Current, 1999–2000: forward to a new regime? *Reports of California Cooperative Oceanic Fisheries Investigations* 41:26–52.

Bograd SJ, Lynn RJ. 2001. Physical-biological coupling in the California Current during the 1997–99 El Niño-La Niña cycle. *Geophysical Research Letters* 28(2):275–8.

Bograd SJ, Checkley DA, Wooster WS. 2003. CalCOFI: a half century of physical, chemical, and biological research in the California Current System. *Deep Sea Research II* 50:2349–53.

Bond NA, Cronin MF, Freeland H, Mantua NJ. 2015. Causes and impacts of the 2014 warm anomaly in the NE Pacific. *Geophysical Research Letters* 42(9):3414–20.

Botsford LW, Lawrence CA. 2002. Patterns of co-variability among California Current chinook salmon, coho salmon, Dungeness crab, and physical oceanographic conditions. *Progress in Oceanography* 53(2–4):283–305.

Botsford LW. 2001. Physical influences on recruitment to California Current invertebrate populations on multiple scales. *ICES Journal of Marine Science* 58(5):1081–91.

Boyd MJ, Mulligan TJ, Shaughnessy FJ. 2002. *Non-Indigenous Marine Species of Humboldt Bay, California.* Sacramento, CA: California Department of Fish and Game.

Bradburn MJ, Keller AA, Horness BH. 2011. *The 2003 to 2008 US West Coast Bottom Trawl Surveys of Groundfish Resources Off Washington, Oregon, and California: Estimates of Distribution, Abundance, Length, and Age Composition. NOAA Technical Memorandum NMFS-NWFSC 114.* Washington, DC: NOAA.

Brand EJ, Kaplan IC, Harvey CJ, Levin PS, Fulton EA, Hermann AJ, Field JC. 2007. *A Spatially Explicit Ecosystem Model of the California Current's Food Web and Oceanography. NOAA Technical Memorandum NMFS-NWFSC-84.* Washington, DC: NOAA.

Brinton E, Townsend A. 2003. Decadal variability in abundances of the dominant euphausiid species in southern sectors of the California Current. *Deep Sea Research Part II: Topical Studies in Oceanography* 50(14–16):2449–72.

Brodeur RD, Ralston S, Emmett RL, Trudel M, Auth TD, Phillips AJ. 2006. Anomalous pelagic nekton abundance, distribution, and apparent recruitment in the northern California Current in 2004 and 2005. *Geophysical Research Letters* 33(22).

Brodeur R, Perry I, Boldt J, Flostrand L, Galbraith M, King J, Murphy J, Sakuma K, Thompson A. 2018. An unusual gelatinous plankton event in the NE Pacific: The Great Pyrosome Bloom of 2017. *PICES Press* 26(1):22–7.

Brophy LS, Greene CM, Hare VC, Holycross B, Lanier A, Heady WN, et al. 2019. Insights into estuary habitat loss in the western United States using a new method for mapping maximum extent of tidal wetlands. *PLoS One* 14(8):e0218558.

Brown JJ. 1994. Treaty rights: twenty years after the Boldt decision. *Wicazo Sa Review* 1:1–6.

Brown JR. 1988. Multivariate analyses of the role of environmental factors in seasonal and site-related growth variation in the Pacific oyster Crassostrea gigas. *Marine Ecology Progress Series* 45(3):225–36.

Brown LR, Moyle PB, Yoshiyama RM. 1994. Historical decline and current status of coho salmon in California. *North American Journal of Fisheries Management* 14(2):237–61.

Brown, J., G.D. Williams, C.J. Harvey, A.D. DeVogelaere, C. Caldow. 2019. *Developing Science-Based Indicator Portfolios for National Marine Sanctuary Condition Reports. Marine Sanctuaries Conservation Series ONMS-19-07.* Silver Spring, MD: NOAA (p. 66).

Bruland KW, Rue EL, Smith GJ. 2001. Iron and macronutrients in California coastal upwelling regimes: Implications for diatom blooms. *Limnology and Oceanography* 46(7):1661–74.

Burge CA, Judah LR, Conquest LL, Griffin FJ, Cheney DP, Suhrbier A, Vadopalas B, et al. 2007. Summer seed mortality of the Pacific oyster, Crassostrea gigas Thunberg grown in Tomales Bay, California, USA: the influence of oyster stock, planting time, pathogens, and environmental stressors. *Journal of Shellfish Research* 26(1):163–73.

Byers JE. 1999. The distribution of an introduced mollusc and its role in the long-term demise of a native confamilial species. *Biological Invasions* 1(4):339–52.

Calambokidis J, Barlow J, Ford JK, Chandler TE, Douglas AB. 2009. Insights into the population structure of blue whales in the Eastern North Pacific from recent sightings and photographic identification. *Marine Mammal Science* 25(4):816–32.

Carlisle AB, Kim SL, Semmens BX, Madigan DJ, Jorgensen SJ, Perle CR, Anderson SD, et al. 2012. Using stable isotope analysis to understand the migration and trophic ecology of northeastern Pacific white sharks (Carcharodon carcharias). *PloS One* 7(2):e30492.

Carlton JT, Thompson JK, Schemel LE, Nichols FH. 1990. Remarkable invasion of San Francisco Bay (California, USA) by the Asian clam Potamocorbula amurensis: introduction and dispersal. *Marine Ecology Progress Series* 66:81–94.

Carr ME, Kearns EJ. 2003. Production regimes in four Eastern Boundary Current systems. *Deep Sea Research Part II: Topical Studies in Oceanography* 50(22–26):3199–221.

Carretta JV, Enriquez L, Villafana CA. 2014. *Marine Mammal, Sea Turtle and Seabird Bycatch in California Gillnet Fisheries in 2012. NOAA Technical Memorandum NMFS-SWFSC-526.* Washington, DC: NOAA.

Carretta JV, Forney KA, Oleson EM, Weller DW, Lang AR, Baker J, Muto MM, et al. 2019. *U.S. Pacific Marine Mammal Stock Assessments: 2018. U.S. Department of Commerce, NOAA Technical Memorandum NMFSSWFSC-617.* Washington, DC: NOAA.

Carretta JV, Oleson EM, Weller DW, Lang AR, Forney KA, Baker JD, Hanson B, et al. 2013. *US Pacific Marine Mammal Stock Assessments. NOAA Technical Memorandum NOAA-TM-NMFS-SWFSC-577.* Washington, DC: NOAA.

Carretta JV, Moore JE, Forney KA. 2017. *Regression Tree and Ratio Estimates of Marine Mammal, Sea Turtle, and Seabird Bycatch in the California Drift Gillnet Fishery: 1990–2015. NOAA Technical Memorandum NOAA-TM-NMFS-SWFSC-568.* Washington, DC: NOAA (p. 83).

Cavanaugh KC, Siegel DA, Reed DC, Dennison PE. 2011. Environmental controls of giant-kelp biomass in the Santa Barbara Channel, California. *Marine Ecology Progress Series* 429:1–7.

Cavole LM, Demko AM, Diner RE, Giddings A, Koester I, Pagniello CM, Paulsen ML, et al. 2016. Biological impacts of the 2013–2015 warm-water anomaly in the Northeast Pacific: Winners, losers, and the future. *Oceanography* 29(2):273–85.

CCIEA (California Current Integrated Ecosystem Assessment Team). 2012. *Draft Annual State of the California Current Ecosystem Report. Agenda Item K.3.a. Supplemental Attachment 1.* Seattle, WA: NOAA CCIEA (p. 20).

CCIEA (California Current Integrated Ecosystem Assessment Team). 2014. *Annual State of the California Current Ecosystem Report. A Report of the NMFS Northwest and Southwest Fisheries Science Centers. 19p. Agenda Item C.1.a. Attachment 1.* Seattle, WA: NOAA CCIEA (p. 20).

CCIEA (California Current Integrated Ecosystem Assessment Team). 2015. *California Current Integrated Ecosystem Assessment (CCIEA) California Current Ecosystem Status Report, 2015. A Report of the CCIEA Team (NOAA Northwest, Southwest and Alaska Fisheries Science Centers) to the Pacific Fishery Management Council.* Seattle, WA: NOAA CCIEA (p. 19).

CCIEA (California Current Integrated Ecosystem Assessment Team). 2016. *California Current Integrated Ecosystem Assessment (CCIEA) California Current Ecosystem Status Report, 2016. A Report of the NOAA CCIEA Team to the Pacific Fishery Management Council. Agenda Item D.1.a. NMFS Report 1.* Seattle, WA: NOAA CCIEA (p. 20).

CCIEA (California Current Integrated Ecosystem Assessment Team). 2017. *California Current Integrated Ecosystem Assessment (CCIEA) California Current Ecosystem Status Report, 2017. A Report of the NOAA CCIEA Team to the Pacific Fishery Management Council. Agenda Item F.1.a. NMFS Report 1*. Seattle, WA: NOAA CCIEA (p. 22).

CCIEA (California Current Integrated Ecosystem Assessment Team). 2018. *California Current Integrated Ecosystem Assessment (CCIEA) California Current Ecosystem Status Report, 2018. A Report of the NOAA CCIEA Team to the Pacific Fishery Management Council*. Seattle, WA: NOAA CCIEA (p. 70).

CCIEA (California Current Integrated Ecosystem Assessment Team). 2019. *California Current Integrated Ecosystem Assessment (CCIEA) California Current Ecosystem Status Report, 2019. A Report of the NOAA CCIEA Team to the Pacific Fishery Management Council, March 7, 2019*. Seattle, WA: NOAA CCIEA (p. 23).

CeNCOOS (Central and Northern California Ocean Observing System). 2016. 2014–2019 *Strategic Plan*. Moss Landing, CA: CeNCOOS Program Office (p. 23).

Chan F, Barth JA, Kroeker KJ, Lubchenco J, Menge BA. 2019. The dynamics and impact of ocean acidification and hypoxia. *Oceanography* 32(3):62–71.

Chasco B, Kaplan IC, Thomas A, Acevedo-Gutiérrez A, Noren D, Ford MJ, Hanson MB, et al. 2017a. Estimates of Chinook salmon consumption in Washington State inland waters by four marine mammal predators from 1970 to 2015. *Canadian Journal of Fisheries and Aquatic Sciences* 74(8):1173–94.

Chasco BE, Kaplan IC, Thomas AC, Acevedo-Gutiérrez A, Noren DP, Ford MJ, Hanson MB, et al. 2017b. Competing tradeoffs between increasing marine mammal predation and fisheries harvest of Chinook salmon. *Scientific Reports* 7(1):15439.

Chavez FP, Ryan J, Lluch-Cota SE, Ñiquen M. 2003. From anchovies to sardines and back: multidecadal change in the Pacific Ocean. *Science* 299(5604):217–21.

Checkley Jr DM, Barth JA. 2009. Patterns and processes in the California Current System. *Progress in Oceanography* 83(1–4):49–64.

Chelton DB, Bernal PA, McGowan JA. 1982. Large-scale interannual physical and biological interaction in the California Current. *Journal of Marine Research* 40(4):1095–125.

Chenoweth M, Landsea C. 2004. The San Diego hurricane of 2 October 1858. *Bulletin of the American Meteorological Society* 85(11):1689–98.

Cheung WW, Brodeur RD, Okey TA, Pauly D. 2015. Projecting future changes in distributions of pelagic fish species of Northeast Pacific shelf seas. *Progress in Oceanography* 130:19–31.

Chew KK. 1984. Recent advances in the cultivation of molluscs in the Pacific United States and Canada. *Aquaculture* 39(1–4):69–81.

Chu D, Thomas R, Clemons J, Parker-Stetter S, Pohl J, Clemons J, Gauthier S. 2016. Application of acoustic technologies to study the temporal and spatial distributions of the Pacific hake (Merluccius productus) in the California Current System. *The Journal of the Acoustical Society of America* 139(4):2173.

Cline TJ, Ohlberger J, Schindler DE. 2019. Effects of warming climate and competition in the ocean for life-histories of Pacific salmon. *Nature Ecology & Evolution* 3(6):935–42.

Cloern JE, Jassby AD. 2012. Drivers of change in estuarine-coastal ecosystems: Discoveries from four decades of study in San Francisco Bay. *Reviews of Geophysics* 50(4).

Cohen AN, Carlton JT, Fountain MC. 1995. Introduction, dispersal and potential impacts of the green crab Carcinus maenas in San Francisco Bay, California. *Marine Biology* 122(2):225–37.

Coleman FC, Figueira WF, Ueland JS, Crowder LB. 2004. The impact of United States recreational fisheries on marine fish populations. *Science* 305(5692):1958–60.

Connell JH. 1972. Community interactions on marine rocky intertidal shores. *Annual Review of Ecology and Systematics* 3(1):169–92.

Connolly TP, Hickey BM, Geier SL, Cochlan WP. 2010. Processes influencing seasonal hypoxia in the northern California Current System. *Journal of Geophysical Research: Oceans* 115:C03021.

Cooley SR, Cheney JE, Kelly RP, Allison EH. 2017. Ocean acidification and Pacific oyster larval failures in the Pacific Northwest United States. *Global Change in Marine Systems* 13:40–53.

Cope JM, DeVore J, Dick EJ, Ames K, Budrick J, Erickson DL, Grebel J, et al. 2011. An approach to defining stock complexes for US West Coast groundfishes using vulnerabilities and ecological distributions. *North American Journal of Fisheries Management* 31(4):589–604.

Cope JM, Haltuch MA. 2012. Temporal and spatial summer groundfish assemblages in trawlable habitat off the west coast of the USA, 1977 to 2009. *Marine Ecology Progress Series* 451:187–200.

Corbosiero KL, Dickinson MJ, Bosart LF. 2009. The contribution of eastern North Pacific tropical cyclones to the rainfall climatology of the southwest United States. *Monthly Weather Review* 137(8):2415–35.

Crozier LG, McClure MM, Beechie T, Bograd SJ, Boughton DA, Carr M, Cooney TD, et al. 2019. Climate vulnerability assessment for Pacific salmon and steelhead in the California Current Large Marine Ecosystem. *PLoS One* 14(7):e0217711.

Cummins PF, Freeland HJ. 2007. Variability of the North Pacific Current and its bifurcation. *Progress in Oceanography* 75(2):253–65.

Daley BA, Scavia D. 2008. *An Integrated Assessment of the Continued Spread and Potential Impacts of the Colonial*

Ascidian, Didemnum sp. A, in US waters. NOAA Technical Memorandum NOS NCCOS 78. Silver Spring, MD: NOAA

Daly EA, Auth TD, Brodeur RD, Jacobson KC. 2019. *Changes in Juvenile Salmon Prey Fields Associated with a Recent Marine Heat Wave in the Northern California Current, Technical Report 15*. Vancouver, Canada, North Pacific Anadromous Fish Commission (pp. 21–74).

Daly EA, Brodeur RD, Auth TD. 2017. Anomalous ocean conditions in 2015: impacts on spring Chinook salmon and their prey field. *Marine Ecology Progress Series* 566: 169–82.

Davis RC, Short FT, Burdick DM. 1998. Quantifying the effects of green crab damage to eelgrass transplants. *Restoration Ecology* 6(3):297–302.

Dawson C, Levin PS. 2019. Moving the ecosystem-based fisheries management mountain begins by shifting small stones: a critical analysis of EBFM on the US West Coast. *Marine Policy* 100:58–65.

Dayton PK. 1971. Competition, disturbance, and community organization: the provision and subsequent utilization of space in a rocky intertidal community. *Ecological Monographs* 41(4):351–89.

de Rivera CE, Ruiz G, Crooks J, Wasson K, Lonhart S, Fofonoff P, Steves B, et al. 2005. *Broad-Scale Non-Indigenous Species Monitoring Along the West Coast in National Marine Sanctuaries and National Estuarine Research Reserves*. Washington, DC: National Fish and Wildlife Foundation.

Defran RH, Weller DW. 1999. Occurrence, distribution, site fidelity, and school size of bottlenose dolphins (Tursiops truncatus) off San Diego, California. *Marine Mammal Science* 15(2):366–80.

DeLong RL, Antonelis GA. 1991. "Impact of the 1982–1983 El Niño on the northern fur seal population at San Miguel Island, California." In: *Pinnipeds and El Niño*. Edited by F Trillmich, KA Ono. Berlin, Germany: Springer (pp. 75–83).

DeMaster DP, Fowler CW, Perry SL, Richlen MF. 2001. Predation and competition: the impact of fisheries on marine-mammal populations over the next one hundred years. *Journal of Mammalogy* 82(3):641–51.

Dengler L, Uslu B, Barberopoulou A, Borrero J, Synolakis C. 2008. The vulnerability of Crescent City, California, to tsunamis generated by earthquakes in the Kuril Islands region of the northwestern Pacific. *Seismological Research Letters* 79(5):608–19.

deReynier, Y. 2012. Making ecosystem-based management a reality: the Pacific fishery management council and the California Current integrated ecosystem assessment. *CalCOFI Reports* 53:81–8.

Di Lorenzo E, Miller AJ, Schneider N, McWilliams JC. 2005. The warming of the California Current System: dynamics and ecosystem implications. *Journal of Physical Oceanography* 35(3):336–62.

Di Lorenzo E, Schneider N, Cobb KM, Franks PJ, Chhak K, Miller AJ, McWilliams JC, et al. 2008. North Pacific Gyre Oscillation links ocean climate and ecosystem change. *Geophysical Research Letters* 35(8).

Dodson S, Abrahms B, Bograd SJ, Fiechter J, Hazen EL. 2020. Disentangling the biotic and abiotic drivers of emergent migratory behavior using individual-based models. *Ecological Modelling* 432:109225.

Dumbauld BR, Kauffman BE, Trimble AC, Ruesink JL. 2011. The Willapa Bay oyster reserves in Washington State: fishery collapse, creating a sustainable replacement, and the potential for habitat conservation and restoration. *Journal of Shellfish Research* 30(1):71–83.

Edwards MS, Estes JA. 2006. Catastrophe, recovery and range limitation in NE Pacific kelp forests: a large-scale perspective. *Marine Ecology Progress Series* 320:79–87.

Edwards MS. 2004. Estimating scale-dependency in disturbance impacts: El Niños and giant kelp forests in the northeast Pacific. *Oecologia* 138(3):436–47.

Emmett RL, Brodeur RD, Miller TW, Pool SS, Bentley PJ, Krutzikowsky GK, McCrae JE. 2005. Pacific sardine (Sardinops sagax) abundance, distribution, and ecological relationships in the Pacific Northwest. *California Cooperative Oceanic Fisheries Investigations Report* 46:122.

EPAP (Ecosystem Principles Advisory Panel). 1999. *Ecosystem-Based Fishery Management Report to Congress*. Silver Spring, MD: National Marine Fisheries Service (p. 62).

Estes JA, Tinker MT, Williams TM, Doak DF. 1998. Killer whale predation on sea otters linking oceanic and nearshore ecosystems. *Science* 282(5388):473–6.

Feely RA, Alin SR, Carter B, Bednaršek N, Hales B, Chan F, Hill TM, et al. 2016. Chemical and biological impacts of ocean acidification along the west coast of North America. *Estuarine, Coastal and Shelf Science* 183:260–70.

Feely RA, Sabine CL, Byrne RH, Millero FJ, Dickson AG, Wanninkhof R, Murata A, Miller LA, Greeley D. 2012. Decadal changes in the aragonite and calcite saturation state of the Pacific Ocean. *Global Biogeochemical Cycles* 26(3).

Ferguson L, Srinivasan M, Oleson E, Hayes S, Brown SK, Angliss R, Carretta J, et al. 2017. *Proceedings of the First National Protected Species Assessment Workshop. NOAA Technical Memorandum NMFS-F/SPO-172*. Washington, DC: NOAA (p. 92).

Fiechter J, Huff DD, Martin BT, Jackson DW, Edwards CA, Rose KA, Curchitser EN, et al. 2015. Environmental conditions impacting juvenile Chinook salmon growth off central California: an ecosystem model analysis. *Geophysical Research Letters* 42(8):2910–7.

Field JC, Baltz KE, Phillips AJ, Walker WA. 2007. Range expansion and trophic interactions of the jumbo squid, Dosidicus gigas, in the California Current. *California Cooperative Oceanic Fisheries Investigations Report* 48:131.

Field JC, Francis RC, Aydin K. 2006. Top-down modeling and bottom-up dynamics: linking a fisheries-based ecosystem model with climate hypotheses in the Northern California Current. *Progress in Oceanography* 68(2–4):238–70.

Field JC, Francis RC. 2006. Considering ecosystem-based fisheries management in the California Current. *Marine Policy* 30(5):552–69.

Fitzgerald D. 1986. *A History of Containerization in the California Maritime Industry: The Case of San Francisco*. La Jolla, CA: UC San Diego.

Fleming AH, Clark CT, Calambokidis J, Barlow J. 2016. Humpback whale diets respond to variance in ocean climate and ecosystem conditions in the California Current. *Global Change Biology* 22(3):1214–24.

Forney KA, Ferguson MC, Becker EA, Fiedler PC, Redfern JV, Barlow J, Vilchis IL, Ballance LT. 2012. Habitat-based spatial models of cetacean density in the eastern Pacific Ocean. *Endangered Species Research* 16(2):113–33.

Foss S. 2008. *Introduced Aquatic Species in the Marine and Estuarine Waters of California. Submitted to the California State Legislature. Appendix C. Introduced and Cryptogenic Species in California by Location*. West Sacramento, CA: California Department of Fish and Game Office of Spill Prevention and Response.

Francis RC, Hare SR, Hollowed AB, Wooster WS. 1998. Effects of interdecadal climate variability on the oceanic ecosystems of the NE Pacific. *Fisheries Oceanography* 7(1):1–21.

Francis RC. 1986. Two fisheries biology problems in West Coast groundfish management. *North American Journal of Fisheries Management* 6(4):453–62.

Fresh KL. 1997. "The role of competition and predation in the decline of Pacific salmon and steelhead." In: *Pacific Salmon & their Ecosystems*. Edited by DJ Stouder, PA Bisson, R Naiman. Boston, MA: Springer (pp. 245–75).

Frissell CA. 1993. Topology of extinction and endangerment of native fishes in the Pacific Northwest and California (USA). *Conservation Biology* 7(2):342–54.

Gibson JR. 1992. Otter Skins, Boston Ships, and China Goods: The Maritime Fur Trade of the Northwest Coast, 1785–1841. Montreal, Canada: McGill-Queen's University Press.

Gibson RN, Barnes M, Atkinson RJ. 2002. Impact of changes in flow of freshwater on estuarine and open coastal habitats and the associated organisms. *Oceanography and Marine Biology: An Annual Review* 40:233–309.

Gislason G, Lam E, Knapp G, Guettabi M. 2017. *Economic Impacts of Pacific Salmon Fisheries*. Vancouver, Canada: Pacific Salmon Commission.

Gleason MG, Newkirk S, Merrifield MS, Howard J, Cox R, Webb M, Koepcke J, et al. 2011. *A Conservation Assessment of West Coast (USA) Estuaries*. Arlington, VA: The Nature Conservancy.

Goericke R, Bograd SJ, Gaxiola-Castro G, Gomez-Valdes J, Hooff R, Huyer AD, Hyrenbach KD, et al. 2004. The state of the California Current, 2003–2004: a rare "normal" year. *California Cooperative Oceanic Fisheries Investigations Report* 45:27.

Goericke R, Venrick E, Koslow T, Sydeman WJ, Schwing FB, Bograd SJ, Peterson WT, et al. 2007. The state of the California Current, 2006–2007: regional and local processes dominate. *California Cooperative Oceanic Fisheries Investigations Report* 48:33.

Grantham BA, Chan F, Nielsen KJ, Fox DS, Barth JA, Huyer A, Lubchenco J, Menge BA. 2004. Upwelling-driven nearshore hypoxia signals ecosystem and oceanographic changes in the northeast Pacific. *Nature* 429(6993):749.

Greene C, Andrews K, Beechie T, Bottom D, Brodeur R, Crozier L, Fullerton A, et al. 2014. *Selecting and Evaluating Indicators for Habitats Within the California Current Large Marine Ecosystem. The California Current Integrated Ecosystem Assessment: Phase III Report. IEA Online Report*. http://www.noaa.gov/iea/Assets/iea/california/Report/pdf/9.Habitat_2013.pdf

Greene CM, Blackhart K, Nohner J, Candelmo A, Nelson DM. 2015. A national assessment of stressors to estuarine fish habitats in the contiguous USA. *Estuaries and Coasts* 38(3):782–99.

Gregr EJ, Christensen V, Nichol L, Martone RG, Markel RW, Watson JC, Harley CD, et al. 2020. Cascading social-ecological costs and benefits triggered by a recovering keystone predator. *Science* 368(6496):1243–7.

Gresh T, Lichatowich J, Schoonmaker P. 2000. An estimation of historic and current levels of salmon production in the Northeast Pacific ecosystem: evidence of a nutrient deficit in the freshwater systems of the Pacific Northwest. *Fisheries* 25(1):15–21.

Griffin D, Anchukaitis KJ. 2014. How unusual is the 2012–2014 California drought?. *Geophysical Research Letters* 41(24):9017–23.

Grosholz ED, Crafton RE, Fontana RE, Pasari JR, Williams SL, Zabin CJ. 2015. Aquaculture as a vector for marine invasions in California. *Biological Invasions* 17(5):1471–84.

Grosholz ED, Ruiz GM, Dean CA, Shirley KA, Maron JL, Connors PG. 2000. The impacts of a nonindigenous marine predator in a California bay. *Ecology* 81(5):1206–24.

Grosholz ED, Ruiz GM. 1996. Predicting the impact of introduced marine species: lessons from the multiple invasions of the European green crab Carcinus maenas. *Biological Conservation* 78(1–2):59–66.

Grosholz ED, Ruiz GM. 1995. Spread and potential impact of the recently introduced European green crab, Carcinus maenas, in central California. *Marine Biology* 122(2): 239–47.

Haigh R, Ianson D, Holt CA, Neate HE, Edwards AM. 2015. Effects of ocean acidification on temperate coastal marine ecosystems and fisheries in the Northeast Pacific. *PLoS One* 10(2):e0117533.

Halpern BS, Kappel CV, Selkoe KA, Micheli F, Ebert CM, Kontgis C, Crain CM, et al. 2009. Mapping cumulative human impacts to California Current marine ecosystems. *Conservation Letters* 2(3):138–48.

Halpern BS, Longo C, Scarborough C, Hardy D, Best BD, Doney SC, Katona SK, et al. 2014. Assessing the health of the US West coast with a regional-scale application of the ocean health index. *PLoS One* 9(6):e98995.

Haltuch MA, Brooks EN, Brodziak J, Devine JA, Johnson KF, Klibansky N, Nash RD, et al. 2019. Unraveling the recruitment problem: A review of environmentally-informed forecasting and management strategy evaluation. *Fisheries Research* 217:198–216.

Haltuch MA, Tolimieri N, Lee Q, Jacox MG. 2020. Oceanographic drivers of petrale sole recruitment in the California Current Ecosystem. *Fisheries Oceanography* 29(2):122–36.

Hamilton S, Baker GB. 2019. Technical mitigation to reduce marine mammal bycatch and entanglement in commercial fishing gear: lessons learnt and future directions. *Reviews in Fish Biology and Fisheries* 29(2):223–47.

Hanak E, Lund JR. 2012. Adapting California's water management to climate change. *Climatic Change* 111(1):17–44.

Hanson MB, Good TP, Jannot JE, McVeigh J. 2019. *Estimated Humpback Whale Bycatch in the US West Coast Groundfish Fisheries. NMFS Report 4.* Seattle, WA: NMFS.

Hard JJ, Kope RG, Grant WS, Waknitz FW, Parker LT, Waples RS. 1996. *Status Review of Pink Salmon from Washington, Oregon, and California. NOAA Technical Memorandum NMFS-NWFSC-25.* Washington, DC: NOAA (p. 131).

Hare SR, Francis RC. 1995. Climate change and salmon production in the Northeast Pacific Ocean. *Canadian Special Publication of Fisheries and Aquatic Sciences* 121:357–72.

Harou JJ, Pulido-Velazquez M, Rosenberg DE, Medellín-Azuara J, Lund JR, Howitt RE. 2009. Hydro-economic models: Concepts, design, applications, and future prospects. *Journal of Hydrology* 375(3–4):627–43.

Harou JJ, Medellín-Azuara J, Zhu T, Tanaka SK, Lund JR, Stine S, Olivares MA, Jenkins MW. 2010. Economic consequences of optimized water management for a prolonged, severe drought in California. *Water Resources Research* 46(5).

Harvey CJ, Gross K, Simon VH, Hastie J. 2008. Trophic and fishery interactions between Pacific hake and rockfish: effect on rockfish population rebuilding times. *Marine Ecology Progress Series* 365:165–76.

Harvey CJ, Reum JC, Poe MR, Williams GD, Kim SJ. 2016. Using conceptual models and qualitative network models to advance integrative assessments of marine ecosystems. *Coastal Management* 44(5):486–503.

Harvey CJ. 2009. Effects of temperature change on demersal fishes in the California Current: a bioenergetics approach. *Canadian Journal of Fisheries and Aquatic Sciences* 66(9):1449–61.

Hauri C, Gruber N, Plattner GK, Alin S, Feely RA, Hales B, Wheeler PA. 2009. Ocean acidification in the California current system. *Oceanography* 22(4):60–71.

Hayes SA, Josephson E, Maze-Foley K, Rosel PE. 2019. *US Atlantic and Gulf of Mexico Marine Mammal Stock Assessments-2018. NOAA Tech Memorandum NMFS-NE-258.* Washington, DC: NOAA (p. 298).

Hayward TL, Cayan DR, Franks PJS, Lynn RJ, Mantyla AW, McGowan JA, Smith PE, Schwing FB, Venrick EL. 1995. The state of the California Current in 1994–1995: a period of transition. *CalCOFI Reports* 36:19–39.

Hayward TL, Cummings SL, Cayan DR, Chavez FP, Lynn RJ, Mantyla AW, Niler PP, et al. 1996. The state of the California Current in 1995–1996: continuing declines in macrozooplankton biomass during a period of nearly normal circulation. *CalCOFI Reports* 37:22–37.

Hayward TL, Mantyla AW, Lynn RJ, Smith PE, Chereskin TK. 1994. The state of the California Current in 1993–1994. *CalCOFI Reports* 35:19–35.

Hazen EL, Jorgensen S, Rykaczewski RR, Bograd SJ, Foley DG, Jonsen ID, Shaffer SA, et al. 2013. Predicted habitat shifts of Pacific top predators in a changing climate. *Nature Climate Change* 3(3):234.

Hazen EL, Palacios DM, Forney KA, Howell EA, Becker E, Hoover AL, Irvine L, et al. 2017. WhaleWatch: a dynamic management tool for predicting blue whale density in the California Current. *Journal of Applied Ecology* 54(5):1415–28.

Hazen EL, Scales KL, Maxwell SM, Briscoe DK, Welch H, Bograd SJ, Bailey H, et al. A dynamic ocean management tool to reduce bycatch and support sustainable fisheries. *Science Advances* 4(5):eaar3001.

Heery E, Cope JM. 2014. Co-occurrence of bycatch and target species in the groundfish demersal trawl fishery of the US west coast; with special consideration of rebuilding stocks. *Fishery Bulletin* 112(1):36–48.

Heinemann D, Gedamke J, Oleson E, Barlow J, Crance J, Holt M, Soldevilla M, Van Parijs S. 2016. *Report of the Joint Marine Mammal Commission-National Marine Fisheries Service Passive Acoustic Surveying Workshop, 16–17 April, 2015, La Jolla, CA. NOAA Technical Memorandum NMFS-F/SPO-164.* Washington, DC: NOAA (p. 107).

Hemery LG, Henkel SK. 2015. Patterns of benthic mega-invertebrate habitat associations in the Pacific Northwest continental shelf waters. *Biodiversity and Conservation* 24(7):1691–710.

Herbold B, Carlson SM, Henery R, Johnson RC, Mantua N, McClure M, Moyle PB, Sommer T. 2018. Managing for salmon resilience in California's variable and changing climate. *San Francisco Estuary and Watershed Science* 16(2).

Herrick Jr SF, Norton JG, Mason JE, Bessey C. 2007. Management application of an empirical model of sardine–climate regime shifts. *Marine Policy* 31(1):71–80.

Hilborn R, Ovando D. 2014. Reflections on the success of traditional fisheries management. *ICES Journal of Marine Science* 71(5):1040–6.

Hilborn RA, Stewart IJ, Branch TA, Jensen OP. 2012. Defining trade-offs among conservation, profitability, and food security in the California current bottom-trawl fishery. *Conservation Biology* 26(2):257–68.

Holbrook SJ, Schmitt RJ, Stephens Jr JS. 1997. Changes in an assemblage of temperate reef fishes associated with a climate shift. *Ecological Applications* 7(4):1299–310.

Holt CA, Punt AE. 2009. Incorporating climate information into rebuilding plans for overfished groundfish species of the US west coast. *Fisheries Research* 100(1):57–67.

Hoover DJ, Odigie KO, Swarzenski PW, Barnard P. 2017. Sea-level rise and coastal groundwater inundation and shoaling at select sites in California, USA. *Journal of Hydrology: Regional Studies* 11:234–49.

Horne PJ, Kaplan IC, Marshall KN, Levin PS, Harvey CJ, Hermann AJ, Fulton EA. 2010. *Design and Parameterization of a Spatially Explicit Ecosystem Model of the Central California Current. NOAA Technical Memorandum, NMFS-NWFSC-104.* Washington, DC: NOAA (pp. 1–140).

Howard MD, Sutula M, Caron DA, Chao Y, Farrara JD, Frenzel H, Jones B, et al. 2014. Anthropogenic nutrient sources rival natural sources on small scales in the coastal waters of the Southern California Bight. *Limnology and Oceanography* 59(1):285–97.

Hsieh CH, Kim HJ, Watson W, Di Lorenzo E, Sugihara G. 2009. Climate-driven changes in abundance and distribution of larvae of oceanic fishes in the southern California region. *Global Change Biology* 15(9):2137–52.

Hughes AR, Williams SL, Duarte CM, Heck Jr KL, Waycott M. 2009. Associations of concern: declining seagrasses and threatened dependent species. *Frontiers in Ecology and the Environment* 7(5):242–6.

Huyer, A, Wheeler, PA, Strub, PT, Smith, RL, Leelier, R, Korso, PM. 2007. The Newport line off Oregon—Studies in the North East Pacific. *Progress in Oceanography* 75:126–60.

Ishimura G, Herrick S, Sumaila UR. 2013. Stability of cooperative management of the Pacific sardine fishery under climate variability. *Marine Policy* 39:333–40.

Jacox MG, Edwards CA, Hazen EL, Bograd SJ. 2018. Coastal upwelling revisited: Ekman, Bakun, and improved upwelling indices for the US West Coast. *Journal of Geophysical Research: Oceans* 123(10):7332–50.

Jacox MG, Hazen EL, Zaba KD, Rudnick DL, Edwards CA, Moore AM, Bograd SJ. 2016. Impacts of the 2015–2016 El Niño on the California Current system: early assessment and comparison to past events. *Geophysical Research Letters* 43:7072–80.

Jamieson GS, Foreman MG, Cherniawsky JY, Levings CD. 2002. "European green crab (Carcinus maenas) dispersal: the Pacific experience." In: *Crabs in Cold Water Regions: Biology Management, and Economics.* Edited by AJ Paul, EG Dawe, R Elner, GS Jamieson, GH Kruse, RS Otto, B Sainte-Marie, TC Shirley, D Woodby. Fairbanks, AK: University of Alaska Sea Grant College Program.

Jamieson GS, Grosholz ED, Armstrong DA, Elner RW. 1998. Potential ecological implications from the introduction of the European green crab, Carcinus maenas (Linneaus), to British Columbia, Canada, and Washington, USA. *Journal of Natural History* 32(10–11):1587–98.

Jannot JE, Somers KA, Tuttle V, McVeigh J, Carretta JV, Helker V. 2002. *Observed and Estimated Marine Mammal Bycatch in US West Coast Groundfish Fisheries, 2002–16.* Washington, DC: NOAA.

Jeffrey KM, Côté IM, Irvine JR, Reynolds JD. 2017. Changes in body size of Canadian Pacific salmon over six decades. *Canadian Journal of Fisheries and Aquatic Sciences* 74(2):191–201.

Jensen CM, Hines E, Holzman BA, Moore TJ, Jahncke J, Redfern JV. 2015. Spatial and temporal variability in shipping traffic off San Francisco, California. *Coastal Management* 43(6):575–88.

Jensen GC, McDonald PS, Armstrong DA. 2002. East meets west: competitive interactions between green crab Carcinus maenas, and native and introduced shore crab Hemigrapsus spp. *Marine Ecology Progress Series* 225: 251–62.

Jensen TC. 1985. United States-Canada Pacific salmon interception treaty: an historical and legal overview. *Environmental Law* 16:363.

Johnson DF, Botsford LW, Methot Jr RD, Wainwright TC. 1986. Wind stress and cycles in Dungeness crab (Cancer magister) catch off California, Oregon, and Washington. *Canadian Journal of Fisheries and Aquatic Sciences* 43(4):838–45.

Johnson OW, Grant WS, Kope RG, Neely K, Waknitz FW, Waples RS. 1997. *Status Review of Chum Salmon from Washington, Oregon, and California. NOAA Technical Memorandum NMFS-NWFSC-32.* Washington, DC: NOAA (p. 279).

Jones HP, Tershy BR, Zavaleta ES, Croll DA, Keitt BS, Finkelstein ME, Howald GR. 2008. Severity of the effects of invasive rats on seabirds: a global review. *Conservation Biology* 22(1):16–26.

Juan-Jordá MJ, Barth JA, Clarke ME, Wakefield WW. 2009. Groundfish species associations with distinct oceanographic habitats in the Northern California Current. *Fisheries Oceanography* 18(1):1–9.

Jurgens LJ, Bonfim M, Lopez DP, Repetto MF, Freitag G, McCann L, Larson K, Ruiz GM, Freestone AL. 2018. Poleward range expansion of a non-indigenous bryozoan and new occurrences of exotic ascidians in southeast Alaska. *Bioinvasions Records* 7(4):357–66.

Kahru M, Kudela R, Manzano-Sarabia M, Mitchell BG. 2009. Trends in primary production in the California Current detected with satellite data. *Journal of Geophysical Research: Oceans* 114:C02004.

Kaldy JE, Lee KS. 2007. Factors controlling Zostera marina L. growth in the eastern and western Pacific

Ocean: comparisons between Korea and Oregon, USA. *Aquatic Botany* 87(2):116–26.

Kämpf J, Chapman P. 2016. "The California Current upwelling system." In: *Upwelling Systems of the World*. Edited by J Kämpf, P Chapman. Cham, Switzerland: Springer (pp. 97–160).

Kaplan I, Ward E. 2017. *Final Report for Phase II of a Spatially-Explicit Ecosystem Model for Quantifying Marine Mammal Impacts on Chinook Salmon in the Northeast Pacific Ocean*. Washington, DC: NOAA.

Kaplan IC, Brown CJ, Fulton EA, Gray IA, Field JC, Smith AD. 2013. Impacts of depleting forage species in the California Current. *Environmental Conservation* 40(4): 380–93.

Kaplan IC, Francis TB, Punt AE, Koehn LE, Curchitser E, Hurtado-Ferro F, Johnson KF, et al. 2019. A multi-model approach to understanding the role of Pacific sardine in the California Current food web. *Marine Ecology Progress Series* 617:307–21.

Kaplan IC, Gray IA, Levin PS. 2013. Cumulative impacts of fisheries in the California Current. *Fish and Fisheries* 14(4):515–27.

Kaplan IC, Horne PJ, Levin PS. 2012. Screening California Current fishery management scenarios using the Atlantis end-to-end ecosystem model. *Progress in Oceanography* 102:5–18.

Kaplan IC, Koehn LE, Hodgson EE, Marshall KN, Essington TE. 2017. Modeling food web effects of low sardine and anchovy abundance in the California Current. *Ecological Modelling* 359:1–24.

Kaplan IC, Leonard J. 2012. From krill to convenience stores: forecasting the economic and ecological effects of fisheries management on the US West Coast. *Marine Policy* 36(5):947–54.

Kaplan IC, Levin P. 2009. "Ecosystem-based management of what? An emerging approach for balancing conflicting objectives in marine resource management." In: *The Future of Fisheries Science in North America*. Edited by RJ Beamish, BJ Rothschild. Dordrecht, Netherlands: Springer (pp. 77–95).

Kaplan IC, Marshall KN, Hodgson E, Koehn L. 2014. *Update for 2014 Methodology Review: Ongoing Revisions to the Spatially Explicit Atlantis Ecosystem Model of the California Current*. Seattle, WA: NMFS (p. 63).

Kaplan IC, Marshall KN. 2016. A guinea pig's tale: learning to review end-to-end marine ecosystem models for management applications. *ICES Journal of Marine Science* 73(7):1715–24.

Kaplan IC, Williams GD, Bond NA, Hermann AJ, Siedlecki SA. 2016. Cloudy with a chance of sardines: forecasting sardine distributions using regional climate models. *Fisheries Oceanography* 25(1):15–27.

Karp MA, Blackhart K, Lynch PD, Deroba J, Hanselman D, Gertseva V, Teo S, et al. 2019. *Proceedings of the 13th National Stock Assessment Workshop: Model Complexity, Model Stability, and Ensemble Modeling. NOAA Technical Memoranda NMFS-F/SPO-189*. Washington, DC: NOAA (p. 49).

Katz J, Moyle PB, Quiñones RM, Israel J, Purdy S. 2013. Impending extinction of salmon, steelhead, and trout (Salmonidae) in California. *Environmental Biology of Fishes* 96(10–11):1169–86.

Kauer K, Bellquist L, Gleason M, Rubinstein A, Sullivan J, Oberhoff D, Damrosch L, Norvell M, Bell M. 2018. Reducing bycatch through a risk pool: a case study of the US West Coast groundfish fishery. *Marine Policy* 96:90–9.

Keller AA, Simon V, Chan F, Wakefield WW, Clarke ME, Barth JA, Kamikawa DA, Fruh EL. 2010. Demersal fish and invertebrate biomass in relation to an offshore hypoxic zone along the US West Coast. *Fisheries Oceanography* 19(1):76–87.

Keller AA, Wallace JR, Horness BH, Hamel OS, Stewart IJ. 2012. Variations in eastern North Pacific demersal fish biomass based on the US west coast groundfish bottom trawl survey (2003–2010). *Fishery Bulletin* 110(2).

Keller AA, Wallace JR, Methot RD. 2017. *The Northwest Fisheries Science Center's West Coast Groundfish Bottom Trawl Survey: History, Design, and Description. NOAA Technical Memorandum NMFS-NWFSC-136*. Washington, DC: NOAA.

Kentula ME, DeWitt TH. 2003. Abundance of seagrass (Zostera marina L.) and macroalgae in relation to the salinity-temperature gradient in Yaquina Bay, Oregon, USA. *Estuaries* 26(4):1130–41.

Kershner J, Samhouri JF, James CA, Levin PS. 2011. Selecting indicator portfolios for marine species and food webs: a Puget Sound case study. *PLoS One* 6(10).

Kilduff DP, Di Lorenzo E, Botsford LW, Teo SL. 2015. Changing central Pacific El Niños reduce stability of North American salmon survival rates. *Proceedings of the National Academy of Sciences* 112(35):10962–6.

King JR, Agostini VN, Harvey CJ, McFarlane GA, Foreman MG, Overland JE, Di Lorenzo E, Bond NA, Aydin KY. 2011. Climate forcing and the California Current ecosystem. *ICES Journal of Marine Science* 68(6):1199–216.

Kittinger JN, Koehn JZ, Le Cornu E, Ban NC, Gopnik M, Armsby M, Brooks C, et al. 2014. A practical approach for putting people in ecosystem-based ocean planning. *Frontiers in Ecology and the Environment* 12(8):448–56.

Klinger T, Chornesky EA, Whiteman EA, Chan F, Largier JL, Wakefield WW. 2017. Using integrated, ecosystem-level management to address intensifying ocean acidification and hypoxia in the California Current large marine ecosystem. *Elementa: Science of the Anthropocene* 5.

Knowlton JL, Donlan CJ, Roemer GW, Samaniego-Herrera A, Keitt BS, Wood B, Aguirre-Muñoz A, Faulkner KR, Tershy BR. 2007. Eradication of non-native mammals and the status of insular mammals on the California Channel

Islands, USA, and Pacific Baja California Peninsula Islands, Mexico. *The Southwestern Naturalist* 52(4):528–41.

Knudsen EE, McDonald D. 2020. *Sustainable Fisheries Management: Pacific Salmon.* Boca Raton, FL: CRC Press.

Knutson P. 1987. The unintended consequences of the Boldt decision. *Cultural Survival Quarterly* 11(2).

Koehn LE, Essington TE, Marshall KN, Kaplan IC, Sydeman WJ, Szoboszlai AI, Thayer JA. 2016. Developing a high taxonomic resolution food web model to assess the functional role of forage fish in the California Current ecosystem. *Ecological Modelling* 335:87–100.

Koehn LE, Essington TE, Marshall KN, Sydeman WJ, Szoboszlai AI, Thayer JA. 2017. Trade-offs between forage fish fisheries and their predators in the California Current. *ICES Journal of Marine Science* 74(9):2448–58.

Kudela RM, Banas NS, Barth JA, Frame ER, Jay DA, Largier JL, Lessard EJ, Peterson TD, Vander Woude AJ. 2008. New insights into the controls and mechanisms of plankton productivity in coastal upwelling waters of the northern California Current System. *Oceanography* 21(4):46–59.

Lackey RT, Lach DH, Duncan SL. 2006. Policy options to reverse the decline of wild Pacific salmon. *Fisheries* 31(7):344.

Lam CH, Kiefer DA, Domeier ML. 2015. Habitat characterization for striped marlin in the Pacific Ocean. *Fisheries Research* 166:80–91.

Large SI, Fay G, Friedland KD, Link JS. 2013. Defining trends and thresholds in responses of ecological indicators to fishing and environmental pressures. *ICES Journal of Marine Science* 70:755–67.

Large SI, Fay G, Friedland KD, Link JS. 2015a. Critical points in ecosystem responses to fishing and environmental pressures. *Marine Ecology Progress Series* 521:1–7.

Large SI, Fay G, Friedland KD, Link JS. 2015b. Quantifying patterns of change in marine ecosystem response to multiple pressures. *PLoS One* 10(3):e0119922.

Lavoie RE. 2005. Oyster culture in North America: history, present and future. *The 1st International Oyster Symposium Proceedings, Oyster Research Institute News* 17:14–21.

Lebon KM, Kelly RP. 2006. Evaluating alternatives to reduce whale entanglements in commercial Dungeness Crab fishing gear. *Global Ecology and Conservation* 18:e00608.

Lederhouse T, Link JS. 2016. A proposal for fishery habitat conservation decision-support indicators. *Coastal Management* 44(3):209–22.

Lees K, Pitois S, Scott C, Frid C, Mackinson S. 2006. Characterizing regime shifts in the marine environment. *Fish and Fisheries* 7(2):104–27.

Leet WS. 2001. *California's Living Marine Resources: A Status Report.* Berkeley, CA: University of California, Division of Agriculture and Natural Resources.

Leising AW, Schroeder ID, Bograd SJ, Abell J, Durazo R, Gaxiola-Castro G, Bjorkstedt EP, et al. 2015. State of the California Current 2014–15: Impacts of the warm water "blob". *California Cooperative Oceanic Fisheries Investigations Reports* 56:1–68.

Lester SE, Costello C, Halpern BS, Gaines SD, White C, Barth JA. 2013. Evaluating tradeoffs among ecosystem services to inform marine spatial planning. *Marine Policy* 38:80–9.

Lester SE, McLeod KL, Tallis H, Ruckelshaus M, Halpern BS, Levin PS, Chavez FP, et al. 2010. Science in support of ecosystem-based management for the US West Coast and beyond. *Biological Conservation* 143(3):576–87.

Levin PS, Breslow SJ, Harvey CJ, Norman KC, Poe MR, Williams GD, Plummer ML. 2016. Conceptualization of social-ecological systems of the California current: an examination of interdisciplinary science supporting ecosystem-based management. *Coastal Management* 44(5):397–408.

Levin PS, Essington TE, Marshall KN, Koehn LE, Anderson LG, Bundy A, Carothers C, et al. 2018. Building effective fishery ecosystem plans. *Marine Policy* 92:48–57.

Levin PS, Francis TB, Taylor NG. 2016. Thirty-two essential questions for understanding the social–ecological system of forage fish: the case of Pacific Herring. *Ecosystem Health and Sustainability* 2(4):e01213.

Levin PS, Holmes EE, Piner KR, Harvey CJ. 2006. Shifts in a Pacific Ocean fish assemblage: the potential influence of exploitation. *Conservation Biology* 20(4):1181–90.

Levin PS, Kaplan I, Grober-Dunsmore R, Chittaro PM, Oyamada S, Andrews K, Mangel M. 2009. A framework for assessing the biodiversity and fishery aspects of marine reserves. *Journal of Applied Ecology* 46(4):735–42.

Lewitus AJ, Horner RA, Caron DA, Garcia-Mendoza E, Hickey BM, Hunter M, Huppert DD, et al. 2012. Harmful algal blooms along the North American west coast region: history, trends, causes, and impacts. *Harmful Algae* 19: 133–59.

Lindegren M, Checkley DM, Rouyer T, MacCall AD, Stenseth NC. 2013. Climate, fishing, and fluctuations of sardine and anchovy in the California Current. *Proceedings of the National Academy of Sciences* 110(33):13672–7.

Lindegren M, Checkley Jr DM, Koslow JA, Goericke R, Ohman MD. 2018. Climate-mediated changes in marine ecosystem regulation during El Niño. Global Change Biology 24(2):796–809.

Link JS. 2018. System-level optimal yield: increased value, less risk, improved stability, and better fisheries. *Canadian Journal of Fisheries and Aquatic Sciences* 75(1):1–6.

Link JS, Marshak AR. 2019. Characterizing and comparing marine fisheries ecosystems in the United States: determinants of success in moving toward ecosystem-based fisheries management. *Reviews in Fish Biology and Fisheries* 29(1):23–70.

Litzow MA, Hunsicker ME, Bond NA, Burke BJ, Cunningham CJ, Gosselin JL, Norton EL, Ward EJ, Zador SG. 2020. The changing physical and ecological meanings of North Pacific Ocean climate indices. *Proceedings of the National Academy of Sciences* 117(14):7665–71.

Lluch-Belda DR, Crawford RJ, Kawasaki T, MacCall AD, Parrish RH, Schwartzlose RA, Smith PE. 1989. Worldwide fluctuations of sardine and anchovy stocks: the regime problem. *South African Journal of Marine Science* 8(1):195–205.

Lockwood BL, Somero GN. 2011. Invasive and native blue mussels (genus Mytilus) on the California coast: the role of physiology in a biological invasion. *Journal of Experimental Marine Biology and Ecology* 400(1–2):167–74.

Lord JP. 2017. Potential impact of the Asian shore crab Hemigrapsus sanguineus on native northeast Pacific crabs. *Biological Invasions* 19(6):1879–87.

Lynch PD, Methot RD, Link JS. 2018. *Implementing a Next Generation Stock Assessment Enterprise. An Update to the NOAA Fisheries Stock Assessment Improvement Plan. NOAA Technical Memorandum NMFS-F/ SPO-183.* Washington, DC: NOAA (p. 127).

Lynham J, Halpern BS, Blenckner T, Essington T, Estes J, Hunsicker M, Kappel C, et al. 2017. Costly stakeholder participation creates inertia in marine ecosystems. *Marine Policy* 76:122–9.

MacCall AD. 2011. "The sardine-anchovy puzzle." In: *Shifting Baselines.* Edited by JBC Jackson, K Alexander, E Sala. Washington, DC: Island Press (pp. 47–57).

MacDonald GM. 2007. Severe and sustained drought in southern California and the West: Present conditions and insights from the past on causes and impacts. *Quaternary International* 173:87–100.

MacWilliams ML, Ateljevich ES, Monismith SG, Enright C. 2016. An overview of multi-dimensional models of the Sacramento–San Joaquin Delta. *San Francisco Estuary and Watershed Science* 14(4).

Mannocci L, Boustany AM, Roberts JJ, Palacios DM, Dunn DC, Halpin PN, Viehman S, et al. 2017. Temporal resolutions in species distribution models of highly mobile marine animals: recommendations for ecologists and managers. *Diversity and Distributions* 23(10):1098–109.

Mantua NJ, Hare SR, Zhang Y, Wallace JM, Francis RC. 1997. A Pacific interdecadal climate oscillation with impacts on salmon production. *Bulletin of the American Meteorological Society* 78(6):1069–80.

Marchesiello P, Estrade P. 2009. Eddy activity and mixing in upwelling systems: a comparative study of Northwest Africa and California regions. *International Journal of Earth Sciences* 98(2):299–308.

Marchesiello P, McWilliams JC, Shchepetkin A. 2003. Equilibrium structure and dynamics of the California Current System. *Journal of physical Oceanography* 33(4):753–83.

Markel RW, Shurin JB. 2020. Contrasting effects of coastal upwelling on growth and recruitment of nearshore Pacific rockfishes (genus Sebastes). *Canadian Journal of Fisheries and Aquatic Sciences* 77(6):950–62.

Marshak AR, Brown SK. 2017. Habitat science is an essential element of ecosystem-based fisheries management. *Fisheries* 42(6):300–300.

Marshall KN, Kaplan IC, Hodgson EE, Hermann A, Busch DS, McElhany P, Essington TE, Harvey CJ, Fulton EA. 2017. Risks of ocean acidification in the California Current food web and fisheries: ecosystem model projections. *Global Change Biology* 23(4):1525–39.

Marshall KN, Stier AC, Samhouri JF, Kelly RP, Ward EJ. 2016. Conservation challenges of predator recovery. *Conservation Letters* 9(1):70–8.

Marshall KN, Levin PS, Essington TE, Koehn LE, Anderson LG, Bundy A, Carothers C, et al. 2018. Ecosystem-based fisheries management for social–ecological systems: renewing the focus in the United States with next generation fishery ecosystem plans. *Conservation Letters* 11(1):e12367.

Mason JE. 2004. Historical patterns from 74 years of commercial landings from California waters. *California Cooperative Oceanic Fisheries Investigations Report* 45:180.

Mate BR, Lagerquist BA, Calambokidis J. 1999. Movements of North Pacific blue whales during the feeding season off Southern California and their southern fall migration. *Marine Mammal Science* 15(4):1246–57.

Matheson K, McKenzie CH, Gregory RS, Robichaud DA, Bradbury IR, Snelgrove PV, Rose GA. 2016. Linking eelgrass decline and impacts on associated fish communities to European green crab Carcinus maenas invasion. *Marine Ecology Progress Series* 548:31–45.

Maxwell SM, Hazen EL, Bograd SJ, Halpern BS, Breed GA, Nickel B, Teutschel NM, et al. 2013. Cumulative human impacts on marine predators. *Nature Communications* 4:2688.

Maxwell SM, Hazen EL, Lewison RL, Dunn DC, Bailey H, Bograd SJ, Briscoe DK, et al. 2015. Dynamic ocean management: defining and conceptualizing real-time management of the ocean. *Marine Policy* 58:42–50.

McCabe RM, Hickey BM, Kudela RM, Lefebvre KA, Adams NG, Bill BD, Gulland FM, et al. 2016. An unprecedented coastwide toxic algal bloom linked to anomalous ocean conditions. *Geophysical Research Letters* 43(19):10–366.

McClatchie S, Goericke R, Koslow JA, Schwing FB, Bograd SJ, Charter RI, Watson WI, et al. 2008. The state of the California Current, 2007–2008: La Niña conditions and their effects on the ecosystem. *CalCOFI Reports* 49:39–76.

McClatchie S, Goericke R, Schwing FB, Bograd SJ, Peterson WT, Emmett R, Charter R, et al. 2009. The state of the California current, spring 2008–2009: cold conditions drive regional differences in coastal production. *CalCOFI Reports* 50:43–68.

McEvoy AF. 1990. *The Fisherman's Problem: Ecology and Law in the California Fisheries, 1850–1980.* Cambridge, UK: Cambridge University Press.

McClatchie S, Jacox MG, Ohman MD, Sala LM et al. 2016. State of the California Current 2015–16: comparisons with the 1997–98 El Niño. *California Cooperative Oceanic Fisheries Investigations*. Data report. 2016 Jan; 57.

McGowan JA, Bograd SJ, Lynn RJ, Miller AJ. 2003. The biological response to the 1977 regime shift in the California Current. *Deep Sea Research Part II: Topical Studies in Oceanography* 50(14–16):2567–82.

McFarlane GA, Smith PE, Baumgartner TR, Hunter JR. 2002. Climate variability and Pacific sardine populations and fisheries. *InAmerican Fisheries Society Symposium* 2002 (pp. 195–214). *American Fisheries Society*.32:195–214.

Mendelssohn R, Schwing FB, Bograd SJ. 2003. Spatial structure of subsurface temperature variability in the California Current, 1950–1993. *Journal of Geophysical Research: Oceans* 108(C3).

McGowan JA, Cayan DR, Dorman LM. 1998. Climate-ocean variability and ecosystem response in the Northeast Pacific. *Science* 281(5374):210–7.

Menge BA, Menge DN. 2013. Dynamics of coastal meta-ecosystems: the intermittent upwelling hypothesis and a test in rocky intertidal regions. *Ecological Monographs* 83(3):283–310.

Miller KA, Aguilar-Rosas LE, Pedroche FF. 2011. A review of non-native seaweeds from California, USA and Baja California, Mexico. *Hidrobiológica* 21(3):365–79.

Miller ML. 1987. Tourism in Washington's coastal zone. *Annals of Tourism Research* 14(1):58–70.

Miller TW, Bosley KL, Shibata J, Brodeur RD, Omori K, Emmett R. 2013. Contribution of prey to Humboldt squid Dosidicus gigas in the northern California Current, revealed by stable isotope analyses. *Marine Ecology Progress Series* 477:123–34.

Mills KL, Laidig T, Ralston S, Sydeman WJ. 2007. Diets of top predators indicate pelagic juvenile rockfish (Sebastes spp.) abundance in the California Current System. *Fisheries Oceanography* 16(3):273–83.

Moffitt EA, Punt AE, Holsman K, Aydin KY, Ianelli JN, Ortiz I. 2016. Moving towards ecosystem-based fisheries management: options for parameterizing multi-species biological reference points. *Deep Sea Research Part II: Topical Studies in Oceanography* 134:350–9.

Morley JW, Selden RL, Latour RJ, Frölicher TL, Seagraves RJ, Pinsky ML. 2018. Projecting shifts in thermal habitat for 686 species on the North American continental shelf. *PLoS One* 13(5):e0196127.

Moore JW, Hayes SA, Duffy W, Gallagher S, Michel CJ, Wright D. 2011. Nutrient fluxes and the recent collapse of coastal California salmon populations. *Canadian Journal of Fisheries and Aquatic Sciences* 68(7):1161–70.

Mueter FJ, Peterman RM, Pyper BJ. 2002. Opposite effects of ocean temperature on survival rates of 120 stocks of Pacific salmon (Oncorhynchus spp.) in northern and southern areas. *Canadian Journal of Fisheries and Aquatic Sciences* 59(3):456–63.

Muething KA. 2020. On the edge: assessing fish habitat use across the boundary between Pacific Oyster Aquaculture and Eelgrass in Willapa Bay, WA. *Aquaculture Environment Interactions* 12:541–57.

Munro G, McDorman T, McKelvey R. 1998. Transboundary fishery resources and the Canada-United States Pacific Salmon treaty. *Canadian-American Public Policy* 1(33):1.

Muto MM, Helker VT, Angliss RP, Boveng PL, Breiwick JM, Cameron MF, Clapham PJ, et al. 2019. *Alaska Marine Mammal Stock Assessments, 2018. NOAA Technical Memorandum NMFS-AFSC-393*. Washington, DC: NOAA (p. 390).

NANOOS (Northwest Association of Networked Ocean Observing Systems). 2017. *Bridge Document—Strategic Operational Plan*. Washington, DC: NANOOS (p. 9).

Neckles HA. 2015. Loss of eelgrass in Casco Bay, Maine, linked to green crab disturbance. *Northeastern Naturalist* 22(3):478–500.

Nehlsen W, Williams JE, Lichatowich JA. 1991. Pacific salmon at the crossroads: stocks at risk from California, Oregon, Idaho, and Washington. *Fisheries* 16(2):4–21.

Nevins, HM, Benson SR, Phillips EM, de Marignac J, DeVogelaere AP, Ames JA, Harvey JT. 2011. *Coastal Ocean Mammal and Bird Education and Research Surveys (BeachCOMBERS), 1997–2007: Ten Years of Monitoring Beached Marine Birds and Mammals in the Monterey Bay National Marine Sanctuary. Marine Sanctuaries Conservation Series ONMS11-02*. Silver Spring, MD: NOAA (p. 63).

NMFS (National Marine Fisheries Service) 2017a. *National Marine Fisheries Service—2nd Quarter 2017 Update*. Washington, DC: NOAA.

NMFS (National Marine Fisheries Service). 2017b. *National Observer Program FY 2013 Annual Report. NOAA Technical Memorandum NMFS F/SPO-178*. Washington, DC: NOAA (p. 34).

NMFS (National Marine Fisheries Service). 2019. Magnuson-Stevens act provisions; fisheries off West Coast states; Pacific coast groundfish fishery; Pacific fishery management plan; amendment 28. *Federal Register* 84(223):63966–92.

NMFS (National Marine Fisheries Service). 2020b. *National Marine Fisheries Service—2nd Quarter 2020 Update*. Washington, DC: NOAA (p. 51).

NMFS (NOAA National Marine Fisheries Service). 2018. *Fisheries Economics of the United States, 2016. NOAA Technical Memorandum NMFS-F/SPO-187*. Washington, DC: NOAA (p. 243).

NMFS (NOAA National Marine Fisheries Service). 2019a. *Status of Stocks 2018: Annual Report to Congress on the Status of U.S. Fisheries*. Washington, DC: NOAA.

NMFS (NOAA National Marine Fisheries Service). 2019b. *Western Regional Implementation Plan for Ecosystem-Based Fisheries Management. NOAA Fisheries Ecosystem-Based Fisheries Management Road Map*. Washington, DC: NOAA (p. 25).

NMFS (NOAA National Marine Fisheries Service). 2020a. *Fisheries of the United States, 2018. NOAA Current Fishery Statistics No. 2018*. Washington, DC: NOAA.

NOAA (National Oceanic and Atmospheric Administration). 2016. Fisheries off West Coast states; Comprehensive ecosystem-based amendment 1; Amendments to the fishery management plans for coastal Pelagic species, Pacific Coast groundfish, U.S. West Coast highly migratory species, and Pacific Coast salmon. Final rule. *Federal Register* 81(64):19054–8.

Norman K, Varney A, Vizek A. 2018. *US West Coast Fishing Communities and Climate Vulnerability in an Ecosystem-Based Management Context*. Washington, DC: NOAA.

Norman KC. 2007. *Community Profiles for West Coast and North Pacific Fisheries: Washington, Oregon, California, and Other US States*. East Seattle, WA: NOAA.

Norton EL, Siedlecki S, Kaplan IC, Hermann AJ, Fisher JL, Morgan CA, Officer S, et al. 2020. The importance of environmental exposure history in forecasting Dungeness crab megalopae occurrence using J-SCOPE, a high-resolution model for the US Pacific Northwest. *Frontiers in Marine Science* 7:102.

Norton J. 1987. *Ocean Climate Influences on Groundfish Recruitment in the California Current. Proceedings of the International Rockfish Symposium, Anchorage, Alaska (AK Sea Grant Rep 87-2)*. Anchorage, AK: University of Alaska (pp. 73–99).

Norton JG, Mason JE. 2005. Relationship of California sardine (Sardinops sagax) abundance to climate-scale ecological changes in the California Current system. *California Cooperative Oceanic Fisheries Investigations Report* 46:83.

NRC (National Research Council). 1996. *Upstream: Salmon and Society in the Pacific Northwest*. Washington, DC: National Academies Press.

NWFSC (Northwest Fisheries Science Center). 2015. *Status Review Update for Pacific Salmon and Steelhead Listed Under the Endangered Species Act: Pacific Northwest*. Washington, DC: NOAA (p. 356).

Oh SJ. 2009. *Role of Science in the Governance of Tuna Fisheries in the Eastern Pacific Ocean*. East Lansing, MI: Michigan State University.

Oreskes N, Finley C. 2007. *A Historical Analysis of the Collapse of Pacific Groundfish: US Fisheries Science, Development, and Management, 1945–1995*. San Diego. CA: University of California.

Paine RT. 1974. Intertidal community structure. *Oecologia* 15(2):93–120.

Parker A, Ollier CD. 2017. California sea level rise: evidence based forecasts vs. model predictions. *Ocean & Coastal Management* 149:198–209.

Parker SJ, Berkeley SA, Golden JT, Gunderson DR, Heifetz J, Hixon MA, Larson R, et al. 2000. Management of Pacific rockfish. *Fisheries* 25(3):22–30.

Peterson WT, Fisher JL, Peterson JO, Morgan CA, Burke BJ, Fresh KL. 2014. Applied fisheries oceanography: ecosystem indicators of ocean conditions inform fisheries management in the California Current. *Oceanography* 27(4):80–9.

Peterson WT, Keister JE. 2003. Interannual variability in copepod community composition at a coastal station in the northern California Current: a multivariate approach. *Deep-Sea Research II* 50:2499–517.

Peterson WT, Keister JE, Feinberg LR. 2002. The effects of the 1997–1999 El Nino/La Nina events on hydrography and zooplankton off the central Oregon coast. *Progress in Oceanography* 54:381–98.

Peterson WT. 2009. Copepod species richness as an indicator of long term changes in the coastal ecosystem of the northern California Current. *California Cooperative Oceanic Fisheries Investigations Report* 50:73–81.

Pfister CA, Berry HD, Mumford T. 2018. The dynamics of Kelp Forests in the Northeast Pacific Ocean and the relationship with environmental drivers. *Journal of Ecology* 106(4):1520–33.

PFMC (Pacific Fishery Management Council). 2013. *Pacific Coast Fishery Ecosystem Plan for the U.S. Portion of the California Current Large Marine Ecosystem*. Portland, OR: PFMC (p. 190).

PFMC (Pacific Fishery Management Council). 2014. *Appendix A to the Pacific Coast Salmon Fishery Management Plan as Modified by Amendment 18. The Pacific Coast Salmon Plan Identification and Description of Essential Fish Habitat, Adverse Impacts, and Recommended Conservation Measures for Salmon*. Portland, OR: PFMC (p. 219).

PFMC (Pacific Fishery Management Council). 2014. *Ecosystem Advisory Subpanel Report on the Atlantis Model Review. Agenda Item H.1.b. Supplemental EAS Report*. Portland, OR: PFMC.

PFMC (Pacific Fishery Management Council). 2017. *Ecosystem Initiatives Appendix to the Pacific Coast Fishery Ecosystem Plan for the U.S. Portion of the California Current Large Marine Ecosystem. Appendix A*. Portland, OR: PFMC (p. 20).

PFMC (Pacific Fishery Management Council). 2018. *Coastal Pelagic Species Fishery Management Plan as Amended Through Amendment 16*. Portland, OR: PFMC (p. 21).

PFMC (Pacific Fishery Management Council). 2018. *Fishery Management Plan for U.S. West Coast Fisheries for Highly Migratory Species as Amended Through Amendment 5*. Portland, OR: PFMC (p. 78).

PFMC (Pacific Fishery Management Council). 2019. *Coastal Pelagic Species Fishery Management Plan as Amended Through Amendment 17*. Portland, OR: PFMC (p. 49).

PFMC (Pacific Fishery Management Council). 2019. *Pacific Coast Groundfish Fishery Management Plan for the California, Oregon, and Washington Groundfish Fishery*. Portland, OR: PFMC (p. 147).

Pinsky ML, Reygondeau G, Caddell R, Palacios-Abrantes J, Spijkers J, Cheung WW. 2018. Preparing ocean governance for species on the move. *Science* 360(6394):1189–91.

Pitcher KW, Olesiuk PF, Brown RF, Lowry MS, Jeffries SJ, Sease JL, Perryman WL, Stinchcomb CE, Lowry LF. 2007. Abundance and distribution of the eastern North Pacific Steller sea lion (Eumetopias jubatus) population. *Fishery Bulletin* 105(1):102–16.

Pomeroy ES. 2003. *The Pacific Slope: A History of California, Oregon, Washington, Idaho, Utah, and Nevada.* Reno, NV: University of Nevada Press.

Powell AM, Xu J. 2011. Abrupt climate regime shifts, their potential forcing and fisheries impacts. *Atmospheric and Climate Sciences* 1(02):33.

PSMFC (Pacific States Marine Fisheries Commission). 2018. *71st Annual Report of the Pacific States Marine Fisheries Commission. 2018 Annual Report.* Portland, OR: PSMFC (p. 90).

Punt AE, MacCall AD, Essington TE, Francis TB, Hurtado-Ferro F, Johnson KF, Kaplan IC, et al. 2016. Exploring the implications of the harvest control rule for Pacific sardine, accounting for predator dynamics: A MICE model. *Ecological Modelling* 337:79–95.

Punt AE. 2011. The impact of climate change on the performance of rebuilding strategies for overfished groundfish species of the US west coast. *Fisheries Research* 109(2–3):320–9.

Radovich J. 1982. The collapse of the California sardine fishery. What have we learned? *CalCOFI Reports* 23:56–78.

Read AJ, Drinker P, Northridge S. 2006. Bycatch of marine mammals in US and global fisheries. *Conservation Biology* 20(1):163–9.

Redfern JV, McKenna MF, Moore TJ, Calambokidis J, Deangelis ML, Becker EA, Barlow J, et al. 2013. Assessing the risk of ships striking large whales in marine spatial planning. *Conservation Biology* 27(2):292–302.

Reeb CA, Arcangeli L, Block BA. 2000. Structure and migration corridors in Pacific populations of the Swordfish Xiphius gladius, as inferred through analyses of mitochondrial DNA. *Marine Biology* 136(6):1123–31.

Reiss CS, Checkley Jr DM, Bograd SJ. 2008. Remotely sensed spawning habitat of Pacific sardine (Sardinops sagax) and Northern anchovy (Engraulis mordax) within the California Current. *Fisheries Oceanography* 17(2):126–36.

Reum JC, McDonald PS, Ferriss BE, Farrell DM, Harvey CJ, Levin PS. 2015. Qualitative network models in support of ecosystem approaches to bivalve aquaculture. *ICES Journal of Marine Science* 72(8):2278–88.

Richards M. 2017. *From Disaster to Sustainability: The Story of the Pacific Groundfish.* Logan, UT: Utah State University.

Richerson K, Holland DS. 2017. Quantifying and predicting responses to a US West Coast salmon fishery closure. *ICES Journal of Marine Science* 74(9):2364–78.

Roemmich D, McGowan J. 1995. Climatic warming and the decline of zooplankton in the California Current. *Science* 267(5202):1324–6.

Ruckelshaus MA, Heppell S. 2004. Chinook Salmon (Oncorhynchus tshawytscha) in Puget Sound. *Species Conservation and Management: Case Studies* 208.

Ruckelshaus MH, Currens K, Fuerstenberg R, Graeber W, Rawson K, Sands NJ, Scott J. 2002a. *Planning Ranges and Preliminary Guidelines for the Delisting and Recovery of the Puget Sound Chinook salmon Evolutionarily Significant Unit. Puget Sound Technical Recovery Team.* Washington, DC: NOAA.

Ruckelshaus MH, Levin P, Johnson JB, Kareiva PM. 2002b. The Pacific salmon wars: what science brings to the challenge of recovering species. *Annual Review of Ecology and Systematics* 33(1):665–706.

Ruesink JL, Lenihan HS, Trimble AC. 2005. Introduction of non-native oysters: ecosystem effects and restoration implications. *Annual Review of Ecology, Evolution, and Systematics* 36:643–89.

Ruiz GM, Carlton JT, Grosholz ED, Hines AH. 1997. Global invasions of marine and estuarine habitats by non-indigenous species: mechanisms, extent, and consequences. *American Zoologist* 37(6):621–32.

Ruzicka JJ, Brink KH, Gifford DJ, Bahr F. 2016. A physically coupled end-to-end model platform for coastal ecosystems: simulating the effects of climate change and changing upwelling characteristics on the Northern California Current ecosystem. *Ecological Modelling* 331:86–99.

Ruzicka JJ, Daly EA, Brodeur RD. 2014. Summer jellyfish blooms in an upwelling ecosystem: Modeling impacts upon fish production and evaluating evidence. *ICES Conference and Meeting* 3662.

Ryan JP, Kudela RM, Birch JM, Blum M, Bowers HA, Chavez FP, Doucette GJ, et al. 2017. Causality of an extreme harmful algal bloom in Monterey Bay, California, during the 2014–2016 northeast Pacific warm anomaly. *Geophysical Research Letters* 44(11):5571–9.

Rykaczewski RR, Dunne JP. 2010. Enhanced nutrient supply to the California Current Ecosystem with global warming and increased stratification in an earth system model. *Geophysical Research Letters* 37(21).

Sage WN. 1946. The Oregon Treaty of 1846. *Canadian Historical Review* 27(4):349–67.

Salcido R. 2011. Siting offshore hydrokinetic energy projects: a comparative look at wave energy regulation in the Pacific Northwest. Golden Gate U. *Environmental Law Journal* 5:109.

Samhouri JF, Andrews KS, Fay G, Harvey CJ, Hazen EL, Hennessey SM, Holsman K, et al. 2017. Defining ecosystem thresholds for human activities and environmental pressures in the California Current. *Ecosphere* 8(6):e01860.

Samhouri JF, Haupt AJ, Levin PS, Link JS, Shuford R. 2013. Lessons learned from developing integrated ecosystem assessments to inform marine ecosystem-based management in the USA. *ICES Journal of Marine Science* 71(5):1205–15.

Samhouri JF, Stier AC, Hennessey SM, Novak M, Halpern BS, Levin PS. 2017. Rapid and direct recoveries of predators and prey through synchronized ecosystem management. *Nature Ecology & Evolution* 1(4):1–6.

Samhouri JF, Levin PS, James CA, Kershner J, Williams G. 2011. Using existing scientific capacity to set targets for ecosystem-based management: a Puget Sound case study. *Marine Policy 35*(4):508–18.

Santora JA, Mantua NJ, Schroeder ID, Field JC, Hazen EL, Bograd SJ, Sydeman WJ, et al. 2020. Habitat compression and ecosystem shifts as potential links between marine heatwave and record whale entanglements. *Nature Communications* 11(1):1–2.

Satterthwaite WH, Andrews KS, Burke BJ, Gosselin JL, Greene CM, Harvey CJ, Munsch SH, et al. 2002. Ecological thresholds in forecast performance for key United States West Coast Chinook salmon stocks. *ICES Journal of Marine Science* 77(4):1503–15.

Savoca MS, Brodie S, Welch H, Hoover A, Benaka LR, Bograd SJ, Hazen EL. 2020. Comprehensive bycatch assessment in US fisheries for prioritizing management. *Nature Sustainability* 30:1–9.

Scales KL, Hazen EL, Jacox MG, Edwards CA, Boustany AM, Oliver MJ, Bograd SJ. 2017a. Scale of inference: on the sensitivity of habitat models for wide-ranging marine predators to the resolution of environmental data. *Ecography* 40(1):210–20.

Scales KL, Schorr GS, Hazen EL, Bograd SJ, Miller PI, Andrews RD, Zerbini AN, Falcone EA. 2017b. Should I stay or should I go? Modelling year-round habitat suitability and drivers of residency for fin whales in the California current. *Diversity and Distributions* 23(10): 1204–15.

SCCOOS (Southern California Coastal Ocean Observing System). 2016. *Strategic Operational Plan 2016–2021.* Silver Spring, MD: IOOS (p. 24).

Schroeder ID, Black BA, Sydeman WJ, Bograd SJ, Hazen EL, Santora JA, Wells BK. 2013. The North Pacific High and wintertime pre-conditioning of California current productivity. *Geophysical Research Letters* 40(3):541–6.

Schroeder ID, Santora JA, Bograd SJ, Hazen EL, Sakuma KM, Moore AM, Edwards CA, Wells BK, Field JC. 2019. Source water variability as a driver of rockfish recruitment in the California Current Ecosystem: implications for climate change and fisheries management. *Canadian Journal of Fisheries and Aquatic Sciences* 76(6):950–60.

Schwantes CA. 1996. *The Pacific Northwest: An Interpretive History.* Lincoln, NE: University of Nebraska Press.

Schwing FB, Bograd SJ, Collins CA, Gaxiola-Castro GI, Garcia JO, Goericke RA, Gomez-Valdez JO, et al. 2002. The state of the California Current, 2001–2002: Will the California Current System keep its cool, or is El Niño looming? *Reports of California Cooperative Oceanic Fisheries Investigations* 43:31–68.

See KE, Feist BE. 2010. Reconstructing the range expansion and subsequent invasion of introduced European green crab along the west coast of the United States. *Biological Invasions* 12(5):1305–18.

Selden RL, Thorson JT, Samhouri JF, Bograd SJ, Brodie S, Carroll G, Haltuch MA, et al. 2020. Coupled changes in biomass and distribution drive trends in availability of fish stocks to US West Coast ports. *ICES Journal of Marine Science* 77(1):188–99.

Sharp GD, McLain DR. 1993. Fisheries, El Niiio-Southern Oscillation and upper-ocean temperature records: an Eastern Pacific example. *Oceanography* 6:13–22.

Shelton AO, Harvey CJ, Samhouri JF, Andrews KS, Feist BE, Frick KE, Tolimieri N, et al. 2018. From the predictable to the unexpected: kelp forest and benthic invertebrate community dynamics following decades of sea otter expansion. *Oecologia* 188(4):1105–19.

Shelton AO, Samhouri JF, Stier AC, Levin PS. 2014. Assessing trade-offs to inform ecosystem-based fisheries management of forage fish. *Scientific Reports* 4:7110.

Sherman K, Henkel S, Webster J. 2013. *Assessing and Addressing Information Needs of Stakeholders Involved in Wave Energy Development and Marine Spatial Planning. NNMREC Report 4.* Corvallis, OR: Northwest National Marine Renewable Energy Center.

Sherman K, DeBruyckere LA. 2018. *Eelgrass Habitats on the U.S. West Coast. State of the Knowledge of Eelgrass Ecosystem Services and Eelgrass Extent.* Portland, OR: PSMFC (p. 67).

Siedlecki SA, Kaplan IC, Hermann AJ, Nguyen TT, Bond NA, Newton JA, Williams GD, et al. 2016. Experiments with seasonal forecasts of ocean conditions for the northern region of the California Current upwelling system. *Scientific Reports* 6(1):1–8.

Simenstad CA, Jay DA, Sherwood CR. 1992. "Impacts of watershed management on land-margin ecosystems: the Columbia River estuary." In: *Watershed Management.* Edited by RJ Naiman. New York, NY: Springer (pp. 266–306).

Simkanin C, Davidson I, Falkner M, Sytsma M, Ruiz G. 2009. Intra-coastal ballast water flux and the potential for secondary spread of non-native species on the US West Coast. *Marine Pollution Bulletin* 58(3):366–74.

Smith SE. 1983. *Energy Related Development in the Columbia River Estuary: Potential, Impacts, and Mitigation.* Washington, DC: NOAA.

Song AM, Temby O, Krantzberg G, Hickey GM. 2017. *Institutional Features of US-Canadian Transboundary Fisheries Governance. Towards Continental Environmental Policy.* Albany, NY: University of New York Press (pp. 156–79).

Sorte CJ, Williams SL, Carlton JT. 2010. Marine range shifts and species introductions: comparative spread rates and community impacts. *Global Ecology and Biogeography* 19(3):303–16.

Srinivasan M, Brown SK, Markowitz E, Soldevilla M, Patterson E, Forney K, Murray K, et al. 2019. *Proceedings of the 2nd National Protected Species Assessment Workshop. NOAA Technical Memorandum NMFS-F/SPO-198*. Washington, DC: NOAA (p. 46).

Steneck RS, Graham MH, Bourque BJ, Corbett D, Erlandson JM, Estes JA, Tegner MJ. 2002. Kelp forest ecosystems: biodiversity, stability, resilience and future. *Environmental Conservation* 29(4):436–59.

Stercho AM. 2006. *The Importance of Place-Based Fisheries to the Karuk Tribe of California: A Socioeconomic Study. Doctoral Dissertation*. Arcata, CA: Humboldt State University.

Stewart IJ, Hamel OS. 2010. *Stock Assessment of Pacific hake, Merluccius Productus,(aka Whiting) in US and Canadian Waters in 2010*. Portland, OR: PFMC.

Stewart JS, Hazen EL, Bograd SJ, Byrnes JE, Foley DG, Gilly WF, Robison BH, Field JC. 2014. Combined climate-and prey-mediated range expansion of Humboldt squid (Dosidicus gigas), a large marine predator in the California Current System. *Global Change Biology* 20(6):1832–43.

Stewart JS, Hazen EL, Foley DG, Bograd SJ, Gilly WF. 2012. Marine predator migration during range expansion: Humboldt squid Dosidicus gigas in the northern California Current System. *Marine Ecology Progress Series* 471:135–50.

Stierhoff K, Zwolinski J, Renfree J, Mau S, Murfin D, Demer D. 2018. *Report on the SWFSC's Collection of Data During the 2015 Joint US-Canada Integrated Acoustic and Trawl Survey of Pacific Hake and Coastal Pelagic Species (SaKe 2015; 1507SH) Within the California Current Ecosystem, 15 June to 10 September 2015, Conducted Aboard Fisheries Survey Vessel Bell M. Shimada*. Washington, DC: NOAA.

Stierhoff KL, Zwolinski JP, Demer DA. 2020. *Distribution, Biomass, and Demography of Coastal Pelagic Fishes in the California Current Ecosystem During Summer 2019 Based on Acoustic-Trawl Sampling. NOAA Technical Memorandum NMFS NOAA-TM-NMFS-SWFSC 626*. Washington, DC: NOAA.

Stock CA, Alexander MA, Bond NA, Brander KM, Cheung WW, Curchitser EN, Delworth TL, et al. 2011. On the use of IPCC-class models to assess the impact of climate on living marine resources. *Progress in Oceanography* 88(1–4):1–27.

Strong DR, Ayres DA. 2016. Control and consequences of Spartina spp. invasions with focus upon San Francisco Bay. *Biological Invasions* 18(8):2237–46.

Sumaila UR, Cheung WW, Lam VW, Pauly D, Herrick S. 2011. Climate change impacts on the biophysics and economics of world fisheries. *Nature Climate Change* 1(9):449.

Sund PN, Blackburn M, Williams F. 1981. Tunas and their environment in the Pacific Ocean: a review. *Oceanography and Marine Biology Annual Review* 19:443–512.

Surma S, Pitcher TJ, Kumar R, Varkey D, Pakhomov EA, Lam ME. 2018. Herring supports Northeast Pacific predators and fisheries: insights from ecosystem modelling and management strategy evaluation. *PloS One* 13(7):e0196307.

Sutherland KR, Sorenson HL, Blondheim ON, Brodeur RD, Galloway AW. 2018. Range expansion of tropical pyrosomes in the northeast Pacific Ocean. *Ecology* 99(10).

Swain DL. 2015. A tale of two California droughts: lessons amidst record warmth and dryness in a region of complex physical and human geography. *Geophysical Research Letters* 42(22):9999–10.

Sydeman WJ, Allen SG. 1999. Pinniped population dynamics in central California: correlations with sea surface temperature and upwelling indices. *Marine Mammal Science* 15(2):446–61.

Sydeman WJ, Thompson SA, Field JC, Peterson WT, Tanasichuk RW, Freeland HJ, Bograd SJ, Rykaczewski RR. 2011. Does positioning of the North Pacific Current affect downstream ecosystem productivity? *Geophysical Research Letters* 38(12).

Sydeman WJ, Thompson SA, Kitaysky A. 2012. Seabirds and climate change: roadmap for the future. Marine Ecology Progress Series 454:107–17.

Sydeman WJ, Mills KL, Santora JA, Thompson SA, Bertram DF, Morgan KH, Wells BK, Hipfner JM, Wolf SG. 2009. Seabirds and climate in the California Current—a synthesis of change. *CalCOFI Report* 50:82–104.

Takada M. 2017. *Economic Effects of Pacific Halibut Closures on Businesses on the North Coast and the Age, Growth, and Reproductive Status of Pacific Halibut in Northern California and Central Oregon*. Arcata, CA: Humboldt State University.

Tallis H. 2009. Kelp and rivers subsidize rocky intertidal communities in the Pacific Northwest (USA). *Marine Ecology Progress Series* 389:85–96.

Tallis HM, Ruesink JL, Dumbauld B, Hacker S, Wisehart LM. 2009. Oysters and aquaculture practices affect eelgrass density and productivity in a Pacific Northwest estuary. *Journal of Shellfish Research* 28(2):251–62.

Tam JC, Link JS, Large SI, Andrews K, Friedland KD, Gove J, et al. 2017. Comparing apples to oranges: common trends and thresholds in anthropogenic and environmental pressures across multiple marine ecosystems. *Frontiers in Marine Science* 4:282.

Teck SJ, Halpern BS, Kappel CV, Micheli F, Selkoe KA, Crain CM, Martone R, et al. 2010. Using expert judgment to estimate marine ecosystem vulnerability in the California Current. *Ecological Applications* 20(5):1402–16.

Tegner MJ, Dayton PK. 2000. Ecosystem effects of fishing in kelp forest communities. *ICES Journal of Marine Science* 57(3):579–89.

Tegner MJ, Dayton PK. 1987. El Niño effects on southern California kelp forest communities. *Advances in Ecological Research* 17:243–79.

Tepolt CK, Darling JA, Bagley MJ, Geller JB, Blum MJ, Grosholz ED. 2009. European green crabs (Carcinus maenas) in the northeastern Pacific: genetic evidence for high population connectivity and current-mediated expansion from a single introduced source population. *Diversity and Distributions* 15(6):997–1009.

Thayer JA, Field JC, Sydeman WJ. 2014. Changes in California Chinook salmon diet over the past 50 years: relevance to the recent population crash. *Marine Ecology Progress Series* 498:249–61.

Thom RM, Haas E, Evans NR, Williams GD. 2011. Lower Columbia River and estuary habitat restoration prioritization framework. *Ecological Restoration* 29(1–2):94–110.

Thomas AC, Strub PT, Brickley P. 2003. Anomalous satellite-measured chlorophyll concentrations in the northern California Current in 2001–2002. *Geophysical Research Letters* 30(15).

Thompson AR, Chen DC, Guo LW, Hyde JR, Watson W. 2017. Larval abundances of rockfishes that were historically targeted by fishing increased over 16 years in association with a large marine protected area. *Royal Society Open Science* 4(9):170639.

Thompson LC, Escobar MI, Mosser CM, Purkey DR, Yates D, Moyle PB. 2011. Water management adaptations to prevent loss of spring-run Chinook salmon in California under climate change. *Journal of Water Resources Planning and Management*. 138(5):465–78.

Thompson, A.R. et al. 2019. State of the California Current 2018–19: a novel anchovy regime and a new marine heat wave? *California Cooperative Oceanic Fisheries Investigations Reports* 60:1–65.

Thomson C. Appendix MS7. *Commercial Fishing Economics Technical Report for the Secretarial Determination on Whether to Remove Four Dams on the Klamath River in California and Oregon.* Santa Cruz, CA: NOAA.

Thorne K, MacDonald G, Guntenspergen G, Ambrose R, Buffington K, Dugger B, Freeman C, et al. 2018. US Pacific coastal wetland resilience and vulnerability to sea-level rise. *Science Advances* 4(2):eaao3270.

Thorson JT, Dorn MW, Hamel OS. 2019. Steepness for West Coast rockfishes: results from a twelve-year experiment in iterative regional meta-analysis. *Fisheries Research* 217: 11–20.

Thorson JT, Stewart IJ, Taylor IG, Punt AE. 2013. Using a recruitment-linked multispecies stock assessment model to estimate common trends in recruitment for US West Coast groundfishes. *Marine Ecology Progress Series* 483:245–56.

Tiner Jr RW. 1984. *Wetlands of the United States: Current Status and Recent Trends*. Washington, DC: United States Fish and Wildlife Service.

Tolimieri N, Haltuch MA, Lee Q, Jacox MG, Bograd SJ. 2018. Oceanographic drivers of sablefish recruitment in the California Current. *Fisheries Oceanography* 27(5): 458–74.

Tommasi D, Stock CA, Pegion K, Vecchi GA, Methot RD, Alexander MA, Checkley Jr DM. 2017. Improved management of small pelagic fisheries through seasonal climate prediction. *Ecological Applications* 27(2):378–88.

Townsend H, Harvey CJ, deReynier Y, Davis D, Zador S, Gaichas S, Weijerman M, et al. 2019. Progress on implementing ecosystem-based fisheries management in the US through the use of ecosystem models and analysis. *Frontiers in Marine Science* 6:641.

Townsend HM, Harvey CJ, Aydin KY, Gamble RJ, Gruss A, Levin PS, Link JS, et al. 2014. *Report of the 3rd National Ecosystem Modeling Workshop (NEMoW 3): Mingling Models for Marine Resource Management, Multiple Model Inference.* Washington, DC: NOAA.

Townsend HM, Link JS, Osgood KE, Gedamke T, Watters GM, Polovina JJ, Levin PS, Cyr EC, Aydin KY. 2008. *Report of the National Ecosystem Modeling Workshop (NEMoW).* Washington, DC: NOAA.

Townsend H, Aydin K, Holsman K, Harvey C, Kaplan I, Hazen E, Woodworth-Jefcoats P, et al. 2017. *Report of the 4th National Ecosystem Modeling Workshop (NEMoW 4): Using Ecosystem Models to Evaluate Inevitable Trade-offs.* NOAA Technical Memorandum NMFS-F/SPO-173. Washington, DC: NOAA (p. 77).

Ulanski S. 2016. *The California Current: A Pacific Ecosystem and its Fliers, Divers, and Swimmers*. Chapel Hill, NC: UNC Press Books.

Venrick EL, Bograd SJ, Checkley DA, Durazo RE, Gaxiola-Castro GI, Hunter JO, Huyer AD, et al. 2003. The state of the California Current, 2002–2003: tropical and subarctic influences vie for dominance. *California Cooperative Oceanic Fisheries Investigations Report* 44:28–60.

Vojkovich M. 1998. The California fishery for market squid (Loligo opalescens). *California Cooperative Oceanic Fisheries Investigations Report* 1:55–60.

Wainwright TC, Emmett RL, Weitkamp LA, Hayes SA, Bentley PJ, Harding JA. 2019. Effect of a mammal excluder device on trawl catches of salmon and other pelagic animals. *Marine and Coastal Fisheries* 11(1):17–31.

Wainwright TC, Weitkamp LA. 2013. Effects of climate change on Oregon Coast coho salmon: habitat and life-cycle interactions. *Northwest Science* 87(3):219–42.

Waldek RD, Chapman J, Cordell J, Sytsma M. 2003. *Lower Columbia River Aquatic Nonindigenous Species Survey.* http://citeseerx.ist.psu.edu/viewdoc/download?doi=10.1.1.494.8789&rep=rep1&type=pdf

Walker R. 2001. Industry builds the city: the suburbanization of manufacturing in the San Francisco Bay Area, 1850–1940. *Journal of Historical Geography* 27(1): 36–57.

Waples RS, Zabel RW, Scheuerell MD, Sanderson BL. 2008. Evolutionary responses by native species to major anthropogenic changes to their ecosystems: Pacific salmon in the Columbia River hydropower system. *Molecular Ecology* 17(1):84–96.

Ward DH, Morton A, Tibbitts TL, Douglas DC, Carrera-González E. 2003. Long-term change in eelgrass distribution at Bahía San Quintín, Baja California, Mexico, using satellite imagery. *Estuaries* 26(6):1529.

Warlick A, Steiner E, Guldin M. 2018. History of the West Coast groundfish trawl fishery: tracking socioeconomic characteristics across different management policies in a multispecies fishery. *Marine Policy* 93:9–21.

Warner MD, Mass CF, Salathé Jr EP. 2012. Wintertime extreme precipitation events along the Pacific Northwest coast: climatology and synoptic evolution. *Monthly Weather Review* 140(7):2021–43.

Watson W, Manion SM, Charter RL. 2007. *Ichtyoplankton and Station Data for Oblique (Bongo Net) Plankton Tows Taken During a Survey of Shallow Coastal Waters of the Southern California Bight in 2004 and 2005. NOAA Technical Memorandum NOAA-TM-NMFS-SWFSC-410*. Washington, DC: NOAA (p. 62).

Wedding L, Yoklavich MM. 2015. Habitat-based predictive mapping of rockfish density and biomass off the central California coast. *Marine Ecology Progress Series* 540:235–50.

Weise MJ, Harvey JT. 2008. Temporal variability in ocean climate and California sea lion diet and biomass consumption: implications for fisheries management. *Marine Ecology Progress Series* 373:157–72.

Weitkamp LA, Wainwright TC, Bryant GJ, Milner GB, Teel DJ, Kope RG, Waples RS. 1995. *Status review of Coho Salmon from Washington, Oregon, and California. NOAA Technical Memorandum NMFS-NWFSC-24*. Washington, DC: NOAA (p. 258).

Welch H, Brodie S, Jacox MG, Bograd SJ, Hazen EL. 2019. Decision support tools for dynamic management. *Conservation Biology* 34(3):589–99.

Wells BK, Huff DD, Burke BJ, Brodeur RD, Santora JA, Field JC, Richerson K, et al. 2020. Implementing ecosystem-based management principles in the design of a salmon ocean ecology program. *Frontiers in Marine Science* 7:342.

Wells BK, Santora JA, Henderson MJ, Warzybok P, Jahncke J, Bradley RW, Huff DD, et al. 2017. Environmental conditions and prey-switching by a seabird predator impact juvenile salmon survival. *Journal of Marine Systems* 174:54–63.

Wells BK, Schroeder ID, Bograd SJ, Hazen EL, Jacox MG, Leising A, Mantua N, et al. State of the California Current 2016–17: still anything but "normal" in the north. *California Cooperative Oceanic Fisheries Investigations Reports* 58:1–55.

Wells BK, Schroeder ID, Santora JA, Hazen EL, Bograd SJ, Bjorkstedt EP, Loeb VJ, et al. State of the California Current 2012–13: no such thing as an "average" year. *California Cooperative Oceanic Fisheries Investigations Reports* 54:37–71.

Weng KC, Boustany AM, Pyle P, Anderson SD, Brown A, Block BA. 2007. Migration and habitat of white sharks (Carcharodon carcharias) in the eastern Pacific Ocean. *Marine Biology* 152(4):877–94.

Whitmire CE, Clarke ME. 2007. *State of Deep Coral Ecosystems of the US Pacific Coast: California to Washington. The State of Deep Coral Ecosystems of the United States NOAA Technical Memorandum CRCP-3*. Washington, DC: NOAA (pp. 109–54).

Wild PW. 1980. Effects of seawater temperature on spawning, egg development, hatching success, and population fluctuations of the Dungeness crab, Cancer magister. *CalCOFI Reports* 21:115–20.

Wilkinson, EB, Abrams K. 2015. *Benchmarking the 1999 EPAP Recommendations with existing Fishery Ecosystem Plans. U.S. Department of Commerce, NOAA Technical Memorandum NMFS-OSF-5* (p. 22).

Williams GH, Blood CL. 2003. Active and passive management of the recreational fishery for Pacific halibut off the US West Coast. *North American Journal of Fisheries Management* 23(4):1359–68.

Wolf PA. 1992. Recovery of the Pacific sardine and the California sardine fishery. *CalCOFI Reports* 33:76–86.

Wonham MJ, Carlton JT. 2005. Trends in marine biological invasions at local and regional scales: the Northeast Pacific Ocean as a model system. *Biological Invasions* 7(3):369–92.

Woodman SM, Forney KA, Becker EA, DeAngelis ML, Hazen EL, Palacios DM, Redfern JV. 2019. ESDM: a tool for creating and exploring ensembles of predictions from species distribution and abundance models. *Methods in Ecology and Evolution* 10(11):1923–33.

Wright BE, Riemer SD, Brown RF, Ougzin AM, Bucklin KA. 2007. Assessment of harbor seal predation on adult salmonids in a Pacific Northwest estuary. *Ecological Applications* 17(2):338–51.

Yamada SB, Peterson WT, Kosro PM. 2015. Biological and physical ocean indicators predict the success of an invasive crab, Carcinus maenas, in the northern California Current. *Marine Ecology Progress Series* 537:175–89.

Yamada SB, Thomson RE, Gillespie GE, Norgard TC. 2017. Lifting barriers to range expansion: the European green crab Carcinus maenas (Linnaeus, 1758) enters the Salish Sea. *Journal of Shellfish Research* 36(1):201–8.

Yoklavich MM, Love MS, Forney KA. 2007. A fishery-independent assessment of an overfished rockfish stock, cowcod (Sebastes levis), using direct observations from an occupied submersible. *Canadian Journal of Fisheries and Aquatic Sciences* 64(12):1795–804.

Yoklavich MM. 1998. *Marine Harvest Refugia for West Coast Rockfish: A Workshop. NOAA Technical Memorandum NOAA-TM-NMFS-SWFSC-255*. Washington, DC: NOAA.

Zabin CJ, Marraffini M, Lonhart SI, McCann L, Ceballos L, King C, Watanabe J, Pearse JS, Ruiz GM. 2018. Non-native species colonization of highly diverse, wave swept outer coast habitats in Central California. *Marine Biology* 165(2):31.

Zeidberg LD, Hamner WM, Nezlin NP, Henry A. 2006. The fishery for California market squid (Loligo opalescens) (Cephalopoda: Myopsida), from 1981 through 2003. *Fishery Bulletin* 104(1):46–59.

Zeidberg LD, Robison BH. 2007. Invasive range expansion by the Humboldt squid, Dosidicus gigas, in the eastern North Pacific. *Proceedings of the National Academy of Sciences* 104(31):12948–50.

Zwolinski JP, Demer DA, Byers KA, Cutter GR, Renfree JS, Sessions TS, Macewicz BJ. 2012. Distributions and abundances of Pacific sardine (Sardinops sagax) and other pelagic fishes in the California Current ecosystem during spring 2006, 2008, and 2010, estimated from acoustic–trawl surveys. *Fishery Bulletin* 110(1):110–22.

Zwolinski JP, Demer DA. 2012. A cold oceanographic regime with high exploitation rates in the Northeast Pacific forecasts a collapse of the sardine stock. *Proceedings of the National Academy of Sciences* 109(11):4175–80.

Zwolinski JP, Emmett RL, Demer DA. 2011. Predicting habitat to optimize sampling of Pacific sardine (Sardinops sagax). *ICES Journal of Marine Science* 68(5):867–79.

CHAPTER 9

The U.S. North Pacific Region

➲ The Region in Brief

- Currently, the North Pacific contains the highest commercial fisheries landings and value compared to the eight other U.S. regional marine ecosystems, with pronounced increases having occurred during fishery expansions in the 1980s.
- Some of the greatest increases in average sea surface temperature have been observed in this region at >2°C since the mid-twentieth century.
- The North Pacific contains the fifth-highest number of managed taxa, including commercially and recreationally important groundfish (e.g., Walleye pollock, Pacific cod, Sablefish, Lingcod, Pacific halibut, rockfishes, Yellowfin sole), cephalopods, King and Tanner crabs, salmon, and Steelhead.
- This region contains a low number and percentage of stocks that are overfished or experiencing overfishing, and a mid-range percentage of stocks of unknown overfished status.
- Human population and population density in this region are lowest in the nation relative to other regions, with values having slowly increased over time. Although less populated than other regions, human activities including fishing, energy exploration and extraction, and coastal tourism exert pressures on marine habitats.
- Among all U.S. regions, the North Pacific has a comparatively mid-range socioeconomic status, with the fifth-highest living marine resources (LMR)-based economy (i.e., establishments, employments, GDP) in the U.S. Additionally, some of the highest contributions of LMRs to total ocean economies are observed in this region, particularly in the Eastern Bering Sea (EBS) and Aleutian Islands subregions.
- In addition to having the cumulatively highest basal ecosystem productivity among U.S. regions, the North Pacific currently contains the highest overall fisheries landings, revenue, and proportions of LMR-dependent jobs in the nation.
- Large expanses of offshore rocky reef, sediment, and hard-bottom habitat, as well as inshore kelp and submerged aquatic vegetation support economically and ecologically important species throughout their life histories.
- Compared across regions, the second-highest numbers of marine mammal stocks (of which approximately one-fourth are strategic stocks) and low to mid-level numbers of threatened/endangered species are found in the North Pacific. Alaskan waters serve as prime feeding grounds for many species of whales, pinnipeds, sea otters, and other marine mammals.
- The North Pacific is the third-highest region nationally in offshore oil production, while also containing no proposed offshore wind energy areas.
- Fisheries management among state and federal entities has been focused upon maintaining good stock status for its relatively newer fisheries and on rebuilding plans for historically overfished or prohibited commercially and recreationally important species. The region has had the second-highest number of fisheries lawsuits in the country and a Fishery Management Council with dominant representation from commercial fishing, with recreational and other interests also represented.

continued

Ecosystem-Based Fisheries Management: Progress, Importance, and Impacts in the United States. Jason S. Link and Anthony R. Marshak, Oxford University Press. © U.S. Department of Commerce, U.S. Government 2021. DOI: 10.1093/oso/9780192843463.003.0009

- Among subregions of the North Pacific, the Gulf of Alaska and EBS-Aleutian Islands contribute most heavily toward fisheries landings, LMR revenue, and LMR employments, while tourism and shipping are most common in the Gulf of Alaska, including Southeast Alaska (aka the Inside Passage) and Prince William Sound. No commercial fisheries occur in the Arctic, where it is banned, with subsistence and some recreational fishing in the Arctic ecosystems (Chukchi and Beaufort seas) also occurring in recent years.
- Overall, a moderate to high degree of ecosystem-based fisheries management (EBFM) progress has been made in the EBS, Aleutian Islands, and Gulf of Alaska in terms of implementation, advancing knowledge of ecosystem principles, examining trade-offs, assessing risks and vulnerabilities, and in beginning to establish and use ecosystem-level reference points for management.
- The North Pacific is excelling in the areas of LMR and socioeconomic status and the quality of its governance system, and is relatively productive, as related to the determinants of successful LMR management.

9.1 Introduction

The U.S. North Pacific fisheries ecosystems bring to mind many nationally popular and iconic fisheries species that include Alaska (Walleye) pollock (*Gadus chalcogrammus*), Pacific cod (*G. macrophalus*), Pacific halibut (*Hippoglossus stenolepis*), Arctic cod (*Boreogadus saida*), Pacific salmon (*Oncorhynchus* spp.), and King (*Lithodes* spp., *Paralithodes* spp.), Snow, and Tanner crabs (*Chionoecetes* spp.). Trawlers, crabbing vessels, marine mammals, glacier-covered mountains, seabirds, cruise ships, oil platforms, and recreational ice fishing adorn its diverse maritime scenery, while plates of King salmon ("Chinook" *O. tshawytscha*), crab legs, farmed oysters, and abundant pollock or cod accompany these Alaskan images. These scenes may also be conjured from afar while consuming Alaskan seafood, given the prominent contributions of its fisheries toward domestic and global seafood supply. Sounds of the North Pacific include those of its many marine species, such as calls, barks, splashes, purrs, and shrieks from whales, seals, otters, and seabirds, together with hums from ship motors and pot haulers, in addition to cultural music of Alaska Native communities and popular American Folk music. Stories by Jack London, including *The Call of the Wild*, tales by James Michener and Kristin Hannah, and more recent non-fiction depictions such as Jonathan Raban's *Passage to Juneau: A Sea and its Meanings* and William McClosky's *Highliners* vividly describe the natural environments and fisheries resources of this region. Additionally, many Alaska Native legends handed down by indigenous tribal communities are rooted in human relations with this productive ecosystem. The marine environments of the North Pacific serve as a major foundation for its indigenous and maritime histories, especially as related to its abundant living marine resources (LMRs) and their overwhelming importance toward the economic livelihood of this region. They are also a major draw for significant tourism and serve as an important resource for offshore energy development and increasingly important mineral extractions.

There are many works that characterize the North Pacific marine ecosystem and its subregions (Feder & Jewett 1981, Walsh & McRoy 1986, Francis & Hare 1994, NRC 1996, Dalsgaard & Pauly 1997, Merrick et al. 1997, Anderson & Piatt 1999, Okey & Pauly 1999, Trites et al. 1999, Aydin et al. 2002, Hunt et al. 2002, Livingston 2002, Olson & Strom 2002, Livingston et al. 2005, Mundy 2005, Kruse et al. 2006, Aydin et al. 2007, Aydin & Mueter 2007, Litzow & Ciannelli 2007, Weingartner et al. 2009, Zador & Gaichas 2010, Gaichas et al. 2011, Hunt et al. 2011, Livingston et al. 2011, Ashjian et al. 2012, 2013, 2014, 2016, Wiese et al. 2012, Harvey & Sigler 2013, Lomas & Stabeno 2014, Moore et al. 2014, Whitehouse et al. 2014, Zador 2015, Dickson & Baker 2016, van Pelt et al. 2016, Siddon & Zador 2018, Zador & Ortiz 2018, Zador & Yasumiishi 2018, Ormseth et al. 2019), its important and major fisheries (e.g., Pacific cod, pollock, halibut, King and Tanner crabs; Bell 1981, Slizkin 1989, Stevens & MacIntosh 1991, Wespestad 1993, Buckley & Livingston 1994, Echeverria 1995, Zhou & Shirley 1998, Clark et al. 1999, Hollowed et al. 2000, Witherell et al. 2000, Barbeaux et al. 2003, Thompson et al. 2003, Kotwicki et al. 2005, Thompson & Dorn 2005, Mattes 2006, Nielsen et al. 2010, Seitz et al. 2011, Siddon et al. 2016, Holsman et al. 2018), its protected species (e.g., whales, seals, Steller sea lion, salmon; Nemoto 1957, Nemoto & Kawamura 1977, Frost & Lowry 1981, Nerini 1984, Frost et al. 1992, Buckland et al. 1993, Olesiuk 1993, York 1994, Merrick et al. 1997, Leatherwood et al. 1998, Mantua et al. 1997,

Sinclair & Zeppelin 2002, Waite et al. 2002, NRC 2003, Mizroch et al. 2004, Rugh et al. 2005, Shelden et al. 2005, Mizroch & Rice 2006, Zerbeni et al. 2006, Urawa et al. 2009, Saulitis et al. 2015, Zerbini et al. 2015), and their contributions toward the marine socioeconomics of this region, especially in how humans use, enjoy, and interact with the marine ecosystems, including their importance to Alaska Native communities (Wolfe & Walker 1987, Haynes & Mishler 1991, Crowell 1999, Richardson & Erickson 2005, Poole & Sepez 2007, Himes-Cornell et al. 2011, Peterson & Carothers 2013, Reedy & Maschner 2014, Young et al. 2014, Link & Marshak 2019). Alaska's resources and its goods and services have played a major role in American History, as illustrated by its evolving importance away from "Seward's Folly" to a lucrative resource for oil, gold, and living resources (Thompson & Freeman 1930, Hunt 1976, Ritter 1993, Crowell 1999, Springer 2011, Otto 2014), a site of conflict between U.S. and Japanese forces during World War II (Garfield 2010), and as a major epicenter for American environmentalism with significant protections of federal lands and first-hand observed consequences from the 1989 Exxon-Valdez oil spill (Hull & Leask 2000, Ross 2000, Peterson 2001, Buklis 2002, Peterson et al. 2003, Loughlin 2013).

What follows is a brief background for this region, describing all the major considerations of this marine ecosystem. Then there is a short synopsis of how all the salient data for those factors are collected, analyzed, and used for LMR management. We then present an evaluation of key facets of ecosystem-based fisheries management (EBFM) for this region.

9.2 Background and Context

9.2.1 Habitat and Geography

For the purposes here, we term the North Pacific region as not implying the entirety of the Northern Pacific or Arctic Oceans, but rather those waters that occur in or near the U.S. Exclusive Economic Zone (EEZ) largely associated with Alaska. The North Pacific makes up the second-largest portion of the United States EEZ (Holsman et al. 2018). It comprises all or portions of five Large Marine Ecosystems (LMEs): the **Gulf of Alaska**, **Aleutian Islands**, **Eastern Bering Sea** (EBS), northern Bering-Chukchi Seas, and Beaufort Sea LMEs, with the latter two part of the **Arctic** (treated herein jointly). These all have connectivity to the eastern portions of the Gulf of Alaska (Canada) and Beaufort Sea, the western Bering Sea (Russia), the California Current, and the rest of the Central Arctic Ocean (Figs. 9.1 and 9.2). The oceanography of this system (Fig. 9.1) is also heavily driven by the Alaska and Alaska Coastal Currents, Alaska Stream, Aleutian North Slope Current, and the Bering Slope Current Currents (Schumacher & Reed 1986, Schumacher et al. 1989, Stabeno et al. 2004, Cummins & Freeland 2007, Stabeno et al. 2016). The Gulf of Alaska has two subregions—**Prince William Sound** and the **Inside Passage** (largely synonymous with Southeast Alaska, but in the fjord-like coastal areas; Fig. 9.2). There are other subregional or LME features in this region, but we emphasize those six emboldened here as exemplary, distinct ecosystems.

The main features of the North Pacific ecosystem include the largest shoreline in the nation that contains expansive rocky intertidal, salt marsh, eelgrass (*Zostera marina*), beach, fjord, and mudflat habitats, which support many ecologically important marine species, in addition to offshore rocky substrates and broad mud/silt shelves, providing habitat to numerous commercially, protected, and recreationally important species (McRoy 1970, Carlson & Straty 1981, Taylor 1981, Vince & Snow 1984, Dean et al. 2000, Murphy et al. 2000, Abookire et al. 2001, Gay & Vaughan 2001, Johnson et al. 2003a, b, Funk et al. 2004, Hayes et al. 2010, Ezer & Liu 2010, Weingartner et al. 2009, Kemp et al. 2013, Logerwell et al. 2005, Pirtle et al. 2017).

9.2.2 Climate and Other Major Stressors

The North Pacific has been subject to substantial regional warming, with some of the highest increases in sea surface temperature (SST) among U.S. regions observed over time (Royer & Grosch 2006, Bond et al. 2015, Praetorius et al. 2018, Zador & Ortiz 2018, Link & Marshak 2019, Cheung & Frolicher 2020). There have been increasingly abundant range-shifting species from the California Current ecosystem to the Gulf of Alaska and from the EBS to the northern Bering-Chukchi and Beaufort seas, and concerns about proliferation of other ocean uses as a result of increasing development in areas that were previously inaccessible (Royer & Grosch 2006, Ruiz et al. 2006, Prowse et al. 2009, Burkett 2011, Coyle et al. 2011, Hunt et al. 2011, Mueter et al. 2011, Day et al. 2013, Hollowed et al. 2013, Sigler et al. 2016, Morley et al. 2018, Nong et al. 2018, Link & Marshak 2019).

More recently, the consequences of ongoing harvest (noted below) coupled with climate change and warming oceanographic events have been documented to affect major fished species, with observed declines in fish apex predator biomass (particularly Pacific cod

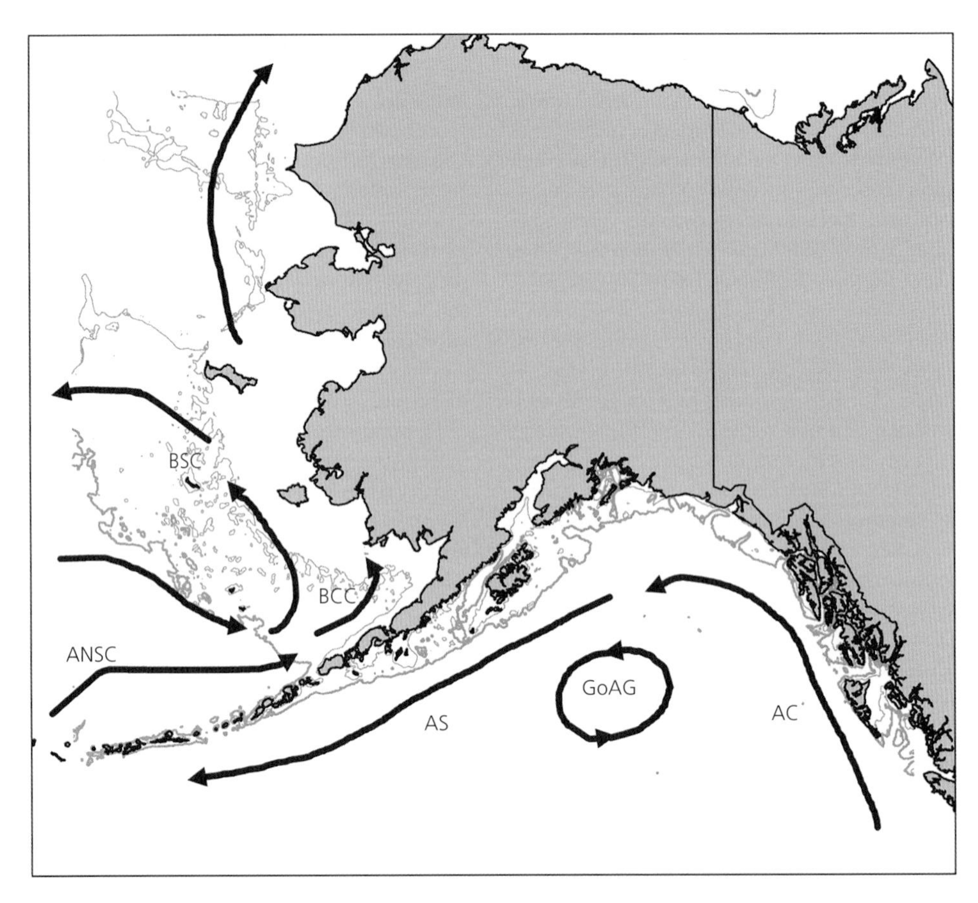

Figure 9.1 Major currents and circulation patterns throughout and encompassing the North Pacific region and subregions. AC = Alaska Current; GoAG = Gulf of Alaska Gyre; AS = Alaskan Stream; ANSC = Aleutian North Slope Current; BCC = Bering Coastal Current (aka Alaska Coastal Current); BSC = Bering Slope Current.

and Arrowtooth flounder (*Atheresthes stomias*), pelagic forage fishes (especially Pacific herring *Clupea pallasii*), also partly caused by the lingering effects of the oil spill in Prince William Sound (PWS) (e.g., Pearson et al. 2012, Ward et al. 2017), and motile epifauna in the Aleutian Islands, EBS, and Gulf of Alaska (Coyle et al. 2011, Mueter et al. 2011, Hollowed et al. 2012, Bond et al. 2015, Siddon & Zador 2018, Zador & Ortiz 2018, Zador & Yasumiishi 2018, Yang et al. 2019, Cheung & Frolicher 2020). Similar below-average or decreasing trends have been observed for seabirds and marine mammal populations over time, especially with lowered numbers of juvenile pinnipeds in recent years (Springer et al. 2003, Trites & Donnelly 2003, Trites et al. 2007, Springer et al. 2008, Siddon & Zador 2018, Zador & Ortiz 2018, Zador & Yasumiishi 2018). Additionally, increased prevalence of harmful algal blooms in the EBS have been thought to affect whale and seal populations, and along with poorer condition of groundfish as prey resources shift; all of which demonstrate the need to consider these factors in ongoing management strategies (Walsh et al. 2011, Lefebvre et al. 2016, Siddon & Zador 2018). Increasing tourism-associated recreational fishing effort, especially throughout the Gulf of Alaska, has been observed over the past decades, causing shifts in fishing allocations for artisanal and local fisheries communities (Colt et al. 2002, Criddle et al. 2003, Pollnac & Pogie 2006, Carothers 2010, Lew & Larson 2012). Despite these stressors this region currently contributes the highest proportions toward domestic fisheries landings and revenue (NMFS 2018a, b, Link & Marshak 2019). Efforts to systematically manage this system and its abundant groundfish resources are also in place, including a two million metric ton total allowable catch for the groundfish species in the EBS and Aleutian Islands and up to

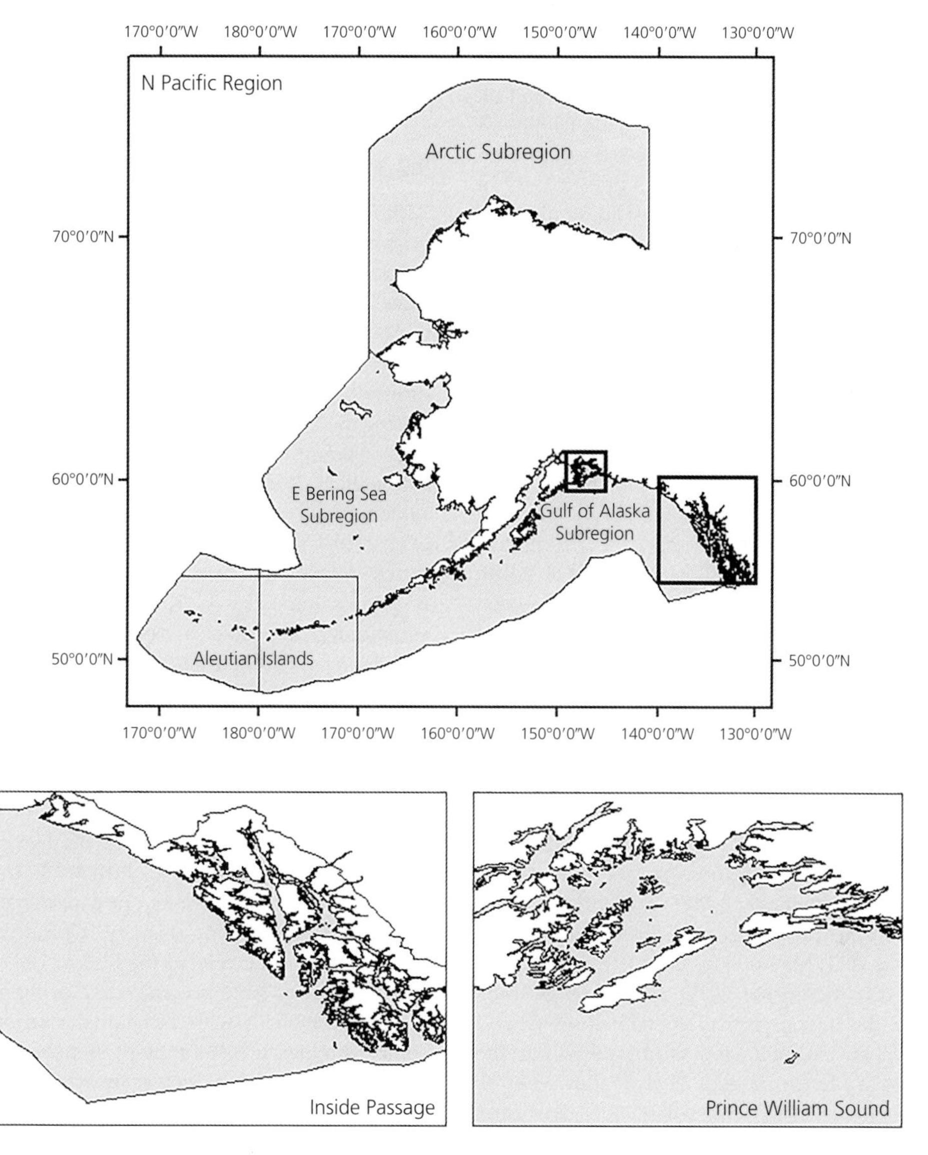

Figure 9.2 Map of the North Pacific region and identified subregions.

800 thousand metric tons in the Gulf of Alaska (Witherell et al. 2000, DiCosimo 2001, Livingston et al. 2011, Link 2018, Ono et al. 2018, NMFS 2020a).

In addition to climate-associated effects, Alaska's marine ecosystems are subject to other stressors, which include habitat loss, increasing coastal and offshore development with associated nutrient loading and pollution, climate-associated sea level rise, ocean acidification, and species range shifts, together with the effects of overfishing and more recent proliferations of exotic species (Hall et al. 1994, Iwata et al. 1994, Kurtz 1994, Hess et al. 1999, Kruse et al. 2000, Musick et al. 2000, Ebbert & Byrd 2002, Wooley 2002, Ruiz et al. 2006, Short et al. 2007, Fabry et al. 2009, Liebezeit et al. 2009, Gorokhovich & Leiserowiz 2011, Hogrefe et al. 2014, Schwörer et al. 2014, Cheung et al. 2015, Laidre et al. 2015, Mathis et al. 2015, Reimer et al. 2017a, Morley et al. 2018). Climate-associated sea ice reduction is responsible for significant habitat loss for many species in this region, especially for marine mammals such as ice seals, walruses, and polar bears (Laidre et al. 2008, Mueter & Litzow 2008, Kovacs et al. 2011, Rode et al. 2014, Laidre et al. 2015, Huntington et al. 2016). Regional warming and human development throughout this region has affected recovery of salmon populations and constrained fisheries productivity in this relatively well-managed system (Hall et al. 1994, Schoen et al. 2017), while effects of the 1989 Exxon-Valdez oil spill and remaining hydrocarbons have continued in PWS and surrounding areas of the Gulf of Alaska (Irons et al. 2000, Peterson 2001, Golet et al. 2002, Peterson et al. 2003, Short et al. 2007, Bowen et al. 2015, Shelton et al. 2017). While losses of seagrasses and kelp beds have not occurred to the degree observed in other, more populated regions, continued effects of climate change and increasing human activities in the North Pacific suggest higher vulnerability of these habitats to stressors (Hogrefe et al. 2014, Krumhansl et al. 2016, Talbot et al. 2016, Filbee-Dexter et al. 2018). Additional impacts in nearshore regions have been observed as a result of human-associated nutrient loading and pollution into essential habitats, and human alterations of Alaskan coastal environments, but their comparative effects have been less consequential than those of oil spills or fisheries overharvesting (Becker 2000, Wooley 2002, Harwell et al. 2010). Emergent impacts of sea level rise have also been observed in this region, although mediated by isostatic rebound in some areas. Ocean acidification has the potential to affect carbonate-producing LMRs, particularly shellfishes and deep corals (Fabry et al. 2009, Gorokhovich & Leiserowiz 2011, Mathis et al. 2015, Long et al. 2013, Punt et al. 2016, Long et al. 2019). Species range shifts are noted in this region, particularly with observed and predicted expansions of lower latitude species into the Gulf of Alaska (de Rivera et al. 2011, Hollowed et al. 2013, Cheung et al. 2015, Morley et al. 2018, Barbeaux & Hollowed 2018), in addition to range contractions of higher latitude crab species (Orensanz et al. 2004). Other range shifts of Alaskan species have been observed as associated with decreases in the "cold pool"—an area of cold bottom water in the EBS shelf that remains following sea ice retreat in the spring—with species such as King crabs and pollock moving farther north (Stabeno et al. 2001, Loher & Armstrong 2005, Grebmeier et al. 2006, Thorson et al. 2017, Alabia et al. 2018, Stevenson & Lauth 2019). Alaska has been subject to severe weather, which though not hurricanes like other regions of the U.S., is still harsh and is predicted to increase with regional warming, and impact its marine ecosystems, LMR-dependent human populations, and regional economies (Lynch et al. 2004, Lynch & Brunner 2007, Rodionov et al. 2007, Mesquita et al. 2010, Maldonado et al. 2013).

9.2.3 Exploitation

Although the large-scale development of Alaskan fisheries occurred later than in other U.S. regions, and few species have been listed in an overfished condition over time, certain consequences of overfishing (e.g., serial depletion of crustacean fisheries in the 1960s-early 1980s, fishing through foodwebs, overfishing of some groundfish and salmon stock depletions), particularly during the 1970s and 1980s, have affected North Pacific marine ecosystems (Armstrong et al. 1998, Goñi 1998, Witherell et al. 2000, Sigler et al. 2003, Clark et al. 2006, Litzow & Urban 2009, DiCosimo et al. 2010, Harwell et al. 2010, Crutchfield & Pontecorvo 2013). This historical overexploitation was compounded by and in many respects trumped by (G. Kruse, pers. comm.) shifts in community composition as a result of the Pacific Decadal Oscillation (PDO), particularly during the late 1970s, which has modified this system between being groundfish and crustacean-dominant over time, and affected annual fisheries population abundances, potential harvest, and species vulnerabilities (Francis & Hare 1994, Mantua et al. 1997, Anderson & Piatt 1999, Hare & Mantua 2000, Benson & Trites 2002, Heymans et al. 2005, Zheng & Kruse 2006, Litzow & Urban 2009, Gaichas et al. 2011). However, since management under the Magnuson-Stevens Fishery Conservation and Management Act extended the EEZ to 200 nmi, North Pacific fisheries have become some of the most well-managed nationally, with catches in this region currently comprising nearly 60% of total U.S. fisheries landings, and low numbers of stocks that are overfished or experiencing overfishing at present (Witherell et al. 2000, Witherell 2004, Litzow & Urban 2009, DiCosimo et al. 2010, Reuter et al. 2010, Link & Marshak 2019). North Pacific salmon populations have remained steady or increased over the past decades due to stricter management actions, with hatcheries also contributing significantly toward enhancement of fisheries biomass (Baker et al. 1996, Kruse et al. 2000, Ruggerone & Irvine 2018). The effects of trawling on seabed habitats are also less intense than in previous decades as a result of enhanced closures, regulations, and cooperation from the fishing industry (McConnaughey et al. 2000, Brown et al. 2005, Amoroso et al. 2018a, b, Siddon & Zador 2018).

The observed consequences of historical overfishing in other U.S. regions, together with the relatively more recent prominence and sustainable management of Alaskan fisheries, has afforded a more connected relationship among stakeholders and managers in the North Pacific (Holland & Ginter 2001, Holt et al. 2008, Ruckelshaus et al. 2008, Fina 2011, Siders et al. 2016, Raymond-Yakoubian & Daniel 2018). A much higher dependence on LMRs toward supporting the livelihoods of local residents, lower human population, and the increasingly observable effects of climate change and other stressors on local resources have facilitated demands for and reception toward more holistic management practices in this region (Callaway et al. 1999, Pollnac & Pogie 2006, Carothers 2010, Lew & Larson 2012, Holsman et al. 2019, Link & Marshak 2019). Increasing awareness of environmental threats to the sustainability of essential habitats, protected species, and commercial fisheries that serve as a major basis for LMR and tourism economies in this region have encouraged greater understanding of their interconnectivity and comprehensive management (Metcalf & Robards 2008, Pinsky & Mantua 2014, Sigler et al. 2016, Dorn et al. 2017, Holsman et al. 2019, Link & Marshak 2019). These concerns were especially magnified following the Exxon-Valdez oil spill and its observed ecological impacts in PWS, some of which have continued in decades since (Hull & Leask 2000, Ross 2000, Peterson 2001, Buklis 2002, Peterson et al. 2003, Loughlin 2013). Anticipated shifts in commercial and recreational fishing effort and concentrations, together with enhanced fisheries conflicts with protected species and expanding marine sectors, suggest that spatial conflicts in the North Pacific are likely to increase as compounded by regional warming (Stram & Evans 2009, Hollowed et al. 2012, 2013, Cheung et al. 2015, Reeves et al. 2012, 2014, Huntington et al. 2015, Laidre et al. 2015). As a result of these ongoing pressures, greater reception toward robust ecosystem-based management frameworks that account for the interconnected effects of human and natural stressors on its socioecological system continues to emerge in this region (Witherell et al. 2000, NPFMC 2007, Zador et al. 2017a, b, Siddon & Zador 2018, Zador & Ortiz 2018, Zador & Yasumiishi 2018, NMFS 2019, NPFMC 2019). Building on these advancements through cumulative consideration of ecosystem-level factors will continue to advance ecosystem-based management (EBM) for the North Pacific.

9.2.4 Invasive Species

With increasing temperatures and human activities in the North Pacific, the proliferation of invasive species within this region has become of increasing concern (Ebbert & Byrd 2002, Ruiz et al. 2006, de Rivera et al. 2011, Schwörer et al. 2014, Verna et al. 2016). Recent risk assessments have identified the Gulf of Alaska as especially vulnerable, given high ship traffic and ballast water exchange in the Ports of Valdez and Drift River Terminal (McGee et al. 2006, Verna et al. 2016). Increasing vulnerability for biological invasions in the Arctic is also anticipated as ship traffic increases in areas of decreasing sea ice (Miller & Ruiz 2014, Verna et al. 2016). Invasive tunicates have been observed in Sitka and other parts of the Inside Passage subregion (Cohen et al. 2011, McCann et al. 2013), while non-native amphipods and other species transferred via ballast water continue to invade Alaskan waters (Ashton et al. 2008). Although biological invasions in the North Pacific ecosystems have not been historically as frequent or concentrated as in other U.S. marine ecosystems, there is great potential for their future spread and ecological effect in this relatively species-poor region as a result of climate-associated thermal expansions of biological invaders (de Rivera et al. 2011, Renaud et al. 2015, Verna et al. 2016). As a result, there are increased concerns regarding expanding European green crab (*Carcinus maenas*) populations along the Pacific coastline into Alaskan waters, changes in fouling communities, and proliferations of sessile invasive fauna in North Pacific ecosystems (e.g., *Didemnum vexillum*, Ascidiacea), especially in the Gulf of Alaska and within PWS (Ebbert & Byrd 2002, Ruiz et al. 2006, McCann et al. 2013). These are also coupled with investigations and monitoring efforts documenting the ecological effects of biological introductions (i.e., rodents, foxes) and invasive plants (e.g., *Elodea* sp.) on commercially and ecologically important marine species such as seabirds and Chinook salmon (*Oncorhynchus tshawytscha*; Ebbert & Byrd 2002, Towns et al. 2011, Luizza et al. 2016).

9.2.5 Ecosystem-Based Management (EBM) and Multisector Considerations

The North Pacific ecosystem is known for balancing trade-offs across its LMRs (and their usage) and between LMR management and other ocean uses (Aydin et al. 2005, Livingston et al. 2005, 2011, Aydin et al. 2007, NPFMC 2007, Hollowed et al. 2011, Zador 2015, Kraska 2016, Tam et al. 2017, Zador et al. 2017a, b, Siddon & Zador 2018, Zador & Ortiz 2018, Zador & Yasumiishi 2018, NPFMC 2019, Watson & Haynie 2018). While fisheries landings and effort are currently

Seal vs. Fish Smackdown: The Trade-Offs Across Conflicting Objectives Between Pinniped and Fishes/Fisheries

—(see also, e.g., Gulf of Maine seals and flatfish, groundfish (Chapter 3); California sea lions and salmon—west coast, Chapter 8)

Some of you may recall the old cage matches in pro wrestling. In one corner, some near-descendant of a Cro-Magnon takes on a steroid-fueled, glistening-skinned, muscle-bound pretty boy with long, flowing, blonde locks. The championship of the free world, or something like that, is at stake. And in this smackdown, only one of them apparently gets to the leave the cage conscious.

It's a bit like what we see for various marine mammals and fish.

Stellar sea lions (*Eumetopias jubatus*) eat forage fish (Gende & Sigler 2006, Womble et al. 2009). They also need to periodically crawl out of the ocean onto land to rest, have babies, and conduct similar life activities. These rookeries provide a sort of home base for these pinnipeds, and as long as they can forage sufficiently within a reasonable distance around their rookeries, all is good.

Groundfish are commercially targeted species in Alaska. They too eat forage fish. Often, the bathymetry and oceanography focus these forage fish around pinniped rookeries (Gende & Sigler 2006). Fisheries that target groundfish thus fish for them around rookeries (Fritz et al. 1995, Witherell et al. 2000). Fisheries also fish for forage fish, again, also around rookeries.

See where this is going?

The competition between sea lions and groundfish, and sea lions and the fishery that targets groundfish and forage, can be intense. To the point that sea lions started to lose out, and their populations declined.

The Magnuson-Stevens Act focuses on ensuring that groundfish are caught at sustainable levels. The Marine Mammal Protection Act, and also the Endangered Species Act, focuses on ensuring that these marine mammals are not experiencing population declines beyond what is viable for the sea lions to persist. The two sets of Acts have, at some level, potentially conflicting goals. Enter the National Environmental Policy Act (NEPA). Essentially NEPA was used to help develop an environmental impact statement to sort out the issue of competition between the two conflicting objectives.

Without acknowledging the conflict, deleterious effects would have continued. The result was a buffer zone around sea lion rookeries (Witherell et al. 2000). Fisheries could still catch groundfish and forage fish, just a little further off the islands. Not everyone liked all the details, there were lawsuits involved, and certainly no interest group got everything they wanted. The aim by the judicial system was to look for a win-win solution.

There are comparable examples for marine mammals and commercially targeted fish all around the country, for instance, Gulf of Maine seals and groundfish, Polar bears and seals vs. forage fish in the Arctic. And even two types of protected species, California sea lions vs. salmon on the West Coast. Among many others. The salient point is not to shy away from and ignore the conflict. Rather the aim is to acknowledge it and attempt to find workable solutions for all interests involved. That is the essence of dealing with trade-offs in EBFM.

References

Fritz LW, Ferrero RC, Berg RJ. 1995. The threatened status of Steller sea lions, Eumetopias jubatus, under the Endangered Species Act: effects on Alaska groundfish fisheries management. *Marine Fisheries Review* 57(2):14–27.

Gende SM, Sigler MF. 2006. Persistence of forage fish 'hot spots' and its association with foraging Steller sea lions (Eumetopias jubatus) in southeast Alaska. *Deep Sea Research Part II: Topical Studies in Oceanography* 53(3–4):432–41.

Witherell D, Pautzke C, Fluharty D. 2000. An ecosystem-based approach for Alaska groundfish fisheries. *ICES Journal of Marine Science* 57(3):771–7.

Womble JN, Sigler MF, Willson MF. 2009. Linking seasonal distribution patterns with prey availability in a central-place forager, the Steller sea lion. *Journal of Biogeography* 36(3):439–51.

c.f.—https://www.eastoregonian.com/opinion/editorials/sea-lions-expose-conflicts-in-laws/article_bc8ea034-fbab-5b94-acb6-315ff84a0b48.html#.Wx_aqssGXgA.linkedin

dominated by commercial interests, greater prominence of recreational fisheries may occur in this region as a result of increasing tourism, human population, and climate-associated range expansions of California Current-associated sportfishes (Colt et al. 2002, Criddle et al. 2003, Pollnac & Pogie 2006, Carothers 2010, Lew & Larson 2012, Cheung et al. 2015, Morley et al. 2018). These increasingly probable fisheries trade-offs also compound current ones among multiple coastal and ocean uses that support Alaska's marine economy, and which co-occur within areas providing habitat to many managed species. Sectors including marine energy, transportation, and tourism (e.g., cruise ships, whale watching, ecotourism) are especially prominent in the North Pacific, while the economic contributions of LMRs are also proportionally highest in this region relative to other U.S. regions (Colt et al. 2002, Criddle et al. 2003, Colt et al. 2007, Dugan et al. 2007, Prowse et al. 2009, Pomeranz et al. 2013, Kraska 2016, Nong et al. 2018, Link & Marshak 2019). Concerns about the effects of these sectors on protected species and their trade-offs with regional fisheries are also among the most nationally prominent, particularly regarding critical habitats for marine mammals, seabirds, and the effects of energy exploration, fisheries bycatch, and marine transportation on the viability of their populations (Fritz et al. 1995, Trites et al. 1997, Kruse et al. 2000, Tasker et al. 2000, Montevecchi 2002, Board 2003, Huntington 2009, Reeves et al. 2012, 2014, Huntington et al. 2015, Laidre et al. 2015). Bycatch of fishes, invertebrates, and seabirds is among some of the highest nationally in this region, particularly in the EBS (Freese et al. 1999, Melvin et al. 2004, Moore et. 2009, Stevenson & Lewis 2010, Benaka et al. 2019, Link & Marshak 2019). This is largely due to the high volume of catches, as the bycatch rates are relatively quite low (Benaka et al. 2019). Increasing ship traffic and expansion of maritime routes into areas with decreasing sea ice may lead to higher proliferation of invasive species, as management of ballast water release has continued to lag in this region (McGee et al. 2006, Chan et al. 2013, Verna et al. 2016). Expansions for marine energy and mineral extraction interests, in addition to potential extension of the Trans-Alaska Pipeline System, have also occurred toward more polar regions of Alaska, with potential consequences to essential habitats, fishing closures, protected species populations, and international resource agreements (Brigham 2010, Hong 2012, Reeves et al. 2012, Levine et al. 2013, Humphries & Hueffman 2014, Kaiser et al. 2016, Nong et al. 2018). To account for these overlaps and potential resource conflicts, more holistic management is necessary to consider the interconnectivity of marine sectors and their cumulative effects on the North Pacific system at large (Duda & Sherman 2002, Kruse et al. 2006, Marasco et al. 2007, Huntington et al. 2015, Siders et al. 2016). Exploring, identifying, and resolving these differences in priorities and goals for the North Pacific is the essence of EBM, and the emerging basis of its regional fishery ecosystem plans (FEPs) and EBFM implementation plans (NPFMC 2007, NMFS 2019, NPFMC 2019). With its many taxa, multiple habitats, large geographic area, extensive shoreline, emerging trade-offs among species and marine sectors, and increasing human pressures, continued implementation and strengthening of ecosystem approaches to marine resource management in the North Pacific remains a priority.

9.3 Informational and Analytical Considerations for this Region

9.3.1 Observation Systems and Data Sources

Data regarding the physical, biological, and socioeconomic components of the North Pacific ecosystem are available through regional observation systems and multiple monitoring programs spanning several decades. The Alaska Ocean Observing System (AOOS) provides oceanographic and atmospheric data from multiple sensor stations throughout coastal and offshore locations (AOOS 2016). The system is a composite of aggregated data from ~20 federal, state, academic, and tribal partnerships and sources, with the majority of observations from coastal sensors maintained by NOAA's National Weather Service (NWS), Center for Operational Oceanographic Products and Services (CO-OPS), and National Data Buoy Center (NDBC), in addition to those maintained by the U.S. Geological Survey (USGS) National Water Information System (NWIS) and Alaska Department of Transportation. Containing sensors that have been collecting physical data at daily to annual scales since the late 1970s, the AOOS portal monitors 43 variables including atmospheric (e.g., air temperature, pressure, precipitation, humidity, winds) and oceanographic (e.g., water temperature, salinity, oxygen, pH, waves, carbon dioxide, aragonite, photosynthetically available radiation, etc.) parameters. This information is reported for the Arctic and Gulf of Alaska, and complemented by long-term monitoring data for sea ice, habitat, and physical parameters in PWS following the Exxon-Valdez oil spill. AOOS priorities also include providing comprehensive information regarding the North Pacific ecosystem, assessing and predicting hazards such as oil spills, harmful algal blooms, and hypoxia, forecasting climatological effects, and enhancing safe maritime practices (AOOS 2016).

Multiple oceanographic and atmospheric data sources have been incorporated toward broader understanding of the North Pacific ecosystem, particularly its overall status (Zador 2015, Siddon & Zador 2018, Zador & Ortiz 2018, Zador & Yasumiishi 2018) and focal components, and applicability to regional ecosystem status reports (ESRs) and integrated ecosystem assessments (IEAs; Levin et al. 2009, 'et al. 2014, Zador 2015, Siddon & Zador 2018, Zador & Ortiz 2018, Zador & Yasumiishi 2018). Data regarding climatological patterns, warming events, and sea surface pressure anomalies have been obtained from satellite-derived SST from NOAA's Optimum Interpolation Sea Surface Temperature (OISST) database, sea level pressure from NOAA's Earth Systems Research Laboratory (ESRL), sea ice data from the National Snow and Ice Data Center (NSIDC), NOAA's Global Ocean Data Assimilation System (GODAS), and applications that include NOAA's Ocean Surface Current Simulator (OSCURS). Information regarding North Pacific climate indices is also maintained at ESRL and applied toward evaluating ecosystem status and projecting seasonal atmospheric and oceanographic trends for the region (Zador 2015, Siddon & Zador 2018, Zador & Ortiz 2018, Zador & Yasumiishi 2018). Information regarding water column dynamics, bottom temperatures, and *in-situ* processes is also available from samplings conducted during NOAA trawl surveys.

Habitat factors and aspects of primary and secondary production are monitored via satellite information and ongoing monitoring programs throughout Alaska's marine ecosystems that account for chlorophyll *a* concentrations, phytoplankton biomass, and their estimated coverage (Arrigo et al. 2008, Coyle et al. 2008, Arrigo & van Dijken 2011, 2015, Brown et al. 2011, Arrigo et al. 2012, Brown & Arrigo 2012, Arrigo et al. 2014, Eisner et al. 2016, Eisner et al. 2017). These include data applied toward conducting habitat risk assessments, and examining the status of nearshore habitats and offshore trawlable and untrawlable substrates, including the amount of area disturbed by bottom tending fisheries gear and status and effects to their associated structural epifauna (McConnaughey et al. 2000, Heifetz et al. 2009, Zador 2015, Amoroso et al. 2018b, Siddon & Zador 2018, Zador & Ortiz 2018, Zador & Yasumiishi 2018). These observations also account for phytoplankton and emergent harmful algal blooms. Zooplankton, jellyfish, and ichthyoplankton abundance and biomass are accounted for in surveys from the NOAA Alaska Fisheries Science Center (AFSC) seasonal monitoring programs and trawl surveys dating back to the early 1970s. Monitoring of forage fish and seabird populations has also continued, with time series information from Middleton Island in the Gulf of Alaska being among the longest source of information throughout Alaskan waters (Piatt et al. 2018, Sydeman et al. 2017, Thompson et al. 2019).

Comprehensive databases for secondary through upper trophic level species include fishery-independent information collected from AFSC seasonal trawl surveys which account for biomass, distribution, abundance, and condition of groundfish and benthic invertebrates, especially crustaceans, while shelf and deep corals, benthic communities, and other sessile epifauna are examined using remotely operated vehicles (ROVs), video, submersible, and acoustic technologies (Swartzman et al. 1992, Kimura & Somerton 2006, Stone 2006, Stone & Shotwell 2007, Heifetz et al. 2009, Von Szalay et al. 2010, Hoff & Britt 2011, Lauth 2011, Siddon & Zador 2018, Zador & Ortiz 2018, Zador & Yasumiishi 2018). The International Pacific Halibut Commission (IPHC) conducts annual coast-wide longline surveys throughout Alaskan waters to determine stock abundance and biomass, which are used to set annual catch limits (Clark 2003, Clark & Hare 2003, 2006, Stewart & Hicks 2019). Nearshore surveys of Alaskan fisheries populations are undertaken through the Alaska Department of Fish and Game (Spalinger 2015). These surveys provide fishery-independent information regarding abundance, biomass, and diet of commercially important fisheries species and composition of species within 3 nautical mi. These data that monitor multiple trophic levels are applied toward broader understanding of ecosystem functioning for the North Pacific and for informing assessments regarding the status of Alaskan fisheries stocks. These surveys are also complemented by NOAA Fisheries fishery-dependent monitoring programs for commercial harvest, landings, revenue, and effort, including fisheries observation data on commercial landings, catches, and bycatch gathered by the North Pacific Observer Program. These observer programs and early commitments to their monitoring and reporting of North Pacific fisheries by NOAA and the NPFMC have been instrumental for data-rich fisheries and effective management (A. Hollowed, pers. comm.). Surveys undertaken by the Alaska Department of Fish and Game account for recreational fishing effort throughout Alaskan waters.

Factors including trends in average size, spatial distributions, aggregate biomass of pelagic and demersal species groupings and their ratios, predator–prey relationships, and mean trophic level, condition indices, multispecies mortality estimates, recruitment predictions, community assessments, and thermal preferences represent many of the ecosystem-level metrics that have been derived from this fishery-dependent and independent survey information (Siddon & Zador 2018, Zador & Ortiz 2018, Zador & Yasumiishi 2018). Socioeconomic information for fisheries and other uses, particularly regarding revenue and employment, are also collected by NOAA Fisheries, the National Ocean Economics Program (NOEP), and NOAA's Office for Coastal Management Economics: National Ocean Watch (ENOW) program. Monitoring programs exist for evaluating the status, recovery, and bycatch of protected species that include pinnipeds and cetaceans, while several monitoring programs for extensive seabird populations in this region are overseen by NOAA, the U.S. Department of Interior, U.S. Geological Service, and Alaska State agencies (Hunt et al. 2000, Ames et al. 2005, Moore et al. 2009, Kuletz et al. 2015, Piatt et al. 2020). In the North Pacific region, all of these aforementioned data sources have been applied toward integrative ecosystem measurements that examine cumulative changes in these metrics over time and their interdependent relationships (Aydin et al. 2007, Hunt et al. 2011, Zador 2015, Zador et al. 2017a, b, Siddon & Zador 2018, Zador & Ortiz 2018, Zador & Yasumiishi 2018).

9.3.2 Models and Assessments

Assessments of commercially important and protected species in the North Pacific are conducted by NOAA Fisheries in collaboration with the NPFMC, the U.S. Fish and Wildlife Service (FWS), IPHC (but see Chapter 11 for international fisheries), and Alaska Department of Fish and Game (Ferguson et al. 2017, Lynch et al. 2018, Srinivasan et al. 2019). Stock assessment models (e.g., index-based, aggregate biomass, statistical catch-at-age and length-at-age) are used to generate estimates of stock status using a multitude of data sources (e.g., landings, catch per unit effort, life history characteristics, fish condition, survey biomass, etc.; Lynch et al. 2018). Stock assessments for North Pacific federally managed species are conducted by stock assessment scientists and vetted through the NPFMC review process (Lynch et al. 2018, Karp et al. 2019). NOAA Fisheries also examines the effects of commercial and recreational fisheries on Alaskan marine socioeconomics through application of the NMFS Commercial Fishing Industry Input/Output and Recreational Economic Impact Models (NMFS 2018a). Data from NOAA Fisheries marine mammal stock abundance surveys are used to conduct population assessments as published in regional marine mammal stock assessment reports and peer reviewed by the Alaska scientific review group (SRG; Muto et al. 2017, 2018, 2019). FWS also prepares stock assessment reports for nearshore and inland threatened and protected species under its jurisdiction such as sea otter, polar bear, and walrus (FWS 2009, 2014). Information collected on protected species is also applied toward 5-year status reviews of those listed under the Endangered Species Act (ESA; FWS-NMFS 2006).

Available data and species assessments have also been applied toward broader ecosystem models for the North Pacific region and its prominent subregions, while efforts to incorporate IEA products (e.g., ESRs) into single and multispecies stock assessments and LMR management decisions are also ongoing (Townsend et al. 2008, 2014, Moffitt et al. 2016, Townsend et al. 2017). Major efforts include modeling food webs, examining multispecies interactions, and end-to-end modeling frameworks (Aydin et al. 2002, 2007, Townsend et al. 2014, 2017). Mass-balance foodweb models for the EBS, Gulf of Alaska, and Aleutian Islands have been developed and incorporated into Stock Assessment and Fishery Evaluation (SAFE) reports for the NPFMC (Aydin et al. 2002, 2005, 2007, Gaichas et al. 2010, Whitehouse et al. 2014). These models have also been developed for Arctic ecosystems, allowing for comparison of ecosystem-level metrics (e.g., productivity to biomass ratios) among polar and subpolar Alaskan ecosystems, and examination of their potential resilience to human activities (Whitehouse et al. 2014). There are several multispecies models used for context, research, and in some cases mortality vectors or matrices in the region (e.g., Mueter & Megrey 2006, van Kirk et al. 2015, Holsman et al. 2016, Thorson et al. 2019). Additionally, coupled multispecies models with a Regional Ocean Modeling System (ROMS) and Nutrient-Phytoplankton-Zooplankton models have been used to produce hindcasts and forecasts for EBS ecosystem dynamics and their effects on Alaskan salmon populations (Aydin et al. 2005, Curchitser et al. 2005).

End-to-end models, including the Forage and Euphausiid Abundance in Space and Time (FEAST) length-based model, also account for climate, oceanography, primary and secondary trophic levels, and fish and fisheries for modeling and predicting population status of fishes that are directly linked to specific zooplankton groups in response to climate projections for the EBS (K. Aydin, pers. comm.; Townsend et al. 2017). These models have also been applied toward management strategy evaluations for Walleye pollock and Pacific cod. Several of these modeling approaches have also been applied toward examining and developing recruitment indices for several species of Gulf of Alaska groundfish, with ongoing incorporation into Ecopath-with-Ecosim (EwE) frameworks. These models have been used to explore the effects of recruitment variability on the Gulf of Alaska ecosystem, and project climate effects on EBS food webs and their fisheries (Hermann et al. 2013, 2016, Ortiz et al. 2016, Townsend et al. 2014, 2017, Hermann et al. 2019). Qualitative modeling of species sensitivities to human and climate-driven changes are also being applied to groundfish in the Gulf of Alaska (Townsend et al. 2019) and for Blue king crab in the Eastern Bering Sea (Townsend et al. 2014). These efforts have built upon other climate-enhanced projection models for Alaskan groundfish species (Ianelli et al. 2016) and dynamic ecosystem modeling efforts for the Gulf of Alaska (Gaichas et al. 2011, 2012).

Over the latter part of the past decade, Models of Intermediate Complexity for Ecosystem Assessments (MICE; Plagányi et al. 2014, Thorson et al. 2019) have been developed. These include CEATTLE (Climate-Enhanced, Age-based model with Temperature-specific Trophic Linkages and Energetics) bioenergetics-based catch-at-age models for the Eastern Bering Sea and Gulf of Alaska (Holsman et al. 2016, Townsend et al. 2017, Hollowed et al. 2020) that examine potential climate-driven changes and evaluate fisheries management trade-offs for multiple species in a food web. These have been applied toward integrated modeling efforts (Alaska Climate Change Integrated Modeling—ACLIM (Hollowed et al. 2020) for use in management strategy evaluations of Eastern Bering Sea groundfishes that predict how climate-driven changes to the

North Pacific may affect future predation, species interactions, and fisheries harvest (A'mar et al. 2009, Ianelli et al. 2011, 2016, Mueter et al. 2011, Wilderbuer et al. 2013, Townsend et al. 2017, Hollowed et al. 2017, 2020). Advancements in applying spatiotemporal MICE ("MICE-in-space") to examine spatial management of Pacific cod and co-occurring species in the Gulf of Alaska have also recently occurred (Thorson et al. 2019). There are other models that inform management discussions in this region (e.g., Swartzman et al. 1992, Buckley & Livingston 1994, Dalsgaard & Pauly 1997, Okey & Pauly 1999, Mueter & Megrey 2006, Kishi et al. 2007, Ezer & Liu 2010, Harwell et al. 2010, Humphries & Huettman 2014, Whitehouse et al. 2014, Luizza et al. 2016, Pirtle et al. 2017, Shelton et al. 2017, Gibson et al. 2018, Rooney et al. 2018, Stockhausen et al. 2018a, Stockhausen et al. 2018b, Holsman et al. 2019b), but most are not directly incorporated into the NPFMC stock assessment review process as noted above.

9.4 Evaluation of Major Facets of EBFM for this Region

To elucidate how well the North Pacific ecosystem is doing in terms of LMR management and progress toward EBFM, here we characterize the status and trends of socioeconomic, governance, environmental, and ecological criteria pertinent for employing an EBFM framework for this region, and examine their inter-relationships as applied toward successful implementation. We also quantify those factors that emerge as key considerations under an ecosystem-based approach for the entire region and for all the subregions of interest. Ecosystem indicators for the North Pacific region related to the (1) human environment and socioeconomic status, (2) governance system and LMR status, (3) natural environmental forcing and features, and (4) systems ecology, exploitation, and major fisheries of this system are presented and synthesized below. We explore all these facets to evaluate the strengths and weaknesses along the pathway:

$$PP \rightarrow B_{\text{targeted, protected spp, ecosystem}} \leftrightarrow L_{\uparrow\text{targeted spp, }\downarrow\text{bycatch}} \rightarrow \text{jobs, economic revenue.}$$

Where PP is primary production, B is biomass of either targeted or protected species or the entire ecosystem, L is landings of targeted or bycaught species, all leading to the other socioeconomic factors. This operates in the context of an ecological and human system, with governance feedbacks at several of the steps (i.e., between biomass and landings, jobs, and economic revenue), implying that fundamental ecosystem features can determine the socioeconomic value of fisheries in a region, as modulated by human interventions.

9.4.1 Socioeconomic Criteria

9.4.1.1 Social and Regional Demographics

In the U.S. North Pacific region, coastal human population values and population densities (Fig. 9.3) have increased from 1970 to present with values currently around 610 thousand residents and ~0.4 people km^{-2}. Human population for this region is lowest in the nation, comprising only 0.2% of United States residents (year 2010), while density values have risen at only ~0.006 people km^{-2} year^{-1}.

Among subregions, human population is more heavily concentrated in areas surrounding the Gulf of Alaska, particularly in the Kenai Peninsula and portions of PWS, where the cities of Anchorage, Homer, and Seward are found. The most populated coastal boroughs in the Gulf of Alaska include Anchorage, Matanuska-Susitna, and Kenai Peninsula, where the human population has increased (~8200 people year^{-1}) over the past 40 years, with currently up to ~536 thousand residents and ~1 person km^{-2} in surrounding areas. Values are somewhat lower in nearby PWS (which includes Valdez-Cordova Borough, AK) where human population has increased by ~5300 people year^{-1} and where currently up to ~357 thousand residents are found. However, PWS currently contains an overall higher coastal population density than the Gulf of Alaska at ~2.1 people km^{-2}. To a lesser extent, human population is also concentrated in the Inside Passage, with inhabitants concentrated in cities including Juneau, Ketchikan, and Sitka, found in eight populated coastal boroughs (i.e., Haines, Juneau, Ketchikan Gateway, Prince of Wales-Hyder Census Area, Sitka, Skagway Municipality, Wrangell City and Borough, and Yakutat, AK). Over the past 40 years, human population in the Inside Passage has increased by ~690 people year^{-1} with currently up to ~66 thousand residents and 0.7 people km^{-2}.

Lower human populations are observed in the Eastern Bering Sea (EBS), Aleutian Islands, and Arctic. Population in the EBS has increased (467 people year^{-1}) to currently ~44.5 thousand residents and ~0.11 people km^{-2}, with most individuals living in Bethel and Kusilvak Census Area, AK. Human population has

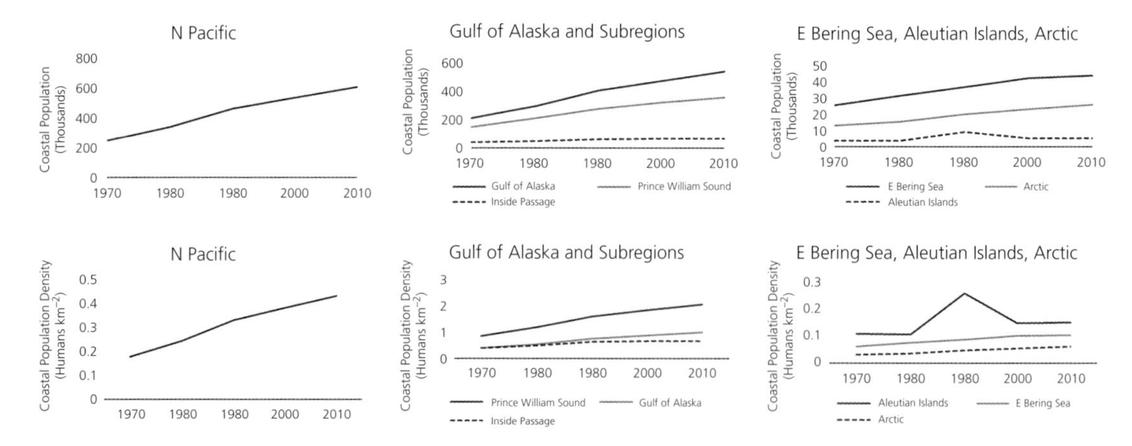

Figure 9.3 Decadal trends (1970–2010) in coastal human population (top) and population density (bottom; humans km^{-2}) for the North Pacific region and identified subregions.

Data derived from U.S. censuses and taken from https://coast.noaa.gov/digitalcoast/data/demographictrends.html.

doubled in the Arctic over the past 40 years (~330 people year^{-1}) at up to ~26.5 thousand residents and 0.06 people km^{-2}. Most individuals reside in Nome and North Slope Borough, AK, with slightly less also living in Northwest Arctic Borough, AK. Lowest human population in the North Pacific occurs in the Aleutian Islands, where values have increased, but at a much lower overall rate (38.3 people year^{-1}). During the 1980s, human population was reported at ~9500 residents, but values have decreased and remained relatively stable in following years to ~5600 residents and 0.15 people km^{-2}. All inhabitants live in the Aleutians West Census Area (AK). Although the eastern portion of the Aleutian Islands also comprises Aleutians East Borough (AK), their inhabitants and fisheries are generally more associated with the EBS and portions of the Gulf of Alaska subregion, as they occur east of 170°W longitude.

9.4.1.2 Socioeconomic Status and Regional Fisheries Economics

Relatively low to mid-range numbers of LMR establishments and employments, and estimated LMR-associated gross domestic product (GDP) values (~$961 million USD), are found throughout the North Pacific region (Fig. 9.4). Fifth- to sixth-highest nationally, the North Pacific contains moderate numbers of reported LMR establishments, employments, and GDP value that have increased over the past decade. These values do not necessarily include current total revenue of fisheries landings (see Fig. 9.6), and are based upon benchmarks of U.S. total GDP conducted by the U.S. Department of Commerce Bureau of Economic Analyses that are currently reflective of the year 2012 (NOAA-OCM 2017). The percent contribution of the North Pacific LMR-associated economy toward its regional ocean economy is highest nationally where LMR establishments comprise up to 13.2% of regional ocean economy establishments and LMR employments have contributed 21–23% of regional ocean employments over time. LMR-associated GDP has increased by ~$470 million since 2005, and has contributed up to 9% toward the regional total ocean economy in recent years.

In nearly all subregions, LMR-associated values have increased over time, with the majority contributing high percentages toward regional ocean economies. The highest numbers of LMR establishments and employments, and GDP value have historically occurred in the Gulf of Alaska subregion (152.3 ± 18.3 standard error (SE), establishments; 2343.9 ± 570.7 SE employments; $171 ± 42.9 million SE, USD), especially following post 2010 increases, with high numbers of LMR establishments also observed in PWS and the Inside Passage. The LMR economy of the Aleutian Islands comprises a major portion of North Pacific LMR employments and GDP value (2222.1 ± 570.7 SE employments; $159 ± 11.3 million SE, USD), with LMRs contributing to nearly 100% of its total ocean employments and GDP, and up to 50% of its total ocean establishments. Similarly, high LMR economic

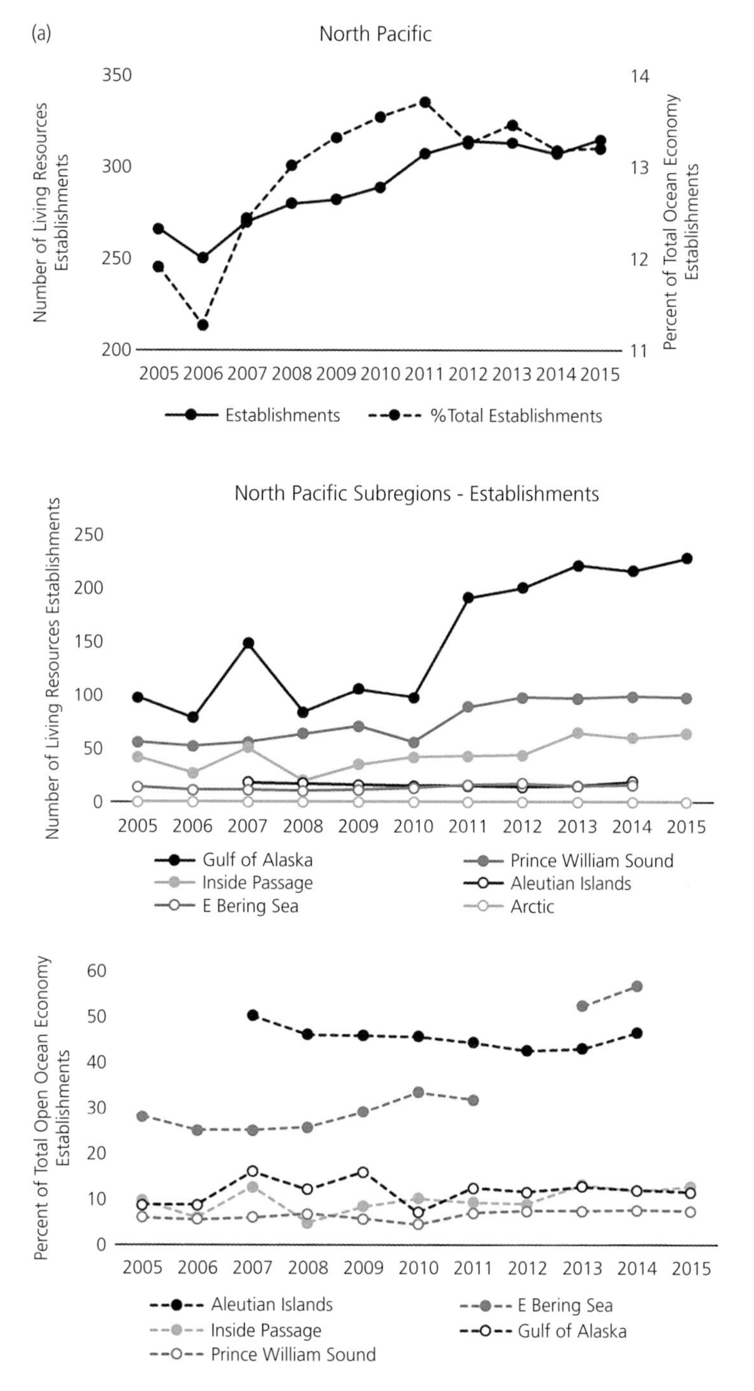

Figure 9.4a Number of living marine resources establishments and their percent contribution to total multisector oceanic economy establishments in the North Pacific region and identified subregions (years 2005–2015).

Data derived from the National Ocean Economics Program.

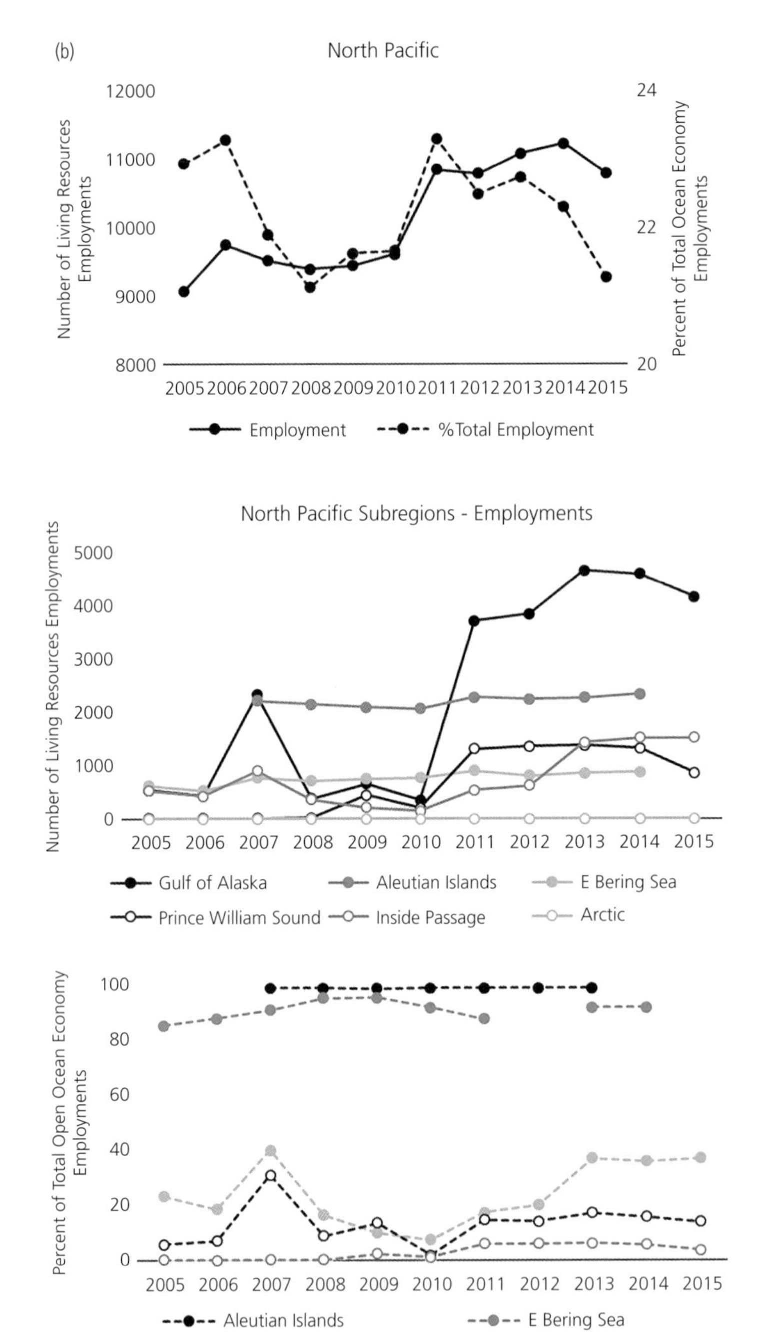

Figure 9.4b Number of living marine resources employments and their percent contribution to total multisector oceanic economy employments in the North Pacific region and identified subregions (years 2005–2015).

Data derived from the National Ocean Economics Program.

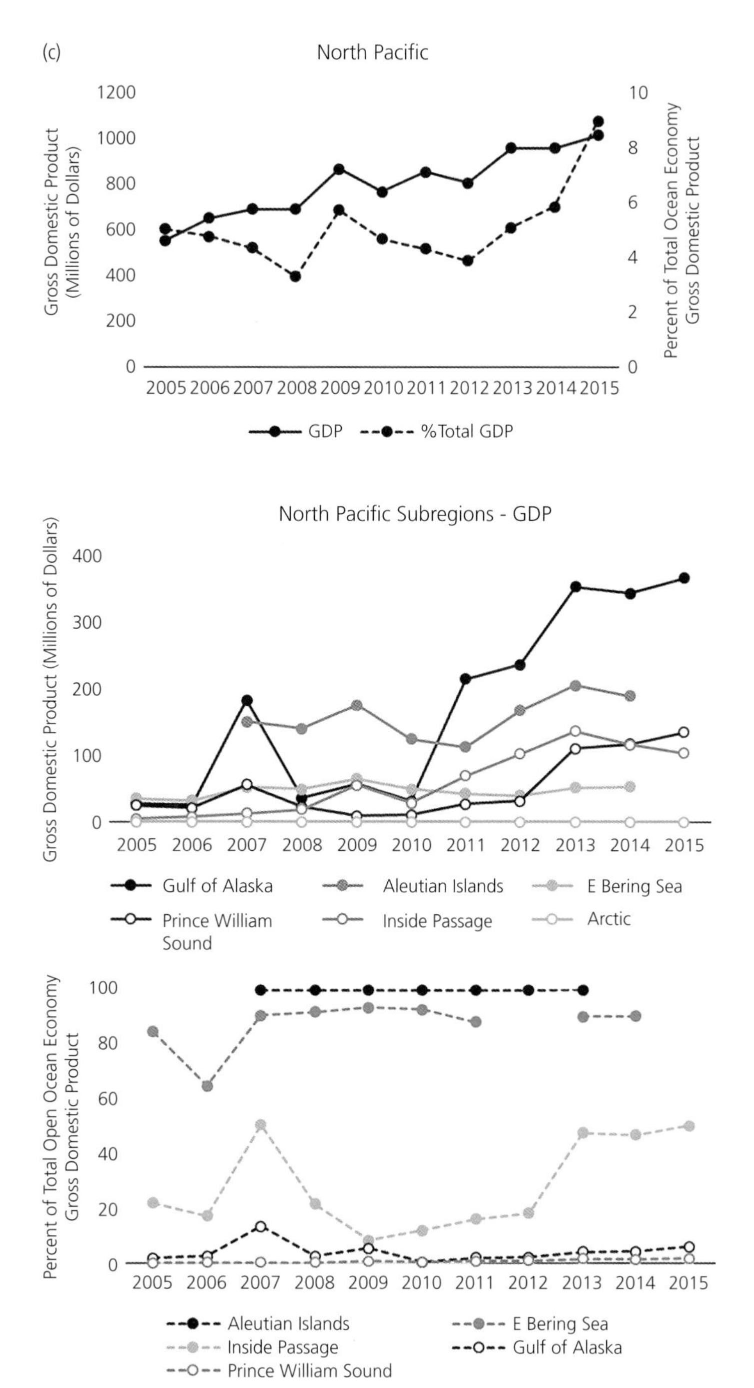

Figure 9.4c Gross domestic product value (GDP; USD) from living marine resources revenue and percent contribution to total multisector oceanic economy GDP for the North Pacific region and identified subregions (years 2005–2015).

Data from the National Ocean Economics Program.

OY Vey: Total Catch Caps, Ecosystem Production, Portfolios, and the Trade-Offs of Dealing with Conflicting Objectives

—(see also every other chapter, especially Chapter 3)

Put all your eggs in one basket. Shun diversification in your investments. Manage for just one or two stocks, while ignoring the broader economy. Disregard overall stock market dynamics. Don't bother with a portfolio of investments. Ignore the full range of risks. Forget about long-term goals. Said no retirement planner or financial advisor ever.

And yet with our overemphasis on population dynamics, reductionist approaches, and a stock-by-stock basis, that's effectively what we do for fisheries.

Alternatively, what if we managed fisheries in marine ecosystems such that: risk of overfishing and ancillary ecosystem impacts were minimized; populations of fishes, catches, and profits were stable; overall value and biomass across all stocks were maximized; bureaucratic oversight and regulatory interventions were minimized; stakeholder disenfranchisement and legal challenges were minimized; and catch and yield were optimized? By adopting systems thinking, particularly using system-level optimal yield, we can move toward these goals.

It is known that for any given patch of the ocean, there can only be so much fish produced (Ryther 1969, Pauly & Christensen 1995, Chassot et al. 2007, 2010, Friedland et al. 2012, Watson et al. 2014, Fogarty et al. 2016, Stock et al. 2017, Link & Watson 2019). Recognizing those limits provides a constraint from which a portfolio of fisheries can be managed, ultimately for an optimal yield (OY) of an entire fishery ecosystem.

One can compare and contrast simulations of a system level vs. a stock-by-stock level approach to estimating OY, and then evaluate *in silico* management performance at aggregate levels. Several instances where such simulation testing has occurred demonstrate improvements across a range of not only yield, but other fisheries objectives (Worm et al. 2009, Link et al. 2010a, b, Fulton et al. 2011, 2019, Kaplan et al. 2012, Fay et al. 2013, 2015, Smith et al. 2015, Jacobsen et al. 2017). A few common outcomes emerge from this collective body of work. One is that the aggregate OY, or more specifically estimates of aggregate maximum sustainable yield (MSY), are on average about 25% lower than summed estimates of single stock MSY in a given system (Worm et al. 2009, Hilborn et al. 2012). Yet when one considers that OY is usually derived and lowered from estimates of MSY, the distinction becomes rather small if not indistinguishable (Restrepo et al. 1998, Restrepo & Powers 1999, Patrick & Link 2015). Another common theme is that the risk of overfishing when managing at an aggregate OY level is much less than managing at component stocks OYs (Worm et al. 2009, Link et al. 2012, Fogarty 2014). For small reductions in overall yield (5–10%), the risk of overfishing any component stock is lessened in many instances by 50–60% (Worm et al. 2009, Hilborn et al. 2012, Link et al. 2012, Fogarty 2014); the economic benefits from a constant revenue stream and not having to implement rebuilding measures likely outweighs any small reduction in yield (Worm et al. 2009, Hilborn et al. 2012, Fay et al. 2013, 2015, Kaplan et al. 2012). Other work has shown that

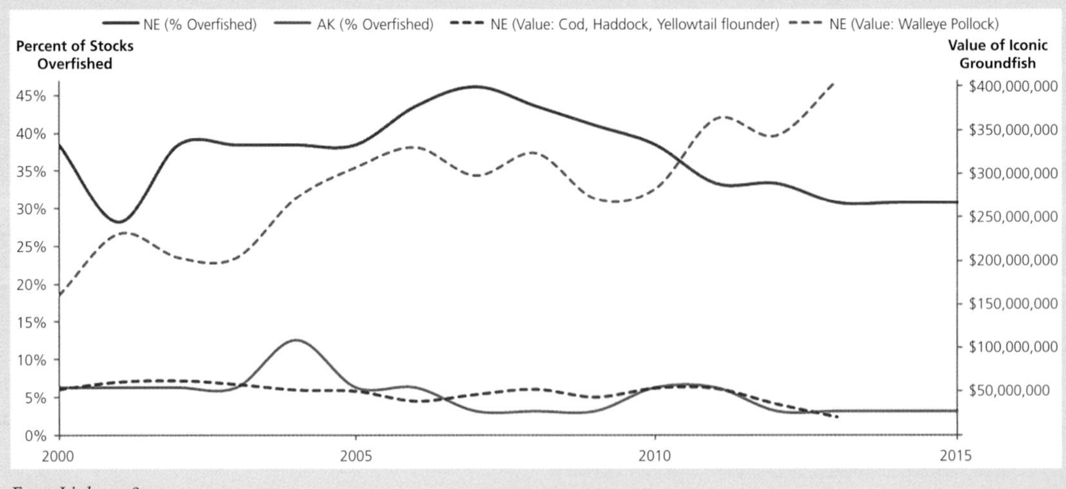

From Link 2018

ecosystem processes and functioning, and hence related OY, is achieved more efficiently when managed at the aggregate or systems level (Jacobsen et al. 2017). Additionally, it has been repeatedly shown that maintaining all stocks in an ecosystem simultaneously at a single-species maximum sustainable or maximum economic yield is an impossibility due to biological and socioeconomic constraints (Au 1973, Crutchfield 1973, May 1975, Pope 1975, 1979, Fukuda 1976, Brown et al. 1976, Larkin 1977, May et al. 1979, Mayo et al. 1992, Collie & Gislason 2001, Walters et al. 2005, Mueter & Megrey 2006, Legovic et al. 2010, Legovic & Gecek 2011, Steele et al. 2011, Fogarty et al. 2012, Heath 2012, Link et al. 2012). Thus, given the energetic constraints of both species production and ecological interactions, coupled with some core tenets of portfolio and hierarchy theory, an aggregate approach is actually more theoretically defensible than attempting to simultaneously maximize all single component stocks or species yield in an entire ecosystem. The theory, simulations, empirical evidence, and contrast in performance all show that adopting an aggregate OY has significant benefits and in many instances is indeed advisable, if not even preferable.

The Eastern Bering Sea in Alaska and Georges Bank/Gulf of Maine in the northeast US have both had a long and important history of fisheries (Witherell et al. 2000, Hollowed et al. 2011, Fogarty & Murawski 1998, Link et al. 2011, respectively). They have broadly similar taxa, high levels of productivity, highly lucrative fisheries, and due to their rich history of scientific studies and long-term data collection, have been used in many comparative ecosystem analyses (e.g., Gaichas et al. 2009, Link et al. 2009, Megrey et al. 2009, Gaichas et al. 2012). There are many similarities in the fisheries context for both ecosystems. Certainly there are differences in oceanography, latitude, regional human populations, regional cultures, and fisheries management measures employed between the two regions. Certainly the nuances of these factors contribute to the different overall value of the fishery in each ecosystem. Yet in a systems context, one such difference is that one fisheries management system has adopted key facets of an aggregate, portfolio, systems-oriented approach, and the other has not.

In one region, Alaska, there is a 2 million metric ton cap that is used to manage the suite of fisheries there; the value of groundfishes in that ecosystem has steadily increased and the number of stocks that are overfished has remained low (<5%). Whereas in the other region, New England, the opposite has occurred; the value of seminal groundfishes has remained relatively flat and low, and the number of stocks overfished is close to 30–35%, with a peak over 40%. This implies that in the region that uses the aggregate estimate of OY as a hard cap for fisheries management the risk of overfishing is lower and the long-term value is higher than the region that has not adopted such an approach.

References

Au DW. 1973. Total sustainable yield from Subareas 5 and 6 based on yield per recruit and primary production consideration. *ICNAF Research Document* 73(10).

Brown B, Breenan J, Grosslein M, Heyerdahl E, Hennemuth R. 1976. The effect of fishing on the marine finfish biomass in the Northwest Atlantic from the Gulf of Maine to Cape Hatteras. *ICNAF Research Bulletin* 12:49–68.

Chassot E, Bonhommeau S, Dulvy NK, Mélin F, Watson R, Gascuel D, Le Pape O. 2010. Global marine primary production constrains fisheries catches. *Ecology Letters 13*(4):495–505.

Chassot E, Mélin F, Le Pape O, Gascuel D. 2007. Bottom-up control regulates fisheries production at the scale of eco-regions in European seas. *Marine Ecology Progress Series* 343:45–55.

Collie JS, Gislason H. 2001. Biological reference points for fish stocks in a multispecies context. *Canadian Journal of Fisheries and Aquatic Sciences* 58(11):2167–76.

Crutchfield JA. 1973. Economic and political objectives in fishery management. *Transactions of the American Fisheries Society* 102(2):481–91.

Fay G, Large SI, Link JS, Gamble RJ. 2013. Testing systemic fishing responses with ecosystem indicators. *Ecological Modelling* 265:45–55.

continued

OY Vey: *Continued*

Fay G, Link JS, Large SI, Gamble RJ. 2015. Management performance of ecological indicators in the Georges Bank finfish fishery. *ICES Journal of Marine Science* 72(5):1285–96.

Fogarty MJ. 2014. The art of ecosystem-based fishery management. *Canadian Journal of Fisheries and Aquatic Sciences* 71(3): 479–90.

Fogarty MJ, Murawski SA. 1998. Large-scale disturbance and the structure of marine systems: fishery impacts on Georges Bank. *Ecological Applications* 8(S1):S6–S22.

Fogarty MJ, Overholtz WJ, Link JS. 2012. Aggregate surplus production models for the demersal fish complex of the Gulf of Maine. *Marine Ecology Progress Series* 459:247–58.

Fogarty MJ, Rosenberg AA, Cooper AB, Dickey-Collas M, Fulton EA, Gutiérrez NL, Hyde KJ, et al. 2016. Fishery production potential of large marine ecosystems: a prototype analysis. *Environmental Development* 17:211–9.

Friedland KD, Stock C, Drinkwater KF, Link JS, Leaf RT, Shank BV, Rose JM, Pilskaln CH, Fogarty MJ. 2012. Pathways between primary production and fisheries yields of large marine ecosystems. *PLoS One* 7(1).

Fukuda Y. 1976. A note on yield allocation in multi-species fisheries. *ICNAF Research Bulletin* 12:83–7.

Fulton EA, Link JS, Kaplan IC, Savina-Rolland M, Johnson P, Ainsworth C, Horne P, et al. 2011. Lessons in modelling and management of marine ecosystems: the Atlantis experience. *Fish and Fisheries* 12(2):171–88.

Fulton EA, Punt AE, Dichmont CM, Harvey CJ, Gorton R. 2019. Ecosystems say good management pays off. *Fish and Fisheries* 20(1):66–96.

Gaichas S, Gamble R, Fogarty M, Benoît H, Essington T, Fu C, Koen-Alonso M, Link J. 2012. Assembly rules for aggregate-species production models: simulations in support of management strategy evaluation. *Marine Ecology Progress Series* 459:275–92.

Gaichas S, Skaret G, Falk-Petersen J, Link JS, Overholtz W, Megrey BA, Gjøster H, et al. 2009. A comparison of community and trophic structure in five marine ecosystems based on energy budgets and system metrics. *Progress in Oceanography* 81:47–62.

Heath M. 2012. Ecosystem limits to food web fluxes and fisheries yields in the North Sea simulated with an end-to-end model. *Progress in Oceanography* 102:42–66.

Hilborn RA, Stewart IJ, Branch TA, Jensen OP. 2012. Defining trade-offs among conservation, profitability, and food security in the California current bottom-trawl fishery. *Conservation Biology* 26(2):257–68.

Hollowed AB, Aydin KY, Essington TE, Ianelli JN, Megrey BA, Punt AE, Smith ADM. 2011. Experience with quantitative ecosystem assessment tools in the northeast Pacific. *Fish and Fisheries* 12:189–208.

Jacobsen NS, Burgess MG, Andersen KH. 2017. Efficiency of fisheries is increasing at the ecosystem level. *Fish and Fisheries* 18(2):199–211.

Kaplan IC, Horne PJ, Levin PS. 2012. Screening California Current fishery management scenarios using the Atlantis end-to-end ecosystem model. *Progress in Oceanography* 102:5–18.

Larkin PA. 1977. An epitaph for the concept of maximum sustained yield. *Transactions of the American Fisheries Society* 106:1–11.

Legović T, Geček S. 2010. Impact of maximum sustainable yield on independent populations. *Ecological Modelling* 221(17):2108–11.

Legović T, Klanjšček J, Geček S. 2010. Maximum sustainable yield and species extinction in ecosystems. *Ecological Modelling* 221(12):1569–74.

Link JS, Bundy A, Overholtz WJ, Shackell N, Manderson J, Duplisea D, Hare J, et al. 2011. Ecosystem-based fisheries management in the Northwest Atlantic. *Fish and Fisheries* 12:152–70.

Link JS, Fulton EA, Gamble RJ. 2010a. The northeast US application of Atlantis: a full system model exploring marine ecosystem dynamics in a living marine resource management context. *Progress in Oceanography* 87(1–4):214–34.

Link JS, Gaichas S, Miller TJ, Essington T, Bundy A, Boldt J, Drinkwater KF, Moksness E. 2012. Synthesizing lessons learned from comparing fisheries production in 13 northern hemisphere ecosystems: emergent fundamental features. *Marine Ecology Progress Series* 459:293–302.

Link JS, Stockhausen WT, Skaret G, Overholtz W, Megrey BA, Gjøsæter H, Gaichas S, et al. 2009. A comparison of biological trends from four marine ecosystems: synchronies, differences, and commonalities. *Progress in Oceanography* 81:29–46.

Link JS, Watson RA. 2019. Global ecosystem overfishing: clear delineation within real limits to production. *Science Advances* 5(6):eaav0474.

Link JS, Yemane D, Shannon LJ, Coll M, Shin YJ, Hill L, Borges MD. 2010b. Relating marine ecosystem indicators to fishing and environmental drivers: an elucidation of contrasting responses. *ICES Journal of Marine Science* 67(4):787–95.

May AW. 1975. *Report of Standing Committee on Research and Statistics. ICNAF Seventh Special Commission Meeting—September 1975.* Dartmouth, Canada: International Convention for the Northwest Atlantic Fisheries.

May RM, Beddington JR, Clark CW, Holt SJ, Laws RM. 1979. Management of multispecies fisheries. *Science* 205:267–77.

Mayo RK, Fogarty MJ, Serchuk FM. 1992. Aggregate fish biomass and yield on Georges Bank, 1960–87. *Journal of Northwest Atlantic Fishery Science* 14:59–78.

Megrey BA, Link JS, Hunt GL, Moksness E. 2009. Comparative marine ecosystem analysis: applications, opportunities, and lessons learned. *Progress in Oceanography* 81:2–9.

Mueter FJ, Megrey BA. 2006. Using multi-species surplus production models to estimate ecosystem-level maximum sustainable yields. *Fisheries Research* 81:189–201.

Patrick WS, Link JS. 2015. Hidden in plain sight: using optimum yield as a policy framework to operationalize ecosystem-based fisheries management. *Marine Policy* 62:74–81.

Pauly D, Christensen V. 1995. Primary production required to sustain global fisheries. *Nature* 374(6519):255–7.

Pope JG. 1975. *The Application of Mixed Fisheries Theory to the Cod and Redfish Stocks of Subarea 2 and Division 3K. ICNAF Research Document 75/IX/126.* Dartmouth, Canada: International Convention for the Northwest Atlantic Fisheries.

Pope J. 1979. *Stock Assessment in Multistock Fisheries, with Special Reference to the Trawl Fishery in the Gulf of Thailand. SCS/DEV/79/19.* Manila, Philippines: FAO

Restrepo VR, Mace PM, Serchuk FM. 1998. *The Precautionary Approach: A New Paradigm or Business as Usual. Our Living Oceans. Report on the Status of US Living Marine Resources. NOAA Technical Memorandum NMFS-F/SPO-41.* Washington, DC: NOAA (pp. 61–70).

Restrepo VR, Powers JE. 1999. Precautionary control rules in US fisheries management: specification and performance. *ICES Journal of Marine Science* 56(6):846–52.

Ruggerone GT, Irvine JR. 2018. Numbers and biomass of natural-and hatchery-origin pink salmon, chum salmon, and sockeye salmon in the north Pacific Ocean, 1925–2015. *Marine and Coastal Fisheries* 10(2):152–68.

Ryther JH. 1969. Photosynthesis and fish production in the sea. *Science* 166(3901):72–6.

Smith L, Gamble R, Gaichas S, Link J. 2015. Simulations to evaluate management trade-offs among marine mammal consumption needs, commercial fishing fleets and finfish biomass. *Marine Ecology Progress Series* 523:215–32.

Steele JH, Gifford DJ, Collie JS. 2011. Comparing species and ecosystem based estimates of fisheries yields. *Fisheries Research* 111:139–44.

Stock CA, John JG, Rykaczewski RR, Asch RG, Cheung WW, Dunne JP, Friedland KD, et al. 2017. Reconciling fisheries catch and ocean productivity. *Proceedings of the National Academy of Sciences* 114(8):E1441–9.

Walters CJ, Christensen V, Martell SJ, Kitchell JF. 2005. Possible ecosystem impacts of applying MSY policies from single-species assessment. *ICES Journal of Marine Science* 62:558–68.

Watson R, Zeller D, Pauly D. 2014. Primary productivity demands of global fishing fleets. *Fish and Fisheries* 15(2):231–41.

Witherell D, Pautzke CP, Fluharty D. 2000. An ecosystem-based approach for Alaska groundfish fisheries. *ICES Journal of Marine Science* 57:771–7.

Worm B, Hilborn R, Baum JK, Branch TA, Collie JS, Costello C, Fogarty MJ, et al. 2009. Rebuilding global fisheries. *Science* 325(5940):578–85.

contributions are found in the EBS (up to 56% of total ocean establishments; 96% of total ocean employments; 93% of total ocean GDP), although with relatively lower numbers of establishments, employments, and GDP value (13.4 ± 0.8 SE establishments; 763 ± 36.6 SE employments; \$46.8 ± 3.1 million SE, USD GDP). Lower percent contributions of LMRs toward subregional total ocean economies are found in the Gulf of Alaska and its subcomponents. However, among these subregions, LMRs in the Inside Passage have contributed most toward its total ocean economy (up to 13% of ocean establishments, 40.2% of ocean employments, and 50.5% of ocean GDP). Due to fisheries closures and limits on its exploitation, no LMR-based economy currently occurs in the Arctic subregion. There is, however, a notable subsistence economy in the Arctic (Fall 2016, Ready 2016, Naves & Fall 2017).

During the 1990s, decreases in the number of reported permitted fishing vessels operating in the North Pacific (Fig. 9.5) occurred, with values highest in the early to mid-1990s at ~25,000 and decreases and fluctuations occurring in the 1990s and 2000s. This was largely due to "rationalization" of the fleet and the fleet being more efficient (Reimer et al. 2017b, Beaudreau et al. 2018). Currently, numbers are around ~5,000 vessels and have remained at this level since 2010.

The North Pacific is highest nationally regarding total revenue (year 2017 USD) of its landed commercial fishery catches (Fig. 9.6). Pronounced increases in commercial landings value have occurred over time for this region, particularly since the 1980s, with peak values observed in 2011 at \$1.89 billion. Among subregions, commercial catch revenues have been relatively

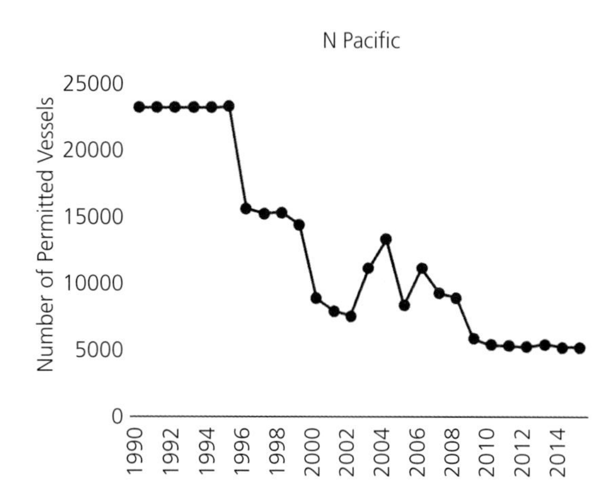

Figure 9.5 Total number of permitted vessels for the North Pacific region over time (years 1990–2015).

Data derived from NOAA National Marine Fisheries Service Council Reports to Congress.

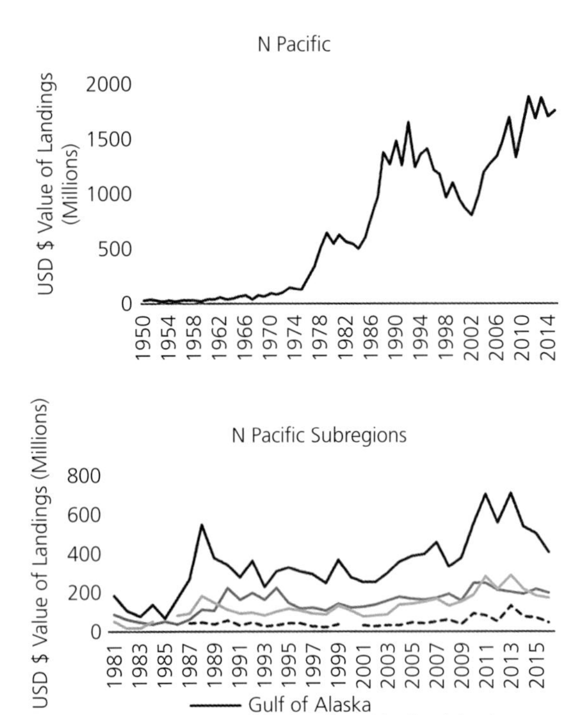

Figure 9.6 Total revenue (year 2017 USD) of landed commercial fishery catches for the North Pacific region (years 1950–2015) and identified subregions (years 1981–2016), over time.

Data derived from NOAA National Marine Fisheries Service commercial fisheries statistics.

steady since 1981, with highest reported commercial catch revenue occurring in the Gulf of Alaska over time and peaking in 2013 for this area and its subcomponents (Gulf of Alaska: $712.3 million; Inside Passage: $291 million; PWS: $132.3 million). In the EBS and Aleutian Islands, values have remained relatively steady, with highest values observed during the early to mid-1990s and later peaking in 2010 at $247.2 million.

Ratios of LMR jobs and commercial fisheries revenue to biomass, and fisheries revenue to primary production (Fig. 9.7), are at mid-level rankings nationally (fifth-highest) in the North Pacific. However, values for jobs to primary production are second-lowest nationally, with highest values among subregions currently observed in the Gulf of Alaska (1.4 jobs million metric tons^{-1}). In the North Pacific, jobs/biomass values have increased over time (up to 0.14 jobs thousand metric tons^{-1}), as have jobs/productivity values (up to 0.9 jobs thousand metric tons^{-1}). Interannual fluctuation has been observed for total revenue/biomass (range: $15.5–$59.8 metric ton^{-1}, with values averaging $43.8 ± 2.4 SE since 2010) with peaks observed during 1990 ($59.8 metric ton^{-1}) and 2008 ($56.3 metric ton^{-1}), while total revenue/productivity values have increased over time (up to $0.32 metric ton^{-1}).

9.4.2 Governance Criteria

9.4.2.1 Human Representative Context

Twenty-seven federal, State, and international agencies and organizations operate within the North Pacific and are directly responsible for fisheries and environmental management (Table 9.1). Fisheries resources throughout state waters (0–3 nautical miles) are managed by the Alaska Department of Fish and Game and Alaska Department of Natural Resources (ADNR), which provide information to the Alaska Board of Fisheries, while those in federal waters are managed by both NOAA Fisheries (Alaska Region) and the North Pacific Fishery Management Council (NPFMC). While Alaska is a member of the Pacific States Marine Fisheries Commission (PSMFC), this organization does not directly manage species in the North Pacific under any fishery management plans (FMPs) and instead serves as a coordinating body for fisheries data and management and conservation issues in this region. Fishery and environmental resources of the North Pacific also fall under several international and multi-jurisdictional agreements, including those of the IPHC, Pacific Salmon Commission (PSC), and Western and Central Pacific Fisheries Commission (WCPFC). Alaskan subregions are subject to these federal, State, and international bodies, while the Arctic is also managed by the Alaska Department of Commerce, Community, and Economic Development (DCCED) and falls under the jurisdiction of several specific and international organizations (Arctic Research Consortium of the US (ARCUS), International Arctic Research Policy Committee (IARPC), and North Atlantic Treaty Organization (NATO)).

Nationally, the North Pacific contains the second-lowest number (i.e., 3) of Congressional representatives, which are concentrated within only one state. When standardized per mile of shoreline and per value of its commercial fisheries (Fig. 9.8), Congressional representation is lowest nationally for the North Pacific. Among subregions, highest state and Congressional representation per mile is found for the Aleutian Islands, Arctic, and PWS. PWS is highest in terms of its Congressional representation in terms of the

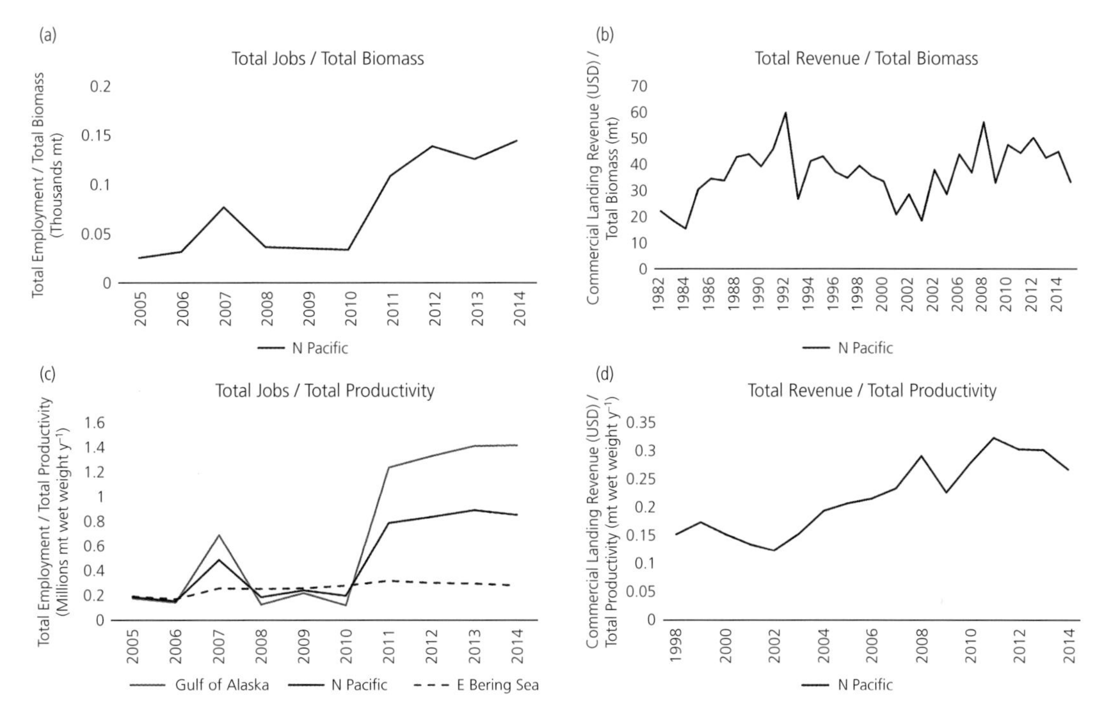

Figure 9.7 Over time, ratios of (a) total living marine resources employments to total fish and invertebrate biomass (thousands of metric tons); (b) total commercial landings revenue (USD) to total biomass (metric tons); (c) total living marine resources employments to total primary productivity (metric tons wet weight year^{-1}); (d) total commercial landings revenue (USD) to total productivity (metric tons wet weight year^{-1}), for the North Pacific region and identified subregions.

value of its fisheries, but values for both metrics still remain comparatively low nationally (Link & Marshak 2019). Historically, composition of the NPFMC has been dominated by members of the commercial fishing sector, with emerging representation from the recreational sector and other representatives occurring during the mid-1990s and early 2000s (Fig. 9.9a) and their expansion and contraction as of the mid-2000s. Current members are at their most balanced among the three major sectors. With eight appointed members, the NPFMC contains a medium number of council members nationally (tied for fifth-highest with the South Atlantic FMC). The Alaska marine mammal SRG is responsible for advising federal agencies on the status of marine mammal stocks covered under the Marine Mammal Protection Act (MMPA) for the North Pacific region. Its composition and membership have remained relatively steady over time (Fig. 9.9b), with individuals from the academic and private institutional sectors making up the majority of its representatives and periodic representation from the tribal community. Higher proportional governmental representation on the SRG has occurred during more recent years.

9.4.2.2 Fishery and Systematic Context

Six current North Pacific-specific FMPs are managed by NOAA Fisheries and the NPFMC for Gulf of Alaska groundfish; Bering Sea-Aleutian Islands groundfish; Bering Sea-Aleutian Islands King and Tanner crabs; region-wide salmon; region-wide scallops; and Arctic resources (Table 9.2). In addition, U.S. and Canadian Pacific halibut (*Hippoglossus stenolepis*) resources are managed through the IPHC. In total, these FMPs have been subject to 318 modifications (i.e., amendments and framework/regulatory amendments), which is fourth-highest for the nation. FEPs exist for the Aleutian Islands and the EBS. These FEPs contain comprehensive information regarding EBS and Aleutian Island fisheries, the habitat and biology of its managed species, social and economic management impacts, and the ecological consequences of conservation and management.

Table 9.1 Federal and State agencies and organizations responsible for fisheries and environmental consulting and management in the North Pacific region and subregions

Region	Subregions	Federal Agencies with Marine Resource Interests	Council	State Agencies and Cooperatives	Multijurisdictional
North Pacific		US Department of Commerce **DoC** (NOAA Fisheries Alaska Fisheries Science Center **AFSC**, Alaska Regional Office **AKRO**); US Department of Defense **DoD** (US Army Corps of Engineers **ACOE**, US Department of the Navy **DoN** Office of Naval Research **ONR**); US Department of Interior **DoI** (US Fish and Wildlife Service **FWS**, National Park Service **NPS**, Bureau of Ocean Energy Management **BOEM**, US Geological Survey **USGS**); US Environmental Protection Agency **EPA**	North Pacific Fishery Management Council **NPFMC**	Alaska Department of Fish & Game **ADFG**; Alaska Department of Natural Resources **ADNR**; Alaska State Legislature; Pacific States Marine Fisheries Commission **PSMFC**; North Pacific Research Board (Together with **DoC**) **NPRB**	International Pacific Halibut Commission **IPHC**; Pacific Salmon Commission **PSC**; Western and Central Pacific Fisheries Commission **WCPFC**
	E Bering Sea, Gulf of Alaska, Aleutian Islands, Inside Passage, Prince William Sound	**DoC** (**AFSC**, **AKRO**); **DoD** (**ACOE**, **DoN-ONR**); **DoI** (**FWS**, **BOEM**, **NPS**, **USGS**); **EPA**	**NPFMC**	**ADFG**; **ADNR**; Alaska State Legislature; **PSMFC**; **NPRB**	**IPHC**; **PSC**; **WCPFC** (Aleutian Islands; portions of E Bering Sea, Gulf of Alaska)
	Arctic	**DoC** (**AFSC**, **AKRO**); **DoD** (**DoN-ONR**); **DoI** (**FWS, NPS, BOEM, USGS**) US Department of State	**NPFMC**	**ADFG**; **ADNR**; Alaska Department of Commerce, Community, and Economic Development **DCCED**; Alaska State Legislature; **NPRB**	North Atlantic Treaty Organization **NATO**; International Arctic Research Policy Committee **IARPC**; Arctic Research Consortium of the US **ARCUS**

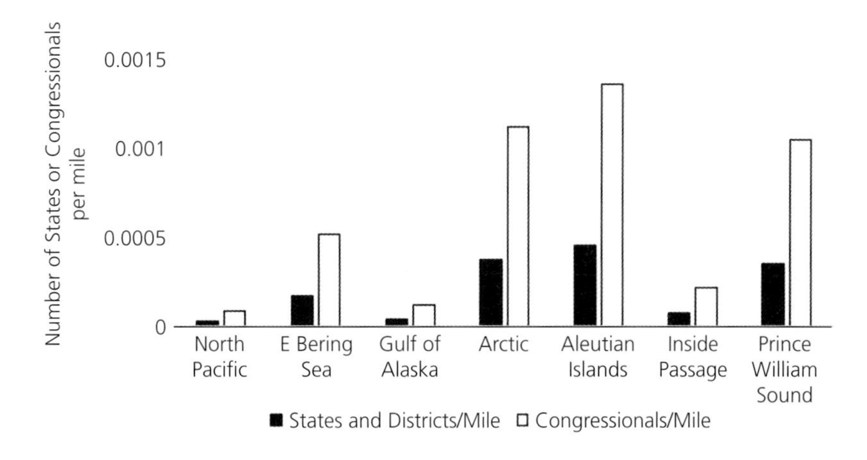

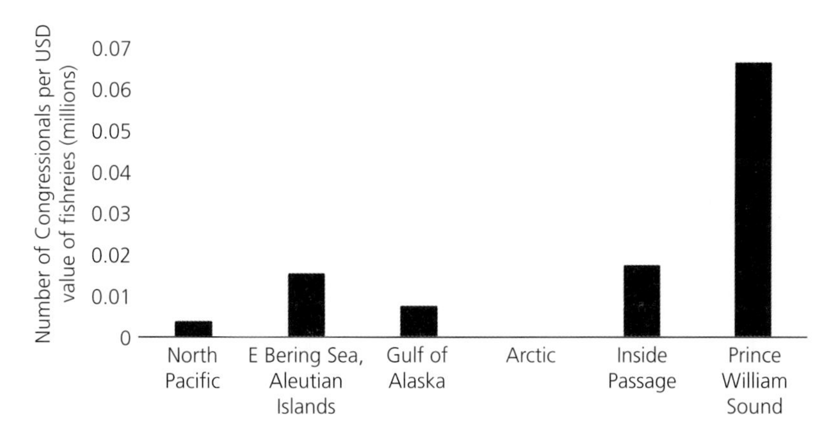

Figure 9.8 Number of states and U.S. Congressional representatives standardized by miles of shoreline (top panel) and as standardized by 2017 USD value of commercial fisheries (bottom panel) for the North Pacific region and its identified subregions.

Ten coastal national parks, monuments/memorials, or historic sites, one NOAA Habitat Focus Area (Kachemak Bay), and one National Estuarine Research Reserve (NERR) are found throughout the North Pacific region (Table 9.3), with the majority occurring in the Gulf of Alaska. Five expansive inshore and offshore regions are also identified as habitat areas of particular concern (HAPCs) in the North Pacific. These include 15 offshore Alaska seamounts, the Bowers Ridge habitat conservation zone, 5 Gulf of Alaska coral habitat protection areas, 10 Gulf of Alaska slope habitat conservation areas, and 6 skate nursery areas. Forty-one named fishing closure zones occur in the North Pacific, while fishing is prohibited throughout the Arctic EEZ in two large specific areas (Arctic Management Area; Arctic National Wildlife Refuge) and restricted in 74 additional portions of the North Pacific (Fig. 9.10). Most of these closures and restricted areas occur in the Gulf of Alaska and EBS subregions, which together contain 38 named closures (~1.7 million km^2) and 67 restricted areas (~2.4 million km^2). Approximately 17% of the North Pacific EEZ contains fishing prohibitions, with the largest prohibited area being the Arctic Management Area (~498,000 km^2). The largest single restricted area in the North Pacific is the Aleutian Islands Habitat Conservation Area (~920,000 km^2), which comprises most of the Aleutian Islands.

9.4.2.3 Organizational Context

Relative to the overall total commercial value of its managed fisheries, the NPFMC budget is lowest (eighth) when compared to other regions (Fig. 9.11), and makes up a small proportion of its overall costs. These relative values increased slightly in the early

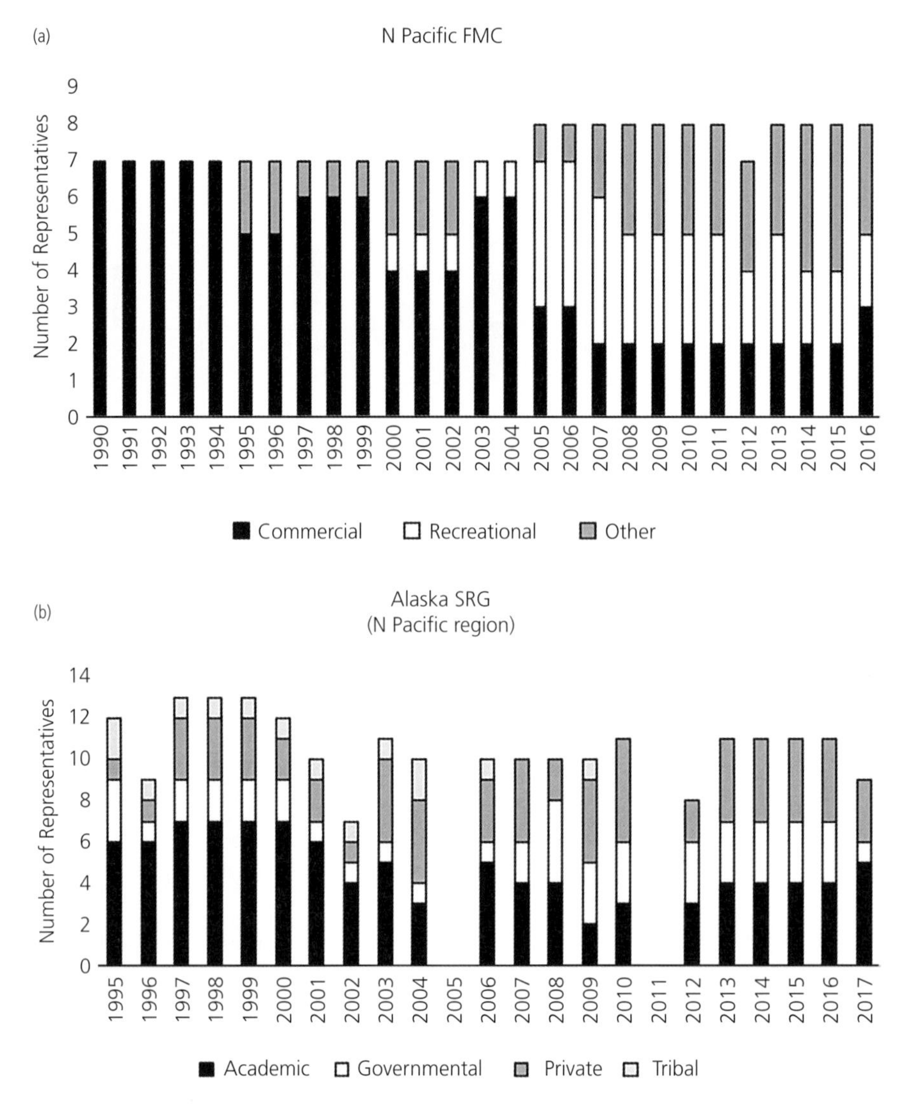

Figure 9.9 (a) Composition of the North Pacific Fishery Management Council (FMC) and the (b) Alaska marine mammal scientific review group (SRG; data unavailable for 2005, 2011) over time.

Data derived from NOAA National Marine Fisheries Service Council Reports to Congress and Scientific Review Group reports.

2000s and have remained relatively stable in more recent years, with current values similar to those in 2007–08. Cumulative National Environmental Policy Act Environmental Impact Statement (NEPA-EIS) actions from 1987 to 2016 in the North Pacific are fourth-highest nationally, and lowest among Pacific Ocean regions, with 206 observed over the past 30 years (Fig. 9.12) and the majority occurring in the Gulf of Alaska (n = 121) and Inside Passage (n = 68). Fourteen fisheries-related lawsuits have occurred in the North Pacific region since 2010, making it the second-most litigious among regions.

9.4.2.4 Status of Living Marine Resources (Targeted and Protected Species)

Nationally, the North Pacific contains the fifth-highest number of managed species. Of the 214 federally managed species in the region, 162 are targeted fishery

Table 9.2 List of major managed fishery species in the North Pacific region, under State, Federal, and international jurisdictions. FMP = Fishery Management Plan

Council/Agency	FMP	FMP Modifications	Major Species/Species Group	Scientific Name(s) or Family Name(s)
North Pacific Fishery Management Council **NPFMC**	Groundfish Gulf of Alaska (141 species)	104 amendments	Walleye pollock	*Gadus chalcogrammus*
			Pacific cod	*Gadus macrophalus*
			Sablefish	*Anoplopoma fimbria*
			Shallow-water flatfish	Northern rock sole *Lepidopsetta polyxystra* Southern rock sole *Lepidopsetta billineta* Butter sole *Isopsetta isolepis* Yellowfin sole *Pleuronectes aspera* Species from family Pleuronectidae
			Deep-water flatfish	Greenland turbot *Reinhardtius hippoglossoides* Dover sole *Microstomus pacificus* Deep-sea sole *Embassichthys bathybius*
			Rex sole	*Glyptocephalus zachirus*
			Arrowtooth flounder	*Astheresthes stomias*
			Flathead sole	*Hippoglossoides elassodon*
			Pacific Ocean perch	*Sebastes alutus*
			Northern rockfish	*Sebastes polyspinus*
			Shortraker rockfish	*Sebastes borealis*
			Dusky rockfish	*Sebastes variabilis*
			Blackspotted rockfish	*Sebastes melanostictus,*
			Rougheye rockfish	*S. aleutianus*
			Thornyhead rockfish	Shortspine thornyhead *Sebastolobus alascanus,* Longspine thornyhead *Sebastolobus altivelis*
			Demersal shelf rockfish	Broadfin thornyhead *Sebastolobus macrochir* Yelloweye rockfish *Sebastes ruberrimus* Species from families Sebastidae and/or Scorpaenidae

Continued

Table 9.2 Continued

Council/Agency	FMP	FMP Modifications	Major Species/Species Group	Scientific Name(s) or Family Name(s)
North Pacific Fishery Management Council **NPFMC**	Groundfish Gulf of Alaska (141 species)	104 amendments	Other rockfish	Sharpchin rockfish *Sebastes zacentrus* Harlequin rockfish *Sebastes variegatus* Silvergray rockfish *Sebastes brevispinis* Redstripe rockfish *Sebastes proriger* Redbanded rockfish *Sebastes babcocki* Species from families Sebastidae and/or Scorpaenidae
			Atka mackerel	*Pleurogrammus monopterygius*
			Skates	Big skate *Beringraja binoculata*, Longnose skate *Raja rhina*, and *Bathyraja* spp.
			Sculpins	Yellow Irish lord *Hemilepidotus jordani* Great sculpin *Myoxocephalus polyacanthocephalus* Bigmouth sculpin *Hemitripterus bolini* Plain sculpin *M. jaok* Species from families Cottidae, Hemitripteridae, Psychrolutidae, and Rhamphocottidae
			Sharks	Pacific spiny dogfish *Squalus suckleyi* Salmon shark *Lamna ditropis* Pacific sleeper shark *Somniosus pacificus* Incidental Apisturus spp., Carcharodon spp., *Cetorhinus* spp., *Hexanchus* spp., *Prionace* spp.
			Squid	*Berryteuthis* spp., *Belonella* spp., *Chiroteuthis* spp., *Eogonatus* spp., *Galiteuthis* spp., *Gonatopsis* spp., *Gonatus* spp., *Moroteuthis* spp., *Onychoteuthis* spp., *Ommastrephes* spp., and Rossia spp.
			Octopus	Giant Pacific octopus *Enteroctopus dofleini* *Benthoctopus* spp., *Japatella* spp., *Octopus* spp., *Opisthoteuthis* spp., and *Vampyroteuthis* spp.

Council/Agency	FMP	FMP Modifications	Major Species/Species Group	Scientific Name(s) or Family Name(s)
North Pacific Fishery Management Council **NPFMC**	Groundfish Bering Sea & Aleutian Islands (148 species)	114 amendments	Pollock	*Gadus chalcogrammus*
			Pacific cod	*Gadus macrophalus*
			Sablefish	*Anoplopoma fimbria*
			Yellowfin sole	*Limanda aspera*
			Arrowtooth flounder	*Atheresthes stomias*
			Kamchatka flounder	*Atheresthes evermanni*
			Rock soles	*Lepidopsetta polyxstra, L. bilineatus*
			Greenland turbot	*Reinhardtius hippoglossoides*
			Alaska plaice	*Pleuronectes quadrituberculatus*
			Flathead sole	*Hippoglossoides elassodon*
			Bering flounder	*H. robustus*
			Other flatfish (flounders, dabs, soles)	Species from Families Pleuronectidae and Paralichthyidae
			Pacific Ocean perch	*Sebastes alutus*
			Northern rockfish	*Sebastes polyspinus*
			Blackspotted rockfish	*Sebastes melanostictus,*
			Rougheye rockfish	*S. aleutianus*
			Shortraker rockfish	*Sebastes borealis*
			Other rockfish	Shortspine Thornyhead *Sebastolobus alascanus* Dusky rockfish *Sebastes variabilis* Species from families Sebastidae and/or Scorpaenidae
			Atka mackerel	*Pleurogrammus monopterygius*
			Skates	Alaska skate *Bathyraja parmifera* *Amblyraja* spp., *Bathyraja* spp., *Beringraja* spp., and *Raja* spp.
			Sculpins	Great sculpin *Myoxocephalus polyacanthocephalus* Threaded sculpin *Gymnocanthus pistillger* Plain sculpin *M. jaok* Warty sculpin *M. verrucosus* Bigmouth sculpin *Hemitripterus bolini* Yellow Irish lord *Hemilepidotus jordani* Species from families Cottidae, Hemitripteridae, Psychrolutidae, and Rhamphocottidae

Continued

Table 9.2 Continued

Council/Agency	FMP	FMP Modifications	Major Species/Species Group	Scientific Name(s) or Family Name(s)
North Pacific Fishery Management Council **NPFMC**	Groundfish Bering Sea & Aleutian Islands (148 species)	114 amendments	Squid	Magistrate armhook squid *Berryteuthis magister* *Belonella* spp., *Berryteuthis* spp., *Chiroteuthis* spp., *Eogonatus* spp. *Galiteuthis* spp., *Gonatopsis* spp., *Gonatus* spp., *Moroteuthis* spp., *Onychoteuthis* spp., and *Rossia* spp.
			Sharks	Pacific sleeper shark *Somniosus pacificus* Spiny dogfish *Squalus acanthias* Salmon shark *Lamna ditropis* Other incidentally captured shark species
			Octopus	Giant Pacific octopus *Enteroctopus dofleini* *Benthoctopus* spp., *Graneledone* spp., *Japatella* spp., *Opisthoteuthis* spp., and *Sasakiopus* spp.
	King and Tanner crabs Bering Sea & Aleutian Islands (5 species)	0 amendments	Red king crab Blue king crab Golden king crab Tanner crab Snow crab	*Paralithodes camtschaticus* *Paralithodes platypus* *Lithodes aequispinus* *Chionoecetes bairdi* *Chionoecetes opilio*
	Salmon (5 species)	10 amendments	Chinook salmon Coho salmon Pink salmon Sockeye salmon Chum salmon	*Oncorhynchus tshawytscha* *Oncorhynchus kisutch* *Oncorhynchus gorbuscha* *Oncorhynchus nerka* *Oncorhynchus ket*
	Scallop (4 species)	15 amendments	Weathervane scallops	*Patinopecten caurinus*
			Other non-commercial scallops	Pink or reddish scallops *Chlamys rubida* Spiny scallops *Chlamys hastata* Rock scallops *Crassadoma gigantea*
	Arctic (3 species)	1 amendment, 2 motions	Arctic cod Saffron cod Snow crab	*Boreogadus saida* *Eleginus gracilis* *Chionoecetes opilio*
International Pacific Halibut Commission **IPHC**—Includes North Pacific and Pacific Regions	Pacific halibut (1 species)	92 Annual Regulations	Pacific halibut	*Hippoglossus stenolepis*

Table 9.3 List of major bays and islands, NOAA habitat focus areas (HFAs), NOAA National Estuarine Research Reserves (NERRs), NOAA National Marine Monuments and Sanctuaries (NMS), coastal national parks, national seashores, and number of habitats of particular concern (HAPCs) in the North Pacific region and subregions

Region	Subregion	Major Bays	Major Islands	HFAs	NERRs	Monuments and NMS	Coastal National Parks	National Sea/Lakeshores	HAPCs
N Pacific	**E Bering Sea**	Bristol Bay, Kuskokwim Bay, Kvichak Bay, Norton Sound	Nunivak Island, Pribilof Islands, St. Lawrence Island						2
	Gulf of Alaska ***Inside Passage*** ***Prince William Sound***	Cook Inlet, Kachemak Bay, Kukak Bay, Iniskin Bay, Icy Bay Berner's Bay Glacier Bay, Lituya Bay William Henry Bay, Yakutat Bay Cochrane Bay, Orca Bay	Kodiak Island	Kachemak Bay, Alaska HFA	Kachemak Bay NERR		Aniakchak National Monument & Preserve, Katmai National Park & Preserve, Lake Clark National Park & Preserve, Kenai Fjords National Park, Wrangel St. Elias National Park & Preserve, Glacier Bay National Park & Preserve, Sitka National Historical Park		3
	Arctic	Goodhope Bay, Prudhoe Bay, Smith Bay					Bering Land Bridge National Preserve, Cape Krusenstern National Monument, Inupiat Heritage Center		
	Aleutian Islands	Akun Bay, Akutan Bay, Unalaska Bay	Fox Islands, Islands of Four Mountains, Andreanof Islands, Rat Islands, Near Islands				Aleutian World War II National Historic Area		

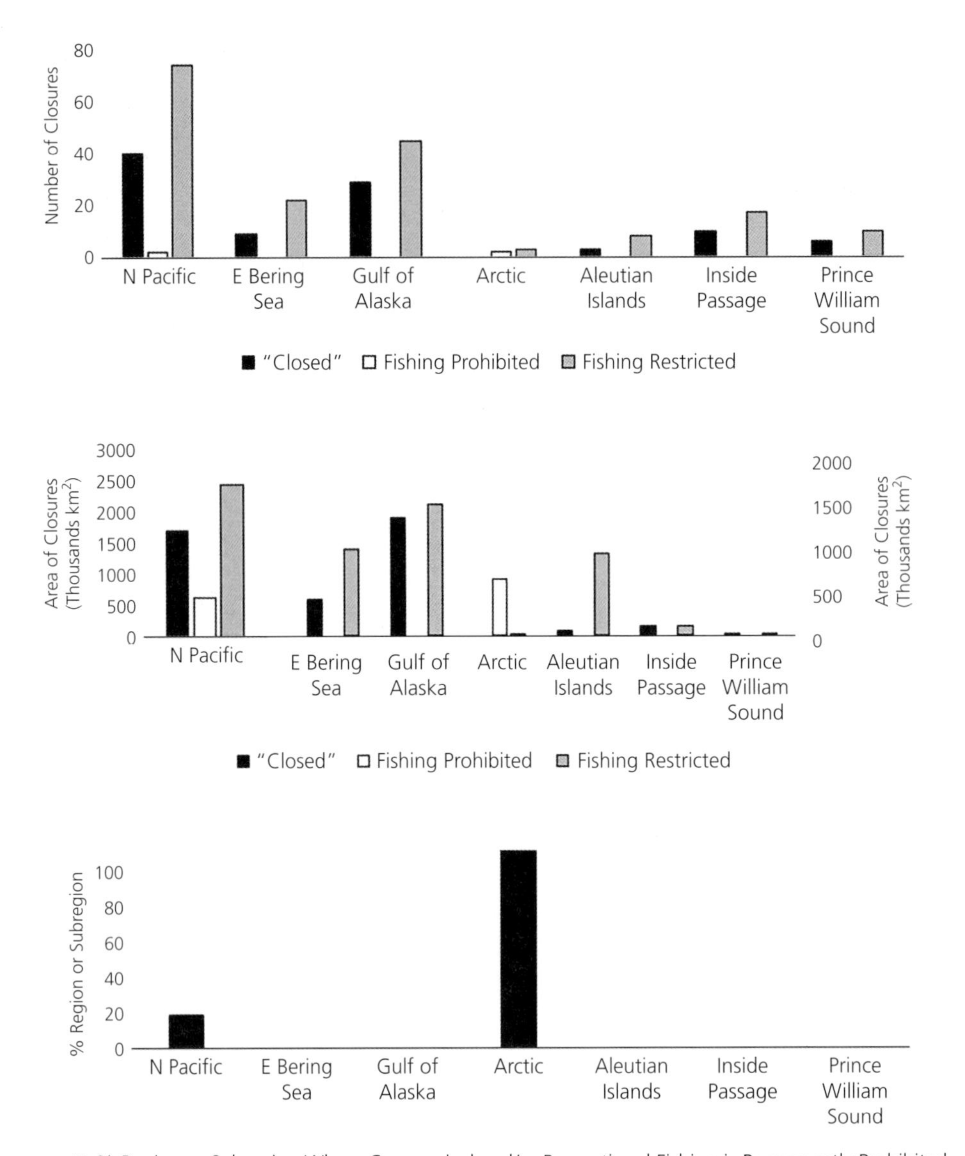

Figure 9.10 Number (top panel) and area (middle panel; km^2) of named fishing closures, and prohibited or restricted fishing areas of the North Pacific region and identified subregions, including percent region or subregion (bottom panel) where commercial and/or recreational fishing is permanently prohibited. Values for North Pacific subregions (middle panel) plotted on the secondary (right) axis.

Data derived from NOAA Marine Protected Areas inventory.

species covered under FMPs, while 16 species (Pacific halibut, Pacific herring, five species of Pacific salmon, Steelhead trout, and nine species of King and Tanner crab) are prohibited from being retained in certain fisheries as bycatch. Overlap in the types of species being managed under corresponding FMPs is also observed among subregions with 162 fishery species managed in the EBS-Aleutian Islands, 150 fishery species managed in the Gulf of Alaska, and 3 identified fishery species managed in the Arctic. Twenty-four species of marine mammals and seabirds are also federally protected under the ESA or MMPA (Fig. 9.13, Tables 9.2, 9.4–5).

As of mid-2017, 65 stocks or stock complexes were federally managed in the North Pacific with 36 listed as a NOAA Fish Stock Sustainability Index (FSSI) stock (Fig. 9.14; NMFS 2017, 2020b).[1] Geographically, stocks

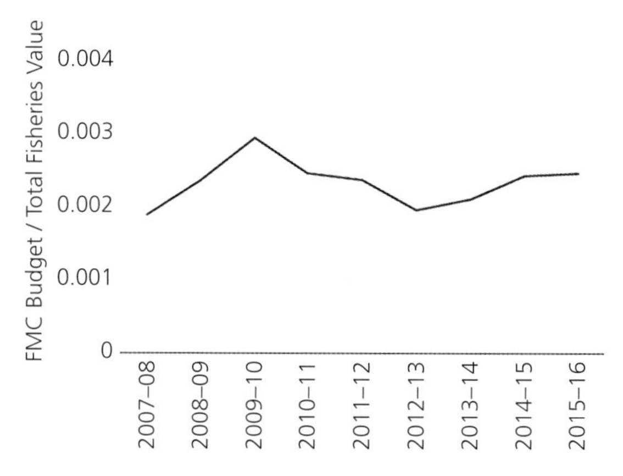

Figure 9.11 Total annual budget of the North Pacific Fishery Management Council (FMC) as compared to the total value of its marine fisheries. All values in USD.

Data from NOAA National Marine Fisheries Service Office of Sustainable Fisheries and NOAA National Marine Fisheries commercial landings database.

are typically managed among three subregions: the EBS-Aleutian Islands (35 total stocks), Gulf of Alaska (25 total stocks), and Arctic (3 total stocks). In addition, three stocks are managed throughout the Alaskan EEZ (**Alaska Coho salmon assemblage**—*Oncorhynchus kisutch*; **Chinook salmon—Eastern North Pacific far north migrating**; **Weathervane scallop**—*Patinopecten caurinus*, **Alaska**). Of all stocks during that period, one stock was listed as experiencing overfishing and was also overfished (**Blue king crab**—*Paralithodes platypus*, **Pribilof Islands**).[2] There were no stocks that remained unclassified as to whether they were experiencing overfishing (0%; lowest nationally) and 29 had unknown overfished status (44.6%; sixth-highest nationally), making the North Pacific mid-range nationally in terms of the proportion of fishery stocks with unknown status.[3] Approximately equal numbers of stocks of unknown overfished status occurred in the EBS-Aleutian Islands (n = 13) and Gulf of Alaska (n = 14), while a higher percentage of Gulf of Alaska stocks had unknown overfished status (51.9%).[4] Fishing mortality in the Arctic, while technically unknown, is expected to be (at or at least near) zero.

Among the protected species of the North Pacific are 49 independent marine mammal stocks, and 15 ESA-

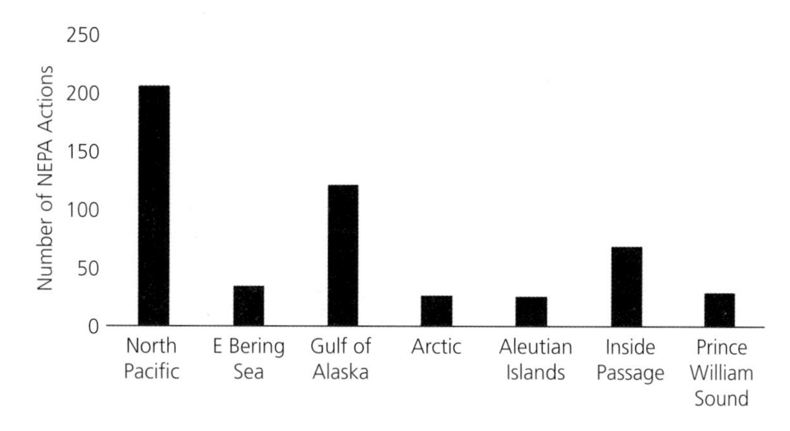

Figure 9.12 Number of National Environmental Policy Act Environmental Impact Statement (NEPA-EIS) actions from 1987–2016 for the North Pacific region and identified subregions.

Data derived from U.S. Environmental Protection Agency EIS database.

[1] As of June 2020, 63 stocks or stock complexes are federally managed in the North Pacific, with 35 listed as FSSI stocks. In 2017, two stocks (Bering Sea/Aleutian Islands squid complex; Gulf of Alaska squid complex) were reclassified under the NPFMC ecosystem component category.

[2] As of June 2020, no North Pacific stocks are experiencing overfishing and two are overfished (Blue king crab, Pribilof Islands and St. Matthew Island stocks).

[3] In June 2020, 26 stocks had unknown overfished status (41.3%).

[4] As of June 2020, 11 stocks in the EBS-Aleutian Islands and 13 stocks in the Gulf of Alaska have unknown overfished status. Between the two subregions, the Gulf of Alaska continues to have a higher percentage of stocks with unknown overfished status (48.1%).

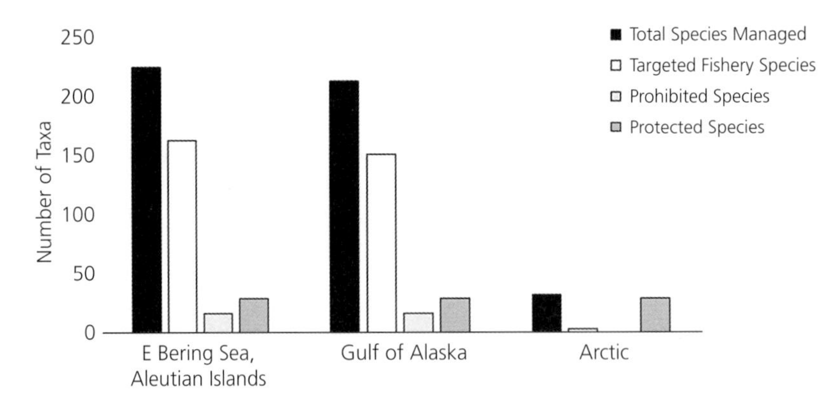

Figure 9.13 Number of managed taxa (species or families) for the identified subregions of the North Pacific.

Table 9.4 List of prohibited fishery species in the North Pacific region, including federal and state jurisdictions. Currently, there are no prohibited fishery species specifically identified in the Arctic Fishery Management Plan.

Council/Authority	Region	Prohibited Species	Scientific Name
International Pacific Halibut Commission **IPHC**; North Pacific Fishery Management Council **NPFMC**; Alaska Regional Office **AKRO**	E Bering Sea/Aleutian Islands (16 species)	Pacific halibut	*Hippoglossus stenolepis*
		Pacific herring	*Clupea pallasii*
		Pacific salmon	*Oncorhynchus* spp.
		Steelhead trout	*Oncorhynchus mykiss*
		King crab	*Lithodes* spp., *Paralithodes* spp.
		Tanner crab	*Chionoectes* spp.
	Gulf of Alaska (16 species)	Pacific halibut	*Hippoglossus stenolepis*
		Pacific herring	*Clupea pallasii*
		Pacific salmon	*Oncorhynchus* spp.
		Steelhead trout	*Oncorhynchus mykiss*
		King crab	*Lithodes* spp., *Paralithodes* spp.
		Tanner crab	*Chionoectes* spp.
	Arctic (0 species)		

listed distinct population segments, of which ~67% are endangered and none have unknown status (Fig. 9.15). Nationwide, high numbers of marine mammal stocks are found in the North Pacific (second-highest nationally; 15.3% of total marine mammal stocks), with approximately one-fourth listed as strategic marine mammal stocks. There are 20.4% of North Pacific marine mammal stocks that have unknown population size, which is second-lowest (seventh-highest) among regions.

9.4.3 Environmental Forcing and Major Features

9.4.3.1 Oceanographic and Climatological Context

Circulation patterns throughout the North Pacific are driven primarily by the Alaska Current, Alaskan Stream, and Alaska Coastal Current (ACC) throughout the Gulf of Alaska (Ladd & Cheng 2016), and the Aleutian North Slope Current, Bering Slope Current, and Bering Coastal Current throughout the EBS and Aleutian Islands subregion (Hurst et al. 2014; Fig. 9.1). The Alaska Current derives from the North Pacific Current as it splits as it approaches North America. The northern limb flows in a counterclockwise pattern in the Gulf of Alaska, while complemented by the inshore ACC that flows along the PWS, Kenai Peninsula, and Kodiak Island subregions, and through some Aleutian Islands passes in the EBS. A gyre containing the Aleutian North Slope Current and Bering Slope Current presents a counterclockwise circulation pattern near the Aleutian Islands, while the ACC promotes connectivity between the Gulf of Alaska and EBS, and continues to move northward into the Chukchi Sea and along the Alaskan Arctic coast (Stabeno et al. 1999). The Bering

Table 9.5 List of protected species under the Endangered Species Act (ESA) or Marine Mammal Protection Act (MMPA) in the North Pacific region

Region	Managing Agency	Protected Species Group	Common Name	Scientific Name
E Bering Sea/Aleutian Islands, Gulf of Alaska, and Arctic (29 species)	North Pacific Fishery Management Council **NPFMC**; Alaska Regional Office **AKRO**	Pinnipeds	Stellar sea lion	*Eumetopias jubatus*
			Harbor seal	*Phoca vitulina*
			Ice seals	Ringed seal *Phoca hispida*
				Ribbon seal *Histriophoca fasciata*
				Spotted seal *Phoca largha*
				Bearded seal *Erignathus barbatus*
			Northern fur seal	*Callorhinus ursinus*
		Whales	Baird's beaked whale	*Berardius bairdii*
			Beluga whales	*Delphinapterus leucas*
			Bowhead whale	*Balaena mysticetus*
			Fin whale	*Balaenoptera physalus*
			Humpback whale	*Megaptera novaeangliae*
			Gray whale	*Eschrichtius robustus*
			Killer whale (Orca)	*Orcinus orca*
			Narwhal	*Monodon monoceros*
			Northern minke whale	*Balaenoptera acutorostrata*
			North Pacific right whale	*Eubalaena japonica*
			Sei whale	*Balaenoptera borealis*
			Sperm whale	*Physeter macrocephalus*
		Dolphins/Porpoises	Dall's porpoise	*Phocoenoides dalli*
			Harbor porpoise	*Phocoena phocoena*
			Pacific white-sided dolphin	*Lagenorhynchus obliquidens*
	U.S. Fish and Wildlife Service **FWS**	Seabirds	Spectacled eider	*Somateria fischeri*
			Steller's eider	*Polysticta stelleri*
		Other species	Sea otter	*Enhydra lutris*
			Polar bear	*Ursus maritimus*
			Walrus	*Odobenus rosmarus*

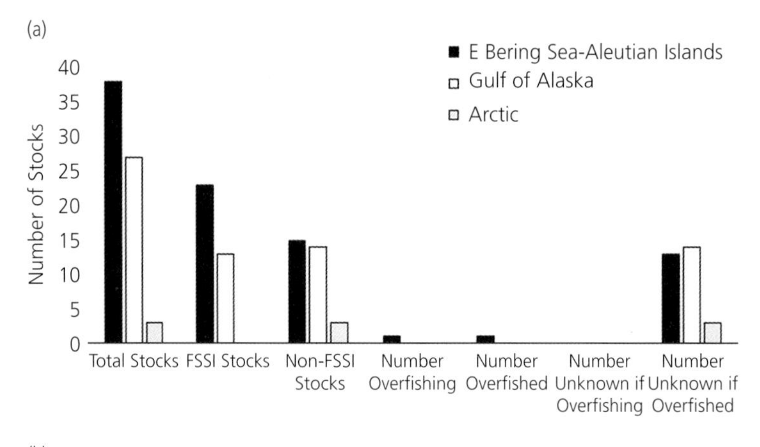

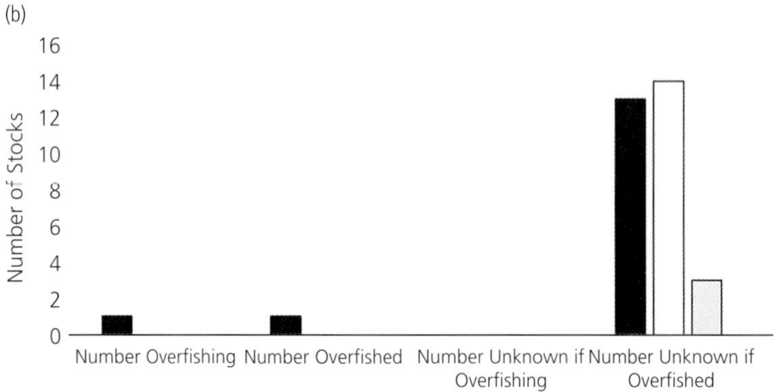

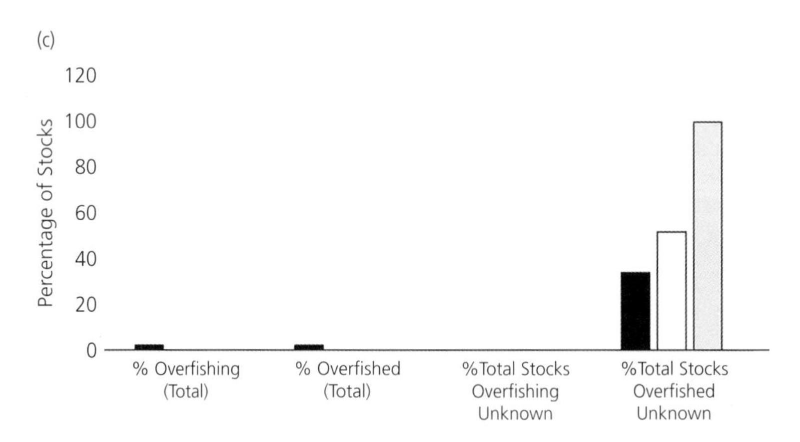

Figure 9.14 For the North Pacific region as of June 2017 (a) Total number of managed Fish Stock Sustainability Index (FSSI) stocks and non-FSSI stocks, and breakdown of stocks experiencing overfishing, classified as overfished, and of unknown status. (b) Number of stocks experiencing overfishing, classified as overfished, and of unknown status. (c) Percent of stocks experiencing overfishing, classified as overfished, and of unknown status.

Data from NOAA National Marine Fisheries Service. Fishing mortality in the Arctic, while technically unknown, is expected to be zero. Note: stocks may refer to a species, family, or complex.

Coastal Current moves northward along the EBS coastline (Hurst et al. 2014). These currents provide means of propagule transport along coastal regions and throughout the North Pacific EEZ. Additional circumpolar circulation patterns are observed in the Arctic, particularly related to the Beaufort Gyre (Brugler et al. 2014).

Clear interannual and multidecadal patterns in average SST have been observed for the North Pacific region (Fig. 9.16), with temperatures varying among

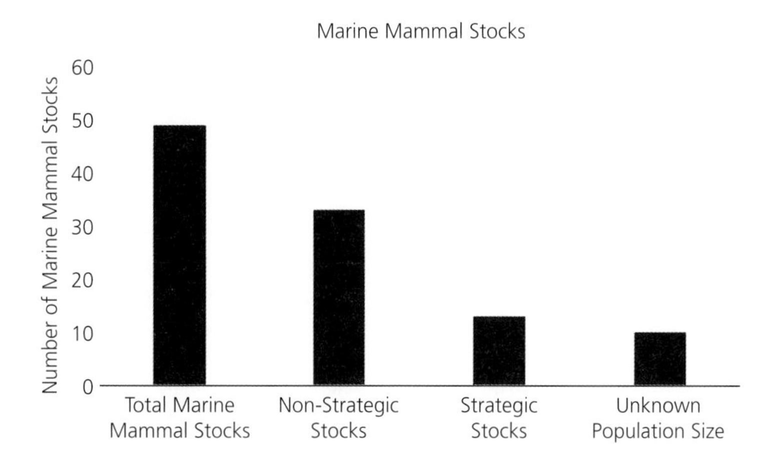

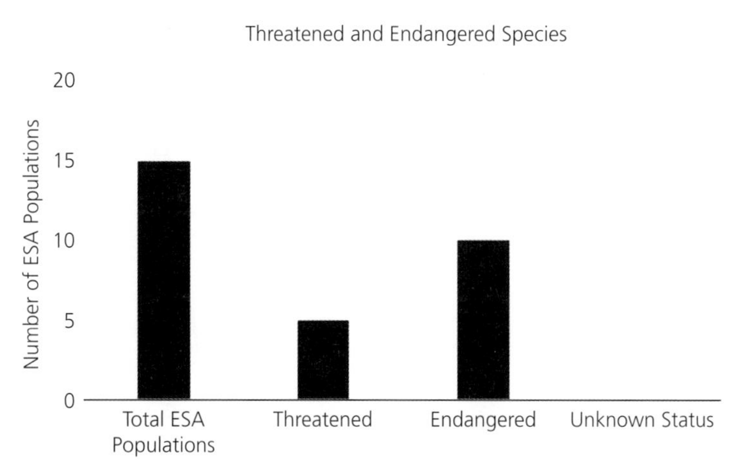

Figure 9.15 Number and status of federally protected species (marine mammal stocks, top panel; distinct population segments of species listed under the Endangered Species Act (ESA), bottom panel) occurring in the North Pacific region.

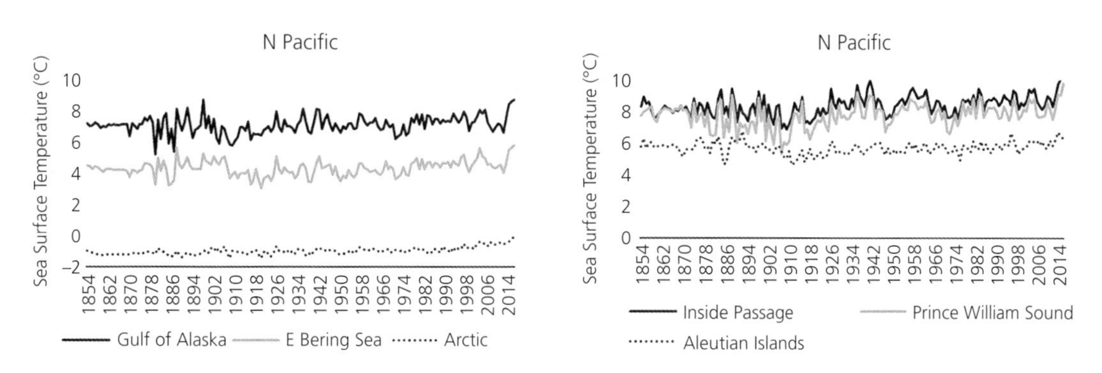

Figure 9.16 Average sea surface temperature (°C) over time (years 1854–2016) for North Pacific subregions.

Data derived from the NOAA Extended Reconstructed Sea Surface Temperature dataset (https://www.ncdc.noaa.gov/).

subregions and averaging 7.1°C ± 0.05 (SE) in the Gulf of Alaska, 4.3°C ± 0.04 (SE) in the EBS, and −0.98°C ± 0.02 (SE) in the Arctic over time (1854–2016). Values have increased by 2.3°C in the Gulf of Alaska, 2.1°C in the EBS, and 1.2°C in the Arctic since the mid-twentieth century. Similar trends have been observed in other subregions, with average temperatures over the 162-year time period also higher in the Inside Passage at 8.4°C ± 0.05 (SE) and PWS at 7.8°C ± 0.06 (SE), and slightly lower in the Aleutian Islands at 5.7°C ± 0.03 (SE). Temperatures in the Inside Passage and PWS have increased by 2.7°C and 2.4°C, respectively, since 1950, while those for the Aleutian Islands have also increased by 1.0°C. Coincident with these temperature observations are Pacific basin-scale climate oscillations (Fig. 9.17), including the El Niño Southern Oscillation Index (ENSO), Multivatiate El Niño Index (MEI), Northern Oscillation Index (NOI), North Pacific Gyre Oscillation Index (NPGO), and PDO, which exhibit decadal cycles that can influence the environmental conditions and ecologies of several U.S. regions, including the North Pacific. Effects of the PDO and North Pacific Index (NPI) on North Pacific LMRs were particularly observed in the 1970s and 1980s, resulting in significant shifts between groundfish and crustacean-dominant communities and their fisheries, and enhancements to salmon populations (Francis & Hare 1994, Mantua et al. 1997, Anderson & Piatt 1999, Hare & Mantua 2000, Benson & Trites 2002, Heymans et al. 2005, Zheng & Kruse 2006, Litzow & Urban 2009, Gaichas et al. 2011). Subsequently, there has been a significant reorientation of the fish community, especially in the EBS, largely due to climate and sea ice extent shifts, species shifting further poleward, and benthic taxa becoming more temperate (Mueter & Litzow 2008, Litzow et al. 2014, 2019).

9.4.3.2 Notable Physical Features and Destabilizing Events and Phenomena

With six prominently identified subregions (including two subcomponents of the Gulf of Alaska), the North Pacific contains 23 major bays, inlets, or sounds

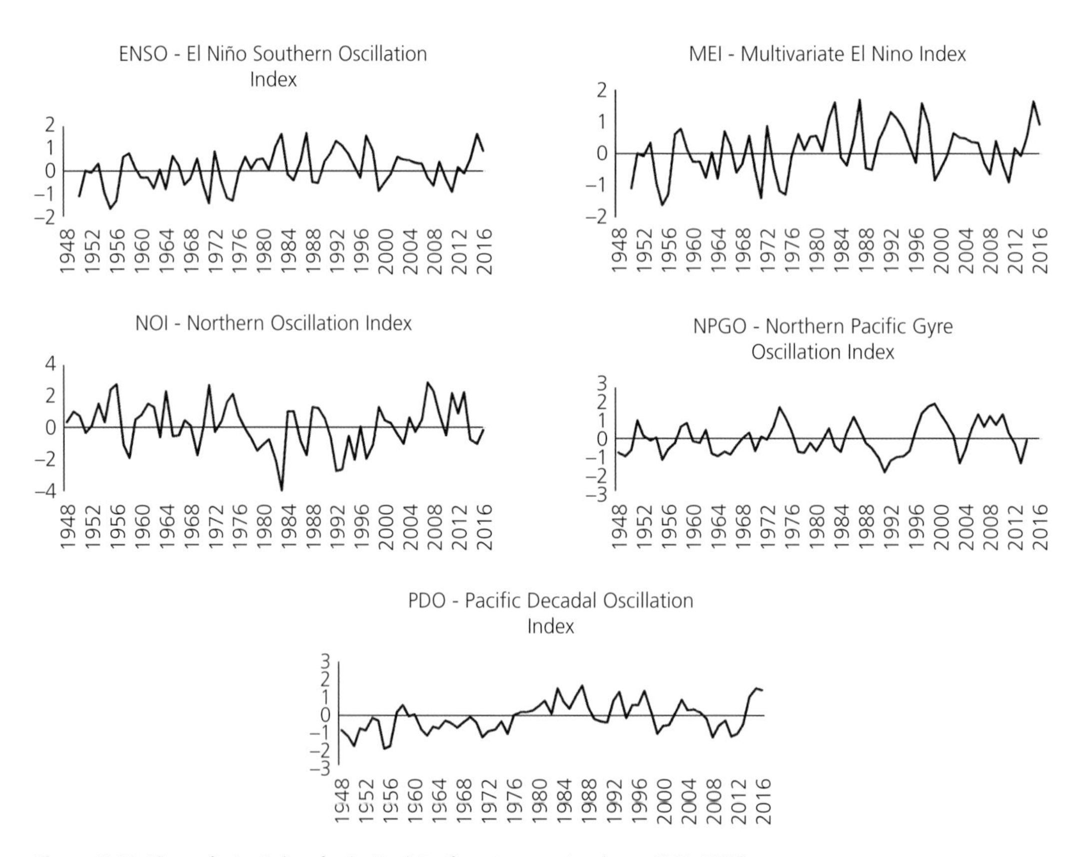

Figure 9.17 Climate forcing indices for the North Pacific region over time (years 1948–2016).

Data derived from NOAA Earth System Research Laboratory data framework (https://www.esrl.noaa.gov/).

(Table 9.3) throughout the Alaskan coast, and 9 major island groupings concentrated throughout its inshore and offshore areas. Proportionally, the North Pacific makes up 27% of the U.S. EEZ (~3.7 million km²), and 34.8% of the national shoreline (~58,400 km), while containing a relatively low overall average depth (~2100 m) among regions and maximum depths of ~7.9 km (Fig. 9.18), which comprises the third-deepest waters nationally.

The Gulf of Alaska, EBS, and Aleutian Islands each contribute significant proportions of national and North Pacific EEZ (Gulf of Alaska: 30.8% Alaska, 8.3% national; EBS: 26.9% Alaska, 7.3% national; Aleutian Islands: 25% Alaska; 6.8% national), with the Gulf of Alaska comprising major components of the Alaskan (70.4%) and national shoreline (24.4%) at ~41,000 km. While the Inside Passage makes up only 2.5% of the Alaskan EEZ (~92,500 km²), it contains 13.6% of the Alaskan shoreline (~23,000 km). PWS is smallest among the identified subregions at ~9,100 km², but contains ~8% of Alaska's shoreline (~4,600 km). The Arctic comprises 17% of Alaska's EEZ (~630,000 km²) and ~7.4% of its shoreline. The deepest average depths of the North Pacific are off the Aleutian Islands

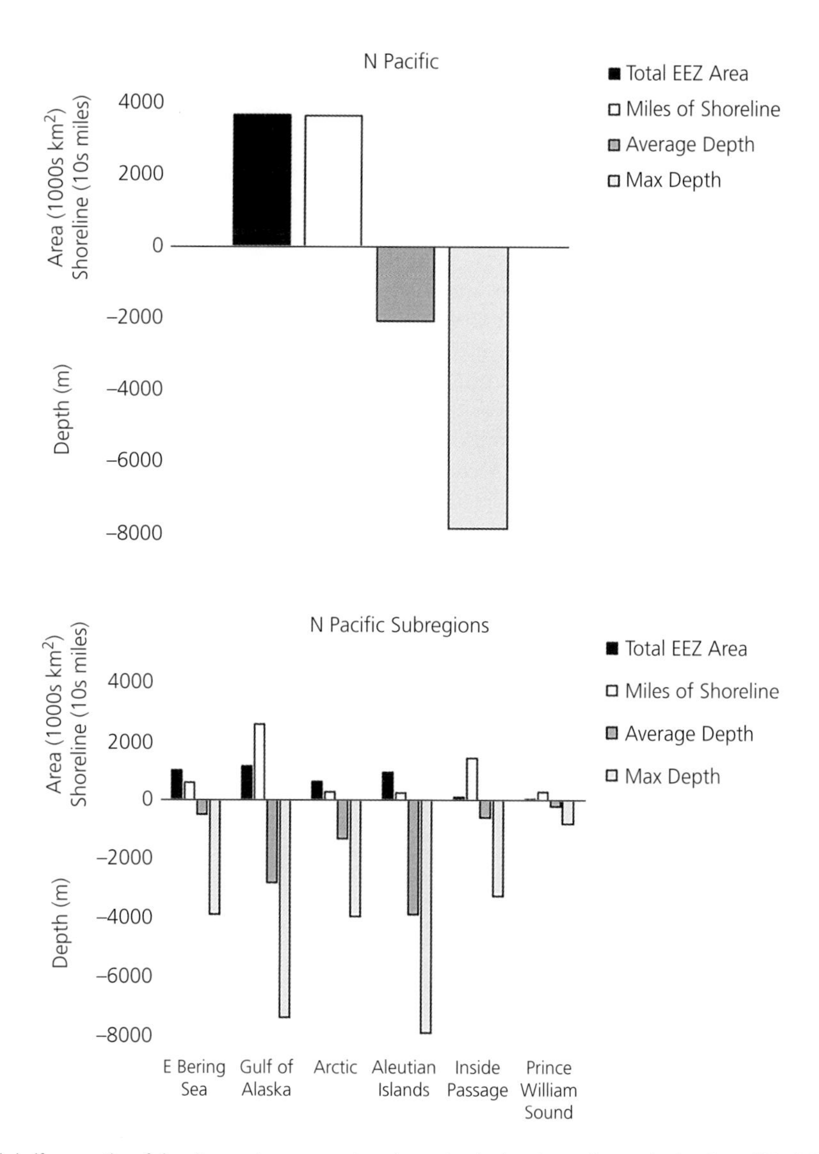

Figure 9.18 Total shelf area, miles of shoreline, and average and maximum depth of marine regions and subregions of North Pacific.

(~3.8 km) and Gulf of Alaska (~2.8 km), with the deepest maximum depths also occurring in these subregions (~7.9 km Aleutian Islands; ~7.4 km Gulf of Alaska).

In the North Pacific region, compared to other regions of the U.S., no hurricanes have been reported over time, though there are hurricane-force winds (Overland & Bond 1993, Deal et al. 2002, Winstead et al. 2006). However, sea ice is concentrated in this region (Fig. 9.19), with highest proportions occurring in the Arctic and EBS. Although values have fluctuated interannually, proportions have generally remained stable in the Arctic subregion, with higher variability observed, especially in recent years, for the EBS. However, declines in sea ice coverage and thickness throughout the Arctic have also accelerated over time (Rothrock et al. 1999, Stroeve et al. 2007, 2008, Comiso et al. 2008, Notz & Stroeve 2016).

9.4.4 Major Pressures and Exogenous Factors

9.4.4.1 Other Ocean Use Context

Ocean uses in the North Pacific are most associated with the tourism/recreation, offshore mineral extraction, and marine transportation sectors (Fig. 9.20). Among sectors, trends for North Pacific Ocean establishments and employments have increased over the past decades. Tourism establishments and employments comprise 67–73% of total North Pacific Ocean establishments and 43–50% of total North Pacific Ocean employments. Following 2011, substantial increases in total ocean establishments were observed, while gradual and steady increases in employments have occurred from 2005 to 2015, with highest values occurring in 2015–2016 at ~2,400 establishments and ~50,000 employments. The GDP from offshore mineral extraction is highest among sectors, contributing 69–91% toward total North Pacific Ocean GDP.

Over the past decades, ocean GDP has oscillated and continued to decline during the 2010s, with most recent estimated values at ~$8.6 billion. Overall, the North Pacific region contributes toward 5.4% of the U.S. total ocean GDP, which is third-lowest nationally. Establishments and employments in the North Pacific offshore mineral extraction sector contribute 6–7% toward North Pacific total ocean establishments and 21–29% toward its total ocean employments. The North Pacific also contributes low percentages (<8%) toward U.S. national marine establishments and employments for all other ocean use sectors, but comprises 8–10% of national offshore mineral extraction employments and GDP, which are each second-highest nationally. Overall, the North Pacific is above average in terms of its marine socioeconomic status, especially as associated with its LMR and marine energy-based economy.

The degree of tourism, as measured by number of dive shops, number of major ports and marinas, and number of cruise ship departure passengers is among mid to low-range nationally in the North Pacific (Fig. 9.21), with nationally lowest numbers of dive shops, and fourth- and fifth-highest numbers of ports and marinas, respectively, mostly concentrated in the

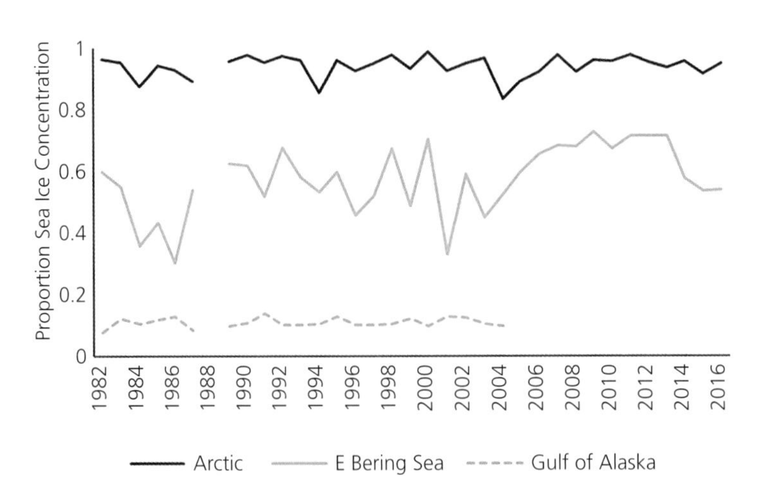

Figure 9.19 Proportional area of sea ice concentration for North Pacific subregions, over time (years 1982–2016).
Data derived from NOAA Earth System Research Laboratory data framework (https://www.esrl.noaa.gov/).

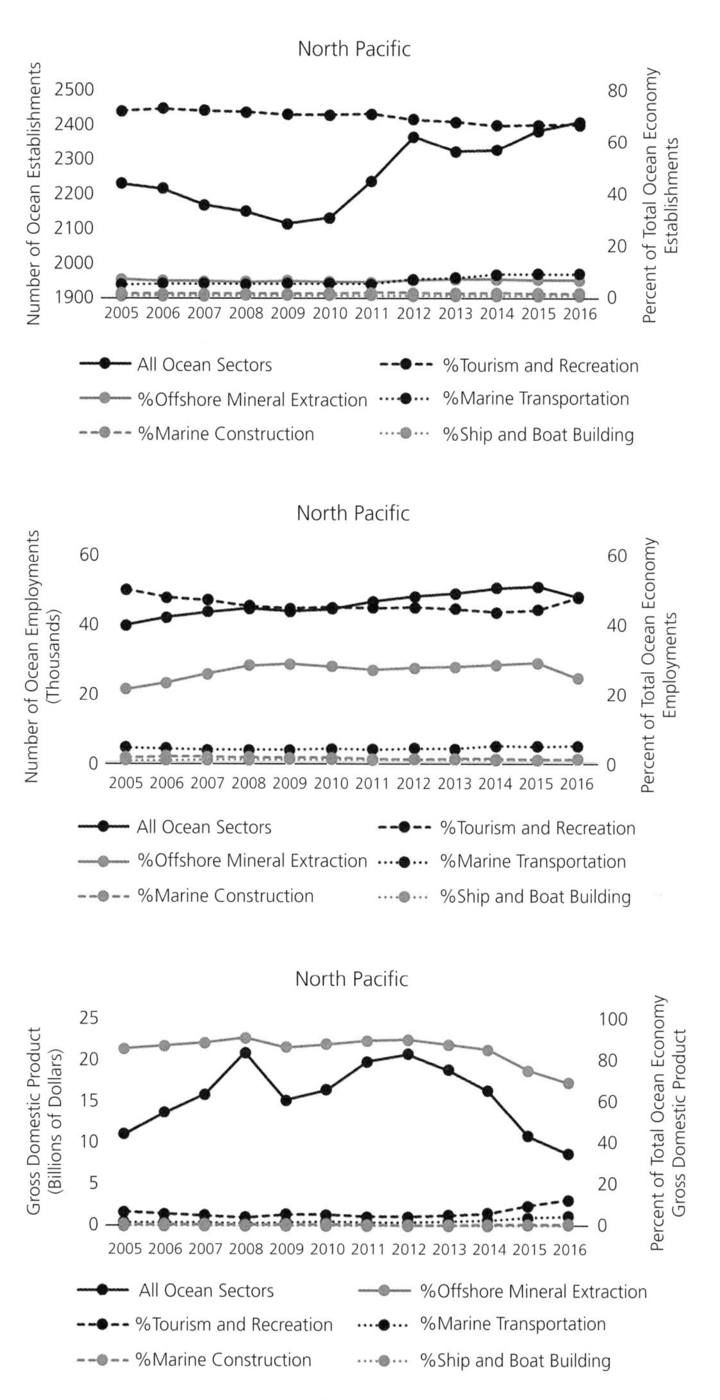

Figure 9.20 Number and percent contribution of multisector establishments, employments, and gross domestic product (GDP) and their percent contribution to the total multisector oceanic economy of the North Pacific region (years 2005–2016).

Data derived from the National Ocean Economics Program.

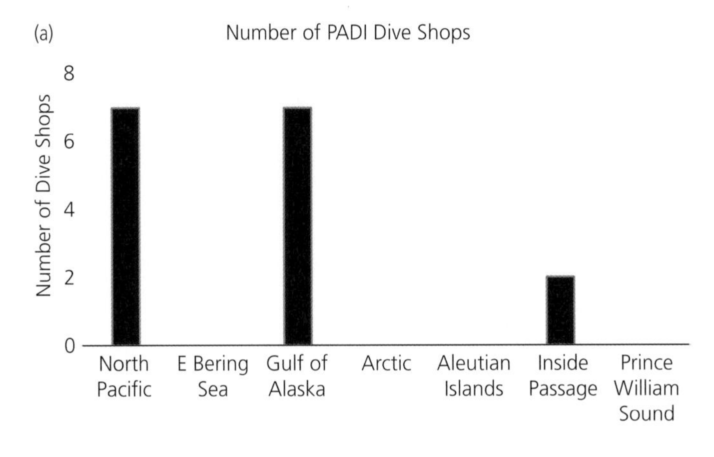

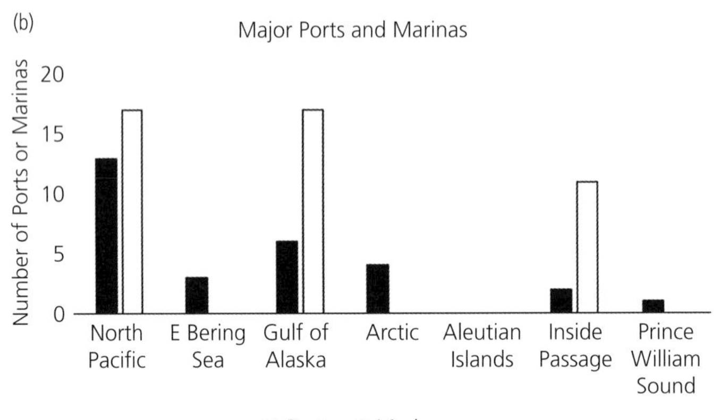

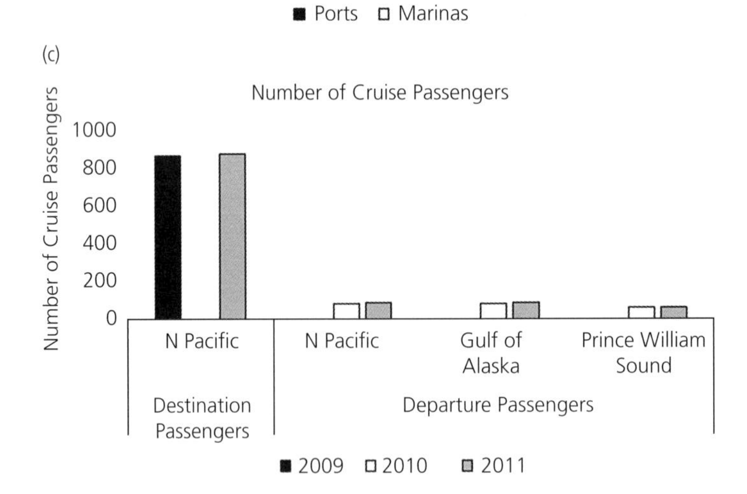

Figure 9.21 Tourism indicators for the North Pacific region and identified subregions: (a) Number of Professional Association of Diving Instructors (PADI) dive shops. Data derived from http://apps.padi.com/scuba-diving/dive-shop-locator/ (b) Number of major ports and marinas. (c) Number of cruise destination (years 2009, 2011) and cruise departure port passengers (years 2010–2011), values in thousands.

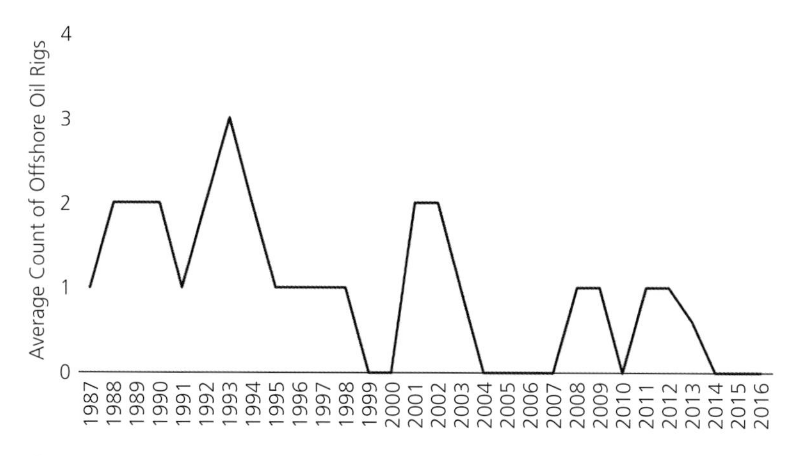

Figure 9.22 Average count of offshore oil rigs for the North Pacific region.

Data derived from Baker Hughes Rotary Rig Counts.

Gulf of Alaska. There are 11 major marinas in the Inside Passage. Cruise destination passengers in the North Pacific are second-highest nationally with ~900 thousand per year as of 2011, while only ~82–85 thousand departure passengers are found in the Gulf of Alaska PWS subcomponent, leaving from Whittier, AK. While very active terrestrially, relatively limited offshore oil production (fourth-highest nationally) in the North Pacific has been observed in terms of the number of oil rigs over time (Fig. 9.22). Peak values (n = 3) occurred in 1993, and oscillating numbers ranging from 0 to 2 have been observed in more recent years. No U.S. Bureau of Ocean Energy Management (BOEM) wind energy sites are found in this region.

9.4.5 Systems Ecology and Fisheries

9.4.5.1 Basal Ecosystem Production

Since 1998, mean surface chlorophyll values throughout subregions of the North Pacific EEZ (Fig. 9.23) have been highest in the Inside Passage (3.5 mg m^{-3} ± 0.18, SE), PWS (2.3 mg m^{-3} ± 0.12, SE), and EBS (1.9 mg m^{-3} ± 0.1, SE). Low and stable values are also found for the Arctic, Gulf of Alaska, and Aleutian Islands with values ranging from 0.7 to 2.1 mg m^{-3} over time. Average annual productivity within the EBS and Gulf of Alaska has remained relatively steady between ~150–170 g C m^{-2} $year^{-1}$ over the same time period, with values historically slightly higher in the EBS and now approximately equal between both subregions. Cumulatively, the North Pacific region contains the highest primary production among U.S. regions.

9.4.5.2 System Exploitation

Nationally, the lowest numbers of landed taxa have been captured in the North Pacific region over time, with values continuing to increase since the late 1970s and peaking at 65 taxa in 1993 and decreasing to ~40 reported taxa $year^{-1}$ since 2000 (Fig. 9.24). These trends were probably mainly due to increased species-specific reporting and fishing of groundfish species during the 1980s, fewer newly reported fisheries since the 1990s, and less regular species-specific reporting of rockfishes after the mid-1990s. Combined surveyed fish and invertebrate biomass for the North Pacific region has remained relatively constant over time (Fig. 9.25), averaging 34 million ± 234,000 (SE) metric tons $year^{-1}$, with increases observed from 1998 onward and peaking in 2003 at 53 million metric tons. More recent values have remained around 30–40 million metric tons, with year 2015 biomass similar to observed peak values.

Values for total U.S. domestic landings/total survey biomass in the North Pacific increased during the 1980s and early 1990s (up to 0.09; Fig. 9.26) with marginal decreases observed over time to current values of ~0.05. These trends are mainly explained by a substantial increase in reported groundfish landings and transition from foreign-dependent and joint venture fisheries to fully domestic ones during the 1980s. Decreases in crab landings after the 1990s have also contributed to more recent trends. Values for total biomass/productivity have remained relatively constant at near 0.003–0.005 for the majority of time recorded, with higher peak values (~0.008) during the early 2000s.

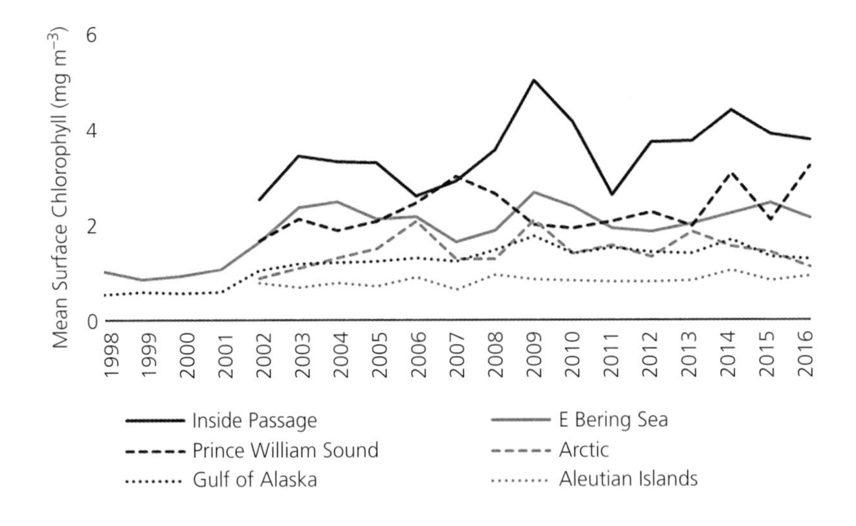

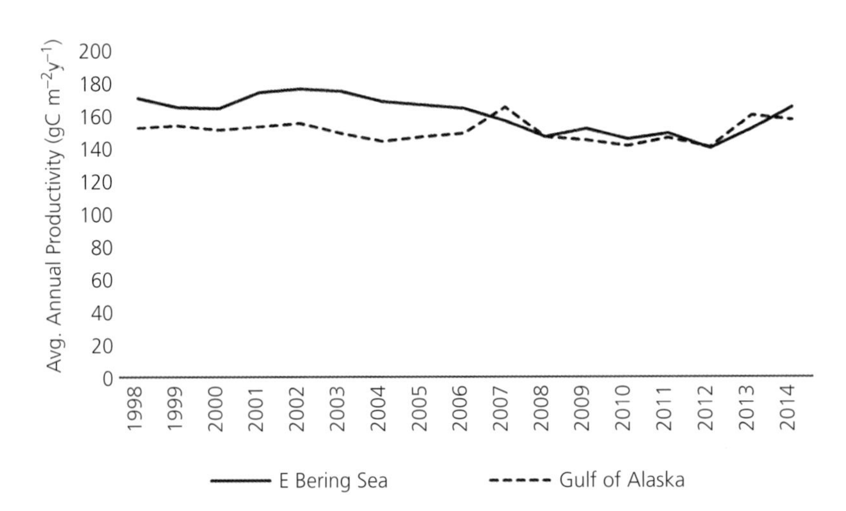

Figure 9.23 Mean surface chlorophyll (mg m^{-3}; top panel) for the North Pacific region and identified subregions, and average annual productivity (grams carbon m^{-2} $year^{-1}$; bottom panel) for the E Bering Sea and Gulf of Alaska subregions over time.

Data derived from NASA Ocean Color Web (https://oceancolor.gsfc.nasa.gov/) and productivity calculated using the Vertically Generalized Production Model (VGPM).

Total landings/productivity values have remained relatively steady (0.0003–0.0004) over time, with higher values observed in more recent years.

9.4.5.3 Targeted and Non-Targeted Resources

Total commercial landings are currently highest in the nation for the North Pacific region (Fig. 9.27), which experienced major increases in the 1980s, and have remained around 2.5 million metric tons since the 1990s. These trends are related to both expansions of Alaskan fisheries (shifting from foreign to domestic emphasis) and PDO and NPI-related shifts in community composition (Francis & Hare 1994, Mantua et al. 1997, Anderson & Piatt 1999, Hare & Mantua 2000, Benson & Trites 2002, Heymans et al. 2005, Zheng & Kruse 2006, Litzow & Urban 2009, Gaichas et al. 2011).

While North Pacific landings have been dominated by commercial fisheries, increases in recreational catches during the early 2000s were also observed. Highest commercial contributions are currently reported in EBS and Aleutian Islands major ports at ~700–800 thousand metric tons since 2011. Similar trends during this time

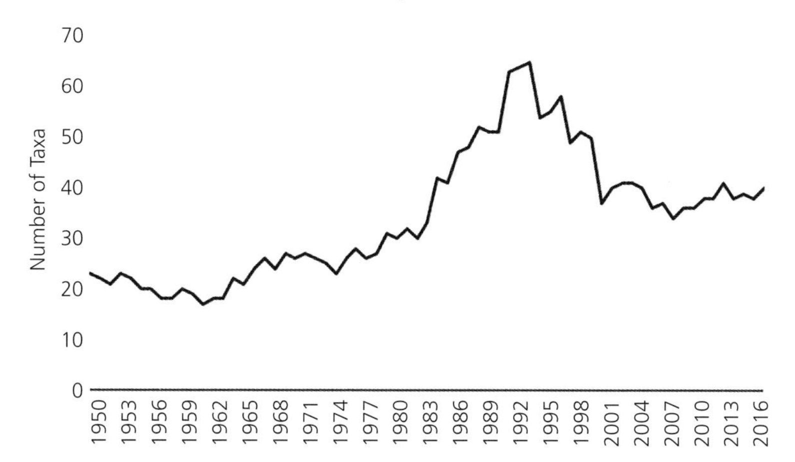

Figure 9.24 Number of taxa reported captured by commercial and recreational fisheries in the North Pacific region.
Data derived from NOAA National Marine Fisheries Service commercial and recreational fisheries statistics.

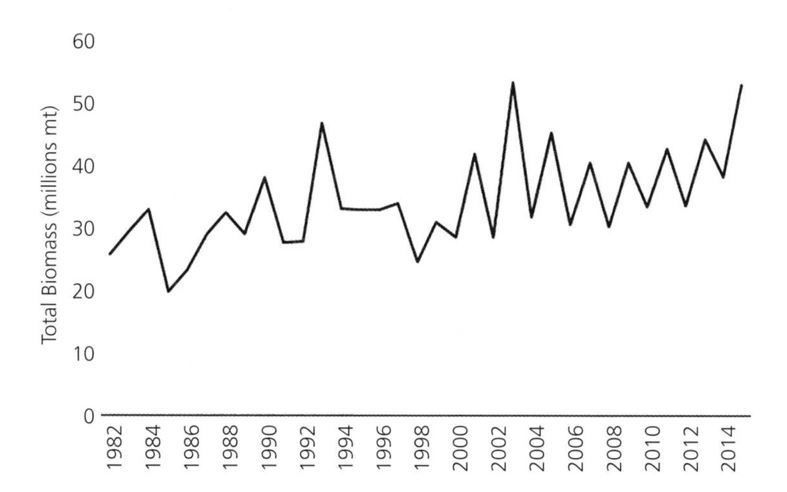

Figure 9.25 Total combined surveyed fish and invertebrate biomass (metric tons; mt) for the North Pacific region, over time.

period have persisted for the Gulf of Alaska at ~300–500 thousand metric tons. In the Gulf of Alaska subcomponents, higher commercial landings are observed in the Inside Passage at ~100 thousand metric tons, while those for PWS are lower at ~40–100 thousand metric tons over time. Trends for these two subregions have remained relatively stable over time, with staggered increases observed since the 2000s. Additionally, no commercial landings are reported (nor allowed) for the Arctic. Increases in recreational and subsistence fisheries catches have been observed for the Gulf of Alaska over time since the mid-1990s, with values up to 1.8 million fishes. Values in the EBS-Aleutian Islands and Arctic have decreased over time from ~3 to 5 thousand fishes during the mid-1990s to current values of ~1200 reported fishes in the Arctic and ~400 in the EBS.

When standardized per square kilometer (Fig. 9.28), North Pacific landings follow very similar patterns as observed for total landings values, with this region currently containing the fourth-highest concentration of fisheries landings over time in the nation. The highest concentrated fishing pressure in the region (up to ~11 metric tons km^{-2}) has been historically observed in and around PWS, while much lower values have been observed in the EBS-Aleutian Islands, Gulf of Alaska, and Inside Passage (up to 0.5 metric tons km^{-2}).

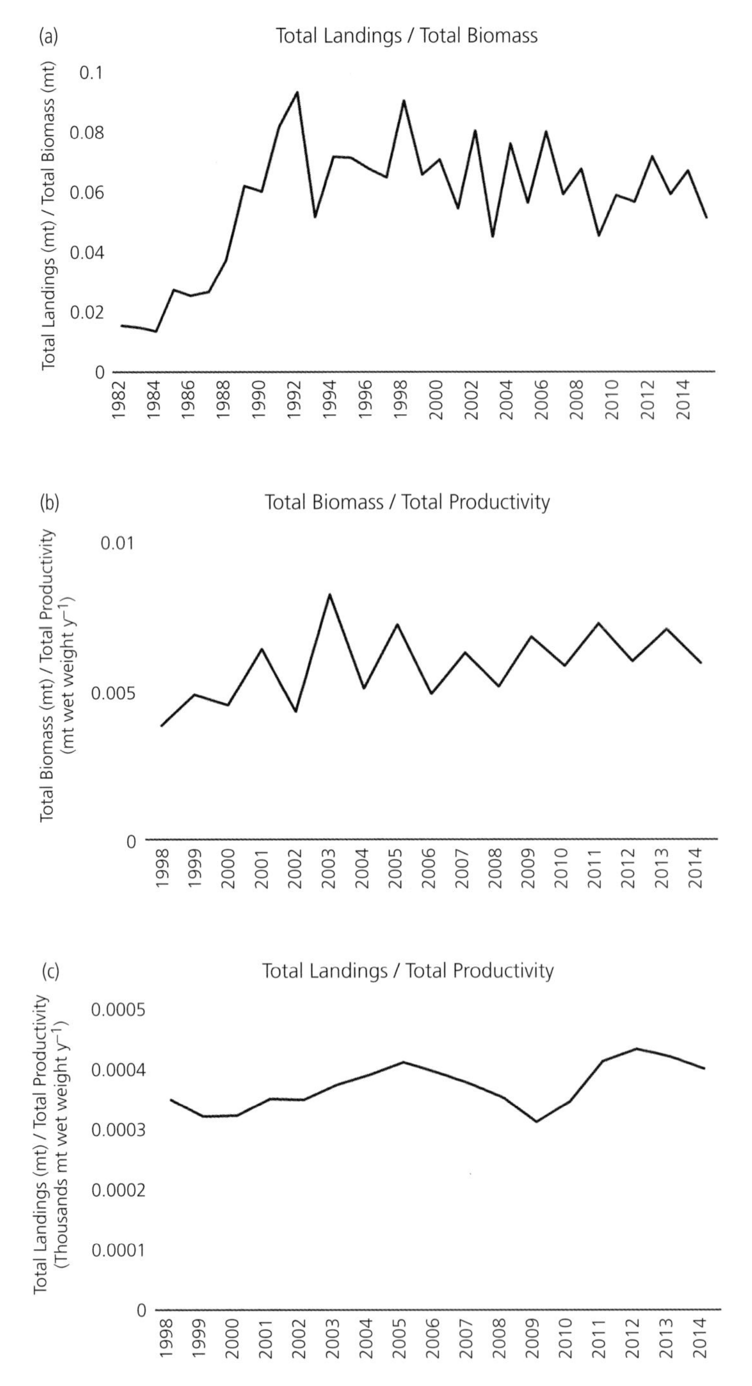

Figure 9.26 Ratios of (a) total commercial and recreational landings (metric tons) to total biomass (metric tons; i.e., exploitation index); (b) total biomass (metric tons) to total productivity (metric tons wet weight year^{-1}); (c) total commercial and recreational landings (metric tons) to total productivity (metric tons wet weight yr^{-1}) for the North Pacific region.

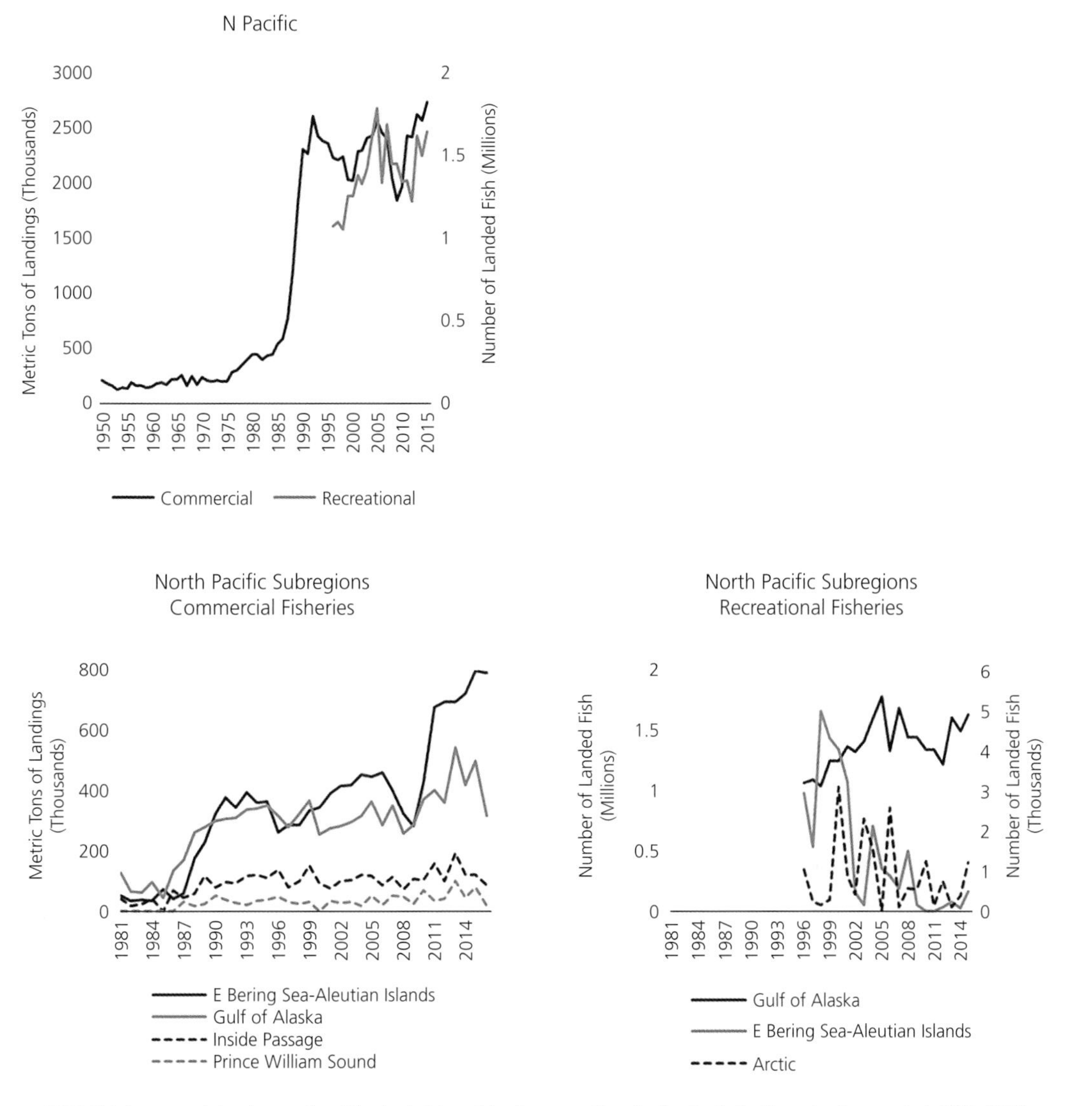

Figure 9.27 Total commercial and recreational (and subsistence) landings over time for the North Pacific region (commerical: 1950–2016; recreational: 1996–2016) and identified subregions (1981–2016). Reported recreational (and subsistence) landings for the North Pacific, and Eastern Bering Sea-Aleutian Islands and Arctic subregions, plotted on the secondary (right) axis.

Data derived from NOAA National Marine Fisheries Service and Alaska Department of Fish and Game commercial and recreational fisheries statistics.

Although values have decreased over time for certain taxa, bycatch continues to persist throughout the North Pacific region (Fig. 9.29). By weight, bycatch is most pronounced for bony fishes and invertebrates, while high by number for seabirds. Nationally, bycatch in this region is second-highest by weight and fourth-highest by number, with decreases observed over time for seabirds. This is largely due to the high volume of catches, as the bycatch rates are relatively quite low (Benaka et al. 2019). Higher bycatch among taxa is observed in the EBS compared to the Gulf of Alaska, except for sharks where values are higher in the Gulf of Alaska and have increased in recent years. For other taxa (i.e., bony fishes, invertebrates, and marine mammals), bycatch values have been more consistent since 2010.

9.5 Synthesis

The North Pacific ecosystem has relatively moderate numbers of managed species (compared to tropical

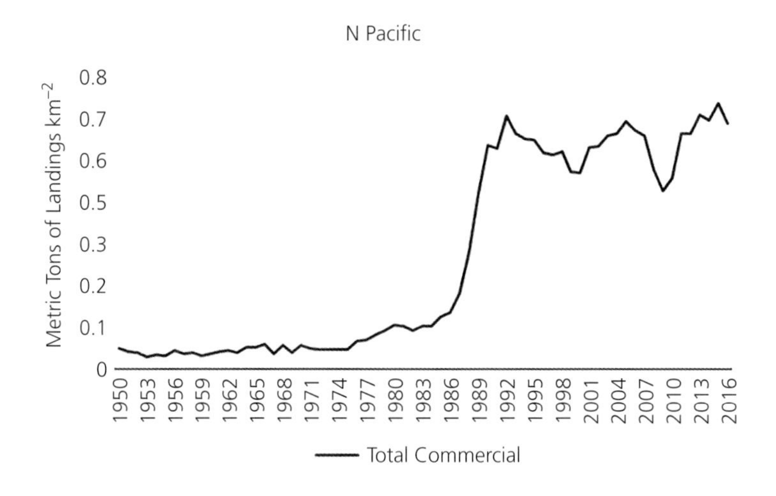

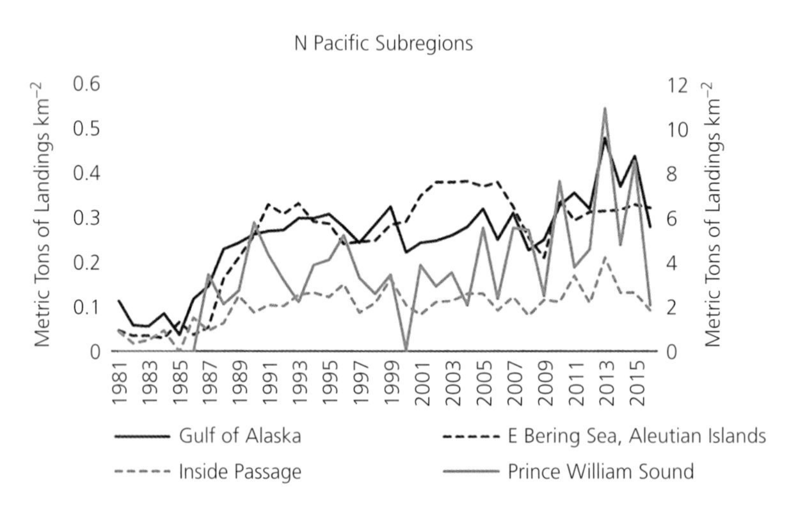

Figure 9.28 Total commercial landings (metric tons) per square kilometer over time for the North Pacific region (1950–2016) and identified subregions (1981–2016). Commercial landings per square kilometer for Prince William Sound plotted on the secondary (right) axis. Commercial data derived from NOAA National Marine Fisheries Service commercial fisheries statistics.

systems), with biota and marine communities that are responding to the consequences of fishing pressure, climate oscillations, and other ocean uses. More recent stressors, including substantial regional warming, associated species shifts, increasing relative human population density, and proliferation of invasive species, are affecting this system and altering its composition, dynamics, and LMR production (Royer & Grosch 2006, Ruiz et al. 2006, Prowse et al. 2009, Burkett 2011, Coyle et al. 2011, Hunt et al. 2011, Mueter et al. 2011, Day et al. 2013, Hollowed et al. 2013, Sigler et al. 2016, Morley et al. 2018, Nong et al. 2018, Link & Marshak 2019). Tourism and other ocean uses, including associated pressures from fishing, recreational, and industrial sectors, continue to affect the North Pacific, while also contributing to its nationally significant LMR and offshore mineral extraction-based economy. Despite continuing and emerging stressors on LMR populations, fisheries landings and revenue in this system remain highest nationally, especially following major regime shifts and increases in landings during the 1980s (Francis & Hare 1994, Mantua et al. 1997, Anderson & Piatt 1999, Hare & Mantua 2000, Benson & Trites 2002, Heymans et al. 2005, Zheng & Kruse 2006, Litzow & Urban 2009, Gaichas et al. 2011). Trends in Alaskan subregions have also remained relatively consistent

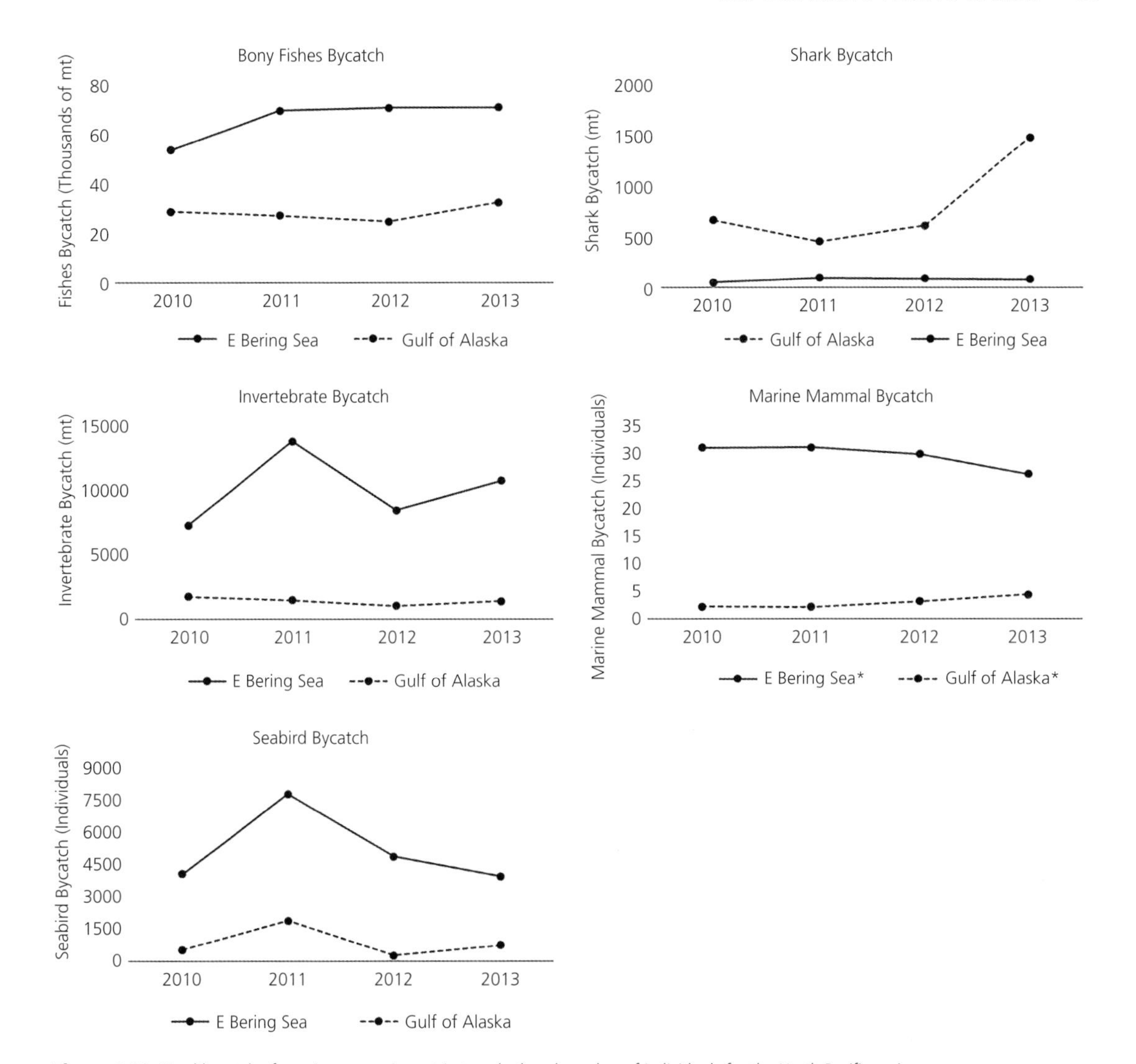

Figure 9.29 Total bycatch of species groups in metric tons (mt) and number of individuals for the North Pacific region.
Data derived from NOAA National Marine Fisheries Service U.S. National Bycatch Report Database System.

over time. As of 2017 the North Pacific contains the nationally lowest number and proportion of stocks that are overfished (n = 1) or experiencing overfishing (n = 1), but also contains moderate numbers of unassessed fisheries stocks, and nationally low numbers of strategic marine mammal stocks (Link & Marshak 2019). While this region has above average socioeconomic status regarding its LMRs (Link & Marshak 2019), it remains low to mid-range nationally in its total contributions toward other ocean uses, while making up ~5% of the U.S. total ocean economy.

As regional warming and oscillations in climate forcing continue to affect this ecosystem and its LMRs (Francis & Hare 1994, Anderson & Piatt 1999, Gaichas et al. 2011, Day et al. 2013, Hollowed et al. 2013, Sigler et al. 2016, Morley et al. 2018, Nong et al. 2018, Link & Marshak 2019), and national seafood demand that is heavily dependent on North Pacific fisheries continues to increase, the continued sustainable production of this system and its marine economy is important to consider. As trade-offs continue to intensify in this region with increasing accessibility to less exploited areas and the development of other ocean uses, EBFM strategies for the North Pacific and its subregions must consider these cumulative factors on its coupled socioecological system. There has been considerable progress toward broader understanding of this ecosystem, in synthesizing advanced ecosystem-level metrics, developing sophisticated models, and establishing quantitative frameworks for estimating system thresholds, and in applying all that information directly into management (Witherell et al. 2000, Aydin et al. 2007, Zador 2015, Tam et al. 2017, Zador et al. 2017a, b, Rooney et al. 2018, Siddon & Zador 2018, Zador & Ortiz 2018, Zador & Yasumiishi 2018). Although some notable progress has been made, expanding management actions that set system-wide control rules to other subregions, species complexes, and trophic levels remains needed. These are especially warranted given past and emerging stressors affecting basal ecosystem productivity, fisheries and protected species biomass, and marine socioeconomics, with differential system responses among subregions to these pressures (Francis & Hare 1994, Armstrong et al. 1998, Peterson et al. 2001, 2003, DiCosimo et al. 2010, Zador & Gaichas 2010, Gaichas et al. 2011, Zador 2015, Rooney et al. 2018, Siddon & Zador 2018, Zador & Ortiz 2018, Zador & Yasumiishi 2018). For example, the resiliency of Arctic systems to multiple stressors is suggested to be less than within the subpolar Alaskan shelf as a result of lower biomass to productivity coupling in this system (Aydin et al. 2007, Zador 2015, Siddon & Zador 2018, Townsend et al. 2018, Zador & Ortiz 2018, Zador & Yasumiishi 2018), for which differing management practices should be applied in response to ongoing human and natural impacts. Currently, the Arctic is closed to all commercial fishing, but climate forcing and emerging other ocean uses still have the potential to affect this system (see Chapter 11 on international agreements).

The North Pacific ecosystem is well known for its high LMR-based socioeconomic status and the significant contribution of LMRs to its subregional marine economies. This is especially related to the overall sustainable management of its fisheries, habitats, and protected resources over recent decades. Recent system exploitation is below suggested sustainable thresholds of 0.1 and is overall second-lowest nationally (Samhouri et al. 2010, Large et al. 2013, 2015, Tam et al. 2017, Link & Marshak 2019). This region is nationally highest in terms of the total contributions of its LMR establishments, employments, and GDP toward its total marine economy, emphasizing their continued importance toward the economic prosperity and cultural identity of this socioecological system. These trends are also fostered by the nationally highest basal ecosystem productivity of its contributing multiple LMEs, which also facilitates high LMR revenue and significant proportional employments as related to the total production and biomass of this system. Governance frameworks are also advanced in this region, especially regarding the multispecies nature of subregional FMPs, concentrated management attention on moderate numbers of managed species, strategic and nationally highest proportional use of fisheries closures, lower numbers of NEPA EIS actions, and cooperative management among the NPFMC and Alaska Board of Fisheries. However, this region is also subject to higher litigation potential in terms of the second-highest number of fisheries-related lawsuits in the nation, has contained a less-balanced FMC dominated by commercial representatives, has the smallest Congressional delegation in the country (but high Congressional attention on its LMRs), and leads in terms of amendments for its FMPs, demonstrating reasonable attention to management challenges.

Priorities for NPFMC EBFM actions include accounting for trophic interactions among protected and fisheries species when setting harvest limits, applying climate vulnerability assessments into management strategies and their evaluations, conserving essential fish habitat and incorporating vulnerability assessments toward its refinement, examining system trade-offs in a changing climate for more robust management strategies, and enacting a 2 million metric ton limit on Eastern Bering Sea-Aleutian Islands groundfish resources (NPFMC 2007, Sigler et al. 2016, Dorn et al. 2017, NMFS 2019, NPFMC 2019). These priorities are also included in a recently released EBFM Implementation Plan for the Alaskan region (NMFS 2019), which focuses on implementing the NOAA Fisheries EBFM Policy and Roadmap (NMFS 2016a, b). Its priorities are specifically focused on initiating and continuing the following activities:

- Continued implementation of the Bering Sea FEP, the Bering Sea and Gulf of Alaska Climate Science Regional Action Plans, the Alaska IEA programs, and other, related NPFMC priorities.
- Continued incorporation of EBFM into NPFMC management actions through the framework created in the Bering Sea FEP.
- Application of the North Pacific Climate Action Module to continue developing climate-enhanced stock assessments and guide incorporation of climate and ecosystem information in fishery management decisions moving forward.
- Continued application of the ACLIM project to advance understanding of fisheries and ecological responses to climate change.

- Continued development of protocols for incorporating local and traditional ecological knowledge into the fishery management process.
- Conduct regional habitat risk assessments for the Eastern Bering Sea ecosystem as related to dynamic species distributions and shifts in multispecies essential fish habitat.
- Continue applying modeling tools that include single-species assessment models coupled to climate drivers, models of technical interactions (gears and bycatch), and multispecies/ecosystem models.
- Continue supporting ESRs, their development as appendices to stock assessments, and the creation of quantifiable metrics relating ecosystem indicators that allow for the development of reference points and delineated thresholds.

Ongoing characterization of North Pacific ecosystem dynamics has allowed for the development of robust multispecies assessments that incorporate ecosystem-level information and account for climate dynamics, which have been applied toward developing system-wide reference points and multispecies approaches to management. These efforts have led to greater understanding of North Pacific ecosystem processes and have stimulated notable management actions, including prohibited directed fishing on forage fish and protected species prey, and the use of ecosystem-based limits on Eastern Bering Sea, Aleutian Islands, and Gulf of Alaska fisheries resources (Witherell 2004, Witherell & Woodby 2005, Livingston et al. 2005, Kruse et al. 2006, Mueter & Megrey 2006, Witherell et al. 2010, Zador 2015, Zador et al. 2017a, b, Siddon & Zador 2018, Zador & Ortiz 2018, Zador & Yasumiishi 2018). These efforts are also complemented by other ecosystem-centric protective measures toward reducing bycatch, applying large-scale protected areas toward species conservation, and preventing benthic habitat damage from fishing practices. These advancements allow for the application of this synthesized information toward continued ecosystem-level planning and risk assessment to bolster current management actions and expand upon their applications. Recurring ESRs that track ecosystem-level indicators, and which enhance the development of socioecological assessments, provide comprehensive information needed to build upon current holistic management practices (Zador & Gaichas 2010, Zador 2015, Siddon & Zador 2018, Zador & Ortiz 2018, Zador & Yasumiishi 2018). Continued application of integrated ecosystem-level models is facilitating more informative and localized assessments that work toward broader understanding of North Pacific subregions and their responses to ongoing stressors, especially regional warming (Zador 2015, Siddon & Zador 2018, Zador & Ortiz 2018, Zador & Yasumiishi 2018, Holsman et al. 2019a, b). All of these advancements are enhancing ecosystem approaches to management in the North Pacific, and providing a framework for its potential expanded application throughout the Alaskan shelf.

Despite these large-scale efforts toward greater scientific understanding and ecosystem-level management of the North Pacific, comprehensive FEPs for the Gulf of Alaska and Arctic still do not exist, while additional ecosystem-based harvest control rules and management measures are needed for salmon and invertebrate fisheries, especially in consideration of their responses to climate change, shifting distributions, and interactions with other species, including trophodynamics. While frameworks have been developed, and efforts toward implementing EBFM for this region are underway, additional actions remain needed toward holistically addressing the multiple stressors affecting this system. Despite having well-managed species and effective protections for their essential and critical habitats, the consequences of climate change continue to significantly affect the productivity and condition of Alaskan fisheries species and protected species populations, and are compounded by environmental stressors. Hence, more comprehensive management strategies that consider cumulative effects on both biological and human communities are warranted, especially as thresholds for exploitation of the North Pacific ecosystem continue to shift as a result of altered productivities. With current advanced understanding of socioecological relationships in the North Pacific, continued application of observed and predicted system-level responses toward establishing and expanding system-level reference points is feasible for broader management of this system.

While human population density is nationally lowest in the North Pacific, and increasing at a low rate, pressures from fishing fleets and developing other ocean use sectors continue to occur, suggesting potential for future resource competition within this region. Although the North Pacific is a geographically large region with currently less concentrated overlap among marine interests than in other regions, increasing national demand on Alaskan resources could lead to significant future exploitation of this region, and potentially concentrate sectors in newly accessible areas. As natural and human-related stressors continue to increase, it is imperative that management measures such as closures and cumulative consideration of stressors on baseline production be used to manage concurrent human activities and limit unsustainable exploitation of the North Pacific. Consideration of system-level interconnected effects from cumulative marine sector impacts, and conflicts among fisheries, tourism, marine transportation, and marine energy development, remain necessary for comprehensive cooperative management strategies, especially in light of species and system-level shifts in response to regional warming and climate forcing. Shifting species compositions and climate-related effects on the condition of economically important species and their prey resources are also likely to affect the viability of commercial fisheries landings and revenue. Increasingly abundant and important sportfishes and recreational fisheries species from lower latitudes may increase conflicts among fishing entities within this commercially dominant fisheries system and economy. With concerns about expanding marine energy and mineral extraction sectors moving into historical commercial fishing grounds (Prowse et al. 2009, Burkett 2011, Nong et al. 2018), systematic management strategies must account for these interacting sectors and their effects on LMRs, their fisheries, and their essential habitats. As is needed in other U.S. regions, management strategies that account for spatially overlapping ocean uses, resource partitioning, multiple species, and the socioeconomic effects of changing environments would be useful in this system. Predictive models that account for these cumulative effects can also enhance multisector management frameworks. With expanding human activities within this system (Brigham 2010, Hong 2012, Reeves et al. 2012, Levine et al. 2013, Humphries & Hueffman 2014, Kaiser et al. 2016, Nong et al. 2018, Link & Marshak 2019), additional non-fishing effects are anticipated, which warrant continued focus toward habitat conservation and ongoing spatial management considerations of the Alaskan EEZ. This is especially necessary in consideration of changing climate, with application toward broadening management and consultation frameworks for considering these emerging issues.

As engagement strategies for ecosystem-level management have been successful in the North Pacific, given a cooperative relationship among scientists, managers, and stakeholders, there is cause for optimism regarding continued application of ecosystem-level management in addressing emerging concerns for the sustainable future of this socioecological system. Although not without contention or limited attention, as observed by the high number of fisheries and LMR-related lawsuits and low numbers of on-site staff throughout a geographically large region, governance frameworks in this area remain very advanced, focused, and targeted toward species complexes. Building on these current frameworks, and applying the wealth of ecosystem-level knowledge toward addressing emerging issues, will strengthen opportunities for advancing ecosystem-based sustainable practices in this region. Given the advanced knowledge of human dimensions and marine socioeconomic effects of ecosystem dynamics in the North Pacific, this region illustrates great potential for continued development of comprehensive management strategies toward preserving livelihoods, ecological integrity, and coastal economies. Despite its unique factors, advances made in coupled understanding of these components also serve as a model for other regions in their progress toward ecosystem-level management.

9.5.1 Progress Toward Ecosystem-Based Fisheries Management (EBFM)

Alaska is excelling in the areas of LMR and socioeconomic status and the quality of its governance system, and is relatively productive, as related to the determinants of successful LMR management (Fig. 9.30; Link & Marshak 2019).

Alaskan fisheries management has continued advancing in approaches that address the cumulative issues affecting its LMRs and fisheries. Through a more systematic and prioritized consideration of all fisheries, pressures, risks, and outcomes, holistic EBFM practices are being developed and implemented across U.S. regions, including Alaska (NMFS 2016a, b, Holsman et al. 2017, 2019a Levin et al. 2018, Marshall et al. 2018, Link & Marshak 2019, NMFS 2019). In accordance with the NOAA Fisheries EBFM Policy and Roadmap (NMFS 2016a, b), an EBFM implementation plan (NMFS 2019) for Alaska's ecosystems was developed through efforts overseen by the NOAA AFSC, NOAA Fisheries Alaska Regional Office (AKRO), and the NPFMC. This plan specifies approaches to be taken toward conducting EBFM and in addressing NOAA Fisheries Roadmap priorities and gaps. Progress for the North Pacific and its major subregions in terms of EBFM Roadmap Guiding Principles and Goals is shown in Table 12.9. Overall, a moderate to high degree of progress has been made in the Eastern Bering Sea, Aleutian Islands, and Gulf of Alaska in terms of implementation, advancing knowledge of ecosystem principles, examining trade-offs, assessing risks and vulnerabilities, and in establishing and using ecosystem-level reference

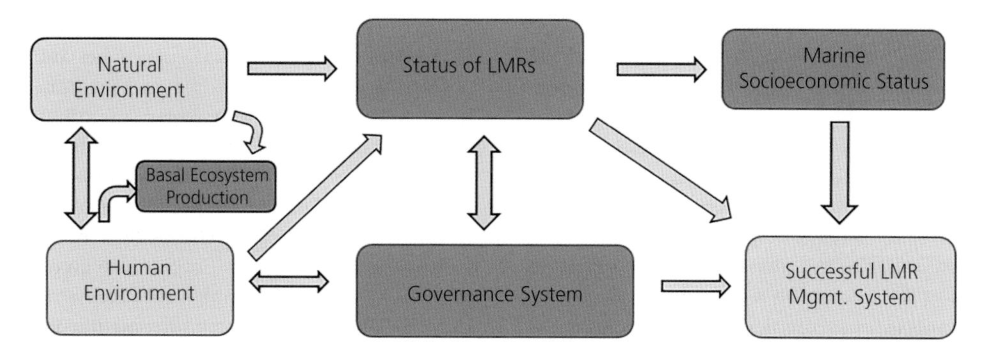

Figure 9.30 Schematic of the determinants and interconnectivity of successful living marine resource (LMR) systems management criteria as modified from Link and Marshak (2019). Those highlighted in blue indicate criteria for which the North Pacific ecosystem is excelling.

points for management. The North Pacific is one of the leading U.S. regions in terms of progress toward these goals. While much information has been obtained and applied toward ecosystem-level calculations, syntheses, and models, continued progress in applying these system-wide emergent properties into regional management frameworks remains necessary.

Progress toward recommendations of the Ecosystem Principles Advisory Panel (EPAP) for the development of FEPs (EPAP 1999, Wilkinson & Abrams 2015) is shown in Table 12.10. FEPs currently exist for two North Pacific sub-regions (Eastern Bering Sea and Aleutian Islands), and a substantial degree of progress has been observed toward characterizing, monitoring, and assessing aspects of ecosystem dynamics in support of this ecosystem approach for fisheries management. More efforts toward assessing the habitat needs for managed species life history stages and examining the ecosystem effects of fishing remain needed. Greater ecosystem-level data and more in-depth syntheses for the Gulf of Alaska and its subregions are also warranted.

While among U.S. regions that are leading in terms of ecosystem status and application of holistic management practices (Patrick & Link 2015, Zador et al. 2017b, Link 2018, Siddon & Zador 2018, Zador & Ortiz 2018, Zador & Yasumiishi 2018, Link & Marshak 2019), continued development of ecosystem-level reference points, assessments, and system-wide thresholds remains needed for the North Pacific ecosystem; research into environmental factors and ecosystem components is also necessary to strengthen current efforts and address emerging issues (Hollowed et al. 2011, Patrick & Link 2015, Zador et al. 2017a, b, Link 2018). Advances toward addressing some of these data gaps and issues have been more concentrated in Alaska in recent years, particularly in accounting for spatial and environmental factors in stock and protected species assessments, foundational understanding of system-wide processes, and applying this information into management frameworks (McConnaughey & Smith 2000, Johnson et al. 2003a, b, Thorson et al. 2016, Pirtle et al. 2017, Somerton et al. 2017, McGowan et al. 2019). Broader understanding and management that accounts for these effects and their interrelations under an ecosystem-based approach is underway, especially through precautionary closures and efforts to examine ecosystem-level linkages in this system, with continued development of management strategies still warranted (Marasco et al. 2007, NPFMC 2007, Stram & Evans 2009, Zador et al. 2017a, b, NPFMC 2019). Progressive efforts such as extensive trawling closures to conserve Alaska's essential and critical habitats and numerous measures to address and reduce the effects of bycatch on targeted and protected species are also extant (NPFMC 1998, Witherell & Woodby 2005, NPFMC 2011, Goetz et al. 2012, Sigler et al. 2012, Rieser et al. 2013, Eich et al. 2016).

9.5.2 Conclusions

Like in the Pacific and Mid-Atlantic, fisheries management in the North Pacific represents some of the most progressive efforts toward holistic LMR management strategies. It additionally serves as a leading example for sustainably managed fisheries through cooperation with engaged stakeholders, including tribal interests (Holland & Ginter 2001, Holt et al. 2008, Ruckelshaus et al. 2008, Fina 2011, Siders et al. 2016, Raymond-Yakoubian & Daniel 2018). Emerging holistic management approaches have been enabled by comprehensive multi-decadal data regarding coastal and offshore fisheries, essential habitats, and climate forcing (McConnaughey & Smith 2000, Gaichas et al. 2011,

Litzow et al. 2014, Zador 2015, Rooney et al. 2018, Siddon & Zador 2018, Zador & Ortiz 2018, Zador & Yasumiishi 2018). Efforts to continue an ecosystem approach in managing the resources of this ecologically significant region are underway to ensure its ongoing sustainable use. Current management frameworks considering the effects of natural stressors and human activities on species complexes are a foundational step toward broader system-wide strategies that account for interacting sectors, multiple trophic levels among species complexes, and socioeconomic feedbacks within the North Pacific. Continued system-wide considerations of the inherent relationships between productivity, biomass, landings, and socioeconomics are needed to maintain this system within needed sustainable thresholds, especially in light of multiple stressors. EBFM provides opportunities that focusing on one stock or species complex, one fishery, or the effects of only the fishing sector would not. Simultaneous controlled harvest of interacting species complexes and fisheries, which account for their trophic interdependence and contributions toward system productivity are essential for optimal management. Overlapping and/or differential fishing effort for these species, in addition to the gear-related effects of their harvest on essential habitats and subsequent ecological consequences to multispecies complexes is partially considered in current North Pacific frameworks, with potential for their expansion. Continued application of portfolio and multi-level management approaches (Edwards et al. 2004, Kellner et al. 2011, Link 2018) can provide for more effective, responsive, and precautionary management strategies that consider the interconnectivity of this ecosystem and its cumulative response to ongoing pressures.

Holistic approaches to management of the North Pacific ecosystem allow for its nationally significant cultural marine heritage and LMR-dependent economy to continue through ecologically sound resource use. Enhancing EBFM in this region advances current fisheries management strategies toward a system-wide focus that accounts for the interdependence of fisheries and protected species complexes, habitat quality, climate, and bottom-up forcing related to productivity thresholds. While substantial progress has been made, continued integration of ecosystem-level factors in consideration of its interacting components remains needed toward setting sustainable limits on resource use and harvest in response to climate and human-related stressors. Applying available, comprehensive information toward the development of system-level reference points and thresholds with respect to basal productivity and biomass, with which dependent LMR status and economies are interconnected, is especially possible in this data-rich and highly productive system where advanced management frameworks are found. Progress is being made in addressing aspects of these concerns through the ongoing development of broader ecosystem-based frameworks that are likewise being applied in other subregions of the North Pacific, and continue to be prioritized for EBFM implementation in this region (NMFS 2019). Merging this information into a more holistic framework that examines these facets together and allows for improved collaboration among multi-sectoral management interests will work toward maintaining the strong socioeconomic and LMR status in the North Pacific as it is subjected to increasing pressures. There is increasing potential for overexploitation of the Alaskan shelf as regional warming and emerging other ocean uses affect its productivity. With foundational metrics having been developed, it is incumbent on managers, stakeholders, and scientists to cooperatively ensure that future fisheries management strategies consider ecosystem-level properties and trade-offs, and account for multiple environmental effects on system productivity and sustainable fisheries harvest. As compounding effects of climate change, fishing pressure, and emergent other ocean uses continue to affect this region, their management throughout this socioecological system can be more effectively addressed by building upon current ecosystem approaches. Applying and expanding comprehensive management strategies in this system and its subregions that consider basal productivities and well-known integrated ecosystem properties will work toward maintaining its LMR and socioeconomic status in light of ongoing pressures. These efforts provide a pathway toward ensuring the continuation of Alaskan maritime culture, including its nationally important LMRs and ocean-based economies, and in preserving the legacy of this historically iconic fisheries ecosystem.

9.6 References

A'mar TZ, Punt AE, Dorn MW. 2009. The impact of regime shifts on the performance of management strategies for the Gulf of Alaska walleye pollock (*Theragra chalcogramma*) fishery. *Canadian Journal of Fisheries and Aquatic Sciences* 66:2222–42.

Abookire AA, Piatt JF, Norcross BL. 2001. Juvenile groundfish habitat in Kachemak Bay, Alaska, during late summer. *Alaska Fishery Research Bulletin* 8(1):45–56.

Alabia ID, García Molinos J, Saitoh SI, Hirawake T, Hirata T, Mueter FJ. 2018. Distribution shifts of marine taxa in the Pacific Arctic under contemporary climate changes. *Diversity and Distributions* 24(11):1583–97.

Ames RT, Williams GH, Fitzgerald SM. 2005. Using Digital Video Monitoring Systems in Fisheries: Application for Monitoring Compliance of Seabird Avoidance Devices and Seabird Mortality in Pacific Halibut Longline Fisheries. Washington, DC: NOAA.

Amoroso RO, Parma AM, Pitcher CR, McConnaughey RA, Jennings S. 2018a. Comment on "Tracking the global footprint of fisheries". *Science* 361(6404):eaat6713.

Amoroso RO, Pitcher CR, Rijnsdorp AD, McConnaughey RA, Parma AM, Suuronen P, Eigaard OR, et al. 2018b. Bottom trawl fishing footprints on the world's continental shelves. *Proceedings of the National Academy of Sciences* 115(43):E10275–82.

Anderson PJ, Piatt JF. 1999. Community reorganization in the Gulf of Alaska following ocean climate regime shift. *Marine Ecology Progress Series* 189:117–23.

AOOS (Alaska Ocean Observing System). 2016. *AOOS FY 2016–2020 Strategic Operations Plan*. Anchorage, AK: AOOS (p. 16).

Armstrong J, Armstrong D, Hilborn R. 1998. Crustacean resources are vulnerable to serial depletion–the multifaceted decline of crab and shrimp fisheries in the Greater Gulf of Alaska. *Reviews in Fish Biology and Fisheries* 8(2):117–76.

Arrigo KR, van Dijken GL. 2011. Secular trends in Arctic Ocean net primary production. *Journal of Geophysical Research* 116 (C9):C09011.

Arrigo KR, van Dijken GL. 2015. Continued increases in Arctic Ocean primary production. *Progress in Oceanography* 136:60–70.

Arrigo KR, Perovich DK, Pickart RS, Brown ZW, van Dijken GL, Lowry KE, Mills MM, et al. 2012. Massive phytoplankton blooms under sea ice. *Science* 336:1408.

Arrigo KR, Perovich DK, Pickart RS, Brown ZW, van Dijken GL, Lowry KE, Mills MM, et al. 2014. Phytoplankton blooms beneath the sea ice in the Chukchi sea. *Deep Sea Research Part II: Topical Studies in Oceanography* 105(Supplement C):1–16.

Arrigo KR, van Dijken G, Pabi S. 2008. Impact of a shrinking Arctic ice cover on marine primary production. *Geophysical Research Letters* 35(L19603):6.

Ashjian CJ, Harvey HR, Lomas MW, Napp JM, Sigler MF, Stabeno PJ, van Pelt TI. 2012. Understanding ecosystem processes in the Eastern Bering Sea. *Deep-Sea Research Part II*. Vol 65–70.

Ashjian CJ, Harvey HR, Lomas MW, Napp JM, Sigler MF, Stabeno PJ, van Pelt TI. 2013. Understanding ecosystem processes in the Eastern Bering Sea II. *Deep-Sea Research Part II*. Vol 94

Ashjian CJ, Harvey HR, Lomas MW, Napp JM, Sigler MF, Stabeno PJ, van Pelt TI. 2014. Understanding ecosystem processes in the Eastern Bering Sea III. *Deep-Sea Research Part II*. Vol 134.

Ashjian CJ, Harvey HR, Lomas MW, Napp JM, Sigler MF, Stabeno PJ, van Pelt TI. 2016. Understanding ecosystem processes in the Eastern Bering Sea III. *Deep-Sea Research Part II*. Vol 109.

Ashton GV, Riedlecker EI, Ruiz GM. 2008. First non-native crustacean established in coastal waters of Alaska. *Aquatic Biology* 3(2):133–7.

Aydin K, Mueter F. 2007. The Bering Sea—a dynamic food web perspective. *Deep Sea Research Part II: Topical Studies in Oceanography* 54(23–26):2501–25.

Aydin KY, McFarlane GA, King JR, Megrey BA, Myers KW. 2005. Linking oceanic food webs to coastal production and growth rates of Pacific salmon (Oncorhynchus spp.), using models on three scales. *Deep Sea Research Part II: Topical Studies in Oceanography* 52(5–6):757–80.

Aydin KY, Lapko VV, Radchenko VI, Livingston PA. 2002. A Comparison of the Eastern and Western Bering Sea Shelf and Slope Ecosystems Through the Use of Mass-Balance Food Web Models. NOAA Technical Memorandum NMFS-AFSC-130. Washington, DC: NOAA (p. 78).

Aydin K, Gaichas S, Ortiz I, Kinzey D, Friday N. 2007. A Comparison of the Bering Sea, Gulf of Alaska, and Aleutian Islands Large Marine Ecosystems Through Food Web Modeling. NOAA Technical Memorandum NMFS-AFSC-178. Washington, DC: NOAA (p. 298).

Baker TT, Wertheimer AC, Burkett RD, Dunlap R, Eggers DM, Fritts EI, Gharrett AJ, Holmes RA, Wilmot RL. 1996. Status of Pacific salmon and steelhead escapements in southeastern Alaska. *Fisheries* 21(10):6–18.

Barbeaux SJ, Hollowed AB. Ontogeny matters: climate variability and effects on fish distribution in the eastern Bering Sea. *Fisheries Oceanography* 27(1):1–15.

Barbeaux S, Ianelli J, Brown E. 2003. *Aleutian Islands walleye pollock SAFE*. Anchorage, AK: North Pacific Fishery Management Council. http://www.afsc.noaa.gov/refm/docs/2003/AIpollock.pdf.

Beaudreau AH, Chan MN, Loring PA. 2018. Harvest portfolio diversification and emergent conservation challenges in an Alaskan recreational fishery. *Biological Conservation* 222:268–77

Becker PR. 2000. Concentration of chlorinated hydrocarbons and heavy metals in Alaska Arctic marine mammals. *Marine Pollution Bulletin* 40(10):819–29.

Bell FH. 1981. *The Pacific Halibut: The Resource and its Fishery*. Anchorage, AK: Alaska Northwest Publishing Company (p. 267).

Benaka LR, Bullock D, Hoover AL, Olsen NA. 2019. U.S. National Bycatch Report First Edition Update 3. NOAA Technical Memorandum NMFS-F/SPO-190. Washington, DC: NOAA (p. 95).

Benson AJ, Trites AW. 2002. Ecological effects of regime shifts in the Bering Sea and eastern North Pacific Ocean. *Fish and Fisheries* 3(2):95–113.

Board OS, National Research Council. 2003. *Decline of the Steller Sea Lion in Alaskan Waters: Untangling Food Webs and Fishing Nets*. Washington, DC: National Academies Press.

Bond NA, Cronin MF, Freeland H, Mantua N. 2015. Causes and impacts of the 2014 warm anomaly in the NE Pacific. *Geophysical Research Letters* 42:3414–20.

Bowen L, Miles AK, Ballachey BE, Bodkin JL, Esler D. 2015. Gulf Watch Alaska Long-term Monitoring Program-Evaluating Chronic Exposure of Harlequin Ducks and Sea Otters to Lingering Exxon Valdez Oil in Western Prince William Sound. Exxon Valdez Oil Spill Restoration Project Final Report (Restoration Project 12120114-Q), Pacific Wildlife Foundation and Centre for Wildlife Ecology, Simon Fraser University, Delta, British Columbia, Canada. Anchorage, AK: US Geological Survey, Alaska Science Center.

Brigham LW. 2010. The fast-changing maritime Arctic. *Proceedings of the US Naval Institute* 136(5):54–9.

Brown EJ, Finney B, Dommisse M, Hills S. 2005. Effects of commercial otter trawling on the physical environment of the southeastern Bering Sea. *Continental Shelf Research* 25(10):1281–301.

Brown ZW, Arrigo KR. 2012. Contrasting trends in sea ice and primary production in the Bering Sea and Arctic Ocean. *ICES Journal of Marine Science: Journal du Conseil* 69(7):1180–93.

Brown ZW, van Dijken GL, Arrigo KR. 2011. A reassessment of primary production and environmental change in the Bering Sea. *Journal of Geophysical Research* 116:C08014.

Brugler ET, Pickart RS, Moore GW, Roberts S, Weingartner TJ, Statscewich H. 2014. Seasonal to interannual variability of the Pacific water boundary current in the Beaufort Sea. *Progress in Oceanography* 127:1–20.

Buckland ST, Cattanach KL, Hobbs RC. 1993. "Abundance estimates of Pacific white-sided dolphin, northern right whale dolphin, Dall's porpoise and northern fur seal in the North Pacific, 1987/90." In: *Biology, Distribution and Stock Assessment of Species Caught in the High Seas Driftnet Fisheries in the North Pacific Ocean. 4–6 November 1991*. Edited by W Shaw, RL Burgner, J Ito. Tokyo, Japan: International North Pacific Fishery Commission Symposium (pp. 387–407).

Buckley TW, Livingston PA. 1994. A Bioenergetics Model of Walleye Pollock (Theragra Chalcogramma) in the Eastern Bering Sea: Structure and Documentation. NOAA Technical Memorandum NMFS-AFSC-37. Washington, DC: NOAA (p. 55).

Buklis LS. 2002. Subsistence fisheries management on federal public lands in Alaska. *Fisheries* 27(7):10–8.

Burkett V. 2011. Global climate change implications for coastal and offshore oil and gas development. *Energy Policy* 39(12):7719–25.

Ca Callaway D, Eamer J, Edwardsen E, Jack C, Marcy, Olrun A, Patkotak M, Rexford D, Whiting A. 1999. "Effects of climate change on subsistence communities in Alaska." In: *Assessing the consequences of climate change for Alaska and the Bering Sea region. Proceedings of a Workshop at the University of Alaska Fairbanks, October 29–30, 1998*. Edited by G Weller, PA Andersen. Fairbanks, AK: University of Alaska Fairbanks.

Carlson HR, Straty RR. 1981. Habitat and nursery grounds of Pacific rockfish, Sebastes spp., in rocky coastal areas of southeastern Alaska. *Marine Fisheries Review* 43(7):13–9.

Carothers C. 2010. Tragedy of commodification: displacements in Alutiiq fishing communities in the Gulf of Alaska. *Maritime Studies* 9(2):95–120.

Chan FT, Bailey SA, Wiley CJ, MacIsaac HJ. 2013. Relative risk assessment for ballast-mediated invasions at Canadian Arctic ports. *Biological Invasions* 15(2):295–308.

Cheung WW, Brodeur RD, Okey TA, Pauly D. 2015. Projecting future changes in distributions of pelagic fish species of Northeast Pacific shelf seas. *Progress in Oceanography* 130:19–31.

Cheung WW, Frölicher TL. 2020. Marine heatwaves exacerbate climate change impacts for fisheries in the northeast Pacific. *Scientific Reports* 10(1):6678.

Clark JH, McGregor A, Mecum RD, Krasnowski P, Carroll AM. 2006. The commercial salmon fishery in Alaska. *Alaska Fishery Research Bulletin* 12(1):1–46.

Clark WG, Hare SR. 2003. *Assessment of the Pacific Halibut Stock at the End of 2003*. Seattle, WA: International Pacific Halibut Commission.

Clark WG. 2003. A model for the world: 80 years of model development and application at the international Pacific halibut commission. *Natural Resource Modeling* 16(4): 491–503.

Clark WG, Hare SR. 2006. *Assessment and Management of Pacific Halibut: Data, Methods, and Policy. Report No. 83*. Seattle, WA: International Pacific Halibut Commission (p. 104).

Clark WG, Hare SR, Parma AM, Sullivan PJ, Trumble RJ. 1999. Decadal changes in growth and recruitment of Pacific halibut (Hippoglossus stenolepis). *Canadian Journal of Fisheries and Aquatic Sciences* 56:242–52.

Cohen CS, McCann L, Davis T, Shaw L, Ruiz GM. 2011. Discovery and significance of the colonial tunicate Didemnum vexillum in Alaska. *Aquatic Invasions* 6(3): 263–71.

Colt S, Dugan D, Fay G. 2007. *The Regional Economy of Southeast Alaska*. Anchorage, AK: University of Alaska Anchorage, Institute of Social and Economic Research.

Colt S, Martin S, Mieren J, Tomeo M. 2002. *Recreation and Tourism in South-Central Alaska: Patterns and Prospects. General Technical Report PNW-GTR-551*. Portland, OR: US Department of Agriculture, Forest Service, Pacific Northwest Research Station (p. 78).

Comiso JC, Parkinson CL, Gersten R, Stock L. 2008 Accelerated decline in the Arctic sea ice cover. *Geophysical Research Letters* 35(1).

Coyle KO, Eisner LB, Mueter FJ, Pinchuk AI, Janout MA, Cieciel KD, Farley EV, Andrews AG. 2011. Climate change in the southeastern Bering Sea: impacts on pollock stocks and implications for the oscillating control hypothesis. *Fisheries Oceanography* 20(2):139–56.

Coyle KO, Pinchuk AI, Eisner LB, Napp JM. 2008: Zooplankton species composition, abundance and biomass on the eastern Bering Sea shelf during summer: the potential of water-column stability and nutrients in structuring the zooplankton community. *Deep Sea Research II* 55:1775–91.

Criddle KR, Herrmann M, Lee ST, Hamel C. 2003. Participation decisions, angler welfare, and the regional economic impact of sportfishing. *Marine Resource Economics* 18(4):291–312.

Crowell AL. 1999. Maritime cultures of the Gulf of Alaska. *Revista de Arqueología Americana* 1(17):177–215.

Crutchfield JA, Pontecorvo G. 2013. *The Pacific Salmon Fisheries: A Study of Irrational Conservation*. New York, NY: RFF Press.

Curchitser EN, Haidvogel DB, Hermann AJ, Dobbins EL, Powell TM, Kaplan A. 2005. Multi-scale modeling of the North Pacific Ocean: assessment and analysis of simulated basin-scale variability (1996–2003). *Journal of Geophysical Research: Oceans* 110(C11).

Dalsgaard J, Pauly D. 1997. Preliminary mass-balance model of Prince William Sound, Alaska, for the pre-spill period, 1980–1989. *Fisheries Centre Research Reports* 5(2):84.

Day RH, Gall AE, Morgan TC, Rose JR, Plissner JH, Sanzenbacher PM, Fenneman JD, Kuletz KJ, Watts BH. 2013. Seabirds new to the eastern Chukchi and Beaufort seas, Alaska: response to a changing climate. *Western Birds* 44(3):174–82.

de Rivera CE, Steves BP, Fofonoff PW, Hines AH, Ruiz GM. 2011. Potential for high-latitude marine invasions along western North America. *Diversity and Distributions* 17(6):1198–209.

Deal RL, Tappeiner JC, Hennon PE. 2002. Developing silvicultural systems based on partial cutting in western hemlock–Sitka spruce stands of southeast Alaska. *Forestry* 75(4):425–31.

Dean TA, Haldorson L, Laur DR, Jewett SC, Blanchard A. 2000. The distribution of nearshore fishes in kelp and eelgrass communities in Prince William Sound, Alaska: associations with vegetation and physical habitat characteristics. *Environmental Biology of Fishes* 57(3):271–87.

Dickson D, Baker MR. 2016. Introduction to the North pacific research board Gulf of Alaska integrated ecosystem research program (GOAIERP): volume I. *Deep Sea Research Part II: Topical Studies in Oceanography* 132:1–5.

DiCosimo J, Methot RD, Ormseth OA. 2010. Use of annual catch limits to avoid stock depletion in the Bering Sea and Aleutian Islands management area (Northeast Pacific). *ICES Journal of Marine Science* 67(9):1861–5.

DiCosimo J. 2001. *Summary of the Gulf of Alaska Groundfish Fishery Management Plan*. Anchorage, AK: North Pacific Fishery Management Council (p. 16).

Dorn MW, Cunningham CJ, Dalton M, Fadely BS, Gerke BL, Hollowed AB, Holsman KK, et al. 2017. *A Climate Science: Regional Action Plan for the Gulf of Alaska. NOAA Tech. Memo. NMFS-AFSC-376*. Washington, DC: NOAA.

Duda AM, Sherman K. 2002. A new imperative for improving management of large marine ecosystems. *Ocean & Coastal Management* 45(11–12):797–833.

Dugan D, Fay G, Colt S. 2007. *Nature-Based Tourism in Southeast Alaska: Results from 2005 and 2006 Field Study*. Anchorage, AK: University of Alaska Anchorage, Institute of Social and Economic Research.

Ebbert SE, Byrd GV. 2002. Eradications of Invasive Species to Restore Natural Biological Diversity on Alaska Maritime National Wildlife Refuge. Turning the Tide: The Eradication of Invasive Species. Cambridge, UK: IUCN Invasive Species Specialist Group (pp. 102–9).

Echeverria TW. 1995. Walleye pollock (Theragra chalcogramma) on the Bering and Chukchi Sea Shelf. *Climate Change and Northern Fish Populations* 121:131.

Edwards SF, Link JS, Rountree BP. 2004. Portfolio management of wild fish stocks. *Ecological Economics* 49(3):317–29.

Eich AM, Mabry KR, Wright SK, Fitzgerald SM. 2016. *Seabird Bycatch and Mitigation Efforts in Alaska Fisheries Summary Report: 2007 through 2015. U.S. Department of Commerce, NOAA Tech.* Memo. NMFS-F/AKR-12 (p. 47).

Eisner LB, Gann JC, Ladd C, Cieciel KD, Mordy CW. 2016. Late summer/early fall phytoplankton biomass (chlorophyll a) in the eastern Bering Sea: Spatial and temporal variations and factors affecting chlorophyll a concentrations. *Deep Sea Research Part II: Topical Studies in Oceanography* 134:100–14.

Eisner LB, Pinchuk AI, Kimmel DG, Mier KL, Harpold CE, Siddon EC. 2017. Seasonal, interannual, and spatial patterns of community composition over the eastern Bering Sea shelf in cold years. Part I: zooplankton. *ICES Journal of Marine Science* 75(1):72–86.

EPAP (Ecosystem Principles Advisory Panel). 1999. *Ecosystem-Based Fishery Management Report to Congress*. Silver Spring, MD: National Marine Fisheries Service (p. 62).

Ezer T, Liu H. 2010. On the dynamics and morphology of extensive tidal mudflats: integrating remote sensing data with an inundation model of Cook Inlet, Alaska. *Ocean Dynamics* 60(5):1307–18.

Fabry VJ, McClintock JB, Mathis JT, Grebmeier JM. 2009. Ocean acidification at high latitudes: the bellwether. *Oceanography* 22(4):160–71.

Fall JA. 2016. 2008. Comprehensive Subsistence Harvest Survey, Emmonak, Togiak, and Akutan, Alaska. Arctic Data Center. Boulder, CO: UCAR/NCAR—Earth Observing Laboratory

Fall JA. 2016. Regional patterns of fish and wildlife harvests in contemporary Alaska. *Arctic* 69:47–64.

Feder HM, Jewett SC. 1981. Feeding Interactions in the Eastern Bering Sea with Emphasis on the Benthos. The Eastern Bering Sea Shelf: Oceanography and Resources, Vol. 2. NOAA Office of Marine Pollution Assessment. Seattle, WA: University Washington Press (pp. 1229–61).

Ferguson L, Srinivasan M, Oleson E, Hayes S, Brown SK, Angliss R, Carretta J, et al. 2017. *Proceedings of the First National Protected Species Assessment Workshop. NOAA Technical Memorandum NMFS-F/SPO-172.* Washington, DC: NOAA (p. 92).

Filbee-Dexter K, Wernberg T, Fredriksen S, Norderhaug KM, Pedersen MF. 2018. Arctic kelp forests: diversity, resilience and future. *Global and Planetary Change* 172(2019):1–14.

Fina M. 2011. Evolution of catch share management: lessons from catch share management in the North Pacific. *Fisheries* 36(4):164–77.

Francis RC, Hare SR. 1994. Decadal-scale regime shifts in the large marine ecosystems of the north-east Pacific: a case for historical science. *Fisheries Oceanography* 3(4):279–91.

Freese L, Auster PJ, Heifetz J, Wing BL. 1999. Effects of trawling on seafloor habitat and associated invertebrate taxa in the Gulf of Alaska. *Marine Ecology Progress Series* 182:119–26.

Fritz LW, Ferrero RC, Berg RJ. 1995. The threatened status of Steller sea lions, Eumetopias jubatus, under the Endangered Species Act: effects on Alaska groundfish fisheries management. *Marine Fisheries Review* 57(2): 14–27.

Frost KJ, Russell RB, Lowry LR. 1992. Killer whales, Orcinus orca, in the Southeastern Bering Sea: recent sighting and predation on other marine mammals. *Marine Mammal Science* 8(2):110–9.

Frost KJ, Lowry LF. 1981. Foods and Trophic Relationships of Cetaceans in the Bering Sea. The Eastern Bering Sea Shelf: Oceanography and Resources, Vol. 2. Seattle, WA: University of Washington Press (p. 825–36).

Funk DW, Noel LE, Freedman AH. 2004. Environmental gradients, plant distribution, and species richness in arctic salt marsh near Prudhoe Bay, Alaska. *Wetlands Ecology and Management* 12(3):215–33.

FWS (Fish and Wildlife Service). 2009. Marine Mammal Protection Act; Stock Assessment Report. *Federal Register* 74(249):69139–43.

FWS (Fish and Wildlife Service). 2014. Marine Mammal Protection Act; Stock Assessment Reports. *Federal Register* 79(249):4696–9.

FWS-NMFS (Fish and Wildlife Service & National Marine Fisheries Service). 2006. *5-Year Review Guidance: Procedures for Conducting 5-Year Reviews Under the Endangered Species Act.* Washington, DC: NOAA (p. 74).

Gaichas S, Gamble R, Fogarty M, Benoît H, Essington T, Fu C, Koen-Alonso M, Link J. 2012. Assembly rules for aggregate-species production models: simulations in support of management strategy evaluation. *Marine Ecology Progress Series* 459:275–92.

Gaichas SK, Aydin KY, Francis RC. 2010. Using food web model results to inform stock assessment estimates of mortality and production for ecosystem-based fisheries management. *Canadian Journal of Fisheries and Aquatic Sciences* 67(9):1490–506.

Gaichas SK, Aydin KY, Francis RC. 2011. What drives dynamics in the Gulf of Alaska? Integrating hypotheses of species, fishing, and climate relationships using ecosystem modeling. *Canadian Journal of Fisheries and Aquatic Sciences* 68(9):1553–78.

Gaichas SK, Odell G, Aydin KY, Francis RC. 2012. Beyond the defaults: functional response parameter space and ecosystem-level fishing thresholds in dynamic food web model simulations. *Canadian Journal of Fisheries and Aquatic Sciences* 69(12):2077–94.

Garfield B. 2010. *Thousand-Mile War: World War II in Alaska and the Aleutians.* Fairbanks, AK: University of Alaska Press.

Gay III SM, Vaughan SL. 2001. Seasonal hydrography and tidal currents of bays and fjords in Prince William Sound, Alaska. *Fisheries Oceanography* 10:159–93.

Gibson GA, Stockhausen WT, Coyle KO, Hinckley S, Parada C, Hermann AJ, Doyle M, Ladd C. 2018. An individual-based model for sablefish: exploring the connectivity between potential spawning and nursery grounds in the Gulf of Alaska. *Deep Sea Research Part II: Topical Studies in Oceanography* 165:89–112.

Goetz KT, Montgomery RA, Ver Hoef JM, Hobbs RC, Johnson DS. 2012. Identifying essential summer habitat of the endangered beluga whale *Delphinapterus leucas* in Cook Inlet, Alaska. *Endangered Species Research* 16(2): 135–47.

Golet GH, Seiser PE, McGuire AD, Roby DD, Fischer JB, Kuletz KJ, Irons DB, et al. 2002. Long-term direct and indirect effects of the "Exxon Valdez" oil spill on pigeon guillemots in Prince William Sound, Alaska. *Marine Ecology Progress Series* 241:287–304.

Goñi R. 1998. Ecosystem effects of marine fisheries: an overview. *Ocean & Coastal Management* 40(1):37–64.

Gorokhovich Y, Leiserowiz A. 2011. Historical and future coastal changes in Northwest Alaska. *Journal of Coastal Research* 28(1A):174–86.

Grebmeier JM, Overland JE, Moore SE, Farley EV, Carmack EC, Cooper LW, Frey KE, et al. 2006. A major ecosystem shift in the northern Bering Sea. *Science* 311(5766):1461–4.

Hall JV, Frayer WE, Wilen BO. 1994. *Status of Alaska Wetlands.* Anchorage, AK: FWS.

Hare SR, Mantua NJ. 2000. Empirical evidence for North Pacific regime shifts in 1977 and 1989. *Progress in Oceanography* 47(2–4):103–45.

Harvey HR, Sigler MF. 2013. An introduction to the Bering Sea project: volume II. *Deep-Sea Research Part II* 94:2–6.

Harwell MA, Gentile JH, Cummins KW, Highsmith RC, Hilborn R, McRoy CP, Parrish J, Weingartner T. 2010. A conceptual model of natural and anthropogenic drivers and their influence on the Prince William Sound, Alaska, ecosystem. *Human and Ecological Risk Assessment* 16(4): 672–726.

Hayes MO, Michel J, Betenbaugh DV. 2010. The intermittently exposed, coarse-grained gravel beaches of Prince William Sound, Alaska: comparison with open-ocean gravel beaches. *Journal of Coastal Research* 26(1):4–30.

Haynes TL, Mishler C. 1991. *The Subsistence Harvest and Use of Steller Sea Lions in Alaska*. Juneau, AK: Alaska Department of Fish and Game, Division of Subsistence.

Heifetz J, Stone RP, Shotwell SK. 2009. Damage and disturbance to coral and sponge habitat of the Aleutian Archipelago. *Marine Ecology Progress Series* 397:295–303.

Hermann AJ, Gibson GA, Bond NA, Curchitser EN, Hedstrom K, Cheng W, Wang M, et al. 2013. A multivariate analysis of observed and modeled biophysical variability on the Bering Sea shelf: multidecadal hindcasts (1970–2009) and forecasts (2010–2040). *Deep-Sea Research Part II: Topical Studies in Oceanography* 94:121–39.

Hermann AJ, Gibson GA, Bond NA, Curchitser EN, Hedstrom K, Cheng W, Wang M, et al. 2016. Projected future biophysical states of the Bering Sea. *Deep-Sea Research Part II: Topical Studies in Oceanography* 134: 30–47.

Hermann AJ, Gibson GA, Cheng W, Ortiz I, Aydin K, Wang M, Hollowed AB, et al. 2019. Projected biophysical conditions of the Bering Sea to 2100 under multiple emission scenarios. *ICES Journal of Marine Science* 76:1280–304.

Hess NA, Ribic CA, Vining I. 1999. Benthic marine debris, with an emphasis on fishery-related items, surrounding Kodiak Island, Alaska, 1994–1996. *Marine Pollution Bulletin* 38(10):885–90.

Heymans JJ, Guénette S, Christensen V, Trites A. 2005. Changes in the Gulf of Alaska ecosystems due to ocean climate change and fishing. *ICES Conference and Meeting* 22:1–31.

Himes-Cornell A, Package C, Durland A. 2011. Improving Community Profiles for the North Pacific fisheries. NOAA Technical Memorandum NMFS-AFSC-230. Washington, DC: NOAA (p. 85).

Hoff GR, Britt LL. 2011. Results of the 2010 Eastern Bering Sea upper Continental Slope Survey of Groundfish and Invertebrate Resources. NOAA Technical Memorandum NMFS-AFSC-224. Washington, DC: NOAA.

Hogrefe K, Ward D, Donnelly T, Dau N. 2014. Establishing a baseline for regional scale monitoring of eelgrass (Zostera marina) habitat on the lower Alaska Peninsula. *Remote Sensing* 6(12):12447–77.

Holland DS, Ginter JJ. 2001. Common property institutions in the Alaskan groundfish fisheries. *Marine Policy* 25(1):33–42.

Hollowed AB, Barbeaux SJ, Cokelet ED, Farley E, Kotwicki S, Ressler PH, Spital C, Wilson CD. 2012. Effects of climate variations on pelagic ocean habitats and their role in structuring forage fish distributions in the Bering Sea. *Deep Sea Research Part II: Topical Studies in Oceanography* 65:230–50.

Hollowed AB, Holsman KK, Haynie AC, Hermann AJ, Punt AE, Aydin K, Ianelli JN, et al. 2020. Integrated modeling to evaluate climate change impacts on coupled social-ecological systems in Alaska. *Frontiers in Marine Science* 6:775.

Hollowed AB, Ito SI, Pinnegar J. 2017. S-CCME workshop W5, "modeling effects of climate change on fish and fisheries". *PICES Press* 25(1):18–23.

Hollowed AB, Planque B, Loeng H. 2013. Potential movement of fish and shellfish stocks from the sub-Arctic to the Arctic Ocean. *Fisheries Oceanography* 22(5):355–70.

Hollowed AB, Aydin KY, Essington TE, Ianelli JN, Megrey BA, Punt AE, Smith ADM. 2011. Experience with quantitative ecosystem assessment tools in the northeast Pacific. *Fish and Fisheries* 12(2):189–208.

Hollowed AB, Ianelli JN, Livingston PA. 2000. Including predation mortality in stock assessments: a case study for Gulf of Alaska walleye pollock. *ICES Journal of Marine Science* 57:279–93.

Holsman KK, Hazen EL, Haynie A, Gourguet S, Hollowed A, Bograd SJ, Samhouri JF, Aydin K. 2019a. Towards climate resiliency in fisheries management. *ICES Journal of Marine Science* 76(5):1368–78

Holsman K, Hollowed A, Ito SI, Bograd S, Hazen E, King J, Mueter F, Perry RI. 2019b. Climate change impacts, vulnerabilities and adaptations: North Pacific and Pacific Arctic marine fisheries. *Impacts of Climate Change on Fisheries and Aquaculture* 6:113.

Holsman KK, Ianelli J, Aydin K, Punt AE, Moffitt EA. 2016. A comparison of fisheries biological reference points estimated from temperature-specific multi-species and single-species climate-enhanced stock assessment models. *Deep Sea Research Part II: Topical Studies in Oceanography* 134:360–78.

Holsman, K, Samhouri J, Cook G, Hazen E, Olsen E, Dillard M, Kasperski S, et al. 2017. An ecosystem-based approach to marine risk assessment. *Ecosystem Health and Sustainability* 3(1):e01256.

Holsman K, Hollowed A, Ito S, Bograd S, Hazen E, King J, Mueter F. 2018. Climate Change Impacts, Vulnerabilities and Adaptations: North Pacific and Pacific Arctic Marine Fisheries. Impacts of Climate Change on Fisheries and Aquaculture: Synthesis of Current Knowledge, Adaptation and Mitigation Options. FAO Technical Paper No. 627. Rome, Italy: FAO.

Holt CA, Rutherford MB, Peterman RM. 2008. International cooperation among nation-states of the North Pacific Ocean on the problem of competition among salmon for a common pool of prey resources. *Marine Policy* 32(4):607–17.

Hong N. 2012. The melting Arctic and its impact on China's maritime transport. *Research in Transportation Economics* 35(1):50–7.

Hull T, Leask LE. 2000. *Dividing Alaska, 1867–2000: Changing Land Ownership and Management*. Anchorage, AK: University of Alaska Anchorage, Institute of Social and Economic Research.

Humphries GR, Huettmann F. 2014. Putting models to a good use: a rapid assessment of Arctic seabird biodiversity indicates potential conflicts with shipping lanes and human activity. *Diversity and Distributions* 20(4):478–90.

Hunt Jr GL, Coyle KO, Eisner LB, Farley EV, Heintz RA, Mueter F, Napp JM, et al. 2011. Climate impacts on eastern Bering Sea foodwebs: a synthesis of new data and an assessment of the oscillating control hypothesis. *ICES Journal of Marine Science* 68(6):1230–43.

Hunt Jr GL, Kato H, McKinnell SM. 2000. *Predation by Marine Birds and Mammals in the Subarctic North Pacific Ocean. PICES Scientific Report No. 14*. Sidney, Canada: North Pacific Marine Science Organization (p. 164).

Hunt WR. 1976. *Alaska: A Bicentennial History*. London, UK: WW Norton & Company.

Hunt Jr GL, Stabeno P, Walters G, Sinclair E, Brodeur RD, Napp JM, Bond NA. 2002. Climate change and control of the southeastern Bering Sea pelagic ecosystem. *Deep Sea Research II* 49:5821–53.

Huntington HP, Daniel R, Hartsig A, Harun K, Heiman M, Meehan R, Noongwook G, et al. 2015. Vessels, risks, and rules: planning for safe shipping in Bering Strait. *Marine Policy* 51:119–27.

Huntington HP, Quakenbush LT, Nelson M. 2016. Effects of changing sea ice on marine mammals and subsistence hunters in northern Alaska from traditional knowledge interviews. *Biology Letters* 12(8):20160198.

Huntington HP. 2009. A preliminary assessment of threats to arctic marine mammals and their conservation in the coming decades. *Marine Policy* 33(1):77–82.

Hurst TP, Cooper DW, Duffy-Anderson JT, Farley EV. 2014. Contrasting coastal and shelf nursery habitats of Pacific cod in the southeastern Bering Sea. *ICES Journal of Marine Science* 72(2):515–27.

Ianelli J, Holsman KK, Punt AE, Aydin K. 2016. Multi-model inference for incorporating trophic and climate uncertainty into stock assessments. *Deep Sea Research Part II: Topical Studies in Oceanography* 134:379–89.

Ianelli J, Hollowed AB, Haynie AC, Mueter FJ, Bond NA. 2011. Evaluating management strategies for eastern Bering Sea walleye pollock (Theragra chalcogramma) in a changing environment. *ICES Journal of Marine Science* 68(6):1297–304.

Irons DB, Kendall SJ, Erickson WP, McDonald LL, Lance BK. 2000. Nine years after the Exxon Valdez oil spill: effects on marine bird populations in Prince William Sound, Alaska. *The Condor* 102(4):723–37.

Iwata H, Tanabe S, Aramoto M, Sakai N, Tatsukawa R. 1994. Persistent organochlorine residues in sediments from the Chukchi Sea, Bering Sea and Gulf of Alaska. *Marine Pollution Bulletin* 28(12):746–53.

Johnson SW, Murphy ML, Csepp DJ, Harris PM, Thedinga JF. 2003b. A Survey of Fish Assemblages in Eelgrass and Kelp Habitats of Southeastern Alaska. NOAA Technical Memorandum NMFS-AFSC 139. Washington, DC: NOAA (p. 39).

Johnson SW, Murphy ML, Csepp DJ. 2003a. Distribution, habitat, and behavior of rockfishes, Sebastes spp., in nearshore waters of southeastern Alaska: observations from a remotely operated vehicle. *Environmental Biology of Fishes* 66(3):259–70.

Kaiser BA, Fernandez LM, Vestergaard N. 2016. The future of the marine Arctic: environmental and resource economic development issues. *The Polar Journal* 6(1):152–68.

Karp MA, Blackhart K, Lynch PD, Deroba J, Hanselman D, Gertseva V, Teo S, et al. 2019. Proceedings of the 13th National Stock Assessment Workshop: Model Complexity, Model Stability, and Ensemble Modeling. NOAA Technical Memoranda NMFS-F/SPO-189. Washington, DC: NOAA (p. 49).

Kellner JB, Sanchirico JN, Hastings A, Mumby PJ. 2011. Optimizing for multiple species and multiple values: tradeoffs inherent in ecosystem-based fisheries management. *Conservation Letters* 4(1):21–30.

Kemp AC, Engelhart SE, Culver SJ, Nelson A, Briggs RW, Haeussler PJ. 2013. Modern salt-marsh and tidal-flat foraminifera from Sitkinak and Simeonof Islands, southwestern Alaska. *The Journal of Foraminiferal Research* 43(1):88–98.

Kimura, DK, Somerton, DA. 2006. Review of Statistical Aspects of Survey Sampling for Marine Fisheries. *Reviews in Fisheries Science* 14:45–283.

Kishi MJ, Kashiwai M, Ware DM, Megrey BA, Eslinger DL, Werner FE, Aita NM, et al. 2007. Effects of advective processes on planktonic distributions in the Kuroshio region using a 3-D lower trophic model and a data assimilative OGCM. *Ecological Modelling*. 202(1–2):105–19.

Kotwicki S, Buckley TW, Honkaletho T, Walters G. 2005. Variation in the distribution of walleye pollock (Theragra chalcogramma) with temperature and implications for seasonal migration. *Fishery Bulletin* 103:574–87.

Kovacs KM, Lydersen C, Overland JE, Moore SE. 2011. Impacts of changing sea-ice conditions on Arctic marine mammals. *Marine Biodiversity* 41(1):181–94.

Kraska J. 2016. "Maritime governance of the US Arctic region." In: *Climate Change and Human Security from a Northern Point of View*. Edited by Heininen L, Nicol H. Waterloo, Canada: Centre on Foreign Policy and Federalism (p. 165).

Krumhansl KA, Okamoto DK, Rassweiler A, Novak M, Bolton JJ, Cavanaugh KC, Connell SD, et al. 2016. Global patterns of kelp forest change over the past half-century. *Proceedings of the National Academy of Sciences* 113(48): 13785–90.

Kruse GH, Funk FC, Geiger HJ, Mabry KR, Savikko HM, Siddeek SM. 2000. Overview of State-Managed Marine Fisheries in the Central and Western Gulf of Alaska, Aleutian Islands, and Southeastern Bering Sea, with Reference to Steller Sea Lions. Regional Information Report 5J00-10. Juneau, AK: Alaska Department of Fish and Game, Division of Commercial Fisheries.

Kruse GH, Livingston P, Overland JE, Jamieson GS, McKinnell S, Perry RI. 2006. *Report of the PICES/NPRB Workshop on Integration of Ecological Indicators of the North Pacific with Emphasis on the Bering Sea*. Sidney, Canada: North Pacific Marine Science Organization.

Kuletz KJ, Ferguson MC, Hurley B, Gall AE, Labunski EA, Morgan TC. 2015. Seasonal spatial patterns in seabird and marine mammal distribution in the eastern Chukchi and western Beaufort seas: identifying biologically important pelagic areas. *Progress in Oceanography* 136:175–200.

Kurtz MD. 1994. Managing Alaska's coastal development: state review of federal oil and gas lease sales. *Alaska Law Review* 11:377.

Ladd C, Cheng W. 2016. Gap winds and their effects on regional oceanography part I: Cross Sound, Alaska. *Deep Sea Research Part II: Topical Studies in Oceanography* 132:41–53.

Laidre KL, Stern H, Kovacs KM, Lowry L, Moore SE, Regehr EV, Ferguson SH, et al. 2015. Arctic marine mammal population status, sea ice habitat loss, and conservation recommendations for the 21st century. *Conservation Biology* 29(3):724–37.

Laidre KL, Stirling I, Lowry LF, Wiig Ø, Heide-Jørgensen MP, Ferguson SH. 2008. Quantifying the sensitivity of Arctic marine mammals to climate-induced habitat change. *Ecological Applications* 18(sp2):S97–125.

Large SI, Fay G, Friedland KD, Link JS. 2013. Defining trends and thresholds in responses of ecological indicators to fishing and environmental pressures. *ICES Journal of Marine Science* 70:755–67

Large SI, Fay G, Friedland KD, Link JS. 2015. Critical points in ecosystem responses to fishing and environmental pressures. *Marine Ecology Progress Series* 521:1–17.

Lauth RR. 2011. Results of the 2010 Eastern and Northern Bering Sea Continental Shelf Bottom Trawl Survey of Groundfish and Invertebrate Fauna. NOAA Technical Memorandum NMFSAFSC-227. Washington, DC: NOAA (p. 256).

Leatherwood S, Reeves RR, Perrin WF, Evans WE. 1988. *Whales, Dolphins, and Porpoises of the Eastern North Pacific and Adjacent Arctic waters: A Guide to their Identification*. New York: Dover Publications (p. 245).

Lefebvre KA, Quakenbush L, Frame E, Huntington KB, Sheffield G, Stimmelmayr R, Bryan A, et al. 2016. Prevalence of algal toxins in Alaskan marine mammals foraging in a changing arctic and subarctic environment. *Harmful Algae* 55:13–24.

Levin PS, Essington TE, Marshall KN, Koehn LE, Anderson LG, Bundy A, Carothers C, et al. 2018. Building effective fishery ecosystem plans. *Marine Policy* 92:48–57.

Levin PS, Fogarty MJ, Murawski SA, Fluharty D. 2009. Integrated ecosystem assessments: developing the scientific basis for ecosystem-based management of the ocean. *PLoS Biology* 7(1):e1000014.

LeVine M, Tuyn PV, Hughes L. 2013. Oil and gas in America's Arctic ocean: past problems counsel precaution. *Seattle University Law Review* 37:1271.

Lew DK, Larson DM. 2012. Economic values for saltwater sport fishing in Alaska: a stated preference analysis. *North American Journal of Fisheries Management* 32(4): 745–59.

Liebezeit JR, Kendall SJ, Brown S, Johnson CB, Martin P, McDonald TL, Payer DC, et al. 2009. Influence of human development and predators on nest survival of tundra birds, Arctic Coastal Plain, Alaska. *Ecological Applications* 19(6):1628–44.

Link JS, Marshak AR. 2019. Characterizing and comparing marine fisheries ecosystems in the United States: determinants of success in moving toward ecosystem-based fisheries management. *Reviews in Fish Biology and Fisheries* 29(1):23–70.

Link JS. 2018. System-level optimal yield: increased value, less risk, improved stability, and better fisheries. *Canadian Journal of Fisheries and Aquatic Sciences* 75(1):1–6.

Litzow MA, Ciannelli L. 2007. Oscillating trophic control induces community reorganization in a marine ecosystem. *Ecology Letters* 10(12):1124–34.

Litzow MA, Mueter FJ, Hobday AJ. 2014. Reassessing regime shifts in the North Pacific: incremental climate change and commercial fishing are necessary for explaining decadal-scale biological variability. *Global Change Biology* 20(1):38–50.

Litzow MA, Urban D. 2009. Fishing through (and up) Alaskan food webs. *Canadian Journal of Fisheries and Aquatic Sciences* 66(2):201–11.

Litzow MA, Ciannelli L, Puerta P, Wettstein JJ, Rykaczewski RR, Opiekun M. 2019. Nonstationary environmental and community relationships in the North Pacific Ocean. *Ecology* 100(8):e02760.

Litzow MA, Mueter FJ, Hobday AJ. 2014. Reassessing regime shifts in the North Pacific: incremental climate change and commercial fishing are necessary for explaining decadal-scale biological variability. *Global Change Biology* 20:38–50.

Livingston PA, Aydin K, Boldt J, Ianelli J, Jurado-Molina J. 2005. A framework for ecosystem impacts assessment using an indicator approach. *ICES Journal of Marine Science* 62(3):592–7.

Livingston PA, Aydin K, Boldt JL, Hollowed AB, Napp JM. 2011. "Alaska marine fisheries management: advances and linkages to ecosystem research." In: *Ecosystem Based Management: An Evolving Perspective.* Edited by A Belgrano, C Fowler. Cambridge, UK: Cambridge University Press (pp. 113–52).

Livingston PA. 2002. *Ecosystem Considerations for 2003.* Anchorage, AK: North Pacific Fishery Management Council (p. 230).

Logerwell EA, Aydin K, Barbeaux S, Brown E, Conners ME, Lowe S, Orr JW, et al. 2005. Geographic patterns in the demersal ichthyofauna of the Aleutian Islands. *Fisheries Oceanography* 14:93–112.

Loher T, Armstrong DA. 2005. Historical changes in the abundance and distribution of ovigerous red king crabs (*Paralithodes camtschaticus*) in Bristol Bay (Alaska), and potential relationship with bottom temperature. *Fisheries Oceanography* 14(4):292–306.

Lomas MW, Stabeno PJ. 2014. An introduction to the Bering Sea project: volume III. *Deep Sea Research Part II: Topical Studies in Oceanography* 109:1–4.

Long WC, Swiney KM, Harris C, Page HN, Foy RJ. 2013. Effects of ocean acidification on juvenile Red King Crab (*Paralithodes camtschaticus*) and Tanner Crab (*Chionoecetes bairdi*) growth, condition, calcification, and survival. *PLoS One* 8(4):e60959.

Long W, Pruisner P, Swiney KM, Foy RJ. 2019. Effects of ocean acidification on the respiration and feeding of juvenile red and blue king crabs (*Paralithodes camtschaticus* and *P. platypus*). *ICES Journal of Marine Science* 76(5): 1335–43.

Loughlin TR. 2013. *Marine Mammals and the Exxon Valdez.* Cambridge, MA: Academic Press.

Luizza MW, Evangelista PH, Jarnevich CS, West A, Stewart H. 2016. Integrating subsistence practice and species distribution modeling: assessing invasive elodea's potential impact on Native Alaskan subsistence of Chinook salmon and whitefish. *Environmental Management* 58(1):144–63.

Lynch AH, Brunner RD. 2007. Context and climate change: an integrated assessment for Barrow, Alaska. *Climatic Change* 82(1–2):93–111.

Lynch AH, Curry JA, Brunner RD, Maslanik JA. 2004. Toward an integrated assessment of the impacts of extreme wind events on Barrow, Alaska. *Bulletin of the American Meteorological Society* 85(2):209–22.

Lynch PD, Methot RD, Link JS. 2018. Implementing a Next Generation Stock Assessment Enterprise. An Update to the NOAA Fisheries Stock Assessment Improvement Plan. NOAA Technical Memorandum NMFS-F/SPO-183. Washington, DC: NOAA (p. 127).

Maldonado JK, Shearer C, Bronen R, Peterson K, Lazrus H. 2013. "The impact of climate change on tribal communities in the US: displacement, relocation, and human rights." In: *Climate Change and Indigenous Peoples in the United States.* Edited by JK Maldonado, C Benedict, R Pandya. Cham, Switzerland: Springer (pp. 93–106).

Mantua NJ, Hare SR, Zhang Y, Wallace JM, Francis RC. 1997. A Pacific interdecadal climate oscillation with impacts on salmon production. *Bulletin of the American Meteorological Society* 78:1069–79.

Marasco RJ, Goodman D, Grimes CB, Lawson PW, Punt AE, Quinn II TJ. 2007. Ecosystem-based fisheries management: some practical suggestions. *Canadian Journal of Fisheries and Aquatic Sciences* 64(6):928–39.

Marshall KN, Levin PS, Essington TE, Koehn LE, Anderson LG, Bundy A, Carothers C, et al. 2018. Ecosystem-based fisheries management for social–ecological systems: renewing the focus in the United States with next generation fishery ecosystem plans. *Conservation Letters* 11(1):e12367.

Mathis JT, Cooley SR, Lucey N, Colt S, Ekstrom J, Hurst T, Hauri C, et al. 2015. Ocean acidification risk assessment for Alaska's fishery sector. *Progress in Oceanography* 136: 71–91.

Mattes LA. 2006. Fishery Management Plan for the Commercial Tanner Crab fishery in the South Peninsula District of Registration Area J, 2007. Report No. 06-70. Anchorage, AK Alaska Department of Fish and Game, Fishery Management. http://www.sf.adfg.state.ak.us/FedAidPDFs/fmr06-70.pdf.

McCann LD, Holzer KK, Davidson IC, Ashton GV, Chapman MD, Ruiz GM. 2013. Promoting invasive species control and eradication in the sea: options for managing the tunicate invader Didemnum vexillum in Sitka, Alaska. *Marine Pollution Bulletin* 77(1–2):165–71.

McConnaughey RA, Mier KL, Dew CB. 2000. An examination of chronic trawling effects on soft-bottom benthos of the eastern Bering Sea. *ICES Journal of Marine Science* 57(5):1377–88.

McConnaughey RA, Smith KR. 2000. Associations between flatfish abundance and surficial sediments in the eastern Bering Sea. *Canadian Journal of Fisheries and Aquatic Sciences* 57(12):2410–19.

McGee S, Piorkowski R, Ruiz G. 2006. Analysis of recent vessel arrivals and ballast water discharge in Alaska: toward assessing ship-mediated invasion risk. *Marine Pollution Bulletin* 52(12):1634–45.

McGowan DW, Horne JK, Parker-Stetter SL. 2019. Variability in species composition and distribution of forage fish in the Gulf of Alaska. *Deep Sea Research Part II: Topical Studies in Oceanography* 165:221–37.

McRoy CP. 1970. Standing stocks and other features of eelgrass (Zostera marina) populations on the coast of Alaska. *Journal of the Fisheries Board of Canada* 27(10):1811–21.

Melvin E, Dietrich K, Van Wormer K, Geernaert T. 2004. *The Distribution of Seabirds on Alaskan Longline Fishing Grounds: 2002 Data Report*. Seattle, WA: Washington Sea Grant Program.

Merrick RL. 1997. Current and historical roles of apex predators in the Bering Sea ecosystem. *Journal of Northwest Atlantic Fishery Science* 22:343–55.

Merrick RL, Chumbley MK, Byrd GV. 1997. Diet diversity of Steller sea lions (Eumetopias jubatus) and their population decline in Alaska: a potential relationship. *Canadian Journal of Fisheries and Aquatic Sciences* 54:1342–8.

Mesquita MS, Atkinson DE, Hodges KI. 2010. Characteristics and variability of storm tracks in the North Pacific, Bering Sea, and Alaska. *Journal of Climate* 23(2):294–311.

Metcalf V, Robards M. 2008. Sustaining a healthy human–walrus relationship in a dynamic environment: challenges for comanagement. *Ecological Applications* 18(sp2):S148–56.

Miller AW, Ruiz GM. 2014. Arctic shipping and marine invaders. *Nature Climate Change* 4:413–6.

Mizroch SA, Rice DW. 2006. Have North Pacific killer whales switched prey species in response to depletion of the great whale populations? *Marine Ecology Progress Series* 310:235–46.

Mizroch SA, Herman LM, Straley JM, Glockner-Ferrari D, Jurasz C, Darling J, Cerchio S, et al. 2004. Estimating the adult survival rate of central North Pacific humpback whales (Megaptera novaeangliae). *Journal of Mammalogy* 85(5):963–72.

Moffitt EA, Punt AE, Holsman K, Aydin KY, Ianelli JN, Ortiz I. 2016. Moving towards ecosystem-based fisheries management: options for parameterizing multi-species biological reference points. *Deep Sea Research Part II: Topical Studies in Oceanography* 134:350–9.

Montevecchi WA. 2002. "Interactions between fisheries and seabirds." In: *Biology of Marine Birds*. Edited by EA Schreiber, J Burger. Boca Raton, FL: CRC Press (pp. 527–57).

Moore JE, Wallace BP, Lewison RL, Žydelis R, Cox TM, Crowder LB. 2009. A review of marine mammal, sea turtle and seabird bycatch in USA fisheries and the role of policy in shaping management. *Marine Policy* 33(3):435–51.

Moore JE, Wallace BP, Lewison RL, Žydelis R, Cox TM, Crowder LB. 2009. A review of marine mammal, sea turtle and seabird bycatch in USA fisheries and the role of policy in shaping management. *Marine Policy* 33(3):435–51.

Moore SE, Logerwell E, Eisner L, Farley EV, Harwood LA, Kuletz K, Lovvorn J, Murphy JR, Quakenbush LT. 2014. "Marine fishes, birds and mammals as sentinels of ecosystem variability and reorganization in the Pacific Arctic region." In: *The Pacific Arctic Region*. Edited by JM Grebmeier, W Maslowski. Dordrecht, Netherlands: Springer (pp. 337–92).

Morley JW, Selden RL, Latour RJ, Frölicher TL, Seagraves RJ, Pinsky ML. 2018. Projecting shifts in thermal habitat for 686 species on the North American continental shelf. *PLoS One* 13(5):e0196127.

Mueter FJ, Bond NA, Ianelli JN, Hollowed AB. 2011. Expected declines in recruitment of walleye pollock (Theragra chalcogramma) in the eastern Bering Sea under future climate change. *ICES Journal of Marine Science* 68(6):1284–96.

Mueter FJ, Litzow MA. 2008. Sea ice retreat alters the biogeography of the Bering Sea continental shelf. *Ecological Applications* 18(2):309–20.

Mueter FJ, Megrey BA. 2006. Using multi-species surplus production models to estimate ecosystem-level maximum sustainable yields. *Fisheries Research* 81(2–3):189–201.

Mueter FJ, Megrey BA. 2006. Using multi-species surplus production models to estimate ecosystem-level maximum sustainable yields. *Fisheries Research* 81(2–3):189–201.

Mundy PR. 2005. *The Gulf of Alaska: Biology and Oceanography*. Fairbanks, AK: University of Alaska Fairbanks (p. 220).

Murphy ML, Johnson SW, Csepp DJ. 2000. A comparison of fish assemblages in eelgrass and adjacent subtidal habitats near Craig, Alaska. *Alaska Fishery Research Bulletin* 7:11–21.

Musick JA, Harbin MM, Berkeley SA, Burgess GH, Eklund AM, Findley L, Gilmore RG, et al. 2000. Marine, estuarine, and diadromous fish stocks at risk of extinction in North America (exclusive of Pacific salmonids). *Fisheries* 25(11):6–30.

Muto MM, Helker VT, Angliss RP, Allen BA, Boveng PL, Breiwick JM, Cameron MF, et al. 2017. *Alaska Marine Mammal Stockassessments, 2016. NOAA Technical Memorandum NMFS-AFSC-355*. Washington, DC: NOAA (p. 366).

Muto MM, Helker VT, Angliss RP, Allen BA, Boveng PL, Breiwick JM, Cameron MF, et al. 2018. *Alaska Marine Mammal Stock Assessments, 2017. NOAA Technical Memorandum NMFS-AFSC-378*. Washington, DC: NOAA (p. 382).

Muto MM, Helker VT, Angliss RP, Boveng PL, Breiwick JM, Cameron MF, Clapham PJ, et al. 2019. *Alaska Marine Mammal Stock Assessments, 2018. NOAA Technical Memorandum NMFS-AFSC-393*. Washington, DC: NOAA (p. 390).

Myers KW, Aydin KY, Walker RV, Fowler S, Dahlberg ML. 1996. Known Ocean Ranges of Stocks of Pacific Salmon and Steelhead as Shown by Tagging Experiments, 1956–1995. (NPAFC Doc. 192.) FRI-UW-9614. Seattle, WA: University of Washington.

Naves LC, Fall JA. 2017. Calculating food production in the subsistence harvest of birds and eggs. *Arctic* 70:86–100.

Nemoto, T. 1957. Foods of baleen whales in the northern Pacific. *Scientific Reports of the Whales Research Institute* 12:33–89.

Nemoto T, Kawamura A. 1977. Characteristics of food habits and distribution of baleen whales with special reference to the abundance of North Pacific sei and Bryde's whales. *International Whaling Commission* 1:80–7.

Nerini M. 1984. "A review of gray whale feeding ecology." In: *The Gray Whale, Eschrichtius Robustus*. Edited by ML Jones, SL Swartz, S Leatherwood. Orlando, FL: Academic Press (pp. 423–50).

Nielsen JL, Graziano SL, Seitz AC. 2010. Fine-scale population genetic structure in Alaskan Pacific halibut (*Hippoglossus stenolepis*). *Conservation Genetics* 11(3): 999–1012.

NMFS (National Marine Fisheries Service). 2016a. Ecosystem-Based Fisheries Management Policy of the National Marine Fisheries Service, National Oceanic and Atmospheric Administration. Washington, DC: NOAA.

NMFS (National Marine Fisheries Service). 2017. *National Marine Fisheries Service—2nd Quarter 2017 Update.* Washington, DC: NOAA.

NMFS (National Marine Fisheries Service). 2016b. *NOAA Fisheries Ecosystem-Based Fisheries Management Roadmap.* Washington, DC: NOAA.

NMFS (National Marine Fisheries Service). 2018a. Fisheries Economics of the United States, 2016. NOAA Technical Memorandum NMFS-F/SPO-187. Washington, DC: NOAA (p. 243).

NMFS (National Marine Fisheries Service). 2018b. *Fisheries of the United States, 2017. NOAA Current Fishery Statistics No. 2017.* Washington, DC: NOAA.

NMFS (National Marine Fisheries Service). 2019. *Ecosystem Based Fisheries Management Alaska Region Implementation Plan.* Washington, DC: NOAA (p. 21).

NMFS (National Marine Fisheries Service). 2020a. Fisheries of the exclusive economic zone off Alaska; Gulf of Alaska; Final 2020 and 2021 harvest specifications for groundfish. *Federal Register* 85(47):13802–30.

NMFS (National Marine Fisheries Service). 2020b. *National Marine Fisheries Service—2nd Quarter 2020 Update.* Washington, DC: NOAA (p. 51).

NOAA-OCM (National Oceanic and Atmospheric Administration-Office of Coastal Management). 2017. *Frequent Questions: Economics: National Ocean Watch (ENOW) Data.* Washington, DC: NOAA (p. 8).

Nong D, Countryman AM, Warziniack T. 2018. Potential impacts of expanded Arctic Alaska energy resource extraction on US energy sectors. *Energy Policy* 119:574–84.

Notz D, Stroeve J. 2016. Observed Arctic sea-ice loss directly follows anthropogenic CO2 emission. *Science* 354(6313): 747–50.

NPFMC (North Pacific Fishery Management Council). 1998. Omnibus Essential Fish Habitat Amendments for Groundfish in the Bering Sea, Aleutian Islands, and Gulf of Alaska, King and Tanner Crab in the Bering Sea and Aleutian Islands, Alaska Scallop, and Alaska Salmon Fish. Washington, DC: NOAA.

NPFMC (North Pacific Fishery Management Council). 2007. *Aleutian Islands Fishery Ecosystem Plan.* Washington, DC: NOAA (p. 190).

NPFMC (North Pacific Fishery Management Council). 2009. *Fishery Management Plan for Fish Resources of the Arctic Management Area.* Washington, DC: NOAA (p. 146).

NPFMC (North Pacific Fishery Management Council). 2011. *Central Gulf of Alaska (GOA) Trawl Sweeps modification Discussion Paper*. February 2011. Agenda D-1(b)(1). Washington, DC: NOAA (p. 11).

NPFMC (North Pacific Fishery Management Council). 2019. *Bering Sea Fishery Ecosystem Plan.* Washington, DC: NOAA (p. 133).

NRC (National Research Council). 1996. *The Bering Sea Ecosystem.* Washington, DC: National Academies Press.

NRC (National Research Council). 2003. *Decline of the Steller Sea Lion in Alaskan waters; Untangling Food Webs and Fishing Nets.* Washington, DC: National Academies Press (p. 204).

Okey TA, Pauly D. 1999. Trophic mass-balance model of Alaska's Prince William Sound ecosystem, for the post-spill period 1994–1996. *Fisheries Centre Research Reports* 7(4).

Olesiuk PF. 1993. Annual prey consumption by harbor seals (Phoca vitulina) in the Strait of Georgia, British Columbia. *Fishery Bulletin* 91:491–515.

Olson MB, Strom SL. 2002. Phytoplankton growth, micro-zooplankton herbivory and community structure in the southeast Bering Sea: Insight into the formation and temporal persistence of an Emiliania huxleyi bloom. *Deep Sea Research Part II* 49(5):5969–90.

Ono K, Haynie AC, Hollowed AB, Ianelli JN, McGilliard CR, Punt AE. 2018. Management strategy analysis for multi-species fisheries, including technical interactions and human behavior in modelling management decisions and fishing. *Canadian Journal of Fisheries and Aquatic Sciences* 75(8):1185–202.

Orensanz JM, Ernst B, Armstrong DA, Stabeno P, Livingston P. 2004. Contraction of the geographic range of distribution of snow crab (Chionoecetes opilio) in the eastern Bering Sea: an environmental ratchet? *CalCOFI Reports* 45:65–79.

Ormseth OA, Baker MM, Hopcroft RR, Ladd C, Mordy CW, Moss JH, Mueter FJ, Shotwell SK, Strom SL. 2019. Introduction to understanding ecosystem processes in the gulf of Alaska, volume 2. *Deep Sea Research Part II: Topical Studies in Oceanography* 165:1–6.

Ortiz I, Aydin K, Hermann AJ, Gibson GA, Punt AE, Wiese FK, Eisner LB, Ferm N, et al. 2016. Climate to fish: synthesizing field work, data and models in a 39-year retrospective analysis of seasonal processes on the eastern Bering Sea shelf and slope. *Deep Sea Research Part II: Topical Studies in Oceanography* 134:390–412.

Otto R. 2014. History of King Crab Fisheries with Special Reference to the North Pacific Ocean: Development, Maturity, and Senescence. King Crabs of the World: Biology and Fisheries Management. Boca Raton, FL: CRC Press (pp. 81–138).

Overland JE, Bond N. 1993. The influence of coastal orography: the Yakutat storm. *Monthly Weather Review* 121(5): 1388–97.

Pearson WH, Deriso RB, Elston RA, Hook SE, Parker KR, Anderson JW. 2012. Hypotheses concerning the decline and poor recovery of Pacific herring in Prince William Sound, Alaska. *Reviews in Fish Biology and Fisheries* 22: 95–135.

Peterson CH, Rice SD, Short JW, Esler D, Bodkin JL, Ballachey BE, Irons DB. 2003. Long-term ecosystem response to the Exxon Valdez oil spill. *Science* 302(5653):2082–6.

Peterson CH. 2001. The "Exxon Valdez" oil spill in Alaska: acute, indirect and chronic effects on the ecosystem. *Advances in Marine Biology* 39:1–103.

Peterson MJ, Carothers C. 2013. Whale interactions with Alaskan sablefish and Pacific halibut fisheries: surveying fishermen perception, changing fishing practices and mitigation. *Marine Policy* 42:315–24.

Piatt JF, Arimitsu ML, Sydeman WJ, Thompson SA, Renner H, Zador S, Douglas D, et al. 2018. Biogeography of pelagic food webs in the North Pacific. *Fisheries Oceanography* 27(4):366–80.

Piatt JF, Parrish JK, Renner HM, Schoen SK, Jones TT, Arimitsu ML, Kuletz KJ, et al. 2020. Extreme mortality and reproductive failure of common murres resulting from the northeast Pacific marine heatwave of 2014–2016. *PLoS One* 15(1):e0226087.

Pinsky ML, Mantua NJ. 2014. Emerging adaptation approaches for climate-ready fisheries management. *Oceanography* 27(4):146–59.

Pirtle JL, Shotwell SK, Zimmermann M, Reid JA, Golden N. 2017. Habitat suitability models for groundfish in the Gulf of Alaska. *Deep Sea Research Part II: Topical Studies in Oceanography* 165:303–21.

Plagányi ÉE, Punt AE, Hillary R, Morello EB, Thébaud O, Hutton T, Pillans RD, et al. 2014. Multispecies fisheries management and conservation: tactical applications using models of intermediate complexity. *Fish and Fisheries* 15(1):1–22.

Pollnac RB, Poggie Jr JJ. 2006. Job satisfaction in the fishery in two southeast Alaskan towns. *Human Organization* 65(3):329.

Pomeranz EF, Needham MD, Kruger LE. 2013. Stakeholder perceptions of indicators of tourism use and codes of conduct in a coastal protected area in Alaska. *Tourism in Marine Environments* 9(1–2):95–115.

Poole A, Sepez J. 2007. Recent and Historic Population Trends in Bering Sea and Aleutian Island Fishing Communities: Hubs and Spokes, Booms and Busts. Annual Meeting March 29–31 2007. New York, NY: Population Association of America.

Praetorius S, Rugenstein M, Persad G, Caldeira K. 2018. Global and Arctic climate sensitivity enhanced by changes in North Pacific heat flux. *Nature Communications* 9(3124).

Prowse TD, Furgal C, Chouinard R, Melling H, Milburn D, Smith SL. 2009. Implications of climate change for economic development in northern Canada: energy, resource, and transportation sectors. *AMBIO: A Journal of the Human Environment* 38(5):272–82.

Punt AE, Foy RJ, Dalton MG, Long WC, Swiney KM. 2016. Effects of long-term exposure to ocean acidification conditions on future southern Tanner crab (Chionoecetes bairdi) fisheries management. *ICES Journal of Marine Science* 73(3):849–64.

Raymond-Yakoubian J, Daniel R. 2018. An Indigenous approach to ocean planning and policy in the Bering Strait region of Alaska. *Marine Policy* 97:101–8.

Ready E. 2016. Challenges in the assessment of inuit food security. *Arctic* 69:266–80.

Reedy K, Maschner H. 2014. Traditional foods, corporate controls: networks of household access to key marine species in southern Bering Sea villages. *Polar Record* 50(4):364–78.

Reeves R, Rosa C, George JC, Sheffield G, Moore M. 2012. Implications of Arctic industrial growth and strategies to mitigate future vessel and fishing gear impacts on bowhead whales. *Marine Policy* 36:454–62.

Reeves RR, Ewins PJ, Agbayani S, Heide-Jørgensen MP, Kovacs KM, Lydersen C, Suydam R. 2014. Distribution of endemic cetaceans in relation to hydrocarbon development and commercial shipping in a warming Arctic. *Marine Policy* 44:375–89.

Reimer JP, Droghini A, Fischbach A, Watson JT, Bernard B, Poe A. 2017a. *Assessing the Risk of Non-Native Marine Species in the Bering Sea. NPRB Project 1523.* Anchorage, AK: University of Alaska, Alaska Center for Conservation Science (p. 39).

Reimer MN, Abbott JK, Haynie AC. 2017b. Empirical models of fisheries production: conflating technology with incentives? *Marine Resource Economics* 32(2):169–90.

Rieser A, Watling L, Guinotte J. 2013. Trawl fisheries, catch shares and the protection of benthic marine ecosystems: has ownership generated incentives for seafloor stewardship?. *Marine Policy* 40:75–83.

Renaud PE, Sejr MK, Bluhm BA, Sirenko B, Ellingsen IH. 2015. The future of Arctic benthos: expansion, invasion, and biodiversity. *Progress in Oceanography* 139:244–57.

Reuter RF, Conners ME, DiCosimo J, Gaichas S, Ormseth O, TenBrink TT. 2010. Managing non-target, data-poor species using catch limits: lessons from the Alaskan groundfish fishery. *Fisheries Management and Ecology* 17(4):323–35.

Richardson J, Erickson G. 2005. *Economics of Human Uses and Activities in the Northern Gulf of Alaska. The Gulf of Alaska: Biology and Oceanography.* Fairbanks, AK: University of Alaska (pp. 117–38).

Ritter H. 1993. *Alaska's History: The People, Land, and Events of the North Country.* Portland, OR: Alaska Northwest Books.

Rode KD, Regehr EV, Douglas DC, Durner G, Derocher AE, Thiemann GW, Budge SM. 2014. Variation in the response of an Arctic top predator experiencing habitat loss: feeding and reproductive ecology of two polar bear populations. *Global Change Biology* 20(1):76–88.

Rodionov SN, Bond NA, Overland JE. 2007. The Aleutian Low, storm tracks, and winter climate variability in the

Bering Sea. *Deep Sea Research Part II: Topical Studies in Oceanography* 54(23–6):2560–77.

Rooney SC, Rooper CN, Laman EA, Turner KA, Cooper DW, Zimmermann M. 2018. Model-Based Essential Fish Habitat Definitions for Gulf of Alaska Groundfish Species. NOAA Technical Memorandum NMFS-AFSC-373. Washington, DC: NOAA.

Ross K. 2000. *Environmental Conflict in Alaska*. Boulder, CO: University Press of Colorado.

Rothrock DA, Yu Y, Maykut GA. 1999. Thinning of the Arctic sea-ice cover. *Geophysical Research Letters* 26(23):3469–72.

Royer TC, Grosch CE. 2006. Ocean warming and freshening in the northern Gulf of Alaska. *Geophysical Research Letters* 33(16).

Ruckelshaus M, Klinger T, Knowlton N, DeMaster DP. 2008. Marine ecosystem-based management in practice: scientific and governance challenges. *BioScience* 58(1): 53–63.

Rugh DJ, Hobbs RC, Lerczak JA, Breiwick JM. 2005. Estimates of abundance of the eastern North Pacific stock of gray whales 1997–2002. *Journal of Cetacean Research and Management* 7(1):1–12.

Ruiz GM, Huber T, Larson K, McCann L, Steves B, Fofonoff P, Hines AH. 2006. *Biological Invasions in Alaska's Coastal Marine Ecosystems: Establishing a Baseline*. Edgewater, MD: Smithsonian Environmental Research Center.

Samhouri JF, Haupt AJ, Levin PS, Link JS, Shuford R. 2014. Lessons learned from developing integrated ecosystem assessments to inform marine ecosystem-based management in the USA. *ICES Journal of Marine Science* 71(5): 1205–15.

Samhouri JF, Levin PS, Ainsworth CH. 2010. Identifying thresholds for ecosystem-based management. *PLoS One* 5(1):e8907.

Saulitis E, Holmes LA, Matkin C, Wynne K, Ellifrit D, St-Amand C. 2015. Biggs killer whale (Orcinus orca) predation on subadult humpback whales (Megaptera novaeangliae) in Lower Cook Inlet and Kodiak, Alaska. *Aquatic Mammals* 41(3):341.

Schoen ER, Wipfli MS, Trammell EJ, Rinella DJ, Floyd AL, Grunblatt J, McCarthy MD, et al. 2017. Future of Pacific salmon in the face of environmental change: lessons from one of the world's remaining productive salmon regions. *Fisheries* 42(10):538–53.

Schumacher JD, Reed RK. 1986. On the Alaska coastal current in the western Gulf of Alaska. *Journal of Geophysical Research: Oceans* 91(C8):9655–61.

Schumacher JD, Stabeno PJ, Roach AT. 1989. Volume transport in the Alaska Coastal Current. *Continental Shelf Research* 9(12):1071–83.

Schwörer T, Federer RN, Ferren HJ. 2014. Invasive species management programs in Alaska: a survey of statewide expenditures, 2007–11. *Arctic* 1:20–7.

Seitz AC, Loher T, Norcross BL, Nielsen JL. 2011. Dispersal and behavior of Pacific halibut Hippoglossus stenolepis in the Bering Sea and Aleutian Islands region. *Aquatic Biology* 12(3):225–39.

Shelden KE, Moore SE, Waite JM, Wade PR, Rugh DJ. 2005. Historic and current habitat use by North Pacific right whales Eubalaena japonica in the Bering Sea and Gulf of Alaska. *Mammal Review* 35(2):129–55.

Shelton AO, Hunsicker ME, Ward EJ, Feist BE, Blake R, Ward CL, Williams BC, et al. 2017. Spatio-temporal models reveal subtle changes to demersal communities following the Exxon Valdez oil spill. *ICES Journal of Marine Science* 75(1):287–97.

Short JW, Irvine GV, Mann DH, Maselko JM, Pella JJ, Lindeberg MR, Payne JR, Driskell WB, Rice SD. 2007. Slightly weathered Exxon Valdez oil persists in Gulf of Alaska beach sediments after 16 years. *Environmental Science & Technology* 41(4):1245–50.

Siddon E, Zador S. 2018. *Ecosystem Status Report 2018 Eastern Bering Sea*. Anchorage, AK: North Pacific Fishery Management Council (p. 229).

Siddon EC, De Forest LG, Blood DM, Doyle MJ, Matarese AC. 2016. Early life history ecology for five commercially and ecologically important fish species in the eastern and western Gulf of Alaska. *Deep Sea Research Part II: Topical Studies in Oceanography* 165(6):7–25.

Siders A, Stanley R, Lewis KM. 2016. A dynamic ocean management proposal for the Bering Strait region. *Marine Policy* 74:177–85.

Sigler MF, Hollowed AB, Holsman KK, Zador S, Haynie AC, Himes-Cornell AH, Mundy PR, et al. 2016. *Alaska Regional Action Plan for the Southeastern Bering Sea: NOAA Fisheries Climate Science Strategy. NOAA Technical Memorandum NMFS-AFSC 336*. Washington, DC: NOAA.

Sigler MF, Lunsford CR, Fujioka JT, Lowe SA. 2003. Alaska Sablefish Assessment for 2003. Stock Assessment and Fishery Evaluation Report for the Groundfish Resources of the Gulf of Alaska as Projected for 2004, November 2003 Plan Team Draft. Washington, DC: NOAA (pp. 229–94).

Sigler MF, Cameron MF, Eagleton MP, Faunce CH, Heifetz J, Helser TE, Laurel BJ, et al. 2012. *Alaska Essential Fish Habitat Research Plan: A Research Plan for the National Marine Fisheries Service's Alaska Fisheries Science Center and Alaska Regional Office. AFSC Processed Rep. 2012-06* (p. 21) Washington, DC: Alaska Fish Science Centre, NOAA.

Sinclair EH, Zeppelin TK. 2002. Seasonal and spatial differences in diet in the western stock of steller sea lions. *Journal of Mammalogy* 83(4):973–90.

Slizkin AG. 1989. *Tanner Crabs (Chionoecetes Opilio, C. Bairdi) of the Northwest Pacific: Distribution, Biological Peculiarities, and Population Structure. Proceedings of the International Symposium on King and Tanner Crabs. Alaska Sea Grant College*

Program Report No. 90-04. Fairbanks, AK: University of Alaska Fairbanks (pp. 27–34).

Somerton DA, McConnaughey RA, Intelmann SS. 2017. Evaluating the use of acoustic bottom typing to inform models of bottom trawl sampling efficiency. *Fisheries Research* 185:14–16.

Spalinger K. 2015. *Operational Plan: Large-Mesh Bottom Trawl Survey of Crab and Groundfish: Kodiak, Chignik, South Peninsula, and Eastern Aleutian Management District-Standard Protocol, 2015–2019. Regional Operational Plan ROP.CF.4K. 2015.20*. Kodiak, AK: Alaska Department of Fish and Game.

Springer AM, Estes JA, Van Vliet GB, Williams TM, Doak DF, Danner EM, Forney KA, Pfister B. 2003. Sequential megafaunal collapse in the North Pacific Ocean: an ongoing legacy of industrial whaling? *Proceedings of the National Academy of Sciences* 100(21):12223–8.

Springer AM, Estes JA, Van Vliet GB, Williams TM, Doak DF, Danner EM, Pfister B. 2008. Mammal-eating killer whales, industrial whaling, and the sequential megafaunal collapse in the North Pacific Ocean: a reply to critics of Springer et al. 2003. *Marine Mammal Science* 24(2):414–42.

Springer E. 2011. "Historical transitions in access to and management of Alaska's commercial fisheries, 1880–1980." In: *World Fisheries: A Social-Ecological Analysis*. Edited by TJ Pitcher. Hoboken, NJ: Blackwell Publishing Ltd (pp. 291–309).

Srinivasan M, Brown SK, Markowitz E, Soldevilla M, Patterson E, Forney K, Murray K, et al. 2019. *Proceedings of the 2nd National Protected Species Assessment Workshop. NOAA Technical Memorandum NMFS-F/SPO-198*. Washington, DC: NOAA (p. 46).

Stabeno PJ, Schumacher JD, Ohtani K. 1999. "The physical oceanography of the Bering Sea." In: *Dynamics of the Bering Sea: A Summary of Physical, Chemical, and Biological Characteristics, and a Synopsis of Research on the Bering Sea*. Edited by TR Loughlin, K Ohtani. Sidney, Canada: North Pacific Marine Science Organization (pp. 1–28).

Stabeno PJ, Bond NA, Kachel NB, Salo SA, Schumacher JD. 2001. On the temporal variability of the physical environment over the south-eastern Bering Sea. *Fisheries Oceanography* 10:81–98.

Stabeno PJ, Bond NA, Hermann AJ, Kachel NB, Mordy CW, Overland JE. 2004. Meteorology and oceanography of the Northern Gulf of Alaska. *Continental Shelf Research* 24(7–8):859–97.

Stabeno PJ, Bell S, Cheng W, Danielson S, Kachel NB, Mordy CW. 2016. Long-term observations of Alaska Coastal Current in the northern Gulf of Alaska. *Deep Sea Research Part II: Topical Studies in Oceanography* 132:24–40.

Stevens BG, MacIntosh RA. 1991. *Cruise Results Supplement, Cruise OH-91-1. A Survey of Juvenile Red King Crabs with Dredge and Beam Trawl in Bristol Bay, Alaska*. Kodiak, AK: NMFS, Kodiak Fisheries Research Center (p. 14).

Stevenson DE, Lewis KA. 2010. Observer-reported skate bycatch in the commercial groundfish fisheries of Alaska. *Fishery Bulletin* 108(2):208–18.

Stevenson DE, Lauth RR. 2019. Bottom trawl surveys in the northern Bering Sea indicate recent shifts in the distribution of marine species. *Polar Biology* 42:407–21.

Stewart I, Hicks A. 2019. *Assessment of the Pacific halibut (Hippoglossus Stenolepis) Stock at the End of 2018. IPHC-2019-AM095-09*. Washington, DC: International Pacific Halibut Commission (p. 26).

Stockhausen WT, Coyle KO, Hermann AJ, Blood D, Doyle MJ, Gibson GA, Hinckley S, Ladd C, Parada C. 2018a. Running the gauntlet: connectivity between spawning and nursery areas for arrowtooth flounder (Atheresthes stomias) in the Gulf of Alaska, as inferred from a biophysical individual-based model. *Deep Sea Research Part II: Topical Studies in Oceanography* 165:127–39.

Stockhausen WT, Coyle KO, Hermann AJ, Doyle MJ, Gibson GA, Hinckley S, Ladd C, Parada C. 2018b: Running the gauntlet: Connectivity between natal and nursery areas for Pacific ocean perch (Sebastes alutus) in the Gulf of Alaska, as inferred from a biophysical individual-based model. *Deep Sea Research Part II: Topical Studies in Oceanography* 165:74–88.

Stone RP, Shotwell SK. 2007. *State of Deep Coral Ecosystems in the Alaska Region: Gulf of Alaska, Bering Sea and the Aleutian Islands. The State of Deep Coral Ecosystems of the United States. NOAA Technical Memorandum CRCP-3*. Washington, DC: NOAA (pp. 65–108).

Stone RP. 2006. Coral habitat in the Aleutian Islands of Alaska: depth distribution, fine-scale species associations, and fisheries interactions. *Coral Reefs* 25(2):229–38.

Stram DL, Evans DC. 2009. Fishery management responses to climate change in the North Pacific. *ICES Journal of Marine Science* 66(7):1633–9.

Stroeve J, Holland MM, Meier W, Scambos T, Serreze M. 2007. Arctic sea ice decline: faster than forecast. *Geophysical Research Letters* 34(9).

Stroeve J, Serreze M, Drobot S, Gearheard S, Holland M, Maslanik J, Meier W, Scambos T. 2007. Arctic sea ice extent plummets in 2007. *Eos, Transactions American Geophysical Union* 89(2):13–4.

Swartzman G, Huang C, Kaluzny S. 1992. Spatial analysis of Bering Sea groundfish survey data using generalized additive models. *Canadian Journal of Fisheries and Aquatic Sciences* 49(7):1366–78.

Sydeman WJ, Piatt JF, Thompson SA, García-Reyes M, Hatch SA, Arimitsu ML, Slater L, Williams JC, et al. 2017. Puffins reveal contrasting relationships between forage fish and ocean climate in the North Pacific. *Fisheries Oceanography* 26(4):379–95.

Talbot SL, Sage GK, Rearick JR, Fowler MC, Muñiz-Salazar R, Baibak B, Wyllie-Echeverria S, Cabello-Pasini A, Ward DH. 2016. The structure of genetic diversity in eelgrass (Zostera marina L.) along the north Pacific and Bering Sea coasts of Alaska. *PLoS One* 11(4): e0152701.

Tam JC, Link JS, Large SI, Andrews K, Friedland KD, Gove J, Hazen E, et al. 2017. Comparing apples to oranges: common trends and thresholds in anthropogenic and environmental pressures across multiple marine ecosystems. *Frontiers in Marine Science* 4:282.

Tasker ML, Camphuysen CJ, Cooper J, Garthe S, Montevecchi WA, Blaber SJ. 2000. The impacts of fishing on marine birds. *ICES Journal of Marine Science* 57(3):531–47.

Taylor RJ. 1981. Shoreline vegetation of the arctic Alaska coast. *Arctic* 1:37–42.

Thompson GG, Dorn MW, Gaichas S, Aydin K. 2003. *Assessment of the Pacific Cod Stock in the Eastern Bering Sea and Aleutian Islands Area. Plan Team for Groundfish Fisheries of the Bering Sea/Aleutian Islands (compiler), Stock Assessment and Fishery Evaluation Report for the Groundfish Resources of the Bering Sea/Aleutian Islands regions.* Washington, DC: NOAA (pp. 127–222).

Thompson SA, García-Reyes M, Sydeman WJ, Arimitsu ML, Hatch SA, Piatt JF. 2019. Effects of ocean climate on the length and condition of forage fish in the Gulf of Alaska. *Fisheries Oceanography* 28(6):658–71.

Thompson WF, Freeman NL. 1930. *History of the Pacific Halibut Fishery.* Vancouver, Canada: International Fisheries Commission.

Thompson GG, Dorn MW. 2005. *Assessment of the Pacific Cod Stock in the Gulf of Alaska. Stock Assessment and Evaluation Report for the Groundfish Resources of the Gulf of Alaska, November 2005, Section 2.* Anchorage, AK: North Pacific Fishery Management Council (p. 155–244).

Thorson JT, Adams G, Holsman K. 2019. Spatio-temporal models of intermediate complexity for ecosystem assessments: a new tool for spatial fisheries management. *Fish and Fisheries* 20(6):1083–99.

Thorson JT, Ianelli JN, Kotwicki S. 2017. The relative influence of temperature and size-structure on fish distribution shifts: a case-study on Walleye pollock in the Bering Sea. *Fish and Fisheries* 18(6):1073–84.

Thorson JT, Rindorf A, Gao J, Hanselman DH, Winker H. 2016. Density-dependent changes in effective area occupied for sea-bottom-associated marine fishes. *Proceedings of the Royal Society B: Biological Sciences* 283(1840):20161853.

Towns DR, Vernon Byrd G, Jones HP, Rauzon MJ, Russell JC, Wilcox C. 2011. "Impacts of introduced predators on seabirds." In: *Seabird Islands: Ecology, Invasion, and Restoration.* Edited by CPH Mulder, WB Anderson, DR Towns, PJ Bellingham. Oxford, UK: Oxford University Press.

Townsend H, Harvey CJ, deReynier Y, Davis D, Zador S, Gaichas S, Weijerman M, Hazen EL, Kaplan IC. 2019. Progress on implementing ecosystem-based fisheries management in the US through the use of ecosystem models and analysis. *Frontiers in Marine Science* 6:641.

Townsend HM, Harvey CJ, Aydin KY, Gamble RJ, Gruss A, Levin PS, Link JS, et al. 2014. *Report of the 3rd National Ecosystem Modeling Workshop (NEMoW 3): Mingling Models for Marine Resource Management, Multiple Model Inference. NOAA Technical Memorandum NMFS-F/SPO-173.* Washington, DC: NOAA.

Townsend HM, Link JS, Osgood KE, Gedamke T, Watters GM, Polovina JJ, Levin PS, Cyr EC, Aydin KY. 2008. *Report of the National Ecosystem Modeling Workshop (NEMoW).* Washington, DC: NOAA.

Townsend HK, Aydin K, Holsman C, Harvey I, Kaplan E, Hazen P, Woodworth-Jefcoats M. et al. 2017. *Report of the 4th National Ecosystem Modeling Workshop (NEMoW 4): Using Ecosystem Models to Evaluate Inevitable Trade-offs. NOAA Technical Memo NMFS-F/SPO-173.* Washington, DC: NOAA (p. 77).

Trites AW, Christensen V, Pauly D. 1997. Competition between fisheries and marine mammals for prey and primary production in the Pacific Ocean. *Journal of Northwest Atlantic Fishery Science* 22:173–87.

Trites AW, Donnelly CP. 2003. The decline of Steller sea lions Eumetopias jubatus in Alaska: a review of the nutritional stress hypothesis. *Mammal Review* 33(1):3–28.

Trites AW, Livingston PA, Mackinson S, Vasconcellos M, Springer AM, Pauly D. 1999. Ecosystem change and the decline of marine mammals in the Eastern Bering Sea: testing the ecosystem shift and commercial whaling hypotheses. *Fisheries Centre Research Reports* 7(1):106.

Trites AW, Livingston PA, Vasconcellos MC, Mackinson S, Springer AM, Pauly D. 1999. *Ecosystem Considerations and the Limitations of Ecosystem Models in Fisheries Management: Insights from the Bering Sea.* Fairbanks, AK: University of Alaska, Ecosystem Approaches for Fisheries Management (pp. 609–20).

Trites AW, Miller AJ, Maschner HD, Alexander MA, Bograd SJ, Calder JA, Capotondi A, et al. 2007. Bottom-up forcing and the decline of Steller sea lions (Eumetopias jubatus) in Alaska: assessing the ocean climate hypothesis. *Fisheries Oceanography* 16(1):46–67.

Urawa S, Sato S, Crane PA, Agler B, Josephson R, Azumaya T. 2009. Stock-specific ocean distribution and migration of chum salmon in the Bering Sea and North Pacific Ocean. *North Pacific Anadromous Fish Commission Bulletin* 5:131–46.

van Kirk KF, Quinn II TJ, Collie JS, A'Mar ZT. 2015. Assessing uncertainty in a multispecies age-structured assessment framework: the effects of data limitations and model assumptions. *Natural Resource Modeling* 28(2):184–205.

Van Pelt TI, Napp JM, Ashjian CJ, Harvey HR, Lomas MW, Sigler MF, Stabeno PJ. 2016. An introduction and overview of the Bering Sea Project: volume IV. *Deep Sea Research Part II: Topical Studies in Oceanography* 134:3–12.

Verna DE, Harris BP, Holzer KK, Minton MS. 2016. Ballast-borne marine invasive species: exploring the risk to coastal Alaska, USA. *Management of Biological Invasions* 7(2):199–211.

Vince SW, Snow AA. 1984. Plant zonation in an Alaskan salt marsh: I. distribution, abundance and environmental factors. *The Journal of Ecology* 1:651–67.

Von Szalay PG, Raring NW, Shaw FR, Wilkins ME, Martin MH. 2010. *Data Report: 2009 Gulf of Alaska Bottom Trawl Survey. NOAA Technical Memorandum NMFS-AFSC-208*. Washington, DC: NOAA.

Waite JM, Friday NA, Moore SE. 2002. Killer whale (Orcinus orca) distribution and abundance in the central and southeastern Bering Sea, July 1999 and June 2000. *Marine Mammal Science* 18:779–86.

Walsh JJ, Dieterle DA, Chen FR, Lenes JM, Maslowski W, Cassano JJ, Whitledge TE, et al. 2011. Trophic cascades and future harmful algal blooms within ice-free Arctic Seas north of Bering Strait: a simulation analysis. *Progress in Oceanography* 91(3):312–43.

Walsh JJ, McRoy CP. 1986. Ecosystem analysis in the southeastern Bering Sea. *Continental Shelf Research* 5:259–88.

Ward EJ, Adkison M, Couture J, Dressel SC, Litzow MA, Moffitt S, Neher TH, Trochta J, Brenner R. 2017. Evaluating signals of oil spill impacts, climate and species interactions in Pacific herring and Pacific salmon populations in Prince William Sound and Copper River, Alaska. *PLoS One* 12(3):e0172898.

Watson JT, Haynie AC. 2018. Paths to resilience: the walleye pollock fleet uses multiple fishing strategies to buffer against environmental change in the Bering Sea. *Canadian Journal of Fisheries and Aquatic Sciences* 75(11):1977–89.

Weingartner T, Eisner L, Eckert GL, Danielson S. 2009. Southeast Alaska: oceanographic habitats and linkages. *Journal of Biogeography* 36(3):387–400.

Wespestad VG. 1993. The status of Bering Sea pollock and the effect of the "Donut Hole" fishery. *Fisheries* 18(3): 18–24.

Whitehouse GA, Aydin K, Essington TE, Hunt GL. 2014. A trophic mass balance model of the eastern Chukchi Sea with comparisons to other high-latitude systems. *Polar Biology* 37(7):911–39.

Wiese FK, Wiseman WJ, Van Pelt TI. 2012. Bering sea linkages. *Deep Sea Research Part II* 65–70:2–5.

Wilderbuer T, Stockhausen W, Bond N. 2013. Updated analysis of flatfish recruitment response to climate variability and ocean conditions in the Eastern Bering Sea. *Deep Sea Research Part II: Topical Studies in Oceanography* 94:157–64.

Wilkinson EB, Abrams K. 2015. *Benchmarking the 1999 EPAP Recommendations with Existing Fishery Ecosystem Plans. NOAA Technical Memorandum NMFS-OSF-5*. Washington, DC: NOAA (p. 22).

Winstead NS, Colle B, Bond N, Young G, Olson J, Loescher K, Monaldo F, Thompson D, Pichel W. 2006. Using SAR remote sensing, field observations and models to better understand coastal flows in the Gulf of Alaska. *Bulletin of the American Meteorological Society* 87:787.

Witherell D, Pautzke C, Fluharty D. 2000. An ecosystem-based approach for Alaska groundfish fisheries. *ICES Journal of Marine Science* 57(3):771–7.

Witherell D, Woodby D. 2005. Application of marine protected areas for sustainable production and marine biodiversity off Alaska. *Marine Fisheries Review* 67(1):1–28.

Witherell D. 2004. *Managing Our Nations Fisheries: Past, Present and Future. Proceedings of a Conference on Fisheries Management in the United States Held in Washington, D.C. Nov. 2003*. Anchorage, AK: North Pacific Fisheries Management Council (p. 256).

Wolfe RJ, Walker RJ. 1987. Subsistence economies in Alaska: productivity, geography, and development impacts. *Arctic Anthropology* 1:56–81.

Wooley C. 2002. The myth of the "pristine environment": past human impacts in Prince William Sound and the Northern Gulf of Alaska. *Spill Science & Technology Bulletin* 7(1–2):89–104.

Yang Q, Cokelet ED, Stabeno PJ, Li L, Hollowed AB, Palsson WA, Bond NA, Barbeaux SJ. 2019. How "The Blob" affected groundfish distributions in the Gulf of Alaska. *Fisheries Oceanography* 28:434–53.

York A. 1994. The population dynamics of northern sea lions, 1975–1985. *Marine Mammal Science* 10(1):38–51.

Young R, Kitaysky A, Carothers C, Dorresteijn I. 2014. Seabirds as a subsistence and cultural resource in two remote Alaskan communities. *Ecology and Society* 19(4).

Zador S, Gaichas S. 2010. *Ecosystem Considerations for 2011. Groundfish Stock Assessment and Fishery Evaluation Report*. Washington, DC: NOAA.

Zador S, Ortiz I. 2018. *Ecosystem Status Report 2018 Aleutian Islands. North Pacific Fishery Management Council Bering Sea and Aleutian Islands Stock Assessment and Fishery Evaluation (SAFE) Report*. Washington, DC: NOAA (p. 130).

Zador S, Yasumiishi. 2018. *Ecosystem Status Report 2018 Gulf of Alaska. North Pacific Fishery Management Council Gulf of Alaska Stock Assessment and Fishery Evaluation (SAFE) Report*. Washington, DC: NOAA (p. 193).

Zador S. 2015. *Ecosystem Considerations 2015 Status of Alaska's Marine Ecosystems. North Pacific Fishery Management Council Ecosystem Considerations*. Anchorage, AK: North Pacific Fisheries Management Council (p. 296).

Zador SG, Gaichas SK, Kasperski S, Ward CL, Blake RE, Ban NC, Himes-Cornell A, Koehn JZ. 2017a. Linking ecosystem processes to communities of practice through commercially fished species in the Gulf of Alaska. *ICES Journal of Marine Science* 74(7):2024–33.

Zador SG, Holsman KK, Aydin KY, Gaichas SK. 2017b. Ecosystem considerations in Alaska: the value of qualitative assessments. *ICES Journal of Marine Science* 74(1):421–30.

Zerbini AN, Baumgartner MF, Kennedy AS, Rone BK, Wade PR, Clapham PJ. 2015. Space use patterns of the endangered North Pacific right whale Eubalaena japonica in the Bering Sea. *Marine Ecology Progress Series* 532:269–81.

Zerbini AN, Waite JM, Laake JL, Wade PR. 2006. Abundance, trends and distribution of baleen whales off Western Alaska and the central Aleutian Islands. *Deep Sea Research Part I* 53:1772–90.

Zheng J, Kruse GH. 2006. Recruitment variation of eastern Bering Sea crabs: climate-forcing or top-down effects? *Progress in Oceanography* 68(2–4):184–204.

Zhou S, Shirley TC. 1998. A submersible study of red king crab and Tanner crab distribution by habitat and depth. *Journal of Shellfish Research* 17:1477–9.

CHAPTER 10

The U.S. Western Pacific Region

➲ The Region in Brief

- The U.S. Western Pacific region composes over half (~51%) of the U.S. Exclusive Economic Zone (EEZ), including multiple remote archipelagos, and extends over much of the Western and Central Pacific Ocean basin.
- The Western Pacific contains the second-highest (among eight regions) number of managed taxa in U.S. waters, including commercially and recreationally important bottomfishes (e.g., emperors, snappers, groupers), pelagic fishes, crustaceans, corals, and coral reef-associated taxa.
- This region contains the seventh-highest domestic commercial fisheries landings and fisheries value in U.S. marine ecosystems (not accounting for all highly migratory species in international ports), with pronounced variability in landings over time.
- This region contains the second-highest number of fish stocks in the nation, a medium number and percentage of which are overfished or experiencing overfishing, and the second-highest percentage of stocks of unknown overfished status.
- The most expansive fishing closures are found in the Western Pacific region, with nearly one-third of its EEZ permanently protected from fishing.
- Human population and population density are second-lowest nationally in this region, with values having steadily increased over time. Human activities in this region including fishing, energy exploration, and high coastal tourism continue to exert heavy pressures on marine habitats.
- The Western Pacific has a relatively mid to low-range socioeconomic status among all U.S. regions. Contributions of living marine resources (LMRs) to total ocean economies are fifth to seventh-highest nationally in this region, particularly in the Hawaii subregion.
- With clearer waters that facilitate the growth of coral reefs, and a large diversity of productivity regimes (e.g., upwellings, oceanographic gyres), basal ecosystem pelagic productivity in the Western Pacific region is among the lowest in the nation.
- Within this region and its subregions, increases in average sea surface temperatures have been observed at >0.6°C since the mid-twentieth century, which are among the lowest observed in U.S. waters.
- Large expanses of fringing and offshore coral reefs support economically and ecologically important species throughout their life histories.
- The third-highest numbers of marine mammal stocks (of which approximately one-sixth are strategic stocks) in the nation and the second-highest numbers of threatened/endangered species are found in this region. In addition, abundant seabird populations are found throughout the Western Pacific.
- The Western Pacific is fourth-highest in the nation in offshore oil production, while also containing one of the lowest numbers of wind energy areas among U.S. regions.
- Fisheries management among state, federal, and tribal entities has been focused upon rebuilding plans and spatial management for historically overfished or prohibited commercially and recreationally important species,

continued

Ecosystem-Based Fisheries Management: Progress, Importance, and Impacts in the United States. Jason S. Link and Anthony R. Marshak, Oxford University Press. © U.S. Department of Commerce, U.S. Government 2021. DOI: 10.1093/oso/9780192843463.003.0010

and is moving toward ecosystem approaches to management through implementing fishery ecosystem plans. Representation on the Western Pacific Regional Fishery Management Council (WPRFMC) has varied over time with oscillating dominance by the recreational/subsistence fishing sector.

- The Hawaiian Archipelago contributes more heavily toward fisheries landings, LMR revenue, and LMR employments, and tourism, while tourism is also pronounced in the Marianas archipelago.
- Overall, significant ecosystem-based fisheries management (EBFM) progress has been made in terms of implementing ecosystem-level planning and advancing knowledge of ecosystem principles. Additionally, some progress has been observed toward assessing risks and vulnerabilities to ecosystems through ongoing investigations into climate vulnerability and species prioritizations for stock and habitat assessments.
- This ecosystem is excelling in facets of its human environment and the quality of its governance system as related to the determinants of successful LMR management.

10.1 Introduction

Thoughts of the U.S. Western Pacific ecosystem bring to mind major fisheries species including Bigeye (*Thunnus obesus*) and Yellowfin (*Thunnus albacares*) tunas called *"ahi,"* Wahoo *"ono"* (*Acanthocybium solandri*), Dolphinfish *"mahimahi"* (*Coryphaena hippurus*), and members of the deep seven bottomfish complex such as Pink snapper *"opakapaka"* (*Pristipomoides filamentosus*)," Long-tail red snapper *"onaga"* (*Etelis coruscans*), and Ruby snapper *"ehu"* (*Etelis carbunculus*"). Visions of tropical sandy beaches, green sea turtles *"honu"* (*Chelonia mydas*), humpback whales (*Megaptera novaeangliae*), Hawaiian monk seals (*Monachus schauinslandi*), snorkeling and diving on coral reefs, surfing enormous waves, sport fishing, exploring atolls, and viewing abundant marine life come to mind when referring to this expansive region. All of these experiences are well accompanied by feasting on some *poke*, *lomi* salmon, and *laulau* or *poi*, ideally served at a Hawaiian luau. Sounds of hula dancing, crashing waves, plucks and strums on ukuleles, traditional sounds of a Chamorro belembaotuyan or a Samoan slit drum, and calls from numerous seabirds along the shores of its many archipelagos punctuate its serene settings, whose origins are told in Polynesian mythology. Hawaiian anthologies, such as those compiled by Mary Kawena Pukui, myths and tales of the Marianas including Louise Stout's *Kalou: A Legend of Saipan*, and novels and poems by Samoan author, Sia Fiegel (e.g., *The Girl in the Moon Circle*) are examples of the diverse literature from this region that vividly depict Pacific Island culture and describe its natural settings. Additional works that take place in this region include James Michener's *Hawaii* and Alan Brennert's *Molokai*. These culturally iconic images are deeply rooted in the marine environments that compose the Western Pacific ecosystem, especially its beaches, coral reef and open-ocean associated habitats, plus their dependent living marine resources (LMRs). These serve as the foundation of its significant tourism-based marine economy and as a major source for the livelihoods of its island communities.

Many studies have characterized major components of the Western Pacific ecosystem (Randall et al. 1993, Richmond et al. 2002, Brainard et al. 2005, Craig et al. 2005, Lundblad et al. 2006, Parrish & Baco 2007, Sabater & Tofaeono 2007, Miller et al. 2008, Ellison et al. 2009, Hoeke et al. 2009, Glazier 2011, Williams et al. 2011, Monaco et al. 2012, Richards et al. 2012, Franklin et al. 2013, Gove et al. 2013, Keener et al. 2013, Heenan et al. 2014, McCoy et al. 2016, PIFSC 2016, Heenan et al. 2017, Gove et al. 2019), its important and iconic fisheries species (e.g., tunas, billfishes, coastal pelagic fishes, deep-slope bottomfishes, precious corals; Yong & Wetherall 1980, Boehlert 1993, Craig et al. 1993, Grigg 1993, Pooley 1993, Hamnett & Pintz 1996, Haight & Dalzell 2000, Dalzell & Boggs 2003, Sabater & Tofaeono 2007, Brodziak et al. 2012, Yau et al. 2016, Oyafuso et al. 2016, 2017, Ault et al. 2018, Langseth et al. 2018), its protected species (e.g., sea turtles, cetaceans, dugongs, monk seals, and corals; Kenyon & Rice 1959, Eckert 1993, Richmond et al. 2002, Eldredge 2003, Baker & Johanos 2004, Antonelis et al. 2006, Goldberg et al. 2008, Ellison et al. 2009, Fulling et al. 2011, Dutton et al. 2014, Martien et al. 2014), and the sociocultural context in which humans use, enjoy, and interact with this marine ecosystem (Maragos 1986, 1993, Ansell et al. 1996, Harrison 2004, Kulbicki et al. 2004, Treloar & Hall 2005, Ingram et al. 2018). This part of the ocean, its resources, and its goods and services have played an important part in American history, including its prominent and strategic roles during American conflicts, and most notably during World War II (Leibowitz 1989, Van Dyke et al. 1996). The main features of the Western

Pacific ecosystem include the importance and connectivity of its coral reefs, high species richness, and the influences of the North and South Equatorial Currents and subtropical gyres (Tabata 1975, Sund et al. 1981, Miller & Crosby 1998, Rohmann et al. 2005, Parrish & Baco 2007, Di Lorenzo et al. 2008, Ellison 2009, Hoeke et al. 2009, Knowlton et al. 2010, Monaco et al. 2012, Gilman et al. 2016, Link & Marshak 2019). A large proportion of the biodiversity and marine economy of this region is associated with the nearshore and offshore coral reef environments, and offshore deep and pelagic habitats, which support ecologically and commercially important species (Haight & Dalzell 2000, Richmond et al. 2002, Turgeon et al. 2002, Dalzell & Boggs 2003, Brainard et al. 2005, Craig et al. 2005, Parrish & Baco 2007, Goldberg et al. 2008, Miller et al. 2008, Williams et al. 2011, Richards et al. 2012, Heenan et al. 2017, Oyafuso et al. 2017, Ault et al. 2018). As in the U.S. Caribbean, more recent investigations into the importance of deep corals, mesophotic reefs, and their diverse flora and fauna have emerged, especially as fisheries and other ocean uses continue expanding into less historically overexploited areas of its insular shelf (Parrish & Baco 2007, Hourigan 2009, Kahng et al. 2010, Wagner et al. 2011, Kane et al. 2014, Wagner et al. 2014, Pyle et al. 2016, Winston et al. 2017).

What follows is a brief background for this region, describing all the major considerations of this marine ecosystem. Then there is a short synopsis of how all the salient data for those factors were collected, analyzed, and used for LMR management. We then present an evaluation of key facets of ecosystem-based fisheries management (EBFM) for this region.

10.2 Background and Context

10.2.1 Habitat and Geography

For the purposes here, we term the Western Pacific region not as that geography *sensu strictu*, but more specifically portions of the U.S. waters and territories therein. The Western Pacific comprises the largest portion of the United States Exclusive Economic Zone (EEZ). The activities of U.S. fisheries certainly span the vast portions of the blue-water, open ocean of the Pacific Ocean basin. Though there is connectivity to international waters of the Western and Central North Pacific, and Central South Pacific with jurisdictions in regional fisheries management organizations (RFMOs, Chapter 11), those waters of the Western Pacific with particular U.S. jurisdiction are found within U.S. territorial waters surrounding these island chains. This includes the Insular Pacific-Hawaiian Large Marine Ecosystem (LME), the proposed Marianas LME, and other island archipelagos not part of any specific LME. The region is notably influenced by local and regional oceanographic processes that occur in large portions of the Pacific Basin, particularly influenced by the North and South Equatorial Currents and North and South Pacific Gyres (Tabata 1975, Carpenter 1998, Parrish & Baco 2007, Hoeke et al. 2010, Bell et al. 2013a, b; Fig. 10.1).

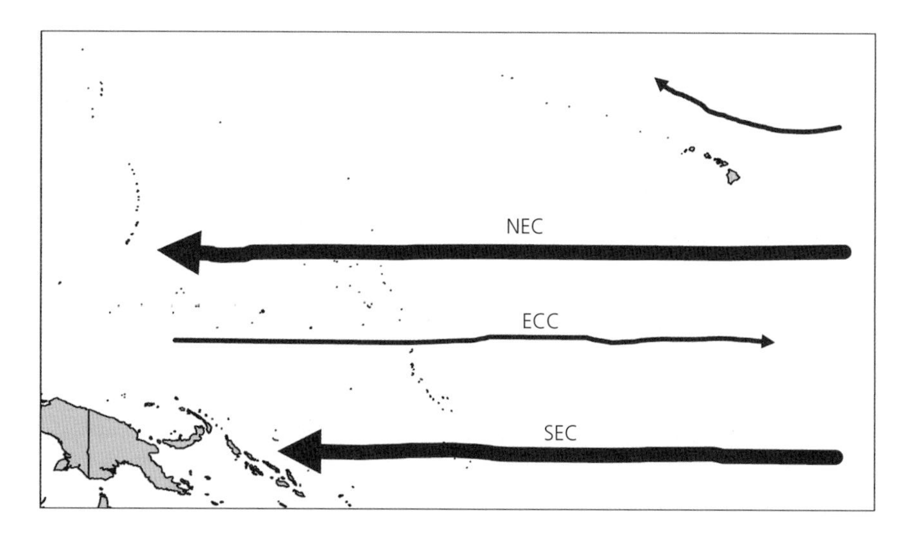

Figure 10.1 Major currents and circulation patterns throughout and encompassing the Western Pacific region and subregions. NEC = North Equatorial Current; SEC = South Equatorial Current; ECC = Equatorial Counter Current.

While overall low concentrations of human population occur in this region, notable domestic and international fisheries landings and economic production are found throughout the Western Pacific here we emphasize two key subregions: **the Hawaiian islands**—Hawaii and **U.S. Pacific island territories** (*American Samoa*¸ *Marianas islands*—Guam and Commonwealth of Northern Marianas islands, and *Pacific Remote Island areas*—Howland, Baker, and Jarvis islands, Kingman Reef, and Johnston, Midway, Wake, and Palmyra Atolls; Fig. 10.2). We acknowledge that we could have separated the territories further, but ultimately wanted to be inclusive in what we considered.

10.2.2 Climate and Other Major Stressors

Compounded with fishing (described next), there are other stressors affecting the Western Pacific ecosystem. These include global climate change and associated regional warming, hurricanes and typhoons, and habitat loss and degradation as a result of coastal development and concentrated human recreational pressures (Maragos 1986, 1993, Hunter 1995, Kodama & Businger 1998, Richmond et al. 2002, Gilman et al. 2006, Chowdhury et al. 2007, Baker et al. 2008, Schroeder et al. 2012, Keener et al. 2013, Jupiter et al. 2014, Lindfield et al. 2014, Williams et al. 2015, Gilman et al. 2016, PIFSC 2016, Shope et al. 2016, Gove et al. 2019). As observed worldwide, increasing water temperatures have led to more frequent coral bleaching events throughout the Western Pacific, causing mortality to reef-building organisms and increased degradation of reef habitats (Glynn 1993, Turgeon et al. 2002, Brainard et al. 2005, Craig et al. 2005, Baker et al. 2008, Hoeke et al. 2009, Heenan et al. 2017, Brainard et al. 2018, Eakin et al. 2019, Vargas-Ángel et al. 2019). Increasing prevalence and susceptibility of bleached corals to disease has also been observed with rising temperatures, particularly in more densely populated regions of the Western Pacific, with susceptibility varying among species (Richmond et al. 2002, Brainard et al. 2005, Craig et al. 2005, Vargas-Ángel 2010, Aeby et al. 2011, Maynard et al. 2015, Shore-Maggio et al. 2018). With continued regional warming and acidification of Western Pacific waters over time, the persistence of its coral reefs and calcified organisms, included harvested shellfish and macroinvertebrates, is of great concern (Carpenter et al. 2008, Bell et al. 2013a, b, Keener et al. 2013, Shinjo et al. 2013, Edmunds et al. 2014, Polovina et al. 2016). Prior to the 2015 bleaching event, and based upon observed rates of coral loss, Western Pacific coral cover was projected to decline by 65–85% over the next 20 years (Bruno & Selig 2007, Hoegh-Guldberg et al. 2011, Keener et al. 2013). However, these values are likely low estimates given more recent bleaching mortalities and temperature projections (I. Williams, pers. comm.). Climate impacts are predicted to significantly affect fisheries biomass for the Western Pacific, particularly for higher trophic levels and the Hawaii longline fishery, suggesting that reductions in fishing mortality may be necessary for future sustainable management (Pratchett et al. 2011, Bell et al. 2013, Howell et al. 2013, Nicol et al. 2013, Kapur & Franklin 2017, Bell et al. 2018, Woodworth-Jefcoats et al. 2019). Alterations to pelagic fish habitats as a result of shifting thermal boundaries and oceanographic properties are also anticipated, with projected shifts of tuna populations toward the central and eastern Pacific and away from current lucrative fishing grounds in the Western North Pacific (i.e., Marshall Islands and the Federated States of Micronesia (FSM); Bell et al. 2011, Keener et al. 2013, Howell et al. 2013, Woodworth-Jefcoats et al. 2017, Bell et al. 2018).

Moderate to severe effects from typhoons and hurricanes have been observed in this region (Kodama & Businger 1998, Saunders et al. 2000, Camargo & Sobel 2005, Link & Marshak 2019). Significant consequences to critical and essential habitats, coastal and inland populations, and state and territorial economies have occurred in recent decades from Hurricanes Iwa (1982), Iniki (1992), and Walaka (2018) in Hawaii, Cyclones Evan (2012) and Gita (2018) in American Samoa, and more recently Typhoon Yutu (2018) in the Mariana Islands, which was a category 5 superstorm (Polhemus 1993, Durocher 1994, Knaff et al. 2000, Keener et al. 2013, Dwyer 2018, Fakhruddin & Schick 2019, Cannizo et al. 2020). With regional warming, changes in storm tracks, frequency, and intensity are anticipated for this region, with tropical cyclone activity predicted to lessen in the Central South Pacific and significantly increase in the Western North Pacific (Emanuel 2005, Keener et al. 2013). While variable results have been projected for the eastern and central Pacific, including Hawaii, cyclone paths are also predicted to shift more toward the Central North Pacific with continued warming (Li et al. 2010, Keener et al. 2013). In addition to the effects from storms, and as likewise observed for the U.S. Caribbean, deforestation and degradation of essential nursery habitats such as mangroves, seagrass beds, and coral reef backreef regions have exacerbated recruitment limitations and affected Western Pacific biological communities

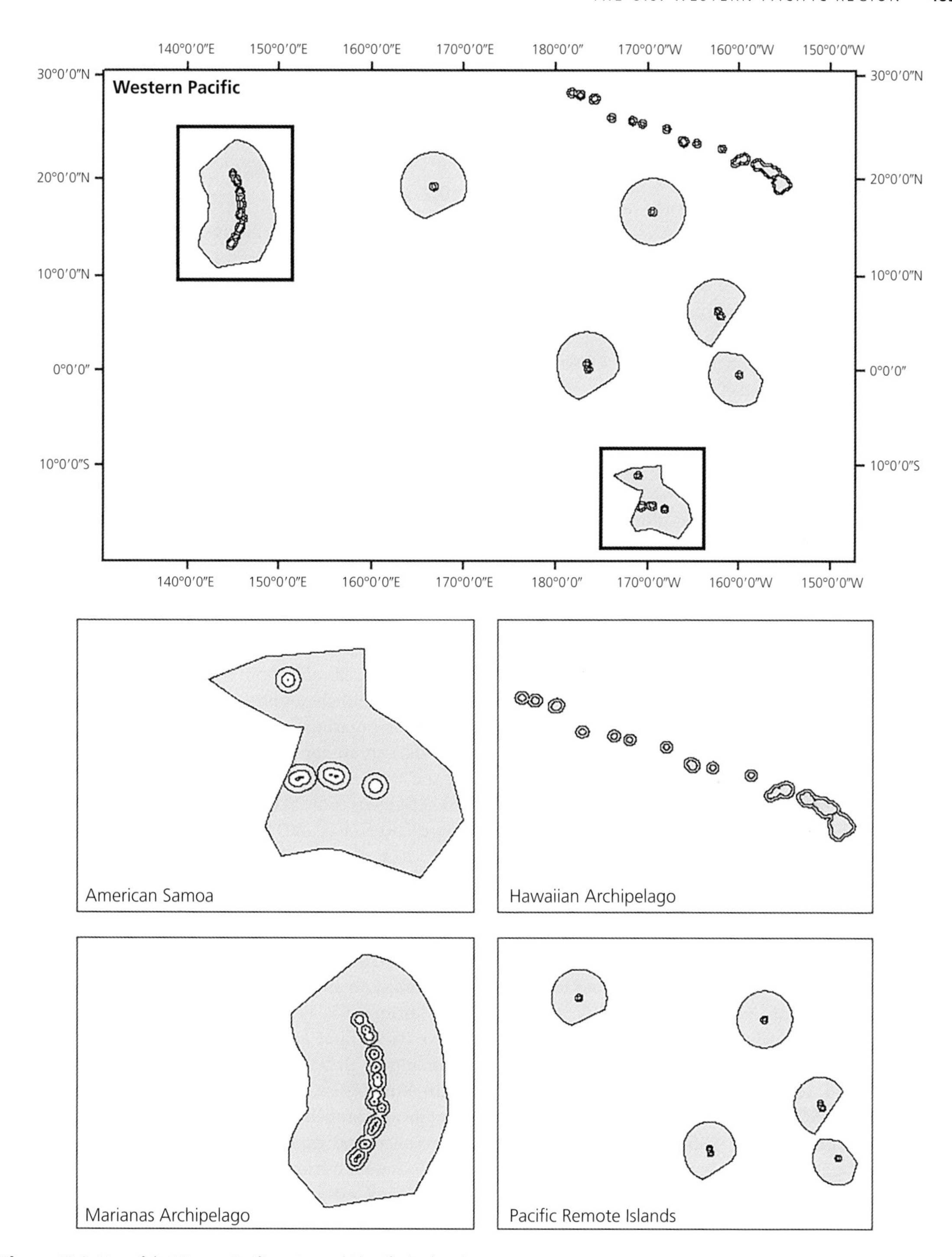

Figure 10.2 Map of the Western Pacific region and identified subregions.

(Maragos 1993, Miller & Crosby 1998, Richmond et al. 2002, Turgeon et al. 2002, Gilman et al. 2006, Baker et al. 2008, Ellison 2009, Keener et al. 2013, Gilman et al. 2016), but note that mangroves are introduced invasive species considered a nuisance in the Hawaiian Islands (Chimner et al. 2006). These impacts are particularly concentrated in populated and urbanized subregions, with increasing coastal development having resulted in high nutrient and sediment runoff in recent decades, compounded with the consequences of nearshore coral reef exploitation (Maragos 1993, Miller & Crosby 1998, Richmond et al. 2002, Turgeon et al. 2002, Craig et al. 2005, Gilman et al. 2006, Ellison 2009, Keener et al. 2013, Gilman et al. 2016).

Further losses of vegetated habitats are anticipated to occur with climate change, sea level rise, and heightened cyclone and storm severity, due to consequential increases in sedimentation, freshwater flow, nutrient loading, and habitat damage (Gilman et al. 2006, Ellison 2009, Waycott et al. 2011, Bell et al. 2013, Keener et al. 2013, Gilman et al. 2016). Increasing sea level may lead to erosion, inundation, and continued loss of Western Pacific mangroves and intertidal flats, especially in highly developed coastal areas that constrain their landward migration and where sedimentation rates are limited (Gilman et al. 2006, 2008, Keener et al. 2013, Gilman et al. 2016, Veitayaki et al. 2017). These factors along with changes in rainfall and local oceanography may lead to increased turbidity and phytoplankton blooms, reducing light availability for corals and seagrass beds, and enhancing the effects of runoff, nutrients, land-based sources of pollution, and destructive practices on Western Pacific nearshore habitats (Miller & Crosby 1998, Short & Neckles 1999, Turgeon et al. 2002, Craig et al. 2005, Porter et al. 2005, Björk et al. 2008, Keener et al. 2013). Cumulative human recreational pressures and heavily concentrated tourism have also significantly affected the quality of these habitats and their associated LMRs, especially in coral reefs, while also increasing recreational fishing pressures on coral reef and pelagic fisheries species (Turgeon et al. 2002, Craig et al. 2005, Porter et al. 2005, Morrison 2012, Zeller et al. 2015, Makaiau et al. 2016). Of particular concern is damage by boats, jet skis, and anchors to mangroves, seagrasses, corals, and their associated communities, physical impacts from highly concentrated SCUBA divers and snorkelers, the effects of sunscreen chemicals on coral species, and the ecological consequences of recreational spearfishing on reef and bottom fishes and their spawning aggregations (Turgeon et al. 2002, Craig et al. 2005, Porter et al. 2005, Morrison 2012, Lindfield et al. 2014, Makaiau et al. 2016). Within Western Pacific offshore benthic habitats, the increasing importance of deep-sea mining also poses emerging threats to its mesophotic and deep coral communities (Rocha et al. 2018).

10.2.3 Exploitation

Concentrated commercial, recreational, and artisanal fishing in the Western Pacific has led to the depletion of several species in pelagic and shallow coral reef habitats, especially surrounding the main Hawaiian islands (MHI; Craig et al. 1993, Smith et al. 1993, Miller & Crosby 1998, Turgeon et al. 2002, Zeller et al. 2006, Sabater & Carroll 2009, Kinch et al. 2010, Williams et al. 2011, Fenner 2014, Nadon et al. 2015, Robinson et al. 2017). For those species that have been assessed, some overfished statuses have continued, while progress toward stock rebuilding from fishing closures and reductions in fishing effort has been observed for other pelagic species (Craig et al. 1993, Miller & Crosby 1998, Stevens et al. 2000, Friedlander et al. 2003, Turgeon et al. 2002, Zeller et al. 2006, Sabater & Carroll 2009, Kinch et al. 2010, Williams et al. 2011, Fenner 2014). Many nearshore fisheries resources surrounding populated areas of Hawaii are currently depleted, with fish and shellfish abundances significantly lower than those surrounding the Northwestern Hawaiian Islands (NWHI) and other protected and less populated areas of the Western Pacific (Smith et al. 1993, Gulko et al. 2000, Stevens et al. 2000, Friedlander & DeMartini 2002, Friedlander et al. 2003, 2007, Grigg et al. 2008, Williams et al. 2008, Friedlander et al. 2014, Nadon et al. 2015). The ecological consequences from fishing, including the loss of apex predators, can have cascading effects on lower trophic levels and diminish overall coral reef integrity. This situation has been observed throughout urbanized areas of the Western Pacific (Friedlander & DeMartini 2002, Friedlander et al. 2003, Gulko et al. 2000, Pandolfi et al. 2005, Friedlander et al. 2007a, b, Williams et al. 2008, Edwards et al. 2014, Nadon et al. 2015, Robinson et al. 2017). A recent assessment found that the bottomfish populations for American Samoa are overfished and experiencing overfishing, while those off Guam are currently overfished, and that Northern Mariana Islands populations are sustainable (Langseth et al. 2019), with all areas experiencing some reductions in catches (Hamnett & Pintz 1996, Zeller et al. 2006, Moffit et al. 2007, Zeller et al. 2007, Brodziak et al. 2012, Yau et al. 2016). Declines in domestic fisheries landings for pelagic species in American Samoa have also been documented over time (Craig et al. 1993, Campbell & Hand 1999, Hamnett & Pintz 1996, Zeller et al. 2006).

As observed for the U.S. Caribbean, data limitations together with historically intensive fishing, differing perceptions by stakeholders regarding the status of local and cross-basin pelagic fisheries, and mixed stakeholder opinion regarding the causes of fishery population declines have led to varied receptions toward fisheries management in the Western Pacific (Gilman 1997, Leung et al. 1998, Allen & Bartram 2008, Glazier 2011, Reiser 2011, Toonen et al. 2011, Williams et al. 2011, 2012, De Santo 2013, Edwards et al. 2014, Weijerman et al. 2015a, b, 2016, 2018, Link & Marshak 2019). Major conservation efforts toward establishing large no-take zones in this region have been implemented in remote areas. Despite little to no negative published impacts of these closures (aka monuments) to fisheries (Chan 2020, Lynham et al. 2020), fishing communities have expressed concern over their large size and perceived restricted access to fisheries resources, or have opposed additional closures near populated regions or traditionally exploited areas (Tissot et al. 2009, Jokiel et al. 2011, Kittinger et al. 2011, Reiser 2011, De Santo 2013, Richmond & Kotowicz 2015, Ayers et al. 2018, Chan 2020, Lynham et al. 2020). Some sharks, tunas, and billfishes have been relatively well-managed based upon available information, while other species continue to experience overfishing or overfished statuses (Craig et al. 1993, Haight & Dalzell 2000, Stevens et al. 2000, Dalzell & Boggs 2003, Dalzell et al. 2008, Link & Marshak 2019). However, several stocks continue to be overfished, subject to overfishing, and many are also unassessed (NMFS 2017a, Link & Marshak 2019, NMFS 2020). Other types of non-commercial fishing (i.e., recreational, cultural, subsistence) and the degree of impact of increasingly important commercial and tourism-related sportfishing activities on pelagic and coral reef fisheries populations remains generally unassessed outside of Hawaiian waters (Craig et al. 1993, Haight & Dalzell 2000, Dalzell & Boggs 2003, Porter et al. 2005, Zeller et al. 2006, 2007, Lindfield et al. 2014, Zeller et al. 2015, PIFSC 2016, Gove et al. 2019). While improving, limited historical interactions among stakeholders, including Pacific Islander communities, and commercial and recreational fishing interests has hindered advancements toward more formalized, large-scale holistic management approaches at the federal level (Gilman 1997, Wright et al. 2006, Jokiel et al. 2011, Levine et al. 2016, PIFSC 2016, Ayers 2018, Kleiber & Kotowicz 2018, Chan & Pan 2019, Gove et al. 2019). This need for improved and ongoing outreach and stakeholder engagement serves as a major priority for fisheries management interests as they move toward implementing ecosystem-based management (EBM) and EBFM. In addition to tenuous relationships, limited resources have also led to data limitations throughout the Western Pacific, with many stocks continuing to remain unassessed, with limited communication of accurate accounting of fishing trips, sites, catches, and harvest by fishers to managers (Gilman 1997, Wright et al. 2006, Allen & Bartram 2008, Jokiel et al. 2011, Kleiber & Kotowicz 2018, Chan & Pan 2019, Link & Marshak 2019). Although efforts to improve ecosystem assessments have been implemented (NMFS 2019b), information regarding status and trends remains needed at broader scales. As observed for several other regions, influence from fishing organizations and their representatives in political and regulatory fora, especially during WPRFMC meetings, remains highly significant, and responses to past proposed closures and monument expansions have been met with resistance. Yet in most cases the proposed closures and expansions were established anyway. Given the historical importance of fishing and seafood to many coastal communities, its complementary demand by tourists, and the pronounced cultural identity, knowledge, and pride associated with fishing and Western Pacific LMRs, sensitivity to perceived U.S. governmental infringements on access to marine resources is of great concern when drafting regional management approaches.

In less populated regions, significant conservation of large portions of the U.S. Western Pacific through the establishment of marine national monuments, marine protected areas, fishing closures, and marine sanctuaries has led to enhanced protections of coral reef fisheries species and the critical habitats of marine mammals, seabirds, and sea turtles (Friedlander et al. 2003, Craig 2006, Mora et al. 2006, Williams et al. 2008, Kittinger et al. 2011, Rieser 2011, Williams et al. 2011, Young et al. 2015, Friedlander et al. 2016, Boyle et al. 2017, Naro-Maciel et al. 2018, Oyafuso et al. 2020). However, these protections have had mixed impacts on pelagic fisheries (Chan 2020, Lynham et al. 2020). Additional fisheries closures and marine protected areas, including Marine Life Conservation Districts (MLCDs) throughout much of the MHI, have marginally contributed to minimizing the effects of fisheries and human impacts on essential fish habitat (EFH), or in conserving reef fish assemblages, ornamental species, or promoting recovery of coral reefs given their small relative areas (Grigg 1993, Friedlander 2001, Friedlander et al. 2003, Williams et al. 2006, Friedlander et al. 2007a, b, Tissot et al. 2009, Friedlander et al. 2010). Further, they are not immune to human effects and overcrowding concerns (Filous et al. 2017), make up a

A Bad Reefer: Coral Reefs, Human Effects, and Responses to Climate Change

—(c.f. chapters 5, 6 & 7)

Imagine soaking in a warm bathtub, with happy colors swirling around you, gentle waters caressing you as you float without a care in the world. Beautiful shapes and forms and hues of life drift by, delighting you by their very audacity of existing in such surprising and unexpected splendor. You close your eyes to take it all in, to cement the memory of such views. Then you open your eyes again, and all you see is bleached-out gray with scum everywhere and little sign of life. The accompanying soundtrack is the proverbial needle scratching a record...what has happened to our imaginary scenario?

One of the most iconic images of the ocean is a tropical coral reef, teeming with sea-life expressing a myriad of colors. Coral reefs are amazing ecosystems, exhibiting a very prominent example of mutualism, where living photosynthetic algal cells (zooxanthellae) symbiotically reside inside coral tissues, providing them with ~90% of their energy needs. These calcareous reefs provide structure and habitat for a wide range of organisms, within which species are highly interconnected. Even the physical properties and chemical processes of the ocean are changed by the presence of these reefs, for example serving as natural breakwaters and producing net organic carbon and calcification (Takeshita et al. 2018). They remain remarkable and unique living habitats. This is true for both iconic tropical coral reefs as well as mostly inaccessible deep-sea corals. Although coral reefs represent only <1% of the world's ocean habitats, they contain over >30% of its biodiversity (Smith 1978, McAllister 1991, Copper 1994, Spalding & Grenfell 1997, Moberg & Folke 1999, Holbrook et al. 2015).

In an ecosystem context, the benefits of coral reefs are too numerous to list exhaustively here. Several summaries have noted these (Moberg & Folke 1999, Lesser et al. 2009, Barbier et al. 2011, Graham & Nash 2013, Micheli et al. 2014, Brander et al. 2015). In a regional economic context, coral reefs also have significant benefits. Various ecosystem goods and services, including protecting shorelines, providing habitat for marine life, preventing erosion, buffering water flows, etc. all have major economic value. That does not include the future benefits from derived bioproducts like new medicines being developed from reef organisms. The world's coral reefs may provide goods and services worth $500 billion each year, with some recent estimates suggesting values up to $9.9 trillion (Costanza et al. 1997, 2014). This is an amazing figure for an environment that covers less than 1 percent of the Earth's surface. The benefits of reefs to a wide variety of living marine resources is rather clear. Many artisanal, ornamental, and commercial fisheries would not exist without corals, whose fisheries are valued at >$100 million. Furthermore, it is difficult to estimate the value of global reef-based tourism, but those figures start in the billions of dollars (Cesar et al. 2003).

As you may have realized by now, coral reefs are important. But they are in serious trouble.

Globally, there are over 800 species of reef building corals (Carpenter et al. 2008), with around 200 species in U.S. territorial waters (Riegl & Dodge 2008, Waddell & Clarke 2008). Most are indeed tropical, but there are many temperate deep-water corals in the U.S. as well. Of these, approximately 83 were petitioned for listing and 25 tropical and subtropical species are listed as endangered or threatened (Waddell &

Clarke 2008, Gregg 2019). Corals are managed and considered not only under the Endangered Species Act, but also the Coral Reef Conservation Act. There are also facets of other LMR legislation (e.g., habitat provisions in the Magnuson-Stevens Act) that afford allowances for corals. In all these mandates, the importance of corals is clearly recognized.

The threats to corals are growing. Climate change is an issue. Not only do we see increased temperature leading to stressors such as increasing prevalence of disease on corals, but the ocean is also increasing in its acidity (i.e., decreasing in pH). An increase in acidity does not bode well for organisms that make calcareous structures. Considering land-use in a watershed is also an issue. From mountains to reefs, the degree of pollution in a watershed can swarm reefs and result in increases to nutrient loading, which can cause other algae to grow that outcompete coral. Additionally, increased runoff associated with coastal development has led to higher terrigenous sedimentation in reefs and smothering of corals. While mostly minimized now, destructive dredging practices still occasionally occur with negative impacts on U.S. corals. Combined, it is estimated that these stressors have led to ~60% of all coral reefs experiencing bleaching (where the symbiotic zooxanthellae leave the coral hosts, stuttering coral growth and survivorship) around the world (Heron et al. 2016, Donner et al. 2017). Continued overfishing has also depleted many coral reef fish and invertebrate populations, including herbivores that serve to limit algal overgrowth on reefs, which further exacerbates effects of coral bleaching.

The issues facing corals are both local and global. Ignoring the issue of coral bleaching is not wise given the multiple benefits and sheer economic volume of coral reefs. This is why in a local ecosystem context the range of land-use trade-offs, mitigation measures, re-colonization and genetic engineering of corals, and related options need to be considered—as an entire system—to elucidate those solutions that are most tractable. At a global level, the scale and scope of climate change is something to consider for reasons more than just for corals. Yet corals represent an important "canary in the coal mine" of our progress on dealing with global change. The degree of mitigation, adaptation, and even evolution of coral reefs requires continued evaluation. Most people have an intuitive understanding that addressing these issues will benefit not only just the corals, but entire ecosystems.

References

Barbier EB, Hacker SD, Kennedy C, Koch EW, Stier AC, Silliman BR. 2011. The value of estuarine and coastal ecosystem services. *Ecological Monographs* 81(2):169–93.

Brander LM, Eppink FV, Schägner P, van Beukering PJ, Wagtendonk A. 2015. "GIS-based mapping of ecosystem services: the case of coral reefs." In: *Benefit Transfer of Environmental and Resource Values.* Edited by RJ Johnston, J Rolfe, RS Rosenberger, R Brouwer. Dordrecht, Netherlands: Springer (pp. 465–85).

Carpenter KE, Abrar M, Aeby G, Aronson RB, Banks S, Bruckner A, Chiriboga A, et al. 2008. One-third of reef-building corals face elevated extinction risk from climate change and local impacts. *Science* 321(5888):560–3.

Cesar H, Burke L, Pet-Soede L. 2003. *The Economics of Worldwide Coral Reef Degradation.* Gland, Switzerland: World Wildlife Federation (p. 23).

Copper P. 1994. Ancient reef ecosystem expansion and collapse. *Coral Reefs* 13(1):3–11.

Costanza R, d'Arge R, De Groot R, Farber S, Grasso M, Hannon B, Limburg K, et al. 1997. The value of the world's ecosystem services and natural capital. *Nature* 387(6630):253–60.

Costanza R, De Groot R, Sutton P, Van der Ploeg S, Anderson SJ, Kubiszewski I, Farber S, Turner RK. 2014. Changes in the global value of ecosystem services. *Global Environmental Change* 26:152–8.

Donner SD, Rickbeil GJ, Heron SF. 2017. A new, high-resolution global mass coral bleaching database. *PLoS One* 12(4).

Graham NA, Nash KL. 2013. The importance of structural complexity in coral reef ecosystems. *Coral Reefs* 32(2):315–26.

Gregg RM. 2019. *Listing of Coral Reef Species Under the U.S. Endangered Species Act.* Climate Adaption Knowledge Exchange (CAKE). https://www.cakex.org/case-studies/listing-coral-reef-species-under-us-endangered-species-act

Heron SF, Maynard JA, Van Hooidonk R, Eakin CM. 2016. Warming trends and bleaching stress of the world's coral reefs 1985–2012. *Scientific Reports* 6:38402.

continued

A Bad Reefer: *Continued*

Holbrook SJ, Schmitt RJ, Messmer V, Brooks AJ, Srinivasan M, Munday PL, Jones GP. 2015. Reef fishes in biodiversity hotspots are at greatest risk from loss of coral species. *PLoS One* 10(5).

Lesser MP, Slattery M, Leichter JJ. 2009. Ecology of mesophotic coral reefs. *Journal of Experimental Marine Biology and Ecology* 375(1–2):1–8.

McAllister DE. 1991. What is a coral reef worth. *Sea Wind* 5(1):21–4.

Micheli F, Mumby PJ, Brumbaugh DR, Broad K, Dahlgren CP, Harborne AR, Holmes KE, et al. 2014. High vulnerability of ecosystem function and services to diversity loss in Caribbean coral reefs. *Biological Conservation* 171:186–94.

Moberg F, Folke C. 1999. Ecological goods and services of coral reef ecosystems. *Ecological Economics* 29(2):215–33.

Riegl BM, Dodge RE. 2008. *Coral Reefs of the USA*. Dordrecht, Netherlands: Springer.

Smith SV. 1978. Coral-reef area and the contributions of reefs to processes and resources of the world's oceans. *Nature* 273(5659):225–6.

Spalding MD, Grenfell AM. 1997. New estimates of global and regional coral reef areas. Coral Reefs 16(4):225–30.

Takeshita Y, Cyronak T, Martz TR, Kindeberg T, Andersson AJ. 2018. Coral reef carbonate chemistry variability at different functional scales. *Frontiers in Marine Science* 5:175.

Waddell JE, Clarke AM. 2008. *The State of Coral Reef Ecosystems of the United States and Pacific Freely Associated States: 2008. NOAA Technical Memorandum NOS NCCOS 73*. Silver Spring, MD: NOAA.

small proportion of MHI coastal waters, and have had minimal effects on Hawaiian coral reefs overall (I. Williams, pers. comm.). Efforts to prevent overfishing of pelagic fisheries species, and minimize bycatch of protected and prohibited species in pelagic longline gears, have also been enacted and led to significant declines in seabird bycatch (Haight & Dalzell 2000, Chapman 2001, Gilman 2006, Ovetz 2007, Wilson & Diaz 2012, Lowe et al. 2016, Ayers et al. 2018). However, progress in the assessment of many fisheries populations in the region has continued to lag due to the lack of reliable information to determine the status of targeted species.

10.2.4 Invasive Species

In addition to these direct effects on shallow coastal habitats, increasing prevalence of marine invasive species, particularly certain reef fishes and vegetated species, has raised concerns regarding their effects on native fauna and flora throughout this region (Friedlander et al. 2002, Smith et al. 2002, Schumacher & Parrish 2005, Demopoulous et al. 2007, Dierking et al. 2009, Ellison 2009, Giddens et al. 2014). Increasing abundance of bluestriped snapper (*Lutjanus kasmira*) on Hawaiian coral reefs has been observed since their introduction from the Marquesas in the late 1950s, with it having become the second most abundant species over hard substrata and dominating trap catches in all habitat types in one Hawaiian bay (Friedlander et al. 2002). This species has been found to compete with native goatfishes (Mullidae) for proximity to sheltering habitats (Schumacher & Parrish 2005). Peacock grouper (*Cephalopholis argus*) were also introduced to Hawaii from French Polynesia as a potentially new fishery, but have also raised concerns regarding their competitive superiority and ecological effects on local reef fish assemblages (Dierking et al. 2009, Meyer & Dierking 2011, Giddens et al. 2014, 2018). Additionally, their association with ciguatera fish poisoning has limited the development of a successful fishery that could reduce their populations, but spearfishing efforts toward their removal appear to be effective (Dierking & Campora 2009, Dierking et al. 2009, Giddens et al. 2014). Several species of algae have been introduced to Hawaii and Guam for commercial agriculture and inadvertently from heavily fouled ships and their ballast waters (Doty 1961, Stimson et al. 2001, Smith et al. 2002, 2004, Vermeij et al. 2009). These include two species of red algae (*Gracilaria salicornia, Kappaphycus* spp.), which have proliferated throughout intertidal and reef habitats since their introductions and expanded their coverage, including overgrowing live corals (Rodgers & Cox 1999, Smith et al. 2002, Conklin & Smith 2005). Efforts to control their spread, including through the introduction of herbivores, suggest that abundance of these invasive algae has decreased in more recent years within particular coastal systems (e.g., Kāne'ohe Bay, Oahu, Hawaii; Neilson et al. 2018, I. Williams pers. comm.). Red mangrove (*Rhizophora mangle*) has also invaded Hawaiian waters, which has altered coastal food webs, macrofaunal community structure, and facilitated additional invasions by exotic species (Demopoulous et al. 2007, D'iorio et al. 2007, Demopoulos & Smith 2010, Siple & Donahue 2013). Meanwhile, continued fishing on herbivores has also allowed for increased proliferation of native and invasive algae (Stimson et al. 2001, Vermeij et al. 2009, Edwards et al. 2014, Lindfield et al. 2014).

10.2.5 Ecosystem-Based Management (EBM) and Multisector Considerations

The Western Pacific ecosystem is known not only for its iconic, tropical species, diverse cultural traditions, and spectacular images, but also the trade-offs that arise across LMRs (and usage thereof) and between LMR management and other ocean uses (Leung et al. 1998, Kulbicki et al. 2004, Glazier 2011, Weijerman et al. 2015a, b, 2016, 2018, Link & Marshak 2019, Oyafuso et al. 2020). In the U.S. Pacific Islands, multiple coastal and ocean uses supporting marine economies co-occur along a mosaic of areas that provide habitat to many managed species and local fisheries, which serve as the foundation for a heavily concentrated marine tourism industry (Leung et al. 1998, Kulbicki et al. 2004, Glazier 2011, Wiejerman et al. 2015a, b, 2016, 2018, Link & Marshak 2019). Tourism and the U.S. military overwhelmingly contribute toward the bulk of this region's marine economy, while extensive fishing effort of reef-associated and pelagic species, and recreational sport-fishing, also contribute toward its LMR-dependent economy. Overlapping expansive fishing effort, ongoing coastal development, recreational boating, and abundant tourism throughout inshore and offshore regions where LMRs migrate and co-occur emphasizes the connectivity of this regional fisheries ecosystem and its other uses. For example, stressors to coral reef habitats occurring through tourism, recreational overuse, and their associated marine pollution compound the effects of fishing in these systems and can affect the long-term viability of fishers and their livelihoods. Through ecosystem-based approaches that directly account for resource conflicts and interconnectivity among fishing interests, coastal and tourism-related development LMR habitat quality, and spatial efforts and practices among diving, fishing, and recreational interests, these factors can be better addressed by being considered concurrently and cumulatively. Exploring, identifying, and resolving these differences among priorities and goals is the essence of EBFM (Craig 2006, Link 2010, Link & Browman 2014, Patrick & Link 2015, Link & Marshak 2019). With diverse taxa among multiple habitats, concentrated trade-offs among tourism and fisheries, historical effects of fishing, and increasingly concentrated human population pressures, the Western Pacific is a priority area for execution of ecosystem-based approaches to management.

In light of these multiple considerations, trade-offs, and dynamics, there is great utility in continuing to employ an ecosystem-based approach to systematically incorporate abiotic, biotic, and human factors into the current Western Pacific fisheries management framework. While efforts toward more comprehensive approaches, including subregional oceanic and island-based frameworks and fishery ecosystem plans (FEPs) have emerged, comprehensive fisheries management has generally remained restricted due to limited resources, continued intensive fishing, inadequate compliance with and enforcement of regulations, and focused interests on certain iconic species and complexes (Leung et al. 1998, Glazier 2011, Toonen et al. 2011, Lindfield et al. 2014, Weijerman et al. 2015a, b, 2016, 2018, Link & Marshak 2019). Fisheries management in the region remains challenging given the lack of sufficient data necessary to assess many fishery species, strong influence of the fishing community in local politics lobbying against legislation (i.e., recreational fishing licenses), government responses that have favored expanding fishing effort into previously unexploited deep-water regions, resistance toward conservative fishing measures, and the underassessed nature of local and subregional fisheries populations (Leung et al. 1998, Williams et al. 2008, Glazier 2011, Toonen et al. 2011, Williams et al. 2011, 2012, Wiejerman et al. 2015a, b, 2016, 2018, Link & Marshak 2019). The multispecies nature of this exploited system, emerging fishing practices targeting multiple trophic levels, inconsistent monitoring programs, transboundary pelagic fisheries management, and at times contentious cooperation among fishing and native communities with management entities create unique challenges for the region (Leung et al. 1998, Williams et al. 2008, Vermeij et al. 2009, Glazier 2011, Toonen et al. 2011, Williams et al. 2011, 2012, Edwards et al. 2014, Weijerman et al. 2015a, b, 2016, 2018, Link & Marshak 2019).

10.3 Informational and Analytical Considerations for this Region

10.3.1 Observation Systems and Data Sources

While considered data-limited overall, a regional observation system and several monitoring programs are available for informing aspects regarding the physical, biological, and socioeconomic components of the Western Pacific ecosystem and its subregions. The Pacific Islands Ocean Observing System (PacIOOS) provides oceanographic and atmospheric information from a network of data buoys, coastal meteorological stations, vessels, sensors, instruments, and radars throughout Hawaii, American Samoa, the Marshall Islands, the Federated States of Micronesia, Palau, Guam and the Northern Marianas islands (PacIOOS 2016). This system represents a composite of aggregated data from federal, state, academic, private, and industrial sources, including automated and manual shore stations, gliders, radar, buoys, *in-situ* water sampling data, ocean acidification sensors, and meteorological stations.

Containing sensors that have been collecting physical data at daily to annual scales since its pilot establishment as the Hawaii Ocean Observing System (HiOOS) program in 2007, this system monitorx multiple physical variables including atmospheric (e.g., air temperature, pressure, precipitation, humidity, winds, etc.) and oceanographic (e.g., water temperature, salinity, turbidity, chlorophyll, oxygen, pH, waves, currents, sea level, surge, carbon dioxide, etc.) parameters. This information is also being applied toward greater understanding of the Western Pacific ecosystem, to assisting in Tiger shark (*Galeocerdo cuvier*) tracking, estimating Hawaiian coral cover, facilitating investigations into the effects of ocean acidification, monitoring water quality, and in enhancing place-based habitat conservation efforts by informing about oceanographic processes within the NOAA West Hawaii Habitat Focus Area (HFA) and Hawaiian Islands Sentinel Site Cooperative. These efforts are also being applied toward producing regional and subregional oceanographic and climatological forecasts, maps, and models for the Western Pacific. Priorities for these observing systems also include assessing and predicting coastal hazards, improving water quality, tracking large predator species, and forecasting the effects of climate variability and change and rising sea level on marine ecosystems (PacIOOS 2016).

Multiple oceanographic and atmospheric data sources have been incorporated in enhanced understanding of the Western Pacific ecosystem, particularly its overall status, focal components, and applicability toward the West Hawaii Integrated Ecosystem Assessment and Ecosystem Status Reports (ESRs) that have been developed since 2016 (PIFSC 2016, Gove et al. 2019). As synthesized in these assessments, data regarding climate and ocean, ecological, and social indicators and ecosystem components have been obtained from several sources. Factors including large-scale climate forcing have been monitored using Pacific decadal oscillation (PDO) index data from the NOAA National Centers for Environmental Information (NCEI) and the Multivariate ENSO Index as monitored by NOAA's Earth System Research Laboratory (ESRL). Additional meteorological and oceanographic data are maintained by the University of Hawaii-Manoa Campus' Rainfall Atlas of Hawaii program (rainfall), the University of Hawaii Sea Level Center (sea level rise), the international Aviso geostrophic currents database (eddy activity and eddy kinetic energy, EKE), NOAA Coral Reef Watch and Division of Aquatic Resources (DAR) (sea surface temperature, thermal stress, bleaching activities), and the University of Hawaii International Pacific Research Center (wave forcing and wave power). Ocean productivity information is also synthesized from NASA Moderate Resolution Imaging Spectroradiometer (MODIS) and Visible Infrared Imaging Radiometer Suite (VIIRS) chlorophyll-a data as a proxy for phytoplankton biomass.

Multi-decadal surveys and monitoring of commercial catches, coral reef ecosystems, Pacific Islands oceanography, and more recent fishery-independent sampling of bottomfish have been undertaken by the NOAA Pacific Islands Fisheries Science Center (PIFSC) and DAR, the latter with an emphasis in the Hawaiian subregion (Brainard et al. 2002, Sundberg & Underkoffler 2011, Williams et al. 2012, Richards et al. 2016, PIFSC 2016, 2017, Lino et al. 2018, Swanson et al. 2019, Gove et al. 2019). Programs investigating reef fish community integrity have focused on factors including total fish abundance and biomass, mean length, species richness, and herbivore and target fish biomass have been examined since 2003 by the DAR West Hawaii Aquarium Project (Tissot et al. 2009, Gove et al. 2019). This program has also monitored trends in aquarium target fish populations, particularly the popular ornamental Yellow tang (*Zebrasoma flavescens*) over time. Factors regarding benthic coral reef community integrity are also monitored, such as hard coral and macroalgal cover and coverage of calcifying and non-calcifying species. These data have also been complemented by studies on spatiotemporal aspects of coral disease throughout West Hawaii (Couch 2014, Couch et al. 2014, PIFSC 2016, Gove et al. 2019).

Ongoing PIFSC-led efforts to survey Western Pacific shallow water coral reef communities also include the Pacific Reef Assessment and Monitoring Program (RAMP), which has mapped and monitored coral reef health, complexity, species diversity, oceanography, bathymetry, and reef fish populations since 2000 (Smith et al. 2011, Ayotte et al. 2011, 2015, Alin et al. 2015, Venegas et al. 2019). Surveys have taken place throughout Hawaii and U.S. Pacific Island territories documenting trends in reef fish size, abundance, and diversity, coral reef dynamics (Becker et al. 2019), bleaching episodes, and the effects of climate, disease, and human-related stressors throughout Western Pacific coral reefs. Collectively these efforts are part of NOAA's National Coral Reef Monitoring Program (NCRMP), which is the primary method to monitor status and trends of biological, climatic, and socioeconomic indicators for U.S. coral reef ecosystems (CRCP 2014). Water column surveys have been undertaken within some portions of the Western Pacific toward characterizing phytoplankton, zooplankton, and ichthyoplankton community composition and abundance, and for examining oceanographic factors that affect fisheries recruitment. Hawaii bottomfish fishery-independent surveys (BFISH) have also been conducted toward monitoring trends in size and abundance of "Deep Seven" bottomfish complex species over time. Data from these programs have been applied toward recent single-species reef fish stock assessments for many previously unassessed species, and toward greater understanding of species life histories. Monitoring and exploration of Western Pacific deep ocean reef habitats has occurred through programs such as the NOAA Deep Sea Coral Research and Technology Program (DSCRTP) and NOAA Campaign to Address Pacific monument Science, Technology, and Ocean Needs (CAPSTONE; Friedlander et al. 2016, Hourigan et al. 2017, Parke 2018, Kennedy et al. 2019). The CAPSTONE program was a three-year mission to explore the deep ocean biology, geology, and oceanography of four Pacific Islands monuments, including the subcomponents of the Pacific Remote Islands Marine National Monument, and the Hawaii Humpback Whale National Marine Sanctuary (NMS). Through this program, ~600 thousand km^2 of ocean floor was mapped and numerous biological and geological samplings took place toward identifying deep ocean coral, sponge, and other invertebrate communities (Kennedy et al. 2019). As a result of these efforts, hundreds of previously undescribed species were imaged and collected throughout this region. These surveys are also complemented by NOAA Fisheries fishery-dependent monitoring programs for commercial and recreational harvest, landings, revenue, and effort, including data on catches and bycatch gathered by the National Observer Program (Ito & Walsh 2011, NMFS 2017b, 2018b, Yau 2018, Ma 2019). Surveys undertaken by DAR and other territorial agencies also account for fishing effort throughout Pacific waters, with Hawaiian recreational effort coarsely estimated through DAR's Hawaii Marine Recreational Fishing Survey (Pooley 1993, Williams & Ma 2013, Nadon et al. 2015). Factors including trends in average size, spatial distributions, aggregate biomass of pelagic and demersal species groupings and their ratios, in addition to community assessments, represent many of the ecosystem-level metrics that have been derived from this fishery-dependent and independent survey information (PIFSC 2016, Lynch et al. 2018, Gove et al. 2019, Karp et al. 2019).

Socioeconomic information for fisheries and other ocean uses, particularly regarding revenue and employment, are also collected by NOAA Fisheries, the National Ocean Economics Program (NOEP), and NOAA's Office for Coastal Management Economics: National Ocean Watch (ENOW) program (Allen & Bartram 2008, Levine & Allen 2009, Hospital & Beavers 2014, Chan & Pan 2017, Kleiber & Leong 2018, NMFS 2018a, Link & Marshak 2019). Data with respect to residential and tourism-related population growth and human land-use patterns have been synthesized through the Hawaii Department of Business, Economic Development, and Tourism, U.S. Census, and NASA Socioeconomic Data and Applications Center (Dator et al. 1999, PIFSC 2016, Gove et al. 2019). Hawaiian coastal development metrics such as shoreline modifications, impervious surface area, and spatial distribution of on-site disposal systems, effluent, and flux have been examined using data from NOAA's Coastal Change Analysis Program (C-CAP), the State of Hawaii Department of Health, and LANDSAT satellite imagery. Factors including invasive algae and introduced fish species are also derived from NOAA's PIFSC, Hawaii's DAR, and the University of Hawaii's Coral Reef Assessment and Monitoring Program (CRAMP) surveys (Jokiel et al. 2001, Brown et al. 2004, Walsh et al. 2013, PIFSC 2016, Gove et al. 2019). Spatial data since 2003 regarding Hawaiian fishing pressure for commercial, non-commercial, and commercial aquarium catch, including per gear type and demersal/pelagic captures, are available from DAR and NOAA Fisheries surveys (Wedding et al. 2018, Gove et al. 2019). The PIFSC bottomfish heritage project has worked toward understanding and documenting fishing practices by the Hawaiian bottomfishing community, including cultural traditions, fishing techniques, exploring fisheries socioeconomics, and building on socioecological studies for the Pacific Islands (Kleiber & Leong 2018, Gove et al. 2019, Ingram et al. 2019, Leong et al. 2019).

Multidecadal monitoring programs exist for evaluating the status, recovery, and bycatch of protected species throughout the Pacific Islands that include surveys of Hawaiian monk seals, sea turtles, and the Hawaiian islands Cetacean and Ecosystem Assessment

Survey for monitoring whales and dolphins, while several monitoring programs for extensive seabird populations in this region are overseen by NOAA, the U.S. Department of Interior, and state and territorial agencies (Citta et al. 2007, Hill et al. 2014, Lopez et al. 2014, Martin et al. 2016, Bradford et al. 2017, Becker et al. 2019). In the Western Pacific region, all of these aforementioned data sources have been applied toward integrative ecosystem measurements that examine cumulative changes in these metrics over time and their interdependent relationships (Levine et al. 2009, PIFSC 2016, Gove et al. 2019).

10.3.2 Models and Assessments

Assessments of commercially important and protected species in the Western Pacific are conducted by NOAA Fisheries in collaboration with the WPRFMC, the U.S. Fish and Wildlife Service (FWS), DAR, and Pacific Islands territorial agencies (Heenan et al. 2014, Ferguson et al. 2017, Lynch et al. 2018, Srinivasan et al. 2019). Several pelagic species stock assessments are conducted by RFMOs, including the Western and Central Pacific Fisheries Commission (WCPFC), the Inter-American Tropical Tuna Commission (IATTC), and the International Scientific Committee (ISC) for tuna and tuna-like species in the North Pacific Ocean (see Chapter 11 on RFMOs). Stock assessment models are used to generate estimates of stock status with a multitude of data sources (e.g., landings, catch per unit effort, life history characteristics, fish condition, survey biomass, etc.). Assessments for Western Pacific species are reviewed thoroughly by scientists at PIFSC or RFMOs with subsequent review via the Western Pacific Stock Assessment Review (WPSAR) process. The WPRFMC Scientific and Statistical Committee, which sets acceptable biological catches (ABCs) for managed stocks, additionally provides input (Lynch et al. 2018, Karp et al. 2019). NOAA Fisheries also examines the effects of commercial and recreational fisheries on west coast marine socioeconomics through application of the NMFS Commercial Fishing Industry Input/Output and Recreational Economic Impact Models (NMFS 2018a). Data from NOAA Fisheries marine mammal stock abundance surveys are used to conduct population assessments as published in regional marine mammal stock assessment reports and peer reviewed by regional scientific review groups (SRGs; Carretta et al. 2019, Hayes et al. 2019, Muto et al. 2019). Information collected on protected species of the Pacific region is also applied toward five-year status reviews of those listed under the Endangered Species Act (ESA).

Available data and species assessments have also been applied toward the development of several broader ecosystem models for the Western Pacific region and its prominent subregions, while efforts to incorporate this information into single and multispecies stock assessments and LMR management decisions are also ongoing (Townsend et al. 2008, 2014, 2017). Major efforts have comprised the development of independent size-based food web and species-based Ecopath with Ecosim (EwE) ecosystem modeling projections of climate and fishing impacts in the Central North Pacific (Kitchell et al. 2002, Parrish et al. 2012, Howell et al. 2013, Townsend et al. 2017, Woodworth-Jefcoats et al. 2019). More recent models are driven by output from NOAA's Geophysical Fluid Dynamics Laboratory (GFDL) coupled climate and biogeochemical Earth System Models (ESM 2.1, ESM2G, and ESM2M), and have been used to examine responses across multiple trophic levels to fishing mortality and climatological-oceanographic processes. This information has been applied toward assessing fisheries yields for the Central North Pacific as related to size and species-specific fishing pressure, and to evaluate climate projections for this region (Townsend et al. 2017, Woodworth-Jefcoats et al. 2019).

More specific modeling efforts also include the development of trade-off analyses for coral reef ecosystems, pelagic fisheries systems, downscaling of climate models to island scales, and conceptual modeling of ecosystem services as related to the West Hawaii IEA. The Guam Atlantis Coral Reef Ecosystem model has been completed (Weijerman et al. 2015, Grafeld et al. 2016, Weijerman et al. 2016a, b, c, Townsend et al. 2017), and applied toward evaluating land- and marine-based management strategies as compounded with climate change. Trade-off analyses for Western Pacific coral reefs have been conducted using EwE platforms and the Coral Reef Scenario Evaluation Tool (CORSET) in examining alternative management strategies under future climate change scenarios (Kapur & Franklin 2017). Global climatological modeling frameworks such as the Climate Model Intercomparison Project (CMIP5, Taylor et al. 2012) have also been incorporated and downscaled toward examining the timing of severe bleaching conditions across the globe, particularly in the Western Pacific (van Hooidonk et al. 2016). EwE modeling efforts have also been applied toward evaluating the efficacy of marine protected areas such as the NWHI Marine National Monument toward recovery of Hawaiian monk seals, including the effects of climate change, PDO climate forcing, and fishing mortality on their populations and in identifying ongoing management actions in light of these factors (Parrish et al. 2012, Weijerman et al. 2017). Data from CMIP5 have also been applied toward examining the effects of climate-related changes in sand, sea surface, and air temperatures on Hawaiian sea turtle populations, with application toward evaluating trade-offs with fisheries. Larval reef fish connectivity has also been examined in the Hawaiian Islands with ocean circulation and dispersion models demonstrating a strong connection from Johnston Atoll to the Hawaiian Islands and a greater flux of recruits to the NWHI from the MHI and high self-recruitment (Kobayashi 2006, Wren et al. 2016). Potential application of these models toward understanding the effects of human use on nearby and distant reef fishes is also being examined (Townsend et al. 2017). More recent modeling applications in the Western Pacific include using conceptual models to identify locally relevant ecosystem drivers, pressures, and their impacts within West Hawaii (Ingram et al. 2018), and the development of an Atlantis model for the MHI. This Atlantis framework is focused on questions regarding monk seals and surrounding coral reef ecosystems. A species-based ecosystem model has been developed to examine species-specific trade-offs within the Hawaiian pelagic longline fishery as related to the impacts of climate forcing (Woodworth-Jefcoats et al. 2019). In addition to these noted examples, there are other models that can inform management discussions in this region and subregion (Gilman 1997, Campbell & Hand 1999, McDole et al. 2012, Bell et al. 2013, Gove et al. 2013, Newman et al. 2015, Weijerman et al. 2015b, Williams et al. 2015), yet few, if any, have been directly incorporated into the WPSAR process.

10.4 Evaluation of Major Facets of EBFM for this Region

To elucidate how well the Western Pacific ecosystem is doing in terms of LMR management and progress toward EBFM, here we characterize the status and trends of socioeconomic, governance, environmental, and ecological criteria pertinent toward employing an EBFM framework for this region, and examine their inter-relationships as applied to its successful implementation. We also quantify those factors that emerge as key considerations under an ecosystem-based approach for the region and for its major subregions of interest when possible. Ecosystem indicators for the Western Pacific region related to the (1) human environment and socioeconomic status, (2) governance system and LMR status, (3) natural environmental forcing and features, and (4) systems ecology, exploitation, and major fisheries of this system are presented and synthesized below. We explore all these facets to evaluate the strengths and weaknesses along the pathway:

$$PP \rightarrow B_{\text{targeted, protected spp, ecosystem}} \leftrightarrow L_{\uparrow\text{targeted spp, }\downarrow\text{bycatch}} \rightarrow \text{jobs, economic revenue.}$$

Where PP is primary production, B is biomass of either targeted or protected species or the entire ecosystem, L is landings of targeted or bycaught species, all leading to the other socioeconomic factors. This operates in the context of an ecological and human system, with governance feedbacks at several of the steps (i.e., between biomass and landings, jobs, and economic revenue), implying that fundamental ecosystem features can determine the socioeconomic value of fisheries in a region, as modulated by human interventions.

10.4.1 Socioeconomic Criteria

10.4.1.1 Social and Regional Demographics

In the U.S. Western Pacific region, coastal human population values and population densities (Fig. 10.3) have increased from 1970 to present with values currently around 1.6 million residents and 41.5 people km^{-2}, respectively. This fishery region is seventh-highest nationally (among eight regions) in human population, comprising 0.5% of United States residents (Year 2010), while density values have risen at ~0.47 people km^{-2} year^{-1}.

Among subregions, human population is most heavily concentrated in the State of Hawaii, within the MHI, particularly in the counties of Honolulu (Oahu), Hawaii (Hawaii), and Maui (Maui). Throughout the MHI, human population has increased over the past 40 years (~14,800 people year^{-1}), with currently up to ~1.4 million residents and ~48 people km^{-2}. Highest human population concentrations are found in Honolulu County (Oahu, MHI) at up to 172.9 people km^{-2} as of 2010, making it the most densely populated area throughout the Western Pacific region. Comparatively lower human populations are found throughout U.S. Pacific Island territories, with highest human concentrations occurring in the Marianas territory of Guam. Human population in Guam has

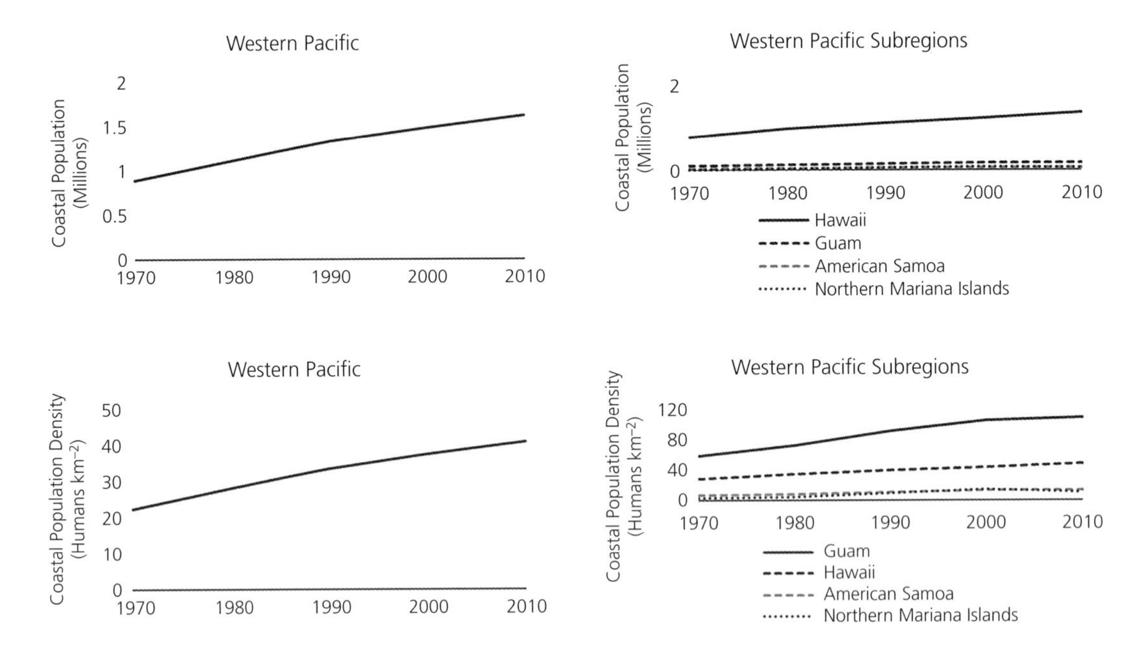

Figure 10.3 Decadal trends (1970–2010) in coastal human population (top) and population density (bottom; humans km^{-2}) for the Western Pacific region and identified subregions.

Data derived from U.S. censuses and taken from https://coast.noaa.gov/digitalcoast/data/demographictrends.html.

increased (~1,850 people $year^{-1}$) to currently ~160 thousand residents and 107.8 people km^{-2}. The population within the Commonwealth of the Northern Mariana Islands (CNMI) has increased (~1,100 people $year^{-1}$) to currently ~54 thousand residents and ~10.5 people km^{-2}, with most individuals living on the island of Saipan. Similar human population values are found in American Samoa, which has increased (707 people $year^{-1}$) to currently ~55.5 thousand residents and 12.7 people km^{-2}, with the majority of the population occurring in the Eastern and Western Districts.

10.4.1.2 Socioeconomic Status and Regional Fisheries Economics

Low to moderate numbers of LMR establishments (third-lowest nationally) and employments (lowest nationally) are found throughout the Western Pacific region. These numbers are only reported for the Hawaii subregion, where they are also presumed to be highest throughout the Western Pacific (142.5 ± 4.2 standard error (SE), establishments; 794.3 ± 8.7 SE employments; \$60.7 ± 2.1 million SE, USD), with marginal decreases observed in recent years (Fig. 10.4). Increased LMR-associated gross domestic product (GDP) values (~68.2 million USD as of 2014) over time have been observed in the Hawaii subregion. The percent contribution of the Western Pacific LMR-associated economy toward its regional ocean economy is seventh-highest nationally where LMR establishments have comprised up to 4.3% of regional ocean economy establishments and LMR employments have contributed only up to 0.8% of regional ocean employments over time. LMR-associated GDP has fluctuated over the past decade with an overall increase of ~\$12.6 million since 2005, and has contributed up to 1.1% toward the regional total ocean economy in recent years. While not reflected in national ocean economic data, total LMR values throughout protected and restored Hawaiian coral reef ecosystems have been estimated at up to \$34 billion (Bishop et al. 2011).

Although reported information on permitted vessels in the Western Pacific has been limited, and values included in Council Reports on fishing vessels have high uncertainty (PIFSC, pers. comms.), some interannual reported estimates are shown (Fig. 10.5). During the 1990s, relatively steady numbers of reported permitted fishing vessels operating in the Western Pacific were observed, which were followed by a near tripling of reported values in the 2000s. Values peaked in the mid to late 2000s at ~12 thousand vessels. During the late 2000s and 2010s, sharp

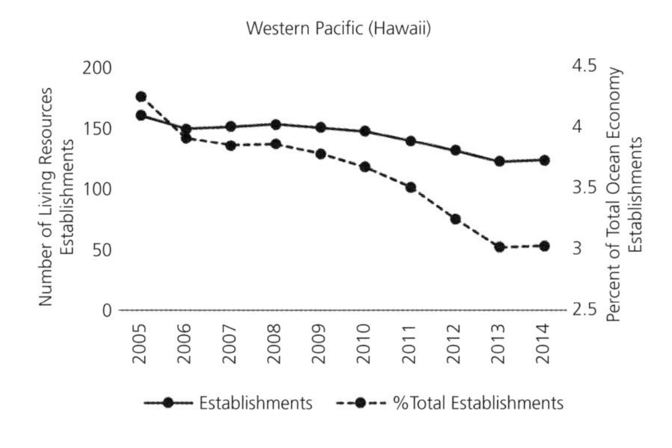

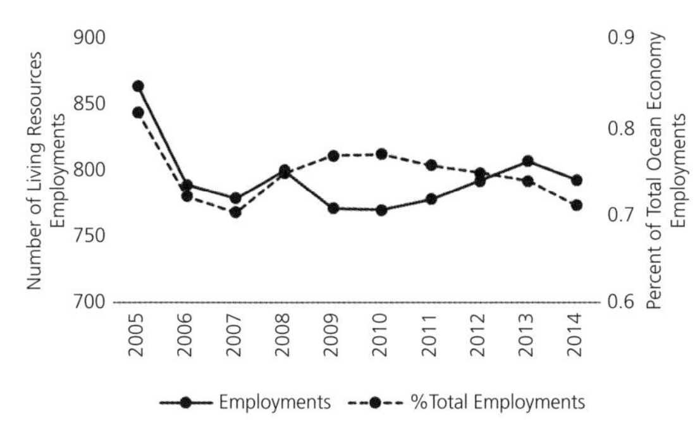

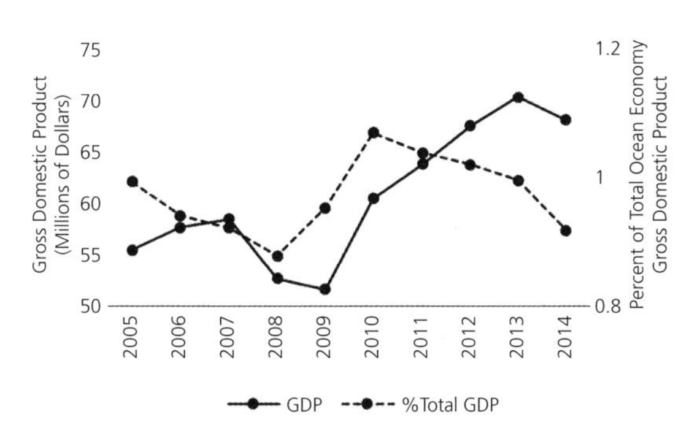

Figure 10.4 Number of living marine resources establishments and employments, and gross domestic product (GDP; USD) value, and their percent contributions to the total multisector oceanic economy in the Western Pacific Hawaii subregion (years 2005–2014).

Data derived from the National Ocean Economics Program.

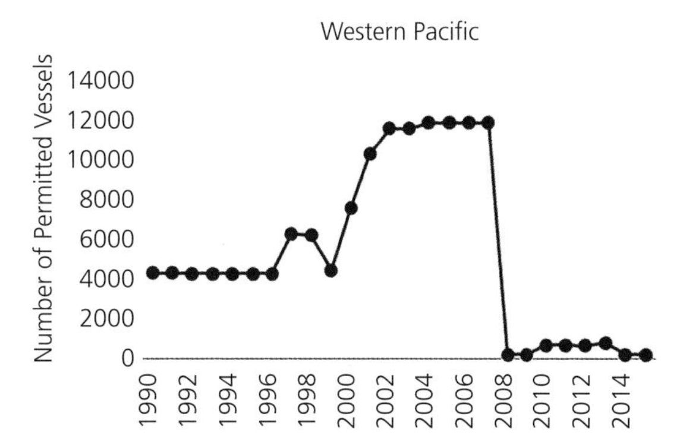

Figure 10.5 Total estimated number of reported permitted vessels for the Western Pacific region over time (years 1990–2015).

Data derived from NOAA National Marine Fisheries Service Council Reports to Congress. Note: It is important to recognize that these values are likely under-reported, are potentially of questionable accuracy (PIFSC, pers. communication), and that summarized trends may not be fully reflective of the Western Pacific fleet.

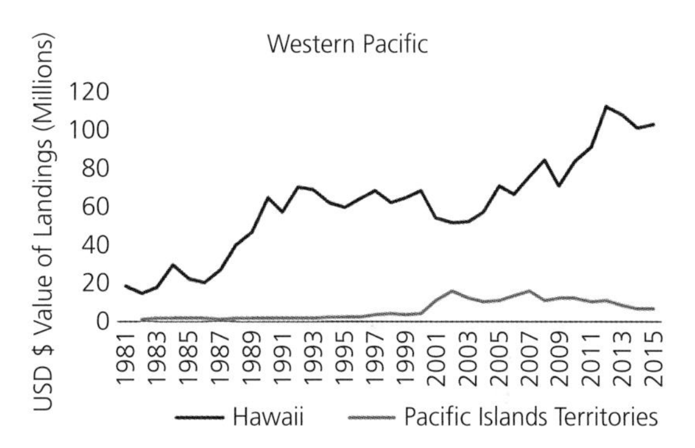

Figure 10.6 Total revenue (year 2017 USD) of landed commercial fishery catches for the Western Pacific Hawaii and grouped U.S. Territory subregions, over time (years 1981–2015).

Data derived from NOAA National Marine Fisheries Service commercial fisheries statistics.

decreases in the number of reported vessels occurred with values currently in the upper hundreds. Overall, reported values are sixth-highest nationally, with the Western Pacific representing on average 7.5% of the total reported vessels across all regions. It is important to recognize that these values are likely under-reported, and that summarized trends may not be fully reflective of the Western Pacific fleet. As observed for other data-limited tropical fisheries (e.g., those in the U.S. Caribbean), these factors reflect the challenges in quantifying fishing patterns, including estimating fleet size throughout this region.

The Western Pacific is seventh-highest nationally in total revenue (year 2017 USD) of its landed domestic commercial fishery catches (Fig. 10.6). These values do not account for all highly migratory species captures occurring in Western Pacific regional waters, a large fraction of which are landed outside Honolulu and U.S. Pacific Island territories in international ports (see Chapter 11 on RFMOs; Craig et al. 2017). While landed highly migratory species are included in NOAA Fisheries statistics for all regions, including the Western Pacific, the numbers and values are limited to those caught by Hawaiian permitted vessels. Therefore, they are underestimates in light of the international jurisdictions of these species and records of capture beyond U.S. waters throughout their range (see Chapter 11 on RFMOs; Craig et al. 2017). Pronounced increases in domestic commercial landings value have occurred over time for this region, particularly during the 1990s and further during the 2000s, with peak values observed for the Hawaii subregion at $112 million in 2012. Within the U.S. Pacific Island territories, domestic values have remained much lower with increases to peak revenue (~$16 million) observed during the 2000s, and current values remaining at $6.3 million. These values do not fully account for recreational and subsistence fisheries, for which data are limited.

Ratios of LMR jobs and domestic commercial fisheries revenue to biomass and primary production for the Western Pacific (Fig. 10.7) are at mid-low rankings nationally (sixth to seventh-highest) in the Western Pacific Hawaii subregion. While remaining steady over time, the Western Pacific (Hawaii) is seventh-highest nationally in value for jobs to primary production (up to 0.4 jobs per million metric tons^{-1}). The Western Pacific (Hawaii) is sixth-highest nationally in jobs/biomass (up to 0.005 jobs thousand metric tons^{-1}) and seventh-highest nationally in domestic revenue/productivity (up to $0.06 metric ton^{-1}), with values having increased for the latter over the past decade. Overall, the Western Pacific (Hawaii) is sixth-highest nationally in terms of domestic fisheries revenue/biomass, with oscillating values (up to $0.03 metric ton^{-1}) over the past decade.

10.4.2 Governance Criteria

10.4.2.1 Human Representative Context

Twenty-four federal, State, and international agencies and organizations operate within the Western Pacific and are directly responsible for fisheries and environmental management (Table 10.1). Fisheries resources throughout Hawaii and U.S. Pacific Island territorial waters (0–3 nautical miles) are managed by individual state and territorial agencies, while those in federal waters (3–200 nm) are managed by both NOAA Fisheries (Pacific Islands region) and the Western Pacific Regional Fishery Management Council (WPRFMC). Western Pacific fishery and environmental resources also fall under several international and multijurisdictional agreements, including those of the Inter-American Tropical Tuna Commission (IATTC), South Pacific Regional Fisheries Management

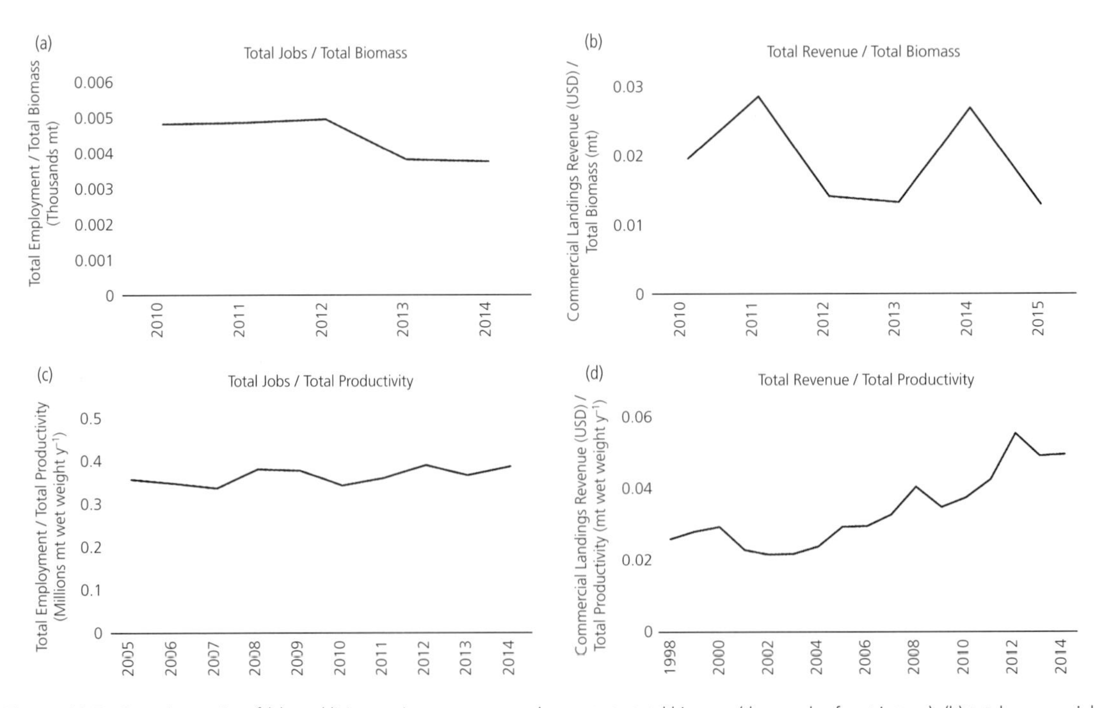

Figure 10.7 Over time, ratios of (a) total living marine resources employments to total biomass (thousands of metric tons); (b) total commercial landings revenue (USD) to total biomass (metric tons); (c) total living marine resources employments to total productivity (metric tons wet weight year^{-1}); (d) total commercial landings revenue (USD) to total productivity (metric tons wet weight year^{-1}) for the Western Pacific Hawaii subregion.

Table 10.1 Federal and State agencies and organizations responsible for fisheries and environmental consulting and management in the Western Pacific region and subregions

Region	Subregions	Federal Agencies with Marine Resource Interests	Council	State Agencies and Cooperatives	Multijurisdictional
Western Pacific		U.S. Department of Commerce **DoC** NOAA Fisheries Pacific Islands Fisheries Science Center **PIFSC**, Pacific Islands Regional Office **PIRO**, NOAA National Marine Sanctuary Program **NMS**; U.S. Department of Interior **DoI** (U.S. Fish and Wildlife Service **FWS**, National Park Service **NPS**, Bureau of Ocean Energy Management **BOEM**, U.S. Geological Survey **USGS**); U.S. Environmental Protection Agency **EPA**	Western Pacific Fishery Management Council **WPFMC**	American Samoa Department of Marine and Wildlife Resources **DMWR**; Guam Division of Aquatics and Wildlife Resources **DAWR**; Commonwealth of the Northern Mariana Islands (CNMI) Division of Fish and Wildlife State of Hawaii Division of Aquatic Resources **DAR**	Commission for the Conservation of Southern Bluefin Tuna **CCSBT**; Coral Reef Task Force **CRTF**; Hawaii Coral Reef Institute **HCRI**; Inter-American Tropical Tuna Commission **IATTC**; International Scientific Committee for Tuna and Tuna-like Species in the North Pacific Ocean **ISC**; International Society for Reef Studies **ISRS**; Secretariat of the Pacific Community **SPC**; South Pacific Regional Fisheries Management Organisation **SPRFMO**; South Pacific Tuna Treaty **SPTT**; Western and Central Pacific Fish Commission **WCPFC**
	Hawaii	**DoC** **PIFSC, PIRO, NMS** **DoI** **FWS**, **NPS**, **BOEM**, **USGS**; **EPA**	**WPFMC**	**DAR**	**CRTF**; **HCRI**; **IATTC**; **ISC**; **ISRS**; **SPC**; **WCPFC**
	Pacific Islands territories *American Samoa* *Guam* *Northern Mariana Islands* *Pacific Remote Island Areas (Howland, Baker, and Jarvis islands, Kingman Reef, and Wake and Palmyra Atolls)*	**DoC** **PIFSC, PIRO, NMS**; **DoI** **FWS**, **NPS**, **BOEM**, **USGS**; **EPA**	**WPFMC**	**DMWR**; **DAWR**; Commonwealth of the Northern Mariana Islands (CNMI) Division of Fish and Wildlife	**CCSBT**; **CRTF**; **IATTC**; **ISC**; **ISRS**; **SPTT**; **SPRFMO**; **SPC**; **WCPFC**

Organisation (SPRFMO), South Pacific Tuna Treaty (SPTT), and the WCPFC.

Within the Hawaiian subregion, the Northwestern Hawaiian Islands are collectively managed as the Papahānaumokuākea Marine National Monument by NOAA, the U.S. Department of Interior FWS, and the State of Hawaii, including its Office of Hawaiian Affairs. This monument also contains the Hawaiian Islands National Wildlife Refuge (administered by FWS), and several state-managed reserves and refuges. Throughout U.S. Pacific Island territories, NOAA manages the Fagatele Bay NMS and Rose Atoll Marine National Monument in American Samoa, the Marianas Trench Marine National Monument in the Marianas archipelago, and the Pacific Remote Islands Marine National Monument.

Nationwide, the Western Pacific contains the third-lowest (sixth-highest) number of congressional representatives (n = 7), which are concentrated within one state and three U.S. territories (Fig. 10.8). Among subregions, the highest numbers of congressionals are found in Hawaii (n = 4), while only one non-voting representative in the House of Representatives each is found for American Samoa, CNMI, and Guam. When standardized per mile of shoreline, congressional representation is second-lowest nationally in the Western Pacific. Historically until the 2000s, composition of the WPRFMC was relatively balanced between members of the commercial and recreational fishing sector, and representatives from other sectors (Fig. 10.9a). From 2000 onward, expanding representation and dominance by the recreational fishing sector was observed until 2010. Current representation is more balanced among sectors, with emerging commercial representation in more recent years. With nine currently appointed members, the WPRFMC contains a medium number of council members nationally (fourth-highest nationally).

The Pacific marine mammal SRG is responsible for advising federal agencies on the status of marine mammal stocks covered under the Marine Mammal Protection Act (MMPA) for the Pacific and Western

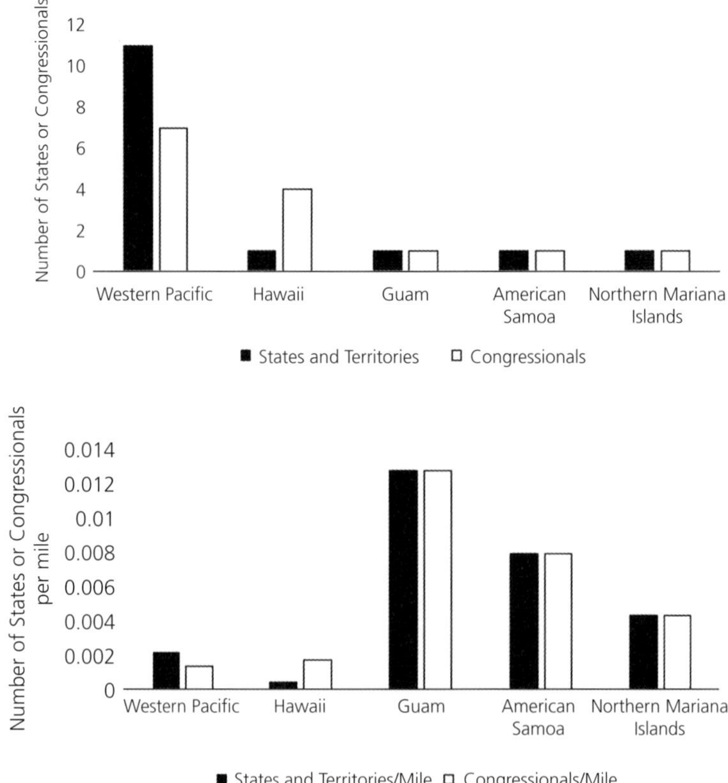

Figure 10.8 Number of states/territories and U.S. Congressional representatives (top panel) and as standardized by miles of shoreline (bottom panel) for the Western Pacific region and its identified subregions (Hawaii and U.S. Pacific Island territories).

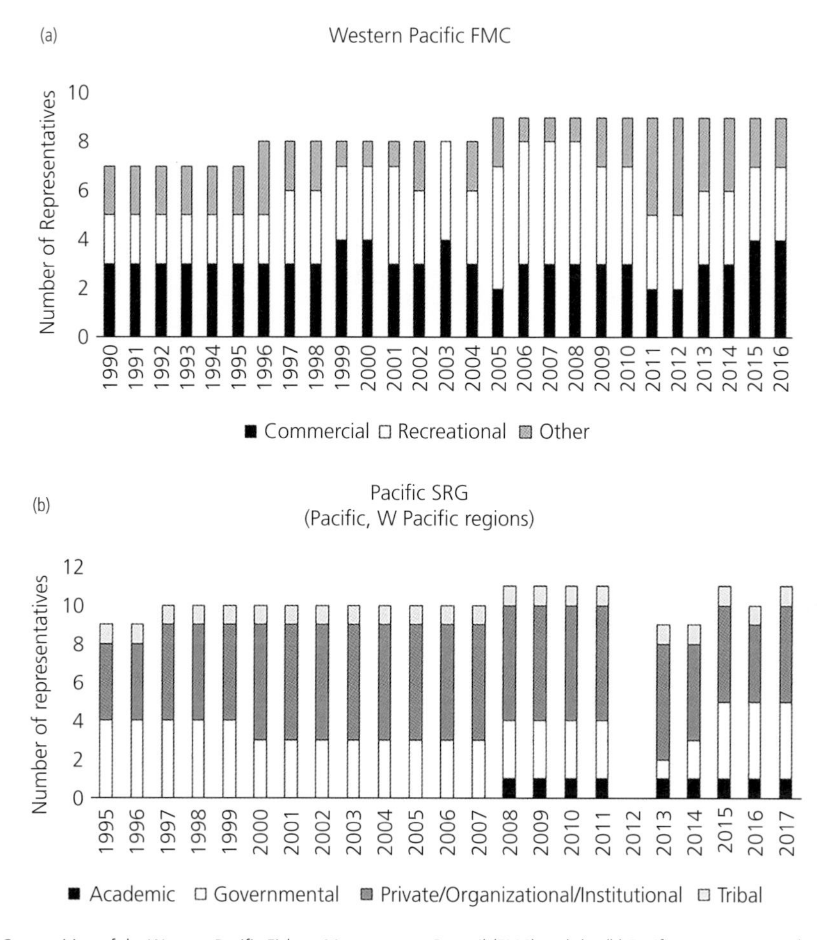

Figure 10.9 (a) Composition of the Western Pacific Fishery Management Council (FMC) and the (b) Pacific marine mammal Scientific Review Group (SRG) over time.

Data derived from NOAA National Marine Fisheries Service Council Reports to Congress and Scientific Review Group reports. SRG composition was unreported for 2012.

Pacific regions. Its composition and membership have remained relatively steady over time (Fig. 10.9b), with individuals from the governmental and private institutional sectors making up the majority of its representatives, and with periodic representation from academics and the tribal community.

10.4.2.2 Fishery and Systematic Context

Five current Western Pacific subregional FEPs are managed by NOAA Fisheries and the WPRFMC for the Hawaiian Archipelago, American Samoa, Mariana Archipelago, Pacific Remote Islands, and pelagic species (Table 10.2). Prior to their development, fishery resources were managed under five region-wide Fishery Management Plans (FMPs) for bottomfish, coral reef ecosystems, crustaceans, precious corals, and pelagic species throughout the entire Western Pacific. In total, these FMPs were subject to 53 amendments, which is second-lowest (seventh-highest) for the nation. All five FEPs have been subject to 21 modifications, the highest nationally. Each subregional FEP contains comprehensive information regarding coastal and offshore fisheries, the habitat and biology of managed species in a given area, and the socioeconomic and ecological impacts of conservation and fisheries management policies.

Nine coastal national parks, monuments/memorials, or historic sites, two NOAA HFAs (West Hawaii, Hawaii; Manell-Geus, Guam), four marine national monuments, one National Estuarine Research Reserve (NERR), and two NMS are found throughout the Western Pacific region (Table 10.3), with the majority occurring in the Hawaiian subregion. Sixty-six inshore and offshore regions are also identified as Habitat Areas of Particular Concern (HAPCs) in the Western

Table 10.2 List of major managed fishery species in the Western Pacific region, under state, federal, and international jurisdictions. FMP = Fishery Management Plan; FEP = Fishery Ecosystem Plan

Council/Agency	FMP/FEP	FEP Modifications	FMP Modifications	Major Species/ Species Group	Scientific Name(s) or Family Name(s)
Western Pacific Fishery Management Council **WPFMC**	Hawaiian Archipelago—Including Midway Atoll (169 species)	3 amendments	14 amendments (original entire W. Pacific FMP)	Bottomfish and seamount groundfish	Species from Families Carangidae (Jacks) and Lutjanidae (Snappers)
			13 amendments (original entire W. Pacific FMP)	Crustaceans	Spiny lobsters *Panulirus* spp., Slipper lobster (Family Scyllaridae), Kona crab *Ranina ranina*, and Deep-water shrimp *Heterocarpus* spp.
			7 amendments (original entire W. Pacific FMP)	Precious corals	*Antipathes* spp., *Corallium* spp., *Gerardia* spp., *Lepidisis* spp., *Narella* spp.
			0 amendments (original entire W. Pacific Coral Reef Ecosystem FMP)	Currently harvested coral reef taxa	Fish species from Families Acanthuridae (Surgeonfishes), Balistidae (Triggerfishes), Carangidae (Jacks), Carcharhinidae (Sharks), Holocentridae (Squirrelfishes), Kuhliidae (Flagtails), Kyphosidae (Rudderfish), Labridae (Wrasses), Mullidae (Goatfishes), Mugilidae (Mullets), Muraenidae (Morays), Polynemidae (Threadfins), Priacanthidae (Big-eyes), Scaridae (Parrotfishes), Serranidae (Groupers/Seabasses), Sphyraenidae (Barracudas), Zanclidae (Moorish Idols). Invertebrate species from Families Octopodidae (Octopuses), Sabellidae (Featherduster Worms), Turbinidae (Snails)
				Potentially harvested coral reef taxa	42 families of fishes, corals, invertebrates, Trumpetfish *Aulostomus chinensis*, Cornetfish *Fistularia commersoni*, Black-lipped pearl Oyster *Pinctada margaritifera*, crustaceans, anemones, zoanthids, bivalves, tunicates, sponges, mollusks, algae, live rock, annelids, all other marine plants, and animals not aforementioned specifically
	American samoa (202 species)	2 amendments	14 amendments (original entire W. Pacific FMP)	Bottomfish	Species from Families Carangidae (Jacks), Lethrinidae (Emperors), Lutjanidae (Snappers), Serranidae (Groupers/Seabasses)
			13 amendments (original entire W. Pacific FMP)	Crustaceans	Spiny lobsters *Panulirus* spp., Slipper lobster (Family *Scyllaridae*), Kona crab *Ranina ranina*, and Deep-water shrimp *Heterocarpus* spp.

Council/Agency	FMP/FEP	FEP Modifications	FMP Modifications	Major Species/ Species Group	Scientific Name(s) or Family Name(s)
Western Pacific Fishery Management Council **WPFMC**	American Samoa (202 species)	2 amendments	7 amendments (original entire W Pacific FMP)	Precious Corals	*Acanella spp., Antipathes* spp., *Calyptrophora* spp., *Corallium* spp., *Gerardia* spp., *Lepidisis* spp., *Narella* spp.
			0 amendments (original entire W. Pacific Coral Reef Ecosystem FMP)	Currently harvested coral reef taxa	Fish species from Families Acanthuridae (Surgeonfishes), Balistidae (Triggerfishes), Carcharhinidae (Sharks), Holocentridae (Squirrelfishes), Kuhliidae (Flagtails), Kyphosidae (Rudderfish), Labridae (Wrasses), Mullidae (Goatfishes), Mugilidae (Mullets), Muraenidae (Morays), Polynemidae (Threadfins), Priacanthidae (Big-eyes), Scaridae (Parrotfishes), Scombridae (Tunas/Mackerels), Siganidae (Rabbitfishes), Sphyraenidae (Barracudas) and invertebrate species from Families Octopodidae (Octopuses), Turbinidae (Snails)
				Potentially harvested coral reef taxa	49 families of fishes, corals, invertebrates, Dog tooth tuna *Gymnosarda unicolor*, Trumpetfish *Aulostomus chinensis*, Bluespotted cornetfish *Fistularia commersoni*, Black-lipped pearl oyster *Pinctada margaritifera*, and crustaceans, anemones, zoanthids, bivalves, tunicates, sponges, mollusks, algae, live rock, annelids, all other marine plants, and animals not aforementioned specifically.
	Mariana Archipelago—Commonwealth of Northern Mariana islands **CNMI** and Guam (213 species)	2 amendments	14 amendments (original entire W. Pacific FMP)	Bottomfish (**CNMI** and Guam)	Species from Families Carangidae (Jacks), Lethrinidae (Emperors), Lutjanidae (Snappers), Serranidae (Groupers/Seabasses)
			13 amendments (original entire W. Pacific FMP)	Crustaceans (**CNMI** and Guam)	Spiny lobsters *Panulirus* spp., Slipper lobster (Family Scyllaridae), Kona crab *Ranina ranina*, and Deep-water shrimp *Heterocarpus* spp.
			7 amendments (original entire W. Pacific FMP)	Precious Corals	*Acanella spp., Antipathes* spp., *Calyptrophora* spp., *Corallium* spp., *Gerardia* spp., *Lepidisis* spp., *Narella* spp.
			0 amendments (original entire W. Pacific Coral Reef Ecosystem FMP)	Currently harvested coral reef taxa	Fish species from Families Acanthuridae (Surgeonfishes), Balistidae (Triggerfishes), Carangidae (Jacks), Carcharhinidae (Sharks), Holocentridae (Squirrelfishes), Kuhliidae (Flagtails), Kyphosidae (Rudderfish), Labridae (Wrasses), Mullidae (Goatfishes), Mugilidae (Mullets), Muraenidae (Morays), Polynemidae (Threadfins), Priacanthidae (Big-eyes), Scaridae (Parrotfishes), Scombridae (Tunas/Mackerels), Siganidae (Rabbitfishes), Sphyraenidae (Barracudas) and invertebrate species from Families Octopodidae (Octopuses), Turbinidae (Snails)
				Potentially harvested coral reef taxa	49 families of fishes, corals, invertebrates, Dog tooth tuna *Gymnosarda unicolor*, Trumpetfish *Aulostomus chinensis*, Bluespotted cornetfish *Fistularia commersoni*, Black-lipped pearl oyster *Pinctada margaritifera*, and crustaceans, anemones, zoanthids, bivalves, tunicates, sponges, mollusks,

Continued

Table 10.2 Continued

Council/Agency	FMP/FEP	FEP Modifications	FMP Modifications	Major Species/ Species Group	Scientific Name(s) or Family Name(s)
Western Pacific Fishery Management Council **WPFMC**	Mariana Archipelago—**CNMI** and Guam (213 species)	2 amendments	0 amendments (originalentire W. Pacific Coral Reef Ecosystem FMP)	Potentially harvested coral reef taxa (cont'd)	algae, live rock, annelids, all other marine plants, and animals not aforementioned specifically
	Pacific Remote Island Areas—Howland, Baker and Jarvis islands, Kingman Reef, and Wake and Palmyra Atolls (143 species)	1 amendment	14 amendments (original entire W. Pacific FMP)	Bottomfish	Species from Families *Carangidae* (Jacks), *Lethrinidae* (Emperors), *Lutjanidae* (Snappers), *Serranidae* (Groupers/Seabasses)
			13 amendments (original entire W. Pacific FMP)	Crustaceans	Spiny lobsters *Panulirus* spp., Slipper lobster (Family Scyllaridae), Kona crab *Ranina ranina*, and Deep-water shrimp *Heterocarpus* spp.
			7 amendments (original entire W Pacific FMP)	Precious Corals	*Antipathes* spp., *Corallium* spp., *Gerardia* spp., *Lepidisis* spp., *Narella* spp.
			0 amendments (original entire W Pacific Coral Reef Ecosystem FMP)	Currently harvested coral reef taxa	Fish species from Families Acanthuridae (Surgeonfishes), Labridae (Wrasses), Mullidae (Goatfishes), Mugilidae (Mullets), Muraenidae (Morays), Priacanthidae (Big-eyes), Scaridae (Parrotfishes), Scombridae (Tunas/ Mackerels), Sphyraenidae (Barracudas) and invertebrate species from Family Octopodidae (Octopuses)
				Potentially harvested coral reef taxa	46 families of fishes, corals, invertebrates, Trumpetfish *Aulostomus chinensis*, Bluespotted cornetfish *Fistularia commersoni*, Black-lipped pearl oyster *Pinctada margaritifera*, and crustaceans, anemones, zoanthids, bivalves, tunicates, sponges, mollusks, algae, live rock, annelids, all other marine plants, and animals not aforementioned specifically.
	Pelagics (32 species)	7 amendments, 1 regulatory amendment	19 amendments (original FMP)	Tunas	Albacore *Thunnus alalunga*, Bigeye tuna *T. obesus*, Yellowfin tuna *T. albacares*, Northern bluefin tuna *T. thynnus*, Skipjack tuna *Katsuwonus pelamis*, Kawakawa *Euthynnus affinis*, other tuna relatives *Auxis* spp., *Scomber* spp., *Allothunus* spp.
				Billfishes (incl. Swordfish)	Striped marlin *Kajikia audax*, Shortbill spearfish *Tetrapturus angustirostris*, Swordfish *Xiphias gladius*, Sailfish *Istiophorus platypterus*, Blue marlin *Makaira mazara*, Black marlin *M. indica*

Council/Agency	FMP/FEP	FEP Modifications	FMP Modifications	Major Species/ Species Group	Scientific Name(s) or Family Name(s)
Western Pacific Fishery Management Council **WPFMC**	Pelagics (32 species)	7 amendments, 1 regulatory amendment	19 amendments (original FMP)	Sharks	Pelagic thresher shark *Alopias pelagicus*, Bigeye thresher shark *A. superciliousus*, Common thresher shark *A. vulpinus*, Silky shark *Carcharhinus falciformis*, Oceanic whitetip shark *C. longimanus*, Blue shark *Prionace glauca*, Shortfin mako shark *Isurus oxyrinchus*, Longfin mako shark *I. paucus*, Salmon shark *Lamna ditropis*
				Other Pelagics	Mahimahi (Dolphinfish) *Coryphaena* spp., Moonfish *Lampris* spp., Wahoo *Acanthocybium solandri*, Family Gempylidae (Oilfish), Family Bramidae (Pomfret)
				Squids	Neon flying squid *Ommastrephes bartamii*, Diamondback squid *Thysanoteuthis rhombus*, Purple flying squid *Sthenoteuthis oualaniensis*

*Note: As of 2019, most Western Pacific species are now listed as "ecosystem component species" (NMFS 2019a). These listings reflect information in Fishery Ecoystem Plans developed prior to this change.

Table 10.3 List of major bays and islands, NOAA Habitat Focus Areas (HFAs), NOAA National Estuarine Research Reserves (NERRs), NOAA National Marine Monuments and Sanctuaries (NMS), coastal national parks, national seashores, and number of habitats of particular concern (HAPCs) in Western Pacific subregions

Region	Subregion	Major Bays	Major Islands	HFAs	NERRs	Monuments and NMS	Coastal National Parks	National Sea/ Lakeshores	HAPCs
W Pacific	**Hawaii**	A'iea Bay, Hanauma Bay, Hilo Bay, Kaneohe Bay, Mamala Bay, Maunalua Bay, Waimea Bay, Yokohama Bay	**8 Main Hawaiian Islands** (Hawai'i, Maui, O'ahu, Kaua'i, Moloka'i, Lāna'i, Ni'ihau, Kaho'olawe), **10 Northwest Hawaiian Islands** (Necker *Mokumanamana*, French Frigate Shoals *Kānemiloha'i*, Gardner Pinnacles *Pūhāhonu*, Maro Reef *Nalukākala*, Laysan *Kauō*, Lisianski *Papaāpoho*, Pearl and Hermes Atoll *Holoikauaua*, Midway *Pihemanu*, Kure *Mokupāpapa*), **119 Islets**	West Hawaii, Hawaii HFA	He'eia NERR	Hawaiian Islands Humpback Whale NMS, Papahānaumokuākea Marine National Monument	World War II in the Pacific National Memorial, Kalaupapa National Historical Park, Haleakala National Park, Hawaii Volcanoes National Park, Pu'uhonua O Honaunau National Historical Park, Kaloko-Honokohau National Historical Park, Puukohola Heiau National Historic Site		36
	American Samoa	Faga'itua Bay, Fagatele Bay, Larsen Bay, Vatia Bay	**5 Islands** (Tutuila, Aunu'u, Ofu, Olosega, Ta'u), **2 coral atolls** (Swains, Rose Atoll)			Fagatele Bay NMS, Rose Atoll Marine National Monument	National Park of American Samoa		12
	Guam and Northern Mariana Islands	Laolao Bay	**16 Islands** (Guam, Saipan, Tinian, Rota, Aguigan, Farallon de Pajaros, Maug islands, Asuncion, Agrihan, Pagan, Alamagan, Guguan, Papaungan, Sarigan, Anatahan, Farallon de Medinilla)	Manell-Geus, Guam HFA		Marianas Trench Marine National Monument,	American Memorial, War in the Pacific National Historical Park		8
	Pacific Remote Islands		Baker Island, Howland Island, Jarvis Island, Johnston Atoll, Kingman Reef, Palmyra Atoll, Wake Island			Pacific Remote Islands Marine National Monument			10

Pacific. These include protected sites for bottomfish and seamount groundfish throughout the Hawaiian Archipelago, and as stipulated in FEPs for American Samoa, the Marianas archipelago, the Pacific Remote Islands, and for pelagic species.

No specifically named fishing closure zones occur in the Western Pacific, while fishing is prohibited in 47 specific areas of the EEZ and restricted in 69 of them (Fig. 10.10). Equal numbers of closures occur in the Hawaii and Pacific Island territories subregions, with the majority of prohibited sites (n = 36) found within U.S. territorial waters and majority of restricted sites (n = 47) occurring in the Hawaii subineegion. These latter two numbers are third and sixth-highest nationally among regions, respectively. However, the area of these prohibited (2.1 million km²) and restricted areas (2.2 million km²) is highest and second-highest nationally, with an overwhelming majority of prohibited and restricted area (84–85%) occurring in waters surrounding U.S. Pacific Island territories. In total, 30.5%

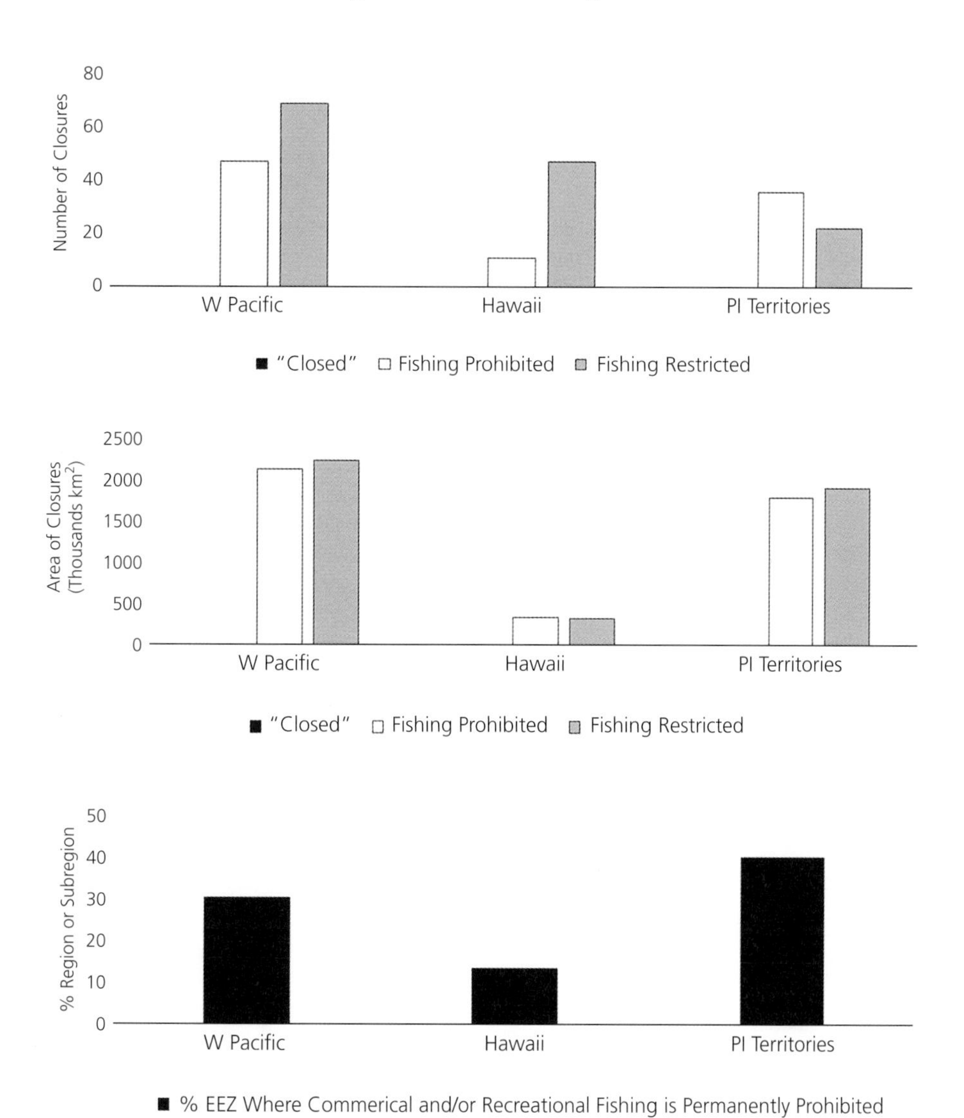

Figure 10.10 Number (top panel) and area (middle panel; km²) of named fishing closures, and prohibited or restricted fishing areas of the Western Pacific region and identified subregions (Hawaii and U.S. Pacific Island, PI, Territories), including percent region or subregion Exclusive Economic Zone (EEZ; bottom panel) where commercial and/or recreational fishing is permanently prohibited.

Data derived from NOAA Marine Protected Areas inventory.

of the Western Pacific is permanently protected from fishing, which is highest nationally, including 13.5% of the Hawaii subregion (~333,900 km²) and 40.1% of U.S. Pacific Island territorial waters (~1.9 million km²). The largest fishing protection sites include the Pacific Remote Islands (1.3 million km²), Marianas Trench (~248,500 km²), and Rose Island National Atoll (~35,000 km²) Marine National Monuments in the Pacific Island territories subregion, and the Papahānaumokuākea Marine National Monument (~364,000 km²) in the Hawaii subregion.

10.4.2.3 Organizational Context

Relative to the total commercial value of its managed fisheries, the WPRFMC budget is second-highest nationally when compared to other regions (Link & Marshak 2019; see Chapter 12, Synthesis, Table 12.4), and makes up a small proportion of the total value of its fisheries (Fig. 10.11). These relative values decreased slightly in the early 2010s and have increased in more recent years, with current values similar to those in 2007–08. Cumulative National Environmental Policy Act Environmental Impact Statement (NEPA-EIS) actions from 1987 to 2016 in the Western Pacific are also second-highest nationally, with 250 observed over the past 30 years (Fig. 10.12) with the majority occurring in the Hawaii subregion (n = 160) and fewer throughout U.S. Pacific Island territories. Six fisheries-related lawsuits have occurred in the Western Pacific region since 2010, making it the fifth-most litigious among regions. These lawsuits have mostly occurred within the Hawaii subregion, with one lawsuit brought by the Territory of American Samoa in 2016.

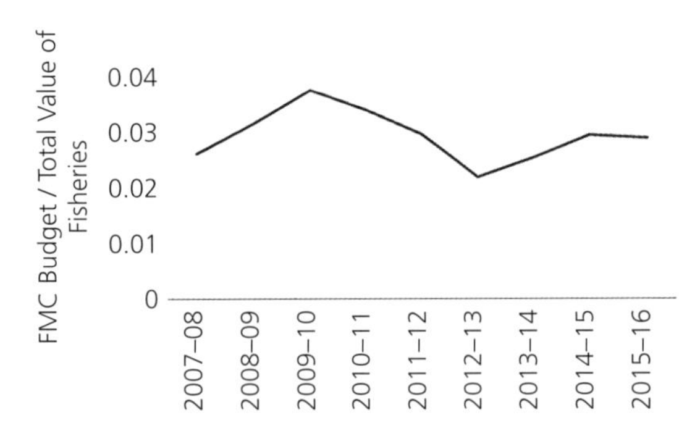

Figure 10.11 Total annual budget of the Western Pacific Fishery Management Council as compared to the total value of its marine fisheries. All values in USD.

Data from NOAA National Marine Fisheries Service Office of Sustainable Fisheries and NOAA National Marine Fisheries commercial landings database.

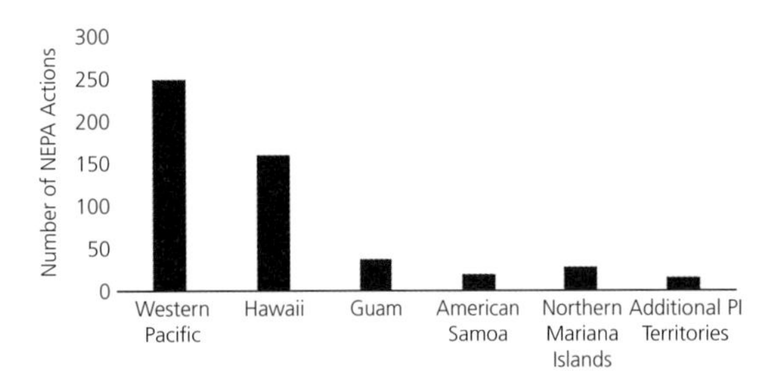

Figure 10.12 Number of National Environmental Policy Act Environmental Impact Statement (NEPA-EIS) actions from 1987 to 2016 for the Western Pacific region (sum of all subregions) and identified subregions (Hawaii and U.S. Pacific Island, PI, Territories).

Data derived from U.S. Environmental Protection Agency EIS database.

10.4.2.4 Status of Living Marine Resources (Targeted and Protected Species)

Nationally, the Western Pacific contains the second-highest number of managed species. Of the 295 federally managed species in the region, 213 are targeted fishery species covered under previous FMPs and current FEPs, while 29 taxa in the Hawaiian subregion and 16 taxa in U.S. Pacific Island territories are prohibited from harvest. The highest numbers of targeted fishery species are found in the Marianas and American Samoa subregions. As of 2019, most Western Pacific species are no longer "federally managed species" and are now considered as "ecosystem component species" instead (NMFS 2019a). The values and listings presented reflect information included in FMPs and FEPs prior to this change. Fifty-three species of marine mammals, sea turtles, fishes, corals, and seabirds are also federally protected under the ESA or MMPA (Fig. 10.13, Tables 10.2, 10.4–5).

During mid-2017, 83 stocks (second-highest nationally) were listed as federally managed in the Western Pacific with only four listed as a NOAA Fish

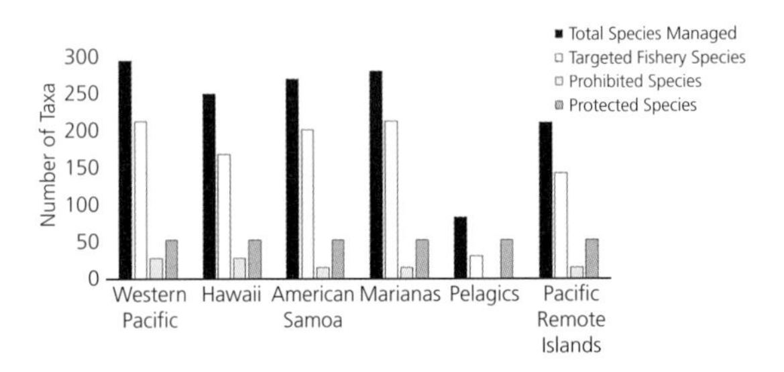

Figure 10.13 Number of managed taxa (species or families) for the entire Western Pacific region and as included for identified fishery ecosystem plan subregions (Hawaii, U.S. Pacific Island territories, and pelagic species).

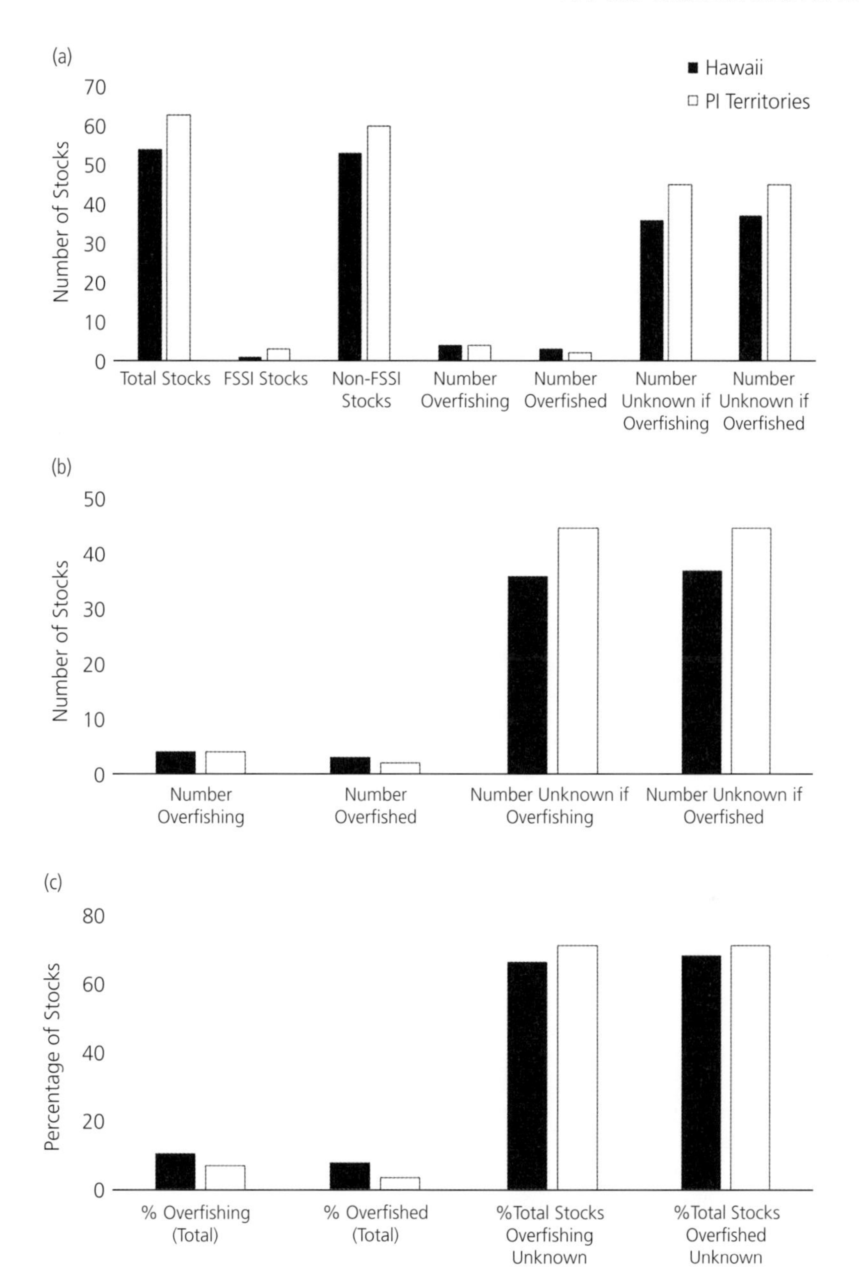

Figure 10.14 For the Western Pacific Hawaii and U.S. Pacific Island (PI) territories subregions as of June 2017 (a) Total number of managed Fish Stock Sustainability Index (FSSI) stocks and non-FSSI stocks, and breakdown of stocks experiencing overfishing, classified as overfished, and of unknown status. (b) Number of stocks experiencing overfishing, classified as overfished, and of unknown status. (c) Percent of stocks experiencing overfishing, classified as overfished, and of unknown status. Note: stocks may refer to a species, family, or complex.

Stock Sustainability Index (FSSI) stock (Fig. 10.14; NMFS 2017a, 2020).[1] At that point in time, of all stocks, four were experiencing overfishing (**Pacific bluefin tuna**—*Thunnus orientalus*; **Swordfish**—

[1] As of June 2020, 57 stocks or stock complexes (now third-highest nationally) are federally managed in the Western Pacific, with six now listed as FSSI stocks. The two newly added FSSI stocks are Green jobfish—*Aprion virescens*, main Hawaiian islands and Spanner (or Kona) crab—*Ranina ranina*, main Hawaiian Islands. Between 2017 and 2020, seven non-FSSI stocks from American Samoa, 14 from the Mariana Archipelago, and five from the Hawaiian Archipelago (Bigeye scad—*Selar crumenophthalmus*; coral reef ecosystem multispecies complex; Mackerel scad—*Decapterus macarellus*; Slipper lobsters—Scyllaridae, Spiny lobsters—*Panulirus* spp.) were reclassified as ecosystem component species.

Table 10.4 List of prohibited fishery species in Western Pacific subregions, including federal and state jurisdictions

Council/Authority	Region	Prohibited Species	Scientific Name
Western Pacific Fishery Management Council **WPFMC**; Pacific Islands Regional Office **PIRO**	Hawaii (29 taxa)	Spectacled Parrotfish *" Uhu uliuli"* (only Maui jurisdiction)	*Chlorurus perspicillatus*
		Ember parrotfish *" Uhu 'ele'ele"* (only Maui jurisdiction)	*Scarus rubroviolaceus*
		Clams, oysters, and other mollusks	Abalone *Haliotis* spp., Cherrystone clam (quahog) *Mercenaria mercenaria*, Coral rock oyster *Crassostrea amasa*, Eastern oyster *C. virginica*, Japanese oyster *C. gigas*, Pearl oyster *Pinctada margaritifera*, Top shell *Trochus* spp.
		Live rock	Any coral reef resource attached to a hard substrate, including dead coral or rock (excluding individual mollusk shells)
		Pink, gold, and black corals (State jurisdiction)	*Antipathes* spp., *Corallium* spp., *Gerardia* spp., *Narella* spp.
		Stony corals	Families Acroporidae, Agariciidae, Astrocoeniidae, Balanophyllidae, Caryophyllidae, Dendrophylliidae, Faviidae, Fungiidae, Pocilloporidae, Poritidae, Siderastreadae
	American Samoa (16 taxa) Marianas (16 taxa) Pacific Remote Islands (16 taxa)	Gold corals	*Calyptrophora* spp., *Gerardia* spp., *Narella* spp.
		Living corals	Families Acroporidae, Agariciidae, Astrocoeniidae, Balanophyllidae, Caryophyllidae, Coralliidae, Dendrophylliidae, Faviidae, Fungiidae, Pocilloporidae, Poritidae, Siderastreadae
		Live rock	Any coral reef resource attached to a hard substrate, including dead coral or rock (excluding individual mollusk shells)

Xiphias gladius; **Striped marlin**—*Kajikia audax*, western/central Pacific; **Bigeye tuna**—*Thunnus obesus*, western/central Pacific) and three were overfished (**Pacific bluefin tuna; seamount groundfish complex, Hancock seamount; Striped marlin, western/central Pacific**).[2] There are 62 stocks which remained unclassified as to whether they were experiencing overfishing (74.7%; highest nationally) and 63 had unknown overfished status (75.9%; second-highest nationally), making the Western Pacific one of the highest nationally in terms of the proportion of fishery stocks with unknown status at that time (NMFS 2017a, Link & Marshak 2019).[3]

Among the protected species of the Western Pacific are 48 independent marine mammal stocks, and 33 ESA-listed distinct population segments (second-highest nationally), of which 65% are endangered and zero have unknown status (Fig. 10.15). Nationwide,

[2] As of June 2020, six Western Pacific stocks are experiencing overfishing (American Samoa bottomfish multispecies complex; Pacific bluefin tuna; Striped marlin—*Kajikia audax*, western and central North Pacific; Swordfish; Yellowfin tuna—*Thunnus albacares*, eastern Pacific; Oceanic whitetip shark—*Carcharhinus longimanus*, western and central Pacific) and six are overfished (American Samoa bottomfish multispecies complex; Guam bottomfish multispecies complex; Pacific bluefin tuna; Striped marlin, western and central North Pacific; Hancock seamount groundfish complex, Hawaii Archipelago; Oceanic whitetip shark, western and central Pacific).

[3] In June 2020, 33 stocks (57.9%) remained unclassified regarding their overfishing status and 34 stocks (59.6%) had unknown overfished status.

Table 10.5 List of protected species under the Endangered Species Act (ESA) or Marine Mammal Protection Act (MMPA) in the Western Pacific region

Region	Managing Agency	Protected Species Group	Common Name	Scientific Name
Hawaii and Pacific Island territories (53 species)	Western Pacific Fishery Management Council **WPFMC**; Pacific Islands Regional Office **PIRO**	Seals	Hawaiian monk seal	*Neomonachus schauinslandi*
		Whales	Blainville's beaked whale	*Mesoplodon densirostris*
			Blue whale	*Balaenoptera musculus*
			Bryde's whale	*Balaenoptera edeni*
			Cuvier's beaked whale	*Ziphius cavirostris*
			Dwarf sperm whale	*Kogia simus*
			False killer whale	*Pseudorca crassidens*
			Fin whale	*Balaenoptera physalus*
			Humpback whale	*Megaptera novaeangliae*
			Killer whale	*Orcinus orca*
			Longman's beaked whale	*Indopacetus pacificus*
			Melon-headed whale	*Peponocephala electra*
			Minke whale	*Balaenoptera acutorostrata*
			Pygmy killer whale	*Feresa attenuata*
			Pygmy sperm whale	*Kogia breviceps*
			Sei whale	*Balaenoptera borealis*
			Short-finned pilot whale	*Globicephala macrorhynchus*
			Sperm whale	*Physeter macrocephalus*
		Dolphins	Common bottlenose dolphin	*Tursiops truncatus*
			Common dolphin	*Delphinus delphis*
			Dall's porpoise	*Phocoenoides dalli*
			Fraser's dolphin	*Lagenodelphis hosei*
			Pacific white-sided dolphin	*Lagenorhynchus obliquidens*
			Risso's dolphin	*Grampus griseus*
			Rough-toothed dolphin	*Steno bredanensis*
			Spinner dolphin	*Stenella longirostris*
			Pantropical spotted dolphin	*Stenella attenuata*
			Striped dolphin	*Stenella coeruleoalba*
		Sea Turtles	Leatherback sea turtle	*Dermochelys coriacea*
			Green sea turtle	*Chelonia mydas*
			Olive Ridley sea turtle	*Lepidochelys olivacea*
			Loggerhead sea turtle	*Caretta caretta*
			Hawksbill sea turtle	*Eretmochelys imbricata*
		Fishes	Scalloped hammerhead shark	*Sphyrna lewini*
		Corals (15 species)		*Acropora* spp., *Euphyllia paradivisa*, *Isopora crateriformis*, *Montipora australiensis*, *Pavona diffluens*, *Porites napopora*, *Seriatopora aculeata*
	U.S. Fish and Wildlife Service **FWS**	Seabirds	Newell's shearwater	*Puffinus auricularis*
			Micronesian megapode	*Megapodius laperouse*
			Short-tailed albatrosses	*Phoebastria immutabilis*
		Other species	Dugongs	*Dugong dugon*

the third-highest number of marine mammal stocks is found in the Western Pacific (15% of total marine mammal stocks), with ~17% listed as strategic marine mammal stocks. Of all Western Pacific marine mammal stocks, 29.2% have unknown population size, which is fourth-highest nationally among regions.

10.4.3 Environmental Forcing and Major Features

10.4.3.1 Oceanographic and Climatological Context

Circulation patterns throughout the Western Pacific EEZ are driven primarily by several major currents and

Watch the Shell, Dude: Dealing with Bycatch of Species with Special Status

Using gear technology and oceanographic predictions of frontal boundaries for species distribution to minimize bycatch —(see also Baleen whales, Odontocetes, Seabirds throughout the book)

Years ago when my (JL) kids were young, I was the heroic father trying to win a stuffed animal for my 4-year-old at the county fair. I kept hitting the wrong thing and ended up winning all these plastic snake and lizard toys that she most definitely did not want. It was kinda frustrating trying to go after one thing and ending up with a fake replica of a reptile instead.

That's a lot like longline bycatch in the central Pacific.

Fishermen target large tunas and billfish and the like. They end up catching turtles. Which is in itself not good, a waste of effort, and frustrating for anyone fishing. But it is also a problem because turtles are endangered and afforded special status. The Endangered Species Act (and Marine Mammal Protection Act) in practice often trump fisheries objectives given the severity of extinction risk. Thus, the bycatch of turtles was potentially going to shut down some of the longline fisheries in the Pacific.

The trade-off was clear—figure out a way to minimize turtle bycatch. The solutions varied. Some involved changing the set depth of the line, whereas others involved toggling lights on the line. Others engineered circle hooks to make it less likely to capture turtles. All excellent ideas and in their own ways effective, particularly when combined and employed together. Gear engineering is a huge but often undersung part of ecosystem-based fisheries management—avoiding bycatch by figuring out how to not catch it is simply just smart.

But it gets better.

Oceanographers were noting that turtles had a slightly different temperature preference than highly migratory fishes. And that lines set in bands of water at certain temperatures had higher by-catches than those that didn't. If there were only a way to predict or even observe those water bands, let the fleet (and everybody) know about them, and see if that could help lower turtle bycatch...Guess what? There was, and is, a way to do that. Hence NOAA Fisheries' "Turtle watch"[4] (Howell et al. 2008, 2015) was born.

This is a great example of considering all facets of the ecology, behavior, species interactions, and oceanography to better manage living marine resources in a way that mitigated a major trade-off. Make no mistake, the conflict between fisheries and bycatch is a challenge. The conflict between fisheries and bycatch that has some sort of protected status is amplified. Finding a solution to minimize that conflict is what EBFM is all about.

Similar efforts have been adopted for baleen whales, odontocetes, seabirds, pinnipeds, and other PET species. Often with a similarly clever incorporation of oceanography

(Grantham et al. 2011, Lewison et al. 2015, Hazen et al. 2017, 2018). More widely adopting such solutions seems both prudent and doable.

[4] https://oceanwatch.pifsc.noaa.gov/turtlewatch.html

References

Grantham HS, Game ET, Lombard AT, Hobday AJ, Richardson AJ, Beckley LE, Pressey RL, et al. 2011. Accommodating dynamic oceanographic processes and pelagic biodiversity in marine conservation planning. *PLoS One* 6(2).

Hazen EL, Palacios DM, Forney KA, Howell EA, Becker E, Hoover AL, Irvine L, et al. 2017. WhaleWatch: a dynamic management tool for predicting blue whale density in the California Current. *Journal of Applied Ecology* 54(5):1415–28.

Hazen EL, Scales KL, Maxwell SM, Briscoe DK, Welch H, Bograd SJ, Bailey H, et al. 2018. A dynamic ocean management tool to reduce bycatch and support sustainable fisheries. *Science Advances* 4(5):eaar3001.

Howell EA, Hoover A, Benson SR, Bailey H, Polovina JJ, Seminoff JA, Dutton PH. 2015. Enhancing the TurtleWatch product for leatherback sea turtles, a dynamic habitat model for ecosystem-based management. *Fisheries Oceanography* 24(1):57–68.

Howell EA, Kobayashi DR, Parker DM, Balazs GH, Polovina JJ. 2008. TurtleWatch: a tool to aid in the bycatch reduction of loggerhead turtles Caretta caretta in the Hawaii-based pelagic longline fishery. *Endangered Species Research* 5(2–3):267–78.

Lewison R, Hobday AJ, Maxwell S, Hazen E, Hartog JR, Dunn DC, Briscoe D, et al. 2015. Dynamic ocean management: identifying the critical ingredients of dynamic approaches to ocean resource management. *BioScience* 65(5):486–98.

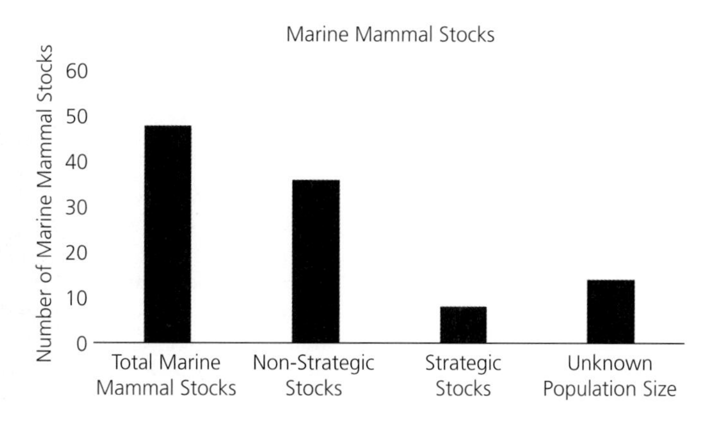

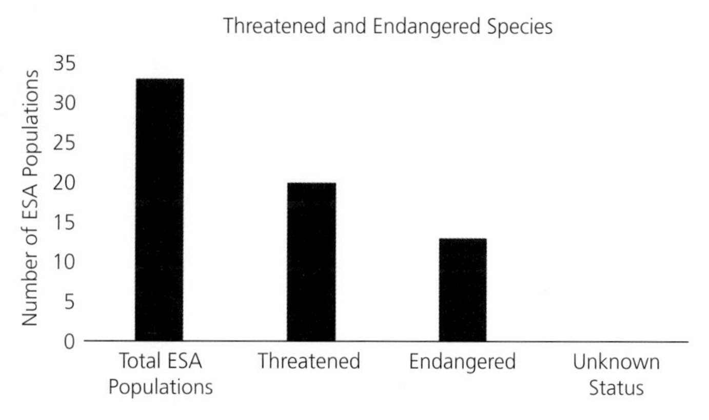

Figure 10.15 Number and status of federally protected species (marine mammal stocks, top panel); distinct population segments of species listed under the Endangered Species Act (ESA, bottom panel) occurring in the Western Pacific region.

components of the North and South Pacific Gyres (Fig. 10.1). Circulation in the Hawaii subregion of the Western Pacific is primarily associated with the clockwise North Pacific Gyre and its southern North Equatorial Current component. The North Equatorial Current continues toward the Marianas archipelago in the Western Pacific, where clockwise circulation influences both Guam and the Northern Mariana Islands. American Samoa and portions of the Pacific Remote Islands are also found within the South Pacific Gyre, as encompassed by the South Equatorial Current, and subject to predominantly counterclockwise circulation patterns. Complex equatorial current systems influence the remaining Pacific Remote Islands, including the Equatorial Undercurrent and the North Equatorial Counter Current.

Clear interannual and multidecadal patterns in average sea surface temperature (SST) have been observed for the Western Pacific region (Fig. 10.16), with temperatures averaging 24.3°C ± 0.03 (SE) over time (1854–2016) in the Hawaii subregion. SST throughout Hawaiian waters has increased by 0.4°C since the mid-twentieth century. Among territorial archipelagos, highest average SST over time has been observed in American Samoa (28.4°C ± 0.02, SE), where temperatures have increased by 0.8°C since the mid-twentieth century. Similar average trends have

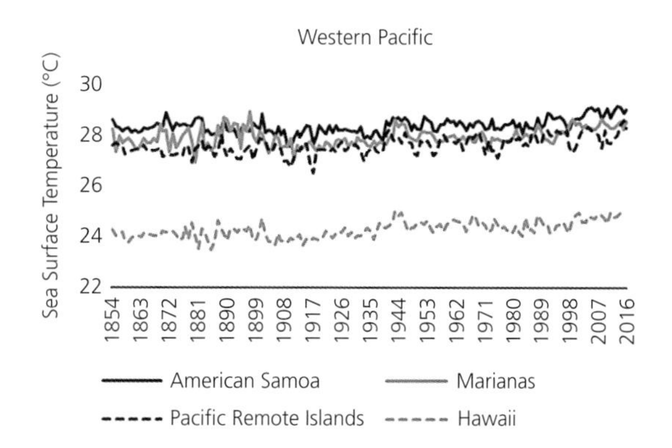

Figure 10.16 Average sea surface temperature (°C) over time (years 1854–2016) for Western Pacific subregions.

Data derived from the NOAA Extended Reconstructed Sea Surface Temperature dataset (https://www.ncdc.noaa.gov/).

occurred for the Marianas(27.9°C ± 0.03, SE) and Pacific Remote Islands Archipelagos (27.6°C ± 0.03, SE), while SST has increased by 0.76°C and 1.1°C, respectively, in those territorial subregions. In addition, SST temperatures between summer and winter range from only 3–5°C for the MHI, but up to 9–11°C for the NW Hawaiian islands (R. Brainard, pers. comm.). In American Samoa, summer–winter temperatures range between 0.17 and 1.8°C on average, while ranges are broader for the Marianas (0.1–4°C) and Pacific remote islands (0.06–1.5°C) archipelagos. Coincident with these temperature observations are Pacific basin-scale climate oscillations (Fig. 10.17), including the El Niño–Southern Oscillation (SOI), Multivariate ENSO Index (MEI), Northern Oscillation Index (NOI), Northern Pacific Gyre Oscillation (NPGO), and PDO, which exhibit interannual and decadal cycles that can influence the environmental conditions and ecologies of several U.S. regions, particularly the Western Pacific.

10.4.3.2 Notable Physical Features and Destabilizing Events and Phenomena

With four prominently identified subregions, the Western Pacific contains 13 major bays (Table 10.3) off the coasts of Hawaii, American Samoa, Guam and the Northern Marianas, as well as the Pacific remote islands and major island groupings throughout these archipelagos. Proportionally, the Western Pacific makes up the largest percentage (~51.9%) of the U.S. EEZ (~7 million km^2), and 4.9% of the national shoreline (~5,100 km), while containing the deepest average depth (~4720 m) among regions and maximum depths of ~10.8 km (Fig. 10.18), making it the deepest of the regions throughout the Pacific basin within which the U.S. EEZ occurs, as well as containing the deepest national waters. The Hawaii subregion (2.6 million km^2) comprises 36.3% of the Western Pacific and 18.8% of the U.S. EEZ, while U.S. Pacific Island territories cumulatively make up 33% of the

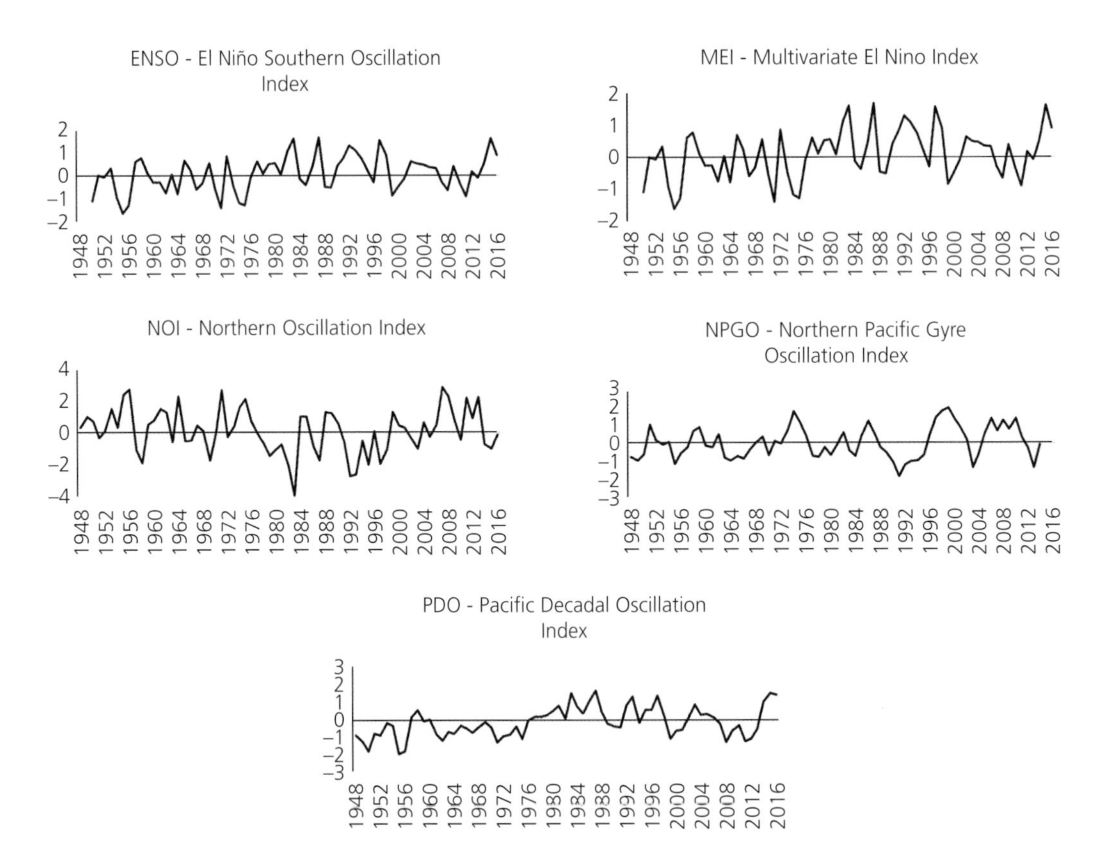

Figure 10.17 Climate forcing indices for the Western Pacific region over time (years 1948–2016).

Data derived from NOAA Earth System Research Laboratory data framework (https://www.esrl.noaa.gov/).

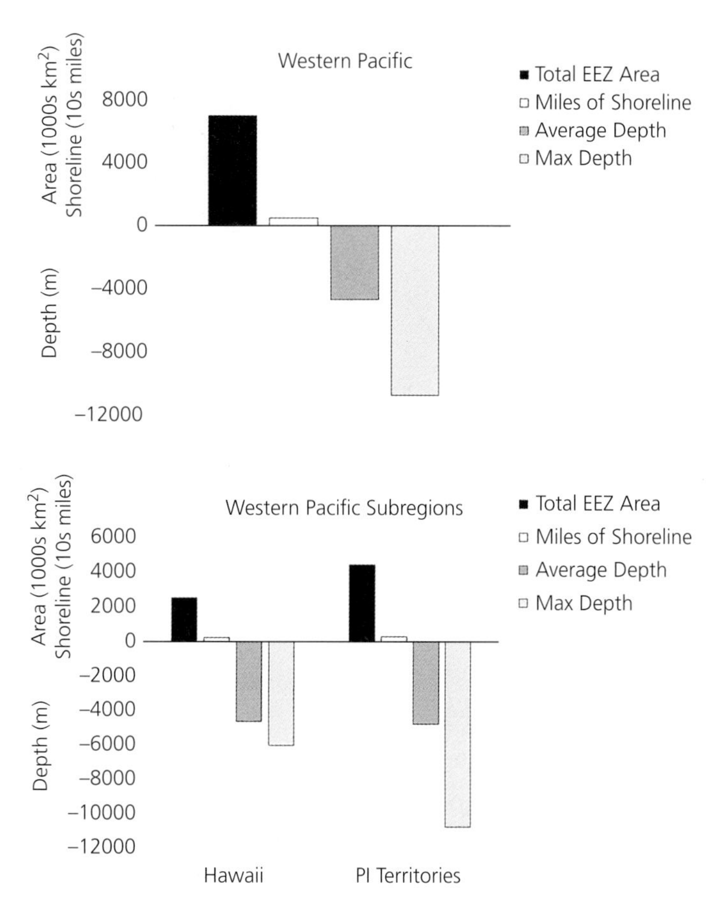

Figure 10.18 Total shelf area, miles of shoreline, and average and maximum depth of marine regions and subregions of Western Pacific.

U.S. EEZ. The Hawaii subregion contains ~2300 km of shoreline, which approaches cumulative values observed throughout all Pacific Island territories (~2800 km). The deepest average and maximum depths (~10.8 km) occur off the Marianas archipelago, given its proximity to the Mariana Trench, while average depths among the Hawaii (~4.6 km) and cumulative territorial subregions (~4.8 km) are similar.

In the Western Pacific region, 1,252 tropical cyclones have been reported over time (Fig. 10.19), with the majority occurring throughout U.S. Pacific Island territories. Frequency of observed and reported storms has increased over time to peak values of 189 tropical cyclones in the 1960s and 166 tropical cyclones in the 1990s, with the majority occurring throughout the Pacific Remote Islands. In more recent years, numbers of storms have decreased to currently 63 reported during the 2010s. Eight tropical cyclones have been reported for Hawaii since 1850, with two storms per decade being observed during the 1950s, 1980s–1990s, and 2010s. Relatively low numbers have been observed in American Samoa, with 24 storms total and up to 4–5 storms per decade since the 1990s. More frequent tropical cyclones have occurred throughout the Marianas archipelago.

10.4.4 Major Pressures and Exogenous Factors

10.4.4.1 Other Ocean Use Context

Other ocean uses, besides fishing, in the Western Pacific (Hawaii) are most obvioiusly represented by the tourism/recreation sector, followed to a much lesser extent by marine transportation and marine construction interests (Fig. 10.20). Among sectors, trends for Western Pacific establishments and employments have shown significant growth over the past decades, with consistent dominance of tourism establishments and employments comprising 91–93% of total Western Pacific establishments and 88–90% of total Western Pacific employments. Increases in total ocean establishments and employments were most prominent

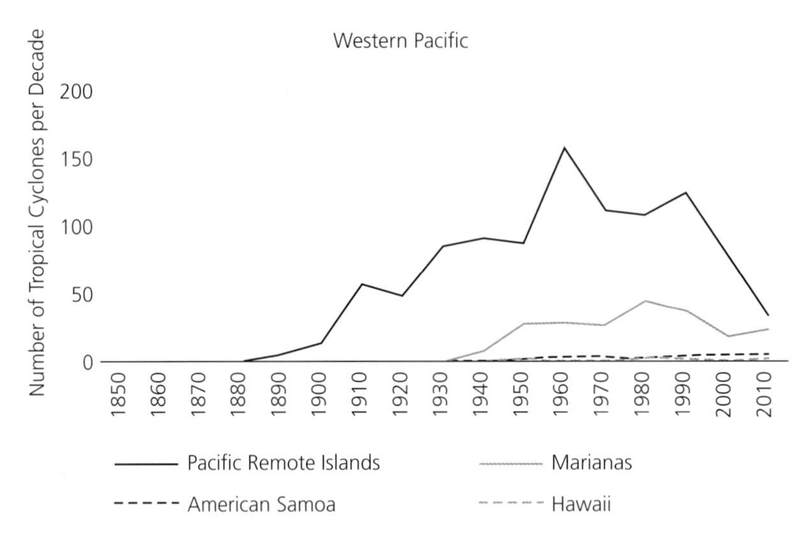

Figure 10.19 Total number of tropical cyclones recorded per decade for the Western Pacific Hawaii and U.S. Pacific Island territories subregions since year 1850.

Data derived from NOAA Office for Coastal Management (https://coast.noaa.gov/hurricanes/).

during the 2010s, with highest values occurring in 2016 at ~4,300 establishments and ~118,000 employments (nationally second-lowest). The GDP from tourism is highest among sectors, contributing 86–89% toward total Western Pacific Ocean GDP. Over the past decades, ocean GDP has steadily increased, with most recent estimated values at ~$8.6 billion. Overall, the Western Pacific only contributes 2.4% of the U.S. total ocean GDP, which is second-lowest nationally. Establishments and employments in the Western Pacific marine transportation sector contribute 2–3% toward its total ocean establishments and comprise 3–6% of its total ocean employments. Employments in the ship and boat building sector contribute 4–5% of Western Pacific total ocean employments. The Western Pacific contributes 3.5% of U.S. tourism establishments, 5.3% of national tourism employments, which are both nationally sixth-highest, and 7.2% of national tourism GDP, which is nationally seventh-highest. It is nationally sixth-highest in its contributions toward ship and boat building establishments, employments, and GDP (1.4% of U.S. ship and boatbuilding establishments, 3.5% of U.S. ship and boatbuilding employments, and 0.4% of U.S. ship and boatbuilding GDP), and marine construction and marine transportation employments and GDP. The Western Pacific is seventh-highest in its contributions to all remaining categories, including toward national marine construction and marine transportation establishments, and offshore mineral extraction establishments, employments, and GDP.

The degree of tourism, as measured by number of dive shops, number of major ports and marinas, and number of cruise ship departure passengers is among mid to low-range nationally in the Western Pacific (Fig. 10.21), with third-highest numbers of dive shops, sixth-highest numbers of ports, and second-highest numbers of marinas (mostly concentrated in Hawaii).

Dive shops are most abundant in Hawaii (n = 64) and the Marianas archipelago (n = 34), while six and two ports are found in Hawaii and Guam, respectively. Cruise destination passengers in the Western Pacific are fourth-highest nationally with ~200,000 per year. However, departure passengers are seventh-highest nationally with the majority leaving from populated ports in Hawaii (i.e., Honolulu) at rates of 126,000 per year. The Western Pacific is fourth-highest nationally in terms of offshore oil production and the number of oil rigs over time, although values have remained low over time (0–2 rigs per year), with occasional increases during the 1990s and 2000s (Fig. 10.22). Two U.S. Bureau of Ocean Energy Management (BOEM) wind energy sites are found in this region, which is lowest nationally.

10.4.5 Systems Ecology and Fisheries

10.4.5.1 Basal Ecosystem Production

Since 1998, mean surface chlorophyll values (Fig. 10.23) throughout the Western Pacific EEZ have been lowest nationally, with values among subregions higher in the Pacific Island territories (0.08 mg m^{-3} ± 0.002, SE). Values have remained relatively steady in the Hawaii subregion

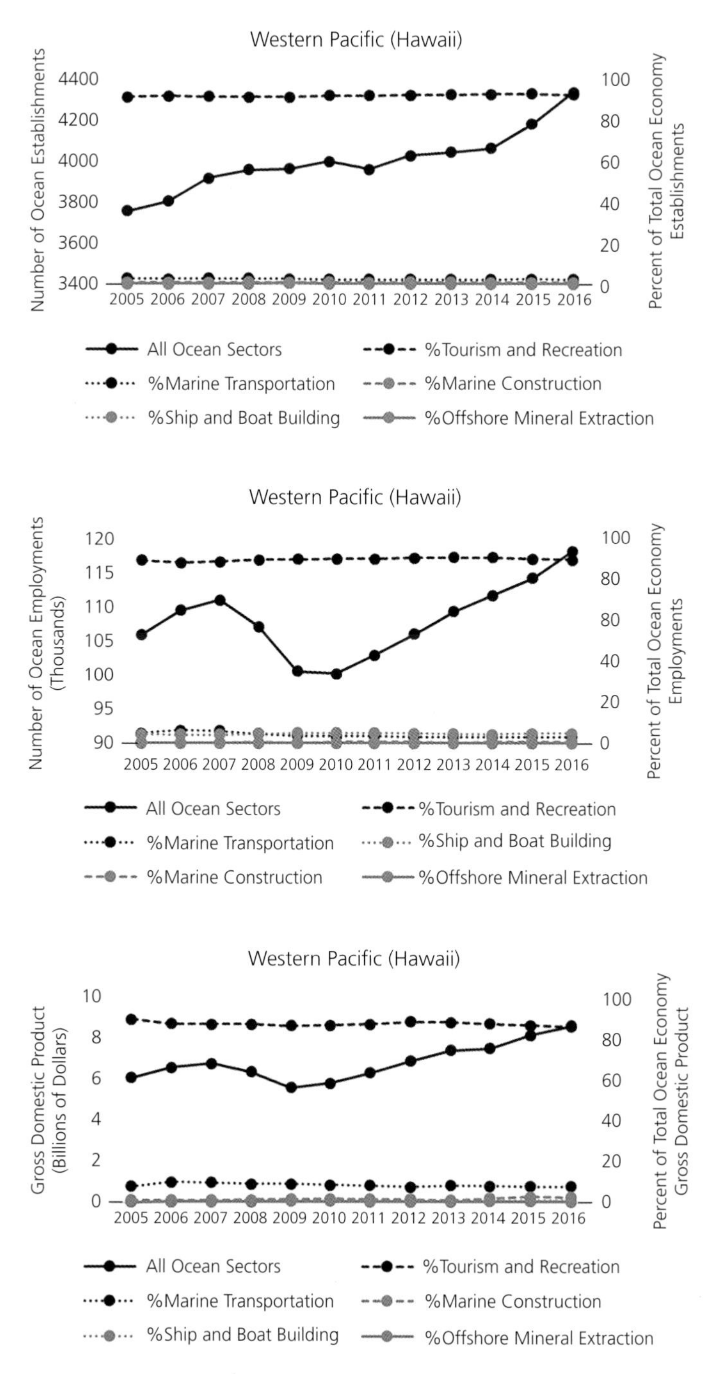

Figure 10.20 Number (primary axis) and percent contribution (secondary axis) of multisector establishments, employments, and gross domestic product (GDP) and their percent contribution to total multisector oceanic economy in the Western Pacific Hawaii subregion (years 2005–2016).

Data derived from the National Ocean Economics Program.

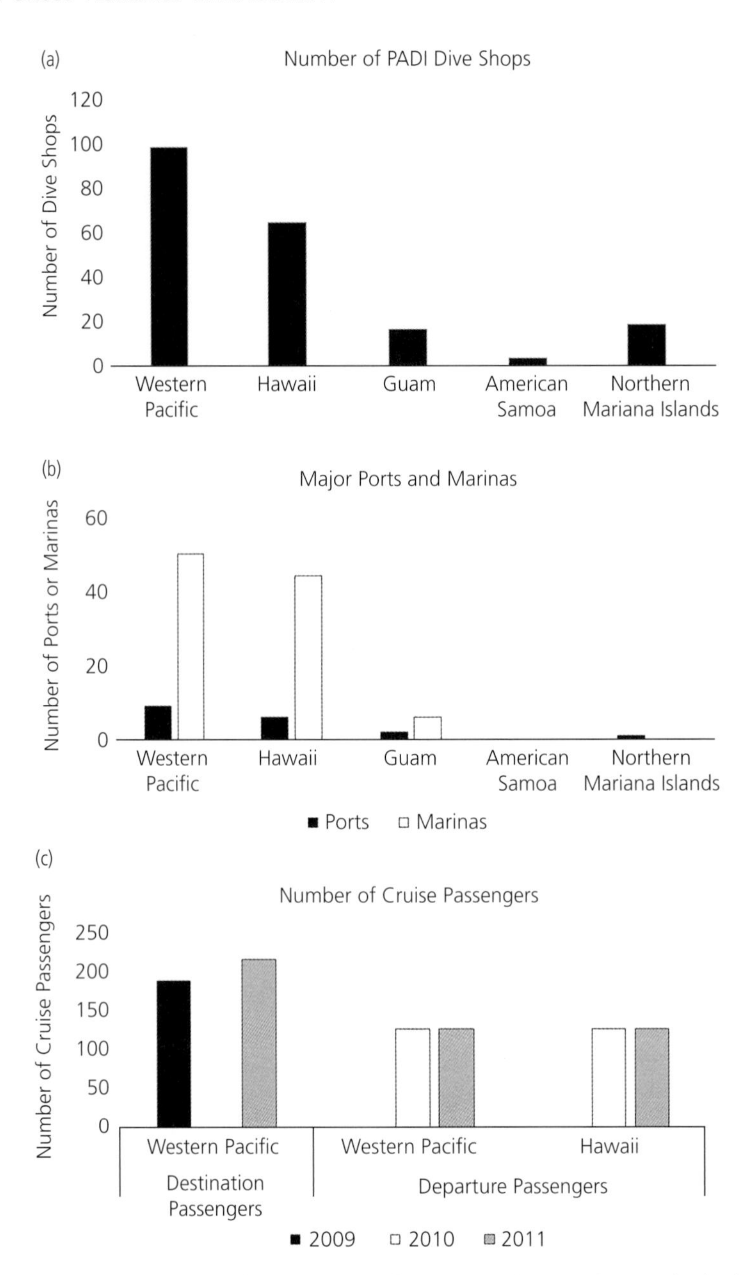

Figure 10.21 Tourism indicators for the Western Pacific region and identified subregions (Hawaii and U.S. Pacific Island territories): (a) Number of Professional Association of Diving Instructors (PADI) dive shops. Data derived from http://apps.padi.com/scuba-diving/dive-shop-locator/. (b) Number of major ports and marinas. (c) Number of cruise destination (years 2009, 2011) and cruise departure port passengers (years 2010–2011); values in thousands.

(0.06–0.07 mg m^{-3}), while slightly increasing over time in the Pacific Island territories throughout the 2000s, and having decreased in more recent years. Average annual productivity in the Hawaii subregion has slightly decreased over time from ~93 to 80 g C m^{-2} $year^{-1}$ (eighth-highest nationally) over the same time period.

10.4.5.2 System Exploitation

Nationally, the eighth-highest numbers of reported landed taxa have been captured in the Western Pacific Hawaii subregion over time, with consistently low values (average: ~16 taxa) during the 1980s–1990s, and a substantial increase occurring in 2002. During

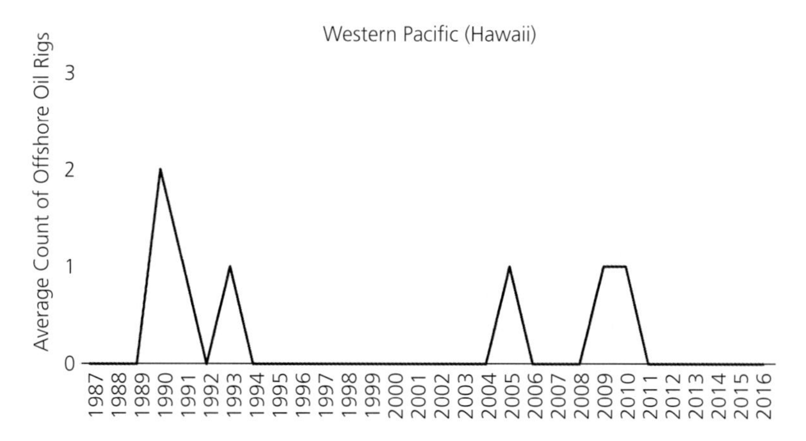

Figure 10.22 Average count of offshore oil rigs for the Western Pacific Hawaii subregion. Data derived from Baker Hughes Rotary Rig Counts.

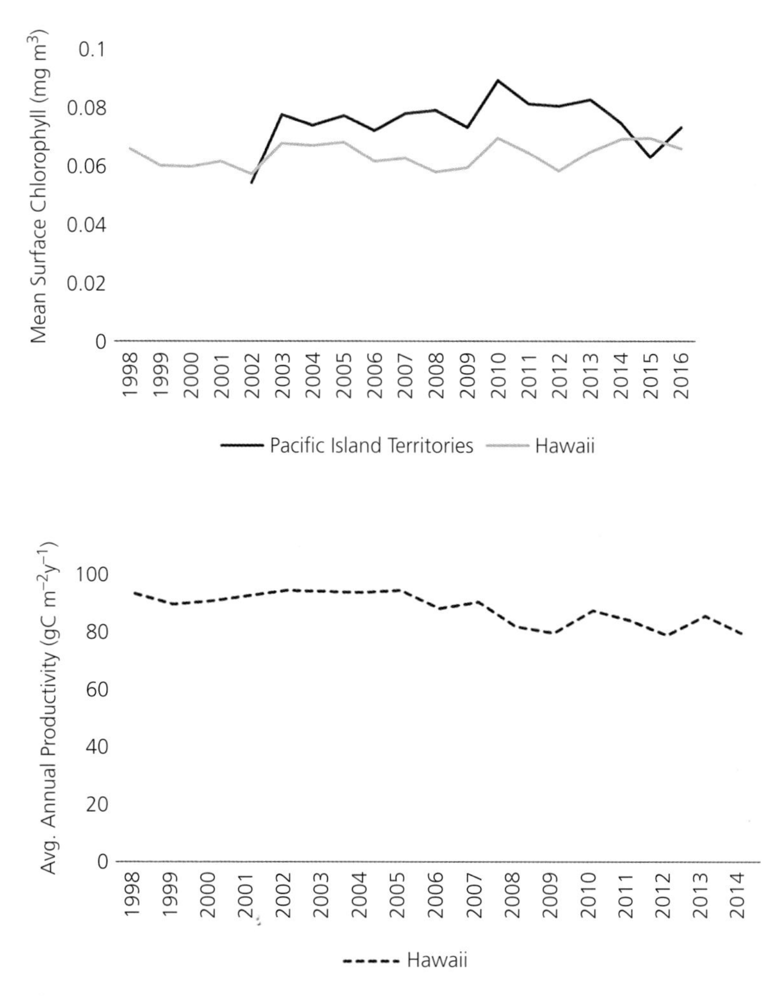

Figure 10.23 Mean surface chlorophyll (mg m^{-3}; top panel) for the Western Pacific region and identified subregions and average annual productivity (grams carbon m^{-2} $year^{-1}$; bottom panel) for the Hawaii subregion over time.

Data derived from NASA Ocean Color Web (https://oceancolor.gsfc.nasa.gov/) and productivity calculated using the Vertically Generalized Production Model (VGPM).

the 2000s and 2010s, values have continued steadily at up to 68 taxa (Fig. 10.24). These increases are related to more species-specific reporting of fisheries catches, including for reef fishes and macroinvertebrates, billfishes, coastal pelagic species, and several shark species. Combined surveyed nearshore fish and invertebrate biomass for the Western Pacific region is highest nationally (Fig. 10.25). Over the past decade, values in the Hawaii subregion have averaged ~186.7 million ± 11.6 million (SE) metric tons year^{-1} and increased to 212.6 million metric tons in 2015. In the Pacific Island territories, cumulative nearshore biomass values have oscillated over the past decade at an average of ~6.2 billion ± 1.03 billion (SE) metric tons year^{-1}, with most current values at 8.3 billion metric tons in 2015. These values do not account for highly migratory species for which total biomass estimates are limited (and are considered more fully in Chapter 11).

Western Pacific values for exploitation (total domestic landings/biomass) have oscillated over the past decade at nationally lowest values, with peaks observed during 2011 (6.3×10^{-6}) and 2014 (5.9×10^{-6}; Fig. 10.26). In the Hawaii subregion, there have been increases in total nearshore biomass/productivity have over time from a minimum value in 2010 (0.07) to a maximum of 0.10 in 2014. While oscillating, total domestic landings/productivity values have increased over time (4×10^{-6} to 1.26×10^{-4}), with peak values observed in 2008, and more recent values remaining at 1×10^{-5}.

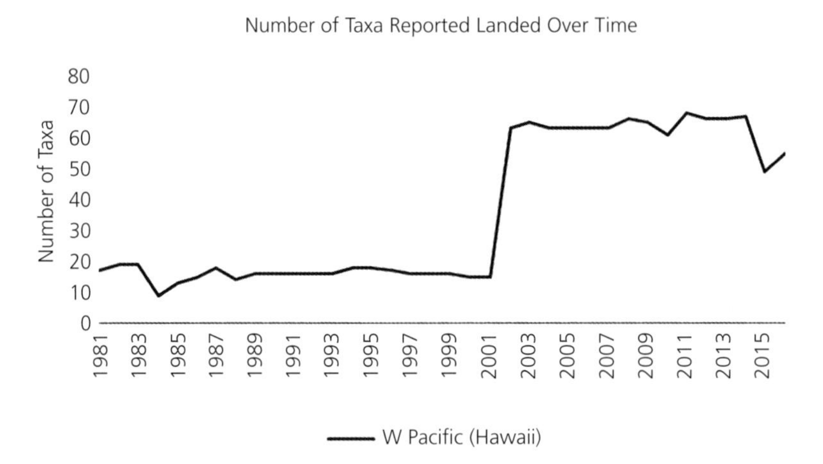

Figure 10.24 Number of taxa reported captured by commercial and recreational fisheries in the Western Pacific Hawaii subregion.
Data derived from NOAA National Marine Fisheries Service commercial and recreational fisheries statistics.

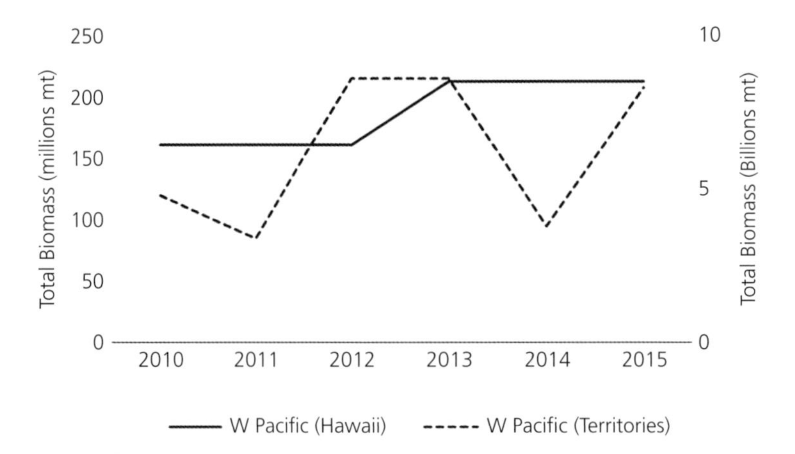

Figure 10.25 Total combined surveyed fish and invertebrate biomass (metric tons; mt) for the Western Pacific region, over time. Data for Pacific Island territories shown on the secondary axis.

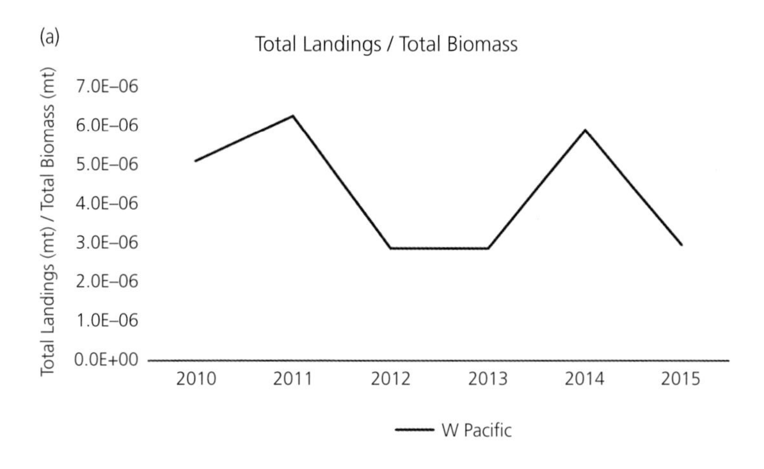

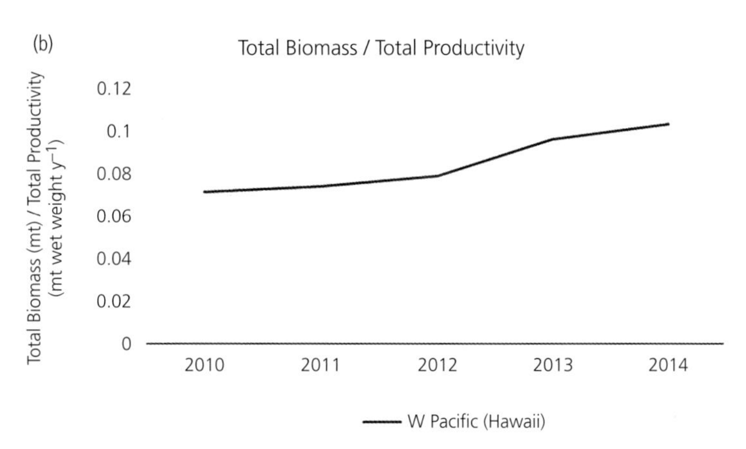

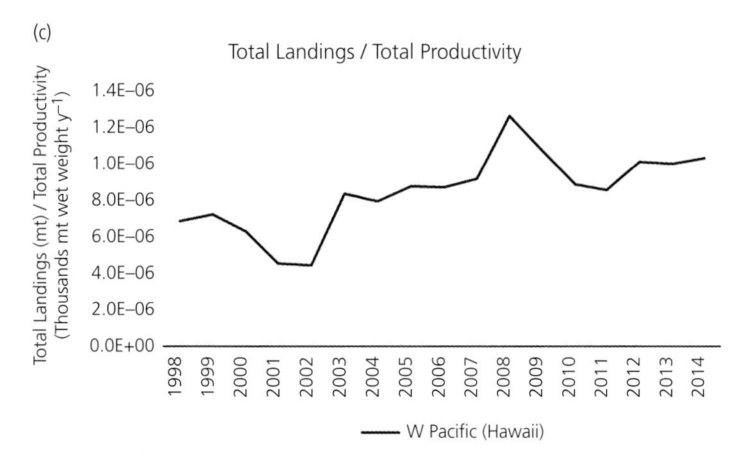

Figure 10.26 Ratios of (a) total commercial and recreational landings (metric tons) to total biomass (metric tons; i.e., exploitation index) for the Western Pacific region; (b) total biomass (metric tons) to total productivity (metric tons wet weight year^{-1}) and (c) total commercial and recreational landings (metric tons) to total productivity (metric tons wet weight yr^{-1}) for the Western Pacific Hawaii subregion.

10.4.5.3 Targeted and Non-Targeted Resources

Total domestic commercial and recreational landings are currently seventh-highest nationally in the Western Pacific region (Fig. 10.27), not accounting for all captures of highly migratory species landed in international ports throughout the Pacific basin, which are not fully reflected in nationally reported fisheries data (see Chapter 11 on RFMOs; Craig et al. 2017). Among subregions, Hawaii domestic commercial landings greatly oscillated in the early 1980s (4800–15,800 metric tons year^{-1}) and increased to peak values (~16,740 metric tons) through the 1990s. Decreases were observed in the early 2000s, following which values have increased to a current value of ~15,700 metric tons in 2015. Domestic commercial landings are much lower in Pacific Island territories, which increased to peak values (~7,500 metric tons) in 2002, and have intermittently decreased over time to current values of ~2400 metric tons. Granted, domestic recreational landings for the Western Pacific region are only estimated for Hawaii, with values over the past decades increasing to peak values (~12,650

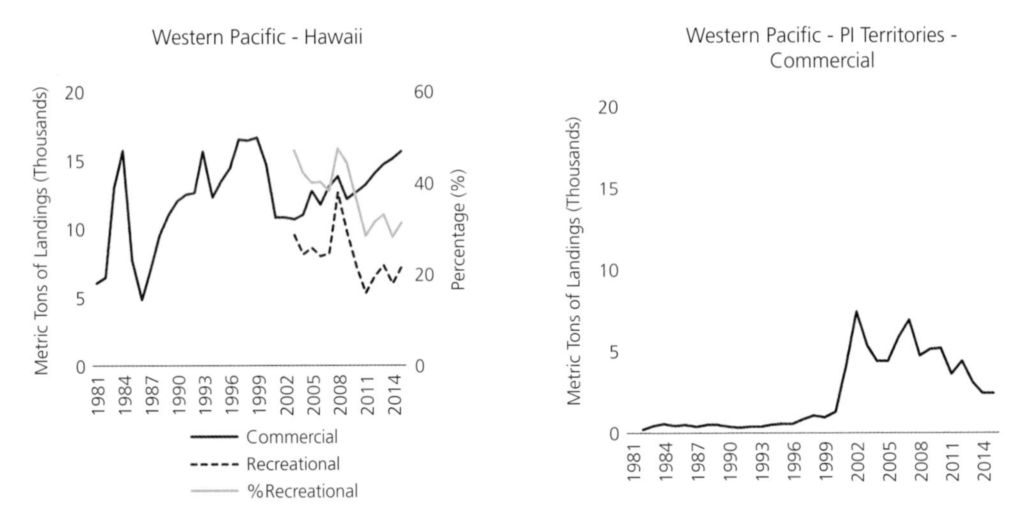

Figure 10.27 Total commercial and recreational landings and percent contribution of recreational landings (%Recr) over time for the Western Pacific Hawaii and U.S. Pacific Island (PI) territories subregions (1981–2015).

Data derived from NOAA National Marine Fisheries Service commercial and recreational fisheries statistics.

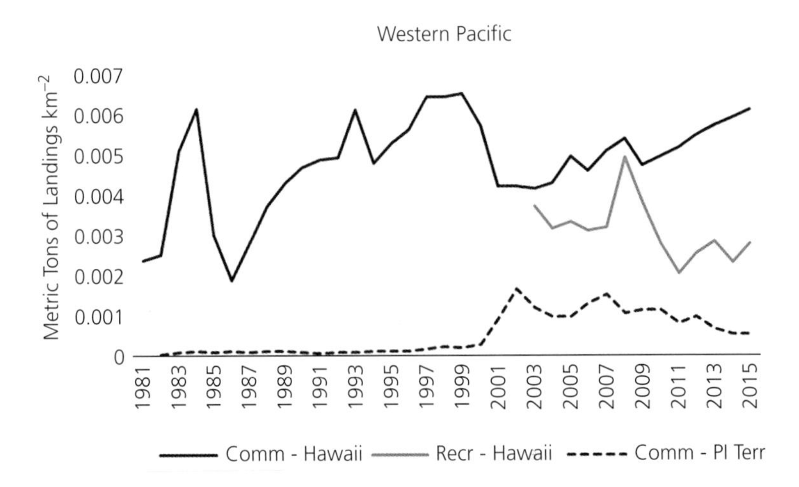

Figure 10.28 Total commercial landings (comm) and recreational (recr; metric tons) per square kilometer over time (1981–2015) for the Western Pacific Hawaii and U.S. Pacific Island territories (PI Terr) subregions.

Data derived from NOAA National Marine Fisheries Service commercial and recreational fisheries statistics.

metric tons) in 2008 and decreasing in the late 2000s and early 2010s to values that have remained around an average of ~6,500 metric tons year^{-1}. Over time, recreational catches in Hawaii have contributed nearly half (up to 48%) of total landings, while currently contributing 31.3% of the total Hawaiian catch in 2015. When standardized per square kilometer (Fig. 10.28), Western Pacific landings follow very similar patterns as observed for total domestic commercial and recreational landings values, with this region containing the lowest concentration of fisheries landings over time nationally. Among subregions, highest concentrated domestic fishing pressure (up to 0.006 metric tons km^{-2}) has been historically observed in Hawaii, while lower values (up to 0.002 metric tons km^{-2}) are observed for Pacific Island territories. However, these values are estimated to be much higher in nearshore areas of Guam and Saipan (M. Parke, pers. comm.).

Bycatch continues to persist throughout the Western Pacific region, with values for most taxa remaining similar or increasing over the four-year period of examination (Fig. 10.29). Values are higher in the Hawaii subregion than in Pacific Island territories (i.e., American Samoa) as likely related to the vast majority of fishing effort being Hawaii's fishery (P. Woodworth-Jefcoats, pers. comm.). By weight, bycatch is most pronounced for bony fishes, while it is not reported for invertebrates. Comparatively among U.S. regions, moderate values of bycatch are observed for sharks, marine mammals, and seabirds. On average, reported

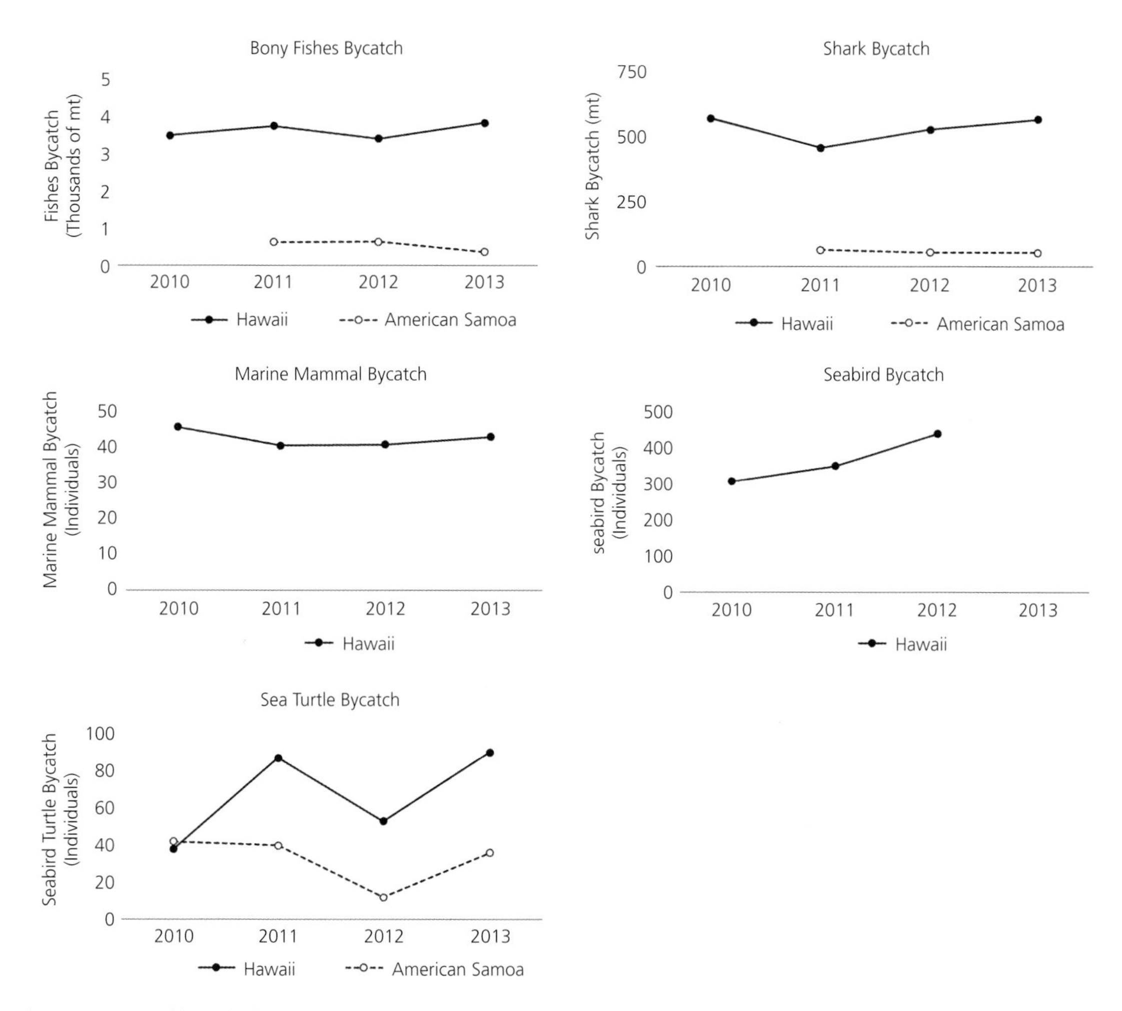

Figure 10.29 Total bycatch of species groups in metric tons (mt) and number of individuals for the Western Pacific Hawaii and American Samoa subregions.

Data derived from NOAA National Marine Fisheries Service U.S. National Bycatch Report Database System.

bycatch in the Western Pacific is lowest among regions by weight and number. For marine mammals and sharks, bycatch values have been more consistent over time, although values for seabird and sea turtle bycatch (number of individuals) increased from 2010–2013 in Hawaii, and slightly decreased in American Samoa.

10.5 Synthesis

The U.S. Western Pacific has been affected by above average natural and human stressors that include the highest frequency and intensity of cyclonic storm activity, intensive fishing, high coastal development, and continually increasing temperatures (Maragos 1986, Maragos 1993, Friedlander et al. 2002, 2003, Camargo & Sobel 2005, Emanuel 2005, Baker et al. 2008, Carpenter et al. 2008, Di Lorenzo et al. 2008, Li et al. 2010, Kittinger et al. 2011, Keener et al. 2013, Edwards et al. 2014, Friedlander et al. 2016, Filous et al. 2017, Veitayaki et al. 2017). Predicted effects from shifting climate and oceanographic forcing are anticipated to compound current anthropogenic impacts on LMRs and their dependent habitats in this region, while also affecting subregional marine economies (Short & Neckles 1999, Camargo et al. 2005, Björk et al. 2008, Di Lorenzo et al. 2008, Gilman et al. 2008, Bell et al. 2011, Pratchett et al. 2011, Waycott et al. 2011, Bell et al. 2013, Howell et al. 2013, Keener et al. 2013, Nicol et al. 2013, Gilman et al. 2016, Bell et al. 2018). Among U.S. marine ecosystems, comparatively mid-ranked LMR status and lower-ranked marine socioeconomics are currently observed in the U.S. Western Pacific (Hamnett & Pintz 1996, Overtz 2007, Sabater & Carroll 2009, Bell et al. 2013, Grafeld et al. 2016, Levine et al. 2016, Chan & Pan 2019, Link & Marshak 2019), with continued vulnerabilities to climate change and overexploitation hindering progress toward more sustainable and productive standings (Bell et al. 2011, Pratchett et al. 2011, Waycott et al. 2011, Shope et al. 2016, Kleiber & Kotowicz 2018). While second-lowest nationally in human population density among regions, the environmental effects of increasingly concentrated fisheries exploitation in the MHI and Pacific Island territories during the 1990s and 2000s, and high coastal development and tourism, have altered top-down and bottom-up forces in this system, with direct effects on LMR production (Maragos 1986, Maragos 1993, Friedlander & DeMartini 2002, Friedlander et al. 2002, 2003, Baker & Johanos 2004, Antonelis et al. 2006, Craig 2006, Friedlander et al. 2007, Grigg et al. 2008, Williams et al. 2008, Friedlander et al. 2010, Williams et al. 2011, Edwards et al. 2014, Heenan et al. 2014, Nadon et al. 2015, Williams et al. 2015, Giddens et al. 2018). These are compounded by the effects of climate change, species invasions, and habitat degradation that continue to affect the Western Pacific. Although monitoring of marine socioeconomics outside of Hawaii remains limited, a 14–16-billion-dollar tourism industry in this region contributes toward ongoing seafood demand, recreational access to marine resources, sportfishing, and complementary human pressures (Tian et al. 2013). These values are also enhanced by multiple international interests in highly migratory species, which contribute heavily toward fisheries landings and revenue throughout this region. While values for these species remain partially underestimated in some national reporting for Western Pacific domestic fisheries landings and revenue (Craig et al. 2017), improved reporting in logbooks, observer records, and at auction has occurred in recent years, which is allowing for more accurate estimates (P. Woodworth-Jefcoats, pers. communication). Refining EBFM strategies for the Western Pacific that also consider the cumulative effects of these natural and human stressors on its coupled socioecological system, particularly their influences on basal ecosystem productivity, habitats, sustainable fisheries landings and biomass, and their ongoing ecological consequences in a changing environment remains a needed next step.

There are several components of the Western Pacific fisheries ecosystem that are performing relatively well, despite its comparatively lower LMR and socioeconomic statuses. In particular, aspects of its human environment and ecosystem-based advances in fisheries governance serve as assets toward more holistic management approaches and ocean planning. Given its overwhelming coral reef and oceanic habitats, basal productivity is lowest nationally (Gulko et al. 2000, Richmond et al. 2002, Brainard et al. 2005, Bruckner et al. 2005, Craig et al. 2005, Porter et al. 2005, Rohman et al. 2005, Goldberg et al. 2008, Williams et al. 2012, Boyle et al. 2017, Link & Marshak 2019). However, coral reef and pelagic biomass for this region is highest nationally, despite low basal pelagic productivity and exploitation, and as fostered by expansive, productive, and protected reef areas (Williams et al. 2012, Link & Marshak 2019). Abundant tourism and marine commerce, low human population density, and expanding energy interests have fostered increasingly diverse ocean uses throughout this region. Efforts toward ensuring more sustainable practices are emerging as concerns continue regarding their environmental impacts. The continued implementation of five subregional FEPs for Hawaii, the Marianas, American Samoa, the Pacific Remote Islands, and pelagic fisheries resources over the past decade (WPRFMC 2010a, b, c, d, e) has served as a foundation toward more comprehensive EBFM, with focused attention away from single-species approaches to broader considerations of multiple species, trophic levels, habitats, fishing fleets and sectors, and subregional fisheries socioeconomics, with a focus on each archipelago. Ongoing protection of significant portions of the Western Pacific EEZ through establishment of marine national monuments and sanctuaries has also served to advance area-based fishing closures. These protections have allowed for the conservation and recovery of ecological integrity within extensive areas, while also limiting multiple human impacts, restricting fishing practices, and focusing attention on promoting and maintaining ecosystem function and integrity (Friedlander et al. 2003, Craig 2006, Mora et al. 2006, Friedlander et al. 2007, 2010, Kittinger et al. 2011, Rieser 2011, Williams et al. 2011, Friedlander et al. 2014, Young et al. 2015, Naro-Maciel et al. 2018). These efforts are complemented by advances in the understanding of ecosystem functioning for components of the Western Pacific ecosystem, modeling of multispecies responses to natural and human stressors, and initiatives toward more specific ecosystem approaches to management (Turgeon et al. 2002, Di Lorenzo et al. 2008, Hoeke et al. 2009, Tissot et al. 2009, Glazier 2011, McClanahan et al. 2011, Howell et al. 2013, Weijerman et al. 2015a, b, Friedlander et al. 2016, Weijerman et al. 2016a, b, c, 2018, PIFSC 2016, Gove et al. 2019, NMFS 2019b). Specifically, these include implementation and continued amendments of its five subregional FEPs, and ongoing efforts to compile information and assess data gaps through additional monitoring of ecosystems. It is important to note that this region needs continuation of efforts toward integrated and interdisciplinary ecosystem monitoring of pelagic and insular systems. Continued efforts to advance and refine modeling frameworks toward addressing and predicting LMR, habitat, and system-wide responses to human, natural, and climatic stressors are facilitating more ecosystem-based approaches for this region. These efforts also enhance information regarding EFH of managed species, and detail anticipated effects of management actions on Western Pacific physical, biological, and social environments.

These priorities are complemented by the completion of an EBFM implementation plan for the U.S. Pacific Islands region (NMFS 2019b), which works to continue these aforementioned priorities within its major subregions, strengthen ecosystem-level planning, and incorporate more specific ecosystem considerations into management actions, including the following:

- Engagement among NOAA Fisheries, WPRFMC, and state and territorial agencies with constituents to identify ecosystem-related information for addressing management needs and knowledge gaps.
- Assist councils, commissions, RFMOs, and other bodies in their development of new, or revision of existing, FEPs or other fisheries plans.
- Advance resources to continue conducting EBFM through the use of advanced sampling technologies and electronic monitoring to improve surveys and data streams, catch and bycatch information, and protected species conservation.
- Continue to collect core oceanographic data, conduct coral connectivity assessments within Pacific monuments and associated archipelagos, and examine coral resilience and recovery to climate change and ocean acidification within the Pacific remote islands.
- Integrate data into annual ESRs and stock assessment and fishery evaluation (SAFE) reports that summarize ecosystem effects on fisheries dynamics.
- Conduct systematic risk assessments of climate impacts on fisheries and protected species, and subregional ecosystems, including marine monuments. These efforts also include advancing habitat science for priority species and evaluating large-scale reef restoration efforts.
- Bolster modeling frameworks for improved understanding of ecosystem factors and ecosystem-level effects on fisheries and LMRs.
- Including ecosystem factors in crafting advice for managed species and their fisheries, and for enhancing species and ecosystem-level assessments.
- Integrating information for management advice across multiple species within ecosystems, including refining essential fish habitat and protected species ESA reviews, coral reef condition indicators, bycatch considerations, and emerging ecosystem-level impacts of aquaculture.
- Explore community well-being and socioeconomic metrics throughout the Pacific Islands, including social adaptive capacities, climate-related vulnerabilities, the roles of cultural values in contributing to Hawaiian marine ecosystems, enhancing socioeconomic monitoring of coral reef communities in Guam, and climate resiliency and adaptation approaches for American Samoan communities.

These efforts especially work to advance current progress toward enhanced characterization and monitoring of Western Pacific ecosystem dynamics, and to address data limitations that have constrained ecosystem-based approaches to management in this region, such as implementing system-wide reference points. They will also enhance identifying and providing quantitative information needed to produce indicators for ecosystem-level assessments, and build on expanding IEA-based efforts that are currently concentrated in western Hawaii (PIFSC 2016, Gove et al. 2019). These efforts have aligned socioeconomic and habitat-based considerations together with advancing understanding of physical and biological ecosystem components for this subregion, while efforts toward expanding and examining community vulnerability assessments to climate change, management practices, and emerging land and ocean use patterns remain priorities for this region (Gilman 1997, Wright et al. 2006, Jokiel et al. 2011, Levine et al. 2016, PIFSC 2016, Ayers 2018, Kleiber & Kotowicz 2018, Chan & Pan 2019, Gove et al. 2019, NMFS 2019b). Although scaling back of monitoring programs and data limitations beyond coral reef ecosystems have severely limited the ability to comprehensively assess many of this region's ecosystems at present, these priorities are working toward providing foundational information to support needed investigations into ecosystem and LMR status and build upon frameworks for broader management strategies. However, it is important to note that at present no milestones have been identified for establishing and using ecosystem-level reference points in the current EBFM plan (NMFS 2019b).

While frameworks and priorities are in place, given limited information applications and compounding natural and human effects on this system, many challenges remain toward advancing ecosystem approaches to management for the Western Pacific. As in other data-limited regions, efforts to mitigate overfishing and to rebuild overfished stocks have remained constrained by limited historical and comprehensive monitoring of fishery populations, fishing effort, or complementary environmental indicators. The U.S. Western Pacific contains high proportions of unassessed stocks in terms of overfishing (highest nationally unassessed) or overfished status (second-highest nationally unassessed; Link & Marshak 2019), all of which have limited progress on broader management practices. While some recent comprehensive investigations into regional fisheries ecosystem dynamics and ecological assessments have occurred, they have generally been restricted to small portions of the Western Pacific, namely the MHI, West Hawaii, Guam, and American Samoa, or within portions of non-populated marine monuments (Richmond et al. 2002, Brainard et al. 2005, Bruckner et al. 2005, Porter et al. 2005, Burdick et al. 2008, Williams et al. 2011, PIFSC 2016, Weijerman et al. 2016c, Gove et al. 2019). Improved understanding of fisheries practices, ecosystem dynamics, and the consequences of emerging human and natural impacts in Guam, the Marianas, American Samoa, and the Pacific Remote Islands remain needed, especially with increasing vulnerability of LMRs to changes in climate forcing, increasing temperatures, ocean acidification, and widespread coral bleaching events. Efforts toward resolving these data gaps and advancing ecosystem-level management frameworks that incorporate broader ecosystem considerations also remain needed, especially in consideration of system-level shifts and thresholds that may be exceeded in Western Pacific subregions. Given the historical and recent overexploitation of this system, especially in urbanized and populated regions, compounded with the effects of increased sedimentation, nutrient loading, recreational pressure, and habitat loss and degradation, the ability to examine and mitigate these impacts as related to system biomass and production, and incorporate the effects of natural and human-derived factors into management strategies is needed. In particular, overall basal ecosystem productivity for this region is expected to decrease with increasing stratification of the water column, due to regional warming, with associated loss of trophic transfer that supports current levels of fisheries harvest (Stock et al. 2017, Gaines et al. 2018). Direct consideration of relationships between fisheries harvest, ecosystem properties, and ocean use is necessary for a sustainable approach to management in the U.S. Western Pacific.

As tourism and residential human population continues to increase and enhance ongoing stressors to this system, the effects of climate-related shifts in LMR distribution and abundance will likely result in intensified resource competition among marine sectors and international interests. Resulting conflicts among commercial fisheries fleets, recreational and Pacific Islander fisheries interests, tourism, marine transportation, and emerging energy industries are anticipated within this region. As such, emerging priorities for the U.S. Western Pacific include greater understanding of Pacific Island fisheries socioeconomics, which should also consider the effects of other ocean uses on these interacting sectors, economies, and dependent coastal communities toward the development of more comprehensive EBM strategies.

Broadening current fisheries management approaches to incorporate expanding and overlapping ocean uses into ones that consider their impacts on human communities and multispecies interactions throughout habitat mosaics systematically addresses many of these ongoing and emerging factors within the Western Pacific. With tourism, LMRs, and military spending contributing significant proportions toward the marine economy of this region, their trade-offs with other expanding sectors would need to be considered in more holistic management frameworks. While the Western Pacific leads nationally in terms of HAPCs and protected areas within its EEZ, with significant portions permanently closed to fishing, there is still a need for more strategic considerations of habitat and spatial management within more populated areas to combat the effects of natural and human-induced stressors. Especially needed for these regions is investment toward improved LMR surveying, more comprehensive fishery-independent and commercial and recreational fishery-dependent data collection (including bycatch), improved data archiving and accessibility, and enhanced communication with managers, stakeholders, and Pacific Islander communities. All of these efforts will be instrumental toward resolving many of the current hurdles facing this ecosystem, especially in the face of climate change and ongoing natural stressors. Fisheries management continues to be challenging in this region, especially given conflicting interests among native fishers and international commercial fleets, differing perceptions regarding the status of many iconic stocks, incomplete information regarding its LMRs, and concern toward accessibility to reef and pelagic fisheries resources in nearby expansive marine monuments. Therefore, efforts that collaboratively address these concerns and build upon current EBFM priorities for the Western Pacific are notably needed.

Although ranked low overall regarding the status of its marine socioeconomics, the Western Pacific leads nationally in terms of aspects of its marine tourism, particularly its significant numbers of visitors to its marine habitats, and moderate cruise ship passenger numbers, all of which contribute heavily to its local economy. While containing some of the highest numbers of managed fisheries species nationally, the nationally lowest numbers of depleted marine mammal stocks, and higher numbers of threatened and endangered species, the Western Pacific is mid-ranked among regions in terms of the status of its LMRs (Link & Marshak 2019). This region contains low total domestic landings, indicators of high system exploitation in urbanized areas, poorly assessed stocks, and continues to exhibit substantial consequences from intensive fishing in its populated subregions (Friedlander et al. 2002, 2003, Porter et al. 2005, Moffitt et al. 2007, Burdick et al. 2008, Williams et al. 2008, 2011, 2012, 2015, Grafeld et al. 2016, Wiejerman et al. 2015b, 2016b, Link & Marshak 2019). Needed advances in data collection and monitoring of this system, its species, and its fisheries remain a significant obstacle to recovery of Western Pacific LMRs and habitats. Nationally, the Western Pacific has been highest in recent years for the percentage of stocks for which overfishing status is unknown (67%) and second-highest for stocks with unassessed overfished status (68.5%). Despite recent management changes that now list most Western Pacific species as "ecosystem component species" (NMFS 2019a), there remains high uncertainty regarding the status of many key economic species in its multispecies fisheries, with urgency for improving this knowledge in light of predicted climatological effects on their populations and resistance from stakeholders to cooperate with fishing restrictions. Despite this, governance in this region is ranked among the highest for all regions (Link & Marshak 2019), with higher congressional representation related to its shoreline and value of fisheries, comparatively higher use of spatial protections and HAPCs, low numbers of lawsuits, a well-balanced fishery management council (FMC), few NEPA-EIS activities, and more strategic use of FEPs. However, many of these factors are also directly related to this region having comparatively low national fisheries value, especially from intensively exploited coral reef fisheries within populated subregions (Friedlander et al. 2002, 2003, Porter et al. 2005, Zeller et al. 2006, Moffitt et al. 2007, Burdick et al. 2008, Williams et al. 2008, 2011, 2012, 2015, Grafeld et al. 2016, Weijerman et al. 2015b, 2016b, Link & Marshak 2019).

10.5.1 Progress Toward Ecosystem-Based Fisheries Management (EBFM)

This ecosystem is excelling in facets of its human environment and the quality of its governance system as related to the determinants of successful LMR management (Fig. 10.30; Link & Marshak 2019).

Western Pacific fisheries management strategies continue advancing toward approaches that address the cumulative issues that affect its LMRs. Through a more systematic and prioritized consideration of all fisheries, pressures, risks, and outcomes, holistic EBFM practices are being developed and implemented

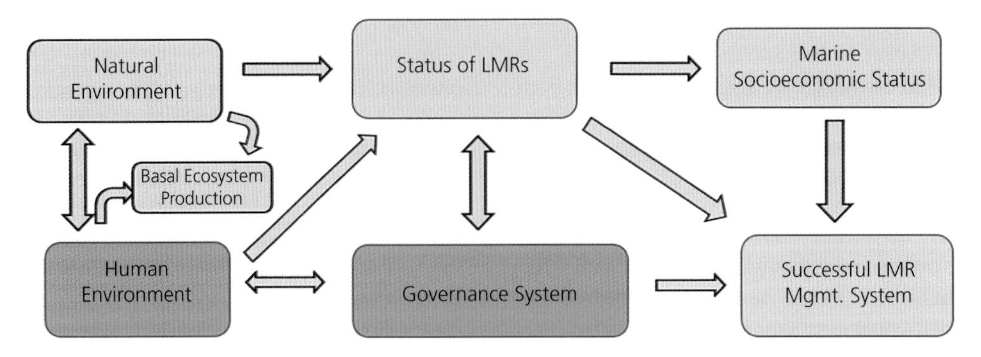

Figure 10.30 Schematic of the determinants and interconnectivity of successful living marine resource (LMR) systems management criteria as modified from Link and Marshak (2019). Those highlighted in blue indicate noteworthy criteria for the Western Pacific ecosystem.

across U.S. regions (NMFS 2016a, b, Levine et al. 2018, Marshall et al. 2018, Link & Marshak 2019). Fisheries management for the Western Pacific by necessity emphasizes not a traditional single-species approach ut rather one that address cumulative issues affecting the LMRs and fisheries of this ecosystem. Over the past few years, efforts to implement EBFM in this region have arisen through refinement of local management strategies. In accordance with the NOAA Fisheries EBFM Policy and Roadmap (NMFS 2016a, b), an EBFM implementation plan (NMFS 2019b) for the Western Pacific has been developed through efforts overseen by the NOAA Pacific Islands Fisheries Science Center (PIFSC), NOAA Fisheries Pacific Islands Regional Office (PIRO), and WPRFMC. This plan specifies approaches to be taken toward conducting EBFM and in addressing NOAA Fisheries Roadmap priorities and gaps. Progress for the Western Pacific and its major subregions in terms of EBFM Roadmap Guiding Principles and Goals is shown in Table 12.9.

Overall, significant progress has been made in terms of implementing ecosystem-level planning and advancing knowledge of ecosystem principles. Additionally, some progress has been observed toward assessing risks and vulnerabilities to ecosystems through ongoing investigations into climate vulnerability and species prioritizations for stock and habitat assessments. However, little or partial progress has been observed toward addressing trade-offs or applying ecosystem-level emergent properties and ecosystem considerations into management frameworks and decision-making.

Progress toward recommendations of the Ecosystem Principles Advisory Panel (EPAP) for the development of FEPs (EPAP 1999, Wilkinson & Abrams 2015) is shown in Table 12.10. Five FEPs has been in effect for the Pacific region since 2010 (WPRFMC 2010a, b, c, d, e), with a substantial degree of progress observed toward characterizing, monitoring, and assessing aspects of its ecosystem dynamics in support of this ecosystem approach, including at the subregional level. As observed in other regions, additional efforts toward assessing the habitat needs for managed species life history stages and examining the ecosystem effects of fishing remain needed. However, indices that account for ecosystem status and uncertainty have been developed for the Hawaii subregion. Greater ecosystem-level data and more in-depth syntheses for Western Pacific subregions are also warranted, while ongoing efforts to enact and build upon ecosystem approaches to governance are required for more effective EBFM.

10.5.2 Conclusions

Building on existing priorities to enact an ecosystem-based approach for the Western Pacific will result in advancements to its overall LMR management, especially in accounting for the cumulative effects of ongoing and increasing human and natural stressors and working to mitigate their impacts. These are especially of concern regarding continued LMR productivity, harvest, and well-being of coral reef-associated biological communities. The local multibillion-dollar tourism industry associated with this region is highly dependent on ensuring a sustainable seafood supply, ecological integrity of coastal and offshore habitats to support diving and other marine recreational activities, and well-managed sport fisheries that go together with healthy ecosystem functioning which are all prioritized in an EBFM context. Management entities must continue to monitor, consider, and incorporate the natural, biological, and socioeconomic factors that affect the degree to which marine resources are sustainably extracted and harvested, especially in a changing, underassessed, and potentially overexploited environment. Continuing with an approach toward exploiting and managing the resources of this ecologically significant region without considering effects on the ecosystem as a whole, or the influences of interacting sectors, historic and ongoing stressors, and governance and socioeconomic feedbacks, limits effectiveness of sustainable practices and ignores the significance of these factors toward LMR production and marine socioeconomic growth. System-wide considerations of the inherent relationships between productivity, biomass, landings, and socioeconomics are needed to refine and maintain this system toward needed sustainable thresholds, while also accounting for the influences from natural and human activities and other ocean uses. EBFM provides opportunities that focusing on one stock or species complex or one fishery would not. Overlapping and differential fishing effort for these species and the effects of different harvesting methodologies among fisheries (e.g., effects of native and tourism-based spearfishing on spawning aggregations of multiple reef fish species) is also considered in these frameworks, while also accounting for their ecological consequences on interacting predator and prey species. As observed for the U.S. Caribbean, given the very high species richness in this system, such an ecosystem-based approach allows for portfolio and multilevel considerations, leading to perhaps more effective management strategies (Edwards et al. 2004, Kellner et al. 2011, Link 2018).

Enhanced holistic approaches to management in the Western Pacific are necessary for sustaining its cultural marine heritage, LMR-dependent tourism-based economy, and future ecological conditions. Strategic EBFM under the umbrella of EBM for this region would refine current fisheries management strategies, incorporate engagement with management bodies from other sectors, and work toward more robust habitat conservation, improved land-use and watershed planning, and reduced nutrient loading. Assessment and rebuilding of overharvested stocks through fishing restrictions and protections, targeted governance, directed resources for monitoring and enforcement of regulations, and consideration of system thresholds in terms of productivity and biomass enhances the potential for refining management of this system and limiting further decline. While some progress is being made in addressing aspects of these concerns and continues to be prioritized for EBFM implementation in this region (NMFS 2019b), advances in providing foundational information and merging it into a more holistic framework that examines these facets together and allows for improved collaboration among management interests will work toward improved sustainability for the Western Pacific and its subregions. While showing signs of overexploitation, quantifiable information and reference points demonstrating the degree to which this system has exceeded thresholds remains needed for both biological and socioeconomic metrics. Additionally, improved abilities to predict the consequences of climate change, regional warming, and acidification throughout this expansive system are required for future viability of its LMRs and marine habitats. Applying available data toward identifying system-wide thresholds remains a major need for this system, as does more focused data collection that parameterizes specific ecosystem models and management actions. This enhanced information, and dedicated resources for its collection, will facilitate future management strategies that can maintain Western Pacific fisheries resources within sustainable limits. As compounding effects of coastal development, habitat loss, climate change, and overfishing continue to affect the Western Pacific, their consequences throughout this socioecological system are more effectively addressed using an ecosystem approach. Applying concrete EBFM actions in this system that consider the limitations of its productivity, and work to resolve its LMR status and dependent socioeconomics, provides a pathway toward ensuring sustainable practices. These efforts will foster continuation of this region's significant LMR-based culture, which is heavily reliant on its expansive marine environments.

10.6 References

Aeby GS, Williams GJ, Franklin EC, Kenyon J, Cox EF, Coles S, Work TM. 2011. Patterns of coral disease across the Hawaiian Archipelago: relating disease to environment. *PLoS One* 6(5):e20370.

Alin S, Brainard R, Price N, Newton J, Cohen A, Hare J, Peterson WT, et al. 2015. Characterizing the natural system: toward sustained, integrated coastal ocean health observing systems to facilitate resource management and decision support. *Oceanography* 28(2):92–107.

Allen SD, Bartram PK. 2008. *Guam as a Fishing Community. Pacific Islands Fisheries Science Center Administrative Report H-08-01*. Washington, DC: NOAA (p. 61).

Ansell AD, Gibson RN, Barnes M, Press UC. 1996. Coastal fisheries in the Pacific Islands. *Oceanography and Marine Biology: An Annual Review* 34(395):531.

Antonelis GA, Baker JD, Johanos TC, Braun RC, Harting AL. 2006. Hawaiian monk seal: status and conservation issues. *Atoll Research Bulletin* 543:75–101.

Ault JS, Smith SG, Richards B, Yau A, Langseth B, Humphreys RL, Boggs CH, DiNardo GT. 2018. *Towards Fishery-Independent Biomass Estimation for Hawaiian Deep 7 Bottomfish. NOAA Technical Memorandum NMFS-PIFSC-67.* Washington, DC: NOAA.

Ayers AL, Hospital J, Boggs C. 2018. Bigeye tuna catch limits lead to differential impacts for Hawaii longliners. *Marine Policy* 94:93–105.

Ayers AL. 2018. *The Commonwealth of the Northern Mariana Islands Fishing Community Profile: 2017 Update. NOAA technical memorandum NMFS-PIFSC 66.* Washington, DC: NOAA.

Ayotte P, McCoy K, Williams I, Zamzow J. 2011. *Coral Reef Ecosystem Division Standard Operating Procedures: Data Collection for Rapid Ecological Assessment Fish Surveys. Pacific Islands Fisheries Science Center Administrative Report H-11-08.* Washington, DC: NOAA (p. 24).

Ayotte P, McCoy K, Heenan A, Williams I, Zamzow J. 2015. *Coral Reef Ecosystem Program Standard Operating Procedures: Data Collection for Rapid Ecological Assessment Fish Surveys. Pacific Islands Fisheries Science Center Administrative Reports H-15-07.* Washington, DC: NOAA (p. 33).

Baker AC, Glynn PW, Riegl B. 2008. Climate change and coral reef bleaching: an ecological assessment of long-term impacts, recovery trends and future outlook. *Estuarine, Coastal and Shelf Science* 80(4):435–71.

Baker JD, Johanos TC. 2004. Abundance of the Hawaiian monk seal in the main Hawaiian Islands. *Biological Conservation* 116(1):103–10.

Becker SL, Brainard RE, Van Houtan KS. 2019. Densities and drivers of sea turtle populations across Pacific coral reef ecosystems. *PLoS One* 14(4):e0214972.

Bell JD, Allain V, Gupta AS, Johnson JE, Hampton J, Hobday AJ, Lehodey P, et al. 2018. *Climate Change Impacts, Vulnerabilities and Adaptations: Western and Central Pacific Ocean Marine Fisheries. Impacts of Climate Change on Fisheries and Aquaculture. FAO Fisheries and Aquaculture Technical Paper No. 627.* Rome, Italy: FOA (p. 305).

Bell JD, Ganachaud A, Gehrke PC, Griffiths SP, Hobday AJ, Hoegh-Guldberg O, Johnson JE, et al. 2013a. Mixed responses of tropical Pacific fisheries and aquaculture to climate change. *Nature Climate Change* 3(6):591.

Bell JD, Johnson JE, Hobday AJ. 2011. *Vulnerability of Tropical Pacific Fisheries and Aquaculture to Climate Change. SPC FAME Digital Library.* Nouméa, New Caledonia: SPC.

Bell JD, Reid C, Batty MJ, Lehodey P, Rodwell L, Hobday AJ, Johnson JE, Demmke A. 2013b. Effects of climate change on oceanic fisheries in the tropical Pacific: implications for economic development and food security. *Climatic Change* 119(1):199–212.

Bishop, RC, Chapman, DJ, Kanninen, BJ, Krosnick, JA, Leeworthy B, Meade, NF. 2011. *Total Economic Value for Protecting and Restoring Hawaiian Coral Reef Ecosystems: Final Report. NOAA Technical Memorandum CRCP 16. Silver Spring, MD: NOAA Office of National Marine Sanctuaries, Office of Response and Restoration, and Coral Reef Conservation Program* (p. 406).

Björk M, Short F, Mcleod E, Beer S. 2008. *Managing Seagrass for Resilience to Climate Change. IUCN Resilience Science Group Working Paper Series No. 3.* Gland, Switzerland: IUCN.

Boehlert GW. 1993. Fisheries and marine resources of Hawaii and the US-associated Pacific Islands: an introduction. *Marine Fisheries Review* 55(2):3–6.

Boyle S, De Anda V, Koenig K, O'Reilly E, Schafer M, Acoba T, Dillon A, et al. 2017. *Coral Reef Ecosystems of the Pacific Remote Islands Marine National Monument: A 2000–2016 Overview.* Washington, DC: NOAA.

Bradford AL, Forney KA, Oleson EM, Barlow J. 2017. Abundance estimates of cetaceans from a line-transect survey within the U.S. Hawaiian Islands exclusive economic zone. *Fishery Bulletin* 115(2):1290142.

Brainard R, DeMartini E, Holzwarth S. 2002. *Coral Reef Ecosystems of the Northwestern Hawaiian Islands: Interim Results Emphasizing the 2000 Surveys.* Honolulu, HI: U.S. Fish and Wildlife Service and the Hawaii Department of Land and Natural Resources (pp. 10–11, 17, 31, 39, 43).

Brainard R, Maragos J, Schroeder R, Kenyon J, Vroom P, Godwin S, Hoeke R, et al. 2005. *The State of Coral Reef Ecosystems of the US Pacific Remote Island Areas. The State of Coral Reef Ecosystems of the United States and Pacific Freely Associated States. NOAA Technical Memorandum NOS NCCOS.* Washington, DC: NOAA (pp. 338–72).

Brainard RE, Oliver T, Venegas R, Heenan A, B Vargas-Ángel, MJ McPhaden, R Rotjan, et al. 2018. Ecological impacts of the 2015/16 El Niño in the Central Equatorial Pacific. *Bulletin of the American Meteorological Society* 99(1):S21–6.

Brodziak JK, O'Malley JM, Richards B, DiNardo GT. 2012. *Stock Assessment Update of the Status of the Bottomfish Resource of American Samoa, the Commonwealth of the Northern Mariana Islands, and Guam. Pacific Islands Fisheries Science Center Administrative Report H-12-04.* Washington, DC: NOAA.

Brown EK, Cox E, Jokiel PL, Rodgers SK, Smith WR, Tissot BN, Coles SL, Hultquist J. 2004. Development of benthic sampling methods for the Coral Reef assessment and monitoring program (CRAMP) in Hawaii. *Pacific Science* 58(2):145–58.

Bruckner A, Buja K, Fairey L, Gleason K, Harmon M, Heron S, Hourigan T, et al. 2005. *Threats and Stressors to US Coral Reef Ecosystems. The State of Coral Reef Ecosystems of the*

United States and Pacific Freely Associated States. Silver Spring, MD: NOAA (pp. 12–44).

Bruno JF, Selig ER. 2007. Regional decline of coral cover in the Indo-Pacific: timing, extent, and subregional comparisons. *PLoS One* 2(8):e711.

Burdick D, Brown V, Asher J, Caballes C, Gawel M, Goldman L, Hall A, et al. 2008. *Status of the Coral Reef Ecosystems of Guam. Bureau of Statistics and Plans, Guam Coastal Management Program IV*. Washington, DC: NOAA.

Camargo SJ, Sobel AH. 2005. Western North Pacific tropical cyclone intensity and ENSO. *Journal of Climate* 18(15):2996–3006.

Campbell HF, Hand AJ. 1999. Modeling the spatial dynamics of the US purse-seine fleet operating in the western Pacific tuna fishery. *Canadian Journal of Fisheries and Aquatic Sciences* 56(7):1266–77.

Cannizzo ZJ, Hutto S, Wenzel L. 2020. Adapting to a changing ocean: experiences from marine protected area managers. *Parks Stewardship Forum* 36(1).

Carpenter KE, Abrar M, Aeby G, Aronson RB, Banks S, Bruckner A, Chiriboga A, et al. 2008. One-third of reef-building corals face elevated extinction risk from climate change and local impacts. *Science* 321(5888):560–3.

Carpenter KE. 1998. *An Introduction to the Oceanography, Geology, Biogeography, and Fisheries of the Tropical and Subtropical Western and Central Pacific. FAO Species Identification Guide for Fishery Purposes. The Living Marine Resources of the Western Central Pacific*. Rome, Italy: FAO (p. 117).

Carretta JV, Forney KA, Oleson EM, Weller DW, Lang AR, Baker J, Muto MM, et al. 2019. U.S. *Pacific Marine Mammal Stock Assessments: 2018. NOAA Technical Memorandum NMFSSWFSC-617*. Washington, DC: NOAA.

Chan HL, Pan M. 2017. *Economic and Social Characteristics of the Hawaii Small Boat Fishery 2014. NOAA Technical Memorandum NOAA-TM-NMFS-PIFSC-63*. Washington, DC: NOAA (p. 107).

Chan HL, Pan M. 2019. *Tracking Economic Performance Indicators for Small Boat Fisheries in American Samoa, Guam, and the Commonwealth of the Northern Mariana Islands*. Washington, DC: NOAA.

Chan HL. 2020. Economic impacts of Papahānaumokuākea marine national monument expansion on the Hawaii longline fishery. *Marine Policy* 14:103869.

Chapman L. 2001. *Bycatch in the Tuna Longline Fishery. 2nd SPC Heads of Fisheries Meeting (Noumea, New Caledonia, 23–27 July 2001)*. Nouméa, New Caledonia: SPC.

Chimner RA, Fry B, Kaneshiro MY, Cormier N. 2006. Current extent and historical expansion of introduced mangroves on Oahu, Hawaii. *Pacific Science* 60(3):377–83.

Chowdhury MR, Chu PS, Schroeder T. 2007. ENSO and seasonal sea-level variability–a diagnostic discussion for the US-Affiliated Pacific Islands. *Theoretical and Applied Climatology* 88(3–4):213–24.

Citta J, Reynolds MH, Seavy N. 2007. *Seabird Monitoring Assessment for Hawai'i and the Pacific Islands. Hawai'i Cooperative Studies Unit Technical Report. HSCU-007*. Honolulu, HI: University of Hawai'i (p. 122).

Conklin EJ, Smith JE. 2005. Abundance and spread of the invasive red algae, Kappaphycus spp., in Kane'ohe Bay, Hawai'i and an experimental assessment of management options. *Biological Invasions* 7(6):1029–39.

Couch C. 2014. *Intrinsic Host and Extrinsic Environmental Drivers of Coral Health and Disease*. Ithaca, NY: Cornell University.

Couch CS, Garriques JD, Barnett C, Preskitt L, Cotton S, Giddens J, Walsh W. 2014. Spatial and temporal patterns of coral health and disease along leeward Hawai'i Island. *Coral Reefs* 33(3):693–704.

Craig M, Bograd S, Dewar H, Kinney M, Lee HH, Muhling B, Taylor B. 2017. *Status Review Report of Pacific Bluefin Tuna (Thunnus Orientalis)*. NOAA Technical Memorandum NMFS-SWFSC-587. Washington, DC: NOAA.

Craig P, DiDonato G, Fenner D, Hawkins C. 2005. *The State of Coral Reef Ecosystems of American Samoa. The State of Coral Reef Ecosystems of the United States and Pacific Freely Associated States. NOAA Technical Memorandum NOS NCCOS 11*. Silver Spring, MD: NOAA (pp. 312–37).

Craig P, Ponwith B, Aitaoto F, Hamm D. 1993. The commercial, subsistence, and recreational fisheries of American Samoa. *Marine Fisheries Review* 55(2):109–16.

Craig RK. 2006. Are marine national monuments better than national marine sanctuaries? US ocean policy, marine protected areas, and the Northwest Hawaiian Islands. *Sustainable Developement Law & Policy* 7:27.

CRCP (NOAA's Coral Reef Conservation Program). 2014. *National Coral Reef Monitoring Plan*. Silver Spring, MD: NOAA Coral Reef Conservation Program (p. 39).

D'iorio M, Jupiter SD, Cochran SA, Potts DC. 2007. Optimizing remote sensing and GIS tools for mapping and managing the distribution of an invasive mangrove (Rhizophora mangle) on South Molokai, Hawaii. *Marine Geodesy* 30(1–2):125–44.

Dalzell P, Boggs CH. 2003. Pelagic fisheries catching blue and striped marlins in the US western Pacific islands. *Marine and Freshwater Research* 54(4):419–24.

Dalzell PJ, Laurs RM, Haight WR. 2008. *Case Study: Catch and Management of Pelagic Sharks in Hawaii and the US Western Pacific Region. Sharks of the Open Ocean: Biology, Fisheries and Conservation*. Oxford, UK: Blackwell Scientific Publications (pp. 268–74).

Dator J, Hamnett M, Nordberg D, Pintz WS. 1999. *Hawaii 2000: Past, Present, and Future*. Honolulu, HI: Office of Planning, Department of Business, Economic Development and Tourism.

De Santo EM. 2013. Missing marine protected area (MPA) targets: how the push for quantity over quality undermines sustainability and social justice. *Journal of Environmental Management* 124:137–46.

Demopoulos AW, Fry B, Smith CR. 2007. Food web structure in exotic and native mangroves: a Hawaii–Puerto Rico comparison. *Oecologia* 153(3):675–86.

Demopoulos AW, Smith CR. 2010. Invasive mangroves alter macrofaunal community structure and facilitate opportunistic exotics. *Marine Ecology Progress Series* 404:51–67.

Di Lorenzo E, Schneider N, Cobb KM, Franks PJ, Chhak K, Miller AJ, McWilliams JC, et al. 2008. North Pacific Gyre Oscillation links ocean climate and ecosystem change. *Geophysical Research Letters* 35(8).

Dierking J, Campora CE. 2009. Ciguatera in the introduced fish Cephalopholis argus (Serranidae) in Hawai'i and implications for fishery management. *Pacific Science* 63(2):193–204.

Dierking J, Williams ID, Walsh WJ. Diet composition and prey selection of the introduced grouper species peacock hind (Cephalopholis argus) in Hawaii. Fishery Bulletin. 2009;107(4):464–76.

Doty MS. 1961. Acanthophora, a possible invader of the marine flora of Hawaii. *Pacific Science* 15(4):547–52.

Durocher JF. 1994. One hotel's story: the Hyatt Kauai. *Cornell Hospitality Quarterly* 35(2):71.

Dutton PH, Jensen MP, Frutchey K, Frey A, LaCasella E, Balazs GH, Cruce J, et al. 2014. Genetic stock structure of green turtle (Chelonia mydas) nesting populations across the Pacific islands. *Pacific Science* 68(4):451–65.

Dwyer C. 2018. *Super Typhoon Yutu,'Strongest Storm Of 2018,' Slams US Pacific Territory*. NPR. https://www.npr.org/2018/10/24/660224741/super-typhoon-yutu-strongest-storm-of-2018-slams-u-s-pacific-territory?t=1619884178383

Eakin CM, Sweatman HPA, Brainard RE. 2019. The 2014–2017 global-scale coral bleaching event: insights and impacts. *Coral Reefs* 38:539–45.

Eckert KL. 1993. *The Biology and Population Status of Marine Turtles in the North Pacific Ocean. NOAA Technical Memorandum NOAA-TM-NMFS-SWFSC-186*. Washington, DC: NOAA.

Edmunds PJ, Adjeroud M, Baskett ML, Baums IB, Budd AF, Carpenter RC, Fabina NS, et al. 2014. Persistence and change in community composition of reef corals through present, past and future climates. *PLoS One* 9(10):e107525.

Edwards CB, Friedlander AM, Green AG, Hardt MJ, Sala E, Sweatman HP, Williams ID, et al. 2014. Global assessment of the status of coral reef herbivorous fishes: evidence for fishing effects. *Proceedings of the Royal Society B: Biological Sciences* 281(1774):20131835.

Edwards SF, Link JS, Rountree BP. 2004. Portfolio management of wild fish stocks. *Ecological Economics* 49(3):317–29.

Eldredge LG. 2003. The marine reptiles and mammals of Guam. *Micronesica* 35(36):653–60.

Ellison JC. 2009. Wetlands of the Pacific Island region. *Wetlands Ecology and Management* 17(3):169–206.

Emanuel K. 2005. Increasing destructiveness of tropical cyclones over the past 30 years. *Nature* 436(7051):686–8.

Fakhruddin BS, Schick L. 2019. Benefits of economic assessment of cyclone early warning systems—a case study on Cyclone Evan in Samoa. *Progress in Disaster Science* 2:100034.

Fenner D. 2014. Fishing down the largest coral reef fish species. *Marine Pollution Bulletin* 84(1–2):9–16.

Ferguson L, Srinivasan M, Oleson E, Hayes S, Brown SK, Angliss R, Carretta J, et al. 2017. *Proceedings of the First National Protected Species Assessment Workshop. NOAA Technical Memorandum NMFS-F/SPO-172*. Washington, DC: NOAA (p. 92).

Filous A, Friedlander AM, Koike H, Lammers M, Wong A, Stone K, Sparks RT. 2017. Displacement effects of heavy human use on coral reef predators within the Molokini Marine Life Conservation District. *Marine Pollution Bulletin* 121(1–2):274–81.

Franklin EC, Jokiel PL, Donahue MJ. 2013. Predictive modeling of coral distribution and abundance in the Hawaiian Islands. *Marine Ecology Progress Series* 481:121–32.

Friedlander AM, Brown E, Monaco ME. 2007. Defining reef fish habitat utilization patterns in Hawaii: comparisons between marine protected areas and areas open to fishing. *Marine Ecology Progress Series* 351:221–33.

Friedlander AM, Brown EK, Jokiel PL, Smith WR, Rodgers KS. 2003. Effects of habitat, wave exposure, and marine protected area status on coral reef fish assemblages in the Hawaiian archipelago. *Coral Reefs* 22(3):291–305.

Friedlander AM, Brown EK, Monaco ME. 2007. Coupling ecology and GIS to evaluate efficacy of marine protected areas in Hawaii. *Ecological Applications* 17(3):715–30.

Friedlander AM, DeMartini EE. 2002. Contrasts in density, size, and biomass of reef fishes between the northwestern and the main Hawaiian islands: the effects of fishing down apex predators. *Marine Ecology Progress Series* 230:253–64.

Friedlander AM, Stamoulis KA, Kittinger JN, Drazen JC, Tissot BN. 2014. Understanding the scale of marine protection in Hawai'i: from community-based management to the remote Northwestern Hawaiian Islands. *Advances in Marine Biology* 69:153–203.

Friedlander AM, Wagner D, Gaymer CF, Wilhelm TA, Lewis NA, Brooke S, Kikiloi K, Varmer O. 2016. Co-operation between large-scale MPAs: successful experiences from the Pacific ocean. *Aquatic Conservation: Marine and Freshwater Ecosystems* 26:126–41.

Friedlander AM, Wedding LM, Brown E, Monaco ME. 2010. *Monitoring Hawaii's Marine Protected Areas Examining Spatial and Temporal Trends Using a Seascape Approach. NOAA Technical Memorandum NOS NCCOS 117*. Silver Spring, MD: NOAA.

Friedlander AM. 2001. Essential fish habitat and the effective design of marine reserves: application for marine ornamental fishes. *Aquarium Sciences and Conservation* 3(1–3):135–50.

Friedlander M, Parrish JD, DeFelice RC. 2002. Ecology of the introduced snapper Lutjanus kasmiva (Forsskal) in the reef fish assemblage of a Hawaiian bay. *Journal of Fish Biology* 60(1):28–48.

Fulling GL, Thorson PH, Rivers J. 2011. Distribution and abundance estimates for cetaceans in the waters off Guam and the Commonwealth of the Northern Mariana Islands. *Pacific Science* 65(3):321–44.

Gaines SD, Costello C, Owashi B, Mangin T, Bone J, Molinos JG, Burden M, et al. 2018. Improved fisheries management could offset many negative effects of climate change. *Science Advances* 4(8):eaao1378.

Giddens J, Friedlander AM, Conklin E, Wiggins C, Stamoulis K, Donovan MK. 2014. Experimental removal of the invasive peacock hind (roi) Cephalopholis argus, in Puakō, Hawai'i: methods for assessing and managing marine invasive species. *Marine Ecology Progress Series* 511:209–21.

Giddens JL, Wiggins C, Friedlander AM, Conklin EJ, Stamoulis KA, Minton D. 2018. Assemblage-level effects of the introduced peacock hind (Cephalopholis argus) on Hawaiian reef fishes. *Environmental Biology of Fishes* 101(2):275–86.

Gilman E, Van Lavieren H, Ellison J, Jungblut V, Wilson L, Areki F, Brighouse G, et al. 2016. *Pacific Island Mangroves in a Changing Climate and Rising Sea. UNEP Regional Seas Reports and Studies No. 179*. Nairobi, Kenya: UNEP.

Gilman E. 2006. *Incidental Capture of Seabirds in Pelagic Longline Fisheries of the Tropical and Subtropical Pacific Islands Region and Draft Pacific Islands Regional Plan of Action for Reducing the Incidental Catch of Seabirds in Pelagic Longline Fisheries*. Honiara, Solomon Islands: Pacific Islands Forum Fisheries Agency.

Gilman EL, Ellison J, Jungblut V, Van Lavieren H, Wilson L, Areki F, Brighouse G, et al. 2006. Adapting to Pacific Island mangrove responses to sea level rise and climate change. *Climate Research* 32(3):161–76.

Gilman EL. 1997. Community based and multiple purpose protected areas: a model to select and manage protected areas with lessons from the Pacific Islands. *Coastal Management* 25(1):59–91.

Gilman EL, Ellison J, Duke NC, Field C. 2008. Threats to mangroves from climate change and adaptation options: a review. *Aquatic Botany* 89(2):237–50.

Glazier E. 2011. *Ecosystem Based Fisheries Management in the Western Pacific*. Hoboken, NJ: John Wiley & Sons.

Glynn PW. 1993. Coral reef bleaching: ecological perspectives. *Coral Reefs* 12(1):1–7.

Goldberg J, Adams K, Albert J, Asher J, Brown P, Brown V, Burdick D, et al. 2008. Status of coral reef resources in Micronesia and American Samoa: 2008. *Status of Coral Reefs of the World* 1:199–212.

Gove JM, Lecky J, Walsh WJ, Ingram RJ, Leong K, Williams ID, Polovina JJ, et al. 2019. *West Hawai'i Integrated Ecosystem Assessment Ecosystem Status Report. Pacific Islands Fisheries Science Center, PIFSC Special Publication, SP-19-001*. Washington, DC: NOAA (p. 46).

Gove JM, Williams GJ, McManus MA, Heron SF, Sandin SA, Vetter OJ, Foley DG. 2013. Quantifying climatological ranges and anomalies for Pacific coral reef ecosystems. *PLoS One* 8(4):e61974.

Grafeld S, Oleson K, Barnes M, Peng M, Chan C, Weijerman M. 2016. Divers' willingness to pay for improved coral reef conditions in Guam: an untapped source of funding for management and conservation? *Ecological Economics* 128:202–13.

Grigg RW, Polovina J, Friedlander AM, Rohmann SO. 2008. "Biology of coral reefs in the Northwestern Hawaiian Islands." In: *Coral Reefs of the USA*. Edited by BM Riegl, RE Dodge. Dordrecht, Netherlands: Springer (pp. 573–94).

Grigg RW. 1993. Precious coral fisheries of Hawaii and the US Pacific Islands. *Marine Fisheries Review* 55(2):50–60.

Gulko D, Maragos J, Friedlander A, Hunter C, Brainard R. 2000. "Status of coral reef in the Hawaiian archipelago." In: *Status of Coral Reefs of the World*. Edited by C Wilkinson. Cape Ferguson, Australia: Australian Institute of Marine Science (pp. 219–38).

Haight WR, Dalzell PJ. 2000. "Satch and management of sharks in pelagic fisheries in Hawaii and the Western Pacific region." In: *Sharks of the Open Ocean: Biology, Fisheries and Conservation*. Edited by MD Camhi, EK Pikitch, EA Babcock. Hoboken, NJ: Blackwell Publishing Ltd.

Hamnett MP, Pintz WS. 1996. *The Contribution of Tuna Fishing and Transshipment to the Economies of American Samoa, the Commonwealth of the Northern Mariana Islands, and Guam*. Honolulu, HI: University of Hawaii, NOAA, Joint Institute for Marine and Atmospheric Research.

Harrison D. 2004. Tourism in Pacific islands. *The Journal of Pacific Studies* 26(1):1–28.

Hayes SA, Josephson E, Maze-Foley K, Rosel PE. 2019. *US Atlantic and Gulf of Mexico Marine Mammal Stock Assessments-2018. NOAA Technical Memorandum NMFS-NE-258*. Washington, DC: NOAA (p. 298).

Heenan A, Ayotte P, Gray AE, Lino K, McCoy K, Zamzow JP, Williams ID. 2014. *Pacific Reef Assessment and Monitoring Program. Data report: Ecological Monitoring 2012–2013: Reef Fishes and Benthic Habitats of the Main Hawaiian Islands, American Samoa, and Pacific Remote Island Areas. PIFSC Data Report DR-14-003*. Washington, DC: NOAA (p. 112).

Heenan A, Williams ID, Acoba T, DesRochers A, Kosaki RK, Kanemura T, Nadon MO, Brainard RE. 2017. Long-term monitoring of coral reef fish assemblages in the Western central pacific. *Scientific Data* 4:170176.

Hill MC, Ligon AD, Deakos MH, Milette-Winfree A, Bendin AR, Oleson EM. 2014. *Cetacean Surveys in the Waters of the Southern Mariana Archipelago (February 2010–April 2014) Pacific Islands Fisheries Science Center, PIFSC Data Report, DR-14-013*. Washington, DC: NOAA (p. 49).

Hoegh-Guldberg O, Ortiz JC, Dove S. 2011. The future of coral reefs. *Science* 334(6062):1494–5.

Hoeke RK, Gove JM, Smith E, Fisher-Pool P, Lammers M, Merritt D, Vetter OJ, et al. 2009. Coral reef ecosystem integrated observing system: in-situ oceanographic observations at the US Pacific islands and atolls. *Journal of Operational Oceanography* 2(2):3–14.

Hospital J, Beavers C. 2014. *Economic and Social Characteristics of Small Boat Fishing in the Commonwealth of the Northern Mariana Islands. Pacific Islands Fisheries Science Center Administrative Report H-14-02.* Washington, DC: NOAA (p. 58).

Hourigan TF, Etnoyer PJ, Cairns SD. 2017. *The State of Deep-Sea Coral and Sponge Ecosystems of the United States. NOAA Technical Memorandum NMFS-OHC-4.* Silver Spring, MD: NOAA (p. 467).

Hourigan TF. 2009. Managing fishery impacts on deep-water coral ecosystems of the USA: emerging best practices. *Marine Ecology Progress Series* 397:333–40.

Howell EA, Wabnitz CC, Dunne JP, Polovina JJ. 2013. Climate-induced primary productivity change and fishing impacts on the Central North Pacific ecosystem and Hawaii-based pelagic longline fishery. *Climatic Change* 119(1):79–93.

Hunter CL. 1995. *Review of Status of Coral Reefs Around American Flag Pacific Islands and Assessment of Need, Value, and Feasibility of Establishing a Coral Reef Fishery Management Plan for the Western Pacific Region.* Honolulu, HI: Western Pacific Regional Fishery Management Council.

Ingram RJ, Leong KM, Wongbusarakum S. 2019. *Bringing Human Well-Being into the West Hawai'i Integrated Ecosystem Assessment Program. Pacific Islands Fisheries Science Center, PIFSC Internal Report, IR-19-005.* Washington, DC: NOAA (p. 43).

Ingram RJ, Oleson KL, Gove JM. Revealing complex social-ecological interactions through participatory modeling to support ecosystem-based management in Hawai'i. *Marine Policy* 94:180–8.

Ito RY, Walsh WA. 2011. *U.S. Commercial Fisheries for Marlins in the North Pacific Ocean. Working Document Submitted to the ISC Billfish Working Group Workshop, 19–27 January 2011, Honolulu, Hawaii, USA, ISC/11/BILLWG-1/02.* Pacific Islands Fisheries Science Center, PIFSC Working Paper, WP-11-003. Washington, DC: NOAA (p. 20).

Jokiel PL, Brown EK, Friedlander A, Rodgers SK, Smith WR. 2001. *Hawaii Coral Reef Initiative Coral Reef Assessment and Monitoring Program (CRAMP) Final Report 1999–2000.* Honolulu, HI: University of Hawaii.

Jokiel PL, Rodgers KS, Walsh WJ, Polhemus DA, Wilhelm TA. 2011. Marine resource management in the Hawaiian Archipelago: the traditional Hawaiian system in relation to the Western approach. *Journal of Marine Biology* 2011:1–16.

Jupiter S, Mangubhai S, Kingsford RT. 2014. Conservation of biodiversity in the Pacific Islands of Oceania: challenges and opportunities. *Pacific Conservation Biology* 20(2):206–20.

Kahng SE, Garcia-Sais JR, Spalding HL, Brokovich E, Wagner D, Weil E, Hinderstein L, Toonen RJ. 2010. Community ecology of mesophotic coral reef ecosystems. *Coral Reefs* 29(2):255–75.

Kane C, Kosaki RK, Wagner D. 2014. High levels of mesophotic reef fish endemism in the Northwestern Hawaiian Islands. *Bulletin of Marine Science* 90(2):693–703.

Kapur MR, Franklin EC. 2017. Simulating future climate impacts on tropical fisheries: Are contemporary spatial management strategies sufficient? *Canadian Journal of Fisheries and Aquatic Sciences* 74(11):1974–89.

Karp MA, Blackhart K, Lynch PD, Deroba J, Hanselman D, Gertseva V, Teo S, et al. 2019. *Proceedings of the 13th National Stock Assessment Workshop: Model Complexity, Model Stability, and Ensemble Modeling. NOAA Technical Memoranda NMFS-F/SPO-189.* Washington, DC: NOAA (p. 49).

Keener V. 2013. *Climate Change and Pacific Islands: Indicators and Impacts: Report for the 2012 Pacific Islands Regional Climate Assessment.* Washington, DC: Island Press.

Kellner JB, Sanchirico JN, Hastings A, Mumby PJ. 2011. Optimizing for multiple species and multiple values: tradeoffs inherent in ecosystem-based fisheries management. *Conservation Letters* 4(1):21–30.

Kennedy BR, Cantwell K, Malik M, Kelley C, Potter J, Elliott K, Lobecker E, et al. 2019. The unknown and the unexplored: insights into the Pacific deep-sea following NOAA CAPSTONE expeditions. *Frontiers in Marine Science* 6:480.

Kenyon KW, Rice DW. 1959. Life history of the Hawaiian monk seal. *Pacific Science* 13(3): 215–52.

Kinch J, Anderson P, Richards E, Talouli A, Vieux C, Peteru C, Suaesi T. 2010. *Outlook Report on the State of the Marine Biodiversity in the Pacific Islands Region.* Nairobi, Kenya: UNEP.

Kitchell JF, Essington TE, Boggs CH, Schindler DE, Walters CJ. 2002. The role of sharks and longline fisheries in a pelagic ecosystem of the central Pacific. *Ecosystems* 5(2):202–16.

Kittinger JN, Dowling A, Purves AR, Milne NA, Olsson P. 2011. Marine protected areas, multiple-agency management, and monumental surprise in the Northwestern Hawaiian Islands. *Journal of Marine Biology* 2011:1–17.

Kleiber D, Kotowicz DM. 2018. *Applying National Community Social Vulnerability Indicators to Fishing Communities in the Pacific Island Region. NOAA Technical Memorandum NMFS-PIFSC 65.* Washington, DC: NOAA.

Kleiber D, Leong K. 2018. *Cultural Fishing in American Samoa. Pacific Islands Fisheries Science Center, PIFSC Administrative Report H-18-03.* Washington, DC: NOAA (p. 21).

Knaff JA, Zehr RM, Goldberg MD, Kidder SQ. 2000. An example of temperature structure differences in two cyclone systems derived from the advanced microwave sounder unit. *Weather and Forecasting* 15(4):476–83.

Knowlton N, Brainard RE, Fisher R, Moews M, Plaisance L, Caley MJ. 2010. Coral reef biodiversity. *Life in the World's Oceans: Diversity Distribution and Abundance* 8:65–74.

Kobayashi DR. 2006. Colonization of the Hawaiian archipelago via Johnston atoll: a characterization of oceanographic transport corridors for pelagic larvae using computer simulation. *Coral Reefs* 25:407–17.

Kodama KR, Businger S. 1998. Weather and forecasting challenges in the Pacific Region of the National Weather Service. *Weather and Forecasting* 13(3):523–46.

Kulbicki M, Labrosse P, Ferraris J. 2004. Basic principles underlying research projects on the links between the ecology and the uses of coral reef fishes in the Pacific. *Challenging Coast, MARE Publication Series* 2004:119–58.

Langseth B, Syslo J, Yau A, Carvalho F. 2019. *Stock Assessments of the Bottomfish Management Unit Species of Guam, the Commonwealth of the Northern Mariana Islands, and American Samoa, 2019. NOAA Technical Memorandum NMFS-PIFSC-86.* Washington, DC: NOAA (p. 177).

Langseth B, Syslo J, Yau A, Kapur M, Brodziak JK. 2018. *Stock Assessment for the Main Hawaiian Islands Deep 7 Bottomfish Complex in 2018, with Catch Projections Through 2022.* Washington, DC: NOAA.

Laughlin Jr SK. 1979. The application of the constitution in United States Territories: American Samoa, a case study. *University of Hawai'i Law Review* 2:337.

Leibowitz AH. 1989. *Defining Status: A Comprehensive Analysis of United States Territorial Relations.* Leiden, Netherlands: Martinus Nijhoff Publishers.

Leong KM, Wongbusarakum S, Ingram RJ, Mawyer A, Poe MR. 2019. Improving representation of human well-being and cultural importance in conceptualizing the West Hawai 'i Ecosystem. *Frontiers in Marine Science* 6:231.

Leung P, Muraoka J, Nakamoto ST, Pooley S. 1998. Evaluating fisheries management options in Hawaii using analytic hierarchy process (AHP). *Fisheries Research* 36(2–3):171–83.

Levine A, Allen S. 2009. *American Samoa as a Fishing Community. NOAA Technical Memorandum NOAA-TM-NMFS-PIFSC-19.* Washington, DC: NOAA (p. 74).

Levine AS, Dillard MK, Loerzel J, Edwards PE. 2016. *National Coral Reef Monitoring Program Socioeconomic Monitoring Component: Summary Findings for American Samoa, 2014. NOAA Technical Memorandum CRCP 24.* Washington, DC: NOAA.

Li T, Kwon M, Zhao M, Kug JS, Luo JJ, Yu W. 2010. Global warming shifts Pacific tropical cyclone location. *Geophysical Research Letters* 37:L21804.

Lindfield SJ, McIlwain JL, Harvey ES. 2014. Depth refuge and the impacts of SCUBA spearfishing on coral reef fishes. *PloS One* 9(3):e92628.

Link JS, Browman HI. 2014. Integrating what? Levels of marine ecosystem-based assessment and management. *ICES Journal of Marine Science* 71(5):1170–3.

Link JS, Marshak AR. 2019. Characterizing and comparing marine fisheries ecosystems in the United States: determinants of success in moving toward ecosystem-based fisheries management. *Reviews in Fish Biology and Fisheries* 29(1):23–70.

Link JS. 2018. System-level optimal yield: increased value, less risk, improved stability, and better fisheries. *Canadian Journal of Fisheries and Aquatic Sciences* 75(1):1–6.

Lino K, Asher J, Ferguson M, Gray A, McCoy K, Timmers M, Vargas-Angel B. 2018. *Ecosystem Sciences Division Standard Operating Procedures: Data Collection for Towed-Diver Benthic and Fish Surveys. Pacific Islands Fisheries Science Center, PIFSC Administrative Report, H-18-02.* Washington, DC: NOAA (p. 76).

Lopez J, Wurth T, Littnan C. 2014. *Report on Hawaiian Monk Seal Survey on Ni'ihau Island, 2014. Pacific Islands Fisheries Science Center, PIFSC Data Report, DR-14-017.* Washington, DC: NOAA (p. 8).

Lowe MK, Quach M, Brousseau KR, Tomita AS. 2016. *Fishery Statistics of the Western Pacific. Volume 29, Territory of American Samoa (2012), Commonwealth of the Northern Mariana Islands (2012), Territory of Guam (2012), State of Hawaii (2012). Pacific Islands Fisheries Science Center administrative report H 16-03.* Washington, DC: NOAA.

Lundblad ER, Wright DJ, Miller J, Larkin EM, Rinehart R, Naar DF, Donahue BT, Anderson SM, Battista T. 2006. A benthic terrain classification scheme for American Samoa. *Marine Geodesy* 29(2):89–111.

Lynch PD, Methot RD, Link JS. 2018. *Implementing a Next Generation Stock Assessment Enterprise. An Update to the NOAA Fisheries Stock Assessment Improvement Plan. NOAA Technical Memorandum NMFS-F/ SPO-183.* Washington, DC: NOAA (p. 127).

Lynham J, Nikolaev A, Raynor J, Vilela T, Villasenor Villaseñor-Derbez JC. 2020. Impact of two of the world's largest protected areas on longline fishery catch rates. *Nature Communications* 979(1):1–9.

Ma H. 2019. *Catch and Effort Estimates for Major Pelagic Species from the Hawaii Marine Recreational Fishing Survey (2003–2018) Pacific Islands Fisheries Science Center, PIFSC Internal Report, IR-19-010.* Washington, DC: NOAA (p. 7).

Makaiau J, Tosatto MD, McGregor M, Kingma E. 2016. *Amendment 4. Fishery Ecosystem Plan for the Mariana Archipelago. Remove the Prohibited Areas for Medium and Large Bottomfish Vessels in the Commonwealth of the Northern Mariana Islands. RIN 0648-BF37.* Honolulu, HI: Western Pacific Fishery Management Council.

Maragos JE. 1986. *Coastal Resource Development and Management in the US Pacific Islands. Island-By-Island Analysis.* St. Thomas, U.S. Virgin Islands: Island Resources Foundation.

Maragos JE. 1993. Impact of coastal construction on coral reefs in the US-affiliated Pacific islands. *Coastal Management* 21(4):235–69.

Marshall KN, Levin PS, Essington TE, Koehn LE, Anderson LG, Bundy A, Carothers C, Coleman F, Gerber LR, Grabowski JH, Houde E, Jensen OP, Möllmann C, Rose K, Sanchirico J, Smith ADM. 2018. Ecosystem-based fisheries management for social–ecological systems: renewing the focus in the United States with next generation fishery ecosystem plans. *Conservation Letters* 11(1):e12367.

Martien KK, Hill MC, Van Cise AM, Robertson KM, Woodman SM, Dolar L, Pease VL, Oleson EM. 2014. *Genetic Diversity and Population Structure in Four Species of Cetaceans Around the Mariana Islands. NOAA Technical Memorandum NOAA-TM-NMFS-SWFSC 536*. Washington, DC: NOAA.

Martin SL, Van Houtan KS, Jones TT, Aguon CF, Gutierrez JT, Tibbatts RB, Wusstig SB, Bass JD. 2016. Five decades of marine megafauna surveys from Micronesia. *Frontiers in Marine Science* 2(116).

Maynard J, Van Hooidonk R, Eakin CM, Puotinen M, Garren M, Williams G, Heron SF, et al. 2015. Projections of climate conditions that increase coral disease susceptibility and pathogen abundance and virulence. *Nature Climate Change* 5(7):688.

McClanahan TR, Graham NA, MacNeil MA, Muthiga NA, Cinner JE, Bruggemann JH, Wilson SK. 2011. Critical thresholds and tangible targets for ecosystem-based management of coral reef fisheries. *Proceedings of the National Academy of Sciences* 108(41):17230–3.

McCoy K, Heenan A, Asher JM, Ayotte P, Gorospe K, Gray AE, Lino K, Zamzow JP, Williams ID. 2016. *Pacific Reef Assessment and Monitoring Program. Data Report: Ecological Monitoring 2015: Reef Fishes and Benthic Habitats of the Main Hawaiian Islands, Northwestern Hawaiian Islands, Pacific Remote Island Areas, and American Samoa. PIFSC data report DR-16-002*. Honolulu, HI: Joint Institute for Marine and Atmospheric Research.

McDole T, Nulton J, Barott KL, Felts B, Hand C, Hatay M, Lee H, et al. 2012. Assessing coral reefs on a Pacific-wide scale using the microbialization score. *PLoS One* 7(9):e43233.

Meyer AL, Dierking J. 2011. Elevated size and body condition and altered feeding ecology of the grouper Cephalopholis argus in non-native habitats. *Marine Ecology Progress Series* 439:203–12.

Miller J, Maragos J, Brainard R, Asher J, Vargas-Ángel B, Kenyon J, Schroeder R, et al. 2008. The state of coral reef ecosystems of the Pacific Remote Island Areas. *The State of Coral Reef Ecosystems of the United States and Pacific Freely Associated States* 73:353–86.

Miller S, Crosby M. 1998. *The Extent and Condition of US Coral Reefs*. Silver Spring, MD: NOAA.

Moffitt RB, Brodziak JK, Flores T. 2007. *Status of the Bottomfish Resources of American Samoa, Guam, and Commonwealth of the Northern Mariana Islands, 2005. Administrative Report H-07-04*. Washington, DC: NOAA.

Monaco ME, Andersen SM, Battista TA, Kendall MS, Rohmann SO, Wedding LM, Clarke AM. 2012. *National Summary of NOAA's Shallow-Water Benthic Habitat Mapping of US Coral Reef Ecosystems. NOAA Technical Memorandum NOS NCCOS 122*. Washington, DC: NOAA.

Mora C, Andréfouët S, Costello MJ, Kranenburg C, Rollo A, Veron J, Gaston KJ, Myers RA. 2006. Coral reefs and the global network of marine protected areas. *Science* 312(5781):1750–1.

Morrison C. 2012. Impacts of tourism on threatened species in the Pacific region: a review. *Pacific Conservation Biology* 18(4):227–38.

Muto MM, Helker VT, Angliss RP, Boveng PL, Breiwick JM, Cameron MF, Clapham PJ, et al. 2019. *Alaska Marine Mammal Stock Assessments, 2018. NOAA Technical Memorandum NMFS-AFSC-393*. Washington, DC: NOAA (p. 390).

Nadon MO, Ault JS, Williams ID, Smith SG, DiNardo GT. 2015. Length-based assessment of coral reef fish populations in the Main and Northwestern Hawaiian Islands. *PLoS One* 10(8):e0133960.

Naro-Maciel E, Arengo F, Galante P, Vintinner E, Holmes KE, Balazs G, Sterling EJ. 2018. Marine protected areas and migratory species: residency of green turtles at Palmyra Atoll, Central Pacific. *Endangered Species Research* 37:165–82.

Neilson BJ, Wall CB, Mancini FT, Gewecke CA. 2018. Herbivore biocontrol and manual removal successfully reduce invasive macroalgae on coral reefs. *PeerJ* 6:e5332.

Newman D, Berkson J, Suatoni L. 2015. Current methods for setting catch limits for data-limited fish stocks in the United States. *Fisheries Research* 164:86–93.

Nicol SJ, Allain V, Pilling GM, Polovina J, Coll M, Bell J, Dalzell P, et al. 2013. An ocean observation system for monitoring the affects of climate change on the ecology and sustainability of pelagic fisheries in the Pacific Ocean. *Climatic Change* 119(1):131–45.

NMFS (National Marine Fisheries Service) 2017a. *National Marine Fisheries Service—2nd Quarter 2017*. Washington, DC: NOAA.

NMFS (National Marine Fisheries Service). 2017b. *National Observer Program FY 2013 Annual Report. NOAA Technical Memorandum NMFS F/SPO-178*. Washington, DC: NOAA (p. 34).

NMFS (National Marine Fisheries Service). 2018a. Fisheries Economics of the United States, 2016. NOAA Technical Memorandum NMFS-F/SPO-187. Washington, DC: NOAA (p. 243).

NMFS (National Marine Fisheries Service). 2018b. *Fisheries of the United States, 2017. NOAA Current Fishery Statistics No. 2017*. Washington, DC: NOAA.

NMFS (National Marine Fisheries Service). 2019a. Pacific island fisheries; Reclassifying management unit species to ecosystem component species. A rule by the National Oceanic and Atmospheric Administration on 02/08/2019. *Federal Register* 84(27):2767–75.

NMFS (National Marine Fisheries Service). 2019b. *Pacific Islands Region Ecosystem-Based Fisheries Management Implementation Plan 2018–2022*. Washington, DC: NOAA (p. 39).

NMFS (National Marine Fisheries Service). 2020. *National Marine Fisheries Service—2nd Quarter 2020 Update*. Washington, DC: NOAA (p. 51).

Ovetz R. 2007. The bottom line: an investigation of the economic, cultural and social costs of high seas industrial longline fishing in the Pacific and the benefits of conservation. *Marine Policy* 31(2):217–28.

Oyafuso ZS, Drazen, JC, Moore CH, Franklin EC. 2017. Habitat-based species distribution models of Hawaiian deep-slope fishes. *Fisheries Research* 195:19–27.

Oyafuso ZS, Leung PS, Franklin EC. 2020. Understanding biological and socioeconomic tradeoffs of marine reserve planning via a flexible integer linear programming approach. *Biological Conservation* 241:108319.

Oyafuso ZS, Toonen RJ, Franklin EC. 2016. Temporal and spatial trends in prey diversity of wahoo (*Acanthocybium solandri*): a diet analysis from the central North Pacific using visual and DNA bar-coding techniques. *Journal of Fish Biology* 88:1501–23.

PACIOOS (Pacific Islands Ocean Observing System). 2016. *Ocean Observation Highlights from the Pacific Islands Region*. Honolulu, HI: PacIOOS (p. 10).

Pandolfi JM, Jackson JB, Baron N, Bradbury RH, Guzman HM, Hughes TP, Kappel CV, et al. 2005. Are US coral reefs on the slippery slope to slime? *Science* 307(5716): 1725–6.

Parke M. 2018. *Deep-Sea Coral Research and Technology Program: Pacific Islands Deep-Sea Coral and Sponge 3-Year Research Wrap-Up Workshop May 23–24, 2018. NOAA Technical Memorandum NMFSPIFSC-78*. Washington, DC: NOAA (p. 30).

Parrish FA, Baco AR. 2007. *State of Deep Coral Ecosystems in the US Pacific Islands Region: Hawaii and the US Pacific Territories. The State Of Deep Coral Ecosystems of the United States. NOAA Technical Memorandum CRCP-3*. Silver Spring, MD: NOAA (pp. 165–94).

Parrish FA, Howell EA, Antonelis GA, Iverson SJ, Littnan CL, Parrish JD, Polovina JJ. 2012. Estimating the carrying capacity of French Frigate Shoals for the endangered Hawaiian monk seal using Ecopath with Ecosim. *Marine Mammal Science* 28(3):522–41.

Patrick WS, Link JS. 2015. Hidden in plain sight: using optimum yield as a policy framework to operationalize ecosystem-based fisheries management. *Marine Policy* 62:74–81.

PIFSC (Pacific Islands Fisheries Science Center). 2016. *West Hawai'i Integrated Ecosystem Assessment: Ecosystem Trends and Status Report. NOAA Fisheries Pacific Science Center, PIFSC Special Publication, SP-16-004*. Washington, DC: NOAA (p. 46).

PIFSC (Pacific Islands Fisheries Science Center). 2017. *Summary of 2016 Reef Fish Surveys Around Kahoolawe Island. Pacific Islands Fisheries Science Center, PIFSC Data Report, DR-17-011*. Washington, DC: NOAA (p. 11).

Polhemus DA. 1993. *A Preliminary Assessment of the Impact of Hurricane Iniki on the Aquatic Insect Faunas on Streams on the Na Pali Coast, Kauai, Hawaii. Hawaii State Department of Land and Natural Resources, Division of Aquatic Resources Assessments of the Impact of Hurricane Iniki on Stream Biota and Ecosystems on Kauai*. Honolulu, HI: Department of Land and Natural Resources.

Polovina JJ, Dreflak K, Baker JD, Bloom S, Brooke SG, Chan VA, Ellgen S, et al. 2016. *Pacific Islands Fisheries Science Center Pacific Islands Regional Action Plan: NOAA Fisheries Climate Science Strategy. NOAA Technical Memorandum NMFS-PIFSC 59*. Washington, DC: NOAA.

Pooley SG. 1993. Hawaii's marine fisheries: some history, long-term trends, and recent developments. *Marine Fisheries Review* 55(2):7–19.

Porter V, Leberer T, Gawel M, Gutierrez J, Burdick D, Torres V, Lujan E. 2005. *Status of the Coral Reef Ecosystems of Guam. University of Guam Marine Laboratory Technical Report No. 113*. Washington, DC: NOAA.

Pratchett MS, Munday PL, Graham NA, Kronen M, Pinca S, Friedman K, Brewer TD, et al. 2011. *Vulnerability of Coastal Fisheries in the Tropical Pacific to Climate Change*. Noumea, New Caledonia: SPC (pp. 167–85).

Pyle RL, Boland R, Bolick H, Bowen BW, Bradley CJ, Kane C, Kosaki RK, et al. 2016. A comprehensive investigation of mesophotic coral ecosystems in the Hawaiian Archipelago. *PeerJ* 4:e2475.

Randall JE, Earle JL, Pyle RL, Parrish JD, Hayes T. Annotated checklist of the fishes of Midway Atoll, northwestern Hawaiian Islands. *Pacific Science* 47(4):356–400.

Richards BL, Williams ID, Vetter OJ, Williams GJ. 2012. Environmental factors affecting large-bodied coral reef fish assemblages in the Mariana Archipelago. *PLoS One* 7(2):e31374.

Richards, BL, Smith SG, Ault JS, DiNardo GT, Kobayashi D, Domokos R, Anderson J, et al. 2016. *Design and Implementation of a Bottomfish Fishery-Independent Survey in the Main Hawaiian Islands. NOAA Technical Memorandum, NMFS-PIFSC-53*. Washington, DC: NOAA (p. 56).

Richmond L, Kotowicz D. 2015. Equity and access in marine protected areas: The history and future of "traditional indigenous fishing" in the Marianas Trench Marine National Monument. *Applied Geography* 59:117–24.

Richmond R, Kelty RU, Craig PE, Emaurois CA, Green AL, Birkeland CH, Davis GE, et al. 2002. "Status of the coral reefs in Micronesia and American Samoa: US affiliated and freely associated islands in the Pacific." In: *Status of Coral Reefs of the World*. Edited by C Wilkinson. Townsville, Australia: Australian Institute of Marine Science (pp. 217–36).

Rieser A. 2011. The Papahanaumokuakea precedent: ecosystem-scale marine protected areas in the EEZ. *Asian Pacific Law and Policy Journal* 13:210.

Robinson JP, Williams ID, Edwards AM, McPherson J, Yeager L, Vigliola L, Brainard RE, Baum JK. 2017. Fishing degrades size structure of coral reef fish communities. *Global Change Biology* 23(3):1009–22.

Rocha LA, Pinheiro HT, Shepherd B, Papastamatiou YP, Luiz OJ, Pyle RL, Bongaerts P. 2018. Mesophotic coral ecosystems are threatened and ecologically distinct from shallow water reefs. *Science* 361(6399):281–4.

Rodgers S, Cox EF. 1999. Rate of Spread of Introduced Rhodophytes Kappaphycus alvarezii, Kappaphycus striatum, and Gracilaria salicornia and Their Current Distribution in Kane'ohe Bay, O'ahu Hawai'i. *Pacific Science* 53(3):232–41.

Rohmann SO, Hayes JJ, Newhall RC, Monaco ME, Grigg RW. 2005. The area of potential shallow-water tropical and subtropical coral ecosystems in the United States. *Coral Reefs* 24(3):370–83.

Sabater MG, Carroll BP. 2009. Trends in reef fish population and associated fishery after three millennia of resource utilization and a century of socio-economic changes in American Samoa. *Reviews in Fisheries Science* 17(3):318–35.

Sabater MG, Tofaeono SP. 2007. Scale and benthic composition effects on biomass and trophic group distribution of reef fishes in American Samoa1. *Pacific Science* 61(4):503–21.

Saunders MA, Chandler RE, Merchant CJ, Roberts FP. 2000. Atlantic hurricanes and NW Pacific typhoons: ENSO spatial impacts on occurrence and landfall. *Geophysical Research Letters* 27(8):1147–50.

Schroeder TA, Chowdhury MR, Lander MA, Guard CC, Felkley C, Gifford D. 2012. The role of the Pacific ENSO applications climate center in reducing vulnerability to climate hazards: Experience from the US-affiliated Pacific islands. *Bulletin of the American Meteorological Society* 93(7):1003–15.

Schumacher BD, Parrish JD. 2005. Spatial relationships between an introduced snapper and native goatfishes on Hawaiian reefs. *Biological Invasions* 7(6):925–33.

Shinjo R, Asami R, Huang KF, You CF, Iryu Y. 2013. Ocean acidification trend in the tropical North Pacific since the mid-20th century reconstructed from a coral archive. *Marine Geology* 342:58–64.

Shope JB, Storlazzi CD, Erikson LH, Hegermiller CA. 2016. Changes to extreme wave climates of islands within the Western Tropical Pacific throughout the 21st century under RCP 4.5 and RCP 8.5, with implications for island vulnerability and sustainability. *Global and Planetary Change* 141:25–38.

Shore-Maggio A, Callahan SM, Aeby GS. 2018. Trade-offs in disease and bleaching susceptibility among two color morphs of the Hawaiian reef coral, Montipora capitata. *Coral Reefs* 37(2):507–17.

Short FT, Neckles HA. 1999. The effects of global climate change on seagrasses. *Aquatic Botany* 63(3–4):169–96.

Siple MC, Donahue MJ. 2013. Invasive mangrove removal and recovery: Food web effects across a chronosequence. *Journal of Experimental Marine Biology and Ecology* 448: 128–35.

Smith JE, Hunter CL, Conklin EJ, Most R, Sauvage T, Squair C, Smith CM. 2004. Ecology of the invasive red alga Gracilaria salicornia (Rhodophyta) on O'ahu, Hawai'i. *Pacific Science* 58(2):325–43.

Smith JE, Hunter CL, Smith CM. 2002. Distribution and reproductive characteristics of nonindigenous and invasive marine algae in the Hawaiian Islands. *Pacific Science* 56(3):299–315.

Smith MK. 1993. An ecological perspective on inshore fisheries in the main Hawaiian Islands. *Marine Fisheries Review* 55(2):34–49.

Smith SG, Ault JS, Bohnsack JA, Harper DE, Luo JG, McClellan DB. 2011. Multispecies survey design for assessing reef-fish stocks, spatially explicit management performance, and ecosystem condition. *Fisheries Research* 109(1):25–41.

Srinivasan M, Brown SK, Markowitz E, Soldevilla M, Patterson E, Forney K, Murray K, et al. 2019. *Proceedings of the 2nd National Protected Species Assessment Workshop. NOAA Technical Memorandum NMFS-F/SPO-198.* Washington, DC: NOAA (p. 46).

Stevens JD, Bonfil R, Dulvy NK, Walker PA. 2000. The effects of fishing on sharks, rays, and chimaeras (chondrichthyans), and the implications for marine ecosystems. *ICES Journal of Marine Science* 57(3):476–94.

Stimson J, Larned S, Conklin E. 2001. Effects of herbivory, nutrient levels, and introduced algae on the distribution and abundance of the invasive macroalga Dictyosphaeria cavernosa in Kaneohe Bay, Hawaii. *Coral Reefs* 19(4): 343–57.

Stock CA, John JG, Rykaczewski RR, Asch RG, Cheung WW, Dunne JP, Friedland KD, et al. 2017. Reconciling fisheries catch and ocean productivity. *Proceedings of the National Academy of Sciences* 114(8):E1441–9.

Sund PN, Blackburn M, Williams F. 1981. Tunas and their environment in the Pacific Ocean: a review. *Oceanography and Marine Biology Annual Review* 19:443–512.

Sundberg M, Underkoffler K. 2011. *Size Composition and Length-Weight Data for Bottomfish and Pelagic Species Sampled at the United Fishing Agency Fish Auction in Honolulu, Hawaii from October 2007 to December 2009. Pacific Islands Fisheries Science Center Administrative Report H-11-04.* Washington, DC: NOAA (p. 34).

Swanson DW, Smith SG, Richards BL, Yau AJ, Ault JS. 2019. *Bottomfish Fishery-Independent Survey in Hawaii: Season and Gear Effects on Abundance. NOAA Technical Memorandum NOAA-TM-NMFS-PIFSC-88.* Washington, DC: NOAA (p. 48).

Tabata S. 1975. The general circulation of the Pacific Ocean and a brief account of the oceanographic structure of the North Pacific Ocean Part I, circulation and volume transports. *Atmosphere* 13(4):133–68.

Taylor KE, Stouffer RJ, Meehl GA. 2012. An overview of CMIP5 and the experiment design. *Bulletin of the American Meteorological Society* 93(4):485–98.

Tian E, Mak J, Leung P. 2013. "The direct and indirect contributions of tourism to regional GDP: Hawaii." In: *Handbook of Tourism Economics: Analysis, New Applications and Case Studies.* Edited by CA Tisdell. Toh Tuck, Singapore: World Scientific Publishing (pp. 523–541).

Tissot BN, Walsh WJ, Hixon MA. 2009. Hawaiian Islands marine ecosystem case study: ecosystem-and community-based management in Hawaii. *Coastal Management* 37(3–4):255–73.

Toonen RJ, Andrews KR, Baums IB, Bird CE, Concepcion GT, Daly-Engel TS, Eble JA, et al. 2011. Defining boundaries for ecosystem-based management: a multispecies case study of marine connectivity across the Hawaiian Archipelago. *Journal of Marine Biology* 2011: 460173.

Townsend HM, Harvey CJ, Aydin KY, Gamble RJ, Gruss A, Levin PS, Link JS, et al. 2014. *Report of the 3rd National Ecosystem Modeling Workshop (NEMoW 3): Mingling Models for Marine Resource Management, Multiple Model Inference. NOAA Technical Memorandum NMFS-F/SPO-149.* Washington, DC: NOAA.

Townsend HM, Link JS, Osgood KE, Gedamke T, Watters GM, Polovina JJ, Levin PS, Cyr EC, Aydin KY. 2008. *Report of the National Ecosystem Modeling Workshop (NEMoW). NOAA Technical Memorandum NMFS-F/SPO-87.* Washington, DC: NOAA.

Townsend HK, Holsman AK, Harvey C, Kaplan I, Hazen E, Woodworth-Jefcoats P, et al. 2017. *Report of the 4th National Ecosystem Modeling Workshop (NEMoW 4): Using Ecosystem Models to Evaluate Inevitable Trade-offs. NOAA Technical Memorandum NMFS-F/SPO-173.* Washington, DC: NOAA (p. 77).

Treloar P, Hall CM. 2005. *Tourism in the Pacific Islands. Oceania: A Tourism Handbook.* Bristol, UK: Channel View Publications (pp. 173–294).

Turgeon DD, Asch RG, Causey B, Dodge RE, Jaap W, Banks K, Delaney J, et al. 2002. *The State of Coral Reef Ecosystems of the United States and Pacific Freely Associated States: 2002.* Washington, DC: NOAA.

Van Dyke JM, Amore-Siah CD, Berkley-Coats GW. 1996. Self-determination for nonself-governing peoples and for indigenous peoples: the cases of Guam and Hawai'i. *University of Hawai'i Law Review* 18:623.

Van Hooidonk R, Maynard J, Tamelander J, Gove J, Ahmadia G, Raymundo L, Williams G, Heron SF, Planes S. 2016. Local-scale projections of coral reef futures and implications of the Paris Agreement. *Scientific Reports* 6(1):1–8.

Vargas-Ángel B. 2010. Crustose coralline algal diseases in the US-Affiliated Pacific Islands. *Coral Reefs* 29(4): 943–56.

Vargas-Ángel B, Huntington B, Brainard RE, Venegas R, Oliver T, Barkley H, Cohen A. 2019. El Nino-associated catastrophic coral mortality at Jarvis Island, central Equatorial Pacific. *Coral Reefs* 38:731–41.

Veitayaki J, Waqalevu V, Varea R, Rollings N. 2017. Mangroves in small island development states in the Pacific: An overview of a highly important and seriously threatened resource. In: *Participatory Mangrove Management in a Changing Climate.* Edited by R DasGupta, R Shaw. Tokyo, Japan: Springer (pp. 303–27).

Venegas RM, Oliver T, Liu G, Heron SF, Clark SJ, Pomeroy N, Young C, Eakin CM, Brainard RE. 2019. The rarity of depth Refugia from coral bleaching heat stress in the Western and Central Pacific Islands. *Scientific Reports* 9(19710).

Vermeij MJ, Smith TB, Dailer ML, Smith CM. 2009. Release from native herbivores facilitates the persistence of invasive marine algae: a biogeographical comparison of the relative contribution of nutrients and herbivory to invasion success. *Biological Invasions* 11(6):1463–74.

Wagner D, Kosaki RK, Spalding HL, Whitton RK, Pyle RL, Sherwood AR, Tsuda RT, Calcinai B. 2014. Mesophotic surveys of the flora and fauna at Johnston Atoll, Central Pacific Ocean. *Marine Biodiversity Records* 7:E68.

Wagner D, Papastamatiou YP, Kosaki RK, Gleason KA, McFall GB, Boland RC, Pyle RL, Toonen RJ. 2011. New records of commercially valuable black corals (Cnidaria: Antipatharia) from the Northwestern Hawaiian Islands at Mesophotic Depths. *Pacific Science* 65(2):249–56.

Walsh W, Cotton S, Barnett C, Couch C, Preskitt L, Tissot B, Osada-D'Avella K. 2013. *Long-Term Monitoring of Coral Reefs of the Main Hawaiian Islands.* Silver Spring, MD: NOAA Coral Reef Conservation Program, Hawai'i Division of Aquatic Resources.

Waycott M, McKenzie LJ, Mellors JE, Ellison JC, Sheaves MT, Collier C, Schwarz AM, et al. 2011. "Vulnerability of mangroves, seagrasses and intertidal flats in the tropical Pacific to climate change." In: *Vulnerability of Tropical Pacific Fisheries and Aquaculture to Climate Change.* Edited by JD Bell, JE Johnson, AJ. Hobday. Noumea, New Caledonia: Secretariat of the Pacific Community (pp. 297–368).

Wedding LM, Lecky J, Walecka HR, Gove JM, Donovan MK, Falinski K, McCoy K, et al. 2018. Advancing the integration of spatial data to map human and natural drivers on coral reefs. *PloS One* 13(9):e0204760.

Weijerman M, Fulton EA, Brainard RE. 2016. Management strategy evaluation applied to coral reef ecosystems in support of ecosystem-based management. *PLoS One* 11(3):e0152577.

Weijerman M, Fulton EA, Janssen AB, Kuiper JJ, Leemans R, Robson BJ, van de Leemput IA, Mooij WM. 2015a. How models can support ecosystem-based management of coral reefs. *Progress in Oceanography* 138:559–70.

Weijerman M, Fulton EA, Kaplan IC, Gorton R, Leemans R, Mooij WM, Brainard RE. 2015b. An integrated coral reef ecosystem model to support resource management under a changing climate. *PLoS One* 10(12):e0144165.

Weijerman M, Gove JM, Williams ID, Walsh WJ, Minton D, Polovina JJ. 2018. Evaluating management strategies to optimise coral reef ecosystem services. *Journal of Applied Ecology* 55(4):1823–33.

Weijerman M, Grace-McCaskey C, Grafeld SL, Kotowicz DM, Oleson KLL, van Putten IE. 2016b. Towards an ecosystem-based approach of Guam's coral reefs: the human dimension. *Marine Policy* 63:8–17.

Weijerman M, Robinson S, Parrish F, Polovina J, Littnan C. 2017. Comparative application of trophic ecosystem models to evaluate drivers of endangered Hawaiian monk seal populations. *Marine Ecology Progress Series* 582:215–29.

Weijerman M, Williams I, Gutierrez J, Grafeld S, Tibbatts B, Davis G. 2016c. Trends in biomass of coral reef fishes, derived from shore-based creel surveys in Guam. *Fishery Bulletin* 114:237–56.

Wilkinson, EB, Abrams K. 2015. *Benchmarking the 1999 EPAP Recommendations with existing Fishery Ecosystem Plans. U.S. Department of Commerce, NOAA Technical Memorandum NMFS-OSF-5*. Washington, DC: NOAA (p. 22).

Williams GJ, Gove JM, Eynaud Y, Zgliczynski BJ, Sandin SA. 2015. Local human impacts decouple natural biophysical relationships on Pacific coral reefs. *Ecography* 38(8):751–61.

Williams I, Zamzow J, Lino K, Ferguson M, Donham E. 2012. *Status of Coral Reef Fish Assemblages and Benthic Condition Around Guam: A Report Based on Underwater Visual Surveys in Guam and the Mariana Archipelago, April–June 2011. NOAA Technical Memorandum NOAA-TM-NMFS-PIFSC-33*. Washington, DC: NOAA (p. 22).

Williams ID, Richards BL, Sandin SA, Baum JK, Schroeder RE, Nadon MO, Zgliczynski B, et al. 2011. Differences in reef fish assemblages between populated and remote reefs spanning multiple archipelagos across the central and western Pacific. *Journal of Marine Biology* 2011:1–14.

Williams ID, Walsh WJ, Miyasaka A, Friedlander AM. 2006. Effects of rotational closure on coral reef fishes in Waikiki-Diamond head fishery management area, Oahu, Hawaii. *Marine Ecology Progress Series* 310:139–49.

Williams ID, Walsh WJ, Schroeder RE, Friedlander AM, Richards BL, Stamoulis KA. 2008. Assessing the importance of fishing impacts on Hawaiian coral reef fish assemblages along regional-scale human population gradients. *Environmental Conservation* 35(3):261–72.

Williams ID, Zamzow JP, Lino K, Ferguson M, Donham E. 2012. *Status of Coral Reef Fish Assemblages and Benthic Condition Around Guam a Report Based on Underwater Visual Surveys in Guam and the Mariana Archipelago, April–June 2011. NOAA Technical Memorandum NMFS-PIFSC-33*. Washington, DC: NOAA.

Williams I, Ma H. 2013. *Estimating Catch Weight of Reef Fish Species Using Estimation and Intercept Data from the Hawaii Marine Recreational Fishing Survey. Pacific Islands Fisheries Science Center Administration Report H-13-04*. Honolulu, HI: NOAA (p. 53).

Wilson JA, Diaz GA. 2012. An overview of circle hook use and management measures in United States marine fisheries. *Bulletin of Marine Science* 88(3):771–88.

Winston M, Taylor B, Franklin EC. 2017. Intraspecific variability in the life history of endemic coral reef fishes between photic and mesophotic depths across the Central Pacific Ocean. *Coral Reefs* 36(2):663–74.

Woodworth-Jefcoats PA, Blanchard JL, Drazen JC. 2019. Relative impacts of simultaneous stressors on a pelagic marine ecosystem. *Frontiers in Marine Science* 6:383.

Woodworth-Jefcoats PA, Polovina JJ, Drazen JC. 2017. Climate change is projected to reduce carrying capacity and redistribute species richness in North Pacific pelagic marine ecosystems. *Global Change Biology* 23(3): 1000–8.

Wren JLK, Kobayashi DR, Jia Y, Toonen RJ. 2016. Modeled population connectivity across the Hawaiian Archipelago. *PLoS One* 11(12):e0167626.

Wright A, Stacey N, Holland P. 2006. The cooperative framework for ocean and coastal management in the Pacific Islands: effectiveness, constraints and future direction. *Ocean & Coastal Management* 49(9–10):739–63.

Yau A, Nadon M, Richards B, Brodziak JK, Fletcher E. 2016. *Stock Assessment Updates of the Bottomfish Management Unit Species of American Samoa, the Commonwealth of the Northern Mariana Islands, and Guam in 2015 Using Data Through 2013. NOAA Technical Memorandum NMFS-PIFSC 51*. Washington, DC: NOAA.

Yau A. 2018. *Report from Hawaii Bottomfish Commercial Fishery Data Workshops, 2015–2016. NOAA Technical Memorandum NOAA-TM-NMFS-PIFSC-68*. Washington, DC: NOAA (p. 105).

Yong MY, Wetherall JA. 1980. *Estimates of the Catch and Effort by Foreign Tuna Longliners and Baitboats in the Fishery Conservation Zone of the Central and Western Pacific, 1965–77. NOAA-TM-NMFS-SWFC-2*. Washington, DC: NOAA.

Young HS, Maxwell SM, Conners MG, Shaffer SA. 2015. Pelagic marine protected areas protect foraging habitat for multiple breeding seabirds in the central Pacific. *Biological Conservation* 181:226–35.

Zeller D, Booth S, Craig P, Pauly D. 2006. Reconstruction of coral reef fisheries catches in American Samoa, 1950–2002. *Coral Reefs* 25(1):144–52.

Zeller D, Booth S, Davis G, Pauly D. 2007. Re-estimation of small-scale fishery catches for US flag-associated island areas in the western Pacific: the last 50 years. *Fishery Bulletin* 105(2):266–77.

Zeller D, Booth S, Pauly D. 2006. Fisheries contributions to the gross domestic product: underestimating small-scale fisheries in the Pacific. *Marine Resource Economics* 21(4):355–74.

Zeller D, Harper S, Zylich K, Pauly D. 2015. Synthesis of underreported small-scale fisheries catch in Pacific island waters. *Coral Reefs* 34(1):25–39.

CHAPTER 11

Regional Fisheries Management Organizations (RFMOs)

- The U.S. participates in transboundary management of migratory and high seas fisheries species as a signatory to 14 major intergovernmental conventions, treaties, and RFMOs throughout the Atlantic and Pacific basins.
- In terms of its highly migratory species (HMS) fisheries (i.e., sharks, tunas, and billfishes), reported U.S. HMS landings and value have been historically greater in the Pacific than in the Atlantic.
- Although U.S. components of the Atlantic and Pacific Ocean basins have cumulatively high socioeconomic status, the U.S. western Atlantic living marine resource (LMR)-based economy, both in terms of reported value and its proportional contribution to total ocean economy, is higher than the combined U.S. western and eastern Pacific LMR economy.
- Among RFMO jurisdictions and Food and Agricultural Organization (FAO) competence areas, U.S. fisheries landings and value are greatest within the areas encompassed by the North Atlantic Fisheries Organization (NAFO), Western and Central Pacific Fisheries Commission (WCPFC), and the Western Central Atlantic Fisheries Commission (WECAFC) competence areas.
- A greater number and larger extent of no-take fishing areas occurs within the Pacific and Antarctic basins than in the Atlantic, while the Antarctic also contains the highest percentage (~10%) of its total area subject to permanent fisheries closures.
- The greatest number of U.S. fisheries stocks subject to RFMO jurisdiction are those under the Pacific Salmon Commission (PSC), while the highest number and proportion of transboundary stocks that are identified as overfished or experiencing overfishing are Atlantic HMS stocks managed by NOAA Fisheries and the International Convention for the Conservation of Atlantic Tunas (ICCAT).
- High proportions of transboundary or multilaterally managed Western Pacific HMS species, and Pacific halibut and Antarctic krill stocks remain unassessed in terms of their overfishing and/or overfished statuses.
- Cumulatively throughout the U.S. Atlantic Exclusive Economic Zone (EEZ), average sea surface temperatures have increased by 0.86°C since the mid-twentieth century. However, a greater increase in temperature has been observed throughout the U.S. Pacific EEZ (1.13°C) during the same time period.
- Other U.S. ocean uses are overwhelmingly greater throughout the cumulative U.S. Atlantic than within the U.S. Pacific in terms of their value and proportional contributions (68–72%) to the total U.S. marine economy.
- Among RFMO jurisdictions, basal ecosystem productivities are highest throughout the Atlantic basin large marine ecosystems (LMEs) subject to ICCAT, U.S. Atlantic HMS fisheries management, and the North Atlantic Salmon Conservation Organization (NASCO). Globally, these RFMO areas also encompass oligotrophic ocean gyres, including those of the North and South Atlantic and North and South Pacific.

Ecosystem-Based Fisheries Management: Progress, Importance, and Impacts in the United States. Jason S. Link and Anthony R. Marshak, Oxford University Press. © U.S. Department of Commerce, U.S. Government 2021. DOI: 10.1093/oso/9780192843463.003.0011

11.1 Introduction

In addition to domestic fisheries resources contained within its Exclusive Economic Zone (EEZ), international, transboundary, and high seas fisheries contribute significantly to U.S. fisheries landings, revenue, and living marine resource (LMR)-based employments. Regional fisheries management organizations (RFMOs) encompass large multinational jurisdictions, including the U.S. EEZ, through which iconic transboundary species and species groups are managed (McDorman 2005, Small 2005, Bensch et al. 2008, Pintassilgo et al. 2010, Aranda et al. 2012, De Bruyn et al. 2013, Gilman et al. 2014, Leroy & Morin 2018). While domestically managing its transboundary and highly migratory species (HMS), the United States is also a participant in 14 several international LMR organizations including RFMOs that encompass major portions of the Atlantic and Pacific basins (Table 11.1). It is also a signatory to international agreements for the continued protection of global cetaceans and efforts toward ongoing conservation of biodiversity (Hourigan 2009, Allen et al. 2010, Cullis-Suzuki & Pauly 2010, Barnard et al. 2012, Brown 2017, Lynch et al. 2018, Nummelin & Urho 2018). Major species of interest include sharks, tunas, and tuna-like billfishes (Teleostei: Scombridae), Pacific halibut (*Hippoglossus stenolepis*), Pacific hake "*Whiting*" (*Merluccius productus*), Pacific salmon (*Oncorhynchus* spp.), Antarctic krill (*Euphausia superba*) and their fisheries, in addition to protected marine mammals and Atlantic salmon (*Salmo salar*). The majority of these managed species undertake large migrations across entire ocean basins, with their fisheries conjuring images of charter boats, massive fishing vessels, historical whaling ships, purse nets, and large fishing rods and reels with heavy lines. Other, nearshore-associated resources including salmon and Pacific groundfishes contribute to major international commercial and recreational fisheries, while also being targeted through subsistence fishing by North American indigenous populations. Such images bring to mind high seas environments chronicled by Herman Melville and Thor Heyerdahl, offshore swordfishing expeditions depicted in Sebastian Junger's *The Perfect Storm*, and legends and tales of Pacific fisheries resources as told by those of the First Nations and in Native American tribal mythology. These jurisdictions and their fisheries ecosystems comprise large water masses subject to large-scale climatological and oceanographic processes, which influence pelagic and benthic habitat quality for multiple fisheries and protected species (Greene & Pershing 2000, Corno et al. 2007, Agostini et al. 2008, Kobayashi et al. 2008, Beaugrand 2009, Bost et al. 2009, Van Houtan & Halley 2011, Tremblay et al. 2011, Greene et al. 2013, Lezama-Ochoa et al. 2016). Similar stressors affecting U.S. regional fisheries ecosystems, especially climatic shifts, regional warming, historical overfishing, and degradation of nearshore and offshore habitats supporting transboundary demersal fisheries species have likewise affected these broader regions and their associated LMRs (Field et al. 2001, Halpern et al. 2009, Gjerde 2011, Miller 2011, Rogers & Laffoley 2011, Cheung et al. 2013, De Bruyn et al. 2013, Wowk 2013, Gutiérrez et al. 2016, Barange et al. 2018).

What follows are descriptions of each of the major taxa groupings and their respective RFMOs that have U.S. involvement and interest. These provide a brief background of each RFMO and set of regional issues. This is followed by an evaluation of the RFMOs relative to their progress toward ecosystem-based fisheries management (EBFM), examining key facets of what amount to large portions of several ocean basins.

11.1.1 Atlantic and Pacific Highly Migratory Species

Significant fisheries landings occur in the United States for species and stocks that extend beyond U.S. boundaries and that are shared internationally. These include both **Atlantic HMS** that are subject to the jurisdictions of NOAA Fisheries and the **International Commission for the Conservation of Atlantic Tunas (ICCAT)**, and **Pacific HMS** that are subject to NOAA Fisheries, the Pacific (PFMC) and Western Pacific Regional Fishery Management Councils (WPFMC), and the **Inter-American Tropical Tuna (IATTC)** and **Western and Central Pacific Fisheries Commissions (WCPFC**; Fig. 11.1).

Since 1990, U.S. Atlantic HMS stocks (including Gulf of Mexico and U.S. Caribbean regions) have been managed by NOAA Fisheries under the authorities of the Magnuson-Stevens Fishery Conservation and Management Act (MSA) and the Atlantic Tunas Convention Act (ATCA; NMFS 2006). These Western Atlantic stocks were originally managed by U.S. regional Fishery Management Councils (FMCs) under separate Fishery Management Plans (FMPs) for sharks, swordfish, billfish, and tunas, which were later carried over to NOAA's jurisdiction under the U.S. Fishery Conservation Amendments of 1990. Their management was later revised and consolidated into a 1999 FMP for sharks, swordfish, and tunas (NMFS 1999),

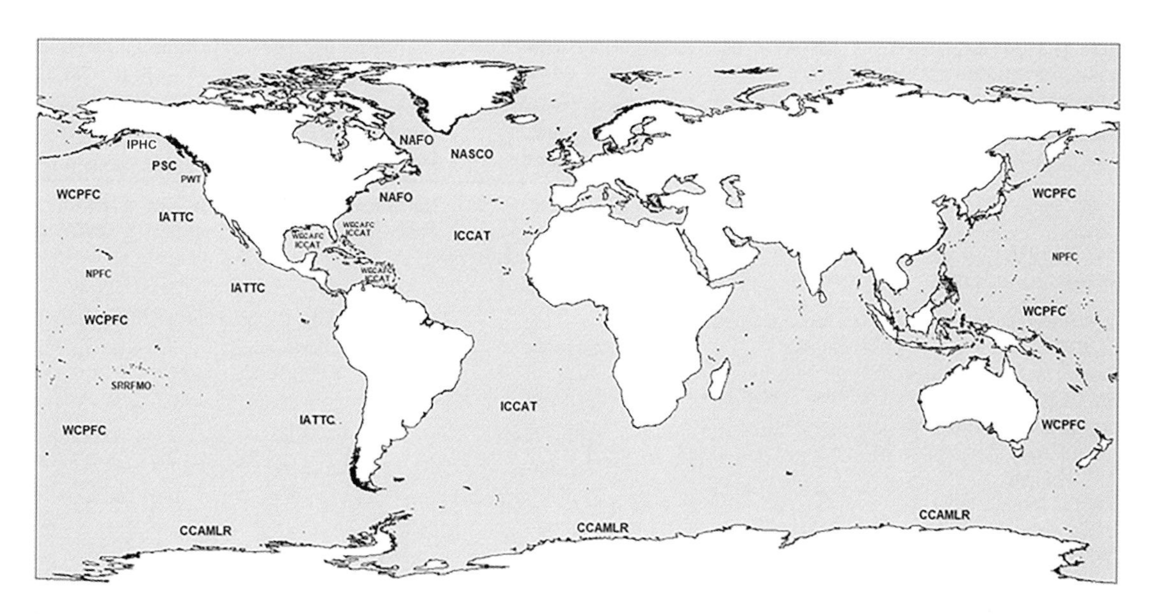

Figure 11.1 Map of global Regional Fisheries Management Organization jurisdictions of which the U.S. is a member. CCAMLR = Convention for the Conservation of Antarctic Marine Living Resources; IATTC = Inter-American Tropical Tuna Commission; ICCAT = International Commission for the Conservation of Atlantic Tunas; IPHC = International Pacific Halibut Commission; NAFO=North Atlantic Fisheries Organization; NASCO = North Atlantic Salmon Conservation Organization; NPFC = North Pacific Fisheries Commission; PSC = Pacific Salmon Commission; PWT = Pacific Whiting Treaty; SPRFMO = South Pacific Regional Fisheries Management Organisation; WECAFC = Western Central Atlantic Fishery Commission; WCPFC = Western and Central Pacific Fisheries Commission.

and a separately amended Atlantic billfish FMP (NMFS 1999). In 2006, all species groups were combined into a single consolidated Atlantic HMS FMP (NMFS 2006), which has since served as the basis for their management. Additionally, Atlantic HMS are managed internationally by ICCAT, which consists of 53 contracting parties and other cooperating entities that fish for tunas and tuna-like species throughout the Atlantic. ICCAT has worked to "cooperate in maintaining the populations of these fishes at levels which will permit the maximum sustainable catch for food and other purposes" (ICCAT 1966). ICCAT also requires countries to collect catch information for HMS species, and to specifically collect and process statistics and data on Atlantic tuna, swordfish, billfish, and some shark fisheries, and serves as a data repository for other internationally managed HMS species (Prince & Brown 1991, Restrepo 2004, NMFS 2006, Babcock et al. 2008, Garibaldi 2012, Herrera & Báez 2018). Stock assessments for most Atlantic HMS are conducted by the ICCAT Standing Committee on Research and Statistics (SCRS), with U.S. domestic shark assessments conducted by NOAA Fisheries through the SouthEast Data, Assessment, and Review (SEDAR) process,[1] which are communicated in annual SCRS reports and Atlantic HMS Stock Assessment and Fishery Evaluation (SAFE) reports (NMFS 2019a). This information is used to produce conservation and management recommendations that include total allowable catches, sharing agreements for member countries, size limits, effort controls, spatiotemporal closures, trade and compliance measures, and monitoring and inspection programs (NMFS 2006, Aranda et al. 2010). As long as the U.S. accepts an ICCAT recommendation, the ATCA provides the U.S. Secretary of Commerce and NOAA with the necessary statutory authority to issue regulations to implement binding ICCAT recommendations to U.S. managed fisheries. However, no regulation promulgated under ATCA may have the effect of increasing or decreasing any allocation or quota to which the U.S. agreed in accordance with an ICCAT recommendation (NMFS 2006). U.S. HMS are subject to those measures agreed upon by ICCAT, while being domestically managed within the U.S. EEZ by the provisions of the NOAA Fisheries FMP and implemented regulations.

[1] http://sedarweb.org/

Table 11.1 List of regional fisheries management organizations (RFMOs) to which the U.S. is a member, and their encompassing U.S. and global large marine ecosystem (LME) geographies

RFMO	Fisheries Managed	U.S. Regions Encompassed	Large Marine Ecosystems (LMEs)/Nations Encompassed
NOAA Fisheries Office of Sustainable Fisheries—Atlantic Highly Migratory Species Division **Atl. HMS**	U.S. Atlantic highly migratory species (HMS)	Gulf of Mexico, U.S. Caribbean, South Atlantic, Mid-Atlantic, New England	**North America**—Caribbean Sea, Gulf of Mexico, Northeast U.S. Continental, Southeast U.S. Continental
Pacific Highly Migratory Species **Pac. HMS** Managed by: NOAA Fisheries West Coast, Pacific Islands Regional Offices **WCRO, PIRO** – Pacific Fishery Management Council **PFMC**, – Western Pacific Fishery Management Council **WPFMC**	U.S. Pacific and Western Pacific HMS	Pacific, Western Pacific	**North America**—California Current, Gulf of Alaska, Insular Pacific Hawaiian
Commission for the Conservation of Antarctic Marine Living Resources **CCAMLR**	All Antarctic marine living resources	All U.S.	**Antarctica LME**, with participation from the European Union and 25 other member countries: **North America**—United States of America **Central/South America**—Argentina, Brazil, Chile, Uruguay **Europe**—Belgium, France, Germany, Italy, the Netherlands, Norway, Poland, Russian Federation, Spain, Sweden, Ukraine, United Kingdom **Africa**—Namibia, South Africa **Asia**—China, India, Japan, Republic of Korea **Oceania**—Australia, New Zealand
Inter-American Tropical Tuna Commission **IATTC**	Eastern Pacific tuna and tuna-like species	Pacific	**North America**—California Current **Central/South America**—Gulf of California, Humboldt Current, Pacific Central American Coastal
International Commission for the Conservation of Atlantic Tunas **ICCAT**	Atlantic tuna and tuna-like species	Gulf of Mexico, U.S. Caribbean, South Atlantic, Mid-Atlantic, New England	**North America**—Caribbean Sea, Gulf of Mexico, Newfoundland/Labrador Shelf, Northeast U.S. Continental, Scotian Shelf, Southeast U.S. Continental, West Greenland Shelf **Central/South America**—Caribbean Sea, East Brazil Shelf, Gulf of Mexico, North Brazil Shelf, Patagonian Shelf, South Brazil Shelf **European Union**—Baltic Sea, Barents Sea, Canary Current, Celtic Biscay Shelf, East Greenland Shelf, Faroe Plateau, Iberian Coastal, Iceland Shelf, Mediterranean Sea, North Sea, Norwegian Sea **Africa**—Benguela Current, Guinea Current
International Pacific Halibut Commission **IPHC**	Pacific halibut in U.S. and Canadian Waters	North Pacific, Pacific	**North America**—California Current, East Bering Sea, Gulf of Alaska
Northwest Atlantic Fisheries Organization **NAFO**	Transboundary fisheries in the western North Atlantic Ocean	Mid-Atlantic, New England	**North America**—Newfoundland/Labrador Shelf, Northeast U.S. Continental, West Greenland Shelf

RFMO	Fisheries Managed	U.S. Regions Encompassed	Large Marine Ecosystems (LMEs)/Nations Encompassed
North Atlantic Salmon Conservation Organization **NASCO**	Atlantic salmon in North Atlantic waters	New England	**North America**—Newfoundland/Labrador Shelf, Northeast U.S. Continental, Scotian Shelf, West Greenland Shelf **European Union**—Baltic Sea, Barents Sea, Celtic Biscay Shelf, East Greenland Shelf, Faroe Plateau, Iberian Coastal, Iceland Shelf, North Sea, Norwegian Sea
North Pacific Fisheries Commission **NPFC**	Transboundary fisheries in the North Pacific Ocean	Western Pacific (excluding American Samoa and Jarvis Island)	**North America**—Insular Pacific Hawaiian
Pacific Salmon Commission **PSC**	Pacific salmon in U.S. and Canadian Waters	North Pacific, Pacific	**North America**—California Current, East Bering Sea, Gulf of Alaska
Pacific Whiting Treaty **PWT**	Pacific hake (Whiting) in U.S. and Canadian Waters	Pacific	**North America**—California Current, Gulf of Alaska
South Pacific regional fisheries management organization **SPRFMO**	Transboundary fisheries in the South Pacific Ocean	Western Pacific—American Samoa and Jarvis Island	**No LMEs included**
Western and Central Atlantic Fisheries Commission **WECAFC**	Transboundary fisheries in the Western and Central Atlantic Ocean	Gulf of Mexico, U.S. Caribbean, South Atlantic	**North America**—Caribbean Sea, Gulf of Mexico, Southeast U.S. Continental **Central/South America**—Caribbean Sea, Gulf of Mexico
Western and Central Pacific Fisheries Commission **WPCFC**	Transboundary fisheries (including HMS) in the western North and Central Pacific Ocean	North Pacific, Western Pacific	**North America**—Aleutian Islands, East Bering Sea, Insular Pacific Hawaiian **Asia**—East China Sea, Gulf of Thailand, Indonesian Sea, Kuroshio Current, Oyashio Current, Sea of Japan, Sea of Okhotsk, South China Sea, Sulu-Celebes Sea, West Bering Sea, Yellow Sea **Oceania**—East Central Australian Shelf, New Zealand Shelf, North Australian Shelf, Northeast Australian Shelf, Southeast Australian Shelf

Domestic U.S. Pacific HMS stocks, including tunas, billfishes, and sharks, are currently managed by the PFMC HMS FMP for those species captured by its west coast fisheries (PFMC 2018), while those captured within the Western Pacific EEZ are managed by the Western Pacific Regional Fishery Management Council (WPFMC) ww Pacific Pelagic Fisheries of the Western Pacific Region Fishery Ecosystem Plan (FEP; WPFMC 2009). The PFMC HMS FMP has been in effect since 2004, with U.S. West Coast HMS also subject to IATTC managerial agreements for the eastern Pacific under the Tuna Conventions Act (Magnuson 1976, Aranda et al. 2010, Rhodes et al. 2014, PFMC 2018). Since its 1949 establishment and later amended Antigua Convention (IATTC 2003), the IATTC and its 21 member states, signatories, and five cooperating non-member states have committed to ensuring "the long-term conservation and sustainable use of the fish stocks covered by this Convention, in accordance with the relevant rules of international law." In addition to management of eastern Pacific HMS species, the IATTC also serves as the Secretariat for the International Dolphin Conservation Program (IDCP), which applies the 1999 Agreement on the International Dolphin Conservation Program (AIDCP) within the eastern Pacific Ocean (IATTC 1999, Hedley 2001, Compeán 2018) and works toward limiting dolphin-associated mortality in tuna-associated fisheries. Historically, all eastern Pacific HMS species have been under the IATTC jurisdiction, with most focus having been on tropical tunas and particularly Yellowfin tuna (*Thunnus albacares*) captured by purse seine, baitboat, and longline fisheries (PFMC 2018). The Antigua Convention promotes an ecosystem approach allowing for the possibility of considering other organisms that interact with Pacific HMS. Scientific staff at IATTC also conduct regular stock assessments on tropical tunas. Eastern Pacific albacore tuna (*Thunnus alalunga*) are also managed under the bilateral 2004 U.S.-Canada Albacore Treaty, allowing each nation's vessels access to the other's jurisdictional waters and ports for fishing, sale, export, and transshipments of albacore resources (PFMC 2018).

Western and central Pacific HMS were originally managed by a WPFMC Pelagic FMP (WPFMC 1986), following which broader geographic FEPs were developed for Western Pacific fisheries resources, including one for pelagic species (WPFMC 2009). This expanded approach compiles ecological, community, and local knowledge into a more comprehensive plan, while also applying this information into enhanced management processes and implementation of ecosystem approaches (WPFMC 2009). In addition, management of major western and central Pacific tuna and tuna-like fisheries has been under the purview of the WCPFC since 2004. Established by the Convention on the Conservation and Management of Highly Migratory Fish Stocks in the Western and Central Pacific Region, the U.S. and 19 other participants have committed to conserve and manage HMS in areas west of 150°W longitude. Together with the WPFMC, efforts toward more ecosystem-based approaches are emerging, especially in regards to improved gear technology, sustainable fishing practices, improved assessments, and sharing knowledge across member bodies (WPFMC 2009). Both the IATTC and WCPFC are members of the International Scientific Committee for Tuna and Tuna-Like Species in the North Pacific Ocean (ISC), which provides scientific information, assessments, and conservation recommendations on northern stocks for the WCPFC Northern Committee. All Atlantic and Pacific HMS stocks are also subject to the 1995 United Nations Agreement on the Conservation and Management of Straddling Fish Stocks and Highly Migratory Fish Stocks, which interprets the duty of nations to cooperate in conservation and management of fishery resources. For U.S. interests, this agreement requires that EEZ management be compatible with management for species under any international agreement (e.g., ICCAT, IATTC). The U.S. also participates in deliberations and decisions of the **United Nations Food and Agriculture Organization (FAO)** in international plans of action toward reducing incidental catch of seabirds in longline fisheries; preventing, deterring, and eliminating illegal, unreported, and unregulated (IUU) fishing of sharks; and for managing HMS fishing capacity. The U.S. has also developed national plans of action toward these priorities, with certain domestic efforts such as west coast seabird avoidance measures consistent with these national and international plans of action (PFMC 2018). The U.S. is also a major player in many aspects of FAO international fisheries policy with representatives on its many fisheries committees, and implements its Code of Conduct for Responsible Fisheries (FAO 1995) and the Ecosystem Approach to Fisheries (Garcia 2003). It participates in data sharing networks, collaborates in intergovernmental efforts, and assists in broader international efforts to prevent, deter, and eliminate IUU fishing.

U.S. and international HMS have been subject to stressors affecting regional marine ecosystems, particularly the effects of overfishing, climate forcing, and regional warming (Meltzer 1994, Maguire 2006, Miller 2007, Cullis-Suzuki & Pauly 2010, Lynch et al. 2011, Lehodey et al. 2013, Muhling et al. 2015, Gilman et al. 2016, Dell'Apa et al. 2018, Lynch et al. 2018, Pentz et al. 2018, Moustahfid et al. 2018, Bell et al. 2018). Overfishing has affected HMS stocks within the Atlantic and Pacific basins, for which U.S. rebuilding plans are in place and ICCAT, IATTC, and WCPFC cooperative conservation measures have been enacted for certain targeted species (Stone et al. 1998, Aranda et al. 2010, Milazzo 2012, De Bruyn et al. 2013, Worm et al. 2013, Juan-Jorda et al. 2015). However, despite these efforts regarding continued overfishing (NMFS 2017, Link & Marshak 2019), selective compliance with laws, incidental captures, and incomplete monitoring and management have hindered ongoing efforts to reach more sustainable levels for some HMS populations and their fisheries (Worm & Vanderzwaag 2007, Cullis-Suzuki & Pauly 2010, Lynch et al. 2011, Juan-Jorda et al. 2015, Boerder et al. 2019). Globally, shark landings have declined by 20% over the past two decades despite progress toward implementation of improved management frameworks (Dulvy et al. 2017). This trend is especially due to continued declines in shark populations for countries with high human coastal populations and limited national-scale management measures outside the U.S. to prevent overfishing (Lam & Sadovy de Mitcheson 2011, Bradley et al. 2014, Davidson et al. 2016, Dulvy et al. 2017; but also c.f. Simpfendorfer & Dulvy 2017). Approximately 10% of the current global shark catch (from at least 33 species) is considered biologically sustainable (Simpfendorfer & Dulvy 2017). High exploitation continues to inhibit the longer recovery times needed for many of these species, while poor monitoring of international catches and limited management has inhibited efforts toward their successful recovery (Camhi et al. 2009, Godin & Worm 2010, Ward-Paige et al. 2012, Worm et al. 2013, Davidson et al. 2016). However, some degree of regional recovery has been observed, such as in the southeastern U.S. (Carlson et al. 2012, Peterson et al. 2017, Simpfendorfer & Dulvy 2017). The ecosystem-level consequences from depleted apex predator populations continue to be suspected in some marine communities, with cascading effects in systems that support other fisheries (Kitchell et al. 1999, Stevens et al. 2000, Cox et al. 2002, Kitchell et al. 2002, Schindler et al. 2002, Hinke et al. 2004, Ward & Myers 2005, Kitchell et al. 2006, Heithaus et al. 2008, Ferretti et al. 2010, Heithaus et al. 2010, Worm et al. 2013, Lascelles et al. 2014, Ortuño Crespo & Dunn 2017, Bornatowski et al. 2018). The consequences of climate change are affecting HMS productivity through physiological stressors, loss and alterations of pelagic habitat, shifts of prey composition and pelagic fisheries (Moustahfid et al. 2018, Bell et al. 2018), and changes in species ranges and migration routes as related to shifting ocean currents and water temperatures (Boyce et al. 2008, Steinacher et al. 2010, Worm & Tittensor 2011, Hazen et al. 2013, Hobday & Evans 2013, Hobday et al. 2013, Lascelles et al. 2014, Robinson et al. 2015, Choy et al. 2016, Horodysky et al. 2016, Lynch et al. 2018). Incidental bycatch of non-targeted pelagic fishes and protected species such as sea turtles and marine mammals have historically been strongly associated with HMS fisheries, with some progress in limiting bycatch effects on their populations (Nitta & Henderson 1993, Lewison et al. 2004, Kerstetter & Graves 2006, Mandelman et al. 2008, Amandè et al. 2010, Curran & Bigelow 2011, Lynch et al. 2011, Huang et al. 2016, Juan-Jorda et al. 2018). Managing these iconic species in consideration of these stressors will require broader multinational management strategies that account for their cumulative effects and the ecosystem-level consequences of historical harvesting practices (Stevens et al. 2000, Cox et al. 2002, Kitchell et al. 2006, Maunder & Harley 2006, Lynch et al. 2011, Worm et al. 2013, Juan-Jorda et al. 2015, Horodysky et al. 2016, Juan-Jorda et al. 2018). Although efforts to implement ecosystem approaches for HMS species are underway (Juan-Jorda et al. 2015, 2018, FAO 2016, 2017, NMFS 2019b), expanded strategies that improve monitoring and assessments, refine multinational limitations toward sustainable management practices, and consider precautionary ecosystem-level reference points are needed to ensure the sustainability of HMS populations (Aranda et al. 2010, Lynch et al. 2011, Juan-Jorda et al. 2015, 2018).

11.1.2 Pacific Halibut and Pacific Whiting

Bilateral agreements between the U.S. and Canada (CA) are in place for two major northeastern Pacific groundfish species, Pacific halibut (*H. stenolepis*) and Pacific hake or "*Whiting*" (*M. productus*). U.S. and Canadian Pacific halibut fisheries are managed through the intergovernmental **International Pacific Halibut Commission (IPHC)**, which has been in effect since the 1923 IPHC Convention (Clark & Hare 2002, Clark 2003, IPHC 2019). Several revisions to this Convention have occurred since the establishment of the IPHC, with the most recent amended protocol (IPHC 1979) extending U.S. and Canadian jurisdictions to 200 miles throughout each respective EEZ. Since this 1979 protocol and the complementary U.S. Northern Pacific Halibut Act of 1982, the joint U.S.-Canadian Pacific Halibut fishery has operated throughout the western EEZs of the U.S. and British Columbia (CA), including southern and western Alaskan waters (Copes & Cook 1982, Wilen & Homans 1998, Clark & Hare 2002, Clark 2003, Williams 2012, IPHC 2019). The IPHC operates under the objectives "to develop the stocks of Pacific halibut in the Convention waters to those levels which will permit the optimum yield from the fishery and to maintain the stocks at those levels" (IPHC 2019). Its functions include dividing Convention waters into allocation areas, establishing open or closed seasons and catch and size limits (including incidental captures) within each area, regulating gear types, licensing of vessels, monitoring the fishery, and enacting spatial closures of nursery grounds (McCaughran & Hoag 1992, Williams & Blood 2003, IPHC 2019).

Prior to the establishment of the IPHC, low Pacific halibut populations were observed during the 1910s–1920s, with initial catch restrictions effective toward population rebuilding (Clark 2003). Winter closures have been effective in regulating and rebuilding the Pacific halibut population over time, with pronounced allocated fishing restrictions occurring during the 1970s-1990s in response to declining landings (Clark 2003, Loher 2011). The Pacific Decadal Oscillation (PDO) has been associated with a decline in growth of Alaskan and Canadian Pacific halibut from the late 1970s to mid-1990s, which also impacted their recruitment (Clark et al. 1999, Clark & Hare 2002, Sullivan 2016). In contrast, decreased halibut recruitment was observed along the U.S. west coast (Clark & Hare 2002, Thom 2018). Size-at-age for Pacific halibut has declined since the 1980s, related to decreasing recruitment and spawning biomass, size-selective fishing practices, and increases in Pacific arrowtooth flounder (*Atheresthes stomias;* a potential competitor) populations (Sullivan 2016, Barnes et al. 2018). In consideration of these climatological, fishing, and ecological effects on halibut life history, more accurate assessment and management methods that are robust to environmental variability and that allow for precautionary harvests have been developed (Clark et al. 1999, Parma 2002, Martell et al. 2013, Barnes et al. 2018). With these more conservative management practices, fishing does not appear to affect recruitment as strongly as climate forcing (Clark & Hare 2002, Beamish 2008). Given their small home ranges and high halibut site fidelity, spatial closures and harvest restrictions appear to be effective for conserving Pacific halibut broodstock and limiting their bycatch (Nielsen et al. 2014, Wallace et al. 2015). However, given the high preference for structured habitat by younger halibut, the effects of fishing gears on benthic habitats may have significant consequences toward their ongoing recruitment (Gibson 1994, Stoner & Titgen 2003). Increasingly frequent hypoxic conditions along the California Current shelf have also affected Pacific groundfish populations, including Pacific halibut and whiting, potentially restricting suitable habitat availability (Keller et al. 2010, Sadorus et al. 2014).

Pacific whiting are subject to the **Pacific Whiting Treaty (PWT)**, which allocates a set percentage of a total U.S. west coast and Canadian harvest quota to American and Canadian fishers (PWT 2004, Ishimura et al. 2005, Hamel & Stewart 2009, Hamel et al. 2015). Currently, the majority of the quota (~75%) is allocated to U.S. fisheries (PWT 2004, Ishimura et al. 2005, Hamel & Stewart 2009, Hamel et al. 2015, Guldin & Anderson 2018). Since its 2003 establishment, the PWT has accounted for the transboundary stock of Pacific whiting through a process where U.S. and Canadian fisheries scientists, managers, and stakeholders determine total catch and apply a percentage formula (PWT 2004, Ishimura et al. 2005, Hamel & Stewart 2009, Hamel et al. 2015, Guldin & Anderson 2018). This agreement formalized a process for Pacific whiting allocations that had been previously conducted since the 1970s through informal joint U.S.-Canadian stock assessments and stock management measures (PWT 2004, Ishimura et al. 2005, Hamel & Stewart 2009, Hamel et al. 2015, Guldin & Anderson 2018). However, past allocations and overfishing resulted in stock declines, with the stock classified as overfished in 2002 (PWT 2004, Ishimura et al. 2005, Hamel & Stewart 2009, Hamel et al. 2015, Guldin & Anderson 2018). Although bilateral conversations had been underway since the 1990s, no formal agreement on percentage shares could be reached until this declaration, with both sides eventually agreeing to the 2003 PWT management and sharing agreement on addressing overfishing. The PWT also formalized scientific and stock assessment collaboration within the Joint Technical Committee, and a scientific review group that reviews the technical committee's assessments (PWT 2004, Ishimura et al. 2005, Hamel & Stewart 2009, Hamel et al. 2015, Guldin & Anderson 2018). Additionally, a joint management committee considers the scientific advice and recommends an overall total allowable catch (TAC), with default U.S. harvest consisting of 73.88% of the TAC.

The west coast Pacific whiting fishery currently represents the largest fishery by weight off the U.S. west coast, with migratory populations accounting for 61% of its pelagic biomass (Ware & McFarland 1995, Agostini et al. 2008, Lomeli & Wakefield 2014, NMFS 2018a). Since the late 1970s, acoustic monitoring has shown a long-term decline in Pacific whiting stock biomass, with estimates continuing to oscillate and remaining at comparatively low values (Helser & Alade 2012). Although no longer classified as overfished or subject to overfishing, low spawning biomass has persisted (Stewart & Hamel 2010, Helser & Alade 2012). Pacific whiting growth rates dramatically increased during the 1970s and decreased in the 1980s as possibly related to increasing water temperatures and lower production in association with the PDO (Helser & Alade 2012). The El Niño Southern Oscillation (ENSO) and increasing temperatures have been shown to favor Pacific Whiting pelagic habitat expansions, resulting in strong year classes (Hollowed et al. 2001, Agostini et al. 2008). However, these have also been related to expanding spawning populations into Canadian waters, which may affect future stock distributions, national allocations, and year-class strength related to anticipated changes in upwelling productivities and zooplankton biomass (McFarlane et al. 2000, Benson et al. 2002, Grover et al. 2002, Beamish 2008). Despite being one of the most data-rich assessments and having a relatively short rebuilding plan, Pacific whiting management challenges have continued due to limited life history information, imprecise surveys, and additional concerns (Helser & Alade 2012, Wetzel & Punt 2016). For instance, high bycatch of overfished rockfishes and other species continues to occur in this fishery (Harvey et al. 2008, Lomeli & Wakefield 2014, 2016).

As observed for other Pacific groundfish throughout the California Current ecosystem (Norton 1987, Oreskes & Finley 2007, Holt & Punt 2009, Punt 2011, Kaplan et al. 2012, 2013), Pacific halibut and whiting have been subjected to multiple stressors that include historical overfishing, climate forcing, regional warming, degradation and compression of essential habitats, and northward expansion of their distributions (Francis et al. 1998, Hollowed et al. 2001, Clark 2003, Stoner & Titgen 2003, PWT 2004, Ishimura et al. 2005, Agostini et al. 2008, Beamish 2008, Hamel & Stewart 2009, Loher 2011, Okey et al. 2014, Hamel et al. 2015, Guldin & Anderson 2018). Trade-offs between the fisheries for these stocks and marine mammals that prey on them are also a significant factor (Best & St-Pierre 1986, Sigler et al. 2009, Peterson & Carothers 2013, Tolmieri et al. 2013, Jannot et al. 2018). While single-species management efforts for both Pacific halibut and Pacific whiting have been quite successful, continued management measures that more broadly incorporate multispecies considerations, ecosystem-level reference points, changing climatic conditions, and which consider these iconic species in context of the California Current and North Pacific ecosystems remain needed (Holt & Punt 2009, Kaplan & Levin 2009, Punt 2011, Kaplan et al. 2012, 2013, Rindorf et al. 2017).

11.1.3 Atlantic and Pacific Salmon

While managed domestically through NOAA Fisheries and the New England Fishery Management Council (NEFMC) as a federally protected species, U.S. Atlantic salmon (*S. salar*) are also subject to the provisions of the multilateral 1983 Convention for the Conservation of Salmon in the North Atlantic Ocean (NASCO 1983, Windsor & Hutchinson 1994, Mills & Piggins 2012). Through this Convention, the intergovernmental **North Atlantic Salmon Conservation Organization (NASCO)** was created to "conserve, restore, enhance, and rationally manage wild Atlantic salmon through international cooperation taking account of the best available scientific information" (NASCO 1983, Windsor & Hutchinson 1994, Mills & Piggins 2012). Since 1984, NASCO parties (Canada, Denmark-Greenland, the European Union, Norway, the Russian Federation, and the U.S.) have functioned to prohibit fishing for Atlantic salmon in most parts of the North Atlantic beyond 12 nautical miles from the coast (Windsor & Hutchinson 1994, Mills & Piggins 2012). These regulatory measures and cooperative state-level management initiatives have resulted in significantly reduced salmon fishing effort throughout the North Atlantic; although local commercial, recreational, and subsistence fishing continues in Greenland (Berg-Hedeholm et al. 2018, Sheehan et al. 2019). Given greatly depleted populations and additional stressors, NASCO efforts also consider precautionary approaches in consideration of fisheries management, protection and restoration of habitat, stock rebuilding, and exogenous effects such as aquaculture (NASCO 1998, 1999, 2005). Despite these efforts, Atlantic salmon populations remain low, with the U.S. Gulf of Maine Distinct Population Segment (DPS) currently classified as endangered under the Endangered Species Act (ESA; Huntsman 1994, Elmen 2003, Jenkins 2003, Robertson 2005, Kircheis & Liebich 2007, VanderZwaag 2015, Hare et al. 2019) and the Kapisillit River, Greenland DPS listed as vulnerable on the International Union for the Conservation of Nature (IUCN) red list (Berg-Hedeholm et al. 2018). With increasing mortality and poor survival of salmon at sea, the International Atlantic Salmon Research Board (IASRB) has been established to improve understanding of causative environmental and human-related factors, including the effects of bycatch, pollution, oceanographic conditions, multispecies interactions, and disease (IASRB 2001a, b, Hansen & Windsor 2006, Hansen et al. 2012).

NOAA Fisheries and regional FMCs manage Pacific salmon (*O. gorbuscha*, *O. keta*, *O. kisutch*, *O. mykiss*, *O. nerka*, *O. tshawytscha*) throughout the U.S. Pacific and Alaskan coasts. However, due to their migratory nature and interception, shared salmon resources are also co-managed with Canada by the 1985 bilateral Pacific Salmon Treaty and its 1999 agreement, as overseen by the **Pacific Salmon Commission (PSC**; PSC 2020**)**. Prior to this treaty, a precursor management organization known as the International Pacific Salmon Fisheries Commission had operated since 1937 as ratified under the Fraser River Convention (IPSC 1945). Together, Canada, Alaska, Washington/Oregon, and

24 treaty tribes throughout Washington, Oregon, and Idaho comprise the Commission, which exists to govern the overall harvest and allocation of salmon stocks that are jointly exploited by the U.S. and Canada (Miller 2003, PSC 2020). A set of conservation and equity principles in Article III state that, "each party shall conduct its fisheries and its salmon enhancement programs so as to: prevent overfishing and provide for optimum production; and provide for each party to receive benefits equivalent to the production of salmon originating in its waters" (PSC 2020). The provisions of the treaty set limits on catches of Pacific salmon throughout transboundary rivers, northern British Columbia and southeastern Alaska boundary areas, within the Fraser and Yukon Rivers, and on southern British Columbia and Washington State Chum salmon (*O. keta*). They also address shared Chinook (*O. tshawytscha*) and Coho salmon (*O. kisutch*) stocks (Miller 2003, PSC 2020). Due to climatic forcing and northward shifts of salmon populations, Alaskan and northern Canadian salmon populations have greatly increased over time, with higher interceptions of British Columbian salmon by Alaskan fishers (Beamish & Bouillon 1993, Hare et al. 1999, Miller 2003). As a result, Canadian fishers were unable to catch agreed-upon harvest ceilings, including for declining southern Chinook and Coho salmon stocks, and increased their fishing effort (Miller 2000, Miller et al. 2001, Miller 2003). These practices in turn affected U.S. west coast fishing interests as a result of perceived Canadian overharvesting of significantly depleted west coast stocks, and resulted in stalled agreements toward accounting for imbalances (Schmidt 1996, Miller 2000, Miller et al. 2001, Miller 2003). Following continued declines of Canadian fall Chinook and Coho stocks, and listings of Pacific Northwest stocks under the ESA, resumed negotiations allowed for the development of the 1999 Pacific Salmon Agreement. This agreement builds upon the 1985 Treaty and replaces its previous ceilings with longer-term fixed harvesting regimes toward conservation and restoration of depressed salmon stocks.

Currently, many identified U.S. salmon populations remain listed as threatened or endangered, including two Chinook salmon ESA-endangered populations (Sacramento River winter-run; Upper Columbia River spring-run), seven threatened populations, two ESA-candidate populations, and two ESA-experimental populations for this species (Williams et al. 2016). One Coho salmon population remains endangered (Central California coast evolutionarily significant unit, ESU), while three are listed as threatened. Endangered populations of Sockeye salmon (*O. nerka*; Snake River ESU) and Steelhead trout (*O. mykiss*; Southern California DPS) occur in U.S. Pacific waters, in addition to one threatened Sockeye population, 11 threatened Steelhead populations, and one ESA-experimental Steelhead population. Two ESA-threatened populations of Chum salmon are found in the U.S. Pacific.

In addition to the effects of overfishing and climatic forcing, Atlantic and Pacific salmon populations have been affected by habitat impediments (i.e., dams), destruction, loss, and degradation of critical habitats, bycatch, and other ocean and riverine uses including aquaculture (Costa-Pierce 2002, Robertson 2005, Welch et al. 2008, Amiro et al. 2009, Limburg & Waldman 2009, Costa-Pierce 2010, Gilman et al. 2014, Nieland & Sheehan 2015, Gibson 2017). Damming of freshwater migration paths has limited access of Atlantic and Pacific salmon to spawning habitats and oceanic juvenile rearing areas, including flooding these locations hat impact habitat quality, and that have led to major population declines over time (Juanes et al. 2004, Holbrook et al. 2009, Weitkamp et al. 2014, Nyqvist et al. 2017). The effects of dams on river flow, water quality, and water temperature have caused significant ecological consequences to salmon populations, their prey, and have constrained spawning potential (Gregory et al. 2002, Angilletta et al. 2008, Keefer et al. 2008, Waples et al. 2008, Olden 2016). Degradation of freshwater habitats through land development and agriculture has led to stream erosion, eutrophication, and increased pollution and toxins that have affected U.S. salmon populations (Waldichuk 1993, Slaney et al. 1996, Cederholm et al. 2001, Kircheis & Liebitch 2007, Arlinghaus et al. 2016, Colvin et al. 2019). Continued bycatch of protected Atlantic and Pacific salmon stocks has constrained rebuilding efforts, while the threats of enhanced disease from and interbreeding with farmed salmon have affected viability of wild populations (Morton & Volpe 2002, Jonsson & Jonsson 2004, Morton et al. 2008, Torrissen et al. 2011, Raby et al. 2012, Teffer et al. 2017). In consideration of these multiple stressors, broader management efforts that consider interconnected climate and environmental effects on U.S. salmon populations, their interactions with co-occurring species within Pacific and Northeast ecosystems, and the effects of land and ocean use practices on these species as a component of broader ecosystem functioning remain necessary (Bisbal & McConnaha 1998, Richards & Maguire 1998, Chittenden et al. 2009, Zhou et al. 2010, Krueger & Zimmerman 2009, Knudsen & McDonald 2019).

11.1.4 Additional Regional RFMOs, Fishery Bodies, and Global Organizations

Notable, broadly encompassing RFMOs and high seas fishery bodies within the Atlantic basin to which the U.S. is a member include the **Northwest Atlantic Fisheries Organization (NAFO)** and **Western Central Atlantic Fisheries Commission (WECAFC**—discussed further below). The U.S. is also a member of several Pacific RFMOs and high seas fishery bodies (discussed further below) including the **North Pacific Fisheries Commission (NPFC)** and the **South Pacific Regional Fisheries Management Organisation (SPRFMO)**.

The U.S. is also a member of several FAO recognized regional fishery bodies (RFBs) that include the North Pacific Anadromous Fish Commission (NPAFC), the North Pacific Marine Science Organization (PISCES), the International Council for the Exploration of the Sea (ICES), and the Convention on the Conservation and Management of the Pollock Resources in the Central Bering Sea (CCBSP) (Marashi 1996, Billé et al. 2016, FAO 2019). All of these entities work toward the conservation and management of fish stocks (Marashi 1996, Sydnes 2002, Billé et al. 2016, Brown 2016, FAO 2019).

NAFO is an intergovernmental fisheries science and management body, which was founded as a successor to the 1949 established International Commission of the Northwest Atlantic Fisheries (Halliday & Pinhorn 1996, NAFO 2009). Since 1979, the NAFO Convention on Cooperation in the Northwest Atlantic Fisheries applies to most fishery resources of the Northwest Atlantic excepting salmon, tunas, and marlins, whales, and sessile species (NAFO 2009). The NAFO Convention Area includes the EEZs of the Northeastern U.S. (Mid-Atlantic and New England regions), Canadian Atlantic waters, western Greenland, and the French territories of St. Pierre and Miquelon. NAFO manages harvest, bycatch, and fishing effort, and protects vulnerable open-ocean marine ecosystems from bottom fishing activities within its regulatory area, which is outside of these EEZs (NAFO 2019). Among its 12 contracting parties are representatives who serve on its Commission, Scientific Council, and their joint working groups for subjects including bycatch and discards, catch estimations, risk-based management, and ecosystem approaches to science, assessment, and fisheries management (NAFO 2009, 2019). NAFO is responsible for fisheries management, control, and enforcement of international North Atlantic fisheries resources to "ensure the long term conservation and sustainable use of the fishery resources of the Convention Area and, in so doing, to safeguard the marine ecosystems in which these resources are found" (NAFO 2009, 2019). The Convention has been amended four times by modifying Convention Area boundaries of its subareas, divisions, and subdivisions, and most recently (in 2017) to incorporate an ecosystem approach to fisheries management (NAFO 2019). As observed throughout U.S. Mid-Atlantic and New England regional ecosystems, major stressors including historically concentrated overfishing, climate-related regional warming, habitat loss, and degradation, coastal development, eutrophication, sea level rise, species range shifts, and invasive species continue to affect NAFO jurisdictional ecosystems and their LMRs (Joseph 1972, Warren & Niering 1993, Auster et al. 1996, Grosholz & Ruiz 1996, Fogarty & Murawski 1998, Chambers et al. 1999, Roman et al. 2000, Orth et al. 2006, Anthony et al. 2009, Nye et al. 2009, Ezer et al. 2013, Nye et al. 2013, Wells et al. 2013, Flanagan et al. 2019).

In addition to NAFO, IPHC, PSC, and HMS RFMO jurisdictions, the U.S. EEZ is significantly encompassed by or shares strong boundaries with three major RFMO convention areas (i.e., those of WECAFC, NPFC, and SPRFMO). Comprising the entire Gulf of Mexico, Caribbean, and U.S. South Atlantic, in addition to the Atlantic coast of South America north of 10°S latitude, WECAFC was established in 1973 by the FAO Council to "promote the effective conservation, management, and development of the area of competence of the Commission, and address common problems of fisheries management and development faced by members of the Commission" (Stevenson 1981, Ehrhardt et al. 2017, FAO 2019). Its 34 member states serve to promote the provisions of the FAO Code of Conduct on Responsible Fisheries and its related instruments, including the precautionary approach, the ecosystem approach to fisheries management, and attention to small-scale fisheries (Ehrhardt et al. 2017, FAO 2019). All LMRs within the Commission competence area fall under WECAFC jurisdiction "without prejudice to the management responsibilities and authority of other competent fisheries and other LMR management organizations or arrangements" (Stevenson 1981, Ehrhardt et al. 2017, FAO 2019). WECAFC consists of a principal governing commission, a scientific advisory group, and 11 working groups on Gulf, Caribbean, and subtropical and tropical Atlantic wide fishery species (Ehrhardt et al. 2017, FAO 2019). As observed throughout U.S. Gulf of Mexico, Caribbean, and South Atlantic regional ecosystems, major stressors including overfishing, climate-related regional warming, hurricanes, habitat loss and degradation, coastal development, sea level rise, species range shifts, high tourism and energy development, and invasive species have significantly affected WECAFC jurisdictional ecosystems and their LMRs (Rogers et al. 1982, Scatena & Larsen 1991, Lugo-Fernández et al. 1994, Winter et al. 1998, Chesney et al. 2000, Lugo et al. 2000, Pandolfi et al. 2005, Sammarco et al. 2006, Caillouet et al. 2008, Cowan et al. 2008, Ryan et al. 2008, Smith et al. 2008, Wilkinson & Souter 2008, Cowan et al. 2011, Karanauskas et al. 2013, Lucas & Carter 2013, Ballantine et al. 2008, Du & Park 2019, Barange et al. 2018).

The NPFC is an intergovernmental organization established by the 2012 Convention on the Conservation and Management of High Seas Fisheries Resources in the North Pacific Ocean, following the initial 1952 establishment of the International North Pacific Fisheries Commission (Johnson et al. 1993, NPFC 2012). The objective of the Convention and NPFC is, "to ensure the long-term conservation and sustainable use of the fisheries resources in the Convention Area while protecting the marine ecosystems of the North Pacific Ocean in which these resources occur" (NPFC 2012). This Convention Area includes high seas waters of the North Pacific Ocean, excluding the U.S. Hawaiian EEZ and encompassing the seaward limit of waters under the jurisdiction of the U.S. Commonwealth of the Mariana Islands (NPFC 2012, FAO 2019). The U.S. and seven other members work toward regional fisheries management of deep-sea fish and protection of marine ecosystems, including all fish, mollusks, crustaceans, and other marine species (excluding sovereign sessile species, catadromous species, marine mammals, reptiles, and seabirds, and marine species already managed by pre-existing international fisheries management bodies) captured within the Convention Area (NPFC 2012). Consequential habitat effects from deep-sea bottom trawling and sea mining, longline, gillnet, and trawl bycatch, and overfishing have been observed throughout NPFC jurisdictional ecosystems with effects on their pelagic and bottomfish stocks (Johnson et al. 1993, Weaver et al. 2011, Koslow et al. 2015, Arias & Marcovecchio 2017, Menezes & Giacomello 2017, Watling & Auster 2017, NPFC/FAO 2018). Concerns about microplastics, marine pollution, piracy, and other emerging ocean uses also continue to affect high seas fisheries and the LMRs that comprise them (McDorman 1994, Stokke 2001, Hughes 2011, Law & Thompson 2014, Blasiak & Yagi 2016, Lusher et al. 2017).

The SPRFMO is another intergovernmental organization that is "committed to the long-term conservation and sustainable use of the fishery resources of the South Pacific Ocean and in so doing safeguarding the marine ecosystems in which the resources occur" (Schiffman 2013, SPRFMO 2015). The 2009 Convention on the Conservation and Management of High Seas Fishery Resources in the South Pacific Ocean led to the establishment of the SPRFMO in 2012 (Schiffman 2013, SPRFMO 2015). This Convention Area includes all high sea regions of the entire South Pacific and accounts for ~25% of the Earth's high seas, while excluding national EEZs including American Samoa and U.S. Western Pacific Remote Islands (Schiffman 2013, SPRFMO 2015). The Scientific Committee plans, conducts, and reviews scientific assessments of the status of fishery resources and provides advice and recommendations to the governing Commission on assessments, management strategies, and the impact of fishing on vulnerable marine ecosystems (VMEs) (SPRFMO 2020) based on the FAO Deep Sea Fisheries Guidelines (DSFG) (FAO 2009). The U.S. and 14 other member states co-manage pelagic and benthic fisheries resources in the Convention Area of which Pacific jack mackerel (*Trachurus symmetricus*), Jumbo flying squid "Humboldt Squid" (*Dosidicus gigas*), Orange roughy (*Hoplostethus atlanticus*), Oreos (Teleostei: Oreosomatidae), Alfonsino (*Beryx decadactylus*), and Bluenose (*Hyperoglyphe antarctica*) are major commercial species (Schiffman 2013, SPRFMO 2015). As fishing methods include purse seining, pelagic and bottom trawling, and bottom longlining, the impacts of overfishing, bycatch, and bottom fishing practices on benthic habitats have been observed throughout SPRFMO jurisdictional ecosystems, with effects on pelagic and bottomfish stocks (Parker et al. 2009, Cullis-Suzuki & Pauly 2010, Hansen et al. 2013, Watling & Auster 2017).

Global organizations which the U.S. is a member include the **Commission for the Conservation of Antarctic Marine Living Resources (CCAMLR)**, which was established by international convention (Convention for the Conservation of Antarctic Marine Living Resources, CAMLR) in 1982 (CCAMLR 1982, Agnew 1997, Constable et al. 2000). The objective of this commission and its convention is to conserve Antarctic marine life in response to increasing commercial interest in Antarctic krill resources and historical overexploitation of other Southern Ocean marine resources (CCAMLR 1982). With 26 member countries (including the European Union) and 11 additional nations acceding to the Convention, CCAMLR produces conservation measures and resolutions to determine the use of Antarctic marine living resources through its management commission and scientific committee. This convention applies to all Antarctic populations of fishes (e.g., Patagonian toothfish—*Dissostichus elieginoides*, *D. mawsoni*), mollusks (particularly gastropods and bivalves), crustaceans (e.g., krill—*E. superba*), and sea birds (e.g., penguins—*Pygoscelis* spp., black-browed albatross—*Thallasarche melanophrys*), while marine mammals are the subject of the International Convention for the Regulation of Whaling and the Convention for the Conservation of Antarctic Seals (CCAMLR 1982, Agnew 1997, Constable et al. 2000, Molenaar 2001). Advances toward practicing an ecosystem-based approach to ensure sustainable harvests have occurred, with consideration of the effects of fishing on other components of the Antarctic ecosystem and the development of ecosystem models in consideration of management strategies in response to climate change (Constable 2002, Guerry 2005, Gascon & Werner 2006, Kock et al. 2007, Cornejo-Donoso & Antezana 2008, Constable 2011, Watters et al. 2013, Dahood et al. 2019, Keith 2018). These also include model-derived buffers on krill harvest to ensure that enough of these dominant zooplankton remain as prey for other taxa in this "wasp-waist" ecosystem, while also accounting for environmental effects on krill populations (Constable et al. 2000, Hewitt et al. 2002, Kock et al. 2007, Atkinson et al. 2009, McBride et al. 2014, Hill et al. 2016). CCAMLR has also worked toward addressing IUU fishing, establishing protected areas in the Southern Ocean, reducing bycatch and seabird mortality, establishing an ecosystem monitoring program (CEMP), and managing vulnerable marine ecosystems (Agnew 1997, De la Mare 2000, Constable 2002, Croxall & Nicol 2004, Miller et al. 2004, Sabourenkov & Appleyard 2005, Constable 2011, Reid 2011). Due to historical overexploitation and its effects on high seas benthic environments, CCAMLR has restricted the use of bottom trawling gear and bottom fishing in high-seas areas of the Convention Area and established protected areas toward reducing their impacts (Monteiro et al. 2010, Wright et al. 2015, CCAMLR 2019). The CAMLR convention is one of several conventions within the Antarctic Treaty System (ATS), which includes management of other ocean uses and other Antarctic issues such as mining, environmental protections, marine pollution, and waste management (Haward 2017).

The U.S. also serves on the intergovernmental **Arctic Council**, which accounts for high seas management and territorial allocations of Arctic marine resources among its eight member countries (Bloom 1999, Correll 2005, Arctic Council 2007). Given a current lack of fishing, fisheries resources have not been directly considered and the Council does not act as a functional RFMO (Molenaar 2012, Pan & Huntington 2016). For the U.S. Arctic, the North Pacific Fishery Management Council (NPFMC) has adopted an Arctic FMP that prohibits fishing in this region at this time (NPFMC 2009; see chapter 9). Recent advancements in managing other ocean uses have occurred, especially regarding emerging marine energy, transportation, and oil and mineral extraction sectors (Molenaar 2012, Hoel 2015, Wegge 2015). These include using the ecosystem approach as proposed in the Arctic Marine Strategic Plan to reduce and prevent pollution, conserve marine biodiversity and ecosystem functions, promote the health and prosperity of Arctic inhabitants, and advance sustainable marine resource use (Arctic Council 2004, Siron et al. 2008). Intergovernmental precautionary measures are in place to abate oil pollution, regulate offshore drilling and expansion in consideration of human health and marine ecosystems, including the Agreement on Cooperation on Marine Oil Pollution Preparedness and Response in the Arctic (Arctic Council 2013, Brigham 2020). Following the release of the Arctic Marine Shipping Assessment Report (Arctic Council 2009), all participatory Arctic states have worked to implement its recommendations that emphasize marine safety, protecting Arctic inhabitants and the environment, and building marine infrastructure (Brigham 2020). In particular, these protections emphasize minimizing the effects of invasive species, preventing future oil spills, and addressing potential encounters with marine mammals (Arctic Council 2009, Brigham 2020).

There are a few other global LMR organizations to which the U.S. is a member. These include the IUCN, the United Nations Convention for Biological Diversity (CBD; non-party), and the International Whaling Commission (IWC).

The IUCN is a membership union of approximately 1300 organizations (e.g., state, governmental, non-governmental, indigenous, scientific/academic institutions) that works "to conserve the integrity and diversity of nature and to ensure that any use of natural resources is equitable and ecologically sustainable" (IUCN 2020). With headquarters in Gland, Switzerland, the IUCN Council governs the Union in between sessions of the World Conservation Congress, which is the general assembly of its members. Among its roles, the Council sets direction and policy guidance for the work of the Union, provides oversight and guidance, and supports the Director General in communicating IUCN objectives to the global community. The Union consists of six commissions that are focused on education and communication; ecosystem management; environmental, economic, and social policy; species survival; environmental law; and protected areas.

Priorities for ecosystem management include the development of a Red List of Ecosystems, ecosystem-based adaptation to climate change initiatives, disaster risk reductions, ecosystem management program activities in islands and drylands, and emerging human activities such as seabed mining (IUCN-CEM 2017).

The CBD has three main objectives that include, "the conservation of biological diversity, the sustainable use of the components of biological diversity, and the fair and equitable sharing of benefits arising out of the utilization of genetic resources" (CBD 1992). In accordance with Article 6 of the Convention, each of the 196 participatory parties develops national strategies, plans, or programs for the conservation and sustainable use of biological diversity and integrates them as far as possible and appropriate (CBD 1992, 2020). This national biodiversity planning includes the development of a National Biodiversity Strategy and Action Plan (NBSAP) that specifically accounts for these approaches among a given party's biomes (CBD 2020). In addition, the CBD has adopted the ecosystem approach as strategy for the integrated management of land, water and living resources that promotes conservation and sustainable use in an equitable way (CBD 2020). With regards to its marine and coastal biodiversity priorities, the CBD has worked to develop conservation and sustainable use programs; implement integrated marine and coastal area management; advance the establishment of marine protected areas in areas beyond national jurisdictions; identify ecologically or biologically significant marine areas (EBSAs); and address anthropogenic impacts such as underwater noise, marine debris, and ocean acidification (CBD 2020).

The IWC was created under the 1946 International Convention for the Regulation of Whaling and sets out specific measures such as catch limits toward regulating whaling and conserving whale stocks (Caron 1995, IWC 2005, Punt & Donovan 2007, Wright et al. 2016). Over time, 89 current countries and five former member nations (Canada, Egypt, Japan, Philippines, and Venezuela) have worked to designate whale sanctuaries, protect calves and accompanying females, and place restrictions on hunting methods. The IWC also coordinates and supports cetacean research and conservation efforts, including prevention of fishing gear entanglements, ship strikes, and establishment of conservation management plans for key populations (Heazle 2004, Reeves et al. 2005, Wright et al. 2016, Vernazzani et al. 2017, IWC 2019). The commission contains six committees for Science, Conservation, Administration, Subsistence Whaling, Humane Whale Killing Methods and Welfare, and Infractions, each with subcommittees and working groups (IWC 2019). Significant effects of historical whaling, fisheries bycatch, climate change, top-down system-level consequences of their depleted populations on marine ecosystems, prey depletions from overfishing, and the effects of marine transportation and noise pollution continue to be observed on global cetacean populations, warranting further ecosystem-based managerial approaches (Alter et al. 2010, Würsig et al. 2002, Wright et al. 2016, Butterworth 2017, Avila et al. 2018, Jeffries 2018).

11.2 Evaluation of Major Facets of EBFM for These Regions

To elucidate how well these RFMOs in Atlantic and Pacific basins are doing in terms of LMR management and progress toward EBFM, here we characterize the status and trends of socioeconomic, governance, environmental, and ecological criteria pertinent toward employing an EBFM framework for the regions associated with these RFMOs. We examine the interrelationships among these factors relative to successful implementation of EBFM. We also quantify those factors that emerge as key considerations under an ecosystem-based approach for each RFMO region, particularly with information for U.S. components of their jurisdictions when available. Ecosystem indicators across RFMOs, particularly U.S. components, related to the (1) human environment and socioeconomic status, (2) governance system and LMR status, (3) natural environmental forcing and features, and (4) systems ecology, exploitation, and major fisheries of this system are presented and synthesized below. We explore all these facets to evaluate the strengths and weaknesses along the pathway:

$$PP \rightarrow B_{\text{targeted, protected spp, ecosystem}} \leftrightarrow L_{\uparrow\text{targeted spp, }\downarrow\text{bycatch}} \rightarrow \text{jobs, economic revenue.}$$

Where PP is primary production, B is biomass of either targeted or protected species or the entire ecosystem, L is landings of targeted or bycaught species, all leading to the other socioeconomic factors. This operates in the context of an ecological and human system, with governance feedbacks at several of the steps (i.e., between biomass and landings, jobs, and economic revenue), implying that fundamental ecosystem features can determine the socioeconomic value of fisheries in a region, as modulated by human interventions.

I'm Hungry: Ensuring There is Adequate Forage for a Range of Trust Species

—(c.f. Chapters 4, 8)

When I was in college, the landlord of our apartment promised to get us a few pizzas for being such good tenants. I got home the night of the pizza party after spending a little extra time in the lab studying. My anticipation of having free pizza was quite high. Yet when I arrived home, the pizza was gone—the landlord only delivered two medium-sized pizzas to begin with, which wasn't that much for college-aged kids. My roommates had invited a few friends over, which added to the demand on the pizza, and ultimately my roommates had not bothered to save me any. I was not pleased.

I had to go forage for another dinner in another location for another type of food. No trivial task for a poor college student. Thankfully, there are ramen noodles...

Whales, dolphins, seals, and birds eat a lot. Tunas and sharks and swordfish and sailfish also eat a lot. They often eat the same thing- small pelagic fishes. They often find schools of these forage fishes and then deplete those "bait balls" by gorge eating (Nakamura 1965, Ménard et al. 2000, Hinke et al. 2004, Potier et al. 2007, Puncher 2015). A lot like my college roommates. Certainly forage fish are quite productive (Alder et al. 2008, Tacon & Metian 2009, Pikitch et al. 2014, Clausen et al. 2018), and certainly productive enough that forage fish support many upper trophic level organisms that feed on them. But they can become locally depleted if too many of them are removed, at which point the sharks and whales and tunas move on.

The challenge is that a lot of these forage fish also support commercially targeted fisheries (Engelhard et al. 2014, Pikitch et al. 2014, Robinson et al. 2014, Essington et al. 2015, Clausen et al. 2018, McClatchie et al. 2018). And these forage fish often have shorter life-spans, making them more susceptible to environmental fluctuations (Gutiérrez et al. 2007, Pethybridge et al. 2013, McClatchie et al. 2017, Siple et al. 2019). But if too many of these forage fish are removed from fishing, and the ocean changes such that it is not conducive to produce as much of these small pelagics, then that has an impact on upper trophic level species that feed on them.

This trade-off has been recognized as one of the more obvious examples of the need for EBFM (Constable et al. 2011, Eero et al. 2012, Essington et al. 2015, Kaplan et al.

continued

I'm Hungry: *Continued*

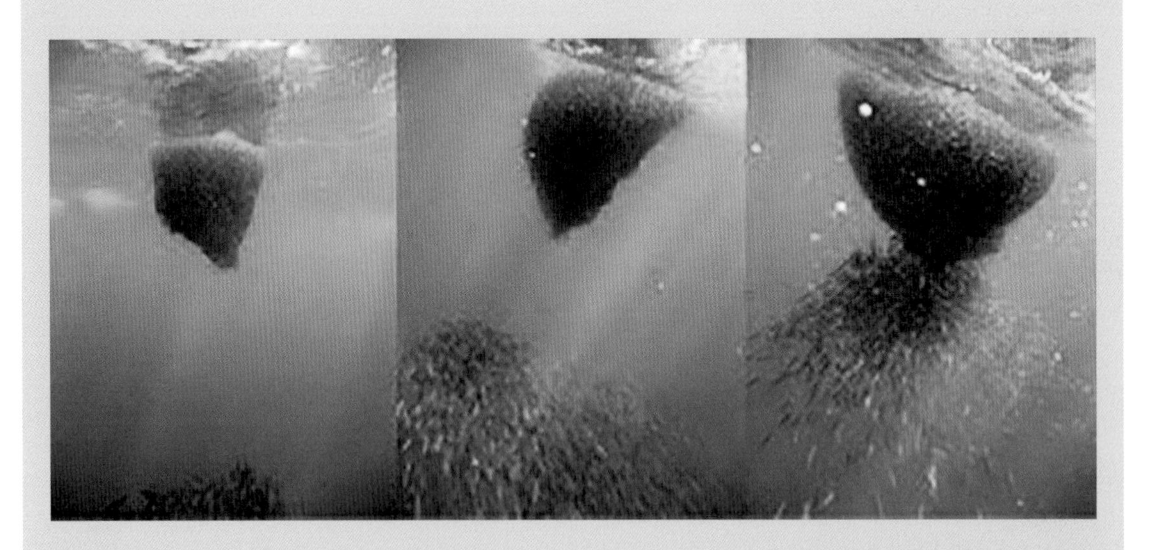

2017, Siple et al. 2019). The challenge is to estimate how much consumption is required by whales, sharks, tunas, birds, etc. to maintain their populations at the desired or mandated sizes, and see if there is sufficient forage biomass to sustain that demand for populations of these upper trophic level species. All while ensuring that any fisheries removals of these forage fishes does not erode the forage capacity to supply that consumptive demand. And while ensuring that any environmental changes to these forage fishes' production does not fall below what can also supply that consumptive demand.

Several laws are in potential conflict here, the Marine Mammal Protection Act and Endangered Species Act on the one hand, and the Magnuson-Stevens Act on the other. And also several international treaties for things like tunas and billfishes and other highly migratory fishes. And treaties that protect some taxa like the International Whaling Commission (IWC) or Convention for the Conservation of Antarctic Marine Living Resources (CCAMLR). The demand for forage highlights the trade-offs and need for EBFM, not only ecologically, but across LMR-oriented mandates.

Several jurisdictions (Pacific Fishery Management Council, PFMC; CCAMLR; North Pacific Fishery Management Council, NPFMC; Constable 2011, NOAA 2016, NPFMC 2019) are beginning to deal with this via forage caps, thresholds or bans that provide a prohibition of catch for certain forage species, or a floor of forage fish biomass below which fishing ceases to ensure other parts of the food web are not deleteriously impacted. Other places (e.g. Atlantic States Marine Fisheries Commission, ASMFC; ASMFC 2008, 2010, Lenfest 2018, Uphoff & Sharov 2018) are beginning to model forage dynamics jointly with recreationally and commercially harvested species to explore the possible solution space to achieve multiple goals among these trade-offs. Continued consideration of how to ensure adequate forage seems like an ongoing mainstay of EBFM.

Reference

Alder J, Campbell B, Karpouzi V, Kaschner K, Pauly D. 2008. Forage fish: from ecosystems to markets. *Annual Review of Environment and Resources* 33:153–66.

ASMFC (Atlantic States Marine Fisheries Commission). 2008. *Multispecies Technical Committee. Update of the Multispecies Virtual Population Analysis*. Washington, DC: ASMFC.

ASMFC (Atlantic States Marine Fisheries Commission). 2010. *Atlantic Menhaden Technical Committee. Atlantic Menhaden Stock Assessment Report for Peer Review. Report 10-2*. Washington, DC: ASMFC.

Clausen LW, Rindorf A, van Deurs M, Dickey-Collas M, Hintzen NT. 2018. Shifts in North Sea forage fish productivity and potential fisheries yield. *Journal of Applied Ecology* 55(3):1092–101.

Constable AJ. 2011. Lessons from CCAMLR on the implementation of the ecosystem approach to managing fisheries. *Fish and Fisheries* 12(2):138–51.

Eero M, Vinther M, Haslob H, Huwer B, Casini M, Storr-Paulsen M, Köster FW. 2012. Spatial management of marine resources can enhance the recovery of predators and avoid local depletion of forage fish. *Conservation Letters* 5(6):486–92.

Engelhard GH, Peck MA, Rindorf A, Smout S, van Deurs M, Raab K, Andersen KH, et al. 2014. Forage fish, their fisheries, and their predators: who drives whom? *ICES Journal of Marine Science* 71(1):90–104.

Essington TE, Moriarty PE, Froehlich HE, Hodgson EE, Koehn LE, Oken KL, Siple MC, Stawitz CC. 2015. Fishing amplifies forage fish population collapses. *Proceedings of the National Academy of Sciences* 112(21):6648–52.

Gutiérrez M, Swartzman G, Bertrand A, Bertrand S. 2007. Anchovy (*Engraulis ringens*) and sardine (*Sardinops sagax*) spatial dynamics and aggregation patterns in the Humboldt Current ecosystem, Peru, from 1983–2003. *Fisheries Oceanography* 16(2):155–68.

Hinke JT, Kaplan IC, Aydin K, Watters GM, Olson RJ, Kitchell JF. 2004. Visualizing the food-web effects of fishing for tunas in the Pacific Ocean. *Ecology and Society* 9(1).

Kaplan IC, Koehn LE, Hodgson EE, Marshall KN, Essington TE. 2017. Modeling food web effects of low sardine and anchovy abundance in the California Current. *Ecological Modelling* 359:1–24.

Lenfest. 2018. *Advancing Ecological Reference Points for Menhaden Using an Ecosystem Model.* Conshohocken, PA: Lenfest Ocean Program (p. 2).

Ménard F, Stéquert B, Rubin A, Herrera M, Marchal É. 2000. Food consumption of tuna in the equatorial Atlantic Ocean: FAD associated versus unassociated schools. *Aquatic Living Resources* 13(4): 233–40.

McClatchie S, Hendy IL, Thompson AR, Watson W. 2017. Collapse and recovery of forage fish populations prior to commercial exploitation. *Geophysical Research Letters* 44(4):1877–85.

McClatchie S, Vetter RD, Hendy IL. 2018. Forage fish, small pelagic fisheries and recovering predators: managing expectations. *Animal Conservation* 21(2018):445–7.

Nakamura EL. 1965. Food and feeding habits of skipjack tuna (Katsuwonus pelamis) from the Marquesas and Tuamotu Islands. *Transactions of the American Fisheries Society* 94(3):236–42.

NOAA (National Oceanic and Atmospheric Administration). 2016. Fisheries off West Coast states; Comprehensive ecosystem-based amendment 1; Amendments to the fishery management plans for coastal pelagic species, Pacific Coast groundfish, U.S. West Coast highly migratory species, and Pacific Coast Salmon. *Federal Register* 81(64):215–8.

NPFMC (North Pacific Fishery Management Council). 2019. *Forage Fish Management. B1 DRAFT Forage Fish Management Summary.* Anchorage, AH: NPFMC (p. 2).

Pethybridge H, Roos D, Loizeau V, Pecquerie L, Bacher C. 2013. Responses of European anchovy vital rates and population growth to environmental fluctuations: an individual-based modeling approach. *Ecological Modelling* 250:370–83.

Pikitch EK, Rountos KJ, Essington TE, Santora C, Pauly D, Watson R, Sumaila UR, et al. 2014. The global contribution of forage fish to marine fisheries and ecosystems. *Fish and Fisheries* 15(1):43–64.

Potier M, Marsac F, Cherel Y, Lucas V, Sabatié R, Maury O, Ménard F. 2007. Forage fauna in the diet of three large pelagic fishes (lancetfish, swordfish and yellowfin tuna) in the western equatorial Indian Ocean. *Fisheries Research* 83(1):60–72.

Puncher GN. 2015. *Assessment of the Population Structure and Temporal Changes in Spatial Dynamics and Genetic Characteristics of the Atlantic Bluefin Tuna Under a Fishery Independent Framework.* Doctoral Dissertation. Bologna, Italy: University of Bologna

Robinson KL, Ruzicka JJ, Decker MB, Brodeur RD, Hernandez FJ, Quiñones J, Acha EM, et al. 2014. Jellyfish, forage fish, and the world's major fisheries. *Oceanography* 27(4):104–15.

Siple MC, Essington TE, Plagányi É. 2019. Forage fish fisheries management requires a tailored approach to balance trade-offs. *Fish and Fisheries* 20(1):110–24.

Tacon AG, Metian M. 2009. Fishing for feed or fishing for food: increasing global competition for small pelagic forage fish. *Ambio* 1:294–302.

11.2.1 Socioeconomic Criteria

11.2.1.1 Social and Regional Demographics

In the U.S., coastal human population values and population densities (Fig. 11.2) have increased from 1970 to present with values currently around 125.6 million residents and ~49.9 people km^{-2}, respectively. Within Western Atlantic U.S. regions, including the northern Gulf of Mexico and U.S. Caribbean, human population has increased (~419.3 thousand people year^{-1}) over the past 40 years, with currently up to ~71.7 million residents and ~89 people km^{-2} (nearly double the 2010 national value) in surrounding areas. Values are cumulatively lower within Pacific regions, including the Western Pacific and Pacific Islands territories, with human population over the past 40 years increasing at ~302.7 thousand people year^{-1}, and current values at ~35.1 million residents and ~20.5 people km^{-2}.

11.2.1.2 Socioeconomic Status and Regional Fisheries Economics

Total U.S. LMR establishments and employments, and LMR-associated gross domestic product (GDP) value (~$10.8 billion USD) remained relatively steady over the past decade, until increases were observed for all three metrics in 2016 (Fig. 11.3). Total U.S. LMR establishments have increased over the past decade by ~2000, and comprise up to ~5.8% of total U.S. ocean economy establishments. Total reported U.S. LMR employments have increased by ~20,000 since 2015,

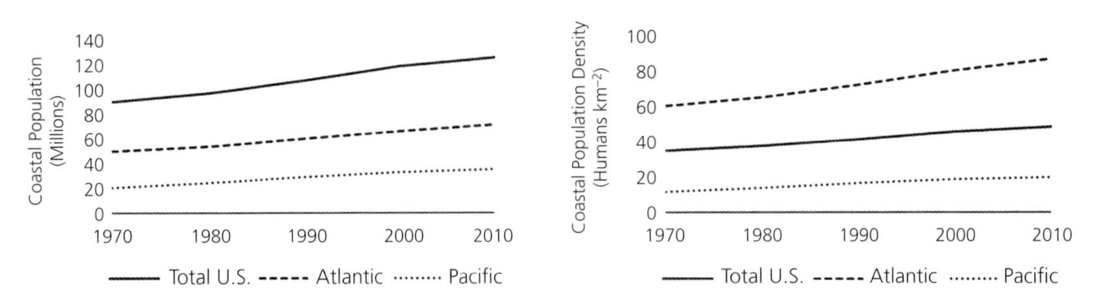

Figure 11.2 Decadal trends (1970–2010) in coastal human population (left) and population density (right; humans km^{-2}) for the entire United States, and its Atlantic and Pacific regions.

Data derived from U.S. censuses and taken from https://coast.noaa.gov/digitalcoast/data/demographictrends.html.

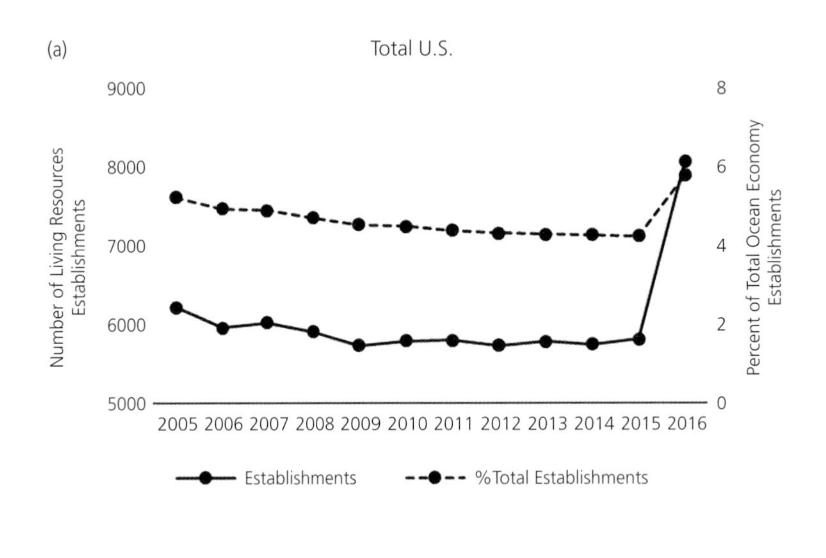

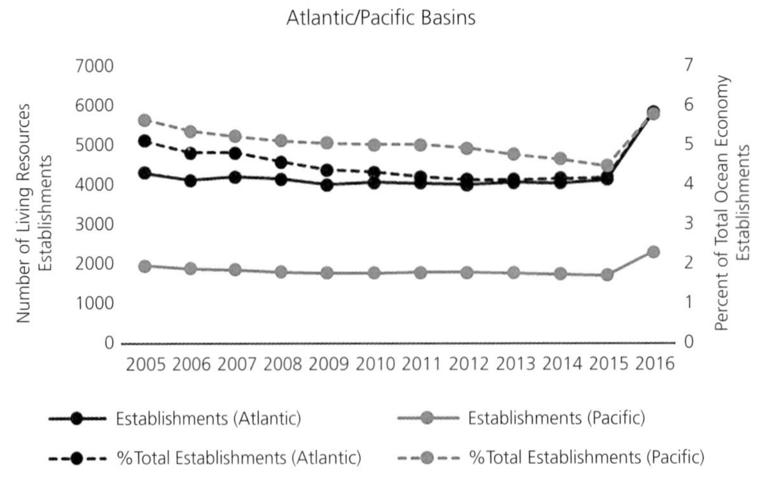

Figure 11.3a Number of living marine resources establishments and their percent contribution to total multisector oceanic economy establishments for the entire United States, and its Atlantic and Pacific regions (years 2005–2016).

Data derived from the National Ocean Economics Program.

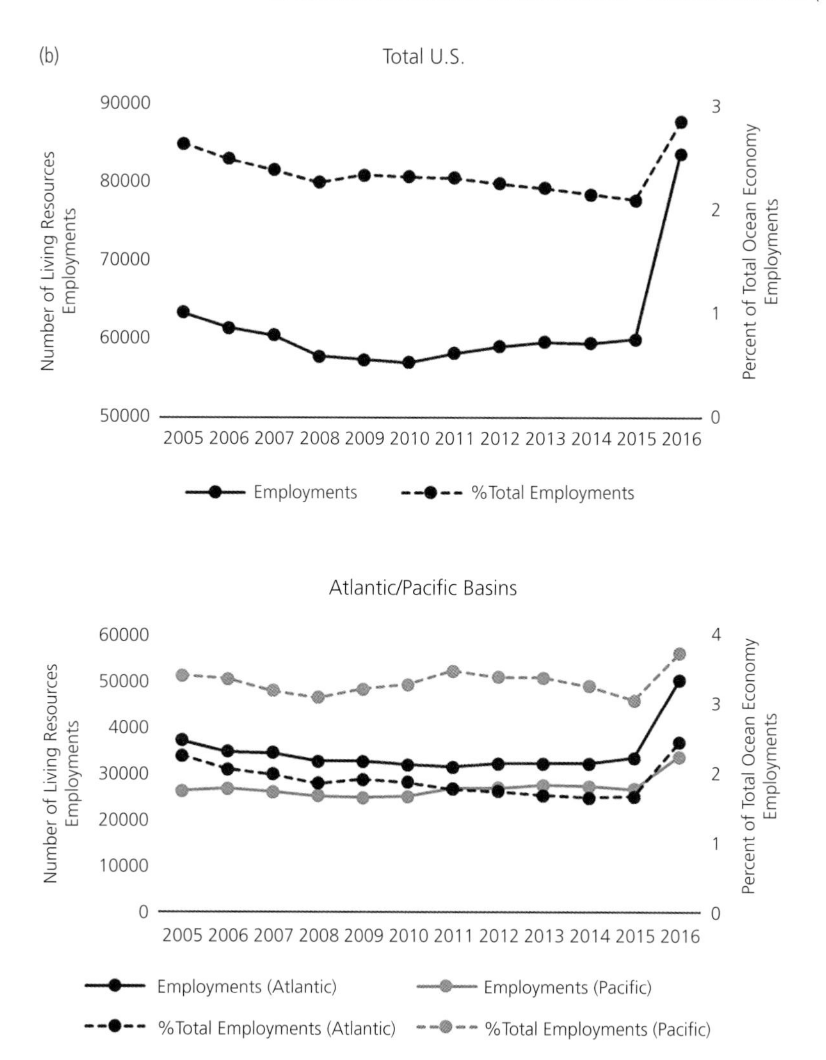

Figure 11.3b Number of living marine resources employments and their percent contribution to total multisector oceanic economy employments for the entire United States, and its Atlantic and Pacific regions (years 2005–2016).

Data derived from the National Ocean Economics Program.

and comprise up to ~2.8% of total U.S. ocean economy employments. LMR-associated GDP has increased by ~$5.4 billion since 2005, and has contributed up to ~3.8% toward the total U.S. ocean economy in recent years.

Cumulatively within U.S. Atlantic and Pacific regions, LMR values have increased over time. The highest numbers of LMR establishments, employments, and GDP value have historically occurred in U.S. Atlantic regions (4216.8 ± 144.6 standard error (SE), establishments; 34,602.4 ± 1487.4 SE employments; $4.1 billion ± 265.8 million SE, USD), with LMRs contributing 4–6% of its total ocean establishments, 2–3% of its total ocean employments, and 2–4% of its total ocean GDP. In U.S. Pacific regions, lower numbers of establishments, employments, and GDP value (1830.4 ± 45.2 SE establishments; 26,916 ± 659.4 SE employments; $2.9 billion ± 143.7 million SE, USD GDP) have occurred over the past decade, but have still increased over that time period. However, LMR-associated economies in the Pacific contribute a higher percentage toward its total ocean economy (up to ~6% of ocean establishments, ~4% of ocean employments, and ~5.3% of ocean GDP).

Among U.S. regional jurisdictions, highest numbers of commercial permits or permitted vessels were observed for those operating within the Atlantic HMS

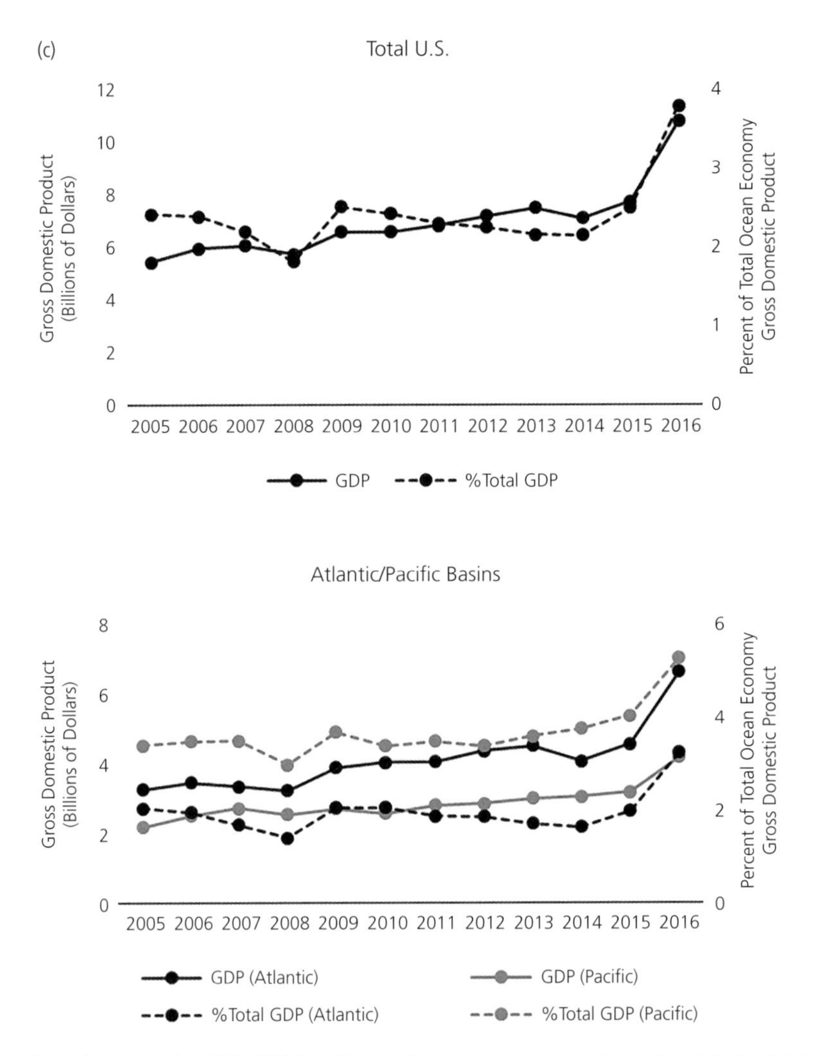

Figure 11.3c Gross domestic product value (GDP; USD) from living marine resources revenue and percent contribution to total multisector oceanic economy GDP for the entire United States, and its Atlantic and Pacific regions (years 2005–2016).

Data from the National Ocean Economics Program.

fishery, and NAFO, WECAFC, and WCPFC jurisdictions (including the overlapping component of the WCPFC with the U.S. North Pacific region) over time (Fig. 11.4). The next highest numbers of permitted vessels are observed for Pacific (California Current) and North Pacific waters, with high numbers of vessels for the Pacific HMS and PSC regions, followed by the NPFC. Lowest numbers are observed for specific Pacific fisheries such as those under limited entry in the IPHC jurisdiction, vessels associated with the PWT, and low reported fishing effort within the SPRFMO jurisdiction. Within recent years, significant increases in the reported numbers of Atlantic HMS permits and IPHC vessels have been observed, while moderate increases in NAFO and WECAFC vessels have also occurred. Atlantic HMS permit increases during 2011 coincided with resumed reporting of commercial charter permits for sharks, billfishes, and swordfish. Declines in the number of reported Pacific HMS, NPFC, and WCPFC vessels have been observed, especially in more recent years, while decreases in PSC and PWT vessels have also occurred since the late 1990s and 2000s. We do not have these data at basin-wide jurisdictions (e.g., for ICCAT, IATTC) or for CCAMLR fleets.

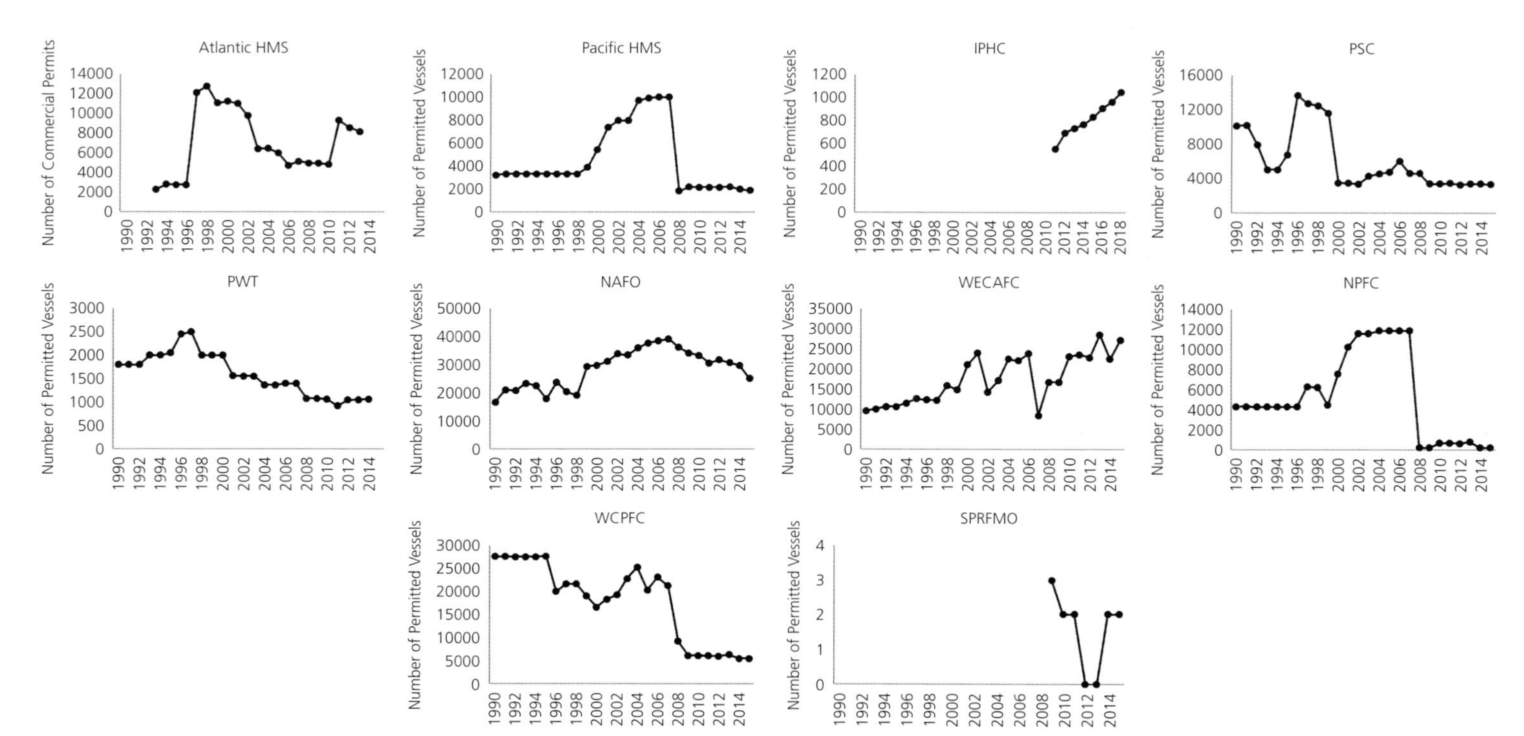

Figure 11.4 Total numbers of reported permits and/or permitted vessels for U.S. components of Atlantic and Pacific highly migratory species (HMS), Pacific halibut (International Pacific Halibut Commission, IPHC), Pacific salmon (Pacific Salmon Commission, PSC), Pacific whiting (Pacific Whiting Treaty, PWT), Northwest Atlantic Fisheries Organization (NAFO), Western and Central Atlantic Fisheries Commission (WECAFC), North Pacific Fisheries Commission (NPFC), Western and Central Pacific Fisheries Commission (WCPFC), and South Pacific Regional Fisheries Management Organisation (SPRFMO) jurisdictions over time (years 1990–2015). Atlantic HMS data reflect total commercial permits.

Data derived from NOAA National Marine Fisheries Service Council Reports to Congress.

Most regions have shown consistent increases in their total revenue (Year 2017 USD) of landed commercial fishery catches, with highest current values associated with NAFO, WECAFC, and PSC jurisdictions (Fig. 11.5). U.S. revenues associated with these management organizations have progressively increased over time, and landings for each of these RFMOs are currently valued at $420 million to $1 billion. Values for PSC-associated U.S. landings significantly increased during the 1970s–1980s, peaking at ~$925 million in 1988, and reemerging in later years to ~$750 million following prior sharp declines. Landings for IPHC, NPFC, and Pacific HMS jurisdictions follow, with increases observed in recent decades, and peak values (~$223 million) and a sharp decline in Pacific HMS revenue having occurred in the early 1980s. Concurrent increases in revenue were observed in later years for Pacific HMS, also coinciding with increased WCPFC revenue. Comparatively lower values have been observed for WCPFC, Atlantic HMS, PWT, and SPRFMO revenues throughout their monitoring periods, with oscillating increases over time observed for Pacific whiting revenues. These data were not readily available for CCAMLR resources.

Over the past two decades, ratios of LMR jobs and commercial fisheries revenue to primary production have remained generally consistent within U.S. EEZs

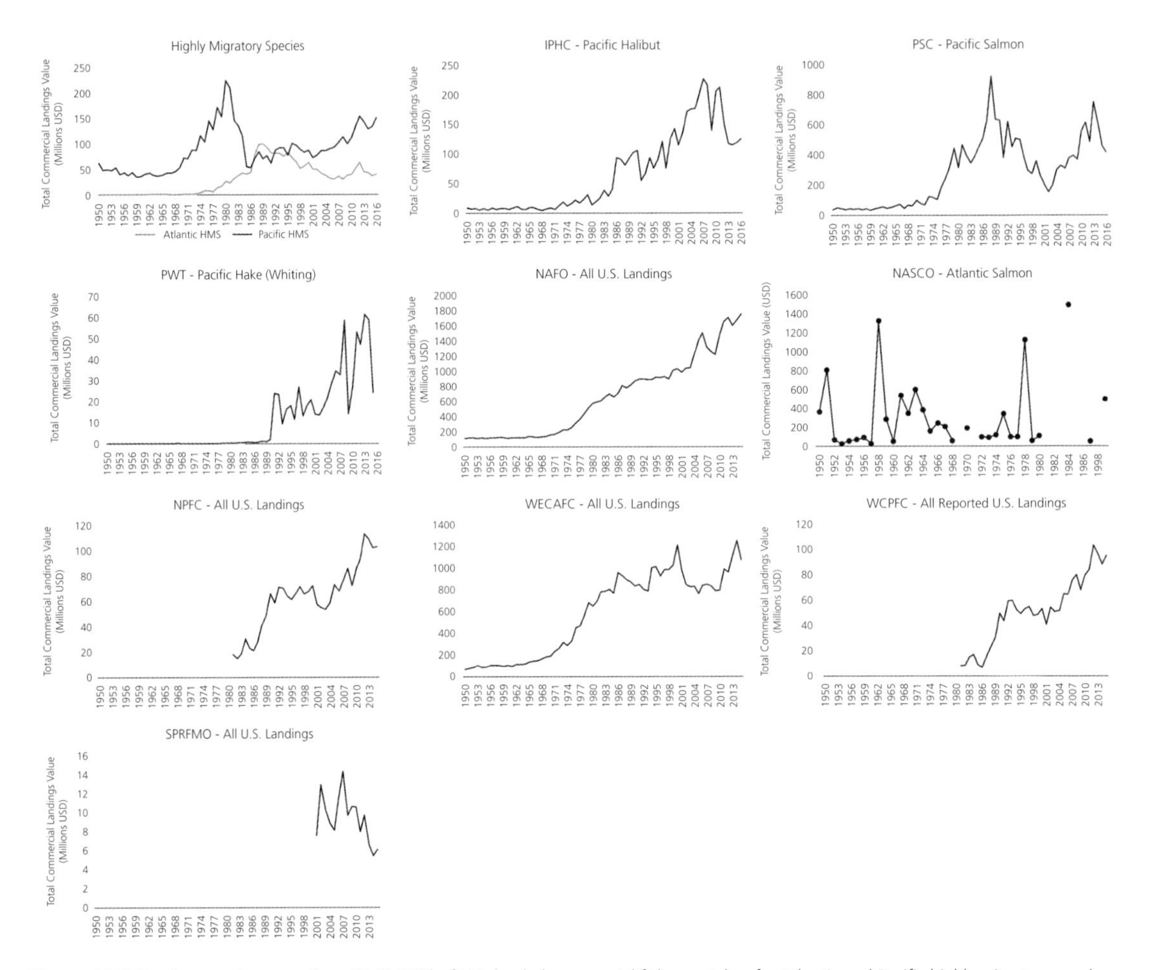

Figure 11.5 Total reported revenue (year 2017 USD) of U.S. landed commercial fishery catches for Atlantic and Pacific highly migratory species (HMS), Pacific halibut (International Pacific Halibut Commission, IPHC), Pacific salmon (Pacific Salmon Commission, PSC), Pacific whiting (Pacific Whiting Treaty, PWT), Northwest Atlantic Fisheries Organization (NAFO), Atlantic salmon (North Atlantic Salmon Conservation Organization, NASCO), Western and Central Atlantic Fisheries Commission (WECAFC), North Pacific Fisheries Commission (NPFC), Western and Central Pacific Fisheries Commission (WCPFC), and South Pacific Regional Fisheries Management Organisation (SPRFMO) over time (years 1950–2016).

Data derived from NOAA National Marine Fisheries Service commercial fisheries statistics.

and throughout RFMO total LME areas (Figs. 11.6–7). Since 2005, ratios within the U.S. EEZ have been highest for NASCO, PWT, and NAFO-associated areas at >15 jobs/million mt wet weight, and <5 jobs/million mt wet weight in all other regions. While NASCO works to limit Atlantic salmon catches in its jurisdiction, other fisheries (e.g., groundfish, scallops, crustaceans) contribute heavily toward local LMR employments where Atlantic salmon also occur. In addition, jobs/production ratios throughout RFMO total jurisdictional areas are consistently highest for WECAFC (average: 2.06 jobs/million mt wet weight), PWT (average: 1.21 jobs/million mt wet weight), IPHC, PSC (average: 1.01 jobs/million mt wet weight), and NAFO (average: 0.61 jobs/million mt wet weight), with all other regions consistently <0.3 jobs/million mt wet weight. In terms of fisheries revenue to primary production ratios within U.S. EEZs, increases across all regions have been generally observed over the past decades, with highest values overwhelmingly occurring within NAFO, and currently at ~$1.50/mt wet weight. Comparatively high values are observed for WECAFC

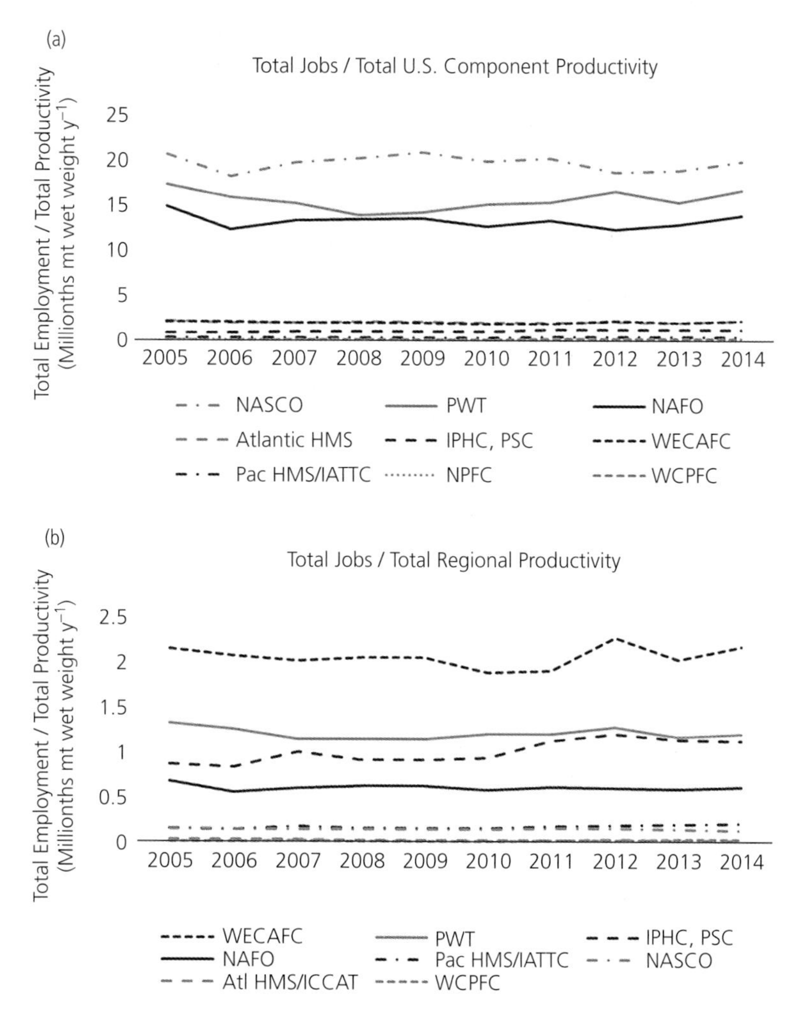

Figure 11.6 Over time, ratios of total U.S. living marine resources employments to (a) total U.S. component productivity (metric tons wet weight year^{-1}) and (b) total summed productivity of encompassing global large marine ecosystems (LMEs) for Atlantic and Pacific highly migratory species (HMS; global productivity for ICCAT; global productivity for Inter-American Tropical Tuna Commission, IATTC), Pacific halibut (International Pacific Halibut Commission, IPHC), Pacific salmon (Pacific Salmon Commission, PSC), Pacific whiting (Pacific Whiting Treaty, PWT), Northwest Atlantic Fisheries Organization (NAFO), Atlantic salmon (North Atlantic Salmon Conservation Organization, NASCO), Western and Central Atlantic Fisheries Commission (WECAFC), North Pacific Fisheries Commission (NPFC), and the Western and Central Pacific Fisheries Commission (WCPFC) jurisdictions.

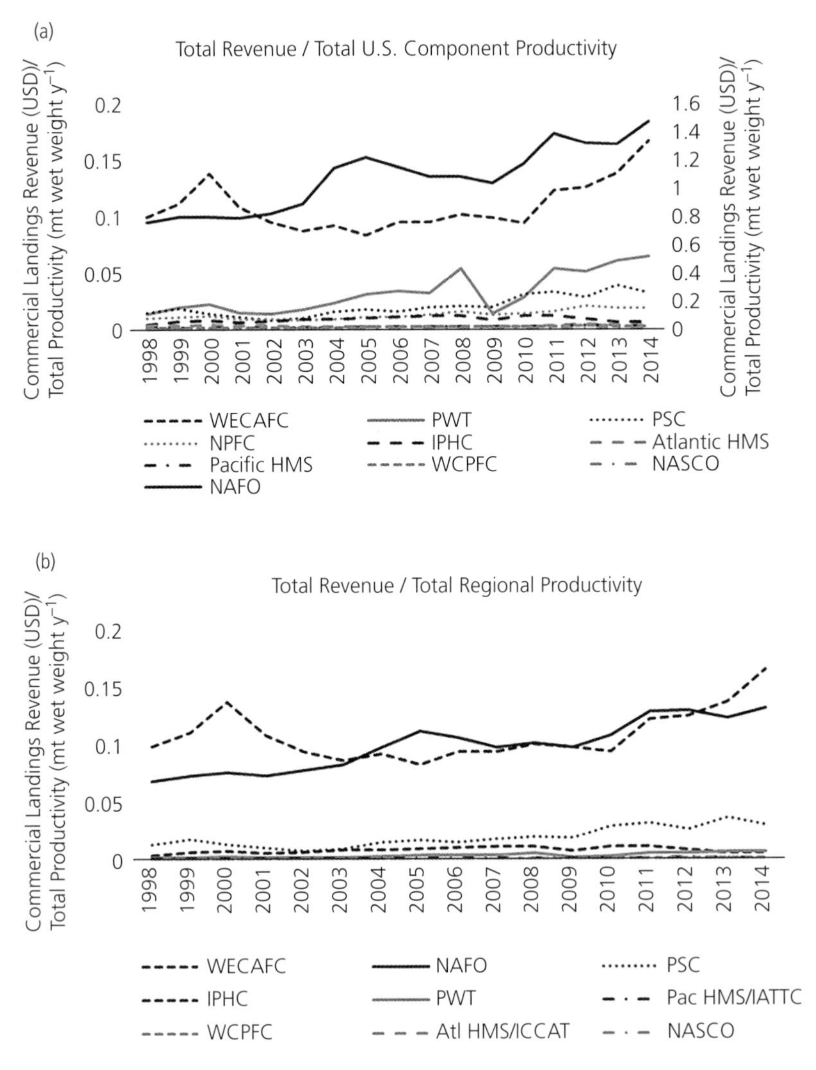

Figure 11.7 Over time, ratios of total U.S. commercial fisheries revenue to (a) total U.S. component productivity (metric tons wet weight year^{-1}) and (b) total summed productivity of encompassing global large marine ecosystems (LMEs) for Atlantic and Pacific highly migratory species (HMS; global productivity for International Commission for the Conservation of Atlantic Tunas, ICCAT; global productivity for Inter-American Tropical Tuna Commission, IATTC), Pacific halibut (International Pacific Halibut Commission, IPHC), Pacific salmon (Pacific Salmon Commission, PSC), Pacific whiting (Pacific Whiting Treaty, PWT), Northwest Atlantic Fisheries Organization (NAFO; plotted on secondary axis for top panel), Atlantic salmon (North Atlantic Salmon Conservation Organization, NASCO), Western and Central Atlantic Fisheries Commission (WECAFC), North Pacific Fisheries Commission (NPFC), and the Western and Central Pacific Fisheries Commission (WCPFC) jurisdictions.

($0.16/mt wet weight), PWT ($0.06/mt wet weight), and PSC ($~0.03/mt wet weight). All other regions are <$0.03/mt wet weight in terms of revenue/production. Throughout RFMO total areas, revenue/production ratios are highest for WECAFC and NAFO (average: $0.10 to $0.11/mt wet weight), for which gradually increasing trends have been observed. For all other RFMO areas, revenue/production values have remained consistently below $0.04/mt wet weight.

11.2.2 Governance Criteria

11.2.2.1 Human Representative Context

For the 14 RFMOs detailed in this chapter, multiple fisheries, nations, and LMEs are encompassed in each of their jurisdictions (Table 11.1). U.S. Atlantic HMS species are directly managed by NOAA Fisheries Office of Sustainable Fisheries Highly Migratory Species Management Division, while U.S. Pacific HMS species are jointly managed by NOAA Fisheries, PFMC, and WPFMC. These species are also accounted for across Atlantic and Pacific basin-wide international boundaries by ICCAT, IATTC, and WCPFC agreements, which together comprise 24 Atlantic and 23 Pacific LMEs. There are 26 member countries that are associated with CCAMLR, with additional multilateral and bilateral agreements for fisheries resources observed within the Atlantic (i.e., NAFO, NASCO, WECAFC) and Pacific (i.e., IPHC, NPFC, PSC, PWT, SPRFMO) basins, which are subject to shared regulation under international fisheries management organizations and the appropriate federal agencies.

11.2.2.2 Fishery and Systematic Context

Multiple species-specific FMPs are managed by NOAA Fisheries and regional FMCs with associated RFMO jurisdictions. Since 2006, 53 species of Atlantic billfishes (5 spp.), sharks (42 spp.), swordfish (1 sp.), and tunas (5 spp.) have been managed under a consolidated Atlantic HMS FMP, which has undergone 14 amendments, and is subject to negotiated agreements through ICCAT and WECAFC (Table 11.2). However, WECAFC does not have binding measures (C. Soltanoff, pers. comm.). Twelve species of eastern Pacific tunas, billfishes, swordfish, and sharks are managed under a single Pacific HMS FMP (with three amendments) and subject to negotiated agreements through the IATTC, while Western Pacific management of these species groups is through a Pelagics FEP, with seven amendments and which accounts for 24 species. Western Pacific pelagic species are also subject to WCPFC, NPFC, and SPRFMO agreements. One species of Antarctic krill (*E. superba*) is managed by CCAMLR, as also detailed in the NOAA Fisheries PFMC Coastal Pelagic Species FMP (with 15 amendments, including Amendment 12 for California Current krill management as influenced by CCAMLR; PFMC 2008). Additionally, many single species or species groups are managed within the Atlantic and Pacific basins, with cooperative agreements overseen by RFMOs. Atlantic salmon are managed by NOAA Fisheries and NEFMC under a single FMP (with four amendments), in coordination with NASCO. U.S. and Canadian Pacific halibut resources are managed through an IPHC FMP with 92 annual regulations, with participation from NOAA Fisheries, NPFMC, and PFMC. Similarly, NOAA Fisheries manages five Alaskan Salmon species under a single NPFMC FMP (10 amendments), and four Pacific salmon species under a PFMC Ocean Salmon FMP (19 amendments), with both regions also subject to U.S.-Canadian PSC bilateral transboundary agreements. NOAA Fisheries and PFMC also participate in the PWT U.S.-Canadian bilateral agreement for management of transboundary Pacific whiting, with U.S. stocks directly managed under the PFMC Pacific Groundfish FMP (28 amendments).

Throughout locations in which U.S. RFMO participation occurs, the highest numbers and areas of partial and complete no-take areas occur within the Pacific basin (Fig. 11.8). This region contains 268 partially protected and 331 completely protected no-take areas that together comprise ~5 million km^2. Within the Atlantic basin, there are 46 partially protected and 71 completely protected no-take areas, comprising a total of ~150 thousand km^2. However, these protected areas make up no more than 2% of the total Atlantic or Pacific basin area. Although no IUCN-designated partial or complete no-take areas are currently identified in Arctic waters, fishing is prohibited throughout the entire U.S. Arctic EEZ and throughout 2.8 million km^2 of the Central Arctic Ocean (Hoag 2017). Antarctic waters contain six complete no-take areas comprising ~2.2 million km^2, and which represent 10% of the total Antarctic basin. Among basins, the largest global no-take areas include the Ross Sea Region Marine Protected Area (~1.6 million km^2) in Antarctica, the Papahānaumokuākea Marine National Monument (~1.5 million km^2) in the Western Central Pacific, and the Monumento Natural Das Ilhas de Trinidade, Martim Vaz e Do Monte Columbia (~68 thousand km^2) in the Atlantic.

11.2.2.3 Status of Living Marine Resources (Targeted and Protected Species)

Across basins, a total of 274 Pacific and 203 Atlantic U.S. RFMO-associated stocks were managed by NOAA Fisheries and regional FMCs as of mid-2017 (Link & Marshak 2019), including 31 Atlantic HMS stocks co-managed with ICCAT, 18 jointly managed PFMC and WPFMC HMS stocks with the IATTC and WCPFC, and 16 WPFMC HMS stocks also co-managed with the WCPFC (Fig. 11.9; NMFS 2017, 2020).[2] These include internationally managed Antarctic krill, and single stocks of Atlantic salmon, Pacific halibut, and Pacific whiting that are subject to bilateral and multilateral international agreements, in addition to 67 stocks of Pacific salmon.

[2] As of June 2020, a total of 253 Pacific stocks and 212 Atlantic U.S. fisheries stocks are managed by NOAA Fisheries and regional FMCs, including 39 Atlantic HMS stocks co-managed with ICCAT, 18 jointly managed PFMC and WPFMC HMS stocks with the IATTC and WCPFC, and 16 WPFMC HMS stocks also co-managed with the WCPFC.

Table 11.2 List of major U.S. managed fishery species (including U.S. managerial bodies and management plans) per U.S. participatory Regional Fisheries Management Organization (RFMO)

Fishery/RFMOs	Fishery Management Plan	FEP Modifications	FMP Modifications	Major Species/ Species Group	Scientific Name(s) or Family Name(s)	Number of Species, Families, and Groupings
Atlantic Highly Migratory Species **Atl. HMS**	**Consolidated HMS Fishery Management Plan**		14 amendments	**Atlantic Billfishes**		5
NOAA Fisheries Office of Sustainable Fisheries—Highly Migratory Species Division **HMS**	Atlantic Billfishes (5 species)			White marlin	*Kajikia albida*	
				Blue marlin	*Makaira nigricans*	
				Sailfish	*Istiophorus platypterus*	
International Commission for the Conservation of Atlantic Tunas **ICCAT**	Atlantic Sharks (42 species)			Roundscale Spearfish	*Tetrapturus georgii*	
	Atlantic Swordfish (1 species)			Longbill spearfish	*Tetrapturus pfluegeri*	
Western and Central Atlantic Fisheries Commission **WECAFC**	Atlantic Tunas (5 species)			**Atlantic Sharks**		
				Large coastal species	*Carcharhinus* spp., *Galeocerdo cuvier*, *Ginglymostoma cirratum*, *Negaprion brevirostris*, *Sphyrna* spp.	11
				Pelagic species	*Alopias vulpinus*, *Carcharhinus longimanus*, *Isurus oxyrinchus*, *Lamna nasus*, *Prionace glauca*	5
				Small coastal species	*Carcharhinus* spp., *Rhizoprionodon terraenovae*, *Sphyrna tiburo*	4
				Prohibited species	*Alopias superciliosus*, *Carcharhinus* spp., *Carcharias taurus*, *Carcharodon carcharias*, *Hexachus* spp., *Isurus paucus*, *Notorynchus cepedianus*, *Odontaspis noronhai*, *Rhincodon typus*, *Rhizoprionodon porosus*, *Squatina demeril*	22
				Atlantic Swordfish	*Xiphias gladius*	1
				Atlantic Tunas		5
				Bluefin	*Thunnus thynnus*	
				Yellowfin	*Thunnus albacares*	
				Bigeye	*Thunnus obesus*	
				Albacore	*Thunnus alalunga*	
				Skipjack	*Katsuwonus pelamis*	

Fishery/RFMOs	Fishery Management Plan	FEP Modifications	FMP Modifications	Major Species/ Species Group	Scientific Name(s) or Family Name(s)	Number of Species, Families, and Groupings
Pacific Highly Migratory Species **Pac. HMS** Managed by: – NOAA Fisheries West Coast Regional Office **WCRO** – Pacific Fishery Management Council **PFMC** Inter-American Tropical Tuna Commission **IATTC**	Highly Migratory Species (13 species)		3 amendments	North Pacific Albacore	*Thunnus alalunga*	
				Other tunas	Yellowfin tuna *Thunnus albacares*, Bigeye tuna *Thunnus obesus*, Skipjack tuna *Katsuwonus pelamis*, Northern bluefin tuna *Thunnus thynnus*	4
				Billfish/ Swordfish	Striped marlin *Tetrapturus audax*, Swordfish *Xiphias gladius*	2
				Sharks	Common thresher shark *Alopias vulpinus*, Pelagic thresher shark *Alopias pelagicus*, Bigeye thresher shark *Alopias superciliosus*, Shortfin mako shark *Isurus oxyrinchus*, Blue shark *Prionace glauca*	5
Pacific Highly Migratory Species **WPac. HMS** Managed by: – NOAA Fisheries Pacific Islands Regional Office **PIRO** – Western Pacific Fishery Management Council **WPFMC** North Pacific Fisheries Commission **NPFC** South Pacific Regional Fishery Management Organization **SPRFMO** Western and Central Pacific Fisheries Commission **WCPFC**	Pelagics Fishery Ecosystem Plan (FEP)	7 amendments, 1 regulatory amendment	19 amendments (original FMP)	Tunas	Albacore *Thunnus alalunga* Bigeye tuna *T. obesus* Yellowfin tuna T. *albacares* Northern bluefin tuna *T. thynnus*, Skipjack tuna *Katsuwonus pelamis* Kawakawa *Euthynnus affinis*, Other tuna relatives Other tuna relatives *Auxis* spp., *Allothunus* spp., *Scomber* spp.	9
				Billfishes (Incl. Swordfish)	Striped marlin *Tetrapturus audax* Shortbill spearfish *T. angustirostris* Swordfish *Xiphias gladius*, Sailfish *Istiophorus platypterus*, Blue marlin *Makaira mazara* Black marlin *M. indica*	6

Continued

Table 11.2 Continued

Fishery/RFMOs	Fishery Management Plan	FEP Modifications	FMP Modifications	Major Species/ Species Group	Scientific Name(s) or Family Name(s)	Number of Species, Families, and Groupings
Pacific Highly Migratory Species **WPac. HMS** Managed by: – NOAA Fisheries Pacific Islands Regional Office **PIRO** – Western Pacific Fishery Management Council **WPFMC** North Pacific Fisheries Commission **NPFC** South Pacific Regional Fishery Management Organization **SPRFMO** Western and Central Pacific Fisheries Commission **WCPFC**	Pelagics FEP	7 amendments, 1 regulatory amendment	19 amendments (original FMP)	Sharks	Pelagic thresher shark *Alopias pelagicus* Bigeye thresher shark *A. superciliousus* Common thresher shark *A. vulpinus* Silky shark *Carcharhinus falciformis* Oceanic whitetip shark *C. longimanus* Blue shark *Prionace glauca* Shortfin mako shark *Isurus oxyrinchus* Longfin mako shark *I. paucus* Salmon shark *Lamna ditropis*	9
Antarctic Krill Managed by: Commission for the Conservation of Antarctic Marine Living Resources **CCAMLR** Additional Participatory Oversight: – NOAA Fisheries West Coast Regional Office **WCRO** – Pacific Fishery Management Council **PFMC**	Coastal Pelagic Species		15 amendments	Krill or Euphausiids	*Euphausia superba*	1

Fishery/RFMOs	Fishery Management Plan	FEP Modifications	FMP Modifications	Major Species/ Species Group	Scientific Name(s) or Family Name(s)	Number of Species, Families, and Groupings
Pacific Halibut Managed by: – NOAA Fisheries Alaska, West Coast Regional Office **AKRO**, **WCRO** – North Pacific Fishery Management Council **NPFMC** – Pacific Fishery Management Council **PFMC** International Pacific Halibut Commission **IPHC**	Pacific halibut (1 species)		92 Annual Regulations	Pacific halibut	*Hippoglossus stenolepis*	
Pacific Salmon (Alaska) Managed by: – NOAA Fisheries Alaska Regional Office **AKRO** – North Pacific Fishery Management Council **NPFMC** Pacific Salmon Commission **PSC**	Salmon (5 species)		10 amendments	Chinook salmon Coho salmon Pink salmon Sockeye salmon Chum salmon	*Oncorhynchus tshawytscha* *Oncorhynchus kisutch* *Oncorhynchus gorbuscha* *Oncorhynchus nerka* *Oncorhynchus keta*	
Pacific Salmon (W Coast) Managed by: – NOAA Fisheries West Coast Regional Office **WCRO** – Pacific Fishery Management Council **PFMC** Pacific Salmon Commission **PSC**	Ocean Salmon (4 species)		19 amendments	Chinook salmon Coho salmon Pink salmon Sockeye salmon	*Oncorhynchus tshawytscha* *Oncorhynchus kisutch* *Oncorhynchus gorbuscha* *Oncorhynchus nerka*	

Continued

Table 11.2 Continued

Fishery/RFMOs	Fishery Management Plan	FEP Modifications	FMP Modifications	Major Species/ Species Group	Scientific Name(s) or Family Name(s)	Number of Species, Families, and Groupings
Pacific Hake (Whiting)	Pacific Groundfish		28 amendments	Pacific whiting	*Merluccius productus*	
Managed by: – NOAA Fisheries West Coast Regional Office **WCRO**						
– Pacific Fishery Management Council **PFMC**						
Pacific Whiting Treaty **PWT**						
Atlantic Salmon	Atlantic salmon		4 amendments	Atlantic salmon	*Salmo salar*	
Managed by: – NOAA Fisheries Greater Atlantic Regional Fisheries Office **GARFO**						
– New England Fishery Management Council **NEFMC**						
North Atlantic Salmon Conservation Organization **NASCO**						

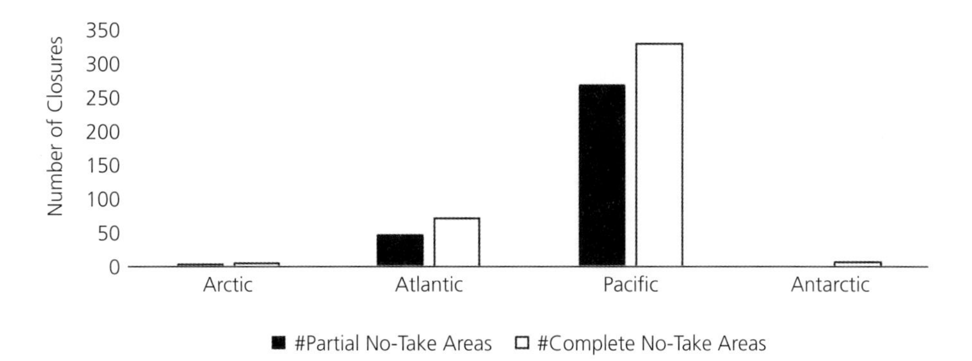

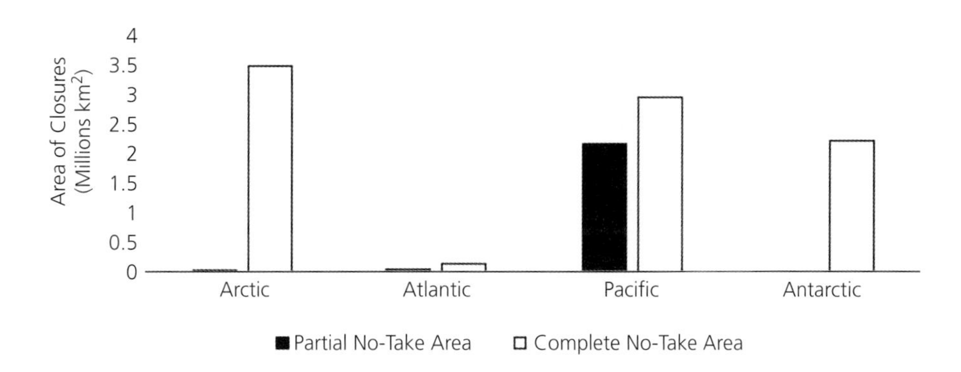

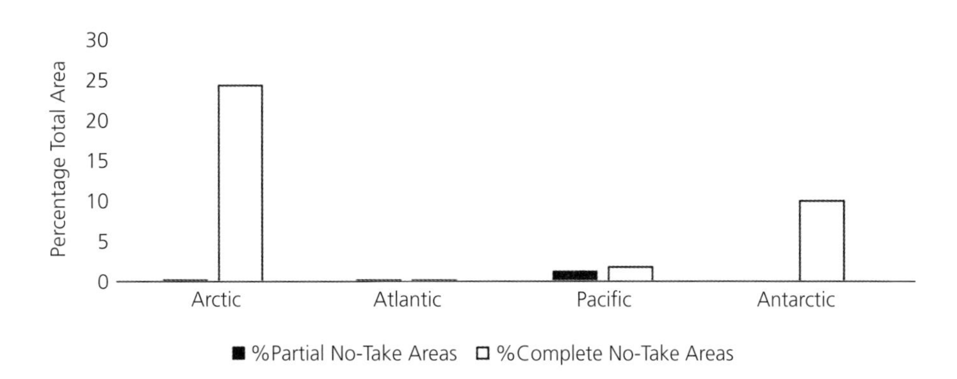

Figure 11.8 Number (top panel) and area (middle panel; km^2) of partial and complete global no-take areas per ocean basin under U.S. jurisdictions, including percent ocean basin (bottom panel) of partial or complete no-take areas.

Data derived from the International Union for the Conservation of Nature-protected areas inventory, NOAA Marine Protected Areas Inventory, and Hoag 2017.

For Atlantic HMS stocks, six stocks were experiencing overfishing (**Atlantic bigeye tuna**—*Thunnus obesus*; **Atlantic blacknose shark**—*Carcharhinus acronotus*; **Atlantic blue marlin**—*Makaira nigricans*; **Dusky shark**—*Carcharhinus obscurus*; **Scalloped hammerhead**—*Sphyrna lewini*; **North Atlantic shortfin mako shark**—*Isurus oxyrinchus*) and nine are overfished (**Atlantic bigeye tuna; Atlantic blacknose shark**; **Atlantic blue marlin**; **Dusky shark; Porbeagle shark**—*Lamna nasus*; **Sandbar shark**—*Carcharhinus plumbeus*; **Scalloped hammerhead**; **North Atlantic shortfin mako shark**; **White marlin**—*Kajikia albidus*).[3] Nine stocks were in an unclassified status (29%).[4] Four Pacific HMS stocks were

[3] These values and species remain the same as of June 2020.
[4] In June 2020, 18 Atlantic HMS stocks were in an unclassified status (46%).

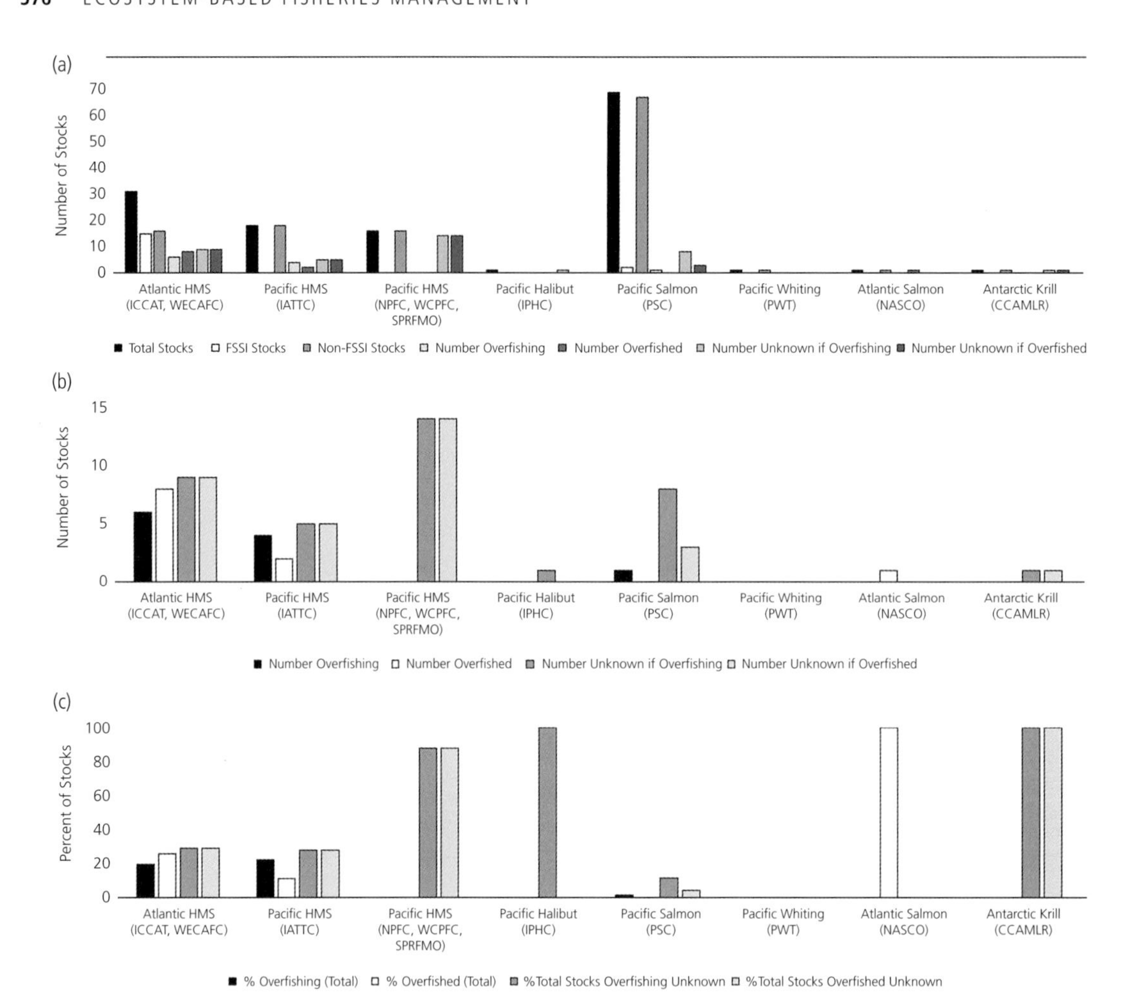

Figure 11.9 For U.S. Atlantic and Pacific highly migratory species (HMS), Pacific halibut, Pacific salmon, Pacific whiting, Atlantic salmon, and Antarctic krill stocks as of June 2017 (a) Total number of managed Fish Stock Sustainability Index (FSSI) stocks and non-FSSI stocks, and breakdown of stocks experiencing overfishing, classified as overfished, and of unknown status. (b) Number of stocks experiencing overfishing, classified as overfished, and of unknown status. (c) Percent of stocks experiencing overfishing, classified as overfished, and of unknown status. Data from NOAA National Marine Fisheries Service. ICCAT—International Commission for the Conservation of Atlantic Tunas; WECAFC—Western Central Atlantic Fisheries Commission; IATTC—Inter-American Tropical Tuna Commission; NPFC—North Pacific Fisheries Commission; WCPFC—Western and Central Pacific Fisheries Commission; SPRFMO—South Pacific Regional Fisheries Management Organisation; IPHC—International Pacific Halibut Commission; PSC—Pacific Salmon Commission; PWT—Pacific Whiting Treaty; NASCO—North Atlantic Salmon Conservation Organization; CCAMLR—Commission for the Conservation of Antarctic Marine Living Resources. Note: stocks may refer to a species, family, or complex.

experiencing overfishing (**Pacific bluefin tuna**—*Thunnus orientalis*; **Western and Central Pacific bigeye tuna**—*Thunnus obesus*; **Western and Central North Pacific Striped marlin**—*Kajikia audax*; **Swordfish**—*Xiphias gladius*) and two were overfished (**Pacific bluefin tuna**; **Western and Central North Pacific Striped marlin**), all of which are co-managed by PFMC, WPFMC, and IATTC.[5] A total of 19 Pacific HMS stocks encompassing IATTC (n = 5) and WCPFC (n = 14) jurisdictions remained unassessed as to their overfished or overfishing statuses (56%) (NMFS 2017).[6] Pacific halibut remained

[5] As of June 2020, five Pacific HMS stocks are experiencing overfishing (Pacific bluefin tuna; Striped marlin, Western and Central North Pacific; Swordfish; Yellowfin tuna—*Thunnus albacares*, Eastern Pacific; Oceantip white shark—*Carcharhinus longimanus*, Western and Central Pacific) and three are overfished (Pacific bluefin tuna; Striped marlin, western and central North Pacific; Oceantip white shark, western and central Pacific).

[6] In June 2020, 21 Pacific HMS stocks encompassing IATTC (n = 3) and WCPFC (n = 18) jurisdictions were unassessed as to their overfished or overfishing statuses (61.8%).

unclassified as to whether it was experiencing overfishing.

Atlantic salmon remains listed as overfished and protected under the ESA, along with 38 Pacific salmon stocks that are also ESA-protected. In mid-2017, only one U.S. Pacific salmon stock was experiencing overfishing (**Puget Sound: Hood Canal Coho salmon**—*Oncorhynchus kisutch*) and none were overfished.[7] Eight U.S. salmon stocks (11.6%) remained unassessed as to whether they were experiencing overfishing and three stocks (4.3%) had unknown overfished status.[8]

U.S. marine mammal stocks include 112 Atlantic and 108 Pacific cetacean stocks that are included under agreement with the IWC. Of Atlantic stocks, 63 are strategic stocks (i.e., those threatened, endangered, declining, and/or depleted stocks for which the level of direct human-caused mortality exceeds the potential biological removal level) and 49 are of unknown population size. These stocks consist of 36 individual whale stocks, 75 dolphin stocks, and 1 porpoise stock. Pacific marine mammal stocks are comprised of 24 strategic stocks and 26 with unknown population size. They consist of 60 whale stocks, 37 dolphin stocks, and 11 stocks of porpoises. Large-scale whaling occurs in Iceland, Norway, and Japan, while subsistence whale hunts on belugas and bowheads continue in the United States. Beluga catches are monitored by the Alaska Beluga Whale Committee, while bowhead hunting by Alaskan indigenous communities is monitored by NOAA and the Alaska Eskimo Whaling Commission (Muto et al. 2017, 2018, 2019). IUCN-listed vulnerable and endangered marine species include 8 Arctic species (75% vulnerable, 25% endangered), 11 Antarctic species (36% vulnerable, 64% endangered), 423 Atlantic species (56% vulnerable, 44% endangered), and 655 Pacific species (71% vulnerable, 29% endangered).

[7] As of June 2020, no Pacific salmon stocks are experiencing overfishing and five are overfished (Chinook salmon—*Oncorhynchus tshawytscha*, California Central Valley: Sacramento River Fall; Chinook salmon, Northern California Coast: Klamath River Fall; Coho salmon, Puget sound: Snohomish; Coho salmon, Washington Coast: Queets; Coho salmon, Washington Coast: Strait of Juan de Fuca).

[8] These values remain the same as of June 2020.

11.2.3 Environmental Forcing and Major Features

11.2.3.1 Oceanographic and Climatological Context

U.S. marine ecosystems included in RFMO jurisdictions are generally defined by their extent and position relative to the Atlantic and Pacific Ocean basins or other large bodies of water (e.g., Gulf of Mexico), and by major currents that characterize and define important features of the U.S. EEZ (Fig. 11.10). These currents include the Alaska Current throughout the North

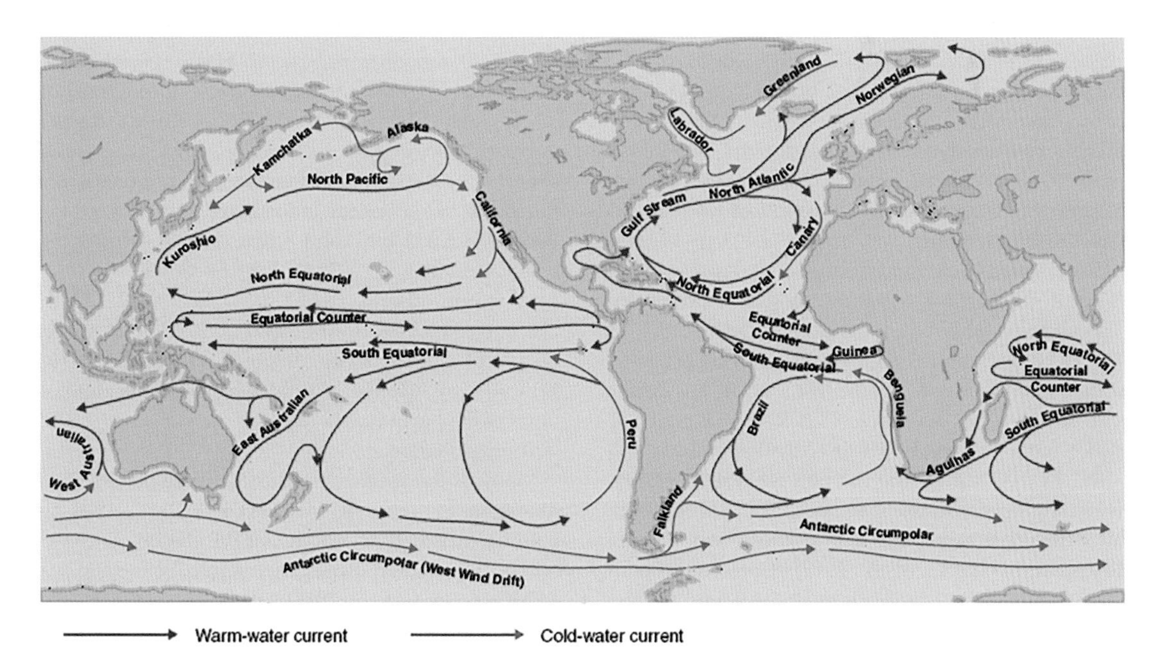

Figure 11.10 Major currents and circulation patterns throughout and encompassing the world's oceans.

Image from Satellite Applications for Geoscience Education (https://cimss.ssec.wisc.edu/sage/oceanography/lesson3/images/ocean_currents2.jpg).

Pacific region and the California Current, which is the major driver for oceanographic dynamics in the Pacific region. Both of these currents are derived from the North Pacific Current as it splits near the U.S. continental landmass. As it moves south, the California Current continues to form the North Equatorial Current, serving as the southern component of the North Pacific Gyre, within which much of the Western Pacific region is found. Similarly, components of the Western Pacific region (i.e., American Samoa and portions of the Pacific Remote Islands) are found within the South Pacific Gyre, as encompassed by the South Equatorial Current. As northern winds regularly move surface waters offshore along its relatively narrow continental shelf, the Pacific coast region is subject to regular upwelling intensities. Broader, relatively shallower continental shelves are found in the Bering Sea region.

The U.S. Caribbean region is influenced by both the South Equatorial and North Equatorial Currents. As the South Equatorial Current passes into the U.S. Caribbean Sea, it becomes the Caribbean Current, which enters the Gulf of Mexico and is responsible for connectivity between these two regions. Within the Gulf of Mexico, the current is commonly referred to as the "Gulf Loop Current," which loops around the Florida peninsula to join the Gulf Stream. the Gulf Stream is also connected to the North Equatorial Current and is the major current for the U.S. Atlantic regions, ultimately merging into the North Atlantic Current as a major component of the North Atlantic Gyre. These major flows are responsible for the thermohaline properties that are associated with U.S. Atlantic marine regions. Broader, relatively shallower continental shelves are also found in the Gulf of Mexico and New England regions.

Clear interannual and multidecadal patterns in average sea surface temperature (SST) have been observed throughout the Atlantic and Pacific U.S. EEZ (Fig. 11.11), with higher temperatures having occurred in the Atlantic (22.9°C ± 0.02, SE) than in the Pacific (21.6°C ± 0.02, SE) over time (1854–2016). Overall SST values have increased by 0.86°C in the Atlantic and 1.13°C in the Pacific since the mid-twentieth century. Increasing temperatures have been observed throughout all major ocean basins during these time periods (Bryndum-Buchholz et al. 2019). Coincident with these temperature observations are Atlantic, Arctic, and Pacific basin-scale climate oscillations (Fig. 11.12). These basin-scale features exhibit decadal cycles that can influence environmental conditions and ecologies of a given region. Specific indices include the AMO—

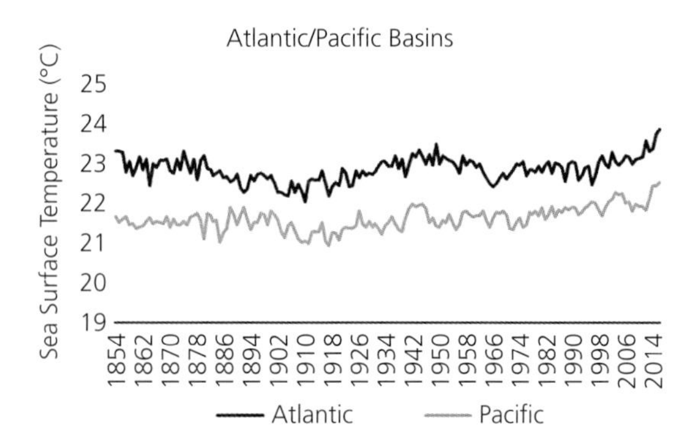

Figure 11.11 Average sea surface temperature (°C) over time (years 1854–2016) for Atlantic and Pacific regions of the U.S. Exclusive Economic Zone (EEZ).

Data derived from the NOAA Extended Reconstructed Sea Surface Temperature dataset (https://www.ncdc.noaa.gov/).

Atlantic Meridional Oscillation Index, AO—Arctic Oscillation Index, ENSO and MEI—Multivariate El Niño index, NAO—North Atlantic Oscillation Index, NOI—Northern Oscillation Index, NPGO—Northern Pacific Gyre Oscillation Index, and PDO. The MEI, NOI, and PDO have exhibited the most pronounced variabilities over time.

11.2.3.2 Notable Physical Features

In terms of proportional contribution of the U.S. EEZ to RFMO jurisdictions, Pacific HMS species under NPFC and WCPFC agreements occur throughout the largest area (10.9 million km^2; Fig. 11.13) and deepest portions (max depth: 10.8 km). Similar U.S. shoreline extent is covered under areas in which Atlantic and Pacific HMS species are landed (Atlantic: ~75,100 km; Pacific: ~76,350 km), and where Pacific salmon and Pacific halibut occur (~72,450 km). Shallower and much smaller portions of U.S. EEZ are subject to NAFO, the PWT, NASCO, WECAFC, and SPRFMO. However, deeper EEZ portions are contained within PSC/IPHC, WECAFC, while the SPRFMO includes deeper portions of the Pacific basin (outside the U.S. Pacific EEZ) in its jurisdiction.

Many of the species captured in these fisheries and managed by these RFMOs occur in areas beyond national jurisdiction (ABNJ). For this reason, SPRFMO, NPFC, NAFO, and others are referred to as deep-seas RFMOs (Ásmundsson 2016). Furthermore, these ABNJ fisheries are prosecuted on the high seas, and even the ones are important to the U.S. are not always included in U.S. landings reports. If considered here, these represent significant portions of ocean basins and are

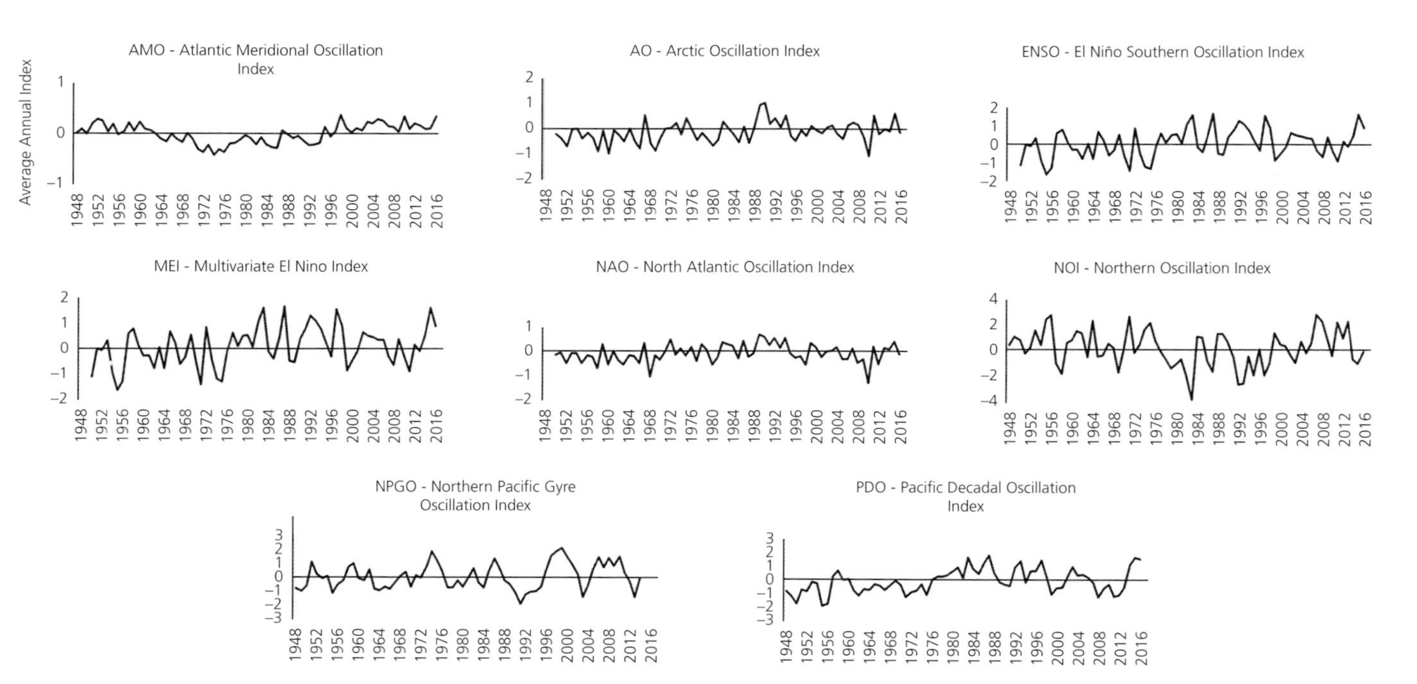

Figure 11.12 Climate forcing indices for major U.S.-associated RFMO jurisdictions over time (years 1948–2016).

Data derived from NOAA Earth System Research Laboratory data framework (https://www.esrl.noaa.gov/).

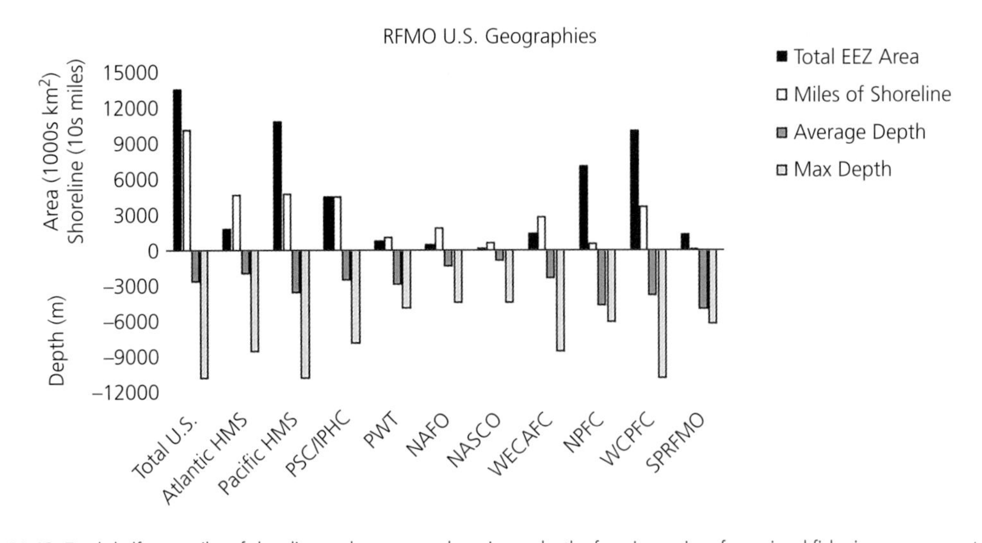

Figure 11.13 Total shelf area, miles of shoreline, and average and maximum depth of marine regions for regional fisheries management organizations (RFMO) comprising the entire United States, U.S. components of Atlantic and Pacific highly migratory species (HMS), Pacific halibut (International Pacific Halibut Commission, IPHC), Pacific salmon (Pacific Salmon Commission, PSC), Pacific whiting (Pacific Whiting Treaty, PWT), Northwest Atlantic Fisheries Organization (NAFO), Atlantic salmon (North Atlantic Salmon Conservation Organization, NASCO), Western and Central Atlantic Fisheries Commission (WECAFC), North Pacific Fisheries Commission (NPFC), Western and Central Pacific Fisheries Commission (WCPFC), and South Pacific Regional Fisheries Management Organisation (SPRFMO) jurisdictions.

often deepwater, open-ocean, oligotrophic ecosystems characterized by high water visibility and major trans-oceanic currents (Ásmundsson 2016).

11.2.4 Major Pressures and Exogenous Factors

11.2.4.1 Other Ocean Use Context

Ocean uses throughout the entire U.S. are comprised mainly by tourism/recreation and offshore mineral extraction sectors (Fig. 11.14a–c). Among sectors, trends for U.S. ocean establishments and employments have shown increases over the past few decades, with consistent dominance of tourism establishments and employments at the national level (81% of total ocean establishments; 72% of total ocean employments) in both the Atlantic and Pacific regions (80–83% of total ocean establishments; 71–73% of total ocean employments), with marginally higher proportions observed in the Pacific. Increases in total ocean establishments and employments were most prominent in the Atlantic during the mid-2000s and in the Pacific during the 2010s, with highest values occurring in 2016 (Atlantic: ~100,000 establishments, ~2 million employments; Pacific: 40,000 establishments, ~900,000 employments). The GDP from offshore mineral extractions has contributed on average 44% toward total national ocean GDP and with peak values at ~\$187 billion. This has typically been the largest sector economically speaking. Yet in 2016, the tourism and recreation sector (including recreational fishing) contributed the greatest proportionally (40%) to national ocean GDP at ~\$113 billion. Similar trends are observed for Atlantic regions, with nearly equivalent contributions most recently observed from the tourism/recreation (2016 value: 36.6%, \$75.3 billion) and offshore mineral extraction sectors (2016 value: 34.5%, \$70.9 billion). Comparatively, highest contributions to Pacific Ocean GDP have been from increasingly important tourism/recreation (2016 value: 47.7%, \$37.9 billion), while decreases in the percent contribution of offshore mineral extractions have occurred in more recent years (2016 value: 11.3%, \$8.9 billion). Overall contributions from the marine transportation sector to total ocean GDP have remained consistent at 23–28% and have ranged from \$18 to 21 billion. Among all sectors, the Atlantic cumulatively leads in terms of its contributions to the U.S. national ocean economy, comprising 68–72% of total ocean establishments, employments, and GDP. Between 80–89% of offshore mineral extraction establishments, employments, and GDP occur within the Western Atlantic. Among the eight major U.S. regions, the Pacific

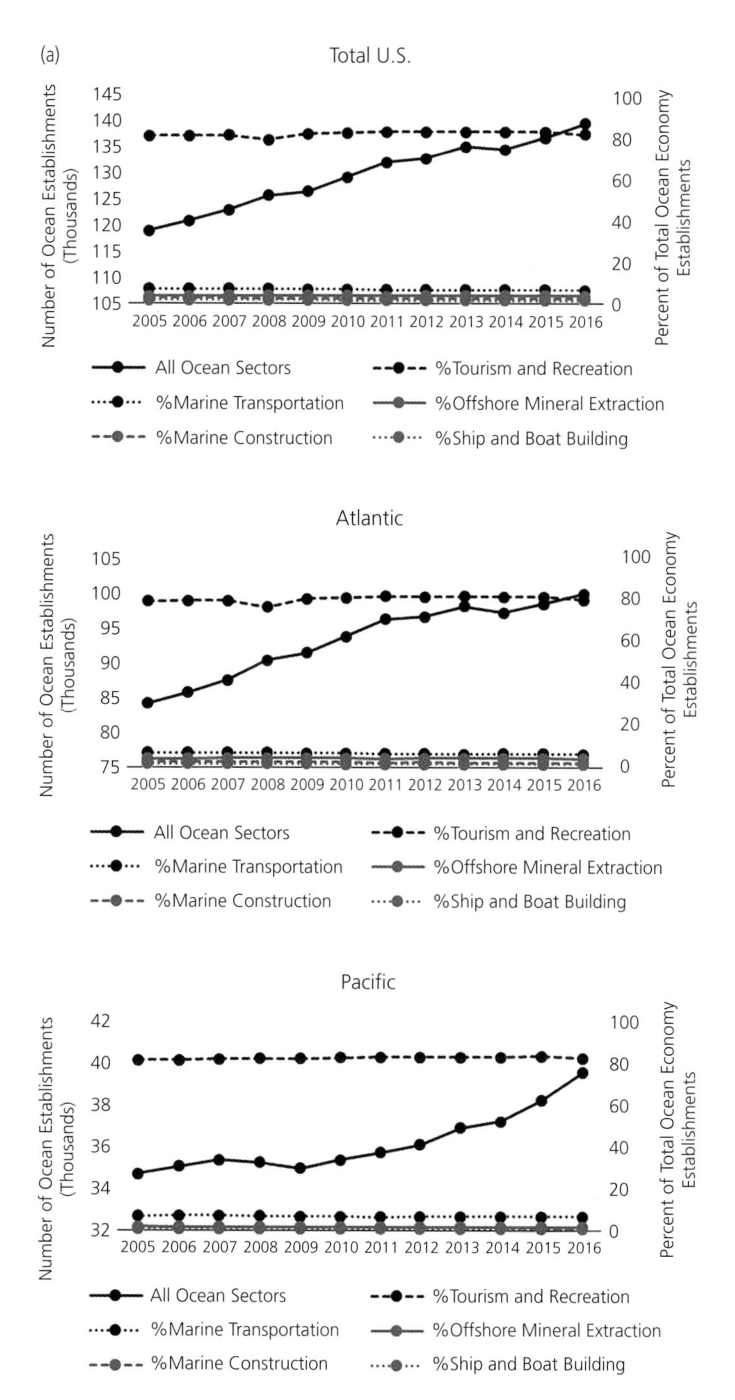

Figure 11.14a Number of total U.S. ocean establishments from all sectors and percent contribution of other ocean uses to total multisector oceanic economy establishments for the entire United States, and Atlantic and Pacific regions (years 2005–2016).

Data derived from the National Ocean Economics Program.

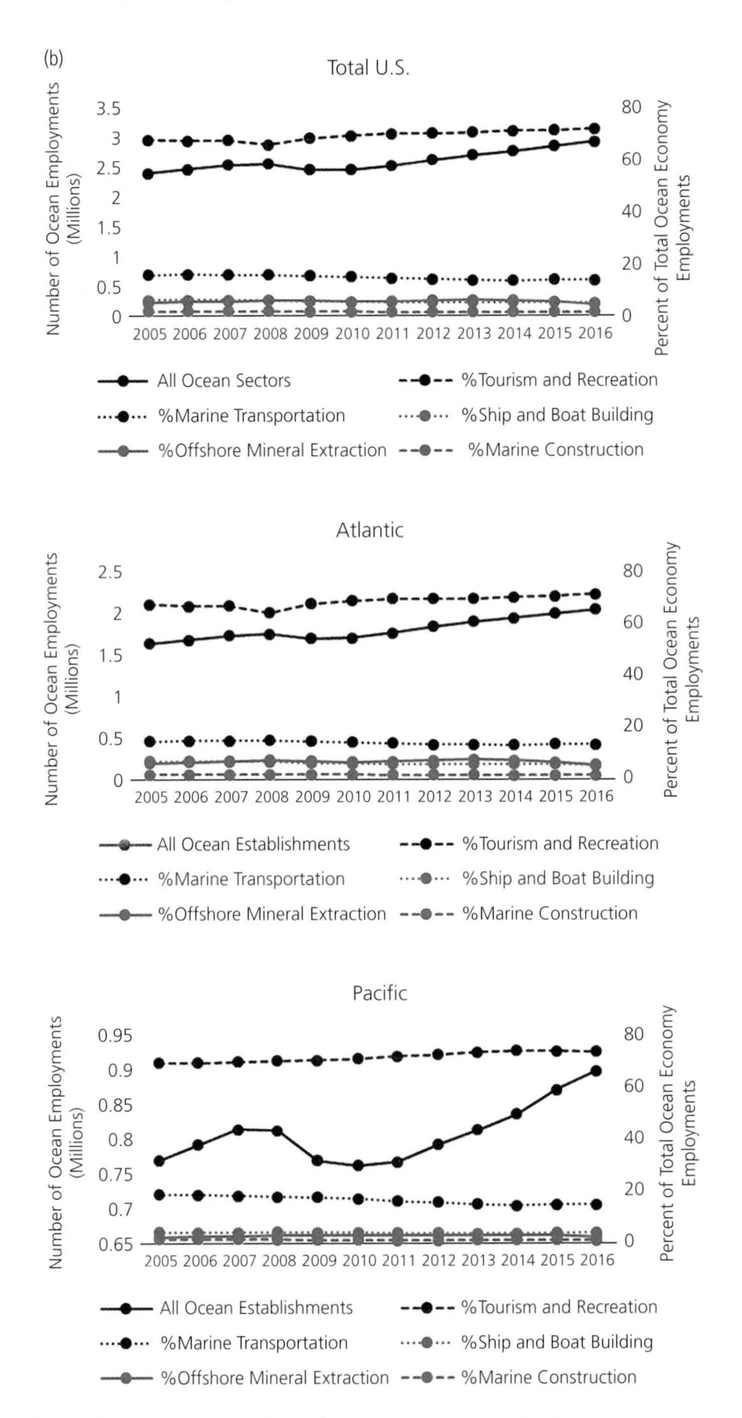

Figure 11.14b Number of total U.S. ocean employments from all sectors and percent contribution of other ocean uses to total multisector oceanic economy employments for the entire United States, and Atlantic and Pacific regions (years 2005–2016).

Data derived from the National Ocean Economics Program.

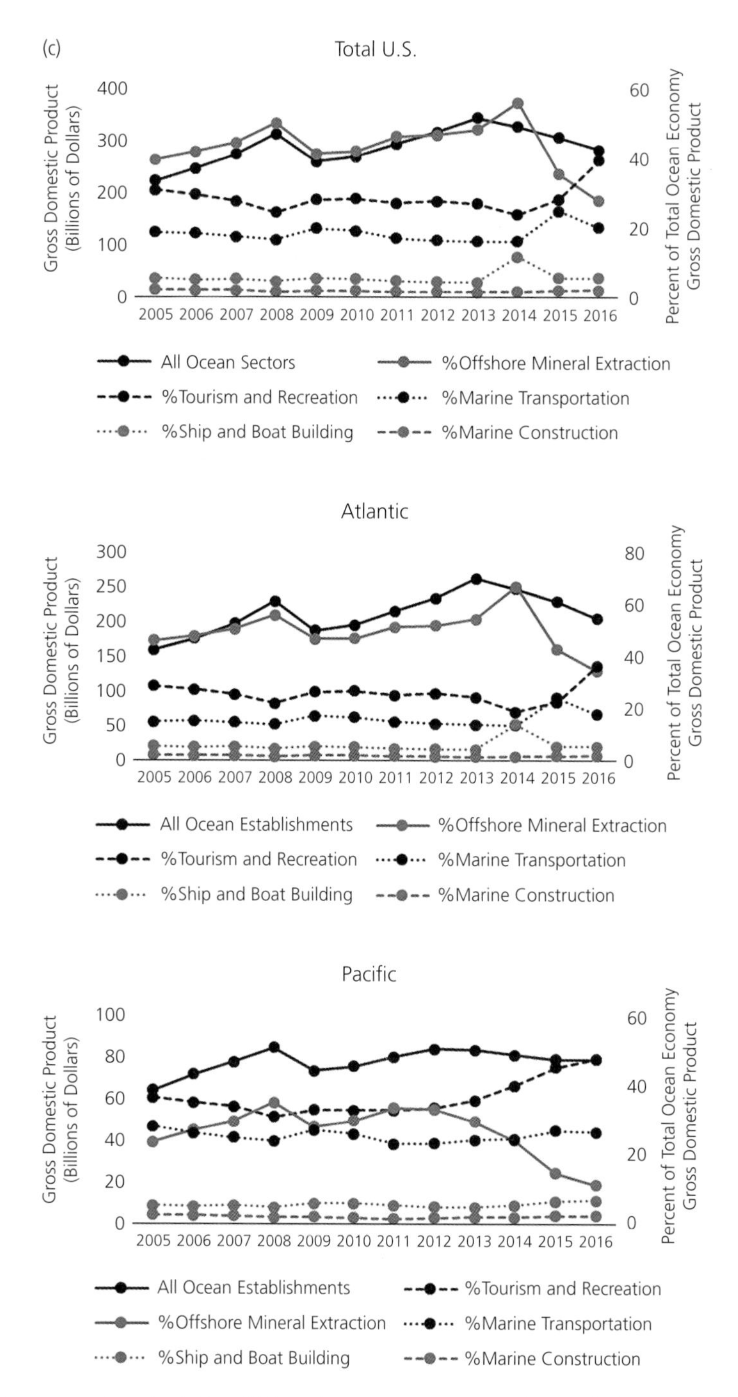

Figure 11.14c Gross domestic product value (GDP; USD) from all sectors and percent contribution of other ocean uses to total multisector oceanic economy GDP for the entire United States, and Atlantic and Pacific regions (years 2005–2016).

Data derived from the National Ocean Economics Program.

(California Current) region still leads in regards to many of these metrics, particularly when compared individually among the five separate regions (Gulf of Mexico, Caribbean, South Atlantic, Mid-Atlantic, New England) that comprise the western Atlantic.

11.2.5 Systems Ecology and Fisheries

11.2.5.1 Basal Ecosystem Production

Since 1998, average annual productivity values have remained generally consistent with marginal decreases over time throughout portions of the U.S. EEZ that have shared or overlapping jurisdictions with RFMOs (Fig. 11.15a). On average, productivities have been highest within waters supporting Atlantic HMS species (917.4 g C m^{-2} $year^{-1}$ ± 12.1, SE), including within WECAFC areas (631.6 g C m^{-2} $year^{-1}$ ± 11.6, SE). Values within areas supporting Pacific HMS species are high (521.6 g C m^{-2} $year^{-1}$ ± 4.8, SE), with similar values observed for PSC/IPHC (432.9 g C m^{-2} $year^{-1}$ ± 3.8, SE) and WCPFC (400.9 g C m^{-2} $year^{-1}$ ± 4.7, SE) jurisdictions. Lowest values occur within NAFO/NASCO, PWT, and NPFC regions. Average annual productivities throughout all LME components of RFMO jurisdictions (Fig. 11.15b) have remained relatively steady over time, with highest values observed throughout Atlantic LMEs under ICCAT (5.3–5.7 kg C m^{-2} $year^{-1}$) and NASCO (2.9–3.1 kg C m^{-2} $year^{-1}$), and Pacific LMEs under WCPFC (2.2–2.4 kg C m^{-2} $year^{-1}$), which are comparatively largest in area. Works by Wroblewski et al. (1988), Polovina and Mitchum (1995), Polovina et al. (2008), Blanchard et al. (2012) and Yen and Lu (2016) have examined trends in productivities at Atlantic and Pacific Ocean basin scales, and all convey the generally consistent nature of production noted here, albeit with changes in spatial and seasonal dynamics noted. More recently and coincident with increasing temperatures, rapid large-scale expansions of low productivity waters have been observed in both the Atlantic and Pacific basins (Polovina et al. 2008), including toward the southeast North Atlantic and historically productive eastern Pacific (Polovina et al. 2008, Yen & Lu 2016). Declines in primary and secondary production near the eastern Indo-Pacific, the northern Humboldt Current (eastern Pacific) and North Canary Current (eastern Atlantic) are anticipated, with similar predicted consequences to fish production in these regions (Blanchard et al. 2012).

11.2.5.2 System Exploitation

Values for total landings/productivity throughout U.S. EEZs that share RFMO jurisdictions have remained generally consistent (Fig. 11.16), with highest values observed for NAFO encompassed areas (0.0005–0.0006) over time and areas subject to PWT (0.0001–0.0003). Values in both RFMO jurisdictions increased during the early to mid-2000s, with oscillations more pronounced in PWT regions. For global LMEs under RFMO jurisdictions, total landings/productivity have been highest in WECAFC (0.00007–0.0001), NAFO (0.00004–0.00005), and PWT (0.00001–0.00003) over time, with greater oscillations observed for WECAFC during the 2000s.

11.2.5.3 Targeted and Non-Targeted Resources

Across RFMO jurisdictions, total U.S. landings have been generally highest within waters subject to WECAFC, NAFO, and PSC agreements (Fig. 11.17). Management actions by RFMOs and within competence areas affect the statuses of their international fisheries, including notable U.S. captures from these locations. Similar to trends observed for the northern Gulf of Mexico, landings for WECAFC doubled from 1950 to 1985, peaking at 1.4 million metric tons, and followed by substantial decreases during the 1990s–2000s to 800 thousand metric tons. U.S. landings within the NAFO Convention Area decreased during the late 1960s from >1 million mt to ~480 thousand mt. During the 1970s–1980s, some marginal increases occurred to ~850,000 mt, and have remained at 600,000–700,000 mt in recent years. Pacific salmon landings under PSC agreements increased during the mid-1970s to ~300,000 mt and have continued oscillating around that level with maximum landings observed in the 2010s at ~480,000 mt. These trends are mostly related to landings occurring in the North Pacific. Increases in landings for Pacific whiting were observed during the late 1980s to 200–260 thousand mt, where they have continued to present.

During the 1950s to early 1980s, Pacific HMS landings remained around 150,000–200,000 mt, then decreased to ~20,000–40,000 mt, where they have remained since. While still greater, these values are similar to landings observed for Atlantic HMS species since 1950. Pacific halibut landings similarly have ranged from 10 to 40 thousand mt over time, with decreases observed from 1950 to 1980 and rebounds to higher values occurring in the 1980s–2000s. In more recent years, declines in reported landings have been observed. Reported U.S. landings for CCAMLR associated species were generally low until the 2000s, when they exploded to ~12,000 mt and have remained around several thousand mt since. Antarctic krill have comprised the majority of these catches, contributing nearly ~99% of reported landings (CCAMLR 2018, 2019).

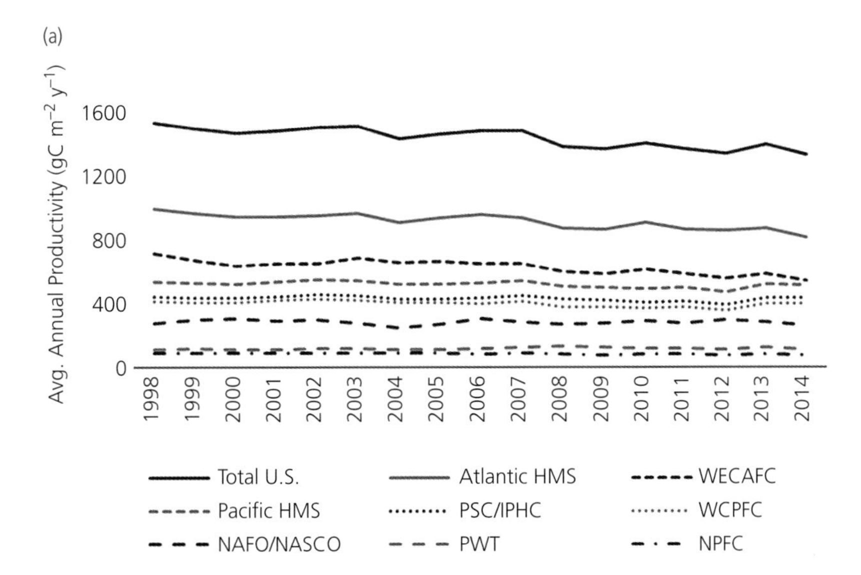

Figure 11.15a Average annual productivity (grams carbon m^{-2} year^{-1}; bottom panel) for the entire United States, and U.S. components of Atlantic and Pacific highly migratory species (HMS), Pacific halibut (International Pacific Halibut Commission, IPHC), Pacific salmon (Pacific Salmon Commission, PSC), Pacific whiting (Pacific Whiting Treaty, PWT), Northwest Atlantic Fisheries Organization (NAFO), Atlantic salmon (North Atlantic Salmon Conservation Organization, NASCO), Western and Central Atlantic Fisheries Commission (WECAFC), North Pacific Fisheries Commission (NPFC), and Western and Central Pacific Fisheries Commission (WCPFC) jurisdictions.

Data derived from NASA Ocean Color Web (https://oceancolor.gsfc.nasa.gov/) and productivity calculated using the Vertically Generalized Production Model (VGPM).

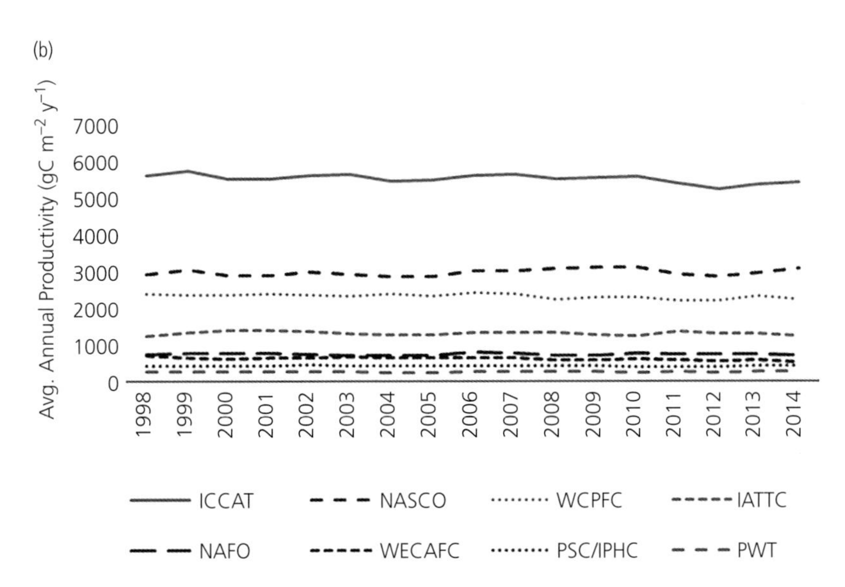

Figure 11.15b Average annual productivity (grams carbon m^{-2} year^{-1}; bottom panel) for all large marine ecosystem (LME) components of the International Commission for the Conservation of Atlantic Tunas (ICCAT), Inter-American Tropical Tuna Commission (IATTC), Pacific halibut (International Pacific Halibut Commission, IPHC), Pacific salmon (Pacific Salmon Commission, PSC), Pacific whiting (Pacific Whiting Treaty, PWT), Northwest Atlantic Fisheries Organization (NAFO), Atlantic salmon (North Atlantic Salmon Conservation Organization, NASCO), Western and Central Atlantic Fisheries Commission (WECAFC), North Pacific Fisheries Commission (NPFC), and Western and Central Pacific Fisheries Commission (WCPFC) jurisdictions.

Data derived from NASA Ocean Color Web (https://oceancolor.gsfc.nasa.gov/) and productivity calculated using the Vertically Generalized Production Model (VGPM).

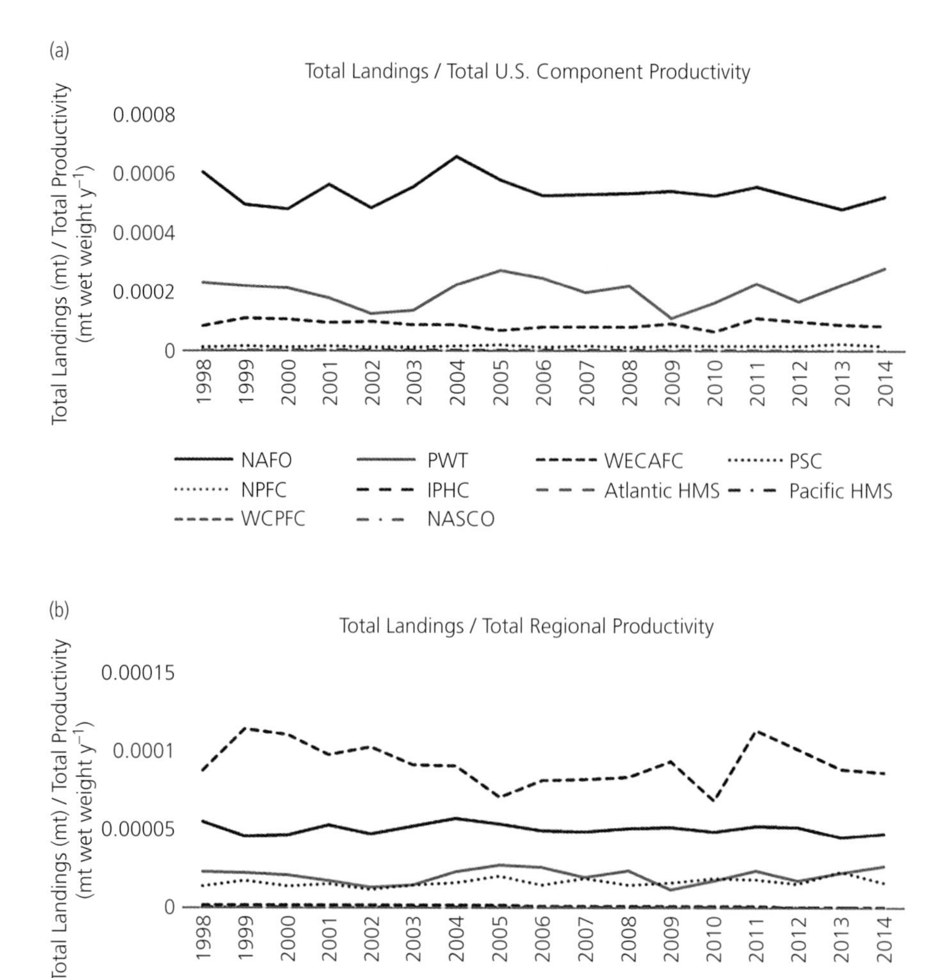

Figure 11.16 Over time, ratios of total U.S. commercial fisheries landings to (a) total U.S. component productivity (metric tons wet weight year^{-1}) and (b) total summed productivity of encompassing global large marine ecosystems (LMEs) for Atlantic and Pacific highly migratory species (HMS; global productivity for International Commission for the Conservation of Atlantic Tunas, ICCAT; global productivity for Inter-American Tropical Tuna Commission, IATTC), Pacific halibut (International Pacific Halibut Commission, IPHC), Pacific salmon (Pacific Salmon Commission, PSC), Pacific whiting (Pacific Whiting Treaty, PWT), Northwest Atlantic Fisheries Organization (NAFO), Atlantic salmon (North Atlantic Salmon Conservation Organization, NASCO), Western and Central Atlantic Fisheries Commission (WECAFC), North Pacific Fisheries Commission (NPFC), and the Western and Central Pacific Fisheries Commission (WCPFC) jurisdictions.

Increases in reported NPFC landings since the 1980s have also been observed toward most recent levels of 20,000–25,000 mt. Lower landings have been observed within SPRFMO regions, with decreases from 5000 to 7000 mt since the late 1990s to 2000–3000 mt in more recent years. With depletion of Atlantic salmon wild populations, NASCO-associated U.S. reported landings have oscillated at low levels at <1 mt over time, with higher values only observed in the 1950s–1960s.

11.2.6 Synthesis

Significant progress has been made toward greater understanding of Atlantic and Pacific ecosystems

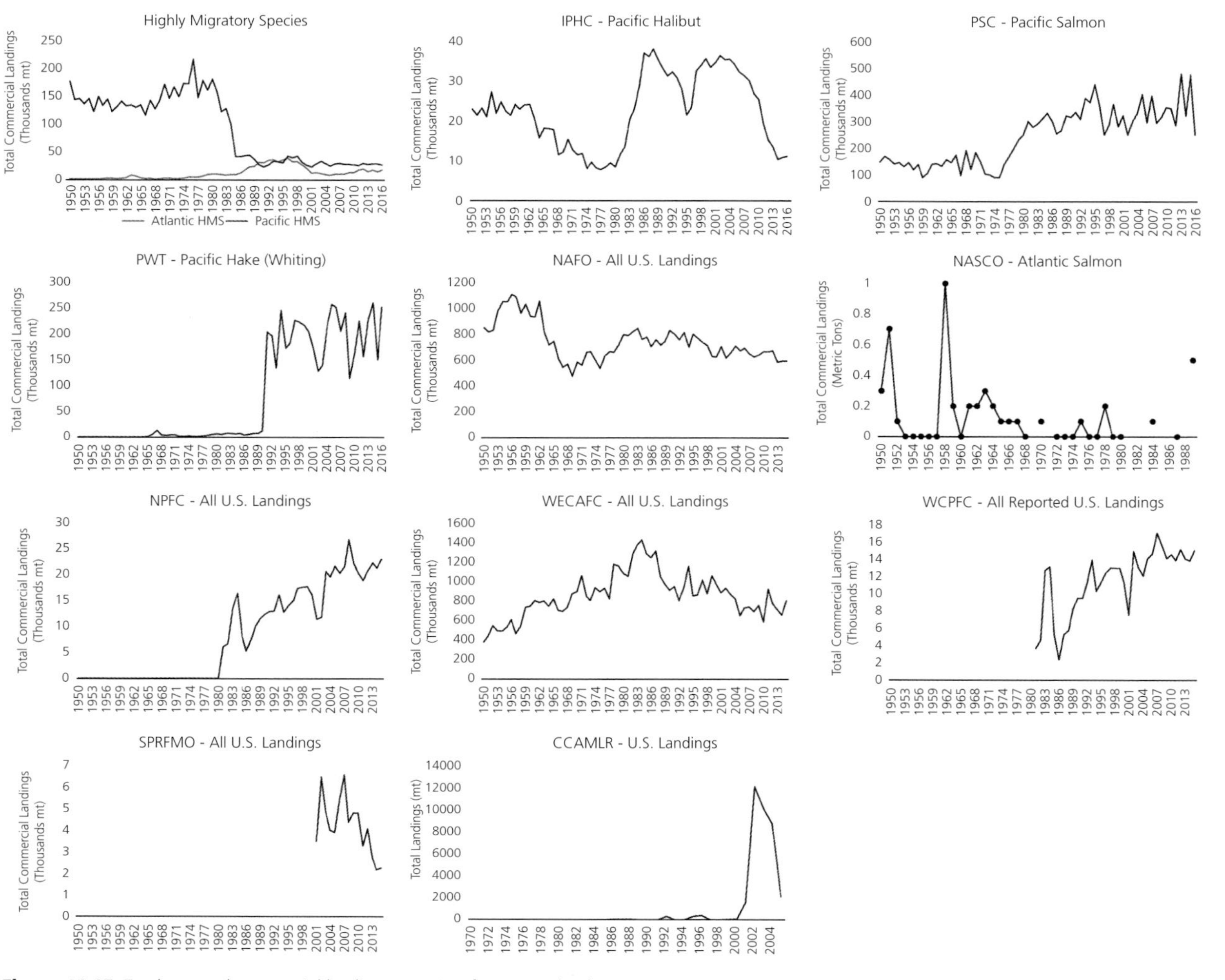

Figure 11.17 Total reported commercial landings over time for Regional Fisheries Management Organizations (RFMO) comprising U.S. components of Atlantic and Pacific highly migratory species (HMS), Pacific halibut (International Pacific Halibut Commission, IPHC), Pacific salmon (Pacific Salmon Commission, PSC), Pacific whiting (Pacific Whiting Treaty, PWT), Northwest Atlantic Fisheries Organization (NAFO), Atlantic salmon (North Atlantic Salmon Conservation Organization, NASCO), Western and Central Atlantic Fisheries Commission (WECAFC), North Pacific Fisheries Commission (NPFC), Western and Central Pacific Fisheries Commission (WCPFC), South Pacific Regional Fisheries Management Organisation (SPRFMO), and Commission for the Conservation of Antarctic Marine Living Resources (CCAMLR) jurisdictions.

Data derived from NOAA National Marine Fisheries Service commercial fisheries statistics, U.N. Fish and Agriculture Organization (FAO) global fish production and NASCO databases, and CCAMLR Statistical Bulletin (CCAMLR 2019).

within RFMO jurisdictions (Agnew 1997, Constable 2002, Cornejo-Donoso & Antezana 2008, Beaugrand 2009, Link et al. 2011, Greene et al. 2013, Brown 2016, Juan-Jorda et al. 2015, Zador et al. 2017a, b, Bornatowski et al. 2018, Harvey et al. 2018, Juan-Jorda et al. 2018, Dahood et al. 2019, Koen-Alonso et al. 2019, Link & Marshak 2019). Particularly in terms of advancing our knowledge of oceanographic, multispecies, and fishing and non-fishing effects on LMRs (Waldichuk 1993, Gibson 1994, Würsig et al. 2002, Hinke et al. 2004, Lewison et al. 2004, Hourigan 2009, Fulton et al. 2014, Nielsen et al. 2014, Muhling et al. 2015, Robinson et al. 2015, Arlinghaus et al. 2016, Lezama-Ochoa et al. 2016, Moffitt et al. 2016, Butterworth 2017, Rindorf et al. 2017, Zador et al. 2017a, b, Avila et al. 2018, Juan-Jorda et al. 2018, Dahood et al. 2019). These factors are being increasingly reflected in the creation of FEPs throughout U.S. regions, the development of ecosystem status reports for ICCAT managed areas, and the availability of some integrated models and management strategy evaluations that examine ecosystem components at the ocean basin scale (Juan-Jorda et al. 2015, 2018, ICCAT 2019, NMFS 2019b). Greater consideration of climatological, oceanographic, and multispecies relationships in assessments and management actions are reflected in approaches by CCAMLR (Constable 2011, Watters et al. 2013), while also in emerging EBFM implementation plans for U.S. managed HMS (NMFS 2019b, c, d), and NAFO ecosystem plans (NAFO 2018, Koen-Alonso et al. 2019). The development of ecosystem report cards by NAFO, considerations of Integrated Ecosystem Approach (IEA) frameworks by ICES, ICCAT, and IATTC, and investigations by scientific working groups of ecosystem-level case studies throughout Atlantic, Pacific, and Antarctic regions are beginning to provide a comprehensive synthesis of information necessary for more holistic management practices (Agnew 1997, Constable et al. 2011, Juan-Jorda et al. 2015, ICES 2018, Juan-Jorda et al. 2018, NAFO 2018, ICCAT 2019, Koen-Alonso et al. 2019). Broader spatial management approaches at the RFMO level (e.g., by IPHC, PSC, IWC, CCAMLR, the Arctic Council, etc.) have contributed to successful rebuilding of some previously overharvested stocks (e.g., Pacific halibut, Pacific salmon, Pacific whiting), and are creating buffers to limit additional human impacts on managed LMRs (Miller 2003, Constable 2011, Helser & Alade 2012, Barnes et al. 2018, Cook et al. 2019). Other intergovernmental efforts have also expanded into minimizing the environmental effects of other ocean uses in polar regions and the high seas, preventing species extinctions, and protecting marine mammals from shipstrikes and bycatch (Arctic Council 2004, 2009, 2013, Constable 2011, Cook et al. 2019). Initiatives by the IUCN and CBD are likewise promoting ecosystem-based approaches that enhance environmental action strategies from its member nations (IUCN-CEM 2017, CBD 2020, IUCN 2020). Notably, commitment to an ecosystem approach that considers broader human and environmental factors when regulating LMRs has been formally adopted by all RFMOs (Garcia 2003). Together these efforts are helping to advance progress toward executing EBFM at the RFMO level, with strengthened potential to develop larger-scale ecosystem-based management (EBM) actions and broader management approaches (Reeves et al. 2005, Small 2005, Karnauskas et al. 2013, Gilman et al. 2014, Nielsen et al. 2014, Wright et al. 2015, Karnauskas et al. 2017, Jeffries 2018).

Challenges remain, however, and additional steps are needed for carrying out EBFM within these larger jurisdictions. Some of this is understandable given that many of these RFMOs are taxa-oriented organizations. While advanced ecosystem-level metrics and quantitative frameworks have been developed throughout major portions of the U.S. EEZ, fewer of these advancements have occurred for fisheries ecosystems within high seas or entire RFMO jurisdictions (Gilman et al. 2014, Juan-Jorda et al. 2015, Moffitt et al. 2016, Juan-Jorda et al. 2018), though these features are beginning to be considered (FAO 2020). These broader developments are warranted given past and emerging stressors that are affecting transboundary LMRs, their fisheries, and their socioecological systems. For example, overfished and unassessed statuses for multiple HMS have continued, which has constrained the recovery of these apex predators. In addition, the effects of climate forcing and increasing temperatures on co-occurring LMRs, their prey, and system productivity continue to affect marine populations and efforts toward more sustainable practices (Gilman et al. 2016, Dell'Apa et al. 2018, Pentz et al. 2018, Barange et al. 2018). Ultimately, these effects radiate throughout the food web, including at basin-wide and transboundary scales. Concerns regarding minimizing and accurately documenting bycatch (especially in HMS fisheries) require ongoing attention, while data limitations for some of these multilaterally managed species (e.g., Pacific whiting, Antarctic predators) continue to hinder more robust management approaches (Constable

2011, Helser & Alade 2012, Wetzel & Punt 2016). Limited alignment of spatial management with concrete ecosystem-focused management priorities, as seen in whale sanctuaries, has also impeded progress toward more comprehensive management actions (Cook et al. 2019). While some positive results have occurred, continued threatened and endangered statuses of protected species, with the addition of newly listed populations, also demonstrates limitations of current management approaches focused on single species (Berg-Hedeholm et al. 2018, Cook et al. 2019, IUCN 2020). Implementing an ecosystem approach with management actions will simultaneously account for basal productivity limits, planktonic and forage prey abundance and distribution, and their bottom-up effects on upper trophic levels when setting harvest limits (Beamish & Bouillon 1993, Hollowed et al. 2001, Parma 2002, Jonsson & Jonsson 2004, Steinacher et al. 2010, Tremblay et al. 2011, Gutiérrez et al. 2016, Libralato et al. 2019, Link & Marshak 2019, Link & Watson 2019, Pranovi et al. 2020). Importantly, trade-offs among fisheries, protected species, other ocean uses, and marine economic development are accounted for in this ecosystem approach. The consequences of overfishing, shifting migration patterns and altered trophodynamics in response to oceanographic processes and climate change, and ongoing effects of pelagic and bottom fishing gears to marine habitats and bycaught species would be integrated into these broader management actions (Hourigan 2009, Gilman et al. 2014, Juan-Jorda et al. 2015, Moffitt et al. 2016, Juan-Jorda et al. 2018). Cumulatively, the U.S. Atlantic and Pacific ecosystems have high LMR-based socioeconomic status (Link & Marshak 2019), including for many transboundary fisheries and protected resources, for which enhanced efforts to reduce overfishing have emerged through multilateral agreements. However, recent system-level exploitations in terms of U.S. EEZ basal productivities remain above suggested sustainable thresholds of 0.0001 (Link & Marshak 2019) for NAFO and Pacific whiting resources, and above this level for U.S. fisheries throughout WECAFC jurisdictional waters. As NAFO's ecosystem approach includes efforts to spatially quantify and consider system exploitation (NAFO 2018, Koen-Alonso et al. 2019), management actions that set sustainable targets with respect to system biomass and productivity are urgently needed. Although LMR-based economies are proportionally high, the continued importance of marine fisheries and protected resources as part of cultural identity is also important (Koen-Alonso et al. 2019, Link & Marshak 2019, Haas et al. 2020).

Basin-scale considerations of system-level interconnected cumulative effects of fishing and non-fishing activities remain necessary as species and system-level shifts occur in response to overexploitation, regional warming, and climate forcing. Shifts in species compositions, migration patterns, and prey concentrations are likely to affect regional viabilities of landings and revenue, leading to conflicts over allocations of shared resources under RFMO jurisdictions. Some of these examples have been previously observed with shifts of salmon and groundfish stocks and fisheries allocations in response to climate forcing (Miller 2000, Miller et al. 2001, Parma 2002, Miller 2003, 2007, Sadorus et al. 2014, Sullivan 2016); these are predicted to intensify for HMS stocks over time (Juan-Jorda et al. 2015, 2018). Thus, establishing the basis to address trade-offs due to these predicted and observed distribution shifts remains an important capability to develop.

With increasing global human population and continued demand for both fish consumption and other ocean resources throughout the Atlantic and Pacific basins, competition and conflict among expanding and developed marine sectors is expected to intensify (Douvere 2008, Curtin & Prellezo 2010, Lewison et al. 2015, Maxwell et al. 2015, Selkoe et al. 2015). Strategic use of closures, cumulative consideration of interacting stressors on baseline production and their ecosystem-level effects, and measures to ensure sustainable limits on system exploitation are needed. Ongoing fishing and human activities in nearshore and offshore environments have also contributed to intermittent declines in their landings and revenue, leading to further resource limitations and competition among international fishing interests (Waldichuk 1993, Angilletta et al. 2008, Amiro et al. 2009, Hansen et al. 2012, Nielsen et al. 2014, Gibson 2017). As global marine sectors continue to expand, especially in terms of marine energy, transportation, and mineral extraction, and they increasingly intersect with commercial fishing areas, management strategies must account for these interactions and their effects on LMRs, fisheries, marine economies, and habitats. Management strategies accounting for increasingly spatially overlapping ocean uses, resource partitioning, multiple species, and socioeconomic effects in changing environments are necessary for continued sustainable practices of marine resource use. Human activities are heavily concentrated in portions of the Atlantic and Pacific basins, and continue to expand into less historically exploited regions, including the Arctic and Antarctic (Constable et al. 2000, Constable 2011, Molenaar 2012, Hoel 2015). Focus on multispecies habitat protections and continued spatial management of EEZs and high seas resources merits further consideration. Although comprehensive information is available for many international LMEs, more robust information is needed regarding other ocean sectors, marine socioeconomics, and the cumulative ecosystem-level effects of marine practices toward fostering more effective management (FAO 2020). These are especially needed as changing climate and emerging practices continue to affect sustainable fisheries harvest potential and the viability of international marine economies.

The execution of precautionary ecosystem approaches remains a central *stated* priority for all RFMOs, especially conserving biodiversity, ensuring sustainable management of marine fisheries resources, and protecting the biological integrity of marine ecosystems (NASCO 1998, Richards & Maguire 1998, NASCO 1999, Shelton & Rice 2002, Mooney-Seus & Rosenberg 2007, De Bruyn et al. 2013, Pan & Huntington 2016). But in practice that has been lagging with few concrete actions with respect to consideration of system thresholds or ecosystem-level reference points (Juan-Jorda et al. 2018, Link 2018). Documenting environmental factors and multispecies interactions affecting stock production has been a focus for some RFMOs (Fromentin et al. 2006, Aranda et al. 2010, Campbell 2010, Juan-Jorda et al. 2015, 2018, NAFO 2018, ICCAT 2019). But again, in practice although some progress has been made scientifically, this has not translated into management advice. While greater scientific understanding of Atlantic, Pacific, and Antarctic fisheries ecosystems is promoting an advanced knowledge base for execution of EBFM, limited implementation of formalized EBFM management actions or ecosystem-level control rules has occurred. While regulatory bodies have adopted the principles in the FAO ecosystem approach, and while RFMO-specific implementation plans, roadmaps, and amended conventions have shifted toward incorporating ecosystem considerations into assessments and management, actual ecosystem-based harvest control rules and management measures remain needed. Limited quantification of ecosystem-level reference points, metrics, and their emerging properties has occurred at these basin scales, and little to no application of these factors or ecosystem-level indicators toward management actions have occurred regionally or within RFMO jurisdictions (Gilman et al. 2014, Moffitt et al. 2016, Juan-Jorda et al. 2015, 2018, Link 2018). Although partial frameworks have been developed within portions of the U.S. EEZ, and EBFM implementation efforts are underway throughout all U.S. regions, additional actions are needed within and among RFMOs in consideration of multiple stressors affecting these systems. Given comprehensive information regarding many U.S. and global marine fisheries ecosystems, the establishment of system-wide control rules that consider production and yields of species complexes and trophic levels can become feasible within current RFMOs to support their recent adoption of ecosystem-based approaches (Libralato et al. 2019, Link & Watson 2019, Pranovi et al. 2020). Additional studies, models, and approaches to examine larger portions of ocean basins, the interactions of their species, and responses to management strategies are allowing for more concrete development of these types of management actions (Constable 2011, Juan-Jorda et al. 2015, Planas 2017, Juan-Jorda et al. 2018, Koen-Alonso et al. 2019). Continued support and expansion of these studies will allow for the development of a more complete EBFM approach through actions that are grounded in regulating harvest with regards to system production, biomass, and the effects of human and environmental stressors on these factors.

An important consideration is that the U.S. is a net seafood importer and thus our direct landings, numbers of vessels, etc. really only reflects a small fraction of our "ecosystem footprint" and the role our nation plays in international fisheries (Kite-Powell et al. 2013, Shamshak et al. 2019). Our international footprint is larger than, perhaps, many American analysts realize. For instance, required catch documentation schemes and eco-labeling designed specifically for the U.S. consumer are huge drivers of international fisheries (Buck 2010, Kite-Powell et al. 2013, Shamshak et al. 2019). For example, MMPA issues and the "Dolphin Safe" label massively affected Pacific HMS ecosystems by simultaneously pushing the U.S. fleet to the Western Pacific and creating a void for fishery aggregating device (FAD)-caught fish from the eastern Pacific that has been filled by fleets from other nations (Joseph 1994, Protopapadakis 2006, NMFS 2012, 2018b, Shamshak et al. 2019). The transition from dolphin sets to FAD sets has had profound ecosystem implications, largely driven by a net loss of U.S. vessels (Joseph 1994, Protopapadakis 2006, NMFS 2012, 2014, Debevec & Taylor 2018, Shamshak et al. 2019). Keeping in mind the indirect impacts, both good and bad, of the U.S. market seems prudent as we consider these interconnected systems.

Efforts to adopt an ecosystem approach in managing the resources of these ecologically significant regions and their historically important LMRs are necessary for their ongoing sustainable use. Increasing consideration of the effects of natural stressors and human activities on species complexes are a foundational step toward broader system-wide strategies that account for interacting sectors, multiple trophic levels among species complexes, and their socioeconomic feedbacks regarding internationally shared marine resources. Continued system-wide considerations of the inherent relationships between productivity, biomass, landings, and socioeconomics are needed to maintain these systems and their iconic LMRs within sustainable thresholds, especially in light of multiple stressors. EBFM provides opportunities that focusing on one stock or species complex, one fishery, or the effects of only the fishing sector would otherwise miss. There remains a continued need for enhanced EBFM in consideration of internationally shared resources and in terms of the role that the U.S. plays in its RFMO memberships. Expanding upon and incorporating domestic and international advances in fisheries management strategies toward broader system-wide focuses, especially in accounting for the interdependence of fisheries and protected species complexes, climate and bottom-up forcing related to productivity thresholds is a logical next step when considering RFMO jurisdictions. Incorporating the determinants of successful LMR management (Link & Marshak 2019) in consideration of system-level production and biomass at the RFMO level will allow for the development of these metrics toward sustainable ecosystem exploitation. These factors ultimately determine the available harvest and socioeconomic potential associated with a given system, and aid in setting management measures that reflect these limits.

11.2.6.1 Progress Toward Ecosystem-Based Fisheries Management (EBFM)

Advances toward EBFM have been occurring throughout several RFMOs, with specific progress of adopting ecosystem considerations occurring in various jurisdictions (Schiffman 2006, Kock et al. 2007, Constable 2011, Link et al. 2011, De Bruyn et al. 2013, Webster 2013, Gilman et al. 2014, Juan-Jorda et al. 2015, Brown 2016, Koen-Alonso et al. 2019, FAO 2020). Over the past few decades, international agreements including the United Nations Fish Stock Agreement, FAO Code of Conduct for Responsible Fisheries, and FAO Ecosystem Approach to Fisheries have increased expectations of RFMOs to incorporate ecosystem considerations into their management (FAO 1995, Garcia 2000, FAO 2003, Garcia & Cochrane 2005, Juan-Jorda et al. 2015, Brown 2016). A subcommittee on ecosystems promotes an ecosystem approach in ICCAT Atlantic HMS management, with priority foci on monitoring and reducing bycatch of targeted and protected species, improving monitoring of HMS interactions with non-target ICCAT species, supporting ecosystem-based research efforts and ecological modeling, and providing ecosystem-level advice to the Commission (Fromentin et al. 2006, Small et al. 2013, Juan-Jorda et al. 2015). Specifically, the subcommittee is working toward developing EBFM goals and objectives, workshops, and ecosystem status reports as outlined in its 2015–2020 Strategic Plan (ICCAT 2015, Juan-Jorda et al. 2015). Recent Atlantic pelagic ecosystem modeling efforts include developing a preliminary food web model to assess *Sargassum* ecosystems and an Ecopath model to test the effects of developing a Gulf of Guinea FAD fishery (ICCAT 2015, 2019, Juan-Jorda et al. 2015). There is increasing use of ecosystem and habitat models such as the spatial ecosystem and population dynamic model (SEAPODYM) and the Apex Predators Ecosystem Model (APECOSM) to investigate spatial dynamics of target species and their responses to climate forcing (Schirripa et al. 2011, Lehodey et al. 2014, Juan-Jorda et al. 2015, Lefort et al. 2015). However, these efforts have had limited impact on formally identifying or protecting habitats for these species, nor have these models been regularly incorporated into the development of management advice (Juan-Jorda et al. 2015, 2018).

Focused efforts toward investigating and evaluating the effects of climate and oceanographic factors, direct and incidental fishing, multispecies trophic interactions, pollution, and other human impacts that affect the abundance, distribution, and migration of ICCAT target species are conducted and supported by this subcommittee (Fromentin et al. 2006, Small et al. 2013, Juan-Jorda et al. 2015, ICCAT 2019). As of 2019, a pilot ecosystem plan has been developed for the tropical ecoregion of the Atlantic Ocean, complemented by ongoing workshops and efforts toward producing ecosystem plans. This includes a case study for the Sargasso Sea (ICCAT 2019). Efforts to produce ecosystem indicators for protected species that interact with HMS fisheries, evaluating the trophic effects of fishing practices, and considering habitat, environmental, and socioeconomic factors are also underway. The IATTC follows a precautionary ecosystem approach under its Antigua Convention, while the WCPFC makes explicit reference to following both ecosystem and precautionary approaches (IATTC 2003, De Bruyn et al. 2013, Juan-Jorda et al. 2015). ICCAT is also developing final resolutions that specifically refer to both these approaches (C. Soltanoff, pers. comm.). During the 1980s–1990s, ICCAT and IATTC scientific programs began to research ecosystem factors including the impacts of tuna fisheries on target and bycatch species, examining forage species, trophic interactions, ecosystem modeling and risk assessments, tracking ecosystem indicators, and incorporating ecosystem considerations into management (Fromentin et al. 2006, Small et al. 2013, Juan-Jorda et al. 2015, ICCAT 2019). Although challenges toward implementation continue, the applications of conceptual ecological modeling that incorporate the IEA (Levin et al. 2009, Dickey-Collas 2014, Juan-Jorda et al. 2015) and ecosystem-based management driver-pressure-state-ecosystem service-response (EBM DPSER) model (Kelble et al. 2013) have been promoted for Atlantic and Pacific HMS (Juan-Jorda et al. 2015). However, at present neither ICCAT nor IATTC have defined or discussed ecosystem-level metrics or thresholds, and have not linked identified ecosystem indicators to operational management (Juan-Jorda et al. 2015, 2018).

Although ICCAT has its own ecosystem efforts, domestically the U.S. is pursuing EBFM in an Atlantic HMS context by addressing the cumulative issues affecting these species and their fisheries (NMFS 2016a, b, Levin et al. 2018, Marshall et al. 2018, Link & Marshak 2019). In accordance with the NOAA Fisheries EBFM Policy and Roadmap (NMFS 2016a, b), an EBFM implementation plan for Atlantic HMS has been developed (NMFS 2019b). This plan specifies approaches to be taken toward conducting EBFM and in addressing NOAA Fisheries Roadmap priorities and gaps. Specifically it advances ecosystem-level planning, knowledge of ecosystem principles, and works toward assessing risks and vulnerabilities of ecosystems through investigations into climate vulnerability and species prioritizations for stock and habitat assessments. Additional priorities include:

- Continued engagement with ICCAT and the SCRS Sub-Committee on Ecosystems in developing best practices for international EBFM efforts for ICCAT and for a dialog across tuna regional fishery management organizations.
- Providing Atlantic HMS EBFM guidance and participating with council and commission ecosystem-related committees as requested.
- Designating an Atlantic HMS management FEP coordinator who would work with relevant management bodies to incorporate Atlantic HMS information into FEPs as appropriate.
- Considering trophic interactions and other ecosystem topics to be included in Atlantic HMS research needs and priorities.
- Continue participation in ICCAT's development of management strategy evaluations.
- Continue working at ICCAT to develop and adopt harvest control rules for internationally managed species.
- Explore including the evaluation of offshore aquaculture as a non-fishing impact on essential fish habitat in amendments to the consolidated Atlantic HMS FMP. Include ecosystem effects as available.
- Encourage consideration of a socioeconomic goal and related indicators to evaluate in the ICCAT ecosystem report card.

As observed throughout the ICCAT management area and despite having relatively more information, little overall progress has been made in applying ecosystem-level properties into management frameworks or exploring system-level trade-offs for U.S. Atlantic HMS. Similarly, limited progress toward these outcomes has occurred for Pacific bilateral fisheries management, despite the development of ecosystem-level calculations, metrics, and models that account for Pacific halibut, Pacific whiting, and Pacific salmon (Witherell et al. 2000, Shelton & Rice 2002, Field & Francis 2006, Zador & Gaichas 2010, Zador 2015, Zador et al. 2017a, b, Harvey et al. 2018, Zador & Ortiz 2018, Zador & Yasumiishi 2018), the recent completion of a U.S. West Coast EBFM implementation plan (NMFS 2019b), and study groups working to account for broader EBM of these transboundary species (Jamieson & Zhang 2005). The IPHC has similarly developed an ecosystem research plan in support of investigations that account for greater environmental influences on Pacific halibut populations, and their roles in Pacific and North Pacific ecosystem processes (Planas 2017).

NAFO has begun to implement facets of an ecosystem approach to management (Link et al. 2011, Rice 2011, Dickey-Collas 2014, Koen-Alonso et al. 2019). Given NAFO's ongoing recognition of the dependence of fish stock production on environmental factors, a Scientific Council Working Group on the Ecosystem Approach to Fisheries Management (WGEAFM; later the Working Group on Ecosystem Science and Assessment, WGESA) was established to provide guidance on ecosystem-related issues (Koen-Alonso et al. 2019).

More recently, this working group has created a Roadmap for Developing an Ecosystem Approach to Fisheries in consideration of long-term ecosystem sustainability (NAFO 2018, Koen-Alonso et al. 2019). Additionally, a joint Fisheries Commission-Scientific Council Working Group on an ecosystem approach to fisheries management was established in 2013. As amended in its 2017 Convention, NAFO has committed to applying an ecosystem approach that "includes safeguarding the marine environment, conserving its marine biodiversity, minimizing the risk of long term or irreversible adverse effects of fishing activities, and taking account of the relationship between all components of the ecosystem" (NAFO 2017, Koen-Alonso et al. 2019). There has been progress in spatially characterizing northwestern Atlantic ecosystems, habitats, and their fisheries with regards to their exploitation rates, and as applied to ecological risk assessments (Koen-Alonso et al. 2019). However, socioeconomic and cultural elements are presently not included, nor have management actions based on ecosystem-level emergent properties been created (Koen-Alonso et al. 2019).

While having examined the ecological effects of fishing on Antarctic predators and key prey species, and modeled potential climate effects on these relationships, CCAMLR CEMP data mostly have been applied to management actions for individual target species (Constable et al. 2000, Constable 2011). For example, krill management now also accounts for natural mortality from predator populations instead of simply the effects of fishing mortality when setting harvest reference points. Previously, a post-fishing target level was developed as half of the unexploited median abundance, which was then revised to 75% of the pre-exploitation median to account for predation (Constable 2011). Catch limits are set to reserve a specific amount of krill production for predators (Hill et al. 2006, Constable 2011, Watters et al. 2013). CCAMLR has specified similar objectives for harvest limits in consideration of Patagonian toothfish and Mackerel icefish (*Champsocephalus gunnari*) biomass, and their effects on the Antarctic ecosystem (Agnew et al. 1998, de la Mare et al. 1998, Constable et al. 2000, Constable 2011). More direct actions for predatory species have yet to be taken, but these issues remain a focus for the CCAMLR Working Group on Ecosystem Monitoring and Management. Proposed actions include developing harvest strategies that account for the vulnerabilities of predator populations to fishing effects based on their foraging concentrations and spatial distributions, applying management strategy evaluations, and considering the ecosystem effects of coastal versus oceanic fishing activities (Constable 2011, Watters et al. 2013). These efforts are also complemented by IWC supported actions to conduct ecosystem modeling and management scenario testing, to examine the partial and cumulative effects of fishing on whales and their prey, and look at trade-offs between whale predation and fishery prey species (Cook 1995, Crespo & Hall 2002, Baker & Clapham 2004, Surma et al. 2014, Surma & Pitcher 2015). While currently limited in terms of their applications to EBM, the creation of IWC whale sanctuaries has allowed for protections of additional species, their habitats, and for management approaches through the development of broader management plans (Cook et al. 2019). While management bodies including NASCO, NPFC, SPRFMO, and WECAFC have adopted the FAO precautionary and ecosystem approaches toward considering broader environmental factors in the management of their fisheries, no concrete management actions have been developed with regard to ecosystem-level reference points or broad system-level trade-offs (Sainsbury et al. 2000, Campbell 2010, Grant et al. 2013, Fulton et al. 2014, Gilman et al. 2014, Moffitt et al. 2016, FAO 2019).

11.2.6.2 Conclusions

Moving forward, other factors including sectorial trade-offs and cumulative effects of human activities on ecosystem dynamics still require incorporation into more holistic conservation and sustainable management efforts. Applying available, comprehensive information toward the development of system-level reference points and thresholds with respect to basal productivity and biomass, with which dependent LMR status and economies are interconnected, is especially possible in systems where management frameworks can be expanded. Progress is being made in addressing aspects of these concerns through the ongoing development of broader ecosystem-based frameworks for some U.S. participatory RFMOs (e.g., NAFO, ICCAT, CCAMLR), for which ecosystem approaches continue to be prioritized and indicators developed. Taking these efforts a step further into holistic system-level approaches to management actions, which examine these facets together and allow for improved collaboration among multisectoral management interests, will work toward maintaining the strong socioeconomic and LMR statuses associated with these U.S. and internationally shared taxa as they are subjected to ongoing and increasing pressures. There is high potential for overexploitation of multiple regions of the Atlantic and Pacific basins, especially as regional warming, significant other ocean uses, and increasing seafood demand affect the viability and productivity of these systems. These biological and socioeconomic metrics have been developed in certain regions, with potential for their application in other geographies and at broader scales. Applying and expanding comprehensive management strategies in these systems and the LMEs that compose them in consideration of their productivities and well-known integrated ecosystem properties will work toward maintaining their LMR and socioeconomic statuses in light of ongoing pressures. These efforts provide a pathway toward ensuring the continuation of internationally important LMRs, global maritime traditions, multilateral ocean-based economies, and internationally iconic fisheries.

11.2.7 References

Agnew DJ. 1997. The CCAMLR ecosystem monitoring programme. *Antarctic Science* 9(3):235–42.

Agnew DJ, Everson I, Kirkwood GP, Parkes GB. 1998. Towards the development of a management plan for the mackerel icefish (Champsocephalus gunnari) in Subarea 48.3. *CCAMLR Science* 5:63–77.

Agostini VN, Hendrix AN, Hollowed AB, Wilson CD, Pierce SD, Francis RC. 2008. Climate–ocean variability and Pacific hake: a geostatistical modeling approach. *Journal of Marine Systems* 71(3–4):237–48.

Allen R, Joseph JA, Squires D. 2010. *Conservation and Management of Transnational Tuna Fisheries*. Hoboken, NJ: John Wiley & Sons.

Alter SE, Simmonds MP, Brandon JR. 2010. Forecasting the consequences of climate-driven shifts in human behavior on cetaceans. *Marine Policy* 34(5):943–54.

Amandè MJ, Ariz J, Chassot E, de Molina AD, Gaertner D, Murua H, Pianet R, Ruiz J, Chavance P. 2010. Bycatch of the European purse seine tuna fishery in the Atlantic Ocean for the 2003–2007 period. *Aquatic Living Resources* 23(4):353–62.

Amiro PG, Brazner JC, Giorno JL. 2009. *Assessment of the Recovery Potential for the Atlantic Salmon Designatable Unit Inner Bay of Fundy: Threats*. Fisheries and Oceans Canada.

Angilletta Jr MJ, Ashley Steel E, Bartz KK, Kingsolver JG, Scheuerell MD, Beckman BR, Crozier LG. 2008. Big dams and salmon evolution: changes in thermal regimes and their potential evolutionary consequences. *Evolutionary Applications* 1(2):286–99.

Anthony A, Atwood J, August PV, Byron C, Cobb S, Foster C, Fry C, Gold A, Hagos K, Heffner L, Kellogg DQ. 2009. Coastal lagoons and climate change: ecological and social ramifications in the US Atlantic and Gulf coast ecosystems. *Ecology and Society* 14:8.

Aranda M, de Bruyn P, Murua H. 2010. A report review of the tuna RFMOs: CCSBT, IATTC, IOTC, ICCAT and WCPFC. *EU FP7 Project* 212188:171.

Aranda M, Murua H, De Bruyn P. 2012. Managing fishing capacity in tuna regional fisheries management organisations (RFMOs): development and state of the art. *Marine Policy* 36(5):985–92.

Arctic Council. 2004. *Arctic Marine Strategic Plan*. Akureyri, Iceland: Protection of the Arctic Marine Environment (PAME) International Secretariat (p. 13).

Arctic Council. 2007. *About Arctic Council*. Tromsø, Norway: Arctic Council Secretariat.

Arctic Council. 2009. *Arctic Marine Shipping Assessment 2009 Report (AMSA)*. Tromsø, Norway: Arctic Council Secretariat.

Arctic Council. 2013. *Arctic Council Observer Manual for Subsidiary Bodies*. Tromsø, Norway: Arctic Council Secretariat.

Arias AH, Marcovecchio JE. 2017. "High seas deep-sea fisheries under the global changing trends." In: *Marine Pollution and Climate Change*. Edited by AH Arias, JE Marcovecchio. Boca Raton, FL: CRC Press (pp. 188–244).

Arlinghaus R, Lorenzen K, Johnson BM, Cooke SJ, Cowx IG. 2016. Management of freshwater fisheries: addressing habitat, people and fishes. *Freshwater Fisheries Ecology* 1:557–79.

Atkinson A, Siegel V, Pakhomov EA, Jessopp MJ, Loeb V. 2009. A re-appraisal of the total biomass and annual production of Antarctic krill. *Deep Sea Research Part I: Oceanographic Research Papers* 56(5):727–40.

Auster PJ, Malatesta RJ, Langton RW, Watting L, Valentine PC, Donaldson CL, Langton EW, Shepard AN, Babb WG. 1996. The impacts of mobile fishing gear on seafloor habitats in the Gulf of Maine (Northwest Atlantic): implications for conservation of fish populations. *Reviews in Fisheries Science* 4(2):185–202.

Avila IC, Kaschner K, Dormann CF. 2018. Current global risks to marine mammals: taking stock of the threats. *Biological Conservation* 221:44–58.

Axelrod M. 2011. Climate change and global fisheries management: linking issues to protect ecosystems or to save political interests? *Global Environmental Politics* 11(3):64–84.

Ásmundsson S. 2016. *Regional Fisheries Management Organisations (RFMOs): Who Are They, What is Their Geographic Coverage on the High Seas and Which Ones Should be Considered as General RFMOs, Tuna RFMOs and Specialised RFMOs?* London, UK: NEAFC (p. 7).

Babcock EA, Nakano H. 2008. Data collection, research, and assessment efforts for pelagic sharks by the International C for the Conservation of Atlantic Tunas. *Sharks of the Open Ocean: Biology, Fisheries and Conservation* 29:472–7.

Ballantine DL, Appeldoorn RS, Yoshioka P, Weil E, Armstrong R, Garcia JR, Otero E, et al. 2008. "Biology and ecology of Puerto Rican coral reefs." In: *Coral Reefs of the USA*. Edited by BM Riegl, RE Dodge. Dordrecht, Netherlands: Springer (pp. 375–406).

Baker CS, Clapham PJ. 2004. Modelling the past and future of whales and whaling. *Trends in Ecology & Evolution* 19(7):365–71.

Barange M, Bahri T, Beveridge MC, Cochrane KL, Funge-Smith S, Poulain F. 2018. *Impacts of Climate Change on Fisheries and Aquaculture. Synthesis of Current Knowledge, Adaptation and Mitigation Options*. Rome, Italy: Food and Agriculture Organization of the United Nations.

Barnard M, Starkey DJ, Holm P. 2012. *Oceans Past: Management Insights from the History of Marine Animal Populations*. London, UK: Earthscan.

Barnes CL, Beaudreau AH, Hunsicker ME, Ciannelli L. 2018. Assessing the potential for competition between Pacific Halibut (Hippoglossus stenolepis) and Arrowtooth Flounder (Atheresthes stomias) in the Gulf of Alaska. *PLoS One* 13(12).

Beamish RJ. 2008. *Impacts of Climate and Climate Change on the Key Species in the Fisheries in the North Pacific. PICES Science Report No. 35*. Sidney, Canada: North Pacific Marine Science Organization (PICES).

Beamish RJ, Bouillon DR. 1993. Pacific salmon production trends in relation to climate. *Canadian Journal of Fisheries and Aquatic Sciences* 50(5):1002–16.

Beaugrand G. 2009. Decadal changes in climate and ecosystems in the North Atlantic Ocean and adjacent seas. *Deep Sea Research Part II: Topical Studies in Oceanography* 56(8–10): 656–73.

Bensch A, Gianni M, Gréboval D, Sanders JS, Hjort A. 2008. *Worldwide Review of Bottom Fisheries in the High Seas*. Rome, Italy: Food and Agriculture Organization of the United Nations.

Benson AJ, McFarlane GA, Allen SE, Dower JF. 2002. Changes in Pacific hake (Merluccius productus) migration patterns and juvenile growth related to the 1989 regime shift. *Canadian Journal of Fisheries and Aquatic Sciences* 59(12):1969–79.

Bell J, Valérie A, Gupta A, Johnson J, Hampton J, Hobday A, Lehodey P, et al. 2018. *Climate Change Impacts, Vulnerabilities and Adaptations: Western and Central Pacific Ocean Marine Fisheries. Impacts of Climate Change on Fisheries and Aquaculture: Synthesis of Current Knowledge, Adaptation and Mitigation Options. FAO Fisheries Technical Paper No. 627*. Rome, Italy: FAO.

Berg-Hedeholm R, Broberg M, Nygaard R, Grønkjær P. 2018. Population decline in the endemic Atlantic salmon (Salmo salar) in Kapisillit River, Greenland. *Fisheries Management and Ecology* 25(5):392–9.

Best EA, St-Pierre G. 1986. *Pacific Halibut as Predator and Prey*. Washington, DC: International Pacific Halibut Commission.

Billé R, Chabason L, Drankier P, Molenaar EJ, Rochette J. 2016. *Regional Oceans Governance. Making Regional Seas Programmes, Regional Fishery Bodies and Large Marine Ecosystem Mechanisms Work Better Together. UNEP Regional Seas Reports and Studies No.197*. Nairobi, Kenya: UNEP.

Bisbal GA, McConnaha WE. 1998. Consideration of ocean conditions in the management of salmon. *Canadian Journal of Fisheries and Aquatic Sciences* 55(9):2178–86.

Blanchard JL, Jennings S, Holmes R, Harle J, Merino G, Allen JI, Holt J, Dulvy NK, Barange M. 2012. Potential consequences of climate change for primary production and fish production in large marine ecosystems. *Philosophical Transactions of the Royal Society B: Biological Sciences* 367(1605):2979–89.

Blasiak R, Yagi N. 2016. Shaping an international agreement on marine biodiversity beyond areas of national jurisdiction: lessons from high seas fisheries. *Marine Policy* 71:210–6.

Bloom ET. 1999. Establishment of the Arctic Council. *American Journal of International Law* 93(3):712–22.

Boerder K, Schiller L, Worm B. 2019. Not all who wander are lost: improving spatial protection for large pelagic fishes. *Marine Policy* 105:80–90.

Bornatowski H, Angelini R, Coll M, Barreto RR, Amorim AF. 2018. Ecological role and historical trends of large pelagic predators in a subtropical marine ecosystem of the South Atlantic. *Reviews in Fish Biology and Fisheries* 28(1):241–59.

Bost CA, Cotté C, Bailleul F, Cherel Y, Charrassin JB, Guinet C, Ainley DG, Weimerskirch H. 2009. The importance of oceanographic fronts to marine birds and mammals of the southern oceans. *Journal of Marine Systems* 78(3):363–76.

Boyce DG, Tittensor DP, Worm B. 2008. Effects of temperature on global patterns of tuna and billfish richness. *Marine Ecology Progress Series* 355:267–76.

Bradley D, Gaines SD. 2014. Extinction risk: counting the cost of overfishing on sharks and rays. *Elife* 3:e02199.

Brigham LW. 2020. "Arctic policy developments and marine transportation." In: *The Palgrave Handbook of Arctic Policy and Politics*. Edited by K Coates, C Holroyd. Cham, Switzerland: Palgrave Macmillan (pp. 393–408).

Brown BE. 2016. Regional fishery management organizations and large marine ecosystems. *Environmental Development* 17:202–10.

Brown BE. 2017. Large marine ecosystem fisheries management with particular reference to Latin America and the Caribbean Sea. *Environmental Development* 22:111–19.

Bryndum-Buchholz A, Tittensor DP, Blanchard JL, Cheung WW, Coll M, Galbraith ED, Jennings S, Maury O, Lotze HK. 2019. Twenty-first-century climate change impacts on marine animal biomass and ecosystem structure across ocean basins. *Global Change Biology* 25(2):459–72.

Buck EH. 2010. *Seafood Marketing: Combating Fraud and Deception. CRS Report for Congress 7-5700, RL34124*. Washington, DC: Congressional Research Service 12).

Butler JR, Wong GY, Metcalfe DJ, Honzák M, Pert PL, Rao N, van Grieken ME, et al. 2013. An analysis of trade-offs between multiple ecosystem services and stakeholders linked to land use and water quality management in the Great Barrier Reef, Australia. *Agriculture, Ecosystems and Environment* 180:176–191.

Butterworth A. 2017. *Marine Mammal Welfare: Human Induced Change in the Marine Environment and its Impacts on Marine Mammal Welfare*. Basel, Switzerland: Springer.

Caillouet Jr CW, Hart RA, Nance JM. 2008. Growth overfishing in the brown shrimp fishery of Texas, Louisiana, and adjoining Gulf of Mexico EEZ. *Fisheries Research* 92(2–3): 289–302.

Camhi MD, Valenti SV, Fordham SV, Fowler SL, Gibson C. 2009. *The Conservation Status of Pelagic Sharks and Rays*. Gland, Switzerland: IUCN Species Survival Commission Shark Specialist Group (p. 78).

Campbell R. 2010. *Identifying Possible Limit Reference Points for the Key Target Species in the WCPFC. Unpublished WCPFC Working Document no WCPFCSC6-2010/MI-IP-01*. Kolonia, Federated States of Micronesia: WCPFC (p. 14).

Carlson JK, Hale LF, Morgan A, Burgess G. 2012. Relative abundance and size of coastal sharks derived from commercial shark longline catch and effort data. *Journal of Fish Biology* 80:1749–64.

Caron DD. 1995. The International Whaling Commission and the North Atlantic Marine Mammal Commission: the institutional risks of coercion in consensual structures. *American Journal of International Law* 89(1):154–74.

CBD (Convention on Biological Diversity). 1992. *Convention on Biological Diversity*. Montreal, Canada: Secretariat of the Convention on Biological Diversity (p. 28).

CBD (Convention on Biological Diversity). 2020. *The Convention on Biological Diversity*. https://www.cbd.int/

CCAMLR (Commission for the Conservation of Antarctic Marine Living Resources). 1982. *Text of the Convention on the Conservation of Antarctic Marine Living Resources Statement by the Chairman of the Conference on the Conservation of Antarctic Marine Living Resources*. Tasmania, Australia: CCAMLR (p. 19).

CCAMLR (Commission for the Conservation of Antarctic Marine Living Resources). 2018. *CCAMLR Statistical Bulletin, Vol. 30*. Tasmania, Australia: CCAMLR.

CCAMLR (Commission for the Conservation of Antarctic Marine Living Resources). 2019. *CCAMLR Commission for the Conservation of Antarctic Marine Living Resources*. Tasmania, Australia: CCAMLR. https://www.ccamlr.org/en

CCAMLR (Commission for the Conservation of Antarctic Marine Living Resources). 2019. *CCAMLR Statistical Bulletin, Vol. 31*. Tasmania, Australia: CCAMLR.

Cederholm CJ, Johnson DH, Bilby RE, Dominguez LG, Garrett AM, Graeber WH, Greda EL, et al. 2001. *Pacific Salmon and Wildlife: Ecological Contexts, Relationships, and Implications for Management. Wildlife-Habitat Relationships in Oregon and Washington*. Corvallis, Oregon: Oregon State University Press (pp. 628–85).

Chambers RM, Meyerson LA, Saltonstall K. 1999. Expansion of Phragmites australis into tidal wetlands of North America. *Aquatic Botany* 64(3–4):261–73.

Chesney EJ, Baltz DM, Thomas RG. 2000. Louisiana estuarine and coastal fisheries and habitats: perspectives from a fish's eye view. *Ecological Applications* 10(2):350–66.

Cheung WW, Sumaila UR. 2013. "Managing multiple human stressors in the ocean: a case study in the Pacific Ocean." In: *Managing Ocean Environments in a Changing Climate: Sustainability and Economic Perspectives*. Edited by KJ Noone, UR Sumaila, RJ Diaz. London, UK: Elsevier (pp. 277–335).

Chittenden CM, Beamish RJ, McKinley RS. 2009. A critical review of Pacific salmon marine research relating to climate. *ICES Journal of Marine Science* 66(10):2195–204.

Choy CA, Wabnitz CC, Weijerman M, Woodworth-Jefcoats PA, Polovina JJ. 2016. Finding the way to the top: how the composition of oceanic mid-trophic micronekton groups determines apex predator biomass in the central North Pacific. *Marine Ecology Progress Series* 549:9–25.

Clark WG. 2003. A model for the world: 80 years of model development and application at the international Pacific halibut commission. *Natural Resource Modeling* 16(4): 491–503.

Clark WG, Hare SR. 2002. Effects of climate and stock size on recruitment and growth of Pacific halibut. *North American Journal of Fisheries Management* 22(3):852–62.

Clark WG, Hare SR, Parma AM, Sullivan PJ, Trumble RJ. 1999. Decadal changes in growth and recruitment of Pacific halibut (Hippoglossus stenolepis). *Canadian Journal of Fisheries and Aquatic Sciences* 56(2):242–52.

Colvin SA, Sullivan SM, Shirey PD, Colvin RW, Winemiller KO, Hughes RM, Fausch KD, et al. 2019. Headwater streams and wetlands are critical for sustaining fish, fisheries, and ecosystem services. *Fisheries* 44(2):73–91.

Compeán G. 2018. Review of management and conservation measures for tropical tunas in the Eastern Pacific Ocean. *Ocean Yearbook Online* 32(1):317–28.

Constable AJ. 2002. CCAMLR ecosystem monitoring and management: future work. *CCAMLR Science* 9:233–53.

Constable AJ. 2011. Lessons from CCAMLR on the implementation of the ecosystem approach to managing fisheries. *Fish and Fisheries* 12(2):138–51.

Constable AJ, de la Mare WK, Agnew DJ, Everson I, Miller D. 2000. Managing fisheries to conserve the Antarctic marine ecosystem: practical implementation of the Convention on the Conservation of Antarctic Marine Living Resources (CCAMLR). *ICES Journal of Marine Science* 57(3):778–91.

Cook D, Malinauskaite L, Roman J, Davíðsdóttir B, Ögmundardóttir H. 2019. Whale sanctuaries—an analysis of their contribution to marine ecosystem-based management. *Ocean & Coastal Management* 182:104987.

Cook JG. 1995. The International Whaling Commission's revised management procedure as an example of a new approach to fishery management. *Developments in Marine Biology* 4:647–57.

Copes P, Cook BA. 1982. Rationalization of Canada's Pacific halibut fishery. *Ocean Management* 8(2):151–75.

Corell R. 2005. Arctic climate impact assessment. *Bulletin of the American Meteorological Society* 86(6):860.

Cornejo-Donoso J, Antezana T. 2008. Preliminary trophic model of the Antarctic Peninsula ecosystem (Sub-area CCAMLR 48.1). *Ecological Modelling* 218(1–2):1–7.

Corno G, Karl DM, Church MJ, Letelier RM, Lukas R, Bidigare RR, Abbott MR. 2007. Impact of climate forcing on ecosystem processes in the North Pacific Subtropical Gyre. *Journal of Geophysical Research: Oceans* 112(C4).

Costa-Pierce BA. 2002. *Ecology as the Paradigm for the Future of Aquaculture. Ecological Aquaculture: The Evolution of the Blue Revolution*. Oxford, UK: Blackwell Science (pp. 339–72).

Costa-Pierce BA. 2010. Sustainable ecological aquaculture systems: the need for a new social contract for aquaculture development. *Marine Technology Society Journal* 44(3):88–112.

Cowan JH, Grimes CB, Patterson WF, Walters CJ, Jones AC, Lindberg WJ, Sheehy DJ, et al. 2011. Red snapper management in the Gulf of Mexico: science-or faith-based? *Reviews in Fish Biology and Fisheries* 21(2):187–204.

Cowan Jr JH, Grimes CB, Shaw RF. 2008. Life history, history, hysteresis, and habitat changes in Louisiana's coastal ecosystem. *Bulletin of Marine Science* 83(1):197–215.

Cox SP, Essington TE, Kitchell JF, Martell SJ, Walters CJ, Boggs C, Kaplan I. 2002. Reconstructing ecosystem dynamics in the central Pacific Ocean, 1952–1998. II. A preliminary assessment of the trophic impacts of fishing and effects on tuna dynamics. *Canadian Journal of Fisheries and Aquatic Sciences* 59(11):1736–47.

Crespo EA, Hall MA. 2002. "Interactions between aquatic mammals and humans in the context of ecosystem management." In: *Marine Mammals.* Edited by H Thewissen, B Würsig, WF Perrin. Boston, MA: Springer (pp. 463–90).

Crowder L, Norse E. 2008. Essential ecological insights for marine ecosystem-based management and marine spatial planning. *Marine Policy* 32(5):772–8.

Croxall JP, Nicol S. 2004. Management of Southern Ocean fisheries: global forces and future sustainability. *Antarctic Science* 16(4):569–84.

Cullis-Suzuki S, Pauly D. 2010. Failing the high seas: a global evaluation of regional fisheries management organizations. *Marine Policy* 34(5):1036–42.

Curran D, Bigelow K. 2011. Effects of circle hooks on pelagic catches in the Hawaii-based tuna longline fishery. *Fisheries Research* 109(2–3):265–75.

Curtin R, Prellezo R. 2010. Understanding marine ecosystem based management: a literature review. *Marine Policy* 34(5):821–30.

Dahood A, Watters GM, de Mutsert K. 2019. Using sea-ice to calibrate a dynamic trophic model for the Western Antarctic Peninsula. *PLoS One* 14(4):e0214814.

Davidson LN, Krawchuk MA, Dulvy NK. 2016. Why have global shark and ray landings declined: improved management or overfishing? *Fish and Fisheries* 17(2):438–58.

De Bruyn P, Murua H, Aranda M. 2013. The precautionary approach to fisheries management: how this is taken into account by Tuna regional fisheries management organisations (RFMOs). *Marine Policy* 38:397–406.

De la Mare WK. 1998. Tidier fisheries management requires a new MOP (management-oriented paradigm). *Reviews in Fish Biology and Fisheries* 8:349–56.

De la Mare WK. 2000. Utilising data from ecosystem monitoring for managing fisheries: development of statistical summaries of indices arising from the CCAMLR ecosystem Monitoring Program. *CCAMLR Science* 7:101–17.

Dell'Apa A, Carney K, Davenport TM, Carle MV. 2018. Potential medium-term impacts of climate change on tuna and billfish in the Gulf of Mexico: a qualitative framework for management and conservation. *Marine Environmental Research* 141:1–11.

Dickey-Collas M. 2014. Why the complex nature of integrated ecosystem assessments requires a flexible and adaptive approach. *ICES Journal of Marine Science* 71(5):1174–82.

Dorn MW, Saunders MW, Wilson CD, Guttormsen MA, Cooke K, Kieser R, Wilkins ME. 1999. *Status of the Coastal Pacific Hake/Whiting Stock in US and Canada in 1998. Canadian Science Advisory Secretariat Research Document No. 99/90*. Ottawa ON: Fisheries and Oceans Canada.

Douvere F. 2008. The importance of marine spatial planning in advancing ecosystem-based sea use management. *Marine Policy* 32(5):762–71.

Du J, Park K. 2019. Estuarine salinity recovery from an extreme precipitation event: Hurricane Harvey in Galveston Bay. *Science of The Total Environment* 670: 1049–59.

Dulvy NK, Simpfendorfer CA, Davidson LN, Fordham SV, Bräutigam A, Sant G, Welch DJ. 2017. Challenges and priorities in shark and ray conservation. *Current Biology* 27(11):R565–72.

Ehrhardt N, Brown JE, Pohlot BG. 2017. *Western Central Atlantic Fishery Commission (WECAFC). Eight Session of the Scientific Advisory Group (SAG). Merida, Mexico, 3–4 November 2017*. WECAFC/SAG/VIII/2017/5. Rome, Italy: FOA.

Elmen J. 2003. Swimming upstream: a legal analysis of listing Atlantic Salmon as an endangered species. *Ocean & Coastal Law Journal* 9:333.

Ezer T, Atkinson LP, Corlett WB, Blanco JL. 2013. Gulf Stream's induced sea level rise and variability along the US mid-Atlantic coast. *Journal of Geophysical Research: Oceans* 118(2):685–97.

FAO (Food and Agricultural Organization of the United Nations). 1995. *Code of Conduct for Responsible Fisheries.* Rome, Italy: FAO (p. 41).

FAO (Food and Agricultural Organization of the United Nations). 2003. *Fisheries Management. The Ecosystem Approach to Fisheries. FAO Technical Guidelines for Responsible Fisheries*, 4 (Suppl. 2). Rome, Italy: FAO (p. 112).

FAO (Food and Agricultural Organization of the United Nations). 2009. *International Guidelines for the Management of Deep-sea Fisheries in the High Seas*. Rome, Italy: FAO (p. 73).

FAO (Food and Agricultural Organization of the United Nations). 2016. *Report of the Joint Meeting of Tuna RFMOs on the Implementation of the Ecosystem Approach to Fisheries Management*. December 12–14, 2016. Rome, Italy: FAO.

FAO (Food and Agricultural Organization of the United Nations). 2017. *Report of the Joint Meeting of Tuna RFMOs on the Implementation of the Ecosystem Approach to Fisheries Management. WCPFC Scientific Committee 13th Regular Session*. WCPFC-SC13-2017/EB-IP-10. Rome, Italy: FAO.

FAO (Food and Agricultural Organization of the United Nations). 2019. *Regional Fishery Bodies (RFB)*. Rome, Italy: FAO.

FAO (Food and Agricultural Organization of the United Nations). 2020. *A Review of the Application of the FAO Ecosystem Approach to Fisheries (EAF) Management by the Regional Fisheries Bodies Responsible for Areas Beyond National Jurisdiction (ABNJ)*. Rome, Italy: FAO.

Ferretti F, Worm B, Britten GL, Heithaus MR, Lotze HK. 2010. Patterns and ecosystem consequences of shark declines in the ocean. *Ecology Letters* 13(8):1055–71.

Field JC, Boesch DF, Scavia D, Buddemeier R, Burkett VR, Cayan D, Fogarty M, et al. 2001. *Potential Consequences of Climate Variability and Change on Coastal Areas and Marine Resources. Climate Change Impacts on the United States: The Potential Consequences of Climate Variability and Change, Report for the US Global Change Research Program*. Cambridge, UK: Cambridge University Press (pp. 461–87).

Field JC, Francis RC. 2006. Considering ecosystem-based fisheries management in the California Current. *Marine Policy* 30(5):552–69.

Flanagan PH, Jensen OP, Morley JW, Pinsky ML. 2019. Response of marine communities to local temperature changes. *Ecography* 42(1):214–24.

Fogarty MJ, Murawski SA. 1998. Large-scale disturbance and the structure of marine systems: fishery impacts on Georges Bank. *Ecological Applications* 8(sp1):S6–22.

Francis RC, Hare SR, Hollowed AB, Wooster WS. 1998. Effects of interdecadal climate variability on the oceanic ecosystems of the NE Pacific. *Fisheries Oceanography* 7(1):1–21.

Fromentin JM, Fonteneau A, Restrepo V. 2006. Ecosystem approach to fisheries: A brief overview and some considerations for its application in ICCAT. *Collective Volume of Scientific Papers* 59(2):682–9.

Fulton EA, Smith AD, Smith DC, Johnson P. 2014. An integrated approach is needed for ecosystem based fisheries management: insights from ecosystem-level management strategy evaluation. *PLoS One* 9(1):e84242.

Garcia SM. 2000. The FAO definition of sustainable development and the Code of Conduct for Responsible Fisheries: an analysis of the related principles, criteria and indicators. *Marine and Freshwater Research* 51(5):535–41.

Garcia SM. 2003. *The Ecosystem Approach to Fisheries: Issues, Terminology, Principles, Institutional Foundations, Implementation and Outlook*. FAO Fisheries Technical Paper. No. 443. Rome, Italy: FAO (p. 71).

Garcia SM, Cochrane KL. 2005. Ecosystem approach to fisheries: a review of implementation guidelines. *ICES Journal of Marine Science* 62(3):311–8.

Garibaldi L. 2012. The FAO global capture production database: a six-decade effort to catch the trend. *Marine Policy* 36(3):760–8.

Gascon V, Werner R. 2006. CCAMLR and Antarctic krill: ecosystem management around the great white continent. *Sustainable Development Law & Policy* 7:14.

Gibson RJ. 2017. Salient needs for conservation of Atlantic Salmon. *Fisheries* 42(3):163–74.

Gibson RN. 1994. Impact of habitat quality and quantity on the recruitment of juvenile flatfishes. *Netherlands Journal of Sea Research* 32(2):191–206.

Gilman E, Passfield K, Nakamura K. 2014. Performance of regional fisheries management organizations: ecosystem-based governance of bycatch and discards. *Fish and Fisheries* 15(2):327–51.

Gilman E, Allain EV, Collette BB, Hampton J, Lehodey P. 2016. "Impacts and effects of ocean warming on pelagic tunas." In: *Explaining Ocean Warming: Causes, Scale, Effects and Consequences*. Edited by D Laffoley, JM Baxter. Gland, Switzerland: IUCN (pp. 255–70).

Gilman E, Passfield K, Nakamura K. 2014. Performance of regional fisheries management organizations: ecosystem-based governance of bycatch and discards. *Fish and Fisheries* 15(2):327–51.

Godin AC, Worm B. 2010. Keeping the lead: how to strengthen shark conservation and management policies in Canada. *Marine Policy* 34(5):995–1001.

Gjerde KM. 2011. "High seas fisheries governance: prospects and challenges in the 21st century." In: *The World Ocean in Globalisation*. Edited by D Vidas, PJ Schei. Leiden, Netherlands: Martinus Nijhoff Publishers (pp. 221–232).

Gjerde KM, Currie D, Wowk K, Sack K. 2013. Ocean in peril: reforming the management of global ocean living resources in areas beyond national jurisdiction. *Marine Pollution Bulletin* 74(2):540–51.

Grafton RQ, Hannesson R, Shallard B, Sykes D, Terry J. 2006. *The Economics of Allocation in Tuna Regional Fisheries Management Organizations (RFMOs). Economics and Environment Network Working Paper EEN0612*. Canberra, Australia: Australian National University (p. 14).

Grant SM, Hill SL, Trathan PN, Murphy EJ. 2013. Ecosystem services of the Southern Ocean: trade-offs in decision-making. *Antarctic Science* 25(5):603–17.

Greene CH, Meyer-Gutbrod E, Monger BC, McGarry LP, Pershing AJ, Belkin IM, Fratantoni PS, et al. 2013. Remote climate forcing of decadal-scale regime shifts in Northwest Atlantic shelf ecosystems. *Limnology and Oceanography* 58(3):803–16.

Greene CH, Pershing AJ. 2000. The response of Calanus finmarchicus populations to climate variability in the Northwest Atlantic: basin-scale forcing associated with the North Atlantic Oscillation. *ICES Journal of Marine Science* 57(6):1536–44.

Gregory S, Li H, Li J. 2002. The conceptual basis for ecological responses to dam removal: resource managers face enormous challenges in assessing the consequences of removing large dams from rivers and evaluating management options. *BioScience* 52(8):713–23.

Grosholz ED, Ruiz GM. 1996. Predicting the impact of introduced marine species: lessons from the multiple invasions of the European green crab Carcinus maenas. *Biological Conservation* 78(1–2):59–66.

Grover JJ, Buckley TW, Woodbury D. 2002. Effects of the 1997–1998 El Niño on early-juvenile Pacific hake Merluccius productus: age, growth, abundance, and diet in coastal nursery habitats. *Marine Ecology Progress Series* 240:235–47.

Guerry AD. 2005. Icarus and Daedalus: conceptual and tactical lessons for marine ecosystem-based management. *Frontiers in Ecology and the Environment* 3(4):202–11.

Guldin M, Anderson CM. 2018. Catch shares and shoreside processors: a costs and earnings exploration into the downstream sector. *Marine Resource Economics* 33(3): 289–307.

Gutiérrez D, Akester M, Naranjo L. 2016. Productivity and sustainable management of the Humboldt Current large marine ecosystem under climate change. *Environmental Development* 17:126–44.

Haas B, McGee J, Fleming A, Haward M. 2020. Factors influencing the performance of regional fisheries management organizations. *Marine Policy* 113:103787.

Halliday RG, Pinhorn AT. 1996. North Atlantic fishery management systems: a comparison of management methods and resource trends. *Journal of Northwest Atlantic Fishery Science* 20.

Halpern BS, Kappel CV, Selkoe KA, Micheli F, Ebert CM, Kontgis C, Crain CM, et al. 2009. Mapping cumulative human impacts to California Current marine ecosystems. *Conservation Letters* 2(3):138–48.

Hamel OS, Ressler PH, Thomas RE, Waldeck DA, Hicks AC, Holmes JA, Fleischer GW. 2015. "Biology, fisheries, assessment and management of Pacific hake (Merluccius productus)." In: *Hakes: Biology and Exploitation*. Edited by H Arancibia. New York, NY: Wiley (pp. 234–62).

Hamel OS, Stewart IJ. 2009. *Stock Assessment of Pacific Hake, Merluccius Productus, (aka Whiting) in US and Canadian Waters in 2009*. Portland, OR: Pacific Fishery Management Council.

Hansen LP, Hutchinson P, Reddin DG, Windsor ML. 2012. Salmon at sea: scientific advances and their implications for management: an introduction. *ICES Journal of Marine Science* 69(9):1533–7.

Hansen LP, Windsor ML. 2006. Interactions between aquaculture and wild stocks of Atlantic Salmon and other diadromous fish species: science and management, challenges and solutions: an introduction by the conveners. *ICES Journal of Marine Science* 63(7):1159–61.

Hansen S, Ward P, Penney A. 2013. *Identification of Vulnerable Benthic Taxa in the Western SPRFMO Convention Area and Review of Move-on Rules for Different Gear Types. SPRFMO Paper SC-01-09*. Canberra, Australia: Department of Agriculture.

Hare JA, Kocik JF, Link JS. 2019. Atlantic Salmon recovery informing and informed by ecosystem-based fisheries management. *Fisheries* 44(9):403–11.

Hare SR, Mantua NJ, Francis RC. 1999. Inverse production regimes: Alaska and west coast Pacific Salmon. *Fisheries* 24(1):6–14.

Harvey CJ, Garfield N, Williams GD, Tolimieri N, Schroeder I, Hazen EL, Andrews KS, et al. 2018. *Ecosystem Status Report of the California Current for 2018: A Summary of Ecosystem Indicators Compiled by the California Current Integrated Ecosystem Assessment Team (CCIEA). NOAA Technical Memorandum NMFS-NWFSC-145*. Washington, DC: NOAA.

Harvey CJ, Gross K, Simon VH, Hastie J. 2008. Trophic and fishery interactions between Pacific hake and rockfish: effect on rockfish population rebuilding times. *Marine Ecology Progress Series* 365:165–76.

Haward M. 2017. Contemporary challenges to the Antarctic Treaty and Antarctic Treaty System: Australian interests, interplay and the evolution of a regime complex. *Australian Journal of Maritime & Ocean Affairs* 9(1):21–4.

Hazen EL, Jorgensen S, Rykaczewski RR, Bograd SJ, Foley DG, Jonsen ID, Shaffer SA, et al. 2013. Predicted habitat shifts of Pacific top predators in a changing climate. *Nature Climate Change* 3(3):234.

Heazle M. 2004. Scientific uncertainty and the International Whaling Commission: an alternative perspective on the use of science in policy making. *Marine Policy* 28(5):361–74.

Hedley C. 2001. The 1998 Agreement on the International Dolphin Conservation Program: recent developments in the tuna-dolphin controversy in the eastern Pacific ocean. *Ocean Development & International Law* 32(1):71–92.

Heithaus MR, Frid A, Vaudo JJ, Worm B, Wirsing AJ. 2020. "Unraveling the ecological importance of elasmobranchs." In: *Sharks and Their Relatives II*. Edited by M Heithaus, JC Carrier, JA Musick. Boca Raton, FL: CRC Press (pp. 627–54).

Heithaus MR, Frid A, Wirsing AJ, Worm B. 2008. Predicting ecological consequences of marine top predator declines. *Trends in Ecology & Evolution* 23(4):202–10.

Helser TE, Alade L. 2012. A retrospective of the hake stocks off the Atlantic and Pacific coasts of the United States: uncertainties and challenges facing assessment and management in a complex environment. *Fisheries Research* 114:2–18.

Herrera M, Báez JC. 2018. *On the Potential Biases of Scientific Estimates of Catches of Tropical Tunas of Purse Seiners Monitored by European Scientists and Catches Reported to the ICCAT and IOTC. IOTC–2018–WPTT20–17*. Rome, Italy: FAO.

Hewitt RP, Watkins JL, Naganobu M, Tshernyshkov P, Brierley AS, Demer DA, Kasatkina S, et al. 2002. Setting a precautionary catch limit for Antarctic krill. *Oceanography* 15(3):26–33.

Hill SL, Atkinson A, Darby C, Fielding S. 2016. Is current management of the Antarctic krill fishery in the Atlantic sector of the Southern Ocean precautionary? *CCAMLR Science* 23:17–30.

Hill SL, Murphy EJ, Reid K, Trathan PN, Constable AJ. 2006. Modelling Southern Ocean ecosystems: krill, the food-web, and the impacts of harvesting. *Biological Reviews* 81:581–608.

Hillary RM, Preece AL, Davies CR, Kurota H, Sakai O, Itoh T, Parma AM, et al. 2016. A scientific alternative to moratoria for rebuilding depleted international tuna stocks. *Fish and Fisheries* 17(2):469–82.

Hinke J, Kaplan I, Aydin K, Watters G, Olson R, Kitchell JF. 2004. Visualizing the food-web effects of fishing for tunas in the Pacific Ocean. *Ecology and Society* 9(1).

Hoag H. 2017. Nations agree to ban fishing in Arctic Ocean for at least 16 years. *Science* 1: aar6437.

Hobday AJ, Arrizabalaga H, Evans K, Nicol S, Young JW, Weng KC. 2015. Impacts of climate change on marine top predators: advances and future challenges. *Deep Sea Research Part II: Topical Studies in Oceanography* 113:1–8.

Hobday AJ, Evans K. 2013. Detecting climate impacts with oceanic fish and fisheries data. *Climatic Change* 119(1):49–62.

Hobday AJ, Young JW, Abe O, Costa DP, Cowen RK, Evans K, Gasalla MA, et al. 2013. Climate impacts and oceanic top predators: moving from impacts to adaptation in oceanic systems. *Reviews in Fish Biology and Fisheries* 23(4):537–46.

Hoel AH. 2015. "Oceans governance, the Arctic Council and ecosystem-based management." In: *Handbook of the Politics of the Arctic*. Edited by G Hønneland, LC Jensen. Cheltenham, UK: Edward Elgar Publishing (p. 265).

Holbrook CM, Zydlewski J, Gorsky D, Shepard SL, Kinnison MT. 2009. Movements of prespawn adult Atlantic salmon near hydroelectric dams in the lower Penobscot River, Maine. *North American Journal of Fisheries Management* 29(2):495–505.

Hollowed AB, Hare SR, Wooster WS. 2001. Pacific Basin climate variability and patterns of Northeast Pacific marine fish production. *Progress in Oceanography* 49(1–4):257–82.

Holt CA, Punt AE. 2009. Incorporating climate information into rebuilding plans for overfished groundfish species of the US west coast. *Fisheries Research* 100(1):57–67.

Horodysky AZ, Cooke SJ, Graves JE, Brill RW. 2016. Conservation physiology of marine fishes fisheries conservation on the high seas: linking conservation physiology and fisheries ecology for the management of large pelagic fishes. *Conservation Physiology* 4(1): cov059.

Hourigan TF. 2009. Managing fishery impacts on deep-water coral ecosystems of the USA: emerging best practices. *Marine Ecology Progress Series* 397:333–40.

Huang HW, Swimmer Y, Bigelow K, Gutierrez A, Foster DG. 2016. Influence of hook type on catch of commercial and bycatch species in an Atlantic tuna fishery. *Marine Policy* 65:68–75.

Hughes J. 2011. *The Piracy-Illegal Fishing Nexus in the Western Indian Ocean*. West Perth, Australia: Future Directions International Strategic Analysis Paper.

Huntsman GR. 1994. Endangered marine finfish: neglected resources or beasts of fiction? *Fisheries* 19(7):8–15.

IASRB (International Atlantic Salmon Research Board). 2001a. *Report of the Inaugural Meeting of the International Cooperative Salmon Research Board Thistle Kensington Park Hotel, London 5–7 December 2001*. Edinburgh, UK: North Atlantic Salmon Conservation Organization (p. 21).

IASRB (International Atlantic Salmon Research Board). 2001b. *Rules of Procedure for the International Atlantic Salmon Research Board*. Edinburgh, UK: North Atlantic Salmon Conservation Organization (p. 2).

IATTC (Inter-American Tropical Tuna Convention). 2003. *Convention for the Strengthening of the Inter-American Tropical Tuna Commission Established by the 1949 Convention Between the United States of America and the Republic of Costa Rica ("Antigua Convention")*. La Jolla, CA: IATTC (p. 21).

IATTC (Inter-American Tropical Tuna Convention). 2017. *Agreement on the International Dolphin Conservation Program*. La Jolla, CA: IATTC (p. 23).

ICCAT (International Convention for the Conservation of Atlantic Tunas). 1966. International Convention for the Conservation of Atlantic Tunas Final Act of the Conference of Plenipotentiaries on the Conservation of Atlantic Tunas. Rio de Janeiro, Brazil: ICCAT (p. 9).

ICCAT (International Commission for the Conservation of Atlantic Tunas). 2015. *Strategic Plan Report 323, Appendix 10. 2015–2020 SCRS Science Strategic Plan*. Rio de Janeiro, Brazil: ICCAT (pp. 323–43).

ICCAT (International Convention for the Conservation of Atlantic Tunas). 2019. *Report of the 2019 ICCAT Sub-Committee on Ecosystems Meeting. Madrid, Spain 8–12 April*. Rio de Janeiro, Brazil: ICCAT (p. 28).

IPHC (International Pacific Halibut Commission). 1979. *Protocol Amending the Convention Between the United States of America and Canada for the Preservation of the Halibut*

Fishery of the Northern Pacific Ocean and Bering Sea. Washington, DC: IPHC (p. 36).

IPHC (International Pacific Halibut Commission). 2019. *International Pacific Halibut Commission Strategic Plan IPHC–2019–SP (2019–23), Seattle, WA.* Washington, DC: IPHC (p. 7).

IPSC (International Pacific Salmon Commission). 1945. *Annual Report 1944.* New Westminster, Canada: IPSC (p. 74).

Ishimura G, Punt AE, Huppert DD. 2005. Management of fluctuating fish stocks: the case of Pacific whiting. *Fisheries Research* 73(1–2):201–16.

IUCN (International Union for the Conservation of Nature). 2020. *Vision and Mission.* Bangkok, Thailand: IUCN. https://www.iucn.org/regions/asia/about/vision-and-mission

IUCN-CEM (International Union for the Conservation of Nature—Commission on Ecosystem Management). 2017. *Mandate 2017–2020.* Bangkok, Thailand: IUCN (p. 3). https://www.iucn.org/sites/dev/files/content/documents/cem_mandate.pdf

IWC (International Whaling Commission). 2019. *International Whaling Commission.* Impington, UK: IWC.

IWC (International Whaling Commission). 2005. International Convention for the Regulation of Whaling, Annual Report of the International Whaling Commission, 2005. Impington, UK: IWC (pp. 157–68)

Jamieson G, Zhang CI. 2005. *Report of the Study Group on "Ecosystem-Based Management Science and its Application to the North Pacific". PICES Scientific Report No. 29.* Sidney, Canada: PICES (p. 77).

Jannot JE, Somers KA, Tuttle V, McVeigh J, Carretta JV, Helker V. 2018. *Observed and Estimated Marine Mammal Bycatch in US West Coast Groundfish Fisheries, 2002–16. NWFSC Processed Report 2018-03.* Washington, DC: NOAA.

Jefferies CS. 2018. International whale conservation in a changing climate: the ecosystem approach, marine protected areas, and the International Whaling Commission. *Journal of International Wildlife Law & Policy* 21(4):239–80.

Jenkins D. 2003. Atlantic salmon, endangered species, and the failure of environmental policies. *Comparative Studies in Society and History* 45(4):843–72.

Johnson DH, Shaffer TL, Gould PJ. 1993. Incidental catch of marine birds in the North Pacific high seas driftnet fisheries in 1990. *International North Pacific Fisheries Commission Bulletin* 53:473–83.

Jonsson B, Jonsson N. 2004. Factors affecting marine production of Atlantic salmon (Salmo salar). *Canadian Journal of Fisheries and Aquatic Sciences* 61(12):2369–83.

Joseph EB. 1972. The status of the sciaenid stocks of the middle Atlantic coast. *Chesapeake Science* 13(2):87–100.

Joseph J. 1994. The tuna-dolphin controversy in the eastern Pacific Ocean: Biological, economic, and political impacts. *Ocean Development & International Law* 25(1):1–30.

Juan-Jordá MJ, Arrizabalaga H, Restrepo V, Dulvy NK, Cooper AB, Murua H. 2015. *Preliminary Review of ICCAT, WCPFC, IOTC and IATTC Progress in Applying Ecosystem Based Fisheries Management. IOTC Technical Report IOTC-2015-WPEB11-40.* Rome, Italy: FAO (p. 70).

Juan-Jordá MJ, Murua H, Arrizabalaga H, Dulvy NK, Restrepo V. 2018. Report card on ecosystem-based fisheries management in tuna regional fisheries management organizations. *Fish and Fisheries* 19(2):321–39.

Juanes F, Gephard S, Beland KF. 2004. Long-term changes in migration timing of adult Atlantic salmon (Salmo salar) at the southern edge of the species distribution. *Canadian Journal of Fisheries and Aquatic Sciences* 61(12):2392–400.

Kaplan IC, Gray IA, Levin PS. 2013. Cumulative impacts of fisheries in the California Current. *Fish and Fisheries* 14(4):515–27.

Kaplan IC, Horne PJ, Levin PS. 2012. Screening California Current fishery management scenarios using the Atlantis end-to-end ecosystem model. *Progress in Oceanography* 102:5–18.

Kaplan IC, Levin P. 2009. "Ecosystem-based management of what? An emerging approach for balancing conflicting objectives in marine resource management." In: *The Future of Fisheries Science in North America.* Edited by RJ Beamish, BJ Rothschild. Dordrecht, Netherlands: Springer (pp. 77–95).

Karnauskas M, Kelble CR, Regan S, Quenée C, Allee R, Jepson M, Freitag A, et al. 2017 *Ecosystem Status Report Update for the Gulf of Mexico. NOAA Technical Memorandum NMFS-SEFSC-706.* Washington, DC: NOAA.

Karnauskas M, Schirripa MJ, Kelble CR, Cook GS, Craig JK. 2013. *Ecosystem Status Report for the Gulf of Mexico. NOAA Technical Memorandum NMFS-SEFSC 653.* Washington, DC: NOAA (p. 52).

Keefer ML, Peery CA, Caudill CC. 2008. Migration timing of Columbia River spring Chinook salmon: effects of temperature, river discharge, and ocean environment. *Transactions of the American Fisheries Society* 137(4):1120–33.

Keith R. 2018. *Climate Change Impacts, Vulnerabilities and Adaptations: Southern Ocean Marine Fisheries. Impacts of Climate Change on Fisheries and Aquaculture: Synthesis of Current Knowledge, Adaptation and Mitigation Options. FAO Fisheries Technical Paper 2018 No. 627.* Rome, Italy: FAO.

Kelble CR, Loomis DK, Lovelace S, Nuttle WK, Ortner PB, Fletcher P, Cook GS, Lorenz JJ, Boyer JN. 2013. The EBM-DPSER conceptual model: integrating ecosystem services into the DPSIR framework. *PLoS One* 8(8):e70766.

Keller AA, Simon V, Chan F, Wakefield WW, Clarke ME, Barth JA, Kamikawa DA, Fruh EL. 2010. Demersal fish and invertebrate biomass in relation to an offshore hypoxic zone along the US West Coast. *Fisheries Oceanography* 19(1):76–87.

Kerstetter DW, Graves JE. 2006. Effects of circle versus J-style hooks on target and non-target species in a pelagic longline fishery. *Fisheries Research* 80(2–3):239–50.

Kircheis D, Liebich T. 2007. *Habitat Requirements and Management Considerations for Atlantic Salmon (Salmo*

Salar) in the Gulf of Maine Distinct Population Segment. Orono, ME: National Marine Fisheries Service (p. 132).

Kitchell JF, Boggs CH, He X, Walters CJ. 1999. *Keystone Predators in the Central Pacific. Ecosystem Approaches for Fisheries Management.* Fairbanks, AK: University of Alaska Sea Grant (pp. 665–83).

Kitchell JF, Essington TE, Boggs CH, Schindler DE, Walters CJ. 2002. The role of sharks and longline fisheries in a pelagic ecosystem of the central Pacific. *Ecosystems* 5(2):202–16.

Kitchell JF, Martell SJ, Walters CJ, Jensen OP, Kaplan IC, Watters J, Essington TE, Boggs CH. 2006. Billfishes in an ecosystem context. *Bulletin of Marine Science* 79(3):669–82.

Kite-Powell HL, Rubino M, Morehead B. 2013. The future of US seafood supply. *Aquaculture Economics and Management* 17(3):228–50.

Knudsen EE, McDonald D. 2019. *Sustainable Fisheries Management: Pacific Salmon.* Boca Raton, FL: CRC Press.

Kobayashi DR, Polovina JJ, Parker DM, Kamezaki N, Cheng IJ, Uchida I, Dutton PH, Balazs GH. 2008. Pelagic habitat characterization of loggerhead sea turtles, Caretta caretta, in the North Pacific Ocean (1997–2006): insights from satellite tag tracking and remotely sensed data. *Journal of Experimental Marine Biology and Ecology* 356(1–2):96–114.

Kock KH, Reid K, Croxall J, Nicol S. 2007. Fisheries in the Southern Ocean: an ecosystem approach. *Philosophical Transactions of the Royal Society B: Biological Sciences* 362(1488):2333–49.

Koen-Alonso M, Pepin P, Fogarty MJ, Kenny A, Kenchington E. 2019. The Northwest Atlantic Fisheries Organization roadmap for the development and implementation of an ecosystem approach to fisheries: structure, state of development, and challenges. *Marine Policy* 100:342–52.

Koslow JA, Auster P, Bergstad OA, Roberts JM, Rogers A, Vecchione M, Harris PT, Rice J, Bernal P. 2015. *Biological Communities on Seamounts and Other Submarine Features Potentially Threatened by Disturbance. United Nations World Ocean Assessment.* Cambridge, UK: Cambridge University Press.

Krueger CC, Zimmerman CE. 2009. *Pacific Salmon: Ecology and Management of Western Alaska's Populations.* Bethesda, MD: American Fisheries Society Symposium.

Lam VYY, Sadovy de Mitcheson Y. 2011. The sharks of South East Asia - unknown, unmonitored and unmanaged. *Fish and Fisheries* 12:51–74.

Lascelles B, Notarbartolo Di Sciara G, Agardy T, Cuttelod A, Eckert S, Glowka L, Hoyt E, et al. 2014. Migratory marine species: their status, threats and conservation management needs. *Aquatic Conservation: Marine and Freshwater Ecosystems* 24(S2):111–27.

Law KL, Thompson RC. 2014. Microplastics in the seas. *Science* 345(6193):144–5.

Lefort S, Aumont O, Bopp L, Arsouze T, Gehlen M, Maury O. 2015. Spatial and body-size dependent response of marine pelagic communities to projected global climate change. *Global Change Biology* 21(1):154–64.

Lehodey P, Senina I, Calmettes B, Hampton J, Nicol S. 2013. Modelling the impact of climate change on Pacific skipjack tuna population and fisheries. *Climatic Change* 119(2013): 95–109.

Lehodey P, Senina I, Dragon AC, Arrizabalaga H. 2014. Spatially explicit estimates of stock size, structure and biomass of North Atlantic albacore tuna (Thunnus alalunga). *Earth System Science Data* 7:169–95.

Leroy A, Morin M. 2018. Innovation in the decision-making process of the RFMOs. *Marine Policy* 97:156–62.

Levin PS, Essington TE, Marshall KN, Koehn LE, Anderson LG, Bundy A, Carothers C, et al. 2018. Building effective fishery ecosystem plans. *Marine Policy* 92:48–57.

Levin PS, Fogarty MJ, Murawski SA, Fluharty D. 2009. Integrated ecosystem assessments: developing the scientific basis for ecosystem-based management of the ocean. *PLoS Biology* 7(1):e1000014.

Lewison R, Hobday AJ, Maxwell S, Hazen E, Hartog JR, Dunn DC, Briscoe D, et al. 2015. Dynamic ocean management: identifying the critical ingredients of dynamic approaches to ocean resource management. *BioScience* 65(5):486–98.

Lewison RL, Crowder LB, Read AJ, Freeman SA. 2004. Understanding impacts of fisheries bycatch on marine megafauna. *Trends in Ecology & Evolution* 19(11):598–604.

Lezama-Ochoa N, Murua H, Chust G, Van Loon E, Ruiz J, Hall M, Chavance P, Delgado De Molina A, Villarino E. 2016. Present and future potential habitat distribution of Carcharhinus falciformis and Canthidermis maculata by-catch species in the tropical tuna purse-seine fishery under climate change. *Frontiers in Marine Science* 3:34.

Libralato S, Pranovi F, Zucchetta M, Monti MA, Link JS. 2019. Global thresholds in properties emerging from cumulative curves of marine ecosystems. *Ecological Indicators* 103:554–62.

Limburg KE, Waldman JR. 2009. Dramatic declines in North Atlantic diadromous fishes. *BioScience* 59(11):955–65.

Link JS. 2018. System-level optimal yield: increased value, less risk, improved stability, and better fisheries. *Canadian Journal of Fisheries and Aquatic Sciences* 75(1):1–6.

Link JS, Bundy A, Overholtz WJ, Shackell N, Manderson J, Duplisea D, Hare J, Koen-Alonso M, Friedland KD. 2011. Ecosystem-based fisheries management in the Northwest Atlantic. *Fish and Fisheries* 12(2):152–70.

Link JS, Marshak AR. 2019. Characterizing and comparing marine fisheries ecosystems in the United States: determinants of success in moving toward ecosystem-based fisheries management. *Reviews in Fish Biology and Fisheries* 29(1):23–70.

Link JS, Watson RA. 2019. Global ecosystem overfishing: clear delineation within real limits to production. *Science Advances* 5(6):eaav0474.

Loher T. 2011. Analysis of match–mismatch between commercial fishing periods and spawning ecology of Pacific halibut (Hippoglossus stenolepis), based on winter surveys and behavioural data from electronic archival tags. *ICES Journal of Marine Science* 68(10):2240–51.

Lomeli MJ, Wakefield WW. 2014. *Examining the Potential Use of Artificial Illumination to Enhance Chinook Salmon Escapement Out a Bycatch Reduction Device in a Pacific Hake Midwater Trawl*. Seattle, WA: National Marine Fisheries Service, Northwest Fisheries Science Center Report.

Lomeli MJ, Wakefield WW. 2016. Evaluation of a sorting grid bycatch reduction device for the selective flatfish bottom trawl in the US West Coast fishery. *Fisheries Research* 183:294–303.

Long RD, Charles A, Stephenson RL. 2015. Key principles of marine ecosystem-based management. *Marine Policy* 57:53–60.

Lucas KL, Carter GA. 2013. Change in distribution and composition of vegetated habitats on Horn Island, Mississippi, northern Gulf of Mexico, in the initial five years following Hurricane Katrina. *Geomorphology* 199:129–37.

Lugo AE, Rogers CS, Nixon SW. 2000. Hurricanes, coral reefs and rainforests: resistance, ruin and recovery in the Caribbean. *AMBIO: A Journal of the Human Environment* 29(2):106–15.

Lugo-Fernández A, Hernández-Avila ML, Roberts HH. 1994. Wave-energy distribution and hurricane effects on Margarita Reef, southwestern Puerto Rico. *Coral Reefs* 13(1):21–32.

Lusher AL, Hollman PCH, Mendoza-Hill JJ. 2017. *Microplastics in Fisheries and Aquaculture: Status of Knowledge on Their Occurrence and Implications for Aquatic Organisms and Food Safety. FAO Fisheries and Aquaculture Technical Paper No. 615*. Rome, Italy: FAO.

Lynch PD, Graves JE, Latour RJ. 2011. *Challenges in the Assessment and Management of Highly Migratory Bycatch Species: A Case Study of the Atlantic Marlins. Sustainable Fisheries: Multilevel Approaches to a Global Problem*. Bethesda, MD: American Fisheries Society (pp. 197–226).

Lynch PD, Methot RD, Link JS. 2018. *Implementing a Next Generation Stock Assessment Enterprise. An Update to the NOAA Fisheries Stock Assessment Improvement Plan. NOAA Technical Memorandum NMFS-F/ SPO-183*. Washington, DC: NOAA (p. 127).

Lynch PD, Shertzer KW, Cortés E, Latour RJ. 2018. Handling editor: Manuel Hidalgo. Abundance trends of highly migratory species in the Atlantic Ocean: accounting for water temperature profiles. *ICES Journal of Marine Science* 75(4):1427–38.

McFarlane GA, King JR, Beamish RJ. 2000. Have there been recent changes in climate? Ask the fish. *Progress in Oceanography* 47(2–4):147–69.

Magnuson WG. 1976. The Fishery Conservation and Management Act of 1976: first step toward improved management of marine fisheries. *Washington Law Review* 52:427.

McDorman T. 2005. Implementing existing tools: turning words into actions, decision-making processes of regional fisheries management organisations (RFMOs). *The International Journal of Marine and Coastal Law* 20(3):423–57.

Maguire JJ. 2006. *The State of World Highly Migratory, Straddling and Other High Seas Fishery Resources and Associated Species*. Rome, Italy: FAO.

Mandelman JW, Cooper PW, Werner TB, Lagueux KM. 2008. Shark bycatch and depredation in the US Atlantic pelagic longline fishery. *Reviews in Fish Biology and Fisheries* 18(4):427.

Marashi SH. 1996. *Summary Information on the Role of International Fishery and Other Bodies with Regard to the Conservation and Management of Living Resources of the High Seas*. Rome, Italy: FAO.

Marshak AR, Link JS, Shuford R, Monaco ME, Johannesen E, Bianchi G, Anderson MR, et al. 2017. International perceptions of an integrated, multi-sectoral, ecosystem approach to management. *ICES Journal of Marine Science* 74(1):414–20.

Marshall KN, Levin PS, Essington TE, Koehn LE, Anderson LG, Bundy A, Carothers C, et al. 2018. Ecosystem-based fisheries management for social–ecological systems: renewing the focus in the United States with next generation fishery ecosystem plans. *Conservation Letters* 11(1):e12367.

Martell S, Stewart IJ, Leaman BM. 2013. *Optimal Harvest Rates for Pacific Halibut. International Pacific Halibut Commission Eighty-Ninth Annual Meeting*. Washington, DC: IPHC.

Maunder MN, Harley SJ. 2006. Evaluating tuna management in the eastern Pacific Ocean. *Bulletin of Marine Science* 78(3):593–606.

Maxwell SM, Hazen EL, Lewison RL, Dunn DC, Bailey H, Bograd SJ, Briscoe DK, et al. 2015. Dynamic ocean management: defining and conceptualizing real-time management of the ocean. *Marine Policy* 58:42–50.

McBride MM, Dalpadado P, Drinkwater KF, Godø OR, Hobday AJ, Hollowed AB, Kristiansen T, et al. 2014. Krill, climate, and contrasting future scenarios for Arctic and Antarctic fisheries. *ICES Journal of Marine Science* 71(7):1934–55.

McCaughran DA, Hoag SH. 1992. *The 1979 Protocol to the Convention and Related Legislation*. Washington, DC: IPHC.

McDorman TL. 1994. Stateless fishing vessels, international law and the UN high seas fisheries conference. *Journal of Maritime Law and Commerce* 25:531.

McFarlane GA, King JR, Beamish RJ. 2000. Have there been recent changes in climate? Ask the fish. *Progress in Oceanography* 47(2–4):147–69.

Meltzer E. 1994. Global overview of straddling and highly migratory fish stocks: the nonsustainable nature of high seas fisheries. *Ocean Development & International Law* 25(3):255–344.

Menezes GM, Giacomello E. 2017. "High seas deep-sea fisheries under the global changing trends." In: *Marine Pollution and Climate Change*. Edited by A Arias, JE Marcovecchio. Boca Raton, FL: CRC Press (pp. 180–236).

Milazzo MJ. 2012. Progress and problems in US marine fisheries rebuilding plans. *Reviews in Fish Biology and Fisheries* 22(1):273–96.

Miller D. 2011. "Climate change and the management of high seas, straddling and migratory fish stocks." In: *The Economics of Adapting Fisheries to Climate Change*. Edited by the Organisation for Economic Co-operation and Development. Paris, France: OECD Publishing.

Miller D, Sabourenkov E, Ramm D. 2004. Managing Antarctic marine living resources: the CCAMLR approach. *The International Journal of Marine and Coastal Law* 19(3):317–63.

Miller KA. 2000. "Pacific salmon fisheries: climate, information and adaptation in a conflict-ridden context." In: *Societal Adaptation to Climate Variability and Change*. Edited by G Yohe, SM Kane. Dordrecht, Netherlands: Springer (pp. 37–61).

Miller KA. 2003. *North American Pacific salmon: A Case of Fragile Cooperation. Norway-FAO Expert Consultation on the Management of Shared Fish Stocks, Bergen, Norway*. Rome, Italy: FAO (pp. 105–22).

Miller KA. 2007. Climate variability and tropical tuna: Management challenges for highly migratory fish stocks. *Marine Policy* 31(1):56–70.

Miller KA, Munro GR, McDorman TL, McKelvey R, Tyedmers P. 2001. *The 1999 Pacific Salmon Agreement: A Sustainable Solution? Canadian-American Public Policy Occasional Paper 2001*. Orono, ME: University of Maine.

Mills D, Piggins D. 2012. *Atlantic Salmon: Planning for the Future the Proceedings of the Third International Atlantic Salmon Symposium—Held in Biarritz, France, October 21–23, 1986*. Berlin, Germany: Springer Science & Business Media.

Moffitt EA, Punt AE, Holsman K, Aydin KY, Ianelli JN, Ortiz I. 2016. Moving towards ecosystem-based fisheries management: options for parameterizing multi-species biological reference points. *Deep Sea Research Part II: Topical Studies in Oceanography* 134:350–9.

Molenaar EJ. 2001. CCAMLR and southern ocean fisheries. *The International Journal of Marine and Coastal Law* 16(3):465–99.

Molenaar EJ. 2012. Current and prospective roles of the Arctic Council system within the context of the law of the sea. *The International Journal of Marine and Coastal Law* 27(3):553–95.

Monteiro S, Vázquez X, Long R. 2010. Improving fishery law enforcement in marine protected areas. *Aegean Review of the Law of the Sea and Maritime Law* 1(1):95–109.

Mooney-Seus ML, Rosenberg AA. 2007. *Regional Fisheries Management Organizations: Progress in Adopting Precautionary Approach and Ecosystem-Based Management*. Rome, Italy: FAO.

Morton A, Routledge R, Krkosek M. 2008. Sea louse infestation in wild juvenile salmon and Pacific herring associated with fish farms off the east-central coast of Vancouver Island, British Columbia. *North American Journal of Fisheries Management* 28(2):523–32.

Morton A, Volpe J. 2002. A description of escaped farmed Atlantic salmon Salmo salar captures and their characteristics in one Pacific salmon fishery area in British Columbia, Canada, in 2000. *Alaska Fishery Research Bulletin* 9(2):102–10.

Moustahfid H, Marsac F, Gangopadhyay A. 2018. *Climate Change Impacts, Vulnerabilities and Adaptations: Western Indian Ocean Marine Fisheries. Impacts of Climate Change on Fisheries and Aquaculture: Synthesis of Current Knowledge, Adaptation and Mitigation Options*. FAO Fisheries Technical Paper No. 627. Rome, Italy: FAO.

Muhling BA, Liu J, Lee SK, Lamkin JT, Roffer MA, Muller-Karger F, Walter III JF. 2015. Potential impact of climate change on the Intra-Americas sea: part 2. Implications for Atlantic bluefin tuna and skipjack tuna adult and larval habitats. *Journal of Marine Systems* 148:1–13.

Muto MM, Helker VT, Angliss RP, Allen BA, Boveng PL, Breiwick JM, Cameron MF, et al. 2017. *Alaska Marine Mammal Stock Assessments, 2016. NOAA Technical Memorandum NMFS-AFSC-355*. Washington, DC: NOAA (p. 366).

Muto MM, Helker VT, Angliss RP, Allen BA, Boveng PL, Breiwick JM, Cameron MF, et al. 2018. *Alaska Marine Mammal Stock Assessments, 2017. NOAA Technical Memorandum NMFS-AFSC-378*. Washington, DC: NOAA (p. 382).

Muto MM, Helker VT, Angliss RP, Boveng PL, Breiwick JM, Cameron MF, Clapham PJ, et al. 2019. *Alaska Marine Mammal Stock Assessments, 2018. NOAA Technical Memorandum NMFS-AFSC-393*. Washington, DC: NOAA (p. 390).

NAFO (Northwest Atlantic Fisheries Organization). 2009. *NAFO Perspectives*. Dartmouth, Canada: NAFO (p. 18).

NAFO (Northwest Atlantic Fisheries Organization). 2017. *Convention on Cooperation in the Northwest Atlantic Fisheries*. Dartmouth, Canada: NAFO (p. 38).

NAFO (Northwest Atlantic Fisheries Organization). 2019. *Conservation and Enforcement Measures 2019*. Dartmouth, Canada: NAFO (p. 181).

NAFO (Northwest Atlantic Fisheries Organization). 2018. *Report of the NAFO Joint Commission-Scientific Council Working Group on Ecosystem Approach Framework to Fisheries Management (WG-EAFFM) Meeting. NAFO/COM-SC Doc*. 18-03. Dartmouth, Canada: NAFO.

Najjar RG, Pyke CR, Adams MB, Breitburg D, Hershner C, Kemp M, Howarth R, et al. 2010. Potential climate-change impacts on the Chesapeake Bay. *Estuarine, Coastal and Shelf Science* 86(1):1–20.

NASCO (North Atlantic Salmon Conservation Organization). 1983. *The Convention for the Conservation of Salmon in the North Atlantic Ocean*. Edinburgh, UK: NASCO (p. 11).

NASCO (North Atlantic Salmon Conservation Organization). 1998. *Agreement on Adoption of a Precautionary Approach*. Edinburgh, UK: NASCO (p. 4).

NASCO (North Atlantic Salmon Conservation Organization). 1999. *Action Plan for Application of the Precautionary Approach*. Edinburgh, UK: NASCO (p. 14).

NASCO (North Atlantic Salmon Conservation Organization). 2005. *Strategic Approach for NASCO's 'Next Steps'*. Edinburgh, UK: NASCO (p. 9).

Nieland JL, Sheehan TF, Saunders R. 2015. Assessing demographic effects of dams on diadromous fish: a case study for Atlantic salmon in the Penobscot River, Maine. *ICES Journal of Marine Science* 72(8):2423–37.

Nielsen JK, Hooge PN, Taggart SJ, Seitz AC. 2014. Characterizing Pacific halibut movement and habitat in a marine protected area using net squared displacement analysis methods. *Marine Ecology Progress Series* 517: 229–50.

Nitta ET, Henderson JR. 1993. A review of interactions between Hawaii's fisheries and protected species. *Marine Fisheries Review* 55(2):83–92.

NMFS (National Marine Fisheries Service). 2006. *Final Consolidated Atlantic Highly Migratory Species Fishery Management Plan*. Silver Spring, MD: NOAA (p. 1600).

NMFS (National Marine Fisheries Service). 2012. *Supplemental Environmental Assessment Implementation of the Western and Central Pacific Fisheries Commission Bigeye Tuna Catch Limits for Longline Fisheries in 2012*. Silver Spring, MD: NOAA (p. 75).

NMFS (National Marine Fisheries Service). 2014. *Supplemental Environmental Assessment for a Rule to Implement Decisions of the Western and Central Pacific Fisheries Commission for Restrictions on the Use of Fish Aggregating Devices in Purse Seine Fisheries for 2015*. Silver Spring, MD: NOAA (p. 87).

NMFS (National Marine Fisheries Service). 2016a. *Ecosystem-Based Fisheries Management Policy of the National Marine Fisheries Service*. Silver Spring, MD: NOAA.

NMFS (National Marine Fisheries Service). 2016b. *NOAA Fisheries Ecosystem-Based Fisheries Management Roadmap*. Silver Spring, MD: NOAA.

NMFS (National Marine Fisheries Service). 2017. *National Marine Fisheries Service—2nd Quarter 2017 Update. Table A. Summary of Stock Status for FSSI Stocks*. Silver Spring, MD: NOAA (p. 53).

NMFS (National Marine Fisheries Service). 2018a. *Fisheries of the United States, 2017. NOAA Current Fishery Statistics No. 2017*. Silver Spring, MD: NOAA.

NMFS (National Marine Fisheries Service). 2018b. Programmatic Environmental Assessment: U.S. *Commercial Fishing for Tropical Tuna in the Eastern Pacific Ocean*. Silver Spring, MD: NOAA (p. 46).

NMFS (National Marine Fisheries Service). 2019a. *2018 Stock Assessment and Fishery Evaluation (SAFE) Report for Atlantic Highly Migratory Species. Atlantic Highly Migratory Species Management Division*. Silver Spring, MD: NOAA (p. 234).

NMFS (National Marine Fisheries Service). 2019b. *Atlantic Highly Migratory Species Ecosystem-Based Fisheries Management Road Map Implementation Plan*. Silver Spring, MD: NOAA (p. 28).

NMFS (National Marine Fisheries Service). 2019c. *Pacific Islands Region Ecosystem-Based Fisheries Management Implementation Plan 2018–2022*. Silver Spring, MD: NOAA (p. 39).

NMFS (National Marine Fisheries Service). 2019d. *Western Regional Implementation Plan for Ecosystem-Based Fisheries Management*. Silver Spring, MD: NOAA (p. 25).

NMFS (National Marine Fisheries Service). 2020. *National Marine Fisheries Service—2nd Quarter 2020 Update*. Silver Spring, MD: NOAA (p. 51).

Norton J. 1987. *Ocean Climate Influences on Groundfish Recruitment in the California Current. Proceedings of the International Rockfish Symposium, Sea Grant Rep 87-2*. Anchorage, AK: University of Alaska (pp. 73–99).

NPFC (North Pacific Fisheries Commission). 2012. *Convention on the Conservation and Management of High Seas Fisheries Resources in the North Pacific Ocean*. Tokyo, Japan (p. 28).

NPFC/FAO. 2018. *VME Workshop. 1st Workshop Report NPFC-2018-WS VME01-Final Report*. Tokyo, Japan (p. 18).

NPFMC (North Pacific Fishery Management Plan). 2009. *Fishery Management Plan for Fish Resources of the Arctic Management Area*. Anchorage, Alaska: NPFMC (p. 146).

Nummelin M, Urho N. 2018. *International Environmental Conventions on Biodiversity*. Oxford, UK: Oxford Research Encyclopedia of Environmental Science.

Nye JA, Gamble RJ, Link JS. 2013. The relative impact of warming and removing top predators on the Northeast US large marine biotic community. *Ecological Modelling* 264:157–68.

Nye JA, Link JS, Hare JA, Overholtz WJ. 2009. Changing spatial distribution of fish stocks in relation to climate and population size on the Northeast United States continental shelf. *Marine Ecology Progress Series* 393:111–29.

Nyqvist D, Nilsson PA, Alenäs I, Elghagen J, Hebrand M, Karlsson S, Kläppe S, Calles O. 2017. Upstream and downstream passage of migrating adult Atlantic salmon: remedial measures improve passage performance at a hydropower dam. *Ecological Engineering* 102:331–43.

Okey TA, Alidina HM, Lo V, Jessen S. 2014. Effects of climate change on Canada's Pacific marine ecosystems: a summary of scientific knowledge. *Reviews in Fish Biology and Fisheries* 24(2):519–59.

Olden JD. 2016. *Challenges and Opportunities for Fish Conservation in Dam-Impacted Waters. Conservation of Freshwater Fishes.* Cambridge, UK: Cambridge University Press (pp. 107–48).

Oreskes N, Finley C. 2007. *A Historical Analysis of the Collapse of Pacific Groundfish: US Fisheries Science, Development, and Management, 1945–1995.* San Diego, CA: University of California.

Orth RJ, Luckenbach ML, Marion SR, Moore KA, Wilcox DJ. 2006. Seagrass recovery in the Delmarva coastal bays, USA. *Aquatic Botany* 84(1):26–36.

Österblom H, Gårdmark A, Bergström L, Müller-Karulis B, Folke C, Lindegren M, Casini M, et al. 2010. Making the ecosystem approach operational—can regime shifts in ecological-and governance systems facilitate the transition? *Marine Policy* 34:1290–9.

Ortuño Crespo G, Dunn DC. 2017. A review of the impacts of fisheries on open-ocean ecosystems. *ICES Journal of Marine Science* 74(9):2283–97.

Pan M, Huntington HP. 2016. A precautionary approach to fisheries in the Central Arctic Ocean: Policy, science, and China. *Marine Policy* 63:153–7.

Pandolfi JM, Jackson JB, Baron N, Bradbury RH, Guzman HM, Hughes TP, Kappel CV, et al. 2005. Are US coral reefs on the slippery slope to slime? *Science* 307(5716): 1725–6.

Parker SJ, Penney AJ, Clark MR. 2009. Detection criteria for managing trawl impacts on vulnerable marine ecosystems in high seas fisheries of the South Pacific Ocean. *Marine Ecology Progress Series* 397:309–17.

Parma AM. 2002. In search of robust harvest rules for Pacific halibut in the face of uncertain assessments and decadal changes in productivity. *Bulletin of Marine Science* 70(2):423–53.

Pentz B, Klenk N, Ogle S, Fisher JA. 2018. Can regional fisheries management organizations (RFMOs) manage resources effectively during climate change? *Marine Policy* 92:13–20.

Peterson CD, Belcher CN, Bethea DM, Driggers III WB, Frazier BS, Latour RJ. 2017. Preliminary recovery of coastal sharks in the south-east United States. *Fish and Fisheries* 18(5):845–59.

Peterson MJ, Carothers C. 2013. Whale interactions with Alaskan sablefish and Pacific halibut fisheries: surveying fishermen perception, changing fishing practices and mitigation. *Marine Policy* 42:315–24.

PFMC (Pacific Fishery Management Council). 2008. *Management of Krill as an Essential Component of the California Current Ecosystem. Amendment 12 to the Coastal Pelagic Species Fishery Management Plan. Environmental Assessment, Regulatory Impact Review, and Regulatory Flexibility Analysis.* Portland, OR: PFMC (p. 114).

PFMC (Pacific Fishery Management Council). 2018. *Fishery Management Plan for U.S. West Coast Fisheries for Highly Migratory Species. As Amended Through Amendment 5.* Portland, OR: PFMC (p. 78).

Pintassilgo P, Finus M, Lindroos M, Munro G. 2010. Stability and success of regional fisheries management organizations. *Environmental and Resource Economics* 46(3):377–402.

Pitcher TJ, Kalikoski D, Short K, Varkey D, Pramod G. 2009. An evaluation of progress in implementing ecosystem-based management of fisheries in 33 countries. *Marine Policy* 33(2):223–32.

Planas JV. 2017. *IPHC Biological and Ecosystem Science Research Plan.* Seattle, WA: IPHC (pp. 10–24).

Polovina JJ, Howell EA, Abecassis M. 2008. Ocean's least productive waters are expanding. *Geophysical Research Letters* 35(3).

Polovina JJ, Mitchum GT. 1995. Impact on biological production in the Central and North Pacific. *Deep Sea Research* 42(10):1701–16.

Pranovi F, Libralato S, Zucchetta M, Anelli Monti M, Link JS. 2020. Cumulative biomass curves describe past and present conditions of large marine ecosystems. *Global Change Biology* 26:786–97.

Prince ED, Brown BB. 1991. Coordination of the ICCAT enhanced research program for billfish. Creel and angler surveys in fisheries management. *American Fisheries Society Symposium* 12:13–18.

Protopapadakis L. 2006. *Ecosystem-Based Management of Pacific Tunas. Doctoral Dissertation.* Durham, NC: Duke University (p. 83).

PSC (Pacific Salmon Commission). 2020. *Treaty Between the Government of Canada and the Government of the United States of America Concerning Pacific Salmon.* Vancouver, Canada: PSC (p. 142).

Punt AE. 2011. The impact of climate change on the performance of rebuilding strategies for overfished groundfish species of the US west coast. *Fisheries Research* 109(2–3):320–9.

Punt AE, Donovan GP. 2007. Developing management procedures that are robust to uncertainty: lessons from the International Whaling Commission. *ICES Journal of Marine Science* 64(4):603–12.

PWT (Pacific Whiting Treaty). 2004. *Agreement with Canada on Pacific Hake/Whiting. Agreement Between the Government of the United States of America and the Government of Canada on Pacific Hake/Whiting.* Washington, DC: NOAA (p. 10).

Raby GD, Donaldson MR, Hinch SG, Patterson DA, Lotto AG, Robichaud D, English KK, et al. 2012. Validation of reflex indicators for measuring vitality and predicting the delayed mortality of wild coho salmon bycatch released from fishing gears. *Journal of Applied Ecology* 49(1):90–8.

Ramirez-Llodra E, Tyler PA, Baker MC, Bergstad OA, Clark MR, Escobar E, Levin LA, et al. 2011. Man and the last great wilderness: human impact on the deep sea. *PLoS One* 6(8):e22588.

Reeves RR, Berggren P, Crespo EA, Gales N, Northridge SP, di Sciara GN, Perrin WF, et al. 2005. *Global Priorities for Reduction of Cetacean Bycatch.* Gland, Switzerland: World Wildlife Fund.

Reid, K. 2011. Conserving Antarctica from the bottom up: implementing UN General Assembly resolution 61/105 in the commission for the conservation of Antarctic marine living resources (CCAMLR). *Ocean Yearbook* 25:131–139.

Restrepo V. 2004. *Estimation of Unreported Catches by ICCAT. Fish Piracy: Combating Illegal, Unreported and Unregulated Fishing*. Paris, France: OECD Publishing (pp. 155–7).

Rhodes A, Stelle WW, Taylor H. 2014. *Environmental Assessment to Revise the United States Commercial Fishery Regulations in Accordance with Inter-American Tropical Tuna Commission Resolution for the Conservation and Management of Pacific Bluefin Tuna in the Eastern Pacific Ocean (C-13-02).* Washington, DC: NOAA.

Richards LJ, Maguire JJ. 1998. Recent international agreements and the precautionary approach: new directions for fisheries management science. *Canadian Journal of Fisheries and Aquatic Sciences* 55(6):1545–52.

Rice J. 2011. Managing fisheries well: delivering the promises of an ecosystem approach. *Fish and Fisheries* 12(2):209–31.

Rindorf A, Dichmont CM, Thorson J, Charles A, Clausen LW, Degnbol P, Garcia D, et al. 2017. Inclusion of ecological, economic, social, and institutional considerations when setting targets and limits for multispecies fisheries. *ICES Journal of Marine Science* 74(2):453–63.

Robertson CT. 2005. *Conservation of Endangered Atlantic Salmon in Maine*. Toronto, ON: University of Toronto.

Robinson LM, Hobday AJ, Possingham HP, Richardson AJ. 2015. Trailing edges projected to move faster than leading edges for large pelagic fish habitats under climate change. *Deep Sea Research Part II: Topical Studies in Oceanography* 113:225–34.

Rogers A, Laffoley DD. 2011. *International Earth System Expert Workshop on Ocean Stresses and Impacts: Summary Workshop Report*. Darby, PA: Diane Publishing.

Rogers CS, Beets J. 2001. Degradation of marine ecosystems and decline of fishery resources in marine protected areas in the US Virgin Islands. *Environmental Conservation* 28(4):312–22.

Rogers CS, Suchanek TH, Pecora FA. 1982. Effects of hurricanes David and Frederic (1979) on shallow Acropora palmata reef communities: St. Croix, US Virgin Islands. *Bulletin of Marine Science* 32(2):532–48.

Roman CT, Jaworski N, Short FT, Findlay S, Warren RS. 2000. Estuaries of the northeastern United States: habitat and land use signatures. *Estuaries* 23(6):743–64.

Ryan KE, Walsh JP, Corbett DR, Winter A. 2008. A record of recent change in terrestrial sedimentation in a coral-reef environment, La Parguera, Puerto Rico: a response to coastal development? *Marine Pollution Bulletin* 56(6): 1177–83.

Sabourenkov EN, Appleyard E. 2005. Scientific observations in CCAMLR fisheries–past, present and future. *CCAMLR Science* 12:81–98.

Sadorus LL, Mantua NJ, Essington T, Hickey B, Hare S. 2014. Distribution patterns of Pacific halibut (*Hippoglossus stenolepis*) in relation to environmental variables along the continental shelf waters of the US West Coast and southern British Columbia. *Fisheries Oceanography* 23(3):225–41.

Sainsbury KJ, Punt AE, Smith AD. 2000. Design of operational management strategies for achieving fishery ecosystem objectives. *ICES Journal of Marine Science* 57(3):731–41.

Sammarco PW, Winter A, Stewart JC. 2006. Coefficient of variation of sea surface temperature (SST) as an indicator of coral bleaching. *Marine Biology* 149(6):1337–44.

Scatena FN, Larsen MC. 1991. Physical aspects of hurricane Hugo in Puerto Rico. *Biotropica* 1:317–23.

Schiffman HS. 2006. Moving from single-spaced management to ecosystem management in regional fisheries management organizations. *ILSA Journal of International & Comparative Law* 13:387.

Schiffman HS. 2013. The South Pacific regional fisheries management organization (SPRFMO): an improved model of decision-making for fisheries conservation? *Journal of Environmental Studies and Sciences* 3(2):209–16.

Schindler DE, Essington TE, Kitchell JF, Boggs C, Hilborn R. 2002. Sharks and tunas: fisheries impacts on predators with contrasting life histories. *Ecological Applications* 12(3):735–48.

Schirripa M, Lehodey P, Prince E, Luo J. 2011. Habitat modeling of Atlantic blue marlin with SEAPODYM and satellite tags. *ICCAT Collective Volume of Scientific Papers* 66:1735–7.

Schmidt Jr RJ. 1996. International negotiations paralyzed by domestic politics: two-level game theory and the problem of the Pacific Salmon Commission. *Environmental Law* 26:95.

Selkoe KA, Blenckner T, Caldwell MR, Crowder LB, Erickson AL, Essington TE, Estes JA, et al. 2015. Principles for managing marine ecosystems prone to tipping points. *Ecosystem Health and Sustainability* 1(5):1–8.

Shamshak GL, Anderson JL, Asche F, Garlock T, Love DC. 2019. US seafood consumption. *Journal of the World Aquaculture Society* 50(4):715–27.

Sheehan TF, Barber J, Dean A, Deschamps D, Gargan P, Goulette G, Maxwell H, et al. 2019. *The International Sampling Program: Continent of Origin and Biological Characteristics of Atlantic Salmon Collected at West Greenland in 2017. Northeast Fisheries Science Center Reference Document 19-12.* Washington, DC: NOAA.

Shelton PA, Rice JC. 2002. Limits to Overfishing: Reference Points in the Context of the Canadian Perspective on the

Precautionary Approach. Ottowa, Canada: Canadian Science Advisory Secretariat.

Sigler MF, Tollit DJ, Vollenweider JJ, Thedinga JF, Csepp DJ, Womble JN, Wong MA, Rehberg MJ, Trites AW. 2009. Steller sea lion foraging response to seasonal changes in prey availability. *Marine Ecology Progress Series* 388:243–61.

Simpfendorfer CA, Dulvy NK. 2017. Bright spots of sustainable shark fishing. *Current Biology* 27(3):R97–8.

Siron R, Sherman K, Skjoldal HR, Hiltz E. 2008. Ecosystem-based management in the Arctic Ocean: a multi-level spatial approach. *Arctic* 1:86–102.

Slaney TL, Hyatt KD, Northcote TG, Fielden RJ. 1996. Status of anadromous salmon and trout in British Columbia and Yukon. *Fisheries* 21(10):20–35.

Small C, Waugh SM, Phillips RA. 2013. The justification, design and implementation of ecological risk assessments of the effects of fishing on seabirds. *Marine Policy* 37:192–9.

Small CJ. 2005. *Regional Fisheries Management Organisations: Their Duties and Performance in Reducing Bycatch of Albatrosses and Other Species.* Cambridge, UK: BirdLife International.

Smith TB, Nemeth RS, Blondeau J, Calnan JM, Kadison E, Herzlieb S. 2008. Assessing coral reef health across onshore to offshore stress gradients in the US Virgin Islands. *Marine Pollution Bulletin* 56(12):1983–91.

SPRFMO (South Pacific Regional Fishery Management Organization). 2020. *8th Meeting of the SPRFMO Commission COMM 8—Inf 02, Preventing Significant Adverse Impacts by Bottom Fisheries on Vulnerable Marine Ecosystems: Important Issues of Scale.* Port Vila, Vanuatu, February 14–18. Wellington, NZ: SPRFMO.

SPRFMO (South Pacific Regional Fishery Management Organization). 2015. *Convention on the Conservation and Management of High Seas Fishery Resources in the South Pacific Ocean.* Wellington, NZ: SPRFMO (p. 42).

Steinacher M, Joos F, Frölicher TL, Bopp L, Cadule P, Cocco V, Doney SC, et al. 2010. Projected 21st century decrease in marine productivity: a multi-model analysis. *Biogeosciences* 7(3):979–1005.

Stevens JD, Bonfil R, Dulvy NK, Walker PA. 2000. The effects of fishing on sharks, rays, and chimaeras (chondrichthyans), and the implications for marine ecosystems. *ICES Journal of Marine Science* 57(3):476–94.

Stevenson DK. 1981. *A Review of the Marine Resources of the Western Central Atlantic Fisheries Commission (WECAFC) Region. FAO Fisheries Technical Papers 211.* Rome, Italy: FAO (p. 132).

Stewart IJ, Hamel OS. 2010. *Stock Assessment of Pacific Hake, Merluccius Productus,(aka Whiting) in US and Canadian Waters in 2010.* Portland, OR: Pacific Fishery Management Council.

Stokke OS. 2001. *Governing High Seas Fisheries: The Interplay of Global and Regional Regimes.* Oxford, UK: Oxford University Press.

Stone RB, Bailey CM, McLaughlin SA, Mace PM, Schulze MB. 1998. Federal management of US Atlantic shark fisheries. *Fisheries Research* 39(2):215–21.

Stoner AW, Titgen RH. 2003. Biological structures and bottom type influence habitat choices made by Alaska flatfishes. *Journal of Experimental Marine Biology and Ecology* 292(1):43–59.

Sullivan JY. 2016. *Environmental, Ecological, and Fishery Effects on Growth and Size-At-Age of Pacific Halibut (Hippoglossus Stenolepis). Doctoral Dissertation.* Fairbanks, AK: University of Alaska Fairbanks.

Surma S, Pakhomov EA, Pitcher TJ. 2014. Effects of whaling on the structure of the southern ocean food web: insights on the "krill surplus" from ecosystem modelling. *PLoS One* 9(12).

Surma S, Pitcher TJ. 2015. Predicting the effects of whale population recovery on Northeast Pacific food webs and fisheries: an ecosystem modelling approach. *Fisheries Oceanography* 24(3):291–305.

Sydnes AK. 2002. Regional fishery organisations in developing regions: adapting to changes in international fisheries law. *Marine Policy* 26(5):373–81.

Teffer AK, Hinch SG, Miller KM, Patterson DA, Farrell AP, Cooke SJ, Bass AL, Szekeres P, Juanes F. 2017. Capture severity, infectious disease processes and sex influence post-release mortality of sockeye salmon bycatch. *Conservation Physiology* 5(1).

Thom B. 2018. *Environmental Assessment and Regulatory Impact Review for 2018 Pacific Halibut Catch Limits in International Pacific Halibut Commission Regulatory Area 2A (Washington, Oregon, and California).* Seattle, Washington: NOAA.

Tolimieri N, Samhouri JF, Simon V, Feist BE, Levin PS. 2013. Linking the trophic fingerprint of groundfishes to ecosystem structure and function in the California Current. *Ecosystems* 16(7):1216–29.

Torrissen O, Olsen RE, Toresen R, Hemre GI, Tacon AG, Asche F, Hardy RW, Lall S. 2011. Atlantic salmon (Salmo salar): the "super-chicken" of the sea? *Reviews in Fisheries Science* 19(3):257–78.

Tremblay JÉ, Bélanger S, Barber DG, Asplin M, Martin J, Darnis G, Fortier L, et al. 2011. Climate forcing multiplies biological productivity in the coastal Arctic Ocean. *Geophysical Research Letters* 38(18).

Trochta JT, Pons M, Rudd MB, Krigbaum M, Tanz A, Hilborn R. 2018. Ecosystem-based fisheries management: perception on definitions, implementations, and aspirations. *PloS One* 13(1):e0190467.

Uphoff Jr JH, Sharov A. 2018. Striped bass and Atlantic menhaden predator–prey dynamics: model choice makes the difference. *Marine and Coastal Fisheries* 10(4):370–85.

Van Houtan KS, Halley JM. 2011. Long-term climate forcing in loggerhead sea turtle nesting. *PLoS One* 6(4):e19043.

VanderZwaag DL. 2015. "Sustaining Atlantic marine species at risk: scientific and legal coordinates, sea of governance challenges." In: *Science, Technology, and New Challenges to*

Ocean Law. Edited by HN Scheiber, J Kraska, MS Kwon. Leiden, Netherlands: Martinus Nijhoff Publishers (pp. 149–64).

Vernazzani BG, Burkhardt-Holm P, Cabrera E, Iñíguez M, Luna F, Parsons EC, Ritter F, et al. 2017. Management and conservation at the International Whaling Commission: a dichotomy sandwiched within a shifting baseline. *Marine Policy* 83:164–71.

Waldichuk M. 1993. Fish habitat and the impact of human activity with particular reference to Pacific salmon. Perspectives on Canadian marine fisheries management. *Canadian Bulletin of Fisheries and Aquatic Sciences* 226: 295–337.

Wallace F, Williams K, Towler R, McGauley K. 2015. *Innovative Camera Applications for Electronic Monitoring. Fisheries Bycatch: Global Issues and Creative Solutions*. Fairbanks, Alaska: University of Alaska Fairbanks (pp. 105–17).

Waples RS, Zabel RW, Scheuerell MD, Sanderson BL. 2008. Evolutionary responses by native species to major anthropogenic changes to their ecosystems: Pacific salmon in the Columbia River hydropower system. *Molecular Ecology* 17(1):84–96.

Ward P, Myers RA. 2005. Shifts in open-ocean fish communities coinciding with the commencement of commercial fishing. *Ecology* 86(4):835–47.

Ward-Paige CA, Keith DM, Worm B, Lotze HK. 2012. Recovery potential and conservation options for elasmobranchs. *Journal of Fish Biology* 80(5):1844–69.

Ware DM, McFarlane GA. 1995. Climate-induced changes in Pacific hake (Merluccius productus) abundance and pelagic community interactions in the Vancouver Island upwelling system. *Canadian Special Publication of Fisheries and Aquatic Sciences* 121:509–21.

Warren RS, Niering WA. 1993. Vegetation change on a northeast tidal marsh: interaction of sea-level rise and marsh accretion. *Ecology* 74(1):96–103.

Watling L, Auster PJ. Seamounts on the high seas should be managed as vulnerable marine ecosystems. *Frontiers in Marine Science* 4:14.

Watters GM, Hill SL, Hinke JT, Matthews J, Reid K. 2013. Decision-making for ecosystem-based management: evaluating options for a krill fishery with an ecosystem dynamics model. *Ecological Applications* 23(4):710–25.

Weaver PP, Benn A, Arana PM, Ardron JA, Bailey DM, Baker K, Billett DS, et al. 2011. *The Impact of Deep-Sea Fisheries and Implementation of the UNGA Resolutions 61/105 and 64/72*. Southampton, UK: National Oceanography Centre.

Webster DG. 2013. International fisheries: assessing the potential for ecosystem management. *Journal of Environmental Studies and Sciences* 3(2):169–83.

Wegge N. 2015. The emerging politics of the Arctic Ocean. Future management of the living marine resources. *Marine Policy* 51:331–8.

Weitkamp LA, Goulette G, Hawkes J, O'Malley M, Lipsky C. 2014. Juvenile salmon in estuaries: comparisons between North American Atlantic and Pacific salmon populations. *Reviews in Fish Biology and Fisheries* 24(3): 713–36.

Welch DW, Rechisky EL, Melnychuk MC, Porter AD, Walters CJ, Clements S, Clemens BJ, et al. 2008. Correction: survival of migrating Salmon smolts in large rivers with and without dams. *PLoS Biology* 6(12):e314.

Wells CD, Pappal AL, Cao Y, Carlton JT. 2013. *Report on the 2013 Rapid Assessment Survey of Marine Species at New England Bays and Harbors*. Boston, MA: Massachussetts Office of Coastal Management.

Wetzel CR, Punt AE, Anderson E. 2016. The impact of alternative rebuilding strategies to rebuild overfished stocks. *ICES Journal of Marine Science* 73(9):2190–207.

Wilen JE, Homans FR. 1998. What do regulators do? Dynamic behavior of resource managers in the North Pacific Halibut fishery 1935–1978. *Ecological Economics* 24(2–3):289–98.

Wilkinson CR, Souter D. 2008. *Status of Caribbean Coral Reefs After Bleaching and Hurricanes in 2005*. Townsville, Australia: Global Coral Reef Monitoring Network.

Williams GH. 2012. *Incidental Catch and Mortality of Pacific Halibut, 1962–2011. International Pacific Halibut Commission Report of Assessment and Research Activities*. Washington, DC: IPHC (pp. 381–98).

Williams GH, Blood CL. 2003. Active and passive management of the recreational fishery for Pacific halibut off the US West Coast. *North American Journal of Fisheries Management* 23(4):1359–68.

Williams TH, Spence BC, Boughton DA, Johnson RC, Crozier LG, Mantua NJ, O'Farrell MR, Lindley ST. 2016. *Viability Assessment for Pacific Salmon and Steelhead Listed Under the Endangered Species Act: Southwest. NOAA Technical Memorandum NMFS-SWFSC-564*. Washington, DC: NOAA.

Windsor ML, Hutchinson P. 1994. International management of Atlantic salmon, Salmo salar L., by the North Atlantic salmon conservation organization, 1984–1994. *Fisheries Management and Ecology* 1(1):31–44.

Winter A, Appeldoorn RS, Bruckner A, Williams Jr EH, Goenaga C. 1998. Sea surface temperatures and coral reef bleaching off La Parguera, Puerto Rico (northeastern Caribbean Sea). *Coral Reefs* 17(4):377–82.

Witherell D, Pautzke C, Fluharty D. 2000. An ecosystem-based approach for Alaska groundfish fisheries. *ICES Journal of Marine Science* 57(3):771–7.

Worm B, Davis B, Kettemer L, Ward-Paige CA, Chapman D, Heithaus MR, Kessel ST, Gruber SH. 2013. Global catches, exploitation rates, and rebuilding options for sharks. *Marine Policy* 40:194–204.

Worm B, Tittensor DP. 2011. Range contraction in large pelagic predators. *Proceedings of the National Academy of Sciences* 108(29):11942–7.

Worm B, Vanderzwaag D. 2007. High-seas fisheries: troubled waters, tangled governance, and recovery prospects. *Behind the Headlines* 64(5):1.

Wowk KM. 2013. "Paths to sustainable ocean resources." In: *Managing Ocean Environments in a Changing Climate. Sustainability and Economic Perspectives*. Edited by RJ Diaz, KJ Noone, UR Sumaila. Amsterdam, Netherlands: Elsevier (pp. 301–48).

WPFMC (Western Pacific Fishery Management Council). 2009. *Fishery Ecosystem Plan for Pacific Pelagic Fisheries of the Western Pacific Region*. Honolulu, HI: WPFMC (p. 249).

Wright AJ, Simmonds MP, Galletti Vernazzani B. 2016. The international whaling commission—beyond whaling. *Frontiers in Marine Science* 3:158.

Wright G, Ardron J, Gjerde K, Currie D, Rochette J. 2015. Advancing marine biodiversity protection through regional fisheries management: a review of bottom fisheries closures in areas beyond national jurisdiction. *Marine Policy* 61:134–48.

Wroblewski JS, Sarmiento JL, Flierl GR. 1988. An ocean basin scale model of plankton dynamics in the North Atlantic: solutions for the climatological oceanographic conditions in May. *Global Biogeochemical Cycles* 2(3):199–218.

Würsig B, Reeves RR, Ortega-Ortiz JG. 2002. "Global climate change and marine mammals." In: *Marine Mammals*. Edited by H Thewissen, B Würsig, WF Perrin. Boston, MA: Springer (pp. 589–608).

Yen KW, Lu HJ. 2016. Spatial–temporal variations in primary productivity and population dynamics of skipjack tuna Katsuwonus pelamis in the western and central Pacific Ocean. *Fisheries Science* 82(4):563–71.

Zador S. 2015. *Ecosystem Considerations 2015 Status of Alaska's Marine Ecosystems*. Anchorage, AK: North Pacific Fishery Management Council Ecosystem Considerations (p. 296).

Zador S, Gaichas S. 2010. *Ecosystem Considerations for 2011. Groundfish Stock Assessment and Fishery Evaluation Report*. Anchorage, AK: North Pacific Fishery Management Council.

Zador SG, Gaichas SK, Kasperski S, Ward CL, Blake RE, Ban NC, Himes-Cornell A, Koehn JZ. 2017a. Linking ecosystem processes to communities of practice through commercially fished species in the Gulf of Alaska. *ICES Journal of Marine Science* 74(7):2024–33.

Zador SG, Holsman KK, Aydin KY, Gaichas SK. 2017b. Ecosystem considerations in Alaska: the value of qualitative assessments. *ICES Journal of Marine Science* 74(1):421–30.

Zador S, Ortiz I. 2018. Ecosystem Status Report 2018 Aleutian Islands. *North Pacific Fishery Management Council Bering Sea and Aleutian Islands Stock Assessment and Fishery Evaluation (SAFE) Report*. Anchorage, AK: North Pacific Fishery Management Council (p. 130).

Zador S, Yasumiishi. 2018. *Ecosystem Status Report 2018 Gulf of Alaska. North Pacific Fishery Management Council Gulf of Alaska Stock Assessment and Fishery Evaluation (SAFE) Report*. Anchorage, AK: North Pacific Fishery Management Council (p. 193).

Zhou S, Smith AD, Punt AE, Richardson AJ, Gibbs M, Fulton EA, Pascoe S, et al. 2010. Ecosystem-based fisheries management requires a change to the selective fishing philosophy. *Proceedings of the National Academy of Sciences* 107(21):9485–9.

CHAPTER 12

An Examination of Progress Toward Ecosystem-Based Management of Living Marine Resources in the U.S.

- In all U.S. marine ecosystems, there has been notable progress toward ecosystem-based fisheries management (EBFM), albeit on different facets for different regions.
- Among all U.S. marine ecosystems, the overall status of living marine resources (LMRs) is highest in the Mid-Atlantic, North Pacific, and Pacific regions and their subregions.
- Among all U.S. marine ecosystems, the overall status of LMR- and ecosystem-associated socioeconomic indicators is highest in the New England, North Pacific, Pacific, Gulf of Mexico, and Mid-Atlantic regions.
- When ranked among U.S. marine ecosystems, highest basal productivities occur in the North Pacific, New England, and Mid-Atlantic, and secondarily in the Gulf of Mexico and Pacific regions.
- When ranked across all indicators (including other ocean uses), the Pacific, Mid-Atlantic, and North Pacific emerge as those regions with currently highest EBFM capacities.
- When ranked among U.S. subregions, highest chlorophyll concentrations occur in Chesapeake Bay, Long Island Sound, Galveston Bay, Matagorda Bay, Mississippi River Delta, Mobile Bay, and the Columbia River.
- When ranked across all indicators (including other ocean uses), Puget Sound-Salish Sea, Chesapeake Bay, Delaware Bay, Hawaii, the Channel Islands, and Columbia River emerge as those subregions with currently highest EBFM capacities.
- When ranked among regional fisheries management organizations (RFMO), highest total productivities were observed in highly migratory species (HMS)-encompassed jurisdictions, and those of the North Atlantic Salmon Conservation Organization (NASCO), International Pacific Halibut Commission (IPHC), Pacific Salmon Commission (PSC), and North Atlantic Fisheries Organization (NAFO).
- When ranked across all indicators and management actions, Pacific and Atlantic HMS management bodies and the Commission for the Conservation of Antarctic Marine Living Resources (CCAMLR) had the relatively highest EBFM capacities, but this is generally low for all RFMOs when compared to the domestic context.
- Overall, more inherently productive marine ecosystems appear to have greater biomass, fisheries landings, proportional LMR-based employments, and fisheries revenue. However, this trend does not necessarily hold in subregions or for RFMO jurisdictions.
- Progress regarding NOAA Fisheries EBFM Road Map action items and Ecosystem Principles Advisory Panel (EPAP) recommendations has been greatest in the Mid-Atlantic, New England, North Pacific, and Pacific regions.

12.1 Introduction

Within this chapter, a cumulative examination of socioeconomic, governance, ecological, and environmental indicators is presented among the eight major United States (U.S.) marine fishery ecosystems, 26 U.S. subregions, and 14 U.S. participatory regional fisheries management organization (RFMO) jurisdictions. Of all indicators examined, 79 regional, 45 subregional, and 55 RFMO jurisdictional indicators were evaluated and compared to gauge relative trends and

Ecosystem-Based Fisheries Management: Progress, Importance, and Impacts in the United States. Jason S. Link and Anthony R. Marshak,
Oxford University Press. © U.S. Department of Commerce, U.S. Government 2021. DOI: 10.1093/oso/9780192843463.003.0012

success toward ecosystem-based management of living marine resources (LMRs). These indicators were related to criteria organized by (1) socioeconomic, (2) governance, (3) environmental forcing and major features, (4) major pressures and exogenous factors, and (5) systems ecology and fisheries, which were examined using extant datasets (c.f. Chapter 2). The majority of these factors were similarly examined by Link and Marshak (2019) toward characterizing and comparing U.S. fisheries ecosystems and determining successful ecosystem-based strategies for LMR management, but here we place them in particular context of specific benchmarks relative to ecosystem-based fisheries management (EBFM).

Among the five indicator categories, a given indicator was examined based upon its current value or cumulative average value and relative standard error over time. To assess cumulative nationwide trends for indicators, the number of regions with values above the total calculated cross-regional mean value for a given indicator (i.e., anomaly) was tabulated following the method of Link and Marshak (2019). Additionally, the number of regions for which relative standard error was greater than 10% (z-score equivalent: 1.645) were tabulated to identify the total highly variable regions for which collective dynamic trends were occurring per indicator. Tabulated values were also averaged among the five indicator categories to gage overall trends per category (Link & Marshak 2019). Current or average values of time-series data as related to relative standard error (i.e., signal to noise ratios) per ecosystem indicator were additionally ranked across U.S. regions (n = 64 applicable indicators), subregions (n = 44 applicable indicators), and RFMO jurisdictions (n = 52 applicable indicators). Rankings were averaged together and within the five indicator categories to examine comparative regional relative success (high, mid, or low) for components of LMR management. Additional details regarding these methodologies are presented in Chapter 2.

In addition to evaluating these metrics, we also benchmarked EBFM progress in each of the eight U.S. regions of interest, their subregions, and RFMO jurisdictions against the Ecosystem Principles Advisory Panel (EPAP) fishery ecosystem plan (FEP) recommendations (EPAP 1999, Wilkinson & Abrams 2015) and NOAA Fisheries EBFM Road Map action items (NMFS 2016a). To reiterate, particularly for the EBFM Road Map, it was understood that these action items against which we benchmarked EBFM progress were done so cognizant that they were options for consideration and were not requirements. Conversely, any common action, or consistent omission, across regions is informative. Thus, this synthesis was not done as a report card or grading exercise, but rather to note common and national areas of progress and impediment, respectively, toward EBFM.

12.2 U.S. Regions

Typically, and as would be expected statistically speaking, 3–4 out of 8 U.S. marine regions were above the mean indicator in most indicator categories (Table 12.1), with average

Table 12.1 Examination per category of ecosystem indicators across U.S. marine regions as related to given cross-regional mean values. Data are presented as the number of regions above a calculated cross-regional mean indicator value (anomaly), the number of highly variable (>10% relative SE over time) regions per indicator, and the overall number (± SE) of above mean and highly variable regions per category. Values for #Congressionals (and standardized), Fishery Management Council (FMC) Composition, and #FMP (Fishery Management Plan) or FEP (Fishery Ecosystem Plan) modifications reflect the number of regions with values below the cross-regional mean value, as indicated by asterisk. EEZ = Exclusive Economic Zone, GDP = Gross Domestic Product, HAPC = Habitat Area of Particular Concern, HFA = NOAA Habitat Focus Area, LMR = Living Marine Resource, NERR = National Estuarine Research Reserve, NEPA EIS = National Environmental Policy Act Environmental Impact Statement, SRG = Scientific Review Group, Var = Variability

	Time frame	Cross-Regional Mean Value	Regions Above Mean (Anomaly)	Regions >10% Variability
Socioeconomic Criteria				
– Avg. Population	1970–2010	11.9 million humans	3	4
– Avg. Population Density	1970–2010	112.9 humans km^{-2}	4	4
– Avg. % LMR Establishments	2005–2016	5.3%	2	0
– Avg. % LMR Employments	2005–2016	4.4%	1	0
– Avg. % LMR GDP	2005–2016	2.6%	3	0
– Avg. #Permitted Vessels	1990–2016	8920.4 vessels	4	3

	Time frame	Cross-Regional Mean Value	Regions Above Mean (Anomaly)	Regions >10% Variability
Socioeconomic Criteria				
– Avg. Fisheries Value	1950–2016	2.9×10^8 USD	4	2
– Avg. Ratio of Total LMR Employments (Jobs)/Total Biomass	2005–2014	1.1	4	1
– Avg. Ratio of Total Value (Revenue) of Commercial Fisheries/ Total Biomass	1968–2016	57.7	3	1
– Avg. Ratio of Total LMR Employments (Jobs)/Total Primary Production	2005–2014	8.0	4	1
– Avg. Ratio of Total Value (Revenue) of Commercial Fisheries/ Total Primary Production	1998–2014	0.51	4	1
Above Mean			*3.2 ± 0.3*	*1.5 ± 0.5*
Governance Criteria				
Human Representative, Fishery/Systematic Context				
– #Organizations for Management	2017	22.5 organizations	3	
– #States	2017	4.9 states	4	
– #Congressionals*	2017	26.6 Congressionals	5	
– #Congressionals/Mile of Shoreline*	2017	0.003 Congressionals/Mile	3	
– #Congressionals/Value of Fishery*	2017	0.103 Congressionals/USD	4	0
– #FMPs (both federal and state)	2017	18.1 FMPs	3	
– #FEPs (current or in development)	2017	2 FEPs	1	
– #FMP Modifications*	2017	215 modifications	4	
– #FEP Modifications*	2017	8.3 modifications	1	
– %EEZ Fishing Permanently Prohibited	2017	8.3%	2	
– #HAPCs	2017	22 areas	3	
– #Protected Areas (i.e., Sanctuaries, NERRs, HFAs, etc.)	2017	15.2 areas	3	
– #Cumulative #NEPA EIS Actions	1987–2017	140 actions	4	
– Avg. FMC Composition—%Commercial*	1990–2016	<47.9%	4	1
– Avg. Number of Members on Marine Mammal SRG	1995–2017	9.7 members y^{-1}	5	0
– Avg. FMC Budget to Fisheries Value	2007–2016	0.029	2	0
– Avg. Lawsuits	2010–2016	1.7 lawsuits y^{-1}	4	7
– Number Lawsuits	2010–2016	9.9 lawsuits	3	
Status of LMRs				
– #Managed Fishery Species	2017	148.5 species	5	
– #Prohibited Species	2017	72.6 Species	3	
– #Protected Species	2017	43.8 Species	6	
– %Stocks Experiencing Overfishing	2017	7.1%	3	
– %Stocks Overfished	2017	8.7%	2	
– %Stocks Unknown Overfishing Status	2017	18.4%	3	
– %Stocks Unknown Overfished Status	2017	43.9%	5	

Continued

Table 12.1 Continued

	Time frame	Cross-Regional Mean Value	Regions Above Mean (Anomaly)	Regions >10% Variability
Governance Criteria				
– %Strategic Stocks	2017	35.7%	3	
– % Marine Mammal Stocks of Unknown Population Size	2017	30.5%	3	
– #Threatened/Endangered Species	2017	24.8 Species	3	
Above Mean			*3.5 ± 0.2*	*1.6 ± 1.4*
Environmental Forcing and Major Features				
– Avg. Sea Surface Temperature	1854–2016	19.0°C	1	0
– Var. Sea Surface Temperature	1854–2016	0.74°C	1	0
– Temperature Increase (°C)	1950–2016	1.1°C	4	
– Avg. Number of Hurricanes per Decade	1850–2016	20.4/decade	2	4
– Avg. Proportion of Sea Ice	1982–2016	0.34	1	0
– Avg. Bottom Water Hypoxia Extent	1985–2017	1600 km^2 bottom water hypoxia	1	0
Above Mean			*2.4 ± 0.7*	*1.0 ± 1.0*
Major Pressures & Exogenous Factors				
Other Ocean Uses				
– Avg. Ocean Economy Establishments	2005–2016	17,244.2 establishments	3	0
– Avg. Ocean Economy Employments	2005–2016	348,332.1 employments	3	0
– Avg. Ocean Economy GDP	2005–2016	$37.1 billion	3	0
– Avg. % Marine Construction Establishments	2005–2016	1.7%	5	0
– Avg. % Marine Construction Employments	2005–2016	1.5%	3	0
– Avg. % Marine Construction GDP	2005–2016	1.7%	4	0
– Avg. % Marine Transportation Establishments	2005–2016	5.6%	5	0
– Avg. % Marine Transportation Employments	2005–2016	11.0%	3	0
– Avg. % Marine Transportation GDP	2005–2016	16.3%	4	1
– Avg. % Offshore Mineral Extraction Establishments	2005–2016	3.2%	2	1
– Avg. % Offshore Mineral Extraction Employments	2005–2016	6.2%	2	1
– Avg. % Offshore Mineral Extraction GDP	2005–2016	22.4%	2	1
– Avg. % Ship & Boatbuilding Establishments	2005–2016	1.2%	3	0
– Avg. % Ship & Boatbuilding Employments	2005–2016	4.6%	4	1
– Avg. % Ship & Boatbuilding GDP	2005–2016	5.7%	3	1
– Avg. % Tourism/Recreation Establishments	2005–2016	82.7%	5	0
– Avg. % Tourism/Recreation Employments	2005–2016	71.9%	5	0
– Avg. % Tourism/Recreation GDP	2005–2016	50.5%	4	1
– Avg. #Oil Rigs	1987–2016	10.8 rigs y^{-1}	1	0
– #Wind Energy Areas	2016	3.4 areas	4	
– #Dive Shops	2016	73.3 shops	5	

	Time frame	Cross-Regional Mean Value	Regions Above Mean (Anomaly)	Regions >10% Variability
Major Pressures & Exogenous Factors				
– #Ports	2016	15.4 ports	3	
– #Marinas	2016	23.6 marinas	2	
– #Cruise Destination Passengers	2009, 2011	941.8 passengers	1	
– #Cruise Departure Passengers	2009, 2011	1289.8 passengers	3	
Above Mean			*3.3 ± 0.2*	*0.4 ± 0.1*
Systems Ecology & Fisheries				
– Avg. Chlorophyll	1998–2016	0.98 mg/m3	3	0
– Avg. Primary Production	1998–2014	215.6 gC m^{-2} y^{-1}	4	0
– Avg. Number of Taxa	1950–2016	92.4 taxa	5	1
– Avg. Total Biomass	1964–2016	1.08×10^9 metric tons	1	1
– Avg. Fisheries Landings	1950–2016	406k metric tons	4	2
– Avg. Total Bycatch (mt)	2010–2013	48.8k metric tons	2	0
– Avg. Total Bycatch (individuals)	1996–2013	369.7k individuals	3	7
– Avg. Ratio of Total Landings/Total Biomass	1964–2016	0.06	3	1
– Avg. Ratio of Total Landings/Primary Production	1998–2014	0.0003	5	0
– Avg. Ratio of Total Biomass/Primary Production	1998–2014	0.019	1	0
Above Mean			*3.1 ± 0.5*	*1.3 ± 0.7*

variability over time per indicator generally low for most regions. The information presented below identifies those regions above a stated cross-regional mean value.

For socioeconomic criteria, indicators were above cross-regional mean values for ~3/8 regions and were generally stable. Human population stressors are strongest in 3–4 regions (Mid-Atlantic, U.S. Caribbean, New England, Pacific), with trends greatly increasing in 4/8 regions since 1970 (North Pacific, South Atlantic, Gulf of Mexico, Western Pacific). Additionally across regions, LMRs contribute >5% to total oceanic establishments for 2/8 regions (New England, North Pacific), with LMRs in only 1/8 regions (North Pacific) contributing greater than the cross-regional mean (4%) toward total oceanic employments. LMRs in New England, the North Pacific, and Pacific each contribute >2.6% toward their respective total oceanic GDPs. In addition, both the North Pacific and New England regions contain notably above-average cross-regional fisheries revenue along with the highest variabilities. Four regions contain an above-average fleet of >8.9k vessels (Mid-Atlantic, North Pacific, New England, Gulf of Mexico). Average integrative relationships among biomass, LMR employments, fisheries revenue, and production are generally steady, with 3–4/8 regions above cross-regional mean values. Above-average relationships related to fisheries revenue/biomass occur in the Gulf of Mexico, Mid-Atlantic, and New England, and in these three regions plus the Pacific with regards to fisheries revenue/production. Likewise, above cross-regional mean values for relationships between LMR employments, biomass, and production are additionally observed for those four regions. Therefore, while many of these criteria are above the national average in the North Pacific, New England, Pacific, Mid-Atlantic, and Gulf of Mexico, the notable LMR-based socioeconomics in these latter four ecosystems are also associated with some of the greatest human population pressures. These pressures are increasing within all regions as coastal populations become more concentrated, and highlight the need to balance marine socioeconomic status with sustainable exploitation. The inherent productivity in these systems is a major factor in their above-average economic standing (Link & Marshak 2019).

Most governance indicators related to human representative, fishery/systematic, and LMR status contexts are above-average values for ~3–4/8 regions (Table 12.1). Additionally, 3/8 regions are greater than the cross-regional mean (>0.003) in terms of congressionals per mile of shoreline (Mid-Atlantic, New England, Pacific) and 4/8 have >0.103 congressionals per value of fisheries (U.S. Caribbean, Mid-Atlantic, South Atlantic, Pacific; Link & Marshak 2019). There are 3/8 regions with >18 state and federal fishery management plans (FMPs) (South Atlantic, New England, Mid-Atlantic), and 4/8 regions (these three and the North Pacific) that have modified their FMPs >190 times. More than one FEP is currently in place within 2/8 regions (North Pacific, Western Pacific), while only the Western Pacific has modified its FEPs >10 times. Conversely, 3/8 regions currently do not have an FEP; others are actively in development. Additionally, the North and Western Pacific (2/8) regions are the only regions with >10% of their Exclusive Economic Zones (EEZs) permanently prohibited from fishing. In addition to the Western Pacific, the U.S. Caribbean, and Pacific (3/8 regions) contain >20 habitat areas of particular concern (HAPCs). Since 2010, four regions experienced an average of > 1.7 lawsuits per year (New England, N Pacific, Pacific, Gulf of Mexico), while 3/10 regions (New England, North Pacific, Pacific) have experienced >10 total lawsuits. Of measured trends since the past 2–3 decades, most are stable while the number of lawsuits per year is highly variable in all regions. Fishery Management Council (FMC) compositions are mostly balanced in 4/8 regions, with commercial fishing representatives making up a cross-regional average of ~38.5% of their total representations as compared to lower representation from recreational fishing and other sectors in less balanced regions. In general, those regions with above-average standardized representation tend to have above-average socioeconomics (particularly the Mid-Atlantic, New England, and Pacific) and/or concentrated human pressures (i.e., Caribbean). Several regions with above-average socioeconomic criteria also have greatest numbers of lawsuits and FMP modifications. Greater numbers of FEPs are currently associated with larger or more remote regional ecosystems, which also have the greatest proportions of their EEZs permanently protected. All regions have some above-average governance metrics, with geographic- or community-based approaches also occurring in the Western Pacific and Caribbean in terms of broader island-based management plans, habitat protections, and shifts toward more balanced FMCs over time.

For LMR status governance indicators, greater than half of all regions (5/8) manage >150 fishery species (U.S. Caribbean, Western Pacific, South Atlantic, Gulf of Mexico, North Pacific; Table 12.1). Most (5/8 regions) manage >45 protected species, while fewer regions (3/8) manage at least 70 prohibited species (mostly corals; Gulf of Mexico, South Atlantic, U.S. Caribbean) or above-average numbers (>24) of threatened or endangered species (Pacific, Western Pacific, South Atlantic). All marine regions have at least 15% of their marine mammals listed as strategic stocks, while 3/8 (Gulf of Mexico, U.S. Caribbean, South Atlantic) have >35% strategic marine mammals and >30% of marine mammal stocks of unknown status. In 3/8 regions (New England, South Atlantic, Western Pacific), >8% (cross-regional mean) of fisheries stocks are identified as experiencing overfishing, with 2/8 also containing above-average proportions of overfished stocks (New England, South Atlantic). Collectively, most regions (5/8) have >10% of fisheries stocks of unknown overfishing status, while three regions (Western Pacific, Pacific, South Atlantic are above the cross-regional average of 18.4%). Additionally, 7/8 regions have >10% of their fisheries stocks with unknown overfished status, with the U.S. Caribbean, Western Pacific, North Pacific, South Atlantic, Gulf of Mexico all above the cross-regional average of 43.9%. These findings reinforce the challenges associated with managing more speciose regions (several of which also have lower economic revenue, representation, and/or limited infrastructure), especially when accounting for the status of fisheries stocks and protected species. Unknown stock status ultimately constrains efforts toward more sustainable LMR management approaches. Despite known stock status, and advancing governance and socioeconomics in some regions, human pressures such as above-average overfishing and LMR population depletions still continue to affect their ecosystems.

The most common natural stressors occurring across regions include temperatures increasing by >1.1°C since 1950 (4/8 regions: North Pacific, U.S. Caribbean, Pacific, New England; Table 12.1), with the N Pacific also experiencing above-average (>0.74°C; 4.4°C) temperature variability over time, and hurricane/typhoon frequency (regularly occurring in 5/8 regions: Western Pacific, Gulf of Mexico, Mid-Atlantic, South Atlantic, U.S. Caribbean). Two of eight regions (Western Pacific, Gulf of Mexico) have experienced at least 20 hurricanes and/or typhoons per decade. Additionally, total ocean economies and other ocean uses are most pronounced in ~4 regions. Three of eight regions (South Atlantic, Pacific, Gulf of Mexico) have above-average total ocean establishments, employments, and GDP, while highest overall concentrations of and economic contributions from other ocean uses occur in the Pacific, Gulf of Mexico, Mid-Atlantic, and South

Atlantic. Together, these findings show that each region is subject to above-average natural stressors (albeit different ones in each region), with above-average thermal and cyclone stressors both affecting the U.S. Caribbean. Thermal pressures are affecting all regions, especially the North Pacific, Pacific, and New England marine economies. In addition, cyclones have major impacts on Gulf of Mexico, South Atlantic, and Western Pacific socioeconomics.

Primary productivities are greater than ~200 gC m^{-2} y^{-1} in 3–4 regions (North Pacific, New England, Mid-Atlantic, Gulf of Mexico) and generally stable, with highest variabilities (~9%) only observed for average chlorophyll values in the Gulf of Mexico and Pacific (Table 12.1). Average total biomass estimates are above 10 million metric tons in 5/8 regions (with the Western Pacific only above the calculated cross-regional mean), with average total fisheries landings above 400,000 metric tons in 4/8 regions (North Pacific, Gulf of Mexico, Pacific, Mid-Atlantic). Integrative relationships among biomass, fisheries landings, and production are generally steady. Above-average relationships for fisheries landings/biomass (i.e., exploitation >0.06) occur in the Gulf of Mexico, New England, and Mid-Atlantic (3/8 regions). Similarly, above-average relationships for landings/production (i.e., exploitation >0.0003) occur in the Pacific, New England, Mid-Atlantic, Gulf of Mexico, and North Pacific (5/8 regions). As observed for socioeconomic factors, the inherent production in a given system can facilitate above-average landings and sustained exploitation levels (Chassot et al. 2010, Link & Marshak 2019), while additional processes (i.e., inverted trophic pyramid) explain the high biomass observed in coral reef-based ecosystems such as the Western Pacific (Williams et al. 2011, 2015, Graham et al. 2017).

When examined for all regions, average rankings across indicator categories were mostly mid-range, with high and low rankings observed for a few particular indicators (Table 12.2). Socioeconomic criteria were ranked highest for the New England, North Pacific, Pacific, and Gulf of Mexico regions. Overall, LMR-based socioeconomic criteria were ranked "high" in New England, where nationally highest LMR GDP proportion, and high fisheries value, numbers of vessels, integrative ratios of LMR employments, and fisheries value to biomass and production occur. Alternatively, the lowest ranked socioeconomics were observed in the U.S. Caribbean, where high human population density and low LMR-based economies and fisheries value were observed.

Highest rankings for human representative and fishery/systematic governance were observed for the South Atlantic, Western Pacific, and U.S. Caribbean, with the former two ranked "mid–high." These were related to the highest numbers of FEPs, HAPCs, and greatest percent EEZ permanently closed to fishing in the Western Pacific. The South Atlantic is tied with New England for the highest number of FMPs, with a relatively high FMC budget relative to the value of its fisheries, a more balanced FMC composition, and high numbers of protected areas and management organizations. Although, FMPs for the region have been frequently modified as well. The U.S. Caribbean contains lower numbers of management organizations, states, National Environmental Policy Act Environmental Impact (NEPA-EIS) actions, and lawsuits, and higher numbers of HAPCs, and greater congressional representation per fisheries value.

Status of LMRs was ranked highest for the Mid-Atlantic, North Pacific, and Pacific regions, with mid–high rankings for these criteria. The lowest number and percentage of overfished stocks, number of stocks of unknown status, and number of threatened or endangered species are found in the Mid-Atlantic. This region also contains the lowest number of managed fishery species. The North Pacific contains the lowest proportions of stocks experiencing overfishing or with unknown overfishing status, is tied in ranking for the lowest number of threatened or endangered species, and is second-lowest proportionally in terms of overfished stocks, strategic marine mammal stocks, and marine mammal stocks of unknown population size.

The Pacific contains the highest number of protected species, lowest proportion of marine mammal stocks of unknown population size, low numbers of prohibited species, and low proportions of stocks that are overfished or experiencing overfishing. However, it is also the region with the highest number of threatened or endangered species. Those regions ranked in the second tier of LMR status were New England, the Gulf of Mexico, and the Western Pacific. Lowest rankings for LMR status were found in the South Atlantic and U.S. Caribbean owing to high or highest numbers and percentages of stocks of unknown status, high percentages of marine mammal stocks with strategic status and unknown population size, high numbers of threatened and endangered species, and lower numbers of protected species.

Environmental forcing and limited natural stressors were generally ranked as mid-range per region with "mid–low" classifications for the Gulf of Mexico and

Table 12.2 Synthesis of average rankings (1–8; ± SE) and classifications (low, mid, high) for ecosystem indicator categories and Living Marine Resource (LMR) management capacity across U.S. marine regions

Average Rankings	Indicators	N Pacific	Pacific	W Pacific	Gulf of Mexico	U.S. Caribbean	S Atlantic	Mid-Atlantic	New England
Socioeconomic Criteria	11	3.1 ± 0.7	3.7 ± 0.6	5.4 ± 0.5	3.6 ± 0.5	7.3 ± 0.6	5.6 ± 0.4	4.2 ± 0.7	2.8 ± 0.5
Governance Criteria									
– *Human Representative, Fishery/Systematic*	18	4.9 ± 0.7	4.4 ± 0.4	3.7 ± 0.5	4.8 ± 0.5	4.0 ± 0.6	3.7 ± 0.4	4.4 ± 0.6	5.4 ± 0.6
– *Status of LMRs*	10	3.5 ± 0.8	3.9 ± 0.9	5.2 ± 0.7	4.8 ± 0.7	5.7 ± 0.6	5.4 ± 0.6	3.2 ± 0.5	4.0 ± 0.8
Environmental Forcing & Major Features	5	4.5 ± 2	3.7 ± 1.2	3.3 ± 1.9	6.8 ± 1.3	6.7 ± 0.9	4.0 ± 1.2	4.7 ± 0.7	5.0 ± 0.8
Major Pressures & Exogenous Factors	10	5.3 ± 0.6	2.4 ± 0.3	5.1 ± 0.6	3.0 ± 0.7	5.7 ± 0.8	2.8 ± 0.4	3.6 ± 0.8	5.0 ± 0.6
Systems Ecology & Fisheries	10	3.7 ± 0.7	3.5 ± 0.6	3.8 ± 1.0	4.5 ± 0.5	5.6 ± 0.9	5.0 ± 0.6	3.7 ± 0.6	4.3 ± 0.5
Total	**64**	**4.2 ± 0.3**	**3.7 ± 0.2**	**4.5 ± 0.3**	**4.4 ± 0.3**	**5.5 ± 0.3**	**4.3 ± 0.2**	**3.9 ± 0.3**	**4.4 ± 0.3**
Classification									
Socioeconomic Criteria	11	Mid–High	Mid–High	Mid	Mid–High	Low	Mid	Mid	High
Governance Criteria									
– *Human Representative Fishery/Systematic*	18	Mid	Mid	Mid–High	Mid	Mid	Mid–High	Mid	Mid
– *Status of LMRs*	10	Mid–High	Mid–High	Mid	Mid	Mid	Mid	Mid–High	Mid
Environmental Forcing & Major Features	5	Mid	Mid–High	Mid–High	Mid–Low	Mid–Low	Mid	Mid	Mid
Major Pressures & Exogenous Factors	10	Mid	High	Mid	Mid–High	Mid	High	Mid–High	Mid
Systems Ecology & Fisheries	10	Mid–High	Mid–High	Mid–High	Mid	Mid	Mid	Mid–High	Mid
Total	**64**	**Mid**	**Mid–High**	**Mid**	**Mid**	**Mid**	**Mid**	**Mid–High**	**Mid**

U.S. Caribbean, owing to concentrated thermal stressors, hurricane frequency, and bottom water hypoxia (Table 12.2). "Mid–high" classifications were observed for the Pacific and Western Pacific. While typhoons are most frequent in the Western Pacific and thermal stressors including coral bleaching have increased in magnitude in recent years (Heenan et al. 2017), average temperatures have remained relatively stable for the entire region over time (26.3 °C ± 0.1 SE) with the lowest rate of increase (0.6°C since 1950) for any region. Additionally, while the Pacific is subject to low hurricane intensity and less variable sea surface temperature (SST), it is third-highest nationally in terms of temperature increase since 1950 (1.3°C).

Major pressures and exogenous factors that account for other ocean uses were ranked highest for the Pacific, South Atlantic, Gulf of Mexico, and Mid-Atlantic regions. In addition to having the second-highest total ocean establishments, employments, and GDP, the Pacific additionally leads in terms of proportional contributions toward its marine transportation, ship and boatbuilding, and offshore mineral extraction economic sectors. Additionally, the Pacific contains the highest number of marinas and high numbers of dive shops, ports, and cruise departure passengers. The South Atlantic contains the highest numbers of dive shops and cruise departure passengers, and high numbers of wind energy areas, ports, and marinas, demonstrating concentrated trade-offs among ocean sectors that can influence LMR socioeconomic status. Several other ocean uses in the Gulf of Mexico are highest nationally, especially total ocean GDP, ports, and oil rigs. This region leads in terms of proportional contributions toward its marine construction, marine transportation, ship and boatbuilding, and offshore mineral extraction sectors. Additionally, the Mid-Atlantic leads nationally in wind energy development and in total ocean economy establishments and employments. While mid-ranked overall, the North Pacific and U.S. Caribbean are ranked relatively lowest for other ocean uses, except cruise destination passengers, and proportional contributions toward offshore mineral extractions and tourism and recreation, respectively.

Related to overall systems ecology and fisheries, ranked productivities were highest in the North Pacific, New England, and Mid-Atlantic, and secondarily in the Gulf of Mexico and Pacific regions (Table 12.2). While average annual productivity was ranked fifth-highest overall for the Pacific region, it is subject to seasonal periods of high productivity from upwelling of nutrient rich deeper waters. Additionally, overall ecological and fisheries indicators were ranked highest in the Pacific, Mid-Atlantic, and North Pacific, owing to higher rankings for total biomass and fisheries landings. The Western Pacific also emerged as fourth-highest as a result of its highest biomass, lower reported average bycatch (by weight and individuals), and lower exploitation rates as related to biomass and production. Overall, more inherently productive regions appear to be more resilient to relative cumulative pressures, as observed in the higher ranked statuses of their LMRs and socioeconomics, including fisheries landings, whereas regions with lower ranked productivities (i.e., Western Pacific, U.S. Caribbean, South Atlantic) also tended to have lower rankings for LMRs, landings, and socioeconomics. While these relationships do not appear to translate directly toward biomass, landings, employments, or fisheries value for all regions in a 1:1 manner, their rankings were generally highest in the four most productive regions (Table 12.3). This is again indicative of support for the general pathway proposed by Link and Marshak

Table 12.3 Synthesis of average rankings (1 to 8; ± SE) and classifications (low L, mid M, high H) for primary production (P), biomass (B), total fisheries landings (L), % Living Marine Resources (LMR) employments (jobs, J), and total value of commercial fisheries landings (V) across U.S. marine regions from 1998 to 2015. The relationship among these relative variables is additionally included

Average Ranking	**N Pacific**	**Pacific**	**W Pacific**	**Gulf of Mexico**	**U.S. Caribbean**	**S Atlantic**	**Mid-Atlantic**	**New England**
Primary Production	1	5	8	4	6	7	2.5	2.5
Systems Ecology and Fisheries								
– Total Est. Biomass	2	3	1	6			4.5	4.5
– Fisheries Landings	1	3	7	2	8	6	4.5	4.5
Socioeconomic Status								
– % LMR Employments	1	2	6	5	8	7	3.5	3.5
–Fisheries Value	1	5	7	2	8	6	3.5	3.5
Classification								
Primary Production	High	Mid	Low	Mid	Mid	Mid	High	High
Systems Ecology & Fisheries								
– Total Est. Biomass	High	High	High	Mid			Mid	Mid
– Fisheries Landings	High	High	Low	High	Low	Mid	Mid	Mid
Socioeconomic Status								
– % LMR Employments	High	High	Mid	Mid	Low	Low	High	High
– Fisheries Value	High	Mid	Low	High	Low	Mid	High	High
Progression PBLJV	**HHHHH**	**MHHHM**	**LHLML**	**MMHMH**	**M—LLL**	**M—MLM**	**HMMHH**	**HMMHH**

(2019) relating productivity, biomass, landings, fisheries value, and LMR-based employments (see Introduction, Chapter 1). Overall, the eight regions have mid-range rankings in terms of their capacity toward successful (i.e., and ecosystem approach to) LMR management, with assets and limitations varying across the different regions reflective of conditions, histories, and stressors in those ecosystems (Table 12.2). The Pacific, Mid-Atlantic, and North Pacific, closely followed by the South Atlantic, New England, and Gulf of Mexico emerge as those regions with currently the highest capacity for consideration of socioeconomic, governance, other ocean uses, ecological, and fisheries-based criteria in their management approaches.

In summary, more productive regional ecosystems tended to have higher LMR status and higher associated socioeconomic conditions. These relationships were additionally amplified or mitigated by the degree of various governance actions and choices. Those regions demonstrating greatest accordance with these trends were the North Pacific, Mid-Atlantic, and New England. Second-tier primary production in the Gulf of Mexico and Pacific also favored comparatively high-ranked fisheries landings and value in these regions. Additionally, both the Gulf of Mexico and South Atlantic are bolstered by the high development of their other ocean uses and tourism-driven economies. Limited primary production in subtropical and tropical regions also appears to be a major driver for their lower LMR and socioeconomic statuses, suggesting greater potential for system overexploitation and the need for longer recovery times of these socioecological systems (SESs).

12.3 U.S. Subregions

Among U.S. marine subregions, ~8–9 of 26 were above the average in most indicator categories (Table 12.4), with average variability also generally low per indicator in most subregions. For socioeconomic criteria, indicators were above cross-regional mean values for ~7/26 subregions and were generally stable. Human population stressors are strongest in and surrounding

Table 12.4 Examination per category of ecosystem indicators across U.S. marine subregions as related to given cross-subregional mean values. Data are presented as the number of subregions above a calculated cross-subregional mean indicator value (anomaly), the number of highly variable (>10% relative SE over time) subregions per indicator, and the overall number (± SE) of above mean and highly variable subregions per category. Values for #Congressionals (and standardized), Fishery Management Council (FMC) Composition, and #FMP (Fishery Management Plan) or FEP (Fishery Ecosystem Plan) modifications reflect the number of subregions with values below the cross-regional mean value, as indicated by asterisk. EEZ = Exclusive Economic Zone, GDP = Gross Domestic Product, HFA = NOAA Habitat Focus Area, LMR = Living Marine Resource, NERR = National Estuarine Research Reserve, NEPA EIS = National Environmental Policy Act Environmental Impact Statement, SRG = Scientific Review Group, Var = Variability

	Timeframe	Cross-Subregional Mean Value	Subregions Above Mean (Anomaly)	Subregions >10% Variability
Socioeconomic Criteria				
– Avg. Population	1970–2010	2.3 million humans	8	8
– Avg. Population Density	1970–2010	109.7 humans km^{-2}	9	8
– Avg. % LMR Establishments	2005–2016	7.8%	7	2
– Avg. % LMR Employments	2005–2016	10.5%	4	0
– Avg. % LMR GDP	2005–2016	10.6%	4	10
– Avg. Fisheries Value	1981–2016	8.03×10^7 USD	8	2
Above Mean			*6.7 ± 0.9*	*5.0 ± 1.7*
Governance Criteria				
Human Representative, Fishery/Systematic Context				
– #Organizations for Mgmt	2017	15.9 organizations	14	
– #States	2017	1.7 states	7	

	Timeframe	Cross-Subregional Mean Value	Subregions Above Mean (Anomaly)	Subregions >10% Variability
Governance Criteria				
– #Congressionals*	2017	5.6 Congressionals	8	
– #Congressionals/Mile of Shoreline*	2017	0.005 Congressionals/Mile	11	
– #Congressionals/Value of Fishery*	2017	$<1.50\times10^{-7}$ Congressionals/USD	7	
– #FMPs (both federal and state)	2017	16.4 FMPs	16	
– #FEPs (current or in development)	2017	1.8 FEPs	2	
– #FMP Modifications*	2017	177.2 modifications	8	
– #FEP Modifications*	2017	10.5 modifications	1	
– %EEZ Fishing Permanently Prohibited	2017	7.9%	5	
– #Protected Areas (i.e., Sanctuaries, NERRs, HFAs, etc.)	2017	3.5 areas	8	
– #Cumulative #NEPA EIS Actions	1987–2017	33.6 actions	8	
– Avg. FMC Composition—%Commercial*	1990–2016	<50.5%	11	0
– Avg. Number of Members on Marine Mammal SRG	1995–2017	9.7 members y^{-1}	20	0
– Avg. FMC Budget to Fisheries Value	2007–2016	0.007	4	0
– Avg. Lawsuits	2010–2016	1.8 lawsuits y^{-1}	15	7
– Number Lawsuits	2010–2016	11.5 lawsuits	15	
Status of LMRs				
– #Managed Fishery Species	2017	127 species	14	
– #Prohibited Species	2017	30 Species	5	
– #Protected Species	2017	43.2 Species	20	
– %Stocks Experiencing Overfishing	2017	5.8%	10	
– %Stocks Overfished	2017	8.9%	6	
– %Stocks Unknown Overfishing Status	2017	13.7%	11	
– %Stocks Unknown Overfished Status	2017	34.9%	13	
– %Strategic Stocks	2017	35.5%	7	
– % Marine Mammal Stocks of Unknown Population Size	2017	28.3%	9	
– #Threatened/Endangered Species	2017	25.8 Species	7	
Above Mean			*9.7 ± 0.9*	*1.8 ± 1.8*
Environmental Forcing & Major Features				
– Avg. Sea Surface Temperature	1854–2016	15.5°C	11	0
– Var. Sea Surface Temperature	1854–2016	0.04°C	12	0
– Temperature Increase (°C)	1950–2016	1.15°C	11	
– Avg. Number of Hurricanes per Decade	1850–2016	6.9/decade	1	16
– Avg. Proportion of Sea Ice	1982–2016	0.43	1	0
Above Mean			*7.2 ± 2.5*	*4.0 ± 4.0*

Continued

Table 12.4 Continued

	Timeframe	Cross-Subregional Mean Value	Subregions Above Mean (Anomaly)	Subregions >10% Variability
Major Pressures and Exogenous Factors				
Other Ocean Uses				
– #Dive Shops	2016	14.5 shops	5	
– #Ports	2016	3.6 ports	12	
– #Marinas	2016	9.9 marinas	7	
– #Cruise Departure Passengers	2009, 2011	449.1 passengers	4	
Above Mean			*7.0 ± 1.8*	
Systems Ecology and Fisheries				
– Avg. Chlorophyll	1998–2016	6.72 mg/m^3	14	1
– Avg. Number of Taxa	1950–2016	65.3 taxa	12	1
– Avg. Fisheries Landings	1950–2016	114.4k metric tons	6	6
Above Mean			*10.7 ± 2.4*	*2.7 ± 1.7*

8–9 subregions (Long Island Sound, Channel Islands, Delaware Bay, San Francisco Bay, Narragansett Bay, Galveston Bay, Tampa Bay, Chesapeake Bay, and Monterey Bay), with noteworthy increases in 8/26 subregions since 1970 (Aleutian Islands, Gulf of Alaska, Tampa Bay, Pacific Island territories, Prince William Sound, Galveston Bay, Arctic, Puget Sound-Salish Sea). Additionally LMRs contribute greater than the cross-subregion average (7.8%) to total oceanic establishments for 7/26 subregions (Aleutian Islands, East Bering Sea, Penobscot Bay, Gulf of Alaska, Columbia River, Inside Passage, Casco Bay), and contribute >10% toward total oceanic employments in 4/26 subregions (Aleutian Islands, East Bering Sea, Inside Passage, Gulf of Alaska). Similarly, LMRs in 4/26 subregions (Aleutian Islands, East Bering Sea, Inside Passage, Puget Sound-Salish Sea) contribute >10% toward their respective total oceanic GDPs, while 10/26 subregions have high variability in terms of these proportions. In terms of total fisheries value, 8/26 subregions contain above-average fisheries revenue (Gulf of Alaska, Mississippi River Delta, Chesapeake Bay, E Bering Sea-Aleutian Islands, Inside Passage, Delaware Bay, Long Island Sound). Generally, subregions with above-average LMR-based socioeconomics are mainly those occurring in the North Pacific, New England, and Mid-Atlantic. Of those locations, greatest human pressures are found in Long Island Sound, Delaware Bay, and Chesapeake Bay.

Most governance indicators related to human representative, fishery/systematic, and LMR status contexts were above average for 9–10/26 subregions (Table 12.4). Additionally, 11/26 subregions have >0.005 (cross-subregion average) congressionals per mile of shoreline (Narragansett Bay, Long Island Sound, Delaware Bay, Mobile Bay, Monterey Bay, Galveston Bay, Penobscot Bay, Tampa Bay, Casco Bay, Matagorda Bay, Channel Islands) and 7/26 have greater than the cross-subregion average ($1.5{\times}10^{-7}$) for Congressionals per value of fisheries (San Francisco Bay, Pacific Islands Territories, Monterey Bay, Tampa Bay, Cape Hatteras-Pamlico Sound, Long Island Sound, Florida Keys). There are 16/26 subregions with >14 state and federal FMPs, which are encompassed in the South Atlantic, New England, Mid-Atlantic, and Gulf of Mexico regions, and 8/26 subregions (Long Island Sound, Casco Bay, Narragansett Bay, Penobscot Bay, Chesapeake Bay, Delaware Bay, Aleutian Islands, East Bering Sea) have had their jurisdictional FMPs modified >180 times. Greater than one FEP is currently in place within 2/26 subregions (Hawaii, PI Territories), while only the Pacific Island territories have modified their FEPs >10 times. Two North Pacific subregions (East Bering Sea, Aleutian Islands) also have individual FEPs. Additionally, only 5/26 subregions (Arctic, PI Territories, Channel Islands, Hawaii, San Francisco Bay) have >8% (cross-subregion average) of their EEZs permanently prohibited from fishing. For several governance indicators, these values apply at the regional scale. While incorporated for examining subregions too, this information is already detailed above under the U.S. regions subheading (Section 12.2). As similarly found among regions, many subregions with above-average congressional representation tend to have above-average socioeconomics and human population stressors. Additionally, relationships among governance indicators for subregions closely align with those observed at the regional level.

The most common natural stressors occurring across subregions include temperatures increasing by >1.15°C since 1950 (11/26 subregions: Inside Passage, Prince William Sound, Gulf of Alaska, East Bering Sea, Puget Sound-Salish Sea, Columbia River, San Francisco Bay, Monterey Bay, Casco Bay, Penobscot Bay, Arctic; Table 12.4), with 12/26 subregions also experiencing above-average (>0.04°C) temperature variability over time. Only 1/26 subregion (PI Territories) has experienced >7 hurricanes per decade, with the majority of these occurring throughout the Pacific Remote Islands and Marianas archipelagos. Of those tourism-based indicators for other ocean uses available at the subregional scale, highest overall concentrations were found for 5/26 subregions in terms of dive shops (Hawaii, Channel Islands, PI Territories, Florida Keys, Chesapeake Bay), 12/26 subregions regarding ports (Mississippi River Delta, Delaware Bay, San Francisco Bay, Gulf of Alaska, Puget Sound-Salish Sea, Hawaii, Arctic, Channel Islands, Columbia River, Galveston Bay, Chesapeake Bay, Long Island Sound), and 7/26 subregions regarding marinas (Hawaii, Puget Sound-Salish Sea, Gulf of Alaska, Chesapeake Bay, Inside Passage, Columbia River). Cruise departure passengers are also above-average in areas surrounding 4/26 subregions (Florida Keys, Channel Islands, Puget Sound-Salish Sea, Galveston Bay). Above-average thermal stressors are most pronounced in North Pacific, Pacific, and New England subregions, particularly the northernmost of the latter two regions. These subregions likewise have above-average marine economies that are vulnerable to these impacts. Above-average cyclone numbers are generally less common throughout subregions, and are mainly concentrated in Pacific Island territories with comparatively low human populations.

Chlorophyll values are above the cross-subregional mean in 14 subregions (Chesapeake Bay, Delaware Bay, Columbia River, Penobscot Bay, Mississippi River Delta, Mobile Bay, Galveston Bay, Matagorda Bay, Cape Hatteras-Pamlico Sound, Long Island Sound, Casco Bay, Puget Sound-Salish Sea, Monterey Bay, San Francisco Bay). Among subregions, average chlorophyll is generally stable, with highest variabilities (~12%) only observed in Narragansett Bay (Table 12.4). Additionally, 12/26 subregions capture above-average numbers of taxa in commercial landings. Average total fisheries landings are above the cross-subregional mean (114,000 metric tons) in 6/26 subregions (Cape Hatteras-Pamlico Sound, Mississippi River Delta, Gulf of Alaska, Chesapeake Bay, East Bering Sea, Channel Islands). Relationships between above-average chlorophyll production and fisheries landings generally hold for Cape Hatteras-Pamlico Sound, and the largest subregions of the Gulf of Mexico (Mississippi River Delta), and Mid-Atlantic (Chesapeake Bay).

Throughout subregions, average rankings across indicator categories were generally mid-range values, with some high and low rankings also observed (Table 12.5). Socioeconomic criteria were ranked highest for the Arctic, Aleutian Island, East Bering Sea, Inside Passage, Gulf of Alaska, Cape Hatteras-Pamlico Sound, and Puget Sound-Salish Sea subregions (Table 12.5). Much of this pattern was related to the occurrence of lower human stressors in many of those subregions. Excluding the Arctic, proportional contributions of LMRs toward total ocean economy are also highest in these subregions, in addition to Long Island Sound. Alternatively, the lowest ranked socioeconomics were observed in more populated regions of San Francisco Bay, Monterey Bay, Tampa Bay, Galveston Bay, and the Channel Islands, where high human population density and comparatively lower LMR-based economies and fisheries value have occurred.

Governance criteria in terms of human representative and fishery/systematic contexts were ranked "mid–high" in Monterey Bay, Hawaii, and the PI Territories (Table 12.5). All other subregions were mid-ranked in terms of these contexts. In these "mid–high" ranked subregions, highest numbers of FEPs, lower numbers of FMPs, greater percentage of EEZ permanently closed to fishing, and low frequency of lawsuits occur. While mid-ranked for all subregions, status of LMRs is highest in Chesapeake and Delaware Bays, the Arctic, Inside Passage, Gulf of Alaska, and Prince William Sound. As observed for the Mid-Atlantic and North Pacific regions, and with information at both regional and subregional scales, low numbers and percentages of overfished or depleted stocks, unassessed stocks, and threatened and endangered species occur in these subregions. Second-tier status was observed for the East Bering Sea and Aleutian Islands and Pacific subregions, while New England, Gulf of Mexico, Western Pacific, and South Atlantic subregions were at the lower end of mid-ranked LMR status.

Environmental forcing and limited natural stressors (i.e., thermal stress and hurricanes) were mostly mid-ranked throughout subregions. Higher rankings were

Table 12.5 Synthesis of average rankings (1 to 26; ± SE) and classifications (low, mid, high) for ecosystem indicator categories and living marine resource (LMR) management capacity across U.S. marine subregions

Average Rankings	Indicators	Arctic	E Bering Sea	Gulf of Alaska	Aleutian Islands	Inside Passage	Prince William Sound	Channel Islands	Columbia River
Socioeconomic Criteria	6	1.5 ± 0.5	2.9 ± 0.6	7.2 ± 1.7	1.9 ± 0.7	6.7 ± 1.7	14.2 ± 2.7	19 ± 1.8	12 ± 1.7
Governance Criteria									
– *Human Representative, Fishery/Systematic*	17	14.9 ± 2.1	16.6 ± 1.9	16.9 ± 2	15.8 ± 1.8	16.6 ± 1.9	16 ± 1.7	11.2 ± 1.3	13.3 ± 1.6
– *Status of LMRs*	10	10.7 ± 10.3	12 ± 6.2	10.7 ± 8	12 ± 6.2	10.7 ± 8	10.7 ± 8	12.3 ± 7.8	12.3 ± 7.8
Environmental Forcing and Major Features	4	14.3 ± 7.3	21.7 ± 2.4	20 ± 4	13.8 ± 5.9	23 ± 3	25 ± 0	14.7 ± 2	17 ± 4
Major Pressures and Exogenous Factors	4	18.2 ± 4.4	19.5 ± 3.1	7.9 ± 2	23.5 ± 1.3	13.7 ± 4.2	19.3 ± 2.8	3.9 ± 1.9	10.3 ± 2.2
Systems Ecology and Fisheries	3	20.8 ± 0.3	15.3 ± 4.4	16.2 ± 5.6	15.7 ± 4.6	15.5 ± 2.8	19.2 ± 0.7	12 ± 5.1	8.3 ± 2
Total	**44**	**13.5 ± 1.6**	**14 ± 1.3**	**13.3 ± 1.3**	**13.2 ± 1.3**	**13.6 ± 1.2**	**15.4 ± 1.1**	**12.2 ± 1.1**	**12.4 ± 1**
Classification									
Socioeconomic Criteria	6	High	High	High	High	High	Mid	Low	Mid
Governance Criteria									
– *Human Representative, Fishery/Systematic*	17	Mid	Mid	Mid	Mid	Mid	Mid	Mid	Mid
– *Status of LMRs*	10	Mid	Mid	Mid	Mid	Mid	Mid	Mid	Mid
Environmental Forcing and Major Features	4	Mid	Low	Low	Mid	Low	Low	Mid	Mid
Major Pressures and Exogenous Factors	4	Mid–Low	Low	High	Low	Mid	Low	High	Mid
Systems Ecology and Fisheries	3	Low	Mid	Mid	Mid	Mid	Low	Mid	High
Total	**44**	**Mid**	**Mid**	**Mid**	**Mid**	**Mid**	**Mid**	**Mid**	**Mid**

Average Rankings		Monterey Bay	Puget Sound	San Francisco Bay	Hawaii	Pacific Islands Terr's	Galveston Bay	Matagorda Bay	Miss River Delta	Mobile Bay
Socioeconomic Criteria	6	21.3 ± 1.5	9.3 ± 2.5	22.2 ± 0.3	11 ± 1.5	13.3 ± 5.9	19.3 ± 1.4	17.5 ± 0.8	13.8 ± 3.2	10.7 ± 1
Governance Criteria										
– *Human Representative, Fishery/Systematic*	17	8.9 ± 1.2	12.2 ± 1.3	10.3 ± 1.8	9.8 ± 1.8	9.9 ± 2.2	12.3 ± 1.6	11 ± 1.6	14.6 ± 1.5	11.2 ± 1.5
– *Status of LMRs*	10	12.3 ± 7.8	12.3 ± 7.8	12.3 ± 7.8	15 ± 8.9	17 ± 8.9	16.1 ± 6.6	16.1 ± 6.6	16.1 ± 6.6	16.1 ± 6.6
Environmental Forcing and Major Features	4	17.5 ± 1.5	18.5 ± 3.5	17 ± 3	13.3 ± 7	13.3 ± 7	12 ± 3	11.7 ± 2.7	13 ± 6	11 ± 5
Major Pressures and Exogenous Factors	4	16.7 ± 4.5	4.8 ± 1.2	7.6 ± 1.9	4 ± 1.9	8.2 ± 3	10.9 ± 3.6	22 ± 1	10.6 ± 4.4	17.4 ± 3.1
Systems Ecology and Fisheries	3	13.3 ± 4.9	9.3 ± 4.6	14 ± 4.5	22 ± 2	25 ± 0	13.3 ± 5	14.7 ± 6	5.5 ± 2	12.2 ± 3.3
Total	**44**	**13.1 ± 1.1**	**11.2 ± 1**	**12.9 ± 1.2**	**11.7 ± 1.2**	**13 ± 1.5**	**14.2 ± 1**	**14.4 ± 1**	**13.7 ± 1.1**	**13 ± 1**
Classification										
Socioeconomic Criteria	6	Low	Mid–High	Low	Mid	Mid	Low	Mid	Mid	Mid
Governance Criteria										
– *Human Representative, Fishery/Systematic*	17	Mid–High	Mid	Mid	Mid–High	Mid–High	Mid	Mid	Mid	Mid
– *Status of LMRs*	10	Mid	Mid	Mid	Mid	Mid	Mid	Mid	Mid	Mid
Environmental Forcing and Major Features	4	Mid-Low	Mid–Low	Mid	Mid	Mid	Mid	Mid	Mid	Mid
Major Pressures and Exogenous Factors	4	Mid	High	High	High	High	Mid	Low	Mid	Mid-Low
Systems Ecology and Fisheries	3	Mid	Mid-High	Mid	Low	Low	Mid	Mid	High	Mid
Total	**44**	**Mid**	**Mid**	**Mid**	**Mid**	**Mid**	**Mid**	**Mid**	**Mid**	**Mid**

Continued

Table 12.5 Continued

Average Rankings	Indicators	Tampa Bay	Cape Hatteras	Florida Keys	Chesapeake Bay	Delaware Bay	Long Island Sound	Casco Bay	Narragansett Bay	Penobscot Bay
Socioeconomic Criteria	6	19.5 ± 1.3	8.8 ± 1.8	10.2 ± 1.6	15.2 ± 2.8	17.7 ± 2.7	13.7 ± 3.9	11.7 ± 1.9	15 ± 3.3	13.3 ± 3
Governance Criteria										
– *Human Representative, Fishery/Systematic*	17	11.7 ± 1.7	10.4 ± 1.5	10.7 ± 1.5	13 ± 2.1	10.6 ± 2.1	17.6 ± 2.1	16.3 ± 2	16.7 ± 2.2	17.2 ± 1.7
– *Status of LMRs*	10	16.1 ± 6.6	18.7 ± 4.6	18.7 ± 4.6	10.5 ± 5.7	10.5 ± 5.7	13 ± 7.5	13 ± 7.5	13 ± 7.5	13 ± 7.5
Environmental Forcing and Major Features	4	11.3 ± 5.8	10.7 ± 6.9	11.3 ± 4.9	12 ± 5.5	13.3 ± 5	17.3 ± 3.8	20.1 ± 1.9	16.3 ± 2.8	18.1 ± 1.8
Major Pressures and Exogenous Factors	4	11 ± 2.3	19.7 ± 1.3	8.6 ± 3.6	6.8 ± 1	8.5 ± 3.4	8.8 ± 1.2	19.7 ± 1.3	17.2 ± 2.2	21 ± 0
Systems Ecology and Fisheries	3	3 ± 0	7.7 ± 3.3	16.3 ± 2.6	2.3 ± 0.9	7.3 ± 0.3	6.7 ± 2.4	14.2 ± 0.4	11 ± 4.2	13.2 ± 0.6
Total	**44**	**13.7 ± 1.1**	**12.8 ± 1.1**	**12.8 ± 1**	**11.3 ± 1.1**	**11.4 ± 1.1**	**14.4 ± 1.3**	**15.3 ± 1.1**	**15.1 ± 1.2**	**15.7 ± 1**
Classification										
Socioeconomic Criteria	6	Low	Mid-High	Mid	Mid	Mid-Low	Mid	Mid	Mid	Mid
Governance Criteria										
– *Human Representative, Fishery/Systematic*	17	Mid	Mid	Mid	Mid	Mid	Mid-Low	Mid	Mid	Mid
– *Status of LMRs*	10	Mid	Mid–Low	Mid–Low	Mid	Mid	Mid	Mid	Mid	Mid
Environmental Forcing and Major Features	4	Mid	Mid	Mid	Mid	Mid	Mid-Low	Low	Mid	Mid-Low
Major Pressures and Exogenous Factors	4	Mid	Low	High	High	High	Mid-High	Low	Mid	Low
Systems Ecology and Fisheries	3	High	High	Mid	High	High	High	Mid	Mid	Mid
Total	**44**	**Mid**	**Mid**	**Mid**	**Mid**	**Mid**	**Mid**	**Mid**	**Mid**	**Mid**

observed for Cape Hatteras-Pamlico Sound, Mobile Bay, Tampa Bay, Florida Keys, and Matagorda Bay. Additionally, "mid–low" and "low" rankings were observed for Penobscot Bay, Puget Sound-Salish Sea, Gulf of Alaska, Casco Bay, East Bering Sea, Inside Passage, Monterey Bay, Long Island Sound, and Prince William Sound, especially as they related to increasing SST, sea ice concentrations, and thermal variability. Tourism-related other ocean uses were ranked highest for the Channel Islands, Hawaii, Puget Sound-Salish Sea, Chesapeake Bay, San Francisco Bay, and Gulf of Alaska, especially related to highest numbers of dive shops, marinas, and cruise passengers. Additional high rankings were observed for the PI Territories collectively (including American Samoa, Guam, and the Commonwealth of the Northern Marianas), Delaware Bay, Florida Keys, and Long Island Sound.

Overall ecological and fisheries indicators examined at the subregional scale were ranked highest in Chesapeake Bay, Tampa Bay, the Mississippi River Delta, Long Island Sound, Delaware Bay, Cape Hatteras-Pamlico Sound, and the Columbia River. In addition, most other regions were mid-ranked overall in terms of their ecological and fisheries status, with Puget Sound-Salish Sea ranked "mid-high" and four regions ranked low. Unlike at the regional scale, inherently productive subregions were not necessarily higher ranked in the statuses of their LMRs or socioeconomics, including fisheries landings. However, these trends were observed for Chesapeake Bay, Long Island Sound, the Mississippi River Delta, the Gulf of Alaska, and in Delaware and Mobile Bays to a lesser extent. These trends may not always follow at the subregional scale, given differential productivities that are subject to more localized stressors and allochthonous inputs, and which may not necessarily translate to higher fisheries production. Concentrated nearshore human effects may also lead to differential LMR and socioeconomic statuses over time. Overall, the 26 subregions have mid-range rankings in terms of their capacity toward successful (i.e., an ecosystem approach to) LMR management, with assets and limitations varying across the different subregions reflective of conditions, histories, and stressors in their ecosystems (Table 12.5). In terms of ranking among all indicators, Puget Sound-Salish Sea, Chesapeake Bay, Delaware Bay, Hawaii, the Channel Islands, and Columbia River Estuary, closely followed by Cape Hatteras-Pamlico Sound, the Florida Keys, and San Francisco Bay, emerge as those regions with currently the highest capacity in consideration of socioeconomic, governance, other ocean use, ecological, and fisheries-based criteria in their management approaches.

In summary, a relationship between ecosystem productivity, LMR status, and associated socioeconomic conditions was not as apparent at the subregional scale. Those subregions demonstrating highest-ranked systems ecology and fisheries were mostly associated with the Gulf of Mexico, Pacific, and South Atlantic regions. Likewise, those subregions with highest overall total rankings were associated with the Pacific, Mid-Atlantic, Western Pacific, and South Atlantic regions. At the subregional scale, socioeconomic and LMR status appear to be greatly influenced by major pressures and system functioning, while also amplified by governance actions.

12.4 U.S. Participatory RFMOs

For RFMO jurisdictions, ~2 out of 12 were above average (Table 12.6), with low average variability over time per indicator. For socioeconomic criteria, indicators were above cross-jurisdictional mean values for ~2/12 regions and were generally stable. Human population stressors were strongest in U.S. Atlantic regions for which highly migratory species (HMS) and jurisdictions for the International Convention for Conservation of Atlantic Tunas (ICCAT) occur. Additionally, U.S. Pacific LMRs contribute higher above-average proportions to total ocean economies above cross-jurisdictional averages of >5% of total oceanic establishments, and >3% for LMR employments and GDP. Permitted vessels were above the cross-jurisdictional average in 4/12 RFMOs (Atlantic HMS-ICCAT, North Atlantic Fisheries Organization—NAFO, Western and Central Pacific Fisheries Commission—WCPFC, Western and Central Atlantic Fisheries Commission—WECAFC). Above-average cross-jurisdictional fisheries revenue additionally occurs in 4/12 RFMOs (WCPFC, NAFO, WECAFC, Pacific Salmon Commission—PSC). Average integrative relationships among LMR employments, fisheries revenue, and production (U.S. contributory and total RFMO) are generally steady, with ~2/12 jurisdictions above cross-regional mean values. Only 1/12 jurisdiction (NAFO) contains above-average relationships related to fisheries revenue/U.S. contributory production, while 2/12 (WECAFC, NAFO) are above average in terms of fisheries revenue/total RFMO production. Three of 12 jurisdictions have above-average LMR employments/U.S. primary production values (North Atlantic Salmon Conservation Organization—NASCO, NAFO, Pacific Whiting Treaty—PWT), while 4/12 jurisdictions (WECAFC, PWT, PSC, International Pacific Halibut Commission—IPHC) have above-average LMR employments/total RFMO production. Highest variability for these ratios

Table 12.6 Examination per category of ecosystem indicators across U.S. participatory regional fishery management organization (RFMO) jurisdictions as related to given cross-jurisdictional mean values. Data are presented as the number of jurisdictions above a calculated cross-jurisdictional mean indicator value (anomaly), the number of highly variable (>10% relative SE over time) jurisdictions per indicator, and the overall number (± SE) of above mean and highly variable jurisdictions per category. Values for #Congressionals (and standardized), #FMP (Fishery Management Plan) or FEP (Fishery Ecosystem Plan) modifications reflect the number of regions with values below the cross-regional mean value, as indicated by asterisk. GDP = Gross Domestic Product, IUCN = International Union for Conservation of Nature, LMR = Living Marine Resource, Var = Variability

	Timeframe	Cross-Regional Mean Value	Regions Above Mean (Anomaly)	Regions >10% Variability
Socioeconomic Criteria				
– Avg. Population	1970–2010	44.3 million humans	1	0
– Avg. Population Density	1970–2010	45.7 humans km^{-2}	1	0
– Avg. % LMR Establishments	2005–2016	4.8%	1	0
– Avg. % LMR Employments	2005–2016	2.6%	1	0
– Avg. % LMR GDP	2005–2016	2.8%	1	0
– Avg. #Permitted Vessels	1990–2015	8959.9 vessels	3	5
– Avg. Fisheries Value	1950–2016	2.3×10^8 USD	4	5
– Avg. Ratio of Total LMR Employments (Jobs)/ Total U.S. Contributory Production	2005–2014	6.1	3	1
– Avg. Ratio of Total Value (Revenue) of Commercial Fisheries/Total U.S. Contr. Prod.	1968–2016	0.143	1	2
– Avg. Ratio of Total LMR Employments (Jobs)/ Total RFMO Production	2005–2014	0.69	4	1
– Avg. Ratio of Total Value (Revenue) of Commercial Fisheries/Total RFMO Production	1998–2014	0.027	2	2
Above Mean			*1.9 ± 0.3*	*1.4 ± 0.6*
Governance Criteria				
Human Representative, Fishery/Systematic Context				
– #States	2017	10.5 states	6	
– #Congressionals*	2017	56.3 Congressionals	3	
– #Congressionals/Mile of Shoreline*	2017	0.003 Congressionals/Mile	5	
– #Congressionals/Value of Fishery*	2017	<6.6×10^{-7} Congressionals/USD	2	
– #FMPs	2017	7.3 FMPs	2	
– #FEPs (current or in development)	2017	1.7 FEPs	3	
– #FMP Modifications*	2017	86 modifications	3	
– #FEP Modifications*	2017	8.6 modifications	2	
– %Complete No-Take Areas	2017	3.0%	1	
Status of LMRs				
– #Managed Fishery Species	2017	99.9 species	4	
– %Stocks Experiencing Overfishing	2017	7.0%	4	
– %Stocks Overfished	2017	15.2%	3	
– %Stocks Unknown Overfishing Status	2017	40.5%	4	
– %Stocks Unknown Overfished Status	2017	36.8%	4	
– %Strategic Stocks	2017	39.2%	1	
– % Marine Mammal Stocks of Unknown Population Size	2017	33.9%	1	

	Timeframe	Cross-Regional Mean Value	Regions Above Mean (Anomaly)	Regions >10% Variability
Status of LMRs				
– #IUCN-Listed Threatened/Endangered Species	2017	274.3 Species	2	
– %IUCN-Listed Endangered Species	2017	40.5%	2	
Above Mean			*2.9 ± 0.3*	
Environmental Forcing and Major Features				
– Avg. Sea Surface Temperature	1854–2016	22.2°C	1	0
– Var. Sea Surface Temperature	1854–2016	0.02°C	1	0
– Temperature Increase (oC)	1950–2016	0.99°C	1	
Above Mean			*1.0 ± 0.0*	*0.0 ± 0.0*
Major Pressures and Exogenous Factors				
Other Ocean Uses				
– Avg. Ocean Economy Establishments	2005–2016	64,818.2 establishments	1	0
– Avg. Ocean Economy Employments	2005–2016	1,310,060 employments	1	0
– Avg. Ocean Economy GDP	2005–2016	$145.3 billion	1	0
– Avg. % Marine Construction Establishments	2005–2016	1.96%	1	0
– Avg. % Marine Construction Employments	2005–2016	1.55%	1	0
– Avg. % Marine Construction GDP	2005–2016	2.00%	1	0
– Avg. % Marine Transportation Establishments	2005–2016	6.73%	1	0
– Avg. % Marine Transportation Employments	2005–2016	15.4%	1	0
– Avg. % Marine Transportation GDP	2005–2016	20.7%	1	0
– Avg. % Offshore Mineral Extraction Establishments	2005–2016	3.0%	1	0
– Avg. % Offshore Mineral Extraction Employments	2005–2016	4.9%	1	0
– Avg. % Offshore Mineral Extraction GDP	2005–2016	38.3%	1	0
– Avg. % Ship and Boatbuilding Establishments	2005–2016	1.2%	1	0
– Avg. % Ship and Boatbuilding Employments	2005–2016	5.2%	1	0
– Avg. % Ship and Boatbuilding GDP	2005–2016	5.7%	1	0
– Avg. % Tourism/Recreation Establishments	2005–2016	82.1%	1	0
– Avg. % Tourism/Recreation Employments	2005–2016	70.3%	1	0
– Avg. % Tourism/Recreation GDP	2005–2016	31.2%	1	0
Above Mean			*1.0 ± 0.0*	*0.1 ± 0.1*
Systems Ecology and Fisheries				
– Avg. U.S. Contributory Primary Production	1998–2014	411.85 gC m^{-2} y^{-1}	4	0
– Avg. Total RFMO Primary Production	1950–2016	1638.6 gC m^{-2} y^{-1}	3	0
– Avg. Fisheries Landings	1950–2016	266.7k metric tons	3	6
– Avg. Ratio of Total Landings/U.S. Contributory Primary Production	1964–2016	9.21×10^{-5}	3	2
– Avg. Ratio of Total Landings/Total RFMO Primary Production	1998–2014	2.11×10^{-5}	2	1
Above Mean			*3.0 ± 0.3*	*1.8 ± 1.1*

was observed for WCPFC and PWT jurisdictional values. Similarly for regions and subregions, above-average human population pressures coincide with RFMO jurisdictions (particularly encompassing the North Atlantic and Pacific basins, and the Gulf of Mexico) where above-average marine socioeconomics are likewise found.

Most governance indicators related to human representative, fishery/systematic, and LMR status contexts are above average for 2–3/12 RFMO jurisdictions (Table 12.6). There are 2/12 jurisdictions with >7 (cross-jurisdictional average) US-associated FMPs (NAFO, WECAFC) and 3/12 with highly modified FMPs (>86 modifications, the cross-jurisdictional average; NAFO, WECAFC, IPHC). U.S. pelagic fisheries resources in the greater Pacific (NPFC, WCPFC) are managed by 1 FEP, with the highest numbers of modifications among jurisdictions. Among basins, only the Antarctic (managed under the Convention for the Conservation of Antarctic Marine Living Resources—CCAMLR) has >10% of its EEZ as a complete no-take area. In general, above-average governance metrics for U.S. transboundary resources tend to occur for jurisdictions encompassing regions with above-average marine socioeconomics.

For LMR status governance indicators, 4/12 jurisdictions encompassing tropical and subtropical environments manage >100 fishery species (WECAFC, WCPFC, NPFC, SPRFMO). Additionally, higher proportions of U.S. strategic and unassessed marine mammal stocks occur throughout U.S. Atlantic jurisdictions. In 4/12 jurisdictions (WCPFC, Inter-American Tropical Tuna Commission—IATTC, Atlantic HMS-ICCAT, NAFO), >7% (cross-jurisdictional average) of U.S. fisheries stocks are identified as experiencing overfishing, with 3/12 also containing above-average proportions of overfished stocks (NASCO, NAFO, Atlantic HMS-ICCAT). Collectively, 4/12 jurisdictions have >40% of U.S. fisheries stocks of unknown overfishing (CCAMLR, IPHC, NPFC, SPRFMO) or overfished (CCAMLR, NPFC, SPRFMO, WECAFC) status. In addition, the Atlantic and Pacific basins contain above-average numbers of International Union for the Conservation of Nature (IUCN)-listed threatened and endangered species, while the Atlantic and Antarctic basins also contain above-average proportions of IUCN-listed endangered species. These findings reinforce the ongoing challenges in accurately assessing speciose tropical environments and for the sustainable management of HMS. In general, the Atlantic basin contains above-average overfishing and overfished statuses of its transboundary populations, while above-average overfishing also continues to affect Pacific HMS. Above-average overfishing and overfished statuses likewise coincide with those jurisdictions having above-average socioeconomics.

The most common natural stressors occurring across RFMO jurisdictions and ocean basins include temperatures cumulatively increasing across U.S. Pacific jurisdictions by >0.99°C since 1950 (Table 12.6). Additionally, above-average (>0.02°C) temperature variability has been observed throughout the U.S. Pacific over time. Total ocean economies and other ocean uses are most pronounced in U.S. Atlantic jurisdictions in terms of establishments, employments, and GDP. Highest overall concentrations of and economic contributions from other ocean uses generally occur in the Pacific for other ocean use establishments and for the tourism/recreation sector. In the Atlantic, proportional values for other ocean use employments and GDP are generally highest, particularly for offshore mineral extractions and the marine construction sector. While above-average thermal stressors are affecting the greater U.S. Pacific EEZ, increasing temperatures and other ocean uses continue to affect transboundary LMRs and marine economies in both the U.S. Atlantic and Pacific.

Primary production from U.S. contributory large marine ecosystems (LMEs) is above-average and stable in 4/12 RFMO jurisdictions (Atlantic HMS-ICCAT, IATTC, IPHC, PSC), while 3/12 RFMO jurisdictions contain above-average stable RFMO primary production from internationally contributing LMEs (Atlantic HMS-ICCAT, NASCO, WCPFC; Table 12.6). Additionally, above-average U.S. fisheries landings occur in 3/12 jurisdictions (WCPFC, WECAFC, NAFO), for which 6/12 jurisdictions (particularly CCAMLR and NASCO) also have variable (>10%) landings. Average integrative relationships among fisheries landings, and production (U.S. contributory and total RFMO) are generally steady and above cross-jurisdictional mean values for 2 to 3 jurisdictions. Above-average relationships for fisheries landings/U.S. contributory production (i.e., exploitation >0.00009) occur for NAFO, PWT, and WECAFC jurisdictions. Similarly, above-average relationships for landings/production (i.e., exploitation >0.00002) occur for WECAFC and NAFO. In general, RFMO jurisdictions with above-average primary production do not necessarily have above-average fisheries landings or exploitation histories at the basin scale.

Throughout RFMO jurisdictions, average rankings across indicator categories were generally mid-range values (Table 12.7). Socioeconomic criteria were ranked highest for NAFO, WECAFC, and PSC jurisdictions. Much of this pattern was related to lower human

Table 12.7 Synthesis of average rankings (1 to 12; ± SE) and classifications (low, mid, high) for ecosystem indicator categories and living marine resource (LMR) management capacity across U.S. participatory regional fishery management organizations (RFMOs)

Average Rankings	**Indicators**	**Atlantic HMS (ICCAT)**	**Pacific HMS (IATTC, WCPFC)**	**IPHC**	**PSC**	**PWT**	**NAFO**	**NASCO**	**WECAFC**	**NPFC**
Socioeconomic Criteria	11	6.4 ± 4	6.1 ± 3.9	6.1 ± 2.2	4.9 ± 1.6	5.5 ± 2.9	2.8 ± 1.9	5.8 ± 3.8	2.9 ± 2.2	6.5 ± 2.7
Governance Criteria										
– *Human Representative, Fishery/Systematic*	9	6.1 ± 1.3	5.8 ± 1.2	5.1 ± 1.3	6.6 ± 0.9	4.5 ± 0.9	8.1 ± 1.5	6.4 ± 1.3	7.6 ± 0.7	4.9 ± 1
– *Status of LMRs*	7	9.1 ± 0.7	8.9 ± 0.9	4.6 ± 1.2	4.8 ± 0.4	3.2 ± 0.5	6.7 ± 1	4.4 ± 1.4	7 ± 1.1	6.6 ± 1.1
Environmental Forcing and Major Features	2	1.5 ± 0.5	1.5 ± 0.5	–	–	–	–	–	–	–
Major Pressures and Exogenous Factors	18	1.4 ± 0.1	1.6 ± 0.1	–	–	–	–	–	–	–
Systems Ecology and Fisheries	5	5.4 ± 1.8	5.2 ± 1.1	6 ± 0.7	4.4 ± 0.6	6 ± 1.4	3.1 ± 1.3	8.2 ± 1.7	4 ± 1.1	8.6 ± 0.8
Total	**52**	**4.7 ± 0.5**	**4.6 ± 0.5**	**5.5 ± 0.5**	**5.2 ± 0.3**	**4.7 ± 0.5**	**5.2 ± 0.7**	**6.1 ± 0.7**	**5.4 ± 0.6**	**6.4 ± 0.5**
Classification										
Socioeconomic Criteria	11	Mid	Mid	Mid	Mid–High	Mid	High	Mid	High	Mid
Governance Criteria										
– *Human Representative, Fishery/Systematic*	9	Mid	Mid	Mid	Mid	Mid–High	Mid–Low	Mid	Mid	Mid–High
– *Status of LMRs*	7	Low	Mid–Low	Mid–High	Mid–High	High	Mid	Mid–High	Mid	Mid
Environmental Forcing and Major Features	2	High	High	–	–	–	–	–	–	–
Major Pressures and Exogenous Factors	18	High	High	–	–	–	–	–	–	–
Systems Ecology and Fisheries	5	Mid	Mid	Mid	Mid–High	Mid	High	Mid–Low	Mid–High	Mid–Low
Total	**52**	**Mid–High**	**Mid–High**	**Mid**	**Mid**	**Mid–High**	**Mid**	**Mid**	**Mid**	**Mid**

Continued

Table 12.7 Continued

Average Rankings	Indicators	WCPFC	SPRFMO	CCAMLR
Socioeconomic Criteria	11	5.5 ± 2.4	7.5 ± 3.9	5.5 ± 3.2
Governance Criteria				
– *Human Representative, Fishery/Systematic*	9	6 ± 1.3	4 ± 0.9	4.2 ± 2
– *Status of LMRs*	7	7.7 ± 1	6.7 ± 1.2	8.2 ± 1.5
Environmental Forcing and Major Features	2	–	–	–
Major Pressures and Exogenous Factors	18	–	–	–
Systems Ecology and Fisheries	5	4.2 ± 0.6	10 ± 0	11 ± 0
Total	**52**	**6 ± 0.6**	**6 ± 0.7**	**7.1 ± 1.1**
Classification				
Socioeconomic Criteria	11	Mid	Mid	Mid
Governance Criteria				
– *Human Representative, Fishery/Systematic*	9	Mid	Mid–High	Mid–High
– *Status of LMRs*	7	Mid	Mid	Mid–Low
Environmental Forcing and Major Features	2	–	–	–
Major Pressures and Exogenous Factors	18	–	–	–
Systems Ecology and Fisheries	5	Mid–High	Low	Low
Total	**52**	**Mid**	**Mid**	**Mid**

stressors in some portions of these jurisdictions, and higher integrated relationships among fisheries value, LMR employments, and production. Alternatively, lower-tier socioeconomic rankings were observed in Atlantic and Pacific HMS jurisdictions, in addition to NPFC and SPRFMO.

Governance criteria in terms of human representative and fishery/systematic contexts were ranked "mid–high" in SPRFMO, CCAMLR, PWT, and NPFC jurisdictions (Table 12.7). Management in these areas contains higher numbers of FEPs, lower numbers of FMPs and amendments, and greater percentages of EEZs permanently closed to fishing. These criteria were ranked "mid–low" for NAFO. Status of LMRs was ranked "high" for the PWT jurisdiction and "mid–high" under NASCO, IPHC, and PSC resources. These trends were mostly due to the low numbers of targeted species upon which management efforts could be focused, and relatively lower proportions of unassessed stocks or species that are overfished or experiencing overfishing. While Atlantic and Pacific salmon stocks are federally protected, they are currently classified as "least concern" on IUCN listings (WCMC 1996, Rand 2011, IUCN 2021). These indicators suggest that conservation efforts have been effective for improving their overall status. In contrasting the status of Pacific HMS and Antarctic LMRs are ranked "mid–low," while Atlantic HMS have "low" rankings, mostly due to continued overfishing, underassessment of stocks, and greater co-occurrence of endangered species.

Environmental forcing and limited natural stressors as based on thermal conditions were accounted for in Atlantic and Pacific HMS jurisdictions throughout the U.S. EEZ, with similar rankings between basins. These pressures are magnified within other jurisdictions, reinforcing these higher rankings in larger areas. Similarly, highest rankings for other ocean uses were found in Atlantic and Pacific HMS areas, where highest values and greatest proportional contributions to U.S. marine economies occur in these broader coastwide jurisdictions. Overall ecological and fisheries indicators were ranked "high" under NAFO and "mid–high" for WECAFC, WCPFC, and PSC jurisdictions. Given their size, highest productivities were observed in HMS-encompassed jurisdictions. However, NASCO, IPHC, PSC, and NAFO to a lesser extent, were also ranked higher among RFMOs in terms of primary production (Table 12.8). While high productivities did not necessarily translate to higher fisheries landings or revenue for most regions, overall fisheries indicators appear to be at least partially related to basal production, with trends strongest for IATTC, WCPFC, PSC, NAFO, and PWT jurisdictions. Other deviations, such as those for Atlantic HMS and ICCAT, appear to be partially related to ongoing overfishing.

In terms of ranking among all indicators, Pacific and Atlantic HMS RMFOs and PWT management is relatively highest overall in terms of EBFM capacity. Despite lower LMR status for HMS, these trends are particularly related to higher economic production from other ocean uses, less concentrated environmental forcing, and higher relative rankings in terms of governance approaches compared to other jurisdictions. However, it is worth noting the many challenges facing HMS management both domestically and at the RFMO level (see RFMO Chapter 11). Additionally, NAFO, PSC, WECAFC, and IPHC were ranked as second-tier jurisdictions where consideration of holistic criteria in management practices is being emergently observed. Although little data were available beyond landings for CCAMLR's jurisdiction, it leads in terms of several enacted ecosystem-level considerations for krill management and consideration of other predators within Antarctic waters (Constable et al. 2000, Constable 2011). Additionally, few common indicators were available at the jurisdictions of the International Whaling Commission (IWC) or Arctic Council to allow for their inclusion in this synthesis.

In summary, a relationship between ecosystem productivity, LMR status, and associated socioeconomic conditions was not as apparent across all RFMO jurisdictions. Despite higher productivities, landings and fisheries values remain quite low for HMS, while moderately productive areas likewise contain differing LMR and socioeconomic statuses. Overall, other factors, including system exploitation and variable data availability prevent determining as strong a relationship between primary production, LMR status, and socioeconomics throughout these broader jurisdictions.

12.5 Progress Regarding EBFM Roadmap Recommendations

Throughout the U.S., and for its regions, subregions, and participatory RFMOs, progress toward completing the six guiding principles of the NOAA Fisheries EBFM Road Map (and associated action items; Table 12.9) has been generally successful in terms of its first two principles: (1) Implementing ecosystem-level planning and (2) Advancing understanding of ecosystem processes. Though we acknowledge that the subregions and RFMOs were not originally the intended geographies of the EBFM Road Map, we include an

Table 12.8 Synthesis of average rankings (1 to 12; ± SE) and classifications (low L, mid M, high H) for primary production (P), biomass (B), total fisheries landings (L), % living marine resources (LMR) employments (jobs, J), and total value of commercial fisheries landings (V) across participatory regional fishery management organization (RFMO) regions from 1998–2015. The relationship among these relative variables is additionally included

Average Ranking	**Atl HMS (ICCAT)**	**Pac HMS (IATTC, WCPFC)**	**IPHC**	**PSC**	**PWT**	**NAFO**	**NASCO**	**WECAFC**	**NPFC**	**WCPFC**	**SPRFMO**	**CCAMLR**
Primary Production— U.S. Component	1	2	3.5	3.5	9	7.5	7.5	5	10	6	–	–
Primary Production—Total RFMO	1	4	6.5	6.5	9	5	2	8	–	3	–	–
Systems Ecology and Fisheries												
–Fisheries Landings	8	5	7	4	6	1	12	2	9	3	10	11
Socioeconomic Status												
–% LMR Employments	2	1	–	–	–	–	–	–	–	–	–	–
–Fisheries Value	8	5	7	4	10	2	11	1	6	3	9	–
Classification												
Primary Production— U.S. Component	High	High	High	High	Low	Mid	Mid	Mid	Low	Mid	–	–
Primary Production—Total RFMO	High	High	Mid	Mid	Low	Mid	High	Mid	–	High	–	–
Systems Ecology and Fisheries												
– Fisheries Landings	Mid	Mid	Mid	High	Mid	High	Low	High	Low	High	Low	Low
Socioeconomic Status												
– % LMR Employments	High	High	–	–	–	–	–	–	–	–	–	–
– Fisheries Value	Low	Mid	Mid	High	Low	High	Low	High	Mid	High	Low	–
Progression PPLJV	**HHMHL**	**HHMHM**	**HMMM**	**HMHH**	**LLML**	**MMHH**	**MHLL**	**MMHH**	**LLM**	**MHHH**	**LL**	**L**

Table 12.9 Summary table of progress (yes, Y; partial, P; no, N) toward the NOAA Fisheries Ecosystem-Based Fisheries Management (EBFM) Policy Guiding Principles, EBFM Roadmap Components, and overarching goals at the national level (in bold; Natl), within U.S. regions (in bold; Pacific, Pac, Gulf of Mexico, GOM, Caribbean, C, South Atlantic, SA, Mid-Atlantic, MA, New England, NE), major subregions (in italics and/or parentheses; Eastern Bering Sea, EBS, Gulf of Alaska, GOA, Arctic, A, Aleutian Islands, AI, Inside Passage, IP, Prince William Sound, PWS, Channel Islands, CI, Monterey Bay, MB, San Francisco Bay, SF, Columbia River, CR, Puget Sound-Salish Sea, PS, Hawaii, H, Pacific Islands Territories, PI, Galveston Bay, GB, Matagorda Bay, Mat, Mississippi River Delta, MRD, Mobile Bay, Mob, Tampa Bay, TB, Cape Hatteras-Pamlico Sound, CH, Florida Keys, FK, Chesapeake Bay, ChB, Delaware Bay, DB, Long Island Sound, LI, Narragansett Bay, NB, Casco Bay, CB, Penobscot Bay, PB), and internationally (in bold; Commission for the Conservation of Antarctic Marine Living Resources, CCAMLR; regional fishery management organizations, RFMOs). Portions in gray indicate non applicability. Progress shown in bold (**Y**, **N**, **P**) indicates that a given action item is highlighted as important in a regional EBFM implementation plan. **I** indicates important in an EBFM implementation plan, but currently no progress reported

EBFM Policy Statement Guiding Principles	EBFM Road Map Components/Overarching Goals	Action Items	Natl	N Pac *EBS (AI)*	*GOA (IP, PWS)*	*A*	Pac *(CI, MB, SF, CR, PS)*	W Pac *H*	*PI*	GOM *(GB, Mat, MRD, Mob, TB)*	C	SA *(CH, FK)*	MA *(ChB, DB)*	NE *(LI, NB, CB, PB)*	CCAMLR	RFMOs
1 – Implement Ecosystem-Level Planning	**1a. Engagement Strategy Goal: Have an EBFM Engagement Strategy for each region**	Establish EBFM Point of Contact at each Regional Office, Fisheries Science Center, and Headquarters Offices	Y	**Y (P)**	**Y**	**Y**	Y	Y	Y	Y	P	**P**	**Y**	**Y**	Y	Y
		Develop National and Regional EBFM engagement strategies	**Y**	**Y**	Y	Y	**Y**	**Y**	**Y**	**Y**	**Y**	**Y**	**Y**	**Y**	N	**Y**
		Develop best practices where there are overlapping jurisdictions.	**N**	**Y**	N	P	**N**	**N**	**N**	**N**	N	N	**Y**	**Y**	Y	**P**
		Develop Standardized EBFM Policy and Road Map Materials for widespread use (e.g. NOAA Fisheries personnel, Sea Grant extension agents)	**P**													
		NOAA Fisheries supports any Ecosystem Plan Development Teams, Ecosystem Committees (or equivalent groups) that Councils establish	**P**	**Y** (Y)	**Y**	**Y**	**Y** (_, _, _, _, Y)	**Y**	**Y**	**Y**	**Y**	**Y**	**Y** (Y, _)	**Y** (_, _, Y, Y)	Y	**Y**
		Continue to explore trade-offs in the context of EBFM issues and relevant statutory mandates	**P**	**Y** (Y)	Y (_, Y)	N	Y (Y, N, Y, Y, Y)	N	N	N	N	N	N (Y, _)	N (_, _, _, Y)	Y	**N**
		Create "X-prize" like competition for visualizing and communicating EBFM	N													
	1b. Fishery Ecosystem Plans (FEPs) Goal: Assist Councils in the development of their FEPs	Establish FEP Coordinator/ Analyst for each NOAA Fisheries Region and in appropriate Headquarters Office	P	**Y** (Y)	**Y**	**Y**	Y	**Y**	**Y**	N	**P**	**P**	**N**	**N**	N	**P**

Continued

Table 12.9 Continued

EBFM Policy Statement Guiding Principles	EBFM Road Map Components/Overarching Goals	Action Items	Natl	N Pac *EBS (AI)*	*GOA (IP, PWS)*	*A*	Pac *(CI, MB, SF, CR, PS)*	W Pac *H*	*PI*	GOM *(GB, Mat, MRD, Mob, TB)*	C	SA *(CH, FK)*	MA *(ChB, DB)*	NE *(LI, NB, CB, PB)*	CCAMLR	RFMOs
1 – Implement Ecosystem-Level Planning	**1b. Fishery Ecosystem Plans (FEPs) Goal: Assist Councils in the development of their FEPs**	Review and develop inventory of existing FEPs and Ecosystem Considerations in FMPs, documenting best practices	P	**I**								**I**	**I**	**I**		
		Assist Councils, as requested, in their development of new, or revision of existing FEPs		**Y** (Y)	**Y**	Y	**Y**	**Y**	**Y**	**Y**	**Y**	**Y**	**Y** (P, _)	**Y**	Y	**Y**
2 – Advance our Understanding of Ecosystem Processes	**2a. Science to Understand Ecosystems Goal: Have robust, innovative Internationally recognized science programs to support management**	Advance resources to conduct EBFM	P	**I**	**I**	**I**	**I**	**I**	**I**	**I**			**I**	**I**		**I**
		Develop capacity for NOAA Fisheries to conduct end-to-end ecosystem studies	**P**	**Y (Y)**	**Y (P, P)**	Y	**Y** (P, P, Y, Y, Y)	Y	Y	N	N	P	**Y** (Y, _)	**Y** (_, Y, Y, Y)	Y	**N**
		Conduct biennial EBFM Science and Management Conference	**P**	**I**	**I**	**I**		**I**								
		Develop and maintain core data and information streams	**P**	**Y (Y)**	**Y** (P, **Y**)	**Y**	**Y** (Y, Y, Y, Y, Y)	**Y**	**Y**	**Y** (Y, Y, Y, Y, Y)	**P**	**Y** (Y, Y)	**Y** (Y, Y)	**Y** (Y, Y, Y, Y)	Y	**Y**
		A national review of the data collection programs on a wide range of disciplines, including but beyond the typical abundance and basic biological data.	**P**													
	2b. Ecosystem Status Reports (ESRs) Goal: Have ESRs for most of our 12 Large Marine Ecosystems (LMEs)	Conduct a national review of existing ESRs to assess Fisheries Science Center (FSC) indicator information needs to identify where ESRs address similar indicators across LMEs	**Y**													
		Establish routine, regular, and dynamic reporting of ESRs for each LME		**Y (Y)**	**Y**	**Y**	**Y** (P, P, P, _, P)	**P**	**N**	**Y** (P, _, P, P, P)	N	**N** (_, P)	**Y** (Y, _)	**Y**	N	**N**

EBFM Policy Statement Guiding Principles	EBFM Road Map Components/Overarching Goals	Action Items	Natl	N Pac EBS (AI)	GOA (IP, PWS)	A	Pac (CI, MB, SF, CR, PS)	W Pac H	PI	GOM (GB, Mat, MRD, Mob, TB)	C	SA (CH, FK)	MA (ChB, DB)	NE (LI, NB, CB, PB)	CCAMLR	RFMOs
3 – Prioritize Vulnerabilities and Risks	**3a. Ecosystem-level Risk assessment Goal: Evaluate majority of main risks, including Climate Change, for most of our 12 LMEs**	Conduct Systematic Risk Assessments for relevant NOAA regional ecosystems	**P**	**Y**	**Y**	**N**	**P**	**N**	**N**	**P**	N	P	**P**	**P**	N	**N**
		Conduct Climate Vulnerability Assessments	**I**	**Y**	N	N	**Y**	**Y**	**Y**	**Y**	N	**Y**	**Y**	**Y**	N	**N**
		Conduct Stock Assessment Prioritization		Y	Y		**P**	N	N	P	P	N	Y	Y	N	N
		Explore protocols for conducting regional habitat risk assessments for those areas known to serve important ecological functions for multiple species groups or will be especially vulnerable or important in the face of climate change	**P**	**Y**	**N**	P	**P**	**Y**	**Y**	N	N	N	**Y** (Y, P)	**Y** (P, P, P, P)	N	N
		Ensure more integrated, systematic risk assessments are used to coordinate regional NEPA analyses	N	P	N	N	N	**N**	**N**	N	N	N	N	P	N	N
	3b. Managed species, Habitats and Communities Risk Assessment Goal: Evaluate risks for all of our managed species	Ensure that factors which impact 800+ US managed species are being considered	**P**	**Y**	**Y**	**P**	**P**	**P**	**N**	P	**N**	**N**	**Y**	**P**	P	**N**
		Conduct Habitat Assessment Prioritization for all NOAA Fisheries regions	**Y**	**Y**	**Y**	**Y**	**Y**	**Y**	**Y**	P	P	P	**Y**	**Y**	N	**P**
		Conduct Fishing Community vulnerability assessments for all NOAA Fisheries regions	**Y**	**Y (Y)**	**Y (Y, Y)**	**Y**	**Y**	**P**	**P**	Y (Y, Y, Y, Y, Y)	**N**	**Y**	**Y** (Y, Y)	**Y** (Y, Y, Y, Y)	N	**P**
4 – Explore and Address Trade-offs within an Ecosystem	**4a. Modeling Capacity Goal: Have sufficient analytical capacity to evaluate a full range of trade-offs**	Assess and bolster ecosystem and LMR modeling needs in each FSC		**Y**	**Y**	P	**P**	**Y**	**P**	N	**N**	**P**	N	P	P	**N**

Continued

Table 12.9 Continued

EBFM Policy Statement Guiding Principles	EBFM Road Map Components/Overarching Goals	Action Items	Natl	N Pac EBS (AI)	N Pac GOA (IP, PWS)	A	Pac (CI, MB, SF, CR, PS)	W Pac H	W Pac PI	GOM (GB, Mat, MRD, Mob, TB)	C	SA (CH, FK)	MA (ChB, DB)	NE (LI, NB, CB, PB)	CCAMLR	RFMOs
4 – Explore and Address Trade-offs within an Ecosystem	4a. Modeling Capacity Goal: Have sufficient analytical capacity to evaluate a full range of trade-offs	Development of an EBFM analytical toolbox that includes ecosystem modeling tools and best practices; data-poor qualitative and semi-quantitative tools; and related decision support tools.	P	I	I			I								
		Encourage and expand the use of multimodal inference	P	P	N	P	P	N	N	N	N	N	P	P	N	N
		Establish suitable review venues and deliberative bodies for ecosystem models and associated information in each FSC region		Y	Y	N	Y	N	N	N	N	N	Y (P, _)	Y	Y	N
	4b. Management Strategy Evaluations (MSEs) Goal: Have MSEs that cover most of our 12 LMEs and Fisheries	Develop functional system-level MSEs		Y	P	N	P	N	N	N	N	N	P	P	N	N
		Hire Full-Time MSE staff		Y	N	N	Y	Y	P	Y	N	N	Y	Y	N	P
		Explore novel Harvest Control Rules (HCRs) and develop associated guidelines, especially to test and explore robust Ecosystem-Level strategies	N	P	N	N	P	N	N	P	N	P	P	P	P	N
		Create "X-prize" like competition for visualizing and communicating complex ecosystem model and MSE outputs	N													
5 – Incorporate Ecosystem Considerations into Management Advice	5a. Ecosystem-level reference points Goal: Establish and use Ecosystem-Level Reference Points	Delineate, evaluate, and explore best practices for estimating and using system-wide or aggregate group harvest limits, ecosystem production measures, and other ELRPs, to inform management decisions.	P	Y	Y	N	N	N	N	N	N	P	N	P	P	N
		Explore best measures of cross-pressure, cumulative impacts in an ecosystem (in conjunction with Guiding Principle 3)		P	N	N	P	N	N	N	N	N	P	P	N	N

EBFM Policy Statement Guiding Principles	EBFM Road Map Components/Overarching Goals	Action Items	Natl	N Pac EBS (AI)	N Pac GOA (IP, PWS)	A	Pac (CI, MB, SF, CR, PS)	W Pac H	W Pac PI	GOM (GB, Mat, MRD, Mob, TB)	C	SA (CH, FK)	MA (ChB, DB)	NE (LI, NB, CB, PB)	CCAMLR	RFMOs
5 – Incorporate Ecosystem Considerations into Management Advice	**5b. Ecosystem considerations for LMRs Goal: Appropriately include ecosystem-level factors in crafting advice for managed species**	Develop Ecosystem-level reference points and Thresholds	N	**Y**	**Y**	**N**	N	N	N	N	N	N	N	P	P	N
		Develop and track fishery stock status indices that denote when ecosystem considerations are used	**Y**	**Y (Y)**	**Y**	N	Y	**Y**	**Y**	Y	N	N (_, P)	**Y** (Y, _)	**Y**	Y	P
		Support consistent and effective implementation of the National Standard 1 Guidelines, which includes guidance on incorporating ecosystem information into stock management.	Y	**I**	**I**			**I**		**I**	**I**	**I**	**I**	**I**		**I**
		Identify best practices for incorporating ecosystem considerations into management decisions.	**P**	**P**	**P**	N	**P**	**P**	**P**	N	N	**N**	**P** (P, _)	**P**	P	**N**
		Establish ecosystem-related TOR for stock assessments (SAs), stock assessment reviews, and support ecosystem-related TOR for status review groups, HCRs, and science and statistical committee (SSC) review processes.	P	**P** (P)	**P**	N	**Y** (_, _, Y, _, _)	N	N	P	N	N	**Y** (Y, _)	**Y**	Y	**N**
	5c. Integrated Advice for other Management Considerations Goal: Systematically evaluate advice provided	Explore protocols for considering ecosystem-level information in EFH reviews, identifying ecosystem-level habitat areas of particular concern, and setting habitat conservation objectives and/or indicators	**P**	**Y (Y)**	**Y**	**P**	**P**	**P**	**P**	N	**P**	**N**	**Y**	**Y**	Y	**N**
		Finalize National Bycatch Reduction Strategy	**Y**					**I**		**I**	**I**	**I**	**I**	**I**		**I**
		Evaluate the ecosystem effects of offshore aquaculture	N	N	N	N	N	**N**	**N**	N	N	N	**N**	**N**		**N**

Continued

Table 12.9 Continued

EBFM Policy Statement Guiding Principles	EBFM Road Map Components/Overarching Goals	Action Items	Natl	N Pac *EBS (AI)*	N Pac *GOA (IP, PWS)*	*A*	Pac *(CI, MB, SF, CR, PS)*	W Pac *H*	W Pac *PI*	GOM *(GB, Mat, MRD, Mob, TB)*	C	SA *(CH, FK)*	MA *(ChB, DB)*	NE *(LI, NB, CB, PB)*	CCAMLR	RFMOs
5 – Incorporate Ecosystem Considerations into Management Advice	**5c. Integrated Advice for other Management Considerations Goal: Systematically evaluate advice provided**	Implement the National Allocation Policy	Y													**I**
		Review long-term protected species recovery and rebuilding plans to ensure they account for the potential effects of near-term and long-term climate change, particularly relating to alterations to food web structure	**P**	P	N	N	**P**	**P**	**P**	**N**	N	N	**P**	**P**	P	**N**
		Conduct Protected Species Climate Vulnerability Analyses		N	N	N	**Y**	**P**	**P**	N	N	N	**P**	**P** (_, _, P, P)	N	N
6 – Maintain Resilient Ecosystems	**6a. Evaluate Resilience Goal: Develop and achieve ecosystem performance measures**	Track Ecosystem-level reference point to assess changes in ecosystem-level resilience	N	**P** (P)	**P**	**P**	N	**P**	**P**	N	**N**	N	**Y** (P, _)	**Y** (_, P, P, P)	P	N
		Evaluate, conduct, and conduct valuation of Ecosystem Goods and Services relative to benchmarks	**N**	**Y** (Y)	**Y**	**Y**	**Y**	N	N	N	**N**	**N**	**Y**	**Y**	N	**N**
		Develop best practices for tradeoff evaluation with respect to overall ecosystem and community resilience and well-being	N	**Y** (Y)	**P**	N	**P**	N	N	N	**N**	**N**	P	P	N	N
		Develop National EBFM Performance measures	**N**													
	6b. Community Well-being Goal: Maintain well-being of coastal communities	Explore community health socioeconomic metrics	**P**	**Y (Y)**	**Y** (Y, Y)	**Y**	**Y**	**N**	**N**	**Y** (P, P, P, P, P)	**N**	**P** (P, P)	**Y** (Y, Y)	**Y** (Y, Y, Y, Y)	N	**N**
		Adopt community vulnerability analyses to a broader range of cumulative factors	**Y**	**Y (Y)**	**Y** (Y, Y)	**Y**	**P**	P	P	P	**N**	P (P, P)	**Y** (Y, Y)	**Y** (Y, Y, Y, Y)	N	**Y**
		Track community health socioeconomic metrics	P	**Y (Y)**	**Y** (Y, Y)	**Y**	**P**	**P**	**P**	N	**N**	N	**Y** (Y, Y)	**Y** (Y, Y, Y, Y)	N	**N**
		Establish National EBFM Coordinator	Y													

evaluation of them here for informational purposes and to glean important insights those other geographies may provide. Among regions and many North Pacific and Western Pacific major subregions, nearly 100% completion has been observed for most ecosystem-level planning actions. Exceptions are found for exploring EBFM-related trade-offs, however. Currently, these have only been explored in the Pacific (West Coast) region and most of its subregions (excluding Monterey Bay), in most North Pacific subregions, and in the Chesapeake and Penobscot Bay subregions. Additionally, the establishment of an FEP coordinator/regional analyst at NOAA Fisheries has only been completed in the North Pacific, Pacific, and Western Pacific regions.

In terms of advancing understanding of ecosystem processes, advanced progress in developing capacity for conducting end-to-end ecosystem studies has occurred in most regions (Pacific, Mid-Atlantic, New England, North Pacific, Western Pacific), with partial developments taking place in the South Atlantic and Gulf of Mexico. Developments in modeling capacity for North Pacific (4/6 subregions), Pacific (3/5 subregions), and both Western Pacific subregions have been observed, in addition to end-to-end modeling capacities existing for Chesapeake Bay and in 3/4 New England subregions. They also partially exist for the Alaskan Inside Passage, Prince William Sound, Channel Islands, and Monterey Bay subregions. The development and maintenance of core ecosystem data and information streams is completely in effect for the overwhelming majority of regions and subregions, while partially underway in the U.S. Caribbean and Inside Passage. Additionally, these data streams exist for some RFMOs and CCAMLR situations and stocks. Regular and repeated preparation of ecosystem status reports (ESRs) is underway in 6/8 major regions (Gulf of Mexico, Pacific, Mid-Atlantic, New England, North Pacific, Western Pacific) and for five specific subregions (East Bering Sea, Gulf of Alaska, Aleutian Islands, Hawaii, Chesapeake Bay). Partial efforts are also underway for the Gulf of Mexico and its subregions, most Pacific subregions (including the incorporation of Integrated Ecosystem Assessments into the Channel Islands Sanctuary Condition Reports; NMS 2019), and the Florida Keys. The U.S. Caribbean and most U.S. subregions currently lack any ESR efforts, while the development of a South Atlantic ESR is in progress (NMFS 2019).

Regarding the four other guiding principles of the EBFM Road Map, less headway has been made, although there has been some progress. Among regions, some progress has occurred for Guiding Principle 3. Prioritizing vulnerabilities and risks in conducting climate/habitat vulnerability assessments and stock assessment prioritizations are the main ways this progress has occurred. Currently, all three of these efforts have been completed for the Mid-Atlantic and New England regions, and the East Bering Sea subregion. In addition, climate vulnerability assessments have been conducted for the Pacific, Gulf of Mexico, South Atlantic, and Western Pacific. Partial completions of stock assessment prioritizations have occurred in the Pacific, Gulf of Mexico, and U.S. Caribbean, and are completed for the Gulf of Alaska. Habitat risk assessment protocols are currently in place for the Mid-Atlantic, New England, and Western Pacific regions, and for Chesapeake Bay, while partial progress has been made in the Pacific region, Arctic, and all other Mid-Atlantic and New England subregions. Prioritized listings of fish stocks that would most benefit from habitat assessments are in place or are being finalized (Gulf of Mexico, Caribbean, South Atlantic) for all regions. Additionally, fishing community vulnerability assessments have been completed in six regions (North Pacific, Pacific, Gulf of Mexico, South Atlantic, Mid-Atlantic, New England) and most subregions. These are partially completed for the Arctic and Western Pacific. Currently, broader factors affecting all U.S. managed species are only being fully considered in Mid-Atlantic risk assessments.

For Guiding Principle 4, exploring and addressing trade-offs within an ecosystem, particularly in terms of modeling capacity, progress has been made throughout regions. At present, several regions (except the Gulf of Mexico, South Atlantic, and U.S. Caribbean) have made partial to significant developments in advancing aspects of ecosystem modeling. The development of functional management strategy evaluations (MSEs), and efforts to explore ecosystem-level harvest control rules (HCRs) have occurred in the Pacific, Mid-Atlantic, New England, and East Bering Sea. In addition, partial to complete MSE development has occurred for the Gulf of Alaska, while hiring of full-time staff for these duties has been observed for 6/8 regions. The incorporation of ecosystem considerations into management advice (Guiding Principle 5), including the establishment, exploration, and use of ecosystem-level reference points (ELRPs), has been limited to partial efforts in the Pacific, New England, Mid-Atlantic, and for major North Pacific subregions, such as the two million metric ton cap for East Bering Sea groundfish (Witherell et al. 2000). Some partial efforts to explore best practices f12r using aggregate limits and ecosystem production measures are also underway for the South Atlantic. Additionally, most regions or subregions have not developed ELRPs or thresholds for management, except in the North Pacific and partial efforts for New England. However, the Atlantic States Marine Fisheries Commission (ASMFC) is currently considering an ELRP for Atlantic menhaden (Buchheister et al. 2017, ASMFC 2018, 2020). The application of ecosystem considerations to evaluate fishery stock status is underway in the Pacific, Gulf of Mexico, Mid-Atlantic, New England, Chesapeake Bay, in major North and Western Pacific subregions, and for CCAMLR stocks. In the Pacific, Gulf of Mexico, Mid-Atlantic, New England, best practices for incorporating ecosystem considerations into stock assessments, status reviews, HCRs, and management decisions are completed or partially completed. Currently, approximately 8% of all stock assessments include some form of ecosystem consideration (Keyl & Wolff 2008, Tyrell et al. 2008, 2011, Lynch et al. 2018, Marshall et al. 2019), with most progress having been made in the northeastern and west coast U.S. regions. Within subregions, these are also completed or partially completed for Chesapeake Bay, East Bering Sea-Aleutian Islands, Gulf of Alaska, San Francisco Bay, and in the Western Pacific. Efforts have yet to be completed for all other action items in consideration of Guiding Principle 5, including examining best practices for addressing cumulative ecosystem impacts, developing ELRPs and thresholds, and their incorporation into management actions.

Only partial progress has been made for most action items related to maintaining resilient ecosystems (Guiding Principle 6), especially in terms of developing ecosystem performance measures. Efforts have been completed in the Pacific region and are partially completed for the Mid-Atlantic (including Chesapeake and Delaware Bays), New England (including Casco and Penobscot Bays), and Western Pacific regions regarding protected species climate vulnerability analyses. Tracking ELRPs to assess changes in ecosystem-level resilience is underway for the Mid-Atlantic (including Chesapeake Bay) and New England (including Narragansett, Casco, and Penobscot Bays) regions and subregions, and partially underway in North Pacific and Western Pacific subregions. Completed and partially completed efforts are also underway for exploring community health socioeconomic metrics in the Pacific, Gulf of Mexico, Mid-Atlantic, New England, South Atlantic, and their subregions. Continued progress for other actions related to maintaining ecosystem resilience is underway among regions, subregions, and in U.S. participatory RFMOs, with most progress having occurred in the Mid-Atlantic, New England, and North Pacific regions.

Overall, greatest progress regarding EBFM Road Map actions has occurred for the Mid-Atlantic (including Chesapeake Bay), New England, North Pacific, and the Pacific (region and subregions). Additionally, continued efforts within the Gulf of Mexico, South Atlantic, Western Pacific, Caribbean, and their subregions are allowing for greater EBFM capacities in these areas following development of their implementation plans.

12.6 Progress Regarding the 1999 EPAP Recommendations for FEPs

In 1999, the EPAP noted steps to determine what would be most effective toward the development of FEPs throughout U.S. regions and subregions as a means of carrying out EBFM. An initial examination of this progress was undertaken by Wilkinson and Abrams (2015), and is updated here for all U.S. regions, subregions, and participatory RFMO jurisdictions. Though we acknowledge that the subregions and RFMOs were not originally the intended geographies of the EPAP, we include an evaluation of them here for informational purposes and to again glean important insights those other geographies may provide. Progress toward completing EPAP recommendations has been moderately successful in the development of regional and subregional FEPs (Table 12.10). Two regions (Pacific, South Atlantic) and five subregions (East Bering Sea, Aleutian Islands, Hawaii, PI Territories, Chesapeake Bay) currently have completed FEPs. Partial FEPs have also been developed for the Mid-Atlantic region, the Channel Islands, and Penobscot Bay. All regions and most subregions have delineated the geographic extent of their ecosystems, with efforts partially completed for San Francisco Bay, the Mississippi River Delta, and Cape Hatteras-Pamlico Sound. Additionally, all regions (excepting partial biological progress for the U.S. Caribbean), the majority of subregions, and CCAMLR have fully characterized the major physical and biological dynamics of their ecosystems.

Completed conceptual foodweb models have been developed for all regions and most subregions. Descriptions of habitat needs for those species comprising the "significant food web" remain needed, with partial descriptions generally existing for most regions and subregions. The least progress has been made in Cape Hatteras-Pamlico Sound, Delaware Bay, most subregions of the Gulf of Alaska and Gulf of Mexico, and for RFMOs in general (excluding CCAMLR). Overall, little of this information has been considered in additional conservation management measures, with the most advanced efforts observed in

Table 12.10 Progress (yes, Y; partial, P; no, N) toward the 1999 recommendations of the Ecosystem Principles Advisory Panel (EPAP) for the development of FEPs within U.S. regions (in bold; Pacific, Pac, Gulf of Mexico, GOM, Caribbean, C, South Atlantic, SA, Mid-Atlantic, MA, New England, NE), major subregions (in italics and/or parentheses; Eastern Bering Sea, EBS, Gulf of Alaska, GOA, Arctic, A, Aleutian Islands, AI, Inside Passage, IP, Prince William Sound, PWS, Channel Islands, CI, Monterey Bay, MB, San Francisco Bay, SF, Columbia River, CR, Puget Sound-Salish Sea, PS, Hawaii, H, Pacific Islands Territories, PI, Galveston Bay, GB, Matagorda Bay, Mat, Mississippi River Delta, MRD, Mobile Bay, Mob, Tampa Bay, TB, Cape Hatteras-Pamlico Sound, CH, Florida Keys, FK, Chesapeake Bay, ChB, Delaware Bay, DB, Long Island Sound, LI, Narragansett Bay, NB, Casco Bay, CB, Penobscot Bay, PB), and internationally (in bold; Commission for the Conservation of Antarctic Marine Living Resources, CCAMLR; regional fishery management organizations, RFMOs). Portions in gray indicate non applicability, "?" indicates unknown progress, and "_" indicates no information

	N Pac			**Pac**	**W Pac**		**GOM**	**C**	**SA**	**MA**	**NE**	**CCAMLR**	**RFMOs**
EPAP Recommendation	*EBS (AI)*	*GOA (IP, PWS)*	*A*	(*CI, MB, SF, CR, PS*)	*H*	*PI*	(*GB, Mat, MRD, MoB, TB*)		(*CH, FK*)	(*ChB, DB*)	(*LI, NB, CB, PB*)		
Develop a Fishery Ecosystem Plan (FEP)	Y (Y)	N	N	Y (P, N, N, N, N)	Y	Y	N	N	Y	P (Y, N)	N (N, N, P, P)	N	N
1a.) Delineate the geographic extent of the ecosystem(s) that occur(s) within Council (commission, RFMO, etc.) authority.	Y (Y)	Y (Y, Y)	Y	Y (Y, Y, P, Y, Y)	Y	Y	Y (Y, Y, P, Y, Y)	Y	Y (P, Y)	Y (Y, Y)	Y (Y, Y, Y, Y)		
1b.) Characterize the biological dynamics of the ecosystem.	Y (P)	Y (N, Y)	P	Y (P, P, Y, P, P)	Y	Y	Y (Y, P, P, P, P)	N	Y (P, Y)	Y (Y, Y)	Y (Y, Y, Y, Y)	Y	N
1c.) Characterize the chemical and physical dynamics of the ecosystem.	Y (Y)	Y (P, Y)	P	Y (P, Y, Y, Y, Y)	Y	Y	Y (Y, Y, Y, Y, Y)	Y	Y (Y, Y)	Y (Y, Y)	Y (Y, Y, Y, Y)	Y	N
1d.) "Zone" the area for alternative uses.	P (?)	?	?	? (P, P, N, N, P)	?	?	?	?	? (N, P)	? (?, N)	?	?	?
2.) Develop a conceptual model of the foodweb.	Y (Y)	Y (N, Y)	P	Y (N, Y, Y, Y, Y)	Y	Y	Y (P, N, P, Y, Y)	P	Y (N, Y)	Y (Y, P)	Y (Y, Y, Y, Y)	Y	N
3a.) Describe the habitat needs of different life history stages for all plants and animals that represent the "significant food web."	P (P)	P	P	P (P, P, Y, Y, P)	P	P	P (P, N, N, N, N)	P	P (N, P)	P (P, N)	P (N, N, N, P)	N	N
3b.) Describe how this information is considered in conservation management measures. [apart from EFH consultations]	N	N	N	N (P, N, Y, Y, P)	N	N	N	N	P (N, P)	N	P	N	N
4a.) Calculate total removals—including incidental mortality	Y (Y)	Y	N	P	N	N	N	N	N	P	Y	Y	N
4b.) Show how this information relates to standing biomass, production, optimum yields, natural mortality, and trophic structure.	Y (Y)	N	N	N	N	N	N	N	N	N	P	P	N
5a.) Assess how uncertainty is characterized.	Y (Y)	Y	N	Y (N, N, Y, Y, P)	Y	N	Y	N	P (N, P)	Y (P, N)	Y	Y	Y
5b.) Include buffers against uncertainty are included in conservation and management.	Y (Y)	Y	N	Y (Y, N, Y, Y, Y)	Y	Y	Y	N	Y (N, Y)	Y (Y, Y)	Y (N, Y, N, N)	Y	N

Continued

Table 12.10 Continued

	N Pac		Pac		W Pac		GOM	C	SA	MA	NE	CCAMLR	RFMOs
EPAP Recommendation	*EBS (AI)*	*GOA (IP, PWS)*	*A*	*(CI, MB, SF, CR, PS)*	*H*	*PI*	*(GB, Mat, MRD, MoB, TB)*		*(CH, FK)*	*(ChB, DB)*	*(LI, NB, CB, PB)*		
6a.) Develop indices of ecosystem health.	Y (Y)	Y (N, Y)	Y	Y (Y, Y, Y, Y, Y)	Y	N	Y (Y, N, Y, N, Y)	N	N (N, Y)	Y (Y, N)	Y (Y, Y, N, N)	Y	N
6b.) Develop these indices as targets for management. [Reference Points, thresholds]	Y (Y)	Y	N	N	N	N	N	N	N	N	N	N	N
7.) Describe available long-term monitoring data and how they are used.	Y (Y)	Y (N, Y)	N	Y (Y, Y, Y, Y, Y)	Y	Y	Y (Y, P, Y, Y, Y)	Y	Y (Y, Y)	Y (Y, Y)	Y (Y, Y, Y, Y)	Y	N
8a.) Assess the ecological, human, and institutional elements of the ecosystem that most significantly affect fisheries.	Y (Y)	Y (N, Y)	Y	Y (Y, N, Y, Y, Y)	Y	Y	Y	Y	Y (N, P)	Y (Y, N)	Y (Y, Y, Y, Y)	Y	N
8b.) Apply this information to areas outside the Council/Department of Commerce (DOC) authority.	Y (Y)	Y (N, Y)	Y	Y (N/A, Y, Y, Y, Y)	Y	P	Y (Y, Y, Y, Y, Y)	N	Y (Y, Y)	Y (Y, Y)	Y (Y, Y, Y, Y)	Y	N
8c.) Included should be a strategy to address those influences in order to achieve both FMP and FEP objectives.	N	N	N	N (P, N, P, N, N)	N	N	N (N, N, P, N, N)	N	N (N, P)	N (Y, N)	N (N, N, N, P)	N	N
Measures to Implement FEPs													
1.) Encourage the councils to apply ecosystem principles, goals, and policies to ongoing activities.	Y (Y)	Y	Y	Y	P	P	N	P	P	Y (Y, _)	Y	Y	N
2.) Provide training to council members and staff.	Y (Y)	Y	N	Y	N	N	N	N	N	N (Y, _)	N	N	N
3.) Prepare guidelines for FEPs.	Y (Y)	Y		Y	P	P	P	P	Y	Y (Y, _)	Y	N	N
4.) Develop demonstration FEPs.	Y (Y)	N	N	N	Y	Y	N	N	Y	N (Y, _)	N	N	N
5.) Provide oversight (support) to ensure development of and compliance with FEPs.	Y (Y)	Y	N	Y	N	N	N	N	N	Y (N, _)	Y	N	N
6.) Enact legislation requiring FEPs.	N	N	N	N	N	N	N	N	N	N (N, _)	N	N	N
Research Required to Support Management													
1.) Determine the ecosystem effects of fishing.	P (P)	P	N	P	N	N	N	N	N	P (N, _)	P	N	N
2.) Monitor trends and dynamics in marine ecosystems (ECOWATCH).	Y (Y)	Y	Y	Y	Y	Y	Y	P	Y	Y (Y, _)	Y	Y	N
3.) Explore ecosystem-based approaches to governance.	Y (Y)	Y	Y	P	N	N	N	N	N	P (N, _)	N	P	N

Pacific subregions, the South Atlantic, Florida Keys, and New England.

Calculations of total removals, including incidental bycatch or mortality, has occurred in the Antarctic, North Pacific, and New England, while having been partially estimated for most regions. This information has also been fully incorporated into assessments of standing biomass, production, optimal yield, natural mortality, and trophic structure for the East Bering Sea, Aleutian Islands, and partially for Antarctica and New England. Uncertainty has been accounted for and included in conservation and management actions for most regions, major North Pacific, and Pacific subregions, and in Antarctica, Hawaii, and Chesapeake Bay. Buffers against uncertainty are generally included in management actions for most regions and subregions to some extent as part of the Magunson–Stevens Act Annual Catch Limit setting process (Lynch et al. 2018). Additionally, the development of ecosystem health indices has occurred for most regions and subregions (excluding the U.S. Caribbean and South Atlantic) as seen in some form of Ecosystem Status Reporting, with their application as management targets having only occurred in the East Bering Sea-Aleutian Islands and Gulf of Alaska. Long-term monitoring data have been collected and described for all regions (including Antarctica) and most subregions, except for Matagorda Bay (partial progress), Inside Passage, and the Arctic. All regions, Antarctica, and most subregions have also assessed ecological, human, and governance elements of their ecosystems that most significantly affect fisheries. Most have also applied these data and assessments to areas outside U.S. Department of Commerce (i.e., NOAA) and FMC authorities. However, complete, concrete strategies that address these influences in terms of (sub)regional FMP and FEP objectives only exist for Chesapeake Bay. Partial strategies have been developed for the Channel Islands, San Francisco Bay, Mississippi River Delta, Florida Keys, and Penobscot Bay.

Measures to implement FEPs in regions, subregions, and RFMO jurisdictions have been completed in the Pacific, Western Pacific, South Atlantic, Mid-Atlantic, New England, North Pacific, and (a revision for) Chesapeake Bay. These measures include encouraging and training Council members and staff to apply ecosystem principles to Council activities, preparing and developing guidelines for FEPs, overseeing FEP compliance, and enacting legislation requiring FEPs. Partial efforts are also underway in the Gulf of Mexico, Caribbean, and Western Pacific for preparing FEP guidelines and encouraging ecosystem applications in Council products. Additionally, research to support ecosystem-level management actions still remains needed for most regions. Partial or completed efforts toward EPAP research recommendations are mostly observed in the Pacific, North Pacific, and Mid-Atlantic. In these locations, greatest progress toward understanding the ecosystem effects of fishing and ecosystem-based approaches to governance has been observed. However, most regions and major subregions have ecosystem-level monitoring programs.

Similar to that observed for EBFM Road Map actions, greatest progress toward completion of EPAP recommendations has occurred in the Mid-Atlantic (including Chesapeake Bay), North Pacific, and Pacific regions. Additionally, efforts to develop FEPs for the Gulf of Mexico, New England, and Caribbean are underway, while EBFM implementation efforts in the Western Pacific and South Atlantic are closely aligned with their FEPs. Continued progress toward enhancing knowledge of these fisheries ecosystems with initial efforts toward applying this information to concrete system-level management actions is ongoing for all regions. Although, much work remains to achieve these goals among all regions, foundational development of these plans and enhanced FEPs is underway through efforts supported by NOAA Fisheries and regional FMCs.

12.7 Issues of Main Concern for U.S. Marine Regions and RFMOs Over the Next Decade

As identified for all regions, subregions, and RFMO jurisdictions, a suite of stressors and management issues continues to occur throughout U.S. marine fisheries ecosystems. These have been identified in detail at differing scales, for which collective system-level approaches will be required. In consideration of those stressors and factors affecting U.S. marine regions, RFMO jurisdictions, and their SESs, the issues identified in Table 12.11 are expected to be of major concern throughout the upcoming decade. When accounting for them, an ecosystem approach allows for their complementary focus and mutual consideration in management strategies moving forward.

Climate and natural stressors are expected to continue affecting all U.S. regions over the next decade, particularly the effects of increased SST, species distribution shifts, and intensified climate oscillations. More frequent marine heatwaves (e.g., "The blob" of 2015) are anticipated throughout the Pacific basin, with consequences to encompassed marine ecosystems. In addition, the effects of sea level rise and ocean acidification are expected to continue affecting coastal environments, calcareous species (i.e., corals, shellfishes), and the habitats comprising them. The effects of climate and other natural stressors are also anticipated to continue affecting the Gulf of Mexico, especially in its continued tropicalization, increasing storm intensity, large-scale hypoxic events, and recurring harmful algal blooms (HABs). Over the next decade, climatological stressors on Atlantic systems are expected to result in continued species composition shifts, with increased tropicalization and losses of major species and their fisheries into more northerly systems. These trends are also anticipated within the Pacific region. The consequences of sea level rise and ocean acidification are also anticipated to intensify in these systems over the latter portion of the decade. Additionally, these climatological effects are expected to foster ongoing coral bleaching events, leading to potential loss of reefs. Similarly, increased warming, species distribution shifts, heightened climate forcing, and system-wide responses to these stressors are also expected at the RFMO and ocean basin scale.

While prevalent throughout all ecosystems, many of the identified major pressures (e.g., nutrient loading, oil spills, habitat loss) are expected to be more pervasive within tropical and subtropical regions over the next decade. In addition, the expansion of fishing and other ocean uses into deeper ocean regions is likely to affect broader portions of their EEZs and the high seas. Increased potential for oil spills is also concerning for the North Pacific, while proliferations of invasive species are also likely to increase within the Pacific region. Additionally, ongoing effects from nutrient loading, habitat loss, and biological invasions are also expected in the Mid-Atlantic over the next decade.

Management issues of concern in the near-term include improving conflict resolution among all interested parties in the fisheries management process. Efforts to more accurately characterize and manage recreational fisheries are also underway, particularly for the Gulf of Mexico and South Atlantic. In addition, addressing the lack of basic fisheries data remains a major concern in the Western Pacific and U.S. Caribbean. Over the next decade, continued stock rebuilding efforts will need to be prioritized in the Gulf of Mexico, U.S. Caribbean, South Atlantic, New England, and for HMS fisheries under RFMO jurisdictions. In addition, continued progress toward incorporating ecosystem considerations into management actions and developing system-level HCRs is anticipated for the North Pacific, Pacific, Mid-Atlantic, and New England regions.

Social issues of concern over the latter part of the next decade include increasing local and external demand for seafood throughout all regions and RFMOs, and increased demand for Gulf of Mexico and North Pacific energy resources. Conflicts among water users are anticipated to intensify in the Pacific region, while changes in spatial fishing effort are most likely to occur in the North Pacific, Mid-Atlantic, and New England regions. Increases in human population are likewise expected to affect all U.S. marine regions and RFMO participatory nations over the next decade, including in the North Pacific.

Throughout all regions, trade-offs among LMRs and marine sectors are anticipated to intensify over the next decade, particularly for commercial and recreational fishing interests, fisheries and protected species, and among fisheries and other ocean uses. In the North Pacific and Pacific regions, conflicts among protected species such as marine mammal populations, seabirds, and salmon populations are expected to continue. Conflicts between North Pacific commercial and subsistence fisheries communities are also expected to intensify over the next decade. Additionally, concerns regarding conflicting interests between the fishing and tourism sectors are anticipated within tropical/subtropical regions and the U.S. west coast. Specific anticipated conflicts among fisheries and other ocean uses include those with aquaculture in the Gulf of Mexico, Mid-Atlantic, and New England. Additionally, conflicts with oil and gas development are anticipated within the North Pacific, Pacific, Gulf of Mexico, and RFMO jurisdictions, while those with wind energy siting are expected to intensify in the Pacific, Mid-Atlantic, and New England regions.

Among regions, many stressors are anticipated to increase over the next decade. Given the degree of progress observed for EBFM within regions, and ongoing efforts to incorporate ecosystem information and ELRPs into specific management practices, there is increasing potential for these stressors to be cumulatively addressed through system-level approaches. However, with increasing climatic and human-associated pressures, more robust management actions are urgently needed that can allow for their mitigations while also working toward fostering more environmentally sound solutions. Sustainably managing ecosystems in consideration of their productivities, regional pressures, trade-offs, and the effects of other ocean uses allows for these factors to be addressed jointly. Continued development of

Table 12.11 Stressors and factors anticipated to affect major U.S. regions (North Pacific, NP, Pacific, P, Western Pacific, WP, Gulf of Mexico, GOM, Caribbean, C, South Atlantic, SA, Mid-Atlantic, MA, New England, NE) and regional fisheries management organizations (RFMOs) over the next 3–5 years (gray; near term), 5–10 years (vertical lines; long term), and 3–10 years (black; persistent over the next decade). Portions in white indicate non applicability. . ENSO = El Niño Southern Oscillation, HABs = Harmful Algal Blooms, SST = Sea Surface Temperature

Criteria	Stressors/Management Factors	NP	P	WP	GOM	C	SA	MA	NE	RFMO
Climate/ Natural Stressors	Changes in species composition									
	Droughts and lack of water (especially for salmonids)									
	Effects of sea level rise and inundation									
	HABs and red tides									
	Heightened effects of climate oscillations (e.g., ENSO)									
	Hypoxia									
	Increased frequency of marine heatwaves									
	Increased SST, coral bleaching events									
	Increased SST, increasing abundance of lower latitude species									
	Increasing effects of ocean acidification									
	Increasing effects of sea ice loss									
	Increasing hurricane/typhoon/storm intensity									
	Ongoing thermal/climate forcing stressors and system-wide responses									
	Shifting of major species and fisheries out of the system									
Major Pressures	Expanded fishing and other ocean uses into deeper oceanic regions									
	Increased nutrient loading, pollution									
	Increased potential for oil spills and oil toxicity									
	Increased proliferation of invasive species									
	Loss of coral and/or oyster reef habitats									
	Loss of wetlands/vegetated coastal habitats									
	Watershed land use practices affecting reef ecosystems									
Management	Conflict resolution challenges									
	Continued stock rebuilding efforts									
	Emerging incorporation of ecosystem considerations into management actions									
	Incorporating recreational fisheries information into management									
	Lack of basic fisheries data									

Continued

Table 12.11 Continued

Criteria	Stressors/Management Factors	NP	P	WP	GOM	C	SA	MA	NE	RFMO
Management	System-level harvest rules									
Social	Changes in spatial fishing effort									
	Conflicts among water users									
	Human population increase									
	Increasing demand for energy resources									
	Increasing local/external and tourism-related seafood demand									
Trade-Offs	Conflicts among protected species									
	Conflicts between commercial and recreational fishing interests									
	Conflicts between commercial and subsistence use fisheries									
	Conflicts between fisheries and protected species conservation									
	Conflicts between fishing and tourism sectors									
	Conflicts with aquaculture									
	Conflicts with oil rigs, related marine energy development, and marine transportation									
	Conflicts with wind farm siting									
	Enhanced other ocean uses and concentrated ocean sectors									
	Expansion of oil/gas production									
	Increasing resource conflicts among marine sectors									

ecosystem-based management actions will remain essential toward cumulatively managing these increasing stressors and their effects on LMRs over the next decades.

12.8 References

ASMFC (Atlantic States Marine Fisheries Commission). 2018. *Ecosystem Reference Point Data Workshop*. Arlington,VA: ASMFC Offices.

ASMFC (Atlantic States Marine Fisheries Commission). 2020a. *ASMFC Atlantic Menhaden Board Adopts Ecological Reference Points*. Arlington,VA: ASMFC Offices (p. 2).

Buchheister A, Miller TJ, Houde ED. 2017. Evaluating ecosystem-based reference points for Atlantic Menhaden. *Marine and Coastal Fisheries* 9(1):457–78.

Chassot E, Bonhommeau S, Dulvy NK, Mélin F, Watson R, Gascuel D, Le Pape O. 2010. Global marine primary production constrains fisheries catches. *Ecology Letters* 13(4):495–505.

Constable AJ, de la Mare WK, Agnew DJ, Everson I, Miller D. 2000. Managing fisheries to conserve the Antarctic marine ecosystem: practical implementation of the convention on the conservation of Antarctic marine living resources (CCAMLR). *ICES Journal of Marine Science* 57(3):778–91.

Constable AJ. 2011. Lessons from CCAMLR on the implementation of the ecosystem approach to managing fisheries. *Fish and Fisheries* 12(2):138–51.

EPAP (Ecosystem Principles Advisory Panel). 1999. *Ecosystem-Based Fishery Management: A Report to Congress*. Silver Spring, MD: NMFS (p. 62).

Graham NA, McClanahan TR, MacNeil MA, Wilson SK, Cinner JE, Huchery C, Holmes TH. 2017. Human disruption of coral reef trophic structure. *Current Biology* 27(2):231–6.

Heenan A, Williams ID, Acoba T, DesRochers A, Kosaki RK, Kanemura T, Nadon MO, Brainard RE. 2017. Long-term monitoring of coral reef fish assemblages in the Western Central Pacific. *Science Data* 4:170–6.

IUCN 2021. The IUCN Red List of Threatened Species. Version 2021-1. https://www.iucnredlist.org.

Keyl F, Wolff M. 2008. Environmental variability and fisheries: what can models do? *Reviews in Fish Biology and Fisheries* 18(3):273–99.

Lenfest. 2018. *Advancing Ecological Reference Points for Menhaden Using an Ecosystem Model*. Conshohocken, PA: Lenfest Ocean Program (p. 2).

Link JS, Marshak AR. 2019. Characterizing and comparing marine fisheries ecosystems in the United States: determinants of success in moving toward ecosystem-based fisheries management. *Reviews in Fish Biology and Fisheries* 29(1):23–70.

Lynch PD, Methot RD, Link JS. 2018. *Implementing a Next Generation Stock Assessment Enterprise. An Update to the NOAA Fisheries Stock Assessment Improvement Plan. NOAA Technical Memorandum NMFS-F/SPO-183*. Washington, DC: NOAA (p. 127).

Marshall KN, Koehn LE, Levin PS, Essington TE, Jensen OP. 2019. Inclusion of ecosystem information in US fish stock assessments suggests progress toward ecosystem-based fisheries management. *ICES Journal of Marine Science* 76(1):1–9.

NMFS (National Marine Fisheries Service). 2016. *NOAA Fisheries Ecosystem-Based Fisheries Management Road Map*. Washington, DC: NOAA (p. 50).

NMFS (National Marine Fisheries Service). 2019. *Ecosystem-Based Fisheries Management Implementation Plan for the South Atlantic*. Washington, DC: NOAA (p. 16).

NMS (Office of National Marine Sanctuaries). 2019. *Channel Islands National Marine Sanctuary 2016 Condition Report*. Silver Spring, MD: NOAA (p. 482).

Rand, P.S. 2011. *Oncorhynchus nerka. The IUCN Red List of Threatened Species 2011: e.T135301A4071001*. https://dx.doi.org/10.2305/IUCN.UK.2011-2.RLTS.T135301A4071001.en.

Tyrrell MC, Link JS, Moustahfid H, Overholtz WJ. 2008. Evaluating the effect of predation mortality on forage species population dynamics in the Northeast US continental shelf ecosystem using multispecies virtual population analysis. *ICES Journal of Marine Science* 65(9):1689–700.

Tyrrell MC, Link JS, Moustahfid H. 2011. The importance of including predation in fish population models: implications for biological reference points. *Fisheries Research* 108(1):1–8.

WCMC (World Conservation Monitoring Centre). 1996. *Salmo Salar. The IUCN Red List of Threatened Species 1996: e.T19855A9026693*. http://dx.doi.org/10.2305/IUCN.UK.1996.RLTS.T19855A9026693.en

Wilkinson EB, Abrams K. 2015. *Benchmarking the 1999 EPAP Recommendations with Existing Fishery Ecosystem Plans. NOAA Technical Memorandum NMFS-OSF-5*. Washington, DC: NOAA (p. 22).

Williams ID, Baum JK, Heenan A, Hanson KM, Nadon MO, Brainard RE. 2015. Human, oceanographic and habitat drivers of central and western Pacific coral reef fish assemblages. *PLoS One* 10(4):e0120516.

Williams ID, Richards BL, Sandin SA, Baum JK, Schroeder RE, Nadon MO, Zgliczynski B, et al. 2011. Differences in reef fish assemblages between populated and remote reefs spanning multiple archipelagos across the central and western Pacific. *Journal of Marine Biology* 2011:1–14.

Witherell D, Pautzke C, Fluharty D. 2000. An ecosystem-based approach for Alaska groundfish fisheries. *ICES Journal of Marine Science* 57(3):771–7.

CHAPTER 13

So What?

If you're reading this chapter, presumably you've stayed with us long enough to wade through a lot of information. Or if you're an executive or resource manager, or just want to know the bottom line first, you may have skipped to the end to see what this all really means. Either way, it is incumbent upon us to provide further context, unpack all the technical details summarized in the last chapter, and attempt to provide some sense of not only what we have learned, but where the discipline/field and practice need to go during the next decade. So what does this all mean?

To summarize, by examining a suite of over 90 indicators for nine major U.S. fishery ecosystem jurisdictions, we have systematically tracked the progress that the U.S. has made toward advancing ecosystem-based fisheries management. In doing so we also document the current status of the management of living marine resources (LMRs) throughout the U.S. These metrics cover a wide range of socioeconomic, governance, environmental forcing, major pressures, systems ecology, and fisheries criteria; this work covers a wide range of longitude, latitude, and parts of major ocean basins, representing over 10% of the world's ocean surface area. In what follows we highlight lessons learned from a national perspective. The overarching theme is that though much work remains, much progress has occurred. Here we provide a short synopsis of several major "so-whats." We do so by posing a series of questions, questions which are routinely presented to us, and then attempt to address them.

13.1 What Have We Actually Accomplished Regarding Ecosystem-Based Fisheries Management (EBFM)?

13.1.1 In Terms of Recognition, Support, and Acceptance of EBFM

Recognition of the importance of EBFM and of the need for broader considerations in LMR management is no longer debated. Nearly everyone recognizes that the ocean is changing beyond anything we have observed in our data (Barange et al. 2014, Busch et al. 2016, IPCC 2019), that the ocean is incredibly dynamic (Steele 1985, 1991, Lewison et al. 2015), and that it is non-linear (Mann & Lazier 1991, Valiela 1995, Milly et al. 2008, Beyragdar Kashkooli & Modarres 2020). Nearly everyone recognizes that we cannot maximize everything at once (Edwards et al. 2004, Sanchirico et al. 2008, Worm et al. 2009, Hilborn et al. 2012, Jin et al. 2016, Hilborn et al. 2020) and that there are legitimate trade-offs that need to be addressed.

In an informal poll one of us (JL) regularly conducts, there is still notable support for EBFM by the vast majority of senior level leaders in NOAA Fisheries. There is also notable support among a wide range of stakeholders for EBFM (Marshak et al. 2017, AORA 2019, Pope & Weber 2019).

Many of NOAA Fisheries strategic plans or related documents (Link et al. 2015, Busch et al. 2016, NMFS 2016a, 2016b, Lynch et al. 2018, Peters et al. 2018 NMFS 2019) and its mission statement clearly contain aspects of EBFM. That clarity and commitment was not as apparent 10, perhaps even 5, years ago. That EBFM has a clear prioritization of how we do business in strategic plans, annual guidance memos, etc. is not a trivial outcome. Many other countries and similar organizations also have relatively recent, comparable commitments to advance EBFM.

We recognize that there is codification and clarification, and mostly agreement, on the main terms of EBFM (Pitcher et al. 2009, Link & Browman 2014, Dolan et al. 2015, Patrick & Link 2015a, Link & Browman 2017, AORA 2019). In polls we have seen (Hutchison et al. 2015, Marshak et al. 2017), the understanding of EBFM and Ecosystem-Based Management (EBM) is converging on common themes. Sure the odd paper arises re-questioning these terms, and sure there are some who continue to remain uninformed on the topic, but as a discipline the vast majority now acknowledge and grasp the common meaning and understanding of EBFM (Yaffee 1996, Pitcher et al. 2009, Link & Browman 2017, Marshak et al. 2017).

Ecosystem-Based Fisheries Management: Progress, Importance, and Impacts in the United States. Jason S. Link and Anthony R. Marshak, Oxford University Press. © U.S. Department of Commerce, U.S. Government 2021. DOI: 10.1093/oso/9780192843463.003.0013

13.1.2 In Terms of Capability, Products, and Accomplishments

Several integrative programs have persisted over the past decade, especially integrated ecosystem assessment (IEA) efforts in the U.S. and EU (Levin et al. 2009, Samhouri et al. 2013, Möllmann et al. 2014, DePiper et al. 2017, ICES 2020). It is not trivial for programs to survive over a decade, and as the products and outputs of these efforts continue to be recognized and increasingly in demand, the maintenance of these programs not only continues, but they are beginning to expand (ICES 2020).[1]

Most U.S. regions now have standard, routine, and expected ecosystem status reports (or similar documents; Slater et al. 2017). These documents do an excellent job of collating much of the information in an ecosystem to provide context for LMR management decisions (Monaco et al. 2021), and in an increasing number of cases even get incorporated directly into those decisions and quantitative analyses that inform them (Dorn & Zador 2020).

There have been several advances in analytical techniques and models, especially formation and growth of formal ecosystem modeling teams and groups. Nearly all of the seven science Centers and Offices in NOAA Fisheries have a modeling team and ecosystem group; 10 years ago there was only 1.5 out of 7. Additionally the number and degree of sophistication of ecosystem models has increased, but more importantly the direct review and use in an LMR management context has notably increased (Townsend et al. 2019). Several regional fishery management organizations (RFMOs), state marine fishery commissions (SMFCs), scientific review groups (SRGs), and fishery management councils (FMCs) have begun asking for advice such as ecosystem reference points for Atlantic menhaden, management strategy evaluations for Atlantic herring, system-level production caps for the Eastern Bering Sea, etc., that only ecosystem models can provide (Townsend et al. 2019, H. Townsend pers. comm.). Additionally, NOAA has begun a broader set of integration across *all* aspects of modeling via its Unified Modeling efforts (Link et al. 2017b), which is helping to increase capacity to model a broader range of issues important to LMRs.

In many regions, we have begun to see widespread activation and use of combined socio-ecological teams. The inherent value of combining these disciplines has been noted elsewhere (Ostrom 2009, Link et al. 2017a, Thébaud et al. 2017), but these cross-disciplinary teams are now more common than not in most regions of the U.S.

[1] https://www.integratedecosystemassessment.noaa.gov/news/10yearsofIEA

13.1.3 In Terms of Organizations and Governance

Changes in or clarification of policy for EBFM have occurred in the past decade (Foran et al. 2016, NMFS 2016a, 2016b, Link 2017, Rudd et al. 2018). Although still more permissive than required (D. Fluharty pers. comm; Link et al. 2018), this set of policy statements does reflect the importance of EBFM and a commitment to implementing it.

There is a growing use of fishery ecosystem plans (FEPs) in the federal fisheries management process (Wilkinson & Abrams 2015, Levin et al. 2018, Marshall et al. 2018). The majority of FMCs have some form of FEP, which means that now the majority of jurisdictions of the U.S. are covered by these FEPs (Wilkinson & Abrams 2015, Link & Marshak 2019). Increasingly these have begun to move beyond a compilation of species-specific concerns or even regional issues to actually codifying priorities and major areas of emphasis for many FMCs.

We see the requisite adjustments to existing governance bodies to begin to take onboard ecosystem issues. That is, there are now an increasing number of EBFM, ecosystem approaches to fisheries management (EAFM) and related committees or subcommittees for many of the FMCs and Commissions.

Many LMR management bodies are now formally adding ecosystem considerations to their standard operating procedures (Pitcher et al. 2009, Skern-Mauritzen et al. 2016, Lynch et al. 2018, Koehn et al. 2020). Changes to terms of reference and ensuring ecosystem-related topics are considered (bycatch, habitat, climate, trophic interactions, etc.), having become much more common now than 10 years ago, both for the stock assessment processes and the FMC Scientific and Statistical Committee (SSC) review processes (Lynch et al. 2018).

Besides these fisheries-oriented organizations, we note the formation of new regional ocean management and EBM bodies that are multisector and extend beyond just the fisheries emphasis (Rosenberg 2009, AORA 2019; various regional ocean councils or organizations that arose from the USCOP 2004; c.f. Christie 2005, Hershman & Russell 2005). Certainly, many of the details remain to be ironed out, particularly with respect to authorities and mandates for these cross-sectoral bodies. However, new bodies and affiliated organizations are arising to address some of these multisector issues (e.g., windfarms and fisheries; aquaculture siting and telecommunication cables, protected species migration and shipping, etc.).

13.2 Has EBFM Actually Improved Conditions in the Ocean, of LMR Populations, and of Fishing Communities? Or at Least Has It Been Able to Identify and Predict Major Sources of Change and Uncertainty to Them?

13.2.1 In Terms of Status of Resources, Ocean, and Fishing Communities

There has been notable progress in terms of improved status of LMRs, both for sustainable fisheries (Hilborn et al. 2015, Hilborn & Ovando 2014, Hilborn et al. 2020, NMFS 2020) and protected resources (Merrick et al. 2004, Roman et al. 2013, Merrick 2018, Valdivia et al. 2019). We discuss whether that is due to single species approaches only, EBFM, or a bit of both below, but that there have been improvements is clear. There has also been notable progress in terms of increased habitat assessment, protection, and restoration, implying improvement in habitat quantity and quality (Lederhouse et al. 2017, Peters et al. 2018, Duarte et al. 2020, Munsch et al. 2020), and improvements to overall emergent properties of ecosystems (Bundy et al. 2012, Link & Marshak 2019, Link & Watson 2019, Pranovi et al. 2020).

Several beginnings of EBFM performance measures have been proposed (Coll et al. 2016, Bundy et al. 2017, Pranovi et al. 2020, K. Osgood pers. comm) that seek to compile the composite status of a given ecosystem, often with assignment of thresholds or limits, or at least qualification into an ordinal scheme depicting the range from good to bad (Fay et al. 2014, Large et al. 2015, Tam et al. 2017, Libralato et al. 2019). The previously noted national and regional status reports serve as both early warning signals and pseudo-report cards (Monaco et al. 2021, Dorn & Zador 2020).

Fishing communities remain vulnerable to a host of economic and environmental changes (Jepson & Colburn 2013, Himes-Cornell & Kasperski 2015, Colburn et al. 2016). And although employments have varied regionally, nationally employments and revenue have been stable and even slightly increasing (c.f. Figure 4 in each chapter; NMFS 2018). The sociological aspects of fisheries relative to EBFM warrant augmented consideration beyond what we briefly highlight here (Charles 2001, Ostrom 2009).

What we have seen, across all regions, which indicates EBFM progress (and to be fair, progress in single species management as well; Hilborn et al. 2020) and which has led to improved living marine resource status (c.f. Chapter 12, Link & Marshak 2019) includes:

- clear fish and protected species stock status identified, even if overfished or overfishing is occurring. Not having status identified tends to indicate broader challenges for a particular social-ecological system;
- relatively stable but attentive management interventions;
- clear tracking of broader ecosystem considerations, e.g., ecosystem status reports (ESRs);
- landings to biomass exploitation rates at typically <0.1;
- areal landings at typically $<1\ \mathrm{t\ km^2\ yr^{-1}}$;
- ratios of landings relative to primary production at typically <0.001;
- and explicit consideration of ecological and particularly socioeconomic factors directly in management.

We also note that more inherently productive marine ecosystems tend to have greater biomass, fisheries landings, proportional LMR-based employments, and fisheries revenue. This generally supports the rubric we posited in Chapter 1 and then throughout each chapter:

$$PP \rightarrow B_{\text{targeted, protected spp, ecosystem}} \leftrightarrow L_{-\text{targeted spp},\downarrow\text{bycatch}} \rightarrow \text{jobs, economic revenue.}$$

again where PP is primary production, B is biomass of either targeted or protected species (or total ecosystem), L is landings of targeted or bycaught species, all leading to the other socioeconomic factors. But it has now been demonstrated that the management interventions can positively, or negatively, influence these connections. Thus, while there may be some limits to what is produced in a given ecosystem and how that ultimately translates into LMR economics, there are interventions that can improve that situation. This does not necessarily mean that just because an ecosystem has low productivity that it is not making progress toward EBFM.

13.2.2 In Terms of at Least Being Able to Identify and Predict Major Sources of Uncertainty

The ability to show options and scenarios among various trade-offs has drastically improved (Fay et al. 2014, Deroba et al. 2019, Feeney et al. 2019, Fulton et al. 2019, Townsend et al. 2020), even if the venues to make decisions about them are still being ironed out. The nascent ability to consider, evaluate and make decisions explicitly acknowledging multiple objectives is not something we were doing ten years ago (Link 2010, Townsend et al. 2019).

The number and scope of risk assessments has grown notably in the past 5 years. There are now climate vulnerability assessments (CVAs) for the vast majority of U.S. large marine ecosystems (LMEs) (e.g., Gaichas et al. 2014, Busch et al. 2016, Hare et al. 2016, Crozier et al. 2019, Spencer et al. 2019). These are based on highly repeatable and intuitive methods, can be based on limited information, and are providing important context of what should be prioritized analytically with respect to this risk. There are also productivity-susceptibility analyses (PSAs; Patrick et al. 2010) for most targeted fishery stocks and that can also be used to gage risk and prioritize efforts. Community vulnerability assessments similarly evaluate the susceptibility of fishing communities to various risks (Jepson & Colburn 2013, Himes-Cornell & Kasperski 2015, Colburn et al. 2016). These CVAs and related risk analyses can be modified to include other factors (Morrison et al. 2015, Lettrich et al. 2019), or the PSAs can at least be updated for the ones done approximately 10 years ago.

The forecasting and prediction skills for a range of taxa-oriented, aggregate taxa, and full ecosystem considerations, under a range of oceanographic, climatological, and human dimension conditions, have expanded in both spatial and temporal resolution and scope (Kaplan et al. 2016, Olsen et al. 2016, Juricke et al. 2018, Hobday et al. 2019, Park et al. 2019). The degree of predictive capacity remains a challenge, as does data needs for further forecasts (Capotondi et al. 2019), but there are more regions and taxa covered than 10 years ago. The ability to see major perturbations coming has improved notably in the past 15–20 years (Libralato et al. 2019, Link & Watson 2019, Park et al. 2019).

13.3 Are There Any Instances that are Good Examples of Fully Implemented EBFM?

Yes and no. No one region is hitting on all the things noted in each of the regional chapters as important for EBFM (c.f. the last figure of each chapter, c.f. Table 1.3 or Table 12.9; i.e., the action items from the EBFM Road Map, NMFS 2016b), or by any other evaluation rubric for EBFM that has been proposed (e.g. the Ecosystem Principles Advisory Panel (EPAP) advice, c.f. Table 12.10; Wilkinson & Abrams 2015), which is not surprising. So, in the sense that no region has fully implemented the entire suite of recommended actions that each region has identified as important, there remains room for progress for EBFM everywhere.

But we note that the regions which tend to rank the highest in many categories (as noted in the prior chapter; c.f. Tables 12.2, 12.3, 12.5, 12.7, 12.8), are instances where we see particularly notable progress toward EBFM. Many of the details were addressed in detail in the prior chapter, but facets of how the Pacific, Mid-Atlantic, and North Pacific have approached the many issues related to EBFM seem to be particularly worth noting.

More so, significant steps have been taken toward EBFM in each region. Some of these may be better categorized as EAFM rather than EBFM (c.f. Fig. 1.3), but the point is that ecosystem considerations are being increasingly included in all regions. The text boxes in each regional chapter were meant to highlight these examples. So in that sense, when we begin to consider forage fishes more broadly, when we deal with bycatch more systematically, when we have system-level caps, when we explicitly consider options between targeted and protected taxa, when we explicitly include environmental conditions, when we systematically restore or consider habitat, when we explicitly incorporate human dimensions, and so on, we are exhibiting significant steps toward EBFM.

13.4 Has Moving Toward EBFM Made a Positive Difference?

Objectively speaking—maybe. Fisheries in the U.S. are some of the most well managed in the world (Pitcher et al. 2009, Hilborn & Ovando 2014, Bundy et al. 2017, Hilborn et al. 2015, Link & Marshak 2019, Hilborn et al. 2020). They are still managed largely on a stock-by-stock basis. So, it is unclear whether the management for a given stock would be improved beyond what we are doing now using current definitions of status. Yet it has been well documented that status determination criteria for stocks are very different when energetic, ecological, or systemic considerations are included (Worm et al. 2009, Rindorf et al. 2017, Link 2018; e.g., multispecies maximum sustainable yield). Thus, with ecosystem-adjusted reference points, the overall status of many stocks may change to portray a different picture.

We do generally know that the management for targeted fish stocks has been improved by considering ecosystem factors (e.g., Keyl & Wolff 2008, Tyrrell et al. 2011). The example from the Gulf of Mexico (Chapter 7) notes the improvements to the situation for Gag grouper by including a harmful algal bloom (HAB) index (Sagarese et al. 2014, 2015, Grüss et al. 2016). The same has been shown for forage stock improvements when including predation (Tyrrell et al. 2011). Including ecosystem information has made a positive impact in the evaluation and status of the stocks where it has been done.

A comparison of similar ecosystems using a system-level cap, or not, showed substantial benefits in the situation where the system-level cap was employed (Link 2018). Certainly there are nuances to this and related studies (Mueter & Megrey 2006, Megrey et al. 2009, Bundy et al. 2012, Fogarty 2014, Patrick & Link 2015b), but the list of benefits that emerged from such an approach were noteworthy. And when enacted, have made a huge difference in terms of many criteria, not least of which was the status of key fish stocks.

But that is emphasizing solely a stock-oriented perspective of how to manage fisheries. We know that fish and fisheries do not operate in a vacuum, and that species and fisheries do indeed interact. Thus, when there have been direct conflicts across different LMRs, EBFM has made a difference. For example, when there are forage needs for commercially targeted or protected species that also support a fishery in their own right (Collie & Gislason 2001, Smith et al. 2011, Tyrrell et al. 2011, Pikitch et al. 2014), EBFM has facilitated a resolution of those conflicting objectives. Or when there are conflicts because gear for targeted species captures too much of unwanted or protected or limiting taxa (Crowder & Murawski 1998, Howell et al. 2015, Lewison et al. 2015, Hazen et al. 2018), adopting a broader perspective has helped to address those multiple objectives. And so on, the salient point being that where there are trade-offs, and there are indeed many, EBFM approaches can result in better outcomes (Fogarty 2014, Kolding et al. 2016, Jacobsen et al. 2017, Link 2018, Fulton et al. 2019). Not necessarily optimal, but better than if the trade-off in objectives were ignored.

Overall, in instances where the EBFM mode of operation has been adopted, there have been documented benefits (e.g., Link 2018) that include:

- risk of overfishing is minimized (i.e., above agreed upon levels, or at least not substantially above levels of natural mortality);
- populations of fishes, catches, and profits are more stable;
- overall value across all stocks is maximized;
- bureaucratic oversight and regulatory interventions are minimized;
- catch and yield are optimized;
- biomass of the resource (in aggregate) is maximized;
- stakeholder disenfranchisement and legal challenges are minimized;
- catch per unit effort is optimized; and,
- risk of ancillary ecosystem impacts is minimized (again, i.e., above agreed upon levels, or at least not substantially above natural levels).

Stated that plainly, it surprises us that EBFM is not adopted more frequently or rapidly than it has been.

13.5 What are Some Areas for Improvement for EBFM in the Next Decade or So?

There has been progress on EBFM the past decade. There also remain some areas for advancement. Here we list them, with limited (to no) commentary, to showcase facets of EBFM that keep arising as needing to be solidified or clarified.

13.5.1 Organizational, Mandate, and Governance Advancements

- The need to move from permissive to required mandates—clearer mandates.
- Better governance venues for trade-off evaluations.
- Absolute certainty by any future, senior level leadership (in NOAA for the US or comparable organizations elsewhere) that EBFM is critically necessary.
- Rearrange budgets/resources to support interdisciplinary work.
- Rearrange organizational structures to allow for more interdisciplinary, integrated teams.
- Clearer protocols and options for management under ocean and climate change.
- Improved coordination across management bodies for species shifting their distributions.
- Better acceptance of and expectation for addressing a changing ocean.
- Increased flexibility in management responses to allow for nimble responses to ecosystem change.

- Better acceptance of and expectation for dealing with trade-offs.
- Increased diversity in stocks targeted to decrease dependence of fishermen and communities on a limited number of stocks.

13.5.2 Technical, Sampling, and Data Advancements

- Better ability to handle real-time data.
- Better ability to handle acute events.
- Expanded/improved data-handling protocols.
- Innovations in advanced sampling methodologies to the point they become operational.
- Better capacity for more integrated data collection.
- Enhanced or augmented sampling, particularly for data-poor regions.

13.5.3 Analytical and Forecast Advancements

- Better use of multispecies and ecosystem analytics, especially for tactical considerations.
- Better use of risk analyses and incorporating that information into management processes, assessments, and decisions.
- Better use of extended stock assessment models, and ESR context incorporated into management decisions.
- More and expanded use and capacity of management strategy evaluations.
- Better use of coupled social-ecological system models.
- Expanded inclusion of human dimensions.
- Better use of ecosystem-level referent points (ELRP) and system-level measures.
- Better predictions and forecasts for changing oceans, and how those changes will impact LMRs.

With the ultimate aim of continued improvement to status of LMRs, along with their associated fisheries and fishing communities.

13.6 What If We Don't Do EBFM?

Ultimately, who cares whether we execute EBFM or not? Why does it matter that all the impacts we have documented herein are happening? What's the downside to ignoring the need and potential benefits of EBFM?

The downside is that we will have continued litigious situations due to mishandling trade-offs by different interested stakeholders with different objectives. We will lose credibility by the lack of transparency when handling these conflicting objectives in isolation, and on a case-by-case, unsystematic manner. We will miss key factors that we should be considering, factors that drive changes in ecosystem production, and hence fish dynamics. Because of that, due to both human induced and environmental changes that impact this production, our advice will be increasingly wrong. We will not be able to triage and prioritize the greatest risks facing LMRs and their associated human communities. We will be caught unaware and unprepared, unable to accommodate rapid changes we already see in the ocean. We will have less stable fish populations, fisheries, management, and the lack of (at least big-picture) transparency that comes with a continued single stock emphasis. The reason we care is that as ocean dynamics and our human uses continue to alter how energy flows through an ecosystem, it may ultimately make it difficult for some stocks to thrive, recover, or even persist. We will continue to remove more than what can be produced from an ecosystem, and just keep shifting that excess harvesting pressure from one species to the next. We will succumb to cumulative impacts that we overlook and that seemingly suddenly pass a tipping point. And so on.

We acknowledge that all of this sounds a bit alarmist, if not even overstated. But in our observations, these are not hypothetical scenarios; we have seen and, in some instances, directly experienced these issues. Hence our desire is to minimize and overcome these challenges moving forward.

When we brief senior leaders in our (and other) marine resource management organizations, all of the above concerns are true but don't always resonant unless we get invited back for conversations on more specific details. What does grab most people's attention is boiling all of the above down in these simple statements, which rapidly capture the risks of not considering EBFM:

In 5–10 years, some species will be leaving a given ecosystem and moving to another jurisdiction or even country (Nye et al. 2009, Pinsky et al. 2013, Cheung et al. 2015, Morley et al. 2018). And that has nothing to do with fishing pressure. For other species it may be 10–20 years, but climate change is impacting the oceans and we see documented shifts in species distribution. Are we prepared for that? Some of these species are iconic for America (or certain regions), so how do we handle this soon coming loss? What are the business rules for handling this situation?

In many areas, the productivity of individual stocks has rapidly changed due to incredibly rapid changing ocean conditions. Yet we often assign decreases in a population to fishing pressure and/or a stock's status. How do we better assign shifts in productivity to these non-fishing phenomena? How do we allow for both stability and changes in biological reference points and status determination criteria? What are the business rules for handling this situation?

In some regions, we have almost what amounts to a "cage-match" between protected species (e.g., salmon and sea lions; Weise & Harvey 2005, Gende & Sigler 2006) or protected and targeted species (e.g., turtles and tunas; Howell et al. 2015) or even different targeted species (e.g., striped bass and menhaden; Garrison et al. 2010, Buchheister et al. 2017, Uphoff & Sharov 2018). How do we decide which species gets to have more catch than the other? Or which gets to have more biomass in the ecosystem? How do we decide which species gets to have less catch? How do we deal with the trade-offs among the viable options that impact all interested parties? What are the business rules for handling these situations?

In all regions, there are extant measures of ecosystem overfishing (Link & Marshak 2019; Link & Watson 2019). Preliminary results show that at least a few regions in the U.S. are experiencing ecosystem overfishing in terms of total composition of catch, across species, and the erosion of what fish can be produced is outstripping the basic production of an ecosystem. Are we able to keep taking more "principal" out of the bank? Preliminary estimates suggest that due to either ecosystem overfishing or single species overfishing, we are taking money off the table and precluding future revenue. Is that an acceptable choice?

Some regions are exhibiting ecosystem underfishing (Link 2018, Link & Watson 2019). That means there could be slightly more fish caught, even with very conservative assumptions. There are both conservation and investment implications from this situation as well. Are we in agreement with leaving money on the table, particularly as issues of community health, economics, and even food security arise?

Individual fish populations can exhibit drastic dynamics; aggregations thereof tend to be much more stable (Schindler et al. 2015, Link 2018). This is often why financial planners invest in mutual funds or aggregated products versus just one stock (Markowitz 1952, Elton & Gruber 1977, Marston 2011). Why aren't we considering this more balanced and nuanced approach to spread the risk across fish stocks and fisheries in a given region?

And similar examples, as seen throughout the regional chapters. When boiling these issues down to economics and social impacts, we demonstrate the impacts of not doing EBFM on the real lives of people who depend on LMRs. And that, we have learned, garners more attention than tracking changes in some second principal component of a detailed suite of indicators. We trust that we have at least partially addressed this question of "So what?" and we ask you the reader to help us continue to do so in the coming years.

13.7 Finis

The final question is this: where will be in 10 years with respect to EBFM? Especially given the anticipated issues that are likely to take priority over the next decade (c.f., Chapter 12, Table 12.11). Integrated, cross-disciplinary perspectives and systematic syntheses such as this work can offer insight into determining both regionally-specific and overarching approaches for successful LMR management. That is, ascertaining progress toward EBFM. Ten years ago the debate shifted from "what and why" of EBFM to "how" we actually do EBFM. Ten years from now we hope that the EBFM debate will have shifted to "how well have we been doing."

13.8 References

AORA. 2019. *Working Group on the Ecosystem Approach to Ocean Health and Stressors*. Lisboa, Portugal: All-Atlantic Ocean Research Alliance.

Barange M, Merino G, Blanchard JL, Scholtens J, Harle J, Allison EH, Allen JI, Holt J, Jennings S. 2014. Impacts of climate change on marine ecosystem production in societies dependent on fisheries. *Nature Climate Change* 4:211–6.

Beyraghdar Kashkooli O, Modarres R. 2020. Is the volatility and non-stationarity of the Atlantic multidecadal oscillation (AMO) changing? *Global and Planetary Change* 189:103160.

Buchheister A, Miller TJ, Houde ED. 2017. Evaluating ecosystem-based reference points for Atlantic Menhaden. *Marine and Coastal Fisheries* 9:457–78.

Bundy A, Bohaboy EC, Hjermann DO, Mueter FJ, Fu C, Link JS. 2012. Common patterns, common drivers: comparative analysis of aggregate surplus production across ecosystems. *Marine Ecology Progress Series* 459:203–18.

Bundy A, Chuenpagdee R, Boldt JL, de Fatima Borges M, Camara ML, Coll M, Diallo I, et al. 2017. Strong fisheries management and governance positively impact ecosystem status. *Fish and Fisheries* 18:412–39.

Busch DS, Griffis R, Link J, Abrams K, Baker J, Brainard RE, Ford M, et al. 2016. Climate science strategy of the US national marine fisheries service. *Marine Policy* 74:58–67.

Capotondi A, Jacox M, Bowler C, Kavanaugh M, Lehodey P, Barrie D, Brodie S, et al. 2019. Observational needs supporting marine ecosystems modeling and forecasting: from the global ocean to regional and coastal systems. *Frontiers in Marine Science* 6.

Charles A. 2001. *Sustainable Fishery Systems.* Oxford, UK: Wiley-Blackwell.

Cheung WWL, Brodeur RD, Okey TA, Pauly D. 2015. Projecting future changes in distributions of pelagic fish species of Northeast Pacific shelf seas. *Progress in Oceanography* 130:19–31.

Christie DR. 2005. Implementing an ecosystem approach to ocean management: an assessment of current regional governance models ocean ecosystem management: challenges and opportunities for regional ocean governance. *Duke Environmental Law & Policy Forum* 16:117–42.

Colburn LL, Jepson M, Weng C, Seara T, Weiss J, Hare JA. 2016. Indicators of climate change and social vulnerability in fishing dependent communities along the Eastern and Gulf Coasts of the United States. *Marine Policy* 74:323–33.

Coll M, Shannon LJ, Kleisner KM, Juan-Jordá MJ, Bundy A, Akoglu AG, Banaru D, et al. 2016. Ecological indicators to capture the effects of fishing on biodiversity and conservation status of marine ecosystems. *Ecological Indicators* 60:947–62.

Collie JS, Gislason H. 2001. Biological reference points for fish stocks in a multispecies context. *Canadian Journal of Fisheries and Aquatic Sciences* 58:2167–76.

Crowder LB, Murawski SA. 1998. Fisheries bycatch: implications for management. *Fisheries* 23:8–17.

Crozier LG, McClure MM, Beechie T, Bograd SJ, Boughton DA, Carr M, Cooney TD, et al. 2019. Climate vulnerability assessment for Pacific salmon and steelhead in the California Current large marine ecosystem. *PLoS One* 14:e0217711.

DePiper GS, Gaichas SK, Lucey SM, Pinto da Silva P, Anderson MR, Breeze H, Bundy A, et al. 2017. Operationalizing integrated ecosystem assessments within a multidisciplinary team: lessons learned from a worked example. *ICES Journal of Marine Science* 74:2076–86.

Deroba JJ, Gaichas SK, Lee MY, Feeney RG, Boelke D, Irwin BJ. 2019. The dream and the reality: meeting decision-making time frames while incorporating ecosystem and economic models into management strategy evaluation. *Canadian Journal of Fisheries and Aquatic Sciences* 76:1112–33.

Dolan TE, Patrick WS, Link JS. 2015. Delineating the continuum of marine ecosystem-based management: a US fisheries reference point perspective. *ICES Journal of Marine Science* 73:1042–50.

Dorn M, Zador S. 2020. A risk table to address concerns external to stock assessments when developing fisheries harvest recommendations. *Ecosystem Health and Sustainability* 6(1):1813634

Duarte CM, Agusti S, Barbier E, Britten GL, Castilla JC, Gattuso JP, Fulweiler RW, et al. 2020. Rebuilding marine life. *Nature* 580:39–51.

Edwards SF, Link JS, Rountree BP. 2004. Portfolio management of wild fish stocks. *Ecological Economics* 49:317–29.

Elton EJ, Gruber MJ. 1977. Risk reduction and portfolio size: an analytical solution. *The Journal of Business* 50:415–37.

Fay G, Link JS, Large SI, Gamble RJ. 2014. Management performance of ecological indicators in the Georges Bank finfish fishery. *ICES Journal of Marine Science* 72:1285–96.

Feeney RG, Boelke DV, Deroba JJ, Gaichas S, Irwin BJ, Lee M. 2019. Integrating management strategy evaluation into fisheries management: advancing best practices for stakeholder inclusion based on an MSE for Northeast US Atlantic herring. *Canadian Journal of Fisheries and Aquatic Sciences* 76:1103–11.

Fogarty MJ. 2014. The art of ecosystem-based fishery management. *Canadian Journal of Fisheries and Aquatic Sciences* 71:479–90.

Foran CM, Link JS, Patrick WS, Sharpe L, Wood MD, Linkov I. 2016. Relating mandates in the United States for managing the ocean to ecosystem goods and services demonstrates broad but varied coverage. *Frontiers in Marine Science* 3:5.

Fulton EA, Punt AE, Dichmont CM, Harvey CJ, Gorton R. 2019. Ecosystems say good management pays off. *Fish and Fisheries* 20:66–96.

Gaichas SK, Link JS, Hare JA. 2014. A risk-based approach to evaluating northeast US fish community vulnerability to climate change. *ICES Journal of Marine Science* 71:2323–42.

Garrison LP, Link JS, Kilduff DP, Cieri MD, Muffley B, Vaughan DS, Sharov A, Mahmoudi B, Latour RJ. 2010. An expansion of the MSVPA approach for quantifying predator–prey interactions in exploited fish communities. *ICES Journal of Marine Science* 67:856–70.

Gende SM, Sigler MF. 2006. Persistence of forage fish 'hot spots' and its association with foraging Steller sea lions (Eumetopias jubatus) in southeast Alaska. *Deep Sea Research Part II: Topical Studies in Oceanography* 53:432–41.

Gruss A, Schirripa MJ, Chagaris D, Velez L, Shin YJ, Verley P, Oliveros-Ramos R, Ainsworth CH. 2016. Estimating natural mortality rates and simulating fishing scenarios for Gulf of Mexico red grouper (Epinephelus morio) using the ecosystem model OSMOSE-WFS. *Journal of Marine Systems* 154:264–79.

Hare JA, Morrison WE, Nelson MW, Stachura MM, Teeters EJ, Griffis RB, Alexander MA, et al. 2016. A vulnerability assessment of fish and invertebrates to climate change on the Northeast U.S. Continental Shelf. *PLoS One* 11:e0146756.

Hazen EL, Scales KL, Maxwell SM, Briscoe DK, Welch H, Bograd SJ, Bailey H, et al. 2018. A dynamic ocean management tool to reduce bycatch and support sustainable fisheries. *Science Advances* 4:eaar3001.

Hershman MJ, Russell CW. 2005. Regional ocean governance in the United States: concept and realty ocean ecosystem management: challenges and opportunities for regional ocean governance. *Duke Environmental Law & Policy Forum* 16:227–66.

Hilborn R, Amoroso RO, Anderson CM, Baum JK, Branch TA, Costello C, de Moor CL, et al. 2020. Effective fisheries management instrumental in improving fish stock status. *Proceedings of the National Academy of Sciences* 117(4):2218–24.

Hilborn R, Fulton EA, Green BS, Hartmann K, Tracey SR, Watson RA. 2015. When is a fishery sustainable? *Canadian Journal of Fisheries and Aquatic Sciences* 72:1433–41.

Hilborn R, Ovando D. 2014. Reflections on the success of traditional fisheries management. *ICES Journal of Marine Science* 71:1040–6.

Hilborn RA, Stewart IJ, Branch TA, Jensen OP. 2012. Defining trade-offs among conservation, profitability, and food security in the California current bottom-trawl fishery. *Conservation Biology* 26(2):257–68.

Himes-Cornell A, Kasperski S. 2015. Assessing climate change vulnerability in Alaska's fishing communities. *Fisheries Research* 162:1–11.

Hobday AJ, Hartog JR, Manderson JP, Mills KE, Oliver MJ, Pershing AJ, Siedlecki S. 2019. Ethical considerations and unanticipated consequences associated with ecological forecasting for marine resources. *ICES Journal of Marine Science* 76(5):9–10.

Howell EA, Hoover A, Benson SR, Bailey H, Polovina JJ, Seminoff JA, Dutton PH. 2015. Enhancing the TurtleWatch product for leatherback sea turtles, a dynamic habitat model for ecosystem-based management. *Fisheries Oceanography* 24:57–68.

Hutchison L, Montagna P, Yoskowitz D, Scholz D, Tunnell J. 2015. Stakeholder perceptions of coastal habitat ecosystem services. *Estuaries Coasts* 38:67–80.

ICES. 2020. *ICES and Ecosystem-Based Management. The Importance and Rationale of EBM to ICES Version 2*. Copenhagen, Denmark: ICES.

IPCC. 2019. *IPCC Special Report on the Ocean and Cryosphere in a Changing Climate.* Geneva, Switzerland: IPCC (p. 755).

Jacobsen NS, Burgess MG, Andersen KH. 2017. Efficiency of fisheries is increasing at the ecosystem level. *Fish and Fisheries* 18:199–211.

Jepson M, Colburn LL. 2013. *Development of Social Indicators of Fishing Community Vulnerability and Resilience in the U.S. Southeast and Northeast Regions NOAA Technical Memorandum NMFS-F/SPO-129*. Washington, DC: NOAA (p. 72).

Jin D, DePiper G, Hoagland P. 2016. Applying portfolio management to implement ecosystem-based fishery management (EBFM). *North American Journal of Fisheries Management* 36:652–69.

Juricke S, MacLeod D, Weisheimer A, Zanna L, Palmer TN. 2018. Seasonal to annual ocean forecasting skill and the role of model and observational uncertainty. *Quarterly Journal of the Royal Meteorological Society* 144:1947–64.

Kaplan IC, Williams GD, Bond NA, Hermann AJ, Siedlecki SA. 2016. Cloudy with a chance of sardines: forecasting sardine distributions using regional climate models. *Fisheries Oceanography* 25:15–27.

Keyl F, Wolff M. 2008. Environmental variability and fisheries: what can models do? *Reviews in Fish Biology and Fisheries* 18:273–99.

Koehn LE, Essington TE, Levin PS, Marshall KN, Anderson LG, Bundy A, Carothers C, et al. 2020. Case studies demonstrate capacity for a structured planning process for ecosystem-based fisheries management. *Canadian Journal of Fisheries and Aquatic Sciences* 77(7): 1256–74.

Kolding J, Jacobsen NS, Andersen KH, van Zwieten PAM. 2016. Maximizing fisheries yields while maintaining community structure. *Canadian Journal of Fisheries and Aquatic Sciences* 73:644–55.

Large SI, Fay G, Friedland KD, Link JS. 2015. Quantifying patterns of change in marine ecosystem response to multiple pressures. *PLoS One* 10:e0119922.

Lederhouse T, T. Marshak, L. Latchford, R. Peters, and K. Latanich. 2017. *Report from the National Essential Fish Habitat Summit. U.S. Dept. of Commerce, NOAA. NOAA Technical Memorandum NMFS-OHC-3* . Washington, DC: NOAA (p. 44).

Lettrich MD, Asaro MJ, Borggaard DL, Dick DM, Griffis RB, Litz JA, Orphanides CD, et al. 2019. *A Method for Assessing the Vulnerability of Marine Mammals to a Changing Climate. NOAA Technical Memorandum NMFS-F/SPO-196*. Washington, DC: NOAA (p. 80).

Levin PS, Essington TE, Marshall KN, Koehn LE, Anderson LG, Bundy A, Carothers C, et al. 2018. Building effective fishery ecosystem plans. *Marine Policy* 92:48–57.

Levin PS, Fogarty MJ, Murawski SA, Fluharty D. 2009. Integrated ecosystem assessments: developing the scientific basis for ecosystem-based management of the ocean. *PLoS Biology* 7:e1000014.

Lewison R, Hobday AJ, Maxwell S, Hazen E, Hartog JR, Dunn DC, Briscoe D, et al. 2015. Dynamic ocean management: identifying the critical ingredients of dynamic approaches to ocean resource management. *BioScience* 65:486–98.

Libralato S, Pranovi F, Zucchetta M, Monti MA, Link JS. 2019. Global thresholds in properties emerging from cumulative curves of marine ecosystems. *Ecological Indicators* 103:554–62.

Link J. 2017. A conversation about NMFS ecosystem-based fisheries management policy and road map. *Fisheries* 42:498–503.

Link J. 2010. *Ecosystem-Based Fisheries Management: Confronting Tradeoffs.* Cambridge, UK: Cambridge University Press.

Link JS. 2018. System-level optimal yield: increased value, less risk, improved stability, and better fisheries. *Canadian Journal of Fisheries and Aquatic Sciences* 75:1–16.

Link JS, Browman HI. 2017. Operationalizing and implementing ecosystem-based management. *ICES Journal of Marine Science* 74:379–81.

Link JS, Browman HI. 2014. Integrating what? Levels of marine ecosystem-based assessment and management. *ICES Journal of Marine Science* 71:1170–3.

Link JS, Dickey-Collas M, Rudd M, McLaughlin R, Macdonald NM, Thiele T, Ferretti J, Johannesen E, Rae M. 2018. Clarifying mandates for marine ecosystem-based management. *ICES Journal of Marine Science* 76(1):41–4.

Link JS, Griffis RB, Busch DS. 2015. NOAA Fisheries Climate Science Strategy. NOAA Technical Memorandum NMFS-F/SPO-155. Washington, DC: NOAA (p. 70).

Link JS, Marshak AR. 2019. Characterizing and comparing marine fisheries ecosystems in the United States: determinants of success in moving toward ecosystem-based fisheries management. *Reviews in Fish Biology and Fisheries* 29:23–70.

Link JS, Thébaud O, Smith DC, Smith AD, Schmidt J, Rice J, Poos JJ, et al. 2017a. Keeping humans in the ecosystem. *ICES Journal of Marine Science* 74(7):1947–56.

Link JS, Tolman HL, Robinson K. 2017b. Earth systems: NOAA's strategy for unified modelling. *Nature* 549:458.

Link JS, Watson RA. 2019. Global ecosystem overfishing: clear delineation within real limits to production. *Science Advances* 5:eaav0474.

Lynch PD, Methot RD, Link JS. 2018. *Implementing a Next Generation Stock Assessment Enterprise. An Update to the NOAA Fisheries Stock Assessment Improvement Plan. NOAA Technical Memorandum TM NMFS-F/SPO-183*. Washington, DC: NOAA.

Mann KH, Lazier JRN. 1991. *Dynamics of Marine Ecosystems: Biological-Physical Interactions in the Oceans.* Oxford, UK: Blackwell (p. 466).

Markowitz H. 1952. Portfolio selection. *Journal of Finance* 7:77–91.

Marshak AR, Link JS, Shuford R, Monaco ME, Johannesen E, Bianchi G, Anderson MR, et al. 2017. International perceptions of an integrated, multi-sectoral, ecosystem approach to management. *ICES Journal of Marine Science* 74:414–20.

Marshall KN, Levin PS, Essington TE, Koehn LE, Anderson LG, Bundy A, Carothers C, et al. 2018. Ecosystem-based fisheries management for social–ecological systems: renewing the focus in the United States with next generation fishery ecosystem plans. *Conservation Letters* 11:e12367.

Marston RC. 2011. *Portfolio Design: A Modern Approach to Asset Allocation.* Hoboken, NJ: John Wiley and Sons Inc.

Megrey BA, Link JS, Hunt Jr GL, Moksness E. 2009. Comparative marine ecosystem analysis: applications, opportunities, and lessons learned. *Progress in Oceanography* 81:2–9.

Merrick R. 2018. Mechanisms for science to shape US living marine resource conservation policy. *ICES Journal of Marine Science* 75:2319–24.

Merrick R, Allen L, Angliss R, Antonelis G, Eagle T, Epperly S. 2004. *Report of the NOAA Fisheries National Task Force for Improving Marine Mammal and Turtle Stock Assessments. NOAA Technical Memorandum NMFS-F/SPO-63*. Washington, DC: NOAA (p. 123).

Milly PCD, Betancourt J, Falkenmark M, Hirsch RM, Kundzewicz ZW, Lettenmaier DP, Stouffer RJ. 2008. Stationarity is dead: whither water management? *Science* 319:573–4.

Möllmann C, Lindegren M, Blenckner T, Bergström L, Casini M, Diekmann R, Flinkman J, et al. 2014. Implementing ecosystem-based fisheries management: from single-species to integrated ecosystem assessment and advice for Baltic Sea fish stocks. *ICES Journal of Marine Science* 71:1187–97.

Monaco ME, Spooner E, Oakes SA, Harvey CJ, Kelble CR. 2021. Introduction to the NOAA integrated ecosystem assessment program: advancing ecosystem based management. *Coastal Management* 49(1).:1–8.

Morley JW, Selden RL, Latour RJ, Frölicher TL, Seagraves RJ, Pinsky ML. 2018. Projecting shifts in thermal habitat for 686 species on the North American continental shelf. *PLoS One* 13:e0196127.

Morrison WE, Nelson MW, Howard JF, Hare JA, Griffis RB, Scott JD, Alexander MA. 2015. *Methodology for Assessing the Vulnerability of Marine Fish and Shellfish Species to a Changing Climate. NOAA Technical Memorandum NMFS-OSF-3*. Washington, DC: NOAA (p. 54).

Mueter FJ, Megrey BA. 2006. Using multi-species surplus production models to estimate ecosystem-level maximum sustainable yields. *Fisheries Research* 81:189–201.

Munsch SH, Greene CM, Johnson RC, Satterthwaite WH, Imaki H, Brandes PL, O'Farrell MR. 2020. Science for integrative management of a diadromous fish stock: interdependencies of fisheries, flow, and habitat restoration. *Canadian Journal of Fisheries and Aquatic Sciences* 77(9):1487–504.

NMFS (National Marine Fisheries Service). 2016a. *Ecosystem-Based Fisheries Management Policy. National Marine Fisheries Service Policy Directive 01-120*. Washington, DC: NOAA.

NMFS (National Marine Fisheries Service). 2016b. *Ecosystem-Based Fisheries Management Policy. National Marine Fisheries Service Policy Directive 01-120-01*. Washington, DC: NOAA.

NMFS (National Marine Fisheries Service). 2018. *Fisheries Economics of the United States, 2016. NOAA Technical*

Memorandum NMFS-F/SPO-187a. Washington, DC: NOAA (p. 243).

NMFS (National Marine Fisheries Service). 2019. *NOAA Fisheries Strategic Plan 2019–2022*. Silver Spring, MD: NOAA (p. 16).

NMFS (National Marine Fisheries Service). 2020. *Status of Stocks 2019. Annual Report to Congress on the Status of U.S. Fisheries*. Silver Spring, MD: NOAA.

Nye JA, Link JS, Hare JA, Overholtz WJ. 2009. Changing spatial distribution of fish stocks in relation to climate and population size on the Northeast United States continental shelf. *Marine Ecology Progress Series* 393:111–29.

Olsen E, Fay G, Gaichas S, Gamble R, Lucey S, Link JS. 2016. Ecosystem model skill assessment. Yes we can! *PLoS One* 11:e0146467.

Ostrom E. 2009. A general framework for analyzing sustainability of social-ecological systems. *Science* 325:419–22.

Park JY, Stock CA, Dunne JP, Yang X, Rosati A. 2019. Seasonal to multiannual marine ecosystem prediction with a global Earth system model. *Science* 365:284–8.

Patrick WS, Link JS. 2015a. Myths that continue to impede progress in ecosystem-based fisheries management. *Fisheries* 40:155–60.

Patrick WS, Link JS. 2015b. Hidden in plain sight: using optimum yield as a policy framework to operationalize ecosystem-based fisheries management. *Marine Policy* 62:74–81.

Patrick WS, Spencer P, Link J, Cope J, Field J, Kobayashi D, Lawson P, et al. 2010. Using productivity and susceptibility indices to assess the vulnerability of United States fish stocks to overfishing. *Fishery Bulletin* 108:305–22.

Peters R, Marshak AR, Brady MM, Brown SK, Osgood K, Greene C, Guida V. 2018. *Habitat Science is a Fundamental Element in an Ecosystem-Based Fisheries Management Framework: An Update to the Marine Fisheries Habitat Assessment Improvement Plan. NOAA Technical Memorandum NMFS-F/SPO-181*. Washington, DC: NOAA (p. 37).

Pikitch EK, Rountos KJ, Essington TE, Santora C, Pauly D, Watson R, Sumaila UR, et al. 2014. The global contribution of forage fish to marine fisheries and ecosystems. *Fish and Fisheries* 15:43–64.

Pinsky ML, Worm B, Fogarty MJ, Sarmiento JL, Levin SA. 2013. Marine taxa track local climate velocities. *Science* 341:1239–42.

Pitcher TJ, Kalikoski D, Short K, Varkey D, Pramod G. 2009. An evaluation of progress in implementing ecosystem-based management of fisheries in 33 countries. *Marine Policy* 33:223–32.

Pope JG, Weber CT. 2019. A parable of compliance issues and their link to EBFM outcomes. *Fisheries Research* 211:51–8.

Pranovi F, Libralato S, Zucchetta M, Anelli Monti M, Link JS. 2020. Cumulative biomass curves describe past and present conditions of large marine ecosystems. *Global Change Biology* 26:786–97.

Rindorf A, Cardinale M, Shephard S, De Oliveira JAA, Hjorleifsson E, Kempf A, Luzenczyk A, et al. 2017. Fishing for MSY: using "pretty good yield" ranges without impairing recruitment. *ICES Journal of Marine Science* 74:525–34.

Roman J, Altman I, Dunphy-Daly MM, Campbell C, Jasny M, Read AJ. 2013. The Marine Mammal Protection Act at 40: status, recovery, and future of U.S. marine mammals. *Annals of the New York Academy of Sciences* 1286:29–49.

Rosenberg AA. 2009. Changing U.S. ocean policy can set a new direction for marine resource management. *Ecology and Society* 14:art6.

Rudd MA, Dickey-Collas M, Ferretti J, Johannesen E, Macdonald NM, McLaughlin R, Rae M, Thiele T, Link JS. 2018. Ocean ecosystem-based management mandates and implementation in the North Atlantic. *Frontiers in Marine Science* 5:485.

Sagarese SR, Bryan MD, Walter JF, Schirripa M, Grüss A, Karnauskas M. 2015. *Incorporating Ecosystem Considerations Within the Stock Synthesis Integrated Assessment Model for Gulf of Mexico Red Grouper (Epinephelus Morio). SEDAR42-RW-01*. North Charleston, SC: SEDAR.

Sagarese SR, Grüss A, Karnauskas M, Walter III JF. 2014. *Ontogenetic Spatial Distributions of Red Grouper (Epinephelus Morio) Within the Northeastern Gulf of Mexico and Spatiotemporal Overlap with Red Tide Events. SEDAR42-DW-04*. North Charleston, SC: SEDAR.

Samhouri JF, Haupt AJ, Levin PS, Link JS, Shuford R. 2013. Lessons learned from developing integrated ecosystem assessments to inform marine ecosystem-based management in the USA. *ICES Journal of Marine Science* 71:1205–15.

Sanchirico JN, Smith MD, Lipton DW. 2008. An empirical approach to ecosystem-based fishery management. *Ecological Economics* 64:586–96.

Schindler DE, Armstrong JB, Reed TE. 2015. The portfolio concept in ecology and evolution. *Frontiers in Ecology and the Environment* 13:257–63.

Skern-Mauritzen M, Ottersen G, Handegard NO, Huse G, Dingsør GE, Stenseth NC, Kjesbu OS. 2016. Ecosystem processes are rarely included in tactical fisheries management. *Fish and Fisheries* 17:165–75.

Slater WL, DePiper G, Gove JM, Harvey CJ, Hazen EL, Lucey SM, Karnauskas M, et al. 2017. *Challenges, Opportunities and Future Directions to Advance NOAA Fisheries Ecosystem Status Reports (ESRs): Report of the National ESR Workshop. NOAA Technical Memorandum NMFS-F/SPO-174*. Washington, DC: NOAA (p. 74).

Smith ADM, Brown CJ, Bulman CM, Fulton EA, Johnson P, Kaplan IC, Lozano-Montes H, et al. 2011. Impacts of fishing low-trophic level species on marine ecosystems. *Science* 333:1147–50.

Spencer PD, Hollowed AB, Sigler MF, Hermann AJ, Nelson MW. 2019. Trait-based climate vulnerability assessments in data-rich systems: an application to eastern Bering Sea fish and invertebrate stocks. *Global Change Biology* 25:3954–71.

Steele JH. 1991. Marine ecosystem dynamics–comparison of scales. *Ecological Research* 6:175–83.

Steele JH. 1985. A comparison of terrestrial and marine ecological systems. *Nature* 313:355–8.

Tam JC, Link JS, Large SI, Andrews K, Friedland KD, Gove J, Hazen E, et al. 2017. Comparing apples to oranges: common trends and thresholds in anthropogenic and environmental pressures across multiple marine ecosystems. *Frontiers in Marine Science* 4:282.

Thébaud O, Link JS, Kohler B, Kraan M, López R, Poos JJ, Schmidt JO, Smith DC. 2017. Managing marine socio-ecological systems: picturing the future. *ICES Journal of Marine Science* 74:1965–80.

Townsend H, Harvey CJ, deReynier Y, Davis D, Zador SG, Gaichas S, Weijerman M, Hazen EL, Kaplan IC. 2019. Progress on implementing ecosystem-based fisheries management in the United States through the use of ecosystem models and analysis. *Frontiers in Marine Science* 6.

Townsend H, Kaplan I, Link J. 2020. *NOAA Fisheries—Virtual Ecosystem Scenario Viewer (VES-V). A Software Tool for Visualizing Complex Data and Model Outputs for Marine Ecosystems. Version 1.37—User Manual*. Silver Spring, MD: NOAA (p. 31).

Tyrrell M, Link J, Moustahfid H. 2011. The importance of including predation in fish population models: implications for biological reference points. *Fisheries Research* 108:1–8.

Uphoff JH, Sharov A. 2018. Striped Bass and Atlantic Menhaden predator–prey dynamics: model choice makes the difference. *Marine and Coastal Fisheries* 10:370–85.

U.S. Commission on Ocean Policy. 2004. *An Ocean Blueprint for the 21st Century. Final Report.* Washington, DC: U.S. Commission on Ocean Policy (p. 676).

Valdivia A, Wolf S, Suckling K. 2019. Marine mammals and sea turtles listed under the U.S. Endangered Species Act are recovering. *PLoS One* 14:e0210164.

Valiela, I. 1995. *Marine Ecological Processes*. New York, NY: Springer (p. 686).

Weise MJ, Harvey JT. 2005. Impact of the California sea lion (Zalophus californianus) on salmon fisheries in Monterey Bay, California. *Fishery Bulletin* 103:685–96.

Wilkinson EB, Abrams K. 2015. *Benchmarking the 1999 EPAP Recommendations with Existing Fishery Ecosystem Plans. NOAA Technical Memorandum NMFS-OSF-5*. Washington, DC: NOAA (p. 22).

Worm B, Hilborn R, Baum JK, Branch TA, Collie JS, Costello C, Fogarty MJ, et al. 2009. Rebuilding global fisheries. *Science* 325:578–85.

Yaffee SL. 1996. Ecosystem management in practice: the importance of human institutions. *Ecological Applications* 6:724–7.

Index

Note to Index: If a user does not know *examples* of bays, islands, monuments, sanctuaries, subregions, they will be found by referring to the Regions

b, f, n and *t* denote *box, figure, footnote and table*

A

D

E

N

O

Q

R

S

Y

Z